OVERDUE FINES:
25¢ per day per item

RETURNING LIBRARY MATERIALS:
Place in book return to remo
charge from circulation re

HISTOCHEMISTRY

Theoretical and Applied

Volume I

FRANÇOIS-VINCENT RASPAIL

1794–1878

The founder of histochemistry

[*Frontispiece*

HISTOCHEMISTRY
Theoretical and Applied

by

A. G. EVERSON PEARSE

M.A. (Cambridge), M.D. (Cambridge & Basel),
F.R.C.P., D.C.P. (London), F.C.Path.

Professor of Histochemistry in the University of London
(Royal Postgraduate Medical School)
Hon. Consultant Pathologist, Hammersmith Hospital
Fulbright Fellow and Visiting Professor of Pathology
in the University of Alabama
Guest Instructor in Histochemistry, University of Kansas
Visiting Lecturer in Histochemistry, Vanderbilt University

THIRD EDITION

VOLUME 1

With 14 Coloured Plates and 159
Black and White Illustrations

LITTLE, BROWN & COMPANY
BOSTON
1968

To all those, histochemists and others, without whose help and without whose work this book could not have been written.

NAM PROPRIUM EST NIHIL.

Hor. Epist. i, II, 23 (apud Feam).

First Edition	1953
Reprinted	1954
Translated into Polish	
Translated into Spanish	
Translated into Russian	
Second Edition	1960
Reprinted	1961
Translated into Russian	1962
Third Edition	1968

Standard Book Number
7000 1326 1

Library of Congress
Catalog Card Number
68-30789

ALL RIGHTS RESERVED

This book may not be reproduced by any means, in whole or in part, without permission. Application with regard to reproduction should be addressed to the Publishers

PRINTED IN GREAT BRITAIN

PREFACE

I HAVE no excuse for the size of this volume, or for having been forced to divide the third edition into two volumes, other than that which I gave in the preface to the second edition.

Histochemistry and Cytochemistry, which together constitute a single subject, have expanded in almost logarithmic manner in the past ten years. It is now clearly beyond the capacity of a single individual to cover the whole subject with first-hand knowledge of all the available techniques. I have therefore been obliged to rely substantially on the testimony of friends and associates. To state this fact is not to avoid responsibility but merely to put on record my inability any longer to follow the principles on which the first and second editions of this book were based.

From the practical point of view the composition of this volume, and its successor, follows that of previous editions in that the technical appendices form an autonomous unit at the end of the text. It is still necessary to warn users of the appendices that many of the methods given are in early stages of development. They will therefore avoid unnecessary labour and disappointment by occasional reference to the text.

It is impossible to be other than impressed by the general progress of Histochemistry and Cytochemistry, as shown by the enormous increase in the number of applications recorded in the literature. Both branches of the subject stand or fall by their usefulness as applied sciences. Their proper recognition by scientists in other disciplines, still far from adequate, depends largely on their success or failure in this respect.

It is indeed difficult to find any subject in the general framework covered by biology, cytology, embryology, histology, pathology or zoology, which has not been influenced significantly by histochemically activated discoveries. Future applications can be envisaged which will certainly eclipse all that has gone before. Once again, remembering that "it is not the beginning but the end that crowneth the work," I can only express the hope that my book will provide the stimulus and the means for further extensions.

A. G. EVERSON PEARSE

The Fortress,
Letchmore Heath,
Hertfordshire

ACKNOWLEDGEMENTS

TWENTY years ago, believing that histochemistry could make a valuable contribution to the development of pathology, my Chief, Professor J. H. Dible, F.R.C.P., suggested that I should undertake this work. Once again I have pleasure in acknowledging his initial responsibility, and his continuing interest.

Many colleagues, past and present, at the Royal Postgraduate Medical School have given sympathetic assistance. I should like to thank especially Professor Sir John McMichael, F.R.S., whose active and practical support, over a long period, has lightened the burden of running an effective Department of Histochemistry on a slender budget.

I have received generous help, in the form of capital grants for equipment and financial assistance for salaries and maintenance, from the British Empire Cancer Campaign, the Muscular Dystrophy Group of Great Britain, the Medical Research Council, and the Wellcome Foundation. The Smith, Kline and French Foundation has also, on occasion, provided timely and generous assistance.

To acknowledge my indebtness to all my associates in the Department of Histochemistry would require me to mention over 150 names. Space allows only a corporate vote of thanks. Special mention must be given, however, to the great volume of technical assistance provided by Mr. J. D. Bancroft.

I have, moreover, derived considerable help from unpublished or prepublication manuscripts provided by friends and associates. Here I refer particularly to Dr. J. H. Tranzer (Chapter 3), Dr. E. P. C. Tock (Chapter 5), Dr. P. J. Stoward (Chapter 10), Dr. T. Lehner (Chapter 11), and Dr. Z. Lojda (Chapter 16). Mr. K. V. Slee and Mr. F. Stead, of the South London Electrical Equipment Company, provided material assistance and advance information necessary for the preparation of Chapter 2, and for this I am especially grateful. Mr. E. J. Warren, Mr. P. Griggs, and Mr. T. Rowe, of Edwards High Vacuum Ltd., performed a similar office in respect of Chapter 3.

During the four years spent in preparation of this volume I have received valuable assistance from my secretaries, notably Mrs. S. Page and Miss Shirley Green. In addition to her other duties in connection with the book, Miss Green has prepared the author index. I am very grateful to her for her untiring efforts on my behalf.

I wish to thank the following authors, editors, and others for permission to use the undermentioned illustrations, some of which have appeared elsewhere in print: Fig. 1, Dr. N. Shimizu; Figs. 2 and 3, Dr. Wayne Thornberg and the Editor, *Journal of Histochemistry and Cytochemistry*; Fig. 4, Mr. R. Perriton and De La Rue Frigistor Ltd.; Fig. 16, Mr. L. J. Wright; Fig. 18, Prof. O. Eränkö; Fig. 22, Dr. W. A. Jensen; Fig. 38, Dr. J. Persijn and the Editor, *Histochemie*; Fig. 69, Dr. H. A. Nielsen; Fig. 70, Prof. R. G. White and the Editor, *British Journal of Experimental Pathology*; Fig. 90, Dr. R. Love; Figs. 106 and 107, Drs. A. S. Cohen, T. Shirahama (Arthritis Laboratory, Boston University Medical Center); Figs. 110 and 111, Dr. W. O. Berg and the Editor, *Acta Pathologica Scandinavica*; Fig. 129, Prof. M. Chèvremont; Figs. 135 and 136, Dr. P. Loustalot; Fig. 138, the Editors, *Biochemical Journal*; Fig. 152, Dr. D. Janigan; Plate Ib, Dr. J. H. Humphrey, F.R.S. and the Editor, *Journal of Immunology*; Plate Id, Prof. C. W. M. Adams and the Editor, *Journal of Pathology and Bacteriology*; Plates IIa and b, Dr. E. A. Tonna and the Editors, *Journal of the Royal Microscopical Society*; Plates IIIa, b, and c, Prof. A. C. Lendrum and the Editor and Publishers, *Journal of Clinical Pathology* (B.M.A. House, Tavistock Square, London, W.C.1); Plates VIIa to f, Dr. J. E. Scott and the Editor, *Histochemie*; Plate VIIIc, Dr. J. D. Hicks and the Editor, *Journal of Pathology and Bacteriology*, Plates IXa and b, Prof. H. Heller and the Editor, *Journal of Pathology and Bacteriology*; Plate IXc, Prof. M. Wolman and the Editor, *Histochemie*; Plate XIIc, Dr. H. J. Stutte and the Editor, *Histochemie*; Plates XIIIa and b, Dr. W. Strauss and the Rockefeller Institute Press (*Journal of Cell Biology*); Plate XIVa, Prof. W. Gössner and the Editor, *Histochemie*; Plate XIVb, Dr. R. Hess and CIBA (Basel). For the reproduction of the portrait of Raspail, which still forms the Frontispiece, I am indebted to the kindness of Miss A.M. Vidal-Hall of the Scientific Office of the French Embassy.

The dedication of this third edition remains as in the previous two. Nevertheless, even more than in the case of previous editions, it is true to say that this volume would never have been completed had I had not been sustained and supported throughout by my wife, Dr. Elizabeth Pearse. In addition to the onerous task of proof-reading she has been solely responsible for the preparation of the subject index. It is impossible to put into words the sum total of my gratitude to her, and I shall not attempt to do it.

CONTENTS

ERRATA

p. vi	line 14 for 'Wellcome Foundation' read 'Wellcome Trust'.
p. vii	line 5 from foot delete second 'had'.
p. 306	line 4 from foot for 'done' read 'due'.
p. 320	(Plate V) transpose Vc and Vd legends.
p. 387	(Plate IX) transpose IXa and IXc legends.
p. 586	second paragraph line 2 for 'nucleic acid' read 'maleic acid'.
p. 709	*(Feigl, Anger and Frehden, 1934).

CHAPTER 1

THE HISTORY OF HISTOCHEMISTRY

THERE is little difference of opinion about the antiquity of histochemistry, "a science as old as histology itself" according to Lison. The schism occurs when the continuity of the old with the newer histochemistry is considered. On the one hand we have the view, put forward by Lison in the first edition of his book, that the modern science differs profoundly from that of 100 years ago. "It is so different," he said, "that one can say it is entirely recent—created *de novo*." On the other hand there is the view, to which I myself subscribe, that the progress of histochemistry has been uninterrupted, and that its aims and principles today are essentially the same as they were in the past. This is not to say that alterations and improvements of the widest variety have not taken place. Histochemistry can sometimes proceed ahead of, and independently of, the sister sciences of biochemistry and chemistry. It continues to depend to a great extent on these two for the development of new techniques and ideas. In several fields, however, histochemical research has preceded and illuminated the way for biochemical studies to follow. An early example of this is the field of pigmented substances where histochemical work in the nineteenth century resulted in the classical biochemical hypotheses on iron metabolism and on the formation of bile pigments. In the present century the successful use of histochemical and histophysical techniques for nucleic acids was directly responsible for initiating biochemical studies, still continuing, which culminated in the concept of the genetic code.

The historical survey given here appeared in shorter form in a review of modern methods in histochemistry (Pearse, 1951). It is designed to give a true impression of the continuity of what is at the same time the youngest and one of the oldest of the biological sciences. Below, as an introduction, appears a brief tabulated history. This is by no means complete, but it forms a skeleton on which to hang the otherwise unconnected individual references which follow.

1800–29 Isolated reports of the investigation of chemical as opposed to morphological structure in tissue preparations. Histochemistry unknown as a separate science.

1830–55 These years saw the beginnings of histochemistry as a science. In its origins it was primarily botanical; for some decades the whole practice of histochemistry in its true sense was in the hands of the botanists. Various works on the subject appeared: Raspail's *Essai de Chimie Microscopique Appliquée à la Physiologie*, 1830; his *Nouveau Système de Chimie Organique*, 1833; Lehmann's *Lehrbuch der physiologischen Chemie*, 1842; Raspail's *New System of Organic Chemistry*, 1834 (translation).

1856–71 Histochemistry, in the case of animal as opposed to plant tissues, was mainly biological chemistry, and most of its methods involved tissue destruction. More textbooks on this aspect of histochemistry began to appear, some of them being translations of earlier works: *Physiological Chemistry* of Lehmann, 1851 (translation); *Traité de Chimie Anatomique* by Robin and Verdeil, 1853; *Chemie der Gewebe des gesammten Thierreichs* of Schlossberger, 1856; *Handbuch der Experimental Physiologie der Pflanzen* of Sachs, 1865; and the *Handbuch der Histologie und Histochemie des Menschen* of Frey, 1867.

1872–98 At this stage histochemistry became divided, part left histology and became attached to physiology and part remained as biological chemistry. Articles and books published during this period included *Ein Beitrag zur Histochemie*, by Miescher, 1874; *Histology and Histochemistry of Man*, by Frey, 1874 (translation); *Traité de Chimie Physiologique* of Lehmann, 1883 (translation); Bunge's *Lehrbuch der physiologischen und pathologischen Chemie*, 1887; and Sachs' *Lectures on the Physiology of Plants*, 1887.

1899–1929 During this period the use of aniline dyes in histology, first described by Bencke in 1862, became widespread. The first quarter of the twentieth century saw the rapid expansion of descriptive histopathology. Histologists became more interested in new dyes and staining techniques and showed less interest in the chemistry of tissue structures. Although diagnostic significance was attached to many of the new colour reactions, no attempt was made to put them on a physical or chemical basis. Morphological studies overwhelmed histochemistry and Mann (1902) was rash enough to say that the study of mammalian micro-anatomy was "almost complete" by 1900. Three contributions of this period described microchemistry and microphysiology; they were Mann's *Physiological Histology*, 1902, Ehrlich's *Encyclopädie der Mikroskopischen Technik*, 1903, and Macallum's *Methoden und Ergebnisse der Mikrochemie*, 1908, all classical works. Prenant (1910) reviewed the general state of histochemistry in a valuable paper, and Molisch's *Mikrochemie der Pflanzen*, 1913, contained much of histochemical and cytochemical interest. Other works published towards the close of this period, which show that histochemistry still flourished, were the *Review of Recent Developments in Histochemistry*, by Parat, 1927; *Animale Histochemie*, by Patzelt, 1928; *Prakticum der Histochemie*, by Klein, 1929; and *Histochemische Methoden*, by Hertwig, 1929.

1930–44 This stage saw the rebirth of histochemistry and its partial return to the domain of histology. The most important work of the period was Lison's *Histochemie Animale* (1936) in which the author proclaimed the new science of histochemistry without tissue destruction. It is impossible to over-estimate the effect of this book upon the progress and practice of histochemistry. It remained, until superseded in 1953 and 1960 by second and third editions of broader scope, the acknowledged bible of histochemists in all fields of the basic sciences. Other excellent, though less important, publications were the *Handbook of Chemical Microscopy*, by Chamot and Mason, 1930; *Histochemische Methoden* of Romeis, 1932; *Die Mikroveraschung als histochemische Hilfsmethode*, by Policard and Okkels, 1932; and Linderstrøm-Lang's *Problems in Histochemistry*, 1936.

1945–58 The first published work of this period was Glick's *Techniques of Histo- and Cytochemistry*, 1949. This dealt somewhat briefly with the theory and practice of histochemistry, as it concerns the histologist. In it was reviewed the entire compass of histochemistry, physiological, physical and histological, and much of the technical information was beyond the scope of any but specialists in small individual fields. After a brief interval there appeared four works largely or wholly devoted to histochemistry. These, in order of appearance, were Gomori's *Microscopic Histochemistry*, 1952; Danielli's *Cytochemistry*, 1953; the first edition of my book, 1953; and Lison's second edition *Histochimie et Cytochimie Animales*, 1953. Bourne's *Functional Histology*, 1953, though not intended as a histochemical treatise, was based largely on the application of histochemical techniques, and Lillie's second edition with its new title *Histopathologic Technic and Practical Histochemistry*, 1954, represented an important addition to histochemical literature though only partly devoted to this subject. A stimulating and critical review of modern histochemistry was given by Vialli in his *Introduzione alla Ricerca in Istochimica*, 1955, in which the merits and demerits of most aspects of the science were discussed. Evidence of the expanding outlook of modern histochemistry was offered by Mellors' *Analytical Cytology*, 1955, a work of many authors concerned with all types of microscopy and their application, and by Eränkö's *Quantitative Methods in Histology and Microscopic Histochemistry*, 1955. These two works stressed particularly the quantitative side of histochemistry, whose future development is closely bound up with progress in this direction. In 1958 appeared the first volume of a very comprehensive treatise, the *Handbuch der Histochemie*, dealing with the general methodology of the subject. The rapidly developing science of histochemistry, which until 1950 had no journals or periodicals devoted to it, acquired in this period no less than seven. If one counted those publications in which histochemical or cytochemical papers formed a substantial part or even a majority of the contributions the number was clearly even larger. The main histochemical and cytochemical journals of this period were, with dates of first publication, *Experimental Cell Research* (1950), *Journal of Histochemistry and Cyto-*

chemistry (1953), *Acta Histochemica* (1954), *Journal of Biophysical and Biochemical Cytology* (1955), *Rivista di Istochimica* (1955), *Annales d'Histochimie* (1956), *Histochemie* (1958).

1959–65 During this period there was again a considerable increase in the number of new books, and new editions, devoted to Histochemistry. A small but effective volume entitled *Histochemical Technique* was produced by Casselman in 1959 and in the following year appeared three new texts. These were McManus and Mowry's *Staining Methods, Histologic and Histochemical*, the second edition of Mellors' *Analytical Cytology*, and the second edition of my own book. In 1962 Burstone's *Enzyme Histochemistry and its Application to the Study of Neoplasms* was published and in the same year Jensen's *Botanical Histochemistry, Principles and Practice*. In the following year appeared Barka and Anderson's *Histochemistry, Theory, Practice and Bibliography*. Further volumes and parts of volumes of the *Handbuch der Histochemie* appeared, spasmodically and in no particular order; these included Vol. II "Polysaccharide" (1962 and 1964), Vol. III "Nucleoproteide" (1959), Vol. V "Lipide" (1964), Vol. VII "Enzyme" (1960, 1962, 1963, 1964). Two further monographs, in the German language, were published in 1964. The first, written by two Hungarian authors, Kiszely and Pósalaky, was entitled *Mikrotechnische und Histochemische Untersuchungsmethoden*. This work was intended largely as an illustrated laboratory handbook as also was the second work, Spannhof's *Einführung in die Praxis der Histochemie*. A third, much augmented, edition of Lillie's classical text appeared in 1965. This was a veritable mine of information, much of which was unobtainable elsewhere. New journals of the period were fewer than in the preceding epoch. The Japanese Histochemical Association commenced, in 1960, the publication of an annual volume of its proceedings in the English tongue. In 1962 the title of the *Journal of Biophysical and Biochemical Cytology* was changed to the *Journal of Cell Biology* without change in the editorial policy giving priority to electron microscopy and electron cytochemistry. The Polish Histochemical Society, founded in 1961, produced the first volume of its journal *Folia Histochemica et Cytochemica* in 1963. This journal is published substantially but not exclusively in English. With the formation of sections for Histochemistry and Cytochemistry (1964) and for Electron Microscopy (1965), the Royal Microscopical Society expanded its journal in order to accept an increased number of papers in these fields. To mark the occasion of the 2nd International Congress of Histochemistry, held in Frankfurt in 1964, there was published a small volume edited by Sandritter entitled *Hundert Jahre Histochemie in Deutschland*. The English version, edited by Sandritter and Kasten, followed shortly after as a significant contribution to the History of Histochemistry.

In a series of papers entitled *Contributions to the History of Microchemistry* Harms (1931–32) gave a detailed account of the histochemical works of the French pharmacist, botanist and microscopist, François-Vincent Raspail (1794–1878), and he concluded, with the support of other competent observers, that Raspail should be regarded as the founder of the science of Microchemistry. These views were carried a stage further by Baker (1943, 1945) in a monograph of the Quekett Microscopical Club and in the first edition of his book on cytological technique. He too considered that Raspail was the real founder of Histochemistry. His claims to that title I regard as unassailable and his portrait therefore forms the frontispiece of this book. It shows Raspail as a young man, at the time when his greatest work was being done, and before he deserted the realms of botany and histochemistry for the field of politics. An excellent account of Raspail's life and works was given by Weiner (1959) in a monograph which concentrates largely on his extensive philanthropies.

The first clear appreciations of the science of microscopic tissue chemistry undoubtedly came from Raspail (1825a and b, 1829). This author, after formulating his four resolutions (quoted in full by Harms, 1931, and partly also by Baker, 1943), settled down to a study of the processes of fertilization in flowers and fruits of the *Graminaceæ*. The most important reaction which he used for this purpose was the iodine reaction for starch, first described by Colin and de Claubry in 1814, and by Stromeyer (1815), and employed in a microscopic study of starch grains by Caventou in 1826. It is not certain whether priority for the histochemical use of iodine solutions should go to Raspail or to Caventou; the point is not of great importance since the latter did not pursue his histochemical studies any further. Raspail, on the other hand, discovered and applied many other histochemical reactions which are still of mportance today. In 1829 he used the xanthoproteic reaction for protein and the hydrochloric acid (furfural) test for carbohydrate which became more widely known as the reaction of Liebermann (1887); the latter is not now employed in animal histochemistry. In applying sulphuric acid to his plant tissues to demonstrate the presence of protein he was, in fact, using the aldehyde method for tryptophan which, modified by many workers since his time, is still applied in histochemistry as the Voisenet-Fürth reaction. According to Reichl (1889) this type of benzylidene condensation reaction was used by Mikosch to demonstrate protein in plant tissues with a mixture of benzaldehyde, sulphuric acid and ferric sulphate.

Raspail is now usually credited with the discovery of micro-incineration (in 1829 according to Baker). He was also the first person to study the pH of protoplasm, using an indicator dye, turnsole, obtained from a species of sun spurge found in the Mediterranean region. This dye, normally blue, turned pink in acid solution.

Almost contemporary with Raspail's discoveries were those of several other botanists who published accounts of true histochemical reactions. Amongst these the work of Mohl (1831) on the iodine reaction, and of Schleiden (1838) on the iodine-sulphuric acid reaction may be mentioned. Apart from botanical work, however, progress in histochemistry was very slow and little work which can be described as histochemical, even in the broadest sense, was recorded until after 1860. Among the oldest published techniques are those for demonstrating iron, and some of these had their origin in the work of Vogel (1845, 1847), who detected iron in the tissues by its conversion to black ferrous sulphide with yellow sulphide of ammonia. In 1867 Perls introduced his Prussian-blue method for demonstrating iron which remains the method of choice up to the present day. Perls was followed in 1868 by Quinke, who used Vogel's sulphide method, and his technique also survives to the present day in practically unmodified form. In 1844 Millon described his reaction for proteins which Hoffmann, in 1853, showed was actually a test for tyrosine. This reaction was not used in histochemistry until 1888 (by Leitgeb), although Payen (1843) had already demonstrated nitrogenous substances in vegetable tissues with mercury proto-nitrate. The first recorded localization of starch in the chloroplast was by Sachs (1887) in his monograph on Plant Physiology.

Pigment histochemistry properly begins with the extensive studies of Virchow (1847) on the products of haemoglobin breakdown in the tissues. He was the first to use the term *hæmatoidin* for the yellow crystalline pigment appearing in areas of extravasation of blood. The term *hæmosiderin* was proposed by Neumann (1888) for an intracellular iron-containing pigment distinct from Virchow's hæmatoidin. Von Recklinghausen (1889) first described hæmofuscin and the term *melanin* was introduced by Langhans although Virchow (1859) had already described a black pigment in the cells of the central nervous system. The characteristics of melanin in tumours were investigated later by Berdez and Nencki (1886).

In 1850, though the result was not published until 1859, Claude Bernard performed his celebrated experiment by the injection into dogs of iron lactate and potassium ferrocyanide, locating the resulting Prussian blue, which developed in the presence of acid, not only in the gastric glands but on the surface of the gastric mucosa. Such *in vivo* techniques really belong to the domain of physiology but this particular example may justly be claimed for histochemistry. In 1850, also, Schulze first demonstrated his chlor-zinc-iodine method for cellulose. A description of this method, which is still in use today, was given by Fürnrohr (1850).

The use of enzymes for tissue digestion was first reported in 1861, by Beale, who used gastric juice in order to remove unwanted tissues from the nerve fibres which he was studying. As such, the technique was really microanatomical, but it developed by the end of the century into a well-recognized technique described as enzymal analysis (Kossel and Mathews, 1898).

Demonstration of the presence of enzymes in tissue began with the work of Klebs (1868) and Struve (1872), both of whom showed that tincture of guaiac gave a blue colour with pus, thus first recording the presence of peroxidases now well known to occur in the granules of the leucocytes. Brandenburg (1900) first demonstrated the peroxidase reaction in the latter site. Cytochrome oxidase was first demonstrated by Ehrlich (1885), though not, of course, under that name. He performed the "Nadi" reaction (Vol. II) *in vivo* by injecting α-naphthol and *p*-phenylenediamine into animals, observing the formation of indophenol blue in situations where "Nadi oxidase" was present.

Besides some of the methods mentioned above, during the period 1856–98 a number of other histochemical methods appeared. Heidenhain, in 1868, showed that ergastoplasm, the deeply basophil substance at the base of secreting gland cells, contained a material which could be precipitated with acetic acid. This is now recognized as ribonucleic acid. In 1870 he described the development of a brown colour in certain cells of the adrenal medulla when these were treated with chromic acid and this phenomenon is now called the chromaffin reaction. According to Lison, its discovery should be attributed to Henle in 1865. The use of enzymes for digestion is recorded during this period by Miescher (1871), who employed pepsin to free nuclei from cytoplasmic material, and by Stirling (1875) who isolated elastic fibres by means of digestion with gastric juice. These two were hardly histochemical techniques in the modern sense, but they serve to illustrate the destructive nature of much of the research into the chemistry of the tissues which was taking place at that time. It was this destructive element which caused Lison to separate the older histochemistry from his new non-destructive science.

In 1873 Miescher isolated nuclear chromatin by making use of its selective affinity for methyl green, and Ehrlich, 1878–79, observed the effects of heat coagulation in increasing the affinity of hæmoglobin for nitro dyes. This last work finds a modern echo in inquiry into the effects of denaturation on the combination of dyes or histochemical reagents with specific groups in the tissue proteins. During this same period (1856–98) the aniline dyes came into general histological use, following their introduction into the field by Bencke (1862). This caused a revolution in the practice of histology and provided a check to the progress of histochemistry. The various dyes were largely used without any attempt to correlate their performance with the chemical nature of the tissue components that were stained, although every endeavour was made to record the correlation of colour with structure in the histological sense. Notwithstanding this criticism, a considerable amount of work was done by a few authors in order to find out how the various stains attached themselves to the tissues. The physical theory of staining was upheld by Witt (1890–91) and particularly strongly by Fischer (1899), who explained all staining on the basis of absorption. Miescher and Ehrlich, and also Knecht (1888), believed that staining was a chemical process. Mann's (1902) comment on all this is interesting. The object of staining, he says, is first to determine morphological facts and second, "to recognize microchemically the existence and distribution of substances which we have been made aware of by macrochemical means." "It is not sufficient," he goes on, "to content ourselves with using acid and basic dyes and speculating on the basic or acid nature of the tissues, or to apply colour radicals with oxidizing or reducing properties; but we must endeavour to find staining reactions which will indicate not only the presence of certain elements such as iron or phosphorus, but the presence of organic complexes such as the carbohydrate group, the nucleins, protamines, and others." These remarks show that Mann, at least, was aware of the problem, but few practical attempts to meet it were made by histologists in general.

Some of the reactions involving the use of aniline dyes were in fact histochemical, although their significance was often unappreciated or wrongly appreciated. In most cases the significance is still not fully understood. A reaction in this last category is the metachromatic staining of amyloid with methyl violet, first described by Cornil (1875). Almost simultaneously Heschl (1875) and Jürgens (1875) were working on general problems of metachromasia. Ehrlich's reaction for mast-cells, using a saturated alcoholic solution of dahlia containing 8 per cent. acetic acid, is essentially similar to modern methods using the thiazine dyes and may well be considered histochemical. Other reactions whose mechanism was poorly understood were the myelin methods of Weigert (1884) and Marchi (1892) and the anilin-violet method of Gram (1884). The chemical theory of staining was strongly supported by Mathews (1898) in his experimental work with albumins and albumoses and important researches into the nucleohistones were conducted by Saint-Hilaire (1898), who noticed, in his attempts to evolve a method for uric acid in the tissues, that nuclei were occasionally stained. He concluded that the presence of histone was responsible for the nuclear reaction. Saint-Hilaire also showed that nucleohistones, precipitated in the tissues by acetic acid, were dissociable by means of dilute solutions of hydrochloric acid, leaving the histones *in situ*. Variations on this theme are widely employed at the present time.

Especially in the second half of the period we are considering, a great deal of work was done on the nature of protoplasm, particularly by Stöhr (1882), who, as the result of coagulation studies, concluded that a protein substance was present in gastric parietal cells, and by Flemming (1882), Kossel (1882, 1886), Altmann (1886, 1889), Schwarz (1887), and Mann (1890). Flemming (1876) had already described a cement substance holding together the fibrils which compose the loose connective tissue bundles. This he considered to be of mucinous nature. Altmann (in 1889) developed his method of fixation by freezing and drying, which has become an important modern tool of histochemistry in the hands of Gersh and his successors. Others continued to work on the chemical nature of staining and much of their work has modern applications. Among works of particular merit are those of Griesbach (1886), who postulated that tissue dye compounds should have properties differing from those of the free radical; of Unna (1887), who tried to confirm this; and of Lilienfeld (1893), who investigated the staining of mucins. Hoyer (1890) demonstrated metachromasia in the cells of the mucous salivary glands with thiazine dyes and, following his discovery, metachromatic methods for mucin became quite popular. Before Hoyer published his paper mucins were usually stained by techniques making use of their strong basophilia. List (1885) was the first to use Bismarck brown for this purpose and the same dye, in alcoholic solution, was used by Hardy and Wesbrook (1895) to stain water-soluble mucoproteins and mucopolysaccharides. In the same year, Heine (1895) made some important observations on the nature of chromatin. He observed that segments of this substance were intensely stained by Millon's reagent and tried, unsuccessfully, to distinguish between nucleoproteins and nucleic acids using mixtures of methyl green and rubin S. Also in 1895, Macallum demonstrated that, after treatment with sulphuric acid, the nuclei were stainable by the usual methods for iron in the tissues. Since this time the question of whether this iron is really present in the nuclei, or adsorbed from elsewhere, has been debated at length.

Enzyme methods of the closing years of the nineteenth century are represented by the contributions of Mall (1891), who investigated the swelling of collagen in various solutions and the action upon it of crude preparations of pepsin and trypsin. Nothing approaching the modern concept of enzymal analysis was achieved at this time. Daddi, in 1896, first used Sudan III for the *in vitro* staining of fat which, after being subsequently ingested by animals, was demonstrated in the tissues by its red colour. Sudan IV was proposed as a fat stain by Michaelis in 1901. This author showed that the staining of fats with Sudan dyes was purely physical, depending on solution of the inert dyes in the fats themselves.

Among methods for revealing inorganic salts in the tissues may be mentioned the techniques of Molisch (1893) who stained tissue-iron by converting it to the red thiocyanate, and of Lilienfeld and Monti (1892), who evolved an ammonium molybdate technique for demonstrating phosphate. This method was modified by Pollacci (1900) and it has been further modified by other workers in the twentieth century. The forerunner of a number of very similar techniques for demonstrating metal salts in the tissues was de Michele's (1891) method for mercury, which he converted to its sulphide by means of H_2S. The method which we still use to demonstrate the presence of calcium in the tissues was described by von Kóssa in 1901; even then it was preceded by the more specific "gypsum" method, described by Schujeninoff (1897), which also survives to the present day.

It is probable that this brief account of the history of histochemistry has failed to give the reader a true sense of the continuity of the science from decade to decade. This is at least partly due to the fact that real continuity was lacking. Nevertheless, it is to be hoped that I have been able to show that the science of histochemistry as formulated by Raspail, and the principles expressed by that great man, continued without serious interruption from the 1820's to the twentieth century. When the nineteenth century came to a close, the majority of histologists were occupied in reaping the rich harvest presented by new developments in the art of staining and few had time to spare for histochemistry. The subject therefore remained for the most part in abeyance, though kept alive by a few practitioners until its revival in the 1930's and its establishment on a modern footing as an independent branch of histology by Lison with his great work *Histochimie Animale*. Since his time expansion has been rapid and progress almost equally so, until now, in the second half of the twentieth century it is to be hoped that the majority of histologists would wish to "recognize micro-chemically" some at least of the substances in the tissues with which they deal.

REFERENCES

ALTMANN, R. (1886). "Studien über die Zelle." Leipzig.
ALTMANN, R. (1889). *Arch. Anat. Physiol. Lpz. Physiol. Abt.*, p. 524.
BAKER, J. R. (1943). *J. Quekett micr. Cl.*, 4 ser., **1**, 256.
BAKER, J. R. (1945). "Cytological Technique." 2nd Ed. Methuen, London.
BARKA, T., and ANDERSON, P. J. (1963). "Histochemistry. Theory, Practice and Bibliography." Hoeber, New York.
BEALE, L.S. (1861). *Arch. Med. Lond.*, **2**, 179.
BENCKE, A. (1862). *Korrespbl. Ver. Gemeinsch. Arbeiten*, **59**, 980.
BERDEZ, I., and Nencki, M. (1886). *Arch. exp. Path. Pharm.*, **20**, 346.
BERNARD, C. (1859). "Leçons sur les Propriétés Physiologiques et les Altérations Pathologiques des Liquides de l'Organisme." Vol. 2. Paris.
BOURNE, G. H. (1953). "An Introduction to Functional Histology." Churchill, London.
BRANDENBURG, K. (1900). *Münch. med Wschr.*, **47**, 183.
BUNGE, G. (1887). "Lehrbuch der physiologischen und pathologischen Chemie." F. C. W. Vogel, Leipzig.
BURSTONE, M. S. (1962). "Enzyme Histochemistry and its Application in the Study of Neoplasms." Academic Press, New York.
CASSELMAN, W. G. B. (1959). "Histochemical Technique." Methuen, London.
CAVENTOU, J. B. (1826). *Ann. Chim. Phys.*, **31**, 337.
CHAMOT, E. M., and MASON, C. W. (1930). "Handbook of Chemical Microscopy." New York.
COLIN, J. J., and DE CLAUBRY, H. G. (1814). *Ann. Chim.*, **90**, 87.
CORNIL, V. (1875). *C. R. Acad. Sci., Paris*, **80**, 1288.
DADDI, L. (1896). *Arch. ital. Biol.*, **26**, 143.
DANIELLI, J. F. (1953). "Cytochemistry." Wiley & Sons, New York; Chapman and Hall, London.
EHRLICH, P. (1878–79). *Verh. physiol. Ges. Berlin*, **20**.
EHRLICH, P. (1885). "Das Sauerstoff-Bedürfniss des Organismus." Berlin.
EHRLICH, P. (1903). "Encyclopadie der Mikroskopischen Technik." Urban and Schwarzenberg, Berlin and Wien.
ERÄNKÖ, O. (1955). "Quantitative Methods in Histology and Microscopic Histochemistry." Karger, Basel and New York.
FISCHER, A. (1899). "Fixierung, Färbung und Bau des Protoplasmas." Leipzig.
FLEMMING, W. (1876). *Arch. mikr. Anat.*, **12**, 434.
FLEMMING, W. (1882). "Zellsubstanz, Kern und Zelltheilung." Leipzig.
FREY, H. (1867). "Handbuch der Histologie und Histochemie des Menschen." Leipzig.
FREY, H. (1874). "Histology and Histochemistry of Man." Trans. Baker, A. F. J. London.
FÜRNROHR, R. (1850). *Flora*, **8**, 641.
GLICK, D. (1949). "Techniques of Histo- and Cyto-chemistry." Interscience, New York.
GOMORI, G. (1952). "Microscopic Histochemistry." Chicago University Press.
GRAM, C. (1884). *Fortschr. med.*, **2**, 185.

GRAUMANN, W., and NEUMANN, K. (Eds.) (1958). "Handbuch der Histochemie", Vol I, Allgemeine methodik. G. Fischer, Stuttgart.
GRIESBACH, H. (1886). *Z. wiss. Mikr.*, **3**, 358.
HARDY, W. B., and WESBROOK, F. F. (1895). *J. Physiol.*, **18**, 490.
HARMS, H. (1931). *Apotheker-Zeitung.*, No. 90, p. 1454.
HARMS, H. (1932). *Ibid.*, Nos. 83, p. 1274; 84, p. 1293; 85, p. 1307; and 86, p. 1324.
HEIDENHAIN, R. (1868). *Stud physiol. Inst.*, *Breslau*, **4**, 88.
HEIDENHAIN, R. (1870). *Arch. mikr. Anat.*, **6**, 368.
HEINE, L. (1895–96). *Hoppe-Seyl. Z.*, **21**, 494.
HERTWIG, G. (1929). "Histochemische Methoden." In von Möllendorff's "Handbuch der mikroskopische Anatomie des Menschen." Berlin.
HESCHL, A. (1875). *Wiener med. Woch.*, **25**, 714.
HOFFMANN, R. (1853). *Ann. Chem. Pharm.*, **87**, 123.
HOYER, H. (1890). *Arch. mikr. Anat.*, **36**, 310.
JENSEN, W. A. (1962). "Botanical Histochemistry, Principles and Practice." W. H. Freeman, San Francisco.
JURGENS, R. (1875). *Virchows Arch.*, **65**, 189.
KISZELY, G., and PÓSALAKY, Z. (1964). "Mikrotechnische und Histochemische Untersuchungsmethoden." Akadémiai Kiadó, Budapest.
KLEBS, E. (1868). *Z. med. Wiss.*, **6**, 417.
KLEIN, G. (1929). "Prakticum der Histochemie." Berlin.
KNECHT, E. (1888). *Ber. deutsch. chem. Ges.*, **21**, 1556.
KÓSSA, J. VON (1901). *Beitr. path. Anat.*, **29**, 163.
KOSSEL, A. (1882). *Hoppe-Seyl. Z.*, **7**, 7.
KOSSEL, A. (1886). *Ibid.*, **10**, 248.
KOSSEL, A., and MATHEWS, A. (1898). *Ibid.*, **25**, 190.
LANGHANS, TH., (1870). *Virchows Arch.*, **49**, 117.
LEHMANN, C. G. (1842). "Lehrbuch der physiologischen Chemie." Leipzig.
LEHMANN, C. G. (1851). "Physiological Chemistry." Trans. Day, G. E. London.
LEHMANN, C. G. (1883). "Traité de Chimie Physiologique." Paris.
LEITGEB, H. (1888). *Mitt. bot. Inst.*, *Graz*, p. 113.
LIEBERMANN, L. (1887). *Zbl. med. Wiss.*, **25**, 321.
LILIENFELD, L. (1893). *Arch. Anat. Physiol.*, *Lpz.* Physiol. Abt., p. 554.
LILIENFELD, L., and MONTI, A. (1892). *Z. wiss. Mikr.*, **9**, 332.
LILLIE, R. D. (1954). "Histopathologie Technic and Practical Histochemistry." 2nd Edn. Blakiston Co., New York (3rd Edn. 1965).
LINDERSTRØM-LANG, K. (1936). *Arch. exp. Zellforsch.*, **19**, 231.
LISON, L. (1936). "Histochimie Animale." Gautier-Villars, Paris.
LISON, L. (1953). "Histochimie et Cytochimie Animales." Gautier-Villars, Paris.
LISON, L. (1960). "Histochimie et Cytochimie Animales." Vols. I and II, Gautier-Villars, Paris.
LIST, J. H. (1885). *Z. wiss. Mikr.*, **2**, 145.
MACALLUM, A. B. (1895). *Quart. J. Micr. Sci.*, **38**, 175.
MACALLUM, A. B. (1908). *Ergebn. Physiol.*, **7**, 552.
MCMANUS, J. F. A., and MOWRY, R. W. (1960). "Staining Methods, Histologic and Histochemical." Hoeber, New York.
MALL, F. (1891). *Abh. sächs. Ges.* (*Akad.*) *Math-phys.*, **17**, 299.
MANN, G. (1890). *Trans bot. Soc.*, *Edinb.*, **18**, 429.
MANN, G. (1902). "Physiological Histology." London.
MARCHI, V. (1892). *Arch. ital. Biol.*, **17**, 191.
MATHEWS, A. (1898). *Amer. J. Physiol.*, **1**, 445.
MELLORS, R. C. (1955). "Analytical Cytology." McGraw Hill, New York.
MELLORS, R. C. (1960). "Analytical Cytology", 2nd Edn. McGraw Hill, New York.
MICHAELIS, L. (1901). *Virchows Arch.*, **164**, 263.
DE MICHELE, S. (1891). *La Riforma Medica*, p. 169.
MIESCHER, F. (1871). *Hoppe-Seyl. med.-chem. Untersuch.*, p. 441.
MIESCHER, F. (1873). *Verh. naturf. Ges.*, *Basel*, **6**, 138.
MIESCHER, F. (1874). *Verh. naturf. Ges.*, *Basel.*, **6**, 138.
MILLON, A. N. E. (1844). *C. R. Acad. Sci.*, *Paris*, **18**, 1041.

MOHL, H. (1831). *Flora*, **15**, 417.
MOLISCH, H. (1893). *Ber. deutsch. bot. Ges.*, **11**, 73.
MOLISCH, F. (1913). "Mikrochemie der Pflanzen." Fischer, Jena.
NEUMANN, E. (1888). *Virchows Arch.*, **111**, 25.
PARAT, M. (1927). *Biol. Rev.*, **2**, 285.
PATZELT, V. (1928). In "Fortschritte der Mikrochemie." Eds. Klein, G., and Strebinger, R. Vienna.
PAYEN, A. (1843). *Mem. Acad. Sci., Paris*, p. 163.
PEARSE, A. G. E. (1951). *J. clin. Path.*, **4**, 1.
PERLS, M. (1867). *Virchows Arch.*, **39**, 42.
POLICARD, A., and OKKELS, H. (1932). "Die Mikroveraschung als histochemische Hilfsmethode." In Abderhalden's "Handbuch der biologischen Arbeitsmethoden." Berlin.
POLLACCI, G. (1900). *Att. Ist. bot. Univ. Pavia*, 2 ser., **6**, 15.
PRENANT, A. (1910). *J. anat. Physiol.*, **46**, 343.
QUINKE, H. I. (1868). Referred to in *Arch. exp. Path. Pharmakol., Leipzig*, **37**, 183, 1895–96.
RASPAIL, F. V. (1825a). *Ann. Sci. nat.*, **6**, 224.
RASPAIL, F. V. (1825b). *Ibid.*, **6**, 384.
RASPAIL, F. V. (1829). *Ann. Sci. Observation*, **1**, 72.
RASPAIL, F. V. (1830). "Essai de Chimie Microscopique Appliquée à la Physiologie." Paris.
RASPAIL, F. V. (1833). "Nouveau Système de Chimie Organique." Paris.
RASPAIL, F. V. (1834). "A New System of Organic Chemistry." Trans. Henderson, W. London.
RECKLINGHAUSEN, F. D. VON (1889). "Tagebuch der 62." Versammlung Dtsch. Naturf. Ärzte, S324, Heidelberg.
REICHL, C. (1889). *Monatschr. Chem.*, **10**, 317.
ROBIN, C., and VERDEIL, F. (1853). "Traité de Chimie Anatomique." J. B. Baillière, Paris.
ROMEIS, B. (1932). "Histochemische Methoden." In Boehm and Oppel's "Taschenbuch der mikroskopischen Technik." Munich.
SACHS, J. (1865). "Handbuch der Experimental Physiologie der Pflanzen." Engelman, Leipzig.
SACHS, J. (1887). "Lectures on the Physiology of Plants." Clarendon Press, Oxford.
SAINT-HILAIRE, C. (1898). *Hoppe-Seyl. Z.*, **26**, 102.
SANDRITTER, W. (Ed.) (1964). "Hundert Jahre Histochemie in Deutschland." F. K. Schattauer, Stuttgart.
SCHLEIDEN, M. J. (1838). *Ann. phys. Chem.*, **43**, 391.
SCHLOSSBERGER, J. E. (1856). "Die Chemie der Gewebe des gesammten Thierreichs." Leipzig.
SCHUJENINOFF, S. (1897). *Z. Heilk.*, **18**, 79.
SCHWARZ, F. (1887). *Beitr. Biol. Pfl.*, **5**, 1.
SPANNHOF, L. (1964). "Einführung in die Praxis der Histochemie." Gustav Fischer, Jena·
STIRLING, W. (1875). *J. Anat. Physiol., Leipzig*, **10**, 185.
STÖHR, P. (1882). *Arch. mikr. Anat.*, **20**, 221.
STROMEYER, H. (1815). *Ann. Phsiol., Leipzig.*, **19**, 146.
STRUVE, H. (1872). *Liebigs Ann. chem. Pharm.*, **163**, 160.
UNNA, P. G. (1887). *Arch. mikr. Anat.*, **30**, 39.
VIALLI, M. (1955). "Introducione alla Ricerca in Istochimica." I.P.L. Milano.
VIRCHOW, R. (1847). *Virchows Arch.*, **1**, 379.
VIRCHOW, R. (1859). *Virchows Arch.*, **16**, 180.
VOGEL, J. (1845). "Pathologische Anatomie des menschlichen Körpers." Leipzig.
VOGEL, J. (1847). "Pathological Anatomy of the Human Body." Trans. Day, G. E. London, p. 346.
WEIGERT, C. (1884). *Fortschr. Med.*, **2**, 190.
WEINER, D. B. (1959). *French Historical Studies*, **1**, 149.
WITT, O. N. (1890–91). "Färberzeitung." Quoted by Mann, G., 1902.

CHAPTER 2

COLD KNIFE AND COLD MICROTOME (CRYOSTAT) METHODS

THE production of fresh sections in large numbers is now a prime requirement not only in histochemical laboratories but in routine histological departments and, increasingly, in laboratories devoted to biological studies.

The advent of the refrigerated cryostat (1951), and of remotely controlled instruments (1954 onwards), brought about considerable changes in the attitude of histologists towards fresh sections. It became possible to cut thin (2 μ) serial sections from most tissues in unlimited numbers and their handling became a matter of great ease. Alternatives to the cryostat were considered in the 2nd Edition of this book (1960) and these, together with improved modern versions, will be considered in this chapter under the heading of cold knife methods. The remainder of the chapter is devoted to the cryostat and its development.

Cold Knife Methods

Freezing microtomes provided with means for cooling the knife blade with an additional CO_2 jet were introduced into histological technique by Schulze-Brauns (1931a and b, 1932). Commercial models became available in the post-war years and they were used by many workers (e.g. Wachstein and Meisel, 1953) for the provision of semi-serial sections for histochemical use. Difficulties attending the use of such microtomes in practice resulted in a search for better means of controlling the large numbers of variable factors which play a part in the production of perfect fresh sections thin enough to allow detailed studies to be made. The various procedures and forms of apparatus which have been developed for this purpose will be discussed below in some detail.

Although the use of standard freezing microtomes with knife-cooling attachments comes under the heading of cold knife methods, it is now customary to use the term to refer to the type of method developed by Adamstone and Taylor (1947, 1948). These authors used a Leitz base sledge microtome with the knife blade cooled continuously by the application of dry ice (solid CO_2), held on the knife surface by retaining fences on either side of the cutting area. With the Adamstone-Taylor technique, as with previous methods using the cold knife principle, curling or rolling of the sections was prevented by gently stroking the section upwards on the knife (with moving knife microtomes) by means of a fine camel-hair brush. The flattened section was lifted from the knife by means of a double-bottomed metal scoop, containing dry ice in its lower half, and transferred to the surface of a clean glass slide. At the moment melting began, and the section commenced to spread on the slide, the two were immersed in a Coplin jar containing either a fixative or some specific reagent in solution. Adamstone and Taylor stressed particularly the necessity to avoid complete melting on the slide if fine structure was to be preserved, and they described the rupture and distortion of cells which occurred if such precautions were not observed. With this in mind, Adamstone (1951) produced an automatic device which immersed the sections in the chosen fixative at precisely the right moment. With a single operator such a device was almost essential and, as the original authors observed, the whole method was much more efficient if two operators were available.

The rapidity with which reagents, including fixatives, can be applied to tissue sections constitutes an important advantage of the Adamstone-Taylor method as does the saving of time by comparison with standard techniques of dehydration and embedding. A severe disadvantage, as pointed out by Taft (1949) is the dependence of the method on atmospheric conditions. Air temperatures over 18° (65° F.) and excessive humidity interfere with its successful performance so that a cool or cold room should be employed if available.

The Adamstone-Taylor method has, in fact, been favourably received on the whole and it is now widely used throughout the world. The full routine, however, which includes the rapid fixation procedures, is very seldom employed. White and Allen (1951) reported a successful modification of the Bausch and Lomb rotary freezing microtome for cold knife sections. They were able to cut serial sections 6–8 μ thick in almost unbroken series. Personal experience of this excellent microtome has shown that such a performance is well within its capabilities. A device for converting rotary microtomes to CO_2-freezing was described by Strike (1962). A most interesting adaptation was described by Shimizu, Kubo and Morikawa (1956), who employed a Schanze sliding microtome, with the knife cooled on either side by blocks of dry ice. In place of the customary camel-hair brush they used a device for preventing the curling of sections (Fig. 1) based on the principle evolved by

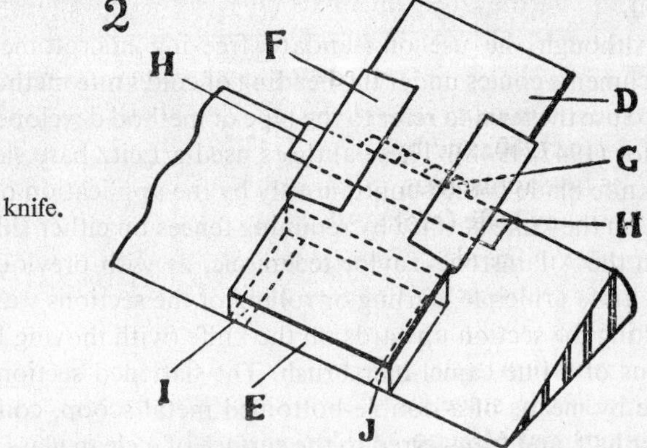

Fig. 1.

D. Glass slide.
E. Cutting edge of knife.
F. Slide holder.
G. Hinge.
H. Hinge clamp.
I.J. Distance pieces.

Linderstrøm-Lang and Mogensen (1938). These devices are now usually described by the term anti-roll plate or guide plate. The whole microtome was placed in a domestic refrigerator (5 cubic feet) whose front was replaced by a wooden door through which passed two gloved armholes for the operator and the CO_2 supply to the microtome. Interior illumination was provided and a small viewing window. With a similar apparatus Thornburg and Mengers (1957) carried out some important studies on optimum conditions for fresh frozen section cutting, using a Spencer sliding microtome incorporated in a modified kitchen type refrigerator. Their apparatus is shown in Figs. 2 and 3. The provision of environmental control in the manner described above certainly overcomes the chief objection to the method of Adamstone and Taylor, that is to say, its sensitivity to atmospheric conditions.

Improved Cold Knife Apparatus

Refrigerated Devices. A considerably greater degree of control of knife and tissue block temperature was obtained by the substitution of compression refrigeration for the customary CO_2-expansion. Batsakis *et al.* (1963) described knife cooler and freezing head attachments for the standard Spencer freezing microtome which were cooled by Freon 12 in a floor-mounted compressor. Commercial equipment of this type has been available for some years but the principle was made obsolete, before it had properly been developed, by the application of thermoelectric devices to cold microtomy.

Thermoelectric Devices. The principle on which these devices function is known as the Peltier effect, after the French watchmaker who first discovered it in 1834. Peltier found that when an electrical current was passed between two different conductors there was either absorption or generation of heat at the junction, depending on the direction of the current. The effect, which is thus easily reversible, is therefore complementary to the so-called Seeback effect (production of an electric current by temperature changes in the junctions between dissimilar conductors). Although the Seeback effect was used for measurement of temperature, proper application of the Peltier effect had to wait until the discovery of new semiconducting thermoelements made possible a great rise in efficiency.

The two above-mentioned effects are described, simply and with clarity, in two contributions by Ioffe (1957, 1958), in "Applications of Thermoelectricity" (Goldsmit, 1960) and by Parrott and Penn (1962).

The first application of thermoelectricity to microtome stage cooling was illustrated by Ioffe (1958) but there was considerable delay in the general development of the idea. Brief descriptions of the use of thermoelectric modules for frozen sectioning were given by Brown and Dilly (1961) and by Hardy and Rutherford (1962). In this year also, as described in Chapter 3, p. 45, the first application of the Peltier effect to freeze drying was made. A water-cooled thermoelectric stage for the Schanze sliding microtome was described by Okamoto and Mizuno (1963) and later (1964) Rutherford *et al.* gave a full description of the application of thermoelectric cooling to both stage and knife, for a variety of freezing microtomes.

Tests carried out in my laboratory with essentially similar water-cooled thermoelements attached to an M.S.E. base sledge microtome indicated that while serial frozen sections of formalin-fixed tissues could readily be obtained, with or without operation of the knife cooler, it was less easy to cut and handle sections from fresh tissue blocks. Some improvement was brought about by the development of a thermoelectric anti-roll plate but trouble was still experienced with condensation of moisture. In my opinion (see below) a cold dry environment is an essential condition for fresh tissue sectioning. A completely thermoelectric assembly (De La Rue Frigistor Ltd.) is shown in Fig. 4, opposite.

Handling of Cold Knife Sections

When it comes to the handling of cold knife sections, there are wide differences of opinion and usage. The two main alternatives are: (1) to pick up the section on a glass slide, prepared or otherwise, or (2) to take up the section with some instrument, cold or otherwise, before transferring it to another situation. Adamstone and Taylor gave the strongest arguments in favour of applying the fixative or other reagent directly to the still frozen sections, as mentioned above. Wachstein and Meisel, on the other hand, found it difficult to fix the still frozen sections to the slides and they therefore transferred them "while thawing" to whatever incubation or staining medium it was intended to employ. Shimizu and his collaborators transferred their sections to slides with a dissecting enedle, removed them from the refrigerator, and allowed them to dry at room temperature. Thornburg and Mengers preferred to use the Adamstone technique with an automatic section dropper installed in their refrigerator by the side of the microtome. Many workers, myself included, developed the habit of carrying cold knife sections directly into incubating media by means of cold section lifters or camel-hair brushes. Further discussion on the handling of sections may be deferred until after the section on cold microtomes which follows, since the problems involved are essentially the same.

Cold Microtome Methods (Cryostats)

In the 2nd Edition of this book (1960) I used the term cold microtome in place of, or as well as, the more customary term cryostat for two reasons. First, the essential difference between the methods of this section and those of the previous one was that the microtome, the knife, and the tissue block were all at the same temperature (—12° to —22°). Secondly, the term fitted into line with the accepted terms cold knife and cold knife sections, so that one could refer to the cold microtome and cold microtome sections. The alternatives, cryostat and cryostat sections, were regarded as perfectly acceptable, but the cryostat (literally a device for maintaining cold, *viz.* a refrigerator) is not

FIG. 2. Domestic refrigerator converted as a microtome chamber with glove holes and lucite window.

FIG. 3. Interior view showing freezing microtome (sliding type) and slide-dropper for instantaneous fixation (Adamstone-Taylor).

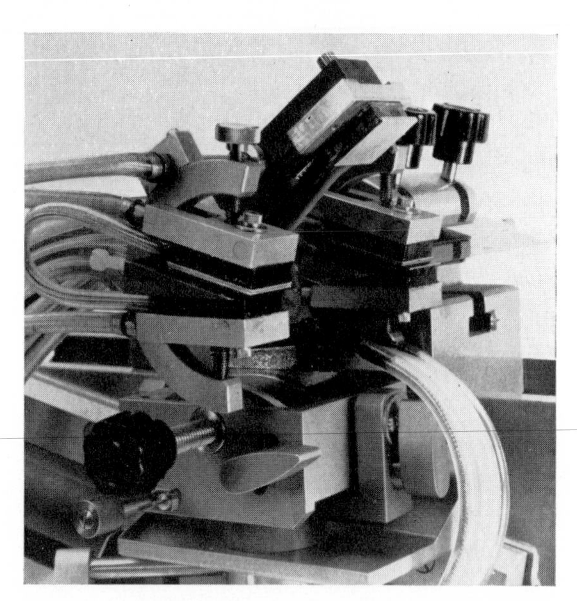

FIG. 4. Fully thermoelectrically controlled apparatus (De La Rue Frigistor Ltd.) attached to a base sledge microtome. Block, knife and anti-roll plate are all provided with cooling modules.

FIG. 5. Commercial cryostat (Dittes, Heidelberg). A Jung rotary microtome is operated by inserting the hands through side ports. The cutting stroke is externally operated by an assistant.

primarily concerned with the production of sections. Moreover, the term is used in cryogenics to describe a number of different devices. By this definition the cold microtome differs from the cold knife with cold environment (Shimizu *et al*, Thornburg and Mengers) simply in the fact that sections are cut without the provision of cold from outside sources and, usually without the provision of a temperature gradient between the knife and the tissues. The cold knife in a cold environment is therefore to be regarded for all practical purposes as a cold microtome. Despite the foregoing arguments the term cryostat has clearly come to stay, both in histochemistry and histology. The companion term cryotomy is much less popular. These matters, and many others concerning the development and use of cryostats, are fully discussed by David and Brown (1966).

The first cryostat was developed in Denmark by Linderstrøm-Lang and Mogensen (1938) for their studies in quantitative histochemistry in which alternate frozen sections were taken for chemical estimation and for histology. In the original cabinet cold was maintained by blocks of dry ice, and from this to the provision of refrigerating coils was a simple step. The microtome was modified by the provision of a glass plate in front of the knife which effectively prevented rolling or curling of sections. No real attempt was made to apply the cold microtome technique thus developed by the Danish authors to histochemical techniques in general. In 1951 Coons and his collaborators (Coons, Leduc and Kaplan) made a number of technical improvements in designing a new cryostat of some 6 cubic feet capacity. This model was made commercially in the United States by the Harris Refrigeration Company of Cambridge, Mass., and in Britain by the Prestcold Company at Cowley, Oxford. The Coons cryostat was originally designed for working temperatures between —16° and —18°, though a temperature as low as —30° could be reached with later models employing more powerful refrigerating machinery. In the sides of the cryostat gloved armholes were provided through which the microtome was operated. The interior was illuminated and provided with a fan (switched off while cutting is in progress) to maintain the circulation of cold air. Below the window a door gave access to the chamber. Coons-type cryostats are still produced in Germany by Dittes, Heidelberg (Fig. 5) and by Phywe A. G., Göttingen. Jung rust-proof rotary microtomes are usually employed although Dittes also manufactures a large glove-box cryostat into which any known microtome can be placed.

The microtome still most commonly used in cryostats in the United States is a rotary machine of the Minot type (International Harris), and this is constructed of stainless steel. Alternative microtomes in the U.S. are the Lipshaw rotary and the Lab-Tek rotary instruments which are constructed from stainless steel and bronze. Normally, rotary microtomes are made with such small tolerances that they are acutely sensitive to cold, and stiffening of the whole mechanism is experienced when these microtomes are employed in a cryostat. They are, in fact, far from ideal and, when operated in "open-top" types of cryostat, must normally be placed on an inclined plane to allow free access to the cutting head. This means that the knife surface becomes nearly horizontal, with consequent loss of the beneficial effects of gravity on the cut section. An exception is the Lab-Tek microtome whose compact design allows it to be used in the normal position even within an open-top cryostat.

The simpler mechanism of the Cambridge Rocking Microtome, and of the Slee Retracting Rocker has made these the microtomes of choice for cryostats designed and manufactured in Britain. The knife edge system of rocker microtomes is insensitive to cold of any degree remotely likely to be atained within the cryostat chamber.

Anti-roll Plates

An essential part of the Coons cold microtome was the ingeniously designed "glass window", applied to the knife to prevent curling up of the sections during cutting. The Coons type of anti-roll plate is illustrated in Fig. 6. It includes adjusting screws which allow height, angle and distance from

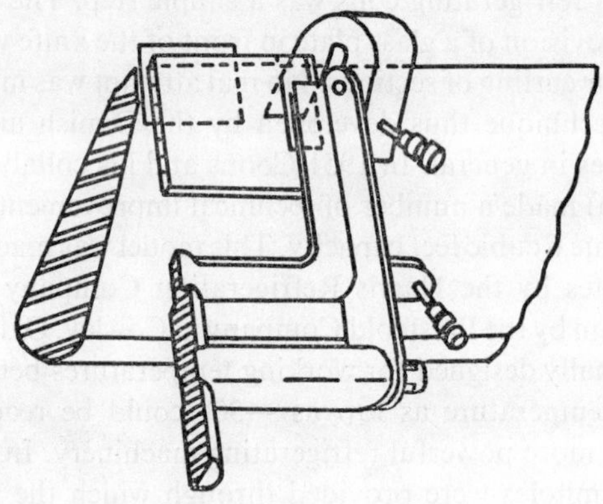

Fig. 6.

the knife edge to be regulated. A rotating pin allows the window assembly to be swung aside after the section has been cut so that it may be picked up on a cold slide kept in readiness in the cryostat. *Correct adjustment of the anti-roll plate is second in importance only to the condition of the knife edge in cutting good thin sections on the cold microtome.*

After carrying out a number of experiments with various materials having a low coefficient of friction, in place of the glass microscope slide usually employed, I found the most suitable material to be polytetrafluoroethylene (Fluon, Teflon). This material had several drawbacks. It was translucent but not transparent and possessed a built-in "memory" which caused the originally flat plate to warp and bend. The advent of the teflon spray and its application to glass anti-roll plates (West, 1962) provides the best possible compromise. It is necessary to renew such plates at regular intervals but their manufacture is the matter of a few minutes work only (see Appendix 2, p. 591). For those who

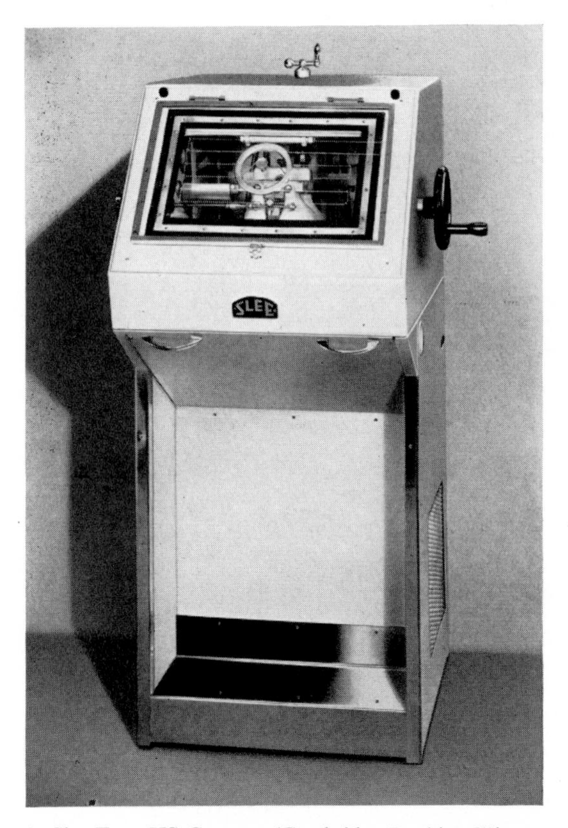

FIG. 8. Slee Type HS Cryostat (Cambridge Rocking Microtome).

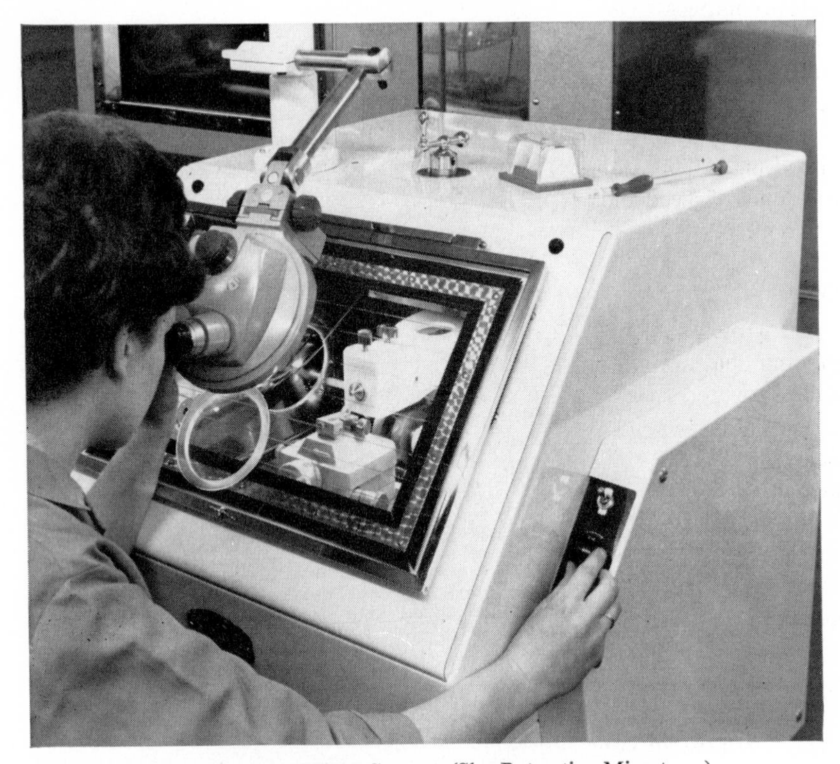

FIG. 9. Slee Type HRM Cryostat (Slee Retracting Microtome).

prefer unbreakable plates there can be little objection to perspex (lucite), as described by Appleton (1964).

More complicated modifications to anti-roll plates, such as those suggested by Iseri *et al* (1961) for the Cambridge Rocking Microtome, are now probably unnecessary. On the other hand, some form of modification is essential in the case of the standard anti-roll device for the International microtome (Fand, 1964). There are still some who advocate the use of the brush in place of the anti-roll plate. This truly constitutes a return to prehistoric technology. On the grounds of speed and efficiency alone, there can be no comparison between the two methods.

Development of the Remotely Controlled Cryostat

As I wrote in the 2nd Edition of this book (1960), comparison of the cold knife and cold microtome methods for the production of large numbers of thin fresh frozen sections for enzyme histochemistry convinced me that only by means of the second could technical efficiency be raised to the point where blocks from all types of tissues could be successfully cut on demand and processed accordingly. I see no reason, at the present time, to alter this view. Bearing in mind the ever-present need to apply the techniques of modern enzyme histochemistry to surgical and autopsy specimens, whose time of arrival could neither be controlled nor predicted, a constant running, remotely controlled, cold microtome and cryostat assembly was designed and produced (in 1954).

The following points were intended to be satisfied:

(1) Low initial cost and constant running with minimum attention.

(2) Instant availability for rapid diagnostic biopsies.

(3) Thermostatic control between —5° and —20°.

(4) Minimum temperature below —20°.

(5) *Automatic* delivery of 8–30 μ *serial* sections directly into incubating media or fixatives.

(6) Production of 4–30 μ *serial* sections, singly or in pairs or triples, for handling on slides or coverslips.

A diagram of the original apparatus is shown in Fig. 7, p. 20. While the essential features remain the same today, there have been a number of modifications in the intervening years. Most important of these is the replacement of the glass or perspex double window, which now has a central 3-inch operating port, by a single layer of perspex containing thin heating wires which carry sufficient current to prevent condensation completely both inside and outside the window.

Many other modifications have culminated in the present design of the standard Slee type H and type HS cryostats, which continue to incorporate the Cambridge Rocking microtome. A full view of one of these cabinets is shown in Fig. 8, facing p. 17. Alternative designs are available containing the Slee Retracting microtome (Fig. 9, facing p. 18), the International microtome and either the Leitz or the M.S.E. Base Sledge microtome (Fig. 10, facing

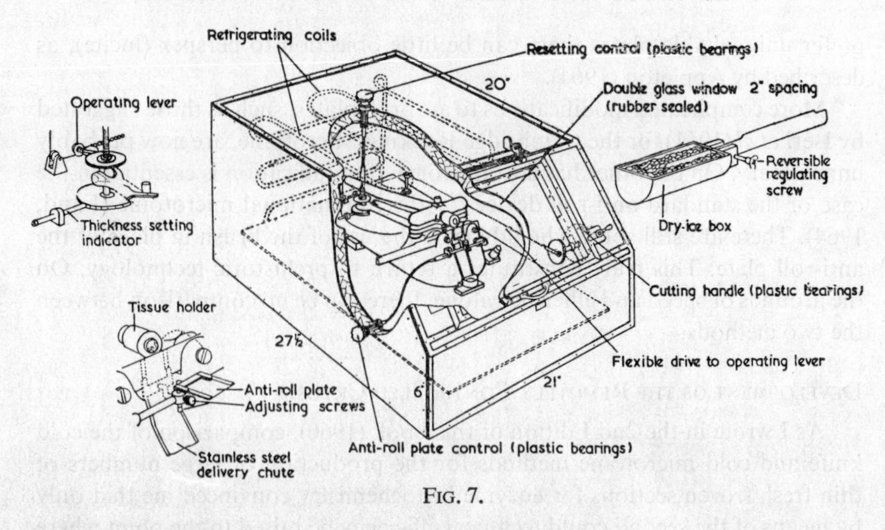

FIG. 7.

p. 21). This last is fitted with a driving motor since the weight of the moving component of the microtome precludes hand operation.

The Slee HR is also available in motorized form (HRM), and, alternatively, with a fine cutting modification of the microtome enclosed in a —70° chamber. This last provides 0·25 to 0·5 μ fresh sections for electron microscopy.

With one or other of this wide range of machines it is possible to cut blocks of almost any type of animal tissue, not excluding bone, from the smallest size up to 10 × 12 cm. Many plant tissues can also be cut in the cryostat without alteration of the customary conditions for animal tissues.

Quenching Techniques

Methods available for initial cooling of the specimens fall into five groups. In the first group are devices (cold metal blocks), cooled by refrigerating coils and situated either inside or outside the cryostat. The second group consists of CO_2-expansion coolers, situated usually on the bench or directly attached to the gas cylinder. The third group of methods is based on dry ice (solid CO_2), either applied directly to the tissue holder or incorporated with some organic solvent as a freezing mixture. The fourth group includes the latest thermo-electrically cooled tissue holders and the fifth comprises the liquid gases, of which liquid N_2 is the safest and most popular. These gases are used alone, or as agents to cool a suitable organic liquid (see Chapter 3, p. 31).

With mammalian tissues the speed of freezing is critical and, although for a few it can be relatively slow, the majority are badly affected if the duration of freezing is more than a few seconds. Of the above-listed methods, those in group one (cold blocks) are too slow for use under any cirumstances. Similarly, removal of heat by thermoelectric devices is at present too slow for effective quenching. Very effective are CO_2-expansion coolers, especially when these incorporate a covering device so that the tissue is instantly smothered by CO_2

FIG. 10. Slee Type MS Cryostat (M.S.E., or Leitz, Base Sledge Microtome; Motorised).

FIG. 11. Stainless steel dry ice box for attachment to Heiffor pattern knives used in Pearse-Slee cold microtome.

snow; somewhat less effective is solid CO_2 itself, applied to the metal tissue holder. Most efficient, but unnecessary for routine and most research purposes is cooling by means of liquid N_2.

The CO_2-expansion freezer is a necessary and useful adjunct to the cryostat. Instruments of this type are described in the manufacturers catalogues (Lab-Tek; International Harris, Slee) and an easily made device was described by Freeman (1963).

OPTIMAL CONDITIONS FOR CUTTING

In their first paper on the analysis of frozen section technique Thornburg and Mengers (1957) pointed out that in frozen tissues water could be regarded as the embedding medium, and sectioning, therefore, as a matter of cutting ice. This conception has obvious limitations. Practical experience leads one to regard the cutting of brain, for instance, much more as a matter of cutting butter. Nevertheless the conception is true enough to form the basis of some interesting theoretical and practical deductions. Thornburg and Mengers found that the optimum values for the temperature of the knife, the chamber and the tissue block varied from tissue to tissue. Cutting was usually possible, however, within a wide range of temperatures.

(1) **Tissue Temperature.** As is the case with ordinary freezing microtomes when the block temperature falls below —45° the tissues are brittle and friable and cannot be cut. Between —40° and —15° block temperature Thornburg and Mengers could cut thin sections, but they noted resistance during the cutting stroke and some crumbling of sections on the knife. At higher temperatures, up to —5°, thin serial sections could be obtained.

With isothermic machines (knife and tissue temperature identical) cutting is possible between —30° and —10°, but at the higher end of the scale the sections tend to be compressed as they pass between the knife and the anti-roll plate. Formalin-fixed blocks can be cut in conventional isothermic cryostats provided that cutting is preceded by thorough washing in water and that the temperature of the cabinet is raised to between —5° and —10°.

(2) **Chamber Temperature.** With the cold environment-cold knife machine chamber temperatures between 0° and —10° were found best. Below —10° the quality of the sections began to deteriorate. The working temperature of most modern cryostats is well below this level; as a rule a constant —20° is maintained. This is doubtless a necessity on account of the relatively high knife temperature in such cabinets. It is a general rule in frozen section cutting that the higher the knife temperature the lower must be the temperature of the block.

Knife Temperature. In Thornburg and Mengers' cryostat the optimum knife temperature was found to be in the region of —50° with the tissue block at —10°. In Coons-type and other cryostats with the tissues held at about —20° a similar knife temperature gives good results with most tissues.

In cryostats of my own design a dry ice box in the form of a hemicylindrical trough can be carried on the knife handle to which it is attached by means of a reversible screw (to allow left-hand or right-hand fitting). This dry ice box is illustrated in Fig. 11. A single charge of pulverized dry ice lasts for about 3 hours (in the early stages it tends to lower the chamber temperature from —20° down to —25° or thereabouts) and the temperature of the actual cutting area is varied by sliding the box in and out on the handle. The provision of this attachment gives knife temperatures between —30° and —60°, but precise control is impossible. Colder than normal knife temperatures are still sometimes required for cutting very thin sections (1–2 microns), and also for low freezing point tissues such as those of marine and terrestrial molluscs. A modified dry ice box suitable for cooling the knife on a Schanze type microtome was described by Ihnuma (1961).

A Theoretical Analysis

In order to provide some interpretation of their results Thornburg and Mengers produced a model for the cutting process suited to their own cabinet and microtome. An adaptation of their diagram is shown in Fig. 12, below. It applies to a cold microtome of the type having accessories for independent cooling of knife, tissue block, anti-roll plate and environment. (The last two are usually kept at identical temperatures.) Making the assumption that fresh-frozen tissue is physically equivalent to ice, Thornburg and Mengers calculated the energy dissipated in cutting a single frozen section. For a block having a cutting resistance of 20 g. per cm. width this is $1{\cdot}96 \times 10^4$ ergs. By converting this figure to calories and dividing by the heat of fusion of ice (73 cals./cm.^{3}) a figure of $6{\cdot}5 \times 10^6$ cm. (650 Å) is obtained for the thickness of the layer of frozen tissue which will be melted by the cutting stroke. The heat generated by the latter can be expected to appear in the region of the cutting edge as a "zone of fusion." Tissues melted in this zone will refreeze as they pass down the two surfaces of the knife, and it is probable that the rate of refreezing at the point A in the diagram is a critical factor for the successful delivery of a perfect section between the knife and the anti-roll plate. A second critical factor

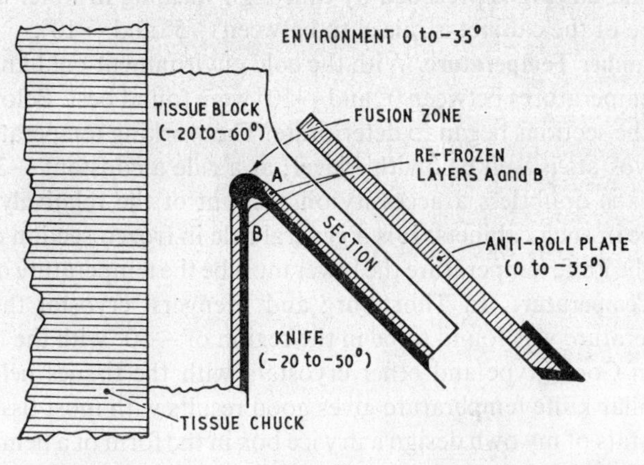

Fig. 12.

here is the rigidity of the section, which must be cold enough or thick enough to withstand frictional resistance to its progress over the knife surface. A clean surface is essential, even a film of condensed vapour is sufficient to increase this resistance beyond the critical point.

The original authors based a number of deductions upon their model of the cutting process. These were that: (1) the thickness of the melting zone increases as the block and knife temperature rise; (2) as the knife temperature is lowered a larger fraction of the energy dissipated in the cutting process is conducted directly into the knife and the thickness of the melting zone is decreased; (3) as the surface temperature of the block is lowered increased pressures are necessary to maintain the melting zone. Thornburg and Mengers predicted on theoretical grounds that conditions must exist under which sections a small fraction of a micron in thickness could be cut from fresh frozen tissues. Experiments with a Cambridge Ultramicrotome, housed in a cryostat with a minimum temperature of —35°, showed that it is indeed possible to cut fresh sections down to approximately 0·2 microns. Subsequent handling of these is extremely difficult; direct mounting on slides or coverslips gives unsatisfactory results. Thornburg and Mengers also predicted that sections cut under optimal thermal conditions would show a surface artifact due to the melting zone. Optimal thermal conditions for cutting are not necessarily equivalent to optimal conditions for tissue preservation and one must add that, since every tissue possesses its own individual optimum cutting and optimum preservation conditions, the only practicable method of determining these is by experiment. The cutting of perfect fresh frozen sections with modern cold microtomes and cryostats is likely to remain, therefore, a skilled technology. This will still be true even with thermoelectrically controlled instruments, such as those described below.

THE THERMOELECTRIC CRYOSTAT

Complete control of the two all-important variables (tissue and knife temperature) can only be obtained by means of thermoelectric modules. Experiments in temperature control were first carried out with equipment of this type, supplied by De La Rue Frigistor Ltd., housed in a standard cryostat, using a cold alcohol circulation to remove heat from the modules. The chamber temperature was found to be unnecessarily low and for almost all purposes 0° to —5° was a suitable range. A thermoelectric temperature controlled anti-roll plate (see Fig. 4) was considered but, in practice, this was also found unnecessary.

A second instrument, constructed on these lines (Pearse and Bancroft 1966), provided the following temperatures:

Chamber	0° to —5°
Tissue Block	—10° to —55°
Knife	0° to —45°

With this instrument considerable variation in cutting optima were found (see Appendix 2, p. 592) but it was possible to find a suitable combination for every type of tissue, including formalin and glutaraldehyde-fixed material.

This latter observation means that the advantages of the cryostat (ease of handling, uniform conditions for serial sectioning) can be transferred to the fixed frozen section technique. For all research purposes standard bench-type freezing microtomes, whether CO_2-cooled or thermoelectric, are rendered obsolete and it is safe to say that they will continue to be employed only in routine work.

A commercial instrument is now in production which incorporates the controlled temperature principle. This is the Frigistor Thermoelectric Cryostat produced by De La Rue Frigistor Ltd., under their patents. It is a compact bench mounted instrument which takes full advantage of current advances in refrigeration technology.

HANDLING OF SECTIONS AFTER CUTTING

With all the cold knife and cold microtome methods described above, handling of the tissue sections after cutting is just as critical as the operations leading up to their production, more so once these have been brought to perfection. A number of different techniques have been evolved by various workers in the field and it must be said at once that no one method is better than all the others. All represent some form of compromise, as will be seen, and the right method for one purpose may well be the wrong one for another.

There are some nine different ways by which cold knife and cold microtome sections can be treated.

(1) Pick up on warm (room temperature) or cold (cryostat temperature) slide or coverslip. Fix by immersion at moment of thawing (Adamstone-Taylor). Alternatively immerse in cold "protective" medium.

(2) Pick up on warm or cold slide or coverslip, thaw, and dry in air.

(3) Pick up on warm or cold slide or coverslip, thaw, dry, and post fix.

(4) Pick up on materials other than glass (e.g. cellulose acetate film, stripping film, etc.).

(5) Transfer or deliver into warm or cold incubating medium (enzyme histochemistry).

(6) Transfer or deliver into warm or cold test reagent solution (especially protein and inorganic histochemistry).

(7) Transfer or deliver into warm or cold fixative.

(8) Freeze dry (for subsequent chemical or biochemical assay).

(9) Freeze substitute.

For the majority of purposes the first three methods are the methods of choice. For histochemical studies on labile enzymes methods 2 or 5 are absolutely necessary. If the enzyme under consideration can withstand fixation, however, methods 1, 3 or 7 may be applied. In the case of soluble enzymes some such procedure is essential unless the fixative is incorporated in the incubating medium (simultaneous fixation). Freeze drying of fresh frozen sections is normally employed only in quantitative histochemistry.

It is important to realize that none of the methods described above is free from objections and that, as a rule, the choice of method must be dictated by practical rather than theoretical considerations. Complete destruction of cells, described by Adamstone (1951) as inevitable unless immediate fixation is applied, is rather more than usually occurs when sections are allowed to thaw and dry unhindered on the slides. This process, however, produces stretching and shrinking artifacts and I believe it should be avoided whenever possible. It is likely that some kind of protein film formation takes place, analogous to that seen in the case of air-dried blood smears, and this may interfere with the subsequent penetration of reagents although it may also act by preventing the escape of soluble proteins and peptides. Thawing of sections in (on the surface of) buffered media was advocated by Wachstein and Meisel (1953). This method produces stretching artifacts in the case of weakly cohesive tissues such as testis, especially if the sections are thinner than 10 μ. In the case of the majority of mammalian tissues residual artifact visible with the light microscope is very small if the sections are subsequently fixed before further handling by the addition of concentrated fixative to the original medium. There is, however, a considerable loss of soluble protein from such sections which is more easily noticed if the protein happens to have the properties of an enzyme.

Probably the technique of section freeze substitution (see Chapter 4, p. 67) has the fewest objections. This is much preferable to the method of fixation at the moment of thawing. Technical details of the maintenance of cryostat and microtome and appropriate methods for handling the sections are described in Appendix 2, p. 591.

Conclusion

The conclusion to the chapter on cold microtomy in the 2nd Edition of this book has been completely confirmed. The cryostat is now an accepted piece of routine apparatus in histological and histochemical laboratories throughout the world. Further developments will presumably centre on improved microtome performance, and on more precise control of cutting conditions.

REFERENCES

Adamstone, F. B. (1951). *Stain Tech.*, **26**, 157.
Adamstone, F. B., and Taylor, A. B. (1947). *Anat. Rec., Suppl.*, **99**, 639.
Adamstone, F. B., and Taylor, A. B. (1948). *Stain Tech.*, **23**, 109.
Appleton, T. C. (1964). *J. roy. micr. Soc.*, **83**, 477.
Batsakis, J. G., Romeika, D. C., and Fahs, G. R. (1963). *Stain Tech.*, **38**, 51.
Brown, R., and Dilly, M. (1961). *J. Physiol.*, **161**, 1 P.
Combs, J. W. (1961). *Stain Tech.*, **36**, 330.
Coons, A. H., Leduc, E. H., and Kaplan, M. H. (1951). *J. exp. Med.*, **93**, 173.
David, G. B., and Brown, A. W. (1966). In "The Interpretation of Cell Structure." Eds. Ross K. F. A. and McGee-Russell, S.M. London, Arnold.
Fand, S. B. (1964). *Stain Tech.*, **39**, 173.
Freeman, C. W. (1963). *Amer. J. clin. Path.*, **39**, 324.

GOLDSMIT, H. J. (1960). "Applications of Thermoelectricity." Methuen, London.
HARDY, W. S., and RUTHERFORD, T. (1962). *Nature*, **196**, 785.
IHNUMA, M. (1961). *Proc. Jap. Histochem. Ass.*, **2**, 55.
IOFFE, A. F. (1957). "Semiconductor Thermoelements and Thermoelectric Cooling." Infosearch, London.
IOFFE, A. F. (1958). *Scientific American*, **199**, 31.
ISERI, O. A., COMBS, J. W., and LAGUNOFF, D. (1961). *Stain Tech.*, **36**, 97.
LINDERSTRØM-LANG, K., and MOGENSEN, K. R. (1938). *C. R. trav. Lab.Carlsberg serie chim.*, **23**, 37.
LUNDY, J. A. (1963). *Stain Tech.*, **38**, 350.
OKAMOTO, M., and MIZUNO, N. (1963). *Stain Tech.*, **38**, 349.
PARROTT, J. E., and PENN, A. W. (1962). *A.E.I. Engineering*, **2**, 202.
PEARSE, A. G. E., and BANCROFT, J. D. (1966). *J. roy. micr. Soc.*, **85**, 385.
RUTHERFORD, T., HARDY, W. S., and ISHERWOOD, P. A. (1964). *Stain Tech.*, **39**, 185.
SCHULTZ-BRAUNS, O. (1931a). *Klin. Woch.*, **10**, 113.
SCHULTZ-BRAUNS, O. (1931b). *Z. allg. Path. path. Anat.*, **50**, 273.
SCHULTZ-BRAUNS, O. (1932). *Ibid.*, **54**, 225.
SHIMIZU, N., KUBO, Z., and MORIKAWA, N. (1956). *Stain Tech.*, **31**, 105.
STRIKE, T. A. (1962). *Stain Tech.*, **37**, 187.
TAFT, E. B. (1949). *Amer. J. Path.*, **25**, 824.
THORNBURG, W., and MENGERS, P. E. (1957). *J. Histochem. Cytochem.*, **5**, 47.
WACHSTEIN, M., and MEISEL, E. (1953). *Stain Tech.*, **28**, 135.
WEST, W. T. (1962). *Stain Tech.*, **37**, 320.
WHITE, R. T., and ALLEN, R. A. (1951). *Ibid.*, **26**, 137.

CHAPTER 3

FREEZE-DRYING OF BIOLOGICAL TISSUES

FROM the histochemical point of view there are many objections to the use of orthodox methods of fixation and embedding and the freeze-drying method was designed with a view to overcoming some of their deficiencies. The chief disadvantages of the ordinary types of fixation are:

(1) Loss of soluble substances, e.g. lipids in fat-solvent fixatives, and proteins, mucoproteins, polypeptides, polysaccharides, biogenic amines, and inorganic materials in aqueous fixatives.

(2) Displacement of cell constituents by diffusion. In this category are placed the so-called streaming artifact and other less obvious types of fixation artifact which include the aggregation of small particles into granules, and the clumping of nucleic acids in the nuclei.

(3) Denaturation of proteins, with alteration in their physical properties and those of the various protein complexes.

(4) Chemical alteration of reactive groups, especially of proteins.

(5) Destruction or inactivation of enzymes by (3) and (4).

The basic principle of freeze-drying is the rapid freezing of tissues at about —160° and their subsequent desiccation at somewhat higher temperatures until all except a small, tightly-bound fraction of their water content has been removed. By this process of rapid freezing the water in the tissues is converted into ice with the minimum formation of tissue-disrupting ice crystals, which form when material is frozen at higher temperatures. The frozen tissues are then dried for a variable period, being maintained during this time at between —35° and —70°. The dried tissues, allowed first to reach room temperature, are usually vacuum-embedded in paraffin or some other wax and cut in the normal manner.

It must be emphasized from the outset that the process of freeze-drying is not a method of fixation, although it is widely referred to as such. It is a means of preparing tissues for microscopical examination which leaves them in such condition that, for the majority of histochemical methods of investigation which are subsequently employed, fixation by one or other of the standard methods is necessary. Because of its obvious relevance for critical cytochemical investigations, by both light and electron microscopy, serious endeavours were made by workers in many disciplines to overcome the difficulties associated with the freeze-drying of tissues.

Until quite recently (1963), in spite of an increasing demand for freeze-dried material for an increasing variety of purposes, the process was sufficiently difficult and tedious to deter all but a few dedicated groups. Although the situation is now transformed, it may be expedient to describe the historical, theoretical and practical aspects of freeze-drying at some length so that those who have but recently entered the field can appreciate the severity of the problems involved.

The History of Freeze-Drying

According to Mann (1902) the preparation of tissues for microscopical research by drying is the oldest of all methods. He describes Leeuwenhoek as having prepared samples of muscle by this method in 1720, carrying them in his pocket and cutting sections from time to time for microscopical examination. The procedure used by Altmann (1890) was more refined than this and involved the maintenance of small portions of tissue *in vacuo* at —20°, over sulphuric acid, in a desiccator. After several days the tissues were embedded *in vacuo* in paraffin wax. Mann himself made use of Altmann's method, employing a mixture of solid CO_2 and alcohol to freeze whole animals; subsequently he broke off small pieces and dried them *in vacuo* at —30°. Bayliss in 1915 used a similar method but dried his tissues at a slightly lower temperature. The method was of little practical use until Gersh (1932) produced the forerunner of all modern freeze-drying apparatus, and his lead was followed by a large number of investigators. Modifications of the original Gersh apparatus or of the technique for using it have been described by Scott (1933), Goodspeed and Uber (1934), Hoerr (1936), Scott and Williams (1936), Packer and Scott (1942), Gersh (1948), Wang and Grossmann (1949), Mendelow and Hamilton (1950), Stowell (1951), Glick and Malmstrom (1952), Treffenberg (1953), Jensen (1954a), Arcadi and Tessar (1954), Moberger, Lindstrøm and Andersson (1954), Lacy and Blundell (1955), Burstone (1956), Caprino (1956), Naidoo and Pratt (1956), Gersh *et al* (1960), Grunbaum and Wellings (1960), Ostrowski and Komender (1960), Seno and Yoshizawa (1960), Benditt *et al* (1961), Branton and Jacobson (1961), and Thaine (1962). The earlier types of apparatus, and the theory and practice of tissue freeze-drying, were ably described in an excellent monograph by Neumann (1952). Some of the above modifications will be more fully considered below, following the section on principles and application. An account of the applications of freeze-drying in biological research was given by Bell (1955) and Cabrini (1955), in an article in Spanish, reviewed the fundamental principles of the process. The technical side has been considered by Kulenkampff (1956), by Mamulina and Orlova (1955), by Hartlieb (1954), and also, more extensively, by Burstone (1962).

The Principles and Practice of Freeze-Drying

The process of freeze-drying for histochemical purposes can conveniently be considered in four subdivisions. These are: (1) the initial freezing or, as it is commonly called, quenching; (2) the subsequent drying; (3) the process of embedding, post-fixation, sectioning and mounting; (4) final treatment before examination.

THE QUENCHING PROCESS

The initial treatment of the tissues by rapid freezing has three main effects. First, it stops all those chemical reactions which are going on in the tissues at the time of their removal from the body which we are accustomed to recognize under the comprehensive term of autolysis. The importance of this effect in critical histochemistry cannot be over-estimated, but the whole benefit of the immediate cessation of chemical reactions in the tissues can only be obtained if material is quenched within a few seconds of the death of the animal from which it is removed. This point is rightly stressed by numerous authors; Bartelmez (1940), for instance, described the development of particle aggregation and of cytoplasmic vacuolation in pieces of uterine epithelium quenched within 30 seconds of removal. It does not mean, however, as is commonly supposed, that the application of freeze-drying to human biopsy and post-mortem material is valueless. This would only be the case if the sole effect of freeze-drying was the prevention of autolysis and sub-mortem change whereas, in fact, this is only a fraction of its function. (By sub-mortem change is meant change of the Bartelmez type whose cause is unknown but which is certainly pre-autolytic.) When the tissues are brought to a solid state, however, the tremendous increase in their viscosity immediately reduces diffusion of substances within them, for practical purposes, to zero. Thus the prevention of sub-mortem change and of diffusion are the two major advantages of the technique.

Its greatest disadvantage is the formation of ice crystals from the unbound water in the cytoplasm of cells and in the tissue spaces, and the subsequent precipitation of materials dissolved in such water at the boundaries of the ice crystals. The latter are of varying size, depending on factors to be discussed below, and it is evident that the smaller they are the more the structure of the freeze-dried sample will reflect the original structure. Conversely, if they are large, resemblance to the original will be small and the artifact will, to all intents and purposes, destroy the structure of the tissue. It is certainly possible to obtain small pieces of tissue in a condition where no ice crystal artifact is visible by light or ultra-violet microscopes. Electron microscopy, however, reveals that ice crystal formation usually occurs, but it is of such size as to be invisible by other means of examination. Gross ice crystal artifact can be conveniently demonstrated by freezing recently formalin-fixed tissues at —20° and subsequently allowing them to thaw before dehydrating and embedding by the usual techniques. The gross disruption which is observed represents the maximum effect of the process we are considering. If this manœuvre is carried out with unfixed tissues, fixed only after thawing, all trace of ice crystals is obliterated. A second, but much less important, artifact of quenching is the development of cracks due to the 2 per cent. contraction in volume which accompanies cooling of the block of tissue.

Bell (1952a), discussing the mechanism of ice crystal formation, stated that in a fixed volume of a solidifying liquid the number of ice crystals formed was directly proportional to the number of nuclei of crystallization and inversely proportional to the rate of growth of the crystals. In order to achieve the smallest crystal size it is therefore necessary to provide a large number of nuclei and a low rate of growth. Luyet and Gehenio (1938) showed that the growth of ice crystals in water was reduced by the addition of gelatin. In a 1 per cent. solution the rate of growth was half that in pure water and in a 3 per cent. solution it was 350 times slower. Thus material in solution in water in the tissues slows the rate of ice crystal growth, depending on its concentration and structure. Many substances can entirely inhibit the formation of crystals so that a paracrystalline solid is obtained on cooling. The physical principles of ice crystal growth were discussed by Meryman (1956). He divided nucleation into two types, homogeneous and heterogeneous, and pointed out that since the former occurs only with extreme degrees of super-cooling the latter is the usual form. Meryman concluded that crystallization is usually extracellular until rapid rates of cooling are achieved and he considered that the rapidity with which destructive ice crystals could grow rendered the thawing process (if this was allowed to occur) as important as freezing. When a crystal nucleus develops and growth continues, loosely bound water is progressively removed from solution until the whole has been converted into ice. This, in effect, is a process of desiccation. There remains always a residuum of tightly bound water (8 to 10 per cent. in mammalian tissues) which never freezes.

More recently the dynamics of freezing in biological materials were dealt with very completely by Rey (1960). He divided the factors concerned into two categories, internal and external. The former include water content, structure and shape of the specimen, and the nature and concentration of solutes and dispersed phases. The external factors include absolute temperature, specific heat of the cold source, and the nature of the thermal contact between the cooled material and the refrigerating system.

As far as we are concerned, in the practical application of the method, the majority of factors influencing crystal growth are not under control. The most important factor, which is the rate of nucleation (i.e. of appearance of new nuclei of crystallization), can be controlled to some extent, however. It varies with the fall in temperature, but up to a certain critical point the process is very slow. This critical point is in the vicinity of —39°. Beyond the critical point the process increases rapidly until the material becomes solid. If a system is cooled rapidly to a temperature at which the rate of nucleation is very great, a non-crystalline structure may be formed. In the case of tissues the presence of dissolved materials has the effect of slowing crystal growth so that the point of rapid nucleation may be reached before the tissue becomes solid.

In practice, therefore, we have to obtain a rapid rate of cooling as has recently been re-emphasized by Meryman (1960). The ability to do this depends on the thermal conductivity of the liquid used for the freezing bath and, if the pieces of tissue are large, it will depend substantially also on their own conductivity. Scott (1933), Hoerr (1936) and Simpson (1941a and b) stressed the necessity for using a cooling liquid of high conductivity and at as low a temperature as possible. A phenomenon called psuedo freeze-drying has been demonstrated in muscle tissue by Luyet and Williams (1962). If the tissue is cooled slowly to —10° part of the water crystallizes while the rest remains as a non-frozen gel. When subjected to a vacuum the ice sublimes (freeze-drying) while the water in the gel evaporates (pseudo freeze-drying) accompanied by marked shrinkage of the tissues. The process must always be considered as a cause for any observed shrinkage.

In his early experiments Gersh used liquid air (—195°) for his quenching bath and although this had a sufficiently low temperature its conductivity was low because of the formation of a layer of vaporized air around the specimen which prevented the transference of heat. In order to overcome this effect Scott (1933) recommended the use of ethyl alcohol cooled to —100° in place of liquid air, but this method had the disadvantage that ethanol is solid at —115° and very viscous at higher temperatures in this region. Hoerr (1936) introduced the method of placing the tissues in crude isopentane cooled to about —165° with liquid nitrogen. Crude isopentane solidifies at —190° and the point at which it is beginning to become viscous is usually employed. Hoerr also used pentane cooled to —120°. Various other fluids have been employed for cooling. Emmel (1946) tried isopentane mixed with dry butane to lower the melting-point of the former and Bell (1952a) proposed a mixture of propane (3 vols.) and isopentane (1 vol.) or, alternatively, pure propane, which has its melting-point at —185°. Some authors, surprisingly continued to use liquid air directly for quenching (Pease and Baker, 1949), and found no advantage in using isopentane. Simpson (1941b) confirmed the excellence of the isopentane technique but maintained that even by this means the resulting freezing was uneven unless very small pieces of tissue were used. He described three zones in the dried tissue caused by artifact during quenching, an outer zone in which ice crystal formation was minimal and preservation of structure excellent, an intermediate zone in which distortion of structure by large crystals was particularly bad, and an inner zone of better but relatively indifferent preservation.

Bell (1952b) recommended Arcton-6 (I.C.I. Ltd), known in the United States as Freon-12, as a safe quenching agent for use with liquid oxygen. This compound, dichlorodifluoromethane, which melts at —158°, is now known as Arcton 12. It was strongly recommended also by Lacy and Davies (1959) and I used it routinely in my own work until 1966. Moberger, Lindstrøm and Andersson (1954) found pure propane cooled in liquid nitrogen to be the best quenching bath. They advised the avoidance of metal holders for quenching and dropped their tissues directly into the bath. Eränkö (1954b), on the other hand, employed a solid bar of copper, cooled to —192° with liquid air. He devised a pair of quenching forceps, solid copper discs with an insulating plastic handle, with which the tissue specimens were picked up. The quenched tissues were then transferred to a cool beaker to await transference into the freeze-drying apparatus. An alternative technique suggested by Zlotnik (1960), which avoids the use of liquid air or liquid nitrogen, employs a mixture of pentane and dry ice. The mixture is contained in a Dewar flask provided with a central perforated zinc well to keep the solid and liquid phases apart. After evaporation under vacuum for an hour the liquid phase reaches —125° and it remains near this level for at least 30 minutes on the bench. Extensive experiments carried out by Rebhun (1965) indicated that, although propane and propylene gave the fastest cooling rates Freon 22 (monochloro-difluoromethane), at 80 per cent., was 20 per cent. more efficient than Freon 12. I therefore now use Freon 22 routinely, as the fastest safe quenching agent.

Rapid freezing with liquid Helium II has been employed by Fernández-Morán (1960) especially for tissues destined for examination by electron microscopy. By comparison with Freon-liquid nitrogen quenched materials, better preservation of fine tissue structure was obtained. Since the improvement is not detectable by light microscopy, however, the present use of liquid Helium II is a restricted one (see Chapter 4, p. 62).

Simpson believed that the structural changes in his intermediate zone were caused, while it was still not solid, by pressure from the hard frozen outer zone. Bell (1952a), on the other hand, considered that it represented a zone in which the temperature of solidification was too high for rapid nucleation to occur but where the rate of crystal growth was high. In the innermost zone, according to this author, the temperature of solidification was too high for rapid nucleation but also too high for the rate of crystal growth to produce the large artifacts of the intermediate zone. This explanation is more satisfactory than that advanced by Simpson. It is supported by the latter's evidence that the dimensions of the three zones could be varied by altering the temperature of the quenching bath. He found that if this was maintained at —90°, only the inner zone remained.

According to Simpson the best practice for tissues required for cytological examination was to regulate the size of the sample and the temperature of the quenching bath so as to obtain entirely outer zone preservation. No means exist for calculating these factors for different tissues and the method of trial and error has therefore been employed. Bell (1952a) conducted a series of experiments into the speed of quenching. By placing a thermocouple within a length of mouse intestine and plunging the whole into cooled isopentane he found that the time taken to reach equilibrium with the bath averaged 10 seconds. (Blank for thermocouple—2 seconds.) The critical temperature of water (—40°) was reached in 4 seconds. Trump *et al* (1964) placed a 30-gauge copper constantan thermocouple in the centre of tissue blocks measuring approximately 5 × 3 × 1 mm. Using dry ice for quenching, the centre reached —79° in less than 2 minutes. With isopentane less than 10 seconds were required to reach —150° and with propane less than 6 seconds to reach —175°.

Agreement between these figures and those of Bell indicates that with such standard methods a surprisingly long period elapses before quenching is complete. It would be of interest to know if these times are substantially reduced by the use of liquid helium II.

THE DRYING PROCESS

The process of drying the solid frozen block of tissue, with which we were left at the end of the previous stage, can conveniently be considered, with Meryman (1960), under three headings. These are (1) the introduction of heat to supply the energy necessary for sublimation, (2) the transfer of water vapour from the ice crystal through the dried shell of the specimen and, (3) the removal of water vapour from the surface of the specimen.

Heating. The water molecules in a block of ice, or in the frozen specimen, still exhibit random thermal motion. This is a function of the temperature and for a water molecule to escape from its environment in the ice block additional energy must be provided. This energy, the latent heat of sublimation, is usually supplied in the form of heat from the environment. The transfer of molecules from ice to vapour removes heat from the environment and, unless this is replaced, the temperature falls. If it falls sufficiently far sublimation ceases. The process of sublimation takes place, therefore, at a rate equivalent to the heat input.

In the older types of freeze-drying apparatus (Fig. 16, p. 38, left and right, above) the sole source of heat was the cooling bath. This was maintained at temperatures between —30° and —78°. When the tissue block temperature fell below that of the cooling bath, heat was transferred by conduction from the bath to the specimen to restore the balance. With instruments of this type the rate of sublimation was very slow. In later types of apparatus (Fig. 16, left and right, below) heat was provided, again by conduction, from an electrically heated surface on which the tissue blocks rested. It was thus applied to the lower surface of the frozen block, while sublimation took place from the opposite (cold) surface. Heat supplied in this way was strangely ineffective and the overall performance of electrical heat conduction driers was hardly superior to that of simpler forerunners.

Many more modern tissue freeze dryers employed the principle of heating by radiation. This could be supplied electrically, as in the Canalco drier shown in Figure 17 (p. 39, left), or from the external environment as in the Fisher dryer on the right of this illustration. With these types of apparatus the rate of sublimation was apparently somewhat higher.

Theoretical reasons for the greater efficiency of radiant heat tissue dryers were set forth by Meryman (1960), who considered that when heat is supplied by conduction to one side of the tissue block it must be transferred to the drying boundary by conduction through the frozen tissue. Since heat flows only when there is a thermal gradient the amount actually supplied to the boundary is determined by the thermal conductivity of the material. According to Meryman (1958) the thermal conductivity of frozen mammalian tissue is 3 per cent.

of that of copper. The conductivity of dried mammalian tissue is not known but, necessarily, it must be well below this figure.

The characteristics of the thermal gradient (T) existing in dryers where heat is supplied solely to the lower surface of the tissue, by conduction, are shown in Figure 13. It will be observed that under these conditions the temperature of the drying boundary must always be lower than that of the heated surface. The temperature of the latter is under direct control of the operator. Its limit is fixed by considerations of the safe maximum temperature for the frozen tissue and this critical factor in the process of freeze drying must now be considered.

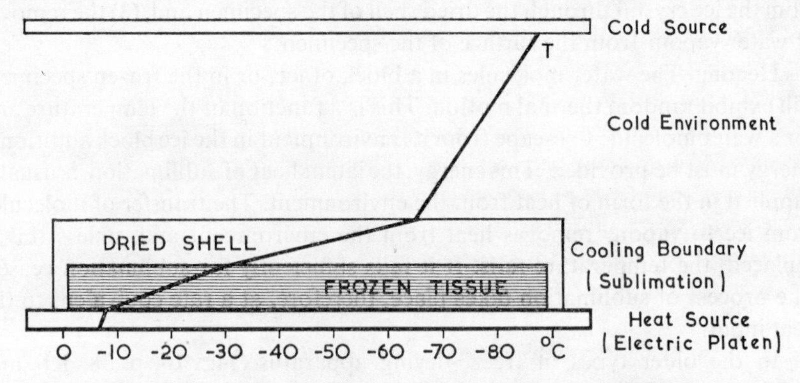

FIG. 13. Thermal Gradient in Conduction-heated Tissue Dryers

Drying Temperature and the Eutectic. In his original studies Altmann used —20°, although he suggested that —30° would be better. Bayliss (1915) dried his tissues at —35°. He was the first to draw attention to consideration of the eutectic point (Greek: ευ easily + τήκειν melt). In general, when a mixture of two liquids is cooled, one of the components will freeze out first, thus concentrating the other. By this process the freezing point is lowered and if cooling is continued the temperature is finally reached where both components freeze out simultaneously and the whole mass then solidifies without a further fall in temperature. This is the eutectic point. The presence of various salts in solution effectively modifies the eutectic point, and Scott (1933) pointed out that the drying temperature originally used by Gersh was not sufficiently low to prevent diffusion of some ions since it was above the eutectic point of many naturally occurring salt systems. Packer and Scott (1942) used a temperature below −54·9°, the eutectic of calcium chloride, since they were particularly interested in the distribution of inorganic materials and since $CaCl_2$ has the lowest eutectic of any naturally occurring salt. It has been observed that the eutectic points of complex mixtures, such as those which occur in the tissues we are trying to dry, are very difficult to calculate. The binary eutectic with the lowest melting point is that of $CaCl_2$ and water as already mentioned; tissues, however, contain complex tertiary and quaternary systems with even lower eutectics.

In spite of these theoretical considerations, practical experience suggested that temperatures between —30° and —40° were about optimal and, although the majority of workers agreed that distinctly better results were obtained at the lower temperature than, say, at —20°, the eutectic theory would indicate that a still lower temperature would be better. An explanation for this discrepancy was offered by Bell (1952a). It is known that in artificial silica sols, quenched and then allowed to thaw slowly, recrystallization takes place if a critical temperature of —55° is maintained. Bell suggested that the critical temperature of recrystallization of tissue proteins might be in the region of —40°, and this would account for the arbitrary selection of this temperature by trial and error. Moberger *et al.* (1954) maintained that temperatures above —60° were unsafe for the preservation *in situ* of highly diffusible ions and stated that at —40° morphological changes due to recrystallization appeared. More recently a method has been described by Stumpf and Roth (1967) in which frozen sections are dried at temperatures below —60°. The process has been called cryosorption pumping and it is claimed that ice crystal and shrinkage artifacts are reduced to a minimum.

According to Rey (1960), although the phenomena observed with binary mixtures do not exist in complex solutions, it is nevertheless permissible to speak of the eutectic temperatures of biological materials. The eutectic temperature is defined as the maximu m temperature at which crystallization can be achieved. Above this temperature the system still contains liquid phases, below it everything exists in the solid form.

Measurement of the eutectic temperature is made by a process described as thermal analysis. In this process (Greaves, 1954) the electrical conductivity of the specimen is measured. As long as liquid is present the current flows without difficulty. When everything is solid the resistivity is extremely high. A sudden increase in resistance signifies the crossing of the eutectic boundary. In order to ascertain the eutectic requirements of a natural substance Rey (1960) recommends (1) Determination of the maximum temperature of complete solidification (Tcs) and (2) Determination by thermal analysis of the minimum temperature of incipient melting (Tim). All that is then necessary is to quench the tissue at a temperature below Tcs and to hold it during drying at or below Tim. Values for Tim for chick embryo extract and horse serum are respectively —17·5° and —34·5°.

It is easy to maintain temperatures much lower than eutectic during drying, but practical requirements dictate that although the temperature must be sufficiently low to prevent melting it must yet be sufficiently high to allow drying within a reasonable period of time. Many methods have been used at one time or another to maintain the frozen state of the tissues at a given level. One means of providing the necessary control of temperature is by means of a compression-expansion refrigerating plant such as as that employed by Gersh (1932) using liquid ammonia and by Harris, Sloane and King (1950) using methyl chloride. Burstone (1956) used a refrigerated box for maintaining his tissues at a selected temperature while drying. The majority of workers, however, used mixtures of solid CO_2 and some organic solvent contained in an insulated vessel surrounding the drying chamber. The original alcohol/CO_2 mixture used by Mann maintained a temperature of —78° and an acetone/CO_2 mixture is stable at about —70°. In the past these two were widely used although the temperatures they give were too low for efficient drying in the shortest possible time. This is because the temperature of the specimen controls its vapour pressure and this in turn controls the rate of evaporation. Ice evaporates ten times more slowly at —60° than at —40°, at which latter point its vapour pressure is $9{\cdot}66 \times 10^{-2}$ mm. Hg. Mendelow and Hamilton (1950) used a slush of solid CO_2 and ethyl oxalate which is stable at —40° as long as solid ethyl oxalate remains in the system. This condition can be maintained by infrequent additions of solid CO_2. Other solvents which have been suggested are phenetole (—34°), nonane (—53°) and isopropyl ether (—60°).

In most modern dryers (with the notable exception of the Thaine, Thieme, and Pearse designs) heat is supplied by radiation alone. Conditions are then very different from those existing in "conduction" driers, as indicated in the diagram (Fig. 14) below. Heat is supplied intermittently to the dried shell, which is a poor conductor, and it passes thence to the cooling (drying) boundary. The main bulk of the frozen tissue remains at a lower temperature than that selected. The gradient through frozen tissue and dried shell is a steep one and the drying boundary is always warmer than the frozen tissue. The environment, always cold in "conduction" dryers may be cold, warm or intermediate in temperature. This will normally depend on the proximity of the cold trap.

In the case of the thermoelectric dryer (see p. 45) the state of affairs is very different, as shown in Figure 15. It can be observed that the heat gradient is nowhere as steep as in the situations shown in Figures 13 and 14. The frozen tissue is maintained within a few degrees of the selected temperature, heat being supplied smoothly and constantly to this end. Heat is likewise supplied to the dried shell whose surface temperature may be in the neighbourhood of 0°. At

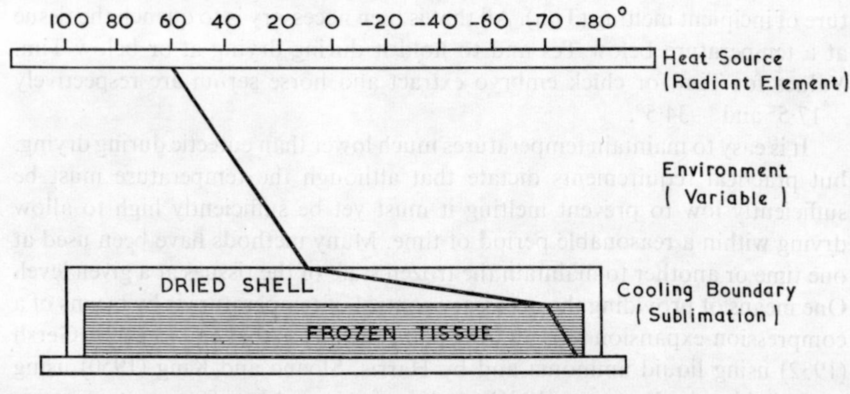

FIG. 14. Thermal Gradient in Radiation-heated Tissue Dryers.

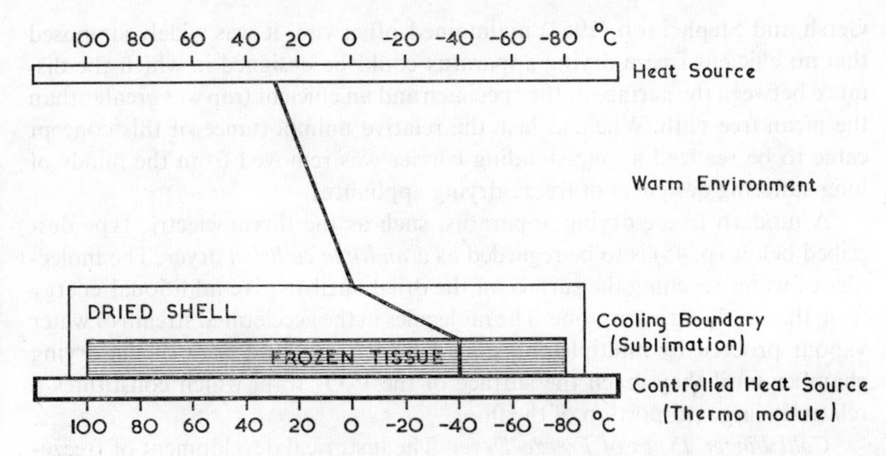

FIG. 15. Thermal Gradient in Warm Environment Thermoelectric Dryers.

a point 5 mm. above the surface the temperature is between +15° and +20° when the room temperature is +23°, so that the environment may be classified as warm. In the Thieme design (p. 45) the environment is likewise warm.

Vapour Transfer. As soon as the process of freeze drying commences a shell of dried tissue appears at the surface of the specimen. As drying proceeds this shell becomes progressively thicker. All water vapour produced by sublimation of ice at the drying boundary must diffuse through this dry barrier layer to reach the tissue surface. The resistance to drying produced by the dried shell is very considerable (for measurements see Stephenson, 1954). The only force available to drive water vapour from the drying boundary to the tissue surface is a concentration gradient (Meryman, 1960). It is thus necessary to produce a high vapour pressure (i.e. a high rate of sublimation) at the boundary and a low vapour pressure at the surface. The latter is obtained by efficient removal of water molecules as soon as they arrive.

Vapour Removal. The removal of water molecules from the surface of the drying specimen depends on the identical force responsible for vapour transfer, that is to say a vapour concentration gradient. In order to provide this it is necessary to have a high vapour pressure at the drying boundary and a low vapour pressure at the surface. The first is directly related to the temperature and thus is limited by considerations discussed above. The second is usually provided by means of a vacuum system containing vapour traps. These may be either cold surfaces or chemical desiccants.

In the early days of tissue freeze drying much was heard of the concept of *molecular distillation* and the *mean free path* (Butler & Bell, 1953). The latter is the distance a molecule can travel before colliding with another molecule. It is dependent on the vacuum and at 10^{-3} Torr* it is about 50 mm. Although

* The International Unit of Vacuum is the Torr. For practical purposes this can be regarded as equivalent to 1 mm. of mercury.

Gersh and Stephenson (1954) maintained otherwise, it was widely supposed that no efficient freeze drying apparatus could be designed in which the distance between the surface of the specimen and an efficient trap was greater than the mean free path. When, at last, the relative unimportance of this concept came to be realized a long-standing barrier was removed from the minds of long-suffering designers of freeze-drying appliances.

A modern freeze-drying apparatus, such as the thermoelectric type described below (p. 45) is to be regarded as a *multiple collision* dryer. The molecules of water reaching the surface of the dried shell receive additional energy from the overlying warm zone. The molecules in the accelerated stream of water vapour proceed by multiple collisions between roof and floor of the drying chamber until they reach the surface of the P_2O_5 trap, which constitutes a relatively large proportion of the floor.

Cold Finger Types of Freeze-Dryer. The historical development of freeze-drying shows clearly the change of emphasis that has taken place. Until fairly recently almost all freeze-dryers embodied an arrangement known as the "cold finger". This finger, of plain or silvered glass, contained liquid N_2 and was introduced into the drying chamber so that the distance between its tip and the specimen was less than 50 mm. The apparatus described by Mendelow and Hamilton (1950), in which a "cold finger" was provided in the shape of a round-bottomed flask, is illustrated in Fig. 16. These authors used an aluminium desiccator as their drying chamber and I experimented with a double-walled brass vessel of similar dimensions in which the outer space contained the ethyl oxalate/CO_2 mixture. While the use of vessels of this sort was fairly

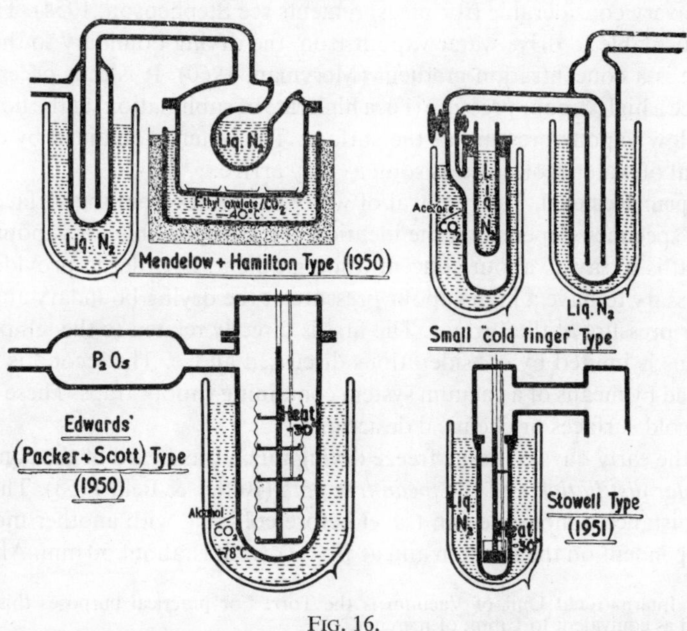

FIG. 16.

satisfactory, the advantages of metal over glass were scarcely so great as claimed by Mendelow and Hamilton, and the difficulty of maintaining a temperature of —40° was increased in the case of these large chambers. In 1951, therefore, I used a simpler form of "cold finger" inserted into the original glass drying chambers. This modification is illustrated diagrammatically in Fig. 16, upper right: the "cold finger" was introduced through the top of the drying tube and the manifold was connected instead to the side of the vessel. These modified tubes could still be immersed in the original Dewar flasks. They suffered from the disadvantage that the volume of liquid N_2 which they contained was small and had to be replenished every $1\frac{1}{2}$ hours. With an apparatus of this sort the average drying time for very small blocks was reduced to 12–24 hours.

An alternative arrangement which also allowed rapid drying and which avoided the complications due to the "cold finger" was the apparatus designed by Stowell (1951), and shown diagrammatically in Fig. 16. In this apparatus the condensing surface was provided by immersing the vessel containing the specimens in a Dewar flask of liquid N_2. The specimens themselves were heated to any desired temperature (Stowell recommended —30°) by means of a heater in their support, and this temperature was indicated by means of a thermocouple and potentiometer.

A conventional type of liquid N_2 cold finger was still employed in the Fisher design (Fig. 17). This was of limited capacity but the drying time was apparently sufficiently short to make refilling a simple task.

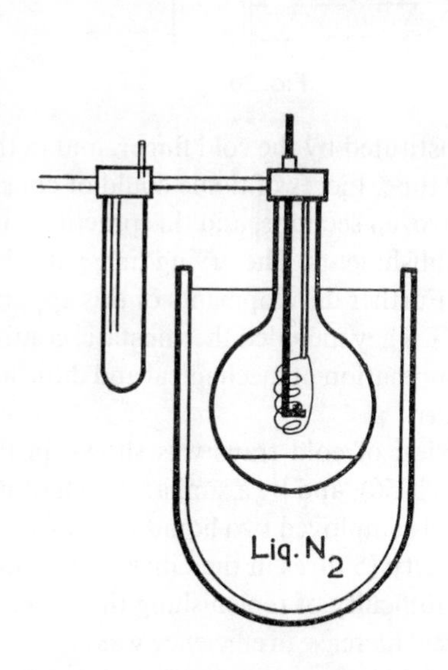

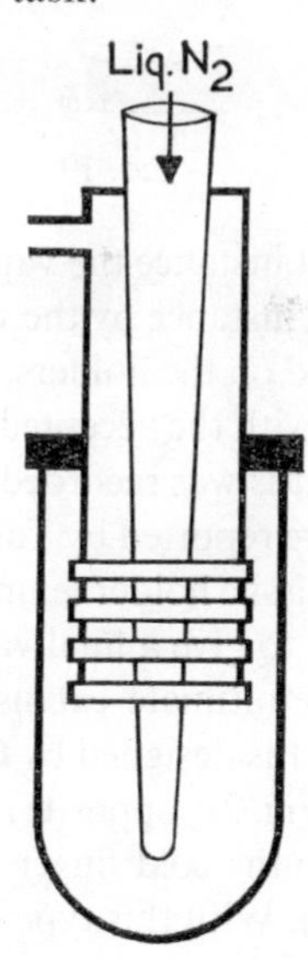

FIG. 17

The small freeze-dryer devised by Eränkö (1954a), and illustrated in Fig. 18, opposite, was intended for use in two different ways. If frozen sections were to be dried they were placed in the holder C (Fig. 19) and the inner tube A was then filled with liquid air. The Dewar flask B contained an ethyl oxalate-CO_2 mixture (—40°). Tissue pieces, on the other hand, were placed in the recesses of the inner tube (A, Figs. 19 and 20), which in this case contained the ethyl oxalate mixture while the Dewar flask, B, contained liquid air. Thus in

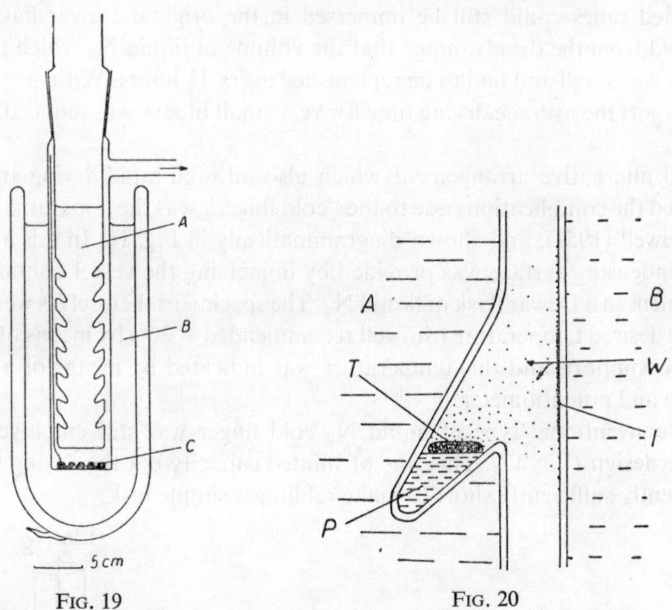

FIG. 19 FIG. 20

the first instance the vapour trap was constituted by the cold finger, and in the second instance by the walls of the outer tube. Pieces of tissue could of course be dried on the holders, as in the case of frozen sections, and this practice conforms with the accepted one in most establishments. The drying time with this apparatus was recorded as 24–48 hours. Further developments of this apparatus were reported by Tallqvist *et al.* (1967). They included thermostatic control of the tissue holder temperature and a combination of mechanical and diffusion pumps to give a final vacuum of 10^{-6} Torr.

The ultimate extension in the provision of cold traps was shown in the apparatus designed by Glick and Bloom (1956), and by a similar design of my own (Fig. 21, opposite). Both these models employed two liquid air traps and in both the cold finger was of large capacity (5 litres in the Glick and Bloom model). With this type of apparatus the difficulty of replenishing the finger at frequent intervals was overcome but no real increase in efficiency was achieved.

In order to test the effectiveness of the cold finger, an apparatus of the above type was fitted with a metal tissue platform whose distance from the finger could be controlled by an external ring magnet. I found that there was a *critical*

Fig. 18. Simple freeze-drying apparatus designed by Eränkö.

Fig. 21. Large "cold finger" type freeze-drying apparatus.

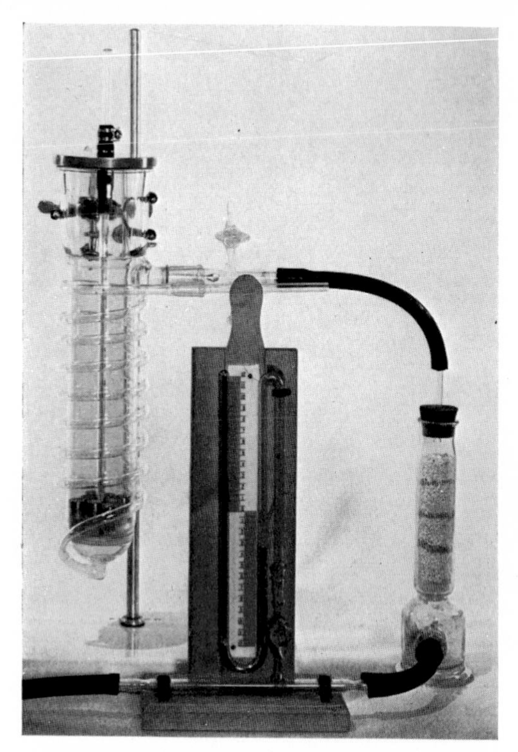

FIG. 22. Moving gas stream type of freeze-drying apparatus designed by Jensen.

distance (3·5 cms) at which radiation of cold from finger to specimen became so large that drying virtually ceased.

Moving-Air Types of Freeze-Dryer. A number of workers (e.g. Treffenberg, 1953) suggested or used a different principle of drying in which a stream of cold gas (air or N_2) passed continuously over the tissue, which was maintained in a moderate vacuum. Jensen (1954a and b) introduced a practical dryer employing this principle after finding that the drying times for most plant tissues, using a conventional cold finger dryer, were excessively long. Modifications of the original instrument resulted in the production of a simple design which was especially suitable for the drying of plant tissues. Jensen's apparatus, which is illustrated in Fig. 22, opposite, consists essentially of a glass drying chamber surrounded by a coil of glass tubing (centre), a desiccant column and a vacuum gauge. The tissues are placed on a fritted glass plate and the chamber is sealed with an "0" ring. Nitrogen from a cylinder is cooled to —30° to —40° by the methyl cellosolve-CO_2 mixture in the Dewar flask and drawn over the tissues by means of a two-stage mechanical vacuum pump at about 15 Torr pressure. The pump is protected from the effects of moisture by the provision of a column of desiccant ($CaCl_2$) inserted into the vacuum line.

With the apparatus shown in Fig. 22 Jensen reported that it was possible to dry 1 mm. sections of *Vicia faba* epicotyl in 4 hours. Root tips required 12–18 hours depending on the size of the tip, while developing stem tips of *Xanthium sp*. dried well in 8–10 hours. Nitrogen was substituted for the air which was used in an earlier apparatus because Jensen and Kavaljian (1957) found that, in the presence of molecular oxygen, easily oxidizable substances such as ascorbic acid could undergo oxidation in spite of the low working temperature in the apparatus.

A slightly different approach was used by Meryman (1959) with the construction of an ingenious moving-air dryer in which a small compressor was used to blow a continuous stream of cold air over the frozen tissues. Drying was thus carried out at atmospheric pressure. Although an average drying time of 8 hours was reported, nothing further has been heard of the device.

An unusual apparatus was described by Branton and Jacobson (1961), of which a diagrammatic illustration is given in Fig. 23. This could be used either as a conventional dryer or as a moving-gas instrument employing nitrogen or CO_2 cooled by a dry ice-alcohol mixture. The authors found that moist gas (moisture 2·6 μg–4·0 μg per litre) dried their specimens faster than dry gas. They suggested that normal freeze-drying made the walls of plant tissues too dry. It is difficult to apply this suggestion to mammalian tissues.

Mechanism of Drying with Moving-Air Dryers. It is necessary to enquire into the factors which contribute to the efficiency of moving-air or moving-gas designs. It appears that the most important effect of the stream of cold gas at —30° to —40° is to maintain the temperature of the tissues at this level during the earlier stages of drying. A second effect, perhaps equally important, is the transport of water molecules away from the surface of the tissues. Jensen and

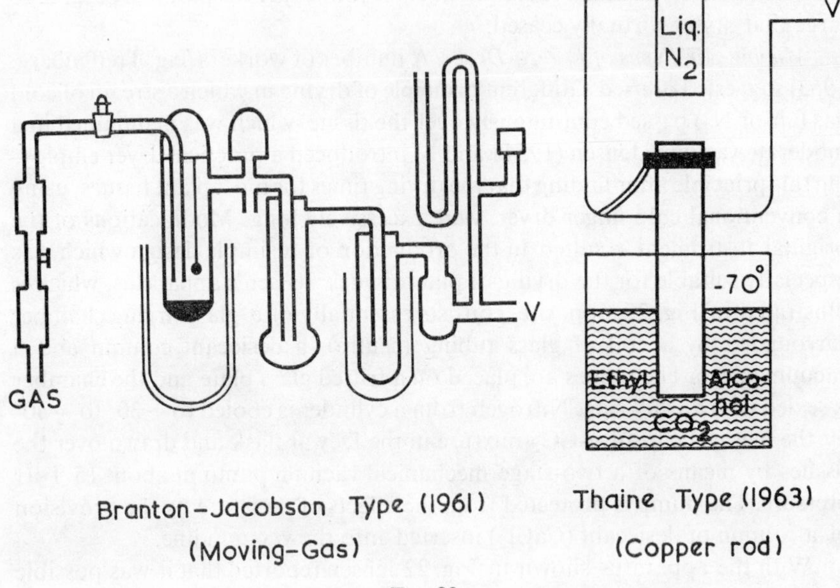

FIG. 23

Kavaljian (1957) considered that a partial vacuum was essential. If the stream of gas was passed over the cold tissues at a positive pressure they found that some drying occurred but not more than 40–60 per cent. of their water could be removed, however long drying was continued. According to Meryman (1960) drying at atmospheric pressure, or presumably in a poor vacuum, was found to be less efficient than high vacuum drying. This was mainly on account of increased resistance to vapour flow in the dried shell of tissue.

Measurement of the Working Pressure. Various types of vacuum gauge have been employed in connection with freeze-drying, the two most popular being the McLeod and the Pirani types. The former measures the volume of gases left in a specified volume of the system by compressing them in a side arm, while the latter is a hot-wire gauge. The Pirani gauge is sensitive to water vapour pressure, but the McLeod gauge is not, and therefore the latter cannot be used to estimate the level to which drying has progressed. Ionization gauges of course give higher accuracy, but costly additional apparatus is necessary which makes them, in most cases, an expensive luxury. They are not usually employed as a permanent fitting in freeze-drying systems as they are subject to contamination and are very difficult to clean. All types of gauge can be used to detect leaks in the apparatus but these can sometimes be estimated by means of the simple and inexpensive Tesla sparking coil. If a spark from this is directed on the glass wall of the system (it cannot be used if the apparatus is entirely of metal) a coloured glow indicates a gas discharge unless the vacuum is better than 10^{-3} Torr when only a greenish fluorescence is visible on the glass. The

Tesla coil can also be used to indicate the exact site of a leak in glass tubing, and in glass-rubber or glass-wax joints.

Detection of the Endpoint of Drying. Hoerr (1936), using a McLeod gauge, claimed that the completion of the drying process was indicated by a fall in indicated vacuum from ·002 to ·0001 Torr. Using an early type of apparatus incorporating a long glass manifold and liquid N_2 traps I observed a similar fall, but considered that it preceded the true dry-point by a long interval. It seems probable that Hoerr's pressure measurements referred only to extraction of air from the system, which of course might be complete before the tissue is dry, although in his case it appeared to be a suitable indication of the dry point. Packer and Scott (1942) used two Pirani gauges, one at either end of the vacuum line, and they claimed that an initial difference between the readings of the two fell to zero when drying was complete. Naidoo and Pratt (1956, 1957) used the stability of the vacuum when cut off from the pump as an accurate indicator of the endpoint of drying.

Only one unexceptionable method for estimating the drying point has been evolved and this is largely of theoretical interest. Jansen (1954) incorporated in his apparatus a torsion balance by which the process of drying could be followed until the endpoint, indicated by constant weight, was reached. Branton and Jacobson (1961) supported their specimens by means of a quartz spring. They could thus follow the loss of water to its conclusion.

Below is a graph of the type obtained by Jansen in his experiments. It shows that a constant weight (W) was reached in 36 hours with only small pressure variations (P) (measured with the pumps turned off) between 10 and 12 millitorr. Temperature variations (T) were simultaneously recorded. Jansen concluded that it was difficult to prevent the specimen from cooling too much by withdrawal of the latent heat of evaporation, and that for an adequate heat supply by radiation the object should be surrounded completely by the radiat-

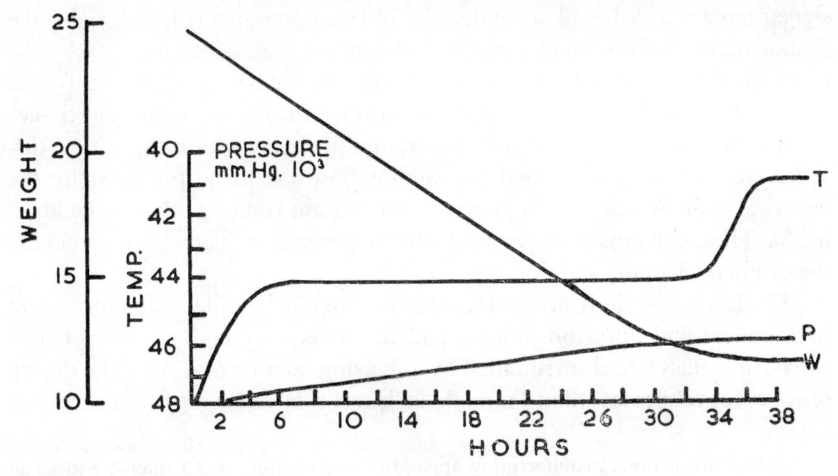

FIG. 24.

ing surface. Alternatively, Jansen suggested fitting the drying tube with a heat-conducting gas at an appropriate pressure.

Jansen objected to Gersh's formula relating the drying rate inversely to the square of shell thickness. He maintained that the slope of the weight and time curve (Fig. 24) did not suggest such a relationship.

Modern Types of Tissue Dryers*

Cold Trap Dryers. Since the second edition of this book appeared (1960) significant changes have taken place in the design of commercial and other freeze-dryers. Their sponsors have all tended to stress shorter drying times although far more important considerations are simplicity of operation and reliability.

The cold finger dryer produced by the Fisher Scientific Co. derives its energy for sublimation by radiation of heat from the environment (Fig. 17, upper right). For the maintenance of specimen temperature (stated to be about —30°) it relies on the latent heat of sublimation. It is doubtful whether the minimum temperature of incipient melting (Tim) is in fact maintained. There is certainly a large temperature gradient across the drying chamber and tissues close to the cold finger will be much colder than those adjacent to the outer wall (room temperature).

The "cold hand" dryer produced by the Canal Industrial Corporation (Fig. 17, upper left) uses radiant heat supplied electrically through a wire spiral surrounding the specimen tray. The wall of the vacuum chamber forms the condensing surface, or cold hand. The Stowell type of apparatus can equally be called a cold hand dryer, as can the Cryo-desiccator of Phipps and Bird Inc. These differ from the Canalco (Meryman) design basically only in supplying heat by conduction to the undersurface of the specimen.

The Thaine tissue dryer (Thaine, 1962; Thaine and Bullas, 1965) was designed particularly for plant materials. In this apparatus (Fig. 23, right) the tissues are carried on top of a copper rod which is maintained at —70° by immersion of its lower end in a freezing mixture. Immediately below the tissues, between them and the copper rod, there is an electrical heating unit. The tissues are in close proximity to the cold finger, and probably within the critical distance, so that additional heat for sublimation has to be provided by the electrical unit. A side arm is provided, to contain resin or other embedding media. These can thus be introduced into the tissue container without breaking the original vacuum.

The Leybold-Elliott design incorporates an efficient high vacuum system (mechanical and diffusion pumps) and the tissues are maintained at below —54°, in a glass vessel surrounded by a freezing mixture at —60°. The drying process, for tissues less than 1 mm. thick, is recorded as lasting for "nearly two days".

* Addresses of firms manufacturing apparatus described in this Chapter are given in Appendix 3, p. 597.

The Thieme (1965) design, shown in Fig. 25, consists of a bath containing a kerosene fraction (Esso Turbo Fuel), which is maintained at −40° by means of a mechanical refrigerator. The cooled fluid is circulated by a separate pump (one per drying unit) up into the base of the tissue container. The latter,

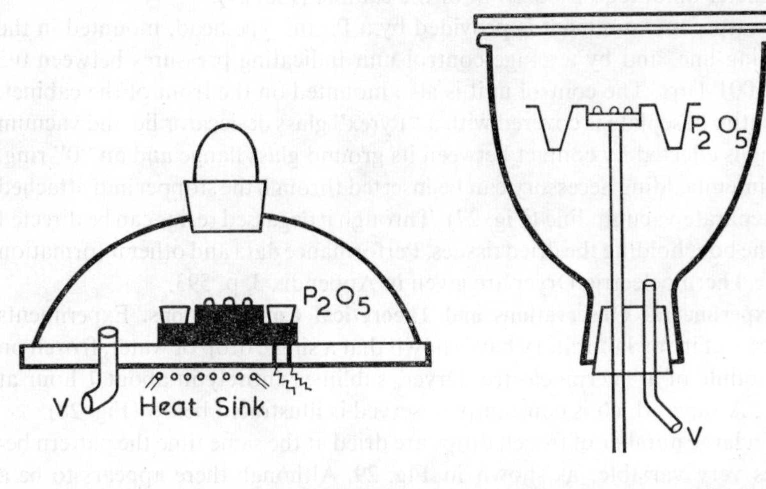

FIG. 25

surrounded by a plastic tray containing P_2O_5, is enclosed in an inverted flask with a removable base. A vacuum line enters, as does the tube carrying the coolant, through a rubber bung closing the neck of the flask.

Examination of Fig. 25 shows that there is an accidental but critically important similarity between the drying chambers of the Thieme apparatus, and of my own design. The significance of this observation will be discussed below, as part of the theoretical treatment of the process of thermoelectric freeze drying.

Thermoelectric Tissue Dryers.* In the original thermoelectric dryer (Pearse, 1963, 1964), which is illustrated diagrammatically in Fig. 25 (left), heat was applied to the specimen both by radiation and by conduction. The vapour gradient was maintained by a vacuum (10^{-3} Torr) and by placing a P_2O_5 trap a short distance away from the drying tissue. The commercial model, produced by Edwards High Vacuum Ltd, is shown in Figs. 26 and 27, facing p. 50.

In this apparatus, as in the original, two thermoelectric modules are mounted in series on a water-cooled base plate. The tissues are dried either in direct contact with the upper surface of the top module, or in the aluminium boat shown in the illustration. The boat is only used if automatic embedding in

* For a discussion of the Peltier effect and thermoelectricity see Chapter 2, p. 15.

wax is required, or if plastic resins are to be introduced through the resin-embedding accessory. A three-sided metal P_2O_5 tray is fitted on the upper surface of the bottom module, and a chrome/alumel thermocouple is inserted into a slot in the top module. The drying temperature can be read on a graduated meter mounted on the front of the cabinet (Fig. 27).

Pressure measurement is provided by a Pirani type head, mounted in the pumping line, and by a gauge control unit indicating pressures between 0·5 and 0·001 Torr. The control unit is also mounted on the front of the cabinet. The entire baseplate is covered with a "Pyrex" glass desiccator lid and vacuum sealing is effected by contact between its ground glass flange and an "0" ring. A resin-embedding accessory can be inserted through the stopper and attached to a separate vacuum line (Fig. 27). Through it degassed resins can be directed into the boat holding the dried tissues. Performance data and other information on the Thermoelectric Dryer are given in Appendix 3, p. 593.

Experimental Observations and Theoretical Considerations. Experiments carried out in my laboratory have shown that a single drop of water, frozen on the module of a Thermoelectric Dryer, sublimes entirely in about 1 hour at —45°. A stage which is constantly observed is illustrated below (Fig. 28).

If a large number of frozen drops are dried at the same time the pattern becomes very variable, as shown in Fig. 29. Although there appears to be a relationship to the situation of the vacuum outlet, and to the P_2O_5 trap, even this is inconstant. There is obviously considerable turbulence of the molecular stream and this, on the whole, is greatest on the side from which heat (light) is radiated on to the surface of the drops.

The "drying shape" of a tissue block depends on the degree of contact with the module. If the block is stuck to the latter with water (ice) the pattern is that shown in Fig. 30 A. If, however, the block is merely in loose contact (as it is in practice) the drying shape changes to that shown in Fig. 30 B.

If a thermocouple, or thermistor, is mounted between two ice drops the temperature which it indicates falls, in the early stages of drying, to about 5° below that of the module (Fig. 31). Subsequently the lower drop of ice evaporates, leaving the drop which is lying on the thermistor unaltered. This indicates the all important effect of heat supplied by the module and the relative

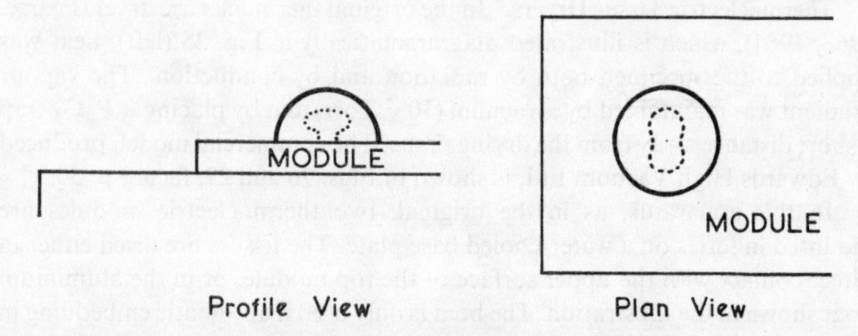

FIG. 28. Freeze-drying water drop, initial and mid-drying stages.

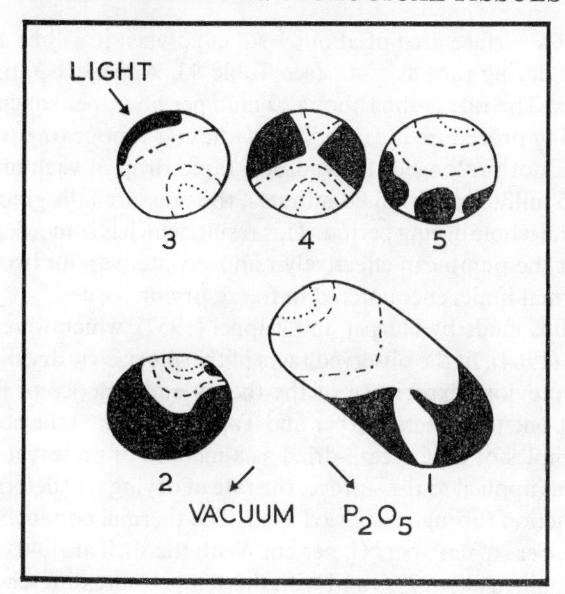

FIG. 29. Variable Drying Pattern of Multiple Water Drops on a Single Module.

unimportance (for sublimation) of heat supplied by radiation from the dome of the drying chamber, even when no dried shell of tissue is present.

As reported by Bancroft (1964), in a thin block of fresh rat kidney weighing 100 mg. there are between 75 and 79 mg. of removable water. When the block has been freeze-dried it will weigh between 21 and 25 mg. Such a block may be

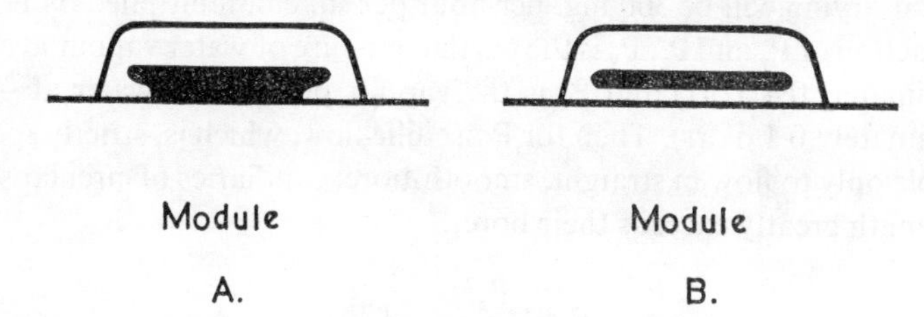

FIG. 30. Dependence of Drying Pattern on Tissue-Module Contact.

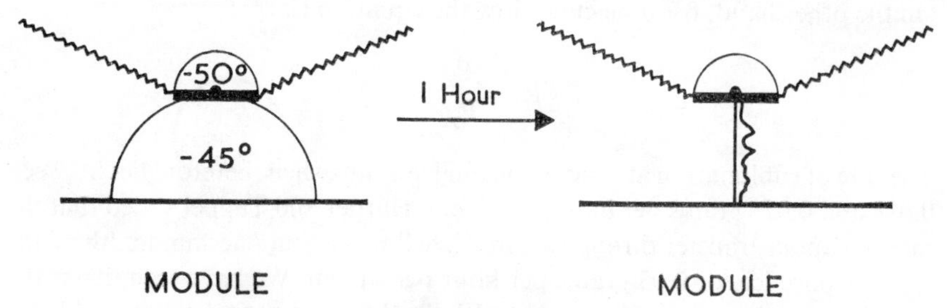

FIG. 31. Two Stages of Drying. Thermistor between two Ice Drops.

cut to present a surface area of about 1 sq. cm., when it will be about 2 mm. thick and the drying time at —40° (see Table 43, Appendix 3, p. 594) will be about 4 hours. The rate is thus about 20 mg., per hour, per sq. cm.

If the drying procedure is carried out without a vapour trap (no P_2O_5) the drying time is not prolonged. Instead of a rapid drop in vacuum to the base level (below 5 millitorr) within 30 minutes, the pressure falls gradually to this level during the whole drying period. This result, which is somewhat surprising, indicates that the pump can effectively remove water vapour from the system up to the normal limits encountered in freeze-drying tissues.

Calculations made by Harper and Tappel (1957), which were later amplified by Rowe (1964), fit the observed facts of thermoelectric drying much more closely than previous experience of the theory and practice of freeze-drying would permit one to expect. Harper and Tappel examined the conditions obtaining in samples of beef, freeze-dried as a method of preservation for food. When heat was applied to the surface, the rate of drying was determined by the rate of conduction through the dried shell. The thermal conductivity was 0·3 cal. per hour, per sq. cm., per °C, per cm. With the shell at +40° C the drying boundary remained at —27° C and with this temperature difference (67° C) the rate of heat conduction was about 200 cals per hour, per sq. cm. Assuming a latent heat of sublimation of 680 calories per gram the rate of sublimation was 0·3g. per hour per sq. cm. (The rate of sublimation of pure ice at —40° C is 3 grams per hour per sq. cm.)

Rowe's calculations were intended to fit the case of a tissue block with the frozen interface maintained at —40° C, assuming the partial pressures of air and water to be zero. According to this author when the dry shell is 1 mm. thick the rate of drying will be 300 mg. per hour per sq. cm. multiplied by R, where R is a function of P_y and P_z. P_y is the partial pressure of water vapour at —27° C (approximately 0·4 Torr) and P_z is the vapour pressure of water at —40° C (approximately 0·1 Torr). Then for Poiseuille flow, which is, strictly speaking, applicable only to flow in straight, smooth-bore, capillaries of circular section, whose length greatly exceeds their bore,

$$R = \frac{P_z^2}{P_y^2} = \frac{1}{16}$$

On the other hand, for molecular flow the equation is,

$$R = \frac{P_z}{P_y} = \frac{1}{4}$$

The rate of sublimation at —40° C through a 1 mm. shell therefore lies between 0·018 and 0·075 grams per hour per sq. cm. Harper and Tappel stated that the rate of vapour transfer through a 1 mm. shell was about one-hundredth of the rate for pure ice, or 0·03 grams per hour per sq. cm. We have seen, from the experimental data given above, that with the thermoelectric dryer the sublimation rate at —40° C for a typical mammalian tissue is 0·02 grams per hour per sq. cm. There is thus general agreement between theory and practice, despite the different conditions allowed for in the theoretical calculations. The experimental data suggest that the 20 calories per hour per sq. cm. required to maintain drying at the observed rate are supplied by conduction from the module and not by radiation through the dried shell.

Embedding, Post-Fixation, Sectioning and Mounting

Embedding in Paraffin Wax. It has been customary in the past, with few exceptions, to embed the dried specimens in paraffin wax. At the end of the drying stage we are left with an almost completely water-free specimen at a temperature of —40° and at 0·001 Torr. There are two principle methods for embedding this in paraffin. If the design of the drying chamber allows, it may contain previously degassed paraffin wax (m.p. 42° to 56°) or, if a drying boat is employed, a ring of wax may be placed in it before drying begins. When the process is complete the chamber is allowed to reach room temperature. While still maintaining some vacuum, heat is applied to the wax (either electrically or by direct application) which then melts and allows the specimens to be impregnated. After a suitable interval the wax is allowed to solidify and air is admitted to the system.The vacuum required is much less than that required for drying (0·1 Torr).

The most usual alternative to this procedure is to allow the dried tissues to come to room temperature while still under vacuum, then to admit air and transfer them from the chamber into a pot of melted wax in a conventional vacuum-embedding bath. Exposure to moist air has no adverse effect but the practice, employed by some earlier authors, of transferring the cold tissues (at —40°) directly into a wax bath at 56° caused appreciable shrinkage. The employment of unsuitable waxes is still the prime cause of destructive artifact at the embedding stage. Suitable waxes are recommended in Appendix 3, p. 595.

For the preparation of rapid paraffin sections for histological (diagnostic) purposes embedding in the drying chamber is clearly the method of choice. If air has been admitted to the tissues, as when the second alternative is employed, there is some evidence that its rapid removal and substitution by hot wax under vacuum causes a certain amount of structural damage. Even if the time of paraffin embedding is kept below 10 minutes, as it should be, there is no doubt that the process can cause both diffusion of lipid and non-polar substances and also a variable amount of shrinkage.

Other Embedding Media. In the past, very few alternative embedding media have been employed for dried mammalian tissues. In order to avoid heating, some workers advised the use of celloidin, either in the usual alcohol-ether solvent or, preferably, in methyl benzoate. It is very much better to follow celloidin by paraffin in a complete double embedding procedure. This technique was employed by Burstone (1962) for many years with excellent results both in terms of preservation of enzymes (his primary objective) and of cytological detail.

Carbowaxes

Embedding in carbowaxes has proved less popular than early experiences suggested and lipid studies, for which such treatment might be considered, are more commonly carried out on fixed tissues (see section on vapour fixation below). Tests carried out with carbowaxes 1500 and 4000, and infiltration at atmospheric pressure, produced extremely bad results both from the point of view of morphology and preservation of chemical activity.

Resins and Liquid Paraffin

There has recently been a great increase in interest in resin embedding for electron microscopy. Attachments are provided for a number of modern freeze-dryers which allow prepolymerized and degassed resins to be introduced into the drying chamber to infiltrate the dried specimens under their original vacuum. If resins such as methacrylate are introduced into the apparatus in this way the resulting contamination is severe and long-lasting. If the final result of the process was rewarding one might accept this but, in fact, methacrylate damage to unfixed freeze-dried tissues is extreme.

In an extensive series of tests carried out in my laboratory, only a few relatively innocuous media for *in vacuo* infiltration could be found. One of these was liquid paraffin (Sp. G. 0·830 to 0·870), which has the advantage of being miscible with paraffin wax.

In my view, the sensitivity of *unfixed* freeze-dried tissues to damage by infiltration *in vacuo* is so high as to make the practice useless. I do not believe that any of the present series of resins employed for electron microscopy can be used in this way.

Post-Fixation (Liquid Media). Simpson (1941b) infiltrated his dried tissue blocks for one week with absolute alcohol in order to remove the last traces of water and to reduce the subsequent shrinkage in paraffin wax. Fixation of this type is seldom if ever employed today. It is possible to infiltrate the dried tissues, in their original vacuum, with an inert medium containing a fixative. For his plant tissues Thaine (1962) used OsO_4 in benzene, followed by embedding in paraffin wax. Alternatively formaldehyde, glutaraldehyde or hydroxyadipaldehyde (see Chapter 4, p. 64) can be extracted from their aqueous solutions into ether or liquid paraffin. The former must be dried over anhydrous sodium sulphate before use. The results produced by these post-fixations in liquid media are better than might be expected, at the light microscope level. When they are followed by embedding for electron microscopy the results are unsatisfactory.

Post-Fixation (Vapour). Introduction of fixatives in vapour form into the drying chamber is generally perfectly feasible, and Lacy and Blundell (1955) incorporated into their Stowell type tissue dryer an additional inlet through

FIG. 26. Close up view of two-stage thermoelectric platen, power supply, temperature recording attachment and O-ring seal.

FIG. 27. Thermoelectric Freeze-Dryer showing complete assembly with resin-embedding accessory.

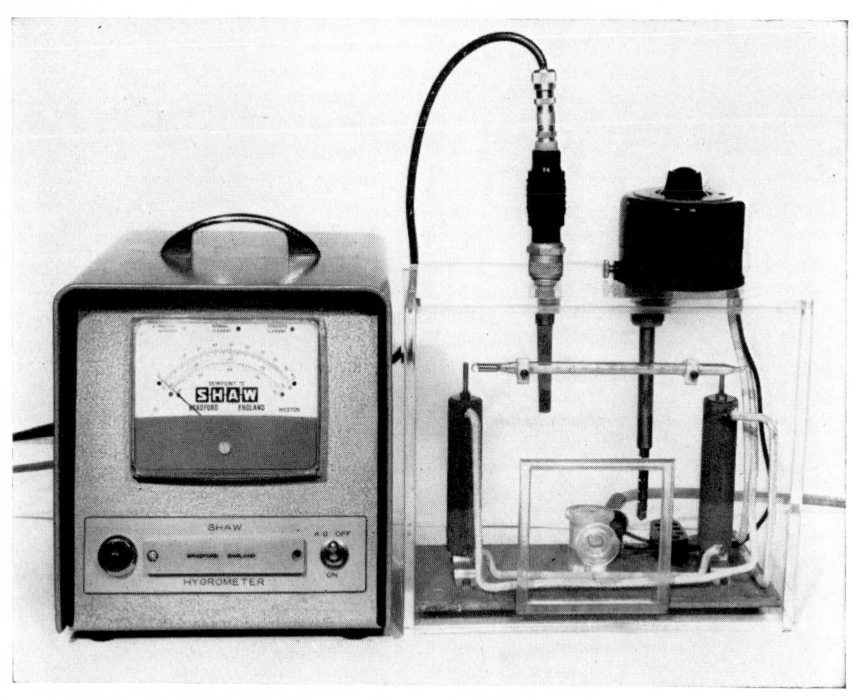

FIG. 32. Simple Thermostatically Controlled Vapour Fixation Chamber.

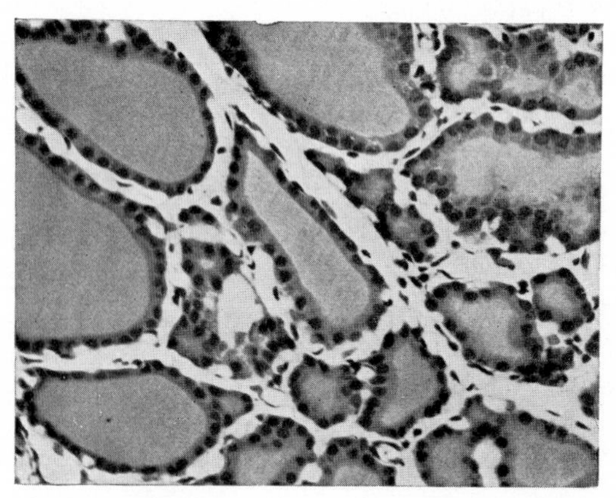

FIG. 33. Freeze-dried, alcohol-fixed, rat thyroid. The evenly stained colloid shows no shrinkage vacuoles. Hæmalum and cosin. × 200.

which formaldehyde vapour could be admitted at the end of the drying process. In many cases, however, the degree of contamination of the apparatus is unacceptable. There is, moreover, no particular advantage over the alternative procedure of removing the dried specimens to a separate apparatus. The latter can be designed specifically to provide the required vapour, at the correct temperature and at the correct level of humidity. Such an apparatus is shown in Fig. 32, opposite. It was designed for formaldehyde vapour (from paraformaldehyde) at 70–80°. In the case of other vapours the temperatures employed for fixation may conveniently be those of the standard histology laboratory incubators (37°, 60°). A closed vessel containing the vapour source, and the tissues resting on a gauze platform, is then all that is required.

The following fixatives and metallic vapours have been tested:

Formaldehyde, 37°, 60°, 70°, 80°.
Glutaraldehyde, 37°, 60°, 70°, 80°.
Acrolein, 37°, 60°.
Osmium tetroxide, 37°, 60°, 70°.
Chromyl chloride, 37°, 60°, 70°.
Chromium hexacarbonyl, 37°, 60°.
Molybdenum hexacarbonyl, 37°, 60°.

Out of this list, as far as light microscopic histochemistry is concerned, by far the most important and successful is the first, used at 50°, 70° or 80°, for 1–3 hours. Its reactivity is then extremely high and preservation of tissue components is correspondingly excellent. The freeze-dried, formaldehyde vapour-fixed, paraffin-embedded block can be used advantageously for almost all histochemical methods except those for enzymes. Wider appreciation of its merits in this respect will certainly produce a technological revolution in histology and histochemistry equivalent to that brought about by the introduction of the cryostat.

Glutaraldehyde, a faster fixative than formaldehyde in aqueous solution, is much slower than the latter in gaseous form. It produces a slower and less adequate fixation, as does acrolein. The rate of penetration of metallic and non-metallic vapours into the dried block appears to be very rapid; OsO_4 and CrO_2Cl_2 are exceedingly fast in this respect.

The chemistry of vapour fixations is discussed in Chapter 5, particularly with reference to formaldehyde (p. 86). Their application to electron microscopy is dealt with in Chapter 32.

Sectioning and Mounting. The most usual practice is to cut the paraffin-embedded tissues on a standard bench type microtome. Until 1960 it was customary to cut material embedded in carbowax, or in low melting-point paraffin wax, on a freezing microtome. The advent of thermoelectric cooling for block holders and knives (see Chapter 2, p. 15) has made it easy to apply suitable cooling to any selected microtome and a base sledge design may well be found best for cutting all types of freeze-dried material.

Floating out provides a further problem. At one time the process was avoided entirely and single sections were applied in the dry state to warm, egg-albuminized, slides and flattened by pressure from a finger. Dry mounting on cold slides was not usually favoured. Mendelow and Hamilton (1950) introduced the useful technique of painting a thick film of melted paraffin on to the exposed face of the block between cutting each section, so that each section is backed by a thick supporting layer of paraffin. The sections are then placed cut face downwards on cold slides and secured by gentle pressure. The floating-out methods used for unfixed freeze-dried sections fall into two categories, those which avoid, and those which involve, fixation of tissue components. In the first category are methods using mercury or light petroleum (petroleum ether). The use of mercury, warmed to about 45°, was suggested by Danielli and first used by Harris, Sloane and King (1950). Sections floated out on this medium flatten well and are easily picked up by allowing the coated side of an albuminized slide to come into contact with them. If light petroleum is used the sections are instantaneously deparaffinized and simultaneously subjected to severe changes by the removal of lipids. Even if the latter are regarded as unimportant, this method should not be employed. (N.B. Mercury vapour is toxic.)

In the second category are methods using 70 per cent. alcohol, 10 per cent formalin and other liquid fixatives. Unfixed sections floated out on these are, of course, subjected to fixation as well as flattening. The use of 70 per cent. alcohol has nothing particular to recommend it, but, if preservation of lipids is particularly desired, 10 per cent. formalin or formol-calcium mixtures may be employed. If paraffin wax is used for embedding, however, its subsequent removal will remove a large proportion of the lipids present in the section. Carbowax sections floated out on formol-calcium mixtures, or on a formol-diethylene glycol mixture such as that described by Rinehart and Abu'l-Haj (1951) (Appendix 1), would be ideal for lipid studies by histochemical techniques of the type described in Chapter 12, but for those inherent difficulties which prevent successful infiltration of properly dried tissues with carbowax. Sections cut from vapour-fixed blocks can usually be floated out on water without loss of tissue components. This refers especially to formaldehyde-treated tissues.

If we now suppose that the material quenched in Section 1, dried in Section 2, and embedded, cut and dry-mounted in Section 3, has reached us in the form of a 6–8 μ section on an albuminized slide with the wax still *in situ*, there remains only the final treatment before it can be brought to the microscope for examination.

Final Treatment before Microscopy. This depends to a large extent on what particular component of the section we intend to examine, and by what means. The various possibilities can be conveniently considered under five headings:

(1) Examination of unfixed material, in inert media, by physical methods.

(2) Examination of unfixed material by micro-incineration or by treatment with buffer extractions, etc.

(3) Examination after post-sectioning fixation by conventional histological and histochemical techniques.

(4) Examination after pre-embedding fixation by physical and chemical techniques.

(5) Examination by autoradiographic techniques.

Examination in Inert Media. Unfixed freeze-dried tissues are particularly suitable for microscopy by phase contrast, and by polarized, ultra-violet or dark-field illumination. Sections can be prepared by freeze-drying for examination by any of these techniques with the minimum of chemical and physical disturbance to tissue components, and their use in histochemistry in combination with freeze-dried material has greatly increased in the past few years. It has been customary for many years to mount sections in glycerine before examination in polarized or ultra-violet light, or by dark-field techniques, and the use of this medium has been extended to freeze-dried sections. It has been applied either after removal of the wax or leaving the wax *in situ*. Although it is an excellent mountant from the physical point of view, since it reduces the amount of scattering of light at refractive index boundaries, it is not so suitable from the histochemical point of view. Many substances in the tissues are soluble in glycerine and those which are not, such as the majority of proteins, undergo marked swelling and consequent distortion. An alternative mountant, much used for studies on tissue fluorescence, is liquid paraffin. This removes the paraffin wax and is sufficiently inert not to alter the tissue proteins appreciably. Still better results are obtained by mounting directly in nonane, as suggested by Bell (1952a), which also removes the wax. This method is particularly suitable for phase contrast microscopy. Mendelow and Hamilton (1950) described a method of mounting in a synthetic medium such as clarite after removal of wax with light petroleum. If the latter is used for removing the wax, care must be taken to avoid drying before application of the clarite or large numbers of air bubbles will remain trapped in the section. The method is more easily carried out with xylene.

Examinations after Extractions, etc. The fact that the proteins of freeze-dried sections remain almost entirely undenatured makes it possible for their solubility in various buffers to be determined. In this type of technique, which was employed by Catchpole (1949) on the pituitary mucoproteins and by Gersh (1949) on thyroid colloid, a series of sections is incubated in equimolar solutions of the buffer at different pH levels. After a suitable period of incubation the sections are fixed in alcohol or formalin and the residual protein is stained either by using an acid dye or by the Millon or other specific reactions for protein. An estimate is then made of the point at which the least extraction of the material under consideration has occurred and this point of minimum solubility is taken as the isoelectric point. Other extractive techniques for proteins are usually performed on sections in which these have been precipitated and denatured but the lipid extraction techniques (see Chapter 12 and Appendix 12) can be performed on unfixed material. Many of these techniques will denature and fix the proteins at the same time as they extract the lipids and it must not be assumed, even with non-fixing extractives such as pyridine, that no alteration in the remaining tissues has occurred. There is evidence (Lovern, 1955) that freeze-drying itself may denature some lipoproteins. He states that β-lipoprotein, for instance, is denatured when its water of hydration, which binds lipid and polypeptide together, is removed by any method. Caspary and Kekwick (1957) found an increase in the sedimentation constant of freeze-dried fibrinogen, and Mayersbach (1957) observed gross differences in the solubilities of normal and freeze-dried proteins after alcohol precipitation. These observations all suggest the need for caution in the interpretation of the results of physical techniques applied to freeze-dried materials. Mayersbach discussed at length the problems of handling freeze-dried sections prior to the application of various histochemical techniques. He drew attention to a number of possible artifacts due to post-fixation and stated clearly that the proteins in freeze-dried tissues could not be regarded as "native".

In studying the distribution of dyes introduced *intra vitam*, freeze-dried material has been found particularly good for the purpose. An embedding medium should be chosen which is soluble in a fluid in which the dye is insoluble, and final examination can be carried out in this same fluid. In this way Catchpole, Gersh and Pan (1950) investigated the distribution of intravitally injected Evans blue in the connective tissues of the rat. For determining the distribution of inorganic substances freeze-dried material is also excellent. Packer and Scott (1942) used micro-incineration methods for calcium and sodium, and these entirely avoid the question of dewaxing and fixation. It is probable that cytochemical methods for inorganic ions, such as those employed by Gersh (1948), allowed too much diffusion of inorganic salts to occur during the reaction for critical localization to be achieved. Even so, the only chance of obtaining reasonably accurate localization is by the use of freeze-dried sections.

Examination after Post-sectioning Fixation. Once again, the choice of fixative for application to the mounted sections will be dictated by the nature of the components we wish to study and the tests which it is proposed to carry out. For purely morphological studies with conventional staining, fixation in 70–80 per cent. alcohol for 16–24 hours has been employed, after rapid removal of wax with light petroleum. This fixative is also useful if histochemical tests for simple proteins and amino-acids are proposed. We know very little of the action of alcohol on freeze-dried proteins, or, for that matter, on native proteins, but in sections the former are not aggregated as are the proteins in undried tissues. Very little morphological difference is observed between unfixed freeze-dried cells, examined by physical means in an inert medium, and similar cells fixed in alcohol and stained histologically or histochemically (Fig. 33).

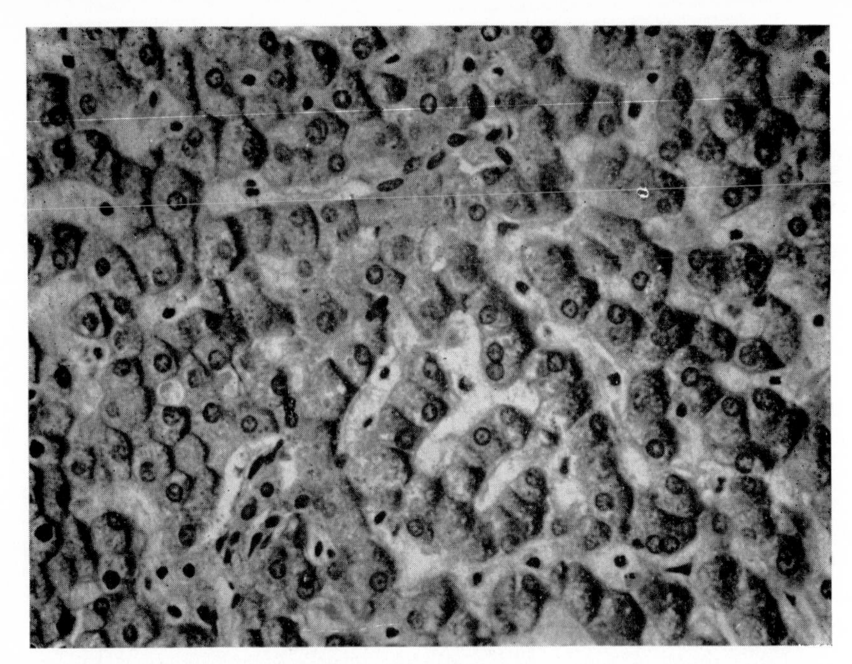

FIG. 34. Rat Liver. Formalin-fixed paraffin section. Best's Carmine, × 370.

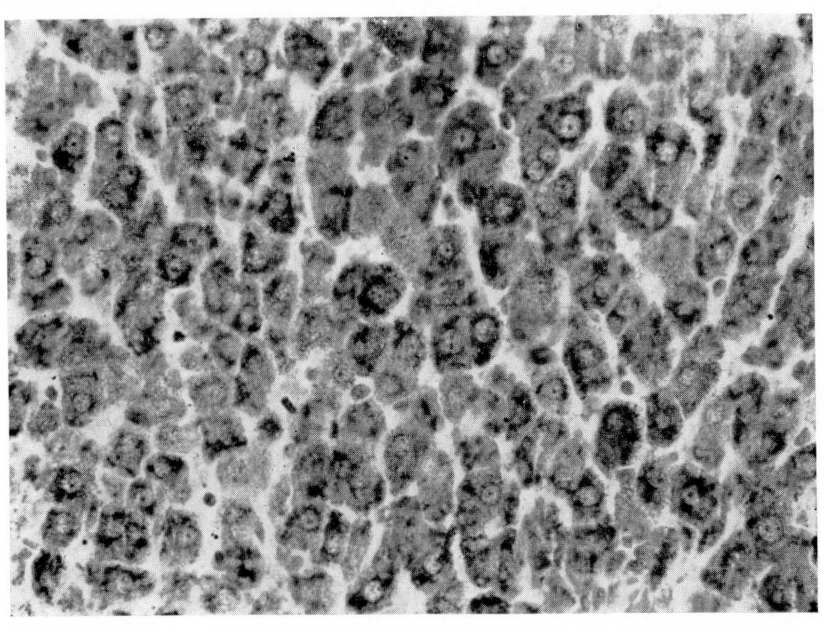

FIG. 35. Rat Liver. Freeze-dried, formaldehyde vapour fixed. Best's Carmine, × 370.

From these points of view, there is therefore little objection to the use of alcohol as a fixative for freeze-dried material. The use of 100 per cent. acetone is similarly free from objection and it may be the best fixative for some enzymes which are more sensitive to alcohol.

In some cases the preservation of enzymes in freeze-dried tissues is particularly good, probably approaching 100 per cent. Doyle (1950) for instance, found that the peptidase activity of rabbit appendix, using alanylglycine as substrate, was unaffected by freeze-drying and also by subsequent embedding in paraffin. If the various histochemical methods for the localization of enzymes are applied to unfixed sections, diffusion artifact is considerable. This is because the enzyme itself diffuses into the incubating medium and produces its products there, the insoluble ones being distributed at random throughout the section. So much material is lost into the incubating fluid from unfixed sections that the general microscopic appearances are very poor. Brief fixation in alcohol or acetone (1–2 hours) is sufficient to reduce the subsequent diffusion of many enzymes without interfering with their function. Even labile enzymes are partly preserved by such treatment, and alkaline phosphatase is particularly well preserved. Acid phosphatase, on the contrary, is less well preserved by freeze-drying and short acetone fixation than by some more conventional techniques (Chapter 16). Reale (1955a and b) compared freeze-dried with cold knife sections in a Gomori-type procedure for alkaline phosphatase. He found the former to be superior in most respects.

It was reported by Naidoo and Pratt (1957) that the specific cholinesterases were very poorly preserved in most tissues after freeze-drying and that in brain they were destroyed. Reasons for these findings were never very clear. The preservation of both specific and non-specific cholinesterases is, in fact, excellent, as reported also by Fredricsson *et al.* (1960). After brief post-sectioning fixation in acetone they can be demonstrated by the majority of the methods described in Chapter 17.

Dehydrogenases are not usually demonstrable histochemically in freeze-dried sections, with or without post-fixation. The complicated problems raised by this observation are by no means solved. Although in the 2nd Edition of this book (1960) it was stated that "not by any means all of the damage can be attributed to the final stages of embedding, mounting and dewaxing", it seems more likely that it is indeed these stages which are responsible for interference with the histochemical demonstration of dehydrogenases.

For studies on mucoproteins such as those of the anterior pituitary gland, after removal of paraffin, fixation by means of alcohol or acetone will be found inadequate. These mucoproteins remain entirely soluble in water after precipitation by alcohol or acetone and other precipitant fixatives, like picric and trichloroacetic acids, leave them in a similar state. A few of the more insoluble mucopolysaccharides and mucoproteins can be preserved by post-fixation with 10 per cent. formalin for 10–60 minutes. It is probable, however, as stated by Hoerr (1936) that no aqueous protein fixative acts fast enough to prevent the prior action of water vapour on the dehydrated tissue. For this reason formalin-alcohol mixtures were at one time recommended for the study of water-soluble mucoproteins. Plate IVa shows a freeze-dried section of pancreas fixed in such a manner and subsequently stained by the periodic acid-Schiff method and toluidine blue.

Glycogen is the only naturally occurring polysaccharide which we need to consider in animal tissues. It is well preserved by the freeze-drying technique and can be demonstrated in unfixed sections by the method of Mancini (1948), referred to below. Other staining techniques for glycogen cannot be applied unless the material is first deparaffinized and then fixed in alcohol, and staining must be carried out with celloidin protection to prevent diffusion of glycogen out of the section. With freeze-drying, the classical streaming artifact is no longer observed (see Fig. 34, facing p. 55). Instead, glycogen is distributed throughout the cytoplasm in diffuse form and in granules generally smaller than those observed with ordinary fixation (see Fig. 35).

Examination after Pre-embedding Fixation. As stated earlier in this chapter, pre-embedding fixation of freeze-dried blocks can be carried out by means of either liquid or vapour. For enzymes only the former offers reasonable preservation. The double embedding method of Burstone (1962) is, in effect, a simultaneous fixation and only a proportion of enzymes survive the procedure. Simpson's (1941) freeze-substitution of the freeze-dried blocks, using dry acetone at —50° for 18 hours, has considerable merit for enzyme preservation. Of the newer fixatives, glutaraldehyde (in anhydrous ether) produces a good morphological result and the paraffin sections can be floated out on water without additional loss of enzymes.

Examination of pre-embedding vapour-fixed material indicates that preservation of mucoproteins and mucopolysaccharides is superior to that achieved by any other technique whatsoever. Plate Vb, facing p. 322 shows the application of the technique to rat submaxillary salivary gland. The mechanism of formaldehyde vapour fixation (Tock and Pearse, 1965) is further discussed in Chapter 5, p. 87.

Preservation of glycogen is equal to that provided by post-fixation. In both cases diastase effectively removes 100 per cent. of the polysaccharide within the usual short incubation period. The nucleic acids are exceptionally well preserved by formaldehyde vapour. This is especially so in the case of RNA and ribonuclease (Chapter 25). requires prolonged incubation to bring about its total removal.

Fixation with chromyl chloride vapour alone produces a brittle section, almost impossible to cut. If it accompanies or follows formaldehyde or osmium tetroxide, chromation of lipids and proteins is effected. These matters are further discussed in Chapter 5, and elsewhere under the appropriate headings.

Examination by Autoradiography. The application of autoradiographic techniques to freeze-dried materials is considered in Chapter 30.

REFERENCES

ALTMANN, R. (1890). "Die Elementarorganismen und ihre Beziehungen zur Zellen." Leipzig.
ARCADI, J. A., and TESSAR, C. (1954). *J. lab. clin. Med.*, **43**, 479.
BANCROFT, J. D. (1964). In "Progress in Medical Laboratory Technique." Vol. 3, p. 41., Ed. F. J. Baker, Butterworths, London.
BARTELMEZ, G. N. (1940). *Anat. Rec.*, **77**, 509.
BAYLISS, W. M. (1915). "Principles of General Physiology." London.
BELL, L. G. E. (1952a). *Int. Rev. Cytol.*, **1**, 35.
BELL, L. G. E. (1952b). *Nature, Lond.*, **170**, 719.
BELL, L. G. E. (1955). In "Physical Techniques in Biological Research." Oster, G., and Pollister, A. W., eds. 1957, Academic Press, Vol. 2.
BENDITT, E. LAGUNOFF, D., and JOHNSON, F. B. (1961). *Arch. Path.*, **72**, 541.
BENSLEY, R. R., and HOERR, N. L. (1934). *Anat. Rec.*, **60**, 251.
BRANTON, D., and JACOBSON, L. (1961). *Ex. Cell Research*, **22**, 559.
BURSTONE, M. S. (1956). *J. nat. Cancer Inst.*, **17**, 49.
BURSTONE, M. S. (1962). "Enzyme Histochemistry and its application to the study of Neoplasms." Academic Press, New York.
BUTLER, L. O., and BELL, L. G. E. (1953). *Nature, Lond.*, **171**, 971.
CABRINI, R. L. (1955). *Rev. Assoc. Med. Argent.*, **69**, 387.
CAPRINO, G. (1956). *Boll. Soc. ital. Sper.*, **31**, 1446.
CASPARY, E. A., and KEKWICK, R. A. (1957). *Biochem. J.*, **67**, 41.
CATCHPOLE, H. R. (1949). *J. Endocrinol.*, **6**, 218.
CATCHPOLE, H. R., GERSH, I., and PAN, S. C. (1950). *J. Endocrinol.*, **6**, 277.
DOYLE, W. J. (1950). *Fed. Proc.*, **9**, 34.
EMMEL, V. M. (1946). *Anat. Rec.*, **95**, 159.
ERÄNKÖ, O. (1954a). *Acta path. Scand.*, **35**, 426.
ERÄNKÖ, O. (1954b). *Acta anat.*, **22**, 331.
FERNÁNDEZ-MORÁN, H. (1960). *Ann. N.Y. Acad. Sci.*, **85**, 689.
FREED, J. J. (1955). *Lab. Invest.*, **4**, 106.
FREDRICSSON, B., FUXE, K., HOLMSTEDT, B., and SJÖQUIST, F. (1960). *Acta morphol., Neerlando-Scand.*, **3**, 107.
GERSH, I. (1932). *Anat. Rec.*, **53**, 309.
GERSH, I. (1948). *Bull. int. Ass. med. Mus.*, **28**, 179.
GERSH, I. (1949). *J. Endocrinol.*, **6**, 282.
GERSH, I., and STEPHENSON, J. L. (1954). In "Theory and Practice of Freeze-Drying." Ed. R. J. Harris. Acad. Press Inc., New York.
GERSH, I., VERGARA, J., and ROSSI, G. I. (1960). *Anat. Rec.*, **138**, 445.
GLICK, D., and MALMSTROM, B. G. (1952). *Exp. Cell Research*, **3**, 125.
GLICK, D., and BLOOM, D. (1956). *Exp. Cell Research*, **10**, 687.
GOODSPEED, T. H., and UBER, F. M. (1934). *Proc. Nat. Acad. Sci., Wash.*, **20**, 495.
GREAVES, R. I. N. (1954). In "Biological Applications of Freezing and Drying." Ed. R. J. C. Harris. Acad. Press, New York, p. 87.
GRUNBAUM, B. W., and WELLINGS, S. R. (1960). *J.Ultrsatruct. Res.*, **4**, 730.
HARPER, J. C., and TAPPEL, A. L. (1957). In "Advances in Food Research." Vol. 7. Eds. E. M. Mrak and G. F. Stewart. Acad. Press, New York, p. 172.
HARRIS, J. E., SLOANE, J. F., and KING, D. T. (1950). *Nature, Lond.*, **166**, 25.
HARTLIEB, J. VON (1954). *Acta histochem.*, **1**, 135.
HOERR, N. L. (1936). *Anat. Rec.*, **65**, 293.
JANSEN, M. T. (1954). *Exp. Cell Research*, **7**, 318.
JENSEN, W. A. (1954a). *Exp. Cell Research*, **7**, 572.
JENSEN, W. A. (1954b). *Stain Tech.*, **29**, 143.
JENSEN, W. A., and KAVALJIAN, L. G. (1957). *Stain Tech.*, **32**, 33.
KULENKAMPFF, H. (1956). *Z. wiss. Mikr.*, **62**, 427.
LACY, D., and BLUNDELL, M. (1955). *J. roy. micr. Soc.*, **75**, 48.
LACY, P. E., and DAVIES, J. (1959). *Stain Tech.*, **34**, 85.
LOVERN, J. A. (1955). "The Chemistry of Lipids of Biochemical Significance." Methuen, London, p. 81.

LUYET, B. J., and GEHENIO, P. M. (1938). *Biodynamica*, **2**, 1.
LUYET, B., and WILLIAMS, R. (1962). *Biodynamica*, **9**, 172.
MAMULINA, V., and ORLOVA, L. V. (1955). *Zh. Obsdich. biol.*, Moskva, **16**, 69.
MANCINI, R. E. (1948). *Anat. Rec.*, **107**, 149.
MANN, G. (1902). "Physiological Histology." Oxford.
MAYERSBACH, H. (1957). *Acta Anatomica*, **30**, 487.
MENDELOW, H., and HAMILTON, J. B. (1950). *Anat. Rec.*, **107**, 443.
MERYMAN, H. T. (1956). *Science*, **124**, 515.
MERYMAN, H. T. (1958). In "Proc. 2nd Symp. Freezing and Drying." Ed. R. J. C. Harris. London.
MERYMAN, H. T. (1959). *Science*, **130**, 628.
MERYMAN, H. T. (1960). *Ann. N.Y. Acad. Sci.*, **85**, 630.
MOBERGER, G., LINDSTRØM, B., and ANDERSSON, L. (1954). *Exp. Cell. Res.*, **6**, 228.
NAIDOO, D., and PRATT, O. E. (1956). *Acta Histochem.*, **3**, 85.
NAIDOO, D., and PRATT, O. E. (1957). *Biochem. J.*, **65**, 10P.
NEUMANN, K. (1952). "Grundriss der Gefriertrocknung." Musterschmidt Gottingen.
OSTROWSKI, K., and KOMENDER, J. (1960). *Folia morphol., Warsaw*, **11**, 171.
PACKER, D. M., and SCOTT, G. H. (1942). *Bull. int. Ass. med. Mus.*, **22**, 85.
PEARSE, A. G. E. (1963). *J. sci. Instrum.*, **40**, 176.
PEARSE, A. G. E. (1964). *Die Kalte*, **2**, 75.
PEASE, D. C., and BAKER, R. F. (1949). *Amer. J. Anat.*, **84**, 175.
REALE, E. (1955a). *Mon. zool. ital.*, **63**, 186.
REALE, E. (1955b). *Mon. zool. ital.*, **63**, 197.
REBHUN, L. I. (1965). *Fed. Proc.*, **24**, S 217.
REY, L. R. (1960). *Ann. N.Y. Acad. Sci.*, **85**, 514.
RINEHART, J. F., and ABU'L-HAJ, S. (1951). *Arch. Path.*, **51**, 666.
ROWE, T. W. G. (1960). *Ann. N.Y. Acad. Sci.*, **85**, 641.
ROWE, T. W. G. (1964).
SCOTT, G. H. (1933). *Protoplasma*, **20**, 133.
SCOTT, G. H., and WILLIAMS, P. S. (1936). *Anat. Rec.*, **66**, 475.
SENO, S., and YOSHIZAWA, K. (1960). *J. biophys. biochem. Cytol*, **8**, 617.
SIMPSON, W. L. (1941a). *Ibid.*, **80**, 173.
SIMPSON, W. L. (1941b). *Ibid.*, **80**, 329.
STEPHENSON, J. L. (1953). *Bull. math. Biophys.*, **15**, 411.
STEPHENSON, J. L. (1954). *Bull. math. Biophys.*, **16**, 23.
STOWELL, R. E. (1951). *Stain Tech.*, **26**, 105.
STUMPF, W. E., and ROTH, L. J. (1967). *J. Histochem. Cytochem.*, **15**, 243.
TALLQVIST, G., TALLQVIST, J., and ERÄNKÖ, O. (1967). *Histochemie*, **8**, 377.
THAINE, R. (1962). *Nature*, **195**, 1014.
THAINE, R., and BULLAS, D. O. (1965). *J. exp. Bot.*, **16**, 192.
THIEME, G. (1965). *J. Histochem. Cytochem.*, **13**, 386.
TRUMP, B. F., GOLDBLATT, P. J., GRIFFIN, C. C., WARAVDEKAR, V. S., and STOWELL, R. E. (1964). *Lab. Invest.*, **13**, 967.
TREFFENBERG, H. (1953). *Arkiv. Zool.*, **4**, 295.
TOCK, E., and PEARSE, A. G. E., (1965). *J. roy. micr. Soc.*, **84**, 519.
WANG, K. J., and GROSSMANN, M. I. (1949). *J. Lab. clin. Med.*, **34**, 292.
ZLOTNIK, I. (1960). *Quart. J. micr. Sci.*, **101**, 251.

CHAPTER 4

FREEZE-SUBSTITUTION OF TISSUES AND SECTIONS

IN order to avoid the need for expensive apparatus for dehydrating quenched tissues Simpson (1941a and b) suggested dissolving the ice in such tissues by placing them in liquid dehydrating agents at low temperatures. He originally recommended methyl cellosolve or ethyl alcohol for the latter purpose, at temperatures between —40° and —78°. There are thus two essential types of freeze-substitution, one in which the quenched tissue is subjected only to dehydration and the other in which it is dehydrated in a fluid which is also a fixative, or which contains an additional fixative. In the latter case the amount of fixation which occurs is governed by the temperature reached during the process. Absolute ethanol, for instance, may be used at —70° without fixing the tissues at all. Ethanol/water mixtures, however, will fix to some extent even at this temperature.

The process which we are considering has been called by many different names by different authors, and even if we allow for translations three or four remain. Lison (1949) used the term *congelation-dissolution*, while Blank, McCarthy and DeLamater (1951) described the process as *non-vacuum freezing-dehydrating*. Persson (1952) preferred *freezing-dehydration in liquid medium*, and Baud (1952) *fixation "par substitution."* Gourévitch (1953), on the other hand, used the general term *congelation-dehydration* to cover freeze-substitution as well as freeze-drying, while Woods and Pollister (1955) were content with *ice-solvent drying*. Peyrot (1956) was content with a straight translation, *congelamento-sostituzione*.

Of the terms quoted above I consider that only two merit final consideration in Chapter 4. These are freeze-substitution and ice solvent drying. The latter is brief, and accurately descriptive of the second half of the process. The former describes accurately the first half of the process and uses to describe the second half a single term which, though less accurate than ice solvent drying, is sufficiently comprehensive to include all variations of technique. It has, moreover, the merit of priority, and for this additional reason I have preferred it to the other.

Below, in Table 1, are set forth in abbreviated form a number of individual variations in the technique of freeze-substitution. It may appear that the differences recorded are not great. Nevertheless it is possible to produce quite extensive variations in the final picture by relatively minor alterations in technique.

Most of the authors quoted below have reported excellent preservation of cytological detail in their material. Woods and Pollister (1955), referring to plant tissues in particular, claimed that the results were fully equal to those

TABLE 1

Techniques of Freeze-Substitution

Author(s)	Quenching	Substituents	Temperature
Simpson (1941)	Isopentane-liquid N_2	Ethanol or methyl cellosolve	−40° to −78°
Lison (1949)	"As for freeze-drying"	Ethanol or Gendre fluid	−40°
Russell *et al.* (1949)	Isopentane-liquid N_2	Ethanol with metal salts	−78°
Blank *et al.* (1951)	Propylene glycol −20°	Propylene glycol	−20°
Persson (1952)	Isopentane-liquid air/N_2	EtOH/MeOH 1:1	−31°
Baud (1952)	Isopentane-liquid N_2	Acetic alcohol	−40°
Gourévitch (1953)	Ethanol −65°	Ethanol	−20°
Woods and Pollister (1955)	Isopentane-liquid N_2	Ethanol	−41° to −45°
Peyrot (1956)	Acetone/CO_2	Rossmann's fluid	−60° to −65°
Hancox (1957)	Isopentane-liquid N_2	*n*-butanol	−38°
Feder and Sidman (1958a)	Acetone/CO_2	Osmic-acetone	−70°
Feder and Sidman (1958b)	Propane-isopropane-liquid N_2	$HgCl_2$-Ethanol	−70°
Fernandez-Moran (1960)	Liquid helium II	Acetone-Ethanol	−80° to −130°
Rebhun (1961)	Freon 12-liquid N_2	Osmic-acetone	−80°
Rebhun and Gagné (1962)	Propane-liquid N_2	Acetone-Ethanol	−108°
Bartl (1962)	Isopentane-liquid N_2	Ethylene glycol monomethacrylate	−65° to −80°

obtained by any chemical fixation and standard dehydration, and Blank *et al.* (1951) showed that with their method not only was cytological detail preserved but that there was no loss of water-soluble substances such as radiophosphorus. They regarded ^{32}P as water-soluble, although presumably only a small proportion of the total amount in their tissues were present in unbound form. Lison (1953), Persson (1952) and Peyrot (1956) have shown that preservation of glycogen is excellent, and I can confirm that the distribution of this substance in freeze-substituted tissues very closely resembles that seen in freeze-dried material. Peyrot (1956) has compared his results (quenching at —80°, substitution at —60°) with those of Lison, and he notes that with the latter's method ice-crystal artifact is sometimes considerable. With the lower temperatures used by Peyrot glycogen was always diffusely intracellular in situation and ice crystals were not visible. This author also noted that intracellular organelles, such as mitochondria, might well be present after substitution by a solvent which would destroy them at room temperature.

Theoretical and Practical Considerations

A few fundamental studies of the mechanism of freeze-substitution have been made and our knowledge of the physics of the process has certainly increased since the publication of the 2nd Edition of this book (1960), in which

I was obliged to describe our understanding of the process as meagre. The earliest critical studies were made by Persson (1952) who used a standard model for his tests, composed of cylindrical slices cut from a bar of 14 per cent. gelatin containing 1 per cent. dextran. The water content of his slices was determined by weighing. It was noted that the dehydration time was shortest when the model blocks were quenched at —170°, and Persson explained this as being due to the production, at low temperatures, of a solid system of the colloids approaching a non-crystalline structure. He supposed that these were more easily dehydrated than the crystalline systems formed at higher temperatures. Bell (1952a), however, remarked that it was not known at what stage the substituent is exchanged for the water in the tissues. It now seems likely that this indeed takes place during the cold phase and not, as once believed, during the warming-up phase.

Studies on the fundamental effects of freeze substitution carried out by Hancox (1957) on various tissues from the rat suggested that protein denaturation was practically absent from tissues quenched at —160° and subsequently dehydrated in *n*-butanol for 3 days at —38°. Using ovalbumin or γ-globulin models Balfour (1961) found that over 90 per cent. of the protein remained soluble after freeze-substitution for 3 days in ethanol or acetone at —70°. Thus, according to this criterion, denaturation of protein was minimal. The effects on immunological reactivity were far greater. The same author showed that the capacity of the protein (γ-globulin) as an antibody was seriously diminished by freeze-substitution and embedding. Only half the expected amount of antigen (human serum albumin) was precipitated by the soluble fraction of the freeze-substituted antibody. According to Balfour the solubility of proteins could be lowered by allowing the substituting fluid to contain a small amount of water, or by adding osmium tetroxide. She found that in both cases immunological reactivity was considerably lowered.

Extraction of Tissue Constituents. The actual losses of material during the substitution phase have been calculated by Ostrowski *et al.* (1962a). These authors showed that nitrogen and phosphorus were lost to the substituting fluid substantially only when the latter was methanol. With ethanol N and P losses were much lower and they were least of all in the case of acetone. These N and P losses apparently were not associated with removal of nucleic acids and proteins but were presumed to involve mainly phospholipids and phosphoproteids. Extraction of lipids may well occur during the cold phase of substitution but this is even more likely during the warming phase and in subsequent procedures.

In a second paper (Ostrowski *et al.*, 1962b) the effects of temperature variation were measured. Only in the case of methanol could improved retention of N and P be shown as a function of lowering the substitution temperature.

It appeared at one time that most enzyme systems were inactivated by freeze-substitution procedures. Birns and Masek (1961) showed that this was by no means true and that alkaline and acid phosphatases, ATP-ase, 5-nucleotidase and aminopeptidase, as well as several dehydrogenases and diaphorases, could be demonstrated even after osmic-acetone substitution. The observations of Balfour (1961) indicate that the main losses of protein (and hence enzyme) occur elsewhere in the procedure, as shown in Table 2, below. This is taken from her results which were obtained with a radioactive labelling method.

TABLE 2

Losses of γ-globulin from Freeze-Substituted Tissue Sections

Medium used for Floating out	Percentage loss
18 % Sodium sulphate	0
Ethanol	1·8
Unbuffered saline	11·0
Buffered saline	26·0
30 % Glucose	31·0
Distilled water	39·0

Further losses of protein (and enzyme) occur every time the section comes into contact with aqueous media, as during staining, incubating and rinsing procedures.

Variables in Freeze-Substitution Procedure. The methods and principles of freeze-substitution were carefully considered by Feder and Sidman (1958b) who subdivided the problem into three main sections: (1) Conditions of Freezing, (2) Conditions of Substitution, (3) Use of Fixatives.

Conditions of Freezing. The most important variables in the first section are clearly the size of the specimen and the nature and temperature of the quenching bath. Although Feder and Sidman were able to obtain good results with whole eyes and intact mouse fœtuses, they admitted a normal limit of 3 mm. diameter. Even in a specimen of such relatively small size the standard of preservation (which depends on freedom from ice crystal formation; see Chapter 3, p. 29) will usually vary from area to area. The smaller the specimen and the faster the removal of the heat therefrom, the better the state of preservation. For most workers the fastest reasonably available quenching bath is a liquid N_2-cooled medium. The use of liquid helium II (Fernandez-Moran, 1959a, b, c, 1960; Bullivant, 1960) is unlikely to become widespread but there is a possibility of obtaining improvement by the technique of powder-coating as suggested by Moline and Glenner (1964). This method avoids the use of organic liquids, with which there is always a chance of explosion. More recent work by Bullivant (1965) suggests that liquid helium II may after all be less effective than previously supposed. His cooling curves were obtained by immersing a 3 mm. sphere of solder, containing a copper-constantan thermocouple, in liquid helium I, liquid helium II, and in propane cooled to $-175°$ with liquid nitrogen. These curves are reproduced in Fig. 36.

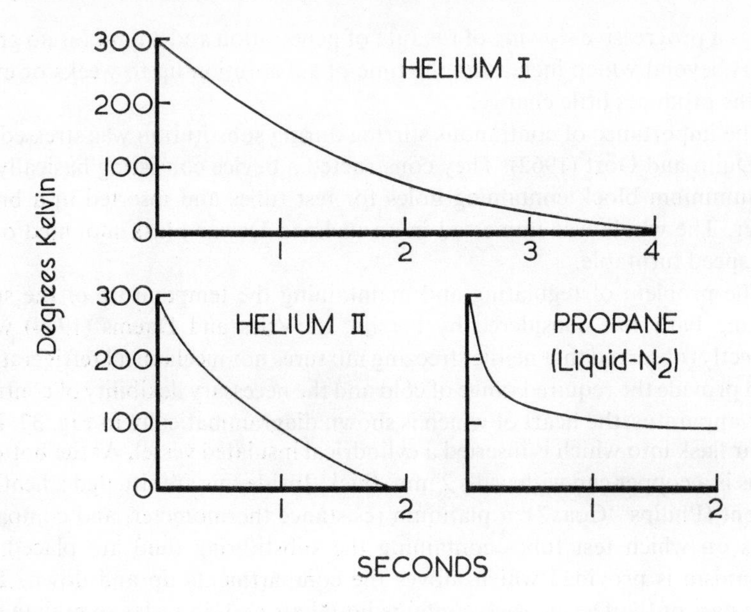

FIG. 36. Cooling Curves for Different Quenching Baths (After Bullivant, 1965)

There is thus substantial agreement with the experimental results of Rinderer and Haenseler (1960) who showed that where substantial temperature gradients exist (as when a specimen at room temperature is quenched in liquid helium II) a film of gas surrounds the specimen and prevents rapid heat transfer. According to Stephenson's (1956) calculations, the rate of cooling required to produce vitrification in tissue samples is in the region of 5000° per second, down to an estimated —100°. Such high rates of cooling are virtually unobtainable, even with very small specimens.

Conditions of Substitution. As shown in Table 1, p. 60, the range of substitution temperatures varies from —20° to —130°. Most work has been done in the middle range from —40° to —70°. Feder and Sidman found structure better preserved following substitution at —70° than at —25° and they attributed their better results to slower growth of ice crystals and the maintenance of temperature well below the melting point of the complex eutectic mixtures of the tissues (Chapter 3, p. 34).

The lower limit of temperature of the solvent (substituting fluid) is provided by increasing viscosity and is in the region of—80° for acetone and—110° for ethanol.

The rate of penetration in ethanol at —70° was stated by Feder and Sidman to be about 0·5 mm. per day, based on the diffusion into the specimen of eosin dissolved in the substituting fluid. Even if this is an accurate indicator of the depth of true substitution, which I doubt, it is my impression that there is always a progressive slowing of the rate of penetration and a limit (at no great depth) beyond which increasing the time of substitution up to weeks or even months produces little change.

The importance of continuous stirring during substitution was stressed by van Duijn and Oort (1962). They constructed a device consisting basically of an aluminium block containing holes for test tubes and inserted in a brass holder. The whole was immersed in an inclined Dewar Flask mounted on a slow speed turntable.

The problem of regulating and maintaining the temperature of the substituting bath was considered by Persijn, de Vries and Daems (1964) who (correctly) observed that neither freezing mixtures nor mechanical refrigerators could provide the required range of cold and the necessary flexibility of control. Their apparatus, the heart of which is shown diagrammatically in Fig. 37, is a Dewar flask into which is inserted a cylindrical insulated vessel. At the bottom of this is a copper can with walls 2 mm. thick. Inside this are situated a heating element (Philips "Coax"), a platinum resistance thermometer, and compartments on which test tubes containing the substituting fluid are placed. A mechanism is provided which moves the compartments up and down. The lower part of the Dewar flask contains liquid air and, in order to maintain a temperature above that of liquid air in the copper can, an equilibrium is established by regulating the input of heat. Needle contacts are set to cut off the supply of heat at any desired temperature and fluctuations are no more than ±3°.

The level of liquid air in the Dewar flask is controlled by a thermistor circuit which initiates refilling from the storage tank as soon as the level falls below the thermistor itself. The whole construction is shown in Fig. 38, opposite. With the aid of such an apparatus it should be possible to carry out critical experiments on the process of freeze-substitution of both tissue blocks and tissue sections (see below). Further information is clearly desirable.

The use of drying agents or molecular sieves in the substituting fluid is by no means universal although the practice of drying the fluid completely by some means or other before use has become so. Balfour (1961) used anhydrous sodium sulphate and found a recognizable improvement (in staining with fluorescent antibody) by comparison with tissues dehydrated in fluids which were incompletely dry. Other workers have used molecular sieves such as Linde type 4A ($\frac{1}{16}$ inch pellets). If the ratio of solvent to specimen is low, as when small test tubes are used as containers, the solvent will ultimately contain an appreciable level of water unless an additional trap is provided.

Use of Fixatives. There are conflicting views upon the degree of fixation (formation of stable chemical bonds between adjacent groups in the tissues) which can be achieved at the temperatures now virtually universally employed in freeze-substitution (below —65°). Inevitably all are agreed that the fixation rate falls with the temperature but the point at which true fixation ceases has not been determined for any of the solvent-fixative media.

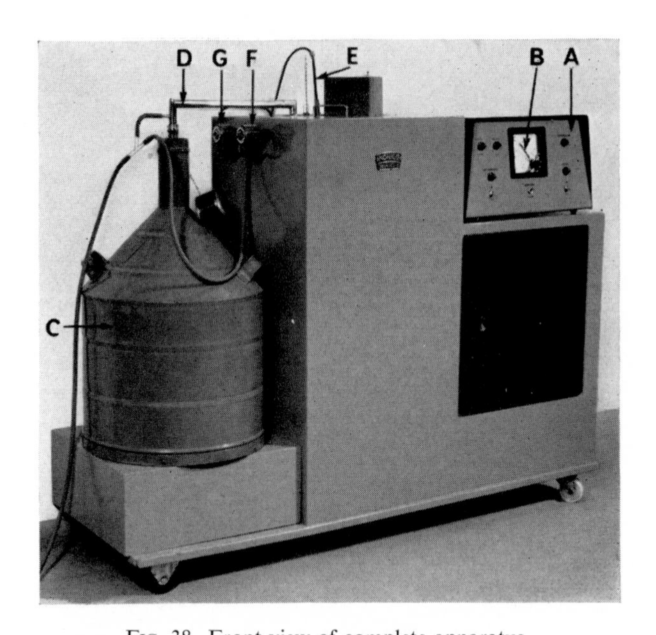

FIG. 38. Front view of complete apparatus.

A and B. Control Panel and Temperature Indicator. C. Liquid Air Tank. D. Connection to Dewar Flask. E. Tube carrying Philips NTC Resistor. F. Electro-magnetically controlled valve. G. Safety valve.

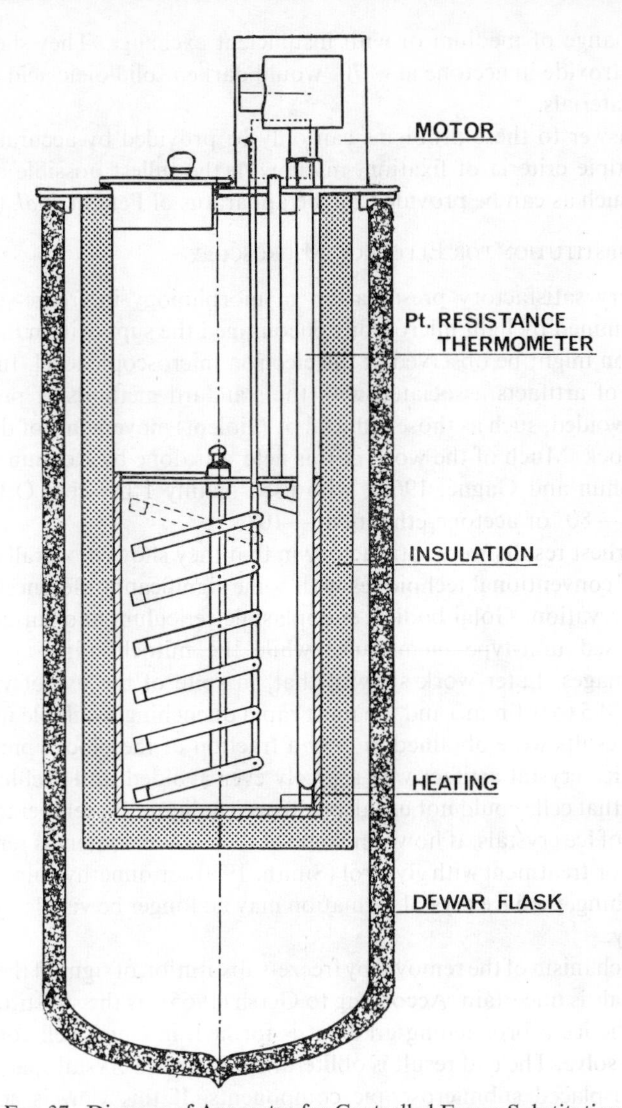

FIG. 37. Diagram of Apparatus for Controlled Freeze-Substitution.

Evidence was provided by Davis *et al.* (1959) that with a variety of anhydrous fixatives (2 per cent. OsO_4, 5 per cent. $HgCl_2$, 2 per cent. trichloracetic acid. between —15° and —75° the rate of insolubilization of proteins substantially exceeded the rate of inactivation of enzymes and antigens. On the other hand Balfour (1961) found complete inactivation of antibody (not antigen) after 1 per cent. OsO_4 in acetone (6 days, —70°) but she observed that $HgCl_2$ possessed hardly any fixing power at —70°. Feder and Sidman (1958b) considered that the fixing agent in the substituting fluid was certainly able to fix at —70°, independently of any effect it might exert if warming was allowed to take place without change of medium or with insufficient exchange. They showed that osmium tetroxide in acetone at —70° would darken solid oleic acid and other reactive materials.

The answer to these problems can only be provided by accurate studies, using multiple criteria of fixation, made with the fullest possible control of variables such as can be provided by the apparatus of Persijn *et al.* (1964).

Freeze-Substitution for Electron Microscopy

The very satisfactory preservation of morphology in freeze-substituted tissues examined by light microscopy encouraged the supposition that similar preservation might be observed at the electron microscopic level. In this case a number of artifacts associated with the standard methods of preparation might be avoided, such as those induced by (violent) movements of the fixative into the block. Much of the work in this field was done by Rebhun (Rebhun, 1961; Rebhun and Gagné, 1962), who used mainly 1 per cent OsO_4 in dry acetone at —80° or acetone-ethanol at —108°.

The earliest results were satisfactory in that they showed overall similarity to those of conventional techniques with some significant differences in membrane preservation. Golgi bodies, endoplasmic reticulum and surface microvilli possessed unit-type membranes while the mitochondria appeared as negative images. Later work showed that, in spite of the use of very small specimens (0·5 to 0·1 mm.) and the most rapid quenching available with liquid N_2, good results were obtained in only a fraction of the blocks prepared. In particular ice crystal artifact was scarcely ever avoided and Rebhun (1965) concluded that cells could not usually be frozen at liquid N_2 temperatures with avoidance of ice crystals. If however some of their water content is removed, as by drying, or treatment with glycerol (Smith, 1950) or dimethylsulphoxide before quenching, then ice crystal formation may no longer be visible by electron microscopy.

The mechanism of the removal by freeze-substitution of signs of the presence of ice crystals is uncertain. According to Gersh (1965), as the substituting fluid dissolves the ice a broadening gradient is formed, in which cell components swell or dissolve. The end result is obliteration of the ice crystal space through shifts in displaced submicroscopic components. If this view is accepted it follows that freeze-substitution for electron microscopy has two limiting features. The first is that solution, diffusion and extraction of cellular components takes place in the diffusing gradient of the substituting fluid. The second is the inevitable deformation and swelling of cellular components, taking place whether ice crystals are initially present or not.

The problems in this field are very great, as they are in the case of freeze-dried tissues, and no complete solution is in sight. There can be no doubt that future advances will be made chiefly by avoidance of ice-crystal damage by improved protective techniques, combined with effective fixation procedures such as those outlined in Appendix 4, p. 598.

SECTION FREEZE-SUBSTITUTION

In a number of laboratories the complete substitution of fresh cryostat sections of tissues was found to be easy and rapid. Subsequent handling of the sections was always a more difficult matter and no great advances were made until the publication by Chang and Hori (1961) of their method of section freeze-substitution. In the original technique 8–10 μ sections were substituted in acetone at —79° for 12 hours or more. They were then usually mounted by a celloidin-coating process (Appendix 4, p. 600), unless required for procedures which were adversely affected by celloidin (stains for nucleic acids and polysaccharides; some techniques for hydrolytic enzymes).

More recently (1965) Chang introduced a number of modifications into his technique, the most important of which was substitution at —20° instead of at —79°. This was based on the reasoning that since penetration of the section by the solvent is rapid at —20° its contained ice crystals would be surrounded by acetone and thus unable to grow. If this were the sole consideration involved, all might be well. Inevitably, however, the higher temperature will involve a higher degree of fixation, increased solution of tissue components in the substituting fluid, and movement within the tissues of small molecular weight substances.

Early claims for section freeze-substitution (Chang and Hori, 1961) suggested that it was a versatile, reliable and practicable method of processing tissues for histochemical research, especially for oxidative enzymes, hydrolytic enzymes, water-soluble isotopes and other chemical constituents. No other preparative technique was considered to preserve so much enzyme activity in sections.

Preservation of Enzymes. Chemical analyses (Hori and Chang, 1962; Chang and Hori, 1962b) showed that, for instance, 70 per cent. of ATP ase remained in acetone substituted tissues, and 76 per cent. acid phosphatase, 92 per cent. alkaline phosphatase, 95 per cent. 5-nucleotidase and 75 per cent. succinate dehydrogenase (but see results of Torack and Markey, Chapter 5, p. 105). Although high levels of enzyme activity are certainly maintained, the final results of applying various standard histochemical enzyme procedures are often less than satisfactory. Presumably this is because a variable amount of the enzyme protein remains soluble and diffusible, depending on the degree and duration of warming permitted during the substituting process.

Preservation of Lipids. One of the limitations of the section freeze-substitution technique, admitted even by its chief supporters, is the solution of lipids in the solvent. In his work on the photoreceptor cells Fernandez-Moran (1959a) noted that the salmon-pink colour of the frozen guinea-pig or frog retina changed after acetone or alcohol freeze-substitution to orange or yellow. Although he attributed this change to lability of photopigment I would suggest solution of the neutral lipid fraction (and its contents) as a more likely cause.

Phospholipids are certainly less readily soluble in cold acetone or alcohol and their preservation can be demonstrated by conventional lipid staining methods, according to Hori and Chang (1962). These authors found improved preservation of mitochondrial and other lipids by the addition of metal salts to the substituting fluid. Among these, uranyl nitrate gave the best results, especially when used in conjunction with chloral hydrate (Appendix 4, p. 598).

Preservation of Nucleic Acids and Proteins. The appearance of nucleic acids (particularly RNA) in freeze-substituted sections closely resembles that seen in freeze-dried tissues according to Chang and Hori (1962a), who suggested that post-fixation was unnecessary. No solubility studies have been made, however. The same authors stated that the Landing and Hall method for histidine (Chapter 6, p. 171) gave good results, again without post-fixation. Barnard (1960), however, found that the coupled tetrazolium reaction after benzoylation (Chapter 6, p. 168) was abolished in smears and sections by freeze-substitution in alcohol. This reaction is now considered to indicate the presence of a labile bond between the histidine residues of the (nucleo) protein and another set of groups in the nucleic acid. Its reversal by freeze-substitution suggests that the effects of the latter may not be so simple as often imagined.

In spite of a number of theoretical and practical advantages over conventional block-substitution, the section technique has not come into general histochemical use. There may be occasions when it provides the best possible preservation of structure and function, always something of a compromise. I consider that for most purposes section freeze-substitution is not superior to alternative methods of preparation and that even for fixation-labile enzyme demonstration it can be replaced by modifications of the newer aldehyde fixation techniques (Chapter 5, p. 74) applied to mounted cryostat sections.

The indications for block freeze-substitution, as far as mammalian tissues are concerned, are still the preservation *in situ* of high mol. wt. polysaccharides, either natural, like glycogen, or artificially introduced like dextran (see Chapter 28). Although the process has been used for retention of water-soluble isotopes and for fluorescent antibody studies, in both cases I prefer one or other of the available alternative procedures. For light microscopic morphology, of both animal and plant tissues, freeze-substitution is an excellent method but it must be appreciated that, even in the case of those tissues for which it works best, it induces a very considerable degree of shrinkage.

REFERENCES

Balfour, B. M. (1961). *Immunology*, **4**, 206.
Barnard, E. A. (1960). *Nature*, **186**, 447.
Bartl, P. (1962). "Proc. Vth Int. Congr. E. M.," Vol. 2, p. P4, Acad. Press, New York.
Baud, C. A. (1952). *Bull. Histol. Tech. Micr.*, 158.
Bell, L. G. E. (1952). *Int. Rev. Cytol.*, **1**, 35.
Birns, M., and Masek, B. (1961). *J. Histochem. Cytochem.*, **9**, 204.
Blank, H., McCarthy, P. L., and de Lamater, E. D. (1951). *Stain Tech.*, **26**, 193.

BULLIVANT, S. (1960). *J. biophys. biochem. Cytol.*, **8**, 639.
BULLIVANT, S. (1965). *Lab. Invest.*, **14**, 1178.
CHANG, J. P. (1965). *J. Histochem. Cytochem.*, **13**, 703.
CHANG, J. P., and HORI, S. H. (1961). *J. Histochem. Cytochem.*, **9**, 292.
CHANG, J. P., and HORI, S. H. (1962a), *Ann. d'Histochimie*, **6**, 419.
CHANG, J. P., and HORI, S. H. (1962b). *J. Histochem. Cytochem.*, **10**, 592.
DAVIS, J., ORNSTEIN, L., TALEPOROS, P., and KOULISH, S. (1959). *J. Histochem. Cytochem.*, **7**, 291.
VAN DUIJN, P., and OORT, J. (1962). *Stain Tech.*, **37**, 116.
FEDER, N., and SIDMAN, R. L. (1958a). *J. Histochem. Cytochem.*, **6**, 401.
FEDER, N., and SIDMAN, R. L. (1958b). *J. biophys. biochem. Cytol.*, **4**, 593.
FERNANDEZ-MORAN, H. (1959a). *Science*, **129**, 1284.
FERNANDEZ-MORAN, H. (1959b). *Rev. mod. Phys.*, **31**, 319.
FERNANDEZ-MORAN, H. (1959c). *J. appl. Phys.*, **30**, 2038.
FERNANDEZ-MORAN, H. (1960). *Ann. N.Y. Acad. Sci.*, **85**, 689.
GERSH, I. (1965). *Fed. Proc.*, **24**, Suppl. 15, p. S233.
GOURÉVITCH, A. (1953). *Bull. Histol. Tech. Micr.*, **3**, 130.
HANCOX, N. M. (1957). *Exp. Cell Res.*, **13**, 263.
HORI, S. H., and CHANG, J. P. (1962). *J. Histochem. Cytochem.*, **9**, 625.
LISON, L. (1949). *C.R. Soc. biol. Paris*, **143**, 115.
LISON, L. (1953). "Histochimie et Cytochimie Animales." Gauthier-Villars, Paris.
LUFT, J. H. (1961). *J. Biophys. Biochem. Cytol.*, **9**, 409.
MOLINE, S. W., and GLENNER, G. G. (1964). *J. Histochem. Cytochem.*, **12**, 777.
OSTROWSKI, K., KOMENDER, J., KOSCIANEK, H., and KWARECKI, K. (1962a). *Experimentia*, **18**, 142.
OSTROWSKI, K., KOMENDER, J., KOSCIANEK, H., and KWARECKI, K. (1962b). *Ibid.*, **18**, 227.
PERSIJN, J-P., DE VRIES, G., and DAEMS, W. TH. (1964). *Histochemie*, **4**, 35.
PERSSON, B. H. (1952). *Acta Soc. med. Upsaliensis*, **57**, 155.
PEYROT, A. (1956). *Riv. Istochim.*, **2**, 197.
REBHUN, L. I. (1961). *J. biophys. biochem. Cytol.*, **9**, 785.
REBHUN, L. I. (1965). *Fed. Proc.*, **24**, Suppl. 15, p. S217.
REBHUN, L. I., and GAGNÉ, H. T. (1961). "Proc. Vth Int. Congr. E. M.," Vol. 2, L2, Acad. Press, New York.
RINDERER, L., and HAENSELER, F. (1960). "Proc. X Int. Congr. Refrigeration," Vol. 1, p. 243. Pergamon, London.
RUSSELL, R. S., SANDERS, F. K., and BISHOP, O. N. (1949). *Nature*, **163**, 639.
SIMPSON, W. L. (1941a). *Anat. Rec.*, **80**, 173.
SIMPSON, W. L. (1941b). *Ibid.*, **80**, 329.
SMITH, A. U. (1950). *Lancet*, **2**, 910.
STEPHENSON, J. L. (1956). *J. Biophys. Biochem. Cytol.*, **2**, 45.
WOODS, P. S. and POLLISTER, A. W. (1955). *Stain Tech.*, **30**, 123.

CHAPTER 5

THE CHEMISTRY OF FIXATION

ATTEMPTS to establish the nature of the chemical and physical changes induced by fixation have received new impetus since the 2nd Edition of this book was published (1960). This has mainly been due to the advent of a new series of aldehyde fixatives, and to a greater appreciation of the importance of vapour fixation. This chapter is divided into two sections. In the first, the chemical aspect of fixation, by formaldehyde and other fixatives, is considered inasmuch as it affects the performance and interpretation of histochemical reactions. In the second, the choice of fixative for a variety of modern histochemical techniques is reviewed.

The best method of fixation for each particular method is of paramount importance and, whatever fixatives are applied as a preliminary to histochemical studies, it is essential to know as precisely as possible their effect on the reactive groups of the various tissue components. The most important of these components are the proteins, and since the classical protein fixative is formalin, the actions of formalin are considered most fully in the section which follows. Full consideration is given also to the other aldehyde and, particularly, dialdehyde fixatives, and other cross-linking agents are considered more briefly. For histochemical purposes alcohol and acetone are the only remaining fixatives of much importance, metallic salts are less often used though they are occasionally essential. Fixation with osmium tetroxide mixtures will be considered in this chapter, where it properly belongs, rather than in Chapter 32, where electron cytochemistry is discussed.

The Chemical Actions of Liquid Fixatives on Tissue Components

ALDEHYDE FIXATIVES

Formalin. Formaldehyde exists in aqueous solutions principally in the form of its monohydrate, methylene glycol, $CH_2(OH)_2$, and as low molecular weight polymeric hydrates or polyoxymethylene glycols, $HO.(CH_2O)_n.H$. Less than 0·1 per cent. of monomeric formaldehyde is present (Walker, 1964). Low formaldehyde concentrations favour methylene glycol and higher concentrations polyoxymethylene glycols. At temperatures below 35° equilibrium is reached only slowly, at least 24 hours being required. This is important when dilute fixatives are made up from concentrated formalin. The Raman spectra of solutions of formaldehyde containing less than 2 per cent. indicate that the carbonyl group is virtually absent, that is to say only methylene glycol is present. Ultraviolet absorption spectra and polarography, however, do indicate that unhydrated formaldehyde monomer is present in aqueous solutions.

The reactions of formaldehyde with tissue proteins are numerous and complex, since it can combine with a number of different functional groups, in many cases forming bridging links between them (see Table 3, p. 73). To this function of forming links between adjacent protein chains formalin owes its success as a polymerizing fixative. According to French and Edsall (1945) the most frequently encountered reaction of formaldehyde is its addition to a compound containing a reactive hydrogen atom with the formation of a hydroxymethyl (methylol) compound.

$$RH + CH_2O \rightleftharpoons R.CH_2(OH)$$

The compound is usually reactive also and it may condense with a further H atom to form a methylene bridge ($—CH_2—$) in the manner illustrated below.

$$R.CH_2(OH) + HR^1 \rightleftharpoons R—CH_2—R' + H_2O$$

These methylene bridges, as the above reaction suggests, are readily ruptured by hydrolysis. Many of the combinations of formaldehyde with tissue proteins are reversible by the simple process of washing or, for instance, by the application of dimedone (see Chapter 13, p. 455). Others are irreversible.

Cross-linking. The essential feature of formaldehyde fixation is thus the formation of cross-links between protein end-groups. Any compound which produces this effect, whether used as a fixative or not, is described as bifunctional. The groups particularly involved in the fixation of proteins by formaldehyde are amino, imino and amido, peptide, guanidyl, hydroxyl, carboxyl, SH and aromatic rings. Methylene bridges may be formed between two similar groups, say NH_2, or between NH_2 and peptide (CONH), or between NH_2 and NH, to quote a few examples. Their occurrence is dependent on the presence of a second suitable group with a suitable spatial relationship to the formaldehyde addition compound formed by the first. Interesting details are available of the action of formaldehyde on some of the commoner proteins. Between pH 6 and 8 formaldehyde reacts with keratin without affecting the S—S links of cystine. In more alkaline solutions, however, it is considered to reduce S—S to two SH groups and subsequently to react with these forming, in some cases, a methylene bridge ($S—CH_2—S$) in place of the original disulphide link (Middlebrook and Phillips, 1942a, b). From experiments in the tanning of casein by formalin at pH 6 Nitschmann and Hadorn (1943, 1944) deduced that a condensation between the ε-amino groups of lysine and the peptide links of another chain was responsible for most of the observed combination with formalin. (Leather chemists, it may be noted, tend to speak of the fixation of formalin by proteins, a reversal of the usual histological terminology.) These same authors (1943) showed that a considerable amount of the formaldehyde remaining bound to protein after 5 hours washing in running water could be removed by further washing (up to 24 days). Even after a period such as this (which is obviously well beyond the upper limit of washing in histological and histochemical practice) some formaldehyde remained in irreversible combination with protein. This was freed by prolonged hot acid hydrolysis and collected for estimation by distillation. A sample of casein washed for 12 days was shown to contain 1·9 per cent. of formaldehyde and a similar sample washed for only $1\frac{1}{2}$ hours contained 2·6 per cent. Obviously something approaching the latter percentage would remain bound to protein in most histological preparations in the block stage at the point of dehydration in alcohol. Neither the amount removed by the further processes of embedding, nor the amount of washing required before the percentage of formaldehyde in the mounted section approaches the lower of the two figures quoted above for bound formaldehyde content, has been determined. Probably very little additional washing of the section is required.

The significance of these observations in histochemical practice is that treatment of proteins with formalin, followed by washing, is likely to leave the majority of active groups in a condition to react with any reagents we may have occasion to use. The Sakaguchi reaction for arginine (Chapter 6, p. 158), for instance, is considerably modified in the presence of formalin, which appears to alter the guanidyl group in such a way as to block the action of hypochlorite upon it. Histochemically, however, the reaction can be performed on formalin-fixed sections, from which the excess formalin has been removed by washing, in a perfectly satisfactory manner. Even if we propose to apply spectrophotometric measurement to make the reaction quantitative, and even if we are uncertain whether a proportion of the guanidyl groups is still blocked in the washed formalin-fixed sections, the application of a standard technique of fixation will allow what will usually be a valid comparison between one tissue and another.

Most of our knowledge of the action of formaldehyde on tissue compounds comes from studies made in connection with the tanning and wool industries, so that collagen and keratin are the two substances we know most about. A great deal of this work has been done with formalin under quite unhistological conditions (e.g. 100° at pH 1, 70° at pH 4, various temperatures at pH 10) and Middlebrook (1949) showed that the amount of formalin irreversibly bound by various proteins dropped sharply, in most cases, when the pH rose above 1·0. This was true, for example, of casein, gelatin and wool but in the case of carboxyhæmoglobin the amount of bound formalin rose to a maximum at pH 4. Middlebrook connected this with the formation of bridges involving the amide groups of asparagine, present in large amounts in hæmoglobin. In histological and histochemical work formalin is nearly always used in buffered solutions at or above the neutral point, and Wolman (1955) suggested that part of the effectiveness of these higher levels was due to the rapid conversion, in neutral and alkaline solution, of the polymerized form of HCHO to the monomer. This explanation seems improbable, even for the usual 4 per cent. formaldehyde solutions where the percentage of hydrated monomer (methylene glycol) will still be very high. The latter is to be regarded as the actual reactant whenever aqueous formaldehyde fixatives are used. The reactions of formalin are set forth below in Table 3.

This list does not exhaust the possible reactions of formaldehyde with groups present in amino-acids and a number of other reactions have been described using pure amino-acids. Such reactions probably do not take place in materials of the type we are dealing with, but the reactions indicated in the Table certainly do occur. It will be observed that the only irreversible reaction is that which takes place with the aromatic hydrogen in tryptophan, tyrosine, phenylalanine and histidine. This reaction occurs very slowly and is often followed by condensation with another reactive group to form a methylene bridge. The phenol formaldehyde polymers are formed in this manner (see Chapter 26. It is worthy of note that when formalin is used as a histological fixative, the process of fixation, as judged by purely histological standards, is not complete until at least 7 days have elapsed.

TABLE 3

Reactions of Formaldehyde
(mainly after French and Edsall, 1945)

Group	Reaction	Remarks
Active hydrogen	$R.H + CH_2O \rightleftharpoons R.CH_2OH$	—
Active hydrogen + condensation	$R.CH_2OH + H.R' = R{-}CH_2{-}R' + H_2O$	Methylene bridge
Amino (1 mol)	$R.NH_2 + CH_2O \rightleftharpoons R.N = CH_2 + H_2O$	—
Amino (2 mols)	$R.NH_2 + 2CH_2O \rightleftharpoons R.N(CH_2OH)_2$	Alkylaminomethanols
Amino (3 mols)	$R.NH_2 + 3CH_2O \rightleftharpoons R.N(CH_2O)_2CH_2 + H_2O$	Triformal
Imino	$R_2 = NH + CH_2O \rightleftharpoons R_2 = N.CH_2OH$	—
Amide	$R.CONH_2 + CH_2O \rightleftharpoons R.CO.NH.CH_2OH$	Methylolamides
Peptide	$-CO{-}NH + CH_2O \rightleftharpoons CO.N(CH_2OH)$	—
Guanidino	Not known	Probably a bridge with NH_2
Hydroxyl (a)	$R{-}OH + CH_2O \rightleftharpoons R.O.CH_2OH$	Hemiacetal formation
Hydroxyl (b)	$2R{-}OH + CH_2O \rightleftharpoons RO.CH_2.OR$	Acetal formation
Carboxyl	$2R{-}COOH + CH_2O \rightleftharpoons (R{-}COO)_2CH_2 + H_2O$	Unimportant in aqueous solution
SH (a)	$R.SH + CH_2O \rightleftharpoons R.S.CH_2OH$	Semithioacetal
SH (b)	$2RSH + CH_2O \rightleftharpoons (R.S)_2CH_2 + H_2O$	Methylene bridge (thioacetal)
Aromatic H	$>CH + CH_2O \rightarrow >C{-}CH_2OH$	Stable C—C bond

It is clear that in using neutral 10 per cent. formalin at room temperature, and even more so at 4°, we are making relatively little use of its capacity to form addition compounds and bridges. Moreover, a high proportion of those initially formed are too labile to withstand washing. Unfortunately we do not know precisely which groups are responsible for the irreversibly-bound formalin fraction of washed tissues, a matter of importance since it is these groups whose reactivity is suppressed. It is likely, however, that aromatic hydrogen and NH_2 groups, by bridge formation, are the two which are mainly involved.

Acrolein. This compound, acrylic aldehyde, $H_2C{=}CH.CHO$, is a bi-functional aldehyde used in the tanning industry which is capable of introducing more cross-links than formaldehyde, under optimum conditions (Bowes, 1963; Cater, 1963). It is unpleasant to use, however, and unstable at alkaline pH levels so that it is little used as a histochemical fixative. The tanning (fixative) effect of acrolein is due to its hydrogen atoms adjacent to the double bond (Gustavson, 1956), since it is abolished by the introduction of ethyl and propyl groups in the 2- and 3-positions of the molecule.

Glutaraldehyde. Glutaric dialdehyde, $(CH_2)_3CHO.CHO$, which is more properly represented by the structural formula,

```
        H2
        C—CHO
       /
   H2C
       \
        C—CHO
        H2
```

was first used in the leather industry as a tanning agent (Seligsburger and Sadlier, 1957; Fein and Filachione, 1957). Its introduction into histochemical use was primarily due to the enterprise of Sabatini, Bensch and Barrnett (1963) who showed that it was a better preservative of structure and enzyme activity than any other of the 9 aldehydes tested. Studies carried out by Bowes (1963) indicated that glutaraldehyde was the most efficient cross-linking agent of

TABLE 4

Cross-links in Collagen treated with Excess Aldehyde
(After Bowes, 1963)

Aldehyde	pH	Mean mol. wt. per cross-linked segment	Cross-links per mol. wt. unit of 10^5g
Control, untreated	—	200,000	0·2
Acrolein	8·0	4,500	11
Acrolein	6·0	9,000	6
Acrolein	4·0	10,500	5
Formaldehyde	8·0	8,900	6
Glyoxal	8·0	5,900	8
Malonic dialdehyde	5·0	12,600	4
Succinic dialdehyde	7·0	5,080	10
Glutaraldehyde	8·0	5,050	10
Glutaraldehyde	6·5	4,240	12
Glutaraldehyde	4·0	7,000	7
Adipic dialdehyde	8·0	24,000	2

those tested. Some reaction was noted even at pH 4·0 while at pH 6·5 up to 12 cross-links per unit molecular weight of 10^5 g were introduced. Assuming that all the lysine and hydroxylysine (of the collagen) were involved the maximum theoretical number would be 16 so that we can assume that unipoint fixation with glutaraldehyde is probably unimportant. Table 4, p. 74, provides a series of relevant comparisons.

It can be seen that when present in excess acrolein is very nearly as efficient as glutaraldehyde, and succinic dialdehyde hardly less so. The performance of the other two dialdehydes is indifferent.

Further experiments were carried out by Bowes (1963) on the effect of varying the concentration of fixative. The concentration used in these experiments corresponded to levels of 0·5 to 1·0 moles per amino group of collagen and 1·0 mole per amino + guanidino group and 1·0 mole per amino + guanidino + amino group. Only the 4 most efficient aldehydes were tested and the results are shown in Table 5.

TABLE 5

Effect of Aldehyde Concentration on Cross-Linking
(After Bowes, 1963)

Aldehyde	pH	Concn. Moles per 10^5g	Mean mol. wt. per cross-linked segment	Cross-links per mol. wt. unit of 10^5g
Control	—	0	200,000	0·2
Acrolein	8·0	17	10,500	5
,,		146	7,800	6
Formaldehyde	8·0	17	26,800	2
,,		34	22,000	2
,,		90	15,500	3
,,		146	15,400	3
,,	5·0	146	32,500	2
Glyoxal	8·0	17	18,500	3
,,		34	6,850	7
Glutaraldehyde	8·0	17	14,000	4
,,		34	9,500	5
,,		90	4,800	10
,,	6·5	17	12,000	4
,,		34	9,550	7
,,		90	7,600	7

By comparison with Table 4 we note that at low concentrations glutaraldehyde is more efficient at pH 8·0 whereas, if present to excess, its performance is better at pH 6·5. Under the usual conditions of fixation there is always excess reagent present but the results given in the two Tables above suggest that with a rapidly acting fixative like glutaraldehyde there may be considerable advantage in using dilute solutions.

Testing the stability of cross-links to boiling water and to acid hydrolysis Bowes (1963) found that those produced by formaldehyde and glyoxal were markedly less stable than those produced by either glutaraldehyde or acrolein. The former is considered to act by bridging the gap between adjacent amino groups while glyoxal can apparently do this only with difficulty. The cross-links of formaldehyde fixation (p. 71) are produced (slowly) as a secondary reaction.

It is clearly not always possible to draw conclusions directly from tanning experiments but the remarkable effectiveness of glutaraldehyde is apparent and it is obvious that we have here a fixative capable of far more variety in application than formaldehyde, for example.

Other Cross-Reacting Fixatives

Chloro-s-triazines. Cyanuric chloride and some of its water-soluble derivatives were tested by Goland and Engel (1963) as fixatives for tissues and for certain exogenous substances introduced into the tissues artificially. These compounds have structural formulæ of the type

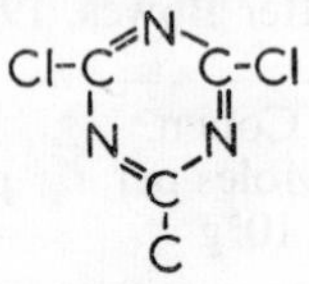

The parent compound is insoluble in water and must be used in a non-polar solvent such a toluene, chloroform, ethyl methyl ketone or acetone. Sodium (dichlorotriazine)-1-naphthalide-5-sulphonate

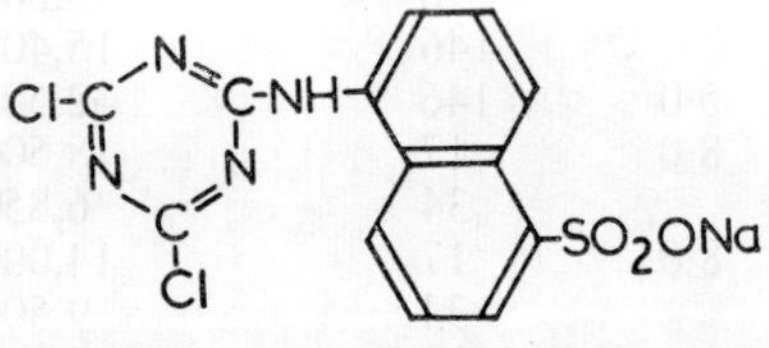

is available commercially, as is 2-amino-4, 6-dichloro-1, 3, 5-triazine and a whole range of dichlorotriazine dyes (Procions, I.C.I. Ltd.). Some of the latter were used by Hess and Pearse (1959), primarily as protein labelling agents. They observed that a certain amount of cross-linking (presumably *via* adjacent amino groups) took place and that this effect was not shown by monochlorotriazinyl compounds. Similarly, Goland and Engel found that monochlorotriazines (Procion and Cibacron dyes) failed to fix cellular components effectively. Further investigation of the properties of cyanuric halides and their derivatives is very necessary.

Carbodi-imides. These compounds, whose general formula is R . N:C:N . R′, react with a carboxyl and an amino group with elimination of water to give a peptide and the corresponding urea (R.NH.CO.NHR′). According to Bowes (1963) the reaction has been used to cross-link soluble proteins but it has not been successful with fibrous proteins like collagen. N.N′-dicyclohexylcarbodiimide is available but, being insoluble in water, it must be used in organic solvents.

Di-isocyanates. Isocyanates and isothiocyanates have been and are used for the introduction of fluorescent labels into proteins (see Chapter 7, p. 188). No cross-linking is intended or obtained in these reactions. With diisocyanates, however, Bowes (1963) succeeded in introducing cross-links into dry collagen. She used the 1:6-hexamethylene and 2:4-toluene derivatives dissolved in pyridine but experiments were also carried out with water-soluble addition compounds. The latter were produced by shaking either of the diisocyanates with a saturated aqueous solution of potassium bisulphite. The hexamethylene derivative introduced 3, and the toluene derivative 4, cross-links per 10^5g. protein.

Once again there is room for experiment with these compounds as histochemical fixatives.

METALLIC IONS AND COMPLEXES

Chromium. There is some modern evidence of the action of the protein-precipitant metals which is worth considering. In histology and histochemistry we are mainly concerned with osmium, mercury and chromium and once again, due to the volume of investigations carried out by the tanning industry, much more is known about the mechanism of fixation with the last of these. Chromium salts have the property of forming complexes with water of the type Cr—O—Cr, and these complexes combine with the reactive groups of adjacent protein chains to bring about a binding effect similar to that of formalin (see Gustavson, 1949, 1956). The main affinity of chromium is for the carboxyl and hydroxyl groups of protein and Bowes and Kenten (1949) found that no chromium is fixed by methylated collagen from solutions of chromium sulphate at pH 4·0. (The effect of methylation is to block carboxyl groups by esterifying them.) Similarly, Gustavson (1940) showed that only small amounts of chromium were taken up by collagen at pH 1; at this pH the carboxyl groups no longer maintain a negative charge and they therefore fail to react. The importance of carboxyl is stressed also by the work of Strakhov (1951), who showed that collagen can fix much more chromium than silk fibroin (which contains fewer COOH and NH_2 groups) and that an ε-caprolactam polymer containing neither of these groups fixed no chromium at all. Green (1953) considers that the primary reaction between the chromium complex and carboxyl groups is followed by the formation of co-ordinates with OH and NH_2 groups. Hexavalent chromium has been shown by Grogan and Oppenheimer (1955) to be bound by egg albumin in decreasing amount as the pH is changed from 4 to 7·35. Gustavson (1949) considered this effect to be due to the larger size of the complexes at acid pH levels and to the increased availability of ionized carboxyl groups.

No histological fixatives containing chromium have a pH lower than 2·9, and the effect of chromium salts in neutral or moderately acid solution will be to form complexes with tissue proteins of the type described. The final effect of these is to break up the internal (salt) links of the protein and to increase the number of basic groups available in a reactive state, thus increasing the acidophilia of the whole structure. Since we have no exact measure of the effect of chromium on the various reactive groups, chromium-containing fixatives, with some specific exceptions, are best avoided in histochemical studies. These exceptions are the use of chromium trioxide as an oxidizing agent in the production of the so-called chromaffin reaction (see Chapter 26.), its possible use for the preservation of glycogen, and the use of $K_2Cr_2O_7$ for lipids and nucleic acids in the process of controlled chromation (see p. 92, and Appendix 5, p. 605).

Mercury. Salts of mercury are much more commonly employed in histological fixatives and in many institutions (my own, for instance) the routine fixative is a mercurial one. Histochemical studies may be called for when no material other than that fixed with mercury (formol-mercury) is available. It is therefore important to consider the effect of salts of this metal on the tissues, and particularly on the proteins of these. In general, Hg^{++} behaves like other metallic ions in combining with the acid groups of proteins, especially carboxyl and hydroxyl, and the phosphoric acid of nucleoproteins. It differs from chromium in not forming complexes capable of binding together adjacent protein chains and also from the majority of metals in one important respect. This is in its selective affinity for thiol (SH) groups. If a very small quantity of a mercuric salt is added to a protein containing reactive thiol groups it wil react with these in preference to any others, and the stability of the mercury-sulphur bond is greater than that between mercury and any other grouping. Hughes (1947) prepared a fraction of serum albumin which contained one SH group per molecule. When this was allowed to react with a mercury salt the resulting protein mercaptide was found to contain $\frac{1}{2}$ an atom of mercury per albumin molecule, suggesting that the unit formed was an albumin dimer. The reactions which take place (modified from Hughes, 1950) are given below.

$$RSH + HgCl_2 \rightleftharpoons RSHgCl + HCl \quad \text{and}$$
$$RSHgCl + RSH \rightleftharpoons (RS)_2Hg + HCl$$

The second reaction is a slow one and probably of little importance in histological fixation. Both reactions are reversible, but while the second can be reversed by any reaction which forms an undissociated mercury complex, the first is only reversed by reagents such as cysteine which form equally stable mercury derivatives.

The part played by metals in the binding of small molecules by protein has been described by Klotz and Loh Ming (1954). Small organic molecules, not ordinarily bound to proteins, form strong complexes if suitable metals are present. The metal acts as a bridge and must be capable of forming co-ordinate bonds with both the protein and the small molecule.

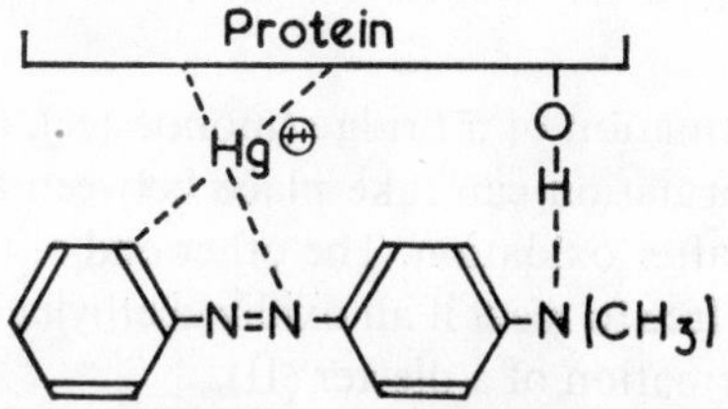

Protein

$Hg^{\oplus}$ O H

$-N{=}N-$ $-N(CH_3)$

The configuration of the co-ordinate links shown in this illustration is not significant. Among the metals which are effective are Hg, Cu, Ag, Ni, Zn, Co and Mn; Ca and Mg are ineffective. It was suggested that protein hydroxyls normally internally bonded could be made available for hydrogen bonding by dyes, in the above manner.

These findings are not immediately applicable to the problem of mercury-fixed proteins in histochemistry, since in ordinary proteins, in contradistinction to Hughes' serum albumin fraction, not all the SH groups present are available for reaction. They do suggest, however, that complete reversal of mercury-binding of tissue groups might be affected by means of reagents like cysteine. For histochemical purposes the important point is whether the reaction of mercury with the various protein groups is reversible by the ordinary processes of embedding, washing, and removal of coarse mercury precipitates by means of iodine and thiosulphate. While specific investigations into this point have not been made, it is certain that some mercury remains bound by the acid groups in nucleoproteins and other proteins and it is probable that some remains bound to SH groups also. It is, therefore, advisable to avoid the use of mercurial fixatives in critical histochemical practice if investigation into the nucleic acids or sulphydryl groups is proposed. Other protein reactions (e.g. Millon, Sakaguchi and tetrazonium) can be performed in spite of the presence of mercury.

Osmium. The so-called osmic acid was employed by Schultze and Rudneff in 1865, and since that time osmium tetroxide (OsO_4) has been in general use as a cytological fixative. Its use in histology has been limited by its poor penetrating qualities, but its current pre-eminence in the preparation of tissues for electron microscopy has brought about a revival of interest in the mechanism by which it acts as a fixative. Our knowledge of this process is still scanty and unsatisfactory in many respects.

Reactions with Lipids. Saturated fats do not react but unsaturated fats reduce OsO_4 with the formation of black compounds containing osmium or its hydroxide, and Criegee (1936) suggested that this is due to oxidation of the double bonds between adjacent carbon atoms.

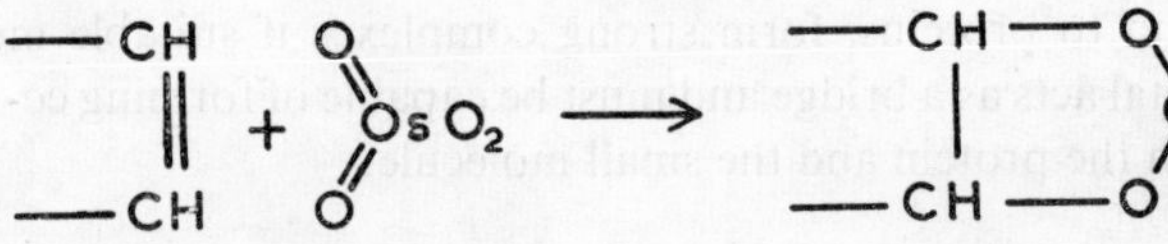

(colourless) (black)

I

This results in the formation of a bridge (monoester), and it has been stated that similar bridge formation can take place between adjacent aliphatic hydroxyls (1:2-glycols) after oxidation. The other end of the molecule shown in the equation above is free to react if an oxidized ethylenic group (diol form) is available, with the formation of a diester (II).

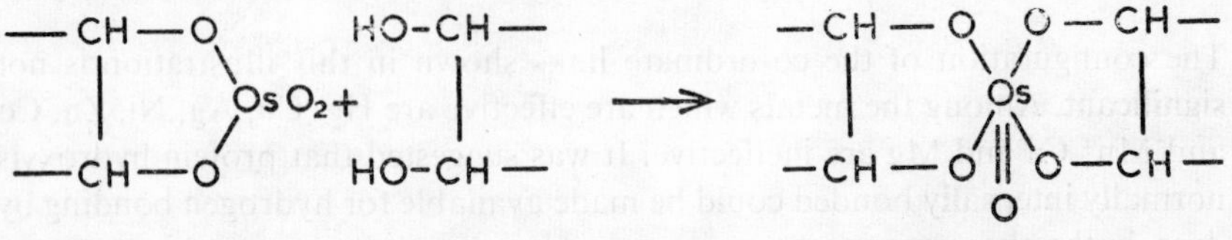

II

No direct evidence exists for the formation of compound II (diester) when lipids from natural sources are fixed with OsO_4. An alternative hypotheses was put forward by Wigglesworth (1957) ,who considered that in place of the second reaction the OsO_2 could form a double co-ordinate linkage directly with the carbon of the double bond (III).

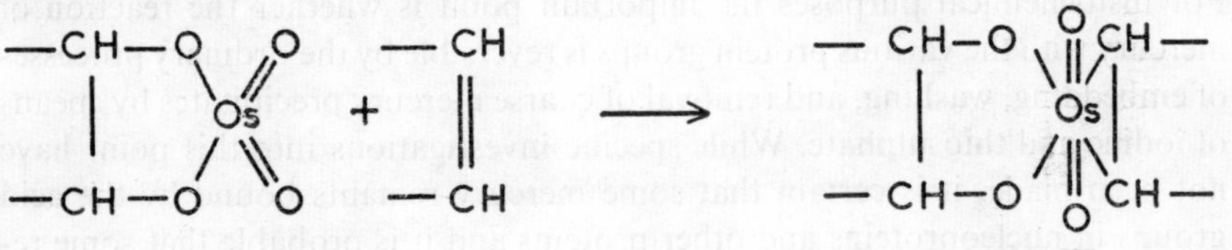

III

In the case of two bonds in adjacent chains in close apposition Wigglesworth considered that double monoester formation could occur, as shown below, with consequent binding of the fatty acid chains (IV).

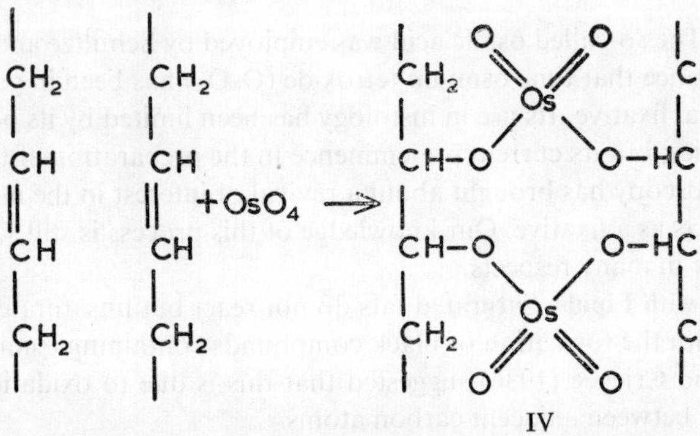

IV

Whatever the nature of the reaction of unsaturated fats with OsO_4 polymerization will occur if, as in triolein, there are a number of unsaturated chains in the molecule. It may also occur when, as in linoleic acid, there are a number of unsaturated bonds in a single chain.

According to Becker (1959) the product of cyclopentene and OsO_4 interaction is a completely insoluble polymer of compound I, for which he proposed the following structure:

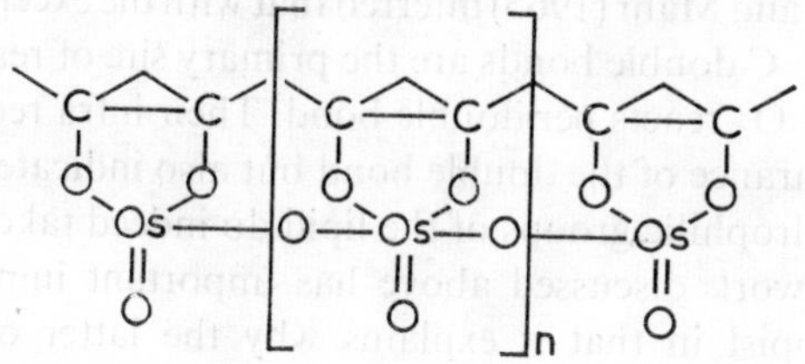

V

Although there is no direct proof of the existence of this compound (V), Becker's hypothesis provides an explanation for the fixation of lipids with only one double bond. Single layers of these could not give rise to either compound II or compound IV.

Finean (1954) used buffered OsO_4 at pH 7·0 to fix sciatic nerve and observed an increase in intensity and a change in relative intensity of the bands in its low angle X-ray diffraction pattern. He suggested that the monoethanoid fatty acids of the myelin sheath were the main site of its action. Alcohol normally extracts the lipids of the myelin sheaths completely, according to Finean. After OsO_4 fixation it no longer does this and it is often assumed that binding of lipid to protein has occurred. This would be unnecessary, however, if the osmium-lipid compound were insoluble in alcohol.

Discussing the mechanism of fixation of lipids with OsO_4 Bahr (1954) had suggested that the reaction of lecithin was a double effect involving both the unsaturated fatty acid and the choline, or cholamine residue in the molecule. Investigations by Khan *et al.* (1961) confirmed the involvement of the polar groups of lecithin as well as its double bonds and further work supporting Bahr's concept was reported by Riemersma and Booij (1962). These authors showed, in the first instance, that the amount of OsO_4 bound by lecithin corresponded to the number of double bonds available (thus confirming Criegee's hypothesis). Secondly, they showed that so-called tricomplex formation (staining of lecithin with the acid dye Brilliant scarlet 3 R after blocking the lipid phosphate groups with uranyl ions) was progressively reversed by reaction with OsO_4. The probable mechanism of tricomplex formation involves binding of the dye by the choline (quaternary ammonium) groups of the lipid after removing interference by its anionic (phosphate) groups. At the same time Riemersma and Booij observed that after reaction with OsO_4 the affinity of lecithin for a basic dye, Brilliant green, was greatly increased. This was considered to be due to reaction with lipid phosphate groups after removal of interference by the quaternary ammonium groups.

Later Riemersma (1963) suggested that hydrated OsO_2, the product of the reaction between lecithin and OsO_4, acts as a negatively charged colloid capable of Coulomb interaction with the quaternary ammonium group. He observed that if this OsO_2 was removed, for example with hydrogen peroxide, the affinity of lecithin for Brilliant scarlet 3 R was fully restored.

From their quantitative and infra red studies of the reaction of OsO_4 with lipids Stoeckenius and Mahr (1965) inferred that with the exception of phosphatidyl serine the C=C double bonds are the primary site of reaction with OsO_4 and that 1 mol. OsO_4 reacts per double bond. Their infra red spectra showed complete disappearance of the double bond but also indicated that secondary reactions with hydrophilic groups of the lipid do indeed take place.

Much of the work discussed above has important implications for the electron microscopist in that it explains why the latter observes that the deposition of osmium in lipid micelles occurs at the site of the polar groups and not at the original site of reaction in the hydrophobic interior of the micelles. This is shown below in Fig. 39, which indicates diagrammatically the two possible sites of osmium deposition.

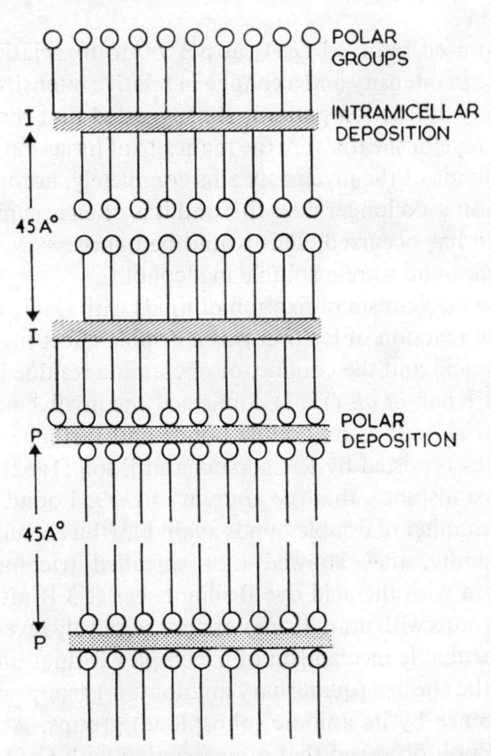

FIG. 39. Sites of Osmium Dioxide Deposition in Lipid Micelles.
(After Riemersma, 1963)

It can be seen that the spacing of the deposits is the same whether they are intramicellar or polar. Stoeckenius and Mahr (1965) observe that the amount of Os in the reaction product of unsaturated lipids is much higher than the amount necessary for a detectable increase in contrast in the E.M. image.

Reaction with Lipoproteins. Quantitative determinations of the OsO_4-lipoprotein interaction were made by Hayes *et al.* (1963), using the X-ray fluorescence technique and serum lipoprotein fractions. These showed that nearly all of the reaction occurred with lipid rather than protein and that as the percentage of the latter rose the uptake of osmium decreased. Recovery of fatty acids from fixed lipoproteins was limited but the results showed that unsaturated fatty acids were preferentially involved in the fixation process.

No estimate was made of the recoverable protein but presumably the negatively charged OsO_2 is able to replace the negatively charged protein groups which are responsible for electrostatic (Coulombic) binding of lipid to protein.

According to Salem (1962) the stability of lipid-protein complexes containing lecithin is mainly due to long range electrostatic forces acting between the positive and negative charged groups of lecithin and similar groups in the constituent amino-acids of the protein. This is shown in Fig. 40, below.

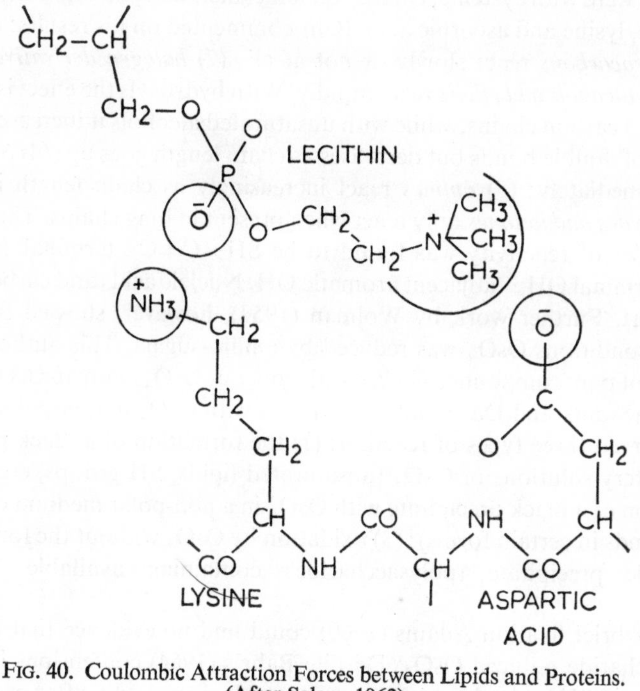

LECITHIN

LYSINE

ASPARTIC ACID

FIG. 40. Coulombic Attraction Forces between Lipids and Proteins. (After Salem, 1962)

Salem estimated the average electrostatic interaction energy between the charged groups of the protein and lipids as 5 kilo-calories per mol. No figures for hydrated OsO_2 are available but presumably its interaction energy considerably exceeds the above value.

Reaction with Protein, Amino-acids, and Sugars. Porter and Kallman (1953) repeated the studies of Fischer, Mann, Baker and others and showed that 2 per cent. OsO_4 formed gels with albumin, globulin and fibrinogen. The clear gel which forms relatively slowly with albumen they regard as an indication of fine micellar or even unimolecular binding. The opaque gel formed with globulin and fibrinogen, on the contrary, is held to indicate some coarser type of binding. Wolman (1955) believed that the bridging which was supposed to occur in the early stages of fixation with OsO_4 was responsible for the excellent preservation of structure, and that with longer fixation damage occurred through over-oxidation. He also stressed the fact that the well-known reversal of tissue acidophilia must be due to blockage or destruction of NH_2 groups.

In a study of the reactions of OsO_4 (and RuO_4) with biologically important substances Bahr (1955) tested hundreds of compounds *in vitro* with OsO_4 at a final concentration of 0·003 M, in water or carbon tetrachloride at 0°, 20° and 50°. The substances tested had a final concentration of 0·6 M. The main positive reactions were with cysteine, triethanolamine, oleic acid, methionine, arginine, ornithine, lysine and ascorbic acid. Bahr commented on his results: (1) *Saturated hydrocarbons* react slowly or not at all; (2) *halogen derivatives* do not react; (3) *alcohols and ethers* react rapidly. With hydroxyls the effect is maximal with 3 to 5 carbon chains, while with unsaturated alcohols it increases with the number of double bonds but decreases as chain length goes up; (4) *SH and SS* react immediately; (5) *amines* react increasingly as chain length increases; (6) *aldehydes and ketones* only react when present in long chains. The approximate order of reactivity was found to be SH$>$C=C$>$terminal $NH_2$$<$SS, CHO, terminal OH$>$ adjacent aromatic OH. Nucleic acids and carbohydrates were inert. Further work by Wolman (1957), however, showed that under certain conditions OsO_4 was reduced by amino-sugars. This author tested a number of pure compounds *in vitro* with aqueous OsO_4, with an OsO_4/oxidant mixture (Swank and Davenport, 1935), and with OsO_4 in non-polar solvents. He observed three types of reaction: (1) the formation of a black precipitate with watery solutions of OsO_4 (unsaturated lipids, SH groups, etc.); (2) the formation of a black precipitate with OsO_4 in a non-polar medium only (NH_2 compounds in certain forms); (3) oxidation by OsO_4 without the formation of a visible precipitate (polysaccharides containing available 1:2-glycol groups).

After brief fixation Adams (1960) could find no evidence that protein or polysaccharide reduced OsO_4. Despite Bahr's (1954) contentions he showed that protein SH and other end groups were blackened only after prolonged (3 day) fixation. He attributed the failure of protein SH groups to reduce OsO_4 to rigidity of the molecule allowing reaction to take place at one position only. Although Wolman's results suggested that absence of a black colour (OsO_2)

could not be taken to indicate the absence of a reaction, Adams pointed out that any bound osmium should have been revealed by treatment with $(NH_4)_2S$ or thiourea. In the event, no such demonstration was made.

Experiments carried out by Wigglesworth (1964) were not in agreement with those of Adams. They indicated that if the salt linkages between protein (histone) and nucleic acid were weakened by exposure to dilute acid, both chromosomes and cytoplasmic nucleic acids would take up large amounts of osmium (1 hour fixation). Tests with blocking agents showed that the chief reactive groups were guanidino (arginine), ε-amino (lysine), imidazole (histidine), and pyrrolidine (proline).

The views expressed by Adams were also opposed by Hake (1965). This author repeated Bahr's series of tests and found that with every amino-acid tested (18) oxidative deamination occurred with reduction of OsO_4. Proteins and peptides reacted similarly, as did SH groups produced in wool fibres by reduction with thioglycollic acid.

In spite of the manifest disagreements recorded in this section, there is no doubt that normal OsO_4 fixation provides no increase in the electron opacity of proteins. These, in fact are substantially leached from the tissues during the process. Dallam (1957) investigated the problem of protein loss from tissue components during the processes of OsO_4 fixation, dehydration, and embedding in methacrylate. In the case of a sample of rat liver mitochondria, for instance, he found a 22·6 per cent. loss of protein during fixation and a further 12·2 per cent. during dehydration. The loss of phospholipid in the fixative could not be calculated for technical reasons, but in the subsequent dehydration stage it was 27 per cent. Not only do these findings indicate the need for caution in interpreting electron micrographs, but also in accepting estimates of the efficiency of any fixative on the basis of histological appearances alone.

Alcohol and Acetone

These two protein precipitants have been used as "fixatives" especially in enzyme histochemistry because, in spite of the morphological disturbances which they create in the tissues, they leave the reactive groups of enzymes to a large extent in their original state. Unfortunately very little is known about the exact effect of alcohol and acetone on proteins. In general, however, denaturation of a protein by precipitation may change the reactivity of its groups in three ways: (1) by rendering parts of the molecule inaccessible to various reagents; (2) by altering the rate at which such reagents diffuse into the protein (not significant in tissue sections); and (3) by bringing into proximity previously separated groups in the protein molecule, thus producing new stereochemical relations. The effects of denaturation are to some extent reversible and many of the original properties of the protein are sometimes regained when the denaturing agent is removed. The mucoproteins of the anterior pituitary gland, for instance, are entirely soluble in water after precipitation by brief treatment with alcohol, and after treatment with acetone the alkaline phosphatase of various tissues is still soluble to some extent in dilute solutions of alcohol (see Chapter 15, p. 499). Other effects are apparently irreversible, such as the partial inactivation of most enzymes. It is not known whether this is due to a chemical effect on their reactive groups or to one of the denaturation factors mentioned above.

The cross linkages which bind protein chains together are of several types, some of these are illustrated in Fig. 41 below.

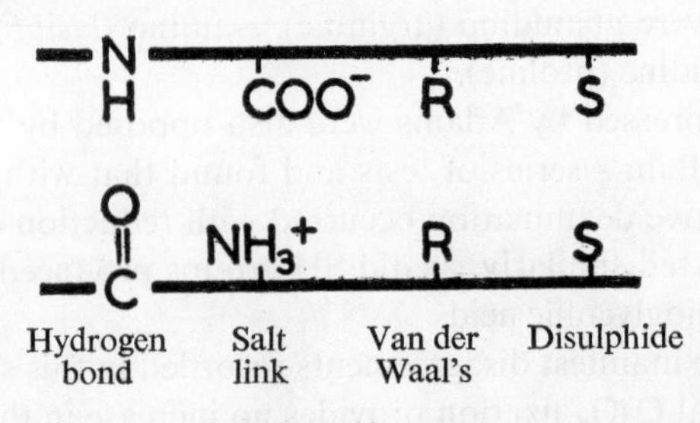

FIG. 41.

It has been suggested that alcohol acts by turning the protein chains outwards (away from each other), breaking the hydrogen bonds and salt links and thus revealing the end groups of the side chains to a varying extent. The precise mechanism, and the extent of liberation of side chains is largely unknown, however. It is discussed at some length by Kauzmann (1959) in an article on the consideration and interpretation of protein denaturation, while the mechanism of denaturation of enzyme proteins has been described by Okunuki (1961).

Although on *a priori* grounds one would imagine that alteration of the reactivity of various groups in the tissues would be greater in the case of the formalin and metal-containing fixatives than with alcohol or acetone, in practice the effect of formalin is not found to be any greater in this respect. Examples of this will be found in the following section on the use of formalin-fixed material for enzyme studies.

THE CHEMICAL ACTIONS OF VAPOUR FIXATIVES

Vapour fixation with volatile fixatives is a practice of long standing in histology and cytology. Its use has been confined mainly to formaldehyde and OsO_4, usually applied to smears or sections at 37° or lower. The motive for employing vapour as opposed to liquid fixatives has been the desire to retain soluble materials *in situ* by converting them to insoluble products before they come into contact with water or non-aqueous solvents.

Application of fixatives and reagents in vapour form to freeze-dried tissues has been recorded occasionally (e.g. Mancini, 1944; van Mullem, 1958) but it was first subjected to serious study by Gersh, Vergara and Rossi (1960).

These authors tested a number of reagents in addition to formaldehyde, such as 1-fluoro-2,4-dinitrobenzene (DNFB), methyl mercuric chloride (MMC) and copper acetyl acetonate. The chief feature of their method was the introduction of the vapour into the drying chamber under its original vacuum.

Selective post-fixation of proteins and polypeptides, in freeze-dried yeast, was produced by Mundkur (1961) using 1,5-difluoro and 2,4-dinitrobenzene. Increased basophilia was demonstrated by subsequent staining with gallocyanin-chromalum (Chapter 9, p. 274). Further studies were reported by Gersh (1964), designed to fix lipids by contact with vapour fixatives *during* the freeze-drying process. The reagents used included ethylamine, ozone, 2,4-dinitrophenylhydrazine and bis-(2-nitro-4-hydrazinophenyl) methane. Ozone was used as a pretreatment, being passed over the specimen at —80° to convert the double bonds of the lipid aliphatic chains to ozonides.

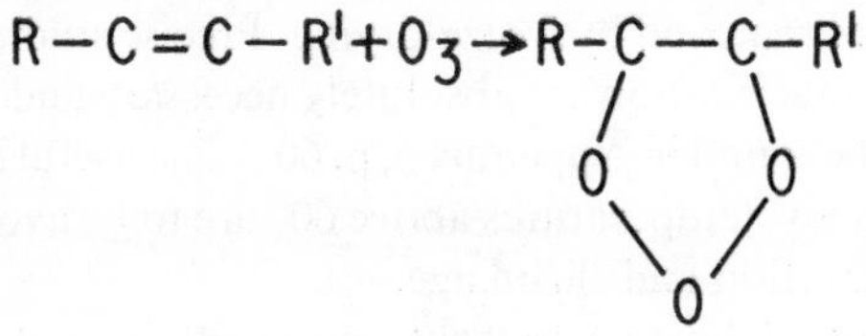

$$R-C=C-R' + O_3 \rightarrow R-C \overset{}{—} C-R' \quad (\text{C–O–O–O–C ozonide ring})$$

Subsequent reaction with phenylhydrazine vapour produced either ketone phenylhydrazones or osazones.

Probably the most important observation in the field of vapour fixation was made by Falck (1962) who described the conversion of catecholamines and 5-hydroxytryptamine in freeze-dried tissues to fluorescent condensation products by treatment with formaldehyde *at elevated temperatures*. Previous work on 5-HT in freeze-dried tissues (Barter and Pearse, 1953, 1955) had employed formaldehyde vapour at room temperature only and previous work on catecholamines (Eränkö, 1951, 1952) was carried out on fresh tissues treated with aqueous formalin. The fluorescent amine methods are fully considered in Chapter 26; their significance here is that they have brought into prominence the ability of highly reactive fixative vapours to capture (render insoluble) highly soluble, low molecular weight compounds. The full range of the latter has not yet been revealed. Obviously it includes smaller protein molecules, polypeptides, nucleic acids and nucleotides, polysaccharides and perhaps di- and monosaccharides also. The products are not usually fluorescent and their identification is not easy because they are *entirely new products* whose reactivity differs totally from that of the parent substance.

Studies carried out by Tock and Pearse (1965), and later reported in full by Tock (1966), clearly showed the effectiveness of "hot" vapour fixation applied to freeze-dried tissues. Comparisons were made between formaldehyde, acrolein, glutaraldehyde, OsO_4, chromyl chloride, cyanuric chloride, and a number of volatile metal derivatives (carbonyls). The most important of these agents are briefly considered below.

Formaldehyde. As Falck and Owman (1965) have rightly stated, "freeze-drying and formaldehyde gas treatment can . . . be recommended for any type of histochemical work where formaldehyde fixation is permissible". The usual source of monomeric formaldehyde is heated paraformaldehyde and, where the moisture content of the environment has been found to be critical, the polymer will probably have been stored before use in an environment with strictly controlled humidity. Even dried paraformaldehyde, it is worth noting, contains 2–4 per cent. water (Walker, 1964).

For trapping small molecules, such as catecholamines, a precise degree of humidity is required but for the fixation of proteins and other tissue components this factor is less important. As shown by Tock and Pearse (1965), formaldehyde gas fixes effectively in the virtual absence of water vapour although possibly it would not do so in the absence of the bound water of the tissues, which is not removed by freeze-drying. Precise studies on the influence of humidity on vapour fixation are absolutely necessary and some information on this point may be found in Appendix 5, p. 605. The useful temperature range is 50°–80° but, usually, temperatures above 60° are to be avoided because they induce heat denaturation and shrinkage.

Other Aldehydes. The two reagents most often used are acrolein and glutaraldehyde. The latter is supplied as a 25 per cent. aqueous solution so that if the vapour is required the aldehyde must first be extracted into ether or liquid paraffin. Glutaraldehyde has little fixing (cross-linking) effect in the absence of water vapour, unlike formaldehyde, and a high temperature (60°) is required for effective fixation in a short period.

Osmium Tetroxide. At 37° the vapour pressure of this reagent is absolutely sufficient. It penetrates very rapidly into freeze-dried blocks of tissue and one hour or less is usually adequate.

Chromyl Chloride. Sufficient vapour pressure is available at 37°. This fixative presumably acts as chromic acid by combination with water vapour and, possibly, with bound water also.

$$2CrO_2Cl_2 + 3H_2O \rightarrow H_2Cr_2O_7 + 4HCl$$

Tissues treated with chromyl chloride vapour alone are rendered extremely brittle and almost impossible to cut after any known embedding process. If it follows formaldehyde, however, this type of post-chromation can be extremely valuable.

Alcohol. Ethanol vapour at 60° has a pronounced denaturating effect on freeze-dried tissues and polysaccharides like glycogen become less soluble than in control tissues. No real indications for the use of ethanol vapour are forthcoming and other alcohols seem not to have been tested.

The Choice of Fixative

In this section it is proposed to discuss the appropriate fixation for tissues required for the various histochemical procedures listed below. Many new methods of fixation have been evolved in recent years, and advances have been made in our knowledge of the ways in which the classical fixatives work and in their application to various problems. In particular, I propose to consider the fixation of glycogen, lipids, nucleoproteins and enzymes. Mucoproteins and mucopolysaccharides will also be dealt with and the choice of fixative to precede treatment with various enzymes, when these are used as histochemical reagents. Inorganic materials, and electron microscopy, will be treated very briefly.

Glycogen

It is now known that a large variety of glycogens occur naturally which are distinguished by varying degrees of polymerization. The larger molecules of the more highly polymerized glycogens are those we have been accustomed to see in tissues fixed by a wide variety of methods as well as the more or less classical alcohol-containing ones. The less highly polymerized glycogens have probably not been successfully fixed in the tissues by any of the processes employed in the past, but have diffused into the fixing fluid. Sedimentation constants from 20 to 120 S were recorded for rabbit liver glycogen (Bridgman, 1942) and from 60 to 300 S for human liver (Polglase *et al.*, 1952). Glycogen from cases of glycogen storage disease is predominantly of the lighter type (Stacey and Barker, 1962) and this is why it can be observed streaming out of cryostat or freeze-dried sections treated with Lugol's iodine.

Histochemically, as demonstrated by Kugler and Wilkinson (1960), it is possible to demonstrate only the so-called "lyo" fraction, that which is soluble in cold trichloracetic acid (see Chapter 10, p. 362). It is the preservation of this "lyo" fraction with which we are concerned, therefore.

A commonly held idea (see Swigert *et al.*, 1960) is that fixation of glycogen occurs not directly but by trapping in protein or other cell components. There is no real reason to suppose that this is true. The effectiveness of 100 per cent. ethanol as a fixative for glycogen suggests that the normal form (the rosette of electron microscopy) contains bound water molecules. Removal of these by dehydration (ethanol, freeze-drying, freeze-substitution) would result in decreased solubility amounting to denaturation. True fixation, on the other hand, might occur with formaldehyde especially when used in vapour form. Working with cellulose (a glucopyranose polymer like glycogen) Heuser (1946) concluded that formal linkages (formals) could connect the OH groups in separate chains. Only a small percentage of linkages was required to modify the properties of cellulose very considerably. The observed changes included diminished solubility in alkali and resistance to direct cotton dyes. Similar changes might be expected to occur in the case of glycogen.

The history of glycogen fixatives is a long one. Standard methods were for a long while carried out at room temperature but Deane, Nesbett and Hastings (1946) suggested an improvement which involved the use of ice-cold picro-alcohol-formalin. Lison and Vokaer (1949) recommended for the same purpose either cold alcohol or a mixture of 96 per cent. alcohol saturated with picric acid (85 parts), 40 per cent. formalin (10 parts), and acetic acid (5 parts). These fixatives were used at a temperature of —73°, produced by an acetone-solid CO_2 mixture, and Lison claimed that they could effectively prevent polarization, that is to say, streaming of the granules to one pole of the cell. Polarization is certainly reduced to a minimum by such methods, It can otherwise only be avoided by the use of freeze-substitution or freeze-drying techniques (see Figs 34 and 35). Alcoholic fixatives are the worst offenders in producing polarization at ordinary temperatures and also in showing a pronounced difference in the amounts of glycogen revealed in sections taken from the surface and from the depth of the block. If acetic acid is used in the fixative the granules are coarser, and with aqueous fixatives, in general, polarization and surface-depth differences are far less than when alcohol is used. Mercuric chloride decreases the amount of glycogen demonstrable by any of the methods discussed in Chapter 10.

More recent data still indicate the overall effectiveness of ethanol fixation. Trott (1961) compared the results obtained with rat liver and 8 different liquid fixatives. Preservation of glycogen (demonstrated with PAS or Best's carmine) varied not only from one fixative to another but also with the duration of fixation. After 24 hours only Rossman's fixative (PAS stain) and 10 per cent. formalin (Best's stain) were inadequate. After 3 months, on the other hand, formalin was inadequate with both stains and the alcoholic fixatives, especially Bouin and Carnoy, were most satisfactory.

Chemical assays carried out by Kugler (1965) on rat liver indicated a clear superiority for absolute ethanol, as shown in Table 6, below, which is taken from his results.

TABLE 6

Preservation of Glycogen by Liquid Fixatives

Fixative	Abs. Alcohol	Gendre	Aq. Bouin	80% Alcohol	Formalin	Carnoy	Alc. Bouin	Susa	Zenker
$\bar{X}$	90	85	82	81	80	79	76	48	37
S	12·2	13·9	10·8	12·3	14·2	15·2	18·2	18·2	18·3
$V=\frac{100S}{\bar{X}}$	13·5	16·4	13·1	15·1	17·7	19·2	23·9	37·9	49·4

$\bar{X}$ = Average percentage of control value.
S = Standard Deviation.
V = Coefficient of Variation.

Parallel histochemical studies confirmed the views of Kugler and Wilkinson (1959) on the threshold values for demonstration of glycogen in tissue sections.

At a level of 0·5 per cent. (wet weight) Susa and Zenker-fixed sections failed to give a positive result. The average threshold for alcoholic fixatives was 0·2 per cent. This agrees fairly well with results obtained by other observers. For instance, Nielson *et al.* (1932) recorded 0·1 to 0·3 per cent. while Deane *et al.* (1946) found that their critical level was 0·08 per cent. Fitzpatrick, Larner and Landing (1948), using mouse liver, observed a somewhat wider spread in their assays, which showed 0·04 to 0·6 per cent.

If liquid fixatives continue to be employed for glycogen studies there is clearly little advantage in using anything except absolute alcohol (and small blocks). Freeze-drying and formaldehyde vapour fixation provides an obvious alternative for research studies but, until assay figures are available, we cannot assume complete fixation. Moreover, as with other small molecular weight components, the chemical and physical reactivity of the glycogen may be entirely altered by this procedure.

Lipids

Only two reagents can be said to fix lipids in the true sense of rendering them insoluble in lipid solvents. These are OsO_4 and chromic acid. Both alter the chemical reactivity of the lipid considerably. Preservation of lipids, without necessarily altering their solubility in organic solvents, can be obtained in a number of ways. The most popular methods are derived from Baker's (1944) fixative, which was designed for the preservation of phospholipids. In his original method this author used formalin together with calcium and cadmium chlorides, and McManus (1946) substituted cobalt nitrate for the more expensive cadmium salt. Berg (1951), in a survey of the lipid-dissolving effects of various solvents on fixed sections, could not find any difference between routine formalin-fixed material and that fixed in formalin with the addition of calcium chloride. This observation confirms Baker's ideas of the mechanism by which calcium and other salts preserve phospholipids. Although the latter are preserved, that is to say prevented from diffusing into the fixing fluid, by Ca, Co and Cd, they are not fixed in the usual sense of the word and are still removable by fat solvents. The action of Ca, Co and Cd ions is presumed to be due to their influence on the formation of complex coacervates of phospholipid with other cellular constituents (protein, mucopolysaccharide, etc.). Coacervates (Latin: *acervus*, a heap or swarm) of the complex type can be regarded as structures resulting from the aggregation of molecules which are held together by intermolecular forces, in the form of a mosaic or lattice of lipid and protein molecules.

Of interest here are the observations of Clayton (1959) who investigated the capacity of various primary (unmixed) fixatives to split (unmask) lipids from lipoprotein complexes. Altogether 16 compounds were tested and the most effective of these was cadmium chloride, followed by mercuric chloride and chloroplatinic acid. Formaldehyde was also moderately powerful in this respect. Reference to Salem's diagram (p. 83) may be helpful.

It is interesting to note that in their extensive studies on fixation of lipids in cryostat sections Tchacarof and Pózalaky (1964) found that the most effective agent was mercuric chloride, either alone or in combination with dichromate or chromic acid.

Two contributions to the better preservation of phospholipids were introduced by Elftman (1954, 1957). The first, a process which its author called *controlled chromation*, was a further refinement in a long line of methods derived from the original method of Weigert (1884) by the work of Smith and Mair (1908), Ciaccio (1909), Dietrich (1910) and Baker (1946). All these were based on the fact that oxidized phospholipids are less soluble in fat solvents than their unoxidized precursors.

Four variables were subjected to control, pH being added to the usual triad of time, temperature and concentration of the reagent. The standard procedure involves fixation in 2·5 per cent. $K_2Cr_2O_7$ at pH 3·5 (acetate buffer) and 56° for 18 hours. Paraffin sections are stained with Sudan black B and with hæmatoxylin (see Appendices 5, p. 605, and 12, p. 690). Chromated lipids are stained by both dyes, non-lipid chromated substances only by hæmatoxylin. This procedure is discussed more fully in the chapter on lipid methods (p. 398).

Elftman's second fixative (dichromate-sublimate) contains 2·5 per cent. $K_2Cr_2O_7$ in 5 per cent. $HgCl_2$. It thus closely resembles Zenker (minus acetic acid) and Helly (minus formalin) and has an initial pH slightly above 3·0. Long fixation (3 days) and paraffin embedding is followed by staining of the sections with Sudan Black B in ethylene or propylene glycol (Appendix 12, p. 691) and mounting in Apathy's medium (Appendix 1, p. 579). Dichromate-sublimate gives good results in mammalian tissues, the mitochondria are well preserved and the Golgi region shows up in many tissues as a circumscribed area of perinuclear staining.

The effects of formalin (as opposed to formaldehyde) on lipids were examined by Heslinga and Deierkauf (1961, 1962) and by Deierkauf and Heslinga (1962). Their results suggested that these should be divided into short-term and long-term actions. Under the former heading the most rapid reaction was with phosphatidyl ethanolamine (presumably with NH_2 groups), followed by the formation of degradation products such as lysophosphatidyl ethanolamine and lysolecithin. One would equally expect formalin to react with the amino groups of phosphatidyl serine (Chapter 12, p. 400) and plasmalogen (Chapter 13, p. 461).

Long-term effects were associated with acid hydrolysis and cannot be regarded as fixation. It is significant, however, that cholesterol, cerebrosides, sulphatides, phosphoinositides and sphingomyelin remained unaffected.

Whether formaldehyde vapour reacts with lipid NH_2 groups is not known with certainty although it may be presumed to do so. According to Walker (1964) formaldehyde only reacts with the —C=C— double bond in the presence of strong acid catalysts. The final products are then 1,3-glycols. It may react with glycerol (Fairbourne *et al.*, 1930) to give glycerol formals. No alteration in solubility could be expected from the reaction with formaldehyde but if glutaraldehyde were to react cross-linking might occur.

A method evolved by Rinehart and Abu'l-Haj (1951) may be mentioned here, although it does not directly concern fixation, because it has to do with the preservation of lipids and lipid coacervates. These authors used one of the polyethylene glycols or carbowaxes which are long chain polymers of the type $HOCH_2(CH_2OCH_2)CH_2OH$, for the dehydration and embedding of tissues after formalin fixation. Lipids, cholesterol and cholesterolesters are insoluble in solutions of carbowaxes in water, and in the carbowaxes themselves. The method gives excellent preservation of lipids and is given in full in Appendix 1, p. 576.

NUCLEOPROTEINS AND NUCLEIC ACIDS

Nucleic acids, like the glycogens, exist in many different states of polymerization (see Chapter 9) and it is certain that any method of fixation induces changes in their physical state. Changes in chemical reactivity are also brought about by the majority of fixatives so that exception can be taken to every method on one or both of these points. From the histochemical point of view formalin is not a good fixative for nucleic acids and nucleoproteins. It blocks a large number of reactive groups so that stainability by both basic and acid dyes is considerably reduced. Improvement occurs when mercury or chromium salts are added but objection must be made to these if histochemical techniques are to be employed. Both salts interfere to some extent with the action of enzymes and lanthanum acetate (Hammarsten *et al.*, 1935; Opie and Lavin, 1946) acts in a similar manner, preventing the action of ribonuclease altogether. Precipitant fixatives, like alcohol and acetic acid, are much more often recommended for the preservation of nucleic acids and Carnoy is the most popular of these. Little exception can be taken to it from the chemical point of view, but White and Elmes (1952a and b) showed that ethyl alcohol rapidly caused irreversible changes in the structure of isolated deoxyribose nucleic acid (DNA) as measured by the ability of the nucleic acid fibres to induce dichroism when stained with a variety of dyes and examined by polarized light. They found methyl alcohol to be much less active in this respect, probably because it is a weaker dehydrating agent than ethyl alcohol. The acid fixatives produce their excellent morphological pictures by precipitating the nucleoproteins and, at the same time, they progressively break the bonds between the nucleic acids and the proteins, increasing the number of acid groups available for reaction and hence the degree of basophilia. From the outset, therefore, the physical state of the nucleoproteins is profoundly altered and a long sojourn in acid fixatives causes extraction first of ribonucleic acid (RNA) and then of DNA. For this reason, prolonged fixation in Carnoy must be avoided.

Differences in fixation have long been known to necessitate different times of exposure to acid hydrolysis in performing the Feulgen reaction (Chapter 9,

p. 254), yet it has been regarded as perfectly permissible to use this technique as a qualitative reaction after almost any fixative. Quantitative estimation of DNA stained by the Feulgen reaction has been carried out by a number of authors and, although the accuracy of some of the earlier results seemed doubtful, direct comparison between biochemical and spectrophotometric results gives good grounds for confidence.

The maximum useful duration of acid hydrolysis after any of the liquid fixatives is in the region of 8 minutes. After freeze-drying and formaldehyde vapour (3 hours, 50°) hydrolysis can be carried out for 1 hour or longer without any reduction in the intensity of the reaction. It is necessary to assume:

(1) that the latter is in no way blocked
(2) that the rapid reversal of the Feulgen reaction, observed in conventionally fixed tissues, does not occur.
(3) that current explanations of the mechanism of the Feulgen reaction (Chapter 9, p. 259) may need revision.

Formaldehyde vapour fixation must be seriously considered for all future studies on the localization and cytophotometric quantitation of nucleic acids.

ENZYMES

The effect of fixation on enzymes continues to be of increasing importance as histochemical techniques for demonstrating them increase in number. Theoretically, the best way of showing enzyme activity by histochemical means would appear to be by the use of fresh frozen sections or tissue blocks. There are considerable objections to both of these in practice, however, especially to the latter. With the cold knife methods (Chapter 2, p. 13) it was technically difficult to obtain good fresh frozen sections of the majority of tissues, but even when these were obtained with greater ease by means of cold microtomes, there remained a further objection. Particularly if incubation with the substrate was prolonged, loss of enzyme and of other protein and non-protein materials into the incubating medium was much greater than in fixed material. This produced not only false localization of enzyme (due to diffusion in the section) but also widespread deposition on the section of the product of enzymic activity in the medium, and general filthiness of the section due to partial dissolution of its components.

The early histochemistry of the phosphatases and esterases was largely carried out with materials fixed in alcohol or acetone, usually at cold-room temperatures (4°), since these two protein precipitants were found to produce only a small loss of activity in the case of alkaline phosphatase. Moreover, they preserved the tissues in such a way as to allow paraffin embedding. Danielli (1946) cut frozen sections of tissues fixed in absolute alcohol and found no significant loss of alkaline phosphatase activity. Up to 75 per cent. loss of activity was caused, however, by the subsequent process of paraffin embedding. Mowry (1949) studied the effect of certain technical procedures on the preservation of alkaline phosphatase for histochemical purposes. He noted no significant differences between acetone and alcohol and no improvement was recorded on buffering these fixatives. Their use in the cold increased the amount of enzyme preserved, and when tissues were fixed in alcohol saturated with sodium α-glycerophosphate preservation of enzyme was improved. Mowry concluded that considerable latitude was possible in the fixation of tissues for alkaline phosphatase studies, but that, on the other hand, the process of embedding did not allow such latitude. The best fixation of enzyme was achieved with acetone or alcohol-fixed blocks cleared in light petroleum (petroleum ether) and embedded *in vacuo* in ordinary or low melting-point (42°–44° C.) paraffin. He recommended that embedding should not occupy longer than 30 minutes. The relative importance of the embedding procedure was later stressed by Dalgaard (1956). This author made model experiments with pieces of gelatine containing serum from dogs with obstructive jaundice. He found that the amount of phenylphosphatase was unaffected by fixation in 80 per cent. ethanol or by dehydration and clearing in methyl benzoate-colloidin (8 hours). The temperature of embedding, however, was critical. At 53° a sharp loss of enzyme occurred and Dalgaard therefore advised embedding in a mixture of soft (42°) and ordinary (55°) paraffin.

There remained practically no recognized alternative to fresh tissues on the one hand and acetone or alcohol-fixed paraffin sections on the other, until Seligman, Chauncey and Nachlas (1951) published their observations on formalin fixation. These were derived from efforts to find a suitable fixative for sulphatase and β-glucuronidase, neither of which could tolerate even the moderate heating required by paraffin embedding after acetone fixation. The authors tested homogenates of rat liver for activity in respect of five different enzymes, the homogenates being prepared from liver slices treated for varying times at varying temperatures with 10 per cent. formalin buffered at pH 7 with phosphate. The following Table, modified from Seligman *et al.* (1951), gives the most interesting results of these experiments.

TABLE 7

Percentage Activity Remaining after Fixation in 10 *per cent. Formalin*

Time	Temperature	Alkaline phosphatase	Acid phosphatase	Esterase	Sulphatase	β-glucuronidase
2 hours	4°	73	79	82	74	86
24 hours	4°	26	55	35	58	87
1 hour	25°	48	58	67	59	100
1 hour	37°	29	41	27	63	105

The blocks used in the preparation of homogenates, from which the above recordings were derived, were washed for 5 minutes only after exposure to the fixative. The authors found no alteration in results if the tissues were washed for 24 hours. This was perhaps somewhat surprising in view of the observations on the action of formalin described in the opening section of this chapter. The results obtained by Seligman *et al.*, compared with those of Stafford and Atkinson (1948), showed that in the case of both alkaline and acid phosphatases greater activity remained after 24 hours in formalin at 4° than with acetone at 4° and subsequent paraffin embedding. Formalin at higher temperatures destroys enzymes much more rapidly. Reis (1956) observed that only 22 per cent. of human placental 5-nucleotidase survived 15 minutes' formalin at 20°, and only 4 per cent. survived 24 hours' fixation. For non-specific alkaline phosphatase the figures were 33 and 17 per cent. respectively.

These figures differ markedly from those published by Novikoff (1952), who tested survival levels in thin slices after cold acetone fixation and paraffin embedding. He found 58 per cent. of alkaline phosphatase and 47 per cent. of 5-nucleotidase surviving. For acid phosphatase and adenosine triphosphatase his figures were 41 per cent. and 1·6 per cent. respectively. The contrast between these results serves to stress the vital importance of temperature in the preservation of enzymes. Acetone continues to be a popular post-fixative for cryostat sections (Novikoff *et al.*, 1960), when it is usually employed at 0°. For blood and bone-marrow smears, however, Kaplow and Burstone (1963) recommended acid-buffered acetone at room temperature (Appendix 5, p. 604).

Further developments in the use of formaldehyde for enzyme studies, particularly for the preservation of acid hydrolases, were reported by Holt and Hicks (1961). These authors found that calcium was an undesirable additive, because of its effects on fine structure at the E.M. level. They recommended 4 per cent. formaldehyde, buffered with 0·067 M phosphate to pH 7·2 and containing 7·5 per cent. sucrose. This was used at 0–2° for 1–24 hours and for light microscopy it was followed by treatment with a gum-sucrose solution. These two procedures (fixation and gum-sucrose) have become standard practice in acid hydrolase cytochemistry (Appendix 5, p. 602).

Graded concentrations (0·7 to 2·0 per cent.) of formaldehyde in Hank's balanced salt solution were used, at 1–4° for very short periods, by Walker and Seligman (1963) for the preservation of bound and so-called soluble dehydrogenases. This work represented a very real step forward, although it was not fully appreciated at the time. A logical extension is contained in the work of Flitney (1966) who found good preservation of soluble dehydrogenases after very brief fixation in dilute buffered glutaraldehyde solutions.

The introduction of glutaraldehyde, and other aldehyde fixatives, by Sabatini *et al.* (1963) was followed by a series of very thorough comparative studies by Janigan (1964, 1965) on the resistance of acid hydrolases to fixation. In his second paper, the activity of acid phosphatase towards four different substrates after fixation with four different aldehydes was recorded. Some of the results are shown in the accompanying histogram (Fig. 42, p. 97).
These data relate to 6 hours' fixation at 0°–2° and they indicate rather regular behaviour towards each of the four substrates. The figures for hydrolysis of AS-TR phosphate are invariably the lowest of the four. Methacrolein, crotonaldehyde and acrolein were also tested but found unsatisfactory both in terms

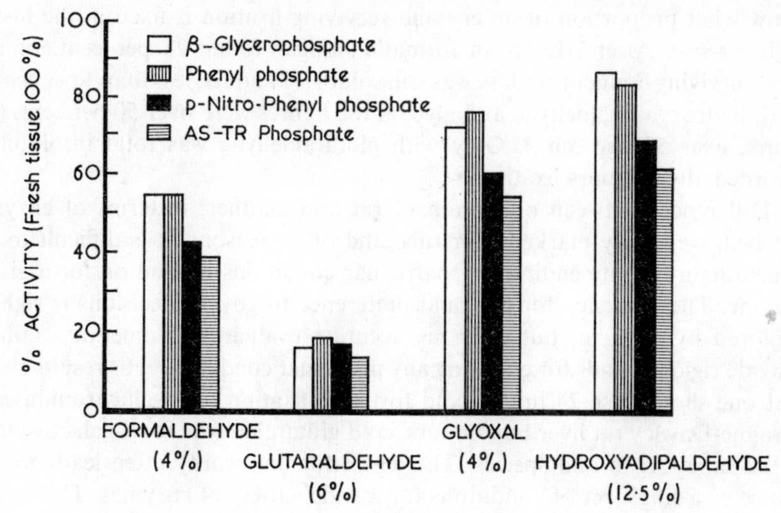

FIG. 42. Survival of Rat Liver Acid Phosphatase after Aldehyde Fixation. (After Janigan, 1965)

of enzyme survival and also of tissue characteristics. Recovery values recorded by Janigan (1965) after 24 hours' cold formalin fixation were close to those recorded by Seligman *et al.* (1951) and by Holt and Hicks (1961).

For the other acid hydrolases (β-glucuronidase, β-galactosidase, *N*-acetyl-β-glucosaminidase and β-glucosidase) the results obtained by Janigan (1964) are summarized in Table 8, below.

TABLE 8

Survival of Acid Hydrolases after Aldehyde Fixation

(Compiled from Janigan, 1964)

Enzyme % survival	4% Formaldehyde (hours)			6% Glutaraldehyde (hours)			12.5% Adipaldehyde (hours)			4% Glyoxal (hours)		
	7	18	24	7	18	24	7	18	24	7	18	24
β-glucuronidase	60	—	45	24	—	12	73	—	68	67	—	59
β-galactosidase	33	—	24	43	—	31	76	—	—	62	—	—
β-glucosidase*	—	56	—	—	32	—	—	82	—	—	98	—
N-acetyl-β-glucosaminidase	—	83	—	—	54	—	—	91	—	—	91	—

* Fixation followed by 24 hours in buffered sucrose.

Although, as can be seen, enzyme recoveries were very high in the case of hydroxyadipaldehyde and glyoxal the tissues themselves were poorly preserved and easily distorted by subsequent procedures (frozen sectioning, paraffin embedding). In terms of practical histochemistry it is often more important to know what proportion of an enzyme surviving fixation is fixed in the histological sense. After 7 hours in formalin Janigan found 28 per cent. of the total surviving β-glucuronidase was still soluble (24 hours, less than 1 per cent.). With hydroxyadipaldehyde and glyoxal the figures were over 50 per cent. (24 hours, over 50 per cent.). Only with glutaraldehyde was total insolubility recorded after 7 hours fixation.

Differences between one strain of rat and another, in terms of enzyme survival, were very marked. For this, and other reasons, it is difficult to be dogmatic in recommending any particular conditions of fixation for a given enzyme. The tendency for personal preference to govern decisions is rightly deplored by Janigan, but even his accurate evaluations sometimes fail to provide rigid grounds for choosing any particular conditions. His results show that one should use 24 hours' cold formalin fixation for β-glucuronidase in Sprague-Dawley rat liver but 7 hours' cold glutaraldehyde for β-galactosidase in the same organ and species. The desire for uniformity often leads to the choice of a single set of conditions for a wide variety of enzymes. This is deplorable, but doubtless the practice will continue.

Formalin containing 0·1 per cent. chloral hydrate was first recommended as a fixative for β-glucuronidase by Baker, Hew and Fishman (1958). Chloral hydrate, $CCl_3CH(OH)_2$, acts as a glycol rather than an aldehyde and, although glycol ethers (cellosolves) have been suggested as fixatives by Shelley and Florence (1961), only a very active agent could be expected to act at so low a concentration. In their comparative studies Hannibal and Nachlas (1959) could find no advantage in its use but Chancey *et al.* (1964) noted a reduction in soluble activity, with full maintainance of the insoluble fraction, after 20 hours' fixation. For acid phosphatase (rat parotid) they recorded 48 per cent. recovery and 1 per cent. soluble activity, a figure which scarcely differs from that of other observers using formalin alone. For β-glucuronidase Manning *et al.* (1963) found about 9 per cent. recovery after either formalin or chloral-formalin while Janigan (1964) found 29 per cent. and 13 per cent. respectively. The evidence for enzyme fixation (or immobilization) by chloral hydrate is not particularly convincing and I consider that its use should be discontinued.

A variety of fixatives has been described for the optimum preservation of particular enzymes. Formalin-sucrose-ammonia (Appendix 5, p. 604) was described by Pearson (1963) especially for cholinesterases, which were well preserved after up to 24 hours' fixation at 0–2°. Monis *et al.* (1965) indicated that although glutaraldehyde was intensely inhibitory towards aminopeptidase (see Chapter 22), a more precise localization could be obtained after its use. They recommended, as others have done for other enzymes, the parallel use of unfixed and glutaraldehyde-fixed sections.

It is difficult, if not impossible, to decide whether enzyme inhibition by successful fixation represents a selective inhibition (lyo- rather than desmo; one isoenzyme rather than another), or the sum of equal degrees of inhibition in various sites or of various species of enzyme.

Broadly speaking, the amount, variety and duration of fixation for any given enzyme must be selected so that the maximum insoluble activity, and no soluble activity whatsoever, are present in the tissue section.

MUCOPOLYSACCHARIDES AND MUCOPROTEINS

The early history of the fixation of mucins is sprinkled with efforts to break away from the use of aqueous formalin, in which and after which many mucopolysaccharides and mucoproteins are still soluble. Holmgren and Wilander (1937) introduced the use of 4 per cent. basic lead acetate as a fixative for acid mucopolysaccharides, when these were to be studied by metachromatic techniques, and Sylvén (1941) used it in combination with formalin. Not all samples of "basic lead acetate" were suitable, but in practice a fresh 4 per cent. solution of pure lead subacetate ($Pb(CH_3CO_2)_2 . PbOH_2$), with or without the addition of an equal volume of 10 per cent. formalin, was usually found satisfactory. The stock solution of subacetate should be made from CO_2-free water and both this and the mixture with formalin should be protected from atmospheric CO_2. If crystals form in tissues fixed in lead acetate mixtures, they can be removed by brief treatment with 0·1 N-HCl. Other authors used 1 per cent. lead nitrate in place of acetate, and Lillie (1954) used an alcoholic 8 per cent. lead nitrate, with or without 10 per cent. formalin, for connective tissue mucins of the type found in umbilical cord and vitreous humour. Dziewiatkowski (1953) used formalin saturated with $Ba(OH)_2$ for preservation of mucopolysaccharides containing ^{35}S, and Mota *et al.* (1956) employed 1 per cent. lead subacetate in 50 per cent. ethanol-acetic acid as a special fixative for mast cells.

Formalin-alcohol mixtures have been popular (Bèlanger, 1963; Quintarelli, 1963) while others have added calcium acetate to formalin, as a cationic precipitant for acidic mucins (Spicer, 1960; Spicer and Jarrels, 1961; Stempien, 1963). In each case an appreciable proportion of tissue mucins remains soluble. For studies of the mucoproteins, exemplified by the anterior pituitary mucoprotein hormones and their precursors, formalin has always been an essential component of whatever fixative is employed. These materials are very soluble in water and partially in 50 per cent. acetone, 70 per cent. alcohol or 50 per cent. dioxane in water (Fevold, 1937). They are not readily precipitated by picric or trichloracetic acids (Meyer, 1945) but absolute alcohol and acetone cause complete precipitation. After such precipitation, however, the mucoproteins of the pituitary gland are still soluble in water and other solvents. In practice, if alcohol or acetone fixation is followed by paraffin embedding, sections cut from such blocks show complete or almost complete loss of mucoprotein. Formalin, on the other hand, is an inadequate fixative for the mucoproteins of many mammalian pituitary glands. These remain water-soluble unless they are subjected to further precipitation in 70–80 per cent. ethanol for 3–6 days before clearing and embedding in paraffin.

Hale (1946) advised the use of Carnoy's fixative for the study of hyaluronic acid by means of his iron method (Chapter 10, p. 349), but it is certainly inadequate for this purpose. It should not be used for mucopolysaccharides in general, despite its employment by Kato and Sirlin (1963) in their radioautographic studies of insect salivary cells.

Comparatively recently Williams and Jackson (1956) introduced two new fixatives for acid mucopolysaccharides. The first was based on a new method of precipitating these compounds with cetylpyridinium salts (Scott, 1955a and b), with which they form highly insoluble complexes. The second depended on the observation that acridines form similarly insoluble complexes. These two fixatives (Appendix 5, p. 604) were superior to formalin and Carnoy, and to formalin-alcohol mixtures, for the preservation of mucopolysaccharides. In a series of comparative tests on human umbilical cord Conklin (1963) found that cetyl pyridinium chloride-formalin gave maximum staining of acid mucopolysaccharide when followed by the Alcian blue and colloidal iron methods. For the preservation of sialic acids Lake (1962) used double fixation, first with 5-aminoacridine-HCl and then with buffered formalin. He claimed better results than with either reagent separately, or combined together.

Fixation in the ordinary sense was avoided by Szirmai (1956) who used cationic dyes in high concentration to obtain precipitation coincident with staining. A similar process was employed by Haust and Landing (1961) for the preservation of the mucopolysaccharides of Hurler's syndrome (Chapter 11, p. 393). As noted by Szirmai (1963) these procedures overrule the greatest disability of the cationic precipitants, their blocking of the anionic groups of the mucin to prevent subsequent staining. An alternative method of overcoming the last objection was the approach used by Kelly *et al.* (1963) who followed fixation in CPC-formalin with selective unblocking procedures using increasing concentrations of KCl (see Chapter 10, p. 336).

The merits of freeze-drying followed by hot formaldehyde vapour for the preservation of all types of mucin have been stressed by Tock and Pearse (1965). There is no doubt about the degree of preservation, which greatly exceeds that of any other type of fixation. The degree of altered reactivity of tissue components requires assessment, however. The reactivity of SO_4 and COOH groups was shown to be unaffected but positive evidence for the total survival of the 1,2-glycols of the mucoproteins was not obtained.

Future studies in mucin cytochemistry will probably accept the obvious compromise and use the selective unblocking techniques in parallel with the absolute fixation technique afforded by freeze-drying and formaldehyde vapour.

Enzymes as Reagents

The necessary prerequisites for any fixative used to prepare tissues for application of various enzymes are (1) that it should preserve without change, movement or loss, the substrate for the enzyme concerned, and (2) that it should leave the substrate in such condition as to be readily attacked by the enzyme. Between these two conditions there is a certain degree of antagonism and, on the whole, those fixatives which satisfy the second condition perform indifferently in the matter of the first. Many fixatives exert a profound effect on the subsequent action of enzymes in digesting or removing various structures. This effect is nearly always to delay or prevent the action of the enzyme entirely. In some cases (e.g. diastase), however, no inhibition is produced by a wide range of fixatives. Stowell and Zorzoli (1947) made a study of the fixation of tissues as it affected the subsequent action of ribonuclease upon them. They used sublimate-alcohol, chrome-sublimate, Zenker-acetic, Zenker-formol, Petrunkevitch, formalin, Susa, Carnoy, Bouin, and alcohol and acetone, and found that fixation of ribonucleic acid (RNA) was adequate with all except Bouin- and Susa-fixed tissues. From Carnoy-fixed material, however, the RNA was entirely removed by buffer controls in the absence of enzyme. In spite of this, Kaufmann *et al.* (1950) particularly recommended Carnoy for the preservation of nucleoproteins and other proteins prior to treatment with various enzymes. Other than for RNA or RNA-protein, fixation of proteins by Carnoy is quite adequate preparation for extractive methods employing enzyme solutions. Stowell and Zorzoli observed that Zenker-acetic and Zenker-formol, and also chrome-sublimate, markedly interfered with the action of ribonuclease; the retention of mercury in the tissues may have a similar inhibitory effect. Chromic, acetic and osmic acid mixtures, such as that of Flemming which was recommended by Kaufmann *et al.* (1948) for the fixation of nucleoproteins, modify protein structures in such a way that subsequent hydrolysis can be achieved only by prolonged digestion with high concentrations of the enzyme.

In most cases formalin is the fixative to which the least objection can be made, although it certainly reduces the action of most enzymes to some extent. An exception is k-toxin collagenase (see Chapter 25) whose action is reduced to a minimum by formalin fixation. The alcoholic fixatives and alcohol itself, as we have already seen, leave many proteins in a water-soluble condition especially if the concentration of electrolytes is high. For this reason, even when controls are used, it is sometimes difficult to determine how much of the removal of substrate is attributable to enzyme action and how much to non-specific removal.

Inorganic Materials

Very little work appears to have been carried out on this aspect of fixation. When a radioactive inorganic ion has been introduced into the tissues with a view to subsequent radioautography, unless the product is considered to be bound to some fixable component, it is usual to prepare sections by freeze-drying with total avoidance of water at all stages. If a soluble inorganic ion (usually cation) is to be demonstrated by chemical methods a similar avoidance of water is recommended.

For studies on calcification, experimentally produced in rat myocardium, Renaud (1959) found that after formalin fixation only grossly macroscopic deposits could be shown. If ethanol (at concentrations of 80 per cent. or higher) was used, however, the finest early deposits were readily demonstrable by alizarin or von Kóssa. This observation accounts for the common failure (well known to pathologists) to demonstrate radiologically demonstrable diffuse calcification in tissues fixed in formalin, especially when the latter is unbuffered.

Electron Microscopy

At one time most of the work in this field concerned variations in the use of OsO_4 mixtures. Subsequently interest arose in a number of other fixatives. Some of these, like $KMnO_4$, were not fixatives in the proper sense and Bradbury and Meek (1960) showed that in this case all fixation was produced by the dehydrating alcohol.

It is perhaps too early in the history of electron microscopy to expect full attention to be paid to the chemical effects of formaldehyde, the newer aldehydes, or even of osmium and chromium. Much of the information given above with respect to light microscopical cytochemical findings will be found useful in the field of electron microscopy. The apparent gulf between the two disciplines is less wide than is often supposed.

REFERENCES

Adams, C. W. M. (1960). *J. Histochem. Cytochem.*, **8**, 262.
Bahr, G. F. (1954). *Exp. Cell Research*, **7**, 457.
Bahr, G. F. (1955). *Exp. Cell Research*, **9**, 277.
Baker, J. R. (1944). *Quart. J. micr. Sci.*, **85**, 1.
Baker, J. R. (1946). *Quart. J. micr. Sci.*, **87**, 441.
Baker, J. R. (1950). "Cytological Technique." 3rd Ed. London.
Ball, J., Bahgat, N. D., and Taylor, G. (1964). *J. Histochem. Cytochem.*, **12**, 737.
Barter, R., and Pearse, A. G. E. (1953). *Nature*, **172**, 810.
Barter, R., and Pearse, A. G. E. (1955). *J. Path. Bact.*, **69**, 28.
Becker, R. (1959). *Diplom. Arbeit. Tech. Hochschule Karlsruhe.*
Berg, N. O. (1951). *Acta Path. Scand., Suppl.*, 90.
Bilanger, L. F. (1963). *Ann. N.Y. Acad. Sci.*, **106**, 364.
Birge, W. J., and Tibbitts, F. D. (1961). *J. Histochem. Cytochem.*, **9**, 409.
Bowes, J. H., and Kenten, R. H. (1949). *Biochem. J.*, **44**, 142.
Bowes, J. H. (1963). "A fundamental study of the mechanism of deterioration of leather fibres." Brit. Leather Manuf. Research Assoc. Report.
Bridgman, W. B. (1942). *J. Amer. chem. Soc.*, **64**, 2349.
Cater, C. W. (1963). *J. Soc. Leath. Trades Chem.*, **47**, 259.
Chauncey, H. H., Kronman, J. H., and Levitt, M. A. (1964). *J. Histochem. Cytochem.*, **12**, 647.
Ciaccio, C. (1909). *Z. allg. Path. path. Anat.*, **20**, 385.
Clayton, B-P. (1959). *Quart. J. micr. Sci.*, **100**, 269.
Conklin, J. L. (1963). *Stain Tech.*, **38**, 56.
Criegee, R. (1936). *Ann. Chem., Pharm.*, **522**, 75.
Dalgaard, J. B. (1956). *J. Histochem. Cytochem.*, **4**, 14.
Dallam, R. D. (1957). *Ibid.*, **5**, 178.
Deane, H. W., Nesbett, F. B., and Hastings, A. B., (1946). *Proc. Soc. exp. Biol., N.Y.*, **63**, 401.
Deierkauf, F. A., and Heslinga, F. J. M. (1962). *J. Histochem. Cytochem.*, **10**, 79.
Dietrich, A. (1910). *Ver. disch. path. Ges.*, **14**, 263.
Dziewiatkowski, D. D. (1953). *J. exp. Med.*, **98**, 119.

ELFTMAN, H. (1954). *J. Histochem. Cytochem.*, **2**, 1.
ELFTMAN, H. (1957). *Stain Tech.*, **32**, 29.
ERÄNKÖ, O. (1951). *Acta physiol. Scand.*, **25**, Suppl. 89, p. 22.
ERÄNKÖ, O. (1952). *Acta anat.*, **16**, Suppl. 17, p. 1.
FAIRBOURNE, A., GIBSON, G. P., and STEPHENS, D. W. (1930). *J. Soc. Chem. Ind.*, **49**, 1069.
FALCK, B. (1962). *Acta physiol. Scand.*, **56**, Suppl. 197.
FALCK, B., and OWMAN, C. (1965). *Acta Univ. Lund.*, Sect. II, No. 7.
FEIN, F. L., and FILACHIONE, E. M. (1957). *J. amer. Leather Chem. Assoc.*, **52**, 17.
FEVOLD, H. L. (1937). *Cold Spr. Harb. Symp. Quant. Biol.*, **5**, 93.
FINEAN, J. B. (1954). *Exp. Cell Research*, **6**, 283.
FISHMAN, W. H., and BAKER, J. R. (1956). *J. Histochem. Cytochem.*, **4**, 570.
FITZPATRICK, T. B., LARNER, J., and LANDING, B. H. (1948)
FLITNEY, E. (1966). *J. Roy. micr. Soc.*, **85**, 353.
FRENCH, D., and EDSALL, J. T. (1945). *Adv. Prot. Chem.*, Vol. II, p. 277.
GERSH, I., VERGARA, J., and ROSSI, G. L. (1960). *Anat. Rec.*, **138**, 445.
GERSH, I. (1964). *Histochemie*, **4**, 322.
GOLAND, P., and ENGEL, M. (1963). *J. Histochem. Cytochem.*, **11**, 751.
GOMORI, G., and CHESSICK, R. D. (1953). *J. Neuropath. exp. Neurol.*, **12**, 387.
GREEN, R. W. (1953). *Biochem. J.*, **54**, 187.
GROGAN, C. H., and OPPENHEIMER, H. (1955). *Arch. Biochem. Biophys.*, **56**, 204.
GUSTAVSON, K. H. (1940). *Svensk. kem. Tid.*, **52**, 75.
GUSTAVSON, K. H. (1949). *Advanc. prot. Chem.*, **5**, 354.
GUSTAVSON, K. H. (1956). "The Chemistry of Tanning Processes." Academic Press Inc., New York.
HAKE, T. (1965). *Lab. Invest.*, **14**, 470.
HALE, C. W. (1946). *Nature, Lond.*, **157**, 802.
HAMMARSTEN, E., HAMMARSTEN, H., and CASPERSSON, T. (1935). *Trans. Faraday Soc.*, **31**, 367.
HAYES, T. L., LINDGREN, F. T., and GOFMAN, J. W. (1963). *J. cell Biol.*, **19**, 251.
HAUST, M. D., and LANDING, B. H. (1961). *J. Histochem. Cytochem.*, **9**, 79.
HESLINGA, F. J. M., and DEIERKAUF, F. A. (1961). *J. Histochem. Cytochem.*, **9**, 572.
HESLINGA, F. J. M., and DEIERKAUF, F. A. (1962). *J. Histochem. Cytochem.*, **10**, 704.
HEUSER, E. (1946). *Paper Trade J.*, **122**, 43.
HOLMGREN, H., and WILANDER, O. (1937). *Z. mikr.-anat. Forsch.*, **42**, 242.
HOLT, S. J., HOBBIGER, E. L., and PAWAN, G. L. S. (1960). *J. biophys. biochem. Cytol.*, 7, 383.
HOLT, S. J., and HICKS, R. M. (1961). *J. biophys. biochem. Cytol.*, **11**, 31.
HUGHES, W. L., Jr. (1947). *J. Amer. chem. Soc.*, **69**, 1836.
HUGHES, W. L., Jr. (1950). *Cold Spr. Harb. Symp. quant. Biol.*, **14**, 79.
JANIGAN, D. T. (1964). *Lab. Invest.*, **13**, 1038.
JANIGAN, D. T. (1965). *J. Histochem. Cytochem.*, **13**, 476.
KAPLOW, L. S., and BURSTONE, M. S. (1963). *Nature, Lond.*, **200**, 690.
KATO, K. I., and SIRLIN, J. L. (1963). *J. Histochem. Cytochem.*, **11**, 163.
KAUFMANN, B. P., GAY, H., and MCDONALD, M. R. (1950)., *Ibid.*, **14**, 85.
KAUFMANN, B. P., MCDONALD, M., and GAY, H. (1948). *Nature, Lond.*, **162**, 814.
KAUZMANN, W. (1959). *Advanc. Protein Chem.*, **14**, 1.
KELLY, J. W., BLOOM, G. D., and SCOTT, J. E. (1963). *J. Histochem. Cytochem.*, **11**, 791.
KHAN, A. A., RIEMERSMA, J. C., and BOOIJ, H. L. (1961). *J. Histochem. Cytochem.*, **9**, 560.
KLOTZ, I. M., and LOH MING, W-C. (1954). *J. amer. Chem. Soc.*, **76**, 805.
KUGLER, J. H., and WILKINSON, W. J. C. (1959). *J. Histochem. Cytochem.*, **7**, 398.
KUGLER, J. H., and WILKINSON, W. J. C. (1960). *J. Histochem. Cytochem.*, **8**, 195.
KUGLER, J. H. (1965). M.Sc. Thesis University of Sheffield.
LAKE, B. D. (1962). Ph.D. Thesis University of London.
LILLIE, R. D. (1954). "Histopathologic Technic and Practical Histochemistry." Blakiston Co. Inc., Philadelphia.
LISON, L., and VOKAER, R. (1949). *Ann. Endocrinol., Paris*, **10**, 66.

MANCINI, R. E. (1944). *Arch. Soc. argent. Anat. norm. patol.*, **6**, 628.
MANNING, J. P., CAVAZOS, L. F., FEAGANS, W. M., and MOSS, R. (1963). *J. Histochem. Cytochem.*, **11**, 383.
MCMANUS, J. F. A. (1946). *J. Path. Bact.*, **58**, 93.
MEYER, K. (1945). *Adv. Prot. Chem.*, Vol. II, p. 249.
MIDDLEBROOK, W. R. (1949). *Biochem. J.*, **44**, 17.
MIDDLEBROOK, W. R., and PHILLIPS, H. (1942a). *Ibid.*, **36**, 294.
MIDDLEBROOK, W. R., and PHILLIPS, H. (1942b). *Ibid.*, **36**, 428.
MONIS, B., WASSERKRUG, H., and SELIGMAN, A. M. (1965). *J. Histochem. Cytochem.*, **13**, 503.
MOTA, I., FERRI, A. G., and YONEDA, S. (1956). *Quart. J. micr. Sci.*, **97**, 251.
MOWRY, R. W. (1949). *Bull. int. Ass. med. Mus.*, **30**, 95.
VAN MULLEM, P. J. (1958). *Acta Histochem.*
MUNDKUR, B. (1961). *Exp. Cell Research*, **21**, 201.
NEWCOMER, E. H. (1953). *Science*, **118**, 161.
NIELSEN, N. A., OKKELS, H., and STOCKHOLM-BORRESON, C. C. (1932). *Acta path. Scand.*, **9**, 258.
NITSCHMANN, H., and HADRON, H. (1943). *Helv. chim. Acta*, **26**, 1075.
NITSCHMANN, H., and HADRON, H. (1944). *Ibid.*, **27**, 299.
NORTON, W. T., GELFAND, M., and BROTZ, M. (1962). *J. Histochem. Cytochem.*, **10**, 375.
NOVIKOFF, A. B. (1952). *Exp. Cell Research*, Suppl. 2, p. 138.
NOVIKOFF, A. B., SHIN, W-Y., and PRUCKER, J. (1960). *J. Histochem. Cytochem.*, **8**, 37.
OKUNUKI, K. (1961). *Advanc. Enzymol.*, **23**, 29.
OPIE, E. L., and LAVIN, G. I. (1946). *J. exp. Med.*, **84**, 107.
PEARSON, C. K. (1963). *J. Histochem. Cytochem.*, **11**, 665.
POLGLASE, W. J., BROWN, D. M., and SMITH, E. L. (1952). *J. Biol. Chem.*, **199**, 105.
PORTER, K. R., and KALLMAN, F. (1953). *Exp. Cell Research.*, **4**, 127.
QUINTARELLI, G. (1963). *Ann. N.Y. Acad. Sci.*, **106**, 339.
REIS, J. L. (1956). Personal Communication.
RENAUD, S. (1959). *Stain Tech.*, **34**, 267.
RIEMERSMA, J. C., and BOOIJ, H. L. (1962). *J. Histochem. Cytochem.*, **10**, 89.
RIEMERSMA, J. C. (1963). *J. Histochem. Cytochem.*, **11**, 436.
RINEHART, J. F., and ABU'L-HAJ, S. K. (1951). *Arch. Path.*, **51**, 666.
SABATINI, D. D., BENSCH, K., and BARRNETT, R. J. (1963). *J. Cell Biol.*, **17**, 19.
SALEM, L. (1962). *Canad. J. Biochem. Physiol.*, **40**, 1287.
SCHULTZE, M., and RUDNEFF, M. (1865). *Arch. mikr. Anat.*, **1**, 298.
SCOTT, J. E. (1955a). *Chem. Indust.*, 168.
SCOTT, J. E. (1955b). *Biochem. Biophys. Acta.*, **18**, 428.
SELIGMAN, A. M., CHAUNCEY, H. H., and Nachlas, M. M. (1951). *Stain Tech.*, **26**, 19.
SELIGSBURGER, L., and SADLIER, C. (1957). *J. amer. Leather Chem. Assoc.*, **52**, 2.
SMITH, J. L., and MAIR, W. (1908). *J. Path. Bact.*, **13**, 14.
SPICER, S. S. (1960). *J. Histochem. Cytochem.*, **8**, 18.
SPICER, S. S., and JARRELS, M. H. (1961). *J. Histochem. Cytochem.*, **9**, 368.
STACEY, M., and BARKER, S. A. (1962). "Carbohydrates of Living Tissues." Van Nostrand, London.
STEMPIEN, M. F. J. (1963). *J. Histochem. Cytochem.*, **11**, 478.
STOECKENIUS, W., and MAHR, S. C. (1965). *Lab. Invest.*, **14**, 458.
STOWELL, R. E., and ZORZOLI, A. (1947). *Stain Tech.*, **22**, 51.
STRAKOV, I. P. (1951). *Zhur. Prikled. Khim.*, **24**, 142.
SWANK, R. L., and DAVENPORT, H. A. (1935). *Stain Tech.*, **10**, 87.
SWIGART, R. H., WAGNER, C. E., and ATKINSON, W. B. (1960). *J. Histochem. Cytochem.*, **8**, 74.
SYLVEN, B. (1941). *Acta chir. scand.*, **86**, Suppl. 66.
SZIRMAI, J. A. (1956). *J. Histochem. Cytochem.*, **4**, 96.
SZIRMAI, J. A. (1963). *J. Histochem. Cytochem.*, **11**, 24.
TCHACAROF, E., and PÓZALAKY, Z. (1964). *Folia Histochem. Cytochem.*, **2**, 123.

TOCK, E. P. C., and PEARSE, A. G. E. (1965). *J. roy. micr. Soc.*, **84**, 407.
TOCK, E. P. C. (1966). Ph.D. Thesis University of London.
TORACK, R. M., and MARKEY, M. H. (1964). *J. Histochem. Cytochem.*, **12**, 12.
TROTT, J. R. (1961). *J. Histochem. Cytochem.*, **9**, 703.
WALKER, D. G., SELIGMAN, A. M. (1963). *J. Cell. Biol.*, **16**, 455.
WALKER, F. J. (1964). "Formaldehyde", IIIrd Edition. Reinhold, New York.
WEIGERT, C. (1884). *Fortsch. Med.*, **2**, 190.
WHITE, J. C., and ELMES, P. C. (1952a). *Nature, Lond.*, **169**, 151.
WHITE, J. C., and ELMES, P. C. (1952b). Personal Communication.
WIGGLESWORTH, V. B. (1957). *Proc. Roy. Soc., B*, **147**, 185.
WIGGLESWORTH, V. B. (1964). *Quart. J. micr. Sci.*, **105**, 113.
WILLIAMS, G., and JACKSON, D. S. (1956). *Stain Tech.*, **31**, 189.
WOLFE, H. J. (1964). *J. Histochem. Cytochem.*, **12**, 217.
WOLMAN, M. (1955). *Int. Rev. Cytol.*, **4**, 79.
WOLMAN, M. (1957). *Exp. Cell Research*, **12**, 231.

CHAPTER 6

PROTEINS AND AMINO-ACIDS

Introduction

IN histological preparations proteins are ubiquitous, but nevertheless histologists sometimes wish to demonstrate the presence of protein in a particular site, and to establish the nature of the predominant amino-acids. They may also be concerned with the separation of materials, identified as protein-containing, into one or other of two main divisions of proteins and into one of the several classes in these two divisions. For histochemical purposes the proteins can be divided into (1) simple proteins, defined by the biochemists as yielding on hydrolysis mainly α-amino-acids and their derivatives, and (2) conjugated proteins, which yield substantial quantities of non-protein substances in addition. Some attempt can be made, therefore, to divide protein material in the tissues into the following classes, though the classification in the Table below is neither complete nor entirely accurate from the biochemical point of view.

TABLE 9

Simple proteins	Conjugated proteins
Albumins	Nucleoproteins
Globulins	Mucoproteins
Albuminoids	Glycoproteins
Globins	Lipoproteins
Histones	Phosphoproteins
Protamines	

The simple proteins can also be separated into fibrous proteins and globular proteins and the first of these divisions contains the collagens, reticulins, keratin, myosin, elastin and fibrin. These are the constructional proteins, or albuminoids of the older terminology, which are insoluble in most aqueous media. The globular proteins, albumins, globulins, globins and histones are soluble in aqueous media and many of them have been crystallized.

General Identification of Protein

The classical colour reactions for the identification of proteins *in vitro* are violently destructive towards tissues, and many efforts have been made to devise innocuous modifications of the original methods. The classical methods which have been employed, with greater or lesser degrees of success, are (1) *Millon's reaction* for tyrosine; (2) *Pauly's diazonium reaction* for tyrosine, trytophan and histidine; (3) *the xanthoproteic reaction* for phenolic compounds; (4) *the Sakaguchi reaction* for arginine; and (5) *the nitroprusside test* for sulphydryl groups. All these reactions determine only a fraction of the constituent amino-acids of any particular protein, nevertheless a positive reaction has been taken as an indication of the presence of protein since free amino-acids do not occur in tissue preparations from animal sources. All these reactions except the last can be applied to fixed tissue sections, but only two are essentially suitable for the simple identification of protein. They are the Millon and the diazonium reactions.

The Millon Reaction

This reaction (Millon, 1849) is due to the presence in the protein molecule of the hydroxy-phenyl group. It is given by any phenolic compound which is unsubstituted in the position *meta* to the hydroxyl group, but such compounds are not found free in the tissues and the only known amino-acid containing the hydroxy-phenyl group is tyrosine.

The original Millon's reagent was made by digesting mercury in nitric acid and diluting the resulting solution with water. Numerous chemical identities occurred in such a mixture and no explanation of its mechanism was initially forthcoming. Following Hoffmann (1853), Meyer (1864) described the essential features when he produced a positive reaction with tyrosine, $HgCl_2$ and nitrite. Nasse (1901) showed that any phenol gives a red colour after contact with mercuric ions followed by nitrite ions.

For histochemical use the reagent was modified by Bensley and Gersh (1933) for use in the cold, and employed in the demonstration of mitochondria in freeze-dried sections. If it is used for paraffin sections the application of moderate heat is essential and the warm reagent must be washed off with dilute nitric acid as soon as the maximum colour has developed. This method is given in Appendix 6 (Fig. 43). Other modifications of the Millon reaction for histochemistry include those of Serra (1946) and Baker (1956). The latter is especially recommended for routine use as being simple and reliable and giving more colour, in my hands, than the other reactions cited. Full details appear in Appendix 6. According to Gibbs (1927) the reaction proceeds in two stages. First, a nitrosophenol is produced by the substitution of NO for H *ortho* or *meta* to the hydroxyl of the phenol. Secondly, Hg^{2+} is incorporated into a new ring, by chelation, which includes the nitrogen of the nitroso group. The new complex is red in colour.

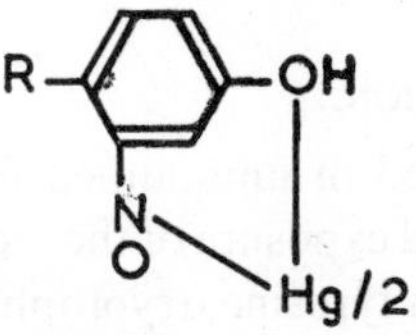

According to Lugg (1937) the nitroso group is ortho to the hydroxyl, but this involves no interference with chelation. I endeavoured, unsuccessfully, to increase the colour of the reaction by treatment with other chelators for mercury. Treatment with diphenylcarbazone, as in the Okamoto reaction (Chapter 12, p. 424), produced a deep purple colour in various tissue components. Blocking reactions, however, showed that mercury is present in large amounts in combination with other reactive groups in addition to nitrosotyrosine.

THE DIAZONIUM REACTION

Diazonium compounds are prepared by the action of nitrous acid in the cold on the salts of primary aromatic amines, sulphanilic acid (*p*-aminobenzene sulphonic acid) being commonly used for this purpose.

$$HO_3S-C_6H_4-NH_2 + HONO + HCl$$

$$\longrightarrow HO_3S-C_6H_4-N{\equiv}NCl + 2H_2O$$

The resulting compounds, acting in alkaline aqueous solutions as diazonium hydroxides, combine with the phenol group of tyrosine, the indole group of tryptophan and the heterocyclic imidazole group of histidine to give coloured products. Other amino-acids present in proteins are known to be able to react with diazonium salts to give products which may be coloured in some instances.

A method for the simple demonstration of proteins in tissue sections, making use of the above principle, is the "coupled tetrazonium reaction" of Danielli (1947). This author observed that the diazonium reaction, as usually employed, was relatively useless because compounds of diazonium hydroxides with the amino-acids tyrosine, tryptophan and histidine were only faintly coloured. He therefore substituted for the original diazonium salt a bis-diazonium salt (diazotized benzidine or *o*-dianisidine) and attached a variety of phenols or amines to the free diazo group of the protein-diazonium compound, making a deeply coloured protein-bisazo-phenol in its place. This reaction, which is considered more fully below and in Chapter 9, is applicable to all kinds of fixed frozen or paraffin sections and also to freeze-dried material. Technical details are given in Appendix 6.

THE XANTHOPROTEIC REACTION.

This reaction is employed in unmodified form using concentrated nitric acid followed by washing and exposure of the section to the fumes of ammonia. It gives a positive result with tyrosine, tryptophan and phenylalanine, and the presence of these can be identified in what remains of the section by a bright orange colour. The method is too destructive to be useful.

The remaining methods listed at the beginning of this section are not used for confirmation of the presence of protein but only to identify the particular amino-acids for which they are specific. They are referred to below in the section on determination of individual amino-acids.

A number of more recently introduced methods, however, have been used for the general identification of proteins.

The Mercury-Bromphenol Blue Method (HgBPB)

This staining method, which is not primarily histochemical, was introduced by Durrum (1950) for the demonstration of protein on filter-paper spots. It was adapted as a general stain for protein by Mazia, Brewer and Alfert (1953). These authors stated that preparations stained by their procedure followed the Beer-Lambert laws and that the amount of dye bound was proportional to the amount of protein over a wide range. The method was employed by Bonhag (1955) for investigating the composition of the ovary of the milkweed bug *Oncopeltus fasciatus*, and some of his modifications are incorporated in the technical details given in Appendix 6. The merits of HgBPB for the demonstration of muscle striations have been pointed out by Menzies (1961) and Kanwar (1960) has re-emphasized the complete lack of specificity of the reaction for proteins. The histochemical uses of this method are necessarily limited and I have not used it myself.

The Oxidized Tannin-Azo Method (Ota)

This method was originally described by Dixon (1959, 1962) for use as a general protein stain. It depends on the reaction of tannic acid (digallic acid) with the tissue proteins, and its subsequent oxidation by periodic acid to a 1,2-quinone. The latter is then coupled with diazotized *o*-dianisidine to produce a salmon-red azo dye (azoquinone). Details of the method are given in Appendix 6, p. 608. The amount of tannic acid bound to protein varies considerably at different pH levels. According to Dixon the tissue NH_2 groups are the main binding site. He found a considerable decrease in staining after treatment with HNO_2 and this finding agrees with those of Bowes and Kenten (1949) but not with the results obtained by Lollar and Kremen (1948). Gustavson (1956), however, points out that under Lollar and Kremens' conditions of deamination only the ε-amino groups of lysine would have been blocked.

The Oxidized Tannin-Oxazine Method (OTO)

This method described by Dixon (1962), is a variation of the foregoing in which the quinone oxidation product of the tissue-bound tannin is converted by reaction with 6-amino-3-dimethylaminophenol into a blue oxazine dye. The original author gives no account of the solubility in water or lipid of either the azoquinone or oxazine product, presuming them to be firmly bound to protein. It is possible that lipid solubility accounts for the staining of mitochondria which was obtained.

Although no specificity or stoichiometry can be claimed for either the OTA or the OTO method, both are simple to perform and may be useful as general oversight methods.

THE ACROLEIN-SCHIFF METHOD

This method was introduced by van Duijn (1961) as a general protein stain. When applied to paraffin sections the controls were Schiff-negative so that the reaction in test sections was entirely attributable to aldehyde groups introduced by the treatment with acrolein ($H_2C{=}CH.CHO$). According to van Duijn the double bond of acrolein could react with SH, aliphatic NH_2 and NH, and imidazoles, leaving a free aldehyde to react with Schiff's reagent. It is customary to consider acrolein fixation of proteins as a property of the double bonds (Gustavson, 1956) and certainly a strong Schiff reaction is given by freeze-dried tissues fixed with acrolein vapour (Chapter 5, p. 88). On the other hand, it is difficult to see how reaction of the double bond of acrolein with, for instance, protein amino groups could give other than unipoint fixation without cross-linking. The compound formed would have the structure $R.NH.CH_2.CH_2CHO$.

The acrolein-Schiff reaction was observed by van Duijn to be completely blocked by acetylation, thus restricting the possible reacting groups to SH, α-amino, ε-amino, NH, OH, guanidino, imidazole, phenol and imino. The participation of hydroxyl groups was rejected from the failure of sugars and glucogen to react and the lack of staining with protamines in model experiments permitted the exclusion of guanidino groups. Other blocking experiments confirmed the reacting groups as SH, NH_2 and imidazole. In the absence of phospholipids the author considered the acrolein-Schiff reaction to be specific for protein.

The method is given in Appendix 6, p. 606.

Physical Differentiation between Classes of Protein

Identification of the simple proteins is achieved by observing their deficiency in those properties which characterize the conjugated proteins. These are subsequently described in the several chapters on conjugated proteins. Some simple protein structures, of particular interest to histologists, are dealt with in Chapter 8.

The chemist separates the various classes of simple protein by making use of their differential solubilities. Albumins, characteristically soluble in pure water; globulins, insoluble in water but soluble in various salt solutions; histones, soluble in water but insoluble in dilute solutions of ammonia; and albuminoids, which are insoluble in all neutral solvents, but soluble in acids and alkalis, constitute the main groups with which the histochemist is concerned. Collagen, elastin and keratin, the so-called scleroproteins, are the chief members of the last group. The globins and histones are known as basic proteins, because they are composed predominantly of the strongly basic amino-acids histidine, arginine and lysine. They differ from each other in that the former contain large amounts of histidine and average amounts of arginine and tryptophan, while the latter contain large amounts of arginine and at most traces of tryptophan. (According to Stedman and Stedman (1947), histones never contain tryptophan.)

The isoelectric points of the various proteins are also significant. At these points many of their physical characteristics, such as solubility, osmotic pressure, viscosity and the property of swelling in water, are reduced to a minimum. The isoelectric point of hæmoglobin, for instance, is between pH 6·7 and pH 6·8, that of serum albumin is pH 4·88 and that of the globulins between pH 5 and pH 7. The extremes are illustrated by pepsin, whose isoelectric point is below pH 1·0 and the protamines, in whose case it is between pH 12 and pH 12·4. The following methods are used to identify various types of protein.

Simple Solubility Tests

Most of the physical properties which divide albumins, globulins, globins and histones cannot be utilized in the study of tissues fixed by ordinary methods; with freeze-dried material, however, it is possible to make solubility studies which give an approximate identification of the tissue protein. Preferably using sections mounted on slides without contact with water (see Chapter 3, p. 52), the solubility of the unfixed protein components can be determined. Water, 1 per cent. NaCl, half and fully saturated ammonium sulphate, dilute solutions of ammonia and various buffers, for instance, are employed as solvents. In this way Gersh (1949) identified the properties of one fraction of thyroid colloid with those of thyroglobulin and Catchpole (1949) similarly identified part of the glycoprotein material in the rat hypophysis with follicle-stimulating hormone. The presence, after various extractions, of residual material giving positive Millon and other reactions may reveal that materials which have been supposed to contain a single protein are in fact composed of several, with widely differing solubilities. This is a method of great promise which should be more widely applied. It must be realized, however, that freeze-dried proteins may differ appreciably in their physical qualities from the original protein of the tissues (see Chapter 3, p. 54).

Isoelectric Point by Minimum Solubility (Appendix 6)

The isoelectric points of proteins, *in vitro* and in freeze-dried preparations, can be determined by estimation of their points of minimum solubility in buffer solutions. For freeze-dried preparations a modification of the method used by Catchpole (1949) has been employed. This is applicable not only to the examination of simple proteins, but also to various conjugated proteins. The maximum amount of stainable protein, in the particular site which is being examined, occurs in the sections treated with buffer solutions in the region of the isoelectric point of that particular protein. If the protein is a simple one it can be demonstrated most conveniently by means of the coupled tetrazonium reaction, or by a staining method such as HgBPB; if it is a muco- or glyco-protein the periodic acid-Schiff reaction (Appendix 10) is more suitable, while nucleoproteins are demonstrated by the Feulgen reaction for deoxyribonucleo-protein or by the Brachet test for ribonucleoprotein (Appendix 9).

ISOELECTRIC POINT BY STAINING AT CONTROLLED pH

On account of their free basic (amino) and acid (carboxyl) groups, proteins are presumed to behave in solution as amphoteric electrolytes which are capable of entering into chemical combinations with acids and bases to form ionizable salts. The dyes used in staining are either acid or basic and dissociate to form negatively or positively charged ions respectively. Below the isoelectric point of a protein acid dyes combine with positively charged protein ions and above this point basic dyes combine with negatively charged protein ions. Thus any structure in the tissues will be stained by a basic dye if the pH is above its isoelectric point and by an acid dye if it is below that point. Methods for determining the isoelectric point of tissue components by staining in solutions of basic and acid dyes, at various pH levels, are based on this assumption.

Pischinger (1927) first attempted to determine the isoelectric point of alcohol-fixed tissue sections, mounted on slides. He used toluidine blue as his basic dye, and first Cyanol extra and later Crystal Ponceau as his acid dye. Zeiger (1930) continued Pischinger's work and investigated the effect of different fixatives on the isoelectric point as determined by controlled pH staining. A different point was obtained in each case. Both Pischinger and Zeiger used strong dye solutions at intervals of about pH 0·2 and plotted the staining intensity of the tissue component against the pH of the solution. They took the "mid-point" of their curves for the basic dye as the isoelectric point of the protein. If their technique is to be considered even approximately accurate, the point at which the curves for the basic and acid dye cross should be taken. Up to this time little consideration had been given to the possible effect of varying the dyes and buffers used, and the concentrations of both. Rawlins and Schmidt (1929), however, had already shown that the dissociation of three different dye-protein compounds, though similar from pH 7 to pH 12, differed widely below pH 6. The important work of Levine (1940), conclusively proved that the staining intensity of a protein in buffered dye solutions was dependent on the interaction of the dye-protein, buffer-protein and dye-buffer systems and that as the type of dye and the type of buffer, or the concentrations of either, were altered, the resultant isoelectric points also varied. Levine considered, however, that even if the true isoelectric point was not obtained, the shift of the apparent isoelectric point with different fixatives might well be significant. He introduced a technique in which all the variable factors were reduced to a minimum. The effect of the rate of staining was eliminated by using dilute (0·5mM) solutions of dyes in dilute (100mM) buffers and by staining until an equilibrium was reached (24 hours). To avoid loss, especially of the basic dyes, when passing through the alcohols during the process of dehydration, Levine (1939) introduced the use of tertiary butyl alcohol in place of ethyl alcohol.

As the result of this work we must conclude that staining at controlled pH does not determine the true isoelectric point. Nevertheless, it remains a useful method for differentiating between various tissue components, especially those which exhibit basophilia, and for quantitative expression of alterations in the basophilia of such components, produced by various experimental manœuvres. Differentiation of mucopolysaccharides from mucoproteins, by a method of controlled pH staining, is referred to in the section dealing with those substances (Chapter 10, p. 341). From Levine's data, it appears that methylene blue and orange G should be the most suitable pair of dyes for general use in controlled pH staining and methylene blue proves entirely satisfactory in practice. Orange G, however, fails in my experience to give a decisive end point to many tissue components at high pH values, although it gives a fairly sharp drop in staining intensity between pH 4 and pH 6 when used upon basophilic tissue components.

The indicator dye Solochrome cyanine RS, which I have used as a complexing reagent for aluminium deposits in the tissues (see Chapter 27, Vol. 2), can be used in acid solutions (pH 1·4 to pH 2·1) as a progressive stain for basic and acidic proteins. At these pH levels it stains nucleic acids blue and basic proteins red, while in fresh or fixed frozen tissues acidic lipids also stain blue (Plate Ic, Fig. 117, facing p. 468). In mammalian and invertebrate tissues the protein component of the mucin takes the stain and shows red, therefore. The acidic (sulphate) groups of most acid mucopolysaccharides apparently have no affinity for Solochrome cyanine RS at acid pH levels. This property is a most convenient one. An exception must be recorded in the case of certain connective tissue mucins, especially those in young connective tissues, which may stain blue rather than red or pink. Changes in the pH at which staining is carried out change red-staining components to blue. The pink (fixed) or red (unfixed) colour of collagen becomes blue at about pH 3·6, for instance. This is not the I.E.P. of collagen, however, which lies between pH 9 and 10. Other red-staining components become blue at different pH levels. The Solochrome cyanine method has not fulfilled early expectations, largely because of a change in the composition of the dye. Other dyes possessing the same property should be sought.

The apparent differences between the isoelectric points of albumins, globulins and albuminoids are insufficient for the controlled pH method to be of much use for distinguishing them. Nevertheless, it can be used to give evidence of the presence of the strongly basic proteins such as globins and histones. On preparations produced in a standard manner, in which all variation of fixation, etc., is avoided, the method can be used to register similarity or dissimilarity between two selected materials. Identical behaviour does not indicate identity; different behaviour indicates difference in structure or composition, but gives no information as to the cause of this difference which is often far from simple. A method for the distinction of basic proteins of the histone type was reported by Hydén (1943), involving staining in a dilute solution of Ponceau 2R, in an acid medium in the presence of a detergent. Later Sokoloff *et al.* (1951), using Fast green and methylene blue as their acid and basic dyes, attempted to distinguish between various types of fibrinoid and collagen and fibrin. They employed the spectrophotometric method of Singer and Wislocki (1948) and found similar "isoelectric points" in every case, so that they were not successful in determining the origin of fibrinoid.

More recently Spicer and Lillie (1961) and Spicer (1962) employed a graded series of alkaline solutions of Biebrich scarlet (C.I. No. 26905) for the identification of basic proteins. Spicer observed that chromatin and nucleoli, fixed in formaldehyde-free fixatives, possessed a selective affinity for the dye at pH $9{\cdot}5 \pm 0{\cdot}4$. Since the effect was abolished by formaldehyde (and also by brief nitrosation or acetylation) Spicer suggested that it was due to the ε-amino groups of the nucleohistones (which may or may not be present in the nucleoli). Staining of the nuclei, and of certain acidic mucins, was metachromatic (orange-red instead of red). Investigating Biebrich scarlet metachromasia Winkelman and Spicer (1963) found that the dye possessed two properties distinct from those of other metachromatic dyes, in being anionic (reacting with cationic chromotropes) and in adhering to Beer's law over a much wider range of concentration than the usual cationic metachromatic dyes. They showed that the mechanism was an electrostatic interaction between the anionic dye and the cationic chromotrope, just as it is between cationic dye and anionic chromotrope in the case of the thiazine dyes (Chapter 10, p. 332).

Identification of Individual Amino-acids

Complete determination of the constituent amino-acids of a particular protein can be achieved by its extraction, purification and the subsequent evaluation of its hydrolysates by chromatography or other methods, Such a manœuvre is beyond the bounds of histochemistry in the restricted sense which is used in this book. Chromatography should, nevertheless, be applicable to the investigation of small quantities of material from freeze-dried preparations, when these have been dissolved in suitable buffer or salt solutions and abstracted from the sections. Thus modern methods of protein analysis can be expected to indicate: (1) which amino-acids are present in a given protein; (2) the percentage amino-acid composition; (3) the minimum mol. wt. of the protein (allowing at least one residue per molecule of the rarest amino-acid present), and (4) the number of amino-acid residues of each sort which go to make up the protein molecule. Modern histochemical methods, on the contrary, can be expected to give simpler and less complete information about a given protein component.

The methods of the previous section will indicate the predominance of acidic groups (aspartic, glutamic, amide-NH_2) or basic groups (arginine, histidine, lysine) in the protein. Methods given in the following section will indicate the presence of α-amino groups (primary amines) and give a relative approximation of their number. This information is roughly parallel to that obtained by demonstrating the acidophilia of a particular component except that acidophilia may be due to primary, secondary or tertiary amines. Table 10, below, will be found useful as an indicator of the range of dissociation of the various ionizable groups in proteins which are responsible for staining with acid and basic dyes. It also indicates the pH level at which reactions for the various end groups should be carried out.

TABLE 10

pH Range of Dissociation of Ionizable Groups
(After Springall, 1954)

Source	Grouping	[———R—H]$^{n+}$	,———R.$^{(n-1)+}$	pH range
Terminal	α-Carboxyl	—COOH	$—CO_2^-$	3–5
Asp. Glut.	β and γ-Carboxyl	—COOH	$—CO_2^-$	3–5
Histidine	Imino	^+NH	N	5–9
Terminal	α-Amino	$^+NH_3$	$—NH_2$	7–9
Lysine	ε-Amino	$^+NH_3$	$—NH_2$	10–11
Tyrosine	Phenolic	OH	O^-	10–12
Cysteine	Sulphydryl	SH	S^-	10–12
Arginine	Guanidino	$C{=}^+NH_2$	C=NH	13

At strongly acid pH levels (pH 1) all the groups in the table exist in the undissociated [———R—H]$^{n+}$ form, while at strongly alkaline pH levels (pH 14) they all exist in the dissociated form [———R]$^{(n-1)+}$. Each group dissociates over a particular short pH range which is characteristic of that group.

Unfortunately the determination of individual amino-acids in tissue sections is limited by the small number of specific procedures available, by the unsatisfactory nature of many of these and by the complexity of some of the others. The behaviour of the amino-acids in denatured proteins in tissue sections differs from their behaviour in native tissue proteins and, even more, from their behaviour in the test-tube. Since the histochemist deals nearly always with amino-acids in the first group he must expect discrepancies when comparing histochemical with biochemical results. The available methods will be considered under the following headings:

(1) Reactions for amino (NH_2) groups.
(2) Reactions for carboxyl (COOH) groups.

(3) Reactions for tyrosine, tryptophan and histidine.
(4) Reactions for cystine and cysteine (SS and SH groups).
(5) Reactions for arginine.
(6) Blocking reactions for protein end-groups.

Reactions for Amino Groups

Ninhydrin. Of the older methods of histochemistry applicable to the study of amines, the best known is the ninhydrin reaction. Ninhydrin reacts with the free NH_2 groups of α-amino-acids, in acid solution, to give a blue-coloured compound, together with carbon dioxide and an aldehyde containing one less carbon atom than the original amino-acid.

$$R-\underset{NH_2}{\overset{H}{C}}-COOH + 2\ (C_6H_4(CO)_2C(OH)_2) \longrightarrow C_6H_4(CO)_2CH{\cdot}N{=}C(CO)_2C_6H_4 + R-\overset{H}{C}{=}O + CO_2$$

According to Slobodian *et al.* (1962) the ε-amino groups of lysine, present in proteins, consistently contribute to the reaction.

During the reaction, which is carried out at 100°, the bluish-violet colour develops rapidly. This is not stable, however, and, in addition, the blue compound is diffusible in the tissues of the section. Since free amino-acids do not occur in ordinary tissue preparations which have been in contact with water, or in freeze-dried material unless contact with water has been specifically avoided, only the α-amino groups of proteins can possibly react. The number of these available for reaction appears to be small and, in practice, little or no colour develops when the method is applied to tissues prepared by the usual histological techniques. Even the modification introduced by Serra and Queiroz Lopes (1945) using ninhydrin in 66mm phosphate buffer at pH 6·98 gave indifferent results on fixed materials, although it gave a satisfactory dark blue colour with freeze-dried sections.

As a colour-giving reaction for NH_2 groups the ninhydrin reaction is now scarcely ever used in histochemical practice. With the increasing use of fluorescence techniques in the future, the capacity of ninhydrin to yield fluorescent products on combination with a number of end groups will probably find application. In alkaline solution, for example, the reaction with guanidino groups produces fluorescent products.

Alloxan. The condensation of alloxan with amino compounds to give a coloured product is largely a biochemical method. Serra (1946), who quotes Winterstein (1933) as saying that the reaction is given by free NH_2 groups and possibly by SH groups also, found that the test was insensitive with fixed materials and that the coloured product was diffusable. The alloxan reaction has been criticized by Romieu (1925) on account of lack of specificity and by Giroud (1929) because of diffusion of colour; Vercauteren (1951) maintained that the reaction product behaved as an anionic dye and stained tissue components having an affinity for such dyes. The alloxan method is, therefore, likely to be of little use in histochemistry.

A number of methods, employing new principles, were suggested for the localization of NH_2 groups in the tissues. Most of these methods were introduced by Danielli (1947, 1950), and they fall into two separate classes, dinitrofluorobenzene and associated reactions, and azomethine reactions.

DNFB and Isocyanate Methods. These methods depend upon treatment of the tissue sections or smears with either DNFB or *p*-nitrophenyl isocyanate, though the latter has scarcely been used in histochemical practice since the original suggestion was made. DNFB, on the other hand, has been much employed as a reagent for protein end groups. Originally introduced by Sanger (1945) as a reagent for N-terminal α-amino groups, it reacts also with the ε-amino groups of lysine and hydroxylysine, the phenolic hydroxyl groups of tyrosine and the SH groups of cysteine. More important still, from the histochemical point of view, it reacts with the imidazole groups of histidine (Porter, 1950), though the precise nature of the reaction is not understood. Investigations carried out by Maddy (1961a) suggested that at pH 7, but not at lower pH values, *im*-dinitrophenylhistidine was formed.

Identification of the NH_2-dinitrophenyl compound can be made by its yellow colour, if this is sufficiently strong, and in this case the reaction of DNFB with tyrosine and histidine can be ignored since the resulting compounds are colourless. The use of monochromatic light at 410 nm, as suggested by Zerlotti and Engel (1962), makes the reaction specific for α-amino, ε-amino and SH groups. The latter can be prevented from reacting by any one of a number of blockade reactions (see p. 172).

Alternatively, after treatment of sections with DNFB, and washing, the NO_2 group of the bound reagent can be reduced, diazotized, and finally coupled to a suitable phenol to produce a deeply coloured product. These points are further considered on p. 167.

Azomethine Methods. The second class of methods described by Danielli made use of the formation of azomethines, or Schiff's bases, when aldehydes condense with primary amines.

$$R.NH_2 + OCH.R' \rightarrow R.N = CH.R' + H_2O$$

In tissue proteins these are the ε-amino groups of lysine, the δ-amino groups of ornithine, and the α-amino groups of terminal amino-acids.

For the histochemical demonstration of NH_2 groups, Danielli suggested the use of several aldehyde compounds, two of which may be cited as illustrating two different visualizing processes. The first compound, terephthalaldehyde, reacts with tissue NH_2 groups as follows:

$$RNH_2 + OHC{-}C_6H_4{-}CHO \longrightarrow RN{=}C{-}C_6H_4{-}CHO + H_2O$$

and the resulting azomethine is treated with fuchsin-SO_2, with which it gives a red colour by virtue of its unattached CHO group. The second type of aldehyde compound suggested by Danielli was *para*-aminobenzaldehyde. It was suggested that this would react with tissue NH_2 groups to form an azomethine, leaving a free amino group in the *para* position. This was treated, as previously described, by diazotization and coupling with a suitable phenol.

As stated in the 2nd Edition of this book (1960) these azomethine methods appeared to offer considerable promise for the accurate localization of NH_2. Of those which I tested, however, only the terephthalaldehyde method gave consistent results with conventionally prepared sections. Structures which stained strongly included nuclei, red cells and voluntary and involuntary muscle. Sections in which the NH_2 group had been destroyed by treatment with nitrous acid showed little difference from those in which the NH_2 groups were intact. This applied to unfixed freeze-dried material (Fig. 44) as well as to fixed, paraffin, sections. Since aldehydes condense *in vitro* with most groups having an active hydrogen, it was presumed that reaction with SH, phenolic OH or indoles was at least possible. The work of Doberneck and Maresch (1952), however, suggests that in tissue sections condensation with indoles will not occur. The results obtained by Akabori and Ohno (1950, 1952) likewise indicate that tyrosine in proteins (i.e. in tissue sections) will not react. Although half mercaptals, R.CH(OH)SR, or mercaptals, $RCH(SR)_2$, are formed when aldehydes condense with thiols (Obata and Yamanishi, 1951), according to Ryzheva (1946) these derivatives are unstable. The possibility must therefore be considered that tissue thiols do not contribute to histochemical reactions involving aldehydes (see section on sulphydryls, p. 138).

A fluorescent azomethine reaction for NH_2 groups was described by Stoward (1963). When salicylaldehyde condenses with tissue amines, in either alkali or acid solution, a green fluorescent anil is formed. Warming to 40° releases the aldehyde and extinguishes the reaction. Details of the alkaline and acid salicylaldehyde methods, regarded by their author as reasonably specific, appear in Appendix 6, p. 611.

3-Hydroxy-2-Naphthaldehyde. A different type o azomethine reaction was proposed by Weiss, Tsou and Seligman (1954) for the demonstration of protein-bound NH_2 groups. Their reagent, 3-hydroxy-2-naphthaldehyde (sometimes written as 2-hydroxy-3-naphthaldehyde)

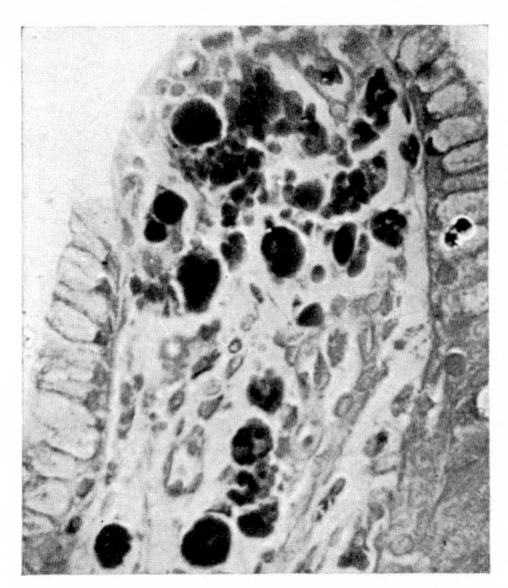

FIG. 43. Human cervix uteri. Intra- and extracellular Russell bodies. Millon reaction. × 400.

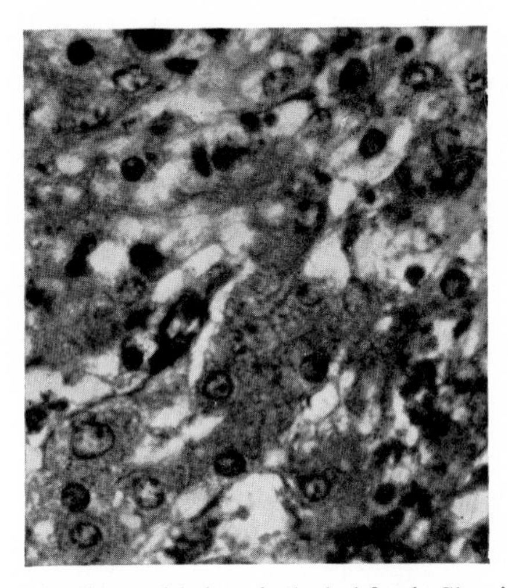

FIG. 44. Human liver, freeze-dried and alcohol-fixed. Showing result of the terephthalaldehyde-leucofuchsin method of Danielli. × 520.

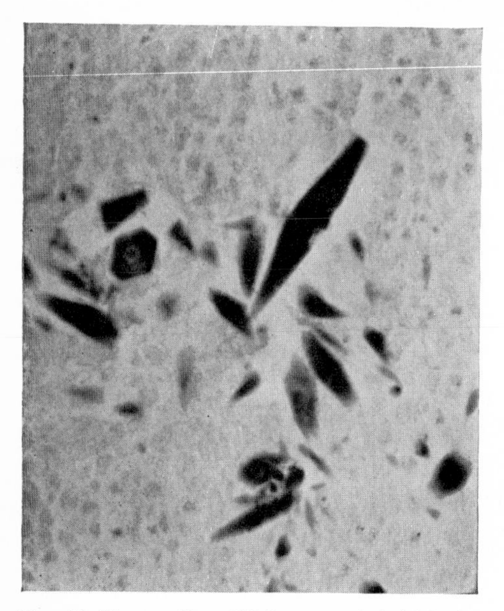

Fig. 45. Human liver. Hydroxy-naphthaldehyde method. × 510.

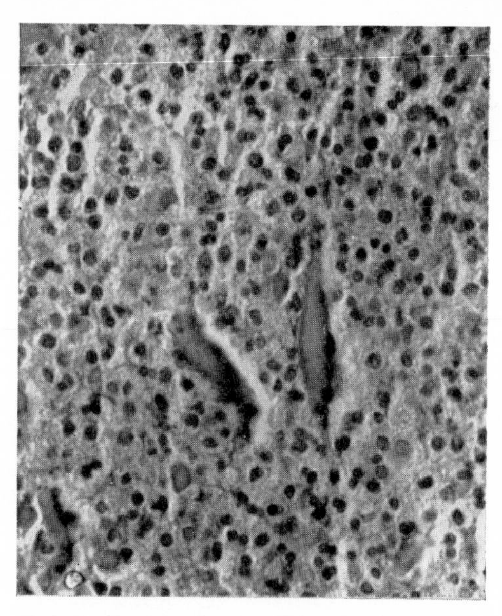

Fig. 46. Human liver as Fig. 45. Paraprotein crystals in a granuloma. Alloxan-Schiff. × 510.

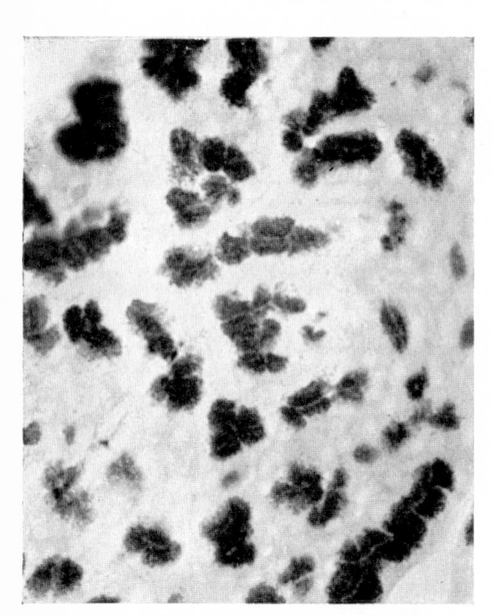

Fig. 47. Mouse pancreas. Intense staining of zymogen granules in the acinar cells. Adams' DMAB-reaction. × 510.

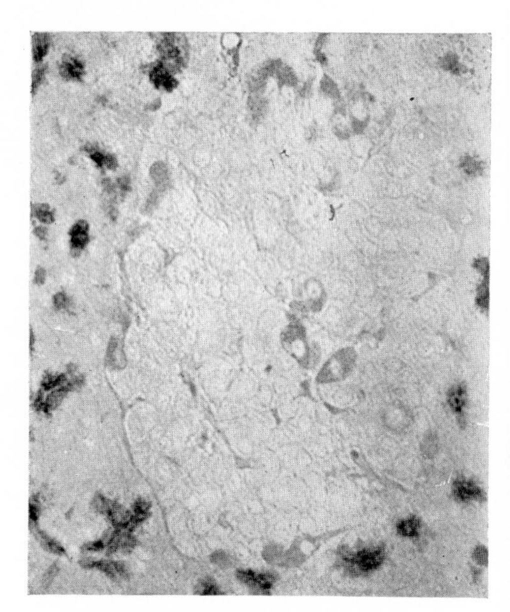

Fig. 48. Rabbit pancreas. Alpha cells in an islet of Langerhans. Post-coupled benzylidene reaction. × 850.

can be prepared from 3-hydroxy-2-naphthoic hydrazide (Appendix 9, p. 648). It is poorly soluble in water and is used, therefore, as a buffered solution in aqueous acetone (pH 8·5). Below this pH level the formation of the pale yellow Schiff's base is considerably reduced. To develop the final colour a second stage is necessary, in which the Schiff's base is coupled with tetrazotized di*ortho*anisidine at pH 7·4. This reaction can be abolished by prior acetylation or deamination and its specificity is therefore high. Fig. 45, opposite, shows the staining of NH_2 groups by this method in formalin-fixed tissue. The original authors noted that formalin suppressed the reaction, and this is true for the majority of tissue components. It is evident, however, that structures containing a large number of available NH_2 groups may still react strongly despite formaldehyde blockade unless this is carried out with hot formaldehyde vapour, applied to freeze-dried blocks of tissue.

Details of this method, which I find very reliable and useful despite its lack of absolute specificity for NH_2 groups, appear in Appendix 6, p. 611.

***O*-Diacetylbenzene.** This reagent was originally used for the histochemical demonstration of α-amino-acids by Voss (1940) and later, for the same purpose, by Dietz (1942). More recently the method has been thoroughly investigated by Wartenberg (1956). The mechanism of the reaction has not been explained, but Winkler (1948) suggested the possibility of a parallel with the formation of phthalocyanins. The equation below shows the condensation of phthalic anhydride with an amino-acid to form a phthalimido acid:

$$C_6H_4\begin{smallmatrix}CO\\CO\end{smallmatrix}\!>O + H_2N-\underset{R}{\underset{|}{CH}}-COOH \longrightarrow$$

$$C_6H_4\begin{smallmatrix}CO\\CO\end{smallmatrix}\!>N-\underset{R}{\underset{|}{CH}}-COOH + H_2O$$

Analysis of the coloured reaction product of the condensation of *o*-diacetylbenzene with ammonia showed that it had a molecular weight of 700 and contained one nitrogen atom.

Wartenberg used *o*-diacetylbenzene as a 2 per cent. solution in 70 per cent. ethanol and found that the reaction was positive in a reasonably short time only if the pH was higher than 8. At this pH level 30 minutes at room temperature produced a deep lilac colour in the sections. The specificity of the reaction is apparently high, since Wartenburg found that it was abolished by acetylation, deamination, benzoylation and by 16 hours' treatment with DNFB. It was weakened, though not reversed, by formalin fixation and reversed by 5 hours' treatment with ninhydrin though not by alloxan for a similar period. Wartenberg claimed, therefore, that *o*-diacetylbenzene reacted only with free NH_2

groups, and it is apparent from his tabular data, and from a study of Table 10, p. 115, that the pH range of the reaction is precisely that at which the dissociation of α-amino groups from $—NH_3$ to $—NH_2$ takes place.

Details of the *o*-diacetylbenzene method are given in Appendix 6, p. 611. A comparative study of this and the foregoing method, made by Goslar (1963), indicated that both possessed a higher specificity for NH_2 groups than the other methods included in the study (acrolein-Schiff, bromphenol blue, ninhydrin-Schiff, coupled tetrazonium).

The König-Sassi Reaction. This reaction was discovered by König (1904a and b) and applied by Sassi (1956) for the detection of amino-acids in chromatograms. Later it was converted into a histochemical reaction by Stoward (1963) who found that when tissue sections were treated with a methanolic solution of pyridine containing a few drops of aqueous cyanogen bromide (NCBr) an intense green fluorescence could be observed which was alcohol-fast and heat stable.

According to König the reaction was due to the formation of cyanine type dyes which are built up of two nitrogen-containing ring systems linked by methine or polymethine chains (König, 1912). Alternative explanations are possible, however. The reaction was used by Donzou and Le Clerc (1955) as the basis of a fluorimetric assay for nicotinamide and these authors suggested that *N*-cyano-3-carbimido-2-pyridone was formed. Later Creyf and Roosens (1959) found that nicotinamide and NCBr reacted in aqueous solution to give not the pyridone derivative but an open ring structure:

$$NC.NH.CH{=}CH.CH{=}\underset{\displaystyle CHO}{\underset{|}{C}}.CONH_2$$

Stoward presumed that in the histochemical König-Sassi reaction pyridine was cleaved similarly to an aldehyde intermediate and that the latter condensed with NH_2 groups of the tissues. Since the reaction can take place in tissue proteins both before and after methylation it seems likely that both primary and secondary amines take part. Details of the König-Sassi reaction are given in Appendix 6, p. 610.

Oxidative Deamination Methods (Ninhydrin-Schiff and Alloxan-Schiff). Two methods in this category were introduced by Yasuma and Ichikawa (1952, 1953) which depended on the demonstration, with Schiff's reagent, of stable tissue aldehydes produced in the course of oxidative deamination with ninhydrin or alloxan (see p. 116). Control sections, not subjected to deamination, were always employed if there was any doubt of the origin of the aldehyde giving rise to the magenta colour of the test section. The mechanism of the two methods is identical and in practice they produce very similar results. I found, however, that the colour developed with the alloxan variant was usually less than that produced by ninhydrin, in parallel sections. Both reactions are blocked by deamination (p. 161) and acetylation (p. 162), applied before the stage of oxidative deamination, and by aldehyde blocking agents (see Chapter 13, p. 453) applied between this stage and treatment with Schiff's reagent. The result of the alloxan-Schiff reaction, applied to paraprotein crystalloids in a granulomatous liver, is shown in Fig. 46, p. 119). Yasuma and Ishikawa specifically stated that 10 per cent. formalin fixation could be employed, as indeed it can, so that in this case formaldehyde blockade of NH_2 groups is apparently reversible.

Doubts as to the specificity of the ninhydrin-Schiff reaction have continued to be expressed since Burstone (1955) found considerable variation in the staining of different proteins, although Fotakis (1960) claimed that its specificity for α-amino-acids was sufficient enough. Puchtler and Sweat (1962) claimed that only free amino-acids (split off by ninhydrin hydrolysis) could yield aldehyde derivatives. These would be soluble in the ninhydrin reagent. Furthermore, they maintained that since glycine, proline and hydroxyproline could not yield aldehydes, as much as 50 per cent. of the amino-acids in collagen would be unreactive.

These views were dismissed by Kasten (1962) as based on misconception of the mechanism of the reaction which, he pointed out, was never supposed to produce colour with α-amino acids such as proline and hydroxyproline. Reviewing the matter on theoretical grounds Glenner (1963) reduced the number of groups which were likely to be reactive to the ε-amino groups of lysine and hydroxylysine and the α-amino groups of terminal γ-glutamyl and β-aspartyl peptides. Certainly only a proportion of the total protein-bound amino groups present in tissue sections is oxidized by ninhydrin.

An attempt to quantitate the reaction was made by Rappay (1963) who observed, as did Puchtler and Sweat, that the colour production was markedly affected by the purity of the reagents employed. The most significant interference came from Cu^{2+} ions which, in a concentration as low as 10^{-6} M, could entirely suppress oxidation by ninhydrin.

Chloramine-T Methods. An oxidative deamination method using Chloramine-T (sodium *p*-toluene sulphochloramine) or hypochlorite was suggested by Chu, Fogelson and Swinyard (1953). The tissue aldehydes were demonstrated by means of Gomori's hexamine silver reagent (Appendix 13, p. 705). Chilled alcohol was found to be the best fixative and formalin was purposely avoided. Burstone (1955) tested the oxidizing capacity of chloramine-T, and of sodium and calcium hypochlorites, finding the last to be only one-third as strong as the other two. He was able to obtain stable aldehydes in the tissues only with chloramine-T oxidation, and these were demonstrated with Schiff's reagent. The reaction was blocked by deamination, but after acetylation the nuclei would still react.

Curtius-Type Reactions. A new histochemical reaction was described by Geyer (1965) for the demonstration of reactive groups in protein side-chains. It was derived from the Curtius (1890, 1894) reaction by which acid azides are converted into amines containing one less carbon atom:

$$R.CON_3 \rightarrow R.N{=}C{=}O \rightarrow R.NH_2$$

The method is more closely derived from the carbobenzoxy technique of Bergmann and Zervas (1936) which was used by Prelog and Wieland (1946) and by Harris and Work (1948) for the preparation of synthetic peptides.

According to Geyer (1965) the acid azide of *p*-nitrobenzoic acid reacts with free amino groups of tissue proteins to give a carbamoyl derivative:

$$R.CON_3 + H_2N.Pr \rightarrow R.CONH.Pr + NH_3$$

The nitro group is then reduced, diazotized, and finally coupled with 1-amino-8-naphthol-3, 6-disulphonic acid to give a red dye. The reaction was observed to be diminished, but not totally blocked, by 24 hours' deamination or acetylation.

The specificity of this method has not been determined and the possibility that the acid azide reacts also with sulphydryl, guanidino and imidazole groups (Herriott, 1947) must still be entertained. For those who wish to test the method details are given in Appendix 6, p. 612.

Reactions for COOH Groups

Acid Anhydride Method. Following the observation by Wiley (1950) that the COOH groups of acylamino-acids could be converted into amido ketones by treatment with acetic anhydride in pyridine, Barrnett and Seligman (1956, 1958) developed a method for the detection of COOH groups in protein. They used paraffin sections fixed in a variety of fixatives, Carnoy being superior to most others, and demonstrated the ketone by means of the Ashbel-Seligman reagent, 2-hydroxy-3-naphthoic acid hydrazide (Chapter 13, p. 453; Appendix 9, p. 648). The reaction was completely prevented by previous methylation of the sections. It was originally considered that only terminal carboxyls having an adjacent α-carbon with an electrophilic N group would contribute to a positive reaction and that the free carboxyls of glutamic and aspartic acids would not do so.

Terminal carboxyls would be expected to be comparatively rare in most proteins. In order to explain the unexpectedly high level of staining obtained with their reaction in certain tissues (such as cartilage) Barrnett and Seligman suggested that this was due to a high concentration of low mol. wt. protein with a high local content of the reactive grouping.

The specificity of the reaction for C-terminal carboxyls was accepted by Gershstein and Svetkova (1960) but questioned by Karnovsky and Fasman (1960) who showed, in experiments with model compounds, that side-chain carboxyls would also react. They considered that all the colour obtained was due to the latter.

The complexity of the problem can be appreciated by studying the alternative mechanisms proposed for the reaction. That most generally accepted provides for an intermediate azlactone stage (I) as described by Levene and Steiger (1927, 1928) and Attenburrow *et al.* (1948):

$$\underset{\text{(C-terminal)}}{R.CONHCH_2COOH} \xrightarrow{(CH_3CO)_2O} \underset{\text{I}}{\text{azlactone } (N{=}C(R){-}O{-}C({=}O){-}CH_2)} \xrightarrow{\text{Base}} (N{=}C(R){-}O{-}C({=}O){-}CH) \xrightarrow{(CH_3CO)_2O} (N{=}C(R){-}O{-}C({=}O){-}CH.COCH_3)$$

$$\rightarrow RCO.NH.CH(COCH_3)(COOH) \rightarrow R.CONHCH_2COCH_3 + CO_2$$

After observing that substitution of NCH_3 for NH prevented azlactone formation Wiley (1950) proposed an alternative mechanism. This involved decarboxylation due to base, caused by the electrophilic nature of the α-amino groups, and an attack of the carbanion thus formed with acetic anhydride. Cornforth and Elliott (1950) and Buchanan *et al.* (1957) objected to this hypothesis since they regarded the H atom on the α-carbon as more important than the H on the nitrogen. Cornforth and Elliott proposed a transitory oxazolium (II) as the intermediate, followed by acetylation on the α-carbon and ring opening and decarboxylation:

$$R'{-}\ddot{N}(C({=}O)R){-}CH_2{-}C({=}O)OH \xrightarrow{(CH_3CO)_2O} \underset{\text{II}}{R'{-}\overset{\oplus}{N}{=}C(R){-}O{-}C({=}O){-}CH_2 \text{ (ring)}}$$

A further alternative was proposed by Karnovsky and Fasman (1960) who suggested that non-ketonic side-chain carboxyls were converted to mixed anhydrides (III) by a reaction of the type:

Anhydride of A + Salt of B $\rightleftharpoons$ Salt of A + mixed anhydride of A and B.

These mixed anhydrides would acetylate naphthoic hydrazide leading either to a dihydrazide (Karnovsky and Fasman) or to compound IV. Either would subsequently couple with the diazonium salt to provide the coloured product of the histochemical reaction. Objections have been made to the fact that acylation of the hydrazide (V) occurs at the same time but, unless it has affinity for particular tissue components, this compound will be removed by the organic solvents of the procedure, in which it is soluble.

$$\underset{\text{(Side-chain)}}{\text{R.CONH.CH.CONHR}'(CH_2)_2C(=O)OH} \xrightarrow{(CH_3CO)_2O} \text{R.CONH.CHCONHR}'(CH_2)_2C(=O)\text{-O-}C(=O)CH_3 \quad \text{III}$$

$$\text{III} \xrightarrow{NH_2NHCO\,\phi} \underset{\text{IV}}{\text{R.CONH.CH.CONHR}'(CH_2)_2C(=O)NH.NHCO\,\phi} + \underset{\text{V}}{CH_3CONH.NHCO\,\phi}$$

Treatment with alkali should destroy mixed anhydrides and Karnovsky and Fasman (1960) found that 0·1 N NaOH at 22° for 15 minutes, following acetic anhydride and pyridine, completely abolished staining even in strongly reactive tissue sites. Methyl ketones derived from C-terminal carboxyls (of tetraglycine) were unaffected by alkali and they continued to react with hydrazide. Distinction between C-terminal and side-chain COOH is thus theoretically possible at the histochemical level, provided that methyl ketones in tissue proteins are similarly resistant to NaOH.

Using his salicyloyl hydrazide-zinc reaction (Chapter 10, p. 324), which produces a blue fluorescence with tissue-bound ketones, Stoward (1963, 1967) observed a strong reaction after acetic anhydride-pyridine treatment. When, however, he produced mixed acid anhydrides in the tissues by acetylation in hot acetic anhydride and 0·1 per cent. H_3PO_4 there was practically no fluorescence. Stoward reasoned, therefore, that if hydrazones were indeed formed from mixed anhydrides their fluorescence was negligible. Several further objections to the mixed anhydride theory were put forward by Stoward (1963).

After abolition of the hydrazide-diazotate reaction by the interposition of NaOH treatment the original COOH groups should be regenerated. A repeat performance of the whole reaction should then be possible; in practice it is not. If this view is accepted the mixed anhydride hypothesis depends on the proposed resistance of methyl ketones to alkali. The evidence for this is unsatisfactory and it is likely, in fact, that they undergo hydrolytic cleaveage:

$$R.CO.NH.CH_2 \vdots CO.CH_3 \xrightarrow{NaOH} R.CO.NHCH_3 + CH_3COONa$$

Reversal of the positive salicyloyl hydrazide reaction, after acetic acid and pyridine, by diazotization and deamination (Appendix 6, p. 613) suggests that an NH group capable of acetylation is essential to the reaction, as proposed by Cornforth and Elliott. Since neither Barrnett and Seligman (1958) nor Karnovsky and Fasman (1960) fully considered the participation of the carbonyl group of acetylated amine or hydroxyl groups in proteins or mucopolysaccharides, Stoward (1963) carried out experiments on this point. He found no evidence that the uronic acid groups of acid mucopolysaccharides could react with hydrazide after acetic anhydride-pyridine, either *in vitro* or in tissue sections. These findings agree with those of Karnovsky and Mann (1961) whose tests on chondromucoprotein indicated that increased staining occurred with an increasing degree of aggregation, supporting the original view of Barrnett and Seligman on staining of cartilage. A puzzling feature of their work was that the results could be obtained with acetic anhydride alone.

It is apparent that in the presence of the NH . R . CH_2 group special conditions are required for the conversion of COOH to methyl ketone. Although no final conclusion can be made, the evidence presented above is perhaps more favourable to the original hypothesis of Barrnett and Seligman (C-terminal COOH → methyl ketone) than to the later hypothesis of Karnovsky and Fasman (Side-chain COOH → mixed anhydride). The reaction, for which full details appear in Appendix 6, p. 628, should be used with confidence in spite of the above considerations. It can be considered to indicate accurately the local concentration of protein-bound carboxyls.

Epoxyether Method. This was the first of a series of methods for the demonstration of protein-bound carboxyls, derived from current methods of peptide synthesis and introduced by Geyer. The first of these (Geyer, 1962a) depends on the reaction of 1-(*p*-biphenyl)-1-methoxyethylenoxide (I) in xylene with protein carboxyl. After subsequent rearrangement of the initial product a ketoester (II) is formed with the release of methanol.

I + R.COOH → II + CH_3OH

The ketoester is then reacted with naphthoic acid hydrazide, as in the acid anhydride method, and finally coupled with a diazonium salt to give a coloured product.

According to the author all available protein-bound carboxyls react positively but those of uronic acid slightly or not at all. Brief details are given in

Appendix 6. Staining closely resembles that given by the acid anhydride method but the reaction has not been sufficiently tested, from the theoretical or practical point of view, for any accurate assessment to be made.

Hydroxamic Acid Methods. These methods for protein-bound carboxyls were suggested by Geyer (1962b). They are based on the reaction of hydroxylamine with carboxyl derivatives to form hydroxamic acids. The latter are demonstrated by forming complexes with ferric iron; these are red in colour (see Chapter 10, p. 359). Suitable carboxyl derivatives were produced by (1) esterification with thionyl chloride in methanol (p. 163); (2) treatment with acetic anhydride in pyridine, which Geyer considered would produce mixed anhydrides; (3) formation of acid chlorides with thionyl chloride.

According to Geyer these hydroxamic acid reactions did not demonstrate uronic acids but "free carboxyl groups of proteins". He disagreed with the interpretations of McComb and McCready (1957), which I had accepted (Pearse, 1960) in formulating a histochemical test for acetylated polysaccharides. In view of the technical, as well as interpretational, difficulties which accompany the use of hydroxamic acid methods I suggest they should not be used in applied histochemistry. Further studies on their specificity may indicate their proper use.

Carbodiimide Method. A method for identification of C-terminal residues was introduced by Khorana (1952, 1953) which depended on the facile condensation of di-*p*-tolycarbodiimide with *N*-acyl peptides to form C-terminal acyl ureas.

$$\text{NH-----CONH.}\underset{|}{\overset{\text{R}}{\text{CH}}}\text{.COOH} + \text{Ø.N:C:N.Ø}$$

$$\downarrow$$

$$\text{NH-----CONH.}\underset{|}{\overset{\text{R}}{\text{CH}}}\text{.C}\begin{matrix}\diagup \text{Ø.N.CO.NH.Ø} \\ \diagdown\!\diagdown \text{O}\end{matrix}$$

These acyl ureas, on treatment with cold alkali, are readily split. In the process the peptide chain is shortened by one residue. The reaction was applied by Geyer (1964) for the histochemical demonstration of free carboxyl groups. Since the usual carbodiimides were insufficiently soluble a new compound, 1-cyclohexyl-3-(2-morpholinyl-(4)-ethyl) carbodiimide, was synthesized for the purpose. The initial reaction medium contained both carbodiimide and naphthoic acid hydrazide and the usual diazonium coupling followed brief acid and alkaline washes.

The results obtained closely resembled those of the acid anhydride method and Geyer declared, on this account, that the side-chain carboxyls of the protein must certainly take part in the reaction. Certain anomalies, like the strong staining of mast cell granules, were not overlooked but they remained unexplained.

Although details of the method appear in Appendix 6, p. 629, the specificity of the reaction must be regarded as far from established. Wider application must await more direct information on this point.

A further application of the carbodiimide reaction was introduced by Geyer (1965). Sections were treated, for 15 hours, in a chloroform, ethanol, tetrahydrofuran bath containing the carbodiimide reagent and *p*-nitrobenzoic acid. The two reagents supposedly combined to give reactive products capable of combination with the free amino groups of the tissue proteins. The bound *p*-nitrophenyl residue was then demonstrated by the usual reduction, diazotisation and coupling sequence.

The overall need for a reaction such as this is very small. It would be necessary to demonstrate that it had some advantage in specificity over the methods for amino groups already described. Failing this, the method must remain one of academic interest only.

Reactions for Tyrosine, Tryptophan and Histidine

Millon Reaction for Tyrosine. This method has already been fully considered (p. 107) as a general reaction for proteins. Its specificity for tyrosine is absolute, as far as mammalian tissues are concerned, although *in vitro* tryptophan also gives a red colour. Pollister and Ris (1947) used the reaction for quantitative spectrophotometry, employing as a "blank" sections treated by the Millon routine without nitrite. They observed that the absorption maxima of the nitroso-mercury-tyrosine compound were 3,550 Å (U.V.) and 4,800 Å (visible). Rasch and Swift (1953) concluded that the Millon reaction should be used for spectrophotometry only if certain conditions, which they specified, were fulfilled. Broadly speaking the method was found useful only for comparing similar proteins.

Tetrazonium Methods. Observations have already been made, earlier in this chapter, on the use of the coupled tetrazonium reaction of Danielli for the demonstration of protein by virtue of its contained amino-acids, particularly tyrosine, tryptophan and histidine. This reaction is based on the combination of amino-acids with tetrazotized benzidine at pH 9·0 and at 4°, and the subsequent coupling of the residual free diazo group with a phenol or aromatic amine. The formulæ which follow express the reaction as it takes place with the aromatic hydroxyl group of tyrosine.

It has been suggested (Pauly, 1904; Gomori, 1952) that diazonium salts will not react with tryptophan, but this view is contradicted by many protein chemists. Frazer and Higgins (1953), for instance, studied the reaction of diazosulphanilic acid with tryptophan, tyrosine and histidine. They showed that ring azo derivatives were formed with the imidazole group of histidine and with the phenolic group of tyrosine. Azotryptophan apparently lacks colour, but this fact has no bearing on the production of a coloured final reaction product in the coupled tetrazonium reaction. According to Barnard (1961) free

R–⟨ ⟩–OH + HON_2–⟨ ⟩–⟨ ⟩–N_2OH ⟶

Protein with tyrosine group   Tetrazotized benzidine

Brown compound
(protein-diazonium

Bright red compound
(protein-bisazo-naphthol)

tryptophan was found to couple with diazobenzene sulphonic acid only to a slight extent, while in *N*-acetyl tryptophan and in gramicidin a considerable degree of coupling occurred.

Zahn, Wollemann and Waschka (1953) found that diazotized 5-chloro-2-methylaniline would react with proline to form a triazene, and with cysteine to form a diazomercaptan, and Gelewitz, Riedemann and Klotz (1954) showed that the reaction of serum albumin with diazosulphanilic acid produced only relatively few reactions with side chains, of which only a third were —N=N— bridges. They suggested that the ε-amino groups of lysine could react, but Frazer and Higgins (*loc. cit.*) denied this possibility.

According to Howard and Wild (1957) the reaction of proteins with diazonium salts certainly occurs with other groups as well as those of tyrosine and histidine. They showed that 2 mols. of diazonium salt combined with phenolic and imidazole groups, and with the ε-amino groups of lysine. One mol. of diazonium salt combined with cysteine (SH) and tryptophan (indole) and 3 mols. with arginine (guanidino).

In view of these findings, particularly those of Howard and Wild, it is necessary to modify our views of the specificity of the coupled tetrazonium reaction for tyrosine, tryptophan and histidine only. Nevertheless, these three amino-acids, and particularly tyrosine and histidine, will produce the major contribution when the reaction is applied to unfixed or suitably fixed tissue sections.

Burstone (1955) emphasized the difficulty of interpreting the results of the coupled tetrazonium reaction. He found in a series of *in vitro* experiments that non-protein substances like filter paper were stained brown by the breakdown products of the tetrazonium salt (the stable bis-diazotate Fast Blue B salt) and noted that even after thorough washing a final coupling with H-acid would produce a deep purple colour. This evidence suggested that some of the colour developed by the reaction might be due not only to amino-acids other than tryptophan, tyrosine and histidine, but to entirely unspecific causes. Using freshly diazotized benzidine I have not found that the colour of either tissue sections or filter paper is other than pale yellow, and, after thorough washing and coupling with H-acid, the latter does not produce a purple colour. Careful attention to technical details will certainly reduce absorbtion artifacts to a negligible level.

The coupled tetrazonium method can be used either as a general reaction for protein or, more usefully, after benzoylation (p. 168) as an indicator of a particular type of bonded histidine component in nucleoproteins. Technical details are given in full in Appendix 6, p. 612.

Diazotization-Coupling Method for Tyrosine

Lillie (1957b) claimed that prolonged treatment of sections with cold HNO_2, following the principles given by Morel and Sisley (1927), resulted in the C-nitrosation of the tyrosine groups of the proteins. The C-nitroso groups were then progressively converted into the quinone oxime tautomer and thence, by further reaction with HNO_2, into diazonium nitrates.

$$R-C_6H_4-OH + HNO_2 \rightarrow R-C_6H_3(N{=}O)-OH + H_2O \rightarrow$$

$$R-C_6H_3(NOH){=}O + 3HNO_2 \rightarrow$$

$$R-C_6H_3(N{=}N-NO_3)-OH + HNO_3 + H_2O$$

These diazonium nitrates can be coupled with amines in alkaline solution to give coloured reaction products. Lillie used 1-amino-8-naphthol-5-sulphonic acid (S-acid) for this purpose. It gives a pinkish red colour in most structures although hair cortex and keratin stain purplish red. As originally described, the D-C reaction was not entirely reliable but further work by Glenner and Lillie (1959) converted the method into one of absolute reliability and specificity. Most of the admitted variability was shown to be due to the photo-sensitivity of diazotyrosine in acid solutions. If nitrosation and diazotization are carried out in the dark loss of diazotyrosine is entirely prevented. Coupling with S-acid, under alkaline conditions, can safely be carried out in the light.

The specificity of the reaction was established by experiments with model proteins (silk fibroin, zein, keratin) known to contain high levels of tyrosine, and by *in vitro* tests on formaldehyde precipitated proteins. Blockade reactions (p. 160) were performed and complete reversal of staining was produced by treatment with acetic anhydride (acetylation) N-bromosuccinimide (bromination) and chloramine-T.

Alternative coupling reactions were proposed. The production of a formazan from benzaldehyde phenylhydrazone was taken to confirm the identity of diazotyrosine but the colour was less than adequate for critical microscopy. Coupling *ortho* to phenolic compounds (phloroglucinol, resorcinol, naphthoresorcinol) produced O, O^1-dihydroxyazo compounds capable of chelating various metal ions.

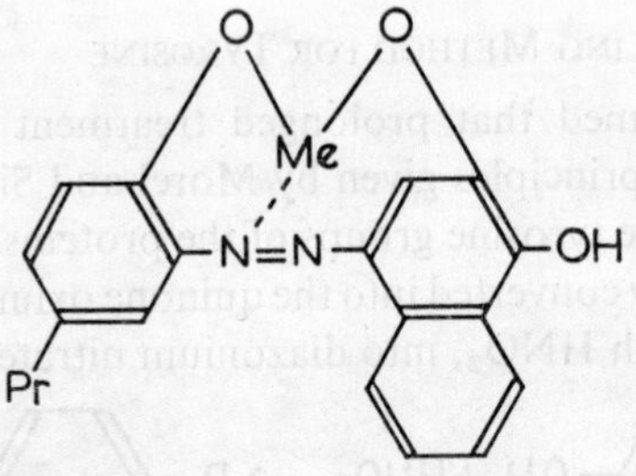

The metal ions which produced coloured chelates were Cu^{2+}, Co^{2+}, Ni^{2+}, VO_2^{2+}, and UO_2^{2+}. Experiments *in vitro* with several amino-acids showed that only diazotyrosine gave rise to a red chelate with naphthoresorcinol and Cu^{2+}, confirming the specificity of the D-C reaction.

Glenner and Lillie suggested the uranyl chelate for electron cytochemical identification of tyrosine but there seems to be no record of its use for this purpose. Glenner and Bagdoyan (1960) described the use of the D-C method for demonstrating pituitary acidophils. Distinction between these cells, which contain high levels of tyrosine, and the other cells of the pars distalis can be accentuated by shortening the period of diazotization. Full details of the method are given in Appendix 6, p. 607.

Nitrosonaphthol Reaction for Tyrosine. A colour reaction for tyrosine was suggested by Gerngross *et al.* (1933) and employed by Udenfriend and Cooper (1952). This depended on a condensation with α-nitroso-β-naphthol (I) in acid solution. The chemistry of the reaction is not well understood but Feigl (1954) believed that the first stage involved oxidation to the α-nitro compound (II).

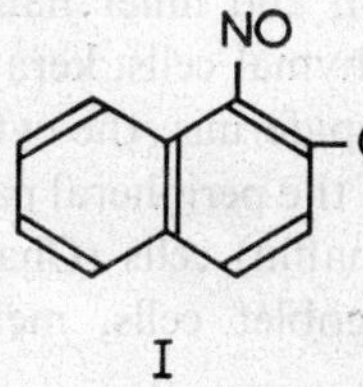

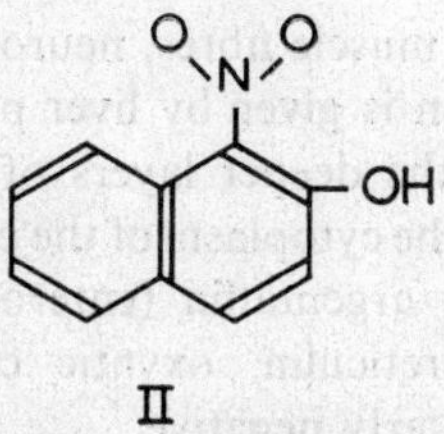

Waalkes and Udenfriend (1957) found that the nitrosonaphthol product of tyrosine was strongly fluorescent (Excitation max. 460 nm; Fluorescence max. 570 nm). When he applied the method to tissue sections Stoward (1963) obtained a greenish-yellow fluorescence which faded rapidly. He reported that the sensitivity of the reaction was much greater than that of the Millon reaction.

Technical details are given in Appendix 6, p. 607.

Dimethylaminobenzaldehyde (DMAB) Methods for Tryptophan. There are at least four biochemical tests for tryptophan: Hopkins-Cole (glyoxylic acid), Voisenet-Rhode or May-Rose (DMAB), Voisenet-Fürth (formol-nitrate) and the Romieu reaction (Blanchetière and Romieu, 1931), employing orthophosphoric acid. The first three are all variations of the historical aldehyde procedure for tryptophan. All these reactions involve the use of strong acids and they are therefore somewhat destructive to the tissues. A modification of the Voisenet-Fürth reaction was reported by Serra and Queiroz Lopes (1945) to be applicable to formalin-fixed paraffin sections, but I found that even when direct mounting in glycerine was employed, as suggested by the original authors, the colour developed was insufficient to allow accurate localization of tryptophan in the tissues. A modification of the May-Rose (1922) reaction, introduced by Bates (1937) and employing sodium nitrate as an oxidizing agent, was used by Kaufmann *et al.* (1946, 1947) in their study of chromosomes, to distinguish histones from tryptophan-containing proteins. I found the amount of colour produced by this reaction similarly inadequate.

Lillie (1956) applied a modification derived from Feigl (1954) using 2·5 per cent. DMAB in a 1:3 mixture of concentrated HCl and acetic acid. He observed that in formalin-fixed tissue the reaction was too insensitive to react with compounds other than those containing large amounts of tryptophan. A positive reaction was obtained with pancreatic zymogen granules which have been shown by Marshall (1954) to contain chymotrypsinogen. This enzyme contains 5·4 per cent of tryptophan.

The DMAB-nitrite Method. A further modification by Adams (1957a) re-introduced, this time successfully, the use of an oxidizing agent to develop the colour of the final product. Following the work of Spies and Chambers (1948a, 1948b) and Portner and Högl (1953), Adams used brief oxidation with sodium nitrite. This method produces an intense blue pigmentation in tissue components containing much tryptophan (fibrin, fibrinoid, Paneth cell granules, chief cell granules, zymogen granules (Fig. 47 and Plate Ia, facing p. 183), thyroid, colloid, muscle fibres, neurokeratin and inner hair-root sheath). A moderate reaction is given by liver parenchymal cells, keratin (after alkaline hydrolysis) and the deeper layers of the epidermis, the cytoplasm of many neurones and by the cytoplasm of the cells of the peripheral parts of argentaffin carcinomas. The argentaffin (enterochromaffin) cells (Chapter 26) do not react; collagen, reticulin, oxyntic cells, goblet cells, melanins and lipofuschins are similarly negative.

In *in vitro* tests Adams found that only 3-indolyl derivatives would form the blue pigment; those which he tested were tryptophan, tryptamine, 5-hydroxytryptamine, and 3-indolyl acetic acid. Indole itself formed a greyish-violet pigment after prolonged oxidation. These results confirm those of Spies and Chambers and of Portner and Högl in suggesting that the DMAB-nitrite reaction takes place in two stages. Adams suggests the following mechanism:

H N NH$_2$ CH.COOH C + H R C O →

Tryptophan Aldehyde

H R H N C NH CH.COOH C + H_2O

β-carboline

The blue pigment (carboline blue), whose structure is apparently unknown, is derived from the *β*-carboline by nitrite oxidation. This concept is supported by the observations of Harvey, Miller, and Robson (1941), who produced such a pigment from the colourless condensation product of tryptophan and formaldehyde (2,3,4,5-tetrahydro-*β*-carboline-4-carboxylic acid), but it fails to explain the formation of a blue pigment from 3-indolyl derivatives having too few C atoms to complete the third ring of the *β*-carboline. The product of DMAB-tryptophan condensation *in vitro* has a red colour, but in the tissues this is seldom seen. I found it only in Russell bodies and in the layer of Huxley and Henle in hair sheaths. The product shows a weak purple fluorescence in UV light and this can be recognized microscopically when the red colour is not visible at all.

The specificity of the Adams DMAB-nitrite reaction is high. Although other amino-acids will react with DMAB through their *α*-amino groups to form Schiff's bases (p. 117) the products are yellow or orange, never red, and they cannot be oxidized to blue pigments. The reaction is entirely prevented by previous oxidation of the tissues with performic acid (p. 166), and also by alkaline persulphate (p. 166), and it is reduced by both chromate and formalin fixation. In the last case interference is slight except with prolonged fixation (e.g. 14 days in 4 per cent. formaldehyde). Technical details are given in Appendix 6, p. 615.

The Rosindole Reaction. Glenner's (1957) rosindole reaction is essentially the same as Adam's DMAB-nitrite reaction except for the use of a mixture of acetic and perchloric acids, containing only a small amount of HCl, as solvent for the DMAB and of acetic-HCl as solvent for the nitrite (Appendix 6, p. 616). The results, in so far as I have compared the two reactions, show an identical distribution of colour. Glenner, however, considers that his rosindole reaction is a benzylidene ($C_6H_5CH{=}$) condensation of the type postulated by Fischer (1911, 1932). According to this author's view pyrrole compounds condense with Ehrlich's reagent in two ways. Either two pyrroles condense with one aldehyde to give a red dyestuff (I), or one pyrrole condenses with one aldehyde (II), again giving a red dye.

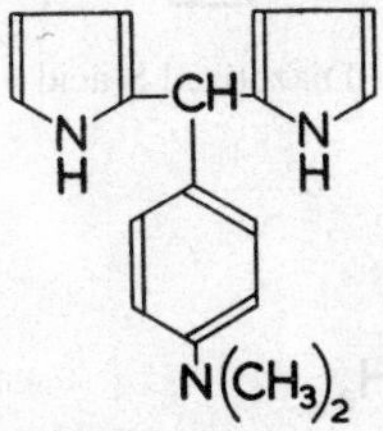

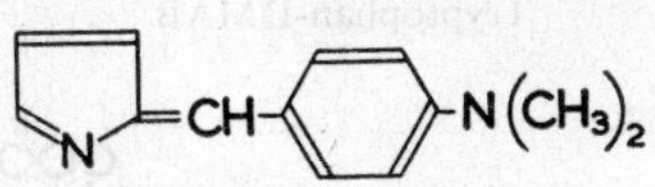

Goessner (1947) has shown that the substance in normal urines which gives a positive reaction is indoxyl and that the condensation which takes place belongs to Fischer's type II. Glenner postulated that in the presence of perchloric acid a molecule of tryptophan condenses with a molecule of DMAB to form a phenylindolyl-methane compound, of Fischer's type II, which is in equilibrium with a leuco compound having a structure analogous to Fischer's type I. Oxidation of the latter produces a stable blue rosindole dye. The histochemical rosindole reaction can be produced by aromatic aldehydes other than DMAB and alternative oxidizing agents can be employed. Among the latter, tested by Glenner, are $KMnO_4$, $FeCl_3$, $K_2Cr_2O_7$ and $K_3Fe(CN)_6$. No combination of alternatives produces any improvement over the standard reagents in the rosindole reaction.

I agree with Glenner that the rosindole reaction is more sensitive than the xanthydrol reaction (see below, p. 135) and that the product is more easily visible. I find, however, that the DMAB-nitrite procedure consistently gives more colour than does Glenner's perchloric acid variant. Although this author refers in his Tables to the positive staining of mucous goblet cells, Adams (1957a) failed to obtain this result, and I cannot reproduce it with either of the methods in mammalian or invertebrate tissues. Since tryptophan can be shown to be present in the mucins derived from these cells it is likely to be present in the goblet cells themselves. Failure to show it must be ascribed to stereochemical factors which interfere with condensation.

The Post-coupled Benzylidene Reaction for Tryptophan. This reaction

(Glenner and Lillie, 1957) is a variant of the original DMAB reactions in which no oxidation step is employed. In place of oxidation the sections are treated with a freshly diazotized amine salt in acetic acid. The original authors tested several compounds and obtained their best results with S-acid (8-amino-1-naphthol-5-sulphonic acid), which gave a stable dark blue product in the tissues. It is assumed that under acid conditions coupling takes place *ortho* to the free dimethyl-amino group of the side chain of the phenylindolylmethane compound:

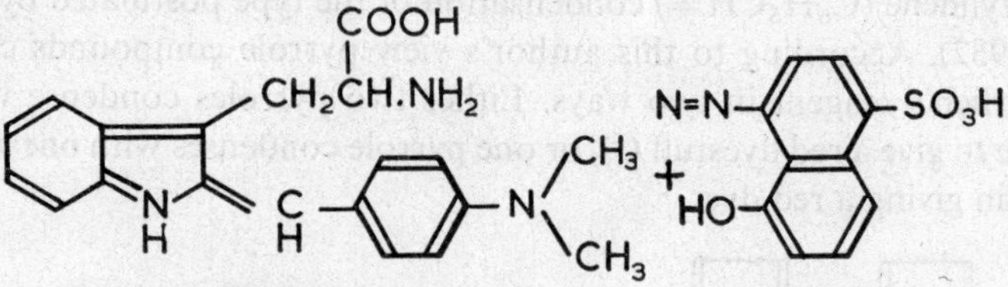

Tryptophan-DMAB + Diazotized S-acid

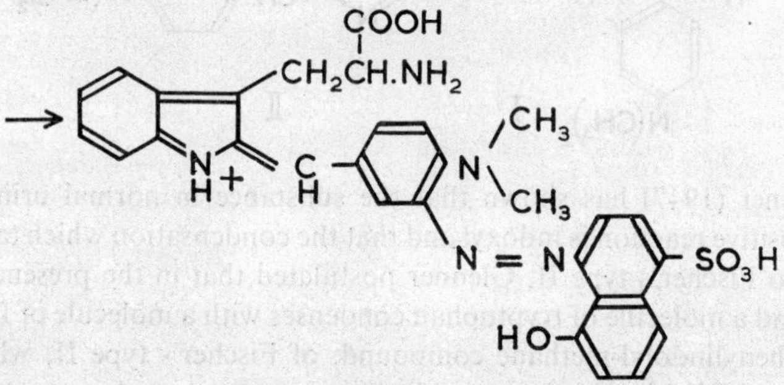

→ Blue azo dye

When benzaldehyde was substituted for DMAB and followed by diazotized S-acid, no increase in colour was produced. Where the same initial procedure was followed by HNO_2 oxidation a dark blue colour developed. This shows that in the post-coupled benzylidene reaction azocoupling certainly takes place.

Glenner and Lillie found that it was possible to couple diazotized S-acid *ortho* to the free amino group of the oxidation product of the rosindole reaction, with some increase in colour. It is not possible to increase the colour of the DMAB-nitrite reaction by this means. The post-coupled benzylidene reaction is to be regarded as entirely specific for indole derivatives. In tissue sections the only reacting compounds likely to be present are tryptophan, tryptamine and 5-hydroxytryptamine, and these give a strong reaction *in vitro* which is slowly reversed by prior treatment with formaldehyde.

The list of tissue components demonstrated by this method is substantially the same as with the two methods described above, but there are a few outstanding points of difference. In particular, by the post-coupled benzylidene and rosindole reactions human pancreatic α-cells are clearly demonstrated

(Fig. 48, facing p. 119), but with the DMAB-nitrite method they are much less well shown. The original authors reported their inability to block the reaction by previous treatment of sections with performic acid, but in my hands this reagent, made according to the instructions given in Appendix 6, p. 620, is quite successful in blocking subsequent reactions of the tissues with DMAB. Alkaline persulphate oxidation has the same effect. Technical details of the method are given in Appendix 6, p. 616, but I consider that for most purposes the more simple DMAB-nitrite method is preferable.

Three further histochemical reactions for tryptophan must be considered. The first of these is the xanthydrol reaction.

Xanthydrol Reaction for Tryptophan. A method for the determination of tryptophan in proteins using xanthydrol in HCl was described by Dickman and Crockett (1956). This was based on earlier work by Illari (1938) and Fearon (1944). These authors showed that pyrroles would react with xanthydrol in acetic acid, giving purplish-red products. According to Fearon the reaction could take place with indoles only if the 3-position was open, and thus tryptophan did not react. Dickman and Westcott (1954), on the other hand, found that xanthydrol in HCl/acetic acid mixtures gave a purple product with both tryptophan and tryptamine.

The Dickman and Crockett method was adapted by Lillie (1957a), and independently by Adams (1957b) for histochemical use. It gives a violet colour with pancreatic zymogen granules in several species, and also with gastric chief cells, salivary duct cells, and in the fibres of the crystalline lens (Lillie). Smooth muscle, cardiac muscle and skeletal muscle fibres give a moderate reaction, goblet cells in the rabbit were just faintly coloured in the original author's hands, but most mucins are negative, as are the argentaffin (enterochromaffin) cells in all species, after formaldehyde fixation. After glutaraldehyde, however, the reaction is strongly positive. As with the DMAB-nitrite reaction, some areas in argentaffin carcinomas (carcinoids) contain cells with a moderately strong reaction in their cytoplasm. Lillie (1957a) suggested two possible mechanisms to explain xanthydrol-tryptophan condensation. The one given below differs from the other only in ignoring the part played by HCl.

In vitro tests which were made by Lillie gave a positive result with tryptophan, tryptamine, indole, skatole and pyrrole. A dark blue-green colour developed with 5-hydroxytryptamine and a deep blue with carbazole. The latter, incidentally, does not react with DMAB. Strong purplish-black colours were obtained with 2-phenyl and 2-methyl-indoles, 2-phenyl and 2-carbethoxytryptophan, and with tetrahydronorharman, pyrogallol, catechol and resorcinol. According to Lillie the reaction was not abolished by treatment of sections with performic acid in the cold. In my hands, using this reagent at room temperature, blocking was complete in 2 hours. The reaction is also blocked by peracetic acid, but benzoylation (p. 168) weakens it only moderately. Iodination with 10 per cent. iodine in methanol (p. 171) also completely prevents the reaction.

The specificity of the xanthydrol method for tryptophan in tissue sections must be regarded as high, but the colour developed, in my hands, is always less than that given by the DMAB-nitrite reaction, which I regard as the more

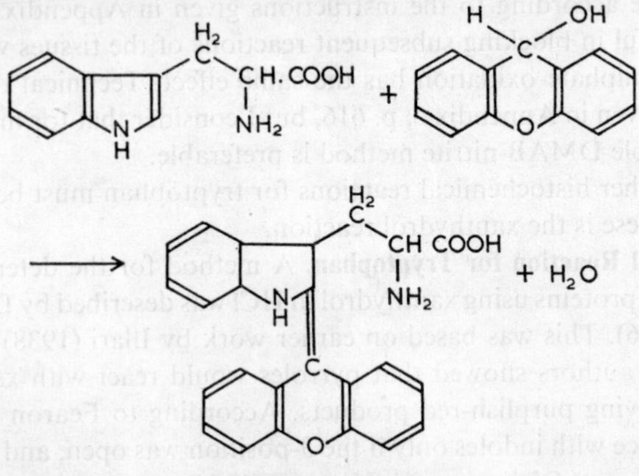

satisfactory of the two. This reaction possesses an additional advantage in that the colour is stable over a long period (years), whereas that of the xanthydrol method fades in a few months.

The Naphthyl-ethylenediamine (NED) Condensation Method for Tryptophan. An entirely different principle has been employed by Bruemmer, Carver and Thomas (1957) in developing a method for tryptophan from the method of Bratton and Marshall (1939) for sulphanilamides. The reagents used by these authors were adapted by Eckert (1943) to produce a red colour with tryptophan.

With the NED condensation method tissue sections are first treated with nitrous acid and then placed in a solution of N-(1-naphthyl)-ethylenediamine dihydrochloride. A violet colour develops in sites containing large amounts of tryptophan. According to Glenner and Lillie (1959) the intensity of the reaction is greatly increased if it is carried out in the dark (c.f. diazotization-coupling, p. 129). The mechanism of the reaction is not so simple as appears at first sight. Primary aromatic amines are diazotized by treatment with HNO_2, as in the original method of Bratton and Marshall, and they can then be coupled with phenols or aromatic amines to form azo dyes. Indoles cannot be diazotized, however, and Bruemmer *et al.* suggest that treatment with HNO_2 produces an N-nitroso compound (cf. Baker's explanation of the Millon reaction, p. 107), and that condensation takes place to link the N of the nitroso group with the free NH_2 group of NED:

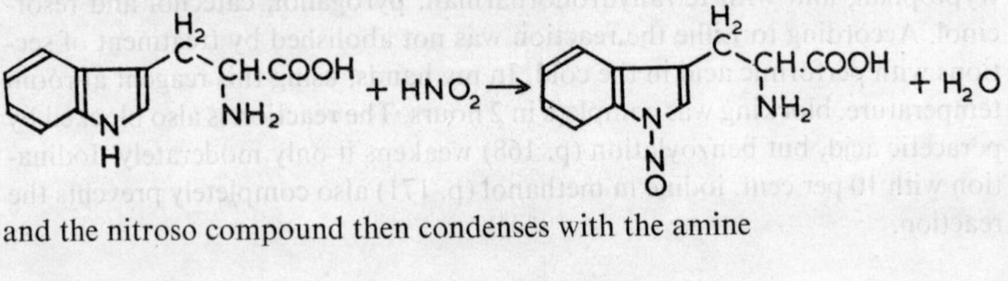

and the nitroso compound then condenses with the amine

$$+ \ RNHCH_2CH_2NH_2 \longrightarrow \ \text{(indole ring)}\text{-}CH_2\text{-}CH.COOH(NH_2),\ N\text{-}N{=}NCH_2CH_2NHR \ + \ H_2O$$

Bruemmer *et al.* found, by infra-red absorption spectroscopy, that the NED-tryptophan condensation product possessed the specific absorption band in the region of 6·3 nm, which is due to the —N=N— configuration.

In place of the NED condensation it might be possible to use the reaction for nitrosamines described by Preussman *et al.* (1964). This involves treatment with 1·25 per cent. diphenylamine in ethanol containing a small amount of palladium chloride.

The original authors carried out specificity tests on a number of compounds *in vitro* and obtained a positive result only with tryptophan, skatole, indole acetic acid, indole butyric acid and isatin. Pyrrole and catechol did not react. No blocking reactions were applied to the histochemical reaction, but the NED condensation in the tissues is abolished by both iodination and by treatment with performic acid. The colour developed is moderately satisfactory (when the reaction is carried out in the dark) but, in view of the observation by Glenner and Lillie (1959) that NED couples also with diazotyrosine, it is necessary to re-examine the whole question of its specificity. Details of the method are given in Appendix 6, p. 617, so that those interested can carry out further investigations. For the present the NED method cannot be recommended for the routine investigation of tryptophan-containing proteins.

Glyceric Aldehyde Condensation Method. A method particularly applicable to formaldehyde-fixed tissues was introduced by Adams (1960). A solution of

$$\text{(indole, NH)}\text{-}CH_2\text{-}CH.COOH(NH_2) \ + \ CH_2OH\text{-}CHOH\text{-}CHO \ \ \text{(I)}$$

$$\downarrow$$

$$\text{(ring)}\text{-}CH_2\text{-}CH.COOH,\ NH\text{-}CH\text{-}CHOH\text{-}CH_2OH \ + \ H_2O$$

(II)

glycerol and ferric chloride dissolved in a sulphuric acid-alcohol mixture is burnt on the section. When the solution is ignited the glycerol is oxidized to glyceric aldehyde (I) and this reacts with protein-bound tryptophan to produce a mauve carboline pigment (II). Both reactions are accelerated by heat and by the rapidly increasing concentration of acids in the burning solvent.

According to Adams the pigment (II) is stable for less than 24 hours. In freeze-dried and freeze-substituted tissues the colour is much less intense than that of the DMAB reaction, as it is in tissues briefly fixed in formalin. A positive reaction is given, however, by sections from blocks fixed for long periods in formalin. These give a negative DMAB reaction. This fact provides the only reason for using glyceric aldehyde condensation (Appendix 6, p. 615).

Dimethylaminocinnamaldehyde as a Tryptophan Reagent. For the demonstration of amines in paper chromatograms Harley-Mason and Archer (1958) used dimethylaminocinnamaldehyde in acid alcohol. Their reagent gave a dark green colour with indoles, violet with skatole and purple with tryptophan. The latter does not react in tissue sections. In glutaraldehyde-, but not formaldehyde, fixed tissues indoles give a positive reaction.

REACTIONS FOR SS AND SH GROUPS (CYSTINE, CYSTEINE)

Histochemical methods for revealing the presence of sulphur-containing amino-acids can be divided into eight groups. First, the older lead and nitroprusside methods and their modifications; secondly, the phenacyl halide methods suggested by Danielli (1950) employing either specific halogen-containing reagents or blocking techniques followed by dinitrofluorobenzene; thirdly, the methods of Bennett (1948) and Yakovlev and Sokolovsky (1955) employing organic mercurials; fourthly, the ferricyanide method of Chevremont and Frederic (1943); fifthly, performic acid oxidation methods (Pearse, 1951; Adams and Sloper, 1955); sixthly, the alkaline tetrazolium methods (Barrnett and Seligman, 1954; Pearse, 1954); seventh, the dihydroxy dinaphthyl disulphide (DDD) method (Barrnett and Seligman, 1952); and lastly, the α-naphthyl maleimide method (Seligman, Tsou and Barrnett, 1954).

The Lead Acetate Method for SS and SH. The principle of this method, adapted from the biochemical test for cystine, is the liberation of sulphur from the tissues by the action of strong alkalis at high temperatures and the subsequent demonstration of this as black sulphide of lead. Alkalis hydrolyse and decompose disulphide cross linkages. Cuthbertson and Phillips (1945) showed that wool cystine breaks down in two ways when treated with sodium hydroxide. Approximately 50 per cent. is converted into lanthionine and sulphur (I) and the remainder is decomposed (II).

Even dilute (0·1 N) NaOH acts slowly in this manner while more concentrated solutions of NaOH (1·0 N) or barium hydroxide (0·385 N), acting for 1 hour at room temperature, produce a complete result, 50 per cent. conversion to lanthionine and 50 per cent. decomposition.

Mann (1902) observed that since the reaction with concentrated solutions

$$
\begin{array}{l}
\begin{array}{ll}
| & | \\
NH & CO \\
| & | \\
CHCH_2S.SCH_2CH \\
| & | \\
CO & NH \\
| & |
\end{array}
\begin{array}{l}
\nearrow \text{I} \quad \begin{array}{l} \text{NH} \qquad\qquad\qquad\quad \text{CO} \\ \text{CH—CH}_2\text{S.CH}_2\text{CH} + \text{S} \\ \text{CO} \qquad\qquad\qquad\quad \text{NH} \end{array} \\[2em]
\searrow \text{II} \quad \begin{array}{l} \text{NH} \qquad\qquad\qquad \text{CO} \\ \text{C—CH}_2 + \text{CH}_2\text{C} + \text{H}_2\text{S} + \text{S} \\ \text{CO} \qquad\qquad\qquad \text{NH} \end{array}
\end{array}
\end{array}
$$

of alkali necessitated almost total destruction of the tissues, its application to histochemistry was not of much use. Nevertheless, using hot concentrated solution of alkali, it is possible to demonstrate the presence of sulphur in keratin at some stage before disintegration of the tissue occurs. The distribution of black lead sulphide is patchy, however, since sulphur is soluble in alkaline solutions, and its localization cannot be accepted as accurate. Theoretically, it should be possible to use more dilute solutions of alkali and still reveal the liberated sulphur *in situ*, unless this has been dissolved away. In practice, I have found that the desired blackening of known sulphur-containing tissues fails to occur when 10 per cent. lead acetate is applied to alcohol or formalin-fixed tissues after the use of alkalis (0·1–2·5 N-NaOH) for various times (1–16 hours) at various temperatures (20°–60°). This method must be regarded as useless and of historical interest only.

Nitroprusside Methods for SH. There are numerous histochemical modifications of the nitroprusside reaction for SH compounds, all of which rely on the production of a purplish-red colour when SH groups come into contact with sodium nitroprusside in the presence of hydroxyl ions (usually NH_4OH) and saturated ammonium sulphate. Nothing is known of the chemistry of the reaction, which according to Lison (1936) was first used in histochemistry by Buffa in 1904. It is given by substances possessing free SH groups such as cysteine and reduced glutathione and has been used for localization of glutathione (Joyet-Lavergne, 1927) and for the free sulphydryl groups of cysteine (Giroud, 1929). These and most other workers have employed frozen sections of fresh tissues in their various researches. Joyet-Lavergne, however, applied the method to the study of alcohol or formalin-fixed sections also, after brief treatment with 2 per cent. trichloroacetic acid to release SH groups from the tissue proteins. Giroud and Bulliard (1932, 1933), suggested that the red colour of a positive reaction might be stabilized by previous treatment of sections with 5 per cent. zinc acetate for a few seconds. Hammett and Chapman (1938), however, experimenting with the nitroprusside reaction on cysteine solutions *in vitro*, found that the presence of zinc acetate decreased the final colour by as much as 75 per cent. These authors evolved a method for application to fresh tissue slices which is given in Appendix 6. In spite of the results obtained by Joyet-Lavergne, I found that the colour developed by the application of his technique, or the modification of Hammett and Chapman, to formalin-fixed frozen and paraffin sections and to alcohol-fixed paraffin sections, was invisible under the microscope. The nitroprusside reaction is, in fact, quite inapplicable to fixed tissues. The colour developed when the method is applied to fresh tissue slices, as Hammett and Chapman have stated, does allow a rough estimate of the relative concentration of SH in different regions. Absolute values for free SH in tissue slices are unobtainable.

Phenacyl Methods for SH (Danielli, 1950). These methods depend on the reaction of thiols with reactive bromides or iodides, such as the phenacyl bromides ($C_6H_5COCH_2Br$). The presence of an aldehyde group does not interfere with the reaction. Thus *p*-bromoacetyl benzaldehyde, one of the compounds suggested by Danielli for the purpose, reacts with SH groups in alkaline solution as follows:

$$CHO-C_6H_4-CO.CH_2\,Br + HS{\cdot}R \rightarrow CHO-C_6H_4-CO.CH_2\,SR + HBr$$

The two other compounds, *p*-iodoacetoxybenzaldehyde and *p*-iodoacetamidobenzaldehyde, would react in a similar manner, it was suggested. In each case, after thorough washing of the tissues, the free aldehyde groups of the combined reagent (thioether) should be available for demonstration by means of Schiff's fuchsin-SO_2 reagent. Although the aldehyde group of these three compounds might react with other groups in the tissues, such a combination would not subsequently be revealed by means of fuchsin-SO_2. Phenacyl bromides and iodides are not known to react with any tissue groups other than sulphydryls, but the method will not demonstrate all the available SH groups unless the phenacyl halogen reacts more readily with SH than does the aldehyde part of the molecule. It is well known that aldehydes do combine with SH groups. I have no experience of this method in practice, and no technical details have been published. One would expect, from the evidence given by Barron (1951), that a satisfactory histochemical method would not be achieved. This author stated that alkylating agents, of which the above are examples, are the least reactive of the commonly used reagents for thiols. "Reactive halogen groups," he said, ". . . cannot be considered suitable reagents for the detection of SH groups in proteins."

A somewhat similar method was described by Savitch and Yakovlev (1957). This depends on the condensation of tissue SH with *p*-nitrobromacetophenone:

$$O_2N-C_6H_4-COCH_2Br + HS.R \rightarrow O_2N-C_6H_4-COCH_2SR + HBr$$

The product is then reduced, diazotized and coupled with H-acid at pH 9·2. The original authors treated control slides in the same manner but without alkylation, in order to avoid the contribution of the D-C method for tyrosine. This precaution was unnecessary because the stage of treatment with nitrous acid is too short.

The method was modified and used by Yakovlev and Nistratova (1958), by Gershtein (1958), and also by Beneš (1962) who found that it produced a strong violet colour in both animal and plant tissues. The strictures of Barron (1951) still apply, however, and it is likely that only a proportion of the available reactive SH groups take part in the reaction. The method, as given by Gershtein (1958) nevertheless appears in Appendix 6, p. 626, since it is the only workable representative of its class.

Blocking Methods for SH Groups (Danielli, 1950). The blocking techniques suggested by Danielli for the histochemical demonstration of SH groups are more complicated than those in the preceding section. The five stages of the method, in order, consist of treatment with:

(1) Phenyl mercuric hydroxide, or chloride in alkaline solution, to block SH groups.
(2) Phenyl isocyanate to block NH_2 and aromatic OH groups.
(3) Performic acid to destroy SS groups.
(4) H_2S or dithioglycerol to unblock the original SH groups.
(5) Dinitrofluorobenzene, reduced, diazotized and finally coupled with a suitable phenol.

Explanation of the theory of these successive stages may be necessary.

(1) methyl or phenyl mercury salts react with thiols in the following manner:

$$C_6H_5-HgOH + HS{\cdot}R \rightleftharpoons C_6H_5-HgSR + H_2O$$

(2) The use of isocyanates for the blocking of NH_2 (and tyrosine OH) groups is based on the reaction below. In alkaline solution amino-acids combine with isocyanates to form substituted hydantoic acids.

$$\underset{\text{Amino-acid}}{H_2N.R.COOH} + \underset{\text{Isocyanate}}{R'N{:}C{:}O} \xrightarrow{\text{KOH}} R'NH.CO.NH.R.COOK + H_2O$$

(3) Toennies and Homiller (1942) showed that performic acid converts

cystine into cysteic acid, breaking the SS bonds and yielding sulphonic acids.

$$\begin{array}{l} S\text{—}CH_2.CH(NH_2).COOH \\ | \\ S\text{—}CH_2.CH(NH_2).COOH \end{array} \xrightarrow{O} \underset{\text{Cysteic acid}}{HO_3S.CH_2.CH(NH_2).COOH}$$

This step prevents the reduction of SS groups to SH by the reagents employed in the fourth stage.

(4) Reversal of the selective blocking of the SH groups by the phenyl mercury salt is obtained by the provision of other SH-containing compounds, such as H_2S or dithioglycerol.

(5) The use of dinitrofluorobenzene, and its reduction, diazotization and coupling, are described in the section on blocking reactions, below, and in Appendix 6.

This method is complicated and I have not succeeded in obtaining a satisfactory result with either fixed or freeze-dried (unfixed) sections.

Blocking Method for SS Groups (Danielli, 1950). This method, suggested by Danielli for the demonstration of tissue SS groups, has three essential stages and consists in treating sections with the following reagents:

(1) Phenyl isocyanate to block SH groups.
(2) Potassium cyanide to reduce SS to SH.
(3) Dinitrofluorobenzene to combine with SH, followed by reduction, diazotization and coupling.

The method, in my hands, failed to give satisfactory results either with formalin or alcohol-fixed material, or with unfixed freeze-dried tissue sections. Since it is relevant to most of the methods in this section, however, it may be appropriate to consider at this point the suggested reduction of SS to SH with potassium cyanide.

Conversion of SS to SH Groups with KCN. Solutions of potassium cyanide have long been employed in histochemistry for the reduction of tissue SS to SH, prior to demonstration of the latter by the nitro-prusside method for instance. Joyet-Lavergne (1927) used 10 per cent. KCN and exposed the tissues for 5 minutes, Lison (1953) recommended 10 per cent. for 5 minutes, and Hammett and Chapman (1938), 5 per cent. for 10 minutes. These last authors, however, stated that SH was destroyed by alkaline solutions and that the alkaline solution of KCN destroyed not only some of the thiol groups initially present in the tissue but also some of those newly converted from disulphide. They observed that cyanide reduction would sometimes establish the presence of SS, but with certainty only when no free SH was present or when the ratio SS to free SH was high. This last condition is present in fixed tissues, but their objection on account of destruction of newly-formed thiol groups remains a valid one.

There are three factors to be considered, the availability of tissue SS groups for reaction with KCN, the speed at which the reaction takes place, and the speed of destruction of thiol groups by the alkaline solution. It is know that the SH groups of free cysteine, and of cysteine combined in various proteins, have different reactivities with specific SH reagents and that denaturation, on the whole, increases rather than reduces this reactivity. As regards the rate of reaction between SS and KCN, Cuthbertson and Phillips (1945) have shown that 85 per cent. of the disulphide-S of sheep's wool is converted by 1 per cent. KCN in excess, acting for 1 hour at 66°. If the reactivity of SS groups in tissue sections is greater than this, as is most likely, then 10 minutes in 10 per cent. KCN should be sufficiently long to achieve a high degree of conversion and yet short enough to minimize destruction of the newly-formed SH groups by alkali. In practice, however, although treatment of sections with KCN in the above manner results in an increased number of reactive SH groups as indicated by both Bennett and ferricyanide reactions (see below), many workers have found it far from satisfactory. I now agree with Lillie (1954) that other methods are preferable.

Comparison of KCN with other Reagents. Barrnett and Seligman (1954) tested a number of agents for their efficiency in reducing tissue SS groups to SH; they were H_2S and its salts, sodium bisulphite, sodium hydrosulphite, sodium thiosulphate, L-ascorbic acid, L-epinephrine, KCN, sodium thioglycollate, cysteine, glutathione, 2-mercaptoethanol, and 2,3-dimercaptopropanol (B.A.L.). Of these they found four particularly useful and efficient; in descending order these are sodium thioglycollate, ammonium sulphide, KCN and hydrosulphite ($Na_2S_2O_4$). The last was effective only at a concentration of 12 molar and at pH 10·5, while KCN (10 per cent.) required a high pH (11–11·5) and a long period of incubation. Moreover, it was pointed out that this reagent produces only one reactive sulphydryl from each disulphide, together with an inert thiocyanate group. Ammonium sulphide was the most powerful reducing agent and it produced particularly strong staining reactions. However, Barrnett and Seligman showed that part of this staining was an artifact due to the creation of new SH groups by condensation of $(NH_4)_2S$ with tissue carbonyl groups. They recommended that this reagent should be used only with suitable controls (carbonyl blocking reagents; see Appendix 13, p. 706), especially when exploring new tissues. Sodium thioglycollate was the best and most specific of the four "short-listed" reducing agents, but even this does not cleave all the tissue disulphide groups. Details of the concentrations used, and of the solvents, etc., will be found in Appendix 6, p. 621.

The Mercury Orange Method for SH (Bennett, 1948, 1951). This method is based on the reaction of mercaptans with organic mercurials like phenyl mercuric chloride, as in the method originally employed by Hellerman *et al.* (1933). These authors also used *p*-chloromercuribenzoate to combine with and block the reactions of SH groups in egg albumin. In an effort to obtain a coloured compound with similar properties to those described above, capable of attachment solely to SH groups, Bennett and Yphantis (1948) synthesized 1-(4-chloromercuriphenylazo)-2-naphthol. This compound is red in colour and it is therefore visible in the tissues if present in sufficient quantity. The reaction with SH groups proceeds as follows:

R·SH + CℓHg–C₆H₄–N=N–(HO-naphthyl) ⇌ R·SHg–C₆H₄–N=N–(HO-naphthyl) + HCℓ

Bennett (1951), recording the use of his reagent for demonstrating tissue thiols, gave it the name red sulphydryl reagent (RSR), but this was subsequently changed to mercury orange. Bennett used small pieces of tissue either fixed in 5 per cent. trichloroacetic acid and dehydrated in alcohol, or prepared by freeze substitution (Appendix 4, p. 598) from liquid N_2 into *n*-propanol or *n*-butanol at —20°. In both cases, before staining, the tissue was teased into small fragments. Mercury orange was employed as a saturated solution in *n*-propanol, toluene, or dimethyl formamide, and staining was usually complete in 2–16 hours. Several variations of the full method are given in Appendix 6.

Specificity of mercury orange for Sulphydryl. Aryl or alkyl mercuric halides have been found to react with thiol groups in native and denatured proteins (Anson, 1941; Hughes, 1950), and there is good evidence in favour of their absolute specificity for SH. With mercury orange, however, there exists the possibility of non-specific attachment to other protein groups in the tissues, or to other non-protein groups. Bennett suggested that four types of binding of mercury orange might occur in the tissues. These are equally pertinent for the newer SH reagents considered at the end of this section.

(1) Specific; by covalent mercaptide linkage to SH.

(2) Non-specific; by hydrogen bonding (Chapter 5, p. 86) to the azo nitrogen or to the oxygen of mercury orange by suitable H groups in the tissues.

(3) Non-specific; by van der Waal's forces (Chapter 5, p. 86) between the aromatic rings of mercury orange and certain tissue components.

(4) Non-specific; by salt linkages between ionized groups in the tissues (such as phosphate) and the Hg^+ or H^+ of mercury orange.

The specificity of mercury orange for SH was established, in the author's view, by the performance of certain critical experiments. It was shown, by substituting phenylazo-2-naphthol for mercury orange, that the presence of the chloromercury group was essential for the reaction. Using teased muscle fibres Bennett found that no colour was taken up from solutions of phenylazo 2-naphthol, whereas, in solutions of mercury orange, the muscle fibres were stained red. If the SH groups of the fibres were blocked by alkylation (p. 172) or by mercaptide formation (p. 172), the reaction with RSR was negative, indicating the necessity of the presence of active SH groups. Bennett prepared a number of solutions designed to compete with molecules bound in tissues by the various types of bonding described above and found that only solutions of mercaptans (β-mercaptoethanol, thioglycollic acid, B.A.L., cysteine) would completely reverse the staining of tissues by mercury orange. Methionine and cystine, which are not mercaptans although they contain sulphur, were quite ineffective.

Bennett's evidence for the specificity of his reaction appeared to be fairly strong, since it took note of most of the possibilities for non-specific attachment to tissue groups. In spite of the results of the author's first critical experiment there remained, however, the possibility of solution in lipid substances to be taken into account. The resemblance of phenylazo-2-naphthol compounds to the Sudan dye Brilliant fat scarlet is very close indeed, as can be seen from a comparison of the formula below with that of mercury orange.

Brilliant Fat Scarlet

Mauri, Vaccari and Kaderavek (1954), after considering and testing a number of reactions for thiols, concluded that only the mercury orange method was sufficiently sensitive and specific. They considered that the insolubility of the reagent in aqueous media was responsible for much of the observed difficulty in staining tissues known to contain SH groups. A water-soluble coloured organic mercurial was therefore synthesized, 1-oxy-2-(*p*-chloromercuriphenylazo)-8-aminonaphthalene-3,6-disulphonic acid. Blocking and other tests suggested that the specificity of this reagent was as high as that of mercury orange.

Horowitz and Klotz (1956) used a different azomercurial, 4-(*p*-dimethylaminobenzeneazo) phenylmercuric acetate, in 0·1 M-glycine buffer at pH 9·0, for estimating SH in proteins. They found that the time of equilibration of their reagent with bovine serum albumin (BSA 6×10^{-6}M, Dye 14×10^{7}M, pH 9, 25°) varied between $4\frac{1}{2}$ and 8 hours. With ovalbumin under the same conditions the time varied from $4\frac{1}{2}$ to 17 hours. These results indicate that mercaptide formation is a slow reaction at room temperature, and in using mercaptide-forming reagents due allowance should be made for this fact. I tested mercury orange in glycine buffer, in which it is moderately soluble, as a histochemical reagent, but only diffuse non-specific staining of the tissues was produced. It was not possible to block this with the usual reagents. This emphasizes the point that different solvents may decrease the specificity of mercaptide-forming reagents to a considerable degree.

Horowitz and Klotz used their reagent (DMAB-phenyl mercuric acetate) in glycine buffer, at the high pH indicated, in order to avoid the possibility of reaction with tissue-bound NH_2 groups. Their whole procedure was adapted for histochemical use by Engel and Zerlotti (1964), who took advantage of the relatively high molar extinction coefficient (25,000) by examining their sections at 430 nm. They were able to block the reaction, which was apparently complete in 30 minutes, with a variety of blocking reagents (pp. 172, 173) and it is clear that mercury orange should be replaced by DMAB-phenyl mercuric acetate, or by one of the newer reagents which have not yet been tested histochemically.

In this category are the two new chromophoric reagents for protein SH groups synthesized by Fai Chong and Liener (1964). The first, 4(*p*-nitrophenylazo)-2-chloromercuriphenol, differs from the second only in having a single nitro group in the *para* position.

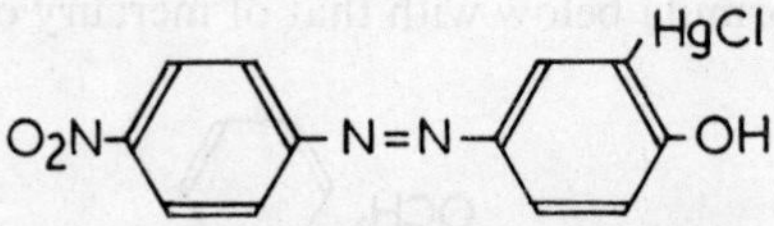

Details of the preparation of these reagents are given in Appendix 6, p. 623.

With all methods which successfully demonstrate SH groups the problem is to determine what proportion of the total available SH is taking part in the reaction. There are two main sources of SH in the tissues, (1) glutathione and (2) protein-bound SH, the latter being either free to react or masked. In Bennett's procedure glutathione is either removed by trichloroacetic acid fixation, or largely preserved by freezing and dehydration in butanol or propanol. Either of these manœuvres will reveal a proportion of the masked SH groups of proteins and enable them to combine with the reagent. Those who investigate the functional aspect of sulphydryl histochemistry will wish to show only free SH groups; investigators of structure desire to show total SH. In practice neither is likely to achieve success, and the final result of the application of a method such as that described in this section is to demonstrate a number of SH groups which lie between the two extremes. Unless the procedure is very carefully standardized, therefore, comparison between one material and another is likely to be open to criticism.

Details of Bennett's method, and also of the method of Engel and Zerlotti, appear in Appendix 6, and their application to problems of diagnostic pathology are considered in Chapter 8.

The Ferric Ferricyanide Method for SH and SS (Chèvremont and Fréderic, 1943; Adams, 1956). This method, according to the authors, can be performed on sections fixed in formol-saline or in Bouin, provided that the duration of fixation is a few hours only. Frozen or paraffin sections may be employed, preferably the former. It can also be applied to unfixed tissues, in freeze-dried or cold microtome sections, or to paraffin sections after fixation in trichloroacetic acid-alcohol. The method depends on the reduction of a fresh solution of ferricyanide, in acid solution at pH 2·4, by sulphydryl groups in the tissues.

$$2[Fe(CN)_6]^{+++} + 2RSH \rightarrow 2[Fe(CN)_6]^{++++} + RS—SR$$

The resulting ferrocyanide combines with ferric iron (provided as ferric sulphate) to give a precipitate of insoluble Prussian blue.

$$4Fe^{+++} + 3K_4[Fe(CN)_6] \rightarrow Fe_4[Fe(CN)_6]_3 + 12K^+$$

Thus, in principle, the Chèvremont method is the same as Schmorl's method for lipofuscin (see Chapter 26) and it gives a positive reaction with the latter substance as well as with melanin and argentaffin cell granules.

Distinction between melanin, lipofuscin, argentaffin granules, and protein structures containing SH is not difficult by ordinary histological techniques. Small quantities of lipid reducing substances attached to protein are by no means easy to detect, however. In my opinion the specificity of the Chèvremont method depends on (1) proof of the absence of non-protein reducing groups, and on (2) the specificity of ferricyanide-reduction, as far as protein is concerned, for SH groups alone. Purity of the reagents used is, of course, essential. Considering the second point first: Anson (1939) stated that denatured proteins which contained no cysteine would not reduce ferricyanide. He showed (1941, 1942) that the extent of the reaction, which is catalysed by copper ions and inhibited by cyanide, depended on the concentration of ferricyanide. When twice the stoichiometric amount of the latter was present, 64 per cent. of the SH groups in denatured egg albumin were oxidized, and with fifty times the amount, 94 per cent. Anson found that partial digestion with trichloroacetic acid made the SH groups of denatured protein more reactive; this recalls Joyet-Lavergne's use of the same reagent. In contradistinction to the above results, which suggest that the specificity of the method should be high, Mirsky and Anson (1936) stated that some protein groups other than SH were oxidized by ferricyanide, notably tyrosine and tryptophan. Herriott (1947) concluded that, in the absence of evidence to the contrary, it might be presumed that the tyrosine groups in some proteins could be oxidized by ferricyanide, and that ferricyanide-oxidation appeared to be non-specific in some instances and ineffective in others. If this is true, a negative result with the histochemical ferricyanide method cannot be taken to indicate absence of SH.

The histochemical interpretation of the ferric ferricyanide reaction was thoroughly considered by Adams (1956). He showed that two pigments are formed when this compound is reduced *in vitro* by a variety of substances. These pigments are Prussian blue and Prussian green. Both are formed during the histochemical performance of the reaction, and the latter forms in small quantities in the medium itself. From this it may possibly be absorbed by tissue components to give rise to non-specific green staining. In any case Adams recommended that the reaction time be limited to 5 minutes and that any green colour be ignored or interpreted with great reserve. This author's theoretical results are also considered in Chapter 26.

Both tyrosine and tryptophan rapidly produce Prussian blue from ferric ferricyanide *in vitro*, but low cysteine, high tryptophan-containing proteins do not give a positive reaction in tissue sections. Even if the specificity of the method for protein-SH is allowed, however, the objection raised in the first point, above, remains valid. It is possible to distinguish lipids, if present in sufficient quantity, by various techniques (see Chapter 12). It is not possible, by direct histochemical means, to exclude the presence of small amounts of lipid.

A different approach to the problem was employed by Yao (1949), who used the Chèvremont method on the eggs of *Drosophila*. He immersed control sections in saturated aqueous mercuric chloride for 1 hour, before treatment with ferricyanide, in order to prevent the reaction of SH groups. Only structures positive in the normally-treated sections and negative in the controls, could then be considered to contain active sulphydryls. A procedure of this kind is essential before the Chèvremont method can properly be used for the demonstration of SH. The mercaptan-forming reagent used by Yao is not the most suitable as the time allowed for the reaction is too short. A saturated solution of phenyl mercuric chloride in 80 per cent. ethanol, or in absolute *n*-butanol, and 48–72 hours exposure, produce a more reliable total blocking of SH groups. Adams (1956) successfully used both types of blocking and also introduced a prior stage of thioglycollate reduction (Appendix 6, p. 621) to convert unreactive SS to reactive SH groups.

For the demonstration of SS groups the Chèvremont method should be employed with a restricted reaction time of 5 minutes, or less, after thioglycollate reduction and with parallel control sections subjected to blocking subsequent to the reduction stage. A blue colour appears in structures containing large amounts of cystine. According to Adams the sensitivity of the method for protein-bound SS is below that of the DDD and alkaline tetrazolium reactions (q.v.). For protein-bound SH the Chèvremont method should also be used with a restricted reaction time, particularly when trichloroacetic acid-fixed sections are employed, and invariably with SH-blocked controls.

Selective Oxidation Methods for SS (Pearse, 1951; Bonomi, 1951). These methods, of which technical details appear in Appendix 6, are based on the selective oxidation of cystine by performic acid. Toennies and Homiller (1942) showed that performic acid reacts only with tryptophan, cystine and methionine among the amino-acids, whereas, according to Nicolet and Shinn (1939) periodic acid attacks tryptophan, methionine, cystine and the α-hydroxy amino-acid serine. In earlier work I found that periodic acid, together with $KMnO_4$, H_2O_2 and perchloric acid, were unsuitable. Only performic and peracetic acids were satisfactory in practice. Theoretical considerations indicated that a major portion of the reaction product from performic acid oxidation of cystine was likely to be cysteic acid (alanine-β-sulphonic acid), and the histochemical problem was resolved into the demonstration of this substance in the

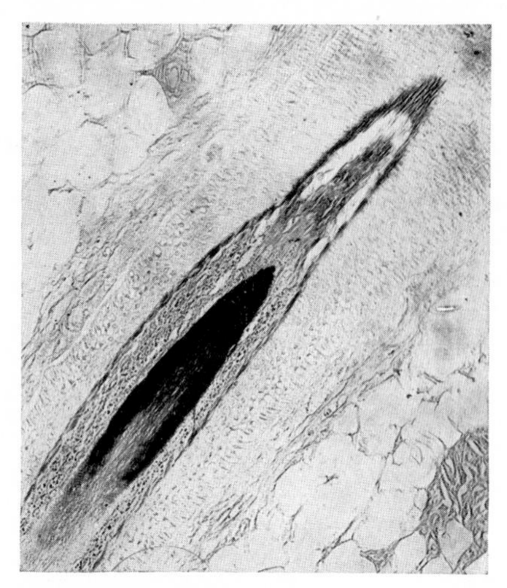

FIG. 49. Longitudinal section of human hair follicle. The outer portion of the hair shaft (cortex and medulla) stains bright magenta. Huxley's layer of the root sheath also stains. Performic acid-Schiff. × 120.

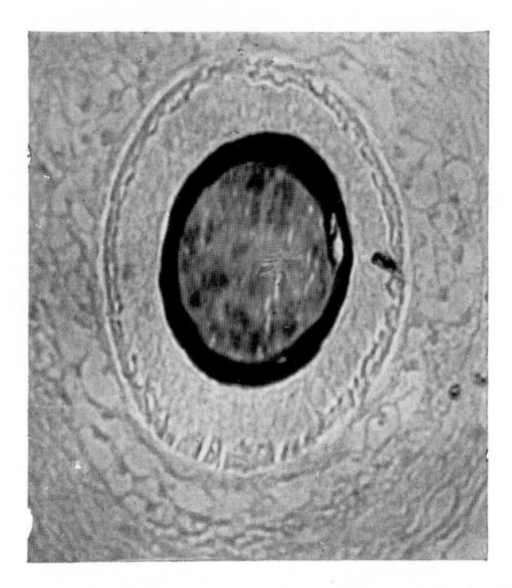

FIG. 50. Transverse section of human hair and root sheath. The cortical layer of hair is stained dark blue, the medulla to a lesser extent. No other structure is stained. Performic acid, 0·5 mM methylene blue at pH 2·6. × 300.

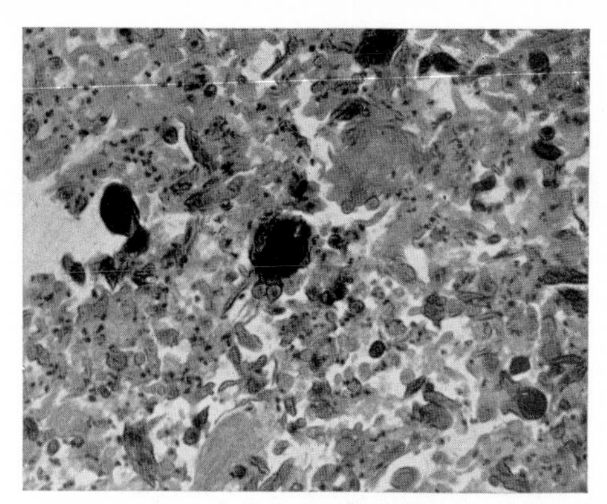

Fig. 51. Keratin masses in a squamous-cell carcinoma, metastasis in a lymph node. Performic acid, cobalt nitrate/sulphide, carmalum. × 105.

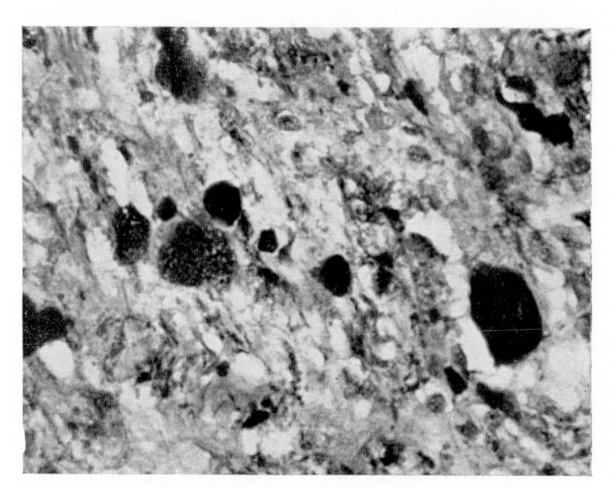

Fig. 52. Human neurohypophysis. Shows large and small masses of neuro-secretory substance. Performic acid—Alcian blue. × 510.

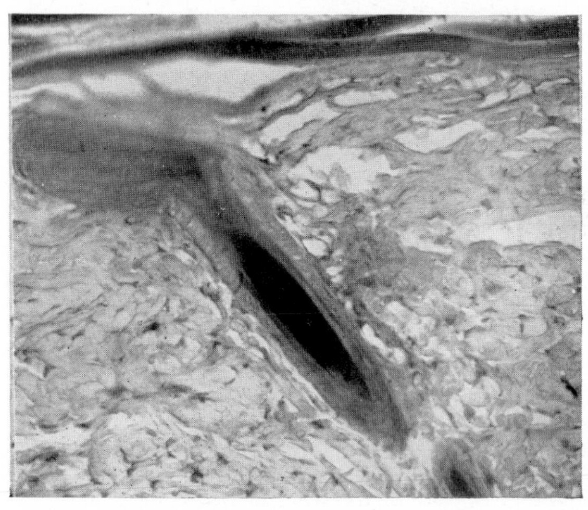

Fig. 53. Rat skin. Positive reaction in hair shaft and epidermal keratin layers Alkaline tetrazolium reaction. × 130.

tissues. A method for the identification of sulphonic acids *in vitro* was suggested by Latimer and Best (1937), which involved the formation of phenylhydrazides, but the colour of these was too faint to be visible in tissue sections. An aryl hydrazide, 2-hydroxy-3-naphthoic acid hydrazide (NAH) was therefore used instead. Final coupling with a diazonium salt was necessary to develop the full colour of the reaction. Using performic oxidation, followed by NAH and diazotized *o*-dianisidine, a positive result was obtained in structures containing keratin.

Since NAH had been usefully employed as an aldehyde reagent (Pearse, 1951), Schiff's solution was automatically suggested as an alternative and from this the performic acid-Schiff procedure was worked out. Alternative methods used to demonstrate the reaction product were staining in dilute aqueous solutions of methylene blue and combination with cobalt salts, subsequently revealed by conversion to black cobalt sulphide. Examples of the application of the three methods are shown in Figs. 49, 50 and 51. Oxidation of tissue with performic acid was shown to give rise to histochemically detectable reaction products particularly in two classes of material. These were keratin, with which we are here concerned, and unsaturated lipids of the phosphatide class. The application of the performic and peracetic acid-Schiff techniques (PFAS, PAAS) to lipids is considered in Chapter 12, p. 437.

SUBSEQUENT DEVELOPMENT OF THE PFA REACTIONS FOR CYSTINE

Of the two alternative reagents originally used in place of Schiff's reagent to demonstrate the sulphonic acid reaction product of performic acid oxidation of cystine-containing structures, methylene blue (2 mM, pH 2·6) was regarded as the most satisfactory and specific. Lillie, Bangle and Fisher (1954) used Azure A and thionin as alternatives, following oxidation by performic and peracetic acids and by bromine in CCl_4. All these variants produced metachromatic staining of keratin but of no other tissue component, so that the methods are inadmissible as reactions for cystine-containing structures in general. A different approach was made by Scott and Clayton (1953). These authors used Lugol's iodine, periodic acid and acidified permanganate followed by Schiff's reagent or by Gomori's (1950) aldehyde fuchsin. A positive result with Schiff's reagent was obtained only after permanganate oxidation, but with aldehyde fuchsin all three types of oxidation were sufficient. Unfortunately, as pointed out by Bangle (1956) and by Braun-Falco (1956), though the sensitivity of aldehyde fuchsin is high, its specificity is low since it combines with aldehydes by non-polar addition as well as with acid radicals by salt linkage.

There is little doubt that performic acid is the most suitable reagent, on both chemical and histochemical grounds, for the oxidation of cystine, and Adams and Sloper (1955, 1956) combined it with an acid solution of the phthalocyanin dye Alcian blue (Chapter 10, p. 344) for their studies on the distribution of the neurosecretory substance (Fig. 52). *The performic acid-Alcian blue method* has a high specificity but a relatively low sensitivity, and it therefore demonstrates only high concentrations of cystine in the tissues. Notwithstanding, there are two important uses of the method in applied histochemistry. These are the demonstration of NSS in various species and the division of the anterior pituitary mucoid (basophil) cells into cystine-rich and cystine-poor groups. Both these applications are further considered in Chapter 8, pp. 239 and 240.

As alternatives to performic and peracetic acids it is possible to use permolybdic and pertungstic acids (Earland *et al.*, 1955), and these can be used histochemically. In the case of permolybdic acid some molybdenum remains in combination with the oxidized keratin, and this suggests a possible development of the method.

THE MECHANISM OF THE PERFORMIC ACID-SCHIFF REACTION

In working out the mechanism of the reaction with Schiff's solution in the case of performic acid-oxidized keratin, it was established that traces of performic acid, or of the formic acid and H_2O_2 from which it was made, could not be responsible for the development of a red colour. It was further shown that a sulphonic acid (*p*-toluene sulphonic acid) produced no colour when added to Schiff's solution *in vitro*. The corresponding sulphinic acid (*p*-toluenesulphinic acid), however, produced a strong but slowly developing reaction under the same circumstances. This finding suggested that the oxidation of cystine in tissue sections produced not only alanine-β-sulphonic acid but also, perhaps as an intermediate stage, alanine-β-sulphinic acid ($HO_2S.CH_2.CH(NH_2)CO_2H$). According to Cecil (1950) the most stable of the intermediate reaction products of cysteine and cystine is the sulphinic acid. *In vitro*, however, oxidation of cystine to cysteic acid was observed to take place without giving a Schiff-positive reaction product.

Experimental evidence (Pearse, 1951) suggested that in performic acid-oxidized sections three possible groups could account for the appearances observed when these were subsequently treated with Schiff's solution, methylene blue, or metal salts. The three groups are sulphonic (SO_3H), sulphinic (SO_2H), and aldehyde. In order to establish responsibility more accurately the reactions concerned were carried out *in vitro* and by a modified Coujard technique (Coujard, 1943). In this method the various reagents are dissolved or suspended in a serum-gelatin mixture which is used to draw lines or bands on the surface of a glass slide. The slides are subsequently dried and exposed to a fixative. They are well washed in running water before use in histochemical reactions. The results of these manœuvres appear in Table 11 (opposite).

It seemed clear that both sulphonic and sulphinic acids might be responsible for the reaction with methylene blue and possibly with Schiff's reagent also, since in this case some colour was produced by sulphonic acid in the Coujard technique. Aldehydes are not responsible for the results with methylene blue but might be responsible for some of the colour both with Schiff and the NAH-dianisidine reagent. Provided that a low pH was employed the methylene blue

TABLE 11

Reactions of Sulphonic, Sulphinic, and Aldehyde Groups

	In vitro		Coujard			
Radicle	Schiff	Ammoniacal silver	Schiff	NAH-dianisidine	Ammoniacal silver	Meth. blue pH 2·6
SO_3H *p*-toluenesulphonic	–	–	faint +	+	–	+
SO_2H *p*-toluenesulphinic	+	–	+	faint +	–	+
CHO phthalaldehyde	+	+	+	+	+	–

technique was found to be more specific than the other two methods for the demonstration of sulphur-containing reaction products. At higher pH levels methylene blue reacts with other groups, such as COOH, produced in the tissues by oxidation.

Münch and Ernst (1965), investigating the use of diamine reagents in place of Schiff's reagent, found that after performic acid the application of dimethyl-*p*-phenylenediamine produced a strong black staining in structures containing SS and SH.

Lillie and Bangle (1954) disagreed with the above hypothesis and suggested that the PAAS (and PFAS) reactions of hair cortex were due to an unsaturated complex of probable lipid nature. They showed that while iodine could oxidize sulphinic acids *in vitro*, it could not prevent subsequent development of the Schiff reaction in oxidized hair. This finding suggested that sulphinic acid was not responsible. The alternative hypothesis of a highly polymerized, partially oxidized lipid cannot be dismissed on account of the fact that a PFAS or PAAS reaction is obtainable after fixation in boiling $CHCl_3/CH_3OH$ since, as Lillie has pointed out, these reactions are positive in the case of ceroid (Chapter 26.) after similar treatment. The observed basophilia, however, is against Lillie's view since the reaction products of oxidized lipids are not basophilic.

The resistance of the basophilia to methylation resembles that of sulphonic rather than sulphinic acids since the latter can be methylated (Field *et al.*, 1961) when lead tetraacetate in methanol acts on disulphides or sulphydryls:

$$RSSR + 3Pb(OAc)_4 + 4CH_3OH \rightarrow 2R.SO.OCH_3 + 3Pb(OAc)_2 + 4AcOH + 2AcOCH_3$$

Findlay (1955a) found that α- and β-keratin membranes prepared from human hair by the method of Alexander and Earland (1950), which involves treatment with peracetic acid in the initial stages, were strongly Schiff-positive without further oxidation. He also found that alternative aldehyde reactions gave negative results with these membranes and with oxidized hair. In the case of the latter the Ashbel-Seligman reagent (Chapter 13, p. 469) gave a red colour rather than the blue which is normally produced by tissue aldehydes. Findlay also found that it was impossible to block the positive Schiff reaction with aldehyde blocking reagents such as cyanide, phenylhydrazine, hydroxylamine and aniline. He considered that the Schiff reaction of peracetic acid-oxidized keratin was probably non-aldehydic and that a relatively intact keratin structure was required for its production. No alternative mechanism for the reaction was forthcoming.

Fractions of isolated wool oxidized with peracetic acid were studied by Earland and Knight (1955). Their two fractions, designated α- and β-keratose, corresponded to Findlay's α- and β-keratins. Neither fraction would form an ammonium salt, although cysteic acid does so readily *in vitro*, and this failure was explained by Alexander, Fox and Hudson (1951) as being due to the formation of sulphonamides rather than sulphonic acids. Weston (1955), however, found no SO_2NH in the infra-red absorption spectra of the two fractions, but SO_3H was definitely present. Earland and Knight suggested, therefore, that strong salt links were formed between the sulphonic acid groups and basic residues ($^+NH_3$) in the tissues and that these prevented the normal reactions of a sulphonic acid from taking place. While these findings do not provide a direct explanation for the PAAS reaction of keratin they suggest the direction in which further inquiries should proceed.

The investigations of Landing and Hall (1956a) added nothing to the solution of the PAAS problem as far as keratin is concerned and the problem must be regarded as still unsettled.

Tetrazolium Methods. Barrnett and Seligman (1952) first suggested the use of ditetrazolium chloride (BT) for the demonstration of SH and SS groups in the tissues. In later papers (1953 and 1954) further details of its application were given, and in the final method celloidin-covered sections were incubated for 8–12 hours in a mixture containing alkali and KCN as well as BT. Because of its higher redox potential (greater ease of reduction) Rogers (1953) preferred to use 2-(*p*-iodophenyl)-3-(*p*-nitrophenyl)-5-phenyl tetrazolium chloride (INT) for the demonstration of SH in frozen sections of skin. He incubated these at pH 7·4 for up to 2 hours and produced a red colour in components containing SH. These conditions are not adequate for the demonstration of total reactive SH groups, and INT forms large crystals, especially in sections which have been stored.

The higher redox potential of Nitro BT produces a positive reaction with SH groups at pH levels as low as 8 and Deguchi (1964) made use of this fact. He assumed that the specificity of the alkaline tetrazolium reaction for SH was made considerably more specific by the substitution of Nitro-BT for BT. Total blocking of the reaction was not obtained, however.

I used three different salts in my early studies of the so-called alkaline tetrazolium reaction (Pearse, 1953, 1954); these were neotetrazolium (NT), BT, and 2,5-diphenyl-3-(4-styrylphenyl)-tetrazolium chloride (M & B 1767). All three were used in 0·1 M-glycine buffers at between pH 11·9 and 12·8 and at 60°. The results with BT were clearly superior to those of the other two salts in that the staining which resulted was non-particulate. Components in the tissues giving rise to a positive alkaline tetrazolium reaction were divided into three groups:

(1) Cystine or cysteine-containing.
(2) Lipid or lipofuscin.
(3) Reducing sugar-containing.

An additional miscellaneous group included such entities as enterochromaffin cell granules and adrenochrome.

Barrnett and Seligman (1954) suggested that reductants such as sugars, adrenalin, and lipid aldehydes would fail to react because they would not be present in paraffin sections. This is not quite true. Protein-bound sugars are retained in any case, and lipid aldehydes also. Adrenalin is preserved, as adrenochrome, by oxidant fixatives, and this substance gives a positive reaction. When the reaction is carried out at pH 12 and at 60° such components undoubtedly react; at 37° the effect of alkaline hydrolysis upon the sugar-containing components is usually insufficient to allow them to stain. Lowering the pH of the reaction to 11·0 has a similar effect, and sugar-containing components largely fail to react. Lipid aldehydes, however, still do so.

Both SS and SH groups have been considered to take part in the reaction. Barrnett and Seligman (1954) believed that under the conditions they employed (pH 11, 37°, KCN) the reaction of sulphur-containing tissues was due to SH groups alone. These were (1) original SH groups present in the tissue, and (2) alkali cyanide-reduced SS groups. According to Cuthbertson and Phillips (1945) at 60° and pH 12 cystine breaks down to form combined lanthionine, combined amino-acrylic acid, H_2S and free sulphur. Beaven and Holliday (1952), on the other hand, considered that SH and SOH groups were formed as intermediate reaction products and that SOH was immediately converted into S or H_2S. At the pH employed in the alkaline tetrazolium reaction (Pearse, 1953) both SH and H_2S, and probably also free sulphur, should reduce tetrazolium salts effectively, and the reaction therefore shows tissue disulphides and sulphydryls without distinction. Findlay (1955b) made a series of experiments to test the validity of the alkaline tetrazolium reaction for tissue sulphydryls. He found that the relative concentration of the reactants made a great difference to the rate and efficiency of the reaction *in vitro*. When the tetrazolium salt was present in excess cysteine produced relatively little reduction between pH 10·8 and 12·6. At this last level, however, provided that the concentration of cysteine was low, the reaction became 100 per cent. efficient. Findlay also compared the DDD and tetrazolium reactions and found a number of inconsistencies. Some of these have not been explained, but his interpretation of the tetrazolium reaction of the nucleoli as due to SS groups is open to the alternative explanation that it is due to reduction by the ribose sugar of RNA. Reversibility by ribonuclease is a point in favour of this second hypothesis.

The low specificity of the alkaline tetrazolium method renders it relatively unsuitable as a method for SS and SH. If it is required for the demonstration of these groups alone then the conditions specified by Barrnett and Seligman should be employed. Their technique is given in Appendix 6, p. 625, and the results are illustrated by Fig. 53, facing p. 149. Since the sensitivity of this method, as opposed to its specificity, is fairly high it may be useful as a control technique for alternative methods.

The Dihydroxy-Dinaphthyl-Disulphide (DDD) Method for SH and SS

This method (Barrnett and Seligman, 1952, 1954; Barrnett, 1953) is based on the use of a reagent, 2,2′-dihydroxy-6,6′-dinaphthyl disulphide, which was synthesized especially for the purpose. The disulphide grouping is the only specific oxidizing agent for sulphydryls and the reagent was presumed to have absolute specificity as far as the first part of the histochemical reaction is concerned. The full mechanism of the reaction is shown in the following series of equations:

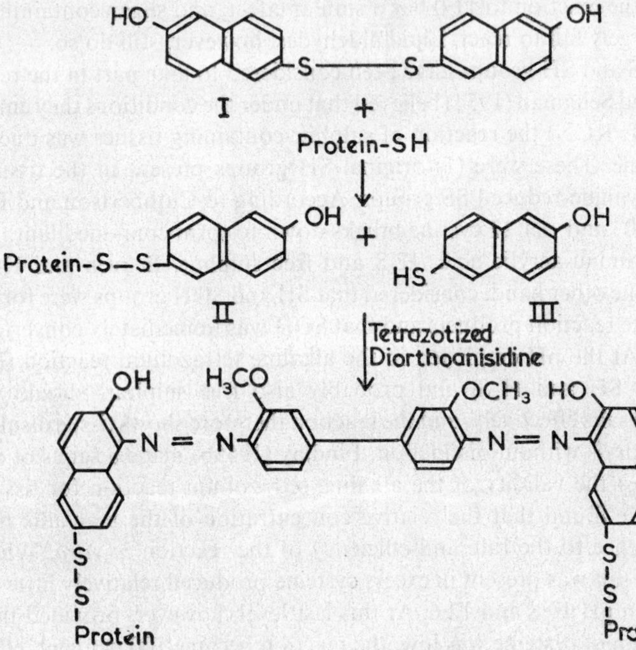

HO OH
S—S
I +
Protein-SH
OH + OH
Protein-S—S HS
II III
Tetrazotized
Diorthoanisidine
OH H_3CO OCH_3 HO
N = N N = N
S S
S S
Protein Protein

In the first stage the reagent (I) combines with protein-SH by splitting to form one protein-naphthyl disulphide (II) and one free naphthyl mercaptan (III). The latter, and any excess of the original reagent adsorbed on the tissues, is removed by washing in a series of alcohols and finally in absolute ether. For the final development of colour the sections are treated with a diazonium salt which combines with the naphthol moiety to form an azo dye. For this purpose

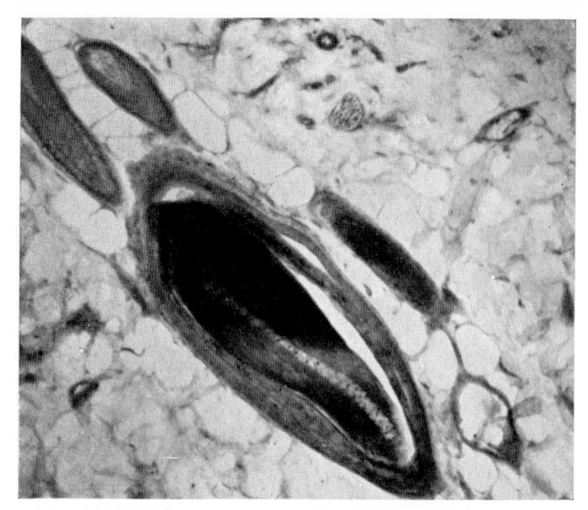

FIG. 54. Rat skin. Positive reaction in hair shaft (blue) and in other structures (reddish purple). DDD reaction. × 130.

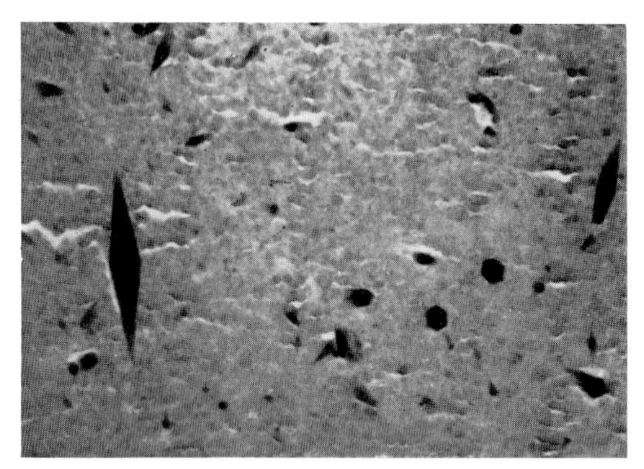

FIG. 55. Human liver, as Figs. 30 and 31. Paraprotein crystals. DDD reaction. × 210.

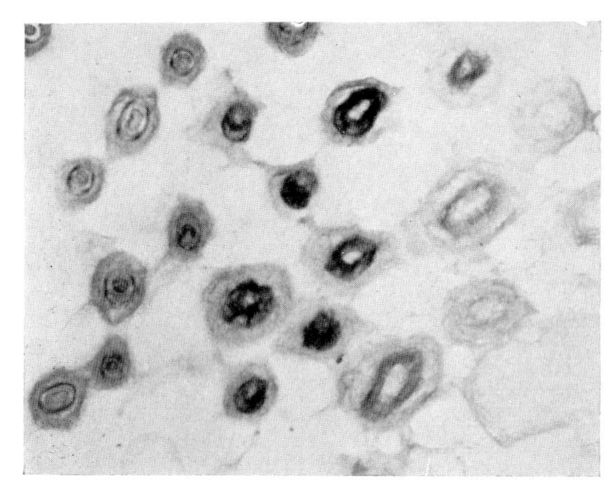

FIG. 56. Rat skin. Strong reaction in hair cortex. HNI—Fast blue B salt. × 210.

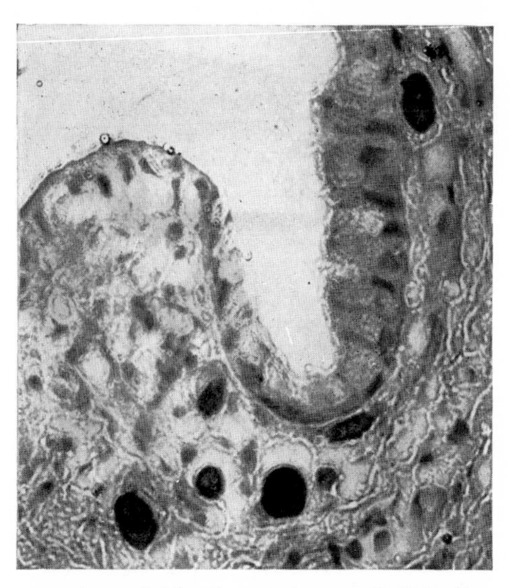

FIG. 57. Human cervix uteri. Simple protein crystalloid bodies containing large amounts of arginine. Sakaguchi reaction (Baker modification). × 400.

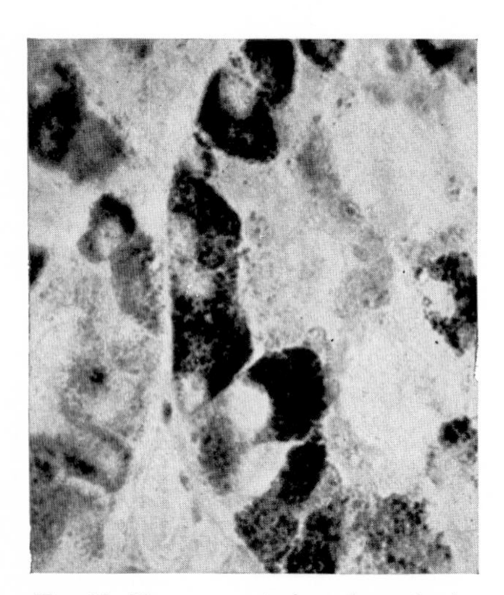

FIG. 58. Human anterior hypophysis. Specific staining of the β-granules after the application of moderate heat followed by 18 hours' benzoylation. Coupled tetrazonium reaction. × 550.

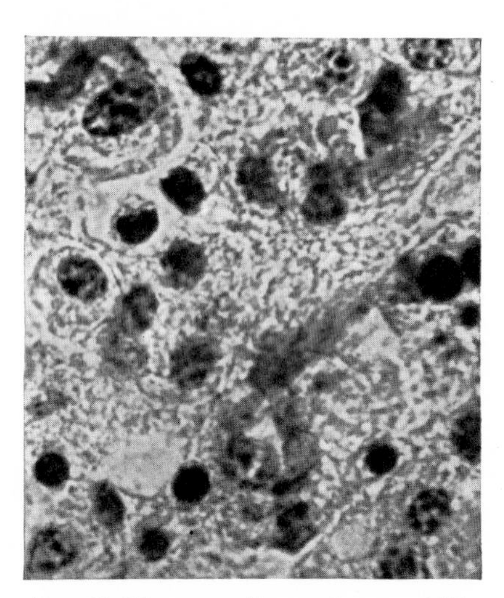

FIG. 59. Plasma cells, and at middle right-hand edge, a group of simple protein Russell bodies. Fourteen hours' benzoylation at 22°. Coupled tetrazonium reaction. × 700.

the original authors used Fast blue B. The results of the method are shown in Fig. 54, and also in Fig. 55, facing p. 156.

The theoretical specificity of the DDD reaction for SH has been confirmed by a rigorous series of control procedures. No colour is produced in the tissues after oxidation of SH groups with iodine (p. 171) or after blocking with specific reagents such as iodoacetate (p. 172) or N-ethyl maleimide (p. 172). Partial failure to block the reaction, often observed when mercurials such as phenyl mercuric nitrate are used, can be attributed entirely to the fact that such blocking is easily reversible. Furthermore, the DDD reaction can be prevented by interposing, before the final stage of colour development, a stage of treatment with excess sulphydryl (glutathione).

Bahr (1957) carried out a series of model experiments to test the specificity of the DDD reaction in which he used the stable diazotate of 4-amino-3,6-dimethyl-4-nitroazobenzene (Fast black salt K) in place of Fast blue B. An alternative which can be used with satisfactory results for the same purpose is the diazotate of 4-benzoylamino-2,5-dimethoxyaniline (Fast Blue RR). By means of a back titration with DDD and ovalbumin Bahr showed that there was a linear relationship between the amount of colour developed in the reaction and the number of protein-bound SH groups available. He considered that the advantage of using Fast black salt K was not confined to the greater amount of colour which developed, but also the absence of a two-coloured result such as occurs when a bis-diazonium salt such as Fast blue B is used.

A discordant note was introduced by the work of Gabler and Scheuner (1966), however. These authors showed, by esterification procedures, that DDD reacts also with carboxyl groups in the tissues.

According to the original authors of the method a blue colour (di-coupling) signifies a high concentration of tissue SH groups, whereas a red or pink colour (mono-coupling) indicates a sparse distribution of reactive groups. Intermediate colours are commonly observed. Only the di-coupling reaction is shown in the final equation for the DDD method (p. 154). A pink blush in connective tissues, which are supposedly sulphydryl free, has been observed by many workers. Such non-specific colouring may result from two causes: (1) a diazonium reaction with histidine and tyrosine residues; and (2) inefficient extraction of unused DDD and of the free reaction product 6-thio-2-naphthol. The second of these is undoubtedly the most important and it is probably responsible for the staining of elastic tissue by the DDD method. Elastic tissue has an affinity for phenols and naphthols which has not been satisfactorily explained but which is not easily reversible by simple extraction with solvents.

Reference has also been made to the lack of contrast obtained with the DDD method. This is largely inevitable with a method of high sensitivity capable of revealing small concentrations of SH in the tissues. By comparison a method of low sensitivity such as Adams' performic acid-Alcian blue (p. 240) produces a sharply contrasting picture which is satisfying to the æsthetic sense of the histologist. The histochemist brought up in this discipline must unlearn some of his acquired discriminatory sense.

I consider that, of all the methods for SS and SH so far discussed in this chapter, the DDD method must form the base line for any investigations in applied histochemistry. I employ it in parallel with chromogenic mercaptide-forming methods, using fresh frozen or freeze-dried sections in this case in spite of the theoretical difficulty of comparing frozen with paraffin sections. The application of SS and SH methods to practical problems is further considered in Chapter 8, p. 238.

Hydrogen Selenide Method for Disulphides

A method for demonstrating disulphide bonds under histochemical conditions was described by Olszewska *et al.* (1967). This was based on a reaction for SS groups in paper chromatograms (Wroński, 1966):

$$RSSR + H_2Se \rightarrow 2RSH + Se$$

The red deposit of selenium is insoluble in water and almost insoluble in ethanol. Since there is no reaction with SH groups the latter can be ignored. Details of the method, including the preparation of H_2Se, are given in Appendix 6. The results achieved are very satisfactory and the use of H_2Se as a vapour "Fixative" for freeze-dried blocks (Chapter 5, p. 86) must have serious consideration.

The α-Napthyl Maleimide and other Maleimide Reactions for SH

Seligman, Tsou and Barrnett (1954) introduced a further series of naphtholic reagents for the demonstration of protein-bound SH groups. Following the use of N-substituted maleimides as specific blocking reactions for SH these authors tested a number of hydroxynaphthyl-*N*-maleimides as histochemical

RS
RSH+O= ... =N– ... –OH ⟶ O= ... =N– ... –OH ⟶
I   II

RS
O= ... =N– ... –OH + CH_3O ... N_2– ... $O.CH_3$ ... N_2 ⟶ [RS ... O ... N ... OCH_3 ... N_2 ... OH]

reagents and found the 1-4 derivative *N*-(4-hydroxy-1-naphthyl) isomaleimide (HNI) to be the best. Trichloroacetic alcohol-fixed sections were reacted with the maleimide at pH 7·5 for 30 minutes, and after extraction of excess reagent the final colour was developed with Fast blue B.

Later the same authors (Barrnett, Tsou and Seligman, 1955) described the synthesis and application of other naphtholic reagents for SH, including 4-hydroxy-1-naphthyl mercuric acetate and 4-(*n*-iodoacetyl)-amino-1-naphthol. Reaction with these two compounds was stated to be slower than with DDD or HNI. Price and Campbell (1957) showed that the sensitivity of HNI for thiols was high and that the reaction proceeded rapidly *in vitro*. In paper chromatograms as little as 0·3 μm-mole could be detected. These authors also found that HNI made by the original method of Tsou, Barrnett and Seligman (1955) contained at least two components and that in some cases it was necessary to separate them chromatographically and to use only the faster component. This does not appear to be necessary in histochemical practice.

The reaction is particularly dependent on pH and should be performed at pH 8·0 or higher, in contradistinction to the blocking reaction with N-ethyl maleimide (NEM), which can be carried out best under acid conditions. A control section treated only with Fast blue B salt is necessary in the case of any tissues containing phenolic substances. This refers particularly to plant tissues.

Of the three reagents mentioned above I have used only HNI. This reacts readily with tissue SH groups and the ultimate picture closely resembles that produced by the DDD method. Apparent differences constantly noted with certain tissue components are presumably due to stereochemical factors. The application of the method to rat skin is shown in Fig. 56, facing p. 156. Price and Campbell also found that the final reaction product (III) could be demonstrated on paper chromatograms by means of its intense red fluorescence (366 nm). This observation can form the basis of a histochemical U.V. method since, fortunately, the reaction product in the tissues gives the same red fluorescence. The resulting increase in sensitivity may make this method the method of choice when tissues containing few reactive SH groups are being investigated.

A new coloured sulphydryl reagent was introduced by Witter and Tuppy (1960). This compound, *N*-(4-dimethylamino-3,5-dinitrophenyl)-maleimide, has been used successfully in protein analysis. It has not apparently been used in cytochemical practice.

Specificity of Maleimide Reactions. Certain objections to the specificity of

R–(imidazole, N, NH) + (N-ethylmaleimide, O, N–Et, O) ⟶ R–(imidazole, N, N–CO.CH=CHCONHEt) I

⟶ OOC.CH=CHCONHEt

II

the NEM reaction have been put forward by Nagamatsu and Fruton (1960). At pH 7 it is agreed that thiols are mainly concerned but reaction with imidazoles is also possible. The latter give rise to acylimidazoles (I) which are subsequently hydrolysed to *N*-ethylmaleimic acid (II).

The latter, according to Gregory (1955) is also produced by the spontaneous hydrolysis of NEM at alkaline pH levels. In acid solution NEM is stable but the reaction with thiols fails to go to completion. This process, if it takes place with HNI or similar chromogenic reagents, will lead to the liberation of naphthylmaleamic acids. If these are absorbed they will give rise to false positive reactions.

A fluorescence method for thiols was suggested by Stoward (1963) which involved reaction of the tissues with NEM in phosphate buffer at pH 7·4, and subsequent treatment with salicyloyl hydrazide followed by zinc acetate.

Details of the HNI reaction, incorporating modifications introduced by Price and Campbell, are given in Appendix 6, p. 626. Details of the fluorescent NEM method are also given.

Reactions for Arginine

The colour reaction for arginine originated by Sakaguchi (1925) is highly specific but the development and preservation of the colour are technically difficult. The reaction was first modified for use in histochemistry by Baker (1944), later by Serra (1944) and, independently, by Thomas (1946). According to the second author, it was to be regarded as specific for guanidine derivatives in which a hydrogen atom of one amino group was substituted by an alkyl or acyl radical. Thomas, however, stated that guanidine derivatives in which a hydrogen atom of one or both amino groups was substituted would react but that those in which both hydrogen atoms of a single amino group were substituted did not do so. Baker (1947) determined that a positive Sakaguchi reaction occurred with compounds having the general structure

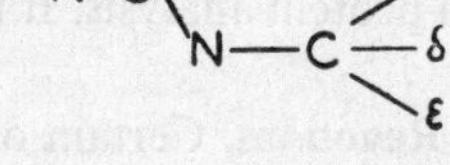

$$\alpha\text{-N=C}\begin{cases}\text{N—}\beta \\ \text{N—C}(\gamma)(\delta)(\varepsilon)\end{cases}$$

where α and $\beta = $ H or CH_3. Such substances include the naturally occurring amino-acids galegine, agmatine and arginine. In practice, a positive reaction is not given by any amino-acids occurring in human tissues other than arginine, nor is it given by guanidine, creatinine or urea.

In the Sakaguchi reaction a red colour is produced when arginine reacts with α-naphthol and hypochlorite or hypobromite in alkaline solution. Hypobromite is more specific than hypochlorite according to MacPherson (1942), who stated that excess hypobromite quickly destroyed the colour already produced. In performing the reaction it is usual to add urea with the α-naphthol to take up the excess hypobromite when this is subsequently added. As pointed out by MacPherson, this manœuvre was only partly effective because although the formation of the coloured compound is instantaneous its destruction by hypobromite begins immediately. Unless the proportion of urea and hypobromite is identical, and unless mixing is efficient, the best results will not be obtained. The original reaction of Thomas, which employed hypochlorite, was performed at room temperature instead of at 0°–4° as in Serra's and the original Sakaguchi method, and this is undoubtedly an advantage where no cold room is available. If one follows MacPherson's view, Serra's is the method of choice because it uses the more selective reagent (hypobromite) and because it avoids the long period in α-naphthol-hypochlorite which characterizes the original Thomas modification. Baker's (1947) modification employs hypochlorite which he prefers to hypobromite for three reasons. First, the validity of the reaction is based on Sakaguchi's work in which he used hypochlorite; secondly, the latter is more easily available; and thirdly, rapid development of colour is not necessary in histochemistry. More recently Thomas (1950) has introduced further refinements into the method which include the provision of tetraethylammonium hydroxide as a stabilizer for the colour developed.

With Serra's method the maximum colour is satisfactory, but this is developed while the section is still immersed in the staining solution after addition of the final portion of hypobromite. Usually, however, before the section can be removed and mounted for inspection considerable loss has occurred. Serra's "stabilization" of the colour, by repeated changes of glycerol and mounting in this medium, takes place too late to prevent this earlier loss of colour, due to excessive action of hypobromite. As MacPherson observes, there is an optimum amount of hypobromite (or hypochlorite), and it is the difficulty of finding this optimum which provides the main problem in performing the arginine reaction on tissue sections.

I have used Serra's method, the original and the newer Thomas' methods, and Baker's method. Of these, the simplest and most convenient is the last. It gives excellent and fairly stable results on tissues fixed in a variety of formalin-containing or alcoholic fluids, and also on freeze-dried material. The application of Baker's method is illustrated in Fig. 57, facing p. 157. Thomas' (1950) method was a great improvement on his original version, but complicated to use. It offers a more permanent preparation than Baker's pyridine-chloroform. Warren and McManus (1951) described the use of 8-hydroxyquinoline (attributed to Sakaguchi) in place of α-naphthol in the histochemical method for arginine. They considered that the stability of the colour developed was greater than with the older methods, but Gilboe and Williams (1956) could find little difference between α-naphthol and oxine in a series of tests.

The oxine reagent of Sakaguchi was also used by Carver, Brown and Thomas (1953) as the basis of an arginine reaction permitting final mounting in a synthetic medium (Permount) with considerable stability of colour. This method is also given in detail in Appendix 6, p. 617. If permanence is regarded as unimportant I still recommend the use of Baker's variant of the α-naphthol-hypochlorite procedure as preferable to the others, either in its original form or using 2,4-dichloro-1-naphthol (McLeish, Bell, La Cour and Chayen, 1957) in place of α-naphthol. This compound gives a strong reddish-orange colour which is sufficiently stable, in an alkaline medium, to allow photometric measurements to be made.

In order to avoid tissue damage from the alkaline reagent Deitch (1961) substituted barium for potassium or sodium hydroxide. She also observed that the presence of an organic amine would prevent fading of the reaction product. For this purpose her sections were mounted in Shillaber's oil (refractive index 1·580) containing 10 per cent. tri-*N*-butylamine.

The nature of the coloured reaction product of the Sakaguchi reaction was investigated by Bhattacharya (1959, 1960) who considered it to be a quinone-imine produced by oxidative coupling between guanidino and phenol, in the manner of an indophenol reaction.

Serra (1946) recommended the arginine reaction as a means of distinguishing basic proteins in the tissues, a distinction which can be achieved by simpler and conventional methods of staining. Nevertheless, in spite of practical drawbacks, when applied to tissue proteins demonstrated by other methods as basic, it affords a means of distinction between arginine-rich histones and arginine-poor globins. This distinction can also be made histochemically by the use of specific methods for tryptophan, since this amino-acid is present in traces, if at all, in histones.

Blocking and Conversion Reactions for Protein End-Groups

A large number of such reactions have been employed or suggested for use in protein histochemistry. Many of them are not specific for a single group but react in varying degrees with others. The reactions are considered below in groups corresponding to those in which the chromogenic end-group reactions have already been considered. In Appendix 6 the methods of application are listed without reference to their specific use but approximately in the order in which they are considered here.

Reactions for Amines (Schiff's base formation, deamination, oxidative deamination, dinitrophenylation, acetylation, benzoylation, nitrobenzoylation, phosphorylation, aryl sulphonation, substituted urea and thiourea formation).

Schiff's Base Formation. The condensation of an aldehyde with a primary amine or amino compound takes place readily, with the formation of a weakly basic compound called variously a Schiff's base, an anil, or an azomethine.

$$R.NH_2 + OCH.C_6H_5 \rightarrow R.N{:}CH.C_6H_5 + H_2O$$

The main example of Schiff's base formation used as a blocking reaction is formalin fixation, and it is thus absolutely necessary to avoid this procedure when total NH_2 groups are to be demonstrated. That the reaction with formaldehyde is incomplete is shown by the positive reaction for NH_2 obtainable by most of the specific chromogenic methods after formalin fixation.

Deamination. Nitrous acid reacts with all types of amino groups in the tissue proteins but most rapidly with the α-amino groups. According to Springall (1954), if one uses the Van Slyke (1929) reagent, the α-amino nitrogen is evolved in 4 minutes, the ε-amino nitrogen from lysine in 30 minutes, and the guanidino-nitrogen of arginine in 6 hours. In histochemical practice nitrous acid mixtures have been used for 1–12 hours at room temperature and, according to Lillie (1954), the acetic acid mixtures were more rapid in action than those based on HCl, in spite of their higher pH levels. Stoward (1963) suggested that such conditions are quite inadequate and he recommended 48 hours in the presence of a strong mineral acid. I agree with this view, which is based on the realization that HNO_2 cannot completely deaminate tissues. Amino groups are removed in two stages which may be called diazotization and deazotization. The latter takes place by reaction with water or alcohol:

$$RNH_3Cl + HNO_2 \rightarrow RN_2Cl + 2H_2O$$

and

$$RN_2Cl + R'CH_2OH \rightarrow RH + N_2 + R'CHO + HCl$$

The second stage of the reaction requires prolonged treatment with water or ethyl alcohol (Appendix 6, p. 613).

Oxidative Deamination. It is possible to carry out this reaction in a number of ways. It forms the first stage of the ninhydrin and alloxan-Schiff methods (p. 116) and is, in fact, seldom used purely as a blocking reaction.

$$R.CHNH_2.COOH + \tfrac{1}{2}\,O_2 \rightarrow R.CO.COOH + NH_3$$

Monné and Slauterback (1951) used both ninhydrin and chloramine-T for blocking by oxidative deamination, and Burstone (1955) showed that the latter would produce what he considered to be Schiff-reactive aldehydes in tissue proteins. Experiments carried out by Stoward (1963) indicated that Burstone's results were partly attributable to the thiosulphate rinse. He substituted washing in alkaline (ammoniacal) alcohol and stated that in his opinion hypochlorite could be used for rapid and complete deamination of tissue sections.

Oxidation. Kahr and Berther (1960) observed that primary amines could be oxidized by hydrogen peroxide in the presence of sodium tunsgtate. Stoward (1963) recommended this procedure as a rapid test for primary amines by examining for oxygen release.

$$R.CH_2.NH_2 \rightarrow R.CH_2NHOH \rightarrow RCH{=}N.OH$$

Dinitrophenylation. This reaction was introduced into protein chemistry by Sanger (1945), and it depends on the reaction of 2,4-dinitrofluorobenzene (DNFB), under mild alkaline conditions, with terminal α-amino groups and with the ε-amino groups of lysine.

$$R.NH_2 + F.C_6H_3(NO_2)_2 \rightarrow RHN.C_6H_3(NO_2)_2 + HF$$

Dinitrophenyl amino-acids are bright yellow, and the colour is adequate if examination of the stained tissue sections is carried out at 430 nm. In practice the method is not used for blocking amino groups, or for their demonstration following addition procedures which are described in the section below on tryptophan, tyrosine and histidine.

Acetylation, Benzoylation, Nitrobenzoylation. These are variants of the same essential process, the Schotten-Baumann reaction, by means of which acyl derivatives of primary and secondary amines and amino compounds are produced. These reactions are not used as blocking methods for NH_2 groups, but *p*-nitrobenzoyl chloride and dinitrobenzoyl chloride have been used by Burstone (1955) for demonstration of acylated groups in tissue proteins after procedures similar to those described below for DNFB.

Phosphorylation. This procedure was introduced into histochemical practice by Landing and Hall (1956a), who suggested that some of the phosphoryl groups introduced into tissue proteins were attributable to esterification of the amino groups. What proportion of the available NH_2 groups is affected by phosphorylation is unknown.

Acyl Sulphonation. This process has been used in protein chemistry for the identification of terminal amino-acid residues. Christensen (1945) used *p*-toluene sulphonyl (tosyl) chloride, and Udenfriend and Velick (1951) employed *p*-iodophenyl sulphonyl (pipsyl) chloride for the same purpose. Under mild alkaline conditions a sulphonamide is formed with free α-amino groups:

$$R.NH_2 + I.C_6H_4.SO_2Cl \rightarrow R.NH.SO_2C_6H_4I + HCl$$

Aryl sulphonation has not been used as a blocking reaction, but might well be used as a sensitive method for demonstrating protein-bound amino groups by substituting the fluorescent compound dimethyl-aminonaphthalene sulphonyl chloride (see Chapter 7, p. 190) for the above reagents.

Lillie (1964a) found that acetic anhydride, in pyridine or alcohol, or tosyl chloride in various solvents, were equally ineffective in preventing the reaction of anionic dyes with tissue amines. Blockade of tissue oxyphilia was effectively produced with an acetic anhydride, glacial acetic acid, sulphuric acid mixture (25:75:0·25) and the effect was reversible by methanolysis (see below).

Methylation. Terner and Clark (1960) reported that prior methanolysis of tissues (methanol/HCl, 60°, 24 hours) would protect primary and secondary amines from subsequent deamination by oxidizing agents (chloramine T). They assumed that the most likely mechanism was alkylation of the amino groups by the small amount of methyl chloride produced in the reaction mixture:

$$CH_3OH + HCl \rightarrow CH_3Cl + H_2O$$
$$RNH_2 + CH_3Cl \rightarrow RNH.CH_3 + HCl$$
$$RNH.CH_3 + CH_3Cl \rightarrow RN(CH_3)_2 + HCl$$

The tertiary amines formed would resist oxidative deamination while retaining their acidophilia.

Substituted Urea Formation. Primary and secondary amines and amino compounds react with aryl isocyanates at room temperature to form substituted ureas. Other compounds such as water, alcohols and phenols, which contain a mobile hydrogen, also react and must be excluded. The process has been used to give fluorescent carbamino-proteins for use in the Coons technique (Chapter 7, p. 181), but not for blocking.

REACTIONS FOR CARBOXYL GROUPS (METHYLATION, HYDROXYALKYLATION)

Methylation. This process can be carried out, in chemical practice, in a number of different ways. Compounds used for the purpose include dimethyl sulphate, methyl iodide, methyl bromide and diazomethane. Methyl halides in alkali are too reactive and destructive for histochemical purposes and it is therefore necessary to examine alternatives. Fraenkel-Conrat and Olcott (1945) described a method for methylating proteins using methanol-HCl mixtures, and this procedure was successfully adopted for histochemical use by Wigglesworth (1952), and later by Fisher and Lillie (1954). While the majority of methylating agents affect other groups, particularly NH_2 and tyrosine-OH, Fraenkel-Conrat and Olcott considered that the methanol method specifically blocked only carboxyl groups:

$$R.COOH + CH_3OH \overset{HCl}{\rightleftharpoons} R.COO.CH_3 + H_2O$$

This reaction is reversible *in vitro* by treatment with HCl, but in the tissues this result can only be achieved with strong oxidizing agents such as $KMnO_4$.

Fisher and Lillie used the method in a study of its effects on tissue metachromasia and basophilia. They found that mild methylation (37°) quickly reversed the metachromasia of most tissue components, and even the basophilia of the nucleic acids was not resistant. They concluded that SO_3H and PO_4 groups were also blocked by the process, which could not therefore be regarded histochemically as specific for COOH. Furthermore, after more drastic methylation (60°) with methanol-HCl the PAS reaction of mucins was considerably reduced. It appeared, therefore, that 1,2-glycol groups might also be affected. Subsequent evidence from a number of sources has indicated that reduction or abolition of the PAS reaction is often due simply to removal of the reacting groups by their extraction from the tissues.

Brenner and Huber (1953) and Bello (1956) claimed that thionyl chloride ($SOCl_2$) in methanol would specifically methylate carboxyl groups *in vitro*, leaving phenols and amines intact. The reagent was employed histochemically by Geyer (1962), as a 4 per cent. solution, at either 20° or 56°. According to Lewis and Boozer (1952) and Boozer and Lewis (1953) thionyl chloride reacts with methanol to form intermediate products (I, II) which rapidly methylate tissue anionic groups:

$$CH_3OH + SOCl_2 \longrightarrow CH_3O{-}SO{-}Cl + HCl$$

$$\underset{II}{CH_3Cl} \leftarrow \underset{I}{CH_3} + SO_2 \rightleftharpoons CH_3OSO + Cl \qquad CH_3 + OSO.Cl$$

In contradistinction to the *in vitro* results, in histochemical use this reagent (Appendix 6, p. 619) rapidly methylates primary amines in 30–60 minutes without affecting thiol groups at all. Tissue basophilia is abolished completely after 4–6 hours and that due to COOH and RNA disappears after 30 minutes. Sulphated mucopolysaccharides are no longer basophilic after 4 hours. Complete extraction of nucleic acid phosphates occurs in 24 hours but, up to 12 hours, extraction of mucopolysaccharides is negligible.

The reagent is a violent one and damage to tissue structure occurs, particularly in the longer treatments. It is difficult, moreover, to establish a precise time for completion of any of the stages of blocking.

Methyl iodide, as a 12·5 per cent. solution in methanol and with or without added sodium carbonate, was used by Terner (1964) as a histochemical alkylating agent. Complete abolition of tissue basophilia was achieved in 18 hours. Longer treatment resulted in the extraction of various protein and mucoprotein constituents of the tissues. Table 12, below, shows the results obtained.

TABLE 12

Alkylation of Tissue End Groups with Methyl Iodide
(after Terner, 1964)

Groups	Alkylation	Time (hours) at 45°
Carboxyl	+	6–18
Phosphate	+	18–24
O-Sulphate	0	24
N-Sulphate	0	24
Amino	+	4
Guanidino	0	24
Tyrosine	+	6
Sulphydryl	+	1
Vic-Glycol	0	24
Tryptophan	0	24

Details of the use of methyl iodide in methanol are given in Appendix 6, p. 619.

Investigating the mechanism of silver staining Peters (1955) applied diazomethane ($N_2{=}CH_2$) in ethereal solution to sections previously "hydrated" in 75 per cent. ethanol. He claimed that selective methylation of carboxyl groups was obtained and that this was responsible for blockage of silver staining. From *in vitro* experiments there is considerable evidence that other protein groups are involved and Kuhn and Baer (1953) showed that in the presence of methanol diazomethane would methylate the hydroxyl groups of carbohydrate. Under histochemical conditions (Appendix 6, p. 619), although the basophilia of proteins and nucleic acids is abolished after 10 hours treatment, the PAS reaction is unaffected.

Of the four main methylating reagents used in histochemical practice the most popular has been methanol/HCl but as an alternative methyl iodide merits serious consideration. The preparation of diazomethane is inconvenient and each batch of reagent lasts only 24 hours. It is therefore unlikely to become universally popular.

Reactions for Tryptophan, Tyrosine and Histidine (Oxidation, Acylation, Benzoylation, Dinitrophenylation). Danielli (1947, 1950) suggested a series of reactions to be used with the coupled tetrazonium reaction in order to distinguish each of the above-mentioned amino-acids. Of these reactions benzoylation and dinitrophenylation have already been mentioned as possible blocking reagents for amines, though they are not much used for this purpose in histochemistry. A third reagent, performic acid, is used here as a specific oxidizing agent for tryptophan.

Differential reagents of the types suggested by Danielli can be used in three ways:

(1) To prevent, by chemical combination, the subsequent staining of a given group. The presence of this group is recognized by deficiency in staining by comparison with unblocked sections.

(2) To convert or destroy the selected group. The use of the word "blocking" is not strictly permissible in this case and the reaction, unlike many of those in the first category, is always irreversible.

(3) To combine with a given group in the tissues, leaving part of its own molecule reactive. The latter is then made visible by some additional process.

The use of DNFB as a blocking reaction falls into the first of the above categories. As can be seen from Table 13, the reagent has a low specificity but at pH 5·5 and below it probably reacts only with SH and phenols and not with protonated amines (Zahn and Traumann, 1954). According to Zuber *et al.* (1955), at 2° only thiols react with DNFB.

TABLE 13

Reactivity of Protein End Groups after Blocking Procedures

End Group	DNFB pH5	DNFB pH8	Performic acid	HNBB*	Benzoylation
Tyrosine	−	−	+	+	−
Tryptophan	+	+	−	−	−
Histidine	+	−	+	+	+
Lysine	+	−	+	+	−
Cysteine	−	−	+	−	−
Arginine	+	+	+	+	−

* Hydroxy-nitro-benzyl bromide

At higher pH levels (pH8 and above), besides SH, NH_2 and phenols, at least three other groups react:

(1) Aliphatic OH (Freudenberg, 1956)
(2) Heterocyclic N (Krönhnke and Leister, 1958)
(3) Imidazoles with broken ring (Zahn and Pfanmüller, 1958).

Performic Acid. The principle of the second type of differential method can be illustrated by the use of selective oxidation with performic acid. It was shown by Toennies (Toennies and Homiller, 1942; Toennies, 1942; Bennett and Toennies, 1942), that, of the commonly occurring amino-acids only cystine, methionine and tryptophan are oxidized by performic acid. After treatment with this acid, therefore, the indole ring of tryptophan fails to react by the coupled tetrazonium method while tyrosine and histidine continue to do so. The presence of tryptophan is inferred from the reduction in colour caused by pre-treatment with performic acid. This compound is produced by mixing hydrogen peroxide and formic acid in suitable proportions (see Appendix 6) and it is used at room temperature.

Persulphate. It has been shown by Boyland, Sims and Williams (1956) that alkaline persulphate can break the pyrrole ring of tryptophan at the 2,3-bond to give a number of acid-labile products:

C.CH₂CH COOH (indole, NH, CH, NH₂) → COCH₂.CH COOH (NH₂) → COOH, NH₂

Using a 2 per cent. solution of $K_2S_2O_8$ in 0·5 N-KOH, I found that 16 hours' incubation at 22° was sufficient to reverse the various specific methods for tryptophan discussed earlier in this chapter. The nature of the reaction product in the tissues has not been elucidated.

Hydroxy-nitrobenzyl bromide. This reagent, 2-hydroxy-5-nitrobenzyl bromide (HNBB) was first described by Koshland, Karkhanis and Latham (1964).

OH, CH_2Br, NO_2

It reacts with tryptophan residues in proteins, and to a much lesser extent with SH groups. According to Horton and Koshland (1965) the *p*-nitrophenyl group confers an absorption spectrum which is sensitive to environment and which absorbs in a region of the spectrum where the protein does not. In regions containing high-tryptophan proteins, such as chymotrypsin, the HNBB-protein can be distinguished by examination at 400 nm.

A chromogenic histochemical reaction can be carried out with Koshland's reagent. After blocking SH groups (see p. 172) and reacting with HNBB, reduction, diazotization and coupling with H-acid follow, as indicated in the DNFB schedule (Appendix 6, p. 614).

Dinitrofluorobenzene. The use of the third type of differential method is illustrated by the reaction between the aromatic hydroxyl group of tyrosine and dinitrofluorobenzene (DNFB). The compound formed by the reaction of these two substances is colourless. However, if the 4-nitro group of the attached DNFB is reduced to NH_2, then diazotized with nitrous acid at 0° and finally coupled in cold alkaline solution with a suitable phenol or aromatic amine, the location of the tyrosine-DNFB compound becomes strongly coloured. Danielli used 1-amino-8-naphthol-3:6-disulphonic acid ("H-acid") as his final coupling agent and I also found it suitable. This reagent gives a rich reddish-purple colour which is insoluble in alcohol or xylene; sections stained with it can, therefore, be dehydrated and mounted in the ordinary manner. The chain of reactions involved in this second type of blocking is thought to proceed as follows:

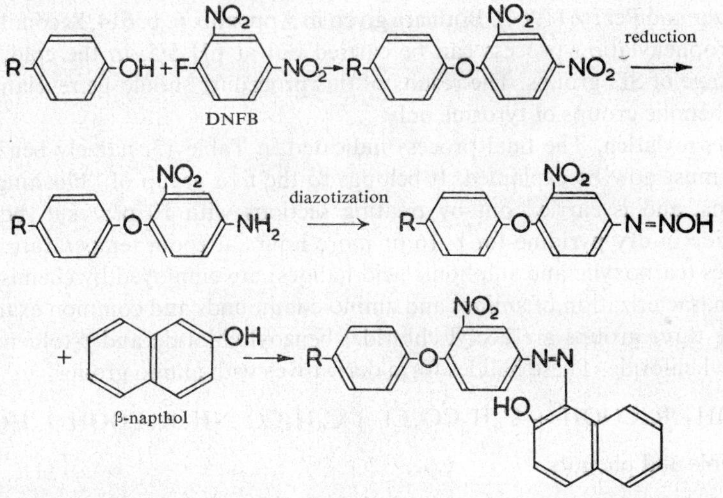

R-C₆H₄-OH + F-C₆H₃(NO₂)-NO₂ (DNFB) → R-C₆H₄-O-C₆H₃(NO₂)-NO₂ —reduction→

R-C₆H₄-O-C₆H₃(NO₂)-NH₂ —diazotization→ R-C₆H₄-O-C₆H₃(NO₂)-N=NOH

+ β-napthol (OH) → R-C₆H₄-O-C₆H₃(NO₂)-N=N-(HO-naphthyl)

Unfortunately, as indicated above, the use of this method for demonstrating tyrosine is complicated by the fact that DNFB reacts with other groups in the tissues such as SH groups, the free α-amino groups of proteins and, to some extent, with the ε-amino groups of lysine (Sanger, 1945, 1950). According to Porter (1950) the imidiazole group of histidine also reacts to give a colourless product with DNFB. Danielli proposed to overcome these problems by using specific agents to block SH and NH_2 groups before treatment with DNFB. For such purposes he suggested naphthoquinone, which blocks both groups, or nitrous acid followed by iodoacetamide. The participation of non-protein components, such as the free NH_2 groups of ethanolamine or serine-containing phospholipids was noted by Maddy (1961a) who reviewed the whole subject of DNFB as a cytochemical reagent in his comprehensive paper.

The original method, in my hands, was quite unsatisfactory when performed upon conventionally fixed tissue sections, although a colour was developed with freeze-dried tissues. Burstone (1955), however, showed that this failure was due to the use of weak reducing agents. The substitution of sodium hydrosulphite for stannous chloride resulted in the strong staining of a number of tissue components such as collagen and elastic tissue. Tranzer and Pearse (1964) found that even with this reagent, as with its forerunners, reduction of nitro groups was incomplete. With titanous chloride ($TiCl_3$), however, complete reduction was obtained in a short period at room temperature. Burstone suggested that the DNFB-H-acid sequence was relatively specific for hydroxyl and primary NH_2 groups and that it could reasonably be employed as such.

Probably there are two main indications for using the DNFB method. First, as a general protein reaction. In this case it is performed at pH 8 and 25°, as indicated by Zerlotti and Engel (1962), with examination of the yellow-stained sections at 410 nm, or by the complete sequence as described by Tranzer and Pearse (1964). Both are given in Appendix 6, p. 614. Secondly, the dinitrophenylation process can be carried out at pH 5·5, in the cold, after blockage of SH groups. The results of this procedure should be revelation of the phenolic groups of tyrosine only.

Benzoylation. The final process indicated in Table 13, namely benzoylation, must now be explained. It belongs to the first group of "blocking" reactions, and is carried out by treating sections with 10 per cent. benzoyl chloride in dry pyridine for 8–16 or more hours at room temperature. Acyl halides (carboxylic and sulphonic acid halides) are employed by chemists for the characterization of amines and amino compounds and common examples in the three groups are acetyl chloride, benzoyl chloride and *p*-toluene sulphonyl chloride. These halides form derivatives with amino groups,

$$NH_2.R.COOH + C_6H_5CO.Cl \rightarrow C_6H_5CO.NH.R.COOH + HCl$$

alcohols and phenols,

$$R.OH + C_6H_5CO.Cl \rightarrow C_6H_5CO.OR + HCl$$

and they therefore block the reaction of the last group with diazonium and tetrazonium salts. The chemist arranges for the hydrochloric acid evolved to be taken up by performing these reactions in strongly alkaline solutions and this method was used by Mitchell (1942) in an earlier histochemical adaptation. Danielli (1947), however, used a 10 per cent. solution of benzoyl chloride in dry pyridine which acted as vehicle and catalyst, absorbing the hydrochloric acid produced by the reaction but disturbing the tissues far less than the strong alkali of the earlier method. Acetyl and *p*-toluene sulphonyl chlorides require high temperatures, if rapid results are required, and the reactions must therefore be conducted in pyridine at 100° under a reflux condenser. In practice, acetic anhydride is usually used in place of acetyl chloride.

$$NH_2.R.COOH + (CH_3CO)_2O \rightarrow CH_3CO.NH.R.COOH + CH_3COOH$$

Tissues firmly attached to slides stand up to such treatment remarkably well but benzoyl chloride in pyridine, since it can be used at room temperature, is clearly the reagent of choice.

As can be seen from Table 13, treatment with benzoyl chloride should prevent the subsequent reaction of lysine, cysteine, arginine, tyrosine, and tryptophan by the coupled tetrazonium method. Danielli (1947) suggested that the last traces of histidine might be difficult to eliminate by benzoylation and later (see Chapter 9, p. 253), he showed that in tissue sections histidine was in fact resistant although it is easily benzoylated *in vitro*. According to Mitchell (1942) and Danielli (1947, 1950), the positive tetrazonium reaction of benzoylated tissues was due to purine and pyrimidine-containing compounds.

Using samples of DNA and DNA-protein, I found (1949) that the purines and pyrimidines of DNA would not react *in vitro* by the tetrazonium reaction (see Chapter 9, p. 252). A survey of the histochemical effects of the reaction, in various tissues, was therefore undertaken in the hope that it might prove possible to attach to it a definite chemical or physical significance.

When paraffin sections of tissues fixed in formalin-containing or alcoholic fixatives, or freeze-dried tissues fixed in absolute alcohol, were used, a large number of structures were found to react by the coupled tetrazonium method after as much as 12–18 hours' benzoylation. As expected, nuclear chromatin, nucleoli and some cytoplasmic ribonucleoprotein continued to react, but, in addition, collagen and reticulin fibres invariably stained as strongly as in the control sections not exposed to benzoylation. This effect was found to be partly due to the action of even mild heat, such as that involved in drying the sections after mounting on slides in the usual manner (see Chapter 8, p. 220). Other structures which continued to stain only after being subjected to mild degrees of heating were fibrin and fibrinoid deposits, the β granules of the anterior pituitary gland (Fig. 58), intracellular gastric, intestinal (Plate IVd), and tracheal mucin, the mucoid granules of the salivary glands, Paneth cell granules, mast-cell granules, Russell bodies (Fig. 59), the ground substance of cartilage and the intercellular substance of arterial walls. If benzoylation was continued for a period longer than 16 hours all the above structures and even the nucleoproteins gradually ceased to react. Alternative treatment with acetic anhydride (10 per cent. in pyridine at 100° for 8 hours) rendered all these structures negative with the exception of nuclei, nucleoli, collagen, reticulin and some cytoplasmic ribonucleoprotein. Even brief acetylation sufficed to render the pituitary β granules negative. Lillie (1964b) noted that of all the acyl procedures which he tested for effectiveness in blocking azo coupling with proteins, acetic anhydride in absolute ethanol was the most effective.

Stoward (1963) recommended using acetic anhydride in the presence of perchloric acid. According to Burton and Prail (1950) perchloric acid ruptures acetic anhydride to give the acetylium radical:

$$H^{+}(ClO_4^{-}) + (CH_3CO)_2O \rightleftharpoons CH_3CO^{+} + CH_3COOH$$

and

$$CH_3CO^{+} + R.NH_2 \rightarrow RNHCOCH_3 + H^{+}$$

Incubation with ribonuclease for only a short period prevented the subsequent reaction of the nucleoli and cytoplasmic ribonucleoprotein; but, on the other hand, after performing a Schneider extraction with hot trichloroacetic acid (see Chapter 9, p. 279, and Appendix 9) so that all Feulgen-positive material was removed from the nuclei, it was still possible to demonstrate the nuclear chromatin by means of the tetrazonium reaction both before and after benzoylation. It may be significant that the mucin of the brush border of the duodenum and free mucin in the stomach, which have been acted upon by enzymes, are not stained by the coupled tetrazonium reaction after benzoylation, although they are still brilliantly stained by the periodic acid-Schiff reaction for mucosubstances (Chapter 10, p. 307). Plates IVc and IVd illustrate this particular point.

Originally (1953, 1960) I considered that the reactive groups which resisted benzoylation did so because of "physical causes, inherent in the spatial arrangement of the polypeptide chains of the proteins containing them, making them susceptible to some kind of rearrangement through mild heating." The experiments of Barnard and Danielli (1956), reviewed by Barnard (1961), pointed strongly to the conclusion that *protective bonds* adjacent to histidine groups were responsible for preventing the latter from reacting with benzoyl chloride, thus allowing them to react subsequently with diazonium salts.

The sensitivity of the protective bonds to hydration has been emphasized by Barnard (1961). Exposure of sections to water after the primary dehydration (by whatever method this is achieved) destroys the protective effect. Barnard presumed that after mild desiccation the structural alteration necessary to form protective bonds reached a stable limit. My own work suggests that heating introduces another factor in producing structural alteration and that additional stable protective bonds can be produced in this way.

The coupled tetrazonium reaction after benzoylation is further considered in Chapters 8 and 9. I believe that it is a most valuable and significant reaction, particularly in indicating the state of nucleoproteins in the cytoplasm and in the nuclei.

Nitrobenzoylation. Burstone (1955) adapted the Schotten-Baumann blocking procedure by substituting *p*-nitrobenzoyl chloride for the unsubstituted acid chloride and using the Danielli sequence (reduction, diazotization and coupling) to render visible the sites of reaction in the tissues. An intense reddish-purple colour was produced in collagen and elastic tissue and a less strong colour in mucins and in the nuclei. No blockage was caused by deamination but, as might be expected, benzoylation reversed the reaction entirely. Erythrocytes do not stain by this method and the results, in general, are difficult to interpret. The reaction is worthy of further study, however, and an elucidation of the mechanism by which gastric parietal and chief cells are distinguished in the dog and guinea-pig, but not in other species, might well be illuminating.

Iodination. Richards and Speakman (1953) investigated the iodination of tyrosine side chains in keratin under various conditions. They found that with 0·1 N-I_2 in aqueous solutions only 75 per cent. of the available reactive groups were iodinated (64 hours, 22°, pH 9). When, on the other hand, they used 0·78 N-I_2 in ethanol for 72 hours at 25° they were able to produce 100 per cent. iodination. Histidine was affected to a negligible degree and tryptophan and proline not at all. Gemmill (1956) studied the iodination of phenolic compounds related to thyroxine, using aqueous solutions of iodine in KI at pH 12·05 (0·05 N-KOH). By varying the concentration of iodine Gemmill found that only at higher concentrations did di-iodination take place.

Landing and Hall (1956c) made use of the fact that iodination (0·3 per cent. I_2 in 0·6 per cent. aqueous KI) at pH 10 for 24 hours will block the subsequent reaction of tyrosine with Millon's reagent. They considered that tryptophan would be blocked by this procedure and that any positive coupled tetrazonium reaction obtained after iodination would be due to histidine. Unfortunately for the specificity of this reaction there is strong evidence that complete iodination of tryptophan does not take place and, in fact, it is possible to obtain a positive Adams' reaction for tryptophan after iodination. Even if this were not so the reaction of other amino acids (see above) would have to be taken into account. According to Lillie (1957b) histidine is also iodinated, but he noted that, according to Ramachandra (1956), the reaction was much slower than with tyrosine. According to Bachmann and Seitz (1961) the specificity of the Landing technique for histidine is increased by following the usual iodination by reduction with potassium metabisulphite. These authors used Fast black K in place of tetrazotized *o*-dianisidine.

Bromination. The use of *N*-haloamides for bromination is known as the Wohl-Ziegler reaction and *N*-bromosuccinimide $(CH_2CO)_2NBr$, is the reagent usually employed. Using a concentration of 0·02 per cent. in 50 per cent. ethanol at pH 4 Glenner and Lillie (1959) produced complete blockage of the DC reaction for tyrosine. Corey and Haefele (1959) showed that proteins containing tyrosine and tryptophan were split by bromination with *N*-bromosuccinimide:

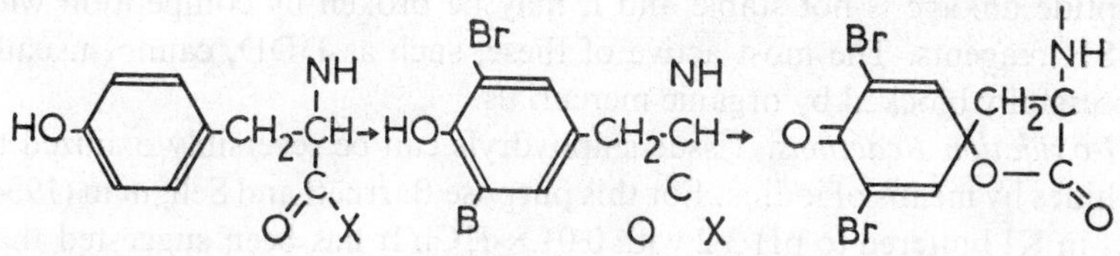

The mechanism, as can be seen, involves cyclization of the molecule after splitting. Bromination must be regarded as an important method for blocking tyrosine groups of proteins provided that it is not required that tryptophan remains reactive.

REACTIONS FOR SH AND SS (Carboxyalkylation, Maleimide block, Mercaptide block, SH-oxidation, SS-reduction)

Carboxyalkylation. Rapkine (1933) described the formation of alkyl carboxy derivatives of thiols by treatment with halogenated fatty acids at alkaline pH levels:

$$Pr—SH + I—CH_2COO \rightarrow Pr—S—CH_2COO + HI$$

The reaction has been observed to occur also with iodoacetamide, iodo-ethyl alcohol and with brominated fatty acids. Iodoacetamide is more reactive than iodoacetate, though the latter is more frequently used in protein and enzyme histochemistry. All existing histochemical SH reactions are blocked irreversibly by prior treatment with iodoacetate.

Maleimide Block. N-ethyl maleimide reacts with tissue SH groups as shown in the equation for α-naphthyl isomaleimide on p. 156. Sheets, Hamilton, De Gowin and King (1956) drew attention to the failure of *N*-ethyl maleimide to react with SH groups in intact erythrocytes. They concluded from their experimental data, that inability to penetrate the red cell envelope was responsible for this. Apart from this circumstance tissue SH groups are entirely prevented from reacting in histochemical tests by prior treatment with *N*-ethyl maleimide. Solutions of this compound (pH 7·4) lose their activity after 48 hours (Gregory, 1955) due to decomposition to a compound of the type OOC.CH=CHCONHEt. At lower pH levels NEM is more stable but blockage of SH is then incomplete.

Mercaptide Block. A large number of compounds containing the mercuric ion have been employed as blocking reagents for SH groups. Among these are $HgCl_2$, phenyl mercuric acetate, methyl mercury iodide, *p*-chloromercuribenzoic acid and *p*-chloromercuriphenyl sulphonic acid. The reaction on which SH blocking by organic mercurials is based is given on p. 141.

In all cases, as pointed out by Horowitz and Klotz (1956), the reaction for SH with mercury compounds is slow, many hours being required, at pH 9·0 and 25°, for equilibration to occur. Apart from this drawback, which makes it difficult to be certain that all reactive SH groups have been blocked, the mercaptide linkage is not stable and it may be broken by competition with other SH reagents. The most active of these, such as DDD, cannot usually be successfully blocked by organic mercurials.

SH-oxidation Reactions. Tissue sulphydryls can be reversibly oxidized to disulphides by means of iodine. For this purpose Barrnett and Seligman (1954) used I_2 in KI buffered to pH 3·2 with 0·01 N-HCl. It has been suggested that at this pH level only SH groups are oxidized. Bennett (1951) used 0·001 M-I_2 in *n*-propanol, which was just as effective though probably less specific in action.

Subsequent reduction with any of the reagents described below restores the original SH groups.

As an alternative to I_2 Bennett used a mixture of H_2O_2 and $FeCl_3$, in water or in *n*-propanol, but this has no advantages over the two iodine solutions mentioned above. Hird and Yates (1961) compared the oxidation rates of protein thiols with iodate, bromate and persulphate. They showed that the first of these is the most rapid. Two extremes of oxidation were postulated.

(1) $$6R.SH + IO_3^{\ominus} \rightarrow 3R.S{-}S.R + I^{\ominus} + 3H_2O$$

(2) $$R.SH + IO_3^{\ominus} \rightarrow R.SO_3H + I^{\ominus}$$

At high thiol concentrations the only product was disulphide but clearly, under histochemical conditions, reaction 2 is possible.

SS-reduction Reactions. A large number of reagents has been employed for reducing SS to SH groups. The strongest of these, namely zinc-HCl and sodium amalgam, are not used in histochemistry. They bring about the reaction

$$R.S{-}S.R \rightarrow R.SH + HS.R$$

That is to say, one cystine molecule gives rise to two cysteine molecules. Cyanides have been employed both by chemists and histochemists for the same purpose, but here the reaction is as follows (Clarke, 1932):

$$R.S{-}S.R \rightarrow R.SH + CN.SR$$

where only one cysteine group per cystine residue is produced. Another reagent in common use for the same purpose is sodium sulphite. According to Cecil and McPhee (1955) the reaction in this case proceeds in the following manner:

$$R.S{-}S.R + SO_3^{--} \rightarrow R.S^- + R.S.SO_3^-$$

and

$$R.S^- + H^+ \rightleftharpoons R.SH$$

In this case also, therefore, only one cysteine group is produced.

Barrnett and Seligman (1954) tested sodium bisulphite ($NaHSO_3$), sodium hydrosulphite ($Na_2S_2O_4$), sodium thiosulphate ($Na_2S_2O_3$), ascorbic acid, L-epinephrine, potassium cyanide, sodium thioglycollate, cysteine, glutathione, 2-mercaptoethanol and 2,3-dimercaptopropanol at various pH levels from 8 to 11·5 at various temperatures from 37° to 50°. They found, as stated earlier in this chapter, that the most effective reagents were thioglycollate, $(NH_4)_2S$, KCN and $Na_2S_2O_4$. Of these the first is now usually employed in histochemical practice as the standard reagent for the purpose.

Probably even thioglycollate fails to cleave all tissue SS groups. For this reason it is impossible to be absolutely certain of the significance of histochemical tests for disulphides.

REFERENCES

ADAMS, C. W. M. (1956). *J. Histochem. Cytochem.*, **4**, 23.
ADAMS, C. W. M. (1957a). *J. clin. Path.*, **10**, 56.
ADAMS, C. W. M. (1957b). Personal communication.
ADAMS, C. W. M. (1960). *J. Path. Bact.*, **80**, 442.
ADAMS, C. W. M., and SLOPER, J. C. (1955). *Lancet*, **1**, 651.
ADAMS, C. W. M., and SLOPER, J. C. (1956). *J. Endocrinol.*, **13**, 221.
AKABORI, J., and OHNO, K. (1950). *Proc. jap. Acad.*, **26**, 39.
AKABORI, J., and OHNO, K. (1952). *Chem. Abstracts*, **46**, 940.
ALEXANDER, P., and EARLAND, C. (1950). *Nature, London*, **166**, 396.
ALEXANDER, P., FOX, M., and HUDSON, R. F. (1951). *Biochem. J.*, **49**, 129.
ANSON, M. L. (1939). *J. gen. Physiol.*, **23**, 247.
ANSON, M. L. (1941). *Ibid.*, **24**, 399.
ANSON, M. L. (1942). *Ibid.*, **25**, 355.
ATTENBURROW, J., ELLIOTT, D. F., and PENNEY, G. F. (1948). *J. chem. Soc.*, 310.
BACHMANN, R., and SEITZ, H. M. (1961). *Histochemie*, **2**, 307.
BAHR, G. F. (1957). *Acta radiol.*, Suppl., 147.
BAKER, J. R. (1944). *Quart. J. micr., Sci.*, **85**, 1.
BAKER, J. R. (1947). *Ibid.*, **88**, 115.
BAKER, J. R. (1956). *Ibid.*, **97**, 161.
BANGLE, R., Jr. (1956). *Amer. J. Path.*, **32**, 349.
BARNARD, E. A. (1961). In "General Cytochemical Methods." Vol. 2, p. 203. Ed. J. F. Danielli, Acad. Press, New York.
BARNARD, E. A., and DANIELLI, J. F. (1956). *Nature, London*, **178**, 1450.
BARRNETT, R. J. (1953). *J. nat. Cancer Inst.*, **13**, 905.
BARRNETT, R. J., and SELIGMAN, A. M. (1952). *Ibid.*, **13**, 215.
BARRNETT, R. J., and SELIGMAN, A. M. (1953). *J. Histochem. Cytochem.*, **1**, 392.
BARRNETT, R. J., and SELIGMAN, A. M. (1954). *J. nat. Cancer Inst.*, **14**, 769.
BARRNETT, R. J., and SELIGMAN, A. M. (1956). *J. Histochem. Cytochem.*, **4**, 411.
BARRNETT, R. J., and SELIGMAN, A. M. (1958). *J. biophys. biochem. cytol.*, **4**, 169.
BARRNETT, R. J., TSOU, K-C., and SELIGMAN, A. M. (1955). *Ibid.*, **3**, 406.
BARRON, E. S. G. (1951). *Advanc. Enzymol.*, **11**, 201.
BATES, R. W. (1937). *Proc. Amer. Soc. biol. Chem., J. biol. Chem.*, **119**, p. vii.
BEAVEN, G. H., and HOLLIDAY, E. R. (1952). *Advanc. Prot. Chem.*, **7**, 320.
BELLO, J. (1956). *Biochem. Biophys. Acta.*, **20**, 426.
BENEŠ, K. (1962). *Biol. plantarum, Prague*, **4**, 61.
BENNETT, H. S. (1948). *Abstr. Amer. Ass. Anat. Rec.*, **100**, 640.
BENNETT, H. S. (1951). *Anat. Rec.*, **110**, 231.
BENNETT, H. S., and YPHANTIS, P. A. (1948). *J. Amer. chem. Soc.*, **70**, 3522.
BENNETT, H. S., and TOENNIES, G. (1942). *J. biol. Chem.*, **145**, 671.
BENSLEY, R. R., and GERSH, I. (1933). *Anat. Rec.*, **57**, 217.
BERGMANN, M., and ZERVAS, L. (1936). *J. Biol. Chem.*, **113**, 341.
BHATTACHARYA, K. R. (1959). *Nature, London*, **184**, 53.
BHATTACHARYA, K. R. (1960). *Arch. biochem. exp. Med.*, **20**, 93.
BLANCHETIÈRE, A., and ROMIEU, M. (1931). *C.R. Soc. Biol., Paris*, **107**, 1127.
BONHAG, P. F. (1955). *J. Morphol.*, **96**, 381.
BONOMI, V. (1951). *Boll. Soc. ital. Sper.*, **27**, 1418.
BOOZER, C. E., and LEWIS, E. S. (1953). *J. amer. Chem. Soc.*, **75**, 3182.
BOWES, J. H., and KENTEN, R. H. (1949). *Biochem. J.*, **44**, 142.
BOYLAND, E., SIMS, P., and WILLIAMS, D. C. (1956). *Biochem. J.*, **62**, 546.
BRATTON, A. C., and MARSHALL, E. K. Jr. (1939). *J. Biol. Chem.*, **128**, 537.
BRAUN-FALCO, O. (1956). *Acta Histochem.*, **2**, 264.
BRENNER, M., and HUBER, W. (1953). *Helv. chim. Acta*, **36**, 1109.
BRUEMMER, N., CARVER, M. J., and THOMAS, L. E. (1957). *J. Histochem. Cytochem.*, **5**, 140.
BUCHANAN, G. L., REID, S. T. THOMPSON, R. E. S., and WOOD, E. G. (1957). *J. chem. Soc.*, 4427.
BURSTONE, M. S. (1955). *Ibid.*, **3**, 32.

BURTON, H., and PRAIL, P. F. G. (1950). *J. chem. Soc.*, 1203.
CARVER, M. J., BROWN, F. C., and THOMAS, L. E. (1953). *Stain Tech.*, **28**, 89.
CATCHPOLE, H. R. (1949). *J. Endocrinol.*, **6**, 218.
CECIL, R. (1950). *Biochem. J.*, **47**, 572.
CECIL, R., and MCPHEE, J. R. (1955). *Ibid.*, **60**, 496.
CHÈVREMONT, M., and FRÉDERIC, J. (1943). *Arch. Biol.*, **54**, 580.
CHRISTENSEN, H. N. (1945). *J. biol. Chem.*, **160**, 75.
CHU, C. H. U., FOGELSON, M. A., and SWINYARD, C. A. (1953). *J. Histochem. Cytochem.*, **1**, 39.
CLARKE, H. T. (1932). *J. biol. Chem.*, **97**, 235.
COREY, E. J., and HAEFELE, L. F. (1959). *J. amer. Chem. Soc.*, **81**, 2225.
COUJARD, R. (1943). *Bull. Histol. Tech. micr.*, **20**, 161.
CORNFORTH, J. W., and ELLIOTT, D. F. (1950). *Science*, **112**, 534.
CREYF, S., and ROOSENS, L. (1959). *Chem. Abstracts*, **59**, 53/6070.
CURTIUS, T. (1890). *Ber.*, **23**, 3023.
CURTIUS, T. (1894). *J. prakt. Chem.*, **50**, 275.
CUTHBERTSON, W. R., and PHILLIPS, H. (1945). *Biochem. J.*, **39**, 7.
DANIELLI, J. F. (1947). *Symp. Soc. exp. Biol.*, **1**, 101.
DANIELLI, J. F. (1950). *Cold Spr. Harb. Symp. quant. Biol.*, **14**, 32.
DEGUCHI, Y. (1964). *J. Histochem. Cytochem.*, **12**, 261.
DEITCH, A. D. (1961). *J. Histochem. Cytochem.*, **9**, 477.
DICKMAN, S. R., and CROCKETT, A. L. (1956). *J. biol. Chem.*, **220**, 957.
DICKMAN, S. R., and WESTCOTT, W. L. (1954). *Ibid.*, **210**, 481.
DIETZ, E. (1942). *Z. mikr. Anat. Forsch.*, **51**, 14.
DIXON, K. C. (1959). *Amer. J. Path.*, **35**, 199.
DIXON, K. C. (1962). *Quart. J. exp. Physiol.*, **47**, 1.
DOBERNECK, H. V., and MARESH, G. (1952). *Zeit. physiol. Chem.*, **289**, 271.
DONZOU, P., and LE CLERC, A. M. (1955). *Anal. chim. Acta*, **12**, 239.
VAN DUIJN, P. (1961). *J. Histochem. Cytochem.*, **9**, 234.
DURRUM, E. L. (1950). *J. Amer. chem. Soc.*, **72**, 2943.
EAGLE, H. (1936). *Proc. Soc. ex. Biol. N.Y.*, **34**, 39.
EARLAND, C., and KNIGHT, C. S. (1955). *Biochem. Biophys. Acta*, **17**, 457.
EARLAND, C., MACRAE, T. P., WESTON, G. J., and STATHAM, K. (1955). *Textile Res. J.*, **26**, 963.
ECKERT, H. W. (1943). *J. biol. Chem.*, **148**, 205.
ENGEL, M. B., and ZERLOTTI, E. (1964). *J. Histochem. Cytochem.*, **12**, 156.
FAI CHONG, S., and LIENER, I. E. (1964). *Nature (Lond.)*, **203**, 1065.
FEARON, W. R. (1944). *Analyst*, **69**, 122.
FEIGL, F. (1954). "Spot Tests," Voll. 2, p. 292, 4th Edition, Elsevier, Amsterdam.
FIELD, L., HOELZEL, C. B., LOCKE, J. M., and LAWSON, J. E. (1961). *J. amer. Chem. Soc.*, **83**, 1257.
FINDLAY, G. H. (1955a). *J. Histochem. Cytochem.*, **3**, 430.
FINDLAY, G. H. (1955b). *Ibid.*, **3**, 331.
FISCHER, H. (1911). *Hoppe-Seyl. Z.*, **75**, 261.
FISCHER, H. (1932). *Ibid.*, **206**, 187.
FISHER, E. R., and LILLIE, R. D. (1954). *J. Histochem. Cytochem.*, **2**, 81.
FOTAKIS, N. S. (1960). *Histochemie*, **2**, 43.
FRAENKEL-CONRAT, H. (1944). *J. biol. Chem.*, **154**, 227.
FRAENKEL-CONRAT, H., and OLCOTT, H. S. (1945). *Ibid.*, **161**, 259.
FRAZER, D., and HIGGINS, H. G. (1953). *Nature, London*, **172**, 459.
FREUDENBERG, F. (1956). *Ber.*, **89**, 258.
GABLER, W., and SCHEUNER, G. (1966). *Acta histochemica*, **23**, 102.
GELEWITZ, E. W., RIEDEMANN, W. L., and KLOTZ, I. M. (1954). *Arch. Biochem. Biophys.*, **53**, 411.
GEMMILL, C. L. (1956). *Ibid.*, **63**, 177.
GERNGROSS, O., VOSS, K., and HERFIELD, H. (1933). *Ber.*, **66**, 435.
GERSH, I. (1949). *J. Endocrinol.*, **6**, 282.
GERSHTEIN, L. M. (1958). In "Histochemical Methods in Normal and Pathological Morphology." Eds. Portugalov, V. V., and Strukob, A. I. Moskva.

GERSHTEIN, L. M., and SVETKOVA, I. V. (1960). *Cytol.*, **2**, 201.
GEYER, G. (1961). *Acta Histochemica*, **13**, 355.
GEYER, G. (1962a). *Ibid.*, **14**, 284.
GEYER, G. (1962b). *Ibid.*, **14**, 1.
GEYER, G. (1962c). *Ibid.*, **14**, 67.
GEYER, G. (1964). *Ibid.*, **19**, 73.
GEYER, G. (1965a). *Ibid.*, **20**, 121.
GEYER, G. (1965b). *Acta Histochem.*, **20**, 130.
GIBBS. H. D. (1927). *J. Biol. Chem.*, **71**, 445.
GILBOE, D. D., and WILLIAMS, J. N. Jr. (1956). *Proc. Soc. exp. Biol.*, **91**, 535.
GIROUD, A. (1929). *Protoplasma*, **7**, 72.
GIROUD, A., and BULLIARD, H. (1932). *Bull. Soc. chim. Biol.*, **14**, 278.
GIROUD, A., and BULLIARD, H. (1933). *Protoplasma*, **19**, 381.
GLENNER, G. G. (1963). *J. Histochem. Cytochem.*, **11**, 285.
GLENNER, G. G. (1957). *J. Histochem. Cytochem.*, **5**, 297.
GLENNER, G. G., and BAGDOYAN, H. E. (1960). *J. Histochem. Cytochem.*, **8**, 138.
GLENNER, G. G., and LILLIE, R. D. (1957). *Ibid.*, **5**, 279.
GLENNER, G. G., and LILLIE, R. D. (1959). *J. Histochem.*, *Cytochem.* **7**, 416.
GOESSNER, W. (1947). *Zbl. allg. Path. path. Anat.*, **85**, 434.
GOMORI, G. (1950). *Amer. J. clin. Path.*, **20**, 665.
GOMORI, G. (1952). "Microscopic Histochemistry." Chicago Univ. Press,
GOSLAR, H. G. (1963). *Histochemie*, **3**, 249.
GREGORY, J. T. (1955). *J. amer. chem. Soc.*, **77**, 3922.
GUSTAVSON, K. H. (1956). "The Chemistry of the Tanning Processes." Acad. Press, New York. p. 161.
HAMMETT, F. S., and CHAPMAN, S. S. (1938). *J. Lab. clin. Med.*, **24**, 293.
HARLEY-MASON, J., and ARCHER, A. A. P. G. (1958). *Biochem. J.*, **69**, 60P.
HARRIS, J. I., and WORK, T. S. (1948). *Biochem. J.*, **46**, 582.
HARVEY, D. G., MILLER, E. J., and ROBSON, W. (1941). *J. chem. Soc.*, 153.
HELLERMAN, L., PERKINS, M. E., and CLARK, W. M. (1933). *Proc. Nat. Acad. Sci. Wash.*, **19**, 855.
HERRIOTT, R. M. (1947). *Advanc. Prot. Chem.*, **3**, 169.
HIRD, F. J. R., and YATES, J. R. (1961). *Biochem. J.*, **80**, 612.
HOFMAN, K. A. (1900). *Ann.*, **312**, 21.
HOFFMANN, R. (1853). *Liebigs Ann.*, **87**, 123.
HOROWITZ, M. G., and KLOTZ, I. M. (1956). *Arch. Biochem. Biophys.*, **63**, 77.
HORTON, H. R., and KOSHLAND, D. E. (1965). *J. amer. chem. Soc.*, **87**, 1126.
HOWARD, A. N., and WILD, F. (1957). *Biochem. J.*, **65**, 651.
HUGHES, W. L., Jr. (1950). *Cold Spr. Harb. Symp. quant. Biol.*, **14**, 79.
HYDÉN, H. (1943). *Acta physiol. scand.*, **6**, Suppl. 17.
ILLARI, G. (1938). *Gazz. chim. Ital.*, **68**, 103.
JOYET-LAVERGNE, P. (1927). *C.R. Soc. Biol., Paris*, **97**, 140.
KAHR, K., and BERTHER, C. (1960). *Ber.*, **93**, 132.
KANWAR, K. C. (1960). *Experientia*, **16**, 355.
KARNOVSKY, M. J., and FASMAN, G. D. (1960). *J. Biophys. Biochem., Cytol.*, **8**, 319.
KARNOVSKY, M. J., and MANN, M. S. (1961). *Histochemie*, **2**, 234.
KASTEN, F. H. (1962). *J. Histochem. Cytochem.*, **10**, 769.
KAUFMANN, B. P., MCDONALD, M. R., GAY, H. (1946). *Carnegie Inst. of Washington Year Book*, **45**, 159.
KAUFMANN, B. P., MCDONALD, M. R., GAY, H., WILSON, K., and WYMAN, R. (1947). *Ibid.*, **46**, 136.
KHORANA, H. G. (1952). *J. chem. Soc.*, 2081.
KHORANA, H. G. (1953). *Chem. Rev.*, **53**, 145.
KÖNIG, W. (1904a). *J. Prakt. Chem.*, **69**, 105.
KÖNIG, W. (1904b). *Ibid.*, **70**, 19.
KÖNIG, W. (1912). *Ibid.*, **86**, 166.
KOSHLAND, D. E., KARKHANIS, Y. D., and LATHAM, H. G. (1964). *J. amer. Chem. Soc.*, **86**, 1448.
KRÖNHNKE, F., and LEISTER, H. (1958). *Ber.*, **91**, 1295.

KUHN, R., and BAER, H. H. (1953). *Ber.*, **86**, 124.
LANDING, B. H., and HALL, H. E. (1956a). *J. Histochem. Cytochem.*, **4**, 382.
LANDING, B. H., and HALL, H. E. (1956b). *Stain Tech.*, **31**, 197.
LANDING, B. H., and HALL, H. E. (1956c). *J. Histochem. Cytochem.*, **4**, 41.
LATIMER, P. H., and BEST, R. W. (1937). *J. Amer. chem. Soc.*, **59**, 2500.
LEVENE, P. A., and STEIGER, R. E. (1927). *J. biol. Chem.*, **74**, 689.
LEVENE, P. A., and STEIGER, R. E. (1928). *Ibid.*, **79**, 95.
LEVINE, N. D. (1939). *Stain Tech.*, **14**, 24.
LEVINE, N. D. (1940). *Ibid.*, **15**, 91.
LEWIS, E. S., and BOOZER, C. E. (1952). *J. amer. Chem. Soc.*, **74**, 308.
LILLIE, R. D. (1954). "Histopathologic Technic and Practical Histochemistry." Blakiston, New York.
LILLIE, R. D. (1956). *J. Histochem. Cytochem.*, **4**, 118.
LILLIE, R. D. (1957a). *Ibid.*, **5**, 188.
LILLIE, R. D. (1957b). *Ibid.*, **5**, 528.
LILLIE, R. D. (1964a). *J. Histochem. Cytochem.*, **12**, 821.
LILLIE, R. D. (1964b). *J. Histochem. Cytochem.*, **12**, 522.
LILLIE, R. D., and BANGLE, R., Jr. (1954). *Ibid.*, **2**, 300.
LILLIE, R. D., BANGLE, R., Jr., and FISHER, E. R. (1954). *Ibid.*, **2**, 95.
LISON, L. (1936). "Histochimie Animale." Gautier-Villars, Paris.
LISON, L. (1953). "Histochimie et cytochimie animales." Gautier-Villars, Paris.
LOLLAR, R. M., and KREMEN, S. S. (1948). *J. amer. Leather Chem. Ass.*, **43**, 452.
LUGG, J. W. H. (1937). *Biochem. J.*, **31**, 1422.
MADDY, A. H. (1961a). In "General Cytochemical Methods." Ed. J. F. Danielli, Academic Press, New York, **2**, 259.
MADDY, A. H. (1961b). *Exp. Cell Research*, **22**, 169.
MADDY, A. H. (1961c). *Exp. Cell Research*, **22**, 181.
MACPHERSON, H. T. (1942). *Biochem. J.*, **36**, 59.
MANN, G. (1902). "Physiological Histology." London.
MARSHALL, J. M., Jr. (1954). *Exp. Cell Research*, **6**, 240.
MAURI, C. E., VACCARI, F., and KADERAVEK, G. P. (1954). *Haematologia*, **38**, 263.
MAY, C. E., and ROSE, E. R. (1922). *J. biol. Chem.*, **54**, 213.
MAZIA, D., BREWER, P. A., and ALFERT, M. (1953). *Biol. Bull.*, **104**, 57.
MCCOMB, E. A., and MCCREADY, R. M. (1957). *Analyt. Chem.*, **29**, 819.
MCLEISH, J., BELL, L. G. E., LA COUR, L. F., and CHAYEN, J. (1957). *Exp. Cell Research*, **12**, 120.
MENZIES, D. W. (1961). *Stain Tech.*, **36**, 285.
MEYER, L. (1864). *Ann. Chim.*, **132**, 156.
MILLON, A. N. E. (1849). *C.R. Soc. Biol., Paris*, **28**, 40.
MIRSKY, A. E., and ANSON, M. L. (1936). *J. gen. Physiol.*, **19**, 451.
MITCHELL, J. S. (1942). *Brit. J. exp. Path.*, **23**, 296.
MONNÉ, L., and SLAUTERBACK, D. B. (1951). *Ark. Zool.*, **1**, 455.
MOREL, A., and SISLEY, P. (1927). *Bull. Soc. Chim.*, **41**, 1217.
MÜNCH, O., and ERNST, B. (1965). *Acta histochem.*, **20**, 125.
NAGAMATSU, A., and FRUTON, J. S. (1960).
NASSE, O., (1901). *Pflug. Arch. ges. Physiol.*, **83**, 361.
NICOLET, B. H., and SHINN, L. A. (1939). *J. Amer. chem. Soc.*, **61**, 1615.
OBATA, Y., and YAMANISHI, T. (1951). *J. agr. chem. Soc. Jap.*, **24**, 226.
OLSZEWSKA, M. J., WROŃSKI, M., and FORTAK, W. (1967). *Folia Histochem. Cytochem.*, **5**, 7.
PAULY, H. (1904). *Z. physiol. Chem.*, **42**, 508.
PEARSE, A. G. E. (1951). *Quart. J. micr. Sci.*, **92**, 393.
PEARSE, A. G. E. (1953). *J. Histochem. Cytochem.*, **1**, 460.
PEARSE, A. G. E. (1954). *J. Path. Bact.*, **67**, 129.
PEARSE, A. G. E. (1960). "Histochemistry, Theoretical and Applied." 2nd Edition, J. & A. Churchill, London, p. 263.
PETERS, A. (1955). *Quart. J. micr. Sci.*, **96**, 84.
PISCHINGER, A. (1927). *Pflüg. Arch. ges. Physiol.*, **217**, 205.
POLLISTER, A. W., and RIS, H. (1947). *Cold Spr. Harb. Symp. quant. Biol.*, **12**, 147.

PORTER, R. R. (1950). *Biochem. J.*, **46**, 304.
PORTNER, C., and HÖGL, O. (1953). *Analyt. Chim. Acta*, **8**, 29.
PRELOG, V., and WIELAND, P. (1946). *Helv. chim. Acta*, **29**, 1128.
PREUSSMAN, R., DAIBER, D., and HENGY, H. (1964). *Nature, Lond.*, **201**, 502.
PRICE, C. A., and CAMPBELL, G. W. (1957). *Biochem. J.*, **65**, 512.
PUCHTLER, H., and SWEAT, F. (1962). *J. Histochem. Cytochem.*, **10**, 365.
RAMACHANDRA, L. K. (1956). *Chem. Rev.*, **56**, 199.
RAPKINE, L. (1933). *C.R. Soc. Biol., Paris*, **112**, 79 and 1794.
RAPPAY, G. (1963). *Nature, Lond.*, **200**, 274.
RASCH, E. M., and SWIFT, H. (1953). *J. Histochem. Cytochem.*, **1**, 392.
RAWLINS, L. M. C., and SCHMIDT, C. L. A. (1929). *J. Biol. Chem.*, **82**, 709.
REMSCHNEIDER, R. (1947). *Gazz. chim. ital.*, **77**, 607.
REMSCHNEIDER, R., and WEYGAND, C. (1955). *Ibid.*, **86**, 201.
RICHARDS, H. R., and SPEAKMAN, J. B. (1953). *Nature, Lond.*, **171**, 751.
ROGERS, G. E. (1953). *Quart. J. micr. Sci.*, **94**, 253.
ROMIEU, P. (1925). *Bull. Histol. Tech. micr.*, **92**, 185.
RYZHEVA, A. P. (1946). *Biokhimya*, **11**, 391.
SAKAGUCHI, S. (1925). *J. Biochem., Tokyo*, **5**, 25.
SANGER, F. (1945). *Biochem. J.*, **39**, 507.
SANGER, F. (1950). *Cold Spr. Harb. Symp. quant. Biol.*, **12**, 142.
SASSI, L. (1956). *Arch. Inst. Pasteur, Tunnis*, **33**, 451 and 461.
SAVITCH, K. V., and YAKOVLEV, V. Y. (1957). *Vopr. med. chim.*, **3**, 121.
SCOTT, H. R., and CLAYTON, B. P. (1953). *J. Histochem. Cytochem.*, **1**, 336.
SELIGMAN, A. M., TSOU, K-C., and BARRNETT, R. J. (1954). *Ibid.*, **2**, 484.
SERRA, J. A. (1944). *Naturwiss.*, **32**, 46.
SERRA, J. A. (1946). *Stain Tech.*, **21**, 5.
SERRA, J. A., and QUEIROZ LOPES, A. (1945). *Port. Acta biol.*, **1**, 111.
SHANKLIN, W. M., and ISSIDORIDES, M. (1960). *Stain Tech.*, **35**, 46.
SHEETS, R. F., HAMILTON, H. E., DE GOWIN, E. L., and KING, R. L. (1956). *J. appl. physiol.*, **9**, 145.
SINGER, M., and WISLOCKI, G. B. (1948). *Anat. Rec.*, **102**, 175.
SLOBODIAN, E., MECHANIC, G., and LEVY, M. (1962). *Science*, **135**, 3502. 441.
SOKOLOFF, L., MUND, A., and KANTOR, T. G. (1951). *Amer. J. Path.*, **27**, 1037.
SPICER, S. S. (1962). *J. Histochem. Cytochem.*, **10**, 691.
SPICER, S. S., and LILLIE, R. D. (1961). *Stain Tech.*, **36**, 365.
SPIES, J. R., and CHAMBERS, D. C. (1948a). *Analyt. Chem.*, **20**, 30.
SPIES, J. R., and CHAMBERS, D. C. (1948b). *J. Amer. chem. Soc.*, **70**, 1682.
SPRINGALL, H. D. (1954). "The Structural Chemistry of Proteins." Acad. Press New York.
STEDMAN, E., and STEDMAN, E. (1947). *Symp. Soc. exp. Physiol.*, **1**, 232.
STOWARD, P. J. (1963). D.Phil. Thesis, University of Oxford.
STOWARD, P. J. (1967). *J. roy. micr. Soc.*, **87**, 247.
TERNER, J. Y. (1964). *J. Histochem. Cytochem.*, **12**, 504.
TERNER, J. Y., and CLARK, E. (1960). *J. Histochem. Cytochem.*, **8**, 184.
THOMAS, L. E. (1946). *J. cell. comp. Physiol.*, **28**, 145.
THOMAS, L. E. (1950). *Stain Tech.*, **25**, 149.
TOENNIES, G. (1942). *J. biol. Chem.*, **145**, 667.
TOENNIES, G., and HOMILLER, R. D. (1942). *J. Amer. chem. Soc.*, **64**, 3054.
TRANZER, J-P., and PEARSE, A. G. E. (1964). *J. Histochem. Cytochem.*, **12**, 325.
TSOU, K-C., BARRNETT, R. J., and SELIGMAN, A. M. (1955). *Ibid.*, **77**, 4613.
UDENFRIEND, S., and COOPER, J. R. (1952). *J. biol. Chem.*, **196**, 227.
UDENFRIEND, S., and VELICK, S. F. (1951). *J. biol. Chem.*, **190**, 733.
VAN SLYKE, D. D. (1929). *Ibid.*, **83**, 425.
VERACAUTEREN, R. (1951). *Nature, Lond.*, **167**, 819.
VOSS, H. (1940). *Z. mikr. Anat. Forsch.*, **49**, 51.
WAALKES, T. P., and UDENFRIEND, S. (1957). *J. lab. clin. Med.*, **50**, 733.
WARREN, T. N., and MCMANUS, J. F. A. (1951). *J. nat. Cancer Inst.*, **12**, 223.
WARTENBERG, H. (1956–57). *Acta Histochem.*, **3**, 145.
WEISS, L. P., TSOU, K.-C., and SELIGMAN, A. M. (1954). *J. Histochem. Cytochem.*, **2**, 29.

WESTON, G. J. (1955). *Biochem. Biophys. Acta*, **17**, 462.
WILEY, R. H. (1950). *Science*, **111**, 259.
WIGGLESWORTH, V. B. (1952). *Quart. J. micr. Sci.*, **93**, 105.
WINKELMAN, J., and SPICER, S. S. (1963). *J. Histochem. Cytochem.*, **11**, 489.
WINKLER, W. (1948). *Chem. Ber.*, **81**, 256.
WINTERSTEIN, A. (1933). In Klein's "Hanbuch der Pflanzeanalyse." Springer, Vienna.
WITTER, A., and TUPPY, H. (1960). *Biochem. Biophys. Acta*, **45**, 429.
WROŃSKI, M. (1966). *J. Chromatog.*, **24**, 480.
YAKOVLEV, V. Y., and SOKOLOVSKY, V. V. (1955). *Dokl. Adad. Nauk.*, **101**, 211.
YAKOVLEV, V. Y., and NISTRATOVA, S. N. (1958). "Histochemical Methods in Normal and Pathological Morphology." Ed. Portugalov, V. V. and Strukov, U. V. Moskva.
YAO, T. (1949). *Quart. J. micr. Sci.*, **90**, 401.
YASUMA, A., and ICHIKAWA, T. (1952). *Nagoya J. med. Sci.*, **15**, 96.
YASUMA, A., and ICHIKAWA, T. (1953). *J. Lab. Clin. med.*, **41**, 296.
ZAHN, H., and PFANMÜLLER, H. (1958). *Biochem. Z.*, **330**, 97.
ZAHN, H., and TRAUMANN, K. (1954). *Z. Naturforsch.*, **96**, 518.
ZAHN, H., WOLLEMANN, B., and WASCHKA, O. (1953). *Hoppe-Seyl. Arch.*, **294**, 100.
ZEIGER, K. (1930). *Z. Zellforsch.*, **10**, 481.
ZERLOTTI, E., and ENGEL, M. B. (1962). *J. Histochem. Cytochem.*, **10**, 537.
ZUBER, H., TRAUMANN, K., and ZAHN, H. (1955). *Z. Naturforsch.*, **106**, 457.

CHAPTER 7

FLUORESCENT ANTIBODY METHODS (IMMUNOFLUORESCENCE)

Introduction (1960). In the first edition of this book (1953) a small subsection of the chapter on proteins and amino-acids dealt with the localization of proteins by the fluorescent antibody method of Coons (Coons, Creech and Jones, 1941; Coons, Creech, Jones and Berliner, 1942). Since this time the method has been increasingly employed as a research tool and the results obtained by various workers are sufficiently important to justify treatment of the method and its applications in a separate chapter.

Introduction (1968). In the interval since the last introduction to this chapter was written the fluorescent antibody technique has been employed with ever increasing frequency for the localization of small concentrations of macromolecular proteins, in many different categories. The number of references to papers in the field of applied immunohistology exceeds those of any other single cytochemical technique by a wide margin. Many technical improvements have been made and the most important of these are fully discussed below. Nevertheless, as its originator remarks (Coons, 1964) it is still true that "as a tool immunofluorescence is simple in concept, but its successful use requires knowledge of immunology and skill in its application".

Several critical works and reviews have appeared since 1960. Chief among these is "Fluorescent Protein Tracing" by Nairn and collaborators (1962). Other works include "Fluorescent Antibody Techniques" (Cherry, Goldman and Carski, 1960); a similarly entitled review (White, 1960); "Immunofluorescent Staining" (Beutner, 1961); "The Fluorescent Antibody Method in Medical and Biological Research (Borek, 1961); "Nuovi problemi nella tecnica dell'anticorpo fluorescente e nelle sue applicazione istochimiche" (Rossi, Zaccheo and Grossi, 1964); Imunofluorescenta (Bals, 1966), and others.

Entitled "The Beginnings of Immunofluorescence", Coon's (1961) historical review should be read widely, not only as an introduction to modern fluorescent antibody techniques, but as an object lesson in the management of research for administrators as well as researchers.

For all the advances which have been made, immunohistology remains primarily a matter of technology. Scarcely less important than technical skill, however, is the suitability of the problem. This point will further be considered towards the end of the chapter.

Principle of the Method

The essentially simple principle of the fluorescent antibody technique is perhaps best illustrated, in the first instance, by describing the steps through which it was developed. Basically it depends on the fact that dye molecules can be attached to a protein without impairment of its function as an antibody. This was first shown by Reiner (1930), who found that the agglutinating titre of pneumococcal antisera was unchanged after coupling with diazotized atoxyl. He suggested that sera labelled in this way might be useful for studying quantitative aspects of the antigen-antibody reaction. A coloured antigen was produced by Heidelberger, Kendall and Soo Hoo (1933), who coupled R-salt-azo-benzidine to ovalbumin, and a year later Marrack (1934) produced colour-labelled antibodies (anti-typhoid and anti-cholera), in the same manner, which he showed would react with the homologous organisms and colour them pink. Coons, in unpublished studies, repeated this work with rabbit anti-pneumococcus serum, but found that although the organisms were agglutinated the colour they developed was less than adequate. He therefore tested the fluorescent labelling method originated by Creech and Jones (1940, 1941). These authors prepared the isocyanates of several aromatic hydrocarbons and coupled them to various proteins to give a highly fluorescent product. They analysed a fluorescent β-anthryl carbamido derivative of rabbit anti-pneumococcus III serum and found that it contained two hydrocarbon groups per molecule of protein. This was sufficient to give an intense (blue) fluorescence even in dilute solution. Their product agglutinated Type III pneumococci only and the clumped organisms showed macroscopic and microscopic fluorescence.

It was soon found, however, that the blue fluorescence of the β-anthryl carbamido-protein could be distinguished only with difficulty from the normal blue autofluorescence of the tissues. In order to overcome this difficulty and to provide fluorescence in a contrasting colour, fluorescein-4-isocyanate was prepared and coupled to the antibody. The product showed the high intensity greenish-yellow fluorescence of fluorescein in high dilution. The essential feature of the fluorescent antibody method, thus developed, was the precipitation of labelled antibody from a concentrated solution placed on a tissue section containing the antigen in reactive form, that is to say substantially undenatured in most cases. Conditions were so arranged that the labelled antibody became firmly attached only to sites containing reactive antigen. The excess of labelled antibody was washed away before examining the section with the fluorescence microscope.

The method of Coons and his collaborators opened up an entirely new field of protein histochemistry by allowing the histochemist to make use of the very high specificity of immune reactions in the localization of antigenic proteins.

TECHNIQUES OF IMMUNOFLUORESCENCE

Direct Localization of Antigen. The original method, depending on the attachment of ambient antibody to fixed location antigen was illustrated graphically by Mellors (1955) in a convenient manner which is reproduced below.

With this method a large amount of work was carried out, but certain defici-

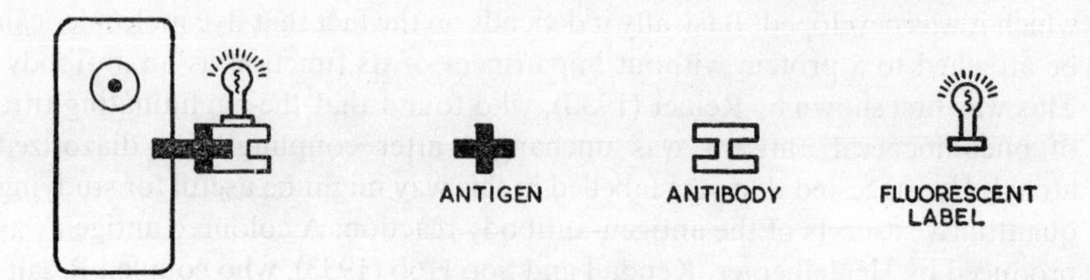

encies were evident. For instance it was virtually incapable of localizing tissue antibodies. To serve this and other purposes, multiple-layer techniques were evolved.

Direct Localization of Antibody. Although there are accounts of the localization of antibody by direct attachment of fluorescent antigen (Willson and Katsh, 1960; Parker, Elevitch and Grodsky, 1963) there are potent reasons why the method is unlikely to be very successful.

Indirect Localization of Antibody. The simplest of these methods made use of the interposition of a layer of antigen between the tissue antibody (fixed location) and the same antibody coupled *in vitro* with fluorescein isocyanate.

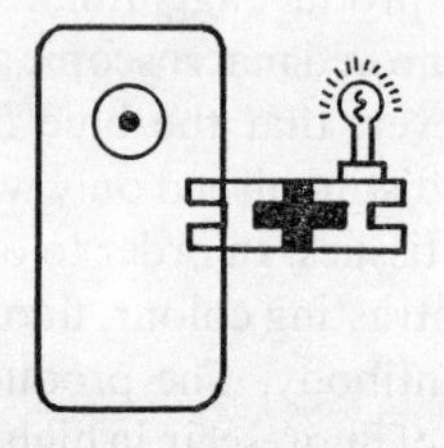

This technique (Coons, Leduc and Connelly, 1955) is now usually called the sandwich technique.

Indirect Localization of Antigen. An alternative application of the sandwich technique was to use fluorescent anti-globulin for the detection of layered specific antibody attached to the antigen in or on the cell (Weller and Coons, 1954). This technique differs from the other in the fact that the intermediate layer is acting as an antibody on the one hand but as an antigen on the other.

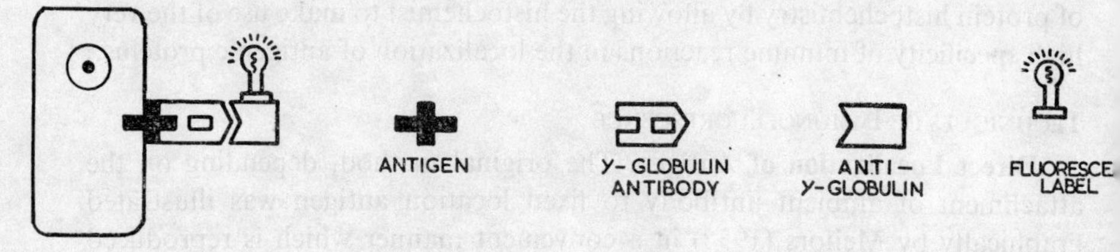

This variety of the sandwich technique is especially useful in cases where the antigen is difficult to obtain and the production of the specific fluorescent

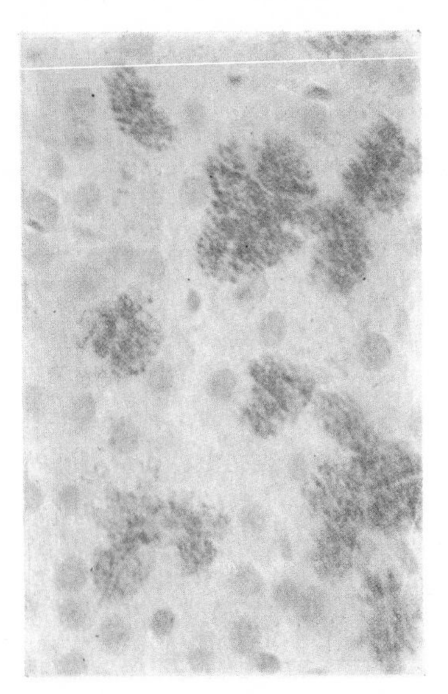

IA. Rat Pancreas, freeze-dried formaldehyde vapour-fixed section. Shows high tryptophan content (blue) in the zymogen granules. DMAB method. × 250

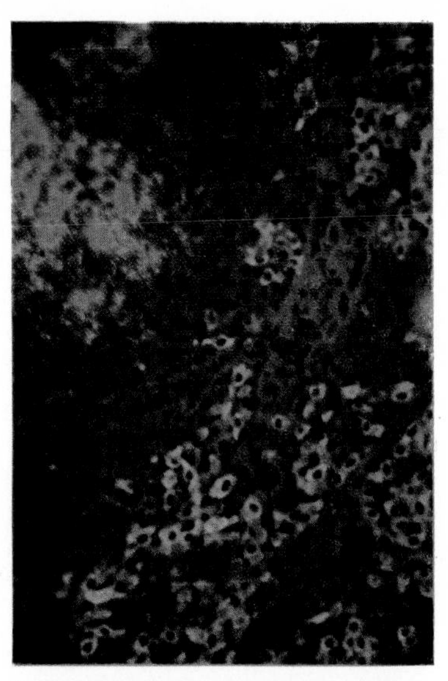

IB. Germinal centre and surrounding area of lymph node from mouse given 3mg [^{125}I] haemocyanin 5 weeks after a priming dose of 10 mg 'cold' haemocyanin. Intense antigen (red dots) deposition in germinal centre and absence of antigen over specific antibody-producing cells (green).

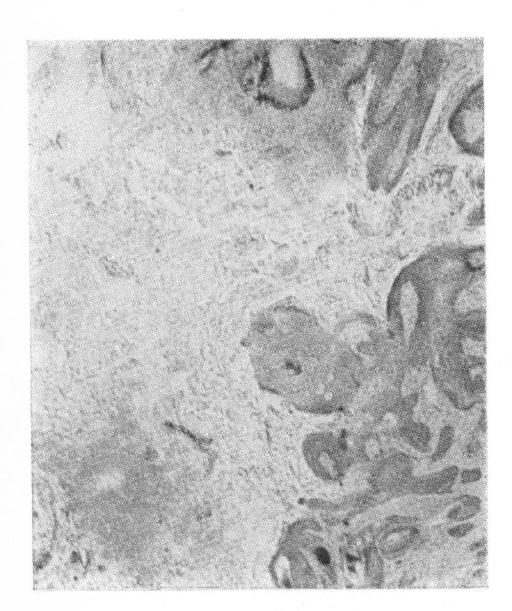

IC. Mouse cervix. Methylcholanthrene-induced squamous cell carcinoma. Growing edge of the tumour. Solochrome cyanine RS. × 124

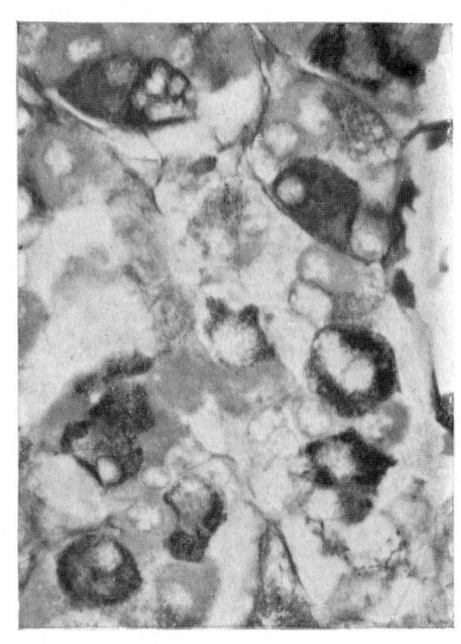

ID. Human adenohypophysis. Distinction between R-cells (magenta), S-cells (blue) and acidophils (orange). Performic acid—Alcian blue, PAS-orange G. × 870

antibody thus difficult to achieve. Its greatest advantage is described below, however. As indicated in the diagram, tissue antigen (AG) combines with 'n' (here 3) molecules of antibody (AB). When the same antibody in turn acts as antigen a further 'n' molecules (here 4) of (fluorescent) antibody can be attached to each antigen molecule.

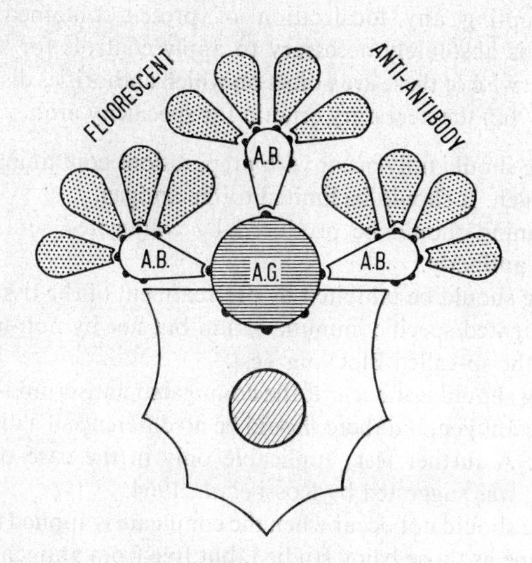

By comparison with the direct technique the sandwich technique thus provides an increased final fluorescence which may be of the order of 3 or 4 times.

Indirect Localization of Complement. A further amplification of the sandwich technique depends on the demonstration of complement bound by an antigen-antibody reaction, using fluorescent anticomplement (Goldwasser and Shepard, 1958). The greatest advantage of this technique is that it requires only a single fluorescent conjugate to demonstrate immune complexes of all types, provided that they bind complement. Although its specificity is uncertain the method, with minor modifications, has been used by a number of workers (Klein and Burkholder, 1959; Müller and Klein, 1959; Vogt and Kochem, 1960; Strauss *et al.*, 1960; Mikhailov and Dashkevich, 1961; Cramer and Langer, 1961; Lachmann *et al.*, 1962; Müller, 1962; Hinuma *et al.*, 1962).

Localization of Immune Complex with Fluorescent Complement. Observing that complement was not inactivated by labelling with fluorescein, Sacchi, Costanzi and Mancini (1962) used this reagent in order to avoid the necessity to prepare specific antiglobulin conjugates. Later Grossi and Zaccheo (1963) successfully demonstrated influenza virus bound to leucocytes, by demonstrating the site of antigen-antibody reaction with fluorescent complement.

There are thus two ways in which complement can be bound in the tissues. Antigen-antibody complexes, occurring as part of some pathological process, may bind autologous complement *in vivo*. On the other hand the tissues may be exposed to heterologous complement during reaction *in vitro* with specific antibody. In the first case demonstration will be by heterologous anti-complement but in the second it is usually homologous.

Controls for Specificity

Before accepting any localization of protein obtained by immunofluorescence it is absolutely necessary to apply controls for specificity. The validity or otherwise of these are points on which authorities disagree. According to Nairn (1962) the necessary criteria for specificity are:

(1) Staining should only occur with preparations containing the appropriate antigen. It should be limited to the antigen.
(2) No staining should be produced by conjugated sera free from the specific antibody.
(3) Staining should be inhibited by pretreatment of the tissue section with unconjugated specific immune serum but not by non-immune serum. This is the so-called Blocking Test.
(4) Staining should not occur if the conjugated antiserum is first absorbed with the antigen, but there should be no difference if a different a ntigen is used. A further test, applicable only in the case of heterologous antigen, was suggested by Rossi *et al.*, 1964.
(5) Staining should not occur when the conjugate is applied to tissues of the same type as those being studied, but free from antigen.

With each of these tests interpretation of the results may be difficult. The blocking test is often regarded as an absolute test but, as indicated by Scott (1960), positive staining will still occur after blocking if incubation is prolonged (over 30 minutes). This is presumably due to exchange between non-fluorescent antibody from the tissue and fluorescent antibody in the staining solution.

Testing for specificity by blocking the antiserum with antigen is usually satisfactory unless the antigen is soluble and the antibody non-precipitating. In this case the complex will be present in the staining medium and the tissue antigen may compete successfully for neutralized antibody.

If an impure antigen has been used for immunization (as is virtually always the case) unwanted antibodies will be present in the antiserum. Absorption of the latter with antigen may block immunological staining but distinction between wanted and unwanted antibodies can scarcely be made.

Non-specific Staining

At one time it was considered that there were at least three important causes of non-specific fluorescent staining in tissue sections, using the fluorescent antibody method. These were:

(1) Unbound dye still present in the conjugated serum.

(2) Non-immunological staining by non-immune protein-dye complexes. (Riggs *et al.*, 1960.)

(3) Non-immunological staining of tissues with immunologically active conjugate—"Protein-protein reaction". (Mayersbach, 1959; Mayersbach and Schubert, 1960; Grossi and Mayersbach, 1964.)

The first cause of non-specific fluorescence, which was certainly an important one, has been eliminated entirely by the techniques of gel filtration, either with Sephadex* or DEAE-cellulose (Zwaan and van Dam, 1961; Goldstein *et al.*, 1961; Öystese, 1962; Wagner, 1962; McDevitt *et al.*, 1963).

The chief reason for the second cause, that is to say denaturation of immune protein during preparation of the conjugate, has been largely eliminated by avoiding excessive reaction with label (Goldstein *et al.*, 1961; Wood *et al.*, 1965). We are left, therefore, with the third cause outstanding. This is the so-called protein-protein reaction, or interaction.

Protein-Protein Interaction. Extensive studies by Mayersbach (1959) and his associates (Mayersbach and Schubert, 1960; Grossi and Mayersbach, 1964), and also by Rossi *et al.* (1964), have drawn attention to electrostatic binding of the fluorescent conjugate with basic proteins in tissue sections. From a different point of view the work of Louis (Louis, 1958; Hughes and Louis, 1959; Louis and White, 1960), which involves the staining of fresh tissue sections with fluorescein-protein conjugates (originally egg-white), certainly supports Mayersbach's hypothesis.

Additional evidence was provided by Allerand and Yahr (1964) who found that the 7S γ-globulin fraction of human serum or C.S.F. was strongly bound to the myelin sheaths and glial cells of the white matter of the brain, in fresh frozen sections. This binding was best demonstrated with fluorescent anti-7S globulin. If antiserum to 19S globulin was used, a similar but less intense fluorescence was observed. The reaction, which was not dependent on complement, was regarded by the authors as an example of protein-protein interaction.

An extensive study of non-specific fluorescence was carried out by Goldstein, Slizys and Chase (1961), who disagreed with the hypothesis of Riggs *et al.* (1960) which attributed the phenomenon to fluorescent non-immune globulins. Goldstein *et al.* found that fluorescent γ-globulins were retained by DEAE cellulose columns and that the strongest binding fractions were those associated with non-specific fluorescence. Methods designed to prevent non-specific staining by absorbing conjugates with dried tissue powders or activated charcoal (Chadwick *et al.*, 1958), or by extraction with ethyl acetate (Dineen and Ada, 1957), were observed to be completely unsatisfactory when applied to globulins having high fluorescein-protein ratios. In the case of globulins with low ratios, there was some evidence that tissue powder absorption decreased non-specific staining.

* Registered Trade Mark. Pharmacia, Uppsala, Sweden.

The importance of the fluorescein-protein ratio was stressed by Grossi and Mayersbach (1964) who indicated that progressive loss of NH_2 groups (by reaction with isothiocyanate) produces a progressive rise in non-specific staining. Mayersbach and Schubert (1960) had already showed that changes in the electrical charge of tissue proteins, induced by fixation or by changing the pH of the buffer used for application of the conjugate, could alter the degree of staining observed. At low pH levels very strong non-specific staining occurred and raising the pH reduced this considerably. Unfortunately, however, high pH levels also have an adverse affect on the antigen-antibody reaction.

Extending the above work Grossi and Mayersbach (1964) demonstrated that the extractive effect of the buffer, by removing predominantly acidic tissue components (e.g. nucleic acids), increased the overall basicity of the tissues. This produced, as expected, a profound increase in non-specific staining. The three worst buffers were veronal-acetate, Tris, and phosphate. No single buffer was recommended, however, but the lowest extractive effect occurred with 0·5 M carbonate-bicarbonate buffer at pH 8·65.

Experimental data provided by Hebert *et al.* (1967) showed that non-specific staining increased in linear fashion as the fluorescein-protein ratio increased. The effect was directly related to the fluorescein isothiocyanate content of the conjugates. The authors indicated that, in their view, provided that conjugates of sufficiently high specific titre could be obtained, dilution was the simplest and most effective way of reducing or eliminating non-specific staining.

Electrophoretic studies made by Curtain (1961) indicated that fluorescein-labelled γ-globulins were heterogeneous. By electrophoresis convection it was possible to isolate most of the non-specific staining material in the faster fractions. These contained an excess of negatively charged fluorescein groups while the slower fractions contained specific-staining antibody. Since electrophoresis convection is a tedious technique Curtain recommended fractionation by chromatography on cross-linked dextran (Sephadex) and ion-exchange cellulose (DEAE cellulose).

Considerable support has accumulated for the DEAE cellulose fractionating procedure (Riggs *et al.*, 1960; Goldstein *et al.*, 1961, 1962; Tokumaru, 1962; Frommhagen and Martins, 1963; McDevitt *et al.*, 1963; Burtin and Buffe, 1963; Rossi *et al.*, 1964). Following the critical studies of Wood *et al.* (1965), it has become almost the only acceptable method for preparing conjugates. Full details are given in Appendix 7, p. 633. Wood and his associates tested a number of different factors affecting the preparation of conjugates and their purification. The effect of pH on the efficiency of coupling was most marked and its maintenance above 9·0 during the 2 hours of the reaction was critical.

As the molecules of FITC were progressively attached, so the solubility, specific activity and isoelectric point of the starting globulin were observed to decrease. When too few FITC molecules were attached to the antibody, light emission from the conjugate was inadequate. If too many were attached brilliant staining was noted but at the expense of marked non-specific protein-protein interaction. Wood *et al.* summed up their views succinctly as follows: "Between these extremes there is a range where bright and specific staining occurs. To obtain such staining it is necessary to remove the undercoupled and overcoupled fractions."

Here we have the whole position of immunofluorescence and non-specific staining set out in a few words. Consideration of the whole of the above section indicates that non-specific staining can be brought to a minimum by the following procedures:

(1) Reduction of tissue non-specific reactive sites (basic groups) by denaturation, fixation, or other special procedure.
(2) Preparation of a "correctly" coupled conjugate.
(3) Efficient U.V. microscopy (see Chapter 29) in order to allow the lowest possible level of labelling to be "correct".
(4) Proper choice of problem.

With regard to this last proposition I have consistently observed that when the answer to the problem is already known with certainty, as in the case of localization of pituitary growth hormone, immunofluorescence gives precise and accurate results (Fig. 60, facing p. 188). When conditions are otherwise, as for instance when studying the production of antibodies at the cellular level, it is often impossible to separate specific and non-specific staining with any degree of confidence.

The greatest degree of non-specific staining is found in fresh-frozen cryostat sections, even after the usual post-fixation in alcohol, acetone, or alcohol followed by acetone (Öystese, 1962). Freeze-dried unfixed and post-fixed sections show the effect much less strongly but post-fixation in efficient and rapid bifunctional fixatives such as glutaraldehyde (Chapter 5, p. 74) soon destroys all specific activity. Alcohol-fixed paraffin embedded sections (Sainte-Marie, 1962) provide an excellent compromise in many cases but, where they can be used, formalin-fixed paraffin sections may give adequate specific fluorescence with no non-specific fluorescence whatsoever.

Tissues which are particularly prone to non-specific reactions should not be regarded as the seat of specific fluorescence until the most rigorous series of control tests has been carried out. Tissues with which most trouble can be anticipated are epithelia, particularly stratified squamous variety and those of glandular origin, whether endocrine or exocrine. Mucins, in general, do not show high levels of protein-protein interaction but histiocytes, and neutrophil and eosinophil leucocytes, are notoriously bad. Voluntary muscle exhibits weak affinities but myocardial fibres react strongly (Rossi *et al.*, 1964). Nervous tissues do likewise (Allerand and Yahr, 1964). Neoplastic tissues react indifferently or not at all, even when derived from strongly reacting precursors. This observation formed the basis of a fluorescent protein test for cancer (King, Hughes and Louis, 1959).

FLUORESCENT TRACERS

The compounds described in this section can be, or have been, applied to the problem of tracing the fate of injected foreign protein antigens. They are equally applicable to labelling of antibodies for use in the fluorescent antibody methods and the particular compounds discussed below are dealt with in this chapter, rather than in Chapter 29, mainly because they have been most extensively used for immunofluorescence.

The following requirements, which should be satisfied by fluorescent compounds designed for use in tracer experiments, were set forth by Chadwick, McEntegart and Nairn (1958b). Such compounds should:

(1) Possess chemical groups capable of reacting directly with proteins to form a stable linkage such as azo, carbamido or sulphonamido.
(2) Not possess groups capable of interfering with the reaction with protein.
(3) Have an intensity of fluorescence (fluorescence yield) of the same order as fluorescein and this should not be diminished appreciably by conjugation.
(4) Give a fluorescence colour of longer wavelength than fluorescein so as to provide better contrast.

To these requirements one might add:

(5) That the compound should be stable and that its solubility in water, and its reactivity, should be such that it can be used for conjugation under conditions leading to the minimum amount of denaturation of protein.

Some of the above requirements cannot be attained. In particular, the intensity of fluorescence of conjugates never approaches that of the free label.

PREPARATION AND CHARACTERISTICS OF FLUORESCENT LABELS

Fluorescein-based Labels. The first successful label for immunofluorescence, and until 1958 the only label, was fluorescein-4-isocyanate. This can be prepared by passing phosgene into a solution of the amine in acetone. The amine, which occurs in two isomeric forms (I and II) is synthesized by the method of Coons, Creech, Jones and Berliner (1942) according to the details given by Coons and Kaplan (1950).

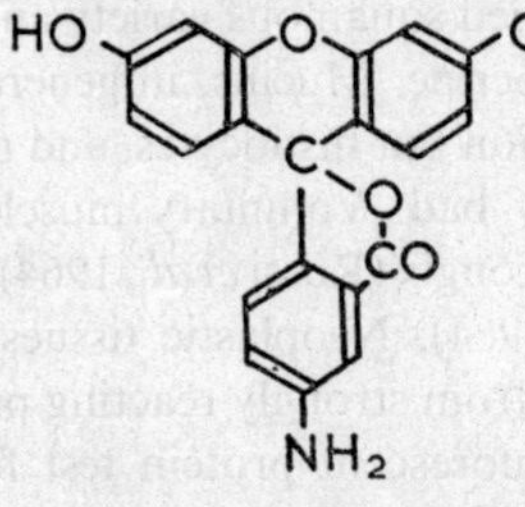

I

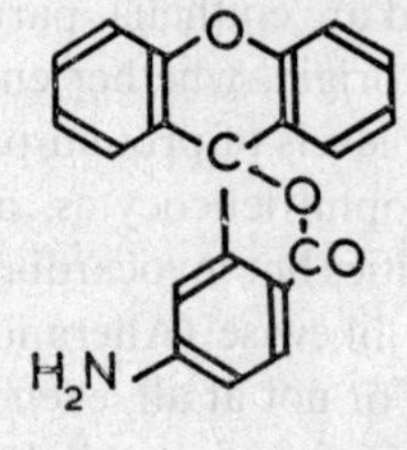

II

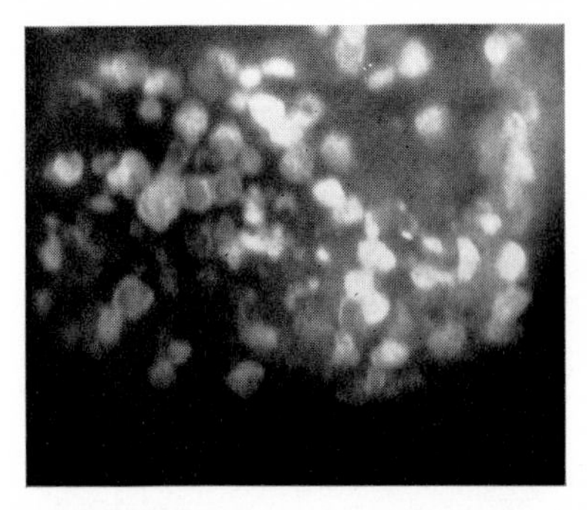

FIG. 60. Unfixed 8μ cryostat section. Anterior hypophysis of a 55-year-old female. Specific fluorescence, confined to the α-cells, after application of anti-growth hormone and fluorescein-labelled anti-globulin (sandwich technique). × 210.

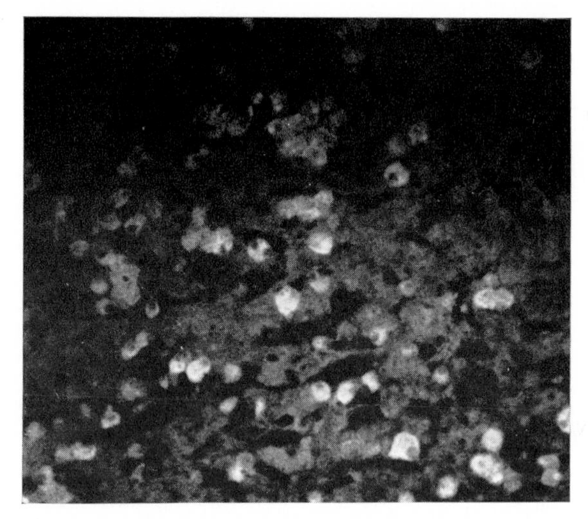

FIG. 68. Unfixed cryostat section. Anterior hypophysis of 28-year-old male (sudden accidental death). Specific anti-ACTH fluorescence confined to R-type mucoid cells. × 190.

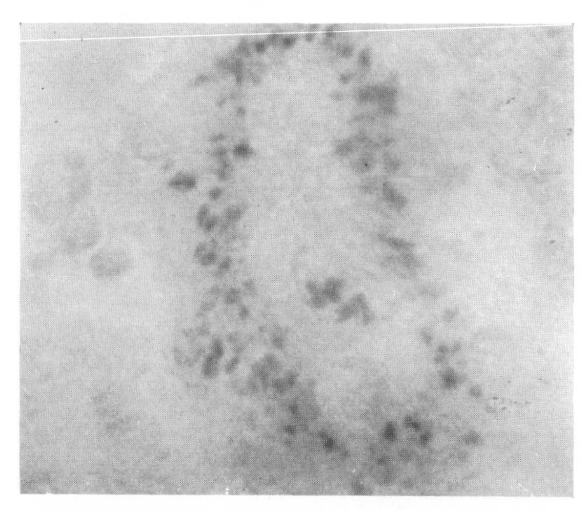

FIG. 66. Mouse kidney. Fresh frozen section. Animal given intraperitoneal injection of ovalbumin labelled with Procion blue HBS, 36 hours previously. Intracellular deposits of the labelled protein can be seen. × 1400.

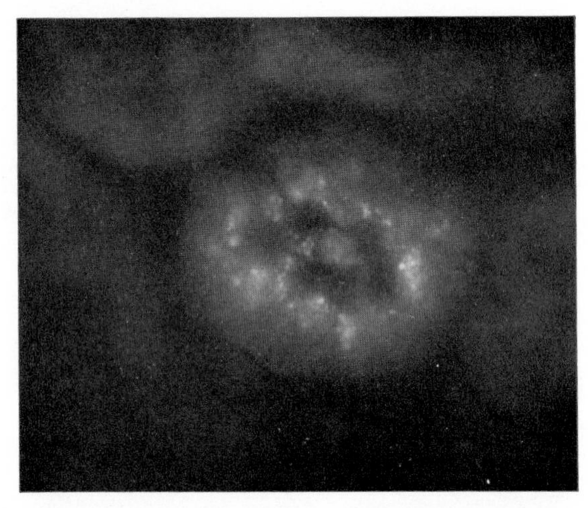

FIG. 67. As Fig. 66 but animal injected with ovalbumin labelled with a fluorescent dichlorotriazene dye, 20 hours previously. Small intracellular protein deposits are brightly fluorescent. × 1400.

Details of the synthesis are given in Appendix 7, p. 630. Fluorescein isocyanate, which is now seldom employed, is available commercially from a number of sources in the U.S. and in Europe.

Although Marshall (1951) reported that sealed ampoules of an acetone solution of the isocyanate would remain stable for a year in the dark, the majority of workers found a much lower stability. A period of 8 weeks was suggested by Coons (1956) as the longest acceptable period. It is worthy of note that Moulton (1956) found that his conjugates had a higher antibody titre when prepared from fresh isocyanate.

The second fluorescein-based label, the isothiocyanate, is now almost universally employed. It was first introduced by Riggs *et al.* (1960) as a stable yellow powder, whose formula appears below.

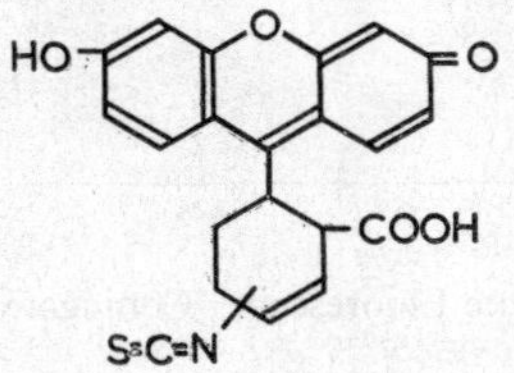

HO — O — =O — COOH — S=C=N

Fluorescin isothiocyanate

Preparative details are given in Appendix 7, p. 631, but fluorescein isothiocyanate (FITC) is widely available as a commercial product.

Observations made by Corey and McKinney (1962) suggested that the purity of FITC samples from commercial sources varied considerably. Some were composed of isomer I and some of isomer II but the majority were mixtures. McKinney *et al.* (1964) found that a sample of FITC Isomer I, prepared by themselves, was equal or superior in FITC content to all the commercial samples tested. Using infra-red assay they found that the worst samples, which were invariably mixtures of the two isomers, contained less than 40 per cent. FITC. Furthermore, these same authors showed that the majority of samples contained an appreciable amount of HCl, in some cases approaching the theoretical level for a monohydrochloride.

The presence of various levels of HCl in FITC samples probably accounts for the observed variation in solubility and, possibly, for variations in stability under storage conditions. Significant deterioration was noted over a long storage period. The fluorescence spectrum of an FITC conjugate is shown in Fig. 61 (Hansen, 1964).

This shows that the useful excitation maximum is at 490 mμ,* with a smaller peak at 320 mμ which is usually employed only when distinction from some autofluorescent tissue component is required. The emission maximum is at 517 mμ, in the blue green region of the spectrum, but this colour is normally modified by the necessary yellow barrier filters. Figures given by Emmart (1958) indicate an emission maximum for fluorescein conjugates of 550 mμ.

*Used in this Chapter as well as the alternative expression (nm).

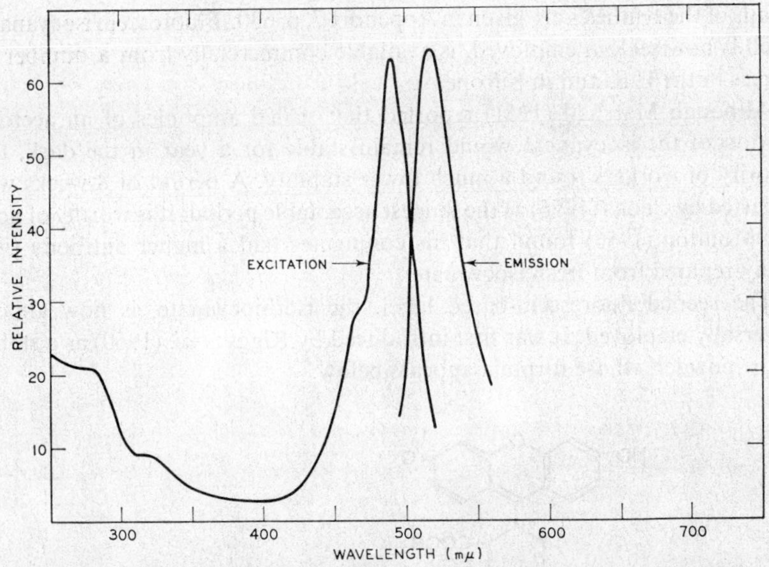

FIG. 61. Fluorescence Spectrum of dilute Fluorescein—Conjugate at pH 7.6

Alternative Fluorescent Labels. The technical difficulties attached to the preparation and purification of fluorescein amines, and objections to phosgenation as the final stage of preparation, naturally resulted in a search for alternative methods of linking a fluorescent label to proteins. Several compounds have been seriously suggested as alternatives to fluorescein isocyanate and isothiocyanate. The first of these, 1-dimethylaminonaphthalene-5-sulphonyl chloride (DANSYL) was used by Weber (1952) and Laurence (1952) in their studies on polarization of fluorescence. This compound is easily prepared by methylation of 1-aminonaphthalene-5-sulphonic acid (DANS) and subsequent treatment with PCl_5 (Appendix 7, p. 631).

NH_2 / SO_3H $\xrightarrow[(CH_3)_2SO_4]{KOH}$ H_3C CH_3 N / SO_3H

$+PCl_5$ → H_3C CH_3 N / SO_2Cl

It couples with protein α-amino groups by covalent bonding to form sulphonamides. Between one and three groups per molecule of protein can be attached in this way and the resulting conjugate gives an intense yellow fluorescence.

Clayton (1954) used this method for labelling antisera prepared against certain embryonic antigens, but, in my original studies (with H. Mayersbach) carried out in 1956, it compared unfavourably with fluorescein isocyanate in two respects. First, a substantial amount of the compound was adsorbed on to the protein which had been labelled by covalent bonding, and prolonged dialysis was necessary before the dialysate ceased to be fluorescent. Secondly, when this point was reached the resulting conjugate was considerably less fluorescent than the fluorescein conjugate prepared from the same protein.

In further extensive studies carried out independently, however, Mayersbach (1958) found that the difficulty of successful dialysation after DANSYL conjugation was confined to whole serum or to albumen fractions. Only short dialysis was necessary with conjugated globulins and the final fluorescence compared favourably with that given by fluorescein isocyanate conjugates, except that the wavelength of the light emitted was closer to that of normal tissue autofluorescence. The effect of this compound on antigen-antibody reactions was tested by Redetski (1958) who found that γ-globulin-DANSYL conjugates contained 3–4 molecules of dye per mole of protein and that this amount of labelling had no effect on the formation of an antigen-antibody complex. Unfortunately, measurements of remaining free dye in the supernatant after precipitation of the protein conjugate do not take into account the amount of adsorbed dye. Redetski's statement that precipitation with alcohol or acetone does not split the protein-DANSYL "bond" must be regarded with caution until proof is offered that such a chemical bond is in fact formed under the usual conditions of conjugation.

Hartley and Massey (1956) investigated the reaction of 1-dimethylamino-naphthalene-5-sulphonic acid with various amino-acids. They found: (1) a very stable bonding with α-amino groups, resistant to both acid and alkaline hydrolysis; (2) a similarly stable bonding with the ε-amino groups of lysine; (3) an acid-labile bonding with cysteine with the probable formation of S-sulphonyl cysteine; (4) a relatively stable bonding with the phenolic group of tyrosine; and (5) formation of an unstable ring-labelled compound with the imidazole group of histidine. These results suggested that theoretically it should be possible to introduce more fluorescent groups per molecule of protein with DANSYL than with fluorescein isocyanate. The optimum conditions for conjugation of DANSYL with γ-globulins have been described by Albrecht and Sokol (1961). When the dye concentration is raised above 2 per cent. of the weight of the protein, considerable precipitation occurs. In Fig. 62, below, is shown the fluorescence spectrum of a DANSYL conjugate (Hansen, 1964). All the maxima are at longer wavelengths than those of the unbound dye.

The excitation maximum is at 250 mμ which is of no value to most fluorescent microscopists since radiation of this wavelength does not pass through glass optics. The second peak, at 313 mμ, provides radiation which is presumably the main source of excitation in practice.

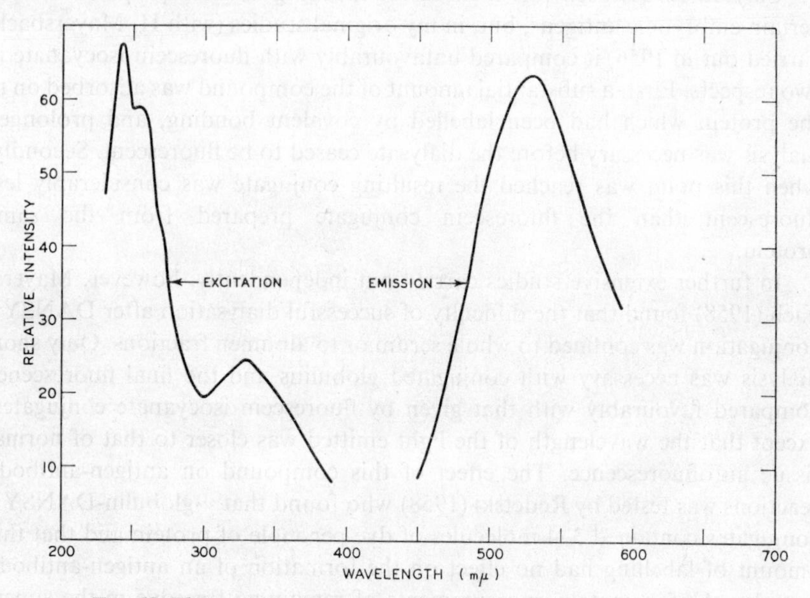

FIG. 62. Fluorescence Spectrum of DANSYL—Conjugate at pH 7.

An alternative to fluorescein, suggested by Clayton and used in the work mentioned above, was described as Nuclear Fast red (benzaldehyde-6-nitro-2-sodium diazotate). As stated, the composition of this compound closely resembles that of Nuclear Fast red salt B or Kernechtrotsalz B (5-nitro-2-amino methoxybenzene diazotate), but not that of Nuclear Fast red (Kernechtrot, Bayer), which is an anthraquinone mordant dye and not a diazonium salt. Neither Kernechtrotsalz or Kernechtrot produce a useful fluorescence when coupled to or adsorbed on proteins, although the second does give a faint yellowish-red fluorescence at times. I can neither understand nor recommend the above method.

Another alternative and contrasting label to fluorescein isocyanate was prepared and used successfully by Silverstein (1957). This author prepared the nitro derivative of the xanthine dye Rhodamine B by condensing two equivalents of *m*-diethylaminophenol with one equivalent of 4-nitrophthalic anhydride. This process is followed by catalytic reduction to the amine, and this product, on treatment with phosgene, yields the isocyanate. The latter is coupled to protein in the usual way. In his original note on the use of rhodamine isocyanate Silverstein gave no details of the purification of conjugates, and Hiramoto *et al.* (1958), who used the tetramethyl compound for the demonstration of antithyroid antibodies in human sera from cases of chronic thyroiditis, also gave no technical details. Satisfactory labelling was apparently difficult to achieve and, since the isocyanate was not stable Riggs *et al.* (1960) prepared the isothiocyanate.

Rhodamine B isothiocyanate

This compound proved more satisfactory and it has been employed chiefly where contrasting fluorescence is required.

The isocyanate of tetramethylrhodamine was used by Hiramoto *et al.* (1958) as a fluorescein substitute, as indicated, but this compound has also been replaced by the more stable isothiocyanate. The latter is commercially available and probably the most popular red fluorescing label for immunofluorescence. The base occurs as two isomers, designated G and R, but the commercially available isothiocyanate (TMRITC) is derived from the latter. Spectral characteristics for the conjugate are given in Fig. 63. (Hansen, 1964), below.

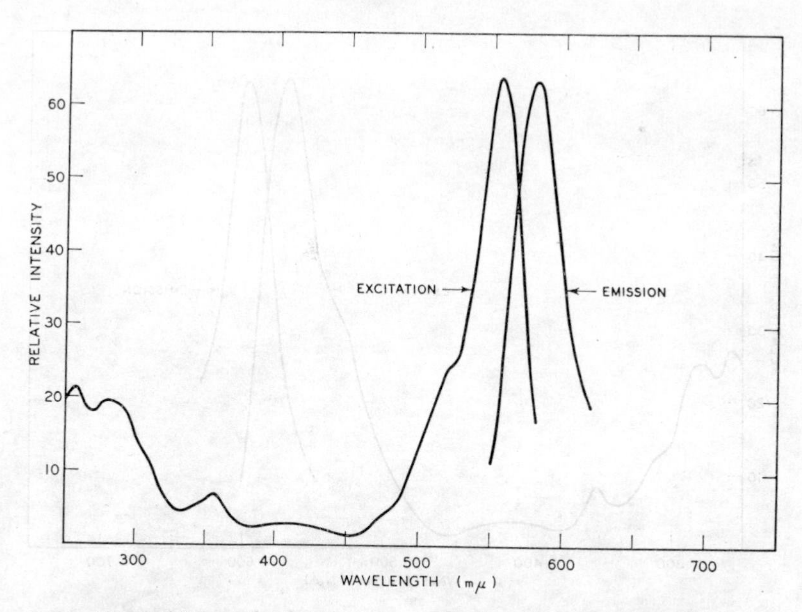

FIG. 63. Fluorescence Spectrum of TMRITC—Conjugate at pH 8

The useful excitation is at the maximum (555 mμ) with the emission maximum at 580 mμ. There is an absorption peak at 517 mμ, but this does not give rise to fluorescence.

A new compound, Lissamine rhodamine sulphonyl chloride was introduced for fluorescent labelling by Chadwick, McEntegart and Nairn (1958a and b). The formula for the parent dye, lissamine rhodamine RB200 (I.C.I. Ltd.) is shown below.

Et_2N $\overset{+}{O}$ NEt_2

SO_3^- (II)

SO_3Na (I)

This dye, which is the disulphonate of Rhodamine B, has two sulphonic acid groups available for conversion to sulphonyl chlorides. Under mild conditions (e.g. chlorosulphonic acid) it is probable that only one of these (I) will be converted, giving a product insoluble in water. Under more severe conditions (PCl_5) both groups (I and II) undergo conversion to sulphonyl chlorides giving a basic dye which is soluble in water but rather easily decomposed. Directions

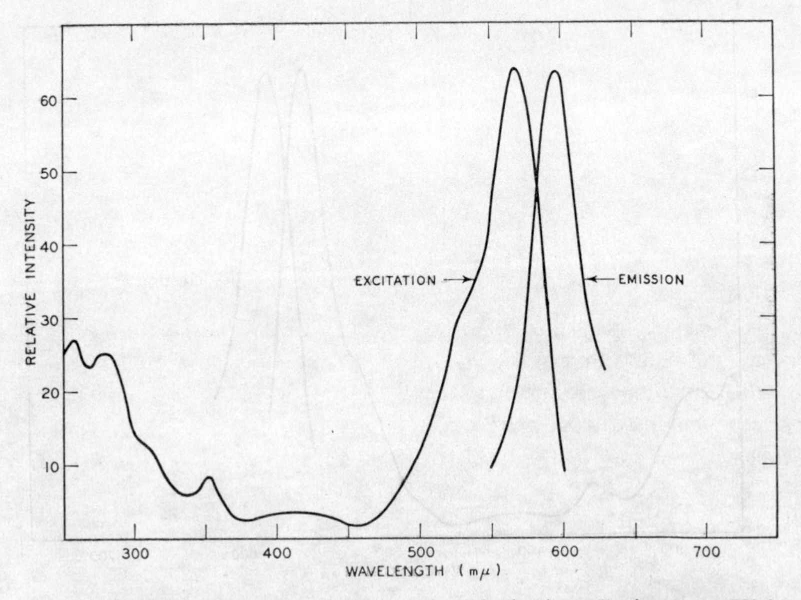

FIG. 64. Fluorescence Spectrum of Lissamine Rhodamine—Conjugate at pH8·0

for the synthesis of lissamine rhodamine disulphonyl chloride are given in Appendix 7, p. 632.

The fluorescence spectrum of Lissamine rhodamine RB 200 conjugate (Hansen, 1964) is shown in Fig. 64.

This compound was successfully employed by Chadwick *et al.* for the labelling of proteins for tracer and antigen-antibody methods.

Excitation occurs at 257, 280, 352 and 568 mμ so that the last is the useful radiation for fluorescence microscopy. The emission maximum is at 597 mμ in the orange region of the spectrum. As observed in the microscope, using an orange-yellow barrier filter, the colour varies from red (high concentrations of the label) to a dirty orange-brown (low concentrations of the label).

Under the synonym sulphorhodamin B (see Harms, 1965) the same dye was used in the form of its sulphonyl chloride by Uehleke (1959) and by Möritz and Ambrosius (1964).

A dye which they called aminorosamine B was synthesized by Borek and Silverstein (1960) and this was used in the form of its isocyanate, isothiocyanate or diazonium derivatives for conjugation with protein. Studies on this dye, and its isothiocyanate, by Hansen (1964) produced unsatisfactory results and it was suggested that this label has very little practical application.

In his search for alternative fluorescent labels Uehleke (1958) found a compound, 3-hydroxypyrene-5,8,10-trisulphonic acid, which he used in the form of its sulphonyl chloride (? disulphonyl chloride) for the preparation of conjugates.

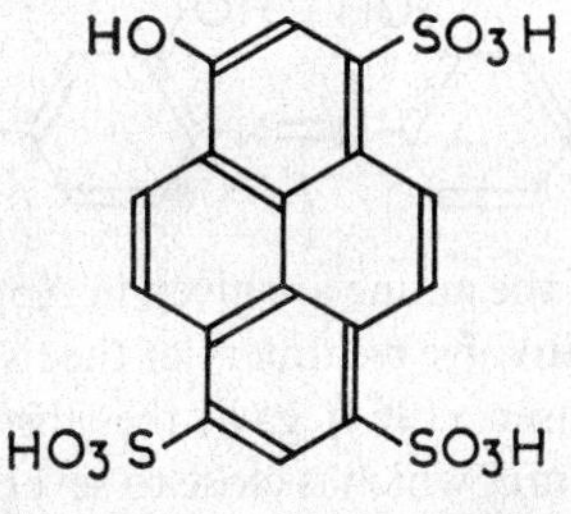

The fluorescence spectra of this dye and of its conjugate were reported by Hansen (1964), and that of the conjugate is shown on p. 196:
Excitation maxima were somewhat different at pH 7 and pH 8 but the emission maximum remained at 515 mμ. Hansen expressed the opinion that this strongly fluorescent label had considerable promise since its (broad) excitation maximum at 374 mμ is in a region where the mercury arc has excellent emission (between 350 and 450 mμ).

Following the discovery by Freeman and White (1956) that azo dyes having hydroxyl groups in the ortho position could form fluorescent chelates with aluminium, Dowdle and Hansen (1959) synthesized 2,2′,4-trihydroxy-4′-

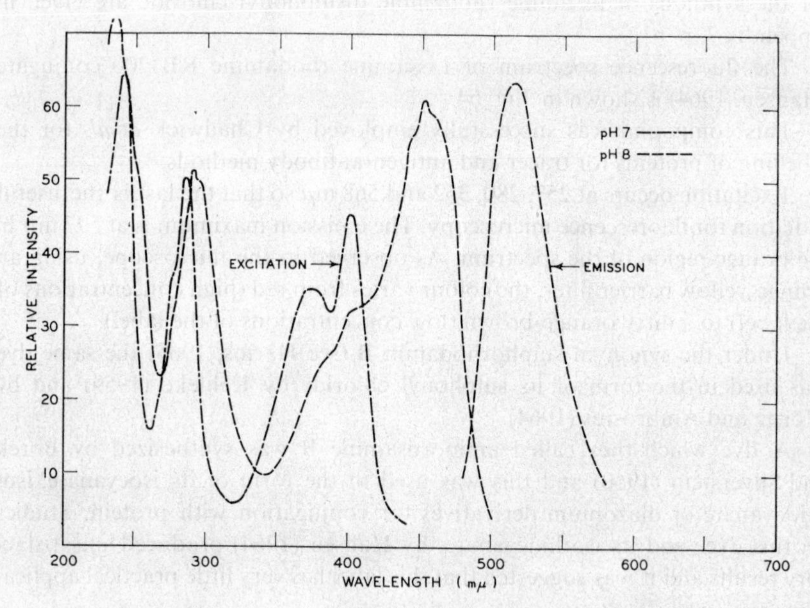

FIG. 65. Fluorescence Spectrum of 3-Hydroxypyrene-5, 8, 10-trisulphonic acid—Conjugate at pH7 and pH8

aminobenzene. The purified amine was reacted with thiophosgene and the isothiocyanate was then used for conjugation.

OH HO
H_2N—⟨ring⟩—N=N—⟨ring⟩—OH

Details of the synthesis of the amine are given in Appendix 7, p. 632. The final stages before microscopy involve treatment of the tissue bound conjugate with aluminium chloride. Hansen (1964) gave the absorption maximum of the chelated conjugate as 365 mμ, which is close to several strong lines of emission of the mercury arc.

A number of similar azo dyes, as metal chelates, have intense fluorescence (see Chapter 29). So far none have proved satisfactory as fluorescent labels for conjugates.

Cellulose Reactive Dyes. A large number of reactive groups are known which will combine rapidly with the hydroxyl groups of cellulose fibres at alkaline pH levels to form stable covalent bonds. Under similar conditions these groups can react with the free amino or hydroxyl groups of proteins. These reactive groups include thiocyano, mono and dichloro-triazinyl, sulphon-β-chloro ethylamido, sulphonfluoride, chloroacetylamino, and γ-chloro-β-hydroxypropyl.

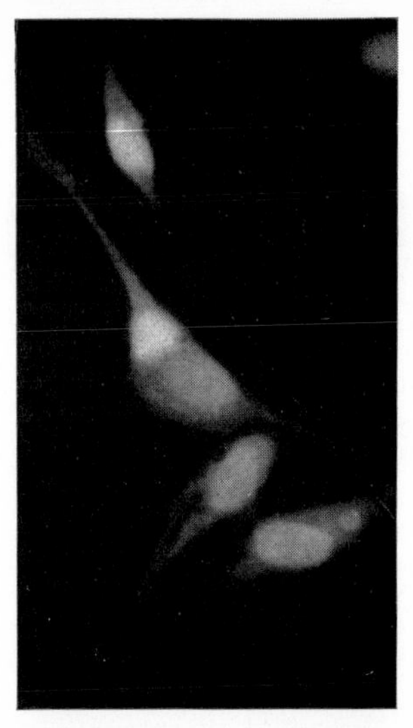

FIG. 69. Tissue culture preparation. Shows localization of Vaccinia virus in the cultured cells. (Fluorescein-labelled antibody.) × 1000.

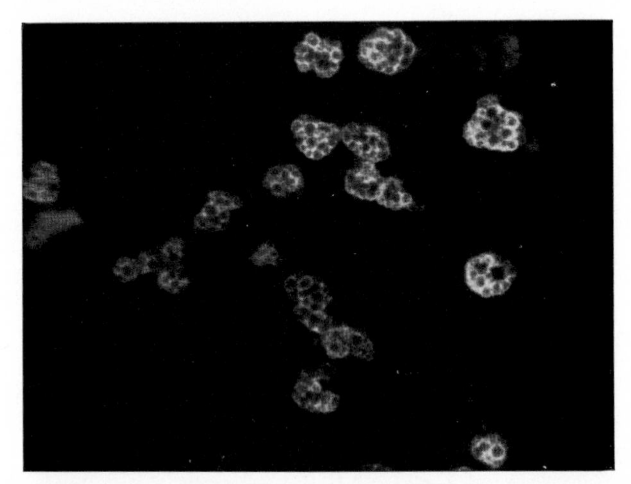

FIG. 70. Mouse spleen. Fresh frozen section from animal immunized with *P. vulgaris* vaccine. Layering technique to show presence of antibody to the organism. (Fluorescein-labelled antibody.) × 600.

A series of coloured or fluorescent compounds containing these groups was tested by Hess and Pearse (1959) in tracer protein experiments and the fluorescent compounds were also tested as labels in antigen-antibody reactions. In these experiments the best results were obtained with dyes (coloured or fluorescent) possessing the mono or dichloro-triazinyl group.

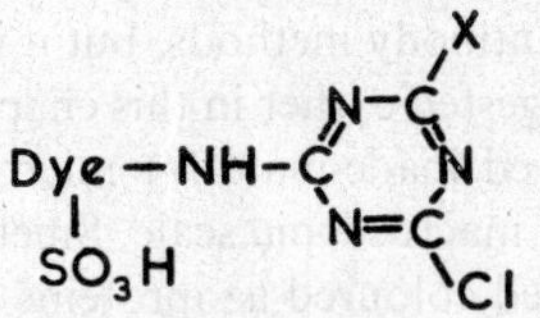

In the above formula Dye-SO_3H represents the sulphonated chromophor (azo, anthraquinone, phthalocyanin, etc.) and X stands for either a chlorine atom or an inactive radical.

The fact that firm chemical linkage results from the conjugation of proteins with mono or dichloro-triazinyl compounds was demonstrated by a method originally described by Vickerstaff (1958). A yellow reactive dye possessing an azo chromophor was coupled to ovalbumin. The dye-protein compound was then reduced with sodium hydrosulphite to a colourless product containing free amino groups (from the split azo linkage). This was then diazotized and coupled with naphthol AS to give a deep red protein-azo dye.

Some of the more highly coloured reactive dyes containing the dichloro-triazine grouping were used to label proteins in tracer experiments in mice. The result given by coupling the dye Procion blue HBS (I.C.I. Ltd.) with ovalbumin are shown in Fig. 66 where the protein droplets in the renal tubules are demonstrated 18 hours after injection of the labelled conjugate. Better results were obtained, however, by using compounds containing a fluorescent chromophor.

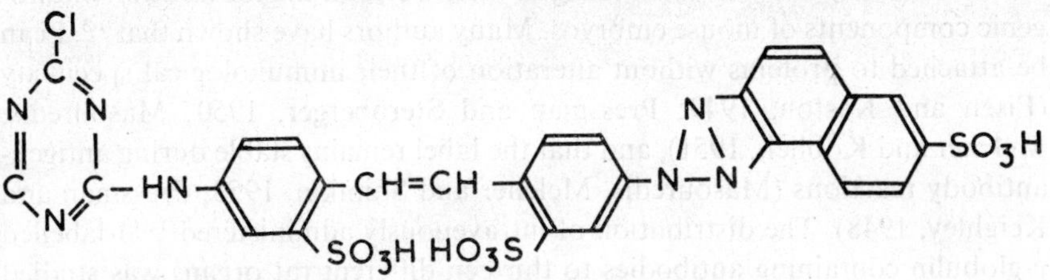

These results are illustrated in Fig. 67, using the same test system described above.

When reactive dyes are coupled with whole serum the result, as with all the fluorescent labelling compounds, is a preferential attachment of the dye to the albumin fraction. Labelling of the γ-globulin fraction may be minimal. If a purified γ-globulin fraction is used, however, labelling with coloured or fluorescent dichloro-triazinyl compounds is quite satisfactory and relatively large amounts of dye can be coupled to the protein if required. In experiments with Procion blue HBS as much as 50 mg. of dye was coupled to 50 mg. of serum albumin (20°; pH 10·0; 1 hour) with recovery, after dialysis, of 2 per cent. of the original weight of dye. It is difficult to determine precisely the amount of dye chemically linked to the protein and the amount which is physically adsorbed.

Alternative Non-fluorescent Labels. Strictly speaking these have no place in a chapter on fluorescent antibody methods, but it is convenient to enumerate them at this point. As suggested earlier in this chapter, the use of *direct colour labelling* has usually proved inadequate for microscopy though it may give satisfactory results on the macroscopic scale. Smetana (1947) and Kruse and McMaster (1949) both used coloured azoproteins as tracer antigens, and the latter showed that diazotized Echtsaure blau and Evans blue could be coupled to proteins in a firm union to give a product visible microscopically when concentrated in such sites as the Kupfer cells.

Of the two alternatives which remain the first, *chromogenic labelling*, was attempted by Mayersbach and Pearse (1956) with little success. A naphthaldehyde conjugated egg white or whole serum was prepared by coupling with 3-hydroxy-2-naphthaldehyde (see Chapter 6, p. 118, and Appendix 6, p. 611) at pH 10. Freeze-dried tissues containing the labelled egg white were treated with a diazonium salt at pH 7·4, but the colour developed (a brownish-red) was too weak to demonstrate any but the most concentrated sites (e.g. Kupfer cells). After chromatographic separation of the labelled whole serum, however, the *albumin* fraction gave a strong black colour on treatment with the diazonium salt suggesting that the principle of chromogenic labelling is at least worthy of further trial.

The final alternative is *isotope-labelling* followed by radioautography. This principle was used by Pressmann, Hill and Foote (1949) as part of a long series of studies on ^{131}I-labelled antisera. Clayton and Feldman (1955) also made use of this method of labelling in their work on the localization of antigenic components of mouse embryos. Many authors have shown that ^{131}I can be attached to proteins without alteration of their immunological specificity (Eisen and Keston, 1949; Pressman and Sternberger, 1950; Masouredis, Melcher and Koblick, 1951), and that the label remains stable during antigen-antibody reactions (Masouredis, Melcher and Shimkin, 1953; Pressman and Keighley, 1948). The distribution of intravenously administered ^{131}I-labelled γ-globulin containing antibodies to thirteen different rat organs was studied by Bale and Spar (1954) and by Spar and Bale (1954). Radio-autographic examination showed the maximum localization of all types of antibody in the renal glomeruli and in the zona reticularis of the adrenal.

Preparation and Application of Fluorescent Antisera

Production of Immune Globulins. As Coons (1954, 1956) emphasized, the production of antisera depends on methods which are largely empirical and, for those who are unfamiliar with immunological technique some assistance will be needed. The necessary methods, and the principles behind them, are dealt with by Kabat and Mayer (1948) and by Boyd (1956). Other articles which may be consulted are those of Wilson and Miles (1955) and Humphrey (1962).

The preparation of the antigen itself may not be entirely straightforward, especially if tissue antigens are being used. Many authors (e.g. Cruickshank and Hill, 1953) have used crude homogenates of single organs or tissue components. These certainly give rise to multiple antibodies. Vaughan and Kabat (1954), for instance, found an antibody in anti-egg albumin sera which was only revealed by the agar diffusion technique. They showed that the specific antigen was present in a concentration of less than 1 per cent. of the total protein and that it was not detectable by the usual methods of purification, such as electrophoresis. The antibody response to such an antigen may often be disproportionately high. Although it has often been stated that precipitating antibodies are necessary for immunofluorescence this is not true, as emphasized by Nairn (1962).

The use of adjuvants is recommended for increasing the antibody response, especially to non-living antigens. For this purpose the method of Freund (Freund and McDermott, 1942; Proom, 1943) has most often been used. Some appropriate schedules for the preparation of antibodies are given in Appendix 7, p. 636.

The production of antibodies to low molecular weight proteins or polypeptides has, in the past, been difficult or even impossible. The successful preparation of antibodies to synthetic octapeptides was reported by Berglund (1965) who adsorbed her antigens on polymethylmethacrylate particles prior to injection. Using this technique Bussolati and Pearse (1967) succeeded in preparing an antibody (non-precipitating) to the polypeptide hormone calcitinin (mol. wt. 3500). Previous attempts, using conventional techniques, had proved abortive.

The most usual procedure for separation of γ-globulins from other serum proteins is by precipitation with ammonium or sodium sulphate. It is necessary completely to remove residual salt from the globulin before labelling. If this is not done there will be interference not only with the latter procedure (Kaufman and Cherry, 1961) but also with the biuret procedure (Gornall *et al.*, 1949). Although there are a number of alternative methods for estimating the protein content of the purified globulin fraction, the most popular is the biuret reaction (Appendix 7, p. 637). It is, of course, necessary to know the protein content in order to couple with the chosen label at the optional proportion.

The concentration of antibody in the globulin may be estimated by a simple procedure and its specificity may be tested by immunoelectrophoresis Details are given in Appendix 7. Alternatively gel diffusion (Ouchterlony plates) is employed.

Storage of immune globulins is usually carried out at 0°–4° and often a preservative is added to prevent bacterial growth. Merthiolate and sodium azide have been used in low concentrations (1:10,000 and 0·015 M, respectively). A hitherto unknown effect of azide was described by Lachmann (1964) who obtained low fluorescein-protein ratios with sera containing azide in spite of efforts to raise the ratio. Removal of the azide restored the situation. In all probability azide and FITC combine to produce an unreactive tetrazole.

Fluorescent Labelling. The mechanism of conjugation of proteins and isocyanates was discussed by Hopkins and Wormall (1933), who showed that 80–90 per cent. of the free amino groups of caseinogen disappeared after treatment with phenyl isocyanate at pH 7–8. Chemical and immunological evidence suggested that combination with the ε-amino groups of lysine was responsible. If this was true (as seems likely) consideration of the pH range of dissociation of this grouping (Table 10, p. 115) would indicate that conjugation should be better at a somewhat higher pH. Coons, in fact, used pH 9 for the purpose. Failure to form phenylcarbamido compounds with zein (a protein deficient in lysine) was noted by Hopkins and Wormall, but Creech and Jones (1941) were able to obtain slight conjugation with this protein. Some terminal α-amino groups were presumably responsible for this effect.

The procedures described above resulted in the attachment of about two molecules of fluorescein per molecule of protein, and attempts to introduce a greater number were not successful. Schiller, Schayer and Hess (1952–53) labelled bovine serum albumin with increasing amounts of fluorescein isocyanate, but found that at a concentration six times that normally employed the number of groups per molecule only rose from 1·4 to 1·9. These authors made a comprehensive study of the physical and immunological properties of fluorescein carbamido-proteins and concluded that they did not differ in size, shape or homogeneity from the uncoupled proteins.

Many variations have been introduced into the technique of conjugation. Coons and Kaplan (1950) added FIC in an acetone-dioxane (1:2) solution to their immune serum fraction containing acetone-dioxane and carbonate-bicarbonate buffer. The temperature was maintained at 0°–2°. Acetone, in the concentration usually employed, was reported to have no effect on immune globulins (Goldstein *et al.*, 1961) and, even after the advent of FITC, it continued to be used as a solvent. Attempts were nevertheless made to avoid organic solvents altogether. Goldman and Carver (1957) impregnated filter paper strips with FIC and, after drying, they used these to introduce the label into the buffered protein medium. Marshall *et al.* (1958) developed a procedure for FITC coupling in which the label was added directly to the carbonate-bicarbonate buffered globulin without any solvent. Conjugation was allowed to proceed for 12–18 hours at 4°. This method found considerable favour and it was adopted by a majority of workers in the field of immunofluorescence (Appendix 7, p. 633).

The use of Celite powder as a carrier for the fluorescent label (Rinderknecht, 1960) provided satisfactory conditions for conjugation and commercial Celite-FITC and Celite-DANSYL became available. Most workers, however, continued to use the technique of Marshall *et al.* Other modifications worthy of mention include the method of dialysis labelling (Clark and Shepard,

1963) by which FITC in buffered aqueous solution was allowed to diffuse through a membrane into the globulin, for 24 hours at 4°.

As indicated earlier in the chapter, the findings of Goldstein *et al.* (1961) showed that the molar fluorescein-protein ratio could not exceed 1·5 without the introduction of non-specific staining. Realization that this was the most important factor in the successful performance of immunofluorescent techniques produced a number of modifications in labelling techniques directed to the precise control of reaction conditions. McKinney *et al.* (1964) examined the effects of temperature, concentration, and pH, on the rate and extent of the reaction between FITC and rabbit γ-globulin. Using pH 9·5, 25°, 2·5 per cent. globulin and 0·05 M buffer, the concentration of FITC was varied from 2·5 to 25 μg/mg. protein. Conjugation was complete in 30 minutes using 6·5 μg/mg. and reproducibility of results was excellent. *From this work it is clear that the fluorescein-protein ratio can be controlled accurately by varying the initial concentration of the label.*

Purification of Conjugates. Crude fluorescein-labelled globulins produce fluorescence in many structures in normal animal tissues, especially in elastic tissue and endothelium of the blood vessels, in eosinophil leucocyte granules, and in a number of other sites. Non-specific "staining" of this sort has to be prevented by purification of the conjugate before use.

At one time the initial purification was invariably carried out by prolonged dialysis in the cold. Although there was evidence that DANSYL conjugates could be freed from non-specific staining effects by this means, neither FIC, FITC or RB 200 conjugates could be so purified. It became customary, following Coons and Kaplan (1950), to absorb non-specific reactants by treatment of the conjugates with dry tissue powders (usually mouse liver). These were denounced as inefficient by Chadwick and Nairn (1960), a view with which most workers at that time agreed. Their abandonment was recommended by McDevitt *et al.* (1963), and modern evidence suggests that tissue powders have no role to play in purification of conjugates.

Extraction of the conjugate with ethyl acetate was suggested by Dineen and Ada (1957) but this was found to cause protein denaturation and its use was soon discontinued. Absorption with powdered activated charcoal was advocated by Chadwick and Nairn (1960), Fothergill and Nairn (1961), and by Nairn (1962) but this procedure also caused large losses of protein. All the above methods (including tissue powder) have entirely been superseded by gel filtration techniques derived from cellulose ion chromatography for proteins (Peterson and Sober, 1956), and from the starch gel chromatographic columns of Lathe and Ruthven (1956).

The only materials in current use are cross-linked dextran (Sephadex G50) and DEAE-cellulose. The Sephadex series was developed by Porath and Flodin (1959) and G50 is the type usually employed. Small molecules are held up in the gel while larger molecules, such as proteins, pass through relatively rapidly. The process of gel filtration was considered by Killander *et al.* (1961), by George and Walton (1961) and also by Zwaan and van Dam (1961).

With DEAE-cellulose larger molecules are also held in the column, from which they can be released by appropriate treatment. Two methods of separation have been used and these are known as gradient elution (Goldstein *et al.*, 1961) and stepwise elution (McDevitt *et al.*, 1963; Wood *et al.*, 1965). The latter procedure is now more usually employed (Appendix 7, p. 634) and the whole reaction mixture, at the end of conjugation is applied to the column.

The principle of stepwise elution is simple. Proteins (globulin conjugates) are applied to the column with a "starting buffer" (often 0·005 M phosphate, pH 8·1). Elution is performed, at a somewhat lower pH, by stepwise increases in the molarity of the buffer (0·01, 0·03, 0·05, and 0·1 M) or by stepwise addition of increasingly concentrated solutions of NaCl in buffer (0·04, 0·1, and 1·0 M). At each step globulins with different degrees of labelling are washed out of the column. Unreacted fluorescein-labelled materials are strongly bound to the column. Lightly coupled globulin is not bound and can therefore be eluted with 0·01 M buffer (Wood *et al.*, 1965). Elution with 0·03 M buffer removes globulin with a higher dye content ("correctly coupled globulin"). Over-coupled globulins are eluted with 0·05 M and 0·1 M buffers.

Concentration of Eluted Fractions. It is usually necessary to reduce the volume of the required globulin fraction after elution from the column. This is done by dialysis against 20 per cent. polyvinyl pyrrolidone or by precipitation with 18 per cent. (w/v) sodium sulphate. There is some evidence to show that the correctly coupled globulin fraction may not be stable in all cases. If there is any question of non-specific staining the selected fraction should be applied to the requisite tissue preparations *as it comes off the column.*

Preparation of Tissues. For many years only two types of tissue section were used for fluorescent antibody studies. These were (1) freeze-dried and (2) fresh frozen sections. The latter were usually produced with a cold microtome or cryostat, but cold knife sections could be used for the purpose. In both cases a necessary prerequisite was satisfied, that the tissue proteins (antigen or antibody) remained substantially unaltered as far as their immunological properties were concerned. Coons and his associates used a modified Linderstrøm-Lang cryostat for the preparation of sections in their extensive studies, while Marshall (1951, 1954), on the other hand, used freeze-dried sections entirely. Mayersbach and Pearse (1956) suggested that freeze-dried deparaffined sections were ideal for immunofluorescence. Their primary (auto) fluorescence corresponds exactly to that seen in fresh frozen tissues, and there is no evidence to suggest that the short exposure to heat during paraffin embedding has any adverse effect on the immunological properties of the proteins. There is no doubt, however, that the physical properties (such as solubility) of many proteins are radically altered by freezing and drying, and Mayersbach (1957b) found that the preservation of freeze-dried DANSYL-labelled protein in liver sections differed radically from that found in frozen sections and depended on the amount of rehydration which was allowed to occur before the fixative was applied.

The most popular method of fixation, for both freeze-dried and cold microtome or cold knife sections, has been ethanol (85–95 per cent.) or absolute methanol. In the case of viruses, whose antigenicity is destroyed by these two fixatives, absolute acetone has been used instead. In contrast with the usual practice Marshall (1954) used buffered formaldehyde in dioxan for his work on the localization of enzymes by the fluorescent antibody technique. This was an indication that some antigens could withstand prolonged formalin fixation and remain antigenic enough to be demonstrated in routine paraffin-embedded sections. Leznoff *et al.* (1960) showed that human pituitary growth hormone could be demonstrated by immunofluorescence in formalin-fixed paraffin sections, as could ACTH (Leznoff *et al.*, 1962). These observations were completely confirmed by Pearse and van Noorden (1963). Since ACTH could not be demonstrated convincingly in fresh sections, after various types of post-fixation, this is an indication that for immunofluorescence studies of low mol. wt. proteins and polypeptides immobilization by chemical as opposed to precipitating fixation may be required. The risk of extinction of antigenic function has to be accepted. Thiede and Choate (1964) reported failure to demonstrate human chorionic gonadotrophin (HCG) in fresh tissues fixed in 10 per cent. formalin. Fresh frozen sections gave strong fluorescence after incubation with anti-HCG.

Paraffin embedding without significant inactivation of many antigens and antibodies can be achieved by the method of Sainte-Marie (1962) which involves fixation in 95 per cent. ethanol, followed by low temperature dehydration and clearing. Bovine γ-globulin, influenza virus A, and antibody contained in mouse spleen are well preserved. The antigenic activity of ovalbumin, on the other hand, is greatly reduced.

Freeze-substituted paraffin-embedded tissues were used by Balfour (1961) in studies already referred to in Chapter 4 (p. 62). In these studies the precipitating power of the soluble fraction of ovalbumin, and also of antibody globulin, are reduced by half.

When a new problem is being considered, the correct course must be to test the preservation of the antigen or antibody concerned by using the most suitable methods:

(1) Freeze-drying, acetone post-fixation.
(2) Fresh frozen sections, acetone post-fixation.
(3) Ethanol fixation, paraffin embedding.
(4) Formalin fixation, paraffin embedding.
(5) Freeze-substitution, paraffin embedding.
(6) Freeze-drying, ethanol vapour fixation (block).

There are a few alternatives which may operate in special cases. Tissue smears or impressions may be fixed solely by air-drying or by subsequent immersion in ethanol or acetone. The possibilities of freeze-drying and vapour fixation of the dried block have not been fully explored.

Fluorescence Microscopy. This differs in no respect from the requirements of ordinary fluorescence microscopy with fluorochromes except that the absolute amount of fluorescent material is small and an intense source of UVL is necessary in order to achieve the maximum degree of excitation. Considerations of the necessary apparatus for microscopy and photomicrography appear in Chapter 29. Although in some earlier studies the carbon arc, or water-cooled mercury vapour arcs, were used for immunofluorescence it is now universal practice to employ the high pressure mercury vapour arc (200–250 watts, 70 atmospheres). The Osram lamp HBO 200 provides illumination in the range 250 to 600 nm (see Chapter 29).

Below 300 nm radiant energy is not transmitted by the glass optics of the microscope and we need not therefore consider it. Normally exciter filters are employed in front of the light source and these restrict the exciting light to between 300 and 400 nm or between 330 and 500 nm. As Hansen (1964) has observed, only a fraction of the energy of the mercury vapour lamp can be used in immunofluorescent studies. Since the strong green line of the mercury arc is close to the emission of fluorescein conjugates (see Fig. 61) it is necessary to use a deep yellow barrier filter with a cut-off of 500 nm. Differentiation of primary autofluorescence and secondary, specific, fluorescence sometimes becomes difficult. A switch to the weaker excitation wavelength (320 nm) may then be helpful.

What is required is a source of radiant energy having uniform intensity between 300 and 500 nm. Such a source is not at present available.

Counterstains. The principle of counterstaining has not received much attention in relation to immunofluorescence. Hiramoto *et al.* (1958) used TMR-labelled antibody as a counterstain for fluorescein and Smith *et al.* (1959) used RB 200-conjugated albumin in the same way. Rhodamine-conjugated papain was employed by Alexander and Potter (1963). Hall and Hansen (1962) described the use of a number of fluorochromes suitable for counterstaining fluorescein-labelled bacteria and tissue sections. The most useful of these, 2-naphthol, 1-(5-chloro-2-hydroxyphenylazo)- is known as Flazo Orange.

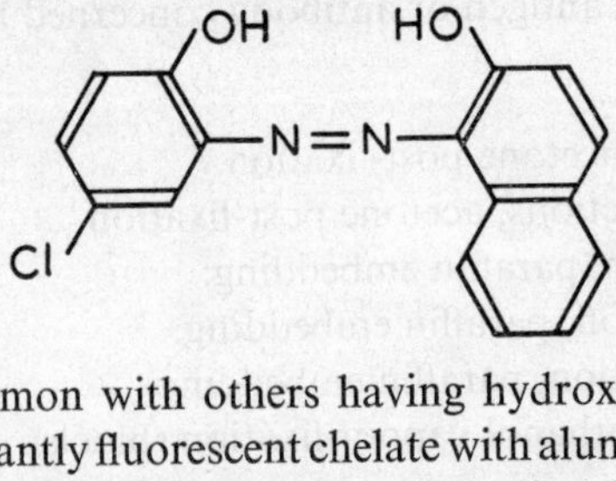

This azo dye, in common with others having hydroxyl groups in the *ortho* position, forms a brilliantly fluorescent chelate with aluminium. The compound is now commercially available in many countries but for those unable to obtain it directions for synthesis are given in Appendix 7, p. 632. The staining procedure is given in the same place.

Effect of pH Changes. Most fluorescent labels emit most efficiently at alkaline pH levels, as already indicated. Below pH 4 fluorescein conjugates are quenched while other labels (TMR, RB 200) continue to fluoresce. This provides a means of distinction in cases where double staining (two fluorescent labels) has been employed (Hiramoto *et al.*, 1964). Details of their method for preparing buffered mounting media appear in this paper.

Applications of the Method

Detection of Foreign Antigens. The greater part of the early work done in this category was carried out on *viruses*. Bacterial infections were largely ignored. More recently many applications to the localization of bacteria, protozoa and their products have been described. An excellent review is that of Cherry, Goldman and Carski (1960). Four early papers described the localization in the tissues of mumps antigens (Coons, Snyder, Cheever and Murray, 1950; Chu, Cheever, Coons and Daniels, 1951; Watson, 1952a and b). The virus was found by Coons and his collaborators in the cytoplasm of the acinar and duct cells of the parotid gland, and in extracellular sites in the brain and elsewhere. Influenza virus infection was investigated by Watson and Coons (1954), and by Liu (1955a and b) in the chick embryo and in the ferret. In the nasal mucosa of the ferret the antigen was found in the cytoplasm of the ciliated cells, and in both species it was observed in the nuclei (see below, p. 206). Studies on human influenza were made by Hers and Mulder (1961).

Varicella virus was studied in infected cells in tissue culture by Weller and Coons (1954). These authors used a layering technique with human convalescent serum as the intermediate and fluorescent anti-human γ-globulin as the demonstrator. The same technique was used by Liu and Eaton (1955) to demonstrate the virus of atypical pneumonia in the bronchial epithelium of infected chick embryos.

Canine infective hepatitis was studied by Coffin, Coons and Cabasso (1953), and canine distemper by Moulton and Brown (1954) and Moulton (1956). The last of these papers dealt with the process of demyelination occurring in distemper and described the localization of specific fluorescence in the cerebellar white matter, where it was principally confined to the astrocyte nuclei. Not surprisingly, some difficulty was experienced in distinguishing astrocytes from other glial cells. The development of vaccinia virus in human epithelial cells was studied by Noyes and Watson (1955), who showed that the antigenic material, at first finely particulate, was distributed in the cytoplasm in close association with the nuclear membrane. Later, collections of antigen were observed in the nucleus. A similar finding was recorded by Nielsen (1966), using whole antivaccinia serum labelled with DANSYL. This is shown in Fig. 69, facing p. 197. Noyes (1955) demonstrated the neurotropic Egypt 101 virus (West Nile Fever) in cells cultured from a squamous carcinoma of the larynx, and also in infected mouse brains. A full list of virus immunofluorescent studies, up to 1960, is given by Nairn (1962), whose later edition should be consulted for more up to date references.

Foreign antigens other than viruses introduced into the organism by injection have been traced by the fluorescent antibody technique in a number of instances. The capsular polysaccharides of the pneumococcus, and of Friedlander's bacillus, were traced in this way by Kaplan, Coons and Deane (1950), and by Hill, Deane and Coons (1950). After intravenous administration in large doses the antigens were found in phagocytic cells throughout the body and also in epithelial cells in many organs. They persisted in the former sites for long periods, but disappeared from most epithelial cells within 3 days. After producing an experimental glomerulonephritis in mice by subcutaneous injection of killed *Proteus mirabilis* Wood and White (1956) found antigenic material from the organism in the affected glomeruli. The polysaccharides of *Cryptococcus neoformans* were found by Eveland *et al.* (1957) in uninfected organs such as kidney and lung. The organisms themselves were easily identified in infected tissues by the fluorescent antibody method.

A number of studies have been devoted to the tracing of foreign protein antigen following their parenteral administration in animals. Egg albumin, bovine serum albumin and human γ-globulin were used by Coons, Leduc and Kaplan (1951) in their work on the fate of intravenously injected foreign protein in the mouse. They found that the initial distribution closely resembled that of injected polysaccharides, except that the proteins were detectable only for a few days at most. Waksman and Bocking (1953) obtained similar results in rabbits following intradermal injection, as did Mayersbach and Pearse (1956) after intraperitoneal injection of egg white in the mouse. More recently Cochrane, Vasquez and Dixon (1957) found the specific antigen (bovine serum albumin) in the lesions of experimental serum sickness, and Wollensak and Seybold (1957) studied the distribution in rat liver and kidney of injected human γ-globulin and serum albumin.

Nuclear Fluorescence. The finding by Coons *et al.* (1951) of specific fluorescence in the nuclei produced a great deal of speculation. Coons considered that the transfer of the foreign protein from the cytoplasm to the nuclei during freezing (quenching) or storage was unlikely. During the cutting of fresh frozen sections with the cold microtome dislocation of tissue components is sometimes noted, but nothing is seen approaching the regular displacement which would be necessary to produce nuclear fluorescence as an artifact. Such an artifact, in any case, would not occur with freeze-dried material. Thawing of the frozen sections on the slides or coverslip certainly produces some distortion, but translocation of cell components is not noted. Non-specific reactions (staining of the nuclei) can be completely ruled out. It is therefore necessary to assume that many foreign proteins, depending no doubt on their molecular weight and other characteristics, can be transferred from the cytoplasm to the nucleus, *in vivo*, in a short space of time.

Evidence in favour of nuclear foreign antigen was given by Crampton and Haurowitz (1950), who found ^{131}I-labelled ovalbumin and bovine γ-globulin in the nuclear fractions of rabbit-liver homogenates. Schiller *et al.* (1952–53) found, however, that the fluorescein marker of fluorescent ovalbumin failed to appear in the nuclei. Mayersbach and Pearse (1956) confirmed this finding and assumed, with the former authors, that differential permeability of the nuclear membrane was the only reasonable hypothesis by which to explain the facts. Later studies by Mayersbach (1957a), however, showed that fluorescein-labelled egg white could be demonstrated in liver cell nuclei for a short period.

A great deal of attention has been focussed on the serum antinuclear factor (ANF), which appears in systemic lupus erythematosus (Holborow *et al.*, 1957; Friou, 1958), and on its demonstration *in situ* by immunofluorescence. Three nuclear patterns were described by Beck (1961) but their significance in relation to the disease, or to any other function, is obscure. The position of ANF was reviewed by Holborow and Johnson (1964).

Detection of Native Antigens. The first application of the fluorescent antibody method to the localization of native antigens was that of Marshall (1951), who prepared rabbit anti-pig ACTH and coupled this with fluorescein isocyanate to localize, in the pig pituitary gland, the sites of origin of this hormone. It was found exclusively in cells classed as basophils by other techniques. Marshall controlled his observations by showing that the fluorescent conjugate would precipitate active ACTH from a commercial preparation, and also that it would not stain any cells in the pituitary glands of other species. Later work by this author (Marshall, 1954) has demonstrated the presence of chymotrypsinogen and procarboxypeptidase in the zymogen granules of ox pancreas.

Many other enzymes have now been studied by immunofluorescent techniques. The localization of glyceraldehyde-3-phosphate dehydrogenase in rat kidney was described by Emmart *et al.* (1963), and that of ribonuclease in mouse and bovine pancreas by Ehinger (1965). Phosphorylase was localized in rabbit skeletal muscle by Dvorak and Cohen (1965), and catalase by Nishimura *et al.* (1964).

The pituitary hormones have been the object of particularly intensive study, as the following list indicates:

Leznoff *et al.* (1960)	GH
Grumbach and Kaplan (1960)	GH
Leznoff *et al.* (1962)	ACTH
Pearse (1962)	GH
Reusser *et al.* (1962)	GH
Pearse and van Noorden (1963)	ACTH
Emmart *et al.* (1963)	Prolactin
Midgeley (1963)	LH
Rennels (1963)	LH, FSH
Robyn *et al.* (1964)	LH
Koffler and Fogel (1964)	LH, FSH
Greenspan and Hargadine (1965)	TSH
Emmart *et al.* (1965)	Prolactin

Midgeley (1966)	LH
Kracht *et al.* (1967)	GH, LH, ACTH, FSH

The localization of human pituitary ACTH in R-type cells (Chapter 8, p. 240) is illustrated in Fig. 68, facing p. 188.

As part of a series of studies under the heading "analytical pathology", Mellors and Ortega (1956) showed that γ-globulin (demonstrated by means of fluorescent rabbit anti-human γ-globulin) was present in the glomeruli in lipid nephroses in childhood and in the same situations in varying types of glomerulonephritis, and in renal amyloidosis. They also found it in periarteritis nodosa. Vasquez and Dixon (1956) also demonstrated the presence of γ-globulin in amyloid deposits by a similar technique. Further studies by Ortega and Mellors (1957) revealed that γ-globulin is normally present in human tissues in the cytoplasm of three distinct cell types. These are: (1) immature and mature plasma cells with basophilic cytoplasm; (2) plasma cells containing Russell bodies; and (3) certain cells in the germinal centres of the lymph nodes (in the cytoplasmic granules only).

Detection of Antibodies to Foreign Antigens. Coons, Leduc and Connolly (1955) demonstrated antibody to ovalbumin or human γ-globulin in plasma cells in the spleen, lymph nodes, ileal submucosa and in the portal spaces of injected rabbits. Lymphocytes never showed any activity. Leduc, Coons an Connolly (1955) investigated the primary and secondary response to diphtheria toxoid and found the former in a few large immature cells in the medulla of lymph nodes draining the injection site. The secondary response was due to the development, supposedly from these immature cells, of colonies of antibody-containing plasma cells. White, Coons and Connolly (1955a) showed that the adjuvant effect of aluminium phosphate was due to delay in absorption of the antigen injected with it from the local site of injection. Prolonged stimulation of cells in the draining lymph nodes was thus possible. These same authors (1955b) injected guinea pigs with an oil-water emulsion of ovalbumin containing a wax fraction from *Mycobacterium tuberculosis*. The wax stimulated a macrophage response with granuloma formation, but the macrophages contained no antibody. A later plasma cell response produced antibody in large amounts.

A study of the contribution of Russell bodies to antibody formation was made by White (1954). Following the injection of rabbits with *Proteus vulgaris* vaccine large numbers of Russell body-containing plasma cells appeared in the spleen. Using the layer technique White showed that these contained antibody, often in the form of a ring (Fig. 70, facing p. 197), but he was unable to show that either Russell bodies or plasma cells contained any antigen.

The relationship between synthesis of anti-ovalbumin and γ-globulin and the distribution of antibody-containing plasma cells was studied by Askonas and White (1956), and the significance of multiple antibody components in the sera of immunized animals was discussed by Askonas, Farthing and Humphrey

(1960). These authors found that antibody to Type III pneumococcal capsular polysaccharide was distributed through the entire range of γ-globulin molecules. A very elegant study was made by McDevitt *et al.* (1966) who demonstrated the localization of foreign antigens (a synthetic multichain polypeptide or *Maia squinedo* hæmocyanin), and the specific antibodies produced by their injection under suitable conditions, in the lymph-nodes of mice. They employed a double technique using radioautography (^{125}I) for the antigens and a sandwich technique (immunofluorescence) for the antibodies. A special photographic method enabled them to record the silver grains (antigen) as red dots on the same colour film as the specific fluorescence (antibody). These results are illustrated in Plate Ib, facing p. 183.

Immunoglobulin Structure. Recently the whole concept of immunoglobulin structure has undergone revision. It is now supposed that individual immunoglobulin molecules are composed of heavy chains and light chains. There are three kinds of heavy chain (γ, α and μ) which represent the class-specific portion of the antibody globulins γG, γA and γM, respectively.

The two light chains, designated $\varkappa$ and λ, are found in all three classes of immunoglobulin. There are therefore 6 polypeptide chain combinations in all: $\gamma\varkappa$, $\gamma\lambda$, $\alpha\varkappa$, $\alpha\lambda$, $\mu\varkappa$ and $\mu\lambda$. Fahey and Goodman (1964) have shown that antibody activity is associated with each of these combinations. Bernier and Cebra (1965) have shown by a double staining technique that, with rare exceptions, the $\varkappa$ and λ chains are mutually exclusive. Similarly α and λ chains are rarely found together.

The main classes of immunoglobulin are therefore:

(1) $\gamma G = 6{\cdot}5$ to 7 S γ-globulin. mol. wt. 150,000; low carbohydrate.

(2) $\gamma A = 7$ to 13 S γ-globulin. (β_2A) mol. wt. 500,000; 9 per cent. carbohydrate.

(3) $\gamma M = 19$ S γ-globulin. (β_2M) mol. wt. 750,000; 10 per cent. carbohydrate.

Very numerous immunofluorescent studies have been made in order to throw light on the mechanism of antibody formation and on the life history of the cells involved. The double staining technique is constantly employed. Clearly the future of immunofluorescence is as bright today as at any time in its 27 years' history.

REFERENCES

ALBRECHT, P., and SOKOL, F. (1961). *Folia microbiol.*, **6**, 49.
ALEXANDER, W. R. M., and POTTER, J. L. (1963). *Immunol.*, **6**, 450.
ALLERAND, C. D., and YAHR, M. D. (1964). *Science*, **144**, 1141.
ASKONAS, B. A., FARTHING, C. P., and HUMPHREY, J. H. (1960). *Immunol.*, **3**, 336.
ASKONAS, B. A., and WHITE, R. G. (1956). *Brit. J. exp. Path.*, **37**, 61.
BALE, W. F., and SPAR, I. L. (1954). *J. Immunol.*, **73**, 125.
BALFOUR, B. M. (1961). *Immunology*, **4**, 206.
BALS, M. G. (1966). "Imunofluorescenta, si applicatiile ei in inframicrobiologie." Ed. Acad. Rep. Soc. Rom., Bucharest.
BECK, J. S. (1961). *Lancet*, **1**, 1203.

BERGLUND, G. (1965). *Nature, London*, **206**, 523.
BERNIER, G. M., and CEBRA, J. J. (1964). *Science*, **144**, 1590.
BERNIER, G. M., and CEBRA, J. J. (1965). *J. Immunol.*, **95**, 246.
BEUTNER, E. H. (1961). *Bact. Rev.* **25**, 49.
BOREK, F. (1961a). *Bull. Wld. Hlth. Org.*, **24**, 249.
BOREK, F. (1961b). *J. org. Chem.*, **26**, 1292.
BOREK, F., and SILVERSTEIN, A. M. (1960). *Arch. Biochem. Biophys.*, **87**, 293.
BOYD, W. C. (1956). "Fundamentals of Immunology." Interscience Publishers Inc., 3rd Ed. New York.
BURTIN, P., and BUFFE, D. (1963). *Proc. Soc. exp. Biol. N.Y.*, **114**, 117.
BUSSOLATI, G., and PEARSE, A. G. E. (1967). *J. Endocrinol.*, **37**, 205.
CEBRA, J. J., and GOLDSTEIN, G. (1965). *J. Immunol.*, **95**, 230.
CHADWICK, C. S., MCENTEGART, M. G., and NAIRN, R. C. (1958a). *J. Immunol.*, **1**, 315.
CHADWICK, C. S., MCENTEGART, M. G., and NAIRN, R. C. (1958b). *Lancet*, **1**, 412.
CHADWICK, C. S., and NAIRN, R. C. (1960). *Immunol.*, **3**, 363.
CHERRY, W. N., GOLDMAN, M., and CARSKI, T. R. (1960). "Fluorescent Antibody Techniques in the Diagnosis of Communicable Disease." Washington, U.S. Government Printing Office.
CHU, T. H., CHEEVER, F. S., COONS, A. H., and DANIELS, J. B. (1951). *Proc. Soc. exp. Biol., N.Y.*, **76**, 571.
CLARK, H. F., and SHEPARD, C. C. (1963). *Virology*, **20**, 642.
CLAYTON, R. M. (1954). *Nature, Lond.*, **174**, 1059.
CLAYTON, R. M., and FELDMAN, M. (1955). *Experientia*, **11**, 29.
COCHRANE, C. G., VASQUEZ, J. J., and DIXON, F. J. (1957). *Amer. J. Path.*, **33**, 593.
COFFIN, D. L. COONS, A. H., and CABASSO, V. J. (1953). *J. exp. Med.*, **98**, 13.
COONS, A. H. (1954). *Ann. Rev. Microbiol.*, **8**, 333.
COONS, A. H. (1956). *Int. Rev. Cytol.*, **5**, 1.
COONS, A. H. (1961). *J. Immunol.*, **87**, 499.
COONS, A. H. (1964). "Proc. 2nd Int. Cong. Histo and Cytochemistry. Eds. Schiebler, T. H., Pearse, A. G. E., and Wolf, H. H. Springer, Heidelberg, p. 107.
COONS, A. H., CREECH, H. J., and JONES, R. N. (1941). *Proc. Soc. exp. Biol., N.Y.*, **47**, 200.
COONS, A. H., CREECH, H. J., JONES, R. N., and BERLINER, E. (1942). *J. Immunol.*, **45**, 159.
COONS, A. H., and KAPLAN, M. H. (1950). *J. exp. Med.*, **91**, 1.
COONS, A. H., LEDUC, E. H., and CONNOLLY, J. M. (1955). *Ibid.*, **102**, 49.
COONS, A. H., LEDUC, E. H., and KAPLAN, M. H. (1951). *Ibid.*, **93**, 173.
COONS, A. H., SNYDER, J. C., CHEEVER, F. S., and MURRAY, E. S. (1950). *Ibid.*, **91**, 31.
COREY, H. S., and MCKINNEY, R. M. (1962). *Anal. Biochem.*, **4**, 57.
CRAMER, H., and LANGER, E. (1961). *Histochemie*, **2**, 176.
CRAMPTON, C. F., and HAUROWITZ, F. (1950). *Science*, **112**, 300.
CREECH, H. J., and JONES, R. N. (1940). *J. Amer. chem. Soc.*, **62**, 1970.
CREECH, H. J., and JONES, R. N. (1941). *Ibid.*, **63**, 1661.
CRUICKSHANK, B., and HILL, A. G. S. (1953). *J. Path. Bact.*, **66**, 283.
CURTAIN, C. C. (1961). *J. Histochem. Cytochem.* **9**, 484.
DINEEN, J. K., and ADA, G. L. (1957). *Nature, Lond.*, **180**, 1284.
DOWDLE, W. R., and HANSEN, P. A. (1959). *J. Path. Bact.*, **77**, 669.
DVORAK, H. F., and COHEN, R. B. (1965). *J. Histochem. Cytochem.*, **13**, 454.
EHINGER, B. (1965). *Histochemie*, **5**, 145.
EISEN, H. N., and KESTON, A. S. (1949). *J. Immunol.*, **63**, 71.
EMMART, E. W. (1958). *Arch. Biochem. Biophys.*, **73**, 1.
EMMART, E. W., BATES, R. W., and TURNER, W. A. (1965). *J. Histochem. Cytochem.*, **13**, 182.
EMMART, E. W., SCHIMKE, R. T., SPICER, S. S., and TURNER, W. A. (1963). *Exp. Cell Res.*, **30**, 460.
EMMART, E. W., SPICER, S. S., and BATES, R. W. (1963). *J. Histochem. Cytochem.*, **11**, 365.

Eveland, W. C., Marshall, J. D., Silverstein, A. M., Johnson, F. B., Iverson, L., and Winslow, D. J. (1957). *Amer. J. Path.*, **33**, 616.
Fahey, J. L., and Goodman, H. (1964). *Science*, **143**, 588.
Fothergill, J. E., and Nairn, R. C. (1961). *Nature, Lond.*, **192**, 1073.
Freeman, D. C., and White, C. E. (1956). *J. Amer. chem. Soc.*, **78**, 2678.
Freund, J., and McDermott, K. (1942). *Proc. Soc. exp. Biol., N.Y.*, **49**, 548.
Friou, G. J. (1958). *J. Immunol.*, **80**, 476.
Frommhagen, L. H., and Martins, M. J. (1963). *J. Immunol.*, **90**, 116.
George, W., and Walton, K. W. (1961). *Nature, Lond.*, **192**, 1188.
George, W. H., and Walton, K. W. (1962). *Nature, Lond.*, **194**, 693.
Goldman, M., and Carver, R. K. (1957). *Science*, **126**, 839.
Goldman, M., and Carver, R. K. (1961). *Exp. Cell Research*, **23**, 265.
Goldstein, G., Slizys, J. S., and Chase, M. W. (1961). *J. exp. Med.*, **114**, 89.
Goldstein, G., Spalding, B. H., and Hunt, W. B. Jr. (1962). *Proc. Soc. exp. Biol., N.Y.*, **111**, 416.
Goldwasser, R. A., and Shepard, C. C. (1958). *J. Immunol.*, **80**, 122.
Gornall, A. G., Bardawill, D. J., and David, M. M. (1949). *J. Biol. Chem.*, **177**, 751.
Greenspan, F. S., and Hargadine, J. E. (1965). *J. Cell Biol.*, **26**, 177.
Grossi, C. E., and Zaccheo, D. (1963). *Riv. istochim. norm. patol.*, **9**, 339.
Grossi, C. E., and Mayersbach, H. (1964). *Acta histochem.*, **19**, 382.
Grumbach, M. M., and Kaplan, S. L. (1960). *Acta Endocrinol.*, **35**, Suppl. **51**, 1101.
Hall, C. T., and Hansen, P. A. (1962). *Z. Bakt.*, **184**, 548.
Hamashima, Y., Harter, J. G., and Coons, A. H. (1964). *J. Cell Biol.*, **20**, 271.
Hansen, P. A. (1964). "Fluorescent Compounds used in Protein Tracing, Absorption and Emission Data." University of Maryland Report.
Harms, H. (1965). "Handbuch der Farbstoffe für die Mikroscopie." Staufen Verlag, Kamp-Lintfort, p. II 99.
Hartley, B. S., and Massey, V. (1956). *Biochim. Biophys. Acta*, **21**, 58.
Hebert, G. A., Pittman, B., and Cherry, W. B. (1967). *J. immunol.*, **98**, 1204.
Heidelberger, M., Kendall, F. E., and Soo Hoo, C. M. (1933). *J. exp. Med.*, **58** 137.
Hers, J. F. P., and Mulder, J. (1961). *Amer. Rev. Res. Dis.*, **83**, 84.
Hess, R., and Pearse, A. G. E. (1959). *Nature*, **183**, 260.
Hill, A. G. S., Deane, H. W., and Coons, A. H. (1950). *J. exp. Med.*, **92**, 35.
Hinuma, Y., Ohta, R., Miyamoto, I., and Ishida, N. (1962). *J. Immunol.*, **89**, 19.
Hiramoto, R., Engel, K., and Pressman, D. (1958). *Proc. Soc. exp. Biol., N.Y.*, **97**, 611.
Hiramoto, R., Bernecky, J., Jurand, J., and Hamlin, M. (1964). *J. Histochem. Cytochem.*, **12**, 271.
Holborow, E. J., Weir, D. M., and Johnson, G. D. (1957). *Brit. med. J.*, **2**, 732.
Holborow, E. J., and Johnson, G. D. (1964). *Arthr. Rheum.*, **7**, 119.
Hopkins, S. J., and Wormall, A. (1933). *Biochem. J.*, **27**, 740.
Hughes, P. E., and Louis, C. J. (1959). *Arch. Path.*, **68**, 508.
Humphrey, J. H. (1962). In: "Immunoassay of Hormones." Ciba Foundation Colloq., Endocrinol., Vol. 14. J. & A. Churchill, London, p. 6.
Kabat, E. A., and Mayer, M. M. (1948). "Experimental Immunochemistry." Thomas, Springfield, Illinois.
Kaplan, M. H., Coons, A. H., and Deane, H. W. (1950). *J. exp. Med.*, **91**, 15.
Kaufman, L., and Cherry, W. B. (1961). *J. Immunol.*, **87**, 72.
Killander, J., Ponten, J., and Roden, L. (1961). *Nature, Lond.*, **192**, 182.
King, E. S. J., Hughes, P. E., and Louis, C. J. (1959). *Cancer*, **12**, 741.
Klein, P., and Burkholder, P. M. (1959). *Schw. Zeit. allg. Path. Bact.*, **22**, 53.
Koffler, D., and Fogel, M. (1964). *Proc. Soc. exp. Biol., N.Y.*, **115**, 1080.
Kohn, J. (1959). *Nature, Lond.*, **183**, 1055.
Kracht, J., Hachmeister, U., Breustedt, H.-J., and Zimmerman, H.-D. (1967). *Materia med. Nordmark*, **19**, 224.
Kruse, D. J. R. (1952). *Biochem. J.*, **51**, 168.
Kruse, H., and McMaster, P. D. (1949). *J. exp. Med.*, **90**, 425.

LACHMANN, P. J. (1964). *Immunol.*, **7**, 507.
LACHMANN, P. J., MULLER-EBERHARD, H. J., KUNKEL, H. G., and PARONETTO, F. (1962). *J. exp. Med.*, **115**, 63.
LATHE, G. H., and RUTHVEN, C. R. J. (1956). *Biochem. J.*, **62**, 665.
LAURENCE, D. J. R. (1952). *Biochem. J.*, **51**, 168.
LAURENCE, D. J. R. (1957). Personal communication.
LEDUC, E. H., COONS, A. H., and CONNOLLY, J. M. (1955). *J. exp. Med.*, **102**, 61.
LEZNOFF, A., FISHMAN, J. GOODFRIEND, L., MCGARRY, E. E., BECK, J. C. and, ROSE, B. (1960). *Proc. Soc. exp. Biol., N.Y.*, **104**, 232.
LEZNOFF, A., FISHMAN, J., MCGARRY, E. E., BECK, J. C., and ROSE, B. (1962). *J. clin. Invest.*, **41**, 1720.
LIU, C. (1955a). *J. exp. Med.* **104**, 665.
LIU, C. (1955b). *Ibid.*, **101**, 677.
LIU, C., and EATON, M. D. (1955). *Bacteriol. Proc.*, p. 61.
LOUIS, C. J. (1958). *Brit. J. Cancer*, **12**, 5.
LOUIS, C. J., and WHITE, J. (1960). *Lab. Invest.*, **9**, 273.
MARRACK, J. (1934). *Nature, Lond.*, **133**, 292.
MARSHALL, J. M., Jr. (1951). *J. exp. Med.*, **94**, 21.
MARSHALL, J. M., Jr. (1954). *Exp. Cell Research*, **6**, 240.
MARSHALL, J. D., EVELAND, W. C., and SMITH, C· W. (1958). *Proc. Soc. exp.Biol., N.Y.*, **98**, 898.
MASOUREDIS, S. P., MELCHER, L. R., and KOBLICK, D. C. S. (1951). *J. Immunol.* **66**, 297.
MASOUREDIS, S. P., MELCHER, L. R., and SHIMKIN, M. B. (1953). *Ibid.*, **71**, 268.
MAYERSBACH, H. (1957a). *Z. Zellforsch.*, **45**, 483.
MAYERSBACH, H. (1957b). *Acta Anatomica*, **30**, 487.
MAYERSBACH, H. (1958). *Acta Histochem.*, **5**, 351.
MAYERSBACH, H. (1959). *J. Histochem. Cytochem.*, **7**, 427.
MAYERSBACH, H., and PEARSE, A. G. E., (1956). *Brit. J. exp. Path.*, **37**, 81.
MAYERSBACH, H., and SCHUBERT, G. (1960). *Acta Histochem.*, **10**, 44.
MCDEVITT, H. O., PETERS, J. H., POLLARD, L. W., HARTER, J. G., and COONS, A. H. (1963). *J. Immunol.*, **90**, 634.
MCDEVITT, H. O., ASKONAS, B. A., HUMPHREY, J. H., SCHECHTER, I., and SELA, M. (1966). *Immunology*, **11**, 337.
MCKINNEY, R. M., SPILLANE, J. T., and PEARCE, G. W. (1962). *J. org. Chem.*, **27**, 3986.
MCKINNEY, R. M., SPILLANE, J. T., and PEARCE, G. W. (1964a). *J. Immunol.*, **93**, 232.
MCKINNEY, R. M., SPILLANE, J. T., and PEARCE, G. W. (1964b). *Analyt. Biochem.*, **7**, 74.
MELLORS, R. C. (1955). *J. Histochem. Cytochem.*, **3**, 284.
MELLORS, R. C., and ORTEGA, L. G. (1956). *Ibid.*, **32**, 455.
MIDGELEY, A. R., Jr. (1963). *Exp. Cell Research*, **32**, 606.
MIDGELEY, A. R. (1966). *J. Histochem. Cytochem.*, **14**, 159.
MIKHAILOV, I. F., and DASHKEVICH, I. O. (1961). *Zh. Mikrobiol.*, **32**, 87.
MÖRITZ, K-V., and AMBROSIUS, H. (1964). *Acta Histochem.*, **19**, 191.
MOULTON, J. E. (1956). *Proc. Soc. exp. Biol., N.Y.*, **91**, 460.
MOULTON, J. E., and BROWN, C. H. (1954). *Ibid.*, **86**, 99.
MÜLLER, F. (1962). *Zbl. Bakt.*, **184**, 361.
MÜLLER, F., and KLEIN, P. (1959). *Dtsch. med. Woch.*, **84**, 2195.
NAIRN, R. C. (1962). "Fluorescent Protein Tracing." Ed. Livingstone, Edinburgh.
NIELSEN, H. A. (1966). Personal communication.
NISHIMURA, E. T., TOBARA, T. Y., and DRURY, J. J. (1964). *Lab. Invest.*, **13**, 69.
NOYES, W. F. (1955). *J. exp. Med.*, **102**, 243.
NOYES, W. F., and WATSON, B. K. (1955). *Ibid.*, **102**, 237.
ORTEGA, L. G., and MELLORS, R. C. (1957). *Amer. J. Path.*, **33**, 614.
ÖYSTESE, B. (1962). *Ver. deutsch. Path. ges.*, **46**, 120.
PARKER, J. V., ELEVITCH, F. R., and GRODSKY, G. M. (1963). *Proc. Soc. exp. Biol., N.Y.*, **113**, 48.
PEARSE, A. G. E. (1962). *Acta Union Int. Cont. Cancer*, **18**, 302.

PEARSE, A. G. E., and VAN NOORDEN, S. (1963). *Canad. Med. Ass. J.*, **88**, 462.
PETERSON, E. A., and SOBER, H. A. (1956). *J. Amer. chem. Soc.*, **78**, 751.
PORATH, J., and FLODIN, P. (1959). *Nature, Lond.*, **183**, 1657.
PRESSMAN, D., HILL, R. F., and FOOTE, F. W., Jr. (1949). *Science*, **109**, 65.
PRESSMAN, D., and KEIGHLEY, G. (1948). *J. Immunol.*, **59**, 141.
PRESSMAN, D., and STERNBERGER, L. A. (1950). *J. Amer. Chem. Soc.*, **72**, 2226.
PRINGSHEIM, P. (1949). "Fluorescence and Phosphorescence." Interscience, New York.
PROOM, H. (1943). *J. Path. Bact.*, **55**, 419.
REDETSKI, H. M. (1958). *Proc. Soc. exp. Biol., N.Y.*, **98**, 120.
REINER, L. (1930). *Science*, **72**, 483.
RENNELS, E. G. (1963). *Colloq. int. C.N.R.S. Paris*, No. 128, p. 203.
REUSSER, F., SMITH, C. G., and SMITH, C. L. (1962). *Proc. Soc. exp. Biol., N.Y.*, **109**, 375.
RIGGS, J. L., LOH, P. C., and EVELAND, W. C. (1960). *Proc. Soc. exp. Biol., N.Y.*, **105**, 655.
RINDERKNECHT, H. (1960). *Experientia*, **16**, 430.
ROBYN, C., BOSSAERT, Y., and HUBIMONT, P-O. (1964). *C.R. Acad. Sci. Paris*, **259**, 1226.
ROSSI, F., ZACCHEO, D., and GROSSI, C. E. (1964). *Riv. Istochim.*, **10**, 53.
SACCHI, R., COSTANZI, and MANCINI, A. M. (1962). *Boll. Soc. ital. Biol. sper.*, **38**, 928.
SAINTE-MARIE, G. (1962). *J. Histochem. Cytochem.*, **10**, 250.
SCHILLER, A. A., SCHAYER, R. W., and HESS, E. L. (1952–53). *J. gen. Physiol.*, **36**, 489.
SCOTT, D. G. (1960). *Immunol.*, **3**, 226.
SILVERSTEIN, A. M. (1957). *J. Histochem. Cytochem.*, **5**, 94.
SMETANA, H. (1947). *Amer. J. Path.*, **23**, 255.
SMITH, C. W., MARSHALL, J. D., and EVELAND, W. C. (1959). *Proc. Soc. exp. Biol., N.Y.*, **102**, 179.
SPAR, I. L., and BALE, W. F. (1954). *J. Immunol.*, **73**, 134.
STRAUSS, A. J. L., SEEGAL, B. C., HSU, K. C., BURKHOLDER, P. M., NASTUK, W. L., and OSSERMAN, K. E. (1960). *Proc. Soc. exp. Biol., N.Y.*, **105**, 184.
TEALE, F. W. J., and WEBER G. (1957). *Biochem. J.*, **65**, 476.
THIEDE, A., and CHOATE, J. W. (1964). *J. Histochem. Cytochem.*, **12**, 17.
TOKUMARU, J. (1962). *J. Immunol.*, **89**, 195.
UEHLEKE, H. (1958). *Z. Naturforsch.*, **13b**, 722.
UEHLEKE, H. (1959). *Schw. Z. Path. Bakt.*, **22**, 724.
VASQUEZ, J. J., and DIXON, F. J. (1956). *J. exp. Med.*, **104**, 727.
VAUGHAN, J. H., and KABAT, E. A. (1954). *J. Immunol.*, **73**, 205.
VICKERSTAFF, T. (1958). *Melliand Textilber.*, **39**, 905.
VOGT, A., and KOCHEM., H. G. (1960). *Z. Zellforsch.*, **52**, 640.
WAGNER, M. (1962). *Zbl. Bakt.*, **185**, 124.
WAKSMAN, B. H., and BOCKING, D. (1953). *Proc. Soc. exp. Biol., N.Y.*, **82**, 738.
WATSON, B. K. (1952a). *Ibid.*, **79**, 222.
WATSON, B. K. (1952b). *J. exp. Med.*, **96**, 653.
WATSON, B. K., and COONS, A. H. (1954). *Ibid.*, **99**, 419.
WEBER, G. (1952). *Biochem. J.*, **51**, 155.
WELLER, T. H., and COONS, A. H. (1954). *Proc. Soc. exp. Biol., N.Y.*, **86**, 789.
WHITE, R. G. (1954). *Brit. J. exp. Path.*, **35**, 365.
WHITE, R. G. (1960). In: "Tools of Biological Research." Ed. Atkins, H. J. B., Blackwell, Oxford, p. 89.
WHITE, R. G., COONS, A. H., and CONNOLLY, J. M. (1955a). *J. exp. Med.*, **102**, 73.
WHITE, R. G., COONS, A. H., and CONNOLLY, J. M. (1955b). *J. exp. Med.*, **102**, 83.
WILLSON, J. T., and KATSH, S. (1960). *Anat. Rec.*, **136**, 327.
WILSON, G. S., and MILES, A. A. (1955). "Topley and Wilson's Principles of Bacteriology and Immunity." 4th Edition, Vol. 2, p. 1253, Arnold, London.
WOLLENSAK, J., and SEYBOLD, G. (1957). *Z. naturforsch.*, **126**, 147.
WOOD, B. T., THOMPSON, S. H., and GOLDSTEIN, G. (1965). *J. Immunol.*, **95**, 228.
WOOD, C., and WHITE, R. G. (1956). *Brit. J. exp. Path.*, **37**, 49.
ZWAAN, J., and VAN DAM, A. F. (1961). *Acta Histochem.*, **11**, 306.

CHAPTER 8

THE HISTOCHEMISTRY OF SOME IMPORTANT SIMPLE PROTEINS

THE purpose of this chapter is to describe the application of modern histochemical techniques to the differentiation of various simple protein structures. In some cases this has been of academic, but in others of strictly practical use. The methods set forth in Table 14, p. 222, were chosen simply to answer the various physical and chemical questions as they arose. They are by no means a comprehensive list of all the available methods but may serve to indicate the manner in which problems concerning these tissue materials may be solved. Although microstructure in the molecular range can only be resolved by the physical techniques of X-ray diffraction and electron microscopy, it is well to remember that both these techniques have their limitations when applied to histological material in its complete state. Between the molecular level at which these physical methods operate and the microscopical level of ordinary morphological histology there lies a realm of submicroscopical structure which can be reached by histochemistry.

COLLAGEN AND RETICULIN

Histologically collagen and reticulin are two easily distinguishable substances, collagen being taken to mean coarse non-branched fibres which are birefringent, red with van Gieson's stain, faintly pink with PAS, blue with sulphation-metachromasia, and yellow or brown with silver impregnation techniques. Reticulin, on the other hand, is taken to mean fine branching fibres which are usually isotropic in paraffin sections, unstained or faintly pink with van Gieson, magenta with PAS, red with sulphation-metachromasia, and black by silver impregnation. On the important question of the relation between these two substances, three views have been expressed by various authors: (a) Collagen and reticulin are one and the same. (2) They are the same chemically but differ physically. (3) They differ both physically and chemically.

The word reticulin, according to Puchtler (1964), is derived from its use by Siegfried (1892) to describe a substance derived from reticulum fibres after removal of their gelatin content. Puchtler suggested that the term reticulin should be reserved for this interfibrillar substance and that reticulum or reticular fibres should be used for the structure known to histologists.

While this view is clearly correct it is probable that pathologists, the group with the strongest interest in the material under discussion, will continue to call it reticulin.

As observed by Lillie (1945, 1952a and b), a wide variety of acid dyes, used at moderately acid pH levels, show essentially the same specificity for the two substances, though van Gieson's method provides an exception. If the concentration of acid fuchsin is raised, however, reticulin fibres can be stained red without difficulty. Methods based on the use of acid dyes therefore indicate a substantial similarity between the two. Mallory and Parker (1927) observed that when certain connective tissue fibres occurred singly they gave the staining reactions of reticulin but that on joining a bundle of fibres they stained as collagen. For this reason these authors considered collagen and reticulin to be identical. Foot (1925, 1927), however, described reticulin as the precursor of collagen and the latter as a chemical substance in the tissue juices which impregnated reticulin fibres. He discarded the unitarian view mainly on evidence derived from his silver staining method, and was unable to demonstrate reticulin fibres arising directly from cell bodies as others claimed to have done. Heringa and Weidinger (1942) and Dublin (1946) also believed that reticulin was "precollagen", in the sense of forming a scaffold, and that collagen was a chemical substance which impregnated reticulin fibres. Lillie (1947), using his periodate-nitric acid-Schiff (periodic acid-Schiff) method, succeeded in obtaining a deep staining of the reticulin network of a variety of organs and the basement membranes in some of those studied. Collagen was only faintly pink. This new evidence was taken to indicate that collagen and reticulin are not chemically identical. Lhotka and Davenport (1950) compared Foot's silver method with the periodic acid-Schiff (PAS) method and found identical histological pictures in a number of tissues. They suggested that the histochemical bonding which occurs in the two procedures is conditioned by the occurrence of similar free radicals in the loci that are stained, and furthermore that these radicals might be aldehydes.

A similar view was expressed by Glegg *et al.* (1953), who showed chromatographically that reticulin contains galactose, glucose, mannose, fucose and ribose in large quantities while collagen contains very small amounts indeed. Bangle and Alford (1954), by means of the anthrone reaction, found 0·55 per cent. of carbohydrate in collagen (commercial hide powder), whereas Windrum *et al.* (1955), with the same reaction, found an average of 4·2 per cent. non-hexosamine carbohydrate in their samples of reticulin. In view of the fact that hide powder contains reticulin and ground substance as well as collagen it is probable that the carbohydrate value for the latter is much lower than 0·55 per cent. In a combined histochemical and chromatographic study Glegg *et al.* (1954) showed that connective tissues in general possessed two carbohydrate fractions, the first rich in glucosaminic acids and the second in aldoses. Most tissues contained more of fraction II, but cartilage contained 14·8 per cent. of its dry weight as fraction I, and 3·5 per cent. as fraction II. The sulphation-metachromasia method of Kramer and Windrum (1953, 1954) (see Chapter 10, p. 338) reflects the high aldose content of reticulin. This component stains red by the method, while collagen remains blue. The absence of galactosamine from reticulin excludes the possibility that it is a complex of collagen-type fibrils with chondroitin, and the absence of uronic acid rules out a similar complex with hyaluronic acid. Chemical investigations carried out on the carbohydrate constituents of reticulin by Velican *et al.* (1965) suggested that there was considerable variation between different organs and also between similar tissues under different physiological conditions and at different ages.

The *in vitro* formation of fibrils by the addition of various acid mucopolysaccharides to solutions of collagen (Gross *et al.*, 1952; Morrione, 1952; Zawisch, 1957) may reflect the necessary *in vivo* conditions for the formation of this component. In the absence of regulator mucopolysaccharides, or of substances capable of acting in the same way, collagen solutions give rise to dispersed gels containing mainly unbanded fibres (i.e. fibres lacking the characteristic 650 Å cross-striations). In the view of Clerici *et al.* (1963), since collagen molecules can assume a number of different paracrystalline states, the word collagen should define the protein and not the filamentous structure. Histologists are unlikely to accept this reservation.

The question of a cementing substance not only in basement membranes, but holding together the individual fibres of connective tissue, has been considered since its original description by Rollett (1871) and by Flemming (1876), who both stated that such a cementing substance occurred and also that it was of a mucinous nature. In more recent years the view arose that the mucinous ground substance of the tissues was the acid mucopolysaccharide hyaluronic acid, present in loose combination with protein substances. The specific enzyme, hyaluronidase, releases this substance from combination in the tissues. McClean (1931, 1936) observed that hyaluronidase had no direct action on collagen fibres, although swelling and splitting of the fibres was noted. This suggested some action on an interfibrillary cement substance. Bensley (1934) reported the observation of a layer of mucoprotein surrounding collagen fibres in tendon, and Robb-Smith (1945), using streptoccal and testis hyaluronidase, produced a separation of the reticulin membrane of muscle from the underlying fibre, which he attributed to removal of a hyaluronic acid-containing cement. He found no action on collagen and reticulin with either of the two hyaluronidases mentioned, or with *Clostridium welchii* filtrates. Day (1947) confirmed the existence of a cementing substance holding together the individual fibrils of loose connective tissue, but did not consider it to be mucinous. He treated unfixed connective tissue with both testicular and streptococcal hyaluronidase without causing any alteration in its properties. A solution of trypsin, however, acting at 20° for 24 hours, caused loss of coherence and elasticity and Day considered this to be due to removal of a proteinous cement-substance. Aikat and Dible (1956), using *Cl. welchii* filtrates as their source of hyaluronidase, showed that collagen bands, staining brown with ammoniacal silver techniques, could be split into individual black-staining reticulin fibrils by its action. This last evidence supports the conception of a mucopolysaccharide-containing cement-substance, depolymerized by hyaluronidase.

PLATE II

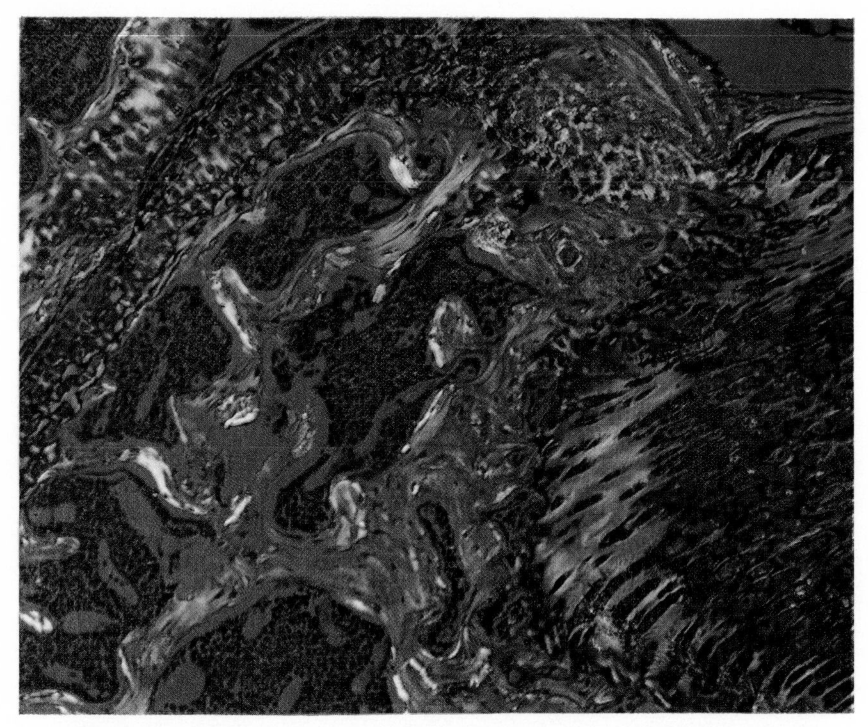

IIA. Mouse Femur (5 weeks old). Stained with toluidine blue O and observed under polarized light and a first-order red retardation plate. Note intensification of interference colours of cortical bone and illumination of the epiphyseal plate and articular cartilage with interference colours of higher order. × 120

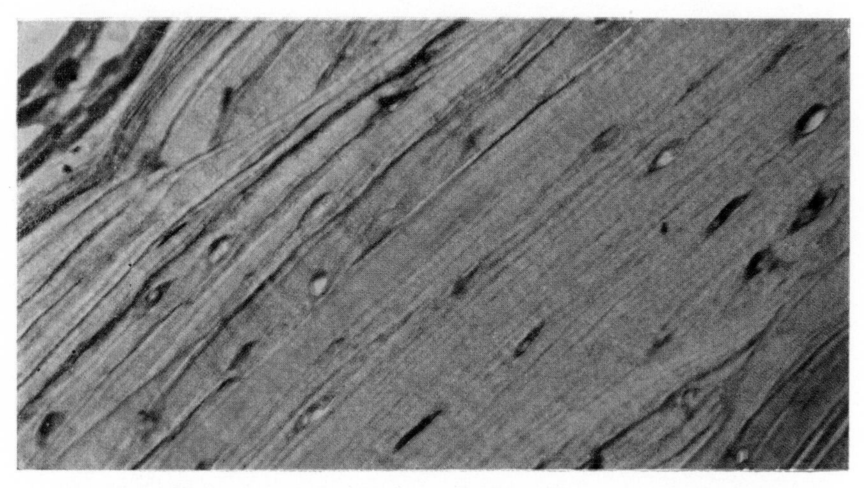

IIB. Mouse Femur (18 months old). Stained with toluidine blue O and observed under polarized light and a first-order retardation plate. Note both fibrous collagen bundles and growth lines (blue) can be seen at the same time. × 487

PHYSICAL CHEMISTRY OF COLLAGEN

Collagen was regarded by Schmitt, Hall and Jakus (1942) as a scleroprotein which when in sol form exists with its individual molecules widely separated. When reticulin fibres appear in this sol they do so by arrangement of the molecules into micellar form, that is to say, into aggregations of molecules orientated in a particular manner. Collagen in gel form was regarded as a further condensation of the scleroprotein molecules. According to Astbury (1940) X-ray diffraction studies indicated that collagen fibres are composed of fully extended chains of C and N atoms held together by lateral bonds of an electrovalent nature. These linkages are comparatively strong but easily penetrated by ions. Later studies of the high-angle X-ray diffraction pattern of collagen fibres, and the interpretation of their results, are given by Cowan *et al.* (1953, 1955) in the proceedings of two valuable symposia, to which the reader is referred. Differences in the pattern given by collagen and reticulin were ascribed by Little and Windrum (1954) to the presence of amorphous material in the latter.

Studies with the electron microscope by Kramer and Little (1953) and by Irving and Tomlin (1954) have shown conclusively that the characteristic fibril of collagen, with its 650 Å cross-striation, is present also in reticulin. Kramer and Little, however, found that whereas collagen is composed of bundles of regularly arranged fibrils, reticulin has a membranous structure in which the fibrils occur in a disorderly manner. They concluded that reticulin differed from collagen in four respects: (1) in having an abundant carbohydrate-rich, amorphous matrix; (2) in having randomly arranged fibrils; (3) in its membranous structure; and (4) in failing to yield gelatin on boiling. They compared collagen and reticulin to rope and linoleum, and this seems to be an excellent analogy, as far as the purely static picture is concerned.

A number of authors, myself included, have stated erroneously that reticulin is not birefringent. Brewer (1957), confirming the earlier observation of von Mollendorf (1932), found that while collagen showed positive form and positive intrinsic birefringence, reticulin possessed positive form and negative intrinsic birefringence. When stained with certain dyes collagen shows dichroism and anomalous colours (see Chapter 11, p. 386), whereas reticulin does not. Brewer suggested that this is because its side chains are not free to combine with dye molecules, and, in confirmation of this thesis, he found that blocking the side chains of collagen, by tanning, prevented the development of dichroism and anomalous colours. Missmahl (1957) also reported some studies on the birefringence of reticulin.

An extensive study was made by Tonna (1964a) of the effects of staining with metachromatic dyes on the birefringence of the fibrous (collagenous) elements of skeletal tissues. A significant increase in the intensity of light passing through bone and cartilage was obtained after staining with Toluidine blue, and to a lesser extent with other thiazine and oxazine dyes. The enhanced birefringence of collagen is clearly due to staining of its associated muco-polyssacharides.

The appearance of the distal epiphysis of a 5-week old mouse femur, stained with Toluidine blue and examined by polarized light and a first-order red retardation plate, is shown in Plate IIa, facing p. 217. By rotating the tissue section through 360° with crossed Nicols a variety of interference colours can be produced. The effect of adding the first-order red retardation plate is to produce a variety of interference colours resulting from the unequal transmission of the various wavelengths of white light by the analyser (Plate IIb, facing p. 217).

In two further papers (1964 b and c) Tonna described the application of his method to problems concerning the arrangement of connective tissues in bone and callus.

BIOCHEMISTRY OF COLLAGEN

Accurate figures for the amino-acid composition of native collagen and of the product of its partial hydrolysis (gelatin), are available and these show a high content, particularly, of glycine, proline and hydroxyproline. Bowes and Kenten (1949) estimated the relative amino-acid contents of collagen and reticulin by judging the intensity of the spots in paper chromatograms of hydrolysates of the two substances. They found the amino-acid distribution to be similar in both except for proline and hydroxyproline in which reticulin was somewhat deficient. Windrum *et al.* (1955), however, made a complete amino-acid analysis of reticulin and compared this with the figures of Bowes *et al.* (1953) for collagen. From a comparison such as this, between human reticulin and bovine collagen, conclusions must be made with circumspection. It seems that the differences between the two, even in respect of proline and hydroxyproline, are scarcely significant.

Jackson (1957) made a biochemical study of the formation and removal of collagen following injection of carrageenin. This is a sulphated polygalactose extracted from the seaweed Irish moss (*Chondrus crispus*), which was shown by Robertson and Schwarz (1953) to produce granulomas rich in collagen when injected subcutaneously into guinea-pigs. Three forms of collagen have been described. One form, which is referred to as acid-soluble collagen, is extractable with dilute acetic acid or with acid buffers. The second form can be extracted from skin with a weakly alkaline phosphate buffer or by means of neutral salts. It is therefore called either alkali-soluble or neutral salt-soluble collagen. These terms are interchangeable (Gross, Highberger and Schmitt, 1955; Jackson and Fessler, 1955). Insoluble collagen, the third form, constitutes the bulk of the collagen of the tissues. Between the first and third forms there is no sharp dividing line. It has been shown, for instance, that treatment with hyaluronidase (Chapter 25) increases the acid-soluble form at the expense of the insoluble form.

In the carrageenin granuloma collagen formation reaches its zenith in 7–9 days and resorption apparently commences as soon as this point is reached.

On the third day Jackson found 40 per cent. of the neutral salt-soluble form, the remainder being insoluble. After the third day the former fell until the fourteenth day, when it began to rise again. The insoluble form, on the other hand, rose to over 80 per cent. on the ninth day and thereafter fell. Comparison with parallel histological studies carried out by Williams (1957) suggested that carrageenin first stimulated the formation of neutral salt-soluble collagen and then, after the fourteenth day, activated a mechanism leading to the removal of collagen.

Jackson showed that (α^{14}-C) glycine was incorporated into the neutral salt-soluble form of collagen with great rapidity, and into acid-soluble and insoluble collagen much more slowly. His view, in opposition to that of Orekhovitch (1955) was that neutral salt-soluble collagen is the precursor of the insoluble form, and that the latter and acid-soluble collagen are very similar. Harkness *et al.* (1954) suggested that acid-soluble collagen was derived from recently laid-down collagen and from the outer layers of larger fibres and reticulin fibres. According to Jackson extraction with 0·2 M-NaCl renders reticulin fibres non-argyrophilic and red with acid fuchsin. After citrate extraction argyrophilic fibres are no longer demonstrable. This suggests that reticulin fibres consist of acid-soluble collagen with a coating of neutral salt-soluble collagen not yet incorporated into the fibre structure.

A tentative scheme for the formation of collagen in developing connective tissue was put forward by Jackson (*loc. cit.*). He considered that neutral salt-soluble collagen was secreted by fibroblasts and transformed into collagen in one of three ways: (1) by formation of new fibrils which are argyrophilic and behave as acid-soluble collagen; (2) by accumulation on fibrils formed as above; and (3) by direct incorporation into pre-existing insoluble collagen fibres. These views are very interesting and, if they can be integrated with the data derived from other disciplines, a great advance in our understanding of the collagen-reticulin enigma will have been made.

Thermal Shortening of Collagen and Reticulin. The shrinkage of collagen fibres on heating in the presence of water is a well-known characteristic and the process has been studied by most of the physicochemical techniques mentioned above. The most prominent changes recorded are (1) loss of resistance to trypsin and (2) loss of the optical properties of the fibre. Gustavson (1956) in his book on the "Chemistry and Reactivity of Collagen" dealt extensively with the whole theme.

If tissues fixed in a variety of aqueous or alcoholic fixatives are subjected to mild heating, subsequent benzoylation no longer prevents the reaction of collagen and reticulin with the coupled tetrazonium reaction (Pearse, 1951). Often the degree of heating necessary to dry the section after mounting on the slide is sufficient to produce this resistance to benzoylation. Exposure to 70°–90° of dry heat, for 5–10 minutes, invariably does so. French and Edsall (1945) stated that formalin fixation increases the temperature at which collagen undergoes thermal shortening from 65°–70° to about 90°. Moreover, the thermal shortening of formalin treated fibres, unlike that of normal fibres, is reversible. The birefringence of collagen in histological sections disappears if the slide is subjected to moderate dry heat (70°–90°) for 30 minutes or more and the colour developed in the tetrazonium reaction after benzoylation becomes deep brownish red instead of purple. These two observations suggest that formalin-fixed collagen can still undergo the same thermal change even when actual visible shortening is prevented by their being fixed on a slide. The process is probably better regarded as a kind of condensation. No difference can be detected in the behaviour of collagen and reticulin, the coarsest and finest fibres staining identically, though the former are deeper coloured in the aggregate. This evidence suggests physical identity, the application of heat being presumed to cause a structural alteration by which the reactive groups of tyrosine, tryptophan or histidine are prevented from combining with benzoyl chloride, but remain reactive to diazonium salts. As described in Chapter 6, many other protein substances, subjected to mild heat, behave in a similar manner so that resistance to benzoylation after heating probably has a physical significance only. The thermal denaturation temperature of newly reconstructed collagen from rat tail tendon is 52° (Gross, 1964) whereas that of collagen in neutral solution is 42°. Gross has suggested that an increased concentration of collagen in the fibril increases the stability of the individual molecules. He rules out cross-linking as being responsible for the effect but it is obviously an important factor after the application of fixatives. The problem of thermal denaturation of collagen has also been considered by Tristam *et al.* (1965).

Further evidence is available which suggests a greater degree of structural identity between collagen and reticulin. This is to be found in studying the effects of mild heat in preventing acetylation of the reactive groups of tyrosine, tryptophan or histidine by acetic anhydride in dry pyridine. Collagen, and to a lesser extent reticulin, remain resistant to acetylation, whereas in many other proteins the specific groups, though resistant to benzoylation, can be successfully acetylated and prevented from giving a positive tetrazonium reaction. Gustavson (1956), using fixation by chromium compounds and condensed naphthalene sulphonic acid as indicators, showed that thermal shrinkage of collagen does not alter the reactivity of its ionic groups (mainly COOH and NH_2). On the other hand, the denatured product combines far more readily with aggregated chromium compounds which react by coordinate valency forces with the OH groups of the collagen. Gustavson suggested, therefore, that the main change in thermal shrinkage is the rupture of hydrogen bond cross-links. How this could affect the resistance of heated collagen to benzoylation and acetylation is not understood.

The behaviour of tissue proteins in the coupled tetrazonium reaction (Appendix 6), with and without benzoylation and with and without mild heating, is shown in Figs. 71, 72 and 73 and also in Figs. 74 and 75. The first three figures illustrate the staining of human renal tubules containing

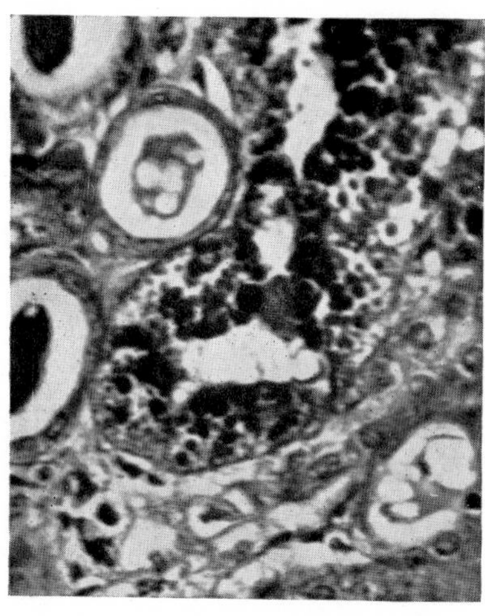

FIG. 71. Formalin-fixed human kidney (5·5 μ paraffin section). All protein structures, including nuclei, cytoplasm, hyaline droplets and tubular casts, are strongly stained. Coupled tetrazonium reaction. × 335.

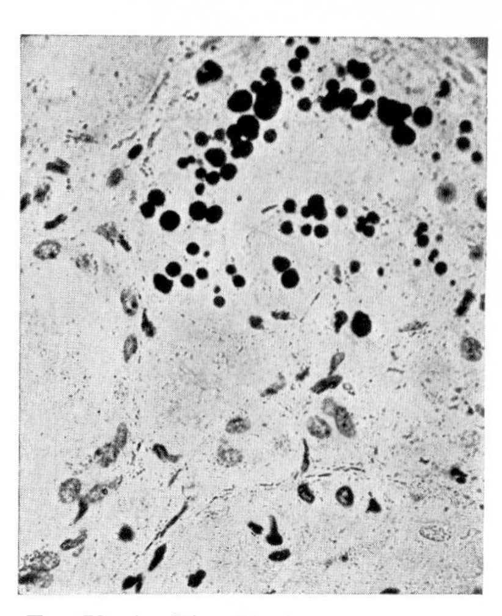

FIG. 72. As Fig. 71, but subjected to 16 hours' benzoylation at 22°. Only the nuclei and some hyaline droplets are stained. The tubule at lower left contains a protein cast. Coupled tetrazonium reaction. × 335.

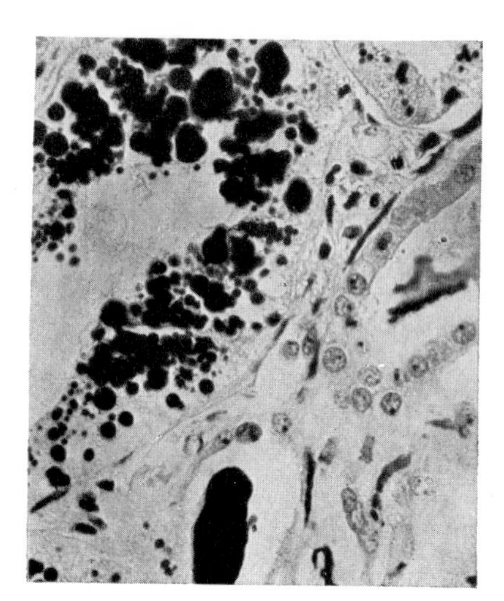

FIG. 73. As Fig. 71, but subjected to 10 minutes' dry heat at 80° before benzoylation. All the hyaline droplets and the protein casts are now stained. Coupled tetrazonium reaction. × 335.

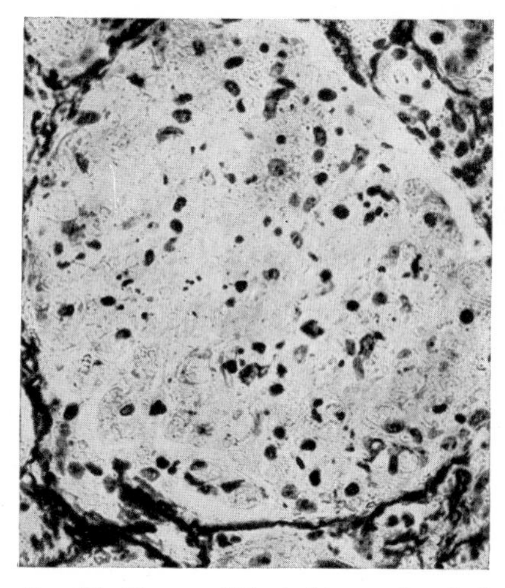

FIG. 74. Human kidney biopsy (cortical necrosis) (5·5 μ formalin-fixed section), showing a single glomerulus. Sixteen hours' benzoylation. Coupled tetrazonium reaction. × 220.

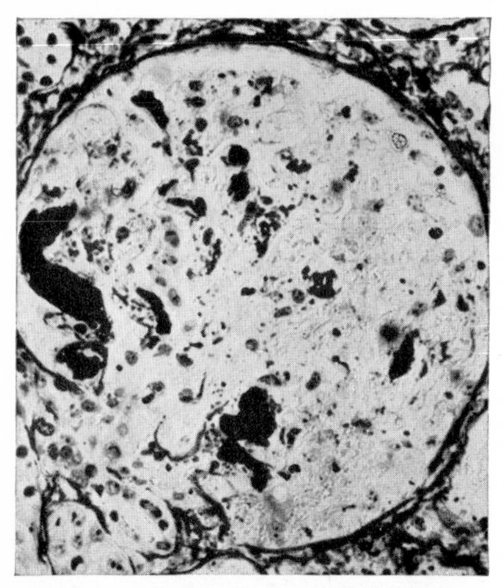

FIG. 75. Serial section to Fig. 74. Subjected to 10 minutes' dry heat at 80° before benzoylation. Fibrin masses in the glomerular tuft are now visible. Coupled tetrazonium reaction. × 220.

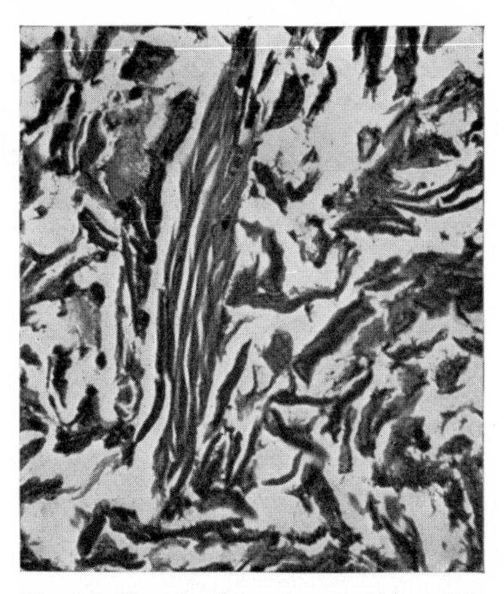

FIG. 78. Keratin flakes in an epidermoid cyst. For comparison with Fig. 79. Hæmalum and eosin. × 350.

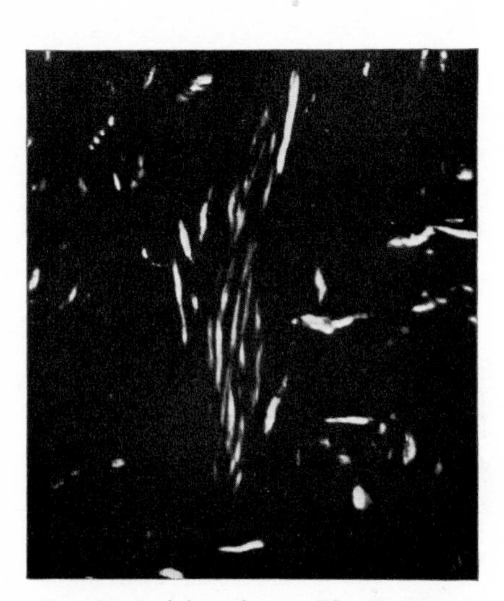

FIG. 79. Serial section to Fig. 78. Photographed by polarized light to show plain birefringence. × 350.

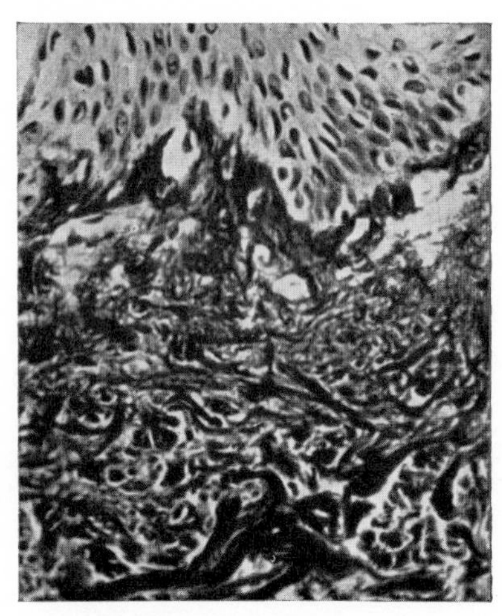

FIG. 80. Human skin (5·5 μ paraffin section). Showing resistance of collagen and reticulin, but not elastic tissue, to the process of benzoylation. Coupled tetrazonium reaction. × 235.

hyaline droplets: (1) by the tetrazonium reaction alone; (2) by the tetrazonium reaction after 16 hours benzoylation; and (3) by the tetrazonium reaction after heating for 10 minutes at 80° (dry) and 16 hours benzoylation. In the first illustration (Fig. 71) all protein-containing structures, including a cast in the lumen of a tubule, are stained with varying degrees of intensity. In the second (Fig. 72), only the nuclei and a few of the hyaline droplets are stained, while a protein cast in the tubule at lower left is colourless. In the third (Fig. 73), all the hyaline droplets are stained and also the protein cast. Figs. 74 and 75 indicate the behaviour of fibrin deposits in the glomerular tuft in a case of cortical necrosis. The first of these two illustrations shows the effect of the tetrazonium reaction after 14 hours benzoylation; only nuclei and the coarser connective tissue fibres are stained. In the second glomerulus, which was treated in parallel with the first after short heating on the warm stage, fibrin deposits are strongly stained.

Lipids of Collagen and Reticulin. A lipid component of human renal reticulin was described by Windrum *et al.* (1955). This was the saturated fatty acid myristic acid, which is apparently closely bound to protein since it resists extraction with boiling $CHCl_3/CH_3OH$. Solvent extracted reticulin, treated in this way, was found to contain about 10 per cent. of myristic acid, calculated on a dry weight basis. The plasmal reaction (Chapter 13, p. 460) indicates that collagen contains a minimal amount of acetal phospholipids, the substances responsible for the positive reaction with fuchsin-sulphurous acid in untreated control sections for the Feulgen and PAS techniques. Although chemical assays reveal a small amount of lipid in white fibrous tissue, which presumably includes any plasmalogens if present, this is not revealed by histochemical procedures. The PAS reaction in the case of collagen, since it is resistant to the action of diastase, can therefore be assumed to indicate protein-bound carbohydrate only. In practice, the amount of colour developed in collagen by this reaction varies from pale to deep pink, even when the times of the various stages are strictly controlled, as they must be in comparative work. One assumes from this that the amount of bound carbohydrate in collagen is variable or that a variable proportion is in a form non-reactive with PAS (e.g. chondroitin sulphate). The degree of reticulin staining with the PAS reaction, and probably therefore its carbohydrate content, vary less than they do in the case of collagen. Windrum and his colleagues showed that their lipid component was not responsible for the PAS reaction of reticulin, and they made several interesting suggestions concerning the role of myristic acid in pathological processes affecting connective tissues.

Immunological Properties of Collagen and Reticulin. Cruickshank and Hill (1953) showed by means of the fluorescent antibody reaction that rabbit anti-rat glomerular serum contained an antibody which would react specifically with reticulin, basement membrane, sarcolemma and neurilemma, throughout the body. There was no reaction with collagen fibres, elastic tissue or cartilage. The suggestion was made, therefore, that the antigenic properties of reticulin might depend on its polysaccharides. This aspect of the problem has been investigated by Milazzo (1957), who endeavoured unsuccessfully to extract a specific haptenic or antigenic fraction from trypsinized reticulin samples. He concluded that the "antigenic component of reticulin is probably integral with its stable glyco-lipo-protein non-collagenous moiety." It could not be decided, from the work of Cruickshank and Hill, whether reticulin and basement membrane were antigenically identical. Scott (1957), however, showed that they were not. He found reticulin antigen in synovia and basement membrane antigen in the arterial media. Glomeruli contained both antigens. Connective tissues must be regarded as antigenically heterogeneous.

TABLE 14

Properties of Various Scleroproteins and Fibrin

Property or test	Frozen or paraffin sections	Collagen	Reticulin	Fibrin	Fibrinoid	Elastic	Keratin
Isoelectric point	..	9–10	..	5·5	..	6·0	..
Mallory PTAH	P	Red-brown	Red, occ. (blue-lined)	Blue	Red-brown to blue	(Coarse) blue (fine) red	Blue
Toluidine blue	F	Variable meta-chromasia	..	Ortho-chromatic	Variable meta-chromasia	Ortho-chromatic	Ortho-chromatic
Sulphation Metachromasia	P	–	+	–	±	–	–
Pseudoplasmal reaction	F	–	–	–	–	+†	–
Lipid (Sudan black)	F	–	–	Variably +	Variably +	–	Outer corneum +
Lipid aldehyde (naphthoic acid)	F	–	–	–	–	+†	–
Birefringence	F	+	+	+ or –*	–	–	+
Autofluorescence	F	–	–	–	–	+‡	–
Gram	P	+	–	+	occ. partly +	–	–
Aldehyde-fuchsin	PF	–	–	–	–	+	–
Tetrazonium	P	Red-purple	Red-purple	Deep red	Deep red	Brownish-gold	Brown
Tetrazonium (benzoylated)	P	Red-purple	Red-purple	Deep red	Deep red	–	–
PAS after diastase	P	Variably + (pink)	Variably +	Occ. + (pink)	++	+ or – §	–
Crystalline pepsin	Unfixed	Rapid	Rapid	Rapid	Less rapid	Nil	Nil
Crystalline trypsin	Unfixed	Slow	Very slow	Rapid	Less rapid	Nil	Nil
Silver methods	P	Brown to yellow	Black	–	Yellow to brown	–	–
Acid hæmatein (Baker)	P	–	–	Pale blue	–	–	–
Sakaguchi reaction	P	Moderate	Moderate	Strong	Strong	Weak	Strong

* Fibrin film is birefringent. Fibrin in the tissues variably so.
† Particularly in rodents.
‡ Arterial and pulmonary elastic tissue especially.
§ Positive in human dermis; weak or negative in most other sites.

At this point is inserted Table 14, which sets forth the various properties both of collagen and reticulin, and of the other scleroproteins which remain to be discussed. It emphasizes the difference between these various scleroproteins and may act as a basis for the correlation of their staining properties with physical and chemical structure.

Conclusions. Very considerable advances continue to be made in our knowledge of the structure of both collagen and reticulin, and the observed differences in staining are reasonably explained by differences in the arrangement of the characteristic component fibrils and of their matrix or cement. Fig. 76 illustrates these points in diagrammatic form.

STRUCTURE OF RETICULIN AND COLLAGEN

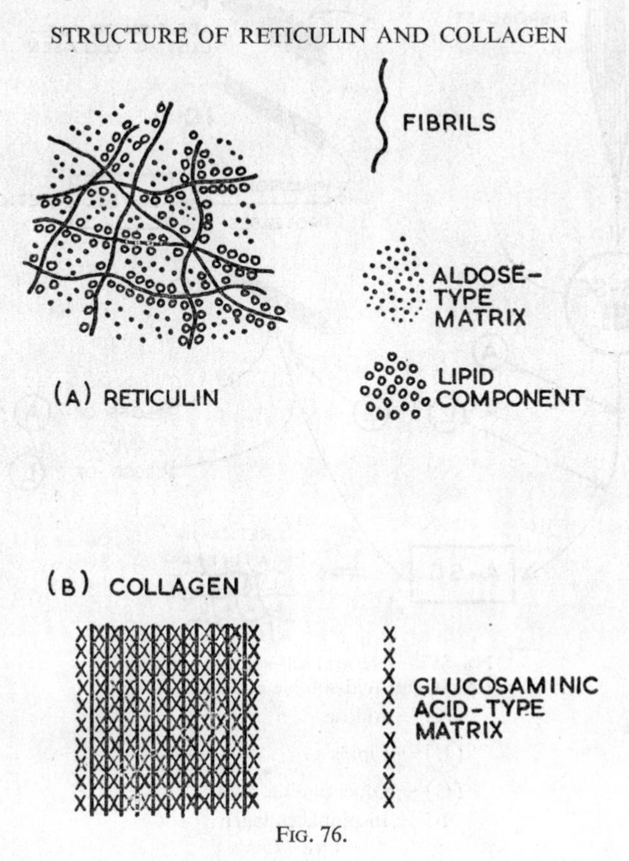

FIG. 76.

If Brewer's (1957) evidence is accepted, the positive form birefringence shown by reticulin indicates that it is made up of longitudinally arranged particles. Since the electron microscope studies of Kramer and Little suggest a random arrangement of fibrils it follows that in most samples of reticulin there is sufficient longitudinal alignment to give the observed birefringence. The negative intrinsic birefringence of reticulin indicates the presence of a component at right angles to the fibre axis, and this is possibly Windrum *et al.*'s myristic acid.

A tentative scheme for the formation of collagen can be drawn up (Fig. 77) which takes into account most of the biochemical factors discussed above. It serves to emphasize the points which still require clarification. The initial stage of collagen formation involves the production of NS-SC by the fibroblast and

BIOSYNTHESIS OF COLLAGEN

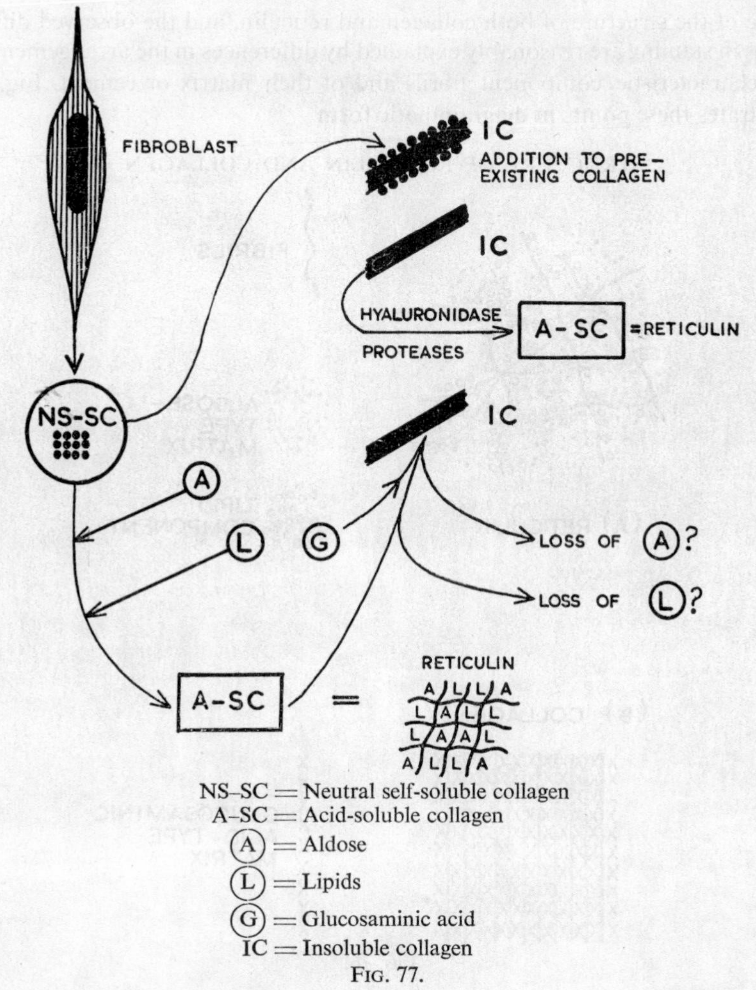

NS–SC = Neutral self-soluble collagen
A–SC = Acid-soluble collagen
(A) = Aldose
(L) = Lipids
(G) = Glucosaminic acid
IC = Insoluble collagen

FIG. 77.

its transformation into A-SC or reticulin. We do not know whether myristic acid is present in the fibroblast and we are fairly certain that the quantity of aldose sugars is small. The addition of these substances to NS-SC, under whatever auspices it takes place, is obviously of vital importance in collagen formation. Although both substances are tightly bound to the protein of reticulin the process appears to be readily reversible and conversion of A-SC to IC must take place by loss of the aldose and lipid fractions of reticulin. It is possible that these are not so prominent a feature of dynamic reticulin, reticulin that is in the process of conversion. The change-over from one type of matrix to another in the conversion of reticulin to collagen cannot at present be explained.

When any of the numerous silver techniques for reticulin are used (Appendix 8, p. 641), many of the fibrils which are stained are certainly fine nerve fibres. This is indicated by the identical pattern obtainable with methods specific for neurofibrils. The reticulin staining obtained by Sen *et al.* (1961) with the bromine-silver method for unsaturated lipids (Chapter 12, p. 440) may have a similar explanation.

ELASTIC TISSUE

I have here purposely used the term elastic tissue rather than elastin because it is the structure, rather than the protein material isolated from it by the chemist, with which we are concerned. As with collagen and reticulin, from the histological point of view, elastic tissue is easily distinguishable by means of a number of "specific" stains, some of which are of considerable antiquity. The most important of these stains are the various resorcin-fuchsin methods associated with the name of Weigert, the orcein method of Tänzer (1891), hæmatoxylin methods such as those of Harris (1902) and Verhoeff (1908), and the latest aldehyde fuchsin method of Gomori (1950). Michaelis (1901) showed that a number of basic dyes could be used in place of fuchsin in the Weigert method and that other phenols might be used instead of resorcin. Michaelis also showed that oxidizing agents such as ammonium persulphate might replace the ferric chloride; he was unable, however, to throw any light on the mechanism of the Weigert stain. Table 15, largely derived from Harms' (1957) review, presents the whole spectrum of elastic tissue staining, historical and present.

TABLE 15

Staining of Elastic Tissues

	Acid Staining
Unna, 1886	0·5 per cent. dahlia in alcoholic HNO_3
Koppen, 1889	5 per cent. crystal violet in 5 per cent. phenol
Tänzer, 1891	1 per cent. orcein in acid alcohol (HCl) pH 1·3
Pranter, 1902	1 per cent. orcein in acid alcohol (HNO_3) pH 0·9
Weigert, 1898	0·25 per cent. resorcin-fuchsin in acid alcohol
Harris, 1902	Mucicarmine with acid alcohol differentiation
Michaelis, 1901	Resorcin and other phenols with fuchsin, safranin, methyl violet, etc.
Verhoeff, 1908	Iron-hæmatoxylin
Matsuura, 1925	Phosphotungstic acid-Congo red
Petry, 1952–53	Phosphomolybdic acid-hæmatoxylin
Gomori, 1950	Aldehyde-fuchsin
	Neutral Staining
Klüver and Barrera, 1953	Alcoholic Luxol fast blue (method only, not application)
	Alkaline Staining
Herxheimer, 1886	Lithium carbonate-hæmatoxylin
Salthouse, 1944	Ammoniacal chlorazol black
Witke, 1951	Alkaline primulin (fluorescence)
Lautsch *et al.*, 1953	Alcoholic chlorazol black-benzazurin

The mechanism by which most of these stains work remains an enigma, but some attempt has been made by Dempsey *et al.* (1952), Weiss (1954) and by Fullmer and Lillie (1956) to explain the nature of orcein staining of elastic tissue. Engle and Dempsey (1954) found that orcein was a mixture of closely related amphoteric dyes which could be separated into four principal coloured fractions. Fractions I and II were cationic, Fraction III anionic, and Fraction IV partly anionic and partly electro-neutral. When these were used to stain formalin-fixed horse ligamentum nuchæ Fractions I, II and III stained certain fibres throughout the pH range up to 9·5, with maximum intensity at pH 3–8·5 for Fraction I, pH 5–7 for Fraction II and pH 2·9 for Fraction III. The authors concluded that orcein staining was relatively insensitive to pH and, therefore, probably due to reaction through non-ionic bonds. Their Fraction II could be used as an acetic acid orcein to stain nuclei blue. Musso (1956), using the technique of distribution chromatography with cellulose or silica gel at pH 11–12, obtained twelve brightly coloured components from samples of orcein. Five of these he was able to crystallize but, because of the small amounts available, none were applied to the staining of elastic tissues. Fullmer and Lillie (1956) were unable to prevent orcein staining of elastic tissue after blocking its carbonyl, hydroxyl, amine and aldehyde groups by a variety of procedures. After such procedures, however, they found that normal collagen would stain with orcein. It was evident, from this work, that polar groups must interfere in some way with orcein staining. Further experiments showed that acetylated or nitrosated (deaminated) orcein could not function as a stain, indicating that the aryl hydroxyl or amine groups of the dye molecule are required for staining. These views are essentially the same as those of Weiss (1954). This author estimated the heat of reaction between elastic tissue and orcein as between −6,000 and −6,500 calories per mole. These figures are consistent with a reaction based on the formation of hydrogen bonds. The question as to which group in elastic tissue takes part in hydrogen bonding with the phenolic groups of the dye is less easily answered. Fractions I and II of Engle and Dempsey, which are responsible for "orcein" staining, carry a positive charge in acid solution. They are therefore repelled by the positively charged side chains of most proteins at such pH levels. Elastin, however, possesses very few positively charged side chains in acid solutions and should not repel the charged orcein molecule.

The mechanism of resorcin-fuchsin staining was considered by Puchtler *et al.* (1961), and subsequently by Puchtler (1964). Depending on the type of pretreatment applied there was positive staining in collagen, reticulum fibres, basement membranes and the ring fibres of blood vessel walls, as well as in elastic fibres. The specificity of resorcin-fuchsin for the latter must clearly be very low.

The involvement of ionic and non-ionic bonds was considered by Goldstein (1962) who found that the staining of elastic tissue by aldehyde-fuchsin, resorcin-fuchsin, Chlorazol black E, and Luxol fast blue was inhibited by urea (saturated solution in the dye bath). Since urea is a powerful H-bonding agent, Goldstein's findings strongly support the theoretical explanations given above.

Chemical Composition. In the foregoing discussion elastic tissue has been assumed to be equivalent to the characteristic protein elastin which it contains. It is possible, of course, that the staining of elastic tissue with any particular dye is due to other components. Elastin differs chemically from collagen in some important respects, chiefly in possessing far less arginine, histidine and lysine and more leucine and valine. It contains no tryptophan but has slightly more tyrosine than collagen (Graham, Waitkoff and Hier, 1949). These differences might be expected to shift the isoelectric point from the alkaline towards the neutral point, which in fact they do. The isoelectric point of collagen is in the region of pH 10·0 and that of elastin about pH 6·0 (Bowes and Kenten, 1949). Leucine residues are known to be lipophilic (Davies and Dubos, 1947) and elastic tissue might, on this account, contain more lipid than collagen. Chemical assays, however, do not reveal any difference in the lipid content of the two substances. The literature concerning the action of proteolytic enzymes on elastic tissue is to some extent contradictory, nevertheless it appears that crude trypsin attacks elastic fibres only very slowly in their native state while crude pepsin acts more rapidly. According to Lansing (1951) the pure enzymes have no effect. Baló and Banga (1950) produced from pancreas a specific elastase which dissolves the elastic tissue of the aorta without hydrolysing it (see Chapter 25), and this enzyme has been shown by Pepler and Brandt (1954) to be identical with chondromucinase. Some authors (Braun-Falco, 1956; Harms, 1957) believe that the effect of elastase in tissues is to remove a chromotropic, mucopolysaccharide-containing cement substance, leaving the fibre (elastin) unaltered. Banga (1951) considered that the enzyme acted on elastin itself, converting it from a fibrous to a globular (soluble) form. During the process an oily layer rises to the top of the incubation medium, and this is said (Dempsey and Lansing, 1954) to have sphingomyelin-like properties.

Elastic fibres from all sources are remarkably insoluble in inorganic or organic solvents, differing in this respect from collagen which dissolves, for instance, in 2 per cent. acetic acid. This difference in reactivity may be connected with the fact that elastic contains 90 per cent. of non-polar amino-acids, and collagen only 50 per cent. Most studies on the property of elastic fibres have been made on material from the ligamentum nuchæ of the ox. Ordinary histological methods of staining suggest that there are wide differences between this material and the elastic fibres of skin, aorta and of the elastic laminæ of the smaller vessels. The elastic fibres of the human dermis are readily stained even with "bad" resorcin-fuchsin mixtures, while the finer fibrils of arterial walls are more difficult to stain satisfactorily. Such differences remain to be explained, but they are most probably of a physical nature. (see below, Oxytalan and Elaunin Fibres). The physical and chemical properties of elastin were comprehensively reviewed by Partridge (1962) who concluded that, despite much conflicting evidence, elastic fibres could be regarded as composed of homogeneous microfibrils of an unique protein. The structure and function of elastin were considered in a monograph by Banga (1966), summarizing her 15 years' work in this field and dealing also with collagen.

The refractility of elastic fibres in tissue sections caused Seki (1934) to propose that they were of high density. The refractive index is, in fact, between 1·47 and 1·54 (Goldstein, 1962) and their refractility in synthetic resins or balsam is due to their having a lower rather than a higher R.I., with respect to the medium.

Electron Microscopy. Electron microscope studies of elastic tissue have not been altogether satisfactory. Gross (1949) observed that tryptic digestion of elastin from various sources released large numbers of filaments, polydisperse in length but constant in diameter. A large proportion of these appeared to be tightly coiled helices composed of two intertwining threads. Franchi and de Robertis (1951), however, repeated the work of Gross and showed that the latter's findings were due to bacterial contamination of the trypsin used for digestion. Lansing *et al.* (1952) studied the process of digestion by elastase and noted the progressive splitting of large elastic fibres into small short ones. In thin sections fixed in buffered osmic acid elastic fibres appear (Dempsey and Lansing, 1954) as a negative image. Only in shadowed preparations could elastic tissue be resolved as an anastomosing, three-dimensional network of fibrils with an average diameter of 200 Å. Collagen fibrils surround the elastic fibres and occasionally penetrate them. Electron microscopic studies by Cox and Little (1962) showed that elastic tissue from various sites in human and animal tissues had the same fine homogeneous structure. No evidence was found to favour a two component concept for elastica.

Fluorescence. Staining with thiazine dyes shows a strict and invariable orthochromasia, but the fibres, though isotropic by polarized light, possess a strong yellowish autofluorescence in ultraviolet light. The significance of these three characteristics is not absolutely clear, but it is worthy of note that the stretched fibres do exhibit birefringence, which is of course never to be observed in histological preparations. Isotropism presumably indicates a random orientation of molecules and autofluorescence may be due to lipid despite its presence in small amounts. According to Partridge (1962) the elastin molecule may consist of two short peptide chains (glycine, proline, alanine, alanine) linked (cross-linked) by a chromophoric nucleus. The latter is responsible for the yellow colour and for the fluorescence. Its identity is still unknown but according to Thomas *et al.* (1963) and Bedford and Katritzky (1963) it may be a pyridinium compound to which peptide chains are linked at four positions.

Cytochemistry. The observed histochemical reactions of elastic tissue are of interest, especially the positive pseudoplasmal reaction. This is due to the presence of oxidized phosphatides whose carbonyl (aldehyde) groups have been demonstrated in elastic tissue by means of the phenylhydrazine and dinitrophenylhydrazine reactions (Albert and Leblond, 1946; Schrek and

McNamara, 1947) as well as by the hydroxynaphthoic acid reagent of Ashbel and Seligman (1949). When this last reaction is employed on formol-fixed frozen sections the internal elastic laminæ of arteries are invariably strongly stained; elastic fibres in the skin, however, usually do not stain, although in both situations a pseudoplasmal reaction occurs. Both Sudan black and Baker's acid hæmatein technique (Chapter 12, p. 422) usually fail to reveal any phosphatide in elastic fibres, although chemical analyses reveal about 1 per cent. of lipid. Hack (1952), moreover, using freeze-dried carbowax-embedded material, has failed to demonstrate acetal phosphatides in elastic tissue. Interpretation of the positive copper phthalocyanin reaction (Chapter 12, p. 426) is difficult. Since nucleic acids and basic amino-acids are definitely absent I feel that it is due to the "sphingomyelin-like" component. Pyridine extraction leaves a positive reaction, as does extraction with acetone and ethanol. This is reversed, however, by boiling $CHCl_3/CH_3OH$.

The weak reaction obtained with various modifications of the Sakaguchi method for arginine reflect the low figures recorded for this amino-acid by chemical assays (Stein and Miller, 1938). With the tetrazonium reaction the colour of elastic tissue in the skin is a golden-brown (weak positive), compared with the strong reddish-purple of collagen, while arterial elastic tissue is even paler. This finding might be due to the absence of histidine though one would have expected the increase in tyrosine, compared with that present in collagen, to have effectively balanced this. Tryptophan is absent from both collagen and elastin. The significance of the tetrazonium reaction here is obviously doubtful. Unless the reactive (tyrosine) groups are much more widely spaced than the reactive (tyrosine, histidine) groups of collagen it is difficult to explain the very pale colour produced in elastic fibres. There is no evidence to suggest that the tyrosine groups are in any way protected from reaction with tetrazonium compounds, in fact the opposite is the case. When the tetrazonium reaction is performed after benzoylation, whether heat has been applied to the sections or not, elastic fibres are always completely colourless. This indicates that their reactive groups are not so arranged as to be resistant to benzoylation, and they are unlikely to be so arranged as to fail to combine with tetrazonium salts, although this possibility still exists.

From histochemical evidence, therefore, we can tentatively suggest that elastic fibres owe their peculiar properties to (*a*) the relative absence of basic amino-acids in the molecule, (*b*) to the wide spacing of the various histochemically reactive amino-acid residues in the polypeptide chain, and (*c*) to the presence of oxidized phospholipids. In what way these factors operate remains a matter for speculation. No single reaction for SO_4 groups is positive in elastic fibres under histochemical conditions, and I cannot, therefore, agree with the views of Harms (1957) on the importance of ester sulphates as an explanation of the observed staining peculiarities of elastin. Moreover, even when the PAS reaction is positive, it has not been shown with certainty that this is due to carbohydrate. Differences between elastic fibres in various situations in the body are more likely to be due to physical than to chemical causes.

Oxytalan and Elaunin Fibres. The first of these, the oxytalan fibre, was first described by Fullmer and Lillie (1958) who found it in regions formerly supposed to contain only collagen. Oxytalan (Greek: *ὀξύς*, acid and *ταλας*, *-νος*, enduring) fibres were so named from their property of resistance to acid hydrolysis. After peracetic acid oxidation they were stained with aldehyde fuchsin, orcein, and resorcin-fuchsin but not with Verhoeff or orcinol-new fuchsin. In a later paper Fullmer (1960) concluded that oxytalan fibres might represent immature or modified elastic tissue, since histochemical tests were unable to distinguish between the two. Fullmer indicated a close relationship between elastic, pre-elastic and oxytalan fibres.

Gawlik (1965) described a further elastic-like fibre which he named elaunin (Greek *ελαυνω*, I stretch). This component stains with orcein, aldehyde fuchsin, resorcin-fuchsin and cresyl violet. It does not stain with Verhoeff or orcinol and new fuchsin. It differs from the oxytalan fibre in not requiring pre-treatment with peracetic acid. Although Gawlik attributed the observed differences between pre-elastic, oxytalan and elaunin fibres to differences in their mucopolysaccharide matrices, his evidence is open to alternative interpretation.

Pathological Changes in Elastic Fibres and Elastotic Degeneration. Elastic fibres are very stable components of the connective tissues and vessels, and (α-C^{14}) glycine studies indicate a very low turnover rate. With advancing age, however, marked changes occur. These take the form of longitudinal splitting, breaking into fragments and ultimately into granules. These changes are associated with chemical changes in amino-acid content (raised glutamic and aspartic acids) and in calcium content (up to 14 per cent.). Changes in acidophilia of aortic elastic tissue, occurring with increasing age, have been considered by Menzies and Roberts (1963) and Menzies *et al.* (1964). The process of senile elastosis was studied by Weidmann (1931) and later by Cooper (1952) and by Lansing *et al.* (1953). Gillman and his associates (1954, 1955) have studied the so-called elastotic degeneration of collagen, and stress the value of Mallory's phosphotungstic acid-hæmatoxylin in distinguishing between normal collagen (reddish-orange) fibrinoid (purple) and elastotic collagen (pale yellow or unstained). According to these authors elastotically degenerated fibres may arise (*a*) by degeneration of pre-existing collagen or (*b*) *de novo* by new formation of abnormal collagen. Changes in elastic fibres are observed in the disorder known as elastosis perforans serpiginosa and these have been described by Greenblatt (1963). Marshall and Lurie (1957), on the other hand, regarded the fibres as normal elastica.

Fibrin

Fibrin is derived from the fibrinogen of the plasma which represents about 4 per cent. of the total plasma proteins. Fibrinogen is converted into fibrin by the addition of small quantities of thrombin, the amount of fibrin formed being independent of the amount of thrombin present, over a wide range (Morrison, 1947). Histologically, two staining methods in particular have been used to demonstrate fibrin. These are the methyl-violet method of Weigert (1887), usually performed as the Gram-Weigert method (Mallory, 1938) and the phosphotungstic acid-hæmatoxylin method (Mallory, 1938). The former requires a fine judgment in differentiation and, of course, stains many other structures, of which the majority are easily distinguished from fibrin. Collagen fibres react quite regularly, however, and these are not so easily distinguishable. The second method gives a deep blue colour with fibrin and although nuclei, muscle, neuroglia and coarse elastic fibres also stain blue, collagen and reticulin are red. Material which is not stained blue by phosphotungstic acid-hæmatoxylin (PTAH) is not fibrin.

Two other staining methods for fibrin have become popular with histopathologists in recent years. These are "Picro-Mallory" and Martius-Scarlet-Blue (MSB). Both are products of long-term studies by Lendrum and his associates, fully described by Lendrum *et al.* (1962). Details of the two methods are given in Appendix 8, pp. 643 and 644. Their application is illustrated in two pictures (Plates IIIa and IIIb, facing p. 238). The problem of staining fibrin is accentuated by changes in its properties which accompany ageing. These have been considered by Lendrum *et al.* (1962) as well as by Lendrum (1963) and Cook (1964). In the earliest stages (4–12 hours) fibrin is difficult to stain by any method. After this period "young" and "middle-aged" fibrin stain easily but the latter has similar staining affinities to elastic tissues. From these it is differentiated by orcein methods. Older fibrins fail to stain by Gram-Weigert or Picro-Mallory at a stage when they can still be shown by PTAH and the MSB method.

At a still later stage, according to Lendrum *et al.*, all stains give the same result as they do with collagen except for a variation of Masson's stain (Masson 44/41). This method, details of which are given in Appendix 8, is less useful for younger fibrins. With the older varieties, as shown in Plate IIIc, facing p. 238, it produces remarkably clear results.

A number of staining techniques which afford distinction between the various stages of age change in fibrin are given by Lendrum, Fraser and Slidders (1964).

The amino-acid content of fibrin, in which the three basic amino-acids are strongly represented, accounts for the strong affinity for acid dyes observed when ordinary histological stains are employed. Human fibrinogen is a large elongated molecule with a length of 700 Å and a molecular weight of 400–700′000 (Edsall, Foster and Scheinberg, 1947) which undergoes no intramolecular rearrangement when changed into fibrin in the process of clotting (Bailey, Astbury and Rudall, 1943). The latter process has been described by Ferry and Morrison (1947), and by Ferry (1948), as the formation of a network whose fine strands consist of chains of fibrinogen molecules joined end to end and occasionally cross-linked by primary chemical bonds formed under the influence of thrombin. These fine strands are joined into coarse strands by secondary cross-linkages which form them into bundles. When the coarse clot is condensed into a film it becomes anisotropic (Ferry, 1948).

Proteolytic enzymes rapidly digest fibrin in unmodified condition and fairly rapidly even after treatment with formalin. Heat treatment, such as exposure to steam at 121° in the presence of 20 per cent. of moisture, increases the resistance of fibrin to enzymic digestion and increases its affinity for both acid and basic dyes (Ferry *et al.*, 1947). These changes, paradoxically at first sight, are attributed to secondary cross-linking (Ferry and Morrison, 1947). It appears that although both heat and formalin prevent a certain number of the charged groups of the protein fibre from reacting with dyes, they increase the availability of all remaining charged groups by some physical modification of structure.

Jaques (1943) studied the reducing capacity of fibrinogen and fibrin for H_2O_2. He found that fibrinogen contained a reducing group with much stronger properties than tyrosine or tryptophan and that conversion to fibrin increased the number of these groups. Of naturally-occurring substances, only cysteine and ascorbic acid possessed a comparable reducing capacity weight for weight. The former is present in insufficient concentration to explain the findings, the latter is not present at all.

Histochemically fibrin possesses properties which correlate well with those determined by physical and chemical investigations. It is invariably orthochromatic and does not exhibit autofluorescence in ultraviolet light. Birefringence is seldom observed in an entire patch of fibrin in the tissues, but often the superficial layers are strongly anisotropic. The Sudan black test for formol-fixed lipid is sometimes positive, though more usually negative, and probably depends on the inclusion of lipid or lipoprotein during the process of clotting. Baker's acid hæmatein method (Chapter 12, p. 422), however, gives a pale blue colour with fibrin, suggesting the presence of phosphatides. The PAS reaction is usually positive, sometimes strongly so. Studies on a series of artificial clots produced from fibrinogen in different ways (Gitlin and Craig, 1957) suggest that included substances may be responsible for this reaction to some extent also. If their clots were formed from fibrinogen and thrombin alone, Gitlin and Craig observed negative reactions with the usual histological stains. If more than 1 g. per cent. of albumin was added, or glutathione at a concentration higher than 0·05 g. per cent., all the histological reactions became positive. Lorand (1950) and Mihályi (1950) both described urea-soluble and urea-insoluble forms of fibrin, and Gitlin and Craig related their positive staining fibrin to the urea-insoluble form, suggesting that some property resident in structural bonds or in cross-linkages was responsible for the observed differences in reactivity.

With the tetrazonium reaction, after heating and before and after benzoylation, both fine fibres and coarse masses of fibrin stain deep brownish-red, suggesting that fibrin exists in a more "condensed" form than do collagen and reticulin, for instance (see Figs. 74 and 75).

The DMAB-nitrite reaction of Adams (Chapter 6, p. 131) is strongly positive with fibrin, which contains 3·5 per cent. of tryptophan, and this reaction is a useful one for demonstrating fibrin deposits in the tissues. It was used by MacMillan *et al.* (1965) to demonstrate accumulation of plasma proteins (including presumably fibrin) in the subendothelial zone of the aortic intima. In most cases the so-called fibrinoid gives an equally strong reaction so that the differential diagnosis between the two substances is not possible by this means. Similarly the alkaline tetrazolium reaction (Chapter 6, p. 152) is intensely positive in both fibrin and fibrinoid.

Summing up, therefore, one may say that the distinction between fibrin and the various other scleroproteins is usually easy. Protein material which is strongly acidophil when stained with acid dyes and blue with Mallory's phosphotungstic acid-hæmatoxylin, which is tetrazonium-positive after mild heat and benzoylation, is almost certainly fibrin. The other reactions are of interest but they are unnecessary for diagnostic purposes. It is necessary to bear in mind the possibility of other constituents of plasma being present as inclusions in a fibrin clot.

Fibrinoids

The term fibrinoid covers a number of different materials with diverse pathogenesis. After its introduction by Neumann (1880) the word was used to describe substances found in the walls of arteriosclerotic vessels which resembled fibrin in their tinctorial behaviour. It was also used to describe changes in the smaller vessels in malignant hypertension, periarteritis nodosa and thrombo-angiitis obliterans. Clark *et al.* (1936) regarded the arteriosclerotic variety as derived from some constituent or constituents of blood plasma and certainly not from necrotic collagen. Grosser (1925) applied the word to the fibrin-like substance deposited in various layers of the placenta. Klinge (1933) described the fibrinoid alteration of collagen in rheumatic fever as being due to hypersensitivity and maintained that a similar mechanism was responsible for its appearance in a number of disorders now considered collectively as collagen diseases.

The type of fibrinoid referred to in Table 14 is that found in areas of collagen necrosis in the rheumatic group of disorders. Its nature is of particular interest at the present time and many theories exist to explain it. Bahrmann (1937) regarded the fibrinoid degeneration of connective tissues as due to necrosis of collagen, while Klemperer *et al.* (1942) regarded the fibrinoid of disseminated lupus and diffuse scleroderma as a coagulation of ground substance. Altschuler and Angevine (1949) went further than this and, after performing a number of histochemical tests, they concluded that fibrinoid of the rheumatic variety was due to precipitation of the acid mucopolysaccharide of the ground substance by its combination with an alkaline (basic) protein. Glynn and Loewi (1952) also used a variety of histological and histochemical methods on rheumatic fibrinoid and considered that the change was essentially due to the deposition of a polysaccharide-containing substance in the altered collagen produced by the initial disorder. These last authors found that the positive PAS reaction of rheumatic fibrinoid could be reversed by means of an enzyme described as pectinase (Chapter 25), but not by hyaluronidase, and they attributed the increased resistance of both fresh and fixed fibrinoid to the action of proteolytic enzymes as being due to protection by polysaccharide.

The results of X-ray diffraction and electron microscope studies (Kellgren *et al.*, 1951) indicated that fibrinoid of rheumatic origin contained no fibrin whatsoever. Material from those parts of a rheumatoid nodule containing large amounts of normal collagen on histological examination also showed normal collagen fibres under the electron microscope, but gave an X-ray diagram of disorientated collagen. In the parts showing advanced fibrinoid change the X-ray diagram of collagen was absent and electron microscopy showed only amorphous material. Intermediate areas between these two were common. Work on the chemistry of connective tissue fibres in rheumatic fever (Consden, Glynn and Stanier, 1952) showed that the material in the central eosinophil core of the rheumatic lesion closely resembled collagen. By chromatographic methods these authors estimated the amounts of hydroxyproline, tryptophan, tyrosine and reducing sugar, in extracts of rheumatic nodules and the normal subcutaneous collagen of the elbow region. The nodules differed only in containing slightly more tyrosine and much more reducing sugar (2·9 as against 1·3 per cent.). Ziff *et al.* (1953) made alkali or trypsin extracts of five rheumatoid nodules which they analysed for hydroxyproline and found 0·0–0·3 mg. per gram of tissue. The collagen content of the nodules, measured by the method of Lowry, Gilligan and Katersky (1941), was between 70 and 159 mg. per gram. Microscopical determination of the collagen/fibrinoid ratio in the nodules gave an estimated fibrinoid content of 17–40 mg. per gram. The maximum amount of collagen present in fibrinoid was estimated to be 2 per cent. by weight, that is to say, a negligible amount.

The nature of the fibrinoid of human placenta was studied particularly by Singer and Wislocki (1948). These authors carried out controlled staining in dilute solutions of orange G and methylene blue at various pH levels. They concluded that the staining characteristics of fibrin and this type of fibrinoid were essentially similar, and the same conclusion was reached by Busanny-Caspari (1952, 1957), who studied the morphogenesis of fibrinoid in placenta and decidua. Other types of fibrinoid have now been the object of critical histochemical studies. Montgomery and Muirhead (1957), using a broad spectrum of histochemical techniques, compared vascular fibrinoid with connective tissue fibrinoid and with fibrin and found the last two tinctorially and histochemically similar. Studies made by Ruiter (1962a and b) indicated that vascular fibrinoid (in cutaneous allergic arteriolitis) was largely composed of fibrin. Gitlin, Craig and Janeway (1957), using the fluorescent antibody technique with rabbit anti-human fibrin, found that the fibrinoid of the collagen diseases (rheumatic arthritis, disseminated lupus erythematosus, dermatomyositis, rheumatic fever, etc.) was partly composed of fibrin. Moreover, they showed that conventional fibrin stains (Mallory-trichome and phosphotungstic acid hæmatoxylin, PAS) were often negative when the fluorescent technique gave a positive result. According to Montgomery and Muirhead (also Muirhead *et al.*, 1956) vascular fibrinoid differed from connective tissue fibrinoid in containing a quantity of cholesterol esters and potassium salts. They believed that the fibrinoid of renal cortical necrosis due to the Schwartzmann phenomenon might originate from smooth muscle and that vascular fibrinoid could have a similar origin. Fibrin and connective tissue fibrinoid gave a strong (Ehrlich-Lison) indole reaction whereas vascular fibrinoid did not. Using an immunofluorescent technique Crawford and Woolf (1960) showed that both hyaline and fibrinoid in vascular lesions were related to fibrin. With a similar technique Fennell *et al.* (1961) found fibrin in the small renal arterial lesions of malignant hypertension. On the other hand Vasquez and Dixon (1958), also using an immunofluorescent method, had stated that the material deposited in the vessel walls in hyperergic fibrinous vasculoses was globulin rather than fibrin.

The greatest weight of evidence at the present time supports the view that fibrinoid normally contains fibrin, more specifically that it is derived from exuded plasma proteins, particularly fibrin (Movat and More, 1957). This, interpretation was supported by the extensive studies of Lendrum *et al.* (1962), who recommended that the term fibrinoid should be abandoned, although this absolutist view was later softened by Lendrum (1963).

Clearly when the term fibrinoid is used, as it certainly will be, it must signify a material composed of plasma protein derivatives including fibrin, which stains as fibrin until the last stages of its evolution. Here, with Lendrum (1963) we can call it pseudo-collagen since it now stains like collagen. The classical changes of chronic fibrinous (or plasmatic) vasculosis are well shown by the renal lesion of diabetes mellitus, which is illustrated in Plate IIId, facing p. 233.

If the plasmatic origin of fibrinoid is accepted it is easy to explain differences in staining from one region to another and between one case and another. Differences may be caused by:

(1) Variation in the proportion of the different (normal) plasma proteins passing through into the lesion.
(2) The presence of abnormal proteins, such as macroglobulins.
(3) The presence of varying amounts of extraneous substances, such as lipids.
(4) Variation inherent in the process of maturation.

Keratin

It is necessary to consider, only briefly, the various properties of keratin which contribute to its histological distinction. This fibrous protein is characterized by its high content of the basic amino-acids arginine, lysine and histidine and of the sulphur-containing amino-acid cystine. Usually easily distinguished on morphological grounds alone, it has a strong affinity for both basic and acid dyes and is impervious to the action of trypsin and pepsin unless the disulphide bonds of cystine, which bind together the individual polypeptide chains of which the molecule is composed, are first broken by some means or other. Keratin in hair differs from that found elsewhere in its particularly high values for cystine, and in its corresondingly lower values for cysteine. Kertain in the stratum corneum and in the flaky contents of epidermoid cysts, as well as that in hair shafts, differs from prekeratin in the other layers of the skin in a physical manner also. In the first three situations it shows simple birefringence by polarized light (see Figs. 78 and 79). This property cannot be used to determine the presence of keratin, in malignant cells for instance, where it exists in isotropic form. The Sakaguchi reaction for arginine is not of use in distinguishing keratin from other materials with a smaller arginine content and the older lead acetate method for cystine only gives a reliable positive reaction in the case of hair shafts, whose identity is unlikely to be in doubt. Rogers (1963), in his studies on hair development, used the Sakaguchi reaction for arginine and a modified Ehrlich DMAB reaction for citrulline. He noted high levels of arginine in the trichohyalin droplets of the inner hair sheath, in both Huxley and Henle layers at the level of the bulb. Higher up, only the Huxley layer gave the reaction and at this level the Henle layer stained for citrulline. Rogers postulated an interaction (NH_3-exchange) between arginine and glutamic acid residues in the polypeptide chains of the inner sheaths, leading to the formation of citrulline (and glutamine). This reaction could explain a number of observed histochemical changes which accompany maturation of hair and skin.

In the identification of an unknown but morphologically suggestive material in the tissues as keratin, the property of resistance to trypsin and pepsin may be of some assistance. Prekeratin, however, with which we are most concerned, is much less resistant than hair and stratum corneum. The specific keratinase present in the intestine of the larval clothes moth, and active in a reducing medium at pH 9·0, has not been purified for histochemical use.

There remain four methods which are of assistance in distinguishing between keratin and other acidophil, fibrous or globular, proteins of high arginine content. The first is the tetrazonium reaction after at least 18 hours benzoylation (with or without previous heating), which is uniformly negative both with keratin in hair and in the stratum corneum (Fig. 80) and also with epithelial pearls and with the intracellular form. The second method, which is sometimes useful, is the performic acid-Alcian blue technique (PFAAB), described in Chapter 6 (p. 155). This method is not a reliable indicator of the presence of prekeratin in tumours of squamous origin although it will invariably give a positive reaction at the point, in epithelial pearls for instance, where the diagnosis would be considered certain on histological grounds alone. The changes which occur during the production of keratin, described below, to some extent explain the failure of this reaction to demonstrate prekeratin. The third and fourth methods, the SH techniques of Bennett (1951) and of Barrnett and Seligman (1954) are considered after the discussion below.

The Process of Keratinization. One of the principal changes which takes place during cornification in normal human skin is the oxidation of SH groups in the lower strata to form SS groups in the stratum corneum. Wilkerson (1934) and Wilkerson and Tulane (1939) found that the keratinization of skin into nails and hair was accompanied by an increase of cysteine and a decrease in methionine. The histidine, lysine and arginine ratios were the same in both cases. Bonting (1950) estimated cysteine, cystine and methionine in the skin of young and adult rats. His findings agreed with those of Klauder and Brown (1936) that adult skin contained only 60 per cent. of the total sulphur present in skin from young animals. During maturation, that is to say during progress from infancy to the adult state, although the percentage of SS groups remained constant the total cystine decreased to 44 per cent. Bonting considered that the cystine lost from the young skin was transferred to the keratin of adult hair. Van Scott and Flesch (1954), however, have pointed out that significant amounts of SH occur in all horny structures and that the SS content of the Malpighian layer of the sole of the foot is the same as that of the stratum corneum.

Mercer (1949) studied the development of human hair by means of birefringence measurements and by X-ray diffraction as well as by histochemical tests. The diagram below (after Mercer, 1949) describes his main findings, which have now been incorporated into a larger work (Mercer, 1963) to which reference may be made.

The hair cortex is divided into four regions. AB is the region of the isotropic bulb, BC the zone of fibrous but unconsolidated prekeratin, CD the

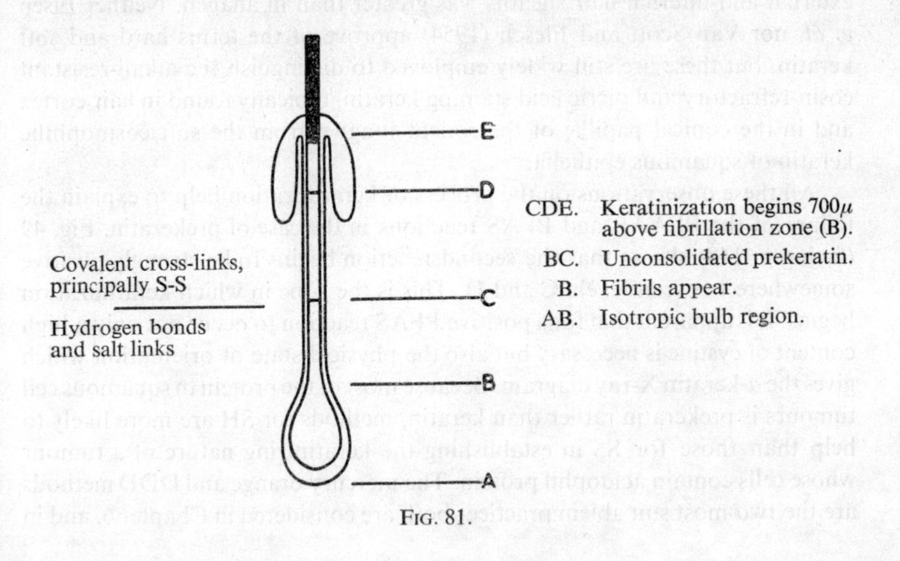

FIG. 81.

zone of progressive hardening and DE the fully hardened zone. Mercer found that 0·1 per cent. trypsin at pH 8 and 40° for 2 hours removed the prekeratin of zone BC, but that the inner root sheath was more resistant and appeared to begin hardening at level B. The fibrils in zone BC were dispersed by a saturated solution of urea and disorientated by warming (90° for 30 seconds in water with a corresponding fall in birefringence. In the whole prekeratinous zone BE, thiols were present in large quantities by the method of Giroud and Bulliard (see Chapter 6, p. 140). Eisen, Montagna and Chase (1953) described the characteristic phases of hair-growth in the mouse and guinea-pig using the mercury orange and DDD methods for sulphydryl groups. The three phases are (1) *Anagen*, (2) *Catagen*, and (3) *Telogen*. During the first there is active cell proliferation, and during this period Bern (1954) observed a high ^{35}S activity in the keratogenous zone 8 hours after the administration of radio-cystine. During the second phase (catagen) proliferation ceases, the hair follicles become shorter and the hair club begins to form. The third phase (telogen) is the resting phase. Details of the distribution of SH and SS groups in human skin were given by Montagna *et al.* (1954) and the whole subject was treated by Montagna (1956) in his excellent book on the "Structure and Function of the Skin."

Eisen *et al.* observed that the results of the mercury orange and DDD reactions were equivalent, but that contrast was sharper with the latter. They found that during telogen SH was confined to the region around the hair club, but that in early anagen it appeared in the thickened epidermis. Later in this phase the greatest concentration of SH occurred in the cortex and cuticle of the keratogenous zone of the hair shaft (BD in Fig. 81), and this accords well with Bern's findings with ^{35}S. During catagen the brush region of the keratinizing hair club contained very large quantities of SH, and the amount in the external and internal hair sheaths was greater than in anagen. Neither Eisen *et al.* nor Van Scott and Flesch (1954) approve of the terms hard and soft keratin, but these are still widely employed to distinguish the alkali-resistant eosin-refractory and picric acid staining keratin, typically found in hair cortex and in the conical papillæ of the rodent tongue, from the soft eosinophilic keratin of squamous epithelia.

All these observations on the process of keratinization help to explain the failure of the PFAAB and PFAS reactions in the case of prekeratin. Fig. 49 (facing p. 148) shows that the second reaction begins to be strongly positive somewhere between levels C and D. This is the zone in which keratinization begins. It is apparent that for a positive PFAS reaction to occur not only a high content of cystine is necessary but also the physical state of orientation which gives the α-keratin X-ray diagram. Because most of the protein in squamous cell tumours is prekeratin rather than keratin, methods for SH are more likely to help than those for SS in establishing the keratinizing nature of a tumour whose cells contain acidophil protein. The mercury orange and DDD methods are the two most suit ablein practice; both are considered in Chapter 6, and in

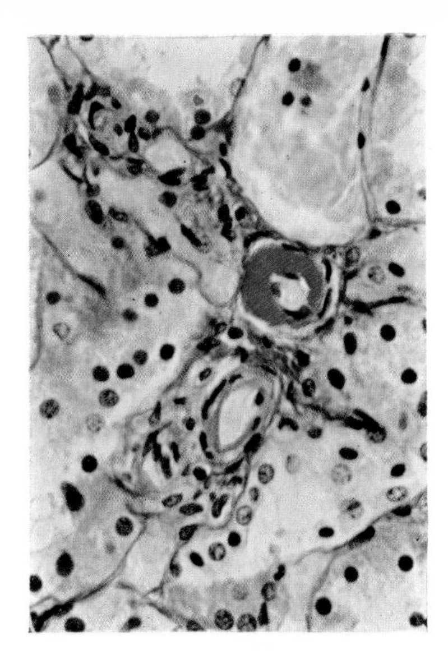

IIIA. Human Kidney. Case of benign hypertension. Formol-mercury fixed, paraffin section. Shows a renal arteriole with fibrin in its wall. Picro-Mallory stain. × 333

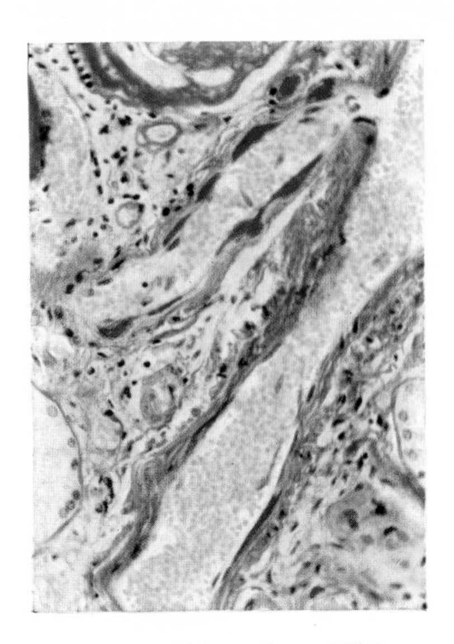

IIIB. Human Kidney. Case of diabetes mellitus. Formol-mercury fixed, paraffin section. Shows an artery with its branch. The latter contains characteristically shaped fibrin "lozenges" which are flat on the luminal side and convex on the other. MSB method. × 166.

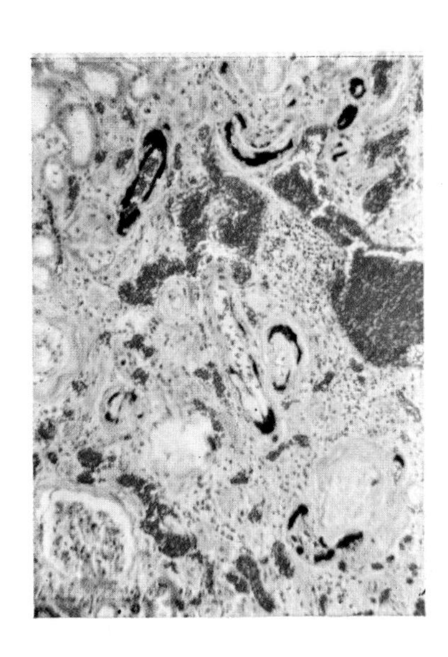

IIIC. Human Kidney. Case of diabetes mellitus. Formol-mercury fixed, paraffin section. The renal arteries are widely affected by infiltration with proteins derived from the plasma. Fibrin appears blue-black. 18/41 Masson technique. × 66

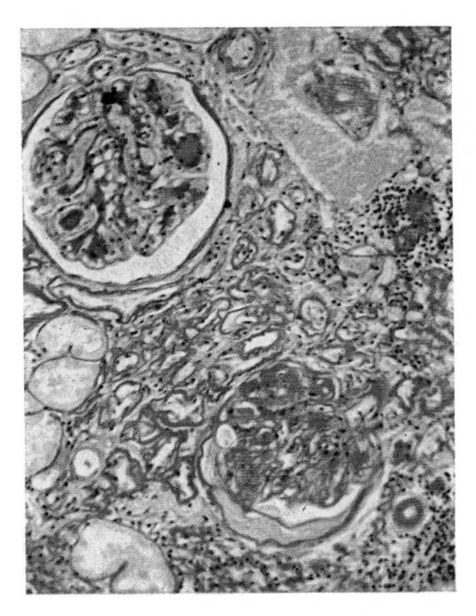

IIID. Human Kidney. Kimmelstiel-Wilson syndrome (Overstained to provide colour contrast). Shows basement membranes (red) and plasmatic vasculosis (reddish-orange). Trichrome-PAS method. × 97

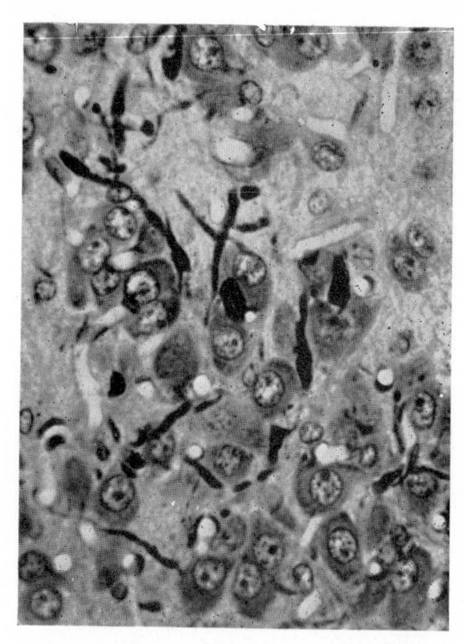

Fig. 82. Rat hypothalamus. Paraventricular nucleus. Shows "axons" charged with "neurosecretory substance". Chrome alum-hæmatoxylin-phloxin. × 370.

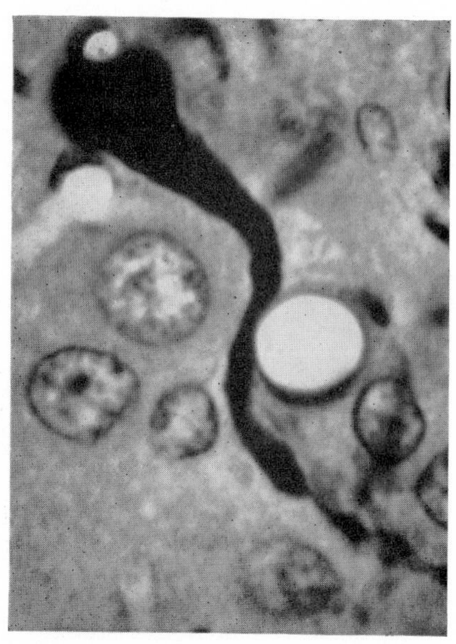

Fig. 83. As Fig. 82. Following incubation with ribonuclease. Chrome alum-hæmatoxylin-phloxin. × 1350.

Appendix 6. There is little to choose between them in the matter of sensitivity, but in either case the investigation is complicated by the absolute necessity for controls.

NEUROSECRETORY SUBSTANCE (NSS)

It is now accepted that the hormones of the posterior lobe of the pituitary gland are produced in the hypothalamus and transported along the axons of the hypothalamo-hypophyseal tract. This hypothesis (Scharrer and Scharrer, 1937) was for a long time opposed to that of Gersh (1939), who held that the hormones were secreted by the pituicytes of the infundibular process. It was not until Bargmann (1949, 1950) successfully modified Gomori's (1939) chrome-hæmatoxylin stain (see Appendix 8, p. 644, and Fig. 82) and applied it to the demonstration of NSS that the Scharrer's hypothesis was confirmed.

Later Dawson (1953) showed that, after mild oxidation, Gomori's (1950) aldehyde fuchsin stain could also demonstrate NSS. Sloper (1955) used a modification of Matsuura's (1925) phosphotungstic acid-Congo red method, which stains the material blue. The mechanism by which these three methods stain NSS is poorly understood and they are in no way specific for the purpose. Chrome hæmatoxylin stains nuclei, lipofuchsin, Nissl substance and the anterior pituitary basophils. The other two stain these cells and also elastic tissue.

The oxytocic and anti-diuretic principles, long known to be present in extracts of the neurohypophysis, were shown by du Vigneaud and his colleagues (Turner, Pierce and du Vigneaud, 1951; du Vigneaud *et al.*, 1953, 1954) to be two distinct octapeptides, both of which contain relatively large amounts of cystine. As a result of this work histochemical techniques for cystine were successfully applied to the demonstration of NSS (Barrnett, 1954; Sloper, 1954), and Adams and Sloper (1955) used the PFAAB method with very good results. This must be considered the histochemical reaction of choice; but it may also be used purely as a staining reaction (Sloper and Adams, 1956).

The stainable NSS was considered by Schiebler (1951, 1952a and b) to be a complex consisting of the posterior lobe hormones and a glycolipoprotein, and by Hild and Zetler (1953) to be a bearer substance. Sloper (1955) and Howe and Pearse (1956), using the dog and the rat respectively, were unable to find either a lipid or a carbohydrate component of the NSS. Both considered it to be a protein or polypeptide containing a relatively large amount of cystine, and both found that it was digested by trypsin and by pepsin. Berezin *et al.* (1955 produced some very interesting results with tissue homogenates. They showed that homogenates of various toad tissues (*Bufo arenarum*, *marinus* and *ictericus*) would remove from formalin-fixed tissues (4 hours, 37°) nothing except NSS and the β-cell granules of the pancreas. An interpretation of these results is awaited. Although the antidiuretic hormone contains arginine as one of its eight amino-acids the Sakaguchi reaction (Chapter 6, p. 158) fails to give a positive reaction with NSS as does the copper phthalocyanin reaction (Chapter

12, p. 426), which stains arginine-containing basic proteins as well as phospholipids and nucleic acids.

Since it is the "classical" method for the purpose, the chrome hæmatoxylin method will probably continue to be widely used. It must be emphasized that the standard procedure (as given for staining lipofuscins in Appendix 26 does not stain NSS. Bargmann's modification given in the Appendix to this chapter must be employed (see Figs. 82 and 83).

Anterior Pituitary Gland Mucoid Cells

The mucoprotein-secreting (PAS+) cells of the adenohypophysis are called mucoid cells, since their contained granules are composed of material falling into this group of the conjugated proteins. The majority of mucoid cells belong to the basophil series of the older terminology, but the terms are not synonymous since there is an overlap into the chromophobes. Adams and Swettenham (1957), using the PFAAB reaction, followed by a PAS-Orange-G routine (Appendix 8, p. 645), showed that the granules of the mucoid cells could be divided into two types on a firm histochemical basis. These two types of granule, designated S and R, are illustrated in Plate Id, facing p. 183. The former contains cystine in relatively high concentration and its carbohydrate moiety that part which is responsible for PAS staining) is removed by treatment with performic acid. The second type of granule contains either no cystine or, at all events, an insufficient amount to give a positive PFAAB reaction. Its carbohydrate component withstands performic acid treatment and is therefore stainable by the PAS routine after such treatment.

Correlation with other terminologies is not easy since the various methods are not equally applicable to the pituitary gland of any one species. The two glands most often studied are those of man and the rat. The results of various staining methods and histochemical reactions, given in Table 16, are derived mainly from Adams and Swettenham (1957).

Since the original method of Halmi (Appendix 8, p. 645) will not work in human glands correlation between this and Adams' method is possible only in the rat. The introduction of a buffered PFAAB method (Swettenham, 1960) has made it possible to use this technique for the rat as well as the human hypophysis. The β-granule of Pearse (1952) and of Wilson and Ezrin (1954) is distinguishable as an R-granule, while the δ-granule of these authors and Pearse's γ-granule are S-granules.

Adams and Pearse (1959) studied the distribution of the two types of mucoid granule in various pathological states and correlated R-type cells with thyrotrophs (Halmi, 1952; Purves and Griesbach, 1951), and S-type cells with gonadotrophs. The S-type cells were regarded as responsible for the production of the cystine-containing protein, isolated by Li *et al.* (1943) and Sayers *et al.* (1943), which was originally but erroneously thought to be ACTH.

Further studies by Pearse and van Noorden (1963a and b), however, indicated that the large purple S cells (S_2) were certainly thyrotrophs. Addi-

TABLE 16

Reactions of S and R-type Pituitary Mucoid Granules

Method	R granule	S granule	Vesicles (vesiculate mucoids)	α-granule
PFAAB–PAS–OG	Magenta++	Blue++	Magenta+	Orange
PFAAB	— or Pale Blue	Blue	—	—
PAS	Magenta +++	Magenta++	Magenta++	—
PFA–PAS	Magenta++	— or Pink	— or Pink	—
PFA–AF–LG–OG (modified Halmi)	Green	Purple	Purple	Orange
CT	Brown++	Brown+	—	Brown+
CT post-benzoylation (Pearse, 1952)	Brown (β) ++	(γ) —	—	—
Wilson-Ezrin	(β) Magenta +++	(δ) Purple	Purple	Orange

PFAAB = Performic acid—Alcian blue
OG = Orange G
AF–LG = Aldehyde Fuchsin—light green
CT = Coupled tetrazonium

tional confirmation came from radioautographic experiments (Deminatti, 1964) which showed that after thiourea administration there was a pronounced uptake of ^{35}S by the thyrotrophs. Subsequently Wolff (1965) showed that after suitable oxidation the thyrotrophs of the rat could be stained electively with pseudoisocyanin. The results were examined by fluorescence microscopy, or by light microscopy. The small blue S cells (S_1) were still regarded as possible gonadotrophs and the R cells were shown by an immunofluorescent technique (see Fig. 68, facing p. 188) to be responsible for ACTH production.

Pancreatic α- and β-cell Granules

The α-granules of the pancreatic islets have been distinguished by means of specific silvering techniques which blacken them but not the β-granules. Of these the Gross-Schultze and Gross-Bielschowsky methods have been consistently employed. The α-granules can also be distinguished by methods like Bensley's aniline acid fuchsin-methyl green method for mitochondria, which stains them red and the β-granules green. Mallory's trichome stain produces red α-granules and blue β-granules, but good differentiation is very dependent on fixation. Recently Glenner and Lillie (1957) have shown that the α-granules in several species can be specifically stained by the post-coupled benzylidene reaction for indoles (Chapter 6, p. 133), and they attribute this to their content of glucagon, which has a high tryptophan content according to Mohnike and Boser (1953). This method of demonstrating the α-granules is more satisfactory than any of the staining methods and I consider that it should be employed in place of them. The effect of fixation on the α and β-granules was investigated by Levene and Feng (1962). They found that both types of granules were preserved by all fixatives containing formaldehyde or dichromate. The best results were obtained with formalin-acetic acid.

Several authors have noted the association of the α-cells with a strong non-specific esterase. This is particularly evident with the α-naphthyl and indoxyl esterase techniques (Chapter 17), and these can therefore be used as functional indicators of the presence of α-cells in those species (man, rat, etc.) in which the reaction is positive.

For distinguishing the β-cell granules of the islets Gomori's chrome-hæmatoxylin phloxin method has been the most popular method. It stains the β-granules dark blue and the α-granules red. Gomori's aldehyde-fuchsin stain, without prior oxidation, has been used as an alternative in recent years. Bangle (1956) found that after peracetic acid, permanganate or bromine oxidation, an acidic substance was produced in the β-cells which would take up a number of basic dyes from solutions having a pH as low as 1·3. These included basic fuchsin, aldehyde-fuchsin, crystal violet and methyl violet 2B. Of these aldehyde-fuchsin was unique in staining the granules more or less specifically without previous oxidation. If aldehyde-fuchsin is to be used for staining pancreatic β-granules it should be more than 8 days old. If it is less mature than this it can be extracted from the granules by treatment with acid alcohol; the mature dye is resistant to such extraction. A modified aldehyde-fuchsin technique for β-granules was given by Sieracki *et al.* (1960).

The β-cells of the pancreatic islets in man and the rat possess a strong acid phosphatase (Goessner, 1957). The azo dye methods for this enzyme (Chapter 16, pp. 559 and 563) are especially convenient for this purpose and the method may be used as a functional indicator for the β-cells, in conjunction with an esterase technique for the β-granules.

This association of acid phosphatase with the production of insulin, a disulphide-linked peptide, has a parallel in the association of this enzyme with the formation of S—S containing proteins and peptides elsewhere. The acid phosphatase of hair follicles, and of the neurones of the supraoptic and paraventricular nuclei, can be cited as instances of this. The mechanism by which this enzyme participates in the formation of S—S linkages is unknown.

Pseudoisocyanin. A new metachromatic reagent for the demonstration of insulin in the β-granules was introduced by Schiebler (1958) and further work with this dye was reported by Schiebler and Schiessler (1959, 1960). The reaction depends on the production of sulphonic acid groups (SO_3H) by oxidation with performic acid or with acidified permanganate. Subsequently the induced SO_3H groups are demonstrated by staining with diethyl-dichloro-pseudoisocyanin. Background structures stain red while the β-granules stain metachromatically deep bluish red (abs. max. 578 nm). Only structures in which closely spaced SO_3H groups are present, or can be induced, have the property of staining metachromatically. The A-chains of insulin can react since their induced SO_3H groups are about 4·5 Å apart. Details of the pseudo-isocyanin reaction are given in Appendix 8, p. 644.

A number of alternative uses have been found for the pseudoisocyanin method. It was used by Wolff (1965) for the demonstration of pituitary thyrotrophs (S-cells) either by examination in visible light or by fluorescence microscopy. Sterba (1964) demonstrated neurosecretory substance (NSS) in a number of vertebrate species, making use of the strong yellow fluorescence of the protein-dye complex.

REFERENCES

ADAMS, C. W. M., and SLOPER, J. C. (1955). *Lancet*, **1**, 651.
ADAMS, C. W. M., and SWETTENHAM, K. V. (1957). *J. Path. Bact.*, **75**, 5.
ADAMS, C. W. M., and PEARSE, A. G. E. (1959). *J. Endocrinol.*, **18**, 147.
AIKAT, B. K., and DIBLE, J. H. (1951). Personal communication.
AIKAT, B. K., and DIBLE, J. H. (1956). *J. Path. Bact.*, **71**, 461.
ALBERT, S., and LEBLOND, C. P. (1946). *Endocrinology*, **39**, 386.
ALTSCHULER, C. H., and ANGEVINE, D. M. (1949). *Amer. J. Path.*, **25**, 1061.
ASHBEL, R., and SELIGMAN, A. M. (1949). *Endocrinology*, **44**, 565.
ASTBURY, W. T. (1940). "Molecular Structure of Fibres of the Collagen Group." London.
BAHRMANN, E. (1937). *Virchows Arch.*, **300**, 342.
BAILEY, K., ASTBURY, W. T., and RUDALL, K. M. (1943). *Nature, London*, **151**, 716.
BALÓ, J., and BANGA, I. (1950). *Biochem. J.*, **46**, 384.
BANGA, I. (1951). *Z. Vit-Horm. Fermentforsch.*, **4**, 49.
BANGA, I. (1966). "Structure and Function of Elastin and Collagen." Akad. Klad., Budapest.
BANGLE, R. Jr. (1956). *Amer. J. Path.*, **32**, 349.
BANGLE, R., Jr., and ALFORD, W. C. (1954). *J. Histochem. Cytochem.*, **2**, 62.
BARGMANN, W. (1949). *Z. Zellforsch.*, **34**, 610.
BARGMANN, W. (1950). *Mikroskopie*, **5**, 289.
BARRNETT, R. J. (1954). *Endocrinology*, **55**, 484.
BARRNETT, R. J., and SELIGMAN, A. M. (1954). *J. nat. Cancer Inst.*, **14**, 769.
BEDFORD, G. R., and KATRITZKY, A. R. (1963). *Nature*, **200**, 652.
BENNETT, H. S. (1951). *Anat. Rec.*, **110**, 231.
BENSLEY, S. (1934). *Ibid.*, **60**, 93.
BEREZIN, A., HADLER, W. A., and TRAMEZZANI, J. H. (1955). *Nature, Lond.*, **176**, 600.
BERN, H. A. (1954). *Ibid.*, **174**, 509.
BONTING, S. J. L., Jr. (1950). *Biochim. Biophys. Acta*, **6**, 183.
BOWES, J. H., ELLIOT, R. C., and MOSS, J. A. (1953). In "Nature and Structure of Collagen." Butterworth, London, p. 199.
BOWES, J. H., and KENTEN, R. H. (1949). *Biochem. J.*, **45**, 281.
BRAUN-FALCO, O. (1956). *Dermatol. Wschr.*, **134**, 1036.
BREWER, D. B. (1957). *J. Path. Bact.*, **74**, 371.
BUSANNY-CASPARI, W. (1952). *Virchow's Arch.*, **322**, 452.
BUSANNY-CASPARI, W. (1957). *Acta Histochem.*, **4**, 304.
CLARK, E., GRAEF, I., and CHASIS, H. (1936). *Arch. Path.*, **22**, 183.
CLERICI, E., BAIRATI, A. Jr., and MOCARELLI, P. (1963). *Experientia*, **18**, 241.
CONSDEN, R., GLYNN, L. E., and STANIER, W. M. (1952). *Proc. Biochem. Soc., p. xix; Biochem. J.*, **50**.
COOPER, Z. (1952). "Problems of Ageing." 3rd Ed. Williams Wilkins, Baltimore.
COWAN, P. M., NORTH, A. C. T., and RANDALL, J. T. (1953). In "Nature and Structure of Collagen." Butterworth, London, p. 241.
COWAN, P. M., NORTH, A. C. T., and RANDALL, J. T. (1955). In "Fibrous Proteins and their Biological Significance." Brown and Danielli, Ed. C.U.P., p. 115.

COX, R. C., and LITTLE, K. (1962). *Proc. roy. Soc. B.*, **155**, 232.
CRAWFORD, T., and WOOLF, N. (1960). *J. Path. Bact.*, **79**, 221.
CRUICKSHANK, B., and HILL, A. G. S. (1953). *J. Path. Bact.*, **66**, 283.
DAVIES, B. D., and DUBOS, R. J. (1947). *J. exp. Med.*, **86**, 215.
DAWSON, A. B. (1953). *Anat. Rec.*, **115**, 63.
DAY, T. D. (1947). *J. Path. Bact.*, **59**, 567.
DEMINATTI, M. (1964). *J. Histochem. Cytochem.*, **12**, 215.
DEMPSEY, E. W., and LANSING, A. I. (1954). *Int. Rev. Cytol.*, **3**, 436.
DEMPSEY, E. W., VIAL, J. D., LUCAS, R. V., Jr., and LANSING, A. I. (1952). *Anat. Rec.*, **113**, 197.
DUBLIN, W. B. (1946). *Arch. Path.*, **41**, 299.
EDSALL, J. T., FOSTER, J. F., and SCHEINBERG, I. H. (1947). *J. Amer. chem. Soc.*, **69**, 2731.
EISEN, A. Z., MONTAGNA, W., and CHASE, H. B. (1953). *J. nat. Cancer Inst.*, **14**, 341.
ENGLE, R. L., and DEMPSEY, E. W. (1954). *J. Histochem. Cytochem.*, **2**, 9.
FENNELL, R. H., Jr., REDDY, C. R. R. M., and VASQUEZ, J. J. (1961). *Arch. Path.*, **72**, 209.
FERRY, J. D., and MORRISON, P. R. (1947). *J. Amer. chem. Soc.*, **69**, 388.
FERRY, J. D., SINGER, M., MORRISON, P. R., PORSCHE, J. D., and KUTZ, R. L. (1947). *Ibid.*, **69**, 409.
FERRY, J. D. (1948). *Advanc. Prot. Chem.*, **4**, 2.
FLEMMING, W. (1876). *Arch. mikr. Anat.*, **12**, 434.
FOOT, N. C. (1925). *Amer. J. Path.*, **1**, 341.
FOOT, N. C. (1927). *Ibid.*, **3**, 401.
FRANCHI, C. M., and DE ROBERTIS, E. (1951). *Proc. Soc. Exp. Biol., N.Y.*, **76**, 515.
FRENCH, D., and EDSALL, J. T. (1945). *Advanc. Prot. Chem.*, **2**, 320.
FULLMER, H. M. (1960). *J. Histochem. Cytochem.*, **8**, 290.
FULLMER, H. M., and LILLIE, R. D., (1956). *Stain Tech.*, **31**, 27.
FULLMER, H. M., and LILLIE, R. D. (1958). *J. Histochem. Cytochem.*, **6**, 425.
GAWLIK, Z. (1965). *Folia Histochem. Cytochem.*, **3**, 233.
GAWLIK, Z., and JAROCINSKA, M. (1964). *Folia Histochem. Cytochem.*, **2**, 257.
GERSH, I. (1939). *Amer. J. Anat.*, **64**, 407.
GERSH, I. (1947). *Fed. Proc.*, **7**, 270.
GILLMAN, T., PENN, J., BRONKS, D., and ROUX, M. (1954). *Nature, Lond.*, **174**, 789.
GILLMAN, T., PENN, J., BRONKS, D., and ROUX, M. (1955). *Arch. Path.*, **59**, 733.
GITLIN, D., and CRAIG, J. M. (1957). *Amer. J. Path.*, **33**, 267.
GITLIN, D., CRAIG, J. M., and JANEWAY, C. A. (1957). *Amer. J. Path.*, **33**, 55.
GLEGG, R. E., EIDINGER, D., and LEBLOND, C. P. (1953). *Science*, **118**, 614.
GLEGG, R. E., EIDINGER, D., and LEBLOND, C. P. (1954). *Ibid.*, **120**, 839.
GLENNER, G. G., and LILLIE, R. D. (1957). *J. Histochem. Cytochem.*, **5**, 279.
GLYNN, L. E., and LOEWI, G., (1952). *J. Path. Bact.*, **64**, 329.
GOESSNER, W. (1957). Habilitationsschrift, Tübingen.
GOLDSTEIN, D. J. (1962). *Quart. J. micr. Sci.*, **103**, 477.
GOMORI, G. (1939). *Amer. J. Path.*, **15**, 497.
GOMORI, G. (1950). *Amer. J. clin. Path.*, **20**, 665.
GORDON, H., and SWEETS, H. H., Jr. (1936). *Amer. J. Path.*, **12**, 545.
GRAHAM, C. E., WAITKOFF, H. K., and HIER, S. W. (1949). *J. biol., Chem.*, **177**, 529.
GREENBLATT, M. (1963). *Arch. Path.*, **75**, 177.
GROSS, J. (1949). *Amer. J. Path.*, **25**, 805.
GROSS, J. (1964). *Science*, **143**, 960.
GROSS, J., HIGHBERGER, J. H., and SCHMITT, F. O. (1952). *Proc. Soc. exp. Biol., N.Y.*, **80**, 462.
GROSS, J., HIGHBERGER, J. H., and SCHMITT, F. O. (1955). *Proc. nat. Acad. Sci. Wash.*, **47**, 1.
GROSSER, O. (1925). *Z. ges. Anat. I.Z. Anat. Entwegsch.*, **76**, 304.
GUSTAVSON, G. H. (1956). "Chemistry and Reactivity of Collagen." Acad. Press Inc., New York.
HACK, M. H. (1952). *Anat. Rec.*, **112**, 275.
HALMI, N. S. (1952). *Stain Tech.*, **27**, 61.

HARKNESS, R. D., MARKO, A. M., MUIR, H. M., and NEUBERGER, A. (1954). *Biochem. J.*, **56**, 558.
HARMS, H. (1957). *Acta Histochemica*, **4**, 314.
HARRIS, H. E. (1902). *Z. wiss. Mikr.*, **18**, 290.
HERINGA, G. C., and WEIDINGER, A. (1942). *Acta necrl. Morph.*, **4**, 291.
HERXHEIMER, K. (1886). *Fortschr. Med.*, **4**, 785.
HILD, W., and ZETLER, G. (1953). *Z. ges. exp. Med.*, **120**, 236.
HOWE, A., and PEARSE, A. G. E. (1956). *J. Histochem. Cytochem.*, **4**, 561.
IRVING, E. A., and TOMLIN, S. G. (1954). *Proc. Roy. Soc. B.*, **142**, 113.
JACKSON, D. S. (1957). *Biochem. J.*, **65**, 277.
JACKSON, D. S., and FESSLER, J. H. (1955). *Nature, Lond.*, **176**, 69.
JAQUES, L. B. (1943). *Biochem. J.*, **37**, 344.
KELLGREN, J. H., ASTBURY, W. T., REED, R., and BEIGHTON, E. (1951). *Nature, Lond.*, **168**, 493.
KLAUDER, J. V., and BROWN, H. (1936). *Arch. Derm. Syph.*, **34**, 568.
KLEMPERER, P., POLLACK, A. D., and BAEHR, G. (1942). *J. Amer. med. Ass.*, **119**, 331.
KLINGE, F. (1933). *Erghn. allg. Path.*, **27**, 1.
KLÜVER, H., and BARRERA, E. (1953). *J. Neuropath. exp. Neurol.*, **12**, 400.
KOPPEN, A. (1889). *Z. wiss, Mikr.*, **6**, 473.
KRAMER, H., and LITTLE, K. (1953). In "Nature and Structure of Collagen." Butterworth, London, p. 33.
KRAMER, H., and WINDRUM, G. M. (1953). *J. clin. Path.*, **6**, 239.
KRAMER, H., and WINDRUM, G. M. (1954). *J. Histochem. Cytochem.*, **2**, 196.
LANSING, A. I. (1951). Josiah Macy Foundation, N.Y., 2nd Conf. Connective Tissues, p. 45.
LANSING, A. I., COOPER, Z. K., and ROSENTHAL, T. B. (1953). *Anat. Rec.*, **115**, Suppl., p. 340.
LANSING, A. I., ROSENTHAL, T. B., ALEX, M., and DEMPSEY, E. W. (1952). *Anat. Rec.*, **114**, 555.
LAUTSCH, E. V., MCMILLAN, G. C., and DUFF, G. L. (1953). *Lab. Invest.*, **2**, 397.
LENDRUM, A. C. (1963). *Canad. med. Ass. J.*, **88**.
LENDRUM, A. C., FRASER, D. S., and SLIDDERS, W. (1964). *Ned. T. Geneesk.*, **108**, 2373.
LENDRUM, A. C., FRASER, D. S., SLIDDERS, W., and HENDERSON, R. (1962). *J. Clin. Path.*, **15**, 401.
LEVENE, C., and FENG, P. (1962). *Quart. J. micr. Sci.*, **103**, 45.
LHOTKA, J. F., and DAVENPORT, H. A. (1950). *Stain. Tech.*, **25**, 129.
LI, C. H., EVANS, H. M., and SIMPSON, M. E. (1943). *J. biol. Chem.*, **149**, 413.
LILLIE, R. D. (1945). *Bull. int. Ass. med. Mus.*, **25**, 1.
LILLIE, R. D. (1947). *J. Lab. clin. Med.*, **32**, 910.
LILLIE, R. D. (1951). *Amer. J. clin. Path.*, **21**, 484.
LILLIE, R. D. (1952a). Josiah Macy Foundation, N.Y., 3rd Conf. Connective Tissues, p. 11.
LILLIE, R. D. (1952b). *Arch. Path.*, **54**, 220.
LITTLE, K., and WINDRUM, G. M. (1954). *Nature, Lond.*, **174**, 789.
LORAND, L. (1950). *Ibid.*, **166**, 694.
LOWRY, O. H., GILLIGAN, D. R., and KATERSKY, E. M. (1941). *J. biol. Chem.*, **139**, 795.
MACMILLAN, R., ADAMS, C. W. M., and IBRAHIM, M. Z. M. (1965). *J. Path. Bact.*, **89**, 225.
MALLORY, F. B. (1938). "Pathological Technique." Saunders, Philadelphia.
MALLORY, F. B., and PARKER, F. (1927). *Amer. J. Path.*, **3**, 515.
MARSHALL, J., and LURIE, H. T. (1957). *Brit. J. Dermatol.*, **69**, 315.
MATSUURA, S. (1925). *Fol. anat. jap.*, **3**, 107.
MCCLEAN, D. (1931). *J. Path. Bact.*, **34**, 459.
MCCLEAN, D. (1936). *Ibid.*, **42**, 477.
MCFARLANE, D. (1944). *Stain. Tech.*, **19**, 29.
MENZIES, D. W., and ROBERTS, J. T. (1963). *Nature, Lond.*, **198**, 1006.
MENZIES, D. W., RYAN, G. B., and ROBERTS, J. T. (1964). *Nature, Lond.*, **203**, 195.

MERCER, E. H. (1949). *Biochim. Biophys. Acta*, **3**, 161.
MERCER, E. H. (1963). "Keratin and Keratinization." Int. Monogr. Pure and Appl. Biol., Pergamon, Oxford.
MICHAELIS, L. (1901). *Dtsch. med. Wschr.*, **27**, 219.
MIHÁLYI, E. (1950). *Acta chem. Scand.*, **4**, 344.
MILAZZO, S. C. (1957). *J. Path. Bact.*, **73**, 527.
MISSMAHL, H. P. (1957). *Z. Zellforsch.*, **45**, 612.
MOHNIKE, G., and BOSER, H. (1953). *Naturwiss.*, **40**, 295.
VON MOLLENDORFF, W. (1932). *Z. Zellforsch.*, **15**, 131.
MONTAGNA, W. (1956). "Structure and Function of the Skin." Acad. Press, New York.
MONTAGNA, W., EISEN, A. Z., RADEMACHER, A. H., and CHASE, H. B. (1954). *J. invest. Derm.*, **23**, 23.
MONTGOMERY, P. O'B., and MUIRHEAD, E. E. (1957). *Amer. J. Path.*, **33**, 285.
MOORE, G. W. (1943). *Bull. inst. Med. Lab. Tech.*
MORRIONE, T. G. (1952). *J. exp. Med.*, **96**, 107.
MORRISON, P. R. (1947). *J. Amer. chem. Soc.*, **69**, 2733.
MOVAT, H. Z., and MORE, R. H. (1957). *Amer. J. clin. Path.*, **28**, 331.
MUIRHEAD, E. E., BOOTH, E., and MONTGOMERY, P. O'B. (1956). *Amer. J. Path.*, **32**, 662.
MUSSO, H. (1956). *Chem. Ber.*, **89**, 1659.
NEUMANN, E. (1880). *Arch. mikr. Anat.*, **18**, 130.
NEUMANN, R. F., and TYTELL, A. A. (1950). *Proc. Soc. exp. Biol., N.Y.*, **73**, 409.
OREKHOVITCH, V. N. (1955). Commun. 3rd Int. Congr. Biochem., Brussels, p. 50.
PARTRIDGE, S. M. (1962). *Advanc. Prot. Chem.*, **17**, 227.
PEARSE, A. G. E. (1951). *J. clin. Path.*, **4**, 1.
PEARSE, A. G. E. (1952). *J. Path. Bact.*, **64**, 791.
PEARSE, A. G. E., and VAN NOORDEN, S. (1963a). *Canad. med. Ass. J.*, **88**, 462.
PEARSE, A. G. E., and VAN NOORDEN, S. (1963b). "Cytologie de L'Adenohypophyse." Colloq. Int. C.R.N.S., Paris, p. 63.
PEPLER, W. J., and BRANDT, F. A. (1954). *Brit. J. exp. Path.*, **35**, 41.
PETRY, G. (1952). *Z. wiss. Mikr.*, **61**, 66.
PETRY, G. (1953). *Ibid.*, **61**, 121.
PRANTER, V. (1902). *Zbl. allg. Path.*, **13**, 292.
PUCHTLER, H. (1964). *Histochemie*, **4**, 24.
PUCHTLER, H. (1964). *J. Histochem. Cytochem.*, **12**, 552.
PUCHTLER, H., BATES, R., and BROWN, J. H. (1961). *J. Histochem. Cytochem.* **9**, *553*.
PURVES, H. D., and GRIESBACH, W. E. (1951). *Endocrinology*, **49**, 244.
ROBB-SMITH, A. H. T. (1945). *Lancet*, **2**, 362.
ROBERTSON, W. VAN B., and SCHWARTZ, B. (1953). *J. Biol. Chem.*, **201**, 689.
ROGERS, G. E. (1963). *J. Histochem. Cytochem.*, **11**, 700.
ROLLETT, A. (1871). In Stricker's "Handbuch der Lehre von den Geweben des Menschen und des Thierreichs." Englemann, Leipzig.
ROSA, C. G. (1953). *Stain tech.*, **28**, 299.
RUITER, M. (1962a). *J. Invest. Dermatol.*, **38**, 85.
RUITER, M. (1962b). *J. Invest. Dermatol.*, **38**, 117.
SALTHOUSE, T. N. (1944). *Ibid.*, **19**, 91.
SAYERS, G., WHITE, A., and LONG, C. N. H. (1943). *J. biol. Chem.*, **149**, 425.
SCHARRER, E., and SCHARRER, B. (1937). *Biol. Rev.*, **12**, 185.
SCHIEBLER, T. H. (1951). *Acta Anat.*, **13**, 233.
SCHIEBLER, T. H. (1952a). *Exp. Cell Research*, **3**, 249.
SCHIEBLER, T. H. (1952b). *Acta Anat.*, **15**, 393.
SCHIEBLER, T. H. (1958). *Naturwiss.*, **45**, 214.
SCHIEBLER, T. H., and SCHIESSLER, S. (1959). *Histochemie*, **1**, 445.
SCHIEBLER, T. H., and SCHIESSLER, S. (1960). *J. Histochem. Cytochem.*, **8**, 312.
SCHMITT, F. O., HALL, C. E., and JAKUS, M. A. (1942). *J. cell. comp. Physiol.*, **20**, 11.
SCHREK, R., and MCNAMARA, W. L. (1947). *Amer. J. clin. Path.*, **17**, 84.
SCOTT, D. G. (1957). *Brit. J. exp. Path.*, **38**, 178.

SEKI, M. (1934). *Folia Anat. Jap.*, **12**, 11.
SEN, P. B., MUKHERJI, M., DEB, C., and MOOKERJEA, C. (1961). *Ind. J. med. Research.*, **49**, 1051.
SIERACKI, J. C., MICHAEL, J. E., and CLARK, D. A. (1960). *Stain Tech.*, **135**, 67.
SIEGFRIED, M. (1892). "Uber die chemischen Eigenshaffen des reticulirten Gewebes." Habilitationschrift, Leipzig.
SINGER, M., and WISLOCKI, G. B. (1948). *Anat. Rec.*, **102**, 175.
SLOPER, J. C. (1954). *J. Anat.*, **88**, 576.
SLOPER, J. C. (1955). *Ibid.*, **89**, 301.
SLOPER, J. C. , and ADAMS, C. W. M. (1956). *J. Path. Bact.*, **72**, 587.
STEIN, W. H., and MILLER, E. G. (1938). *J. biol. Chem.*, **125**, 599.
STERBA, G. (1964). *Acta histochem.*, **17**, 268.
SWETTENHAM, K. (1960). *J. clin. Path.*, **13**, 256.
TÄNZER, L. (1891). *Monatsch. prakt. Derm.*, **12**, 394.
THOMAS, J., ELSDEN, D. F., and PARTRIDGE, S. M. (1963). *Nature*, **200**, 651.
TONNA, E. A. (1964a). *Anat. Rec.*, **149**, 559.
TONNA, E. A. (1964b). *J. Roy. micr. Soc.*, **83**, 307.
TONNA, E. A. (1964c). *Anat. Rec.*, **150**, 349.
TRISTRAM, G. R., WORRALL, J., and STEER, D. C. (1965). *Biochem. J.*, **95**, 350.
TURNER, R. A., PIERCE, J. G., and DUVIGNEAUD, V. (1951). *J. biol. Chem.*, **191**, 21.
UNNA, P. G. (1886). *Monatsch. prakt. Derm.*, **5**, 243.
VAN SCOTT, E. J., and FLESCH, P. (1954). *Science*, **119**, 70.
VASQUEZ, J. J., and DIXON, F. J. (1958). *Arch. Path.*, **66**, 504.
VELICAN, D., RADU, S., and VELICAN, C. (1965). *Rev. Roum. méd. int.*, **2**, 243.
VERHOEFF, F. H. (1908). *J. Amer. med. Ass.*, **50**, 876.
DU VIGNEAUD, V., LAWLER, H. C., and POPENOE, E. A. (1953). *J. Amer. chem. Soc.*, **75**, 4880.
WEIDMAN, F. D. (1931). *Arch. Derm. Syph.*, **24**, 954.
WEIGERT, C. (1887). *Fortschr. Med.*, **5**, 228.
WEIGERT, C. (1898). *Zbl. allg. Path.*, **9**, 284.
WEISS, J. (1954). *J. Histochem. Cytochem.*, **2**, 21.
WILKERSON, V. A. (1934). *J. biol. Chem.*, **107**, 377.
WILKERSON, V. A., and TULANE, V. J. (1939). *Ibid.*, **129**, 477.
WILLIAMS, G. (1957). *J. Path. Bact.*, 73, 557.
WILSON, W. D., and EZRIN, C. (1954). *Amer. J. Path.*, **30**, 891.
WINDRUM, G. M., KENT, P. W., and EASTOE, J. E. (1955). *Brit. J. exp. Path.*, **36**, 49.
WITKE, F. (1951). *Mikroskopie*, **6**, 316.
WOLFF, H. H. (1965). *Histochemie*, **4**, 388.
ZAWISCH, C. (1957). *Acta anat.*, **29**, 143.
ZIFF, M., KANTOR, T., BIEN, E., and SMITH, A. (1953). *J. clin. Invest.*, **32**, 1253.

CHAPTER 9

NUCLEIC ACIDS AND NUCLEOPROTEINS

Introduction

THE first conjugated proteins with which we have to deal are the nucleoproteins, which are combinations of a basic protein with various highly polymerized polynucleotides (nucleic acids). Two types of nucleic acid are found in animal and plant tissues, the deoxypentose nucleic (deoxyribonucleic, thymonucleic) acid of the nuclei and the pentosenucleic (ribonucleic, plasmonucleic) acid of the cytoplasm, and they differ in many chemical and physical respects. However, both are precipitable from alkaline solutions by hydrochloric acid, and ribonucleic acid is also precipitated by acetic acid. Both form salts with alkaline earths or heavy metals and on hydrolysis they yield phosphoric acid, purine and pyrimidine bases, and a carbohydrate or carbohydrate derivative. Histochemical methods for their demonstration in the tissues have been based on reactions for all these constituents. According to the older, tetranucleotide, theory deoxyribonucleic acid (DNA) was regarded as a polymer made up of four distinct units (nucleotides) and containing a total of four PO_4 radicals, one residue each of two purine bases (adenine and guanine), one each of two pyrimidines (thymine and cytosine), and four molecules of the pentose sugar deoxyribofuranose.

New concepts, described below, have modified this theory so that it is now untenable. Chargaff (1955a) first clearly proposed that the only recognizable sub-unit of structure in DNA was the mononucleotide. If repeating polynucleotide units recur in the nucleic acid chain there is, as yet, no way of demonstrating them. In spite of great differences in the nucleotide composition of nucleic acids certain features occur with regularity (Chargaff, 1950, 1951, 1955b). In most cases (1) The sum of the purine nucleotides is equal to that of the pyrimidine nucleotides. (2) The ratio of adenine to thymine equals 1. (3) The ratio of guanine to cystosine equals 1. (4) The number of 6-amino groups (adenine + cytosine + methylcytosine) equals that of 6-keto groups (guanine + thymine). Elson and Chargaff (1954) showed that this last feature, with uracil taking the place of thymine, applies also to some of the pentose nucleoproteins. Some authors (e.g. Allen, 1954), besides dismissing the tetranucleotide theory, regarded such terms as highly polymerized and depolymerized in respect of DNA as unjustified. The majority of authors, however, continued to use them.

Ribonucleic acid (yeast) differs from DNA only in yielding four molecules of ribofuranose in place of the deoxy sugar and one of uracil in place of the thymine of deoxyribonucleic acid. Since the identity of the pentose sugar has not been determined as ribose except in yeast nucleic acid, it is strictly more

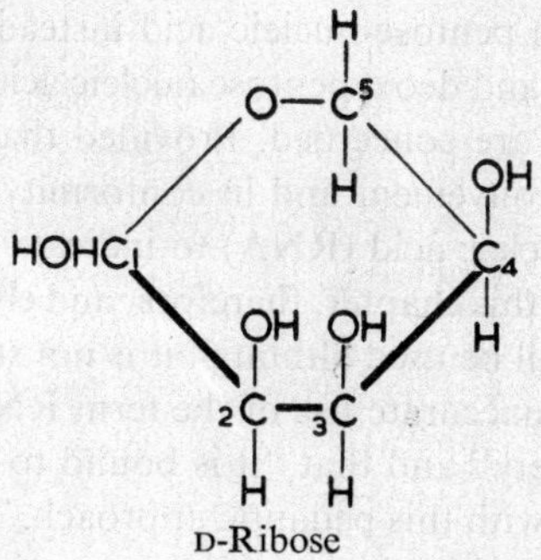

D-Ribose

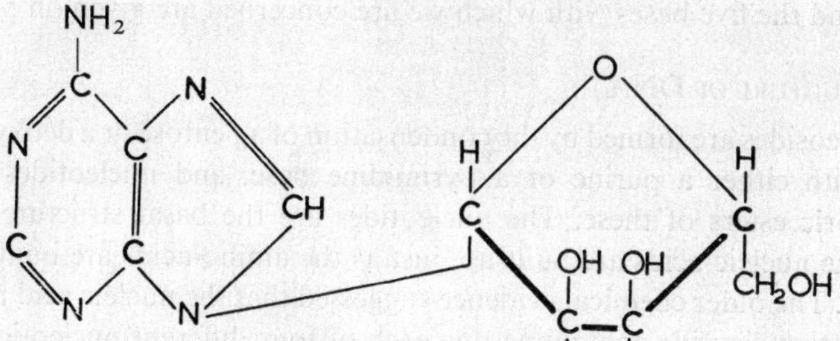

Adenosine (Nucleoside)
9-β-D-Ribofuranosidoadenine

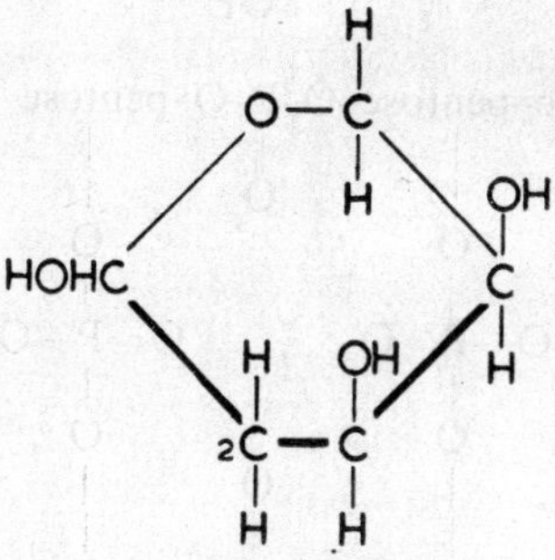

D-2-Deoxyribose

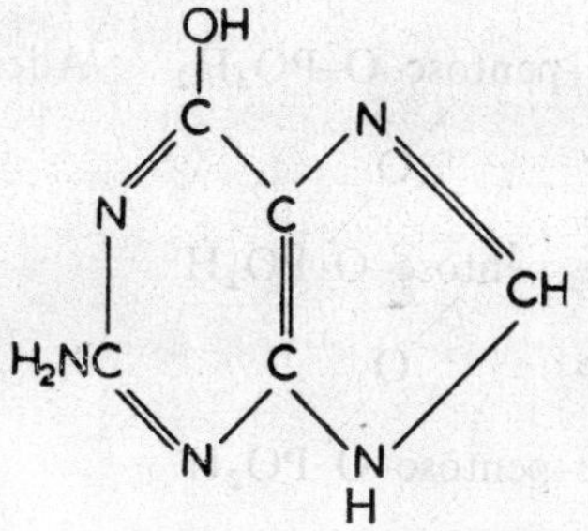

Guanine
2-amino-6-oxypurine

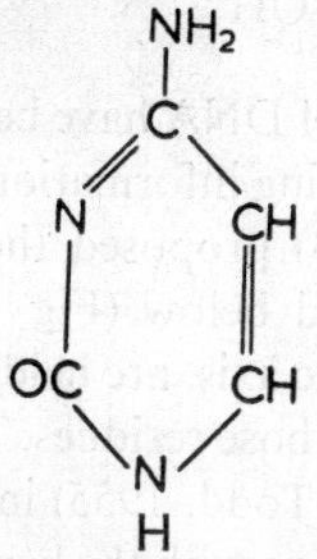

Cytosine
6-amino-2-oxypyrimidine

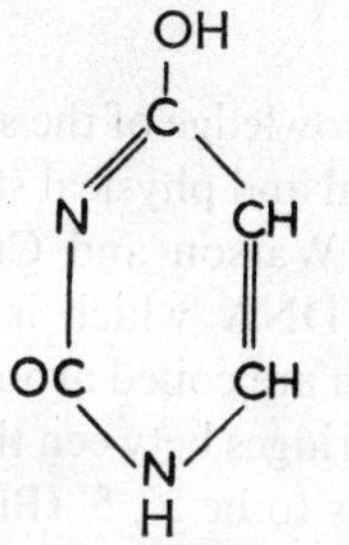

Uracil
2 : 6-dioxypyrimidine

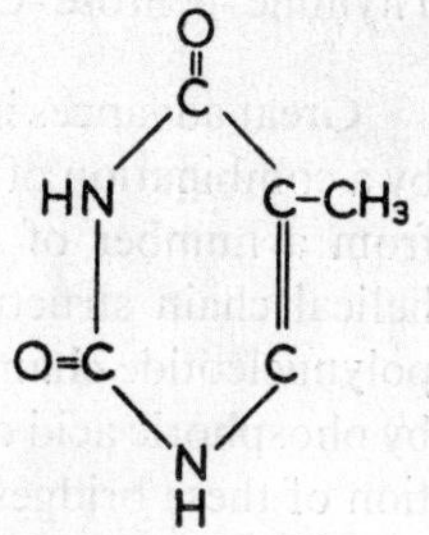

Thymine
5-methyi-uracil

accurate to use the term pentose-nucleic acid instead of ribonucleic acid for the cytoplasmic variety, and deoxypentose nucleic acid for the nuclear variety, as far as animal tissues are concerned. Provided that this fact is recognized (Davidson, 1965), it is convenient and in conformity with common usage to employ the term ribonucleic acid (RNA) to indicate the pentose polynucleotide from any source. In this chapter, therefore, and elsewhere in this book, the term ribonucleic acid will be used although it is not strictly accurate. Kanwar (1960) objects that the inaccurate use of the term RNA is "a callous indifference to scientific propriety" and that "it is bound to create unnecessary confusion". I cannot agree with this pedantic approach. The formulæ for the two sugars and the five bases with which we are concerned are given on page 249.

The Structure of DNA

Nucleosides are formed by the condensation of a pentose or a deoxypentose sugar with either a purine or a pyrimidine base, and nucleotides are the phosphoric esters of these. The nucleotides are the basal structures out of which the nucleic acids are built up, just as the amino-acids are built up into proteins. The older chemical evidence suggested that the nucleic acid molecule was a tetranucleotide containing one each of four different nucleotides. Two different structures were proposed for nucleic acids, the open chain structure of Levene and Sims (A), and the cyclic structure of Takahashi (B):

(*A*)

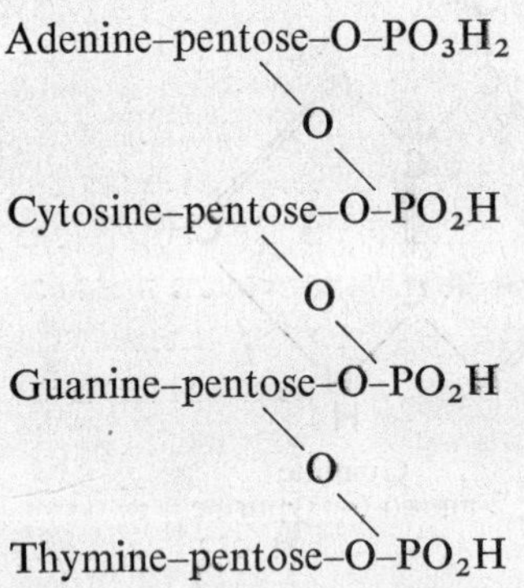

Adenine–pentose–O–PO_3H_2
O
Cytosine–pentose–O–PO_2H
O
Guanine–pentose–O–PO_2H
O
Thymine–pentose–O–PO_2H

(*B*)

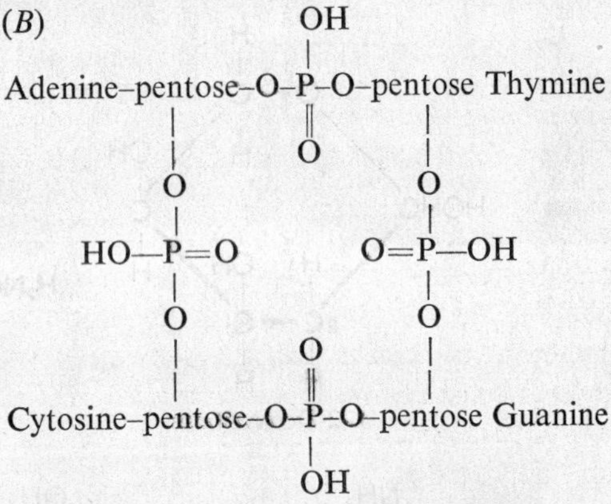

```
                        OH
                        |
Adenine–pentose–O–P–O–pentose Thymine
              |         ||          |
              |         O           |
              O                     O
              |                     |
        HO—P=O                  O=P—OH
              |                     |
              O                     O
              |         O           |
              |         ||          |
Cytosine–pentose–O–P–O–pentose Guanine
                        |
                        OH
```

Great advances in our knowledge of the structure of DNA have been made by a combination of chemical and physical studies. Using information derived from a number of sources Watson and Crick (1953) proposed the double helical chain structure for DNA which is illustrated below (Fig. 84). The polynucleotide chains, which are coiled around a single axis, are held together by phosphoric acid diester bridges between the deoxyribose residues. The position of these bridges appears to be 3′, 5′ (Brown and Todd, 1955) in the case of calf thymus DNA, and the spiral is a right-handed one with the bases on the inside and the phosphates on the outside. The two chains are held together by the purine and pyrimidine bases, which are oriented perpendicular to the fibre axis. They are joined in pairs, a single base from one chain being hydrogen-bonded to a single base from the other chain. One of the pair must be a purine and the other a pyrimidine for bonding to occur. Wilkins *et al.* (1953) regarded the Watson and Crick model as a simplification, while agreeing that it fitted most of the known facts. Their X-ray diffraction studies revealed no differences between samples of crystalline DNA from various species although the ratio of the bases varied widely. From these studies they deduced the helical structure of DNA and showed that a major part of the helix has one sharply defined diameter. Two coaxial helices of 18 Å diameter, one half pitch apart, surround a central helix of mean diameter 10 Å. For each turn of the helix there are 11 nucleotides whose shape resembles that of a rod inclined to the helical axis.

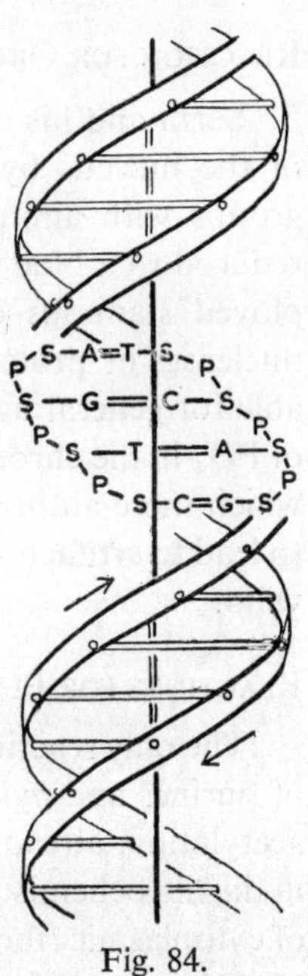

Fig. 84.

A symposium of authors (Feughelman *et al.*, 1955) agreed that the exact configuration of Watson and Crick was incorrect but found it possible to build a model containing the Watson and Crick hydrogen-bonding which gave the required Fourier transform.

These authors also considered the evidence for the combination of DNA with protein in the formation of nucleoproteins. In synthetic DNA-protamines the ratio between the numbers of basic amino-acids and phosphate groups may be varied but in sperm heads the ratio is always close to unity. This is assumed to indicate chemical combination. Protamine can be regarded as a polyarginine chain with repeating 4-arginine units. Nucleic acid presumably combines with poly-L-arginine by joining the four arginine guanidinium groups to the phosphate groups of the helix. For the combination of basic polypeptides with nucleic acids regular spacing of the phosphate groups is evidently necessary. Basic dyes (see below, p. 267) do not combine with mononucleotides; they are bound to nucleic acids not only by ionic bonds to PO_4 groups but also because they lie with their surface in close contact with the sugar surface of the nucleic acid molecule. It is apparently quite feasible for short side chains to the main polynucleotide chain to occur as branches at triply esterified phosphate groups. These can lie in the grooves of the helix without interfering with the packing of the nucleic acid molecules and can terminate as secondary phosphate groups. These will provide additional binding sites for basic dyes.

Histochemically, the difference between the pyrimidines of the two nucleic acids is not detectable, but the difference between the two sugar residues is all-important. A third point of difference, the degree of polymerization of the constituent nucleic acid chains, is also of great histochemical importance. As stated in Chapter 5, the demonstration of this property is influenced by the type of fixative employed in the preparation of the tissues. This point has therefore to be considered in the interpretation of results, especially those obtained by the staining and extraction techniques described below.

General Reactions for Nucleic Acids

Reaction for Organic Phosphate (Serra and Queiroz Lopes, 1945)

Serra and his co-author demonstrated the phosphate radical in the DNA of the nucleus by hydrolysis and subsequent fixation of the released PO_4 groups with ammonium molybdate. The resulting phosphomolybdate was reduced to a blue compound with benzidine instead of the more usually employed stannous chloride. They employed either enzymic hydrolysis with nucleases or prolonged hydrolysis with N-HCl. The method is not very suitable for general use in histochemistry but serves to demonstrate the presence of PO_4 in the chromosomes. It may be objected that the prolonged hydrolysis, which these authors employed in the method finally evolved, would be likely to lead to artifacts due to diffusion of phosphate ions. This criticism is certainly valid.

Reactions for Purines and Pyrimidines

The only reaction which has been used for the histochemical demonstration of purines and pyrimidines is the tetrazonium reaction after benzoylation or acetylation, already referred to at some length in Chapter 6. It was first used in the histochemistry of nucleic acids by Mitchell (1942) for the demonstration of cytoplasmic ribonucleoprotein in tumour cells and subsequently by Danielli (1947) for staining chromatin in the chromosomes. Disregarding for the moment the other aspects of the tetrazonium reaction, it is clear that it stains particularly strongly nuclear chromatin, nucleoli, and such ribonucleoprotein as is not removed during the process of benzoylation or acetylation. Only the interpretation of the reaction is open to question.

Both Mitchell and Danielli assumed that the development of a colour with tetrazotized benzidine, after the action of acetic anhydride or benzoyl chloride, was not due to histidine, tryptophan or tyrosine. They considered that it was most likely to be due to purine and pyrimidine groups, although Danielli felt that there might well be other substances in the tissues which would react with diazonium and bis-diazonium salts. Mitchell (1942) stated that adenine, guanine and uracil could react with diazonium salts *in vitro* and Danielli claimed that both DNA and RNA were able to react *in vitro* with tetrazotized benzidine. My own results (reported in the first edition of this book) with purified DNA-protein and with DNA almost free from protein, deposited in crystals of sodium chloride and fixed on slides with methanol, showed that any positive staining by the coupled tetrazonium reaction was due to the presence of protein since the pure DNA crystals completely failed to react. In view of this evidence it was stated categorically that the staining of nuclei by the tetrazonium reaction is not due to the purines and pyrimidines of their constituent nucleic acids. The positive reaction obtained by Burstone (1955) with DNA was almost certainly due to contamination with protein.

An attempt was therefore made to determine the nature of tetrazonium staining of the nuclei, before and after benzoylation, by further experiment. The application of mild heat (see Chapter 6, p. 169) is not necessary to develop the resistance of nuclear chromatin to benzoylation, though if such heat is used the degree of resistance is increased, just as it is in certain simple protein components of the tissues. It seemed, therefore, probable that the reaction of chromatin after benzoylation was due to tyrosine, tryptophan or histidine groups, protected from benzoylation by the physical state of the whole nucleoprotein. This belief was supported by the fact that if the nuclei were subjected to extractions of the type described below in this chapter, and also in Chapter 25, which remove all nucleic acids, the protein structure which remained gave a positive tetrazonium reaction and therefore presumably contained one or other of the amino-acids.

Barnard and Danielli (1956), and later Barnard (1961), reported comprehensive studies of the coupled tetrazonium reaction in relation to nucleic acids. The first authors claimed that the reaction was specific for "protein in a special form of combination with nucleic acid." Observing that benzoylation *should* block the tetrazonium staining of all tissue components they endeavoured by fractionation studies to determine the cellular distribution of the reactant responsible for its failure to do so, that is to say of the protein-bound group X in the structure

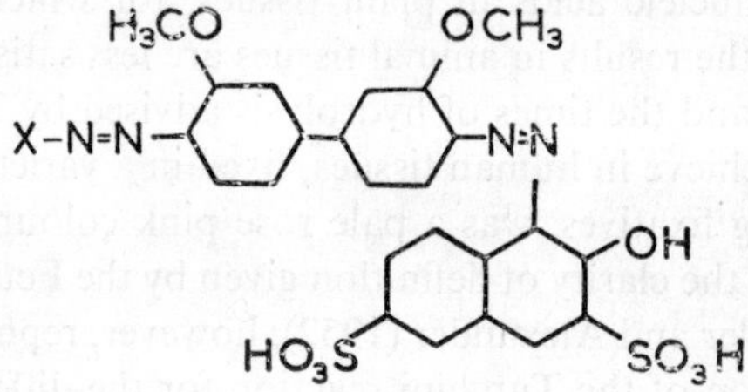

H_3CO OCH_3

X–N=N– –N=N

OH

HO_3S SO_3H

For this purpose they used chick erythrocyte nuclei coupled with Fast blue B salt (tetrazotized diorthoanisidine) and then with 2-naphthol-3,6-disulphonic acid to confer water solubility. Solvent extraction and trichloroacetic acid extraction showed that neither the lipid nor the nucleic acid fractions contained any of the dye component. Extraction of the residue with alkali yielded an intensely coloured macromolecular compound which gave a typical amino-acid mixture after reduction of the dye and hydrolysis. This process was achieved, not without difficulty, by refluxing in 15 mM-cetyltrimethyl ammonium bromide in 0·3 N-NaOH. Paper chromatography of the dye fraction in two solvents gave one main spot suspected to be aminohistidine. Barnard and Danielli suggested that a positive coupled tetrazonium reaction after benzoylation is normally limited to histidine residues in a nucleoprotein fraction and that failure to benzoylate histidine in tissue sections is due to "protective bonds" which exist only after dehydration.

It is now clear that the reaction cannot be used for the identification of nucleic acids but that it remains useful for demonstrating histidine-protection not only in nucleoproteins but in many other types of protein.

As shown by the work of Barnard and Bell (1960) histidine groups are not alone in exhibiting the protective state. These authors found that arginine groups of nucleoproteins, but not of cytoplasmic proteins, were also resistant to benzoylation. This resistance is unaffected by hydration or dehydration and it is regarded as due to stable salt linkages.

REACTIONS FOR DEOXYRIBOSE AND RIBOSE (TURCHINI)

Turchini *et al.* (1944) developed a method for DNA and RNA depending on mild acid hydrolysis followed by reaction with phenyl (or methyl) trihydroxyfluorone. The formula for the phenyl compound is given below:

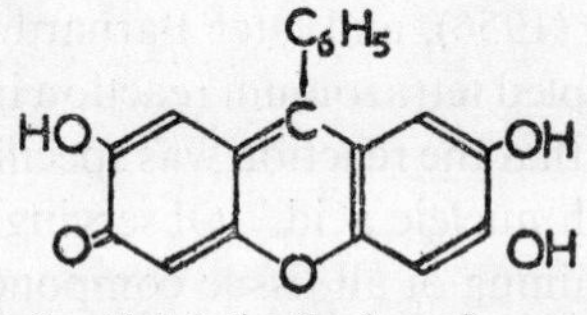

C_6H_5 HO OH O O OH

9-phenyl-2:3:7-trihydroxyfluorone

Although the method has the advantage that distinct and different colours are given by the two nucleic acids in plant tissues, for which the method was primarily evolved, the results in animal tissues are less satisfactory. Using the phenyl compound and the times of hydrolysis advised by Turchini *et al.*, the best that I could achieve in human tissues, fixed in a variety of alcoholic and formalin-containing fixatives, was a pale rose pink colour in the nuclei, nowhere approaching the clarity of definition given by the Feulgen reaction, described below. Backler and Alexander (1952), however, reported the successful use of a modification of the Turchini reaction for the differential staining of DNA and RNA. Using the methyl substituted fluorone, DNA stained violet to blue black and RNA yellow to red. With this modification violet-stained nuclei can be obtained regularly but the results are dependent on the fixative employed and, in general, the times of hydrolysis and subsequent staining are longer than usually suggested.

Reactions Specific for Deoxyribonucleic Acid

THE FEULGEN (FEULGEN-SCHIFF) REACTION

The reaction was introduced by Feulgen and Rossenbeck (1924) as a specific test for thymonucleic acid (DNA). Its historical aspects were well reviewed by Kasten (1964). The reaction depends on the treatment of fixed tissues by mild acid hydrolysis (with N-HCl at 60°) which Feulgen showed could release aldehyde groups from the deoxypentose sugar of DNA. Following hydrolysis the tissues are washed and transferred to a solution of Schiff's reagent (fuchsin-sulphurous acid) which reacts with the exposed aldehyde groups to produce a purple dye in the nuclear chromatin alone (Fig. 85). Bauer (1932) worked out the optimum times of acid hydrolysis, which depend on the fixative employed, and his investigations were extended to plant tissue by Boas and Biechele (1932) and Hillary (1939). A chemical explanation of the reaction was produced by Stacey *et al.* (1946). These authors claimed that a relatively gentle acid hydrolysis would transform deoxyribose into ω-hydroxylævulinic aldehyde ($HOCH_2.CO.CH_2.CH_2.CHO$), and they suggested that this labile aldehyde was responsible for the purple colour given with Schiff's reagent in the Feulgen test. In consequence of the explanation offered by Stacey *et al.*, the specificity of the test, or more strictly speaking its specificity for the localization of DNA, became the object of increasing doubt. In a later section, the modern history of the Fuelgen reaction is discussed. The arguments for and against it are put forward at some length and they serve to illustrate the complexity of the problem. For those who are interested solely in applying the reaction, it may be well to state here that its specificity for DNA is able to survive the conflict reported below, and that it may be used with confidence for the purpose.

Some Important Results of the Feulgen Reaction. Use of the Feulgen reaction for DNA has resulted in a great increase in our knowledge of the distribution of this substance in various types of cell. Darlington (1942, 1947) studied particularly the chromosomes and pointed out the vital importance of the reaction in making chemical processes in the nuclei structurally visible. Stowell (1942, 1945) and Stowell and Cooper (1945) used a modified Feulgen reaction, together with a system of photometric recording, to estimate the amount of DNA in a variety of normal and neoplastic tissues. By means of Stowell's method it is possible to measure mean amounts of substances per unit volume and per cell and thus to correlate quantitative results with morphological data. Further knowledge of the function of the nucleic acids in nuclear division has come from the studies of Thorell and Caspersson (1941), Thorell (1944), and Hydén (1943). All these authors used ultraviolet absorption techniques as well as the Feulgen reaction and from their work developed the concept of nucleolar associated chromatin. Caspersson (1947) considered that this material was responsible for the change in malignant cells which gives rise to the production of large quantities of protein. Davidson and Waymouth (1946) showed that in the nucleus of the rat liver cell the peripheral part of the nucleolus contained DNA and the central part RNA. Important information on the constancy of nuclear DNA has been derived from the studies of Vendrely and Vendrely (1948), Ris and Mirsky (1949), Pasteels and Lison (1950), and Swift (1950). The procedure used by the last author has been employed by a substantial number of workers in this field. Correlated chemical and quantitative cytochemical studies have been carried out in many cases but the problem of

DNA in heterogeneous cell populations can be dealt with *only* by cytochemical means. Ris and Mirsky, for instance, showed that rat liver contained three classes of nuclei in which the DNA content (intensity of the Feulgen reaction) was close to 1:2:4. The polyploidy thus demonstrated was shown in other tissues by Pasteels and Lison and by Swift. These studies are further considered, from the cytophotometric angle, in Chapter 31.

From the point of view of the histologist the Feulgen reaction is less useful, in practice, than theoretical considerations might indicate. It is often necessary to apply it solely in order to be able to state with confidence that a given basophilic inclusion contains no DNA. For this purpose it is the only convenient and acceptable reaction. Greater sensitivity for the reaction may be expected when improved fixation (free-drying, formaldehyde vapour) is routinely employed, together with acid hydrolysis under conditions producing the maximum number of reactive aldehydes without loss of apurinic acid (see below).

The Specificity of the Feulgen Reaction. Most of the doubts concerning the specificity of the Feulgen reaction were raised, and later sustained, by the researches of Stedman and Stedman (1943a and b, 1944, 1947, 1950). Other workers, including Serra (1943), Choudhuri (1943) and Carr (1945), also offered evidence against the specificity of the reaction. Stedman and Stedman (1943a) isolated an acidic protein containing 9 per cent. arginine, 5 per cent. histidine, 11 per cent. lysine and a considerable amount of tryptophan, from various nuclear sources. The whole protein was acidic, despite its considerable content of the basic amino-acids, because of its very large content of glutamic acid. This protein, which they called chromosomin, was originally regarded by its discoverers as the principal component of the chromosomes chiefly on account of its staining properties with methylene blue and with the hæmatoxylins. With the latter dye it assumed the "typical blue colour familiar to histologists." It is interesting to note that in tissues fixed by formaldehyde vapour, after freeze drying, there is a strong reaction for COOH groups in the nucleoli and in some of the nuclear chromatin. This procedure (Chapter 5, p. 86) blocks all NH_2 groups and reduces the solubility of the nuclear proteins to a very low figure.

Stedman and Stedman denied that the Feulgen reaction could localize DNA because they found that the red dye, produced by the addition of fuchsin sulphurous acid to the acid hydrolysis product of DNA, was soluble in water and could be taken up from aqueous solution by chromosomin. Choudhuri (1943) stained chromosomes in plant cells with the red dye produced as above, which he termed "developed nucleal stain". The Stedmans argued, therefore, that this red dye acted as an ordinary basic stain, that it was diffusible in the tissues, and that it stained the chromosomes by attachment to the acidic groups of chromosomin. It would thus be sufficient for DNA to be present anywhere in the nucleus, and not necessarily in the chromosomes, for a positive Feulgen test to occur. Callan (1943) suggested that the Stedmans' view could be refuted by the experiments of Mazia and Jaeger (1939) who used nuclease to reverse the positive Feulgen reaction in *Drosophila* salivary gland chromosomes, and by the ultraviolet absorption experiments of Caspersson (1936), who found the spectra of purines and pyrimidines in the chromosomes. He also cited the micro-incineration experiments of Norberg (1943), who showed that these structures contained nearly all the phosphate in the nucleus. Stedman and Stedman (1943b), however, considered that Mazia and Jaeger's results confirmed their own point of view in that deoxyribonuclease,while destroying the staining capacity of the chromosomes towards aceto-carmine, did so without affecting their integrity. The reason why the basophilia of chromosomin was apparently reversed by treatment with nuclease was not explained by these authors. The reversal of Feulgen staining brought about by deoxyribonuclease might be taken to support rather than refute the argument of Stedman and Stedman, since, in the absence of DNA, no soluble aldehyde-leucofuchsin will be produced. Dobson (1946) and Brachet (1947), however, showed that pure DNA, fixed in Zenker and embedded in agar, gave a strong Feulgen reaction.

The exchange of views continued with Caspersson (1944) and Stedman and Stedman (1944) presenting their respective arguments on adjacent pages in the same journal. Throughout these often confusing arguments the real point at issue remained the same, the solubility or insolubility of the red reaction product produced in the Feulgen test. Baker and Sanders (1946) argued that if the aldehyde formed from DNA by acid hydrolysis could diffuse in the tissues it would also diffuse into the surrounding fluid, and Danielli (1947) criticized the explanation of the Feulgen reaction given by Stacey *et al.* on the grounds that a substance of such small molecular weight as ω-hydroxy-lævulinic aldehyde would certainly be washed out of the tissues during the reaction. Carr (1945) attacked the specificity of the reaction on three grounds. First, he suggested that chromosomes could regenerate the colour of Schiff's reagent by simple absorption in the manner of a column of alumina, and secondly, that selectivity for the nucleus depended on the destruction of cytoplasmic components, which would otherwise react, by the acid hydrolysis. Thirdly, he suggested that excess of SO_2 did not block the reaction with Schiff's solution as it should do if this was due to the formation of aldehyde groups. Modern knowledge of the mechanism of the Schiff reaction (see Chapter 13, p. 447) provides a complete explanation and refutation of these views.

Stowell (1946a) reviewed the evidence for and against the specificity of the Feulgen reaction and concluded that it was relatively specific when properly controlled. Both this author and Dobson (1946) took pains to refute the experimental evidence of Carr, Dobson suggesting that Carr's first point was answered by the results, quoted above, which were obtained with nucleases. Stowell obtained far less satisfactory results with nuclease and did not draw any conclusions from its use. Carr's second point was answered more satisfactorily by Dobson, who proved, by weighing, that acid hydrolysis caused an insignificant loss of material from fixed tissue. The third point remained unanswered by Dobson or by Stowell, although Feulgen (1927) had shown that no colour developed with Schiff's reagent if the aldehyde groups responsible for the reaction were blocked with ammoniacal silver solutions or with bromine, and Brachet (1947) later showed that sodium bisulphite in a non-acid medium had the same effect. I observed that dimedone or hydroxylamine hydrochloride, both of which combine with aldehydes to form unreactive compounds, would also block the reaction of hydrolysed nuclei with Schiff's reagent.

After these further exchanges between the orthodox school and their opponents, the point at issue still remained the solubility or insolubility of the Feulgen reaction product. More work was required to throw light upon the still unsettled problem. Sibatani (1950) described a remarkable and unexplained effect of solutions of histones, or other proteins, in enhancing the colour developed in the Feulgen reaction with extracted DNA. The stage at which they were added did not influence the effect. He believed, therefore, that the dye produced in the reaction was diffusible and that its absorption was not confined solely to chromosomin, as postulated by the Stedmans, but occurred with other proteins as well. Di Stefano (1948) studied the effect of Feulgen hydrolysis on the composition of cell nuclei fixed in acetic alcohol, by means of photometric measurements. He concluded that after 12 minutes' hydrolysis only the two purine groups were lost, while a further 12 minutes' hydrolysis removed all the phosphorus and histone of the nuclei. Stedman and Stedman (1950), however, by direct chemical estimations of isolated nuclei fixed in acetic alcohol, found that 10 minutes' hydrolysis removed all the histone and 27 per cent. of the phosphorus. These same authors also found that the hydrolysis fluid, separated from nuclei, gave an intense positive reaction with Schiff's solution. They concluded that since the form of the chromosomes was unaltered by a 10 minute hydrolysis, histones could not be present therein and, therefore, that nucleic acid combined as nucleohistone must necessarily be extra-chromosomal. These conclusions do not seem to follow from their experimental results although the latter do indicate that the structural form of the chromosome can be maintained in the absence of nucleic acids and histones. Kaufmann *et al.* (1948, 1950), as the result of experiments with enzymic and other extraction techniques on fixed material in tissue sections, concluded that chromosomes represented an integrated fabric in which no single protein or nucleic acid could be regarded as the primary structural component.

Experiments carried out by Kasten (1956) suggested that chromosomin played no part in the development of a positive Feulgen reaction. This author prepared a modified Schiff's reagent by bubbling SO_2 into 1 per cent aqueous acid fuchsin. This dye is a mixture of trisulphonated triphenylmethanes and these could hardly be expected to combine with an acidic protein like chromosomin. In practice, Kasten was able to obtain Feulgen-type staining in hydrolysed nuclei with his acid fuchsin reagent and this result was not achieved in unhydrolysed tissues. The possibility that the reconstituted acid dye might be staining the basic (histone) protein of the nuclei has to be considered. The negative result obtained after various extraction procedures for DNA renders this explanation unlikely and we must conclude that Kasten's work provides further evidence against the chromosomin hypothesis. There is evidence, however, from the work of Chayen and Norris (1953) that *false* localization of DNA may be produced by acid hydrolysis. These authors found, in the root meristem cells of *Vicia faba*, some cytoplasmic particles removable by DNase but not by RNase. These were not stained *in situ* by the Feulgen technique but their DNA was shown to be transferred to the nucleus.

Further Work on the Mechanism of the Feulgen Reaction. The original explanation offered by Stacey *et al.* (1946) was generally accepted, and its acceptance was responsible for much of the now historical argument given in the preceding section. Later, however, Overend and Stacey (1949) followed up the original work by studying the reaction of a large number of compounds in the Feulgen test. As a result, they explained the latter as follows: acid hydrolysis first breaks the sugar linkages engaged in polymeric bonding and secondly ruptures the glycoside linkages between the sugars and the purine bases. The deoxyribose compounds (apurinic acids) thus revealed are described as being attached through PO_4 linkages in the main nucleic acid chain where they are firmly held in the furanose form, with exposed aldehyde residues. Overend and Stacey suggested that the arrangement at carbon atom 1 of the furanose sugar (illustrated below), after removal of the base, was that of a potential aldehyde and that this group (HO.CHO) would react as an aldehyde with Schiff's and other reagents. The ribose sugars, with an OH at carbon atom 2 in place of the second hydrogen atom of the deoxy sugars (see formula for D-ribose, p. 249), are not hydrolysed by normal HCl and therefore do not react in the Feulgen test. The exact mode of attachment of the fuchsin-sulphurous acid molecule to the sugar aldehydes has not been ascertained (see Chapter 13, p. 449). Probably two potential sugar aldehydes react with one molecule of fuchsin-sulphurous acid to give the red product on which the test depends.

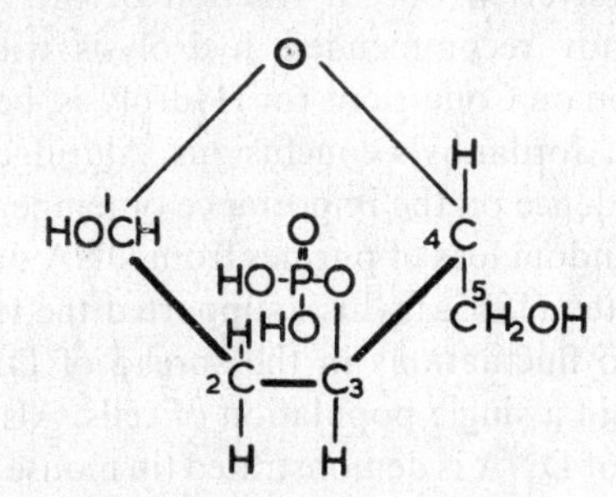

2-Deoxyribofuranose-3-phosphate

Using Fisher-Hirschfelder atomic models based on van der Waals volumes Lessler (1951, 1953) found that the S to S distance in fuchsin-sulphurous acid lies between 10 and 11 Å. X-ray diffraction studies indicate that the alternate nucleotides of DNA are about 10·2 Å apart. If indeed one molecule of fuchsin-sulphurous acid condenses with two molecules of aldehyde these must be situated on alternate nucleotides.

An alternative explanation for the Feulgen hydrolysis is suggested by the successful fluorescent assay for DNA developed by Kissane and Robins (1958). This was based on the formation of a quinaldine by a Döbner-Miller condensation of the hydrolysed DNA with 3,5-diaminobenzoic acid. Condensation takes place only with aldehydes containing an α-methylene group. Thus it is necessary to postulate that after removal of the purine bases the structure at C_2 is CH_2CHO, as indicated in the equation below

O HOH O CH_2OH 2 3 $O=P-OH$ O− ⇌ HO CHO O C CH_2OH $O=P-OH$ O−

The fluorescent quinaldine reaction was tested under cytochemical conditions by Stoward (1963), who was unable to obtain a positive result.

Li and Stacey (1949) showed that the loss of intensity of the Feulgen reaction by over-hydrolysis could be attributed to the lability of sugar aldehyde attachments and to instability of the furanose form itself. Jordanov (1963), however, found that the chief factor in the loss of apurinic acids was the high temperature (60°) at which hydrolysis was carried out. In his experiments the intensity of Feulgen staining was as much decreased by treatment with water at 60° as by N-HCl at 60°. Observing that the standard Feulgen hydrolysis, for 10–15 minutes, converted only a fraction of the available DNA into apurinic acids Jordanov recommended hydrolysis with 5N-HCl at room temperature (see section on Conditions for Hydrolysis, below). Although they disagreed with most of Jordanov's conclusions, Aldridge and Watson (1963) fully supported his evidence on the importance of temperature.

Evidence for the random loss of purines from DNA during acid hydrolysis was presented by Kasten (1965a) which supported the idea that this process does not contribute to fluctuations in the spread of DNA values (Feulgen Cytophotometry) within a single population of cells. Also supported was the view that a single type of DNA is demonstrated (in mouse liver) by the Feulgen reaction. On the other hand Böhm and Sandritter (1966) showed that the time of hydrolysis considerably influenced the dye content of normal and tumour cells. They suggested that two types of DNA existed which differed in acid sensitivity and which gave atypical hydrolysis curves.

From all the foregoing work it is clear that hot acid hydrolysis of nuclei causes two things to occur almost simultaneously. First, the purine bases are rapidly and completely removed and aldehyde groups are produced in the uncovered deoxyribose groups which remain: secondly, the histones and apurinic acids are progressively removed. Thus the Feulgen reaction is always a compromise between these two actions. After short hydrolysis the first predominates and appliction of Schiff's reagent at this point results in a high level of staining of the chromosomes. As hydrolysis proceeds the second action gains momentum and the Schiff reaction increases in the hydrolysis fluid, while it decreases in the chromosomes. Finally, the second overwhelms the first and the chromosomes cease to react. It is significant that the optimum time of hydrolysis for the material used in di Stefano's and the Stedmans' (1950) experiments was in the region of 6 minutes. The fact that over-hydrolysis produces a negative Feulgen reaction has been known since its earliest days. The realization, long overdue, that high temperature hydrolysis is undesirable should lead to considerable improvements in DNA cytochemistry.

The Feulgen Reaction in Practice. With ordinary histological materials the Feulgen reaction can be applied after almost any fixative except Bouin, with which excessive hydrolysis occurs during fixation. A list of the optimum times for hydrolysis with various fixatives, determined by Bauer (1932), is given in Appendix 9, p. 648. For ordinary work the modified Schiff's solution of de Tomasi (1936) has been widely used and for small materials, such as smear preparations, that of Rafalko (1946). A similar preparation, made by bubbling SO_2 gas through a solution of basic fuchsin, was employed by Itikawa and Ogura (1954). This found considerable favour with workers using Schiff's reagent for routine purposes on account of its ease of production and good stability. Coleman (1938) introduced the use of activated charcoal to remove residual colour in Schiff's reagent prepared by de Tomasi's method and this was considered a worthwhile modification. Lison (1936) advised the addition of 0·2 ml. of acetaldehyde per 100 ml. to the prepared reagent, standing for 30 minutes and finally adding a further 20 ml. of N-HCl and 1 g. of sodium sulphite. This modification had no advantages, however, over the simpler method of Coleman, using activated charcoal. Lhotka and Davenport (1947, 1949a) applied a modified Feulgen technique to the staining of tissues in the block. Thin slices of tissue were fixed in equal parts of 5 per cent. sulphosalicylic acid and saturated aqueous picric acid for 48 hours, they were then washed and treated with Schiff's reagent for 24–48 hours. Structures other than nuclei, including myelin, elastic fibrils and cartilage, were also stained red by this procedure. Lhotka and Davenport (1949b) investigated the deterioration of Schiff's reagent under varying conditions. They found that two factors were involved, first an irreversible colour change caused by exposure to air and secondly, a partly reversible change caused by a shifting of the chemical equilibrium between sulphite and dye (Karrer, 1950). They concluded that Schiff's reagent could be prepared in large quantities and that, provided it was stored between 1° and 5° in tightly stoppered bottles, it would retain its staining efficiency for 6 months or longer. It is preferable, I think, to make up sufficient reagent for one month's use only, and to store it under the conditions advised by Lhotka and Davenport. In this way, if it is kept in the dark and used in the dark in a sealed container, deterioration is minimal and false reactions due to this factor are avoided. Fading of Feulgen-stained preparations has often been noted, and many times reported. That this depends on the particular batch of Schiff's reagent was stated by de la Torre and Salisbury (1962). These authors found a fading of 14·6 per cent. in 30 days in Feulgen-stained bovine spermatozoa, using a fresh Schiff reagent. With an older batch of reagent, prepared in the same way, no fading was recorded over a period of 38 days.

Alexander *et al.* (1950) gave details of a rapid method of preparing Schiff reagent involving the simultaneous addition to the basic fuchsin solution of sodium hydrosulphite ($Na_2S_2O_4$, dithionite) and activated charcoal. Thorough shaking decolorized the fuchsin in the absence of the usual acid (20 per cent. HCl) but, since escess SO_2 was not present, oxidation had to be prevented by layering with xylene. Sections were subjected to hydrolysis in the usual way and placed in the reagent for 2 hours. A new dithionite-leucofuchsin reagent was described by Kodousek (1965), prepared by adding both sodium dithionite and sodium metabisulphite to an acid solution of basic fuchsin. As alternatives to the latter thionin and Azure I were sometimes employed.

A rapid method for the preparation of Schiff's solution was also given by Barger and DeLamater (1948). It employed thionyl chloride ($SOCl_2$) in place of the more usual sulphite compounds, and this was considered to break down to give a steady supply of SO_2.

$$SOCl_2 + HOH \rightarrow SO_2 + 2HCl$$

This method yields a remarkably stable solution whose only drawback appears to be that with certain tests, such as the performic acid or peracetic acid-Schiff tests for unsaturated lipids (Chapter 12, p. 437), colour is developed less readily than with less stable modifications. The de Tomasi-Coleman and the thionyl chloride methods for preparing Schiff's solution are given in Appendix 9, p. 647, together with the procedure of Itikawa and Ogura. The last two are recommended for general routine use.

It has sometimes been suggested that basic fuchsin should be used instead of Schiff's reagent for demonstrating the aldehydes produced by the Feulgen hydrolysis (DeLamater, 1948; Arzac, 1950). The disadvantage of this modification lies in the fact that, unless a long staining period is employed, the aldehyde-*basic fuchsin* linkages (Schiff bases) cannot withstand alcoholic dehydration. This is in sharp contrast to the extreme stability of the aldehyde-*fuchsin-sulphurous acid* compound in all the usual solvents. Basic fuchsin is not recommended as a substitute for Schiff's reagent in any of the tests in which the latter is customarily employed.

Alternative Basic Dyes in Preparation of Schiff's Reagent. A dilute aqueous

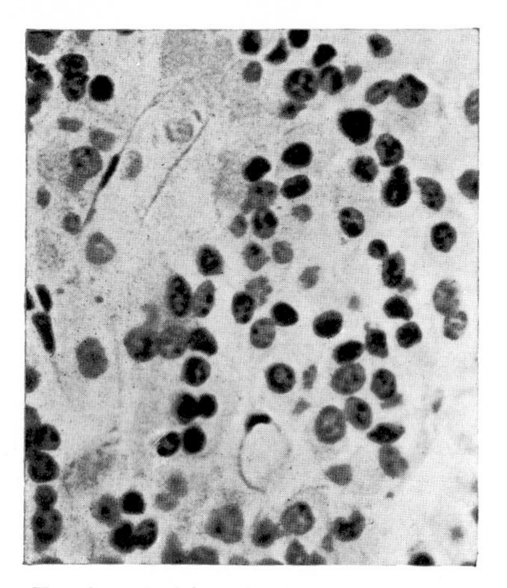

FIG. 85. Nuclei stained by the Feulgen method for thymonucleic acid (Schiff's reagent after N-HCl at 60° for 8 minutes). Only chromatin stains purple. × 520.

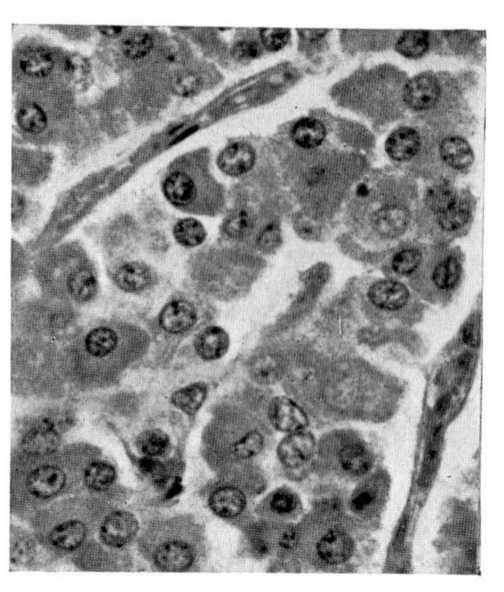

FIG. 86. Serial section to Fig. 85. Nuclei stained by 2-hydroxy-3-naphthoic acid and diazotized *o*-dianisidine after N-HCl at 60° for 8 minutes. Chromatin purple, cytoplasm pink. × 520.

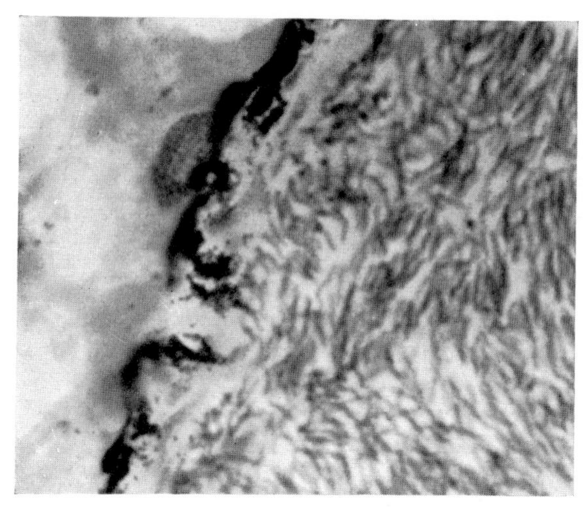

FIG. 87. Snail (*Helix aspersa*). Testis. Shows (right) green stain heads of spermatozoa and (left) pigment layer with supporting cells containing much red-stained RNA. Methyl green—pyronin. × 930.

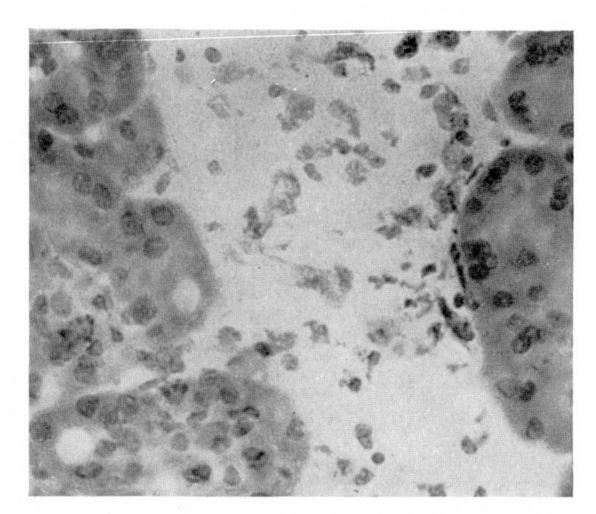

FIG. 88. Rat pancreas. Acinar cells contain much RNA (red). Connective tissue and acinar cell nuclei bright green. Methyl green—pyronin. × 400.

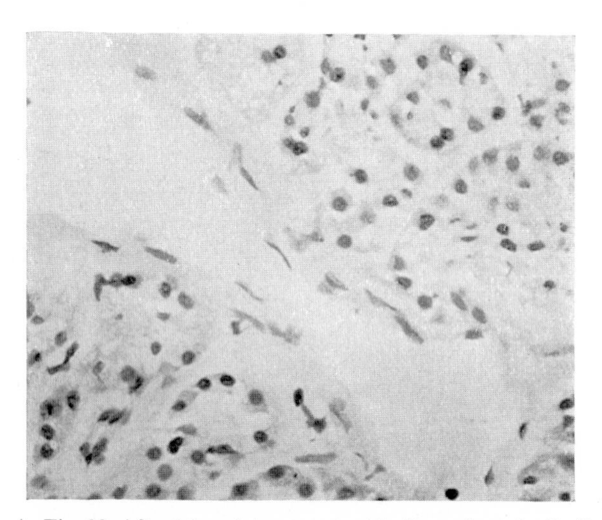

FIG. 89. As Fig. 88. After 1 hour's treatment with ribonuclease in distilled water. Only the nuclei are stained (green). Methyl green—pyronin. × 400.

solution of thionin, decolourized with SO_2, was used by Östergren (1948) in place of Schiff's reagent for hydrolysed plant nuclei and De Lamater *et al.* (1955) noted that thiazine and ozazine dyes would produce substitutes for Schiff's reagent. Van Duijn (1956) used a modification of De Lamater's thionyl chloride-thionine in which the dye concentration was originally reduced to 0·05 per cent. The improved results were certainly due to the *tert*-butanol water solvent and this was subsequently found to provide a stable solution when the higher concentration of thionin was used. This method stained hydrolysed nuclei blue, but unhydrolysed nuclei also stained to some extent, probably because acid hydrolysis was produced by the staining bath itself. These alternative methods have largely been used for double staining techniques and they have little application at present in routine histochemistry.

Conditions for Hydrolysis in the Feulgen Reaction. Alternative reagents for this purpose are fairly numerous. With a single exception all have been in aqueous solution. Brachet and Quertier (1963) proposed and employed N-HCl in absolute ethanol in a variation of the Feulgen reaction designed to reveal cytoplasmic DNA in oocytes. According to Cowden (1965) the alcoholic Feulgen method was not found satisfactory. Its cytoplasmic effects were attributed, probably correctly, to a pseudoplasmal reaction due to the mercurial fixative. If evidence from experiments with tritiated thymidine is accepted, however, it remains true that non-Feulgen positive DNA occurs in the cytoplasm of *Amœba*, *Lilium* species and in certain cells of *Gryllus* (Lima de Faria, 1962).

Widström (1928) used citric acid and De Lamater *et al.* (1950) used N-H_2SO_4 or HNO_3. Di Stefano (1952) preferred perchloric acid and Hashim (1953) phosphoric acid. Later additions to the list include 5 N-HCl and 5 N-HNO_3 (Itikawa and Ogura, 1954), trichloroacetic acid (Bloch and Godman, 1955) and bromine in carbon tetrachloride (Barka, 1956). The last alternative is the most interesting although the author considers that the mechanism is the same as that of the standard Feulgen reaction. According to Roberts and Friedkin (1958) bromine, in the form of bromine-water, could oxidize the thymine residues of DNA to acetol (I) as shown in the equation below:

Br. water

CH_3
|
C=O
|
CH_2OH (I)

The free acetol was condensed with *o*-aminobenzaldehyde (II) to give a fluorescent compound, 3-hydroxyquinaldine (III).

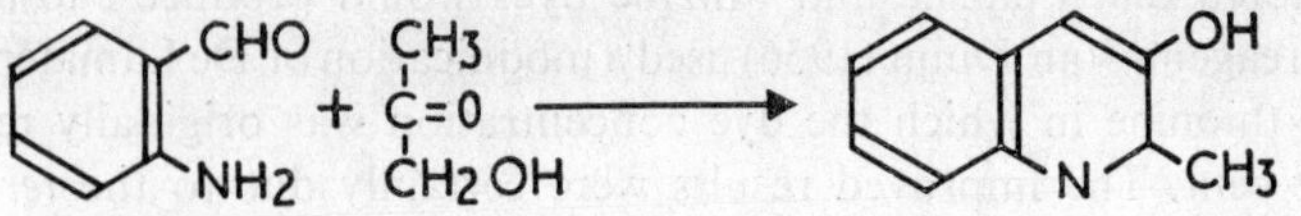

$$\text{(II)}\ C_6H_4(CHO)(NH_2) + CH_3{-}CO{-}CH_2OH \longrightarrow \text{(III)}$$

An attempt was made by Stoward (1963) to perform this reaction at the histochemical level, using salicyloyl hydrazide in place of *o*-aminobenzaldehyde, but without success. Further attempts should be made.

The reaction of RNA with bromine was described by Kanngiesser (1959). He found that uracil, cytosine and guanine all reacted to give bromo-derivatives but that adenine did not react. At the cold temperatures employed by Kanngiesser 5-bromouracil, 5-bromocytosine and 8-bromoguanine were apparently stable.

The Plasmal Reaction. Interference with the performance of the Feulgen reaction occurs through the so-called plasmal reaction of Feulgen and Voit (1924). This is due to the presence in the tissues of labile acetal phospholipids, whose aldehyde groups are easily revealed by mild hydrolysis or oxidation, after which they react with Schiff's reagent to give the usual red dye. In paraffin sections this interference is minimal but it is much greater in frozen sections. Its importance in frozen sections is presumably the reason why the subject appears frequently in the continental literature, but hardly at all in English and American work. Acetal phospholipids (plasmalogens) may be removed from sections by prolonged immersion in 90 per cent. alcohol or by treating the sections with a solution of hydroxylamine hydrochloride in sodium acetate (Appendix 13, p. 706) which converts them to oximes. These acetal phospholipids are considered more fully in Chapter 13.

ALTERNATIVE METHODS FOR NUCLEAR ALDEHYDES

The Feulgen-naphthoic Acid Hydrazide Reaction. Complications due to the use of Schiff's reagent for revealing aldehydes produced in the nuclei by mild acid hydrolysis can be avoided by the use of a different reagent for this purpose. Danielli (1947) used 2:4-dinitrophenylhydrazine, which stained the chromosomes yellow. He concluded that this result confirmed those given by Schiff's reagent. Better results are obtained by using 2-hydroxy-3-naphthoic acid hydrazide (Pearse, 1951). This reagent, first synthesized by Franzen and Eichler (1908), was used by Camber (1949) for the localization of ketosteroids in the adrenal cortex. It combines with aldehydes as well as ketones and the resulting compound, which is dark yellow in colour, is coupled with diazotized *o*-dianisidine in alkaline solution to give a purplish-blue compound. By this method the localization of chromatin staining in the nuclei is precisely similar to that shown by the true Feulgen reaction (Figs. 85 and 86). Cytoplasmic detail is visible in Fig. 86 because the hydrazide reagent combines with tissue proteins by virtue of other groups in its molecule and the subsequent treatment imparts a pinkish colour to these protein hydrazides. Although this does not provide a complete confirmation of the specificity of the Feulgen-Schiff reaction for the location of DNA, if the appearances in the latter were due to diffusion of the soluble dye product (aldehyde-fuchsin-sulphurous acid) different appearances might be expected to occur with the use of so different a compound as the naphthoic acid hydrazide. The solubility of the aldehyde-hydrazide compound doubtless differs from that of the aldehyde-fuchsin-sulphurous acid compound and, while the latter is a basic dye with affinity for acid substances (such as nucleic acids), the former combines with tissue proteins by virtue of the hydroxynaphthoic acid part of the molecule and here probably acts as an acid "dye".

Silver Methods. Several accounts have appeared describing the use of alkaline silver solutions for the demonstration of aldehydes produced in DNA by acid hydrolysis (Bretschneider, 1949; Winkle *et al.*, 1953). Although Gomori (1946) maintained that only aldehydes induced in glycogen and mucins could be demonstrated by silver methods Korson (1964) showed that, especially after hydrolysis with molar citric acid, hexamine (methenamine)-silver solutions gave excellent results with DNA. Very similar findings were reported by de Martino *et al.* (1965) who used the conventional acid hydrolysis (N-HCl, 60°).

Fluorescence Methods. Hydrolytically induced aldehydes in DNA can be demonstrated with so-called fluorescent Schiff reagents, of which a number have been described. According to Kasten (1959) acridine-orange (AO) and Auramine O both function adequately but considerable doubt exists as to whether the mechanism is the same as in the case of fuchsin-sulphurous acid (see Chapter 13, p. 447). Probably it is best to regard these dyes, in the presence of SO_2 and HCl, as pseudo-Schiff reagents.

The effect of acid hydrolysis on the AO fluorescence of DNA has attracted a considerable degree of attention. Schümmelfeder *et al.* (1957) first proposed that the observed shift (orthochromatic green to metachromatic orange or red) was due to depolymerization. In agreement with this view Bradley and Wolf (1959) considered that the steric rigidity of double stranded DNA would prevent the "stacking" of dye molecules required to give metachromatic staining. The depolymerization theory is opposed by the evidence of Joshi and Korgaonkar (1959) that X-irradiation of DNA produces the metachromatic shift.

An alternative theory was provided by Steiner and Beers (1959), who proposed that AO metachromasia was due to binding of dye by the purine and pyrimidine bases. The latter might be more accessible to AO after conversion of DNA to apurinic acid.

Recent reports (Paolillo, 1964a; Roschlau and Reinke, 1964; Roschlau, 1965) have thrown additional light on the mechanism. Roschlau (1965) showed that the metachromatic shift occurred after as little as 1 minute's treatment with N-HCl at 37°, particularly in the nuclei of necrotic cells and of some tumours, and also in mitotic chromosomes. These conditions will produce much less than the maximum amount of apurinic acid but they might be sufficient to produce a minor structural change in the DNA molecule (such as breaking protein-NA bonds).

Since the demonstration by Rigler (1964) that in the AO-DNA complex the dye is monomeric, and in AO-RNA dimeric, the way is open for proper interpretation of the above phenomena (see p. 271, below, and Chapter 29).

Stoward (1963) suggested the use of the salicyloyl hydrazide sequence (Chapter 6, p. 124 and Appendix 10, p. 661) as an alternative to fluorescent Schiff reagents. This produces a bright blue fluorescence in the nuclei which resists alcoholic dehydration.

Miscellaneous Methods

There are several biochemical methods for DNA, depending on the presence of 2-deoxyribose, which have a high degree of specificity. These are the carbazole and diphenylamine reactions of Dische (1930, 1944) and the tryptophan-perchloric acid condensation method of Cohen (1944), which was employed for the demonstration of DNA in normal and pathological sera by Seibert *et al.* (1948). The carbazole reaction, which Schneider (1948) considers specific for the pyrimidine-bound aldehyde of DNA, and the diphenylamine reaction are both carried out in acid solutions with heating to 100°, the former in 85 per cent. sulphuric acid and the latter in glacial acetic acid. They have not been successfully modified for histochemical use. The method of Cohen depends on the condensation of a secondary amine (tryptophan) with deoxyribose freed from purine and pyrimidine groups by treatment with hot perchloric acid. This condensation takes place slowly even in the test tube and cannot be produced *in situ* in tissue sections since the removal of DNA by hot perchloric acid takes place too rapidly.

Hydrazine-Benzaldehyde-Schiff (HBS). Following the description by Takemura (1958) of a method for cleaving pyrimidines (uracil and thymine), using anhydrous hydrazine at elevated temperatures, Smith and Anderson (1960) produced an adaptation for histochemical use. The reaction takes place, in the absence of water, as indicated in the equation below:

```
      CH3                                  CH3
 H     |     O                        H     |     O
  \C=C\C=                              \C=C\C=
   |     |   + H2NNH2 ——→               \    /    + H2NCO.NH2
   N     N                              -N—N
 H/ \C/   \H                           H     \H
     ||
     O
```

While it is clear that *in vitro* hydrazinolysis of DNA produces apyrimidinic acid (polydeoxyribose phosphate with purine bases intact) it appears that this compound still has the urea residue attached (in the 1-position). Treatment with benzaldehyde removes the urea and reforms the carbonyl group. The latter is then demonstrated by means of Schiff's reagent. The final colour of the nuclei is blue rather than magenta, for reasons which are not yet clear.

The authors suggested that the HBS reaction might be modified to give reliably quantitative data and that it might be so used in sequential combination with the Feulgen reaction so that the purine/pyrimidine ratios of individual nuclei or chromosomes could be determined.

Clearly the HBS reaction is worthy of further study and application, particularly since there is a paucity of new cytochemical reactions in the nucleic acid field. Details of the method are given in Appendix 9, p. 650.

Staining of DNA and RNA with Basic Dyes

General Considerations

The primary PO_4 groups of the nucleic acids have a pK in the region of 2, according to Jordan (1952), and above pH 2 they therefore carry a negative charge and combine with cationic groups of the basic dyes. It is usually considered that union of the latter with nucleic acids is by salt linkages but recent evidence, considered below, suggests that hydrogen bonding may be involved. In the tissues the nucleic acids are present as nucleoproteins and a proportion of their PO_4 groups is bound to protein (protamine, histone). Displacement of the latter frees phosphate groups for combination with basic dyes. The ratios of arginine and lysine to PO_4 are constant (Butler *et al.*, 1955) but the total amount of basic amino-acid is usually 10 per cent. less than the total amount of phosphate so that some of the latter remains unbound to histone. Knoblock and Vendrely (1956) found a constant ratio (5·0) between DNA and arginine in fish erythrocyte nuclei but in the sperm of some species (*Trutta, Esox*) it was much lower (1·6). In *Cyprinus* and *Tinca* the ratio in sperm was the same as that of the somatic nuclei.

A number of investigators have studied the problem of dye binding by nucleic acids in solution and in tissue sections. Hermann *et al.* (1950), using solutions and sections of muscle, found that between pH4 and pH6 the binding of *toluidine blue* was a function of the RNA content. A series of investigations, particularly on the properties of nucleic acids in nerve cells, and of ribonucleoproteins (RNP) in mitochondria, was carried out by Schabadasch (1957, 1960). Alterations in the I.E.P. of RNP are due, according to this author, to precise functional changes in the cell. Kurnick and Mirsky (1950) reported stoichiometric precipitation of polymerized DNA by *methyl green* and Kurnick (1950a) obtained a similar result *in vitro* with crystal violet. Irvin and Irvin (1949, 1952) studied the binding of *aminoacridines* and found that under certain conditions depolymerization resulted in decreased binding of the dye. Similar reports have been made for 5-amino-acridine (Lawley, 1956a) *quinolines* (Parker, 1949), methyl green, *ethyl green* and *malachite green*

(Kurnick, 1950a and c). On the other hand Lawley (1956b) and Cavalieri and Angelos (1950) reported a decreased uptake of *rosanilin* by depolymerized DNA. Lawley also found that metal cations (Mg, Ba) could compete with dyes for binding sites but that their effect was greater (30×) for rosanilin than for 5-aminoacridine. This observation agrees with those of De Bruyn *et al.* (1953) who suggested that the acridines (particularly 2,8-diamino-acridine) possess NH_2 groups correctly spaced for hydrogen bonding with the PO_4 groups of DNA. These are approximately 7-8 Å apart whereas the NH_2 groups of rosanilin are 10 Å apart.

It is evident that relatively minor changes can alter very greatly the uptake of basic dyes by nucleic acids.

The Ribonuclease, Methyl Green-pyronin Method (Brachet)

The standard method for demonstrating RNA in both nucleus and cytoplasm is the indirect method evolved by Brachet (1940a, 1942, 1944), which depends on the specific depolymerization of RNA by the enzyme ribonuclease. This treatment does not affect DNA in nuclear chromatin or elsewhere. Brachet showed that the methyl green of Pappenheim's (1899) mixture was an elective stain for DNA in chromatin, while the pyronin G of this mixture had an elective affinity for RNA in both nucleolus and cytoplasm. With Brachet's method of two parallel sections, therefore, one is exposed to ribonuclease and then both are stained with methyl green-pyronin. Alternatively, staining may be carried out with 1 per cent. aqueous toluidine blue, with or without counterstain, or with 1 per cent. toluidine blue in 95 per cent. ethanol which is recommended by Kurnick (1952) as a rough screening test for nucleic acids. Material staining blue with toluidine blue or red with pyronin, and removable by treatment with ribonuclease, is considered to be RNA; that which is not so removable is not. The Methyl green-Pyronin method is probably the method of choice for investigating the RNA content of the majority of tissues, although it does not always allow critical appreciation of cytological detail. If this is required, separate sections must be stained by some wholly cytological method or with toluidine blue followed by a few seconds staining with 2 per cent. orange G in 5 per cent. phosphotungstic acid. This last method shows both RNA and cytological detail. If other types of nucleic acid extraction are used (see below) staining with toluidine blue is preferable to the use of the Methyl green-Pyronin method.

Some Important Results of the Methyl Green-Pyronin Method. The use of this method, especially in the hands of its originator, greatly increased our knowledge of the relationship between the two nucleic acids in the nucleus and RNA in the cytoplasm. Most of the important papers of Brachet on this subject are given in his book (1950), a second edition of which appeared in 1958. Two papers which may be mentioned specifically (1940a, 1940b) refer to the use of the method for determining RNA in various tissues of the frog. Brachet's work was directly responsible for the realization that RNA was intimately

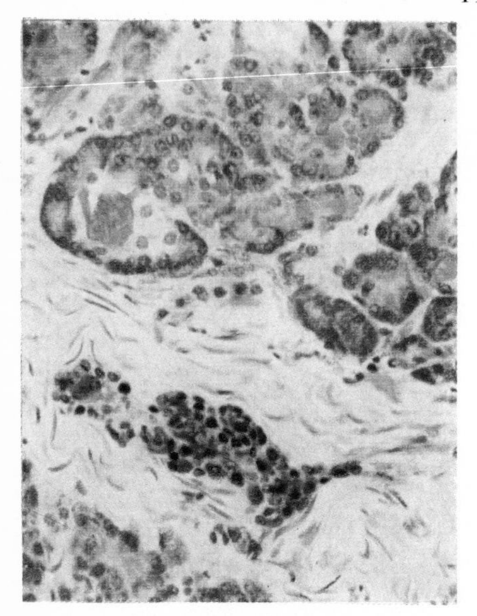

IVA. Human pancreas, freeze-dried formalin-alcohol-fixed section. periodic acid-Schiff, toluidine blue. G mucosubstance in acini CG mucin in duct. × 525.

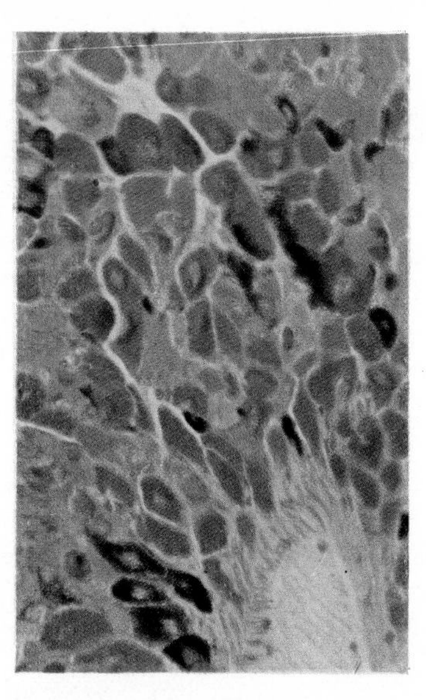

IVB. *Helix aspersa*, Salivary Gland. Freeze-dried formaldehyde vapour-fixed section. Shows DNA (green to bluish-green) and RNA (red) distribution in the different cell types composing the gland. × 250.

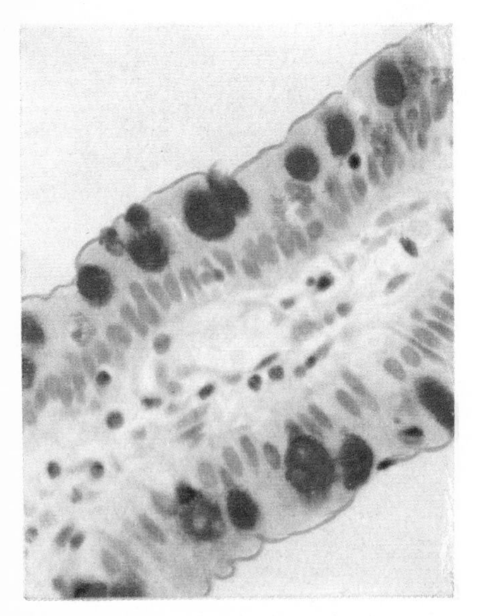

IVC. Human duodenal villus (cf. Plate IVD) SG-mucin in goblet cells and in the brush border. × 440

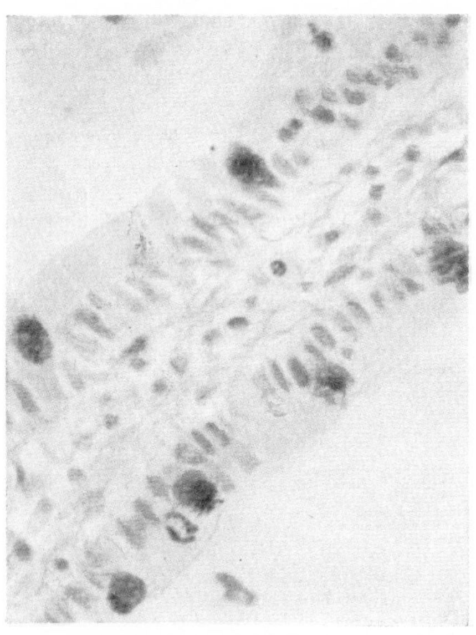

IVD. Human duodenal villus (cf. Plate IVC) Sixteen hours' benzoylation at 22°. Coupled tetrazonium reaction. × 440

concerned with protein synthesis and it was thus connected with the electron microscopic discovery of the ribosomes. It was the direct forerunner, also, of a great mass of work on the different forms of RNA (messenger, transfer, ribosomal) and their participation in protein synthesis. Following Brachet's lead protein-RNA relationships were studied in a wide variety of animal and plant tissues, by many different workers. Desclin (1940), for instance, studied the hypophyses of rats and guinea-pigs and found RNA in the basophils and Dustin (1942) found RNA in mammalian and reptilian reticulocytes. Biesele (1944) added some important findings in the case of skin tumours produced in mice by the action of methyl cholanthrene. Roskin and Ginsberg (1944) applied the method to protozoa, particularly to plasmodia and trypanosomes, and Dempsey and Wislocki (1945) examined human placental syncytium. Mottram and Selbie (1945) examined the distribution of RNA in bean roots and in Rous sarcomas in chickens, *inter alia*. Dempsey and Singer (1946) found ribonuclease-resistant material in the Z line of mammalian muscle. Stowell (1946b) and Deane (1946) investigated RNA in various human tumour cells and in normal human liver cells, respectively, and numerous authors have applied the method to bone marrow smears (Brachet, 1942; van den Berghe, 1942; White, 1947). Smears of the peripheral blood were also stained by the methyl green-pyronin method (Perry and Reynolds, 1956) and it was used for further studies in cancer cytology (Hopman, 1956). Laverack (1955) recommended the technique for application to frozen sections and it is certainly one of the better methods for studying nucleic acids in unfixed or post-fixed (cold microtome) sections. Barka *et al.* (1953) used the method on tissue cultures of chick embryo liver prepared by the hanging drop method. Their investigations showed a close correlation between the capacity of the cells to produce RNA and their capacity to proliferate in tissue culture.

The papers quoted above represent only a fraction of those published in which the Brachet method has been used. They are mainly those in which the method was the only important one employed. The volume of papers on the subject alone indicates its importance and the method remains today as standard histochemical practice for all investigations involving the distribution of the nucleic acids.

Theoretical and Practical Considerations of the Methyl Green-Pyronin Technique. Kaufmann *et al.* (1948), using sections or smears fixed in acetic alcohol and stained with methyl green-pyronin, found that the chromosomes were blue-lavender in colour, while the nucleoli, and the cytoplasm of cells containing RNA, were red. After ribonuclease no component colourable with pyronin remained, so that the cytoplasm and nucleoli were colourless while the nuclei were green. This suggested that RNA was present in the chromosomes. If sections were treated with pepsin, and subsequently stained with methyl green-pyronin, chromosomes, nucleoli and cytoplasm all stained more intensely with pyronin than in control sections. After successive treatment with pepsin and ribonuclease, chromosomes, cytoplasm and nucleoli no longer stained red, indicating that ribonuclease had removed the pyronin-stainable material. Kaufmann *et al.* suggested that RNA was present in cells, either free or combined as ribonucleoprotein, and that its affinity for dyes was influenced by the state of association. Tables of the staining patterns of chromosomal nucleic acids and proteins, using Feulgen, Methyl green-pyronin and Fast green in combination with enzymal analysis and acid extractions, are given by Kaufman *et al.* (1960).

It was first suggested by Kurnick (1947, 1949, 1950a and b) that the difference in stainability of chromosomes and cytoplasmic nucleoprotein was a matter of the degree of polymerization. This author, using the same two dyes together and separately, to stain DNA and RNA *in vitro*, found that highly polymerized DNA stained with pyronin about one-fifth as intensely as RNA and that DNA-histone stained about one-sixth as intensely. DNA which had been depolymerized, however, stained as strongly as RNA with pyronin. Using the two dyes together, RNA stained pale pink, depolymerized DNA a similar colour, and polymerized DNA green. Kurnick concluded, therefore, that pyronin distinguished only different states of polymerization of the nucleic acids rather than chemical differences between the two acids.

Vercauteren (1950), while agreeing with Kurnick's conclusions, considered that less drastic changes in structure than depolymerization could cause DNA to lose its affinity for methyl green. Using unfixed smears of thymus nucleohistone he found that 5 minutes' treatment with N-HCl at 60° would destroy its affinity for methyl green while enhancing that for pyronin. Short immersion in veronal-acetate buffer at pH 1·6, or in neutral solutions of formalin, had a similar effect. Vercauteren claimed that the stereochemical factor in the nucleic acid molecule which accounted for its high affinity for methyl green was the presence of negatively charged phosphate residues at a distance corresponding to that between two possible sites for positive charges on the methyl green molecule.

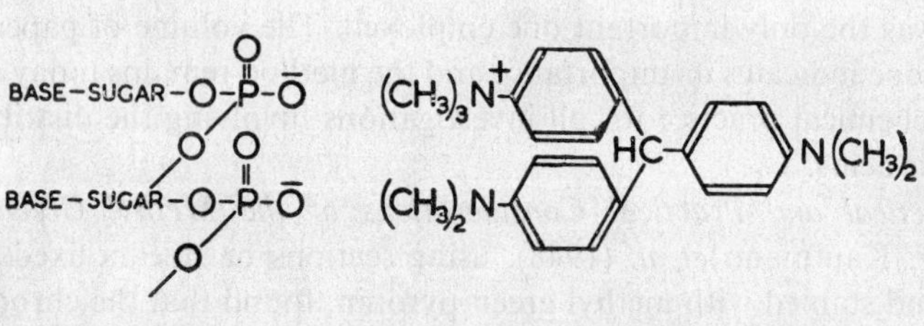

He suggested that the effect of dilute acids at low temperatures was sufficient to break certain weak (hydrogen) bonds in the DNA molecule, causing a coiling of the molecule to occur. This coiling would alter the distance between adjacent PO_4 residues which would thus no longer fit the methyl green molecule.

Opposition to Kurnick's hypothesis came mainly from Taft (1951b) and Alfert (1952). The former treated tissues at pH 3 and pH 11·7 and at 1° and 24° to produce depolymerization of DNA, without affecting methyl green staining. It is probable, however, that depolymerization of this type, which is due merely to rupture of hydrogen bonds (Gulland and Jordan, 1947), is reversible. Nuclear staining with methyl green was considered by Alfert to be at least partly due to blocking of the stainable groups in RNA by proteins and he considered also that the staining of DNA by this dye was subject to variations in degree due to varying amounts of protein binding. There is, however, no essential disagreement between these findings and Kurnick's observations, and most workers are prepared to agree that under carefully controlled conditions methyl green staining is selective for "polymerized" DNA. As pointed out by Godman and Deitch (1957), however, even when methyl green is used with the precautions advised by Kurnick (1955b) it cannot by itself afford unequivocal evidence of depolymerization of DNA. Godman and Deitch showed that acetylation, which blocks the positively charged ε-amino groups of lysine and the guanidyl groups of arginine, restored the methyl green staining of LE bodies in systemic lupus to levels equivalent to those derived from Feulgen staining. The results obtained by Rosenkranz and Bendich (1958) suggested that the specificity of methyl green for DNA depended upon the maintenance of the double stranded condition, and that the latter was disrupted by heating and by other factors. Experiments carried out by Goldstein (1961), on the other hand, were more easily explained on the basis of the spacing of reactive sites for the dye molecule.

In addition to methyl green-pyronin other contrasting pairs of dyes were selected. In every case DNA was stained by the larger molecule and RNA by the smaller. Strong and cogent objections to Goldstein's hypothesis were put forward by van Duijn (1962) and it is clear that explanations of differential NA staining, based on density differences, must be supplemented by additional evidence if cytochemical specificity is claimed. Using methyl green alone Cowden (1965) produced specific staining of cytoplasmic RNA in oocytes and he postulated a special state of the ribonucleoproteins to account for his observations. When methyl green and pyronin-Y were used together the oocyte RNA stained only with the latter.

Using his critical electrolyte concentration CEC method (Chapter 10, p. 348) Scott (1967) has introduced new concepts to explain Kurnick's original findings. Applying the CEC method to a number of polynucleotides and their combination with pyronin, methyl green, malachite green, crystal violet and Alcian blue, Scott found that all these dyes, except the last, bound polynucleotides more strongly than would be expected if only electrostatic bonds were present. Pyronin and other planar monovalent cationic dyes (toluidine blue, acridine orange, 5-aminoacridine, thioflavine T. etc.) were observed to react best with nucleic acids having freely accessible purine and pyrimidine bases, as in single-stranded molecules without extensive secondary structure. Thus they bind strongly to RNA or denatured DNA. Non-planar dyes (methyl green, malachite green), on the other hand, were shown to bind much less strongly to such substrates but much more so to DNA with intact double helical structure (Fig. 84).

Although it provides no complete answer to the relative specificities of methyl green and pyronin for DNA and RNA Scott's work indicates that the two really important variables in the technique are staining time and salt concentration. Scott indicated that 16 hours in 0·15 per cent. methyl green and 0·25 per cent. pyronin in 50 mM-sodium acetate (pH 5·6) with 2M-$MgCl_2$ had been found to give good results.

Many methods for standardizing the performance of the methyl green pyronin stain appeared in the years immediately following Brachet's re-introduction of the method. One, described by Taft (1951a and b), depended on repeated chloroform extractions of a 0·5 per cent. solution of methyl green in 0·1 M-acetate buffer, to which was subsequently added 0·2 per cent. pyronin B. Dehydration was carried out in a mixture of 3 parts of tertiary butyl alcohol and 1 part ethanol, and clearing and mounting as usual. A second method for preparing the methyl green-pyronin stain (Trevan and Sharrock, 1951) is given in Appendix 9, p. 651 and a third (Kurnick, 1955b) is given immediately after this. The method, in which chloroform-washed pyronin-Y is used, is more selective for RNA than other existing methods. A valuable discussion of the theory and practice of methyl green (and methyl green-pyronin) staining appears in Kurnick's (1955a) review of the histochemistry of the nucleic acids, and some useful information on the selectivity of various pyronins for RNA in plant tissues is given by Paolillo (1964b).

The type of result given by the methyl green-pyronin method, with and without prior ribonuclease digestion, is shown in Figs. 87, 88, and 89. It is my impression that far superior results are obtainable by the use of freeze-dried and formaldehyde vapour-fixed materials (Chapters 3, p. 50, and 5, p. 86). The coloured illustration (Plate IVb, facing p. 269) indicates the superior preservation of both nucleic acids obtained by this procedure. Technical and theoretical points concerning the use of ribonuclease are discussed in Chapter 25.

Toluidine Blue-Molybdate Methods (Love)

In a long series of publications Love (Love, 1957, 1962; Love and Liles, 1959; Love and Suskind, 1961; Love and Walsh, 1963; Love and Rabotti, 1963; Love *et al.*, 1964) reported the development and application of staining methods capable of distinguishing 9 types of ribonucleoprotein. These methods are based on the use of preparations wet-fixed in formol sublimate and subsequently stained in aqueous toluidine blue, followed by ammonium molybdate.

The theoretical basis of the TBM method, though not yet fully substantiated, is the progressive blocking of the amino groups of the nucleoproteins and the consequent progressive unmasking of nucleic acid phosphate groups which can bind cationic dyes. At a certain point in this process it is considered that molybdate treatment induces "polymerization" and intensely metachromatic staining with toluidine blue. When complete blockage of protein amino groups is achieved (as by treatment with HNO_2) the metachromatic

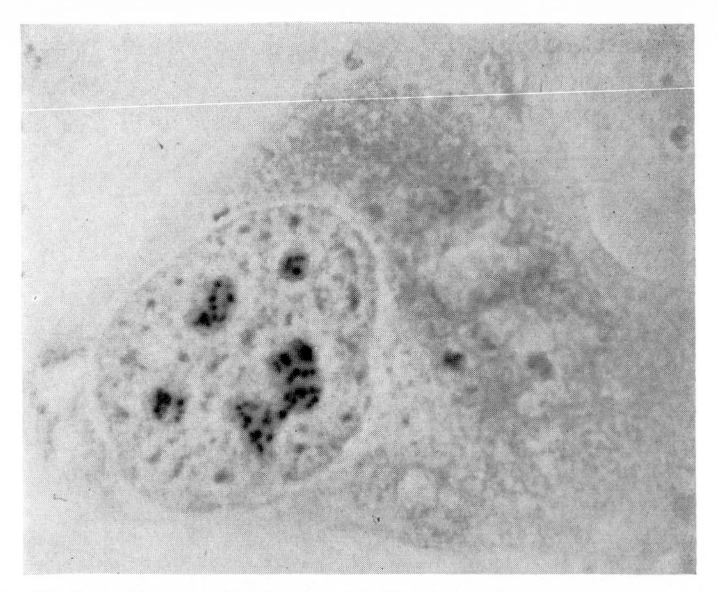

Fig. 90. Spherical nucleolini (black) embedded in the lightly staining (grey) pars amorpha of the nucleolus. Love's TBM, method B. × 2200.

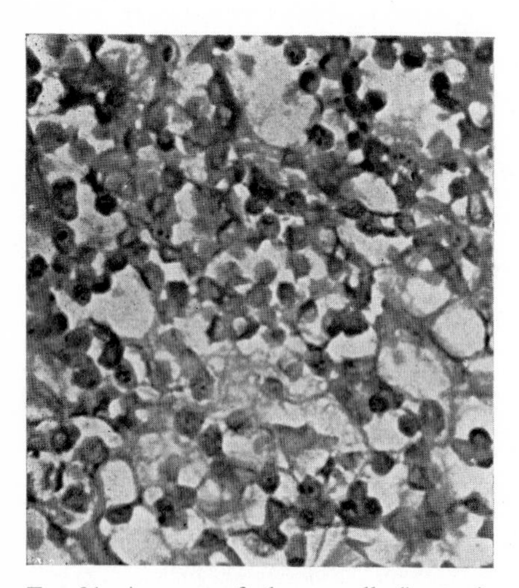

Fig. 91. A group of plasma cells (human), showing the small affinity of their nuclei for acid dyes. Light green S.F. × 520.

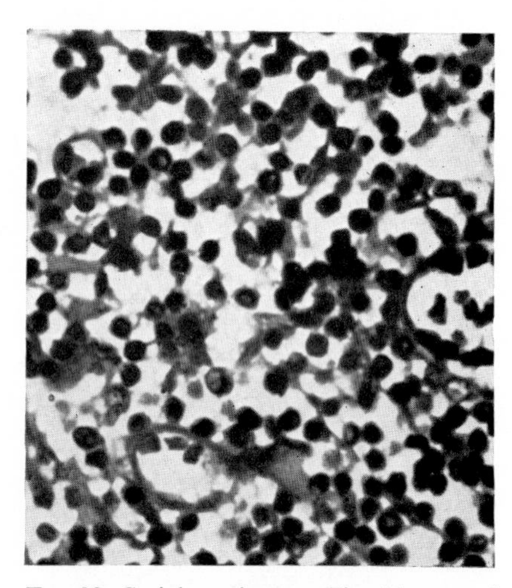

Fig. 92. Serial section to Fig. 91, treated with 5 per cent. trichloroacetic acid at 90° for 15 minutes (Schneider extraction). Stained and photographed as Fig. 91. Light green S.F. × 520.

nucleic acid dye complexes are not formed. Details of the TBM method are given in Appendix 9, p. 655 and metachromatically stained nucleolini are shown in Fig. 90, opposite. These structures are subject to early alteration in virus infections (Love *et al.*, 1965; Love and Fernandes, 1965) and they may perhaps be useful as diagnostic markers for such infections.

Using a modified Giemsa procedure Mironescu (1965a and b) demonstrated nucleolini in unfixed touch preparations but his method appears to be less sensitive than TBM for differentiating RNA-containing components of nuclei and cytoplasm.

Considerations of Toluidine Blue Metachromasia of RNA. The metachromasia of nucleic acids with toluidine blue was noted many years ago (Lison and Mutsaars, 1950; Ghiara, 1953). More recent studies by Lamm *et al.* (1965) and by Feder and Wolf (1965) have strongly supported the original hypothesis of Bradley and Wolf that the capacity of nucleic acids to form dye aggregates (stacking coefficient) is responsible for metachromatic staining. Lamm *et al.* (1965) showed that different types of fixation could influence this capacity. Formaldehyde, for instance, increased the stacking coefficient of RNA, as did other aldehyde fixatives. Feder and Wolf (1965) found that optimal preservation of structure as well as the greatest differential toluidine blue staining of DNA and RNA was produced by acrolein fixation and ester-wax embedding. They suggested that DNA-protein and RNA-protein in the tissues differ more in their conformation than do free DNA and RNA in solution. They suggested, further, that quantitative studies of altered dye spectra in stained tissues have great promise as sources of new information with regard to the structure of the components of living matter (see also Chapter 31).

Quantitative Histochemical Estimation of Nucleic Acids

Three methods have been described for the quantitative estimation of nucleic acids, two applicable solely to DNA and the third to total nucleic acid irrespective of type.

The first method, which has already been mentioned briefly, was that of Stowell (1942) who described a relatively simple apparatus for the photometric measurement of the amount of colour developed in nuclei stained by the Feulgen reaction. A similar method was used by Pollister and Ris (1947), by Leuchtenberger (1958), and by many others who have followed their lead. Measurements were made of the transmission of light through stained and unstained sections and through portions of the glass slides away from the sections in both cases. The method depends on a proportional correlation between the amount of the stain present and the substance for which it is specific, a relationship which has now been elucidated in the case of the Feulgen reaction by the DNA-film studies of Persijn and van Duijn (1961).

A method depending on the application of spectrophotometry to nuclei stained with methyl green was described by Kurnick (1949). Tissues for examination were immersed in 30 per cent. sucrose, as described by Hogeboom *et al.* (1948), before fixation in neutral formalin. This process yielded nuclei which were homogeneous except for the nucleoli. After removal of histones by treatment with N-HCl, sections were stained for 14 hours at 4° in a 0·25 per cent. solution of methyl green in 0·2 M-acetate buffer at pH 4·1, the solution being chloroform-washed before use (see Appendix 9). After staining, the sections were mounted in buffer and the coverslips ringed with paraffin wax. Photometry was carried out at 645 nm. Using this method the author found $0{\cdot}6 - 0{\cdot}8 \times 10^{-8}$ mg. DNA per nucleus in calf thymocytes, values in close agreement with those found in chemical analysis and by densitometry of photographic plates of thymus nuclei photographed at several different wavelengths in the ultraviolet.

Kurnick criticized the method of Pollister and Leuchtenberger (1949), who made a photometric assay of sections of mouse liver stained with methyl green, on account of the high staining temperature employed (56°) and because of their use of alcohol for differentiation. These criticisms were valid but Kurnick's method itself was also open to criticism. Much depended on the exact ratio in which methyl green combines with the nucleic acid. In calculations made from his original figure (dye-P ratio 1:10) certain differences were present between the theoretical and the observed coefficients of extinction. Kurnick suggested that these discrepancies might be due to the combination of DNA with proteins other than histones, which would consequently fail to be removed by treatment with N-HCl, but found that in practice competing protein was not a problem. Further investigation of the stoichiometry of methyl green staining revealed that the dye-P ratio of 1:10 was given by heptamethylpararosanilin (C.I. 42590) and that the ratio 1:13, quoted in an addendum to the paper, was produced by the closely related dye hexamethylethylpararosanilin (C.I. 42585).

The Gallocyanin-Chromalum Method. The use of the chromalum lake of gallocyanin as a basic dye was proposed, in a long series of papers, by Einarson (1935, 1936, 1949). Later (1951) the same author proposed a quantitative method using this dye. Gallocyanin is an oxazine dye which acts in aqueous solution as a weakly acid stain.

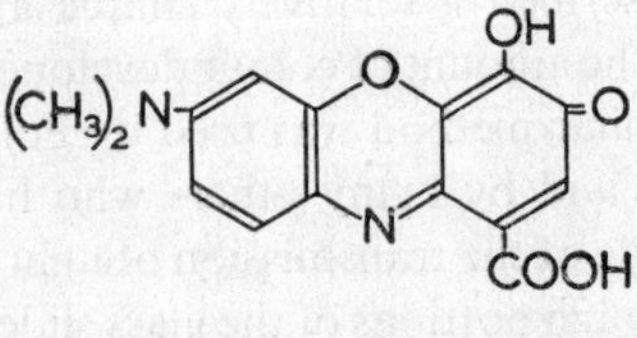

OH
$(CH_3)_2N$— O — =O
N
COOH

Einarson supposed that when mixed with chromalum the dye formed three salts. These he referred to as lake-cation [gallocyanin-$Cr(H_2O)_4$], lake hydroxide [gallocyanin-$Cr(H_2O)_4OH$] and lake sulphate [gallocyanin-$Cr(H_2O)_4]_2SO_4$. In his view the red-coloured lake cation combined with the phosphate groups of the nucleic acids to form a dark blue lake-tissue salt. Harms (1965) considered Einarson's formulæ for lake-cation and lake-sulphate improbable. He also indicated that metal-complex formation involving the carboxyl and *para*-hydroxyl groups was unlikely since it takes place with related dyestuffs having $CONH_2$ in place of COOH. Complexing by the two adjacent hydroxyl groups was regarded as the most likely explanation for the formation of the chromium lake of GC. Sandritter *et al.* (1963) investigated the influence of heating time on the production of the red component, finding an increase up to beyond 30 minutes. The red component, regarded by Harms (1965) as the decarboxylated dye, galloviolet, was considered to play little part in the staining of nucleic acids. It has a strong affinity for acid mucopolysaccharides, producing pseudo-metachromatic (red) staining.

Einarson (1951) stated that the reaction between gallocyanin-chromalum-lake-cation and the nucleic acids was sufficiently specific for use with photometric or densitometric estimations of the degree of basophilia. "From the light transmitted the relative values of the quantity of nucleic acid may be deduced." Sandritter *et al.* (1954) reported that *in vitro* 1 molecule of gallocyanin was bound by 15 phosphorus atoms in "polymerized" RNA and by 23 atoms in heated RNA. The stoichiometry of DNA was not reported by these authors, who did not give any information about the application of their findings. Diefenbach and Sandritter (1954), however, stated that GC combined stoichiometrically with DNA, the ratio GC to nucleic acid being 1:4·7. Later Sandritter *et al.* (1963) reported that, in spite of its low specificity for DNA, GC staining after ribonuclease could certainly be employed for cytophotometric measurements. The theoretical nucleic acid phosphorus: GC ratio (1:1) was not achieved; the ratio observed was, in fact, 1:0·6 or roughly 2 to 1.

For the simple staining of nucleic acids gallocyanin-chromalum (GC) possesses two features of considerable value. It is a progressive stain, and it withstands alcoholic dehydration and clearing in xylene. Staining can be carried out at any pH between 0·8 and 4·3, but it is particularly at the lower levels, between pH 1·5 and 1·75, that specific staining of nucleic acids is at its highest and non-specific staining is negligible. The latter is considered to be due to attachment of the lake-sulphate to acid groups in the tissues through its dimethylamine grouping, $(CH_3)_2N$—. The specificity of the GC method for nucleic acids was investigated by Stenram (1953, 1954) who found that he could still obtain staining of Nissl substance in nerve cells after ribonuclease extraction although sections stained with toluidine blue or methylene blue showed no staining at all. This result occurred using GC at pH 1·0 or at any pH up to pH 4·0. Stenram suggested that the non-specific staining might be due to a separate (removable) component, and subsequent work by de Boer and Sarnaker (1956) appeared to confirm this. These authors extracted from aqueous solutions of gallocyanin a component which gave bright blue staining of Nissl bodies with no staining of other cytoplasmic components. Criticisms of the GC method by Stenram, and also by Terner and Clark (1960) were answered in part by Pakkenberg (1962) who emphasized, once again, the absolute necessity to use pure reagents, and alcoholic fixation.

A report by Berube *et al.* (1966), however, clearly showed that when the three fractions of de Boer and Sarnaker (1956) are prepared from pure samples of the dye all three give identical absorption spectra. There is thus no need for purification procedures before the chelation stage with chrome-alum. Berube *et al.* indicated, however, that the principal component derivable from this stage is a single chelate compound. They recommended its purification and use as a 3 per cent. solution in N-H_2SO_4.

Mayersbach (1956) considered the GC method to be more reliable than methyl green-pyronin and he preferred it on that account, for the demonstration of RNA. I do not consider that it can in any way take the place of the methyl green-pyronin method but recommend it as a method of sufficiently high specificity, involving no differentiation, which is relatively insensitive to differences in fixation. Details of the GC method as a progressive stain and instructions for its preparation are given in Appendix 9, p. 653.

Extraction Techniques for Nucleic Acids

A number of extraction techniques for nucleic acids, derived from chemical procedures in the first instance, have been applied to histochemistry. Most of the methods extract both acids, but at different rates, and they vary in the degree of specific removal of one or both nucleic acids which they are able to achieve.

Sodium Chloride Solutions

It has long been known that certain concentrations of sodium chloride remove the basophilic material from the cytoplasm while leaving the nuclei intact. Levene (1901) used sodium or ammonium chloride solutions to extract nucleic acids from cells and his lead has been followed, after a long interval, by numerous workers in this field. Mirsky and Pollister (1942, 1943, 1946), for instance, found that ribonucleoprotein and deoxyribonucleoprotein could be extracted from liver by 0·15 M (0·85 per cent.) and 1·0 M-NaCl, respectively. The removal of these substances was thought to be due to the depolymerizing effect of salt solutions on nucleohistones. Opie and Lavin (1946) found that 1·0 M and 0·5 M-solutions of sodium chloride did not affect cytoplasmic basophilia but that some diminution occurred in 0·33 M and 0·2 M concentrations; 0·17 M-NaCl (0·95 per cent.) achieved almost complete removal of RNA from sections fixed in acetic-alcohol-formalin, or in formalin alone, the sections being immersed for 5 hours at 37° or for 2 hours at 56°. White (1950), using human liver smears wet-fixed in Susa, demonstrated these actions cytologically by staining the smears, previously extracted with saline, in aniline blue-orange G mixtures at pH 2–3. He found that with the lower concentration of NaCl the blue-staining cytoplasmic ribonucleoproteins were replaced by protein with an affinity for orange G. After 1·0 M-NaCl solutions a similar effect was noticeable in the nuclei.

PERCHLORIC ACID

After initial work by Ogur and Rosen (1949), who used perchloric acid for the extraction of both nucleic acids from onion root tip material, Erickson *et al.* (1949) evolved a method for the extraction of RNA from alcohol-fixed sections. They employed 10 per cent. aqueous perchloric acid at 4° for periods of between 4 and 18 hours, and found that the longest period invariably removed all cytoplasmic basophilia, using 1 per cent. aqueous toluidine blue as an indicator. These authors also used 10 per cent. hot perchloric acid (70°) for up to 20 minutes to extract both DNA and RNA from sections. The method was employed by Sulkin and Kuntz (1950) on Zenker-fixed mammalian tissue sections, followed by Mallory's phloxin-methylene blue stain. Erickson and his co-workers, and Sulkin and Kuntz, both considered that perchloric acid could be substituted for ribonuclease in the histochemical detection of ribonucleic acid. The former authors stated that they obtained "comparable," and the latter "identical," results using ribonuclease on the one hand and cold perchloric acid on the other. Seshachar and Flick (1949), however, employed perchloric acid in varying strengths to extract RNA from protozoan cells fixed in 1:3 acetic acid-alcohol, and concluded that only under rather rigid conditions could perchloric acid imitate ribonuclease. They found that 2 per cent. perchloric acid, in the cold, had no effect on the intensity of the Feulgen reaction, but that it appeared to depolymerize DNA progressively as judged by the progressive decrease in staining with the methyl green component of Pappenheim's stain. Parallel with the decrease in methyl green staining a rise in affinity for pyronin occurred in the structures concerned. As already noted, Kurnick (1947) believed that this affinity for pyronin indicated the presence of nucleic acid in a lower degree of polymerization. He based his assumption on the fact that pyronin stains only RNA and depolymerized DNA *in vitro.* Seshachar and Flick found that 10 per cent. perchloric acid, in the cold, removed RNA very rapidly; moreover, it progressively diminished the intensity of the Feulgen reaction and the affinity of chromatin for methyl green. The accumulated results, obtained by these authors, appear in Table 17, below.

TABLE 17

Per cent. $HClO_4$	Temp. °C	Time (hours)	Result	
			Cytoplasm	Nucleus
2	5	3, 10, 18, 24, 72	Progressive diminution of basophilia from 10 hours	Feulgen unchanged. Methyl green diminished from 18 hours
10	5	3, 10, 18, 24, 72	No basophilia	Feulgen diminished. Methyl green nil at 72 hours
5–10	35	18	No basophilia	Feulgen and methyl green very faint
5–10	70	20	No basophilia	Feulgen and methyl green nil

My experience (1960), on human and animal tissues fixed in various fixatives, agreed with that of Seshachar and Flick. I did not consider that parallel results could be obtained with ribonuclease and perchloric acid. Ribonuclease, freed from proteolytic activity, removed only RNA from sections. Cold perchloric acid extraction certainly removed all RNA, but at the same time it produced a result which I described (1960) as depolymerization of DNA. This effect (alteration of nuclear methyl green-pyronin staining from green to red) indicates some structural alteration which in modern terms is expressed as destruction of the double-stranded structure of DNA. Protein, polysaccharide and lipoprotein were also observed to suffer progressive removal.

Re-examining the validity of cold perchloric acid extraction, for the specific removal of RNA from sections prepared for both light and electron microscopy, Aldridge and Watson (1963) found that this procedure extracted only minimal amounts of protein from acetic-ethanol fixed tissues and none after acrolein fixation. It extracted (depolymerized and depurinized) RNA but did not depolymerize DNA. The latter was slowly depurinized, however, and Aldridge and Watson suggested that this was accompanied by collapse of the secondary structure of the molecule and hence by rapidly progressive failure to bind methyl green.

It may be significant that in freeze-dried, formaldehyde-vapour treated, tissues cold perchloric acid removes RNA (pyroninophilic) very rapidly but is prevented from reversing the affinity of nuclear chromatin for methyl green for much longer than in comparably treated routine preparations.

Results obtained with cold perchloric acid by Kasten (1965b) indicated that if washing in water followed the acid treatment (as it must in the case of cytochemical preparations) then considerable loss and displacement of proteins and glycogen took place. Lipids were not affected.

Investigations on the effects of fixation were made by Goessner (1954) who found, as did Franz *et al.* (1954) and Koenig and Stahlecker (1951, 1952), that this process had a considerable effect on the amount of stainable RNA after various times of extraction.

It is clear, from Aldridge and Watson's results, that fixation in some way preserves the secondary structure of DNA and that the more effective fixation can be made the longer will the affinity of the nuclei for methyl green withstand varius mineral acid extractions. It is clear also that cold perchloric acid is an acceptable alternative to ribonuclease provided that the final method applied to the material is insensitive to the changes which it induces in RNA and DNA (or that the observer is indifferent to these alterations). I agree with Kasten (1965b) that ribonuclease is the reagent of choice for the selective removal of RNA in critical cytochemical studies.

Bile Salts

Henry and Stacey (1943), in their experiments on the nature of the Gram-positive complex in micro-organisms, employed solutions of bile salts to remove what was subsequently demonstrated to be magnesium ribonucleate from the Gram-positive bacterial cytoskeletons. They used 2 per cent. aqueous sodium cholate at 60° in the presence of oxygen. Two to three hours' incubation was sufficient to remove the RNA from the majority of cells. No explanation for this effect of bile salts appears to have been offered, but it has been stated that it is not due to lowering of surface tension. Foster and Wilson (1951, 1952) studied the effect of bile-salt extraction on the Gram-positive staining of the pituitary β granules, which they considered might be due to the presence of RNA. With this process they were able to destroy both Gram-positiveness and basophilia in the β granules, although they were able to reverse only their basophilia by means of hydrolysis with ribonuclease. The bile-salt extraction technique might become a more useful one in histochemistry when something more is known of the types of materials removed and of the optimum duration of incubation. Little further work on this subject has been reported, however, with the exception of a chemical study by Kay and Dounce (1953).

Trichloroacetic Acid

Cohen (1944) used trichloroacetic acid for the extraction of DNA and Schneider (1945), with whose name the process is usually associated, employed a 5 per cent. aqueous solution at 90° for 15 minutes to extract nucleic acids from the tissues. Both deoxyribonucleic and ribonucleic acids are removed by this procedure, which was applied by Kaufmann *et al.* (1950) to onion root tips and by White (1950) to bone marrow smears. Both these authors showed that Schneider extraction leaves behind a protein component of the nucleus which can be demonstrated by its increased affinity for acid dyes. Kaufmann *et al.* used fast green and White a mixture of aniline blue and orange G at pH 2–3. The latter showed that the globin of the erythrocytes, and to a lesser extent the intranuclear histones, retained orange G, while the plasma proteins and cytoplasmic proteins, including those coupled to RNA, stained more slowly with aniline blue. Figs. 91 and 92 illustrate the effect of Schneider extraction on plasma cell nuclei, Helly-fixed, stained with 1 per cent. aqueous light green SF.

Other Methods

Constantin Saint-Hilaire (1898), in his researches on nucleohistones, used water to extract these substances from unfixed tissues. Mann (1902) showed that water would not remove nucleic acids from fixed tissues but that if acetic acid-fixed material was treated for a short while with dilute hydrochloric acid nucleohistones were decomposed, leaving the histone *in situ.* More prolonged extraction removed the histones also. Hydrolysis in N-HCl at 60° for 10 minutes can usefully be employed to remove purines from smears fixed in acetic-alcohol or from sections fixed in alcoholic fixatives. Slightly longer incubation may be necessary in the case of formalin and longer still if sublimate or chromates have been used. More recently, Sevag *et al.* (1940) and Davidson and Waymouth (1944) have used dilute HCl for nucleoprotein extraction, and normal HCl at 37° for 3 hours has been employed by Dempsey *et al.* (1950) as a means of reversing the tissue basophilia due to nucleoproteins of both types. This last procedure leaves the tissues in good condition for subsequent staining, unlike the more violent trichloroacetic acid method, but they do not stand up so well to further procedures involving heat or to analysis by treatment with enzymes. Caspersson *et al.* (1935) used malonic acid for the extraction of nucleic acids and formic acid has also been employed for this purpose. Neither of these techniques has any advantage for use in histochemistry.

BUFFER SOLUTIONS

Brachet (1940a) noted that incubation in water at 70° progressively diminished the cytoplasmic basophilia due to RNA and subsequently numerous observers (e.g. White, 1947; Stowell and Zorzoli, 1947) described the extraction of RNA by various buffers, especially on the alkaline side of neutrality. To avoid these effects it is now customary to use depolymerizing and other enzymes in glass distilled water in order to escape the action of electrolytes, and at 37° to avoid the extractive effect of water at higher temperatures.

THE HISTOCHEMISTRY OF THE GRAM STAIN

Introduction

In 1884 Christian Gram described his aniline-gentian violet and iodine method of staining which has since been subjected to a large number of variations (the earlier ones listed by Hucker and Conn, 1923). It is in bacteriology, of course, that the method finds its most important application in the division of micro-organisms into the two classes of Gram-positive and Gram-negative. As a histochemical reaction, difficulties in standardizing the technique and in interpretation of the results detract largely from its usefulness. Nevertheless, since great advances have recently been made in determining the mechanism of the Gram stain with reference to bacterial cells, it is pertinent to enquire how much these mechanisms apply to the staining of tissue components by the same method.

In histological practice the term "Gram stain" means the use of aqueous solutions of one of the methyl violet dyes, followed by the application of iodine in KI and differentiation by means of acetone, alcohol, aniline or aniline-xylene mixtures. The latter, which is employed in the eosin-Gram Weigert modification, is preferred by histologists on account of the slower differentiation which it allows. A variety of structures in tissue sections can be shown to be Gram-positive by the Weigert modification (Appendix 9, p. 654), some inconstantly, some constantly. Amongst these are fibrin, nuclear chromatin, hyalin droplets in the renal tubules, the heads of spermatozoa, mucin granules, collagen fibres and the β granules of the anterior pituitary gland. Between these structures and the Gram-positive bacteria there exists a quantitative difference at least, in that the latter have usually far greater resistance to decolorization by acetone or alcohol.

Theoretical Aspects of the Gram Stain

Older Theories

Bacteriologists have frequently endeavoured to find a satisfactory explanation for the mechanism of the Gram stain and many theories have been put forward, some chemical, some physical and some a combination of both. Schumacher (1926) believed that a crystal-violet-iodine-cell complex was formed in Gram-positive but not in Gram-negative cells, while Stearn and Stearn (1924) considered that the lower isoelectric point of the cytoplasm of the former cells, and their consequent stronger basophilia, was the main factor concerned. Iodine was relegated to the role of an oxidizing agent increasing the basophilia of the cytoplasmic proteins and thus their affinity for the basic dye methyl violet.

The factor of cell membrane permeability as responsible for the reaction has also been strongly supported. Burke (1922) and Burke and Barnes (1928, 1929) stated that the dye-iodine complex which was precipitated in both types of cell would pass readily only through the membrane of the Gram negatives; in this way they substantially agreed with the views of Benians (1920). Kaplan and Kaplan (1933) also believed that the dye-iodine precipitate was formed in both Gram-positive and negative cells and that on subsequent treatment with alcohol it was dissociated and dissolved out of the latter. Other workers have regarded the Gram-positive property as due to the possession of a Gram positive layer of material coating an otherwise Gram-negative cytoplasm (Gutstein, 1925; Churchman, 1927, 1929).

The Hypothesis of Henry and Stacey

Strong evidence in favour of the views of Gutstein and Churchman was produced by Henry and Stacey (1943), who found that they could render Gram-positive organisms Gram-negative by a variety of methods. Chief of these was the extraction from the washed cells of a material, shown to be magnesium ribonucleate, by treatment with bile salts. There was no evidence, however, to suggest that solely an outer layer was affected. The Gram-negative cytoskeleton which remained could be restored to its original condition by recombination with ribonucleate, provided it was maintained in a reduced condition with formaldehyde or some other mild reducing agent. On the basis of this work Henry and Stacey suggested that the dye-retaining material in Gram-positive organisms was a high molecular complex formed by the combination of a reduced basic protein with magnesium ribonucleate. Bartholomew and Umbreit (1944) removed what they regarded as the outer layer of Gram-positive organisms with ribonuclease, and supported the former authors' conception. They considered more fully the nature of the bond between the ribonucleate and the material of the cytoskeleton which acts only in its reduced form and suggested that sulphydryl groups might be responsible. Three points supported this reasoning; first, the autoxidizable nature of SH groups, which would only react in reduced form, secondly the effect of formalin which is known to prevent oxidation of sulphydryls, and thirdly the essential action of iodine, which is also known to react with sulphydryls, in the performance of the Gram stain. Panijel (1950, 1951), after a long series of experiments concerning the role of ribonucleoprotein in the Gram reaction, concluded that the Gram-positive structure produced by the magnesium ribonucleate-protein complex could be reproduced by a certain molecular disposition of the protein in the absence of RNA.

In later experimental work Henry, Stacey and Teece (1945) obtained from *Clostridium welchii*, and from yeast cells, nucleoproteins of an unusual type which when fixed on a slide were intensely Gram-positive. Material with these characteristics could not be obtained from Gram-negative organisms. The nucleic acid which they extracted was found to be entirely of the ribose variety while the protein, free from pentose and phosphorus, contained 5·5 per cent. of arginine and SS linkages in addition. Neither the nucleic acid, nor the protein alone, were Gram-positive, but on reforming the complex in the presence of magnesium ions and formaldehyde at pH 5·0, the Gram-positive property was restored. When typical nucleates were made by adding nucleic acids to various basic proteins some of these, particularly histonedeoxyribonucleate, were found to retain methyl violet in some measure, but none so strongly as the natural substance from Gram-positive organisms. The authors considered that in the natural nucleoprotein the linkage of protein to nucleic acid was not of the simple electrovalent type, and that the mechanism of dye retention was more fundamental than its mere combination with a basophilic salt-like linkage. This work suggested that the cell membranes had nothing to do with the presence or absence of Gram-positive staining in bacteria, but that the latter depended on the unusual chemical or physical nature of the nucleoprotein present in organisms which reacted positively. Although Henry and Stacey (1943) had been able to "replate" both intact and crushed bacterial cells with magnesium ribonucleate, Mittwer *et al.* (1950) could not repeat this work on crushed cells and they therefore considered that the intact cell membrane was still an important factor.

Larose and Fischer (1953) examined the phenomenon of reversal of Gram staining in bacteria. They considered that the mechanism was similar to that previously found in wool fibres (Fischer and Larose, 1952a) which was thought to be due to oxidation of SH groups to SS groups. Larose and Fischer explained Bartholomew and Mittwer's finding that UV irradiation would restore Gram positive staining as due to reduction of SS to SH and SOH. Libenson and McIlroy (1955), however, disagreed entirely with this hypothesis and stated that SH groups did not take part in the Gram reaction of bacteria. Mitchell and Moyle (1950) showed that while gelatin was Gram-negative formaldehyde-treated gelatin was strongly Gram-positive. Later (1954) they produced an entirely different hypothesis after finding that RNA and DNA levels in the two groups of bacteria (Gram-positive and Gram-negative) did not differ significantly while Gram-positive organisms contained larger (x2) amounts of lipid phosphate and organic phosphate. They suggested that the Gram reaction was due to the presence of a lipid polyglycerophosphate-protein in the cell envelope. Shugar and Baranowska (1957) were able to make Gram-positive the "peripheries" of different organisms by treatment with protein. If the organisms were then ruptured, these "peripheries" retained their staining properties. These authors concluded that the extraction of a substance from bacteria with loss of Gram-positive staining could not be accepted as proof of causation. With this view, at least, one is able to agree whole heartedly.

The Role of Dyes in the Gram Reaction

Before we consider the application of the knowledge gained from this experimental work on bacteria to the histochemical use of the Gram reaction, there remains to be discussed some recent evidence concerning the role of the dyestuff used and the mechanism of its combination with iodine. Bartholomew and Mittwer (1950) tested a large number of dyes for suitability in the Gram stain. They found that any dye which gave a precipitate with iodine would give differentiation between Gram-positive and Gram-negative organisms, and of the eight dyes which did so in a satisfactory manner, seven are chlorides or acid sulphates of basic triphenylmethanes with the general formula:

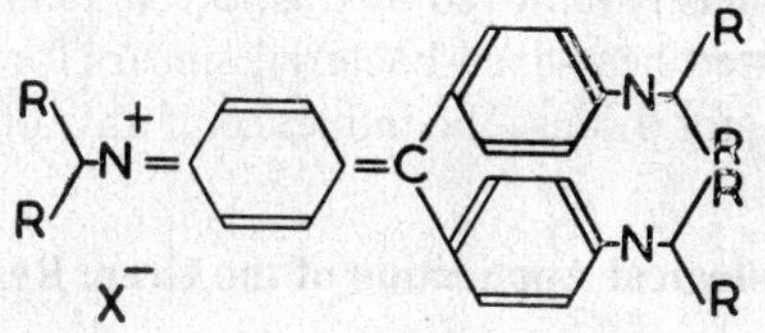

Here R may be H, NH_2, CH_3 or C_2H_5. The seven dyes are brilliant green and malachite green (diamino-triphenyl methanes), crystal violet, methyl violet B, ethyl violet, Hoffmann's violet, and basic fuchsin. The eighth dye, Victoria blue R, is a diphenylnaphthylmethane. It seems possible. therefore, that at one end of the dye molecule the basic groups react with the acid groups of the nucleoprotein, leaving the basic groups at the other end to react with iodine to form the protein-dye-iodine complex long ago conceived as being formed in the Gram reaction. There are other possibilities, however, and these will be referred to below. When treatment with the dye and later with aqueous iodine is followed by differentiation with alcohol or some other solvent, the protein-dye-iodine complex is dissociated and dye and iodine are extracted from all but Gram-positive organisms. In a later paper (the third in their series) Bartholomew and Mittwer (1951) reported on the ability of twenty-nine dye samples to replace crystal violet in the Gram procedure. None proved to be efficient substitutes. Nevertheless, it is important to note that Bartholomew *et al.* (1949) demonstrated that after the methyl esterification of acid carboxyl groups in bacterial cells, a Gram reaction could still take place if a suitable acid dye, such as acid fuchsin, was employed. Iodine forms a precipitate with acid fuchsin and is evidently able to do so after combination of the latter with methylated protein groups produced in the bacterial cell in the above procedure.

Iodine Substitutes

Investigating various substitutes for iodine in the Gram reaction, Mittwer *et al.* (1950) showed that bromine, picric acid and mercuric iodide, all of which give dense precipitates with crystal violet, produced comparable results to those obtained by using iodine. A large number of reagents which gave equally good precipitates *in vitro*, however, would not allow performance of the Gram reaction. Mittwer *et al.*, therefore, considered that precipitate formation alone was not sufficient and that the critical factor was the relative insolubility of the dye-precipitates in alcohol or water, and presumably in other solvents. In addition to the factors of precipitate formation and the relative insolubility of these precipitates, a third factor was found. This was the permeability of iodine through the membrane of the bacterial cell. In aqueous solution iodine was found to permeate both Gram-positive and Gram-negative cells equally freely. In alcoholic solution, however, penetration into Gram-positive cells was very slow. Conversely, once deposited in the cells from aqueous solutions, iodine was relatively resistant to extraction from the Gram-positives by 95 per cent. alcohol, but easily removed by this solvent from the Gram-negatives. The above authors used heat-fixed bacterial smears for their experiments on permeability but did not discuss the influence, if any, of this factor.

Histological Application of the Gram Reaction

Transferring these findings to the Gram reaction as applied to tissue sections, it is at once obvious that in most of the structures listed earlier in the chapter as Gram-positive, ribonucleoprotein is unlikely to be present and the presence of a cell-membrane, in the usual sense of the word, can be excluded. We are left only with the formation of the dye-iodine precipitate, and its low solubility in various solvents, as essential factors for the histological performance of the Gram reaction. Obviously some other factor or factors must be present and recent experimental work, though it does not solve the problem, at least points the way to a better understanding. Foster and Wilson (1951) have shown that whereas the normal Gram reaction, considered by McLetchie (1944) to be specific for the purpose, stains only the β granules of the anterior hypophysis, an alcoholic solution of the crystal violet-iodine precipitate stains only the acidophil α granules. Thus conversion of the basic triphenylmethane dye into an acid dye by means of iodine results in complete alteration of its properties. Furthermore, I have observed that if Gomori's (1950) paraldehyde-basic fuchsin stain for elastic tissue (Appendix 8, p. 639) is followed by Gram's iodine and then by alcoholic differentiation, acidophil structures such as red cells become "Gram-positive." Since fuchsin can take the place of methyl violet in the true Gram reaction, this manœuvre is essentially a Gram reaction performed with a fuchsin whose chemical properties have been altered by combination with aldehyde. Although Gomori did not explain the mechanism of his stain, it appears that the addition of aldehyde causes the basic dye basic fuchsin to behave to some extent as an acid dye with affinity especially for the NH_2 groups of the tissues. Paraldehyde is acetaldehyde trimer and in acid solution, as in Gomori's mixture, it breaks down slowly to yield acetaldehyde. This presumably forms a compound with the basic groups of fuchsin. Iodine is able to form a precipitate with the aldehyde-fuchsin complex *in vitro*, and it can then be shown that all the free aldehyde groups are present in the supernatant, the precipitate being composed of fuchsin and iodine only. When the reaction between aldehyde-fuchsin and iodine is performed on stained sections, subsequent differentiation with alcohol gives the apparent Gram effect in structures which previously stained with the aldehyde fuchsin. Since Bartholomew and Mittwer (1950) found that acid dyes, with the exception of acid fuchsin and aniline blue, gave no hint of Gram differentiation when used as substitutes for crystal violet, this paradoxical effect of aldehydefuchsin requires some explanation.

The positive Gram reaction of tissue components and cells has continued to absorb the interest of a number of groups of investigators. Most of the work reported has concerned keratin and its derivatives but Monné (1955a and b) and Monné and Borg (1954) have examined the nature of the positive reaction given by the egg envelopes of a number of parasitic nematodes. The views of Fischer and Larose (1952b) on the role of SH groups in the Gram reaction of keratin have already been mentioned. Some authors (Harms, 1958) appeared to agree with them while others (Johnson *et al.*, 1957) could find no evidence of the participation of SH. The abolition of the positive reaction by deamination, also observed by Panijel (1950) in the case of *Ascaris* sperm, certainly suggests that reactive protein groups other than SH are concerned but no other evidence, as far as I know, implicates the NH_2 groups in this way. Theoretically one would expect that the basic NH_2 groups would compete with the basic CH_3 groups of the dye for acid binding sites. The latter would then be released by deamination. Probably the nitroso or hydroxyl groups produced by deamination interfere in some way. Gianni (1952) and Gianni and Della Torre (1952) ascribe the positive Gram reaction to the presence of adjacent COOH groups on the same molecule (aspartic and glutamic acids) and their view agrees with that of Panijel (1951). The latter, maintaining that acid groups are essential, showed that those of the dicarboxylic amino-acids were sufficient. Monné, moreover, determined that the Gram-positive reaction of the fibrous protein of his helminth ova was due to COOH groups although he also implicates acid polysaccharides and phenolic substances.

To my mind, the only explanation which will fit the facts is that iodine is able to displace the triphenylmethane dyes from combination with acid groups whatever their nature, reacting with some, or all, of the basic side chains to give the usual iodine-triphenylmethane precipitate. I therefore consider that in the histochemical performance of the Gram reaction the initial acidic protein-crystal violet complex is broken by iodine, which combines with the basic groups at both ends of the triphenylmethane molecule to form a relatively alcohol-insoluble crystal violet-iodine precipitate *in situ.* With subsequent differentiation this precipitate is less easily removed from certain structures than from others, two main factors being concerned. First, more of the crystal violet-iodine complex is formed where there was initially more crystal violet in combination with protein and secondly, certain physical barriers offer resistance to the extraction of crystal violet-iodine by alcoholic or other differentiation. Among the physical barriers may be mentioned the presence of a lipid or lipoprotein membrane, such as is probably present in Russell bodies, and the arrangement of protein molecules, having the necessary reactive acid groups, in the form of a close network. Such a state is often referred to as polymerized. In this context the similarity of results obtained by the coupled tetrazonium reaction after benzoylation (Chapters 6 and 8), and the Gram reaction using aniline-xylene differentiation, is particularly striking. Figs. 93 and 94, opposite, illustrate this point. The first shows intracellular mucin in the human stomach mucosa stained by the tetrazonium reaction after 16 hours benzoylation; the second shows the Gram reaction on similar material. Gastric and intestinal mucin granules, salivary gland mucoid granules, pituitary β granules, Russell bodies and collagen, to mention a representative few, are all stainable by either method, but the two methods do not give exactly parallel results, for two reasons. First, the coupled tetrazonium reaction does not depend on the presence of a sufficient number of reactive basophil (acid) groups, as does the Gram reaction. For this reason a variety of Gram-negative, acidophilic, structures are able to give a positive tetrazonium reaction when the physical state of their aromatic amino-acids (most probably histidine) affords protection from benzoylation. Secondly, from some structures such as reticulin, which are also positive by the tetrazonium reaction after benzoylation, the crystal violet-iodine complex is too easily removed for them to behave as Gram-positive. Mild heat, which has so interesting and profound an effect on the intensity with which various structures react by the tetrazonium method after benzoylation has decidedly less effect on the degree of Gram-positiveness manifested by tissue structures. In some cases, however, the effect is still quite noticeable.

In addition to the initial and insuperable objection to the use of the Gram stain as a histochemical reaction, because of the impossibility of controlling differentiation adequately, further conditions must now be imposed. If the

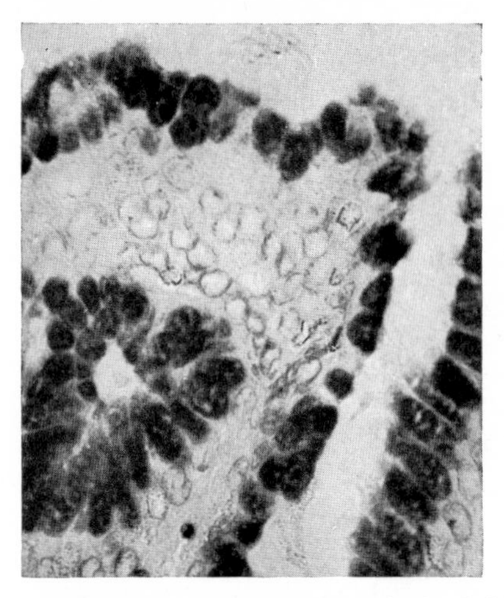

FIG. 93. Human stomach mucosa (5·5 μ paraffin section). Subjected to 16 hours' benzoylation at 22°. Intracellular mucin granules are strongly stained. Coupled tetrazonium reaction. × 400.

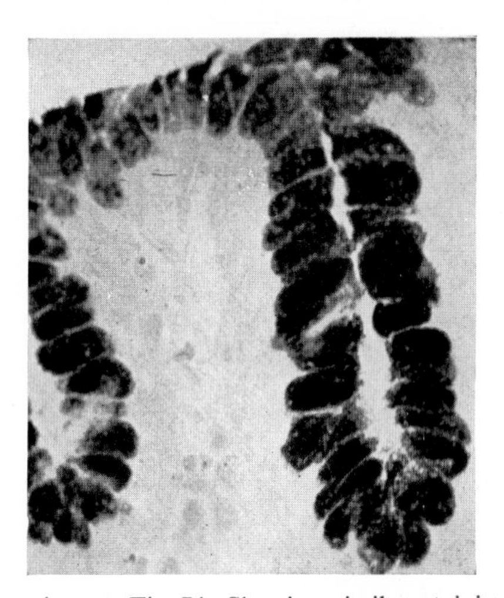

FIG. 94. Similar section to Fig. 74. Showing similar staining of intracellular mucin. Eosin-Gram-Weigert method. × 400.

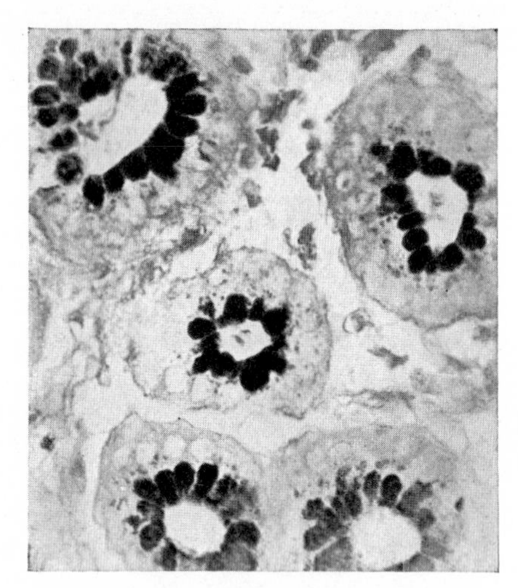

FIG. 96. Human gastric mucosa, oxidized with 1 per cent. aqueous HIO_4 for 10 minutes at 22°, showing granules of mucin strongly stained. Phenylhydrazone-formazan method. × 400.

Gram reaction alone is used, and a particular tissue component is found to be Gram-positive, the only conclusions which can legitimately be drawn are: (1) that reactive acid groups are present, and (2) that either the constituent molecules are in some way condensed or polymerized, or else a physical barrier such as a lipid or lipoprotein membrane is present. Although this information may be useful by itself, further confirmation must be obtained by more histochemically-reliable methods if the Gram reaction is to achieve its full degree of usefulness.

Demonstration of Nucleohistones

For many years the standard methods for staining the basic proteins of nucleus and cytoplasm have been the alkaline fast green method (Alfert and Geschwind, 1953) and the naphthol yellow S method (Deitch, 1955). As an alternative for high arginine containing histones the Sakaguchi reaction (Chapter 6, p. 158) has been used. Recently several alternative staining techniques for histones have been offered. Most of these have involved simple substitution of other dyes in place of fast green, including amido black 10 B (Geyer, 1960), alkaline eosin or bromophenol blue (Bloch and Hew, 1960), and Biebrich scarlet (Spicer, 1962 and see Chapter 6, p. 114). Black and Ansley (1964) described an ammoniacal-silver method and Jobst and Sandritter (1964), noting that after treatment with metaphosphoric acid the basic proteins of the tissues could be stained with basic dyes, evolved a metaphosphoric acid-gallocyanin sequence for the purpose.

Of all the above-mentioned methods only the alkaline fast green (or similar) methods can stand critical examination. It is essential to use these methods after formalin fixation (Cowden, 1966) or, preferably, after formaldehyde vapour fixation of freeze-dried tissues. Nucleic acids must first be removed by one or other of the special methods available. An excellent account is given by Davenport and Davenport (1965).

According to Phillips (1962) histones are soluble in 98 per cent. formic acid and Luck *et al.* (1958) indicated that the dinitrophenyl derivatives of histones are partially soluble in dimethyl formamide. Unfortunately these qualities are altered by effective fixation so that they are of little cytochemical significance.

Extra N-terminal groups (alanine, lysine, glycine) appear when histone solutions are kept at pH 7 to 8. This is attributed to the activities of a proteinase with the properties of chymotrypsin (inhibited by DFP), according to Phillips and Johns (1959). Alterations in stored or post-mortem tissues may be due to this effect.

REFERENCES

ALDRIDGE, W. G., and WATSON, M. L. (1963). *J. Histochem. Cytochem.*, **11**, 773.
ALEXANDER, J., MCCARTY, K. S., and ALEXANDER-JACKSON, E. (1950). *Science*, **111**, 13.
ALFERT, M. (1952). *Biol. Bull.*, **103**, 145.
ALFERT, M., and GESCHWIND, I. I. (1953). *Proc. Nat. Acad. Sci. Wash.*, **39**, 991.
ALLEN, F. W. (1954). *Ann. Rev. Biol.*, **23**, 99.

ARZAC, J. P. (1950). *Stain Tech.*, **25**, 187.
ATKINSON, W. B. (1952). *Ibid.*, **27**, 153.
BACKLER, B. S., and ALEXANDER, W. F. (1952). *Ibid.*, **27**, 147.
BAKER, J. R., and SANDERS, F. K. (1946). *Nature, Lond.*, **158**, 129.
BARGER, J. D., and DELAMATER, E. D. (1948). *Science*, **108**, 121.
BARKA, T. (1956). *J. Histochem.*, **4**, 208.
BARKA, T., TORO, I., and POSALAKI, Z. (1953). *Acta morph. Acad. Sci. Hung.*, **3**, 437.
BARNARD, E. A. (1961). *Gen. cytochem. Methods*, **2**, 203.
BARNARD, E. A., and BELL, L. G. E. (1960). *Nature, Lond.*, **187**, 508.
BARNARD, E. A., and DANIELLI, J. F. (1956). *Nature, Lond.*, **178**, 1450.
BARTHOLOMEW, J. W., EVANS, E. E., and NIELSON, E. D. (1949). *J. Bact.*, **58**, 347.
BARTHOLOMEW, J. W., and MITTWER, T. (1950). *Stain Tech.*, **25**, 103.
BARTHOLOMEW, J. W., and MITTWER, T. (1951). *Ibid.*, **26**, 231.
BARTHOLOMEW, J. W., and UMBREIT, W. W. (1944). *J. Bact.*, **48**, 4567.
BAUER, H. (1932). *Z. Zellforsch.*, **15**, 225.
BENIANS, T. H. C. (1920). *J. Path. Bact.*, **23**, 401.
BERUBE, G. R., POWERS, M. M., KERKAY, J., and CLARK, G. (1966). *Stain Tech.*, **41**, 73.
BIESELE, J. J. (1944). *Cancer Res.*, **4**, 737.
BLACK, M., and ANSLEY, H. R. (1964). *Science*, **143**, 693.
BLOCH, D. P., and GODMAN, G. C. (1955). *J. Biophys. Biochem. Cytol.*, **1**, 17.
BLOCH, D. P., and HEW, H. C. Y. (1960). *J. Biophys. Biochem., Cytol.*, **8**, 69.
BOAS, F., and BIECHELE, O. (1932). *Biochem. Z.*, **254**, 467.
DE BOER, J., and SARNAKER, R. (1956). *Med. Proc. S.A.*, **2**, 218.
BÖHM, N., and SANDRITTER, W. (1966). *J. Cell. Biol.*, **28**, 1.
BRACHET, J. (1940a). *C.R. Soc. Biol., Paris*, **133**, 88.
BRACHET, J. (1940b). *Ibid.*, **133**, 90.
BRACHET, J. (1942). *Arch. Biol., Paris*, **53**, 207.
BRACHET, J. (1944). "Embryologie Chimique." Leige.
BRACHET, J. (1947). *Symp. Soc. exp. Biol.*, **1**, 207.
BRACHET, J. (1950). "Chemical Embryology," 1st Edn. New York.
BRACHET, J. (1958). "Chemical Embryology," 2nd Edn. New York.
BRACHET, J., and QUERTIER, J. (1963). *Exp. Cell Research*, **32**, 410.
BRADLEY, D. F., and WOLF, M. K. (1959). *Proc. nat. Acad. Sci. Wash.*, **45**, 944.
BRETSCHNEIDER, L. H. (1949). *Proc. kon. ned. Wet.*, **52**, 301.
BROWN, D. M., and TODD, A. R. (1955). In "The Nucleic Acids." Vol. I, p. 40.
DE BRUYN, P. P. H., FARR, R. S., BANKS, H., and MORTHLAND, F. W. (1953). *Exp. Cell Res.*, **4**, 174.
BURKE, V. (1922). *J. Bact.*, **7**, 159.
BURKE, V., and BARNES, M. W. (1928). *Ibid.*, **15**, 12.
BURKE, V., and BARNES, M. W. (1929). *Ibid.*, **18**, 69.
BURSTONE, M. S. (1955). *J. Histochem. Cytochem.*, **3**, 32.
BUTLER, J. A. V., DAVISON, P. F., and PHILLIPS, D. M. (1955). *Brit. Emp. Cancer Campaign, 33rd Annual Rpt.*, p. 41.
CALLAN, H. G. (1943). *Nature, Lond.*, **152**, 503.
CAMBER, B. (1949). *Ibid.*, **163**, 285.
CARR, J. G. (1945). *Ibid.*, **156**, 143.
CASPERSSON, T. (1936). *Arch. physiol. scand.*, Suppl., 8.
CASPERSSON, T. (1944). *Nature, Lond.*, **153**, 499.
CASPERSSON, T. (1947). *Symp. Soc. exp. Biol.*, **1**, 127.
CASPERSSON, T., HAMMARSTEN, E., and HAMMARSTEN, H. (1935). *Trans. Faraday Soc.*, **31**, 367.
CAVALIERI, L. F., and ANGELOS, A. (1950). *J. Amer. Chem. Soc.* **72**, 4686.
CHARGAFF, E. (1950). *Experientia*, **6**, 201.
CHARGAFF, E. (1951). *Fed. Proc.*, **10**, 654.
CHARGAFF, E. (1955a). In "Fibrous Proteins and their Biological Significance." p. 32. Cambridge Univ. Press.
CHARGAFF, E. (1955b). In "The Nucleic Acids." Vol. I, p. 307. Acad. Press Inc.
CHAYEN, J., and NORRIS, K. P. (1953). *Nature*, Lond., **171**, 472.

CHOUDHURI, H. C. (1943). *Nature, Lond.*, **152**, 475.
CHURCHMAN, J. W. (1927). *J. exp. Med.*, **46**, 1007.
CHURCHMAN, J. W. (1929). *J. Bact.*, **18**, 413.
COHEN, S. J. (1944). *J. biol. Chem.*, **156**, 691.
COLEMAN, L. D. (1938). *Stain Tech.*, **13**, 123.
COWDEN, R. R. (1965). *Histochemie*, **5**, 441.
COWDEN, R. R. (1966). *Histochemie*, **6**, (In press).
DANIELLI, J. F. (1947). *Symp. Soc. exp. Biol.*, **1**, 101.
DARLINGTON, C. D. (1942). *Nature, Lond.*, **149**, 66.
DARLINGTON, C. D. (1947). *Symp. Soc. exp. Biol.*, **1**, 252.
DAVENPORT, R., and DAVENPORT, J. C. (1965). *J. Cell Biol.*, **25**, 319.
DAVIDSON, J. N. (1965). "The Biochemistry of Nucleic Acids." Methuen, London, p. 13.
DAVIDSON, J. N., and WAYMOUTH, C. (1944). *Biochem. J.*, **38**, 39.
DAVIDSON, J. N., and WAYMOUTH, C. (1946). Quoted by Davidson, J. N. (1947). *Symp. Soc. exp. Biol.*, **1**, 77.
DEANE, H. W. (1946). *Amer. J. Anat.*, **78**, 227.
DEITCH, A. D. (1955). *Lab. Invest.*, **4**, 324.
DELAMATER, E. D. (1948). *Stain Tech.*, **23**, 161.
DELAMATER, E. D., MESCON, H., and BARGER, J. D. (1950). *J. Invest. Derm.*, **14**, 133.
DELAMATER, E. D., SCHAECHTER, M., and HUNTER, M. E. (1955). *J. Histochem. Cytochem.*, **3**, 16.
DEMPSEY, E. W., and SINGER, M. (1946). *Endocrinology*, **38**, 270.
DEMPSEY, E. W., SINGER, M., and WISLOCKI, G. B. (1950). *Stain Tech.*, **25**, 73.
DEMPSEY, E. W., and WISLOCKI, G. B. (1945). *Amer. J. Anat.*, **76**, 277.
DEMPSEY, E. W., WISLOCKI, G. B., and SINGER, M. (1946). *Anat. Rec.*, **96**, 221.
DESCLIN, L. (1940). *C. R. Soc. Biol., Paris*, **133**, 457.
DIEFENBACH, H., and SANDRITTER, W. (1954). *Acta Histochem.*, **1**, 5.
DISCHE, Z. (1930). *Mikrochemie*, **8**, 4.
DISCHE, Z. (1944). *Proc. Soc. exp. Biol. N.Y.*, **55**, 217.
DOBSON, E. O. (1946). *Stain Tech.*, **21**, 103.
DUSTIN, P. (1942). *Le Sang.*, **15**, 193.
EINARSON, L. (1935). *J. Comp. Neurol.*, **61**, 101.
EINARSON, L. (1936). *Acta psychiat., Kbh.*, **13**, 861.
EINARSON, L. (1949). *Acta orthopæd. scand.*, **19**, 27.
EINARSON, L. (1951). *Acta path. scand.*, **28**, 82.
ELSON, D., and CHARGAFF, E. (1954). *Nature, Lond.*, **173**, 1037.
ERICKSON, R. O., SAX, K. O., and OGUR, M. (1949). *Science*, **110**, 472.
FEDER, N., and WOLF, M. K. (1965). *J. Cell Biol.*, **27**, 327.
FEUGHELMAN, M., LANGRIDGE, R., SEEDS, W. E., STOKES, A. R., WILSON, H. R., HOOPER, C. W., WILKINS, M. H. F., BARCLAY, R. K., and HAMILTON, L. D. (1955). *Nature, Lond.*, **175**, 834.
FEULGEN, R. (1927). *Z. phys. Chem.*, **165**, 215.
FEULGEN, R., and ROSSENBECK, H. (1924). *Ibid.*, **135**, 203.
FEULGEN, R., and VOIT, K. (1924). *Pflüg. Arch. ges. Physiol.*, **206**, 389.
FISCHER, R., and LAROSE, P. (1952a). *J. Bact.*, **64**, 435.
FISCHER, R., and LAROSE, P. (1952b). *Canad. J. Med., Sci.*, **30**, 86.
FOSTER, C. L., and WILSON, R. R. (1951). *Nature, Lond.*, **167**, 528.
FOSTER, C. L., and WILSON, R. R. (1952). *Quart. J. micr. Sci.*, **93**, 147.
FRANZ, F., WARDEN, I., MAYER-ARENDT, J. (1954). *Naturwiss.*, **7**, 165.
FRANZEN, H., and EICHLER, T. (1908). *J. prakt. Chem.*, **78**, 157.
GABLER, W. (1965). *Acta Histochem.*, **21**, 387.
GEYER, G. (1960). *Acta Histochem.*, **10**, 286.
GHIARA, F. (1953). In: "Convegno di Genetica." Suppl. La Ricerca Scientifica, p. 137.
GIANNI, A. (1952). *Boll. Soc. ital. Sper.*, **31**, 148.
GIANNI, A., and DELLA TORRE, B. (1952). *Ibid.*, **31**, 312.
GODMAN, G. C., and DEITCH, A. D. (1957). *J. exp. Med.*, **106**, 575.
GOESSNER, W. (1954). *Z. wiss. Mikr.*, **61**, 377.
GOLDSTEIN, D. J. (1961). *Nature, Lond.*, **191**, 407.

GOMORI, G. (1950). *Stain Tech.*, **25**, 81.
GOMORI, G. (1946). *Amer. J. clin. Path.*, **10**, 177.
GRAM, C. (1884). *Fortschr. Med.*, **2**, 377.
GULLAND, J. M., and JORDAN, D. O., (1947). *Symp. Soc. Exp. Biol.*, **1**, 56.
GUTSTEIN, M. (1925). *Zbl. Bakt.*, **94**, 145.
HARMS, H. (1958). "Farbstoffe für die Mikroskopie." Springer.
HARMS, H. (1965). "Handbuch der Farbstoffe dür die Mikroskopie." Staufen-Verlag, Kamp-Lintfort. p. II/137.
HASHIM, S. A. (1953). *Stain Tech.*, **28**, 27.
HENRY, H., and STACEY, M. (1943). *Nature, Lond.*, **151**, 671.
HENRY, H., STACEY, M., and TEECE, E. G. (1945). *Ibid.*, **156**, 720.
HERMANN, H., NICHOLAS, J. S., and BORICIOUS, J. K. (1950). *J. biol. Chem.*, **184**, 321.
HILLARY, B. B. (1939–40). *Bot. Gaz.*, **101**, 276.
HOGEBOOM, G., SCHNEIDER, W. C., and PALADE, G. (1948). *J. biol. Chem.*, **172**, 619.
HOPMAN, B. C. (1956). *Mikroskopie*, **10**, 251.
HUCKER, G. J., and CONN, H. J. (1923). *N.Y. Agr. exp. Sta. Tech. Bull.*, **93**.
HYDÉN, H. (1943). *Acta physiol. scand.*, **6**, suppl. 17.
IRVIN, J. L., and IRVIN, E. M. (1949). *Science*, **110**, 426.
IRVIN, J. L., and IRVIN, E. M. (1952). *Fed. Proc.*, **11**, 235.
ITIKAWA, O., and OGURA, Y., (1954). *Stain Tech.*, **29**, 9.
JOBST, K., and SANDRITTER, W. (1964). *Histochemie*, **4**, 277.
JOHNSON, P. L., HOFFMANN, H., and ROLLE, G. K. (1957). *J. Histochem. Cytochem.*, **5**, 84.
JORDAN, D. D. (1952). *Prog. Biophys.*, **2**, 51.
JORDANOV, J. (1963). *Acta. histochemica*, **15**, 135.
JOSHI, V. N., and KORGAONKAR, K. S. (1959). *Nature, Lond.*, **183**, 400.
KANNGIESSER, W. (1959). *Hoppe Seyl. Z.*, **316**, 146.
KANWAR, K. C. (1960). *The Microscope*, **12**, 245.
KAPLAN, M. L., and KAPLAN, L. (1933). *J. Bact.*, **25**, 309.
KARRER, P. (1950). "Organic Chemistry." 2nd Ed. Elsevier, New York.
KASTEN, F. H. (1956). *J. Histochem. Cytochem.*, **4**, 310.
KASTEN, F. H. (1959). *Histochemie*, **1**, 466.
KASTEN, F. H. (1964). *Acta Histochem.*, **17**, 88.
KASTEN, F. H. (1965a). *J. Histochem. Cytochem.*, **13**, 13.
KASTEN, F. H. (1965b). *Stain Tech.*, **40**, 127.
KAUFMANN, B. P., GAY, H., and MCDONALD, M. R. (1950). *Cold. Spr. Harb. Symp. quant. Biol.*, **14**, 85.
KAUFMANN, B. P., GAY, H., and MCDONALD, M. R. (1960). *Int. Rev. Cytol.*, **9**, 77.
KAUFMANN, B. P., MCDONALD, M. R., and GAY, H. (1948). *Nature, Lond.*, **162**, 814.
KAY, E. R. M., and DOUNCE, A. L. (1953). *J. Amer. Chem. Soc.*, **75**, 4041.
KISSANE, J. M., and ROBINS, E. (1958). *J. biol. Chem.*, **233**, 184.
KNOBLOCH, A., and VENDRELY, R. (1956). *Nature, Lond.*, **178**, 261.
KODOUSEK, R. (1965). *Acta Histochem.*, **21**, 150.
KOENIG, H., and STAHLECKER, H. (1951). *J. Nat. Cancer Inst.*, **12**, 237.
KOENIG, H., and STAHLECKER, H. (1952). *Proc. Soc. exp. Biol. N.Y.*, **79**, 159.
KORSON, R. (1964). *J. Histochem. Cytochem.*, **12**, 875.
KURNICK, N. B. (1947). *Cold Spr. Harb. Symp. quant. Biol.*, **12**, 141.
KURNICK, N. B. (1949). 1st Int. Cong. Biochem. Abstr., Cambridge, p. 264.
KURNICK, N. B. (1950a). *Exp. cell Res.*, **1**, 151.
KURNICK, N. B. (1950b). *J. gen. Physiol.*, **33**, 243.
KURNICK, N. B. (1950c). *Arch. Bioch.*, **29**, 41.
KURNICK, N. B. (1952). *Stain Tech.*, **27**, 233.
KURNICK, N. B. (1955a). *Int. Rev. Cytol.*, **4**, 221.
KURNICK, N. B. (1955b). *Stain Tech.*, **30**, 213.
KURNICK, N. B., and MIRSKY, A. E. (1950). *J. gen. Physiol.*, **33**, 265.
LAMM, M. E., CHILDERS, L., and WOLF, M. K. (1965). *J. cell Biol.*, **27**, 313.
LAROSE, P., and FISCHER, R. (1953). *Science*, **117**, 449.
LAVERACK, J. O. (1955). *Quart. J. micr. Sci.*, **96**, 29.
LAWLEY, P. D. (1956a). *Biochim. Biophys. Acta.*, **22**, 451.

Lawley, P. D. (1956b). *Ibid.*, **21**, 481.
Lessler, M. A. (1951). *Arch. Biochem. Biophys.*, **32**, 42.
Lessler, M. A. (1953). *Int. Rev. Cytol.*, **2**, 231.
Leuchtenberger, C. (1958). *Gen. Cytochem. Methods*, **1**, 219.
Levene, D. (1901). *J. Med. Res.*, **6**, 135.
Lhotka, J., and Davenport, H. A. (1947). *Stain Tech.*, **22**, 139.
Lhotka, J., and Davenport, H. A. (1949a). *Ibid.*, **24**, 127.
Lhotka, J., and Davenport, H. A. (1949b). *Ibid.*, **24**, 237.
Libenson, L., and McIlroy, A. P. (1955). *J. Infect. Dis.*, **97**, 22.
Li, Chong-Fu, and Stacey, M. (1949). *Nature, Lond.*, **163**, 538.
Lima de Faria, A. (1962). *Prog. Biophys.*, **12**, 281.
Lison, L. (1936). "Histochemie Animale." Gautier-Villars, Paris.
Lison, L., and Mutsaars, W. (1950). *Quart. J. micr. Sci.*, **91**, 309.
Longley, J. B. (1952). *Stain Tech.*, **27**, 161.
Love, R. (1957). *Nature, Lond.*, **180**, 1338.
Love, R. (1962). *J. Histochem. Cytochem.*, **10**, 227.
Love, R., Clark, A. M., and Studzinski, G. P. (1964). *Nature, Lond.*, **203**, 1384.
Love, R., and Fernandes, M. V. (1965). *J. cell Biol.*, **25**, 529.
Love, R., and Liles, R. N. (1959). *J. Histochem. Cytochem.*, **7**, 164.
Love, R., and Rabotti, G. (1963). *J. Histochem. Cytochem.*, **11**, 603.
Love, R., Studzinski, G. P., Clark, A. M., and Tressan, E. R. (1965). *J. nat. Cancer Inst.*, **35**, 55.
Love, R., and Suskind, R. G. (1961). *Exp. cell Res.*, **24**, 521.
Love, R., and Walsh, R. J. (1963). *J. Histochem. Cytochem.*, **11**, 188.
Luck, J. M., Rasmussen, P. S., Sutake, K., and Tsvetikov, A. N. (1958). *J. biol. Chem.*, **233**, 1407.
Mann, G. (1902). "Physiological Histology." Oxford.
Margatroyd, L. B. (1963). *Mikroscopie*, **18**, 257.
Margatroyd, L. B. (1963). *Mikroscopie*, **18**, 285.
de Martino, C., Capanna, E., Civitelli, M. V., and Procicchiani, G. (1965). *Histochemie*, **5**, 78.
Mayersbach, H. (1956). *Acta Histochem.*, **3**, 128.
Mazia, D., and Jaeger, L. (1939). *Proc. Nat. Acad. Sci., Wash.*, **25**, 456.
McLetchie, N. G. B. (1944). *J. Endocrinol.*, **3**, 329.
Mirsky, A. E., and Pollister, A. W. (1942). *Proc. Nat. Acad. Sci., Wash.*, **28**, 344.
Mirsky, A. E., and Pollister, A. W. (1943). *Biol. Symp.*, **10**, 241.
Mirsky, A. E., and Pollister, A. W. (1946). *J. gen. Physiol.*, **30**, 117.
Mironescu, S. (1965a). *Acta Histochem.*, **20**, 366.
Mironescu, S. (1965b). *Acta Histochem.*, **21**, 228.
Mitchell, J. S. (1942). *Brit. J. exp. Path.*, **23**, 296.
Mitchell, P., and Moyle, J. (1950). *Nature, Lond.*, **166**, 218.
Mitchell, P., and Moyle, J. (1954). *J. gen. Microbiol.*, **10**, 533.
Mittwer, T., Bartholomew, J. W., and Kallman, B. J. (1950). *Stain Tech.*, **25**, 169.
Monné, L. (1955a). *Ark. Zool.*, **7**, 559.
Monné, L. (1955b). *Ibid.*, **9**, 93.
Monné, L., and Borg, K. (1954). *Ibid.*, **6**, 555.
Mottram, J. C., and Selbie, F. R. (1946). *Brit. J. exp. Path.*, **26**, 377.
Norberg, B. (1943). *Acta physiol. scand.*, Suppl. 14.
Ogur, M., and Rosen, G. (1949). *Fed. Proc.*, **8**, 234.
Opie, E. L., and Lavin, G. I. (1946). *J. exp. Med.*, **84**, 107.
Östergren, G. (1948). *Hereditas*, **34**, 510.
Overend, W. G., and Stacey, M. (1949). *Nature, Lond.*, **163**, 538.
Pakkenberg, H. (1962). *J. Histochem. Cytochem.*, **10**, 367.
Panijel, J. (1950). *Biochim. Biophys. Acta.*, **6**, 79.
Panijel, J. (1951). "Les Problemes de l'Histochimie et la Biologie Cellulaire," Hermann, Paris.
Paolillo, D. J. (1964a). *Acta Histochem.*, **18**, 276.
Paolillo, D. J. (1964b). *Acta Histochem.*, **18**, 283.

Pappenheim, A. (1899). *Virchows Arch.*, **157**, 19.
Parker, F. S. (1949). *Science*, **110**, 426.
Pasteels, J., and Lison, L. (1950). *C.R. Soc. Biol. Paris.*, **230**, 780.
Pearse, A. G. E. (1951). *J. clin. Path.*, **4**, 1.
Perry, S., and Reynolds, J. (1956). *Blood*, **11**, 1132.
Persijn, J-P., and van Duijn, P. (1961). *Histochemie*, **2**, 283.
Phillips, D. M. P. (1962). *Prog. Biophys.*, **12**, 213.
Phillips, D. M. P., and Johns, E. W. (1959). *Biochem. J.*, **72**, 538.
Pollister, A. W., and Leuchtenberger, C. (1949). *Proc. Nat. Acad., Sci., Wash.*, **35**, 111.
Pollister, A. W., and Ris, H. (1947). *Cold Spring Harbour, Symp. Quant. Biol.*, **12**, 147.
Rafalko, J. S. (1946). *Stain Tech.*, **21**, 91.
Rigler, R. Jnr. (1964). Proc. 2nd Int. Congr. Histochem. Springer, Heidelberg, p. 233.
Ris, H., and Mirsky, A. E. (1949). *J. gen. Phys.*, **33**, 125.
Roberts, D., and Friedkin, M. (1958). *J. biol. Chem.*, **233**, 483.
Roschlau, G. (1965). *Histochemie*, **5**, 396.
Roschlau, G., and Reinke, J. (1964). *Acta Histochem.*, **18**, 328.
Rosenkranz, H. S., and Bendich, A. (1958). *J. Biophys. Biochem. Cytol.*, **4**, 663.
Roskin, G. I., and Ginsberg, A. S. (1944). *C.R. Acad. Sci., U.R.S.S.*, **43**, 122.
Saint-Hilaire, C. (1898). *Hoppe-Seyl. Z.*, **26**, 102.
Sandritter, W., Diefenbach, H., and Krantz, F. (1954). *Experientia*, **10**, 210.
Sandritter, W., Kiefler, G., and Rick, W. (1963). *Histochemie*, **3**, 318.
Schabadasch, A. (1957). *Dokl. Akad. Nauk., U.S.S.R.*, **144**, 658.
Schabadasch, A. (1960). *Ann. Histochem.*, **5**, 225.
Schneider, W. C. (1945). *J. biol. Chem.*, **161**, 293.
Schneider, W. C. (1948). *Cold. Spr. Harb. Symp. quant. Biol.*, **12**, 169.
Schumacher, J. (1926). *Zbl. Bact.*, **98**, 104.
Schümmelfeder, N., Ebschner, K., and Krogh, E. (1957). *Naturwiss.*, **44**, 467.
Scott, J. E. (1967). *Histochemie*, **9**, 30.
Seibert, F. B., Pfaff, M. L., and Seibert, M. V. (1948). *Arch. Biochem.*, **18**, 279.
Serra, J. A. (1943). *Bol. Soc. broteriana*, **17**, 203.
Serra, J. A., and Queiroz Lopes, A. (1945). *Port. Acta. biol.*, **1**, 111.
Seshachar, B. R., and Flick, E. W. (1949). *Science*, **110**, 659.
Sevag, M. G., Smolens, J., and Lackman, D. B. (1940). *J. biol. Chem.*, **134**, 523.
Shugar, D., and Baranowska, J. (1957). *Biochim. Biophys. Acta.*, **23**, 227.
Sibatani, A. (1950). *Nature, Lond.*, **166**, 355.
Smith, S. W., and Anderson, P. N. (1960). *Anat. Rec.*, **138**, 179.
Spicer, S. S. (1962). *J. Histochem. Cytochem.*, **10**, 691.
Stacey, M., Deriaz, R. E., Teece, E. G., and Wiggins, L. F. (1946). *Nature, Lon.* **157**, 740.
Stearn, E. N., and Stearn, A. E. (1924). *J. Bact.*, **9**, 497.
Stedman, E., and Stedman, E. (1943a). *Nature, Lond.*, **152**, 267.
Stedman, E., and Stedman, E. (1943b). *Ibid.*, **152**, 503.
Stedman, E., and Stedman, E. (1944). *Ibid.*, **153**, 500.
Stedman, E., and Stedman, E. (1947). *Symp. Soc. exp. Biol.*, **1**, 232.
Stedman, E., and Stedman, E. (1950). *Biochem. J.*, **47**, 508.
Stefano, H. S. Di (1948). *Chromosome*, **4**, 282.
Stefano, H. S. Di (1952). *Stain Tech.*, **27**, 171.
Steiner, R. F., and Beers, R. F. (1959). *Arch. Biochem. Biophys.*, **81**, 75.
Stenram, U. (1953). *Exp. Cell Res.*, **4**, 383.
Stenram, U. (1954). *Acta Anat.*, **20**, 36.
Steward, P. J. (1963). D. Phil. Thesis, Oxford.
Stowell, R. E. (1942). *J. Nat. Cancer Inst.*, **3**, 111.
Stowell, R. E. (1945). *Cancer Res.*, **5**, 283.
Stowell, R. E. (1946a). *Stain Tech.*, **31**, 137.
Stowell, R. E. (1946b). *Cancer Res.*, **6**, 426.
Stowell, R. E., and Cooper, Z. K. (1945). *Ibid.*, **5**, 295.
Stowell, R. E., and Zorzoli, A. (1947). *Stain Tech.*, **22**, 51.

SWIFT, H. (1950). *Physiol. Zool.*, **23**, 169.
SULKIN, N. M., and KUNTZ, A. (1950). *Proc. Soc. exp. Biol., N.Y.*, **73**, 413.
TAFT, E. B. (1951a). *Stain Tech.*, **26**, 205.
TAFT, E. B. (1951b). *Exp. Cell Res.*, **2**, 312.
TAKEMURA, S. (1958). *Biochim. Biophys. Acta.*, **29**, 447.
TERNER, J. Y., and CLARK, G. (1960). *Stain Tech.*, **35**, 167.
THORELL, B. (1944). *Acta med. scand.*, **117**, 334.
THORELL, B., and CASPERSSON, T. (1941). *Chromosoma*, **2**, 132.
TOMASI, J. A. DE (1936). *Stain Tech.*, **11**, 137.
TORRE, L. DE LA, and SALISBURY, G. W. (1962). *J. Histochem. Cytochem.*, **10**, 39.
TREVAN, D. J., and SHARROCK, A. (1951). *J. Path. Bact.*, **63**, 326.
TURCHINI, J., CASTEL, P., and KIEN, K. V. (1944). *Bull. Tech. Histol. micr.*, **21**, 124.
VAN DEN BERGHE, L. (1942). *Acta biol. Belg.*, **2**, 390.
VAN DUIJN, C. Jnr. (1962). *Nature, Lond.*, **193**, 999.
VAN DUIJN, P. (1956). *J. Histochem. Cytochem.*, **4**, 55.
VENDRELY, R., and VENDRELY, C. (1948). *Experientia*, **4**, 434.
VERACAUTEREN, R. (1950). *Enzymologia*, **14**, 134.
WATSON, J. D., and CRICK, F. H. C. (1953). *Nature, Lond.*, **171**, 737.
WHITE, J. C. (1947). *J. Path. Bact.*, **59**, 223.
WHITE, J. C. (1950). *Proc. Biochem. Soc., Biochem., J.*, **47**, 16.
WIDSTRÖM, G. (1928). *Biochem. Z.*, **199**, 298.
WILKINS, M. H. F., SEEDS, W. E., STOKES, A. R., and WILSON, H. R. (1953). *Nature, Lond.*, **172**, 759.
WINKLE, VAN Q., RENOLL, H. W., GARVEY, J. S., PALIK, E. S., and PREBUS, A. F. (1953). *Exp. Cell Res.*, **5**, 38.

CHAPTER 10

CARBOHYDRATES AND MUCOSUBSTANCES

SINCE the second edition of this book was written (1960) biochemists have made very considerable advances in their understanding of the complex structure of carbohydrate and carbohydrate-protein macromolecular complexes. Histochemists, for their part, have made several attempts to produce a classification both accurate and acceptable in biochemical terms. Nevertheless it must be appreciated, as pointed out by Spicer, Leppi and Stoward (1965), that histochemical methods presently available provide relatively little chemical information. For some time to come, therefore, it will be necessary to keep biochemical and histochemical classifications of mucosubstances and carbohydrates apart.

Although they cannot determine the individual components of complex naturally-occurring mucins with any chemical precision, histochemists must nevertheless have a clear picture of the structure of the carbohydrates which are responsible for the specific histochemical reactions of the various mucins.

In its 1953 and 1960 editions the classification of mucosubstances given in this text was substantially based on the original plan devised by Meyer (1938) and subsequently modified by him (1945, 1953). The present classification is derived from the work of many authors (Meyer, 1953; Kent and Whitehouse, 1955; Jeanloz, 1960; Stacey and Barker, 1962; Gottschalk, 1962, 1966).

For the majority of carbohydrate biochemists the expressions mucopolysaccharide and mucoprotein are out of date. For Stacey and Barker (1962), for instance, mucopolysaccharides were protein-carbohydrate complexes whose reactions were predominantly polysaccharide while mucoproteins were similar complexes whose reactions were predominantly protein. As Gottschalk (1966) says, however, "each author defines mucopolysaccharides and mucoproteins in his own way resulting in a great confusion of the true meaning, if any, of these terms". Jeanloz (1960) proposed a bold solution to the problem, namely complete omission of the prefix "muco" from the chemical nomenclature of carbohydrates. It is clear that this move has already been accepted by the majority of workers in the field. I propose, therefore, to follow Jeanloz' general plan, introducing some modifications prompted by more recent publications on the subject.

Biochemical Classification of Carbohydrates

In Jeanloz' original (1960) classification of macromolecular complexes containing carbohydrate components (oligo- or polysaccharides) there were five divisions: (1) Pure polysaccharides; (2) Compounds containing a carbohydrate component attached to a polypeptide component through a "weak" link (salt link hydrogen bond); (3) Compounds containing a carbohydrate component attached to a polypeptide component through a "strong" link (covalent linkage); (4) Compounds containing carbohydrate and lipid components; (5) Compounds containing carbohydrate, lipid and polypeptide components.

Evidence for the existence of weak links (Division 2) was derived from the relative ease with which the two components could be separated by treatment with dilute alkali. It is now certain that strong links are the rule in both Division 2 and Division 3. Gottschalk (1962) defined the former as polysaccharide-protein complexes in which the carbohydrate is a homo- or heteropolysaccharide such as chitin, hyaluronic acid or heparin, characterized by a small repeating unit and a high degree of polymerization. In Division 3, on the other hand, the carbohydrate is of relatively low molecular weight and lacks a short repeating structure.

The classification which appears below is essentially a compromise, for the reason given above. The examples given for each division are in no sense a complete list.

(I) GLYCANS (POLYSACCHARIDES OR OLIGOSACCHARIDES)

(*a*) **Homoglycans (One monosaccharide component)**
- Glycogen
- Starch
- Cellulose
- Dextran
- Galactan (Helix)

(*b*) **Homopolyaminosaccharides**
- Chitin

(*c*) **Homopolyuronosaccharides**
- Pectic acid
- Alginic acid

(*d*) **Heteroglycans (Two or more monosaccharide components)**

(1) *Glycosaminoglycans*
- Keratosulphate
- Sialoglycans

(2) *Glycosaminoglucuronoglycans*
- Hyaluronic acid
- Chondroitin-4-sulphate (chondroitin sulphate A)
- Chondroitin-6-sulphate (chondroitin sulphate C)
- Dermatan sulphate (chondroitin sulphate B)
- Heparin

(II) POLYSACCHARIDE-PROTEIN COMPLEXES
- Chondroitin sulphate-protein
- Hyaluronic acid-protein
- Chitin-protein

(III) Glycoproteins and Glycopolypeptides

Ovomucoid
Salivary gland mucoid (Sialoglycoproteins)
Serum glycoproteins (including immunoglobulins)
Blood Group Specific Substances

(IV) Glycolipids

Cerebrosides
Gangliosides

(V) Glycolipid-protein Complexes

Ox Brain mucolipid
Strandin

Individual Glycans

Glycogen is the main carbohydrate reserve store in animal tissues. It is also found in a number of different micro-organisms and in the plant kingdom, but also in the primitive thallophytes. There are a number of different glycogens of varying degrees of polymerization. All are composed of chains of α-glucopyranose (D-glucose) units, linked α-1, 4′ except at points of branching where the linkage is α-1, 6′, and they are readily hydrolysed to glucose by boiling with dilute acids.

Starch, in various forms, is the main carbohydrate reserve substance of the plants. On hydrolysis the starches yield only D-glucose but they are composed of two different glucans, one a linear molecule called amylose and the other a branched molecule called amylopectin. The amylose-amylopectin ratio varies from case to case. The straight chain maltose polymer (amylose) consists of about 20 glucose residues joined α-1,4′, while the branched chains consist of 24 to 30 glucose residues with an α-1,6′ linkage joining one chain to another.

Cellulose is the main constituent of the cell walls of land plants, and, as such, it is more plentifully distributed than any of the other polysaccharides. Like the two preceding polysaccharides cellulose yields only glucose on total hydrolysis. It consists mainly of linear chains of β-D-glucose residues linked together through β-1,4′ linkages.

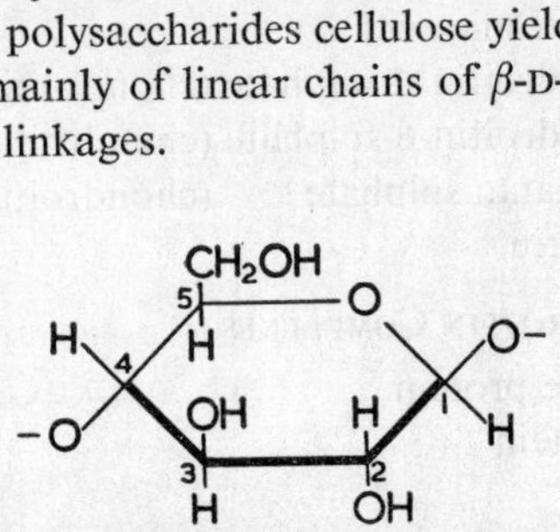

X-ray diffraction analysis has shown that the chains are held in a parallel arrangement with respect to each other and form a crystal lattice with a fibre axis of 10·3 Å (Preston, 1964).

Dextran. The dextrans are a group of polysaccharides containing a backbone of D-glucose units which are predominantly linked α-D-(1 $\rightarrow$ 6). They are produced from sucrose by the activity of bacteria belonging to the *Lactobacteriaceae*. Dextrans have been used extensively in medical practice as plasma substitutes. For this purpose the size of the polymer has usually been kept to a molecular weight between 50′000 and 100′000.

Galactan. The albumin glands of *Helix pomatia* contain a polygalactan whose backbone is composed of D-galactose residues linked β-1:3, to which are attached side-chains of galactose residues linked β-1:6. Similar polysaccharides are present in many other gastropods.

Chitin is a polysaccharide which is widely distributed in the invertebrates and in lower forms of plant life (fungi). In the invertebrates it is the principal structural component of the exoskeleton or shell, hence its name *Χιτὼν*, a tunic or coat of mail. It is most abundant in arthropods, annelids and molluscs and also in the insects. Pure chitin, however, does not occur in nature; it is always mixed with calcium carbonate, or protein, or both.

Structurally chitin is essentially cellulose in which the hydroxyl group on the C_2 carbon atoms are replaced by —$NHCOCH_3$. It consists of linear chains of 2-acetamido-2-deoxy-D-glucose units joined by β-1,4 links.

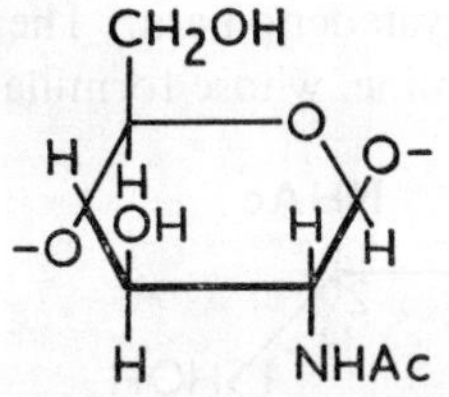

According to Dweltz (1960) the X-ray diffraction fibre pattern of lobster tendon chitin reveals a well-defined orthorhombic lattice with a fibre axis of 19·13 Å.

Pectic Acid. The pectin group of polysaccharides is found in the cell walls and intercellular layers of all plant tissues (Whistler and Smart, 1953) and the main representative, pectic acid, is polygalacturonan. This is a linear polymer composed of α-1,4′-linked D-galacturonic acid units which may be fully or partially esterified. When not fully esterified the carboxyl groups are involved in salt formation with metallic ions, chiefly, calcium and magnesium.

Alginic Acid. This is a linear polymer composed of D-mannuronic acid units which are linked β-1,4′. It is the chief structural polysaccharide of the brown algæ, hence its name.

Keratosulphate. This glycosaminoglycan (heteropolyaminosaccharide) was first isolated by Meyer *et al.* (1953) from bovine cornea but it also occurs in human aorta (Buddecke, 1960) and elsewhere, together with other polysaccharides. As shown by the structural formula below

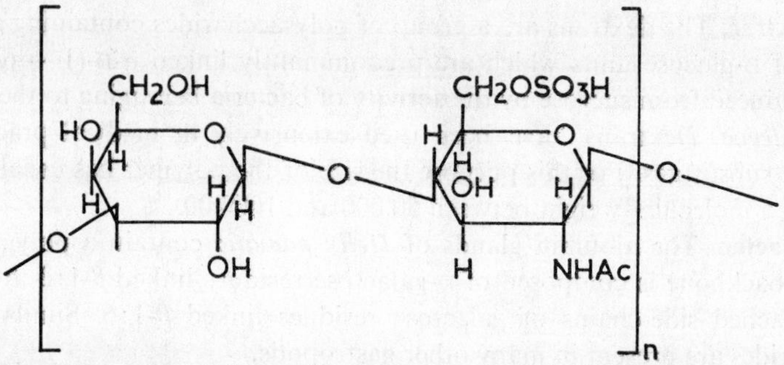

the repeating unit consists of D-galactose and *N*-acetyl-D-glucosamine residues joined by β-D-1,4′-galactosidic links. The repeating units were shown by Rosen *et al.* (1960) to be joined by β-D-1,3-glucosaminidic links and Hirano *et al.* (1961) confirmed that the sulphate is at C-6 of glucosamine.

Sialoglycans. A special family of heteropolyaminosaccharides comprises the neuraminic acids. The naturally-occurring neuraminic acids are *N*-acylated and are known as sialic acids. Discovered in bovine submaxillary gland mucin by Blix (1936), their name is derived from the Greek σιαλον, saliva. In animal tissues sialic acids are found as constituents of both glycoproteins and glycolipids, particularly of the salivary glycoproteins and of the pseudomucins of ovarian cystadenomata. The disaccharide *O*-α-D-sialyl (2 → 6)-*N*-acetyl-D-galactosamine, whose formula is shown here,

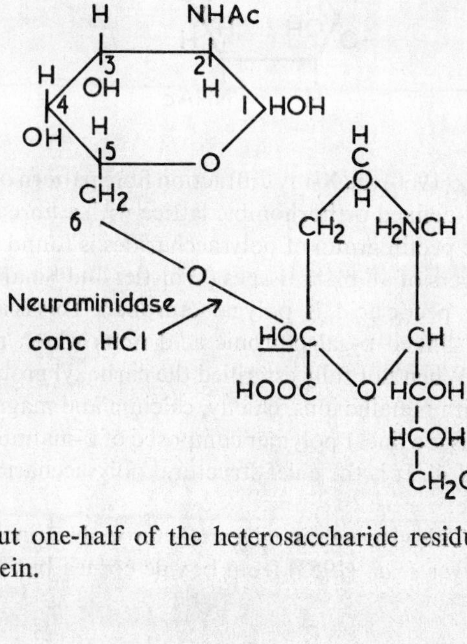

constitutes about one-half of the heterosaccharide residue of submaxillary gland glycoprotein.

The nonulosaminic acid is readily split off from the remaining glycoprotein by the enzyme neuraminidase (see Chapter 25) or by acid hydrolysis. From this observation it has been concluded that in many glycoproteins the carbohydrate moiety is branched and that the sialyl residue is always terminal (Gottschalk, 1956). However, it is not certain that this conclusion is true (Jeanloz, 1963).

Hyaluronic Acid is a polysaccharide composed of equimolar quantities of D-glucuronic acid and 2-acetamido-2-deoxy-D-glucose which occurs widely in connective tissues, in the umbilical cord (Wharton's jelly), in synovial fluids, and in the walls of blood vessels. It was first isolated by Meyer and Palmer (1934) from the vitreous humour of the eye. Hyaluronic acid from this source has been separated into fractions with molecular weights ranging from $7{\cdot}7 \times 10^4$ to $1{\cdot}7 \times 10^6$ (Laurent *et al.*, 1960). The major product of the action of testicular hyaluronidase (Chapter 25) on hyaluronic acid is a tetrasaccharide.

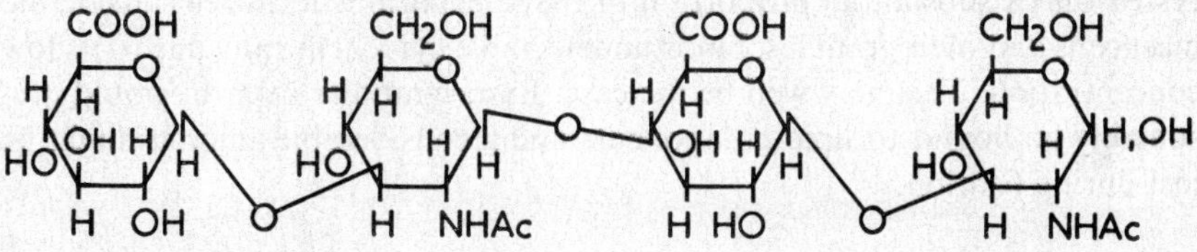

It is now established with certainty that the repeating unit is the disaccharide which forms the middle portion of the tetrasaccharide.

Chondroitin-4 and 6-Sulphates. Formerly known as Chondroitin sulphates A and C, respectively, these two compounds are closely similar and can sensibly be considered together. The repeating unit is a disaccharide, (1 → 4)*O*-*β*-D-glucopyranosyluronic acid – (1 → 3)-2-acetamido-2-deoxy-*O*-sulpho-*β*-D-galactopyranose. The only difference between the two is the position of the sulphate group.

The molecular weight of the two chondroitin sulphate complexes is in the region of 50′000–90′000. They are found in cartilage, cornea, blood vessel walls and in certain human tumours, particularly chondrosarcomata and chordomas.

Dermatan Sulphate. Originally known as chondroitin sulphate B, this compound was isolated by Meyer and Chaffee (1941) from pig skin. The repeating unit is (1 → 4)-*O*-*α*-L-idopyranosyluronic acid -(1 → 3)-2-acetamido-2-deoxy-4-*O*-sulpho-*β*-galactopyranose. Histochemical methods clearly cannot distinguish dermatan sulphate from the two chondroitin sulphates. As far as normal tissues are concerned, dermatan sulphate occurs in gastric mucosa (pig), in tendon, connective tissue, aorta, lung and, as indicated above, it is the chief glucosamino-glucuronoglycan in skin.

Heparin. The full structure of heparin is still not established. It is probably a polymer of D-glucuronic acid and 2-deoxy-2-sulphoamino-D-glucose, both residues contain *O*-sulphate groups but the precise degree of sulphation varies.

The structure of the basic tetrasaccharide repeating unit of heparin is shown below.

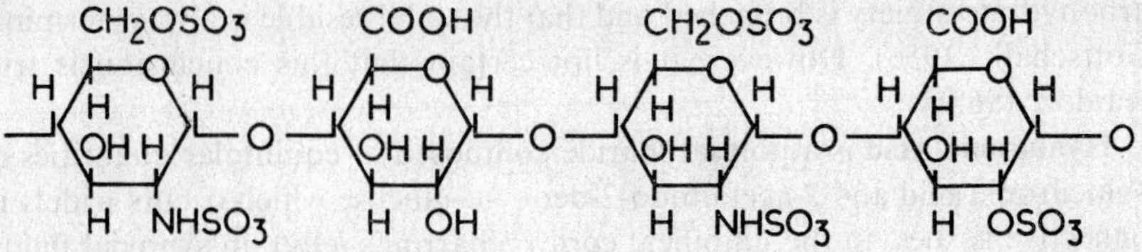

CH_2OSO_3 COOH CH_2OSO_3 COOH

H NHSO$_3$ H OH H NHSO$_3$ H OSO$_3$

All the glycosidic linkages are shown with the α-(1 $\rightarrow$ 4)-configuration but it is possible that some of the glucuronidic linkages may have a (1 $\rightarrow$ 6) structure, or the β-configuration.

Heparin is very widely distributed in mammalian tissues, being stored in the mast cells and, perhaps, synthesized by them also. Jacques (1961) has suggested that a substantial proportion of tissue heparin is localized outside the mast cells and their granules, but is undetectable cytochemically due to its low concentration. This may well be the case. Extra-granular heparin would presumably be bound to lipid components and a considerable amount might be lost during fixation.

Individual Polysaccharide-protein complexes

Acidic Glycosaminoglucuronoglycans (Acid mucopolysaccharides). If cartilage is extracted with 10 per cent. calcium chloride, instead of with alkali, the viscous preparation which results contains a substantial amount of protein, as well as chondroitin sulphate. The molecular weight of this chondroitin sulphate-protein complex may be in the region of 4×10^6. The nature of the linkage between the protein and the polysaccharide is still unknown. There is some evidence that the hydroxyl group of serine is involved (Anderson *et al.*, 1963).

Hyaluronic acid occurs either in soluble form, as in the vitreous and in synovial fluid, or in the form of a protein complex (Ogston and Stanier, 1950). The complex contains 25–30 per cent. protein but it is very easily dissociated. Once again, the mode of attachment of hyaluronic acid to protein is unknown.

Individual Glycoproteins

Ovomucoid. This protein comprizes 12 per cent. of egg-white protein and its carbohydrate moiety (about 25 per cent.) consists of *N*-acetyl-D-glucosamine, D-mannose, D-galactose and *N*-acetyl-neuraminic acid. The last represents about 1 per cent. of the total carbohydrate but in some birds, such as the emu (Melamed, 1966) it may be as high as 10·5 per cent.

Salivary Gland Mucoid. This is an important "type" mucosubstance for the histochemist and it may be helpful, therefore, to describe its characteristics in some detail. Very full accounts are given by Gottschalk (1966) and by Pigman and Gottschalk (1966) and Table 18, which appears below is derived from the latter. According to Gottschalk (1966) the fact that the total number of

TABLE 18

Amino-acid and carbohydrate in ovine submaxillary gland glycoprotein

Component (Amino-acid or sugar)	Grams per 100 g.	Moles per mole (1×10^6)
Alanine	5·08	570
Arginine	3·31	190
Aspartic Acid	4·10	308
Cysteic Acid	0·89	53
Glutamic Acid	6·17	419
Glycine	5·97	795
Histidine	0·30	19
Isoleucine	1·90	145
Leucine	2·92	223
Lysine	1·50	103
Methionine (sulphone)	0·49	27
Phenylalanine	1·96	119
Proline	5·60	486
Serine	7·57	720
Threonine	7·00	588
Tryptophan	0	0
Tyrosine	0·68	38
Valine	4·20	359
Ammonia (amide)	0·046	27
N-Acetylneuraminic acid	22·5 to 25·0	728 to 809
Galactosamine	13·0 to 14·5	729 to 810
Galactose	0·3	17
Mannose	0·15	8
Fucose	0·4	24
Glucosamine	Trace	—

glycine, serine and threonine residues is nearly half the total number may well be significant. The carbohydrate is an equimolar mixture of *N*-acetylneuraminic acid and *N*-acetylgalactosamine, with a few additional sugar residues. The structure of a segment of submaxillary gland mucoid (Gottschalk, 1960) is shown below.

This is the structure which should be intact if our efforts at fixation (Chapter 5, p. 99) have been successful. We cannot be sure that the spacing indicated remains 7·23 Å and, particularly after formaldehyde vapour fixation, it is probable that cross links from the carbohydrate moiety to adjacent protein chains are formed.

The labile α-linkage certainly remains labile, however, in the majority of cases. The β-linkage joining the disaccharide to the seryl and threonyl residues of the protein chain is cytochemically more stable. It can be broken by reductive

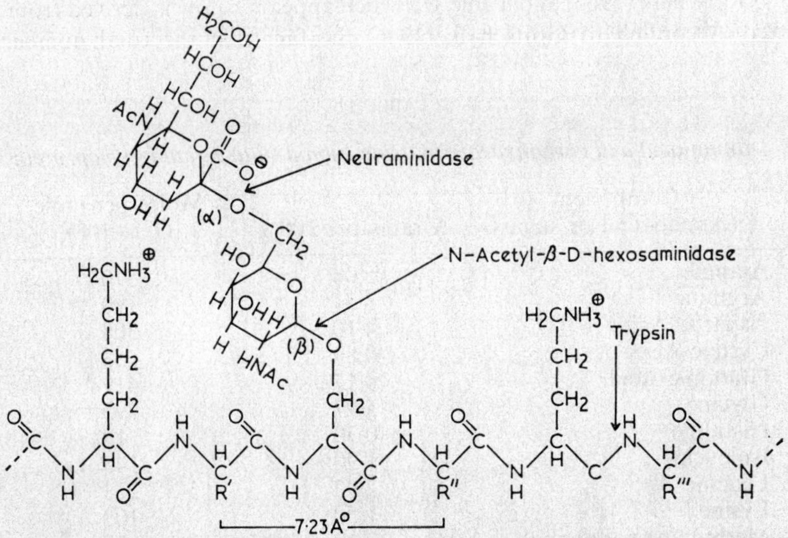

Fig. 95. Structure of the Disaccharide Unit of Submaxillary Mucoid

cleavage with lithium borohydride or, enzymically, by the action of *β-N*-acetylgalactosaminidase.

Immunoglobulins. These are a group of glycoproteins produced in the anti-body-forming cells in response to antigenic stimuli. The function of the carbo-hydrate fraction is not known since the protein remaining after its removal is immunologically competent. As in other fields there is confusion over the nomenclature of the immunoglobulins (Press and Porter, 1966). They are now usually referred to as IgG, IgM, and IgA (Immunoglobulins G, M and A) and some further descriptive details are given in Table 19, below. In the normal serum of all species 90 per cent. of the total immunoglobin is IgG. This has a relatively low carbohydrate content by comparison with the other two types (Hexose 1·02 per cent., Hexosamine 1·2 per cent., Sialic acid 0·03 per cent.; Chaplin *et al.*, 1965). The carbohydrate content of IgM was reported by these authors as hexose 4·8 per cent., hexosamine 3·0 per cent., and sialic acid 1·06 per cent.

TABLE 19

The Immunoglobulins

Immunoglobulins	Synonyms	Mol. Wt.	CHO Content g/100g. protein
IgG	7-Sγ, γ	150′000	2·5
IgM	19-Sγ, β2M, γ1M	1′000′000	10
IgA	γ1A, β2A	150′000	5–10

α_1-Acid Glycoprotein. In addition to the immunoglobins an important carbohydrate-containing serum protein is α_1-acid glycoprotein. Formerly known as seromucoid and, subsequently, as orosomucoid it is the most soluble of all the serum proteins. It has a high carbohydrate content which includes mannose 6·9 per cent., D-galactose 6·5 per cent., L-fucose 1·2 per cent. and *N*-acetylneuraminic acid 10·8 per cent.

As its name suggests this glycoprotein migrates with the α_1-globulin (Weimer *et al.*, 1950). Because of the neuraminic acid content the isoelectric point is low (pH 1–2) and, as expected, it rises to pH 5·0 after treatment by neuraminidase (Popenoe and Drew, 1957). A closely related glycoprotein, fetuin, is found in fœtal calf serum (Deutsch, 1954).

Blood Group Polysaccharides. The antigens responsible for the specificity of the main ABO blood groups are found in varying amounts in different tissues, as well as on the erythrocytes themselves. In addition to the A and B antigens there are others, such as the H substance, found in group O individuals, and Le^a (Lewis system). All the blood group specific substances are characterized by a very high content of carbohydrate (fucose 16–20 per cent., hexosamine 24–36 per cent., sialic acid 1–5 per cent), which is presumed to be covalently linked to the moiety which contains the amino-acid. The carbohydrate accounts for 80–90 per cent. of the whole molecule.

Individual Glycolipids

The cerebrosides, represented chiefly by phrenosin and kerasin, are the principal members of this group. They occur in tissues outside as well as within the central nervous system. These substances contain no phosphoric acid but yield on hydrolysis one molecule of fatty acid, one of sphingosine,

$$CH_3(CH_2)_{12}CH{:}CH.CH(NH_2).CHOH.CH_2OH$$

and one of sugar. The glycolipids are more fully considered in Chapter 12.

Individual Glycolipid-protein Complexes

These compounds are extracted from animal brains by various procedures. They include strandin and ox brain mucolipid but need not concern us further, in this chapter.

Histochemical Classification of Carbohydrates (excluding polysaccharides)

Continuing correspondence in histochemical journals, some of it a good deal less acrimonious than one would wish, testifies to the urgent concern of histochemists with the need for their descriptive terms for mucosubstances to be realistic and up-to-date.

Part of the difficulty, not restricted to the field of mucosubstances, is that these substances in occur the tissues as mixtures of often very different individual components. Since it is the business of histochemists to localize, as well as to identify, their main hopes in the matter of differentiation of mucosubstances must be based on the use of existing, not very specific, techniques, on differential extractions, and on the development of new and more specific methods.

It is not possible, at the present time, to make any classification on a histochemical basis without using the term mucopolysaccharide. *The only possible way in which mucosubstances can be defined histochemically is in terms of their response to the application of localizing reagents used under rigorously defined conditions.* For this reason the classification presented here is closely based on that of Spicer *et al.* (1965).

These authors suggested that tissue mucosubstances should be named by (1) stating the site in which they are found and (2) subtyping them as well as possible as neutral mucosubstances, mucopolysaccharides, sulphomucins and sialomucins. These names are to be given on the basis of histochemical tests for the presence or absence of *vic*-glycols, uronic acids, sulphate groups and sialic acid carboxyls. These components are designated by the letters G, U, S and C respectively. It is suggested that further subdivision of these subtyped mucosubstances can be achieved by means of the following histochemical reactions:

(1) Affinity for basic dyes such as azure A.
(2) Affinity for Alcian blue.
(3) Lability with respect to testicular hyaluronidase.
(4) Lability with respect to *Vibrio choleræ* neuraminidase.

These properties are to be signified by the letters B, A, T and N, respectively. I have replaced the S (sialidase) of Spicer *et al.* by N (neuraminidase) chiefly in order to avoid having two letters S in the abbreviated description of a given mucin, each having a different significance. After B and A numbers are placed to indicate the lowest pH at which a positive reaction is obtained.

In cases where any particular property, described above, is acquired only after some preliminary manœuvre, the latter is signified by an abbreviation in brackets as, for example, (Sap) for saponification. The suffix ± is suggested as an indication of a slow or feeble response as, for example, towards enzymic digestion. More precisely, the prefix G indicates a positive PAS reaction after 10 minutes oxidation of sections with 1 per cent. aqueous periodic acid. It does *not*, and could not, indicate the presence or absence of periodate reactive *vic*-glycols in the strict biochemical sense.

As in every other branch of histochemistry, there is a threshold concentration per unit area of the reactive group, below which a negative result will occur. These thresholds produce the black and white responses of histochemistry which have to be compared with the continuous grey spectrum of the biochemist.

Below is set forth a working classification of mucosubstances which is substantially that of Spicer *et al.* (1965). Further comments and explanations follow.

(I) NEUTRAL MUCOSUBSTANCES

Neutral glycoproteins
Immunoglobulins
Fuco-mucoids
Manno-mucoids
} (All PAS-positive) (Periodate-reactive)

SYMBOL. G-mucosubstance

(II) ACID MUCOSUBSTANCES

A. Sulphated

(1) *Connective Tissue Mucopolysaccharides*. (PAS-negative) (Periodate unreactive)

(*a*) Stable to testicular hyaluronidase

(i) Alcianophilic in 1·0M-$MgCl_2$-keratan sulphate, heparin.

SYMBOL. S-mucopolysaccharide A(1·0$MgCl_2$)

(ii) Alcianophilic in 0·7M-$MgCl_2$-dermatan sulphate.

SYMBOL. S-mucopolysaccharide A(0·7$MgCl_2$)

(*b*) Labile to testicular hyaluronidase

(i) Alcohol-resistant azurophilia (0·02 per cent.) at or above pH 2·0—chondroitin sulphates in cartilage.

SYMBOL. S-mucopolysaccharide B2·0T

(ii) Alcohol-resistant azurophilia (0·02 per cent.) at or above pH 4·0—chondroitin sulphates in vascular walls.

SYMBOL. S-mucopolysaccharide B4·0T

(2) *Epithelial Sulphomucins* (testicular hyaluronidase-resistant)

(*a*) PAS-negative

(i) Sulphate esters on *vic*-glycols.

SYMBOL. VGS-mucin

(ii) Sulphate esters not on *vic*-glycols.

SYMBOL. S-mucin

(*a*) Alcohol-resistant azurophilia (0·02 per cent.) at or above pH 2·0.

SYMBOL. S-mucin B2·0

(*b*) Alcohol-resistant azurophilia (0·02 per cent.) at or above pH 4·5.

SYMBOL. S-mucin B4·5

B. Non-sulphated

(1) *Hexuronic acid-rich mucopolysaccharides*—hyaluronic acid.

SYMBOL. U-mucopolysaccharide

(2) *Sialic acid-rich mucosubstances*

(*a*) Connective Tissue Sialomucins

SYMBOL. C-mucopolysaccharide

(*b*) Epithelial Sialomucins

(i) Labile to Neuraminidase, PAS-positive, metachromatic with Azure A.

SYMBOL. CG mucin BN

(ii) Slowly digested by Neuraminidase

(*a*) PAS-positive.

SYMBOL. CG-mucin N±

(*b*) PAS-negative.

SYMBOL. C-mucin N±

(iii) Stable to Neuraminidase.

(*a*) Rendered metachromatic and susceptible to enzyme by prior saponification.

SYMBOL. S-mucin (Sap)BN

(*b*) Stable to Neuraminidase after saponification.

(i) PAS-positive.

SYMBOL. GC-mucin

(ii) PAS-negative.

SYMBOL. C-mucin

It is possible to use the above classification in two ways, by reference numbers or by symbols, i.e. as II Alai or as S-mucopolysaccharide A(1·0MgCl$_2$). I believe that most histochemists will prefer the second alternative and I have therefore used it throughout this chapter. Table 20, opposite, is appended as an illustration of some histochemically distinct types of mucosubstance, giving their symbols and localization as they should be designated according to the above classification.

Rosan and Saunders (1965) strongly recommended that histochemists should fall into line with biochemical nomenclature, a practice which I support whenever possible. It seems clear, however, that in the matter of mucosubstances this policy is permissible only when a component whose biochemical identity is known (or half known), such as heparin, is localized in a distinct organelle (the mast-cell granule). Eventually histochemistry may develop methods which will allow much greater precision in delineating exact chemical structure. At present differences between one mucosubstance and another, particularly where comparison between species is concerned, may well be done to the physical state of the component rather than to its chemistry. As always in histochemical practice, it is the difference which is significant even when no explanation is forthcoming.

TABLE 20

Symbol Equivalents of Some Characteristic Mucosubstances

Symbol	Localization and Description of Tissue Mucosubstance
G-mucosubstance	Gastric surface epithelia Thyroid Colloid (Man, Guinea pig, Rabbit) Coagulating Gland Fluid (Rat, Mouse)
S-mucopolysaccharide A(1·0$MgCl_2$)	Cornea and Mast Cells
S-mucopolysaccharide B2·0T	Cartilage Ovarian Follicle Fluid
S-mucopolysaccharide B4·0T	Aorta, Heart valves, Renal papilla, Some areas of Cartilage
VGS-mucin	Sublingual glands (Hamster), Glossal mucous glands (Rabbit), Colonic Goblets (Rabbit)
S-mucin B2·0	Colon (Guinea pig)
S-mucin B4·5	Exorbital Lacrimal Gland (Mouse)
SG-mucin B2·0	Glossal mucous glands (Rat, Mouse) Recto-sigmoid colonic goblets (Rat, Mouse)
SG-mucin B4·5	Duodenal Goblets (Rat, Mouse) Pyloric Glands (Rat, Mouse)
U-mucopolysaccharide	Ganglion cysts of Synovia (Man), Vitreous (Man), Cock's comb. Oestrogen-treated cervix uteri (Mouse)
C-mucopolysaccharide	? Cartilage
CG-mucin BN	Sublingual glands (Mouse, Hamster, Guinea pig)
CG-mucin N±	Vaginal epithelium (Pregnant Mouse)
C-mucin N±	Rectosigmoid mucous cells (Mouse)
S-mucin (Sap) BN	Sublingual Gland (Rat)
GC-mucin	Sublingual Gland (Man, Monkey)
C-mucin	Mammary gland secretion (Mouse)

Oxidation Methods for Mucosubstances

THE PERIODIC ACID-SCHIFF REACTION

Principles. The whole of the modern histochemistry of the polysaccharides, mucopolysaccharides and mucoproteins is bound up with the periodic acid-Schiff reaction (PAS), and it is therefore necessary to consider at this point the principles on which this reaction is based. Periodic acid (HIO_4) is an oxidizing agent which was employed by Malaprade (1928, 1934) for the chemical estimation of glycols ($CH_2OH.CHOH$) and by Nicolet and Shinn (1939, 1941) for amino-substituted alcohols ($CH_2OH.CHNH_2$) such as those which occur in the amino-acids serine, threonine and hydroxylysine. It was first employed as a reagent for polysaccharides by Jackson and Hudson (1937). Its use in histology was first described by McManus (1946) for the demonstration of mucin, and by Schabadasch (1947) for glycogen. At much the same time the method was elaborated by Lillie (1947a and b) and Hotchkiss (1948) so that it soon came to be used for demonstrating a wide variety of polysaccharide substances in the tissues. McManus (1948a) regarded the method as primarily of histological importance but later (1948b) accepted Hotchkiss's view that it could be used legitimately in histochemistry. The importance of the method in carbohydrate histochemistry is emphasized by the number of papers describing work in which it has been employed, often as the sole reaction, and by a number of excellent reviews on the subject which should be consulted by those particularly interested in the matter (Hashim, 1952; Gedigk, 1952; Graumann, 1952, 1956; Bangle and Alford, 1954; McManus, 1954; Lillie, 1953, 1954; Hale, 1957; Leblond *et al.*, 1957; McManus, 1961).

Periodic acid is an oxidant which breaks the C—C bonds in various structures, where these are present as 1:2-glycol groups (CHOH—CHOH), converting them into dialdehydes (CHO.CHO). The 1:2 glycols (*vic*-glycols) may exist in either *cis* or *trans* forms and there is considerable evidence (Bobbitt, 1956) that oxidation of *cis*-orientated hydroxyls is more rapid than that of *trans*-hydroxyls. According to Dimler (1952) secure *trans*-glycols are not oxidized by periodic acid but apparently the conformation of the hydroxyl groups can be altered during oxidation. The relative rate is correlated with the formation of a cyclic complex in which the C—O bonds of the glycols are rotated to become nearly coplanar (Honeyman and Shaw, 1959). The exact course of the reaction is unknown but it is considered to proceed by way of the singly charged anionic complex of the diol with periodic acid, as originally suggested by Criegee.

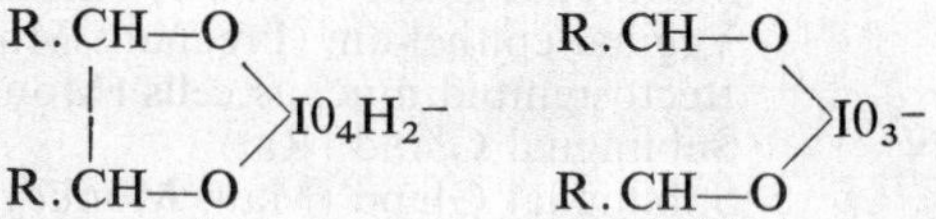

$$\begin{array}{c} R.CH{-}O \\ | \\ R.CH{-}O \end{array}\!\!>IO_4H_2^- \qquad \begin{array}{c} R.CH{-}O \\ \\ R.CH{-}O \end{array}\!\!>IO_3^-$$

$$\text{I} \qquad\qquad\qquad \text{II}$$

This complex (I) is easily dehydrated to (II) and the latter decomposes unimolecularly into the final reaction products R.CHO and IO_3^- by a process analogous to internal rearrangement (Buist and Bunton, 1954; Nevell, 1959).

The equivalent amino or alkylamino derivatives of 1:2-glycol or its oxidation product (CHOH.CO) are also attacked and converted into dialdehydes. The particular property of periodic acid, which renders it immeasurably superior to other reagents commonly used in histochemistry for oxidation of C—C bonds ($KMnO_4$, H_2CrO_4, H_2O_2), is that, under the conditions usually employed, it does not further oxidize the resulting aldehydes and these can, therefore, be localized by combination with Schiff's reagent to give a substituted dye which is red in colour. Carbonyl groups are said to be oxidized to carboxylic groups and thereby prevented from subsequent reaction with Schiff's reagent. The reaction of periodic acid with the dehydroglucose residue of a polysaccharide molecule is illustrated below. The formation of the dehydroglucose residue here shown is *trans* and rotation is assumed to have occurred during oxidation. It should be noted that the red dyestuff formed by union of fuchsin-sulphurous acid with dialdehyde is a new compound and not, as previously supposed, re-oxidized fuchsin.

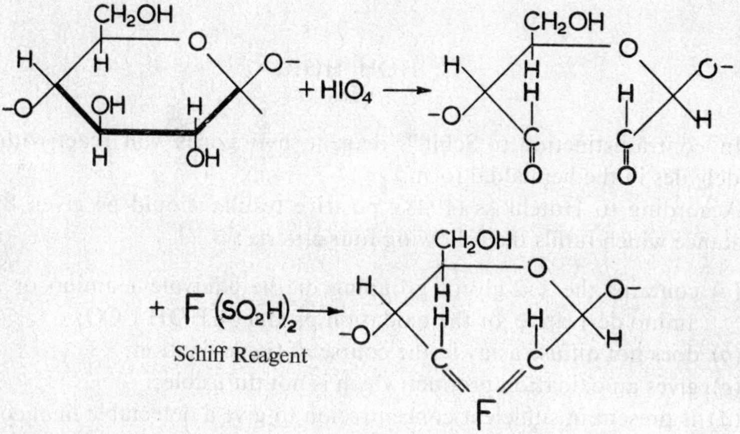

CH_2OH ... $+ HIO_4 \longrightarrow$ ... CH_2OH

$+ F(SO_2H)_2 \longrightarrow$ ... CH_2OH

Schiff Reagent

The amount of colour developed by the reaction is dependent primarily on the amount of reactive glycol structure present in the tissues and Glegg *et al.* (1952) consider that the reactive groups concerned are those of the hexose sugars glucose, galactose and mannose and of the methylpentose sugar fucose. This view is questioned by Dahlqvist *et al.* (1965) who consider that alterations in the structure of PAS-positive substances may alter the intensity of colour produced in the reaction equally as much as alteration in their amounts. The possible part played by hexosamines and hexuronic acids is ignored, but Hooghwinkel and Smits (1957) found that the hexosamine groups of CSA and hyaluronic acid were intact after oxidation with periodic acid. They noted at the same time that although the hexuronic acids were oxidized no aldehydes could be demonstrated. It was therefore concluded that "the substances responsible for a positive PAS reaction must be entirely or partly built up of other carbohydrates than hexosamine and hexuronic acid."

There are other considerations which prevent the colour developed by the PAS reaction from reflecting precisely the number of *vic*-glycols oxidized. Schiff's reagent is relatively insensitive to some aldehydes, owing to variable rates of colour formation (Sawicki, 1962). It will not react with dialdehydes (III) which have undergone hydration (IV and VI) and internal cyclization to hemialdals (V) (Guthrie and Honeyman, 1959; Guthrie, 1961).

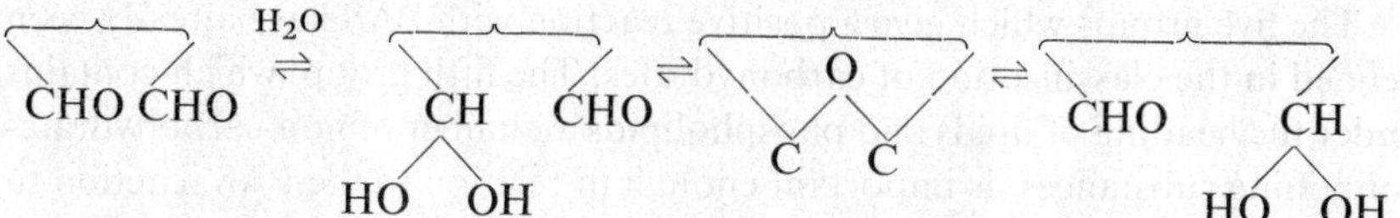

CHO CHO $\underset{}{\overset{H_2O}{\rightleftharpoons}}$ CH(HO)(OH) CHO $\rightleftharpoons$ C–O–C $\rightleftharpoons$ CHO CH(HO)(OH)

III IV V VI

Reaction of hemialdals with alcohols, as may occur if alcoholic periodic acid is used, may result in the formation of hemialdal monoglycosides (VII).

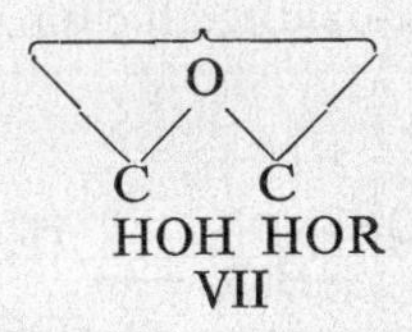

In contradistinction to Schiff's reagent, hydrazines will react with allo-dialdehydes in the hemialdal form.

According to Hotchkiss (1948), positive results should be given by any substance which fulfils the following four criteria:

(*a*) contains the 1:2-glycol grouping or the equivalent amino or alkyl-amino derivative, or the oxidation product CHOH—CO;
(*b*) does not diffuse away in the course of tissue fixation;
(*c*) gives an oxidation product which is not diffusible;
(*d*) is present in sufficient concentration to give a detectable final colour.

A large number of substances contain the 1:2-glycol grouping for which the reaction is specific and should give a positive result, **but compounds in which the glycol groups are substituted do not do so.** In the latter class are ribonucleic acid, in which the glycol group is substituted by phosphoric acid, and serine, threonine and hydroxylysine when these are combined in the polypeptide chains of proteins. Hotchkiss stated that those naturally-occurring animal substances which gave a positive result in practice were monosaccharides, polysaccharides, mucoproteins, glycoproteins, phosphorylated sugars, cerebrosides and inositol-containing lipids. He considered that after the use of ordinary aqueous fixatives only substances of high molecular weight remained in the tissues in sufficient quantity to give a positive result. Such substances were polysaccharides, hyaluronic acid, mucoproteins and mucins.

By the work of Leblond and his associates (Glegg *et al.*, 1952; Glegg, Eidinger and Leblond, 1953, 1954) the status of the PAS reaction in histochemistry was greatly clarified. In Table 21 I have listed, in various groups, the types of material which may be expected to give a positive PAS reaction. In this table there is no reference to the type of preparation employed (fresh sections, formalin-fixed frozen sections, paraffin sections, etc.) but the individual components listed therein will react positively if they are present in concentrations above a certain minimum level, and if the degree of oxidation employed has been adequate (see p. 314).

The five groups which give a positive reaction with PAS have already been defined in the classification of carbohydrates. The fifth group, which contains under the heading of lipids and phospholipids a number of non-carbohydrate-containing substances, is important enough in relation to the PAS reaction to be considered here in this chapter as well as in Chapter 12. The main substance

TABLE 21

PAS-positive Tissue Compounds

I	Polysaccharides (Glycans)		Glycogens Starches Cellulose
II	Glucosaminoglycans and Glucosaminoglucuronoglycans		Sialoglycans
III	Glycoproteins and Glycopeptide	Salivary gland mucoid Seromucoid, seroglycoid Gastric mucoid Chorionic gonadotrophin Pituitary gonadotrophins (FSH, LH). Pituitary TSH Blood group A substance Thyroglobulin	Fractions of: serum albumin serum globulin collagen
IV	Glycolipids	Gangliosides Cerebrosides { Phrenosin Kerasin	Inositol phosphatides
V	Unsaturated lipids and phospholipids	Sphingomyelin	Lipofuscins Cardiolipin Ceroid

in the phospholipid group is sphingomyelin, whose final hydrolysis products are sphingosine, fatty acids, phosphoric acid and choline. Since sphingosine contains a primary acylated amine adjacent to a hydroxyl group it is capable of reacting with HIO_4 to give an aldehyde and thus a positive PAS reaction. Verne (1929) showed that many "pure" fats, among them triolein, butter fat, and lecithin, would given a typical aldehyde reaction with Schiff's solution after mild oxidation. Evidence on the production of aldehydes from lipids and phospholipids by periodic acid oxidation of tissue sections cannot yet be considered conclusive, but Hack (1949) showed that the following lipid substances, *in vitro*, react with Schiff's reagent after oxidation with periodic acid: kerasin and phrenosin (cerebrosides) phosphatidylethanolamine and phosphatidylcholine, acetal and inositol phosphatides, and ganglioside. Obviously we require to know much more about the reaction of periodic acid with lipids other than glycolipids. Wolman (1950) carried out some useful investigations in this respect. By means of a Coujard type of technique (see Chapter 6, p. 150), this author studied the effect of oxidation by HIO_4, $KMnO_4$ and Cr_2O_3 on pure samples of cerebroside and sphingomyelin, and on a lecithin-kephalin mixture. He also used commercial samples of saturated and unsaturated fatty acids. The effect of oxidation was revealed by 10 minutes' treatment with Schiff's reagent. Wolman concluded that unsaturated lipids stained while saturated lipids did not; sphingolipids stained positively whether they contained a carbohydrate moiety or not. Later (1956) Wolman confirmed some of these results and showed that the positive reaction of sphingomyelin was dependent on the strength and duration of oxidation. If a 1 per cent. solution of HIO_4 was employed, and an oxidation time of less than ten minutes, practically no staining occurred. Although the *N*-acetylated *vic*-aminohydroxyl group is normally stable to periodic acid the molecule of sphingomyelin must be considered an exception. Carter *et al.* (1947, 1951) have shown that sphingomyelin to the left of the dotted line gives a positive PAS reaction.

$$\begin{array}{l} \qquad\qquad\qquad\qquad\qquad\qquad\qquad\qquad\qquad\qquad\qquad\quad OH \\ \qquad\qquad\qquad\qquad\qquad\qquad\qquad\qquad\qquad\qquad\qquad\quad | \\ CH_3(CH_2)_{12}CH{=}CH{-}CH{-}CH{-}\vdots{-}CH_2{-}O{-}P{-}OCH_2CH_2N.(CH_3)_3 \\ \qquad\qquad\qquad\qquad\quad | \qquad\quad | \qquad\qquad\qquad\qquad\quad \| \\ \qquad\qquad\qquad\qquad OH \quad NH{-}\vdots{-}COCH_3 \quad\; O \end{array}$$

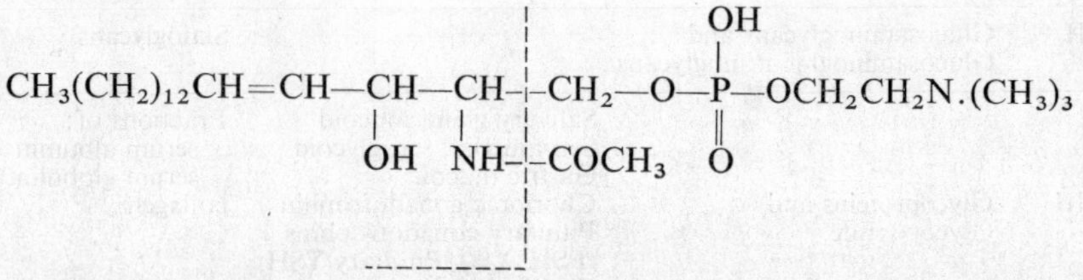

From the practical point of view, therefore, there is no doubt that lipid substances in the tissues, present as free lipids or as lipo-protein complexes, which can reasonably be considered to contain no glycolipid, can react positively by the PAS technique.

In paraffin sections, however, they are either no longer present or present in such small quantities that they do not react. Leblond *et al.* (1957) considered that once glycogen has been removed, only one class of materials gave a positive PAS reaction. These were referred to as carbohydrate-protein complexes and they are synonymous with the glycoproteins of the classification employed in this chapter. Glegg *et al.* (1954) found that these complexes contained, as their main carbohydrate components, the hexose sugars glucose, mannose and galactose, together with the methylpentose sugar fucose, and various hexosamines. The intensity of PAS staining in paraffin sections was found by Leblond *et al.* (1957) to correspond closely with the values for hexose sugars found in tissue extracts and they stressed the fact that their Fraction I of these extracts, which contained all the acid mucopolysaccharides present in the original, was PAS-negative. Fraction II, containing the above-mentioned hexoses, fucose, hexosamines and sialic acid, was strongly PAS-positive.

It is clear, from the work of these and other authors that the glucosaminoglucuronoglycans (acid mucopolysaccharides) assumed by myself and others to be capable of giving a positive though weak reaction with PAS, in fact do not react at all. The weak PAS reaction of intensely metachromatic tissue components containing acid mucopolysaccharides is thus due to carbohydrate-protein complexes of the type mentioned above. The *in vitro* tests carried out by Braden (1955) on various acid mucopolysaccharides confirmed this impression, as did those of Hooghwinkel and Smits (1957) who found that both chondroitin-4-sulphate and hyaluronic acid were PAS-negative.

Hyaluronic acid should be attacked between C_2 and C_3 but, according to Stacey and Barker (1962), acid conditions favour 3:6 lactone formation which confers resistance to periodic acid oxidation. Chondroitin sulphates react after adequate oxidation time (Wolfrom *et al.*, 1952), such as 60 hours or more, and

Stoward (1963) found that after 24 hours' oxidation they gave a moderately strong fluorescence with the salicyloyl hydrazide reaction (p. 324). With the standard 10 minutes' oxidation, therefore, the PAS reaction can be assumed to be negative with both hyaluronic acid and chondroitin sulphates.

The PAS reactivity of sialic acids must also be considered here. As indicated by the investigations of Blix *et al.* (1956), ketosidically bound *N*-acetylneuraminic acid consumes 2 mols HIO_4. It is probable that the three contiguous hydroxyls (C_7 to C_9) are responsible since participation of the *N*-acetylamino-hydroxyl link is ruled out by the work of Warren (1959). The PAS-reactivity of glycoproteins (α_1-acid glycoprotein) was stated by Montreuil and Biserte (1959) to be entirely due to sialic acid residues. This view conflicts, to some extent, with that of Clegg and Leblond, quoted above.

Further Considerations of the PAS Reaction

Oxidation. In carrying out the PAS technique it is usual to employ the double hydrate of periodic acid (H_5IO_6) which is easily obtained from most manufacturers of organic chemicals. Lillie (1947b) employed a modification in which periodic acid was produced from 1 per cent. sodium periodate (Na_3IO_5) by the action of dilute nitric acid (0·5 ml. of 70 per cent. acid per 100 ml.). This method had the advantage that a fresh periodic acid solution was produced from the stable periodate at the time of use, but it is less useful now that periodic acid of reputable quality is available. Within wide limits (0·5 to 2·5 per cent. HIO_4) the strength of the solution of periodic acid makes no difference to the final result. The majority of workers use 1 per cent., and this has become the standard concentration. The concentrations used by biochemists in elucidating the structure of glycoproteins and glycopeptides are of the same order (0·005M to 0·1M). The importance of carrying out oxidations in the presence of excess periodic acid has been pointed out by Eylar and Jeanloz (1962). Under histochemical conditions an excess will be invariably present.

In histochemical practice little notice has been taken of the observations of Head (1950) and Head and Hughes (1952). These authors showed that light had considerable effect on the specificity of periodate oxidation, due in part to the production of ozone by photochemical decomposition of metaperiodate. With fresh solutions of periodic acid, and short oxidation times, it is probable that no lack of specificity will be observed under histochemical conditions.

Solvent. Only two solvents are commonly used for periodic acid in histochemistry; they are water and ethyl alcohol. A 70 per cent. solution of the latter, buffered with 0·2M sodium acetate, was recommended by Hotchkiss (1948) for use whenever there was any question of the solubility in water of the polysaccharide under investigation, or of its oxidation product. He suggested that, where such precautions were judged unnecessary, aqueous solutions, which are more rapid and vigorous in action, might be used instead. At the present time the buffered alcoholic solution of Hotchkiss is infrequently employed. Even for staining glycogen (q.v.), for which it was originally intended, it has been found to give far less staining than the conventional 0·5 per cent. aqueous solution, after treatment with Schiff's reagent. For staining the very water soluble dextrans Mowry (1952, Mowry and Millican, 1952) used periodic acid in 90 per cent. ethanol. With this combination the time of oxidation must be extended to 120 minutes or more.

While no advantage can be anticipated, in terms of rate of reaction, from using non-aqueous solvents it is possible that selective oxidation of certain types of mucosubstances can be obtained in this way.

Time of Reaction. The time of exposure of sections to the acid is especially important. Dempsey, Singer and Wislocki (1950) showed that exposure to 1 per cent. aqueous periodic acid for 1 hour at 37° resulted in a great increase of tissue basophilia as judged by the depth of methylene blue staining. The ability of the oxidized tissues to stain with this dye at low pH values indicated that the reactive groups dissociated as strongly as some mineral acids, and the authors therefore suggested that the increased basophilia might be due to the formation of sulphonic acids by oxidation of tissue sulphydryl groups. Support for this view was derived from the structures in which the increased basophilia was especially manifest, including epidermis, hair shafts and follicles, and the endothelia of blood vessels. If Hotchkiss was correct in his assertion that periodic acid has a negligible tendency to oxidize aldehydes to carboxylic acids, and if it can be shown that the latter are not sufficiently strongly dissociated to combine with methylene blue at low pH values, then the assumption of Dempsey *et al.* that sulphonic acids are responsible for the increased tissue basophilia after periodic acid oxidation is probably correct. However, as these authors admitted, the increased basophilia produced by oxidation with potassium permanganate is comparable with that produced by HIO_4 and this similarity can be shown to exist whether methylene blue, or Schiff's solution, is used as the final indicator of the reaction products. Permanganate is known to break some bonds between adjacent carbon atoms and to oxidize these progressively to carboxylic acids *via* carbonyl groups, but this action is not restricted to 1:2-glycols as it is stated to be with periodic acid. Lillie (1951) made a comparison of the various types of oxidation used in histochemistry with Schiff's solution as final indicator. The differences were in some cases profound. Lhotka (1952a and b) tested the effect of different times (and temperatures) of oxidation. He claimed that whereas the standard 2 minutes' exposure to periodic acid would oxidize *cis*-glycols readily *trans*-glycols would be affected much more slowly. Using gelatin blocks containing isomers of cyclohexane (1952c) Lhotka found that *cis*-glycols would react with Schiff's reagent after 10 seconds oxidation and that the reaction was complete after 60 seconds. The *trans*-glycols gave a colour after 30 seconds and oxidation was complete after 2 minutes. In practice these differences have not proved sufficient basis for distinction between the two types of glycols by the PAS method. McManus (1956) maintained that 0·5 per cent. aqueous HIO_4 (0·022M, pH 2·1) would rapidly oxidize α-glycols, α-hydroxyaldehydes and α-hydroxyketones and diketones, quoting Fleury and Courtois (1950) as his authority. According to these authors (Fleury *et al.*, 1949) α-aminoalcohols, α-aminoaldehydes and α-aminoketones require a neutral or slightly alkaline medium (pH 7–8). According to Dyer (1956), however, α-glycols and α-aminoalcohols will be nearly completely oxidized in 5–10 minutes if excess HIO_4 is present, as it is when the reaction is performed histochemically. Longer than 10 minutes is required for the oxidation of α-hydroxyaldehydes, α-hydroxyketones, α-ketoaldehydes and diketones and their amino derivatives.

It is recommended that the standard time of 10 minutes' oxidation be employed for the PAS reaction, unless there are strong reasons for longer oxidation. In such cases the short (10 min.) oxidation should form the control and the longer (1–6 hour) oxidation the test. Differences observed cannot, at present, be interpreted in terms of quantitative or qualitative chemistry.

Influence of pH. The optimum pH for cleavage of *vic*-glycols lies between 3 and 4·5. At more acid levels non-specific oxidations proceed more rapidly and over-oxidation occurs more readily (Neuberger, 1941). At pH values above 5 overoxidation occurs to a very marked extent (Bobbitt, 1956; Dyer, 1956).

Temperature. If used at high temperatures for long periods, periodic acid probably oxidizes aldehydes to carboxylic acids, and perhaps even oxidizes groups other than the specific 1:2-glycol group in a similar manner. If this is true the effect shown by Dempsey *et al.* is likely to be due to carboxylic acids as well as to the oxidation products of sulphydryls and disulphides. Whatever the true explanation of the observations made by these authors, their importance is undeniable. They indicate that in the histochemistry of polysaccharide-containing substances periodic acid solutions must be used for short periods only.

An increase of only a few degrees rapidly increases the rate of oxidation, and probably affects its specificity for the conversion of 1:2-glycol groups to dialdehydes in the two ways suggested above. **In the histochemical use of the PAS reaction for substances in Groups I, II, and III, therefore, nothing higher than room temperature should be employed and the time of oxidation should be restricted to 10 minutes.**

Suggestions have beeen made (Bobbitt, 1956; Cantley *et al.*, 1963) that more selective cleavage of *vic*-glycols is obtained at 0–4°, since the reaction proceeds quite readily at this temperature.

Reducers. The use of a reducing rinse was first suggested by Hotchkiss (1948) for the purpose of removing periodate or iodate remaining combined in the tissues after the periodic acid bath, since iodates and periodates restore the colour to Schiff's solution. Hotchkiss' rinse contained KI and $Na_2S_2O_3$, in 60 per cent. alcohol with 0·02 N-HCl, and it acted as an iodide-thiosulphate solution in which the maximal amount of acid compatible with the thiosulphate was incorporated. It reduced both periodates and iodates when in the acid state needed to be reacidified when its acidity fell. An objection to its use, suggested by McManus (1948b), was that it decreased the amount of aldehyde available for colouring the Schiff reagent and thus decreased the brilliancy of the stain. Hotchkiss himself suggested that prolonged rinsing in 70 per cent. alcohol or in water might be adequate to remove trapped periodate and iodate. McManus preferred to avoid either type of rinse since he could show that control sections placed in 0·5 per cent. sodium periodate or potassium iodate for 5 minutes gave a negative Schiff reaction. There is no doubt that the aldehydes formed by the periodic acid oxidation are slowly destroyed by Hotchkiss' reducing rinse. Though at one time (1953) I considered the reducing rinse essential for the histochemical, as opposed to the histological, use of the PAS reaction, I have long since abandoned this view. If a reducing rinse is used it is reasonable to follow McManus (1954) and to designate this by the inclusion of the letter R to make the complete reaction PARS, but for most purposes the standard PAS method (Appendix 10, p. 660) should replace it.

Schiff's Reagent. The preparation of Schiff's solution has been considered in Chapter 9 and methods are given in Appendix 9. It may be stressed once more that various batches of the reagent vary considerably in performance. Some workers prefer to dilute the original solution with an equal part of distilled water, slowing its action and allowing greater flexibility. I prefer to test each new batch with a known control section, containing a large amount of the substance for which the PAS reaction is to be employed, and to calculate the optimum time of immersion from this. For human mucoprotein materials an average time of 5–6 minutes is usual with the de Tomasi variant of Schiff's solution. As Stoward (1963) has observed, Schiff-type reagents are not reliable indicators of dialdehyde formation. He has suggested that hydrazine-type reagents should take their place. These reagents are discussed below, p. 324.

Sulphite Baths. In the Feulgen reaction the Schiff bath is followed by two or three washes in sulphite water before dehydation and mounting are carried out. Both Hotchkiss and McManus employed these baths in their techniques for the PAS reaction and Lillie (1947, 1948) also employed them in his version. McManus (1948a) noticed that washing in running water for 5–10 minutes after the last sulphite bath enhanced the final colour considerably. For routine use in histology or histochemistry, I consider that the sulphite baths are unnecessary and that their place may be taken by the 10-minute wash in running water.

Interferences. The interference produced by periodate and iodate has already been considered above. That produced by the presence of acetal phosphatides (plasmalogens) has been mentioned in Chapter 9, and these substances are fully considered in Chapter 13. This type of interference is considerably greater in frozen than in paraffin sections. If unoxidized sections, treated with Schiff's reagent, indicate the presence of a considerable amount of plasmal aldehyde, a number of expedients may be employed for its blocking or removal. These are considered in Chapter 13, in the section on blocking techniques for aldehydes.

Methods for Increasing the Specificity of the Reaction. It was suggested by

Gersh (1949) and by McManus and Cason (1950) that confirmation of the fact that a given PAS reaction was due to the presence of 1:2-glycol groups might be obtained by acetylation of these groups and the consequent blocking of their oxidation by periodic acid.

$$\begin{array}{c} R \\ | \\ H-C-OH \\ | \\ H-C-OH \\ | \\ R \end{array} + (CH_3CO)_2O \rightarrow \begin{array}{c} R \\ | \\ H-C-O-C(CH_3)=O \\ | \\ H-C-O-C(CH_3)=O \\ | \\ R \end{array}$$

Gersh himself used 2 per cent., 10 per cent., or 99 per cent. acetic acid, or acetic anhydride, for this purpose remarking that, although it was unusual for simple carbohydrates to form compounds with acetic acid, the negative PAS reaction after this treatment was understandable only on such a basis. He observed that if the acetylated sections were subsequently treated with 0·1 N-HCl the 1:2-glycol group was restored and reaction with periodic acid could once more take place. I employed 10 per cent. acetic anhydride in dry pyridine for 1 hour at room temperature to effect acetylation of the 1:2-glycol groups responsible for the PAS reaction (Pearse, 1950b) but, as pointed out by Lillie (1954), the strength of the acetic anhydride and the time of incubation were inadequate for complete acetylation of hydroxyl groups though probably adequate for NH_2 groups. Lillie used 40 per cent. acetic anhydride in anhydrous pyridine for 1–24 hours at 25° or for ½ to 6 hours at 58°, and this formula is recommended. McManus and Cason (1950) used dilute alkali (0·1 N-KOH) to reverse the acetylation process and concluded that any material giving a colour with PAS, obliterated by acetylation and restored by subsequent treatment with alkali, possessed numerous 1:2-glycol groups and was likely to be carbohydrate. They supposed that acetylated amino and substituted amino groups would be less easily deacetylated than the hydroxyl groups of unsubstituted 1:2-glycols and that acetylation followed by deacetylation, HIO_4, and Schiff's reagent would distinguish 1:2-glycols from α-amino alcohols. As pointed out by Hale (1957) this is not true, since periodic acid will react as soon as the adjacent OH is deacetylated. Weak NaOH or KOH have usually been employed for deacetylation. Lillie (1954) recommended treatment with an alcoholic ammonia for the same purpose (see Appendix 10, p. 660). Hale (1953) considered that acetylation blocked all potentially PAS-positive substances and that deacetylation restored their activity. I considered, and still consider with Lison (1953), that the whole process is of academic interest rather than practical utility.

Objections to the Specificity of the Reaction. Jeanloz (1950) raised objections to the use of the PAS reaction for identification of polysaccharide structure on the grounds that the presence of two adjacent free OH groups (1:2-glycol) in the carbon chain bore no relationship to the positive reaction. He stated that numerous sugars possessing such groups, cellobiose and methyl α-glucopyranoside for instance, gave a negative result, while the consumption of HIO_4 by substances which react positively varied from 1 molecule per hexose residue in the case of starch, glycogen and cellulose to 0·1 to 0·4 molecules per hexose residue in the case of hyaluronic acid and chitin. McManus and Hoch-Ligeti (1952), however, using a modification of the Hotchkiss spot test, obtained a positive PAS reaction with cellobiose, and hyaluronic acid and chitin have now been shown to be PAS-negative. It is clear that consumption of HIO_4 need not be related in any way to the PAS reaction since the latter will be negative however much or little HIO_4 is involved unless the reaction product is a stable non-diffusible aldehyde. The work of Dahlqvist *et al.* (1965) is relevant here. These authors found, in their *in vitro* tests, very little correlation between the amount of periodate consumption and of PAS reactivity except in so far as they observed that substances which consumed HIO_4 also reacted positively.

Compounds containing pentose sugars with unsubstituted glycol groupings may give a positive result in fixed tissues. The methylpentose sugar fucose has already been shown to do this (Leblond *et al.*, 1957). Compounds such as adenosine-5-phosphoric acid (muscle adenylic acid) and adenosine triphosphate, in which the 1:2-glycol of D(-) ribose is free, give a positive result with the PAS reaction *in vitro*. In frozen sections especially, the A bands of voluntary muscle are strongly stained with PAS even when the glycogen they contain has been removed. Since hexose-containing glycoproteins are not known to be present in the situation, despite the assurances of Hotchkiss that substances of low molecular weight are removed during fixation, dehydration and embedding, it is possible that the reacting substance is protein-bound adenylic acid. Bearing these possible exceptions and lipids of various classes in mind, a positive PAS reaction must be considered to be due to the presence of hexoses. The part played by hexosamines appears to be unimportant and hexuronic acids can be regarded as non-contributory.

It must be emphasized that differences in the PAS reactions of given structures, reported by different workers, are as much due to variation in technique as to variation in the structures themselves. Paneth cell granules and blood and tissue eosiniphil granules are often reported as staining positively. The former undoubtedly vary depending on the species from which they are derived and on the type of fixative employed (Lillie, 1950). In human intestine they can be made positive or negative by varying the solvent used for the periodic acid and the time of oxidation. On the other hand, the reported differences in the case of human Russell bodies are equally due to variation in these structures themselves since they may be composed of strongly or weakly staining mucoprotein or glycoprotein, or of PAS-negative simple protein. Only a proportion of mast cells, demonstrable as such by their metachromasia, stain by the PAS method, and in many of these the colour developed is weak. Gomori (1952) considered that this was due to the fact that fully substituted (sulphated) heparin is PAS-negative (Jorpes *et al.*, 1948). Heparin monosulphate, however, was observed by these last authors to be PAS-positive *in vitro*.

Alternative Oxidizing Agents. Lead tetraacetate ($PbAc_4$) was first used by Criegee (1931, 1948) for chemical estimations on 1:2-glycols. It was also extensively used by Perlin (1955) for elucidation of the structure of reducing disaccharides. Each position of the glycosidic linkage is associated with a characteristic oxidation pattern. A number of authors applied the reagent to the histochemical study of carbohydrates (Crippa, 1951; Glegg *et al.*, 1952; Shimizu and Kumamoto, 1952; Jordan and McManus, 1952; Lhotka, 1952d; Hashim and Acra, 1953; Graumann, 1953a and b). Originally the reagent was used as a 1 per cent. solution in acetic acid but later workers added sodium or potassium acetate following the finding by Criegee and Buchner (1940) that cleavage of 1:2-glycols was greatly accelerated by the presence of these compounds. Hashim and Acra (1953) found that strong solutions of acetic acid caused swelling of the tissues and slowed the rate of the reaction. They therefore diluted the solution with water or organic solvents (benzene, toluene). Casselman (1954) made a comparative study of the various methods employing $PbAc_4$. He found that starch and glycogen did not give consistently positive or negative results. Their reaction depended on the conditions of oxidation. The least active, but more specific reagent, leaving starch and glycogen unstained, was the simple solution in acetic acid (glacial). If acetate was added both substances tended to give a weak positive reaction. Dilution with water increased the rate of oxidation and produced a regularly positive reaction with starch and glycogen.

It was noted by Criegee (1935) that, in addition to 1:2-glycols, $PbAc_4$ would readily oxidize α-hydroxy acids (CHOH.COOH) whereas HIO_4 would more readily oxidize α-hydroxyaldehydes and ketones (CHOH.CHO,CHOH.CO). Lhotka (1957) therefore, suggested that the difference between PAS and PA,$PbAc_4$,S should indicate the amount and location of α-hydroxy acids. He was unable to observe any difference, however, in practice. In a previous paper (1952c), following the observations of Hockett and McClenahan (1939), Bell and Baldwin (1941) and Hockett *et al.* (1943), Lhotka had shown that with $PbAc_4$ in acetic acid the respective oxidation times for *cis* and *trans*-glycols were 15 and 30 minutes (see formula for β-D-glucose on p. 296; this is written as a *trans*-glycol with the hydroxyls in the 2- and 3-positions on opposite sides of the ring). He suggested that distinction between the two might be made by these means but little use seems to have been made of this principle. Owing to overlap in the middle of the range it would be necessary to confine oxidation to less than 5 minutes to be reasonably certain that mainly *cis*-glycols were reacting. This period of oxidation is quite inadequate; Glegg *et al.* (1952) used 1–4 hours and Crippa (1951) 30 minutes.

The reactions of boric acid with polysaccharides were recorded by Deuel *et al.* (1948). They noted that cyclic glycols with the *cis*-configuration formed complexes with boric acid much more easily than did *trans*-glycols. One boric acid molecule combines with one or two glycol molecules. Staple (1955, 1957) was unable to use this reaction to block the oxidation of glycols with $PbAc_4$

but he found that the presence of 0·1 to 0·01M-boric acid during oxidation would prevent subsequent reaction with Schiff's reagent only in certain structures. The effect was better demonstrated with alcoholic than with aqueous Schiff's reagent. It seems that Staple's results are similar to those obtained by the interposition of various aldehyde-blocking reagents after HIO_4 oxidation and that, as with these, interpretation of the effect is complicated by steric factors which are difficult to assess.

Other oxidizing agents which have been employed histochemically for cleavage of 1:2-glycols include sodium bismuthate (Lhotka, 1952e), manganese acetates (Lhotka, 1953) and phenyl iodosoacetate (Lhotka, 1954). The first of these, which was used by Rigby (1949, 1950) in chemical studies, apparently acts by forming a cyclic compound with the glycol which then splits to form 2 carbonyl compounds and bismuth pentoxide. The reaction takes place in aqueous or organic media but Lhotka found that the best medium was 20 per cent. orthophosphoric acid. His results were essentially similar to those produced by HIO_4 and $PbAc_4$. Zonis and Pesina (1950) and Zonis and Kornilova (1950) showed that the tri- and tetraacetates of manganese would split 1:2-glycols to produce aldehydes and Lhotka (1953) applied this reaction to the histochemical demonstration of glycols. He stated that these reagents could not replace HIO_4 and $PbAc_4$ but that they might prove useful as supplementary reagents if their specificity became better understood. In further studies devoted to the testing of alternative oxidants Lhotka (1954) used phenyl iodosoacetate prepared by the method of Pausacker (1953). He found that a 0·02M solution in acetic acid produced the optimal degree of oxidation in the minimum time. Owing to the fact that it is necessary to synthesize both this and the manganic acetate reagents they have not been further tested and no valid opinion as to their value in carbohydrate chemistry can be given at the present time. A number of alternative oxidants used in polysaccharide chemistry have not been tested in histochemical practice. Among these are the Midas reaction (H_2O_2 and OsO_4 in *tert*-butanol) and perbenzoic acid in $CHCl_3$ at 0°.

The standard PAS reaction demonstrates the G-mucosubstances, and the SG-, CG-, and GC-mucins of the Spicer classification given on p. 305. These include the neutral and acid glycoproteins, gastric surface mucin and thyroglobulin, the mucins of various mucous glands and of the intestine, and the epithelial sialomucins. Some of these are shown in the illustrations in Plates IVa and c facing p. 269, and Vb opposite.

A method based on the condensation of an arylamine with the aldehydic products of periodic acid oxidized mucosubstances was reported by Lillie *et al.* (1961). Further elucidation of the reaction was given by Lillie (1962). It was shown that *m*-aminophenol produced a more rapid and more complete blockage of the Schiff reactivity of periodic acid-engendered aldehydes than more usual methods, such as acetic-aniline (p. 457). Furthermore, the bound aminophenol could then be demonstrated by reaction at pH 8 with fresh or stabilized diazonium salts. For this purpose Lillie *et al.* used the stable diazo-

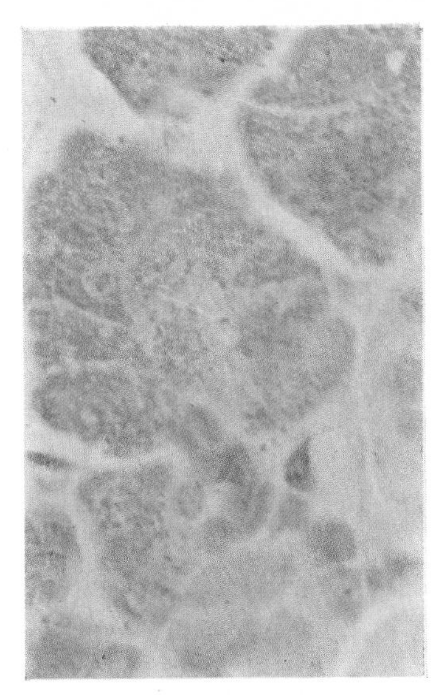

VA. Rat Salivary Gland. Freeze-dried, formaldehyde vapour-fixed section. Alcian blue - haematoxylin staining. Fine (blue) granules fill the cytoplasm of the acinar cells. × 870

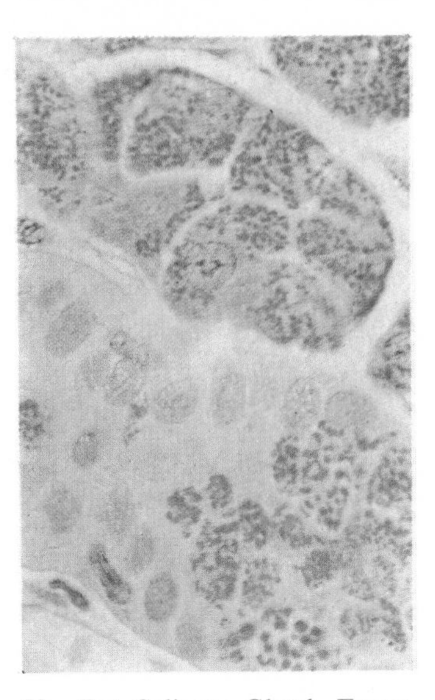

VB. Rat Salivary Gland. Freeze-dried, formaldehyde vapour-fixed section. PAS-haematoxylin staining. Shows PAS-reactive mucosubstances (GC-mucin) in coarse granular form in convoluted tubules (right) and in fine granules in the acini (left). × 870

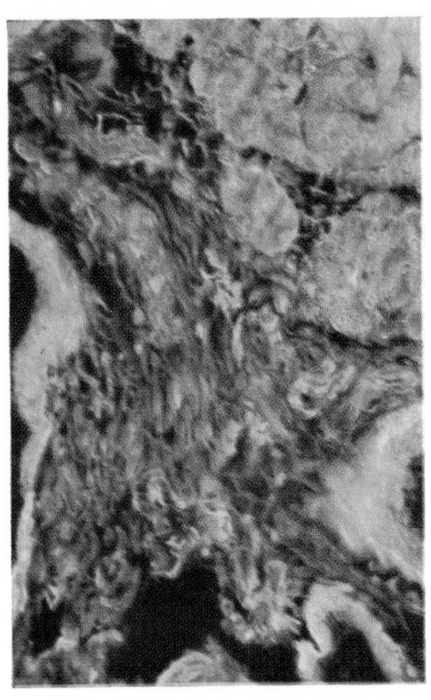

VC. Rat Colon. Goblet cell mucins in freeze-dried formaldehyde vapour-fixed section. Treated by (1) Methylation (2) Periodic acid oxidation (3) Acriflavine. × 170.

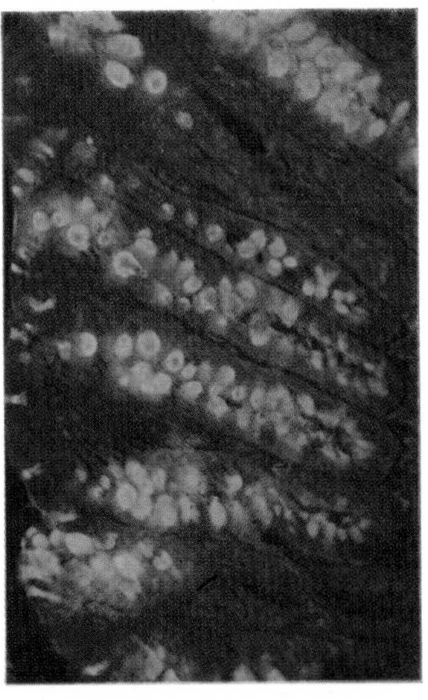

VD. Rat Salivary Gland. Freeze-dried, formaldehyde vapour-fixed section. PA-Coriphosphine method. Shows sulphated mucosubstance (in mast cells) red. Acinar, periodate-reactive, mucosubstances appear yellow. × 170

tate of 4(*p*-nitrophenylazo)-2,5-dimethoxyaniline (Fast Black Salt K; p. 516) and produced a black reaction product on a red to pink background.

A series of tests was carried out by Lillie (1962) to determine whether the Schiff's base formed in the reaction was an azomethine (I) or a secondary amine (II).

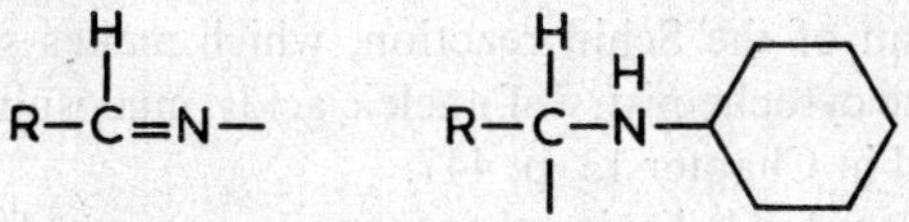

It was observed that nitrosation (1 N-$NaNO_2$ and 1 N-Acetic acid, equal parts, 10 mins, 25°), following *m*-aminophenol condensation, weakened any subsequent reaction with diazonium salts. On reacting the nitroso compound (III) with a Griess-Feigl reagent (*p*-nitroaniline), (IV), a red or orange azo-dye (V) was formed at sites containing mucosubstances. This sequence of tests indicates that the Schiff base is a secondary amine and not an azomethine.

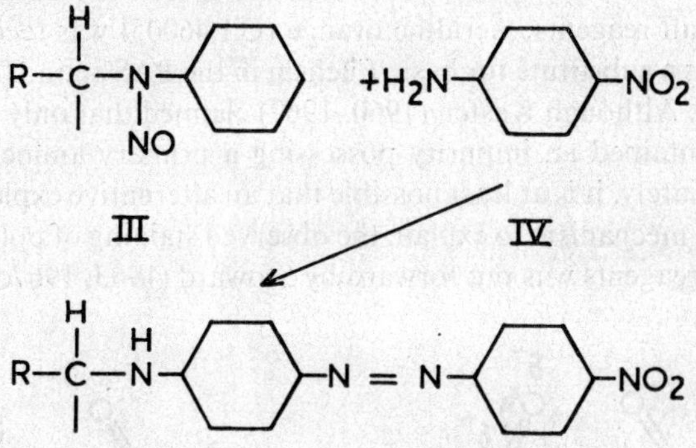

The periodic acid-Schiff base method (called Black Periodic by its authors) has not been used to any great extent. It has not been shown that it affords any distinction of types of mucosubstance though it would be expected to show differences when compared with the standard PAS reaction.

In many cases, particularly when potentially fluorescent aromatic amines are required to condense with tissue aldehydes to form fluorescent anils, it is impossible to form amine-aldehyde complexes in tissue sections. When aldehydes react with potentially fluorescent amines there is necessarily a loss of reactivity of the amine groups. Their contribution to the resonance energy of the compound, and hence to its fluorescence, is thus lost. It seems probable that chemically active amines capable of condensation with CHO groups will not give rise to highly fluorescent anils. Stoward (1963) reported, however, that he could obtain fluorescent tissue anils with 5-aminosalicylic acid and with *p*-aminohippuric acid.

Periodic Acid—PseudoSchiff Reactions

The mechanism of the Schiff reaction, which makes such an important contribution to the cytochemistry of nucleic acids, mucosubstances and lipids, is fully considered in Chapter 13, p. 447.

There are reasons for believing that a number of so-called Schiff reagents, formed by adding a mineral acid and SO_2 to aqueous solutions of basic dyes of the aminoacridine series (Kasten, 1960), combine with aldehydes produced by HIO_4 oxidation by a mechanism different from that of the true Schiff reaction. Three such dyes, in particular, have been used for the demonstration of induced polyaldehydes by fluorescence microscopy. They are Acridine yellow (C.I.46025), Coriphosphine O (C.I.46020) and Benzoflavine (C.I.46065).

According to Craig and Short (1945) the amine groups of monoaminoacridine dyes do not take up protons, except in a very acid medium, and the amine groups of the diaminoacridines are even more inert. Hence neither can form true Schiff reagents. Acridine orange (C.I.46005) was recommended by Betts (1961) as a substitute for basic Fuchsin in the PAS stain. This dye has no amine groups. Although Kasten (1960, 1962) claimed that only batches of the dye which contained an impurity possessing a primary amine group would react appropriately, it is at least possible that an alternative explanation exists.

A possible mechanism to explain the observed staining of polyaldehydes by pseudoSchiff reagents was put forward by Stoward (1963, 1967c). This author

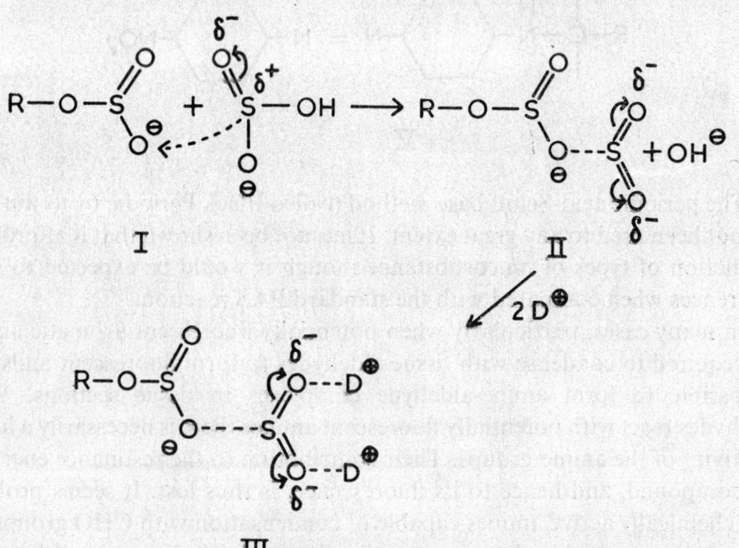

I

II

$2\,D^{\oplus}$

III

found that pseudoSchiff reagents stained not only induced polyaldehydes but also strongly basophilic sites containing either PO_4^- or SO_4^-. For the latter he proposed that in solutions containing both SO_2 and HCl the SO_2 would be adsorbed at its electrophilic S atom on to PO_4^- or SO_4^- (I) by way of an intermediate complex (II) having two centres of comparatively high electron density on the two oxygen atoms of the SO_2 moiety.
Basic dyes (D^+) could then form salt-like links at each of these oxygen atoms, as indicated in the equation (III).

Mechanism of the Reaction between Aldehydes and PseudoSchiff Reagents

Because of its obvious theoretical importance, the explanation offered by Stoward (1963) for the polyaldehyde-pseudoSchiff reaction is given below. It is suggested that the SO_2 is attached either to the ring or to the amino nitrogen atoms of acridine dyes by a salt linkage. Positive charges are carried on the ring and amino N atoms, as shown below on the two canonical forms of the resonance hybrid (I, II) of the basic ion of the 2:8 diaminoacridine dyestuffs.

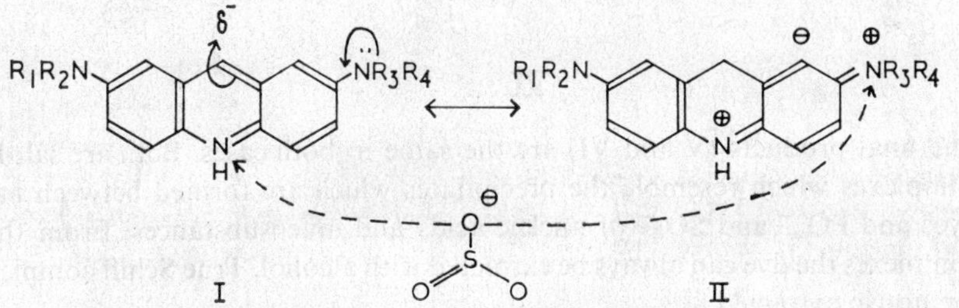

I II

$R_{1,2,3,4}$ = H or CH_3

The negatively charged sulphite ion neutralizes the electron-deficient centres on the dye to form the intermediate III, which is presumed to ionize to IV.

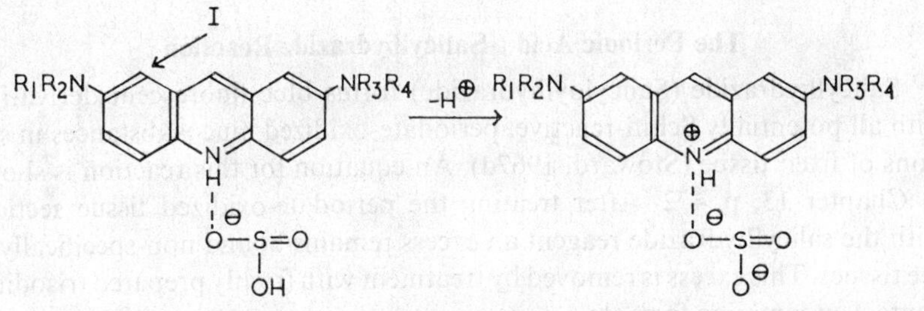

III IV

It is considered unlikely that the O_2 atom of the SO_2 is attached to the nuclear N of the dye by a covalent linkage. The instability of this linkage is illustrated by the reduction of acridine-N-oxide with sodium metabisulphite (Lehmstedt and Klee, 1936). The dye-SO_2 complexes probably react with aldehyde by one of two routes:

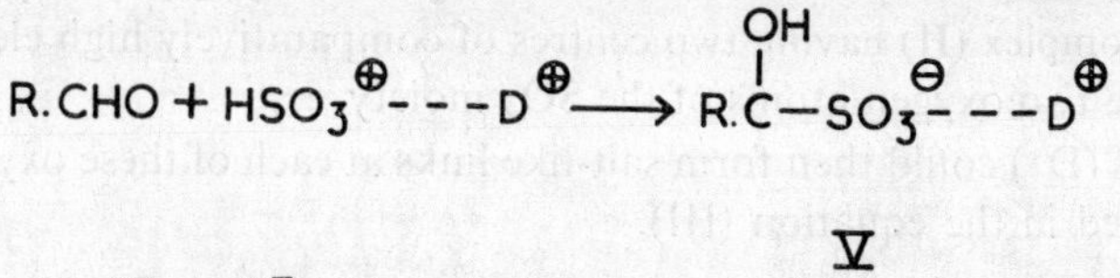

$$R.CHO + HSO_3^{\oplus}---D^{\oplus} \longrightarrow R.\overset{OH}{\overset{|}{C}}-SO_3^{\ominus}---D^{\oplus}$$

V

$$R.C(=O^{\delta-}\leftarrow H-O^{\delta-})(H)(\delta^{+}) + S(=O)(O^{\ominus}) \longrightarrow R.\overset{OH}{\overset{|}{\underset{H}{C}}}-O-S(=O)-O^{\ominus} \xrightarrow{D^{\oplus}} R.\overset{OH}{\overset{|}{CH}}-O-S(=O)-O^{\oplus}--D^{\oplus}$$

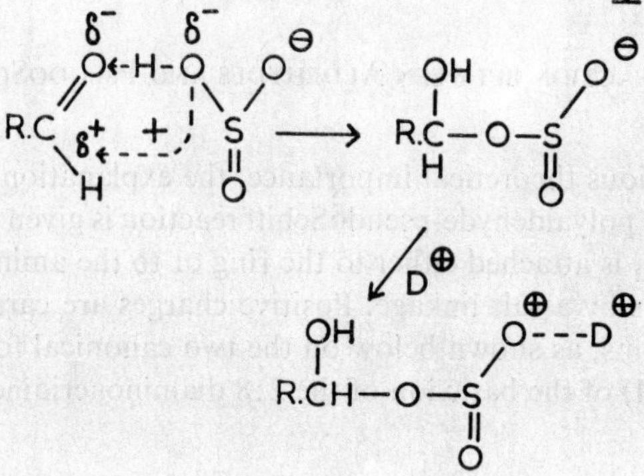

VI

The final products (V and VI) are the same in both cases. Both are salt-like complexes which resemble the precipitates which are formed between basic dyes and PO_4^- and SO_4^- of nucleic acids and mucosubstances. From these complexes the dye can always be extracted with alcohol. True Schiff complexes are not so extracted.

The three dyes mentioned at the beginning of this section can be used as fluorescent pseudoSchiff reagents for the demonstration of tissue polyaldehydes. Details are given in Appendix 10, p. 664, and the results of the application of two of them are shown in Plates Vc and Vd, facing p. 320.

The Periodic Acid—Salicylhydrazide Reaction

Salicylhydrazide (Salicyloylhydrazide) forms blue fluorescent derivatives with all potentially Schiff-reactive, periodate-oxidized mucosubstances in sections of fixed tissue (Stoward, 1967d). An equation for this reaction is shown in Chapter 13, p. 472. After treating the periodate-oxidized tissue sections with the salicylhydrazide reagent an excess remains bound non-specifically to the tissues. This excess is removed by treatment with freshly prepared trisodium pentacyanoammine ferroate.

According to Stoward (1967d), with certain exceptions mucosubstance salicylhydrazones emit a bluer and more intense fluorescence after being treated with a solution of an aluminium salt. Zinc salts, on the other hand, quench the fluorescence emission of almost all mucosubstance salicylhydrazones. In some cases the autofluorescence of tissue proteins is such as to make distinction from mucosubstance salicylhydrazone fluorescence very difficult. Stoward found that aluminium-chelating dyes of the Solochrome black series could be used to "quench" tissue protein autofluorescence by converting it to a red fluorescence, easily distinguishable from the blue of the salicylhydrazones. Details of the periodic acid-salicylhydrazide reaction, and of its various modifications, are given in Appendix 10, p. 661. The result obtained, in the case of rat salivary gland, is shown in Plate VIIIb, facing p. 355.

Periodic Acid—Phenylhydrazone-Formazan Reactions

The Phenylhydrazone Formazan Reaction

This reaction was introduced by Seligman, Gofstein and Rutenberg (1949) as part of the synthesis of an iodized tetrazolium salt and it was stated by Ashbel and Seligman (1949) to be specific for aldehydes as opposed to ketones. In the first and second editions of this book, therefore, it appeared in the chapter on Aldehydes and Ketones. Since 1960 it has taken on new significance and it is more fitting to deal with the subject at this point, rather than in Chapter 13.

According to Ashbel and Seligman the phenylhydrazones of aldehydes would couple with diazonium salts in pyridine to form water insoluble, coloured, formazans. The phenylhydrazones of ketones, because of the absence of replaceable hydrogen, would not undergo the reaction which was presumed to proceed as follows:

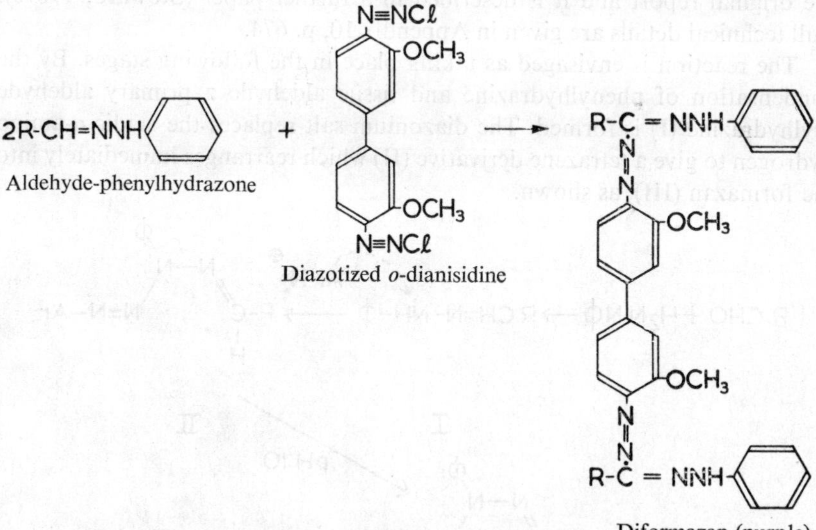

2R-CH=NNH-C_6H_5 + (N≡NCl, OCH_3 / OCH_3, N≡NCl) → R-C(=NNH-C_6H_5)-N=N-(…OCH_3 / OCH_3…)-N=N-C(R)=NNH-C_6H_5

Aldehyde-phenylhydrazone

Diazotized *o*-dianisidine

Diformazan (purple)

The negative results obtained in formalin-fixed sections of adrenal cortex when this method was applied suggested that free aldehydes were not present.

Like the original authors I could not succeed in making the method work with formalin-fixed adrenals but, in paraffin sections exposed to periodic acid, the sites which were normally coloured by means of Schiff's reagent were coloured purplish-red by the use of phenylhydrazine followed by the stable diazotate of dianisidine in watery pyridine. The original (1953) illustration (Fig. 96, facing p. 287) shows granules of mucosubstances in the gastric mucosa stained in this manner after oxidation with periodic acid. I was unable (1953, 1960) to show aldehydes, as revealed in DNA by acid hydrolysis, by the phenylhydrazone-formazan method using a number of diazonium salts. At the time, I did not consider that the negative formazan reaction in the adrenal cortex offered conclusive proof of the absence of aldehydes. Great differences in reactivity were noted between the various aldehydes in the tissues and the substances with which they combined and the negative formazan reaction with the aldehydes engendered by Feulgen hydrolysis was presumed to be due to the protection of the hydrogen atom from substitution by the diazotate.

The PA-Phenylhydrazone-Formazan Reaction

A revival of interest in this reaction was created by the report of Stoward and Mester (1964) that whereas only some of the periodic acid-produced dialdehydes from *vic*-glycols would react with Schiff's reagent, all would react with aromatic hydrazines. The latter included salicyloyl hydrazide and 2-hydroxy-3-naphthoic acid hydrazide. Stoward and Mester used the formazan reaction successfully to show differences between aldehydes, ketones and hemialdals or hemiacetals. The method has been developed considerably since the original report and it is described in a further paper (Stoward, 1967e). Full technical details are given in Appendix 10, p. 674.

The reaction is envisaged as taking place in the following stages. By the condensation of phenylhydrazine and tissue aldehyde a primary aldehyde arylhydrazine (I) is formed. The diazonium salt replaces the acidic α-amino hydrogen to give a tetrazene derivative (II) which rearranges immediately into the formazan (III), as shown.

$$\mathrm{R.CHO + H_2N.N\phi \longrightarrow \underset{I}{R.CH{=}N{-}NH{-}\phi} \xrightarrow{Ar\,N_2^{\oplus}} \underset{II}{R{-}C(H){=}N{-}N(\phi){-}N{=}N{-}Ar}}$$

$$\mathrm{II \xrightarrow{pH\ 10} R.C(=N{-}N(\phi){-}H)(N{=}N{-}Ar)}$$

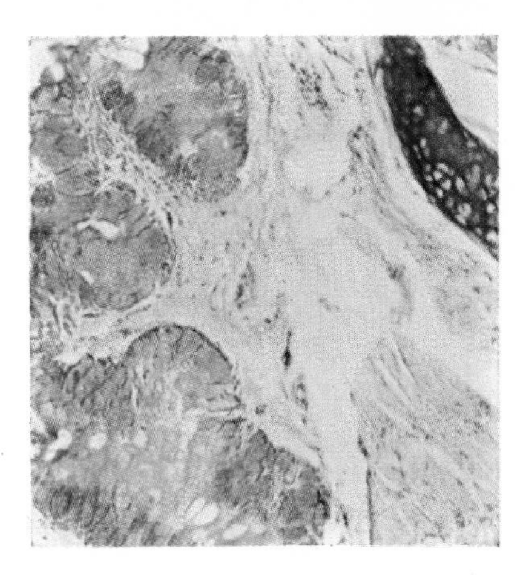

VIa. *Sceloporus undelatus* (lizard), foregut. Blue staining of sulphated epithelial mucins. Alcian blue—Neutral red. × 65.

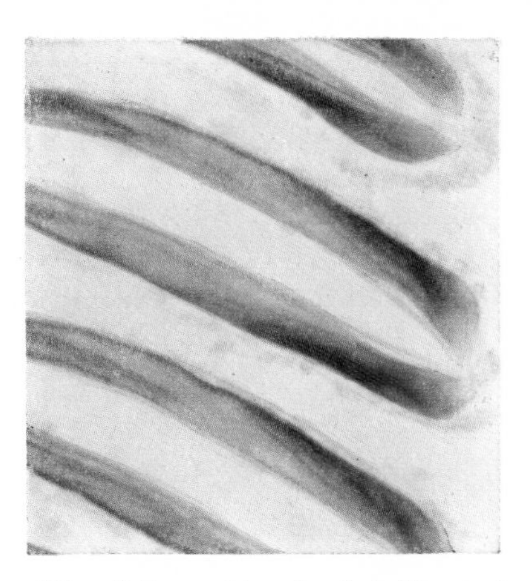

VIb. *Triton rubicundus* (whelk) ctenidium. The gill plates stain blue over the outer two-thirds of their length. In the absence of SH groups and reducing lipids this reaction indicates the presence of reducing sugars (chitin). Alkaline tetrazolium reaction. × 270.

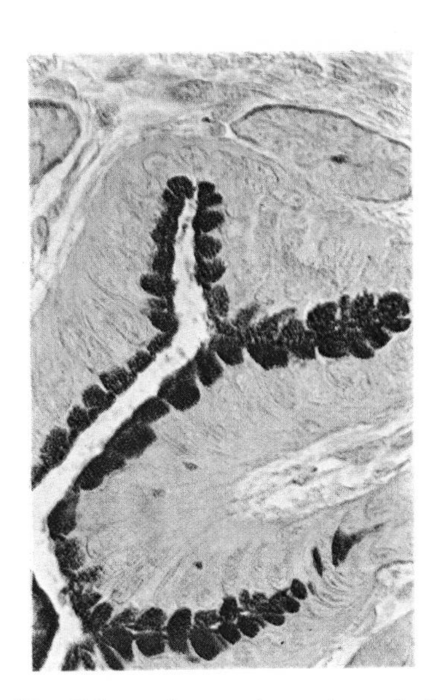

VIc. *Salamandra s. salamandra*, pyloric region of stomach. Freeze-dried, formaldehyde vapour-fixed preparation. TDMBF method, counterstained with methyl green. × 260

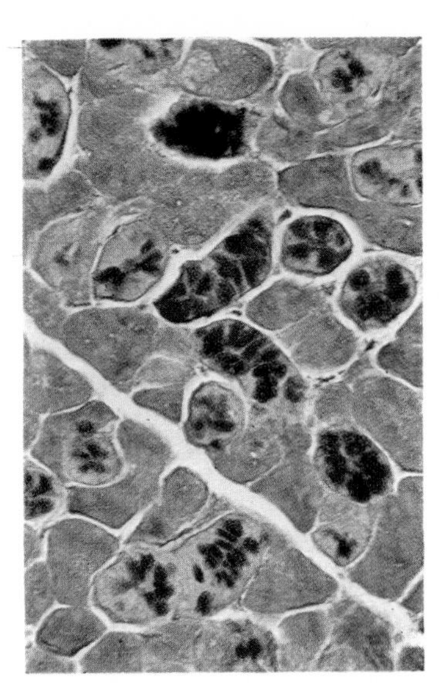

VId. Rat Salivary Gland. Freeze-dried, formaldehyde vapour-fixed preparation. Shows periodate-reactive mucosubstance in the convoluted ducts, stained black by the ANSA modification of the TDMBF method. × 170

Arylhydrazones cannot couple in this way unless they possess a replaceable hydrogen or if their α-amino hydrogen is not acidic. Consequently ketone phenylhydrazones, and practically all aldehyde acid hydrazones, cannot be converted into formazans.

Failure to obtain the expected histochemical result was attributed to the degradation of the aldehyde phenylhydrazones soon after their formation by the strong, hot solutions of phenylhydrazine acetate which Pearse (1953, 1960) used. Such degradation has been observed *in vitro* (Barry andMitchell,1954). By leaving the periodate-oxidized tissues in lower concentrations of phenylhydrazine in acetic acid, for only 3–7 minutes at room temperature, Stoward and Mester (1964) produced yellow to yellowish-brown formazans in oxidized mucosubstances after treatment with freshly diazotized aniline.

Since the colours of the diphenyl formazans produced with diazotized aniline were too pale to provide adequate contrast Stoward (1967b) investigated a number of alternatives. He soon found that freshly tetrazotized 3,3′-dimethoxybenzidine (TDMB) gave rise to much more highly coloured derivations. The original stable diazotate recommended by Ashbel and Seligman (1949) was Fast Blue B, which is derived from this base, and its stabilized zinc double salt was used by Pearse (1953) to produce Fig. 96.

It is evident, nevertheless, that reliable and reproducible results can only be obtained if the diazonium solutions used for coupling with aldehyde phenylhydrazones are free of metals. Stoward (1967b) gives directions for the synthesis of tetrazotized 3,3′-dimethoxybenzidine fluoroborate (TDMBF) which provides a stable and acceptable form of Fast blue B. It does, however, give rise to one difficulty. This is due to staining of the background in various shades of yellow or brown often sufficiently dark to mask the colour of the reacted mucosubstance. The difficulty is due essentially to the uptake of TDMBF by the carbohydrate-free proteins of the connective tissue. Stoward (1967a and c) found that the only way to prevent uptake of diazonium salts by tissue proteins was by prior methylation with methanolic thionyl chloride (see Chapter 6, p. 163 and in this chapter, p. 329). He emphasized that methylation did not affect the PAS-reactivity of tissue mucosubstances.

Mono and Diformazan Formation. Theoretically the mucosubstance phenylhydrazone formazans formed in the tissues could be either mono- or diformazans, or both. Because the diformazans *in vitro* are black, and those formed in tissue sections red, brown or yellow, it is reasonable to assume that the latter are monoformazans. Confirmation of this point comes from the observation that they can successfully be postcoupled with 1-amino-8-naphthol-4-sulphonic acid (ANSA) to give a differently coloured product.

The colours of mucosubstance phenylhydrazone formazans were observed by Stoward (1967b) to differ widely from one site to another. As expected, fixation affected the result and this was presumed to be due to varying degrees of preservation. Methylation also affected the strength and final colour of the reaction. This was particularly evident in the case of glycogen which did not form a coloured product at all with the methylation-formazan-ANSA sequence although its PAS reactivity was undiminished by methylation. In Table 22, below, is given a list of mucosubstances and their products in the various PA-phenylhydrazone formazan reactions.

TABLE 22

Phenylhydrazone-formazan Reactivity of Hamster Tissues

Histological Site and Tissue[1]	Histochemical[2] Classification	$PhN_2ØN_2$[3]	Meth.[4] $PhN_2ØN_2$	Meth.[5] $PhN_2ØN_2$– ANSA
Tongue:				
Mucous glands	SG-mucinB1·5A1·0	3Br	3PkBr	4P
Serous glands	SG-mucinBr·0	3Br	±Br	2BrR
Subdermal collagen	G-mucinB±(20$MgCl_2$)	1–2PkBr	±Pk	2PkBr
Tracheal Cartilage:	SG-mpsB1·5A1·0T	1Br	1Br	1PkV
intercellular lacunar	SG-mpsB1·5A1·0T (0·6$MgCl_2$)	2Br	2Br	2PkV
Sublingual Gland:				
mucous	GC-mucinB1·5A2·5N	3PkP	3BrPk	3V
Submandibular Gland:				
mucous	CSG-mucinB3·5A2·5	3PBr–4RBr	3BrPk	30rP
Stomach:				
Fundus, surface	G-mucin	2–3Br	2–3BrP	3PBr
Pylorus, surface	G-mucin	3Br	3PBr	—
Pylorus, glands	SG-mucinB3·5A1·0 (0·2$MgCl_2$)	4Br	2PBr	—
Duodenum:				
sup. Brunner	SG-mucinB1·5A1·0	4BrR	3BrR	4PR
deep Brunner	G-mucin	3BrP	1–2V	3V
Jejunum-Ileum: goblet	CG-mucinB3·5A1·0	3–4Br	3Br	3Pbr
Ilio-Colon, junct:				
deep mucous cell	CSG-mucinB1·5A1·0	2YBr	2–3YBr	2–3P
surface cell	CSG-mucinB1·5A1·0	3YBr	3–4PkBr	3–4P
goblet	SG-mucinB3·5A1·0	4–5Br	±PBr	1–2PkBr
Placenta: glycogen	Glycogen	4RBr	—	—
Vagina: epithelium	SG-mucinB3·5A1·0 (0·2$MgCl_2$)	1–3Br	3BrP	—

Notes:

(1) Pregnant Syrian hamster tissues, fixed in neutral formalin.
(2) Classification of Spicer *et al.* (1965); see p. 305.
(3) Phenylhydrazine in phosphate buffer, pH5·0, 25°, 3 minutes, followed by TMBDF.
(4) Prior methylation (p. 329) followed by (3).
(5) Prior methylation, followed by (3), followed by ANSA sequence (p. 327).

Colours: Br = brown; Or = orange; P = purple; Pk = pink; R = red; V = violet; Y = yellow.

Intensity: Arbitrary scale 0–5.

Examples of the results of the PA-phenylhydrazone-formazan reaction, and the ANSA modification, are shown in Plates VIc and VId, facing p. 326. Details of the various technical procedures appear in Appendix 10, pp. 674 and 675.

With one exception, all mucosubstances that are potentially periodic acid-Schiff positive give rise to $PhN_2ØN_2$ formazans, according to Stoward (1967a) This exception is the mucin of the mucous cells of the colon. This is PAS-positive but it gives rise to no formazan. Stoward suggested that this was due to the presence in the carbohydrate moiety of large numbers of hydroxyl groups. He drew an analogy with periodic acid-oxidized glycogen which is both Schiff-positive and phenylhydrazone-formazan positive. After methylation, however, PA-oxidized glycogen remains Schiff-positive but will no longer give rise to formazans. The process of methylation, and its significance in mucosubstance histochemistry, is considered below.

Methylation

If fixed tissues containing mucosubstances are treated with methanolic thionyl chloride (see Chapter 6, p. 163 and Appendix 6, p. 619) for 4 hours at room temperature, nucleic acid phosphate groups, protein and sialomucin carboxyl groups, and the sulphate groups of most sulphomucins are apparently esterified. In sulphomucins containing both carboxyl and sulphate groups it is possible that only the sulphate is esterified. Methanolic thionyl chloride is distinguished from hot methanolic-HCl by the fact that it esterifies SO_4^- groups without desulphation or extraction, in the majority of cases. A few sulphated mucins, however, such as those in the goblet cells of the small intestine, are desulphated completely. After saponification (Appendix 10, p. 660) some of the reactions for free SO_4^- are restored.

Mechanism of the Methanol-Thionyl Chloride Reaction

Although much is still not understood concerning the mechanism of this reaction the views put forward by Stoward (1967a) are summarized here. The methylchlorosulphinic acid ester (I; $R{=}CH_3$) is thought to be the first product formed when thionyl chloride is added to methanol (Stähler and Schirm, 1911; Voss and Blanke, 1931; Carré and Libermann, 1933). The chlorosulphinate intermediate (I) will then react with more alcohol to form the sulphite (II). If excess thionyl chloride is added (more than half a molecular equivalent) this will be converted back to the chlorosulphinic acid ester (I) (Gerrard, 1944).

$$ROH + SOCl_2 \rightarrow \underset{\text{I}}{R.OSOCl} + HCl$$

$$ROH + ROSOCl \rightleftharpoons \underset{\text{II}}{(RO)_2SO} + HCl$$

$$(RO)_2SO + SOCl_2 \rightarrow 2ROSOCl$$

Once formed, the chlorosulphinate (I) may possible dissociate into the alkyl chloride (III) and SO_2 (Boozer and Lewis, 1953).

$$ROSOCl \xrightarrow{Cl^-} \underset{\text{III}}{RCl} + SO_2 + Cl^-$$

It is not known which of the various intermediates is responsible for methylating polyanionic substances in tissue sections. The important fact is that fixed tissues after 4 hours' treatment with methanolic thionyl chloride at room temperature will no longer take up diazotized aromatic amines at pH 9 or basic fluorescent dyes such as coriphosphine or acridine orange (see p. 342, below).

The intensity of the PAS reaction is totally unaffected by methanolic thionyl chloride treatment.

CARBOXYLATION

Treatment of fixed mucosubstances with phthalic anhydride (PTA) was suggested by Shackleford (1963). This compound probably reacts with hydroxyl groups in the tissues to produce free carboxyls whose reactions can be studied histochemically.

$$C_6H_4{<}^{CO}_{CO}{>}O + HO\text{-}R \longrightarrow C_6H_4{<}^{C(=O)\text{-}O\text{-}R}_{C(=C)\text{-}OH}$$

This carboxylation reaction is blocked by prior acetylation (Appendix 6, p. 613); it restores the alcianophilia of tissues in which this has been reversed by prior methylation. PTA carboxylation produces an intense orthochromatic basophilia for thiazine dyes, especially in epithelial mucins. Since it does not block the PAS reaction it must be presumed that PTA reacts with hydroxyls other than those of *vic*-glycols, such as the hydroxyls of serine and threonine. If this can be shown to be the case, the significance of the PTA reaction will be much greater than it is at present.

Metachromatic Methods for Mucosubstances

Histochemically metachromasia can be defined as the staining of a tissue component so that the absorption spectrum of the resulting tissue-dye complex differs sufficiently from that of the original dye, and from its ordinary or orthochromatic tissue complexes, to give a marked contrast in colour. As such, it may be produced by a wide variety of dyes (a full list was given by Lison, 1935b). Most histochemical work on metachromasia has been carried out using dyes of the thiazine group and most of our information on the mechanism of metachromasia refers, similarly, to dyes of this group.

Two invariable criteria for metachromasia are:

(1) A shift of the absorption maximum of the dye towards shorter wavelengths;

(2) A decrease in the molar extinction coefficients at the absorption maximum.

In the case of the thiazine dyes this means a change of colour from the orthochromatic blue, through violet, to red.

TABLE 23

Absorption Spectra of Metachromatic Dyes
(After Kelly, 1956)

Dye	Absorption Maxima (nm)	
	Orthochromatic	Metachromatic
Toluidine blue	630	480–540
Azure A	620	480–530
Methylene blue	665	570
Cresyl blue	625	530
Crystal violet	590	510
Basic fuchsin	543	510
Thionine	597	557

With the thiazines the mechanism underlying the production of metachromasia is better understood than with the unrelated triphenylmethane dyes (methyl violet) or azines (neutral red). The metachromasia of the triphenylmethanes is important histochemically only in the case of amyloid, and the reactions of this substance are dealt with in Chapter 11. On account of its practical simplicity thiazine dye metachromasia was extensively employed as a histochemical method but there is now some falling off in the number of applications reported in the literature. A great volume, of paper at least, has been covered by articles and reviews dealing with the mechanism of the reaction and it may be well to consider the latter at this point. Reference should be made, by those especially interested in metachromasia, to the origin papers of Sylvén (1941, 1945, 1954), Kramer and Windrum (1955), Kelly (1956), Schubert and Hamerman (1956), Scheibe and Zanker (1958), and Ghiara (1959).

Mechanism of the Metachromatic Reaction. Michaelis (1902) suggested that metachromasia with the thiazine dyes was due to the formation of tautomers of the dyes employed which possessed absorption spectra differing from the original and Holmes (1928) showed that this explanation was probably true in the case of the oxazine dye brilliant cresyl blue. More modern work by Michaelis, in collaboration with Granick (1945), introduced the conception of polymer formation. The monomeric form of the dye was considered to be blue or violet, the dimers and trimers progressively more violet and the polymers red or pink. Polymerization of the substrate, with which the dye combines, induced polymerization of the dye, and hence metachromasia. Contributions to this theory were made by Bank and Bungenberg de Jong (1939), Rabinowitch and Epstein (1941), Sheppard and Geddes (1944a and b), Vickerstaff and Lemin (1946), Massart *et al.* (1951) and Sylvén and Malmgren (1952).

Later contributions have come from Pal and Schubert (1963), Schoenberg and Moore (1964), and Graumann *et al.* (1966). The influence of the molecular structure of the dyes (thiazines and oxazines) on their metachromatic properties was considered by Taylor (1961).

Following Michaelis (1947) it is possible to recognize two varieties of metachromasia in histochemistry. This author showed that toluidine blue has an absorption spectrum with three bands, alpha, beta and gamma. The monomeric alpha form is blue, and dimeric beta form violet and the polymeric gamma form is red. The production of γ-(red) metachromasia with toluidine blue in tissue sections is due to predominance, in or on the substance stained metachromatically, of the γ form of the dye. The purple β-metachromasia is not due to predominance of the β band since, as Schubert and Hamerman have pointed out, this actually disappears with formation of the γ band. It is presumably due to the simultaneous presence of the γ- and α-forms of the dye. Provided that this is recognized there is no harm in using the terms γ- and β-metachromasia to refer to the red and purple types respectively, as far as tissue sections are concerned.

Sylvén (1954, 1959) regarded metachromasia as "a special type of orderly dye aggregation characterized by the formation of new intermolecular bonds between adjacent dye molecules." The molecules of the substrate he regarded as centres of orientation whose free anionic groups would attract the polar groups of the dyestuff, as postulated by Sheppard (1942) and as indicated in Fig. 97, below, left. This can be considered as the primary requirement for metachromasia.

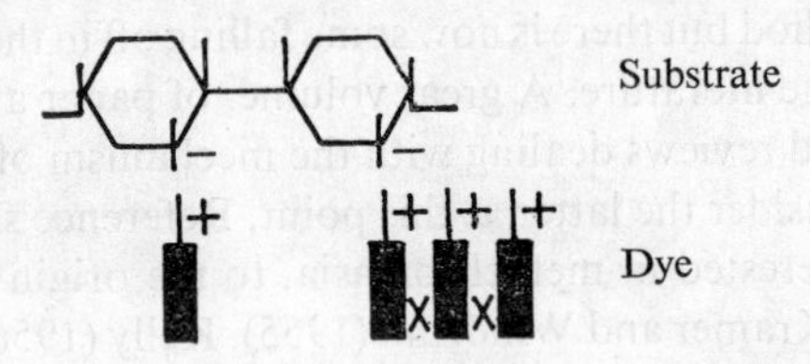

FIG. 97. X = Water molecule.

Aggregation of the dyestuff, the secondary requirement for metachromasia, is shown in the right-hand part of the diagram. It is favoured by high dye concentrations, and by high pH levels which favour the maintenance of a higher degree of hydration. Sylvén first stressed the fact that the presence of water is essential for metachromatic interaction.

For the first stage of metachromasia to take place a certain minimum surface density of negative charges on the substrate is necessary. In hyaluronic acid, where the distance between COOH groups is about 10·3 Å, no metachromasia occurs in dilute aqueous solutions. In the case of pectic acid, a straight chain uronic acid polymer, the intercharge distance between COOH groups is about 5 Å and this is enough to produce a weak metachromasia. Glucosaminoglucuronoglycans containing alternate SO_3H and COOH groups would be expected to have an intercharge distance of about 4 and 6 Å and to show a moderate degree of stable metachromasia. With the introduction of two or three SO_3H groups per disaccharide unit the intercharge distance is reduced to less than 4 Å and this results in a strong and stable metachromasia.

Sylvén considered that the spectral shifts which constitute metachromasia were due to the formation of new bonds, with higher bonding energy, between the aggregated dye molecules. Because of their low energy and reversibility these have been considered to be secondary valency bonds (van der Waal's forces), but their true nature is not understood. According to Pal and Schubert (1962) and Mukherjee and Ghosh (1963), they were to be regarded as hydrophobic. For their occurrence certain minimal energy requirements must be satisfied. Complete metachromasia over a substrate, without admixture of orthochromatic components, is first reached when the γ-band prevails. This occurs at about 520 nm which corresponds to about 8 cal/mol, according to calculations based on Planck's formula. Table 24, below gives some of Sylvén's calculations which serve to illustrate this important conception.

TABLE 24

Energy requirements for production of metachromasia

Absorption maxima (nm)	E (cal./mol.) Δ = E max − 615 nm	Maximum absorption of purified compounds
600	1·0	—
580	2·8	RNA
560	4·5	Pectinate
540	6·4	Chondroitin sulphate
520	8·5	KSCN, detergents
500	10·6	Heparin
480	13·0	Trisulphonated HA

The conception of polymerization of the dye, which is inherent in the above-mentioned work, is strongly held by the majority of workers in the field.

The question of polymerization of the substrate is another matter. Sylvén and Malmgren (1952) found that metachromasia was more easily produced when their substrate (S-free hyaluronic acid) was present in gel form. Concentrated gels and dried films showed marked γ metachromasia though solutions were devoid of this effect. It was observed that the introduction of more COOH, PO_4 or SO_3H groups produced metachromasia even in the sol state, and this agrees with the observations of Walton and Ricketts (1954) who tested the precipitation of toluidine blue *in vitro* by dextran polymers and their sulphated and carboxylated analogues. They found that the amount of dye precipitated was a linear function of the concentration of the polymer and its SO_3H content. The degree of polymerization had no effect on the metachromasia of the precipitates and Walton and Ricketts concluded that if, under histochemical conditions a tissue component showed a fall in its metachromatic reaction this could mean (1) Decrease in concentration of acidic polysaccharide (glucosaminoglucuronoglycan) or (2) Loss of acidic groups by protein binding (see below), or (3) Disruption of the molecular structure of the polysaccharide or degradation to diffusible fractions. It did *not* mean depolymerization of the polysaccharide. Walton and Ricketts also found that one molecule of the dye reacted with one acidic group of the substrate which agreed with the later conception of Michaelis (1950) that dye was distributed in monomolecular fashion over the surface of the substrate. Schubert and Hamerman (1956) pointed out that Walton and Ricketts, in their work on dextrans, were measuring the amount of dye precipitated and not metachromatic activity and that the two were not necessarily the same. They also pointed out, and this should always be remembered, that there is no method known by which the degree of metachromasia can be measured since the dyes concerned do not obey Beer's law in aqueous solutions.

To sum up; from the theoretical point of view metachromasia signifies *only* the presence of free electronegative surface charges of a certain minimum density. The older theory of specific groups cannot be sustained. Negative metachromasia *may* mean bound or masked electronegative surface charges, or the absence of these from the material being examined.

Metachromasia of Fluorescent Dyes. The problem of the metachromasia of fluorescent dyes is more complicated. Most work has been carried out with dyes of the acridine series (acridine yellow, acridine orange, coriphosphine, benzoflavine). All glucosaminoglucuronoglycans *in situ* take up basic fluorescent dyes but, with the exception of acridine orange and coriphosphine, their fluorescence is usually quenched. Artificially sulphated mucosubstances, *in vitro*, form strongly fluorescent complexes with most aminoacridines but in sulphated tissue sections (p. 337) they do not do so. According to Stoward (1963) the ionic charge density on the SO_4^- of mucopolysaccharides is so high that it distorts the symmetry of the π electron system of the dye and thus destroys an essential prerequisite for fluorescence. It should be noted that fully sulphated glycogen, stained *in situ* with coriphosphine fluoresces in a dull red brown only.

All acridines adsorbed to glucosaminoglucuronoglycans may undergo photolytic decomposition to non-fluorescent products after exposure to UVL. The least susceptible dye is coriphosphine (Stoward 1967b).

The metachromasia of fluorescence of the acridine dyes was investigated by Kuyper (1962) who calculated the fluorescence and absorption spectra for a number of these dyes at different pH levels and temperatures. From his results Kuyper concluded that the hypothesis of specific aggregation of the dye was doubtful. This view was directly refuted by Scheibe and Zanker (1962). It conflicts, also, with the evidence produced by Pal (1965) who showed that aggregation of fluorescent basic dyes gave rise to metachromatic spectra and to quenching of their fluorescence.

The Practice of the Metachromatic Reaction. *Fixation.* This subject has already been discussed in Chapter 5, p. 99, where some selective fixatives for acid mucopolysaccharides have been listed. Many studies on metachromasia are still carried out on paraffin sections of tissues fixed in the standard medium used in the department concerned and it is not possible to be dogmatic in this matter. It must be emphasized, despite the opposite views of a number of authorities, that fixation is still of considerable consequence and that lesser degrees of metachromasia can easily be abolished by fixation and subsequent embedding. I would like to enter a plea for the extended use of fresh, thin, cold microtome sections in studies on metachromasia. These should be used in pairs, one section totally unfixed and the other rapidly fixed in one of the preferential fluids such as Lillie's alcoholic lead nitrate (p. 99). In this way a check can be made on water-soluble substances which may be lost even during the brief staining involved. The use of freeze-dried formaldehyde vapour-fixed sections must also be given due consideration since this process affords the highest degree of preservation of soluble mucosubstances.

Concentration of the Dye. The ratio of dye to substrate, or chromotrope as it is often called, has considerable influence on the production of metachromasia (Levine and Schubert, 1952; Schubert and Levine, 1953, 1955; Michaelis, 1950) and if insufficient dye is present metachromasia will not occur (Sylvén and Malmgren, 1952). The usual concentration of dye is of the order of 0·1 per cent. but even 0·01 per cent. is adequate (Kramer and Windrum, 1955). Following the example of Sylvén (1941, 1945) the solvent now commonly used is 30 per cent. alcohol. Chromatographic and spectrophotometric investigations carried out by Ball and Jackson (1953) drew attention to the unsuitability of some samples of toluidine blue for use as metachromatic stains. Most samples of toluidine blue perform adequately and contain, besides the metachromatic component, only non-staining fractions. Other dyes, such as thionin, contain different staining fractions which have different colours. A metachromatic effect with such dyes may be due to differential staining.

Duration of Staining. For the dye concentrations usually employed a time of 5–20 minutes is appropriate but with very dilute solutions (0·001 per cent.) times of up to 4 days have been used. Kramer and Windrum pointed out that overstaining of strongly metachromatic tissue components could lead to an appearance of orthochromasia. Thus the staining time should not only be related to the concentration of dye but also to the type of tissue component being studied.

Post-Treatment of Sections. In the first edition of this book it was stated, in strong terms, that "alcoholic dehydration was capable of removing all metachromasia" and that the process of differentiation thus involved should be avoided by examination of the sections in water, after staining. Exception was taken to this statement, or rather to the manner in which it was stated, and so it was restated in the second edition in different terms.

The conception of Sylvén (1941, 1945) that alcohol-resistant metachromasia is "true" and alcohol-labile metachromasia is "false" must be regarded as a definition, which can be accepted or rejected. Kramer and Windrum (1955) supported this definition and performed experiments to prove it. They contended that alcohol *never* reverses metachromasia due to ester sulphates. Others (Baker, 1951; Hess and Hollander, 1947; Landsmeer, 1951; and Wislocki and Singer, 1950) considered that alcohol could abolish metachromasia and, on the other hand, Persson (1953) noted that the metachromasia of sulphate-free hyaluronate was not abolished by alcohol.

My own objections were and are merely to the use of the words true and false in describing the effects of alcoholic dehydration. I did not and do not disagree with Sylvén's contention that alcohol-fast metachromasia is most likely to be due to the presence of sulphate esters. With Kelly (1956) I still feel that it is essential to examine sections first in water, or in the solvent in which they were stained. After noting, and if necessary recording photographically, the metachromasia present, the section can be mounted in a medium such as Apathy's syrup or dehydrated and mounted in a synthetic resin. According to Kramer and Windrum these two procedures have similar effects. An alternative procedure for retention of toluidine blue metachromasia was suggested by Izard (1964) who mounted his wet preparations directly in prepolymerized butyl methacrylate. The effects on metachromasia of various solvents were described by Pal (1965).

In animal tissues strong γ-metachromasia in water-mounted sections may be due to ester sulphates or to materials containing other acidic groups· β-metachromasia in such tissues is due to the presence of the weaker acidic groups (PO_4; COOH) in high local concentration. At pH 5, high local concentrations of side-chain polypeptide carboxyls give γ-metachromasia. Alcohol-fast or Apathy-stable γ-metachromasia is largely due to ester sulphates but it has not yet been shown that some of the metachromasia of these is not extinguished by differentiation. Alcohol-stable β-metachromasia is normally exhibited only by nucleoproteins as first observed by Wislocki, Bunting and Dempsey (1947).

Ionic Competition. Lison (1953), in the second edition of his book, agreed that his original idea that stable metachromasia was the sole property of higher esters of sulphuric acid could not be maintained. He pointed out, however, that where the metachromasia of polysaccharides was due to carboxyl groups it was easily reversed by treatment with 1 per cent. NaCl and suggested that this might form the basis of a differential method. Sylvén (1954) also commented on the ease with which small concentrations of salt would abolish or *prevent* a metachromatic reaction, and Ishizuka (1955) also described the effects of neutral salts on metachromasia. More recent work on this subject was reported by Arnold (1966) who used NaCl, Na_2SO_4 and $CaCl_2$ in concentrations from 10^{-5}M to 10^{-2}M. This author concluded that their only effect was to disrupt the build-up of γ-metachromasia by metal cation competition for the anionic groups of the chromotrope.

Protein Competition. It was conclusively demonstrated by Kelly (1955), Hamerman and Schubert (1953) and French and Benditt (1953) that the addition of soluble proteins would prevent metachromasia. Competition between the dye and the protein for the available acid groups of acidic polysaccharides is markedly modified by pH. In the case of histone, elevation of the pH favours the dye and lowering of the pH favours the protein, but Noguchi (1956) showed that anionic polysaccharides combine with proteins even on the alkaline side of the I.E.P., where both have negative net charge. This principle has not been applied to the problem of increasing the specificity of the metachromatic reaction.

SULPHATION-INDUCED METACHROMASIA

Metachromasia in polysaccharides after sulphation was first shown by Bignardi (1940a and b). Following the demonstration by McManus and Mowry (1952) that treatment of tissue sections with sulphuric acid would induce hæmatoxylin-staining of basement membranes Kramer and Windrum (1953, 1954) developed sulphation-metachromasia as a histochemical technique. Several methods of sulphation were employed: (1) Conc.H_2SO_4, followed by washing in water; (2) Chlorosulphonic acid in pyridine (Ricketts, 1952); (3) Sulphuric acid-acetic anhydride (Schrauth, 1932); (4) Pyridine sulphur trioxide (Baumgarten, 1926); (5) Chlorosulphonic-acetic-chloroform; (6) Sulphuric acid-acetic anhydride-ether.

It may be convenient, at this point, to consider the merits of some of the different methods of sulphation listed above, together with some newer methods.

Concentrated Sulphuric Acid. In spite of its histochemical priority this reagent cannot be recommended for *O*-sulphation. It has a number of concomitant and deleterious actions:

(1) Extraction of nucleic acids
(2) Degradation of aldehydes
(3) Sulphation and hydrolysis of glycogens
(4) Hydrolysis of glycosidic links of various glycans
(5) Oxidation of primary hydroxyls to carboxyls
(6) Sulphation of hydroxyls of amino-acids.

Although there is evidence (Sutur, 1944) that the hydroxyls of simple alcohols can be sulphated, as can those of phenols and proteins, by conc. H_2SO_4 *in vitro* there is no evidence that carbohydrates respond in the same way. On the contrary, Rice and Fishbein (1956) indicated that they are converted to hydroxy-furfuraldehydes.

Pyridine-SO_3 Complexes. These were originally called chlorosulphonic acid-pyridine and subsequently pyridine-SO_3. Both preparations contain pyridine-SO_3 with such contaminants as pyridine-SO_3H, pyridine-Cl and hydrochloric acid. Chlorosulphonic acid in pyridine will sulphate, *in vitro*, not only the hydroxyls of mucosubstances but also those of protein residues. When carbohydrates are treated with pyridine-sulphur trioxide complex (in excess pyridine) a number of side reactions occur. Some of the hydroxyl sulphonic acid esters initially formed may be replaced by chloride or pyridinium ions or they may be eliminated during the process of intramolecular ether formation (Reynolds and Kenyon, 1951).

Chlorosulphonic Acid in Acetic Anhydride. This Kramer and Windrum reagent displaces secondary hydroxyl groups and substitutes chloro (Roberts, 1957; Guiseley and Ruoff, 1961). Chlorosulphonic acid alone, *in vitro*, preferentially sulphates primary hydroxyl groups.

Sulphuryl Chloride Vapour. This reagent was introduced into histochemical practice by Lewis and Grillo (1959), for rapid sulphation of glycogen. According to the authors isolated neutral and acid mucopolysaccharides could also be sulphated very rapidly. Neutral mucopolysaccharides *in situ* in tissue sections were not rendered basophilic, however, with the sole exception of glycogen. No explanation for this difference is available.

Sulphuryl chloride is an active reagent which gives rise to a number of side reactions. In tissue sections it attacks the thiol and disulphide groups of the proteins, either by oxidation or by chlorosulphation.

$$RSH + ClSO_2Cl \rightarrow R.SSO_2Cl + HCl$$

and the thiosulphuryl chlorides so formed can react with water or alcohols to give thiosulphonates.

$$RS.SO_2Cl + R'OH \rightarrow R.SOO_2OR + HCl$$

The complexity of reactions which this reagent affords militate against its histochemical use without appropriate caution. The fact is, however, that it performs as the original authors suggested, whatever reason may be.

Dioxane-SO_3 Complex. This reagent is made by distilling sulphur trioxide into a cooled chloroform/dioxane mixture. Mono- and di-adducts are formed (compare with pyridine-SO_3, described above).

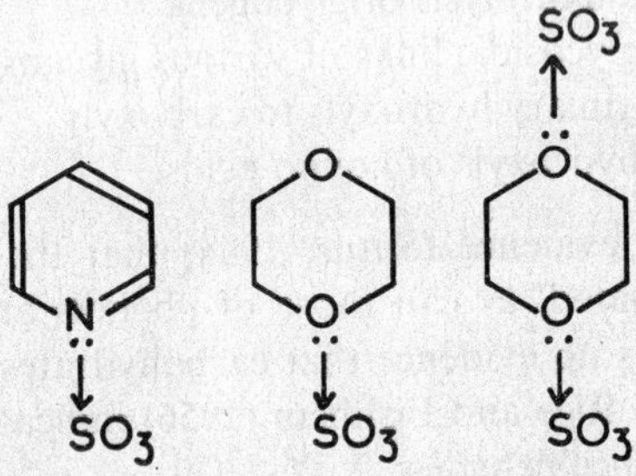

SO3 SO3 SO3 SO3

Stoward (1963) found that dioxane-SO_3, suspended in chloroform, produced some γ-metachromasia in liver after 3–5 minutes. Better results were obtained using dimethyl formamide (DMF) as solvent for the adduct but 10–15 minutes treatment was required to develop the full intensity.

Dimethylformamide-SO_3 Complex. This reagent was described by Coffey *et al.* (1949) and used by them (1948, 1949) for sulphation of leuco vat dyes. It is made by distilling SO_3 into DMF, and is soluble in excess DMF. Stored at

0° it is stable for 1–2 months and it is used at room temperature. All available mucosubstance hydroxyls are sulphated but so are NH_2 and OH groups of proteins.

Stoward (1963) considered that the dipolar complex

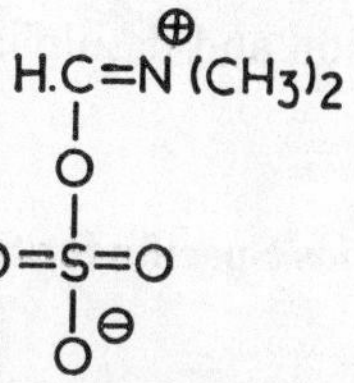

$$\begin{array}{c} \overset{\oplus}{\text{H.C=N}}(CH_3)_2 \\ | \\ O \\ | \\ O{=}S{=}O \\ |\ominus \\ O \end{array}$$

was the best reagent for sulphating OH groups in tissue sections, provided that the simultaneous sulphation of amino groups could be avoided, or selectively reversed. He attempted a number of methods for blocking NH_2 prior to sulphation, but found that each had a serious defect.

(1) Conversion by diazo-deazo reaction (p. 161).
Extracted mucosubstances excessively during reaction.

(2) König-Sassi (p. 120) or salicylaldehyde (p. 118) reactions to block amines.
Both produced steric interference with sulphation of mucosubstance hydroxyls.

(3) Oxidative deamination (p. 161), using HClO or tungstate/H_2O_2.
These methods did not block a sufficient number of amine groups. Because of the failure of these amine-blocking techniques it was necessary to consider their removal *after* sulphation. A number of methods for desulphation are available and these will now be considered.

Desulphation Methods

Acetylative Desulphation. This method was used by Wolfrom and Montgomery (1950) for desulphation of chondroitin sulphate. They treated this product with almost absolute H_2SO_4 at $-10°$, followed by acetylation with excess acetic anhydride. In tissue sections this method fails to reduce γ-metachromasia of sulphated mucosubstances.

Reduction with Metal Hydrides. Bollinger and Elrich (1952) and later Grant and Holt (1960) reported the slow removal of sulphate groups by complex metal hydrides and Rees (1961) described desulphation of chondroitin-4-sulphate, *in vitro*, by potassium borohydride in alkali. In spite of these reports, sulphated mucosubstances are not desulphated by borohydride reduction.

Methanolic-HCl. According to Charles and Scott (1936) the *N*-sulphates of natural sulphated mucopolysaccharides (heparin)are hydrolyzed, *in vitro*, by nitrous or acetic acids. Dilute solutions of mineral acids remove *N*-sulphates in 30 minutes while *O*-sulphates require more than 12 hours (Wolfrom, Shen and Summers, 1953). Extending this principle Kantor and Schubert (1957) used 0·06M-methanolic HCl for the desulphation of chondroitin sulphate and Danishefsky *et al.* (1960) used it for heparin. Chondroitin sulphate B (dermatan sulphate) was desulphated by means of mature (aged) solutions of acetyl chloride in methanol (Stoffyn and Jeanloz, 1960). After considerable experiment Stoward (1963) concluded that 1 per cent. HCl in methanol at 22° would remove only *N*-sulphates from tissue sections. If temperatures higher than 22° were used, however, *O*-sulphates could be methylated or removed.

The techniques of sulphation and desulphation are given in detail in Appendix 10, pp. 666 and 667.

Reactions Specific for Sulphates

The Bracco-Curti Method

This method, introduced into histochemistry by Bracco and Curti (1953) depends on the reaction between tissue sulphate and benzidine. The insoluble salt so formed is demonstrated by oxidation to benzidine blue by oxidation with potassium dichromate. The method was criticized by Geyer (1962b) who indicated that its efficiency was restricted by the method of colour development which resulted in loss of specificity and accuracy of localization. There is no doubt that this last point is a valid one (localization) and Geyer suggested that the oxidation step should be omitted and replaced by inspection under UV light, taking advantage of the ultra-violet absorption of the tissue-bound benzidine salts.

The original method was used by Tock and Pearse (1966) who found that its sensitivity was low, so that thicker than normal sections could be used to advantage. No fault was found, however, with respect to its specificity. The results are illustrated in Fig. 98, p. 350 and details are given in Appendix 10, p. 676.

The Tetrazonium-Sulphonate Method

Based on a similar principle to the foregoing reaction is the tetrazonium sulphonate method of Geyer (1962a). Tissue sulphates, or sulphonic acid groups produced by oxidation or sulphation, react with tetrazotized benzidine at pH 2·3:

$$R.SO_3H + Cl{-}\underset{\underset{N}{|||}}{N}{-}R' \rightarrow R.SO_3{-}\underset{\underset{N}{|||}}{N}{-}R' + HCl$$

After washing out excess of the tetrazonium salt the sulphate-bound tetrazotate is coupled with 1-naphthol at pH 9·4. No reaction with phenols or other tissue aromatic groups occurs at pH 2·3 (see Chapter 6, p. 127). The resulting reddish-violet azo dye is soluble in organic solvents and the final product is therefore mounted in a watery medium. By comparison with the Bracco-Curti method more colour is developed and it is not necessary to employ thicker than normal sections.

Details of this reaction are given in Appendix 10, p. 676.

THE BARIUM-RHODIZONATE METHOD

This method, which was introduced by Stempien (1963), was stated to be based on the affinity of barium ions for tissue sulphate. The sites of uptake were subsequently demonstrated by forming the reddish-brown rhodizonate complex *in situ*. Most authors who have described their attempts to demonstrate uptake of Ba^{2+} by tissue components have indicated a complete lack of success.

Basic Dye Methods

THIAZINES

Methylene blue extinction (MBE). From the earliest days of histological staining it was customary to use the basophilia of the acid mucopolysaccharides as the basis of a qualitative procedure (e.g. the hæmatein and carmine of the older mucin stains, which depend upon it). Using the method originated by Pischinger (1926, 1927), Dempsey and Singer (1946), Dempsey *et al.* (1947), and Dempsey, Singer and Wislocki (1950) used a technique for making quantitative estimation of the degree of basophilia registered by tissue compounds. They stained sections with 0.5mM methylene blue at various pH levels for 24 hours, mounting them with minimum dehydration, and measured (usually spectrophotometrically) the amount of dye bound by various tissue components. I employed this method, without spectrophotometry, using as end point the virtual extinction of binding capacity for methylene blue (Pearse, (1949a, 1950a). The MBE method appeared to make possible a comparative estimation of the degree of basophilia manifested by various structures and it was found useful for distinguishing between acid mucopolysaccharides and mucoproteins. Dempsey and Singer (1946) found that mast-cell granules and the ground substance of cartilage were able to bind methylene blue at pH 2·0, whereas the binding capacity of tracheal mucus was extinguished at pH 3·0 and that of thyroid colloid ribonucleic acid at pH 4. Using the same solutions as the above authors I found, however, that the RNA of plasma cells could still bind methylene blue at pH 2·6. Friedenwald (1947), who had shown that the distinction between metachromatic and orthochromatic staining of tissues could be exaggerated by staining in the presence of high concentrations of magnesium, also showed that at pH 1·5 nucleic acids failed to stain if the solution was saturated with magnesium chloride, whereas cartilage and mast-cell granules still stained. He suggested that this was due to blockage of the phosphate groups of the nucleic acids by magnesium.

Alternative dyes to methylene blue have found favour in recent years. Spicer (1960) advised and used 0·02 per cent. solutions of azure A in HCl-phosphate or phosphate-citrate buffers, at various selected pH levels. It was supposed that staining of mucosubstances with this dye between pH 1·5 and 4·0 might be due to their carboxylic acid and sulphate groups, and that the latter could be partially or entirely masked by proteins.

For characterization of mucosubstances according to the classification given on p. 305, Spicer's azure A series is recommended in place of the methylene blue series but technical details of both are given in Appendix 10. The need to use a single buffer to cover the whole pH range (for MB or azure A extinctions) has probably been exaggerated. When there is good reason for its use, an excellent buffer is that recommended by Lewis (1962) which is based on mixtures of formic acid and sodium acetate. This buffer is described in Appendix 1, p. 585.

There are a number of theoretical and practical objections to the use of MBE and azure A techniques for specific identification and distinction of carboxyl or sulphate groups, or even for the identification of mucosubstance carboxyls as opposed to protein carboxyls. Although according to Springall (1954) the latter are considered not to be dissociated below pH 4, the pKa of some of the constituent amino-acids of proteins is 10^{-2} so that 50 per cent. of their carboxyls will be dissociated at pH 2. In his studies on tissue basophilia Stoward (1963) found that although by visible light basic dyes were not apparently taken up by tissues at pH levels in the range of 2 to 3, there was diffuse uptake of coriphosphine (see below) by various tissues with low content of mucosubstances. This occurred even after removal of RNA phosphate. Especially notable among the tissues which took up the dye was smooth muscle whose acidic amino-acids (lysine, leucine, glutamic acid, aspartic acid) have pKa values from 1·8 to 2·36. At pH 2·7, therefore, the percentage dissociation of their constituent carboxyls should be between 69 and 87 per cent. By staining below pH 2 the uptake of basic dyes by carboxyl groups is largely suppressed but, even at pH 1·5, Stoward found that muscle proteins would stain with coriphosphine.

ACRIDINES

Trypaflavin. This dye (Acriflavine; C.I.46000) was used by Takeuchi (1962) and by Holländer (1964) for the demonstration of sulphate esters in tissue sections. It was used by the second author in two different ways. First as a fluorescent basic dye in 1:20′000 solution in citrate-HCl buffer at pH 2·5 and, secondly, by forming the reddish-brown dimethylaminobenzaldehyde complex *in situ*. The two methods are given in Appendix 10, p. 676. The second gives better results than the cresyl violet method with the sulphatides of metachromatic leucodystrophy.

Acridine orange. A combination of acridine orange staining and salt elution was recommended by Saunders (1964) who observed that the dye complexes with different mucosubstances were labile to different concentrations of sodium chloride. In the case of heparin, for instance, the complex was resistant to 1·6M NaCl whereas with hyaluronate it was labile to 0·2M NaCl. Details of the technique appear in Appendix 10, p. 668, but it should be observed that there are severe limitations to this approach (see Schubert and Hamerman, 1956).

Coriphosphine. This dye (Basic yellow; C.I.46020) was considered by Stoward (1963, 1967b) to be the best fluorescent indicator for SO_4 groups (used at pH 1·5 and pH 5) after removal of PO_4 (see p. 334, above).

Monazo Dyes

Maxilon blue RL. This is a basic monazo dye belonging to the Maxilon series of dyes developed by Geigy for dyeing polyacrylonitrile fibres. These dyes have the general formula:

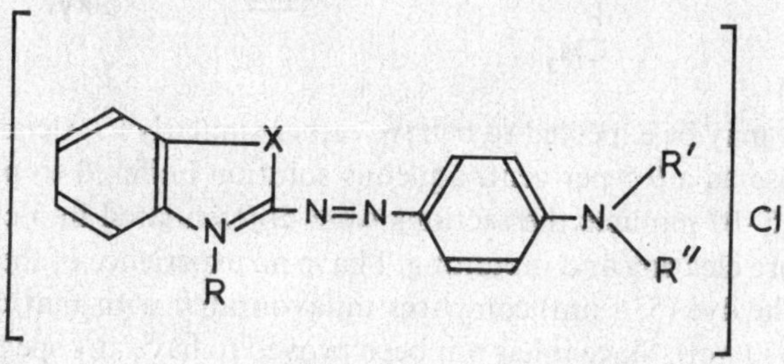

According to de Almeida (1960) Maxilon RL possesses the property of staining acid mucopolysaccharides metachromatically (red) whereas it stains other basophilic tissue components, including nuclei, bright blue. The mechanism of this metachromasia is unknown. Details of the technique are given in Appendix 10.

Triphenylmethane Dyes

Aldehyde Fuchsin. This dye complex was introduced by Gomori (1950) as an elastic stain (see Chapter 8, p. 225). It was soon observed that it stained a variety of acid mucosubstances (Abu'l Haj and Rinehart, 1952; Scott and Clayton, 1953; Halmi and Davies, 1953) and further evidence was provided by Spicer and Meyer (1960) that sulphated mucins were preferentially stained.

The mechanism of staining was investigated by Bangle (1954) and by Elftman (1959) and, more recently, by Ortman *et al.* (1966). Spectrophotometric studies carried out by these last authors indicated that the active dye molecule was pararosaniline and confirmed the well-known impression that aldehyde-fuchsin is not a stable product. In the frog pituitary Ortman *et al.* showed that staining was due to combination of carboxyl groups (present particularly in the thyrotrophs) with an intermediate *meta*-stable species formed in the dye solution.

Aldehyde-fuchsin should probably not be used as a specific stain for mucosubstances. It may be useful as one component of a combined mucin staining procedure (Aldehyde fuchsin-Alcian blue). Details of such a combined procedure are given in Appendix 10, p. 669.

Astrazone Dyes

Experimental Cyanine Red. This cationic astrazone dye, produced by Dupont, was introduced by Quay (1957) for staining acid mucopolysaccharides and nucleic acids. Its formula

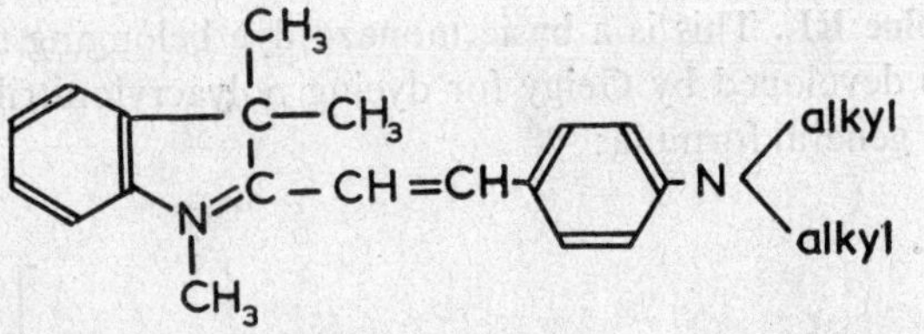

shows that it may be expected to behave rather similarly to Alcian blue. Quay advised its use as a 0·5 per cent. aqueous solution buffered to pH 2·9. After staining for 5–10 minutes the sections were differentiated in 3 changes of *n*-butanol before clearing and mounting. I have no experience of the method but the E^{max} of the dye (536 nm) compares unfavourably with that of the CuPC series (690–670 nm). Since it has not been proved to have any special qualifications as a stain for mucosubstances, this dye has made little impression in histochemical practice.

Phthalocyanin Dyes

Alcian Blue. The use of dilute solutions of Alcian blue 8GS as a specific stain for mucins was introduced by Steedman (1950). This author, using mainly material of other than human origin, showed that the dye possessed great advantages over other dyes for the staining of mucin, the chief of these advantages being its rapidity and ease of application. Alcian blue is a water-soluble copper phthalocyanin precursor of the insoluble dye Monastral Fast blue, into which it is converted by treatment with alkalis (Haddock, 1948).

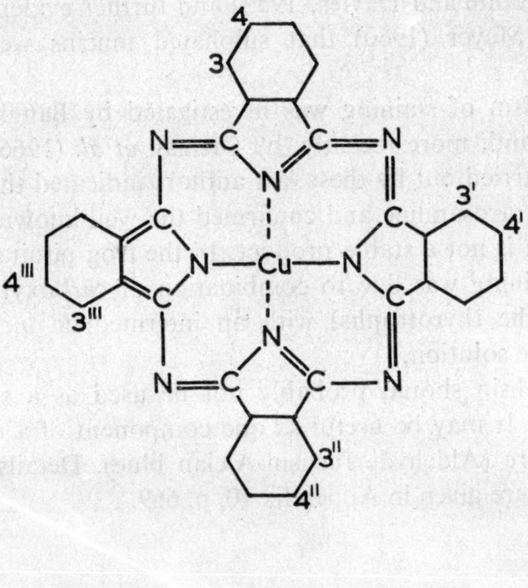

The copper phthalocyanins (CuPC) have a basic structure of the type shown above, and as such they are highly coloured water-insoluble pigments. Various types of solubilizing groups are introduced into the molecule in the preparation of water-soluble dyes, the three most commonly employed being sulphonic acid, carboxylic acid and chloromethyl. It is theoretically possible to substitute 1, 2, 3 or all four of the benzene rings in the molecule. The sulphonyl groups enter in the 3-positions but less is known about the positions taken by COOH and CH_2Cl. Alcian blue is recorded as belonging to a group of chloromethyl-substituted CuPC dyes in which the chloromethyl groups have been reacted with thiourea or alkylthioureas to give isothiouronium derivatives which are very soluble in water.

$$\mathrm{CuPC}\left[CH_2{-}S{-}C\begin{matrix}\diagup NH \\ \diagdown\!\!\diagdown NH_2 + Cl^-\end{matrix}\right]_x$$

According to the original author, Alcian blue did not distinguish chondroitin sulphuric acid complexes from mucoitin sulphuric acid complexes, and it therefore stained cartilage and mucin equally. In practice it was observed to stain acid mucopolysaccharides of epithelial and connective tissue mucin, but not stain the majority of mucoproteins. Its selectivity for mucins depended on the use of short staining; with longer staining nearly every tissue component became coloured. The original reaction, which was used by Vialli (1951) to demonstrate acidic groups after oxidation of tissues with chromic acid, was modified by Lison (1953) and Mowry (1956) by staining in acid solution. Lison used 0·5 per cent. Alcian blue in 0·5 per cent. acetic acid (30 minutes) and Mowry 0·1 per cent. of the dye in 3 per cent. acetic acid (30 minutes, pH 2·7 to 3·0). In fixed tissues and paraffin sections Alcian blue in acid solution stained connective tissue mucins and most epithelial mucins with usually negligible staining of the background proteins. In fresh tissues some of the nuclei stained even if the time of staining was kept to a minimum. This point was considered by Salthouse (1961) who showed that after treatment with certain enzymes (notably β-glucuronidase and hyaluronidase) nuclear chromatin could be stained with Alcian blue (0·1 per cent. in 2 per cent. acetic acid).

In 1957, or thereabouts, the original Alcian blue 8 GS was replaced by a new Alcian blue (8 GX) and the older dye ceased to be available. It was soon observed that 8 GX was more soluble than 8 GS and less resistant to decolourization. Its staining qualities were altered and new conditions became necessary. The work of Mowry (1960) culminated in new Alcian blue routines which gave results superior to those of the original technique. These routines are fully described in Appendix 10, p. 673.

Two alternative dyes of similar type to Alcian blue are Alcian green 3BX and Alcian green 2GX. The former stains bluish-green and the latter in a bright apple-green shade. The performance of the three dyes, as tested on paraffin sections of a variety of mucin-containing mammalian tissues and on unfixed and acetic-ethanol or alcoholic lead nitrate-fixed invertebrate tissues, is essentially very similar. Alcian green 2GX, however, sometimes stains known acid mucopolysaccharides in low concentrations when the other two dyes may fail. Alcian green staining of *Helix aspersa* mucins is shown in Fig. 99, p. 350.

In Germany an equivalent dye of unpublished composition, known as Astrablau was used in place of Alcian blue for mucin staining (Pioch, 1957). The staining properties of Astrablau, particularly with respect to mast cells, were fully described by Bloom and Kelly (1960). These authors used the dye at 0·1 per cent. in 0·6 to 0·7 N HCl (pH 0·2). They stressed the fact that over-staining was not a serious problem. More recently Mowry and Emmel (1966) have used National Fast Blue 8XM (National Aniline) in place of Alcian blue.

By the use of a strong counterstain the staining of connective tissues by Alcian blue can be overlaid, leaving only the connective tissue mucins stained blue. This procedure is recommended for studying the latter and details are given in Appendix 10. Lison (1954) recommended chlorantine fast red 5B as a suitable counterstain; others are tartrazine and neutral red or carmalum. A new oxazine dye, Darrow red, was recommended by Powers *et al.* (1960). Staining with an acid solution of Alcian blue can conveniently be followed by counterstaining with acid Solochrome cyanine R.S. (pp. 113 and 435). This combination gives brilliant nuclear detail and distinctive and different qualities of staining in the various mucopolysaccharides and mucoproteins. The use of Alcian blue is shown in Plate VIa, facing p. 326, where the highly sulphated mucins of the foregut of the lizard *Sceloporus undelatus* are seen to be brilliantly stained. Here the counterstain is neutral red. Another excellent counterstain is the PAS routine, with or without nuclear counterstaining, or iron hæmatoxylin alone (Plate Va, facing p. 320).

Mechanism of Staining with Alcian blue. The specificity of Alcian blue staining was, from its inception, the subject of considerable debate. Much of the difficulty was cleared up by Spicer (1960), and by Lev and Spicer (1964), who showed that at pH 2 the dye reacted mainly with uronic acid groups of mucosubstances whereas at pH 1 it could be made specific for sulphated mucopolysaccharides alone. Earlier work had indicated that Alcian blue staining was prevented by prolonged methylation (Chapter 6, p. 162), which would certainly remove *O*-sulphates, and increased by sulphation *or* by oxidation with chromic acid. Mild methylation (see p. 329) which does not remove *O*-sulphates, blocked Alcian blue uptake by cartilage (chondroitin-4-sulphate) but demethylation with KOH restored staining. On the other hand, the staining of sulphated gastric mucins is not reversed by mild methylation although it is if this procedure is followed by reduction with lithium aluminium hydride in hot dioxane (Stoward, 1963). This procedure is considered to reduce uronic acids to primary alcohols. Rizzoli (1955) had indicated that treatment with acetic anhydride in pyridine increased Alcian blue staining of the tissues. This would presumably convert C-terminal COOH groups to C=O, provided they were adjacent to NH_2 groups.

It was observed by Zugibe (1963) that hyaluronic acid was strongly stained *in vitro* by Alcian blue at pH 2·2. Glucuronic acid and galacturonic acids, *per contra*, did not stain, presumably because of their conversion in acid solution to lactones (Hirsch, 1952).

According to Spicer (1960) the staining of uronic acids could be attributed to the formation of amide bonds. This could scarcely occur unless the amino groups of the dye were spontaneously dealkylated. Alternative hypotheses (Stoward, 1963) suggested that staining could be due to hydrogen bonding of the unionized carboxyl to the copper and phthaloyanin nitrogen to form a stable six-membered structure.

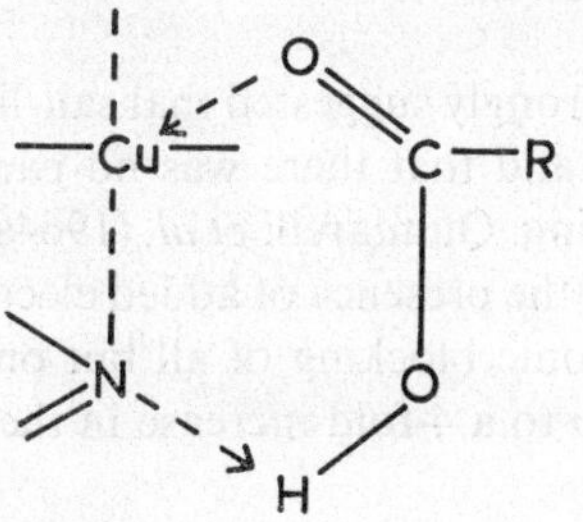

or to binding with the thiouronium group;

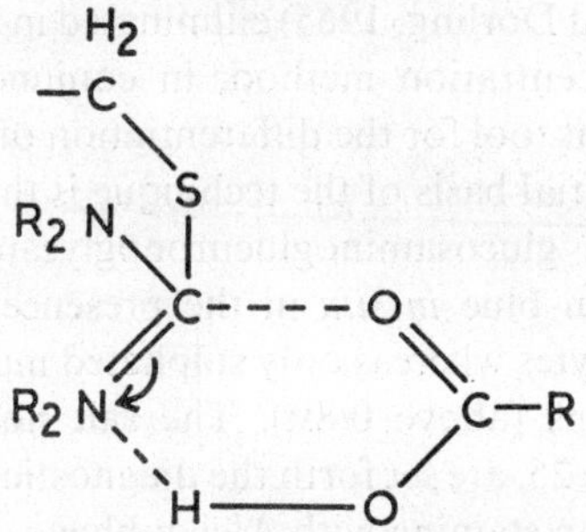

The hydrogen bonding theory seemed the more likely of these two hypotheses, especially in view of the opinion of Venkataraman (1952) that dyes are bound in this manner to COOH and OH groups of textile fibres.

An entirely different explanation of Alcian blue staining was advanced by Goldstein (1961, 1962) who deduced that there was an inverse relationship between the ionic weight of a basic dye and the "density" of its stainable substrate. According to this hypothesis the least dense tissue components, in terms of weight rather than charge (acid mucosubstances), take up dye cations whose ionic weight exceeds 1000 (Alcian blue = 1342). The denser nucleic acids are better stained by dyes whose cationic weight is less than 280.

According to Yamada (1964) Alcian blue staining of sulphated polysaccharides, though attributable mainly to their acidic groups, was affected by the presence of hydroxyl groups.

In developing the critical electrolyte concentration method Scott (Scott, Quintarelli and Dellovo, 1964) carried out a number of electrophoretic experiments with Alcian blue 8GX. In free solution the mobilities indicated the expected presence of basic groups with a pK of at least 7 (isothiouronium). Since, under the conditions normally used in staining, Alcian blue carries at least 2 and possibly 4 positive charges Scott *et al.* postulated that it would combine by salt linkages with polyanionic substances. In fact, Alcian blue was observed to combine with and to precipitate polyanions in the same way as cationic detergents (Scott, 1960). As with cationic polyanion complexes, the Alcian blue complexes were soluble in strong salt solutions and staining was completely blocked by prior treatment of sections with cetyl pyridinium chloride.

These observations strongly suggested that salt-linkage was the mechanism of Alcian blue staining, and that there was no particular specificity for any particular anionic grouping. Quintarelli *et al.* (1964a) showed that Alcian blue staining was increased in the presence of added electrolytes (HCl, NaCl, KCl). They assumed extra-anionic blocking of all but one positive charge per dye molecule, resulting in up to a 4-fold increase in the amount of dye bound by tissue anionic sites.

Critical Electrolyte Concentration. An important series of papers by Scott and his associates (Quintarelli *et al.*, 1964b; Scott *et al.*, 1964; Quintarelli and Dellovo, 1965; Scott and Dorling, 1965) culminated in the establishment of the critical electrolyte concentration method, in conjunction with Alcian blue staining, as an important tool for the differentiation of acidic mucosubstances in the tissues. The essential basis of the technique is the observation that both sulphated mucins and glucosaminoglucuronoglycans containing carboxyl groups will bind Alcian blue *in situ* in the presence of low concentrations (below 0·3M) of electrolytes whereas only sulphated mucosubstances will do so at higher concentrations (above 0·8M). The salt now usually employed is $MgCl_2$. Below, in Table 25, are set forth the diagnostic implications for critical electrolyte concentration staining with Alcian blue.

The results of the application of the critical electrolyte concentration technique are shown in Plate VIIa, b, c, d, e, f, opposite. Details are given in Appendix 10, p. 673.

Extending their studies Scott and Dorling (1966) have used the method to divide cationic dyes into two main classes. These are Group I dyes (acidophilic) whose charge is diffused by resonance (as in the imino-quinones) or "concealed" in bulky organic groupings (as in quaternaries). Group II dyes (carboxyl or phosphate seeking) are usually chelates of Al^{3+} or other metal ions (hæmatoxylin, aluminon) in which the positive charge is highly localized. In the presence of electrolytes Group I dyes are displaced more easily from carboxyl

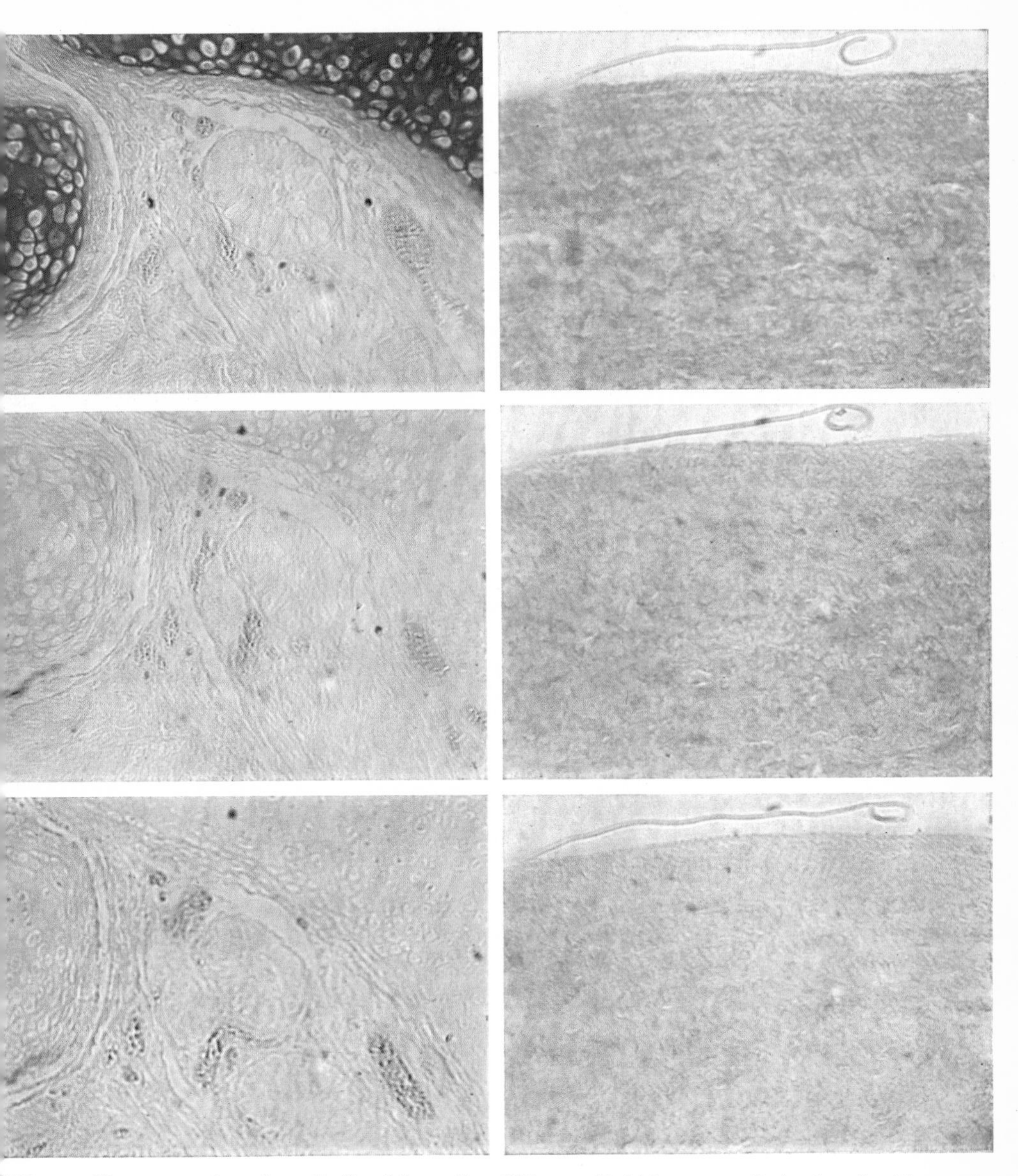

VIIA,B,C. Human new born lung. Buffered formalin-fixed. Stained with fresh 0.05 per cent Alcian blue 8GX in 3 per cent acetic acid (pH 2.5). Concentrations of $MgCl_2$ were 0.5 M, 0.7 M, and 1.0 M, respectively.

VII D,E,F. Rabbit cornea. Stained as for Figs. a, b, and c. Concentrations of $MgCl_2$ were 0.5 M, 0.7 M, and 1.0 M, respectively.

and phosphate groups than from ester sulphates while Group II dyes are more easily displaced from the latter.

TABLE 25

Alcianophilia of Acid Mucosubstances

Alcianophilia		Molarity of $MgCl_2$	Mucosubstances COOH	Mucosubstances SO_4
+	at	0	++	
—	at	0·2		
+	at	0·2 to 0·3	++	++
—	at	0·6		
+	at	0·6	+	++
—	at	0·8		
+	at	0·8 and above	—	++

METAL ION BINDING METHODS

Colloidal Iron. The colloidal iron method, introduced by Hale (1946), depends on the affinity of free acidic groups in the tissues for colloidal Fe^{+++} at pH 2. After washing, bound ferric iron is demonstrated by means of the Prussian Blue reaction. This gives a particularly easily visible and generally clean result. Originally Hale maintained that this method would demonstrate non-sulphated hyaluronic acid as well as sulphated mucopolysaccharides in Carnoy-fixed tissues. The specificity of the method seemed doubtful and Lillie and Mowry (1949) investigating the absorption of iron by tissue sections using M/100 ferric chloride in 1 per cent. mannitol (pH 5) found that mucin, collagen and reticulin were selectively impregnated. In similar experiments (1951) I could not find that the affinity of the tissues for iron was limited to the acid mucopolysaccharides. Because of this I noted in the first edition of this book that the usefulness of the method was restricted. Lison (1953) observed that nucleoproteins and other proteins were stained by the method and Lillie (1954) agreed that the method was not selective for acid polysaccharides. Mancini (1950), however, and Gomori (1952) found that staining was largely confined to those structures which gave metachromasia with toluidine blue. Davies (1952), using various preparations of hyaluronic acid on slides, found that all gave an intense reaction. He found also that a positive reaction was given by fibrin, gelatin, casein and even by peptone. Braden (1955) confirmed these results and disagreed with Immers (1954) whose experiments had upheld the specificity of the reaction. The specificity of iron staining in general was discussed at length by Wigglesworth (1952). Following Fischer and Hultzsch (1938), this author suggested that there is no doubt that iron would be bound by any monoesters or diesters of phosphoric acid present in fixed tissues and available for reaction. If the ionic weight is raised, as in the case of colloidal iron, it is probable that binding will no longer take place. In the case of binding to protein (amino-acid) groups there is considerable lack of agreement. Bechtold (1928) found that constant amounts of iron were firmly held by proteins and he considered that this was due to electrostatic absorption to the amino and carboxyl groups of the protein. Other authors consider that chemical combination between protein and iron takes place. Wigglesworth believed that with ferric iron the initial linkage was an electrostatic bond with the free carboxyl of the protein. Subsequently chelate complexes of the type shown below would be formed.

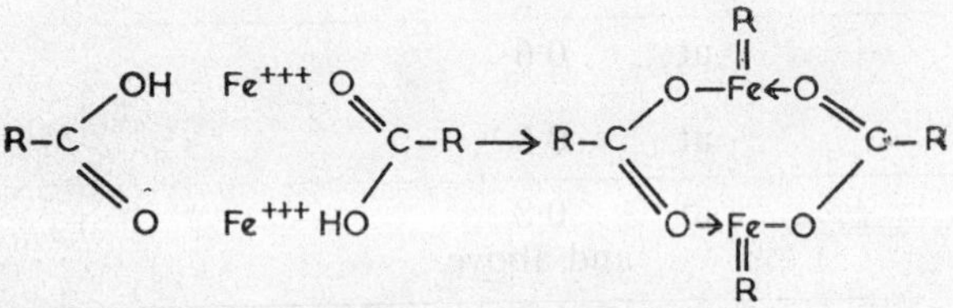

$$R{-}C(OH){=}O \quad Fe^{+++} \quad Fe^{+++} \quad O{=}C(HO){-}R \longrightarrow R{-}C\begin{matrix} O{-}Fe(=R){\leftarrow}O \\ O{\rightarrow}Fe(=R){-}O \end{matrix}C{-}R$$

Wigglesworth emphasized the fact that the formation of complexes of this kind must depend ultimately on steric factors and these have not been critically considered as yet. According to Sidgwick (1950) the affinity of Fe^{+++} for COOH is so great that calcium oxalate will dissolve in solutions of $FeCl_3$. *Per contra* oxalic acid will remove protein-bound iron from hæmosiderin and ferritin (see Chapter 27) and this effect might possibly be used as a method of differentiating various types of iron staining.

Many modifications of the method have been used successfully to stain acid mucopolysaccharides; two early ones were those of Hudack *et al.* (1949) and Ritter and Oleson (1950). The latter's method has been extensively used. Romanini (1951), Romanini and Giordano (1952) and Morone (1952) used Hale's method for studies on various mucopolysaccharides and the former authors regarded it as a refinement of the ferric chloride technique of Benazzi-Lentati (1941, 1942). Rinehart and Abu'l Haj (1951) introduced a new method for making the reagent and claimed that with this the specificity for acid polysaccharides was increased. These authors, as well as Lillie and Mowry, noted that fewer tissue components stained when the pH of the reagent was lowered and in fact the pH of their reagent is below pH 2. Thus, provided the staining solution was strongly acid, it was considered that only SO_3H and PO_4 groups would be sufficiently strongly dissociated to bind ferric iron and, if nucleic acids were excluded (as is easily done) the specificity of colloidal iron for sulphated mucopolysaccharides became much higher.

In spite of these and similar theoretical considerations many observers

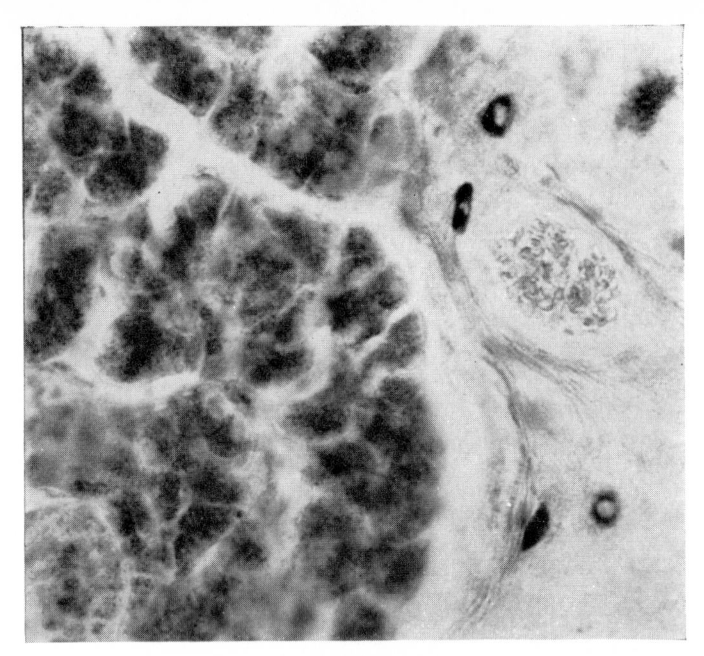

FIG. 98. Rat Salivary Gland. Freeze-dried formaldehyde vapour-fixed section. Shows sulphated mucins in the acinar cell granules. Convoluted ducts essential negative. Very strong staining of mast-cell heparin. Bracco-Curti reaction. ×370.

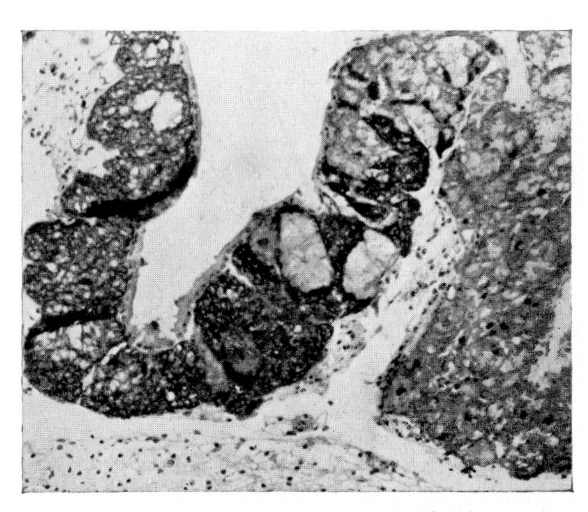

FIG. 99. Snail (*Helix aspersa*). Mucous secreting glands of two types are shown Alcian green—hæmalum. × 126.

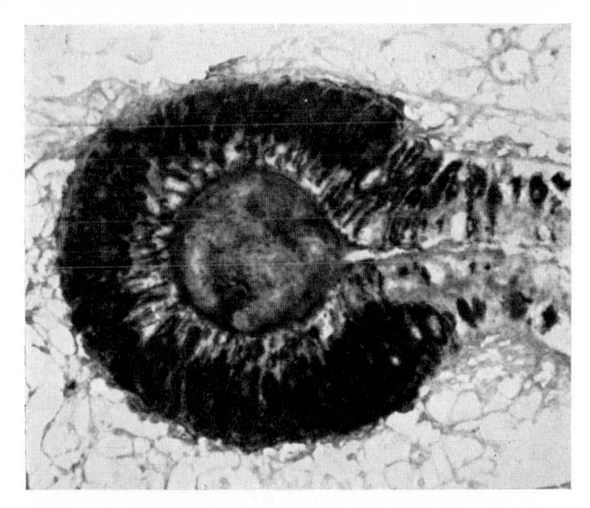

FIG. 100. Snail (*Helix pomatia*). Diverticulum of intestine. Four μ fresh frozen section. Shows intense staining of intracellular mucins. Dialysed Iron method. × 130.

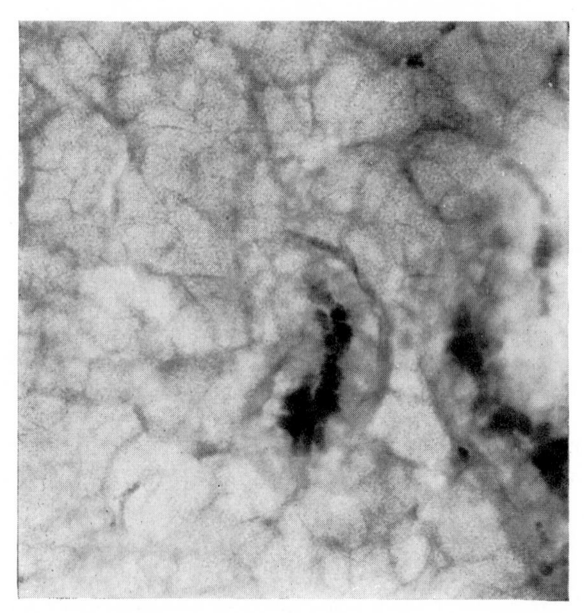

FIG. 101. Rat Salivary Gland. Freeze-dried formaldehyde vapour-fixed section. Shows a moderately strong reaction in the (granular) sialomucins in the convoluted tubules. Ravetto's Orcinol Method. × 370.

found that colloidal iron (CI) could stain uronic acid-containing mucosubstances. Zugibe (1963), for instance, found that hyaluronic acid gave a positive reaction even at low pH levels. It was reported by Gasic and Gasic (1962) that the positive CI reaction of the coating of tumour cells was reversible by prior treatment with neuraminidase (Chapter 25). They concluded that CI staining depended on galactosamine-linked neuraminic acid residues on the cell surface. While this may be so in a particular case, it is clear that CI is bound by both COOH and SO_4 groups at pH 1·8 (the recommended pH; see Appendix 10, p. 671).

Using the new (1951) reagent Abood and Abu'l Haj (1956) studied the interneuronal and intra-neuronal mucopolysaccharide in the sciatic nerve of bull-frogs and other species. They extracted from these by chemical means a substance regarded as non-sulphated hyaluronic acid. Histochemically the CI method (Rinehart and Abu'l Haj modification) gave weak staining of the material in formalin-fixed paraffin sections. Treatment with testis hyaluronidase reversed this staining both in axoplasm and in the neurilemmal sheath. Ghiara (1954, 1956) compared the CI reaction with the reaction of Benazzi-Lentati (1941, 1942) in which a dilute solution of ferric chloride (2–4 drops of a 10 per cent. solution in 10 ml. of distilled water) is applied to the tissues for 10 minutes and followed, after washing, by the Prussian Blue reaction. He obtained better results with Hale's technique and showed that raising the pH of the staining solution to pH 3 made the method much less selective. Lowering the pH or lowering the concentration of the colloidal iron both produced a greater sensitivity for strongly acidic polysaccharides.

Wolman (1956a), observing that at low pH levels only strongly acidic groups should be blocked by the colloidal iron reagent, proposed to follow the full method with a method capable of demonstrating weakly acidic groups such as those of the uronic acids. For this purpose he found a positively charged colloidal gold solution to be satisfactory and the combined technique which he evolved was given the name BiCol. According to the original author this method (1) allowed the differentiation of hyaluronic acid and CSA in skin, arterial wall, umbilical cord, etc.; (2) it showed the presence of weakly acidic compounds in young cartilage, which was followed later by fixation of sulphate; (3) it demonstrated intercellular material in the brain and a weakly acidic component in myelin and (4) it showed that a weak acid was secreted by the gastric parietal cells. I applied the method to a study of the mucins of *Helix aspersa*, using cold microtome sections fixed briefly in acetic ethanol as well as paraffin sections after several special fixatives, and found that it was relatively easy to perform and that it gave a clear-cut distinction between blue and reddish-brown-stained materials. In my hands nearly all protein structures not stained blue appeared reddish-brown and it seemed possible that the specificity of the colloidal gold part of the reaction for weakly acidic polysaccharides was low. Wolman (1956b) suggested that myelin contains an acid polysaccharide complex and he regarded his findings with the Bi-Col method as supporting this concept. The blue interneuronal material which he demonstrated was identified as "ground substance". According to Wolman hyaluronic acid should and did stain brown by the Bi-Col method whereas according to Abood and Abu'l Haj it stained blue, albeit weakly. The latter authors' acceptance of hyaluronidase-lability as indicating the presence of hyaluronic acid is open to question. Since they were able to identify hyaluronic acid in their nerve extracts, however, and to show that this was not sulphated, the discrepancy between their results and those of Wolman could not be resolved.

The original Bi-Col method was modified by Wolman (1961) and this modification is given in Appendix 10, p. 671. A further simplification reported by Módis *et al.* (1964) used a modified technique for preparing the colloidal gold solution. This technique is also given in the Appendix.

An interesting application of a modified CI reagent was reported by Berenbaum (1955) who obtained strong staining of the capsules of *Str. pneumoniae* and *Sgr. pyogenes* using saccharated iron oxide (ferrivenin, Benger) instead of colloidal iron. The majority of organisms showed a peripheral blue-stained layer which was more marked in the mucoid phase. Berenbaum's modification is applicable only to smears since the ferrivenin reagent stains all tissue components very strongly when used on tissue sections.

Hale's method was modified by Müller (1955–56a) who substituted a solution known as "ferrihydrozydsol" for colloidal iron. This reagent is prepared by adding 12 ml. of a 32 per cent. solution of $FeCl_3$ to 750 ml. of boiling distilled water. A colloidal ferric hydroxide results which is stable for some months. This reagent found considerable favour and it is now the standard component of the majority of CI methods. The original details have been supported by considerable theoretical additions (Müller, 1964). Geyer (1956–57) compared Müller's method with the original method of Hale and found that it produced a more distinct picture without blue colouring of the cytoplasm of renal tubular cells. Nuclei, which stained by Hale's method, were unstained by the modification.

I concluded (1960) that although the specificity of Hale's method for acid polysaccharides, and for individual types within this group, was less precise than one would desire, the reaction was a useful one in applied histochemistry. No significant reason can be found to change this opinion. In tissues such as brain, where a large part of the material is non-reactive a clear-cut demonstration of (pathological) acidic polysaccharide can be obtained. Where most of the protein material is reactive it may be necessary to use a more dilute staining solution and to lower the pH. In the case of fresh cold microtome sections (Figs. 100 and 105, facing pp. 351 and 389) these conditions produce excellent results and such sections are particularly recommended for use with the method. If any of the mucosubstances are soluble, or subject to redistribution by liquid fixation techniques, freeze-dried formaldehyde vapour-fixed sections should be used. The results (Tock and Pearse, 1965) are very satisfactory.

Ferric Ions. A fluorescent method for mucins was introduced by Hicks and Matthaei (1958), based on their observation that pretreatment of sections with ferric chloride or iron alum would prevent normal fluorescent staining with acridine orange (AO) while producing a bright orange fluorescence in acid mucopolysaccharides (Plate VIIIc, facing p. 355).

The mechanism of this reaction was not elucidated by the original authors but it has implications for a number of techniques based essentially on iron mordanting, including the foregoing colloidal iron methods. The addition of ferric ions to an aqueous medium produces a number of different products:

$$\underset{\text{I}}{Fe^{3+}} + H_2O \rightleftharpoons \underset{\text{II}}{FeOH^{++}} + H^+$$

$$Fe^{3+} + 2H_2O \rightleftharpoons \underset{\text{III}}{Fe(OH)_2^+} + 2H^+$$

$$Fe(OH)_2^+ + H_2O \rightleftharpoons Fe(OH)_3 + 2H^+$$

$$2Fe^{3+} + 2H_2O \rightleftharpoons \underset{\text{IV}}{Fe_2(OH)_2^{4+}} + 2H^+$$

There is no evidence that ferric monoions (I) form complexes with AO, to explain the reaction of Hicks and Matthaei, and their affinity for N is small. The suggestion was made by Stoward (1963) that the dye replaced Fe^{3+} initially adsorbed on to acid mucins. He was able to displace such adsorbed Fe^{3+} with AO or coriphosphine, preventing the subsequent development of a Prussian blue reaction.

The apparently more specific fluorescence in sections mordanted in iron alum below pH 2 may be due to (1) more specific uptake of iron by acid mucosubstances, (2) changes in the structure of the ferric ions themselves below pH 2, and (3) proportional affinity of iron between the tissue substrate and the basic dye used to render the adsorbed iron visible.

The equilibria given above for Fe^{3+} ions in aqueous solutions were deduced by Bray and Hersey (1934) and studied in detail by Rabinowitch and Stockmayer (1942). The final reaction was suggested by Hëdstrom (1953) and by Mulay and Selwood (1954). Above pH 2 acidic groups in the tissues will apparently take up iron principally in forms II, III and IV but the strength of binding will not be very strong as the ions carry a low net charge. Since none of them has the structure I they cannot form a red thiocyanate. Only $Fe(CNS)^{++}$ is red while $Fe(CNS)_3$ and $Fe(CNS)_6$ are colourless (Schlesinger and Valkenburgh, 1931). The complex $Fe(CNS)^{++}$ is weaker than those formed by ions bound to acidic groups in the tissues. This is particularly the case with Fe^{3+} bound to PO_4^- groups which have an OH group adjacent to the anionic centre. This arrangement promotes chelation and is responsible for the well-known series of iron-phosphate complexes.

In all the foregoing discussion there is no explanation for the Prussian blue reaction of hydrated ionic species of Fe. It seems that the reaction is probably due to the acid pH since H ions can displace adsorbed II and IV, finally converting them to I which will then react with $Fe(CN)_6$. As the pigment (Prussian blue) precipitates the equilibrium moves to the left until all species of iron are converted to I.

Aluminium. A method claimed to be specific for sulphated mucopolysaccharides was described by Heath (1962). This depended on staining with dilute solutions of basic dyes (Nuclear Fast red or methylene blue) in 5 per cent. $Al_2(SO_4)_3$. Control tests carried out with methylation, saponification and sulphation, indicated that the presence of SO_4 was essential.

Once again the mechanism of this reaction is obscure. Tests with Solochrome black indicate no uptake of Al^{3+} by acid mucosubstances in the tissues and the presence or absence of this ion seems hardly to affect the degree of staining or fluorescence obtained with basic dyes. Heath's observation that oxidized keratin reacts may be countered by the observation that it does so without the intervention of any metallic mordant (Chapter 6, p. 150).

Diamine Methods

Periodic Acid-Diamine (PAD)

These methods should properly have been considered under the heading of periodic acid-Schiff base reactions earlier in the chapter (p. 321). They are derived from the observations of Spicer and Jarrels (1961) that in aqueous solutions *N,N'*-dimethyl-*p*-phenylenediamine rapidly formed orange-brown condensation products with periodate-engendered aldehydes. With acid groups of mucopolysaccharides, on the other hand, it gave rise to black complexes after prolonged reaction (up to 72 hours). Further work (Spicer, 1961, 1965) indicated that the *m*-diamine, in the presence of the *p*-isomer, could be used to distinguish two types of mucopolysaccharides. The first type failed to stain after standard periodate oxidation while the second was unaffected. Chromatographic investigations showed that highly purified *N,N'*dimethyl-*m*-phenylenediamine failed to stain acidic mucins whereas samples contaminated with the *p*-isomer did so readily.

In the original method solutions of the *meta* isomer, buffered to pH 6, remained colourless for several days and failed to stain muco-substances. Solutions of the *para* isomer developed a reddish-purple colour and these stained acidic mucosubstances in shades of grey. A solution containing both isomers in the proportion of 6:1 became purple-blue in colour and gave rise to a strong and selective staining of acid mucosubstances, after differentiation in acid alcohol. Finally, it was found that solutions of the two diamines at pH 3·4 to 4·0 would stain selectively without differentiation in acid alcohol.

The nature of the cationic (basic) dye(s) produced by oxidation of the diamines is not known, but their development can be accelerated by the addition of a number of oxidizing agents. The most useful (Spicer, 1965) appears to be $FeCl_3$ and two variants of the original method have been evolved which

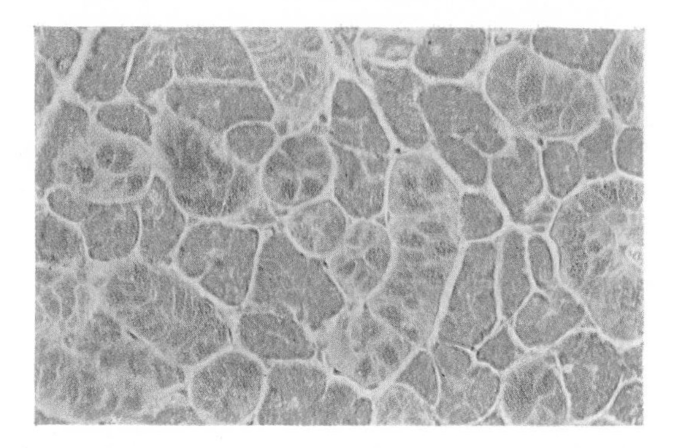

VIIIA. Rat Salivary Gland. Freeze-dried, formaldehyde vapour-fixed section. Granular mucosubstances in the convoluted ducts are stained black. Acinar mucosubstance unstained. PAD method, Methyl green counterstained. × 170

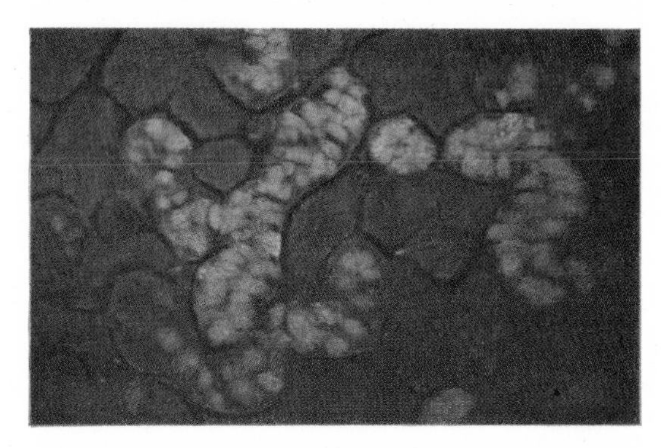

VIIIB. Rat Salivary Gland. Freeze-dried formaldehyde vapour-fixed section. Blue fluorescent periodate-reactive mucosubstances in granular form in the convoluted tubules. PA-Salicylhydrazide method. × 170

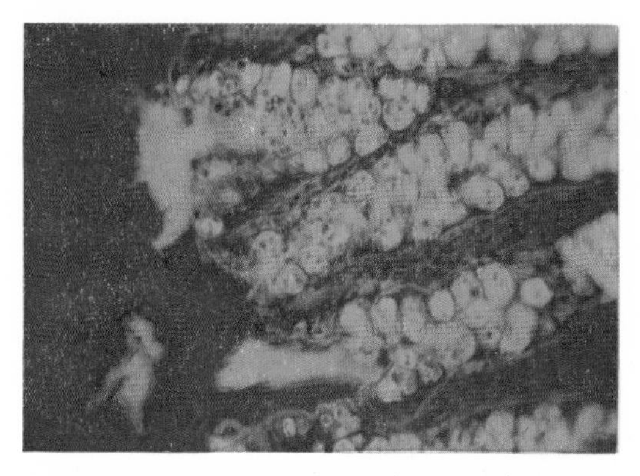

VIIIC. Human colon. Bright orange fluorescence of goblet cell and luminal mucins. Iron hæmatoxylin acridine orange. × 190

make use of this reagent. These are the low iron diamine (LID) and high iron diamine (HID) methods. The first stains most sialomucins and sulphomucins and the second only sulphated mucosubstances. Both LID and HID are best used as combined techniques with (followed by) Alcian blue staining. Details of the mixed diamine stain, and of its two iron-containing variants, are given in Appendix 10, p. 662. The type of result obtainable is shown in Plate VIIIa, opposite.

The effect of periodate oxidation on the diamine reaction cannot be interpreted in chemical terms (Münch and Ernst, 1964). There is no apparent reason why some mucosubstances retain their basophilia while others lose it. Spicer (1965) presumes that condensation of the diamine with periodate-engendered aldehyde groups may bring the positive charges of the disubstituted amine into such proximity to the negatively charged residues of the mucosubstance as to neutralize their attraction for positively charged dye molecules. Blockage of basophilia by the periodate *meta*diamine sequence is therefore interpreted on the basis of *vic*-hydroxyls closely associated with sulphate or carboxyl groups in the mucosubstance concerned. Resistance to the periodate *meta*diamine sequence (as in cartilage and mast cells) is then consistent with the absence of oxidizable *vic*-glycols. Expressed in another way, periodate-*meta*diamine insensitive mucosubstances should be PAS negative and PAD sensitive materials should be PAS positive.

The affinity of nuclear chromatin for Azure A after Feulgen hydrolysis (p. 254) is abolished by *meta*diamine but, according to Münch and Ernst (1965) the aldehyde groups of hydrolysed DNA cannot be demonstrated by the diamine method. These authors showed, on the other hand, that sulphonic acid groups from performic acid-oxidized disulphides and sulphydryls were well shown by the diamine sequence.

The place of the diamine methods has not yet been established. It is likely that with improved interpretation they will prove more and more useful for distinguishing various types of mucosubstances *in situ.*

Ion-Association Techniques

Cetyl Pyridinium-ferric Thiocyanate

This method, described by Zugibe and Fink (1966a and b) is based on the property of quaternary salts to form insoluble complexes with polyanions (Scott, 1955). Complexes formed *in situ* with cetyl trimethyl ammonium bromide or cetyl pyridinium chloride are colourless. By treatment with ferric thiocyanate the colourless compounds are converted into red ion association compounds. These are soluble in organic solvents but insoluble in water.

Zugibe and Fink (1966a) were successful in identifying acid mucosubstances (S-mucopolysaccharide A(1·0$MgCl_2$); S-mucopolysaccharide B2·0T or B4·0T) in freeze-dried (unfixed) or in cryostat (unfixed) sections. Formalin fixation produced "extraneous staining" for which they were unable to account,

Infra-red spectral studies showed that the type of complex formed, in the case of both quaternary compounds, was the hexathiocyanatoferroate III.

An extension of the original method by Zugibe and Fink (1966b) used the critical electrolyte concentration technique developed by Scott (1960) (see p. 348, above). Both methods are given in Appendix 10, p. 669.

"Specific" Methods for Sialic Acid and Uronic Acids

Dische Reaction for Uronic Acids

This test is based on the observation by Dische (1947) that treatment of uronic acids with concentrated H_2SO_4 gives a violet colour which fades on dilution with water. Stoward (1963) adapted the test as a histochemical reaction. He found that by taking sections through rising concentrations of H_2SO_4 to the concentrated acid heated to 70°, a purple colour was produced in sites containing uronic acids. Taking the sections back to water, examination by UVL revealed a strong blue fluorescence in the same sites. (See Appendix 10, p. 677.)

The chemistry of the fluorescent product is unknown. It may possibly be related to the primary degradation product, 5-carboxyfurfural (Neukom and Hui, 1959). On the other hand, it may be a hydrated furfuraldehyde or even 2′5-furandicarboxylic acid (Yoden and Tollens, 1901).

Resorcinol-Nitrite Reaction for Sialic Acid

A direct histochemical method for sialic acid using Bial's reagent was described by Diezel (1957) but this method would not work unless the dye orcein was substituted, as it often was, for the required orcinol. A modification of the resorcinol assay technique of Svennerholm (1957) was used by Shear and Pearse (1963) as a specific method for sialic acid. Although *in vitro* the method had a high degree of sensitivity, its application to formaldehyde vapour-fixed cryostat sections was complicated by the fact that any residual formaldehyde remaining in the section would give rise to the moderately soluble dye "resorcinol red". This dye, which was assumed to be the product of the specific reaction with sialic acid, is formed by oxidation (nitrite) of the anhydro condensation product of resorcinol with aldehyde.

Resorcinol Red

Although the precise localization obtained with the resorcinol-nitrite reaction could not be produced by using resorcinol red as a stain, sufficient doubt remains with regard to accuracy of localization to make it impossible to recommend the method.

MODIFIED BIAL REACTION FOR SIALIC ACID

The first truly successful application of Bial type reagents for the histochemical identification of sialic acids was introduced by Ravetto (1964). Using Svennerholm's orcinol reagent, containing $CuSO_4$, and formalin-fixed frozen sections, this author indicated that he obtained a positive result in sites containing sialic acid. Using formaldehyde vapour fixed freeze-dried sections Tock and Pearse (1965) obtained satisfactory results on material from rodent salivary glands. These results are shown in Fig. 101, p. 351.

THIOBARBITURIC ACID METHOD FOR SIALIC ACID

This method is derived from the now standard assay procedure for sialic acid (Warren, 1959; Aminoff, 1959). It depends on the use of periodate oxidation, after hydrolysis of the *N*-acetyl grouping, to produce a chromogen capable of reacting with thiobarbituric acid. Using Warren's reagent Stoward (1963) found that a diffuse fluorescence was induced in the tissues and this he attributed tentatively to their sialic acid content. Until more proof is available this method cannot be used as a reliable indicator for sialic acid. Technical details are given in Appendix 10, however, so that those who are interested may follow up this promising beginning.

ANISIDINE REACTION

The possibility of developing a fluorescent reaction on the basis of the assay technique of Cerbulis and Zittle (1961) should be examined. This method depends on the hydrolysis of *N*-acetylneuraminic acid in methanolic phosphoric acid, in the presence of *p*-anisidine. Brief details are given in Appendix 10, purely to indicate the basis for further developments.

Miscellaneous Dye Methods

THE EVANS BLUE METHOD

Gersh and Catchpole (1949) and later Catchpole, Gersh and Pan (1950) used intravital staining with Evans blue in their studies of connective tissue mucopolysaccharide-protein complexes (referred to in their terminology as glycoproteins). This dye forms a complex with plasma proteins and Catchpole and Gersh considered that it also formed a complex with the water soluble ground substance of connective tissue when this had been altered, by natural or other processes, to its more "fluid" depolymerized state. The method involved the injection of 1·25 per cent. Evans blue intravenously to rats (1 ml. per 100 g. body-weight) and killing the animals some 10 minutes later. In order to avoid loss of the water-soluble Evans blue-tissue mucopolysaccharide complex, fixation by freeze-drying was employed. According to these authors an "increase in depolymerization" of connective tissue mucopolysaccharides was also indicated by an increase in the degree of staining achieved by the PAS technique. If as in chondroitin sulphate the unit (disaccharide) is built up into blocks of $n = 60$ (1,3-linkage) only the two end glycols should be oxidized by HIO_4. If the linkage between the units is 1,3- or 1,2- no glycol is available for oxidation. If the linkage is 1,4-, periodic acid can cleave the glycol in the 2,3-position. Hydrolysis of bound CSA to individual disaccharides will therefore increase the amount of PAS staining provided the disaccharides also remain bound before and during oxidation.

Odd Methods for Polysaccharides

The Okamoto Method for Glucose

This method is based on the insolubility of Barium-glucose and Barium-lactose in alcohol. Thin slices of fresh tissue are fixed in methyl alcohol, saturated with barium hydroxide, for about 24 hours at 0°–4° and, after processing, the precipitate of Ba-sugar is transformed into a silver precipitate. Gomori (1952), in quoting the method, felt that the accuracy of localization with this method was unlikely to be very high, but Müller (1955–56a and b) made experiments on filter-paper and concluded that the reaction should be capable of localizing glucose in the tissues. When the method was tested on mouse tissues a strongly positive reaction was obtained in the liver cells but not in the kidney.

Further model experiments were carried out by Stiller (1965), using glucose-6-phosphate and fructose-1,6-diphosphate. He found that only unphysiological concentrations of these sugar phosphates could be demonstrated by the Okamoto method. It must be concluded that the reaction is of small interest from the histochemical view, despite reported satisfactory results after injection of intravenous glucose (Lange, 1966).

Methods for Pectin

The standard staining method for pectin, introduced by Mangin (1893) depends on its affinity for ruthenium red (ammoniacal ruthenium trichloride). Some workers have considered the method to have histochemical status and Carré and Horne (1927) showed that only pectic acid or pectates would give the reaction. When esterified, (pectin) pectic acid would not react. Bonner (1936) pointed out that the reagent would stain other substances containing free carboxyl groups and therefore most uronic acid-containing polysaccharides. Klein (1932) maintained that no specific reaction for pectin existed at that time.

Krajčinović *et al.* (1954–55) introduced a new reaction supposedly specific for pectins. This depends on the formation of an addition compound between pectic acid and benzidine. After diazotization of the free NH_2 group which remains at one end of the molecule an azo dye is formed by treatment with β-naphthol in alkaline solution. Details of this method are given in Appendix 10, p. 679.

A specific hydroxylamine-ferric chloride reaction for pectin was described by Reeve (1959). According to the author methyl esters of pectin react with hydroxylamine in alkaline solutions to produce hydroxamic acids. After acidification these react with $FeCl_3$ to give red products. Prior methylation (p. 162) increased the intensity of the reaction which clearly resembles the reaction described in the 2nd Edition of this book (p. 263) and below.

Reactions for Acetylated Pectin

A reaction for the chemical estimation of acetyl groups in pectins was described by McComb and McCready (1957), based on earlier work by McCready and Reeve (1955). This depended on the production of pectin hydroxamic acids by treatment with hydroxylamine and the formation of the red ferric complexes of II and III by treatment with ferric perchlorate.

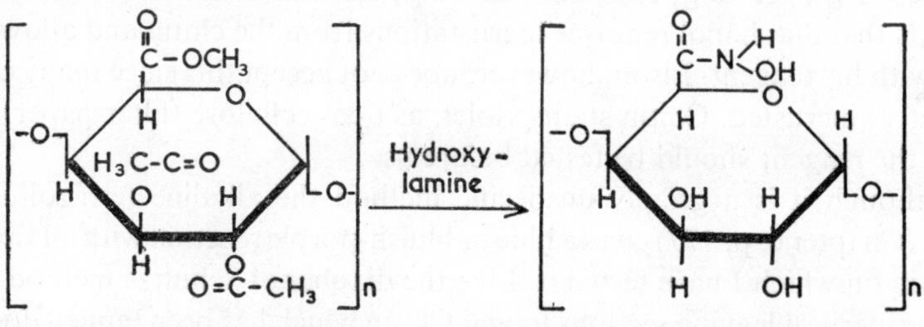

I Acetylated Pectin —Hydroxylamine→ II Pectin Hydroxamic acid

$$+ 2n\ CH_3CONHOH + n\ CH_3OH$$

III Acetyl hydroxamic acid

This reaction can be performed histochemically on plant tissues though any colour in the sections must be due solely to pectin hydroxamic acid (II). It can also be carried out on mammalian tissues after acetylation of their polysaccharides in the usual manner. Fig. 104, facing p. 389, shows the positive reaction obtained in carbohydrate-protein compounds treated in this way. The specificity of the histochemical reaction, if any, remains to be determined for both plant and animal tissues.

Germanic Acid Method for Polyvalent Alcohols

According to Feigl (1954) germanic acid, like boric acid, forms complex compounds with polyvalent alcohols such as glycerol, mannitol and glucose. Treatment of unfixed, cold microtome, sections with germanium dioxide in alkaline solution results in the formation of germanium complexes in materials (mucins) where hexose sugars are presumably present. After acid washing the presence of bound germanium can be demonstrated by means of an acid alcoholic solution of 9-phenyl-2,3,7-trihydroxy-6-fluorone. This forms a bright red absorption compound which is insoluble in water and alcohol and apparently stable. Fig. 102, facing p. 390 shows the result of this method applied to the mucins of the snail *Helix aspersa.*

Methods for Chitin

The occurrence of this polysaccharide has already been mentioned (p. 297, above). The oldest method for demonstration of the various chitins employs the Schultze, chlor-zinc-iodine reagent after treatment with warm KOH (50 per cent.) and washing with alcohol. Chitin stains reddish violet. This reaction works much better in the test tube than under histochemical conditions. Lison remarks (1953) à propos such methods that "les réactions de la chitine . . . n'étaient guère commodément réalisables à cause de leur brutalité." A better variation (Schultze, 1922) depends on prior treatment of sections with diaphanol (chlorodioxyacetic acid) followed by the chlor-zinc-iodine reagent (Iodine 6·1 g.; KI 10 g; zinc chloride 60 g; distilled water 14 g.) The author suggests that diaphanol removes incrustations from the chitin and allows it to react with his reagent. Lison, however, does not accept this since many chitins are not so incrusted. Chitin stains violet, as does cellulose (filter-paper), with which the reagent should be tested before use.

Although it is a grossly unspecific method the alkaline tetrazolium reaction (Chapter 6, p. 127) gives a blue or bluish-purple reaction with all varieties of chitin on which I have tested it. Like the diaphanol-Schultze method it has also the merit of leaving sections intact. Chitin which has been tanned does not give a positive reaction but chitin which is free from phenolic substances (and also from SH groups and reducing lipids) reacts strongly. Plate VIb shows a positive reaction in the ctenidium of *Triton rubicundus* in which the gill plates stain deep blue at one end, tailing off to a fainter blue at the other end.

The histochemical specificity of the alkaline tetrazolium reaction for sugars (leaving aside considerations of non-saccharide reductants) is difficult to determine. According to Avigad *et al.* (1961) tetrazoles in alkaline aqueous solutions at pH 12·4 and at 37° are reduced by sugars manifesting enediol isomerism but not by sugars in which such isomerism is impeded by the presence of a substituent in a carbinol group vicinal to the carbinyl group. Disaccharides in which the glycosidic linkage is through the 2-position do not reduce tetrazoles.

As an alternative to the above methods chitin can be effectively sulphated by one of the methods suggested by Kramer and Windrum (1954) and given in Appendix 10, p. 666. The resulting chitin sulphate can sometimes be demonstrated by the metachromatic reaction or better by staining with an acid solution of one of the Alcian dyes (pH 3·6).

Arguments concerning the presence or absence of positive staining in chitin by the PAS, Alcian blue and colloidal iron techniques (Runham, 1961, 1962; Salthouse, 1962) have not been resolved in a satisfactory manner. It seems that when a positive reaction occurs it may be due to associated mucosubstances.

ALKALINE DMAB METHOD FOR AMINO-SUGARS

It was observed by Ehrlich (1901) that certain mucins and mucoids gave a purple colour when treated with warm solutions of dilute alkali and subsequently with *p*-dimethylaminobenzaldehyde. Aminoff *et al.* (1952), described the reaction as characteristic of *N*-acetylhexosamines and they showed that it depended not only on the pH at which the reaction was carried out but also on the nature of the buffer. The strongest reaction was observed with borate at pH 9·8.

This reaction can be applied histochemically but the colours given by the *N*-acetyl sugars of the majority of mucins are very weak. The reaction is negative with insect chitin and it cannot be recommended in its present form.

Suggested Investigation Procedures for Mucosubstances

From the mass of new investigational tools which have been described in the chapter some readers may have difficulty in making a choice of methods. While it is impossible to give precise instructions the following section may provide some general indications.

Where possible it is strongly recommended that investigations on mucosubstances be carried out on tissues fixed with formaldehyde vapour after freeze-drying.

Acid mucosubstances should be characterized by staining in azure A solutions at pH 1·5, 3·5 and 4·0 (p. 342) and by an Alcian blue 8GX stain at pH 2·5 and pH 1·0 (p. 346). This should be followed by Alcian blue staining in the presence of 0·6M magnesium chloride (p. 348).

Cytochemically, periodate-reactive mucosubstances should be demonstrated by the PAS reaction with the standard 10-minute oxidation and distinction between periodate-reactive and periodate-unreactive mucosubstances should be made by means of the PAD method (p. 354). The reactive substances are coloured brown and the unreactive ones black by this technique.

Sulphomucins may be demonstrated by the LID (p. 355) and HID (p. 355) techniques. The former stains most sialo and sulphomucins grey to black (after 18 hours' incubation). If the HID method is followed by Alcian blue at pH 2·5 highly sulphated mucins are stained black while non-sulphated mucins (and perhaps weakly acidic sulphated mucins) stain blue. Sites containing both types are stained blue and black. With the aldehyde fuchsin—Alcian blue pH 2·5 sequence sulphomucins are stained purple or bluish-purple whereas sialomucins and other acidic, non-sulphated mucosubstances stain blue.

Sialomucins can be confirmed by decrease in basophilia towards azure A and Alcian blue after treatment with *Vibrio cholerae* neuraminidase (100 units/ml.) for 18 hours at 39° (Spicer and Warren, 1960).

Sulphomucins can be blocked by esterifying their SO_4 groups with methanolic thionyl chloride (p. 162). As an indicator of this effect reversal of purple staining with the aldehyde fuchsin-Alcian blue pH 2·5 routine is normally employed. Restoration of sulphate group reactivity is achieved by saponification with barium hydroxide in 70 per cent. ethanol (Lev and Spicer, 1965).

The most satisfactory method for demonstration of sulphomucins is by radioautographic detection of 35S uptake (see Chapter 30.)

The phenylhydrazone-Formazan reaction, and the Formazan-ANSA sequence undoubtedly indicate significant variations in the structure of the periodate-oxidized saccharide units of the mucosubstances. Until further work has been done interpretation of the results of these two techniques will be speculative. I consider that they should be employed as ancillary methods in any complete mucosubstance studies.

The Cytochemical Demonstration of Glycogen

Physiochemistry of Glycogen

The concept of glycogen being bound to protein, and therefore less easily extractable, led Willstätter and Rohdenwald (1934) to describe a complex called desmoglycogen. Meyer and Jeanloz (1943) proved that glycogen was not chemically bound to protein but, after mechanical inclusion in coagulated protein, glycogen of high molecular weight cannot be extracted by washing in hot water. This bound fraction was considered by Prins and Jeanloz (1948) to be the equivalent of Willstätter's desmoglycogen.

Molecular weight determinations on glycogens were carried out by Bell *et al.* (1948), using sedimentation and osmotic pressure measurements. These produced the following figures: horse muscle glycogen, $2{\cdot}9 \times 10^6$; rabbit muscle glycogen, $2{\cdot}6 \times 10^6$; human muscle glycogen, $2{\cdot}4 \times 10^6$ and rabbit liver glycogen, $4{\cdot}4 \times 10^6$. Ultracentrigation studies of the latter (Bridgman, 1942) indicated sedimentation constants with a spread from 20 to 120 S. Normal human liver (Polglase *et al.*, 1952) contains two major components, one 60–100 S and the other 150–300 S.

The method used for extracting glycogen has a great influence on the mol. wt. (Stetten *et al.*, 1956, 1958), acid, and particularly alkaline, extractions cause breakdown of the glycogen molecules. The most popular extraction technique has been the use of trichloroacetic acid (TCA) (Bloom *et al.*, 1951). Using conc. TCA these authors showed that in the case of rat liver 85 per cent. of the total glycogen was extractable whereas in skeletal muscle the figure was 55 per cent. Kugler and Wilkinson (1959, 1960) have produced strong evidence that only lyo-glycogen, defined as free, labile and cold TCA-soluble, can be demonstrated by cytochemical methods. Desmo-glycogen, defined as bound, residual and cold TCA-insoluble, is probably undetectable by these methods. These criteria refer, of course, to glycogen in unfixed or native samples of tissue.

The question of the cytochemical threshold for glycogen has been investigated by a number of authors, and opinions differ to some extent. Nielsen *et al.* (1932) gave a figure of 0·1 to 0·3 per cent., Deane *et al.* (1946) quoted 0·08 per cent. and Fitzpatrick *et al.* (1948) 0·04 to 0·6 per cent. Later investigations (Kugler, 1965) indicating a figure of 0·2 per cent. (wet weight) were obtained by carrying out parallel assay and histochemical staining on samples of myocardium. Wittels (1963) nevertheless quoted a figure of 0·5 per cent. for human fœtal and neonatal myocardium.

Correlation with Electron Microscopy

According to Bondareff (1957) glycogen particles in electron micrographs could be divided into three orders of size: third order, 3000 Å; second order, 1500 Å; first order, 130 Å. Certainly the large rosette form is always present whenever glycogen can be shown easily by light microscopic techniques but, according to Kugler (1965) at about 0·25 per cent. it is impossible to distinguish glycogen particles from ribonucleoprotein.

Enzymal Analysis

Whichever of the methods described below is employed for demonstrating glycogen, control sections must always be employed. In the past these were incubated with saliva (Bauer, 1933; C. M. Bensley, 1939) or with malt diastase (Lillie and Greco, 1947; Bunting and White, 1950). The debranching enzyme, diazyme, is strongly recommended by Kugler (1965) in place of either of the foregoing. It is usually accepted that any of the three, correctly used, will remove all stainable glycogen from tissue sections and that what remains is not glycogen. It is not possible to remove glycogen from araldite embedded material with diastase, diazyme or saliva.

Methods for Glycogen

Iodine Methods

The staining of glycogen with aqueous solutions of iodine, though certainly the oldest method, is still employed in modern histochemistry. The technique was used by Claude Bernard (1877) in his classical researches on the liver and also by Paul Ehrlich (1883) for the demonstration of glycogen in diabetic livers. Solutions of iodine in potassium iodide (such as Lugol's) are usually employed and the main practical difficulty is the production of permanent sections. Lison (1936) gave a selection of three methods designed with this object. That of Langhans (1890) modified by Carleton (1938) employed dehydration of paraffin sections in saturated iodine in absolute alcohol, followed by clearing in essence of origanum, xylene, and mounting in balsam. Gage's (1917) method used fixation in acetic alcohol containing iodine. Paraffin sections were subsequently treated with an alcoholic iodine solution containing potassium iodide, without removal of the paraffin. After drying, the sections were deparaffinized in xylene and mounted in balsam. Driessen (1905) fixed his material in alcohol and stained paraffin sections with phenolic xylene containing Lugol's iodine, vigorously agitated at the time of preparation. Subsequently the sections were cleared in phenolic xylene and in xylene before mounting in balsam. Of the three methods mentioned, Carleton's is probably the best. Iodine methods have the drawback that their specificity is not very high, since a similar brown colour is produced by various protein constituents and by amyloid. The use of saliva or diastase controls removes this objection. Mancini (1944) used the iodine method (iodine vapour) to demonstrate glycogen in cartilage cells fixed by the freeze-drying method. The vapour has not been used, apparently, for "fixation" of glycogen in freeze-dried blocks.

The Bauer-Feulgen Method

The original method described by Bauer (1933) employed hydrolysis and oxidation of the polysaccharide glycogen with 4 per cent. chromic acid for a short period, and demonstration of the resulting polyaldehydes with Schiff's solution. This method has certain disadvantages which preclude its use for accurate work although it is still used in modern histochemistry (Schabadasch, 1937; C. M. Bensley, 1939; Bunting and White, 1950). The chief disadvantage is the non-specific colouring occurring in various tissue components which do not contain glycogen. Among these mucin and thyroid colloid, which certainly contain no glycogen, have been shown to react strongly (Wallraf and Bechert, 1939; Bignardi, 1940; Dempsey and Wislocki, 1947). Although the use of diastase or diazyme controls reduces this objection it remains true that small amounts of glycogen are difficult to appreciate by Bauer's method, which also suffers from the non-specific staining of plasmalogens, inherent in the use of Schiff's reagent, in the same way as do the Feulgen nuclear reaction and the PAS technique. Good histological results can be obtained, and the method can still be used, especially after formaldehyde vapour fixation of freeze-dried tissues.

Silver Methods

Several methods have been devised for the demonstration of glycogen by means of silver complexes. They all employ acid hydrolysis and oxidation to reveal aldehyde groups in the polysaccharide chain prior to their combination with silver. Silver methods have the advantage that the end product is metallic silver which, being black, is easily visible under the microscope and is especially easy to photograph. The methods so far evolved may be divided into two classes, depending on whether or not additional reduction by means of formalin follows the silver bath. In the first group, in which formalin reduction is employed, the first to be described was the method of Mitchell and Wislocki (1944). These authors used acid potassium permanganate as oxidizing agent followed by bleaching in alcoholic oxalic acid-silver nitrate before treatment with ammoniacal silver oxide, and formalin. A second method in this group was that described by Arzac (1947) who also used $KMnO_4$, followed by ammoniacal silver carbonate and formalin reduction. Gomori (1946) criticized the method of Mitchell and Wislocki and his objections apply equally to the method of Arzac. He considered that Mitchell and Wislocki's method was unselective because the correct pre-treatment for the liberation of aldehyde groups was not employed and because the employment of formalin as a reducing agent stained collagen and reticulin as well as specific polysaccharide. Gomori introduced a different technique employing hydrolysis and oxidation with chromic acid (5 per cent.) as in the Bauer method. He did not advance any reason for regarding chromic acid, rather than any other oxidizing agent, as the correct method for liberating aldehydes. Hydrolysis was followed by staining in a solution of a hexamine-silver nitrate complex buffered to alkaline pH with sodium borate. By this method, which belongs to the second group of silver methods for glycogen, mucin also stained strongly.

For all these three methods collodion protected slides were recommended since it was supposed that a thin film of this material would prevent loss of glycogen from the sections which otherwise occurs during hydrolysis and oxidation. As Lille and Greco (1947) showed, in the case of diastase control sections, collodion must be applied after and not before incubation with the enzyme, since it prevents the action of the latter to a variable extent. There is little evidence in favour of the use of collodion films and, in fact, they are seldom employed today.

Arzac and Flores (1949) investigated the different effects of various reagents on the staining of glycogen by silver methods. They found that 10 per cent. chromic acid (20–30 minutes) was the best but specifically excepted periodic acid which was not available to them at that time. Lillie (1948) showed that this reagent could precede staining with silver complexes in the demonstration of glycogen. Arzac and Flores also studied the effect of formalin and of the type of silver solution employed. Of the latter, ammoniacal silver carbonate was found easy to prepare, stable, and rapid in action. In order to increase the rate of reduction of silver the solutions were used at 50° (for 1 hour). Gomori's hexamine-silver and a piperazine-silver complex were also tried but found to be slower in action than the ammoniacal carbonate. Formalin reduction was found to widen the range of substances giving a positive reaction, independently of the type of hydrolysis and silver complex employed. Reticulin was invariably stained. However, sharper images and quicker results were obtained in the case of glycogen.

It must be obvious that Gomori's (1946) objection to formalin reduction is a valid one if the silver techniques for glycogen are to be regarded as having histochemical significance. Gomori's original method is recommended and this and the two solutions evolved by Arzac and Flores to take the place of hexamine-silver appear in Appendix 10.

The Periodic Acid-Schiff Method

The PAS method reveals glycogen particularly clearly and it has been used for this purpose by Marchese (1947), Wislocki, Rheingold and Dempsey (1949), Gibb and Stowell (1949), and by McManus and Findley (1949), among others. Although it was with the specific idea of preserving glycogen in the tissues that Hotchkiss introduced his buffered alcoholic solutions there is no doubt that these reduce the rate of HIO_4 oxidation so much that part of the glycogen bound in the tissues is not stained afterwards by fuchsin-sulphurous acid. It is therefore essential to use watery solutions of HIO_4, preferably on paraffin sections after fixation in a suitable manner (Chapter 5, p. 89). The method is less satisfactory for formalin-fixed frozen sections, and cold microtome sections.

The Lead Tetra-acetate-Schiff Method

This method, introduced by Shimizu and Kumamoto (1952) as a substitute for the PAS reaction, appears to possess no advantages over the latter for the demonstration of glycogen. Casselman (1954) found that the demonstration of glycogen by this method was dependent on the conditions of oxidation. If the $PbAc_4$ was dissolved in anhydrous solvents no staining occurred. If aqueous sodium acetate, or water alone, were present, strong reactions were obtained. The reaction can be used as an alternative to PAS, to which it gives practically equivalent results provided that fresh solutions of the reagent are used.

Best's Carmine Method

Although the use of such an empirical method cannot be reconciled with the strict histochemical point of view, the ammoniacal carmine solution of Best (1906) cannot be dispensed with. It remains in current use by histochemists as well as by histologists (Lison and Vokaer, 1949). According to Baker (1945), the active principle of the natural dye carmine is carminic acid which, at a pH on the alkaline side of its isoelectric point (4·0–4·5), is negatively charged and behaves like an acid dye. Although we know that the carminic acid in Best's solution is present in this form, its staining of glycogen depends on physical and stereochemical factors which at present we do not understand. Basing his opinion on the abolition of staining by urea Goldstein (1962) suggested that the mechanism was by hydrogen bonding on to 1,2-glycol groups. It is not necessary to use celloidin-embedded sections as originally advised by Arnold (1908), but double-embedding of tissue blocks in celloidin and paraffin can be employed. Mullen (1944) described a rapid method of staining glycogen with Best's solution using alcohol-fixed paraffin sections and covering with celloidin in the above-mentioned manner. Differentiation may be carried out with Best's differentiator (aqueous methyl and ethyl alcohols) or simply with absolute methyl alcohol.

FLUORESCENT METHODS

The salicyloyl hydrazide method (Stoward, 1967d) applied after oxidation with periodic acid, gives a bright greenish blue fluorescence with glycogen and Kugler (1965) suggested that its sensitivity might perhaps be higher than any of the light microscopic methods in current use.

Opinions as to which of the methods listed above is the most suitable vary considerably. A recent evaluation of the various methods by Carpenter, Polonsky and Menten (1951) stressed the difficulty of ascribing specificity for localization of glycogen to any of them, on account of the large number of factors operating against such specificity. For routine use, I consider that paraffin sections, fixed in whatever fixative is habitually used, should be stained with Best's carmine. Control diastase-treated sections should be employed in parallel. If an accurate study of the glycogen distribution in any particular tissue is proposed, some improved fixation such as that suggested by Lison and Vokaer (Chapter 5), followed by Best's stain, or by PAS, is a suitable combination. If circumstances allow, freeze-dried formaldehyde vapour fixed paraffin sections should be used, in conjunction with PAS and Best's stain. The iodine methods will not generally be employed because of the difficulty of obtaining good results with ordinarily fixed material and of obtaining permanent mounts; their only real use is for freeze-dried material and for fresh frozen cold microtome sections. The silver methods, which have considerable advantages for photography, are not otherwise especially to be recommended.

The Demonstration of Ascorbic Acid

THEORETICAL CONSIDERATIONS

For the demonstration of ascorbic acid in the tissues only one recognized method exists. This is the acid silver nitrate technique of Bourne (1933a and b, 1936). Attempts have been made to transfer methods of biochemical estimation to the histochemical demonstration of the substance (Giroud *et al.*, 1939) but these have not been successful. Bourne's original method was based on the observation by Svirbely and Szent-Györgi (1933) that vitamin C in the adrenal cortex would reduce silver nitrate solutions, and the very similar method of Giroud and Leblond (1934, 1935a) was also based on this fact. The reduction of acid silver nitrate without the action of light or heat was thought to be specifically the property of ascorbic acid. However, ascorbic acid may be present either in reduced or oxidized form, and it is only the former that will give a positive reaction with acid silver nitrate. Efforts to obviate this have been based on pre-treatment of the fresh frozen sections with H_2S, by saturating their container with this gas and by subsequently exposing the tissues to a vacuum in order to remove the excess of H_2S not used in the process. The following reaction takes place:

L-Ascorbic acid (reduced form) $\underset{+2H\ (H_2S)}{\overset{(H_2O_2)\ -2H}{\rightleftharpoons}}$ dehydroascorbic acid (oxidized form)

From the outset strong criticism was levelled at the specificity of the acid silver reaction. Harris and Ray (1933) drew attention to the fact that in the liver and adrenal medulla, both rich in ascorbic acid by chemical estimation, no black deposit of silver was produced by the Bourne technique. Glick (1935) also doubted the specificity of the reaction and used instead a refinement of the 2:6-dichlorphenolindophenol method of Birch, Harris and Ray (1933), upon tissue sections 20 μ thick cut with a rotary microtome. With this technique Glick and Biskind (1935, 1936a, b and c) and Biskind and Glick (1936) made a study of the distribution of ascorbic acid in the anterior and posterior hypophysis, corpora lutea, small intestine, thymus and adrenal. These studies have been continued, using refinements of the same technique, by Glick *et al.* (1953) and by Bahn and Glick (1954). King (1936) likewise maintained that silver nitrate was reliable neither as a qualitative nor as a quantitative index for vitamin C in the tissues. On the other hand, Giroud and Leblond (1936) defended the specificity of the silver nitrate test and claimed that negative reactions in scorbutic tissues could be reversed in a few hours by the administration of ascorbic acid. A positive reaction, according to these authors, could be taken as an indication of the presence of ascorbic acid while a negative reaction, which might be due to factors inhibiting the reduction of silver nitrate, did not necessarily indicate absence of ascorbic acid.

Barnett, Bourne and Fisher (1941), while using the acid silver nitrate technique to study developing bone, observed a heavy deposit of silver in the trabeculæ which blackened by the action of light during further preparation of the sections. These authors found that this type of silver precipitation, which differed morphologically from the small granules produced by ascorbic acid, could be removed by subsequent treatment of the tissues with 5 per cent. ammonia for 10–15 minutes in the dark. The inclusion of this modification was therefore considered to increase the specificity of the reaction. Further criticism came from Bauer (1943) who concluded from his experiments that either vitamin C granules could be demonstrated by a variety of methods or that silver granules could be produced in the cytoplasm by the acid silver technique which had no relation to ascorbic acid. Barnett and Fisher (1943) conducted a series of experiments with the staining of mixtures of gelatine, ascorbic acid, and olive oil or ground glass, by the usual silver method. They concluded, despite their earlier work, that it was unjustifiable to infer the whereabouts of ascorbic acid from the situation of silver precipitate obtained by the method, and that the precipitation of acid silver nitrate took place at various interfaces. The tendency of the Golgi apparatus to stain by the method was explicable on this basis. Bourne (1944, 1950), however, continued to support the specificity of the method partly on his own (1933a) evidence that the disposition of mitochondria in the pituitary and adrenal stained with Janus green was identical with the distribution of silver-stained granules and partly on the work of Giroud (1938) who found complete similarity in the morphology and dimensions of mitochondria and of the silver-stained granules. Chayen (1953) also found ascorbic acid in the mitochondria, using the method on plant tissues, and he introduced a number of modifications designed to increase its specificity. Jensen and Kavaljian (1956), however, found that in *Allium cepa* and *Vicia faba* the silver deposits appeared as cytoplasmic particles 0·2 to 0·6 μ in diameter. These were localized predominantly in the vicinity of the cell walls, in the root tip cells and also in the elongating cortical parenchyma.

Lison (1953) maintained that there was no doubt that the cytochemical localization of ascorbic acid by the technique of Giroud and Leblond corresponded to that given by chemical estimations. He did not believe, however, that the intracellular localization could be regarded as precise. Clara (1952, 1953) gave a list of materials other than ascorbic acid which are present in cells and tissues and which reacted with the acid $AgNO_3$ reagent. Amongst these were melanin, adrenochrome, enterochromaffin substance, pancreatic α-cell granules and granules of neurosecretory substance in the hypothalamus. Hagen (1954) carried out estimations of ascorbic acid in particulate and non-particulate fractions prepared by ultracentrifugation of sucrose dispersions of ox adrenal and dog liver. He found that ascorbic acid was entirely or almost entirely in the non-particulate fractions and concluded that the localization of reduced silver particles could not indicate the localization of ascorbic acid within the cell. Eränkö (1954) made a number of experiments to check the reliability of the method. He found that a clear solvent zone diffused into fresh adrenal tissue at a higher speed than silver nitrate and suggested, for this reason, that sections should be much better than slices. Certainly freeze-dried sections, as used by Eränkö, are excellent for the purpose and fresh cold microtome sections can be used equally well. Ascending paper chromatography was carried out by Eränkö with 30 μ adrenal sections placed on the paper at varying distances from the base line. In every case at least 20 per cent. of the ascorbic acid in the sections diffused out with the solvent front before the $AgNO_3$ arrived. He agreed with Bourne (1950), however, that reduction of $AgNO_3$ by ascorbic acid is so rapid that once reduced, diffusion of the silver precipitate would not occur. It is evident, however, that even with thin sections mounted on slides or coverslips some diffusion will take place before this event occurs.

Application of the Silver Nitrate Technique

There are two principal ways of applying the acid silver nitrate technique for ascorbic acid: (1) by application of the reagent to unfixed frozen sections, and (2) by fixation of freshly excised blocks of tissue with alcoholic acid silver nitrate. Deane and Morse (1948) found this last variant to be the most successful, and it was also employed by Bacchus (1950), who placed rat adrenal glands in a dark vial containing 5 per cent. $AgNO_3$ at pH 2 to 2·5 within 1 minute of extirpation. This author found a reducing substance present in nearly all the cells of the cortex which he considered mainly attributable to ascorbic acid. Two types of cortical cells were recognized, one with "ascorbic acid" granules aggregated peripherally close to the cell membrane, the other with aggregation in the perinuclear region. Whichever of the above variants is employed, it is possible to produce small black silver-stained granules in a wide variety of tissues, not only in the cells but also in tissue spaces and in the connective tissues. Macrophages in the supporting tissues of the adrenal cortex and pituitary gland constantly appear full of granules by the acid silver method.

It is not necessary, therefore, to decide whether the method works, but whether its specificity for ascorbic acid is absolute, relative or doubtful. Several substances besides ascorbic acid are known to reduce acid silver nitrate, the phosphate deposits observed by Barnett, Bourne and Fisher (1941), melanin granules (Bizzozero, 1908; Giroud and Leblond, 1936) and the various materials mentioned by Clara (1952) which have been listed above. Phosphates will obviously give little trouble in most tissues and in any case they are prevented from interfering in the final result by the modification introduced by Barnett *et al.* Melanin granules, similarly, cause little confusion. There is a significant common factor present, however, in three of the tissue components which are reported as positively staining with acid silver in the fresh state. These components are the mitochondria (Giroud and Leblond, 1935b), the Golgi apparatus (Hirsch, 1938) and the large granules of the adrenal and pituitary macrophages. All of these contain lipid, which is probably phosphatide. It has been shown by Alsterberg (1941) that phosphatides will cause the precipitation of silver from silver nitrate solutions and that the optimum pH for this reaction is in the acid range. There is no direct evidence that such lipid substances are responsible for the ascorbic acid reaction and these speculations do not exclude the possibility that, as the chief supporters of the reaction contend, ascorbic acid is present in all three situations. They do give rise to legitimate grounds for suspicion that the method has low specificity and that accurate intracellular histochemical localization of ascorbic acid by its means is not possible. Danielli (1946) put forward a reasoned criticism of the specificity of the method with respect to localization. According to his view the ascorbic acid and silver ions and probably their initial reaction products are highly diffusible. Consequently one would expect the silver precipitate, if localized at all, to be in the mixing zone in the cell cortex if the reaction product is indiffusable and on any suitable adsorbing surface if it is diffusable. For these reasons he considered that the acid silver technique could not show the *in vivo* distribution of ascorbic acid unless this was fixed chemically to an "indifferent body", and either the reaction product was not diffusible or else had a high affinity for fixed ascorbic acid. It is possible that these conditions, especially the first, may be fulfilled, but they have not been shown to exist by experimental means.

Since no other technique exists, the frozen section method of Bourne, together with the modification of Barnett *et al.* and the block staining technique by Bacchus (1950), appear in Appendix 10.

REFERENCES

Abood, L. G., and Abu'l Haj, S. K. (1956). *J. Neurochem.*, **1**, 119.
Abu'l Haj, S. K., and Rinehart, J. F. (1952). *J. Nat. Cancer Inst.*, **13**, 232.
Alsterberg, G. (1941). *Z. Zellforsch. A.*, **31**, 364.
Aminoff, D. (1959). *Virology*, **7**, 355.
Aminoff, D., Morgan, W. T. J., and Watkins, W. M. (1952). *Biochem. J.*, **51**, 379.
Anderson, B., Hoffman, P., and Meyer, K. (1963). *Biochem. Biophys. Acta.*, **74**, 309.
Arnold, E. (1908). *Arch. path. Anat.*, **193**, 1975.
Arnold, M. (1966). *Histochemie*, **6**, 1.
Arzac, J. P. (1947). *Analecta med.*, **8**, No. 2, p. 9.
Arzac, J. P., and Flores, L. G. (1949). *Stain Tech.*, **24**, 25.
Ashbel, R., and Seligman, A. M. (1949). *Endocrinology*, **44**, 565.
Avigad, G., Zeliksson, R., and Hestrim, S. (1961). *Biochem. J.*, **80**, 57.
Bacchus, H. (1950). *Amer. J. Physiol.*, **163**, 326.
Bahn, R. C., and Glick, D. (1954). *J. Histochem. Cytochem.*, **2**, 103.
Baker, J. R. (1945). "Cytological Technique," 2nd Edn., Methuen, London.
Baker, J. R. (1946). *Quart. J. micr. Sci.*, **87**, 441.
Baker, J. R. (1951). "Cytological Technique," 3rd Edn., Methuen, London.
Ball, J., and Jackson, D. S. (1953). *Stain Tech.*, **28**, 33.
Bank, O., and Bungenberg de Jong, H. G. (1939). *Protoplasma*, **32**, 489.
Bangle, R., and Alford, W. C. (1954). *J. Histochem. Cytochem.*, **2**, 62.
Bangle, R. Jr. (1954). *J. Histochem. Cytochem.*, **2**, 291.
Barnett, S. A., Bourne, G., and Fisher, R. B. (1941). *Nature, Lond.*, **147**, 542.
Barnett, S. A., and Fisher, R. B. (1943). *J. exp. Biol.*, **20**, 14.
Barry, V. C., and Mitchell, P. W. D. (1954). *J. chem. Soc.*, 4020.
Bauer, H. (1933). *Z. mikr.-anat. Forsch.*, **33**, 143.
Bauer, K. F. (1943). *Z. wiss. Mikr.*, **59**, 142.
Baumgarten, P. (1926). *Chem. Ber.*, **59**, 1166.
Bechtold, H. (1928). *Biochem. Z.*, **199**, 451.
Bell, D. J., and Baldwin, E. (1941). *J. Chem. Soc.*, p. 125.
Bell, D. J., Gutfreund, H., Cecil, R., and Ogston, A. G. (1948). *Biochem. J.*, **42**, 405.
Benazzi-Lentati, G. (1941). *Boll. Zool.*, **12**, 135.
Benazzi-Lentati, G. (1942). *Arch. ital. Anat. Embriol.*, **47**, 417.
Berenbaum, M. C. (1955). *J. clin. Path.*, **8**, 343.
Bensley, C. M. (1939). *Stain Tech.*, **14**, 47.
Bernard, C. (1877). "Leçons sur le Diabète." Paris.
Best, F. (1906). *Z. wiss. Mikr.*, **23**, 319.
Betts, A. (1961). *Amer. J. clin. Path.*, **36**, 240.
Bignardi, C. (1940a). *Boll. Sci. ital. sper.*, **15**, 593.
Bignardi, C. (1940b). *Att. Soc. nat. Modena.*, **71**, 59.

BIRCH, T. W., HARRIS, L. J., and RAY, S. N. (1933). *Biochem. J.*, **27**, 590.
BISKIND, C. R., and GLICK, D. (1936). *J. biol. Chem.*, **113**, 27.
BIZZOZERO, E. (1908). *Münch. med. Wschr.*, **55**, 2140.
BLIX, G. (1936). *Z. physiol. Chem.*, **240**, 43.
BLIX, G., TISELIUS, A., and SVENNSSON, H. (1941). *J. Biol. Chem.*, **137**, 485.
BLIX, G., GOTTSCHALK, A., and KLENK, E. (1957). *Nature*, **179**, 1088.
BLIX, G., LINDBERG, E., ODIN, L., and WERNER, I. (1956). *Acta Soc. med. upsal.*, **61**, 1.
BLOOM, W. L., LEWIS, G. T., SCHUMPERT, M. Z., and SHEN, T. (1951). *J. biol. Chem.*, **188**, 631.
BLOOM, G., and KELLY, J. W. (1960). *Histochemie*, **2**, 48.
BOBBITT, J. M. (1956). *Advanc. Carb. Chem.*, **11**, 1.
BOLLINGER, H. R., and ULRICH, P. (1952). *Helv. chim. Acta.*, **35**, 93.
BONDAREFF, W. (1957). *Anat. Rec.*, **129**, 97.
BONNER, J. (1936). *Bot. Rev.*, **2**, 475.
BOOZER, C. E., and LEWIS, E. S. (1953). *J. amer. chem. Soc.*, **75**, 3182.
BOURNE, G. H. (1933a). *Nature, Lond.*, **131**, 874.
BOURNE, G. H. (1933b). *Aust. J. exp. Biol. Med.*, **11**, 261.
BOURNE, G. H. (1936). *Anat. Rec.*, **66**, 369.
BOURNE, G. H. (1944). *Nature, Lond.*, **153**, 254.
BOURNE, G. H. (1950). *Ibid.*, **166**, 549.
BRACCO, M., and CURTI, P. M. (1953). *Biol. Latina*, **6**, 412.
BRADEN, A. W. H. (1955). *Stain Tech.*, **30**, 19.
BRAY, W. C., and HERSEY, A. V. (1934). *J. amer. chem. Soc.*, **56**, 1899.
BRIDGMAN, W. B. (1942). *J. amer. chem. Soc.*, **64**, 2349.
BRIMACOMBE, J. S., and WEBBER, J. M. (1964). "Mucopolysaccharides, Chemical Structure, Distribution and Isolation." Elsevier, Amsterdam.
BRUNET, P. (1952). *Science*, **116**, 126.
BUDDECKE, E. (1960). *Z. physiol. Chem.*, **318**, 33.
BUIST, G. J., and BUNTON, C. A. (1954). *J. chem. Soc.*, 1406.
BUNTING, H., and WHITE, R. F. (1950). *Arch. Path.*, **49**, 590.
CANTLEY, M., HOUGH, L., and PITTET, A. O. (1963). *J. chem. Soc.*, 2527.
CARLETON, H. M. (1938). "Histological Technique," 10th Edn., Oxford, p. 171.
CARPENTER, H. M., POLONSKY, B., and MENTEN, M. L. (1951). *Arch. Path.*, **51**, 480.
CARRÉ, P., and HORNE, G. (1927). *Ann. Bot.*, **41**, 193.
CARRÉ, P., and LIBERMANN, D. (1933). *Bull. Soc. chim. Franc.*, **53**, 1050.
CARTER, H. E., GLICK, F. J., NORRIS, W. P., and PHILLIPS, G. E. (1947). *J. biol. Chem.*, **170**, 285.
CARTER, H. E., and HUMISTON, C. G. (1951). *J. biol. Chem.*, **191**, 727.
CASSELMAN, W. G. B. (1954). *Quart. J. micr. Sci.*, **95**, 323.
CATCHPOLE, H. R., GERSH, I., and PAN, S. C. (1950). *J. Endocrinol.*, **6**, 277.
CERBULIS, J., and ZITTLE, C. A. (1961). *Analyt. Chem.*, **33**, 1131.
CHAPLIN, H., COHEN, S., and PRESS, E. M. (1965). *Biochem. J.*, **95**, 256.
CHARLES, A. F., and SCOTT, D. A. (1936). *Biochem. J.*, **30**, 1927.
CHAYEN, J. (1953). *Int. Rev. Cytol.*, **2**, 77.
CLARA, M. (1952). *Mikroskopie*, **7**, 387.
CLARA, M. (1953). In "Vitamins and Hormones," **6**, 12.
COFFEY, S., DRIVER, G. W., FAIRWEATHER, D. A. W., and IRVING, F. (1948). *Brit. Pat.*, **610**, 117.
COFFEY, S., DRIVER, G. W., FAIRWEATHER, D. A. W., and IRVING, F. (1949). *Chem. Abstracts*, **43**, 3205.
COURTOIS, J., FLEURY, P., and GRANDCHAMP, M. (1949). *Bull. Soc. chim. Franc.*, **31**, 567.
CRAIG, D. P., and SHORT, L. N. (1945). *J. chem. Soc.*, 419.
CRIEGEE, R. (1931). *Ber deutsch chem. Ges.*, **64**, 260.
CRIEGEE, R. (1935). *Chem. Abstr.*, **29**, 6820.
CRIEGEE, R. (1948). "Newer Methods of Preparative Organic Chemistry." Interscience, New York.

CRIEGEE, R., and BÜCHNER, E. (1940). *Ber deutsch chem. Ges.*, **73**, 563.
CRIPPA, A. (1951). *Boll. Soc. ital. sper.*, **27**, 599.
DAHLQVIST, A., OLSSON, I., and NORDEN, A. (1965). *J. Histochem. Cytochem.*, **13**, 423.
DANIELLI, J. F. (1946). *Nature, Lond.*, **157**, 755.
DANISHEFSKY, I., EIBER, H. B., and CARR, J. J. (1960). *Arch. Biophys. Biochem.*, **90**, 114.
DAVIDSON, E. A., and MEYER, K. (1955). *J. Amer. Chem. Soc.*, **77**, 4796.
DAVIES, D. V. (1952). *Stain Tech.*, **27**, 65.
DE ALMEIDA, D. F. (1960). *Stain Tech.*, **35**, 129.
DEANE, H. W., NESBETT, F. B., and HASTINGS, A. B. (1946). *Proc. Soc. exp. Biol. N.Y.*, **63**, 407.
DEANE, H. W., and MORSE, A. (1948). *Anat. Rec.*, **100**, 127.
DEMPSEY, E. W., BUNTING, H., SINGER, M., and WISLOCKI, G. B. (1947). *Ibid.*, **98**, 417.
DEMPSEY, E. W., and SINGER, M. (1946). *Endocrinology*, **38**, 270.
DEMPSEY, E. W., and WISLOCKI, G. B. (1947). *Anat. Rec.*, **98**, 417.
DEMPSEY, E. W., SINGER, M., and WISLOCKI, G. B. (1950). *Stain Tech.*, **25**, 73.
DEUEL, H., NEUKOM, H., and WEBBER, F. (1948). *Nature, Lond.*, **161**, 96.
DEUTSCH, H. F. (1954). *J. biol. Chem.*, **208**, 669.
DIEZEL, P. B. (1957). In "Cerebral Lipidoses." Ed. Cummings, J. N., Blackwell, Oxford.
DIMLER, R. J. (1952). *Advanc. Carb. Chem.*, **7**, 96.
DISCHE, Z. (1947). *J. biol. Chem.*, **167**, 189.
DRIESSEN, A. (1905). *Z. allg. Path. Anat.*, **16**, 129.
DWELTZ, N. E. (1960). *Biochem. Biophys. Acta*, **44**, 3, 416.
DYER, J. R. (1956). In "Methods of Biochemical Analysis." Ed. D. Glick, Interscience, New York.
DYER, J. R. (1956). "Methods Biochemical Analysis," **3**, 111.
EHRLICH, P. (1883). *Z. klin. Med.*, **6**, 33.
EHRLICH, P. (1901). *Deutsch. med. Woch.*, **15**, 434 and 498.
ELFTMAN, H. (1959). *J. Histochem. Cytochem.*, **7**, 98.
ERÄNKÖ, O. (1954). *J. Histochem. Cytochem.*, **2**, 167.
EYLAR, E. H., and JEANLOZ, R. W. (1962). *J. biol. Chem.*, **237**, 1021.
FEIGL, F. (1954). "Spot Tests," Vol. I, p. 107. Elsevier, Amsterdam.
FEULGEN, R., and VOIT, K. (1924). *Pflüg. Arch. ges. Physiol.*, **206**, 389.
FISCHER, F. G., and HULTZSCH, K. (1938). *Biochem. Z.*, **299**, 104.
FISHER, E. R., and LILLIE, R. D. (1954). *J. Histochem. Cytochem.*, **2**, 81.
FITZPATRICK, T. B., LARNER, J., and LANDING, B. H. (1948). *Bull. Int. Assoc. med. Mus.*, **28**, 96.
FLEURY, P. F., and COURTOIS, J. (1950). 8th Internat. Conseil, Solvay Conseil Chem. Brussels.
FLEURY, P. F., COURTOIS, J. E., and GRANDCHAMP, M. (1949). *Bull. Soc. chim. franc.*, 88.
FOLCH, J., and SPERRY, W. M. (1948). *Ann. Rev. Biochem.*, **17**, 147.
FRENCH, J. E., and BENDITT, E. P. (1953). *J. Histochem. Cytochem.*, **1**, 321.
FRIEDENWALD, J. S. (1947). In article by Michaelis, *Cold Spr. Harb. Symp. quant. Biol.*, **12**, 141.
GAGE, S. H. (1917). *J. comp. Neurol.*, **27**, 451.
GASIC, G., and GASIC, T. (1962). *Nature*, **196**, 170.
GEDIGK, P. (1952). *Klin. Woch.*, **30**, 1057.
GERRARD, W. (1944). *J. chem. Soc.*, **85**.
GERSH, I., and CATCHPOLE, H. R. (1949). *Amer. J. Anat.*, **85**, 457.
GEYER, G. (1956–57). *Acta Histochem.*, **3**, 118.
GEYER, G. (1962a). *Acta Histochem.*, **14**, 26.
GEYER, G. (1962b). *Acta Histochem.*, **14**, 307.
GHIARA, F. (1954). *Arch. Zool. ital.*, **39**, 17.
GHIARA, F. (1956). *Arch. Sci. biol.*, **40**, 192.

Ghiara, G. (1959). *Histochemie*, **1**, 274.
Gibb, R. P., and Stowell, R. E. (1949). *Blood*, **4**, 569.
Giroud, A. (1938). "L'acide ascorbique dans la cellule et les tissus." Protoplasma, Monograph, Berlin.
Giroud, A., Giro, E., Rabinowicz, M., and Hartmann, E. (1939). *Bull. Soc. Chim. biol. Paris*, **21**, 1021.
Giroud, A., and Leblond, C. P. (1934). *Arch. Anat. micr. Morph. exp.*, **30**, 1035.
Giroud, A., and Leblond, C. P. (1935a). *Ibid.*, **31**, 111.
Giroud, A., and Leblond, C. P. (1935b). *Bull. Histol. Tech. micr.*, **12**, 49.
Giroud, A., and Leblond, C. P. (1936). *Nature, Lond.*, **138**, 247.
Glegg, R. E., Clermont, Y., and Leblond, C. P. (1952). *Stain Tech.*, **27**, 277.
Glegg, R. E., Eidinger, D., and Leblond, C. P. (1953). *Science*, **118**, 614.
Glegg, R. E., Eidinger, D., and Leblond, C. P. (1954). *Ibid.*, **120**, 839.
Glick, D. (1935). *J. biol. Chem.*, **109**, 433.
Glick, D., and Biskind, G. R. (1935). *Ibid.*, **110**, 583.
Glick, D., and Biskind, G. R. (1936a). *Ibid.*, **113**, 427.
Glick, D., and Biskind, G. R. (1936b). *Ibid.*, **114**, 1.
Glick, D., and Biskind, G. R. (1936c). *Ibid.*, **115**, 551.
Glick, D., Alpert, M., and Stecklein, H. R. (1953). *J. Histochem. Cytochem.*, **1**, 326.
Goldstein, D. J. (1961). *Nature*, **191**, 407.
Goldstein, D. J. (1962). *Stain Tech.*, **37**, 79.
Goldstein, D. J. (1962). *Quart. J. micr. Sci.*, **103**, 477.
Gomori, G. (1946). *Amer. J. clin. Path.*, **16**, 347.
Gomori, G. (1950). *Amer. J. clin. Path.*, **20**, 665.
Gomori, G. (1952). Microscopic Histochemistry. Chicago University Press.
Gottschalk, A. (1956). *Biochim. biophys. Acta*, **20**, 560.
Gottschalk, A. (1960). *Nature*, **186**, 949.
Gottschalk, A. (1962). *Persp. Biol. Med.*, **5**, 327.
Gottschalk, A. (Ed.) (1966). "Glycoproteins; Their Composition, Structure, and Function." Elsevier, Amsterdam.
Grant, D., and Holt, A. (1960). *J. chem. Soc.*, 5026.
Graumann, W. (1952). *Anat. Anz.*, **99**, 19.
Graumann, W. (1953a). *Z. wiss. Mikr.*, **61**, 361.
Graumann, W. (1953b). *Mikroscopie*, **8**, 218.
Graumann, W. (1956). *Acta Histochem.*, **3**, 326.
Graumann, W., Arnold, M., and Gleissner, U. (1966). *Acta histochem.*, **23**, **276.**
Guiseley, K. B., and Ruoff, P. M. (1961). *J. org. Chem.*, **26**, 1248.
Guthrie, R. D., and Honeyman, J. (1959). *J. chem. Soc.*, 2441.
Guthrie, R. D. (1961). *Advance. Carb. Chem.*, **16**, 105.
Hack, M. H. (1949). Personal communication quoted by Gersh, 1949.
Haddock, N. H. (1948). *Research*, **1**, 15.
Hagen, P. (1954). *Biochem. J.*, **56**, 44.
Hale, A. J. (1953). *Quart. J. Micr. Sci.*, **94**, 303.
Hale, A. J. (1955). *J. Histochem. Cytochem.*, **3**, 421.
Hale, A. J. (1957). *Int. Rev. Cytol.*, **6**, 193.
Hale, C. W. (1946). *Nature, Lond.*, **157**, 802.
Halmi, N. S., and Davies, J. (1953). *J. Histochem. Cytochem.*, **1**, 447.
Hamerman, D. J., and Schubert, M. (1953). *J. gen. physiol.*, **37**, 291.
Harris, L. J., and Ray, S. N. (1933). *Biochem. J.*, **27**, 2006.
Hashim, S. A. (1952). *Acta anat.*, **16**, 355.
Hashim, S. A., and Acra, A. N. (1953). *Stain Tech.*, **28**, 1.
Head, F. S. H. (1950). *Nature, Lond.*, **165**, 236.
Head, F. S. H., and Hughes, G. (1952). *J. chem. Soc.*, 2046.
Heath, I. E. (1962). *Quart. J. micr. Sci.*, **103**, 457.
Hëdstrom, B. O. A. (1953). *Arkiv. Kem.*, **6**, 1.
Herlant, M. (1943). *Arch. Biol., Paris*, **54**, 225.
Hess, M., and Hollander, F. (1944). *J. Lab. clin. Med.*, **29**, 321.
Hess, M., and Hollander, F. (1947). *Ibid.*, **32**, 905.

Hicks, J. D., and Matthaei, E. (1958). *J. Path. Bact.*, **75**, 373.
Hirano, S., Hoffman, P., and Meyer, K. (1961). *J. org. Chem.*, **26**, 5064.
Hirsch, C. G. (1938). "Form und Stoffweschel der Golgi-Korper." Protoplasma, monograph, Berlin.
Hirsch, P. (1952). *Rec. Trav. chim.*, **71**, 999.
Hockett, R. C., and McClenahan, W. S. (1939). *J. Amer. Chem. Soc.*, **61**, 1667.
Hockett, R. C., Dienes, M. T., and Ramsden, H. E. (1943). *Ibid.*, **65**, 1474.
Honeyman, J., and Shaw, C. J. G. (1959). *J. chem. Soc.*, 2451.
Holczinger, L. (1956/57). *Acta Histochem.*, **3**, 19.
Holftreter, J. (1947). *J. Morph.*, **80**, 25.
Holländer, H. (1964). *Histochemie*, **3**, 387.
Holmes, W. C. (1928). *J. Amer. Chem. Soc.*, **50**, 1939.
Hooghwinkel, G. J., and Smits, G. (1957). *J. Histochem. Cytochem.*, **5**, 120.
Hotchkiss, R. D. (1948). *Arch. Biochem.*, **16**, 131.
Hudack, S., Blunt, J. W., Higbee, P., and Kearn, G. M. (1949). *Proc. Soc. exp. Biol. N.Y.*, **72**, 526.
Immers, J. (1954). *Exp. Cell Research*, **6**, 127.
Ishizuka, Y. (1955). *J. med. Univ., Kyoto*, **57**, XX.
Izard, J. (1964). *J. Histochem. Cytochem.*, **12**, 486.
Jackson, E. L., and Hudson, C. S. (1937). *J. Amer. Chem. Soc.*, **59**, 2049.
Jacques, L. B. (1961). *Canad. J. Biochem. Physiol.*, **39**, 643.
Jeanloz, R. W. (1950). *Science*, **111**, 289.
Jeanloz, R. W. (1960). *Arthritis and Rheumatism*, **3**, 233.
Jeanloz, R. W. (1963). *Advanc. Enzymol.*, **25**, 433.
Jensen, W. A., and Kavaljian, L. G. (1956). *J. Biophys. Biochem. Cytol.*, **2**, 87.
Jordan, R. H., and McManus, J. F. A. (1952). *J. Nat. Cancer Inst.*, **13**, 228.
Jorpes, J. E., Weiner, B., and Aberg, B. (1948). *J. biol. Chem.*, **176**, 277.
Kantor, T. G., and Schubert, M. (1957). *J. Amer. Chem. Soc.*, **79**, 152.
Kasten, F. (1960). *Histochemie*, **1**, 466.
Kasten, F. (1962). (See Chapter 9 refs.)
Kelly, J. W. (1955). *Arch. Biochem. Biophys.*, **55**, 130.
Kelly, J. W. (1956). Protoplasmatologia, Vol II, "The Metachromatic Reaction." Springer, Vienna.
Kent, P. W., and Whitehouse, M. W. (1955). "Biochemistry of the Aminosugars." Butterworth, London.
King, C. G. (1936). *Physiol. Rev.*, **16**, 38.
Klein, G. (1932). "Handbuch der Pflanzenanalyse," Vol. 2, p. 5. Vienna.
Krajčinovič, M., Puric, V., and Krajčinovič, M. (1954–55). *Acta Histochemica*, **1**, 76.
Kramer, H., and Windrum, G. M. (1953). *J. clin. Path.*, **6**, 239.
Kramer, H., and Windrum, G. M. (1954). *J. Histochem. Cytochem.*, **2**, 196.
Kramer, H., and Windrum, G. M. (1955). *Ibid.*, **3**, 227.
Kugler, J. H., and Wilkinson, W. J. C. (1959). *J. Histochem. Cytochem.*, **7**, 398.
Kugler, J. H., and Wilkinson, W. J. C. (1960). *J. Histochem. Cytochem.*, **8**, 195.
Kugler, J. H. (1965). M.Sc. Thesis, University of Sheffield.
Kuyper, Ch. M. A. (1962). *Histochemie*, **3**, 46.
Landsmeer, J. M. F. (1951). *Act. physiol. pharmacol. neerl.*, **2**, 112.
Lange, W. (1966). *Acta histochemica*, **24**, 320.
Langhans, C. (1890). *Virchows Arch.*, **120**, 28.
Laurent, T. C., Ryan, M., and Pietruskiewicz, A. (1960). *Biochim. biophys. Acta*, **42**, 476.
Leblond, C. P., Glegg, R. E., and Eidinger, D. (1957). *J. Histochem. Cytochem.*, **5**, 445.
Lehmstedt, K., and Klee, H. (1936). *Ber.*, **69**, 1514.
Lev, R., and Spicer, S. S. (1964). *J. Histochem. Cytochem.*, **12**, 39
Lev, R., and Spicer, S. S. (1965). *Amer. J. Path.*, **46**, 23.
Levine, A., and Schubert, M. (1952). *J. Amer. Chem. Soc.*, **74**, 91.
Lewis, P. R., and Grillo, T. A. I. (1959). *Histochemie*, **1**, 391.
Lewis, P. R. (1962). *Ibid.*, **2**, 423.

LHOTKA, J. F. (1952a). *Anat. Rec.*, **112**, 422.
LHOTKA, J. F. (1952b). *Ibid.*, **112**, 455.
LHOTKA, J. F. (1952c). *Nature, Lond.*, **170**, 751.
LHOTKA, J. F. (1952d). *Stain Tech.*, **27**, 213.
LHOTKA, J. F. (1952e). *Ibid.*, **27**, 259.
LHOTKA, J. F. (1953). *Ibid.*, **28**, 245.
LHOTKA, J. F. (1954). *Ibid.*, **29**, 295.
LHOTKA, J. F. (1957). *Ibid.*, **32**, 275.
LILLIE, R. D. (1947a). *Bull. int. Assoc. Med. Mus.*, **27**, 23.
LILLIE, R. D. (1947b). *J. Lab. clin. Med.*, **32**, 910.
LILLIE, R. D. (1948). "Histopathologic Technic," 1st Ed. Blakiston, New York.
LILLIE, R. D. (1950). *Anat. Rec.*, **108**, 239.
LILLIE, R. D. (1951). *Stain Tech.*, **26**, 123.
LILLIE, R. D. (1953). *J. Histochem. Cytochem.*, **1**, 353.
LILLIE, R. D. (1954a). "Histopathologic Technic and Practical Histochemistry," 2nd Ed. Blakiston, New York.
LILLIE, R. D. (1954b). *J. Histochem. Cytochem.*, **2**, 127.
LILLIE, R. D. (1956). *Ibid.*, **4**, 479.
LILLIE, R. D. (1962). *J. Histochem. Cytochem.*, **10**, 303.
LILLIE, R. D., and GRECO, J. (1947). *Stain Tech.*, **22**, 67.
LILLIE, R. D., and MOWRY, R. W. (1949). *Bull. int. Assoc. med. Mus.*, **30**, 91.
LILLIE, R. D., GILMER, P. R., and WELSH, R. A. (1961). *Stain Tech.*, **36**, 361.
LISON, L. (1935a). *C.R. Soc. Biol., Paris*, **118**, 821.
LISON, L. (1935b). *Bull. Histol. Tech. micr.*, **12**, 279.
LISON, L. (1936). "Histochimie Animale." Paris.
LISON, L. (1953). "Histochemie et Cytochimie Animales." Gautier Villars, Paris, p. 534.
LISON, L. (1954). *Stain Tech.*, **29**, 131.
LISON, L., and VOKAER, R. (1949). *Ann. Endocrinol., Paris*, **10**, 66.
MCCOMB, E. A., and MCCREADY, R. M. (1957). *Analyt. Chem.*, **29**, 819.
MCCREADY, R. M., and REEVE, R. M. (1955). *Agricult. Food Chem.*, **3**, 260.
MCMANUS, J. F A. (1946). *Nature, Lond.*, **158**, 202.
MCMANUS, J. F. A. (1948a). *Amer. J. Path.*, **24**, 643.
MCMANUS, J. F. A. (1948b). *Stain Tech.*, **23**, 99.
MCMANUS, J. F. A. (1954). In "Connective Tissue in Health and Disease." Ed. Asboe-Hanson, Munksgaard, Copenhagen, p. 31.
MCMANUS, J. F. A. (1956). *Nature, Lond.*, **178**, 914.
MCMANUS, J. F. A. (1961). *Gen. Cytochem. Methods*, **2**, 171.
MCMANUS, J. F. A., and CASON, J. E. (1950). *J. exp. Med.*, **91**, 651.
MCMANUS, J. F. A., and FINDLEY, L. (1949). *Surg. Gynec. Obstet.*, **89**, 616.
MCMANUS, J. F. A., and HOCH-LIGETI, C. (1952). *Lab. Invest.*, **1**, 19.
MCMANUS, J. F. A., LUPTON, C. H. Jr., and GRAHAM, L. S. Jr. (1951). *Anat. Rec.*, **110**, 75.
MCMANUS, J. F. A., and MOWRY, R. W. (1952). *Lab. Invest.*, **1**, 208.
MALAPRADE, L. (1928). *C.R. Soc. Biol., Paris*, **186**, 625.
MALAPRADE, L. (1934). *Bull. Soc. chim., Fr.*, **1**, 833.
MANCINI, R. E. (1944). *Arch. Soc. argent. Anat. norm. patol.*, **6**, 628.
MANCINI, R. E. (1950). *Rev. Soc. argent. Biol.*, **26**, 139.
MANGIN, L. (1893). *C.R. Soc. biol., Paris*, **116**, 53.
MARCHESE, S. (1947). *Atti Soc. lombarda Sci. med. biol.*, **2**, 1. Quoted by McManus, 1948b.
MASSART, L., COUSSENS, R., and SILVER, M. (1951). *Bull. Soc. chim. biol.*, **33**, 514.
MELAMED, M. D. (1966). In "Glycoproteins, Their Composition, Structure and Function." Ed. Gottschalk, A. Elsevier, Amsterdam, p. 317.
MEYER, K. (1938). *Cold Spr. Harb. Symp. quant. Biol.*, **6**, 91.
MEYER, K. (1943). *Advanc. Enzymol.*, **3**, 109.
MEYER, K. (1945). *Advanc. Prot. Chem.*, **2**, 249.
MEYER, K. H. (1953). In "Some Conjugated Proteins," Symposium Rutgers Univ., N.Y. Ed. W. H. Cole.

MEYER, K., and PALMER, J. W. (1934). *J. biol. Chem.*, **107**, 624.
MEYER, K., and SMYTH, E. M. (1937). *J. biol. Chem.*, **119**, 507.
MEYER, K., SMYTH, E. M., and PALMER, J. W. (1937). *Ibid.*, **119**, 73.
MEYER, K., and CHAFFEE, E. (1941). *J. biol. Chem.*, **138**, 491.
MEYER, K. H., and JEANLOZ, R. W. (1943). *Helv. chim. Acta*, **26**, 1784.
MEYER, K., LINKER, A., DAVIDSON, E. A., and WEISSMANN, B. (1953). *J. biol. Chem.*, **205**, 611.
MICHAELIS, L. (1902). "Einfürhung in die Farbstoffchemie für Histologen." Berlin.
MICHAELIS, L. (1947). *Cold. Spr. Harb. Symp. quant. Biol.*, **12**, 142.
MICHAELIS, L. (1950). *J. Phys. Coll. Chem.*, **54**, 1.
MICHAELIS, L., and GRANICK, S. (1945). *J. Amer. chem. Soc.*, **67**, 1212.
MITCHELL, A. J., and WISLOCKI, G. B. (1944). *Anat. Rec.*, **90**, 261.
MÓDIS, L., SÜVEGES, I., and FÖLDES, I. (1964). *Acta histochemica*, **19**, 343.
MONTREUIL, J., and BISERTE, G. (1959). *Bull. Soc. chim. Biol.*, **41**, 959.
MORRISON, R. W., and HACK, M. H. (1949). *Amer. J. Path.*, **25**, 597.
MORONE, G. (1952). *Atti Soc. oftal. Lombarda*, **7**, 312.
MOWRY, R. W. (1952). *J. Nat. Cancer Inst.*, **13**, 230.
MOWRY, R. W. (1956). *J. Histochem. Cytochem.*, **4**, 407.
MOWRY, R. W. (1960). *J. Histochem. Cytochem.*, **8**, 323.
MOWRY, R. W. (1963). *Ann. N.Y. Acad. Sci.*, **106**, 402.
MOWRY, R. W., and MILLICAN, R. C. (1952). *Amer. J. Path.*, **28**, 522.
MOWRY, R. W., and EMMEL, V. M. (1966). *J. Histochem. Cytochem.*, **14**, 799.
MUKHERJEE, P., and GHOSH, A. K. (1963). *J. phys. Chem.*, **67**, 193.
MULAY, L. N., and SELWOOD, P. W. (1954). *J. Amer. Chem. Soc.*, **76**, 6207.
MULLEN, J. P. (1944). *Amer. J. clin. Path.* (*tech. sect.*), **14**, 9–10.
MÜLLER, G. (1955–56a). *Acta Histochem.*, **2**, 68.
MÜLLER, G. (1955–56b). *Ibid.*, **2**, 73.
MÜLLER, G. (1964). *Acta histochemica*, **17**, 61.
MÜNCH, O., and ERNST, B. (1964). *Acta histochemica*, **18**, 51.
MÜNCH, O., and ERNST, B. (1965). *Ibid*, **20**, 125.
NEUBERGER, A. (1941). *J. chem. Soc.*, **50**.
NEUKOM, H., and HUI, P. (1959). *Chim. Schw.*, **13**, 390.
NEVELL, T. P. (1959). *Chem. Ind.*, 567.
NICOLET, B. H., and SHINN, L. A. (1939). *J. Amer. chem. Soc.*, **61**, 1615.
NICOLET, B. H., and SHINN, L. A. (1941). *J. biol. Chem.*, **139**, 687.
NIELSEN, M. A., OKKELS, H., and STOCKHOLM-BORRESON, C. C. (1932). *Acta Path. Scand.*, **9**, 258.
NOGUCHI, H. (1956). *Biochim. Biophys. Acta.*, **22**, 459.
OGSTON, A. G., and STANIER, J. E. (1950). *Biochem. J.*, **46**, 364.
OKAMOTO, K., KADOTA, I., and AOYAMA, Z. (1948). *Taishitzu. Gaku Zasshi*, **14**, 35.
ORTMAN, R., FORBES, W. F., and BALASURBRAMANIAN, A. (1966). *J. Histochem. Cytochem.*, **14**, 104.
OTTENSTEIN, B., SCHMIDT, G., and THANNHAUSER, S. J. (1948). *Blood*, **3**, 1250.
PAL, M. K. (1965). *Histochemie*, **5**, 24.
PAL, M. K., and SCHUBERT, M. (1962). *J. Amer. chem. Soc.*, **84**, 4384.
PAL, M. K., and SCHUBERT, M. (1963). *J. phys. Chem.*, **67**, 1821.
PAUSACKER, K. H. (1953). *J. Chem. Soc.*, p.107.
PEARSE, A. G. E. (1949a). *J. clin. Path.*, **2**, 81.
PEARSE, A. G. E. (1949b). *J. Path. Bact.*, **61**, 195.
PEARSE, A. G. E. (1950a). *Ibid.*, **62**, 351.
PEARSE, A. G. E. (1950b). *Stain Tech.*, **25**, 95.
PEARSE, A. G. E. (1951). *J. clin. Path.*, **4**, 1.
PERL, E., and CATCHPOLE, H. R. (1950). *Arch. Path.*, **50**, 233.
PERLIN, A. S. (1955). *Analyt. Chem.*, **27**, 396.
PERSSON, B. H. (1953). *Acta Soc. med. upsal.*, **58**, Suppl. 2.
PIGMAN, W., and GOTTSCHALK, A. (1966). In "Glycoproteins; Their Composition, Structure and Function." Elsevier, Amsterdam, p. 434.
PIOCH, W. (1957). *Virchows Arch.*, **330**, 337.
PISCHINGER, A. (1926). *Z. Zellforsch.*, **3**, 169.

PISCHINGER, A. (1927). *Pflüg. Arch. ges. Physiol.*, **217**, 205.
POLGLASE, W. J., BROWN, D. M., and SMITH, E. L. (1952). *J. biol. Chem.*, **199**, 105.
POPENOE, E. A., and DREW, R. M. (1957). *J. biol. Chem.*, **228**, 673.
POWERS, M. M., CLARK, G., DARROW, M. A., and EMMEL, V. M. (1960). *Stain Tech.*, **35**, 19.
PRESS, E. M., and PORTER, R. R. (1966). In "Glycoproteins; Their Composition, Structure and Function." Ed. Gottschalk, A., Elsevier, Amsterdam, p. 395.
PRESTON, R. D. (1964). *Endeavour*, **23**, 153.
PRINS, P. A., and JEANLOZ, R. W. (1948). *Ann. Rev. Biochem.*, **17**, 67.
QUAY, W. B. (1957). *Stain Tech.*, **32**, 175.
QUINTARELLI, G., SCOTT, J. E., and DELLOVO, M. C. (1964a). *Histochemie*, **4**, 86.
QUINTARELLI, G., SCOTT, J. E., and DELLOVO, M. C. (1964b). *Ibid*, **4**, 99.
QUINTARELLI, G., and DELLOVO, M. C. (1965). *Histochemie.*, **5**, 196.
RABINOWITCH, E., and EPSTEIN, L. F. (1941). *J. Amer. chem. Soc.*, **63**, 69.
RABINOWITCH, E., and STOCKMAYER, W. H. (1942). *J. Amer. chem. Soc.*, **64**, 335.
RAVETTO, C. (1964). *J. Histochem, Cytochem.*, **12**, 306.
REES, D. A. (1961). *J. chem. Soc.*, 5168.
REEVE, R. M. (1959). *Stain Tech.*, **34**, 209.
REITZ, H. C., FERRELL, R. E., FRAENKEL-CONRAT, H., and OLCOTT, H. S. (1946). *Ibid.*, **68**, 1024.
REYNOLDS, D. D., and KENYON, W. O. (1951). *J. Amer. chem. Soc.*, **72**, 1587.
RICE, F. A. H., and FISHBEIN, L. (1956). *J. Amer. chem. Soc.*, **78**, 3731.
RICKETTS, C. R. (1952). *Biochem. J.*, **51**, 129.
RIES, E. (1935). *Z. Zellforsch.*, **22**, 523.
RIGBY, W. (1949). *Nature, Lond.*, **164**, 185.
RIGBY, W. (1950). *J. chem. Soc.*, p. 1907.
RINEHART, J. F., and ABU'L HAJ, S. K. (1951). *Arch. Path.*, **52**, 189.
RITTER, H. B., and OLESON, J. J. (1947). *Ibid.*, **43**, 330.
RITTER, H. B., and OLESON, J. J. (1950). *Amer. J. Path.*, **26**, 639.
ROBERTS, R. W. (1957). *J. Amer. chem. Soc.*, **79**, 7175.
ROMANINI, M. G. (1951). *Acta anat.*, **13**, 256.
ROMANINI, M. G., and GIORDANO, A. R. (1952). *Mikroskopie*, **7**, 26.
ROSAN, R. C., and SAUNDERS, A. M. (1965). *J. Histochem. Cytochem.*, **13**, 518.
ROSEN, O., HOFFMAN, P., and MEYER, K. (1960). *Fed. Proc.*, **19**, 147.
RUNHAM, N. W. (1961). *J. Histochem. Cytochem.*, **9**, 87.
RUNHAM, N. W. (1962). *J. Histochem. Cytochem.*, **10**, 504.
SALTHOUSE, T. N. (1961). *Stain Tech.*, **36**, 63.
SALTHOUSE, T. N. (1962). *J. Histochem. Cytochem.*, **10**, 109.
SAUNDERS, A. M. (1964). *J. Histochem. Cytochem.*, **12**, 164.
SAWICKI, E. (1962). In "Microchemical Techniques." Interscience, New York, p. 59.
SCHABADASCH, A. L. (1937). *Bull. Biol. Med. exp. U.S.S.R.*, **4**, 13.
SCHABADASCH, A. L. (1947). *Proc. nat. acad. Sci., U.S.S.R. Ser. biol. No.* 6, p. 745.
SCHEIBE, G. S., and ZANKER, V. (1958). *Acta histochem.*, Suppl. 1, p. 6.
SCHEIBE, G. S., and ZANKER, V. (1962). *Histochemie*, 3, 122.
SCHLESINGER, H. I., and VALKENBURGH, H. B. van (1931). *J. Amer. chem. Soc.*, **53**, 1212.
SCHOENBERG, M. D., and MOORE, R. D. (1964). *Biochim. Biophys. Acta*, **83**, 42.
SCHRAUTH, W. (1932). *Chem. Abstracts*, **26**, 3082.
SCHUBERT, M., and LEVINE, A. (1953). *J. Amer. chem. Soc.*, **75**, 5842.
SCHUBERT, M., and LEVINE, A. (1955). *Ibid.*, **77**, 4197.
SCHUBERT, M., and HAMERMAN, D. (1956). *J. Histochem. Cytochem.*, **4**, 159.
SCHULTZE, P. (1922). *Biol. Z.*, **42**, 388.
SCOTT, H. R., and CLAYTON, B. P. (1953). *J. Histochem. Cytochem.*, **1**, 336.
SCOTT, J. E. (1955). *Chem. Ind.*, 168.
SCOTT, J. E. (1960). "Methods of Biochemical Analysis," **8**, 145.
SCOTT, J. E., DORLING, J., and QUINGARELLI, G. (1964). *Biochem. J.*, **90**, 4P.
SCOTT, J. E., QUINTARELLI, G., and DELLOVO, M. C. (1964). *Histochemie*, **4**, 73.
SCOTT, J. E., and DORLING, J. (1965). *Histochemie*, **5**, 221.
SCOTT, J. E., and DORLING, J. (1966). *J. Histochem. Cytochem.*, **14**, 801.

SEIBERT, P. B., PFAFF, M. L., and SEIBERT, M V. (1948). *Arch. Biochem.*, **18**, 279.
SEITELBERGER, F. (1947). In "Cerebral Lipidoses." Blackwell, Oxford, p. 77.
SELIGMAN, A. M., GOFSTEIN, R., and RUTENBERG, A. M. (1949). *Cancer Res.*, **9**, 366.
SHACKLEFORD, J. M. (1963). *Ann. N.Y. Acad. Sci.*, **106**, 572.
SHEAR, M., and PEARSE, A. G. E. (1963). *Nature*, **198**, 1273.
SHEPPARD, S. E., and GEDDES, A. L. (1942). *Rev. mod. Phys.*, **14**, 303.
SHEPPARD, S. E., and GEDDES, A. L. (1944a). *J. Amer. chem. Soc.*, **66**, 1995.
SHEPPARD, S. E., and GEDDES, A. L. (1944b). *Ibid.*, **66**, 2003.
SHIMIZU, N., and KUMAMOTO, T. (1952). *Stain Tech.*, **27**, 97.
SIDGWICK, N. V. (1950). "The Chemical Elements and their Compounds." Oxford.
SPICER, S. S. (1960). *J. Histochem. Cytochem.*, **8**, 18.
SPICER, S. S. (1961). *Amer. J. clin. Path.*, **36**, 393.
SPICER, S. S. (1965). *J. Histochem. Cytochem.*, **13**, 211.
SPICER, S. S. and LILLIE, R. D. (1959). *J. Histochem. Cytochem.*, **7**, 123.
SPICER, S. S., LEPPI, T. J., and STOWARD, P. J. (1965). *J. Histochem. Cytochem.*, **13**, 599.
SPICER, S. S., and MEYER, D. B. (1960). *Tech. Bull. Reg. Med. Technol.*, **30**, 53.
SPICER, S. S., and WARREN, L. (1960). *J. Histochem. Cytochem.*, **8**, 135.
SPICER, S. S., and JARRELS, M. H. (1961). *J. Histochem. Cytochem.*, **9**, 368.
SPRINGALL, H. D. (1954). "The Structural Chemistry of Proteins." Acad. Press, N.Y.
STACEY, M., and BARKER, S. A. (1962). "Carbohydrates of Living Tissues." Van Nostrand, London.
STÄHLER, A., and SCHIRM, E. (1911). *Ber.*, **44**, 319.
STAPLE, P. H. (1955). *Nature, Lond.*, **176**, 1125.
STAPLE, P. H. (1957). *J. Histochem. Cytochem.*, **5**, 472.
STAPLE, P. H. (1958). *Nature, Lond.*, **178**, 288.
STEEDMAN, H. F. (1950). *Quart. J. micr. Sci.*, **91**, 477.
STEMPIEN, M. F. Jr. (1963). *J. Histochem. Cytochem.*, **11**, 478.
STETTEN, M. R., KATZEN, H. M., and STETTEN, D. (1956). *J. biol. Chem.*, **222**, 587.
STETTEN, M. R., KATZEN, H. M., and STETTEN, D. (1958). *Ibid.*, **232**, 475.
STILLER, D. (1965). *Acta histochemica*, **22**, 46.
STOFFYN, P. J., and JEANLOZ, R. W. (1960). *J. biol. Chem.*, **235**, 2507.
STOWARD, P. J. (1963). *D. Phil. Thesis.* Oxford University.
STOWARD, P. J. (1967a). *J. roy. micr. Soc.*, **87**, 77.
STOWARD, P. J. (1967b). *Ibid.*, **87**, 215.
STOWARD, P. J. (1967c). *Ibid.*, **87**, 237.
STOWARD, P. J. (1967d). *Ibid.*, **87**, 247.
STOWARD, P. J. (1967e). *Ibid.*, **87**, 407.
STOWARD, P. J., and MESTER, L. (1964). *Nature*, **204**, 488.
STRUVE, A. (1955). Quoted by Lubs, H. A. (Ed.), "The Chemistry of Synthetic Dyes and Pigments." Reinhold, New York.
SUTUR, C. M. (1944). In "Organic Chemistry of Sulphur." Wiley, New York.
SVENNERHOLM, L. (1957). *Biochim. Biophys. Acta.*, **24**, 604.
SVIRBELY, J. L., and SZENT-GYÖRGI, A. (1933). *Biochem. J.*, **27**, 279.
SYLVÉN, B. (1941). *Acta chir. scand.*, **86**, Suppl., 66.
SYLVÉN, B. (1945). *Acta radiol., Stockh.*, Suppl., 59.
SYLVÉN, B. (1954). *Quart. J. micr. Sci.*, **95**, 327l.
SYLVÉN, B. (1959). *Acta Histochem.*, **6**.
SYLVÉN, B., and MALMGREN, H. (1952). *Lab. Invest.*, **1**, 413.
TAKEUCHI, J. (1962). *Stain Technol.*, **37**, 105.
TAYLOR, K. B. (1961). *Stain Tech.*, **36**, 73.
TISELIUS, A. (1937). *Biochem. J.*, **31**, 1464.
TOCK, E. P. C., and PEARSE, A. G. E. (1965). *J. roy. micr. Soc.*, **84**, 519.
VENKATARAMAN, K. (1952). "The Chemistry of Synthetic Dyes." Acad. Press, New York, p. 1262.
VERNE, J. (1929). *Ann. Physiol., Paris*, **5**, 245.
VICKERSTAFF, T., and LEMIN, D. R. (1946). *Nature, Lond.*, **157**, 373.
VIALLI, M. (1951). *Boll. Soc. ital. Sper.*, **27**, 597.
VOSS, W., and BLANKE, E. (1931). *Annal. Chem.*, **485**, 258.

WALLRAF, J., and BECHERT, H. (1939). *Z. mikr. anat. Fossch.*, **45**, 510.
WALTON, K. W., and RICKETTS, C. R. (1954). *Brit. J. exp. Path.*, **35**, 227.
WARREN, L. (1959). *J. biol. Chem.*, **234**, 1971.
WEIMER, H. E., MEHL, J. W., and WINZLER, R. J. (1950). *J. biol. Chem.*, **185**, 561.
WHISTLER, R. C., and SMART, C. L. (1953). "Polysaccharide Chemistry." Acad. Press, New York.
WIGGLESWORTH, V. B. (1952). *Quart. J. micr. Sci.*, **93**, 105.
WILLSTÄTTER, R., and ROHDENWALD, M. (1934). *Hoppe-Seyl. Z.*, **225**, 103.
WISLOCKI, G. B., BUNTING, H., and DEMPSEY, E. W. (1947). *Amer. J. Anat.*, **81**, 1.
WISLOCKI, G. B., BUNTING, H., and DEMPSEY, E. W. (1949), *Blood*, **4**, 562.
WISLOCKI, G. B., and SINGER, M. (1950). *J. Comp. Neurol.*, **92**, 71.
WISLOCKI, G. B., RHEINGOLD, J. J., and DEMPSEY, E. W. (1949) *Blood*, **4**, 562.
WITTELS, B. (1963). *Arch. Path.*, **75**, 127.
WOLFROM, M. L., and MONTGOMERY, R. (1950). *J. Amer. chem. Soc.*, **72**, 2859.
WOLFROM, M. L., MADISON, R. K., and CRON, M. J. (1952). *J. Amer. chem. Soc.*, **74**, 1491.
WOLFROM, M. L., SHEN, T. M., and SUMMERS, C. G. (1953). *Ibid.*, **75**, 1519.
WOLMAN, M. (1950). *Proc. Soc. exp. Biol. N.Y.*, **75**, 583.
WOLMAN, M. (1956). *Stain Tech.*, **31**, 241.
WOLMAN, M. (1956a). *Bull. Res. Council, Israel*, 6E, 27.
WOLMAN, M. (1956b). *J. Histochem. Cytochem.*, **4**, 195.
WOLMAN, M. (1961). *Stain Tech.*, **36**, 21.
YAMADA, K. (1964). *J. Histochem. Cytochem.*, **12**, 327.
YODEN, A., and TOLLENS, E. (1901). *Ber.*, **34**, 3446.
ZONIS, S. A., and KORNILOVA, Y. (1950). *Zhur. Obschchen. Klim.*, **20**, 1252.
ZONIS, S. A., and PESINA, A. G. (1950). *Ibid.*, **20**, 1180.
ZUGIBE, F. T. (1963). *J. Histochem. Cytochem.*, **11**, 35.
ZUGIBE, F. T., and FINK, M. L. (1966a). *J. Histochem. Cytochem.*, **14**, 147.
ZUGIBE, F. T., and FINK, M. L. (1966b), *Ibid.*, 153.

CHAPTER 11
APPLIED MUCOSUBSTANCES

In the 2nd Edition of this book (1960), under the heading "Applied Muco-proteins", only amyloid and amyloidosis were considered. This topic remains the principal component of the chapter, but it has now been extended by the addition of other disorders in which mucosubstances are produced and deposited in abnormal amounts. The two most important of these are gargoylism (Pfaundler-Hurler syndrome) and Whipple's disease. Brief consideration is given to the participation of mucosubstances in physiological and pathological processes.

The History and Occurrence of Amyloid

The deposition of a hyaline homogeneous substance in the walls of blood vessels and in the connective tissues was observed by many workers before it was described by Rokitansky (1842) under the comprehensive term lardaceous disease. Virchow (1860), in 1853, named the carbohydrate component of the substance amyloid because of its starch-like behaviour when stained with iodine. The chemical composition of amyloid was extensively discussed by Friedreich and Kekulé (1859) who concluded that the iodine reaction was not due to "cholestearin" and that the main constituent was protein. They excluded carbohydrates. The much later studies of Eppinger (1922) and Hass (1942) showed that amyloid was indeed a carbohydrate-containing protein. Virchow considered that pre-amyloid in the circulating plasma was transformed into a gel after crossing the vessel walls. "Ich bin jetzt vielmehr geneigt anzunehmen dass das Blut in dieser Krankheit eine chemische Veränderung in seinen gelösten Bestandteilen eifahren hat, als dass es die pathologische Substanz in Körperlicher Form enthält." My comment (1960) was "Nothing has occurred in the last 100 years which radically alters this historic conception." A rather similar comment was made by Letterer (1960), discussing the evolution of amyloid. "Nach hundert Jahren sind wir ein Stück weitergekommen."

A very complete review of the early history of amyloid was presented by Puchtler and Sweat (1966). This paper clears up many of the misunderstandings, some of them repeated by me in previous editions, derived from incomplete study or imperfect translation of the original works.

The deposition of amyloid was divided by Reimann and Eklund (1935) into primary and secondary amyloidosis on the basis of the absence or presence of an exciting cause for its production and on the normality or otherwise of its distribution and on its staining characteristics. Primary amyloid, for instance, almost invariably involved the heart and in two-thirds of the cases the liver, kidney, spleen and adrenals were not affected. In the secondary form of amyloidosis the heart was free from deposits, but liver, spleen, adrenal and kidney were invariably involved. Two subsidiary types of amyloid, in the classification of Reimann and Eklund, were tumour-forming amyloid and amyloid associated with multiple myeloma. A more simple classification was proposed by King (1948):

(1) **typical,** divided into two classes: (*a*) associated with other disease and (*b*) not associated with other disease;
(2) **atypical,** occurring in single or multiple foci and divided, as above, into two classes.

Reviewing the whole matter Symmers (1956a and b) indicated that in his experience the differences between generalized amyloidosis with, and without, known causation were not greater than the differences between individual cases in a single category. In his very comprehensive review Cohen (1965) agreed with Symmers and accepted the latter's classification: (1) Generalized amyloidosis with predisposing cause (secondary), (2) Generalized amyloidosis without known cause (primary), (3) Localized amyloidosis.

It was to be expected that removal of the chronic suppurative disorders, and tuberculosis, by modern chemotherapy would reduce the incidence of amyloidosis. On the contrary, it continues to rise. At the Royal Postgraduate Medical School between 1935 and 1962 there were 8758 autopsies and 91 cases of amyloidosis. This represents an incidence of 1·04 per cent. Secondary amyloidosis accounted for 83 cases, the remaining 8 were regarded as primary.

The breakdown of secondary cases was as follows:

Tuberculosis	27 cases
Sepsis	24 cases
Rheumatoid Arthritis	13 cases
Carcinoma	7 cases
Hodgkin's Disease	2 cases
Myeloma	3 cases
Others	7 cases

The term sepsis includes bronchiectasis, osteomyelitis, renal and other suppurative conditions.

Amyloidosis has long been considered a disorder of protein metabolism because it is characterized by the abnormal deposition of protein complexes in various organs. Several observations have suggested that amyloid deposits are at least partly derived from proteins circulating in the blood under abnormal conditions (Richter, 1956). It was shown by Luetscher (1940) and by Johansson (1949), as well as by a number of other authors, that the serum protein pattern is abnormal in amyloidosis and in multiple myelomatosis abnormal levels of serum globulin and amyloid coexist. Moreover, the incidence of amyloidosis is high in hyper-immunized horses used for the production of antisera. Castillo (1935) reviewed very thoroughly the relationship of protein metabolism to the precipitation of amyloid and this was considered also by Teilum *et al.* (1951)

and by Pirani (1951). A comparative study of amyloid deposits in several different species was made by Hass, Huntington and Krumdieck (1943).

A most interesting hypothesis of the cause of amyloidosis was that of Teilum (1956), which was the result of a long series of experimental and pathological studies on the subject. This author postulated that amyloid protein was synthesized by mesenchymal and reticuloendothelial cells in excessive quantities when the normal mechanism controlling this function ceased to operate. It was further postulated that persistent antigenic stimulation and other forms of chronic stress might cause breakdown of the controlling mechanism. This would account for the very high incidence of histologically demonstrable amyloid deposits in rheumatoid arthritis (Reece and Reynolds, 1954; Teilum and Lindahl, 1954). Extensive histological and immunological studies made by Rask-Nielsen *et al.* (1960) offer support for Teilum's hypothesis.

The role of the plasma cell in amyloid production was first suggested by Teilum (1948), who postulated that it was responsible for the abnormal production of globulin in allergic states. My own studies (Pearse, 1949) on plasma cells containing the so-called Russell bodies showed that these consisted for the most part of mucoprotein and that this mucoprotein was manufactured by the cells rather than absorbed from elsewhere. Additional data on the histochemistry of Russell bodies were presented by Bangle (1963). It was suggested by Lundin *et al.* (1954) that plasma cell proliferation was a general reaction promoted by a pituitary factor which Schelin *et al.* (1954) identified with growth hormone (somatotrophin). It had previously been shown by Cavallero (1953) that injection of this hormone into rats produced plasma cell proliferation. There is little doubt that the plasma cell and other mucoprotein-secreting reticulum cells are mainly responsible for production of the protein (mucoprotein) component of amyloid.

Artificial Production of Amyloid

Several techniques have been employed for the production of amyloidosis in animals. In mice, feeding with a casein-rich diet or repeated injections of casein have been used successfully by a number of workers and injections of ribonucleate have also been employed. The latter method was first used by Letterer (1926, 1934) and the production of amyloid was attributed to the formation of abscesses which were produced by the unsterile RNA injections. Goessner *et al.* (1951) used a similar method for the preparation of experimental amyloidosis. Richter (1954), however, produced amyloid in rabbits with buffered RNA passed through a Seitz filter before use. Injections of Freund's adjuvant were used by Rothbard and Watson (1954) to produce artificial amyloid by stimulating the production of antibodies to various antigens. Uotila, Perasolo and Vapaavuori (1955) tested the effect of intercurrent injections of somatotrophin and TSH on caseinate amyloidosis in mice. They found that TSH caused a rise in production of amyloid, but that somatotrophin was without effect. This observation is not in keeping with some of the findings reported earlier in this chapter, where somatotrophin was found to stimulate plasma cell proliferation.

A very complete list, and description, of the various types of experimentally induced amyloidosis appears in Cohen's (1965) review. This should be consulted by those wishing to familiarize themselves with the literature in this field. Recent work by Janigan (1965), and Janigan and Druet (1966), has made an important contribution to our understanding of the mechanism of amyloidosis, and particularly to our concepts of the role of antigenicity. By converting proteins to azo-proteins, Janigan and Druet showed that increasing antigenicity would convert non-amyloidogenic into amyloidogenic protein and decrease the time necessary for a given protein to produce amyloid. They found the following times: azo-gelatin 28–32 days, casein 22 days, azo-albumen 20 days, azo-casein 10 days or, with large amounts, 6–8 days. The substitution of protein hydrolysates for the origin proteins resulted in failure to produce amyloid.

Early work with immunofluorescent techniques (Mellors and Ortega, 1956; Vasquez and Dixon, 1956) had indicated that γ-globulins were present in amyloid. Since this did not, of itself, prove that amyloid is an antigen-antibody complex immunofluorescent techniques were extended to the demonstration of the presence of complement (Vogt and Kochem, 1960). Later studies (Lachmann *et al.*, 1962) indicated that amyloid consistently contained β_1C-globulin, and frequently (two out of three cases) 7S γ-globulin. These authors concluded that there were three alternative explanations: (1) antibody is bound to antigens normally present at the site of amyloid deposition; (2) antigen-antibody complexes unrelated to the tissue site are bound by an immune process or by adsorption; (3) γ-globulin aggregates are bound to the site by some unknown mechanism. Cohen and his associates (Cohen and Paul, 1963; Paul and Cohen, 1963; Cathcart and Cohen, 1966) have produced a considerable volume of evidence that the fibrous protein amyloid does not contain γ-globulin. They consider that when present the latter is merely adsorbed.

Protein Patterns in Amyloidosis

A detailed study of the electrophoretic patterns of sera in cases of amyloidosis was made by Bohle, Hartmann and Pola (1950). These authors found a transient rise of α and β-globulins followed by a more persistent rise in γ-globulins. Wagner (1955) investigated two cases of amyloid and found in one a raised serum α1-globulin fraction associated with a mucopolysaccharide and in the second raised α2- and β-globulin fractions, which were again associated with an abnormal mucopolysaccharide. Changes in the plasma proteins had been considered non-specific by Letterer and Schneider (1953), but Wagner's findings suggested otherwise. They suggested, furthermore, that association of the protein or mucoprotein (globulin) component of amyloid with the mucopolysaccharide component might take place in the bloodstream rather than at the site of deposition.

The composition of amyloid extracts has been investigated by several groups of workers. Calkins and Cohen (1958) found characteristics resembling those of an α-globulin. Benditt *et al.* (1962) produced urea extracts and found five different moieties on ultracentrifugation. These had sedimentation constants of 1, 6, 9, 16 and 23. The S-1 component was present in all four fractions and it was found to contain very high levels of tryptophan and tyrosine. This component is presumed to be responsible for the histochemical protein reactions described below. Mild preparative techniques were used by Muckle (1964) prior to immuno-electrophoresis. He found a variety of types of protein in different amyloids and concluded that some of these could be derived from those normally circulating in the blood plasma.

Studies of the amino-acid content of serum protein, liver protein, amyloid, hyaline and collagen were made by Letterer, Gerok and Schneider (1955). These showed that tyrosine, absent from hyaline, was present in moderate amounts in amyloid. Tryptophan was unfortunately not measured. The total number of lipophilic groups (Gly. Ala. Val. Leuc.) was comparatively high, but the number of basic groups was lower than might have been expected. Figures for cystine and methionine were similar to those published elsewhere for human γ-globulin, but for most of the other amino-acids the relationship was not particularly close. In a sample of human liver amyloid Giles and Calkins (1955) found 1·5 per cent. of hydroxyproline, 3·8 per cent. of glycine and 4·1 per cent. of tyrosine. They concluded that less than 10 per cent. of the dry weight of their material was composed of collagen. Other tests showed that their sample of amyloid contained 82·8 per cent. of water, indicating the presence of a hydrophilic protein. The total carbohydrate content was over 4 per cent. of the dry weight, 1·86 per cent. being neutral sugar, 1·55 per cent. hexosamine and 0·6 per cent. hexuronic acid. It is not possible to tell from these figures what percentage of the carbohydrate content was to be ascribed to internal components of the mucoprotein and how much to associated mucopolysaccharide. Treatment of amyloid with 5 per cent. HCl in methanol at 100° enabled Klenk and Faillard (1955) to extract from it the methylglycoside of neuraminic acid (see Chapter 12, p. 402). From a phenolic extract of amyloid liver Schmitz-Moormann (1961) isolated a mucoprotein containing 6·3 per cent. *N*-acetylneuraminic acid, 7·5 per cent. galactosamine, 6·3 per cent. glucuronic acid, 8·2 per cent. hexoses and 8·9 per cent. amino-acid nitrogen. It is difficult to compare these figures with more recent ones given by Cohen (1966), because of the different methods employed for fractionation.

The Histochemistry and Ultrastructure of Amyloid

Congo Red Staining

Of the two "specific" staining methods for amyloid (methyl violet and Congo red) the Congo red method (Bennhold, 1922) has provided considerably more information about structure than the other. When used in conjunction with simple light microscopy the specificity of the Congo red method is low. It stains most tissues before differentiation, and even after this process eosinophil granules, enterochromaffin granules, Paneth cell granules, and elastic fibres retain some of the dye.

Increased specificity for amyloid is afforded by staining from solution in 80 per cent. ethanol saturated with NaCl (Puchtler *et al.*, 1962). It is suggested that increased staining in the presence of electrolytes indicates non-ionic binding of Congo red to amyloid. The similarity, in terms of staining quality, between cellulose and amyloid has been emphasized by a number of workers (Puchtler *et al.*, 1962; Gueft and Ghidoni, 1963). An investigation of the binding of thirty direct cotton dyes was carried out by Puchtler *et al.* (1964). These authors considered that the dyes were adsorbed on to amyloid by hydrogen bonding between the hydroxyl groups of the polysaccharide chains, and suitable groups in the dye.

Dichroism

Structures which contain groupings which absorb light anisotropically exhibit the phenomenon of dichroism when viewed by polarized light. Cellulose fibres dyed with Congo red absorb green light very strongly if the plane of vibration of the light is parallel to the fibre axis, but not if the plane of vibration is perpendicular to the axis. Stated another way, parallel orientated dye molecules will show different absorption coefficients towards plane polarized light, according to whether the plane of polarization is parallel to or at right angles to the axis of the molecules. In this case the inference is that the long dye molecules are lying along the fibre axis. This is shown below, diagrammatically.

The whole matter of orientation of dyes on fibres was most ably dealt with by Vickerstaff (1954) ,who pointed out that the dichroic effect is greatest at low dye concentrations, due possibly to preferential absorption on the most highly orientated parts of the fibre. Amyloid dyed with Congo red exhibits positive

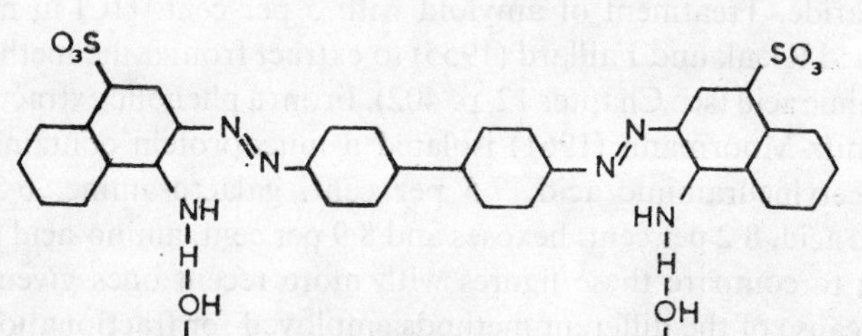

Cellulose

birefringence and dichroism and examination by polarized light should always be carried out if there is any doubt as to the identity of any Congo red-stained tissue component. Polarization optical studies on amyloid were carried out by Divry (1927), Ladewig (1945), Romhany (1949), Pfeiffer (1953), Missmahl and Hartwig (1953) and Dieselben (1954). Many earlier studies on the same lines

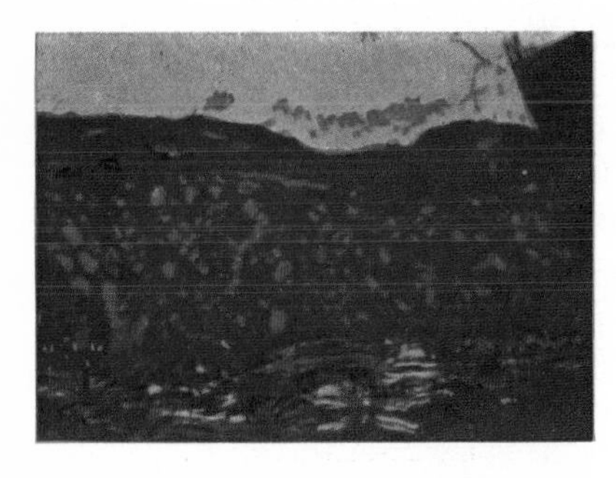

IXA. Congo red crystals examined under polarized light. The colour changes from red, through yellow, to green, in accordance with the thickness of the crystals. × 300

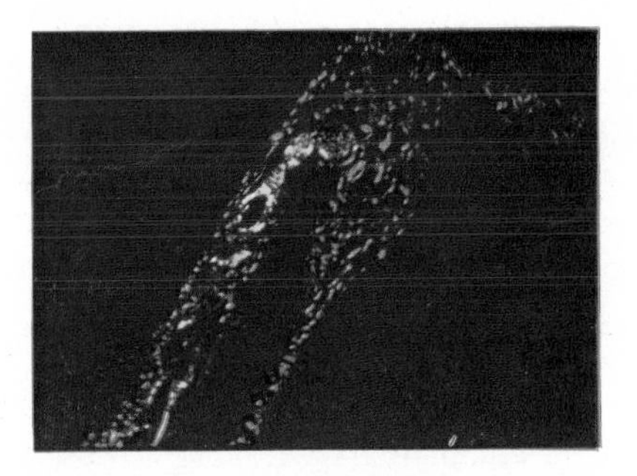

IXB. Human Primary Amyloidosis. Arterial Wall. Stained with Masson Trichrome and Congo red. Photographed by polarized light. (Red) amyloid deposited along collagen bands. × 150

IXC. Human Liver. Case of Familial Mediterranean Fever. Peri-reticulin amyloidosis in intima and beneath the endothelium of a venule. × 150

are referred to in the bibliography of these last three authors. Somewhat later Diezel (1957) reported on the dichroism of amyloid stained with Chlorantin fast red or with Evans blue. Seitelberger (1958) found that amyloid stained by the coupled tetrazonium reaction gave anomalous colours when viewed by polarized light. He pointed out that Congo red and Evans blue are both diazo dyes and that the reaction product of the coupled tetrazonium reaction (see p. 128) is also a disazo dye. This effect may be regarded as characteristic, but not specific, for amyloid.

Further work by Missmahl (1957, 1958, 1959) indicated that the green anisotropic colour of Congo red-stained amyloid was absolutely specific and this work culminated in the evolution of a scheme for differentiation of amyloid into perireticulin and pericollagen types (Heller *et al.*, 1964). According to these last authors every case of amyloid could be grouped in one of these two categories. (See section on Electron Microscopy of Amyloid, below.) The **perireticulin amyloidoses** are: secondary amyloidosis (except multiple myeloma), the hereditary primary amyloidosis of familial Mediterranean fever and the hereditary primary amyloidosis of Muckle and Wells (1962). (Plate IXb, facing p. 387.) The **pericollagen amyloidoses** are: amyloidosis secondary to multiple myeloma, "classical" primary amyloidosis, hereditary amyloidosis (Portuguese type) and localized amyloidosis. (Plate IXa, opposite.)

According to Wolman and Bubis (1965) experiments carried out on Congo red crystals (Plate IXc, facing p. 387), and on Congo red deposits polished in a single direction with a glass wheel, indicated that the green polarization colour depends primarily on the near perfect alignment of the dye particles. The green polarization colour was not present in too thick or too thin sections of amyloid-containing tissues stained with Congo red. Between 5 and 10 μ the correct conditions were found, that is to say, retardation of approximately half a wavelength of red light. Shirahama and Cohen (1966) found green birefringence in thick osmium-fixed epoxy-embedded Congo red-stained sections (0·5 to 2·0 μ). Presumably at this thickness the retardation is again approximately half a wavelength of red light.

AUTOFLUORESCENCE AND SECONDARY FLUORESCENCE

As noted by Cohen *et al.* (1959), amyloid is naturally fluorescent. This property presumably reflects both the high tryptophan and tyrosine content (see below) and the organized and orientated structure. Cohen *et al.* (1959) also reported that Congo red-stained amyloid exhibited a pink fluorescence and Rask-Nielsen *et al.* (1960) suggested that this method was valuable in the detection of small quantities of amyloid. Puchtler and Sweat (1965) agreed with this usage.

The use of an alternative fluorochrome, Thioflavine T (C.I.49005) was proposed by Vassar and Culling (1959) and recommended in further papers (1961, 1962). Amyloid stained with Thioflavine T (Appendix 11, p. 685) fluoresces greenish-yellow when viewed in blue (ultra-violet) light. Strong support for the specificity and sensitivity of the method was given by Hobbs and Morgan (1963) who indicated that false negative stainings with Congo red and methyl violet were invariably positive with the Thioflavine T method. Personally, I am convinced that there are not many cases of amyloid which cannot be demonstrated by Congo red (with fluorescence and polarization microscopy) or by methyl violet (with methyl green differentiation). On the other hand materials giving Thioflavine T fluorescence do not always show the characteristic fibrillar pattern in E.M. studies (see below). There is thus a danger of false positive staining with Thioflavine T, and McAlpine and Bancroft (1964) showed that moderate fluorescence occurred in hyalinized collagen. A very full list of tissue sites giving fluorescence when stained with Thioflavine T was provided by Burns *et al.* (1967), using both blue and UV light. Only with the latter did they find it possible to distinguish amyloid with confidence, by virtue of its relatively specific lime-green to blue fluorescence. The sensitivity of the method for amyloid is, however, undoubted and the results obtained (Plate Xa, facing p. 392) are so satisfactory that its popularity is assured.

Many fluorochromes (and non-fluorochromes) produce different fluorescent colours in amyloid deposits. The non-fluorochromes may be acting as internal filters. Of a wide variety of fluorochromes tested, the most specific, in my hands, has been Phosphine 3R (C.I.46045). Like Thioflavine T, this dye stains the nuclei but their fluorescence colour (copper yellow) is easily distinguishable from the bright clear yellow of amyloid deposits. Distinction from elastic fibres (greenish white) is particularly clear when Phosphine 3R is used (Plate Xb, facing p. 392).

When using fluorochromes (or other dyes) for staining amyloid it must be remembered that *any* fixation may influence the result very strongly. Whatever fixation is employed, unfixed cryostat sections should always be used in parallel in any critical studies on amyloid.

Chemical Composition

Protein. As might be expected from the results of amino-acid assay, histochemical methods for the basic amino-acids give only weakly positive results. Tests for basic protein, such as chromotrope 2R at pH 2 (Wagner, 1957) or acid Solochrome cyanine, are also weakly positive. Although sulphur-containing amino-acids are present in amyloid, tests for these, in my hands (1953), were negative or at most faintly positive. Thompson *et al.* (1961), however, reported that amino-acids and polypeptides containing disulphide linkages were a constant and significant component of amyloid. On the other hand, histochemical tests (1958) indicated the presence of a considerable quantity of tryptophan in samples of human secondary amyloid and at the time this had not been recorded chemically. The application of the DMAB-nitrite method (Chapter 6, p. 131) to an unfixed cold microtome section of kidney containing amyloid deposits is illustrated in Fig. 103). This indicates, at a rough estimate, that amyloid contains about the same amount of tryptophan as fibrin.

The DC method for tyrosine (Chapter 6, p. 129) indicates a high level of tyrosine and this agrees with amino-acid analyses, which now consistently show high tryptophan and tyrosine content and very low levels of cystine, methionine and hydroxyproline.

Solubility studies carried out by Newcombe and Cohen (1965) showed that isolated human amyloid fibrils were somewhat soluble in Sörensen's 0·1M glycinate in the acid range (pH 2–4) but much more soluble in the alkaline range (pH 7–12). This type of solubility indicates predominantly acidic side groups and the point of minimum solubility (pH 4·5) agrees very well with previous histochemical studies on the I.E.P. of amyloid. This was reported by Carnes and Forker (1956) as pH 4·5, and by Goldberg and Deane (1960) as near pH 5·0.

Carbohydrate. The positive PAS reaction of amyloid reflects the high carbohydrate content found in assays, but does not indicate how much of the staining is due to hexoses combined with proteins and how much is due to complexing with polysaccharide. Curiously enough, the alkaline tetrazolium reaction, which in the case of most mucins runs parallel to the PAS reaction, is almost invariably negative in the case of amyloid. The so-called paramyloid reacts readily, though it is, if anything, less PAS-positive. The ferric hydroxamic reaction (Chapter 10, p. 359) gives results which agree closely with those of the PAS reaction, provided the method is applied to unfixed sections. This is shown in Fig. 104.

The majority of assays carried out on amyloid have indicated that carbohydrate is invariably present. The amounts recorded have varied sharply with the method of preparation of the extracts. In the case of highly purified samples of the fibrils the total is about 1–3 per cent. (dry weight) (Giles and Calkins, 1955; Benditt *et al.*, 1962; Pirani *et al.*, 1964). Extensive studies by Schmitz-Moormann (1961, 1964a and b) indicated the presence in amyloid of a glycoprotein containing neuraminic acid, and two other sialic acid-free polysaccharides. The first of these was removable by extraction with cold phenol. Amyloid in tissue sections, treated in this way, was observed to lose its methyl violet metachromasia *and* its dichroism when stained with Congo red.

Cohen (1966) expressed the opinion that amyloid fibrils probably contain no carbohydrate but that the latter is present in the ground substance surrounding the fibrils.

The classical red metachromasia of amyloid with methyl violet has long been ascribed to the presence of acid mucopolysaccharides although, at least in fixed tissues, these are often difficult to demonstrate histochemically. An investigation into the nature of the dye-chromotrope bond in methyl violet-amyloid metachromasia was carried out by Carnes and Forker (1954, 1956). These authors found that methyl violet was bound metachromatically by amyloid at pH levels as low as 1·6, whereas the (non-metachromatic) binding of toluidine blue was abolished at pH 4·5. Differences were found in the spectral absorption curves for amyloid and cartilage (chondroitin sulphate). These and the above-mentioned dye-binding experiments suggested that amyloid possesses few strong anionic binding sites, and that methyl violet metachromasia depends on bonding of a different type from that associated with toluidine blue metachromasia of acid polysaccharides. In fresh cold microtome sections Hale's reaction (Chapter 10, p. 349) is strongly positive. This is illustrated in Fig. 105, facing p. 390. Staining with Alcian blue or Alcian green is also quite strong in such unfixed materials though in formalin-fixed frozen sections heating may be necessary before the dye can combine with the acidic mucosubstance.

The methyl violet method (Plate Xc, facing p. 392) was for many years regarded as the only specific stain for amyloid and the only means of distinction between true amyloid and other hyaline proteins which might resemble it superficially. The method suffers from impermanence; fading takes place with most samples of amyloid in a relatively short time, especially if paraffin sections are employed. With fresh sections, even if these are briefly fixed in alcoholic fixatives such as Wolman's acetic-ethanol, methyl violet metachromasia is more brilliant, more extensive and more permanent than it is in material from the same case prepared by conventional fixing and paraffin embedding. I consider that fresh, cold microtome, sections should be employed for all studies on amyloid, natural or experimental, particularly as there is in this case no prospect of loss by solubility in buffer or aqueous solutions.

Enzymal Analysis

Amyloid is particularly resistant to attack by proteolytic enzymes. Studying this aspect of amyloidosis, Missmahl (1950) reported that samples of secondary amyloid would retain many of their staining properties after digestion with pepsin or after passage through the gastro-intestinal tract of the mouse. Stainability with Congo red and its attendant dichroism, for instance, withstood both these procedures, but methyl violet metachromasia (see below) was more labile. Wagner (1957) tested the effect on amyloid of incubation with trypsin, pepsin, pectin esterase, streptokinase, streptodornase and streptococcal and testicular hyaluronidase. Only very high concentrations of either pepsin or trypsin (10 mg./ml.) had any effect on protein staining or on mucopolysaccharide. High concentrations of testicular hyaluronidase decreased the staining of mucopolysaccharides only very slightly. Windrum and Kramer (1957) likewise observed that partial peptic digestion of secondary amyloid deposits changed their staining properties from orthochromatic to metachromatic with toluidine blue or azure A. Subsequent digestion with testicular hyaluronidase removed most of the metachromatic material, suggesting that it was a polysaccharide of the chondroitin-sulphuric acid type. Arvy and Sors (1959) considered, contrary to Windrum and Kramer, that methyl violet staining of amyloid was a true metachromasia. The effect of collagenase and hyaluronidase on isolated amyloid fibrils was investigated by Cohen and Calkins (1964), who showed that the structural integrity of the fibril was unaltered. This was re-

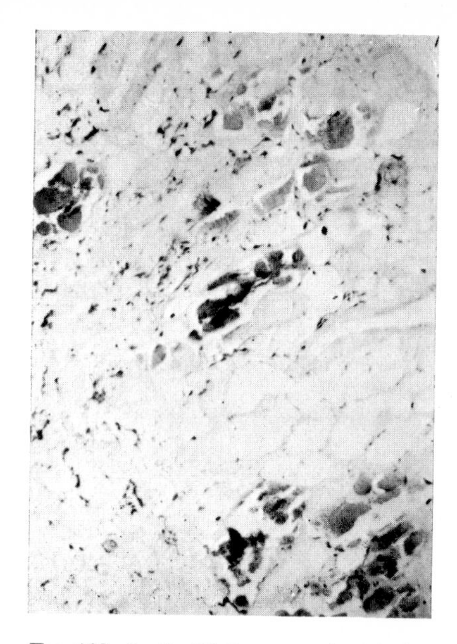

FIG. 102. Snail (*Helix aspersa*). Mucin-staining cells in the loose connective tissues of the foot. Germanate method. × 130.

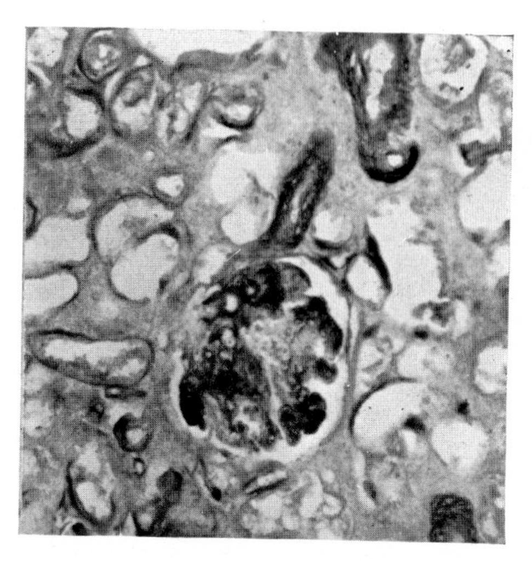

FIG. 103. Human kidney. Amyloidosis. Fresh frozen section. Shows strongly reacting deposits of amyloid in the glomerulus and elsewhere. Adams' DMAB—reaction. × 130.

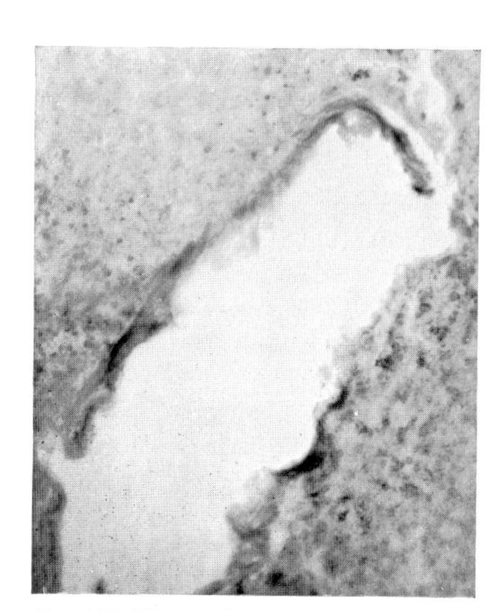

FIG. 104. Human liver. As Fig. 108. Ferric hydroxamate reaction. × 130.

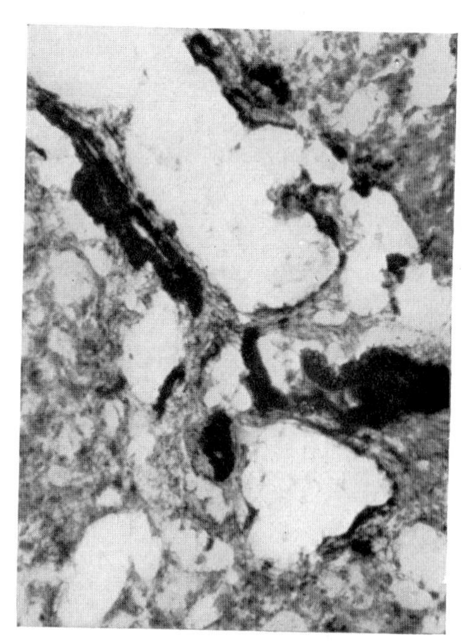

FIG. 105. Human liver. As Figs. 104 and 108. Dialysed Iron reaction. × 510.

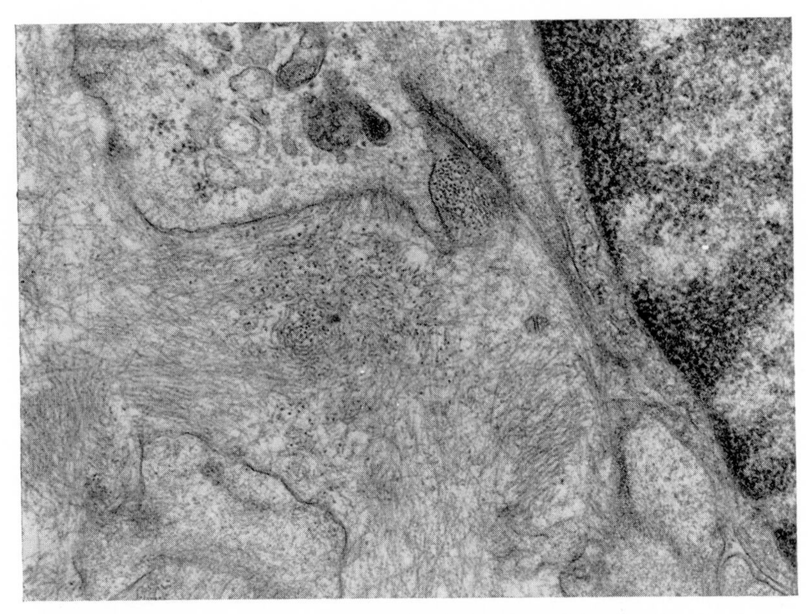

FIG. 106. Rabbit Kidney. Experimentally induced amyloidosis. *Fixation* 1 per cent. 0^s0_4 in phosphate buffer. Epon embedded, uranyl acetate and lead citrate stained. × 32,000.

garded as evidence against either collagen or hyaluronic acid as a major component of amyloid.

Resistance to enzyme attack is not usually used as a diagnostic criterion in the applied histochemistry of amyloid.

Electron Microscopy. The fibrillar ultrastructure of amyloid was confirmed, once and for all, by Cohen and Calkins (1959a and b). It has been found in all types of amyloid tested (Fig. 106) and thus the electron microscope must be considered the absolute arbiter of the presence or absence of amyloid. Combined studies, using "thick" resin-embedded sections stained with Congo red and examined with the polarizing microscope, and electron microscopy will become routine practice in pathology.

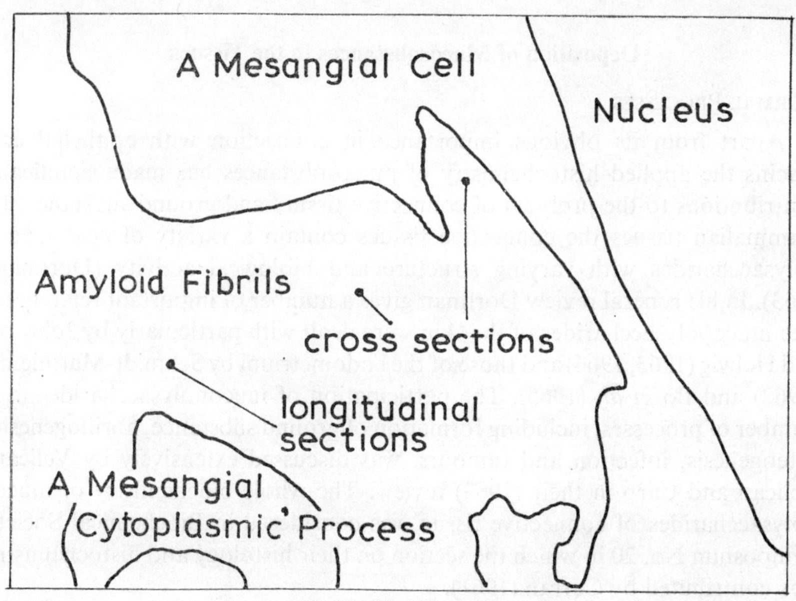

Fig. 106 (line diagram)

The possibility of using old histological material, formalin-fixed and paraffin-embedded, for E.M. studies has been pointed out by Lehner, Nunn and Pearse (1966).

As described by Cohen and Calkins (1959a), Cohen, Weiss and Calkins (1960), and by Pirani *et al.* (1964) amyloid fibrils were 75–100 Å in length and 50–300 Å wide (Fig. 106). Further work by Shirahama and Cohen (1965) has shown that the amyloid fibril is composed of 1–8 laterally aggregated filaments, each about 75 Å wide and showing beading with a periodicity of 100 Å (Fig. 107). They suggest that the beads may represent end-to-end aggregated subunits or, less likely, that there may be a helical arrangement in each filament with 100 Å periodicity.

Diagnosis. The diagnosis of amyloid during life is now frequently made by means of biopsies. Various sites and organs have been recommended; among these are skin, liver, kidney, gum, rectum, bone marrow and small intestine. Certainly rectal biopsy is one of the most reliable in generalized amyloidosis. In localized cases, where the presence of amyloid in a single organ is suspected, this organ (e.g. heart) can be made the object of a special biopsy.

Pathogenesis. An excellent summing up of the current position is given by Cohen (1965). Other important contributions are those of Strukov *et al.* (1963) and of Teilum (1964), whose two-plane cellular theory has received much support from histochemical investigations.

The whole question of paramyloidosis was very well reviewed by Krücke (1959) and its pathology was discussed by Diezel (1961). These two papers should be consulted for full references on the subject.

Deposition of Mucosubstances in the Tissues

Normal Processes

Apart from its obvious importance in connection with epithelial cell mucins the applied histochemistry of mucosubstances has made significant contributions to the problem of connective tissues and ground substance. In mammalian tissues the connective tissues contain a variety of acid mucopolysaccharides with varying structure and biological activity (Dorfman, 1963). In his general review Dorfman gives a number of important references. The mucopolysaccharides of the skin were dealt with particularly by Johnson and Helwig (1963, 1964) and those of the endometrium by Schmidt-Matthiesen (1962) and Bo *et al.* (1965). The participation of mucopolysaccharides in a number of processes, including formation of ground substance, fibrillogenesis, osteogenesis, infection and tumours, was discussed extensively by Velican, Velican and Carp in their (1963) review. The whole biochemistry of mucopolysaccharides of connective tissue was considered in Biochemical Society Symposium No. 20 in which the section on their histology and histochemistry was contributed by Curran (1961).

The participation of mucosubstances in calcification forms an important facet of their applied histochemistry. Important contributions in this field have come from Irving (1959, 1960), from Kasavina and Zenkevich (1961), Partridge and Elsden (1961), van den Hooff (1964), Stockwell and Scott (1965), and Mathews (1967). Konetzki *et al.* (1962) considered the sequential accumulation of calcium and acid mucopolysaccharides in nephrocalcinosis, induced by Vitamin D. This problem had already been considered by Scarpelli *et al.* (1960).

Carbohydrate-containing proteins form an important part of the metabolic structure of the major blood vessels. Neutral, as well as acidic mucosubstances are apparently present, as reported by Bertensen (1960). The majority of papers on the subject, however, refer more or less exclusively to acidic substances (Cain and Pfob, 1962; Böttcher and Klynstra, 1963; Zugibe, 1963; Morin and Bernick, 1963; Manley and Kent, 1963; Sirek *et al.*, 1964).

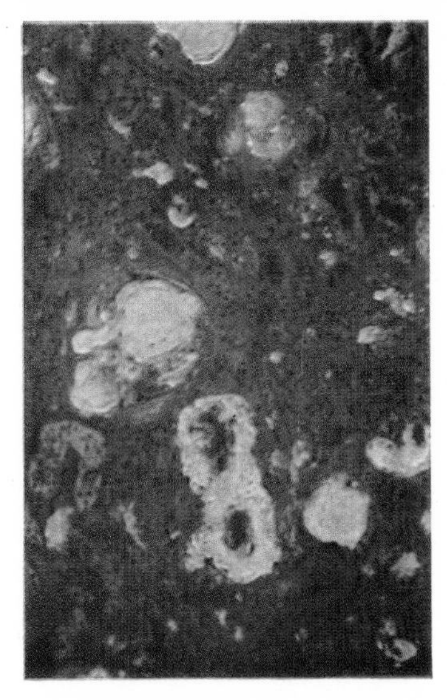

XA. Human Kidney. Case of Renal Amyloidosis. Fresh frozen (cryostat) section. Shows bluish-white flourescence of amyloid deposits. Thioflavin T. method. × 210.

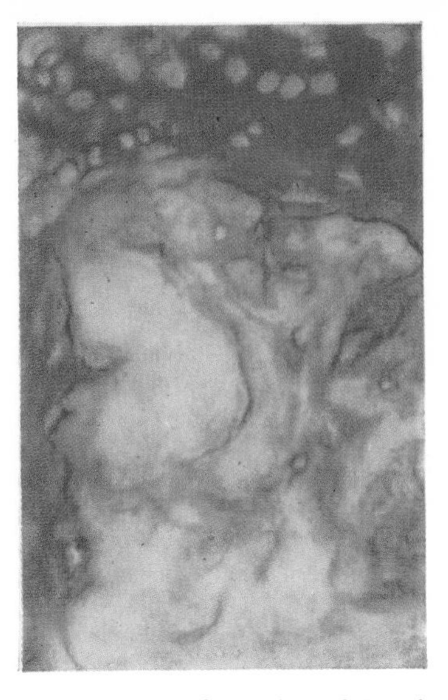

XB. Human Kidney. Case of Renal Amyloidosis. Fresh frozen (cryostat) section. Amyloid deposit, flouresce whitish-yellow, nuclei orange. Phosphine 3R method. × 245

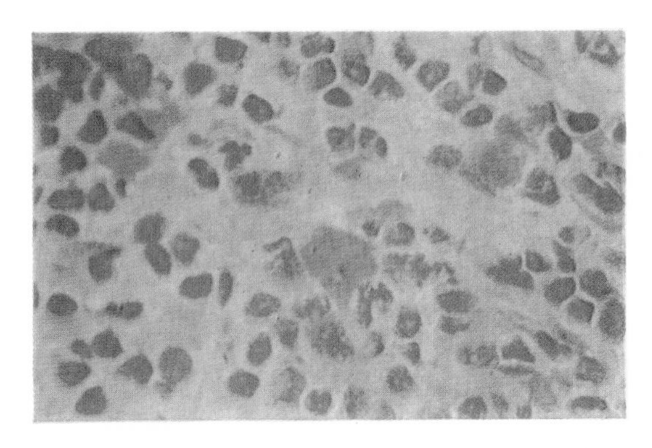

XC. Lymph-node from case of human amyloidosis. Fresh frozen (cryostat) section. Shows red amyloid deposits and green nuclei. Methyl green technique. × 610

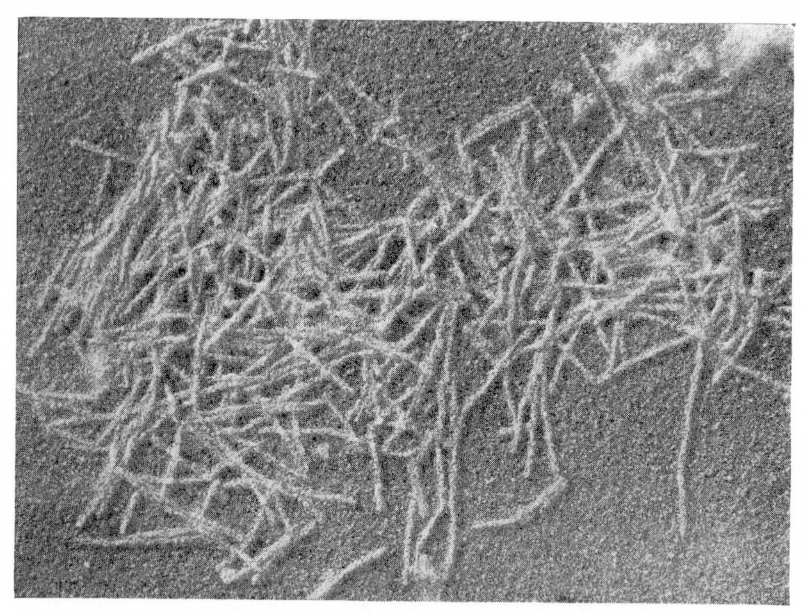

FIG. 107. Amyloid Fibrils. Isolated by sucrose density centrifugation (Cohen, 1966), from a case of human secondary amyloidosis. Shadow casting with platinum-palladium. × 50,000.

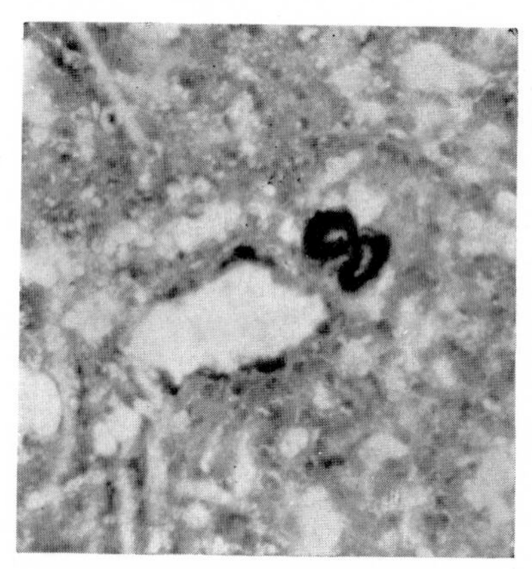

FIG. 108. Human liver. Amyloidosis. Fresh frozen section. Amyloid deposits in the wall of a large sinusoid. Congo red—hæmalum. × 130.

PATHOLOGICAL PROCESSES

Colonic Histiocytosis. It is necessary to mention this condition here because, at frequent intervals, it is described and regarded as pathological. Rowlands and Landing (1960) found histiocytes containing polysaccharide in the colons of two children out of 126 necropsy and surgical biopsy specimens. They reported that the histochemical findings in these two cases were the same as those of Whipple's disease. Discussing the diagnosis of the latter by rectal biopsy Fleming *et al.* (1962) indicated that PAS-positive and mucicarmine-positive material was rarely found in histiocytes as a normal finding. Enzinger and Helwig (1963) further discussed the histochemistry of muciphages and Whipple cells. On the other hand, Fisher and Hellstrom (1964) identified ceroid (Chapter 26) pigment in rectal histiocytes in one in 100 operation specimens. There is still considerable confusion in this field. Azzopardi and Evans (1966) considered that this was partly due to the frequent coincidence of lipofuscin and mucin in the same muciphages. These authors provided a table for the differential diagnosis of Whipple's disease and colonic histiocytosis. They also illustrated the ultrastructure of the two conditions.

Whipple's Disease. This disorder, first described by Whipple (1907) and long considered to be a lipodystrophy, has been given an altered status since the observation of Black-Schaffer (1949) that the stored material was a glycoprotein. Some authors (Kampmeier and Peterson, 1950) regarded it as a form of collagen disease but more modern investigations do not support this notion. Hellwig *et al.* (1961) still continued to favour the hypothesis that Whipple's disease is a disorder of lipid storage. Enzinger and Helwig (1963) examined 15 cases at necropsy and described 3 basic tissue alterations: (1) Deposition of a strongly PAS-positive substance in macrophages; (2) Accumulation of a lipid-containing material in lymph spaces and lymphatics; and (3) Fibrosis of the regional lymph nodes.

Ultrastructural studies (Kurtz *et al.*, 1962, Cohen, 1964) have shown that in the intestinal macrophages of Whipple's disease there are bacilliform inclusions 200–250 mμ in diameter and 1500 mμ in length. These dimensions are precisely those of a typical small rod-shaped bacterium but there is, as yet, no proof that the inclusions are indeed microorganisms.

Gargoylism (Hurler's Disease). This familial disorder affects a number of different tissues, but most importantly cartilage and bone. It has therefore been regarded as a chondrodystrophy, or lipochondrodystrophy. An abnormal material is stored in large histiocytes, particularly in liver, spleen and intestine, and also in liver cells and nerve cells. The histochemistry of this material has been investigated by a number of workers (Schnabel, 1961; Lagunoff *et al.*, 1962; Wolfe *et al.*, 1964). All the organs of a 4½-year-old boy were examined by Schnabel who found storage of gangliosides, phospholipids and acid mucopolysaccharides. Lagunoff *et al.* confined their studies to the mitral valve, in two cases, and found two different types of storage cells. One of these contained glycolipid and the other acid mucopolysaccharide. Wolfe *et al.* studied two patients with Hurler's syndrome, one with normal intelligence and the other mentally retarded. They found only acid mucopolysaccharide storage in the first and both acid mucopolysaccharide and glycolipid storage in the second case. They recommended rectal biopsy, with suitable precautions to ensure adequate fixation of water-soluble mucosubstances (see Chapter 5, p. 99), as a diagnostic procedure.

Complete bibliographies of both Whipple's disease and Hurler's syndrome are provided by Wolman (1964), and the histochemistry of growing tibial cartilage, obtained from a case of the latter, was recorded by Bona *et al.* (1966). Inclusions of mucopolysaccharide in the lymphocytes were described by Mittwoch (1961). This observation has obvious diagnostic importance.

Cystic Fibrosis. This hereditary disorder occurs in about 1 out of every 600 births. A highly viscid mucin is secreted by the tracheobronchial glands, salivary glands, pancreas, and by the intestinal mucosa. The changes in carbohydrate histochemistry were outlined by Shackleford and Bentley (1964). Briefly they are, (1) Synthesis of acid mucopolysaccharides in large amounts by cells which normally secrete little or none. (2) Changes in the nature of the secreted mucins leading to their inspissation.

Changes in the chemical structure of the mucosubstances have been suggested by Dische *et al.* (1962) as the main cause of the disorder, and the application of modern techniques of carbohydrate histochemistry may be expected to contribute further to elucidation of the whole problem.

REFERENCES

ARVY, L., and SORS, C. (1959). *Acta Histochemica*, **6**, 77.
AZZOPARDI, J. G., and EVANS, D. J. (1966). *J. clin. Path.*, **19**, 368.
BANCROFT, J. D. (1963). *Stain Tech.*, **38**, 336.
BANCROFT, J. D. (1967). *J. med. Lab. Tech.*, **24**, 309.
BANGLE, R. A. (1963). *Amer. J. Path.*, **43**, 437.
BENDITT, E. P., LARGUNOFF, D., ERIKSEN, N., and ISERI, O. A. (1962). *Arch. Path.*, **74**, 323.
BENNHOLD, E. (1922). *Munchen. med. Woch.*, **2**, 1537.
BERTENSEN, S. (1960). *Nature*, **187**, 411.
BLACK-SCHAFFER, B. (1949). *Proc. Soc. exp. Biol., N.Y.*, **72**, 225.
BO, W. J., SMITH, S., REITER, R., and PIZZARELLO, D. J. (1965). *J. Histochem. Cytochem.*, **13**, 461.
BOHLE, A., HARTMANN, F., and POLA, W. (1950). *Virchows Arch.*, **319**, 231.
BONA, C., STĂNESCU, V., STREJA, D., and IONESCU, V. (1966). *Acta Histochem.*, **23**, 231.
BÖTTCHER, C. J. F., and KLYNSTRA, F. B. (1963). *Lancet*, **2**, 439.
BURNS, J., PENNOCK, C. A., and STOWARD, P. J. (1967). *J. Path. Bact.*, **94**, 337.
CAIN, H., and PFOB, H. (1962). *Virch. Arch.*, **335**, 240.
CALKINS, E., and COHEN, A. S. (1958). *J. clin. Invest.*, **37**, 882.
CARNES, W. H., and FORKER, B. R. (1954). *J. Histochem. Cytochem.*, **2**, 469.
CARNES, W. H., and FORKER, B. R. (1956). *Lab. Invest.*, **5**, 21.
CASTILLO, P. A. (1935). *Arch. Med. int.*, **1**, 635.
CATHCART, E. S., and COHEN, A. S. (1966). *J. Immunol.*, **96**, 239.
CAVALLERO, C. (1953). *Acta allergol.*, Suppl. III, **6**, 178.

COHEN, A. S. (1964). *J. Ultrastr. Research.*, **10**, 124.
COHEN, A. S. (1965). *Int. Rev. exp. Path.*, **4**, 159.
COHEN, A. S. (1966). *Lab. Invest.*, **15**, 66.
COHEN, A. S., CALKINS, E., and LEVENE, C. I. (1959). *Amer. J. Path.*, **35**, 971.
COHEN, A. S., and CALKINS, E. (1959a). *Nature*, **183**, 1202.
COHEN, A. S., and CALKINS, E. (1959b). *Arthr. Rheum.*, **2**, 70.
COHEN, A. S., WEISS, L., and CALKINS, E. (1960). *Amer. J. Path.*, **37**, 413.
COHEN, A. S., and PAUL, W. E. (1963). *Nature*, **197**, 193.
COHEN, A. S., and CALKINS, E. (1964). *J. cell. Biol.*, **21**, 481.
CURRAN, R. C. (1961). In "The Biochemistry of Mucopolysaccharides of Connective Tissue." Cambridge Univ. Press, p. 24.
DIESELBEN, A. (1954). *Deutsch. Z. Nervenheilk*, **171**, 173.
DIEZEL, P. B. (1957). "Die Stoffwechselstorungen der Sphingolipoide." Springer, Berlin.
DIEZEL, P. B. (1961). *Regensburger Jahrbuch artz. Fortbild.*, **9**, 1.
DISCHE, Z., PALLAVICINI, C., CIZEK, L. J., and CHEIN, S. (1962). *Ann. N.Y. Acad. Sci., Wash.*, **93**, 489.
DIVRY, P. (1927). *J. Belg. Neurol. Psychiat.*, **27**, 642.
DORFMAN, A. (1963). *J. Histochem. Cytochem.*, **11**, 2.
ENZINGER, F. M., and HELWIG, E. B. (1963). *Virch. Arch.*, **336**, 238.
EPPINGER, H. (1922). *Biochem. Z.*, **127**, 107.
FERNANDO, J. C. (1961). *J. Inst. Sci. Tech.*, **7**, 1.
FISHER, E. R., and HELLSTROM, H. R. (1964). *Amer. J. clin. Path.*, **42**, 581.
FLEMING, W. H., YARDLEY, J. H., and HENDRIX, T. R. (1962). *New Engl. J. Med.*, **267**, 33.
FRIEDREICH, N., and KEKULÉ, A. (1859). *Virch. Arch.*, **16**, 50.
GILES, R. B. Jr., and CALKINS, E. (1955). *J. clin. Invest.*, **34**, 1476.
GOESSNER, W., SCHNEIDER, G., SIESS, M., and STEGMANN, H. (1951). *Virchows Arch.*, **320**, 326.
GOLDBERG, A. F., and DEANE, H. W. (1960). *Blood*, **16**, 1708.
GUEFT, B., and GHIDONI, J. (1963). *Amer. J. Path.*, **43**, 837.
HASS, G. (1942). *Arch. Path.*, **34**, 92.
HASS, G., HUNTINGTON, R., and KRUMDIECK, N. (1943). *Ibid.*, **35**, 226.
HELLER, H., MISSMAHL, H. P., SOHAR, E., and GAFNI, J. (1964). *J. Path. Bact.*, **88**, 1.
HELLWIG, C. A., WEINER, R. G., and WILKINSON, P. N. (1961). *Arch. Path.*, **72**, 274.
HOBBS, J. R., and MORGAN, A. D. (1963). *J. Path. Bact.*, **86**, 437.
HOOFF, A. VAN DEN (1964). *Acta anat.*, **57**, 16.
IRVING, J. T. (1959). *Arch. Oral Biol.*, **1**, 89.
IRVING, J. T. (1960). *Clin. Orthop.*, **17**, 92.
JANIGAN, D. T. (1965). *Amer. J. Path.*, **47**, 159.
JANIGAN, D. T., and DRUET, R. L. (1966). *Amer. J. Path.*, **48**, 1013.
JOHANSSON, G. A. (1949). *Klin Woch.*, **27**, 68.
JOHNSON, W. C., and HELWIG, E. B. (1963). *Amer. J. clin. Path.*, **40**, 123.
JOHNSON, W. C., and HELWIG, E. B. (1964). *J. invest. Derm.*, **42**, 81.
KAMPMEIER, R. H., and PETERSON, J. C. (1950). *Trans. Amer. clin. Climatol. Assoc.*, **61**, 248.
KASAVINA, B. S., and ZENKEVICH, G. D. (1961). *Clin. chim. Acta.*, **6**, 874.
KING, L. B. (1948). *Amer. J. Path.*, **24**, 1905.
KLENK, E., and FAILLARD, H. (1955). *Hoppe-Seylers Z.*, **299**, 191.
KONETZKI, W., HYLAND, R., and EISENSTEIN, R. (1962). *Lab. Invest.*, **11**, 488.
KRÜCKE, W. (1959). *Ergeb. Inn. Med.*, **11**, 299.
KURTZ, S. M., DAVIS, T. D., and RUFFIN, J. M. (1962). *Lab. Invest.*, **11**, 653.
LACHMANN, P. J., MÜLLER-EBERHARD, H. J., KUNKEL, H. G., and PARONETTO, F. (1962). *J. exp. Med.*, **115**, 63.
LADEWIG, P. (1945). *Nature*, **156**, 81.
LAGUNOFF, D., ROSS, R., and BENDITT, E. P. (1962). *Amer. J. Path.*, **41**, 273.
LEHNER, T., NUNN, R. E., and PEARSE, A. G. E. (1966). *J. Path. Bact.*, **91**, 297.
LENDRUM, A. C. (1951). In "Recent Advances in Pathology," 2nd Ed. J. and A. Churchill, London, p. 535.

Letterer, E. (1926). *Beit. path. anat.*, **75**, 486.
Letterer, E. (1934). *Virchows Arch.*, **293**, 34.
Letterer, E. (1949). *J. path. Bact.*, **61**, 496.
Letterer, E. (1960). *Arch. de Vechi*, **31**, 303.
Letterer, E., Gerok, W., and Schneider, G. (1955). *Virchows Arch.*, **327**, 327.
Letterer, E., and Schneider, G. (1953). *Plasma*, **1**, 263.
Lundin, P. M., Schelin, U., Pellegrini, G., and Mellgran, J. (1954). *Acta path. Scand.*, **35**, 339.
Lynch, M. J., and Inwood, M. J. H. (1963). *Stain Tech.*, **38**, 259.
Luetscher, J. A., Jr. (1940). *J. clin. Invest.*, **19**, 313.
Manley, G., and Kent, P. W. (1963). *Brit. J. exp. Path.*, **44**, 635.
Mathews, M. B. (1967). In "The Chemical Physiology of Acid Mucopolysaccharides." Ed. G. Quintarelli, Little Brown, Boston.
McAlpine, J. C., and Bancroft, J. D. (1964). *J. clin. Path.*, **17**, 213.
Mellors, R. C., and Ortega, L. G. (1956). *Amer. J. Path.*, **32**, 455.
Missmahl, H. P. (1950). *Virchows Arch.*, **318**, 518.
Missmahl, H. P. (1957). *Z. wiss. Mikr.*, **63**, 133.
Missmahl, H. P. (1958). *Klin. Woch.*, **36**, 29.
Missmahl, H. P. (1959). *Deutsch. ges. Inn. Med.*, **56**, 439.
Missmahl, H. P., and Hartwig, M. (1953). *Ibid.*, **324**, 489.
Mittwoch, U. (1961). *Nature*, **191**, 1315.
Morin, R. J., and Bernick, S. (1963). *Amer. J. Path.*, **43**, 337.
Muckle, T. J. (1964). *Nature*, **203**, 773.
Muckle, T. J., and Wells, M. (1962). *Quart. J. Med.*, **31**, 235.
Newcombe, D. S., and Cohen, A. S. (1965). *Biochim. Biophys. Acta*, **104**, 480.
Partridge, S. M., and Elsden, D. F. (1961). *Biochem. J.*, **79**, 26.
Paul, W. E., and Cohen, A. S. (1963). *Amer. J. Path.*, **43**, 721.
Pearse, A. G. E. (1949). *J. clin. Path.*, **2**, 81.
Pearse, A. G. E. (1963). *Acta neuropath.*, Suppl. 2, p. 100.
Pfeiffer, H. H. (1953). *Exp. Cell Research*, **4**, 181.
Pirani, C. L. (1951). *Lancet*, **2**, 166.
Pirani, C. L., Bestetti, A., Catchpole, H. R., and Meskauskas, M. (1964). *Arthr. Rheum.*, **7**, 338.
Puchtler, H., Sweat, F., and Levine, M. (1962). *J. Histochem. Cytochem.*, **10**, 355.
Puchtler, H., Sweat, F., and Kuhns, J. G. (1964). *J. Histochem. Cytochem.*, **12**, 900.
Puchtler, H., and Sweat, F. (1965). *J. Histochem. Cytochem.*, **13**, 693.
Puchtler, H., and Sweat, F. (1966). *J. Histochem. Cytochem.*, **14**, 123.
Rask-Nielsen, R., Clausen, J., and Cnristensen, H. E. (1960). *J. nat. Cancer Inst.*, **25**, 315.
Rask-Nielsen, R., Christensen, H. E., and Clausen, J. (1960). *J. nat. Cancer Inst.*, **25**, 315.
Reece, J. M., and Reynolds, T. B. (1954). *Amer. J. med. Sci.*, **228**, 554.
Reimann, H. A., and Eklund, C. M. (1935). *Amer. J. med. Sci.*, **190**, 88.
Richter, G. W. (1954). *Amer. J. Path.*, **30**, 239.
Richter, G. W. (1956). *J. exp. Med.*, **104**, 847.
Rokitansky, C. (1842). "Handbuch der speciellen pathologischen Anatomie." Vol. 3, Vienna.
Romhany, G. (1949). *Schw. Z. Path. Bact.*, **12**, 253.
Rothbard, S., and Watson, R. F. (1954). *Proc. Soc. exp. Biol. N.Y.*, **85**, 133.
Rowlands, D. T., and Landing, B. H. (1960). *Amer. J. Path.*, **36**, 201.
Scarpelli, D. G., Tremblay, G., and Pearse, A. G. E. (1960). *Amer. J. Path.*, **36**, 331.
Schelin, U., Hesselsjö, R., Paulsen, F., and Mellgren, J. (1954). *Acta path. Scand.*, **35**, 503.
Schmidt- Matthiesen, H. (1962). *Acta Histochem.*, **13**, 129.
Schmitz-Moormann, P. (1961). *Virch. Arch.*, **334**, 95.
Schmitz-Moormann, P. (1964a). *Z. phys. Chem.*, **338**, 63.
Schmitz-Moormann, P. (1964b). *Z. phys. Chem.*, **339**, 85.

SCHNABEL, R. (1961). *Virch. Arch.*, **339**, 379.
SEITELBERGER, F. (1958). *Naturwiss*, **45**, 40.
SHACKLEFORD, J. M., and BENTLEY, H. P. (1964). *J. Histochem. Cytochem.*, **12**, 512.
SHIRAHAMA, T., and COHEN, A. S. (1965). *Nature*, **206**, 737.
SHIRAHAMA, T., and COHEN, A. S. (1966). *J. Histochem. Cytochem.*, **14** 725.
SIREK, O. V., SCHILLER, S., and DORFMAN, A. (1964). *Biochim. Biophys. Acta*, **83**, 148.
STOCKWELL, R. A., and SCOTT, J. E. (1965). *Ann. rheum. Dis.*, **24**, 341.
STRUKOV, A. I., SEROV, V. V., and PAVLIKHINA, L. V. (1963). *Virch. Arch.*, **336**, 550.
SYMMERS, W. STC. (1956a). *J. clin. Path.*, **9**, 187.
SYMMERS, W. STC. (1956b). *J. clin. Path.*, **9**, 212.
TEILUM, G. (1948). *Amer. J. Path.*, **24**, 389.
TEILUM, G. (1956). *Amer. J. Path.*, **32**, 945.
TEILUM, G. (1964). *Acta path. Scand.*, **61**, 21.
TEILUM, G., ENGBAECK, H. C., HARBOE, N., and SIMONSEN, M. (1951). *J. clin. Path.*, **4**, 301.
TEILUM, G., and LINDAHL, A. (1954). *Acta med. Scand.*, **149**, 449.
THOMPSON, S. W., GEIL, R. G., and YAMANAKA, H. S. (1961). *Amer. J. Path.*, **38**, 737.
UOTILA, U., PERÄSOLO, O., and VAPAAVUORI, M. (1955). *Acta Path. Scand.*, **37**, 322.
VASQUEZ, J. J., and DIXON, F. J. (1956). *J. exp. Med.*, **104**, 727.
VASSAR, P. S., and CULLING, C. F. A. (1959). *Arch. Path.*, **68**, 487.
VASSAR, P. S., and CULLING, C. F. A. (1961). *Amer. J. clin. Path.*, **36**, 244.
VASSAR, P. S., and CULLING, C. F. A. (1962). *Arch. Path.*, **73**, 59.
VELICAN, C., VELICAN, D., and CARP, N. (1963). "Histochimia şi Fiziopatologia Mucopolizaharidelor." Ed. Acad. Rep. Pop. Rom., Bucharest.
VICKERSTAFF, T. (1954). "The Physical Chemistry of Dyeing," 2nd Ed. Oliver and Boyd, Edinburgh.
VIRCHOW, R. (1860). "Cellular Pathology." Trans., London, p. 371.
VOGT, A., and KOCHEM, H. G. (1960). *Z. Zellforsch.*, **52**, 640.
WAGNER, B. M. (1955). *Arch. Path.*, **60**, 221.
WAGNER, B. M. (1957). In "Analytical Pathology." McGraw-Hill, New York.
WHIPPLE, G. H. (1907). *Bull. Johns Hopk. Hosp.*, **18**, 392.
WINDRUM, G. M., and KRAMER, H. (1957). *Arch. Path.*, **63**, 373.
WOLFE, H. J., BLENNERHASSET, J. B., YOUNG, G. F., and COHEN, R. B. (1964). *Amer. J. Path.*, **45**, 1007.
WOLMAN, M. (1964). In "Handbuch der Histochemie," Band V. Lipide, Fischer, Stuttgart, pp. 154 and 220.
WOLMAN, M., and BUBIS, J. J. (1965). *Histochemie*, **4**, 351.
ZUGIBE, F. T. (1963). *J. Histochem Cytochem.*, **11**, 35.

CHAPTER 12

LIPIDS, LIPOPROTEINS AND PROTEOLIPIDS

BEFORE presenting any classification of lipids as a basis for the discussion of their histochemistry, it is necessary first to deal with the nomenclature to be employed. In the case of no other class of substances is there such wide variation in the significance attached to the various descriptive terms. First of all we have to consider the words lipoid, lipid, lipine and lipin.

Cain (1950), in his review of the histochemistry of the substances we are considering, used the word lipoid in the sense in which it was used by Baker (1946a) to mean "all those substances that can be extracted from tissues by several or all the usual fat solvents, and are insoluble or only colloidally soluble in water." The term "lipine" he restricted to substances containing N as well as C, H and O, such as lecithin, kerasin, and allied compounds. In this chapter, instead of the word "lipoid", the word "lipid" is used, and this covers the whole of the large group of naturally occurring fat-like substances which are insoluble in water but soluble in such solvents as benzene, chloroform, acetone, ether, etc. The word "lipoid", if it is used at all, and the word "lipin", are used to refer to the substances classed as "lipines" by Cain. Since the ending "ine" has a definite chemical significance, it has been thought better to avoid use of the word lipine altogether.

Much of the lipid present in mammalian tissues is associated with protein. The so-called masked lipids of the histologist belong to this category although in some cases the phenomenon of unmasking may be due to variation in the degree of dispersion of lipid droplets. Lipoproteins can be defined as substances which behave as proteins, from which lipids can be extracted and proteolipids (Rossiter, 1955) are substances which behave as lipids, from which proteins can be obtained.

Classification of Lipids

According to Lovern (1955) there was no field of biochemistry in which there was less uniformity than the lipids. This is not a promising start for the histochemist, who is bound to adhere to some classification if his work is to be of any significance in the wider sense of the term. The basic classification used below is derived from Bloor (1943).

All naturally occurring lipids belong to one of three main groups: (1) derivatives of non-cyclic, usually straight chain, hydrocarbons; (2) derivatives of cyclopentenophenanthrene (steroids); and (3) derivatives of isoprene (terpenes and carotenoids). Only substances in the first and second groups are considered in this chapter, and they can conveniently be placed in the following sub-divisions:

SIMPLE LIPID ESTERS

These are esters of fatty acids with alcohols and they include the fats, oils and waxes. Fats are neutral esters of glycerol with saturated or unsaturated fatty acids, while the oils are similar substances but liquid instead of solid at normal temperatures. The only substances dealt with in this chapter, which may be included in the class of waxes, are the cholesterol esters. These are esters of various fatty acids with the steroid alcohol cholesterol.

COMPOUND LIPIDS

The compound lipids consist of a fatty acid, an alcohol (usually glycerol) and one or several additional groups. They may be subdivided into (*a*) phospholipids (lecithins, kephalins and sphingomyelins), (*b*) glycolipids (cerebrosides and gangliosides) and (*c*) sulpholipids.

Phospholipids. On hydrolysis the phospholipids yield fatty acids, glycerol (or some other alcohol), phosphoric acid and a base such as choline, ethanolamine, or serine. They are to be regarded as esters of phosphatidic acid, or phosphatides. The chief members of this group are the lecithins and kephalins whose structural formulæ are given below.

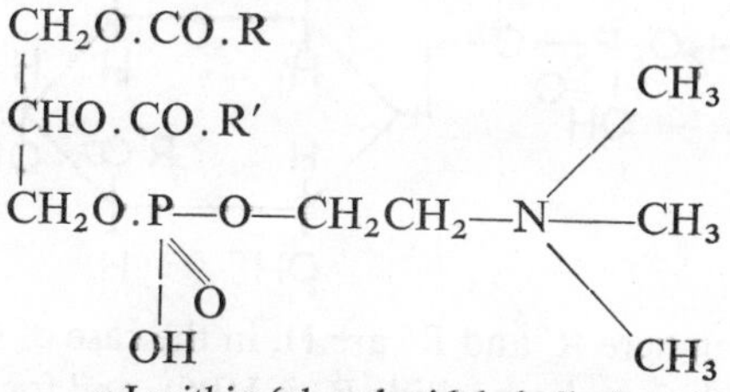

```
CH2O.CO.R
|                                       CH3
CHO.CO.R'                              /
|                                     /
CH2O.P—O—CH2CH2—N———CH3
      |\\                   \
      | O                    \
     OH                       CH3
```

α-Lecithin (phosphatidyl choline)

Lecithin is usually the predominant phospholipid component of animal tissues (Dawson, 1966), in many tissues it makes up about half the total phospholipid fraction (Ansell and Hawthorne, 1964). The removal of phosphoryl choline from lecithin gives rise to a D-2,3 diglyceride but phosphatidyl choline contains not two distinct fatty acids, as suggested by the formula, but a mixture. The 2-position is considered to be largely esterified by unsaturated fatty acids and the 1-position by saturated ones (Tattrie, 1959).

The kephalins are of similar constitution to the lecithins except that the organic base is aminoethylalcohol (ethanolamine), or alternatively serine, instead of choline. In both cases the fatty acids are R.COOH and R′COOH, in the formulæ.

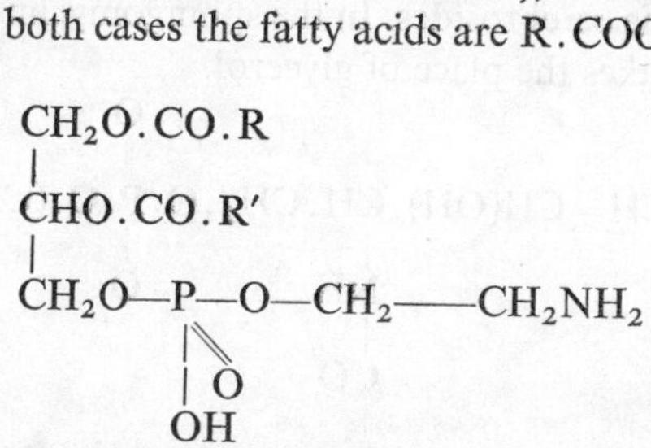

```
CH2O.CO.R
|
CHO.CO.R'
|
CH2O—P—O—CH2——CH2NH2
       |\\
       | O
      OH
```

α-Kephalin (phosphatidyl ethanolamine)

$$
\begin{array}{l}
CH_2O.CO.R \\
| \\
CHO.CO.R' \\
| \\
CH_2O.P{-}O{-}CH_2{-}CH.COOH{-}NH_2 \\
\quad\quad\;\; | \;\backslash\backslash O \\
\quad\quad\; OH
\end{array}
$$

α-Kephalin (phosphatidyl serine)

Phosphatidyl choline and phosphatidyl ethanolamine possess both basic and acidic groups, and electrometric titration data and dielectric constant measurements indicate that the former compound possesses a Zwitterion structure and the latter sometimes a Zwitterion structure and sometimes the form of a neutral molecule. Phosphatidyl serine possesses one basic to two acid groups and hence reacts as an acid.

The phosphoinositides have the same basic composition as the phosphatidyl compounds described above.

CH₂O.P(=O)(OH)—O— [inositol ring with substituents: OH, HO, H, H, H, H, R'O, OR'', OH, H]

In the formula given here R′ and R″ are H, in the case of monophosphoinositide. In the case of diphosphoinositide R″ is HPO_3 and for triphosphoinositide both R′ and R″ are HPO_3. All three compounds are found in the nervous system where di- and triphosphoinositide are present in much higher concentrations than in other tissues. They are selectively localized in the myelin sheath. The configuration of the phosphatidyl group, and the distribution of fatty acids, in phosphatidylinositol was discussed by Brockerhoff (1961). Rat liver inositide was shown to contain a diarachidonyl inositide.

The sphingomyelins, which contain sphingosine in combination with fatty acid, choline and phosphoric acid, occur in brain and in other organs rich in phosphatides. They do not exactly fit into the latter category, having properties more like those of the cerebrosides. In the sphingomyelins, and in the cerebrosides, sphingosine takes the place of glycerol.

$$
CH_3.(CH_2)_{12}.CH{=}CH(OH).\underset{\substack{|\\ NH\\ |\\ CO\\ |\\ R}}{CH}.CH_2.O.\overset{\substack{O\\ \|}}{\underset{\substack{|\\ O}}{P}}.O.(CH_2)_2.N.(CH_3)_3
$$

It will be noted that choline is present here as a phosphoryl ester, as in the case of lecithin, and not as a salt. A mixture of fatty acids is obtained from the sphingomyelins on hydrolysis and these differ from those found in the glycerophosphatides chiefly in being much more saturated. The chief acid is the saturated C_{24} derivative lignoceric acid.

Phospholipids are thus characterized by having a hydrophilic polar "head" and a hydrophobic hydrocarbon "tail". They are therefore described as amphipathic. Molecules which are amphipathic are surface active and will orientate at the interface between two phases of different dielectric constant, e.g. lipid and water (Dawson, 1966).

Folch and Sperry (1948) have proposed that the terms lecithin, kephalin and sphingomyelin be used only for the crude fractions and that the three groups of compound ester phosphatides be designated: (1)phosphoglycerides (lecithin and kephalin), (2) phosphoinositides (lipid containing phosphoric acid and inositol such as the diphosphoinositide of brain), and (3) phosphosphingosides (sphingomyelin). The disadvantage of this classification is that the acetal phosphatides cannot be fitted into it.

The acetal phosphatides, which are always found mixed with the ester phosphatides in varying quantities (up to 12 per cent.), are similar to them in composition but yield higher aldehydes (stearic or palmitic aldehydes) on hydrolysis, in place of fatty acids. Because their histochemistry is closely connected with that of aldehydes in general the acetal phosphatides are considered, together with other substances containing these groups, in the following chapter.

The mitochondrial lipid cardiolipin, or polyglycerol phospholipid (Daw-

```
     n         n          n          n
      \         \          \          \
       CH2       CH2        CH2        CH2
      /         /          /          /
   H2C       H2C        H2C        H2C
      \         \          \          \
       C=O       C=O        C=O        C=O
      /         /          /          /
     O         O          O          O
     |         |          |          |
     C ------ CH2 H2C ---------------C
    /                                 \
 H2C                                   CH2
    \                                 /
     O      O       H       O        O
      \    /  \     C     /  \      /
       O=P      C   O   C      P=O
         |      H2      H2     |
         O                     O
```

son, 1957) at first sight appears not to belong to the phosphatidyl series. As the above formula indicates, however, it can be regarded as consisting of two of the phosphatidyl units which exist in lecithin, esterified to the two primary alcohol groups of a glycerol molecule. As suggested by Benson and Strickland (1960) and by Ansell and Hawthorne (1964), the most suitable chemical name for cardiolipin is diphosphatidyl glycerol.

Glycolipids. The cerebrosides contain fatty acids, a carbohydrate (usually glucose or galactose) and a complex alcohol such as sphingosine ($CH_3.(CH_2)_{12}.CH = CH.CHNH_2.CHOH.CH_2OH$), but no phosphoric acid. The sugar is usually D-galactose but sometimes it is D-glucose and the fatty acid may be unsubstituted or it may contain a hydroxyl group in the α-position. In the latter case it may be called a phrenosin whereas the former type is known as a kerasin. The cerebrosides have the following general formula:

```
CH3.(CH2)12.CH=CH.CH(OH).CH.CH2.O.CH ────┐
                           |         |      │
                           NH        CH(OH) │
                           |         |      │
                           CO        CH(OH) O
                           |         |      │
                           R         CH(OH) │
                                     |      │
                                     CH ────┘
                                     |
                                     CH2OH
```

It is apparent that they bear a close resemblance to sphingomyelin.

The gangliosides (Klenk, 1942) are cerebroside-like substances which are found chiefly in the ganglion cells of the nervous system. They yield on hydrolysis 1 mole each of fatty acid, sphingosine and neuraminic acid, and 3 moles of galactose. Small proportions of glucose accompany the latter. The characteristic component of the gangliosides is the hypothetical amino-sugar neuraminic acid, whose derivatives are found in many mammalian tissues in combined form. The naturally occurring derivatives are known as sialic acids (Blix, Gottschalk and Klenk, 1957). The commonest substituent is acetyl and Gottschalk (1955) proposed the formula given below for *N*-acetylneuraminic acid. It may be regarded as the product of aldol condensation between an *N*-acetylhexosamine and pyruvic acid.

```
                CH·OH
               /     \
            CH2       HC·NHAc
   HO        |        |
     \       |        |
      C ─────┘        CH
     / \             /
HO2C    ───── O ────┘
                      |
                    H·C·OH
                      |
                    H·C·OH
                      |
                    CH2OH
```

N—acetylneuraminic acid

The gangliosides are localized chiefly in the grey matter of the C.N.S., as mentioned above, although Brante (1951, 1952) found small quantities in white matter. The ganglioside which accumulates in the neurons in familial infantile amaurotic idiocy has now been found to be a special type, called Tay-Sachs' ganglioside or GM2 (Svennerholm, 1966). In late infantile amaurotic idiocy the normal major monosialoganglioside, GM_1, is increased. In the leucodystrophies of globoid cell and metachromatic types the ganglioside pattern is also disturbed, with a pronounced increase in levels of the minor mono- and disialogangliosides.

Constitutional formulae for the brain gangliosides were given by Klenk and Gielen (1960) and the whole subject very ably reviewed by Svennerholm (1964).

Cerebrosides, gangliosides, sphingomyelins and "kephalin B" were included by Carter *et al.* (1947) under the common heading of sphingolipids. This is a useful sub-classification to the main one used here which emphasizes the common factor (sphingosine) present in all the above-mentioned lipids. The first two are glycosphingosides and the last two phosphosphingosides. "Kephalin B" is the name given by Brante (1949) to a sphingomyelin-like compound which, like sphingomyelin itself is resistant to mild hydrolysis with N-KOH. The isolation of this compound from cerebral lipids was described by Svennerholm and Thorin (1960).

The table of synonyms given below is derived from Edgar (1956).

TABLE 25

Nomenclature of Lipids

Sulphatides = sulpholipids = sulphur-containing lipids.
Phosphatides = phospholipids = phosphorus-containing lipids.
Glycerol phospholipids = lecithins and kephalins = mono-amino-phospholipids = phosphoglycerides and phosphoinositides (inositol-containing kephalins) = KOH-hydrolysable phospholipids.
Sphingolipids = all sphingosine-containing lipids.
Phosphosphingosides = all phosphorus-containing sphingolipids = sphingomyelins and "kephalin B" = KOH-resistant phospholipids.
Sphingomyelins = diamino-phospholipids.
Glycosphingosides = all hexose-containing sphingolipids = cerebrosides and gangliosides = glycolipids (Glycolipids often called galactolipids since nearly all the hexose in *normal* tissues is galactose).

DERIVED LIPIDS

Under the heading of derived lipids are classified, *inter alia*, the various fatty acids which can be derived from simple and compound lipids by hydrolysis. These fatty acids are of two types, unsaturated or saturated, depending on the presence or absence of double bonds in their molecule. Stearic and palmitic are the chief saturated fatty acids and oleic acid the chief unsaturated fatty acid occurring in animal fats. The sterols (cholesterol) are usually placed in the class of derived lipids. In this chapter we have to consider histochemical methods for distinguishing cholesterol and its esters, neutral fats, fatty acids, phosphatides and cerebrosides. In the following chapter we shall consider methods for the ketosteroids and acetal phosphatides.

Cain (1950), in his review of the subject, emphasized the fact that lipids are really less difficult to analyse histochemically than carbohydrates or proteins. He suggested that the lack of histochemical studies on lipids was due to a general mistrust of the techniques, and his independent studies of the latter went some way to remedy the situation. A feature of lipid histochemistry is its comparative freedom from problems of location. Many lipids occur in droplets, sharply separated from the surrounding aqueous cytoplasm. That this appearance is not fixation artifact is shown by their demonstration in living cells as well as in freeze-dried material. According to Cain the histochemist should be prepared to recognize four possible groups of lipids:

(1) Lipids detectable as such in living and fixed tissues.
(2) Lipids detectable as such in fixed tissues only (but present in living tissues).
(3) Lipids detectable as such in fixed tissues only (not present in living tissues).
(4) Lipids detectable only after special processes additional to fixation.

The methods for detecting and analysing lipids are divisible broadly into two large sections, physical and chemical.

Physical Methods for Lipids

Solubility in Organic Solvents (Extractive Methods)

An important characteristic of the lipids is their ability to dissolve in each other. The rules of solubility are good only for pure substances and pure solvents. Lison (1936, 1953) pointed out the difficulties attached to the analysis of lipids by reference to their solubilities, maintaining that the prolonged extraction techniques of the chemists could not be imitated by histochemical procedures. These strictures must be borne in mind when applying the methods described below. It must also be realized that *any* fixation will modify the results of extractive techniques. In this respect the drying which takes place when cryostat sections are exposed to air is of paramount importance.

The performance of the various tests for lipids after the application of known lipid solvents and comparison with unextracted materials, has long been customary. Four extractive techniques must be mentioned as a preliminary stage to the consideration of extractive techniques as a whole, in relation to lipid histochemistry.

Ciaccio's Method II. This method, which was evolved by Ciaccio (1934), must be distinguished from his method I for phospholipids, referred to later in this chapter. It consisted essentially of acetone extraction of formalin-fixed blocks followed by staining either with Sudan III or with the Smith-Dietrich procedure. Previous authors had found that simple acetone extraction of blocks, followed by staining for lipids with Sudan III, gave unsatisfactory results. Ciaccio added the refinement of short extraction in ascending strengths of acetone (70, 80 and 90 per cent.) containing 1 per cent. cadmium nitrate, followed by prolonged extraction in acetone containing 2 per cent. of a saturated solution of cadmium nitrate in absolute alcohol. Passing thence to 50 per cent. acetone in water by numerous stages, all in the presence of cadmium, Ciaccio postchromed his material for 2–3 days in acetone-dichromate, embedded in paraffin and stained with Sudan III. Lison (1936) noted that better results were obtained by means of Sudan black B but remarked (1953) that although the method was logical it had been found unsatisfactory in the hands of a number of workers. I agreed with Lison that the main fault in Ciaccio's method lay in the fact that when present in mixtures phospho- and glycolipids are to a considerable extent soluble in acetone and acetone-cadmium mixtures. Ciaccio (1956) introduced modifications of his original method designed to increase its specificity. I have not tested these, however.

Baker's Pyridine Extraction Method. In order to distinguish between lipin and non-lipin substances staining by his acid hæmatein method, Baker (1946b) employed an extraction technique using pyridine at 60° on material fixed in weak Bouin. The extraction stage was followed by postchroming and by the full acid hæmatein routine. The special weak Bouin fixative was necessary in order to permit total extraction of phosphatide by the pyridine. This technique can be criticized for employing a different fixative to that used for the test sections which are not subjected to extraction. This does not necessarily invalidate the interpretation put upon it by the original author. There are, however, a number of recent observations which make exact interpretation difficult. These are considered below.

Keilig Extraction Methods. Working on unfixed blocks of human brain, Keilig (1944) used the regular series of extractions given below. Blocks of tissue not more than 3 mm. thick were treated with the solvents shown in the first column with the results indicated in the second.

(1) Cold acetone	Removes glycerides, cholesterol, cholesterides and ketosteroids.
(2) Hot acetone	Removes cerebrosides.
(3) Hot ether	Removes lecithins and kephalins.
(4) Hot chloroform/methanol	Removes all lipids.

Each extractive stage was performed over a period of 24 hours with at least three changes of the solvent. All but the first should be performed in some kind of refluxing apparatus. Keilig herself used a Soxhlet apparatus for her studies on brain; I used a simple apparatus consisting of a reflux condenser over a flask heated by an electric light bulb in a metal housing. The correct temperature was obtained by altering the distance between bulb and flask. Lhotka (1955) modified the standard Soxhlet apparatus by adding a water-jacket around the extraction chamber so as to keep the temperature of tissues undergoing extraction below limits considered damaging. After extraction the blocks of tissue are carried rapidly through descending strengths of alcohol to water and frozen sections are cut and stained with Sudan black B. If any of the Sudan dyes are used for final staining it is essential that no formalin and no mordant of any kind be used after extraction. In practice the results are not usually as clear cut as suggested in the above list, although removal of the various substances occurs as stated. With the phosphatides, for instance, although extraction by hot acetone fails to remove them, a variable dispersive effect is noted so that localization is impossible and the general effect is dirty. This is not a serious drawback since accurate localization will have been made by other methods. As an alternative to this procedure, if phosphatides are the object of study, the extracted blocks may be postfixed in formalin, postchromed, and frozen sections subjected to a Smith-Dietrich type of staining (see p. 421).

Okamoto-Ueda Extraction Methods. In order to distinguish between the various lipids stained by their mercuric nitrate-diphenyl-carbazone method Okamoto and his associates (Ueda, 1952) performed a series of extractions on tissue sections *after* the staining procedure. The latter, which is fully considered on p. 424, was presumed to stain all phospholipids and glycosphingosides bluish-purple. After 48 hours' treatment with pyridine at 0°–4° glycosphingosides were said to be removed. Subsequent treatment with ether at 22° for 48 hours was considered to extract, in addition, all phospholipids except sphingomyelins. These methods were employed by Diezel (1954) in his studies on the primary lipidoses, in a slightly different manner. He performed the extractions first and the staining afterwards, without suggesting any different interpretation.

Extraction Techniques in General. With all extraction techniques using organic solvents it is necessary to be aware of deviations from the accepted rules of solubility derived from biochemical estimations. Edgar and Donker (1957) performed a valuable service in checking by chemical methods the amount of the various lipid fractions extracted from tissue sections by procedures of the Ueda type. They found very considerable differences between fresh and formalin-fixed sections. From the latter 80–90 per cent. of the total glycosphingoside was removed by 48 hours' treatment with cold pyridine. The percentage of ganglioside resistant to removal by pyridine was observed to be greater in cases of amaurotic idiocy than in diffuse sclerosis, in both of which there is an accumulation of these substances in the brain. Pyridine treatment also dissolved some of the KOH-hydrolysable and some of the KOH-resistant phospholipids from formalin-fixed sections. Brante (1949), however, showed that formalin causes part of the original KOH-hydrolysable phospholipid to disappear and that it converts another part of this fraction into a KOH-resistant lipid. This means that in formalin-fixed tissues this last fraction consists not only of phosphosphingosides but is a mixture of lipids of unknown constitution.

Edgar and Donker observed that treatment with ether for 48 hours at room temperature had no effect whatsoever on the glycosphingoside content of the sections (as expected). Moreover, except in cases of Niemann-Pick's disease it had no material effect on the phosphosphingosides. In fresh material cold pyridine extraction did not decrease the phosphosphingoside content of sections, and surprisingly, it failed to dissolve the KOH-hydrolysable phospholipids although lecithins and kephalins are supposed to be soluble in pyridine. In agreement with theoretical expectations much of the KOH-hydrolysable phospholipid was dissolved by treatment with ether but a considerable quantity (10–40 per cent.) remained insoluble. On the other hand, Edgar and Donker found that a considerable part of the KOH-resistant phospholipid, which is theoretically insoluble in ether, was dissolved by this treatment. They considered that this anomaly might be due to Brante's "kephalin B" which is soluble, like other kephalins, in wet ether. If this is so, the ether-soluble part of the KOH-resistant fraction in fresh tissues can be identified with "kephalin B." Edgar and Donker concluded from their results (1) that Okamoto-Ueda extractions are unsuitable for application to fresh tissues. (2) That in the case of formalin-fixed tissues a material *totally* unaffected by cold pyridine is unlikely to be a glycosphingoside. (If a material is partly dissolved, however, the residual part cannot be identified as phospholipid since it may belong to the insoluble fraction of the glycosphingosides.) (3) That ether extraction cannot be assumed to distinguish between phospholipid and sphingomyelins. Table 26, below, gives a list of Edgar and Donker's observations with the theoretical results side by side for comparison.

In the case of Baker's hot pyridine extract the results on formalin-fixed (or weak Bouin-fixed) tissues may also be confusing. Some work carried out by Almeida and Pearse (1958) on the development of myelin in rabbit brain

TABLE 26

Theoretical and Observed Solubilities of Tissue Section Lipids

Lipid Fraction	Pyridine			Ether		
	Expected	Fixed	Fresh	Expected	Fixed	Fresh
Cerebrosides	Sol.	10–15% Insol.	70% Insol.	Insol.	Insol.	20–30% Sol.
Glycosphingosides						
Gangliosides	Sol.	10–15% Insol.	70% Insol.	Insol.	Insol.	20–30% Sol.
Sphingomyelins	Insol.	20–30% Sol.	Insol.	Insol.	Insol.	50–70% Sol.
KOH-resistant phospholipids						
"Kephalin B"	Not known	20–30%	Insol.	Insol.	Insol.	50–70% Sol.
Kephalins	Sol.	20–70% Insol.	Insol.	Sol.	Insol.	20–50% Insol.
KOH-hydrolysable phospholipids	Sol.					
Lecithins	Sol.	20–70% Insol.	Insol.	Sol.	Insol.	20–50% Insol.

illustrates this point. Up to the 11th post-natal day all acid hæmatein-positive lipids were rendered negative by hot pyridine. After this day, however, there was an increasing amount of residual staining in the myelin sheaths which could be demonstrated also by means of Sudan black B staining and by copper phthalocyanin but not by the Feyrter method. There was no reason to doubt the lipid nature of the residual material since it was extractable with chloroform-methanol solutions. If Baker is correct in his assumption that acid hæmatein does not stain cerebrosides the residual material might be regarded as phosphosphingoside. This substance, however, does not stain with alcoholic solutions of copper phthalocyanin and we found, in *in vitro* tests, that a commercial sample of cerebroside was strongly stained with acid hæmatein. Cerebrosides have been shown by many authors to be strongly positive by the very similar Smith-Dietrich procedure (Edgar, 1956; Lennert and Weitzel, 1952; Landing and Freiman, 1957). In any case, extraction with hot pyridine has been shown to be partially ineffective.

The scepticism expressed by Lison and by Edgar (1956) and by myself, with regard to the use of lipid extraction techniques in histochemistry is still to be upheld in the light of modern experience. This does not mean, however, that such techniques should not be attempted though wherever possible they should be controlled by studies of the type carried out by Edgar and Donker. I believed, and still believe, that a series of extractions should be devised for application in the first instance to unfixed cold microtome sections. These could be secondarily post-fixed in formalin, dichromate or chromic acid as an additional means of differentiation. With the cold microtome the handling of alternate sections for histochemical treatment and for chemical assay is a matter of great ease and convenience.

Certainly in the early stages of the development of "certified" extractions for histochemical use it will be necessary to follow the normal practices of lipid chemistry but *without* adopting their conclusions. An excellent survey of extraction techniques was given by Lovern (1955) in the second chapter of his book on the "Chemistry of the Lipids". In view of its well-known effect in liberating protein-bound lipids ethanol and ethanol mixtures will probably have to be avoided as fixatives for fresh tissue sections destined for lipid studies. Since cold acetone is a poor solvent for phosphatides and sphingolipids it may be useful as a substitute, and in order to depress their solubility still further electrolytes such as magnesium chloride may be added to the acetone. Formalin and formol-calcium mixtures may also be used but even with thin sections fixation with this and the various dichromate mixtures is likely to be slow. As Lovern (1955) remarked, there is an immense volume of published material on lipid extraction techniques and many possible variations of solvent mixtures and sequences. Clearly only the fringe of the problem has been approached by histochemists and some resolute and painstaking studies combining histochemical and chemical methods in alternate serial cold microtome sections are absolutely necessary.

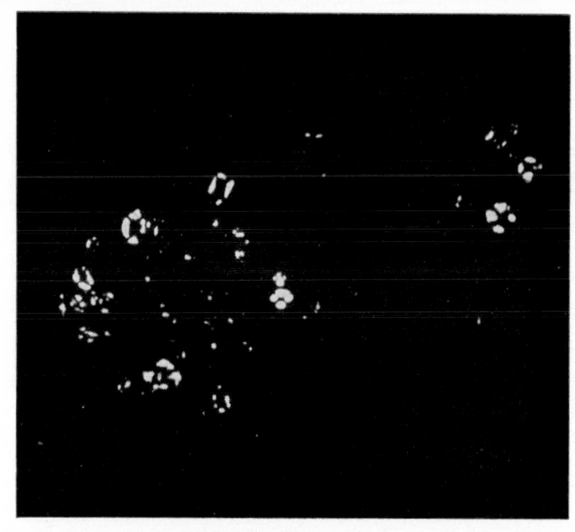

FIG. 109. Myelin sheaths photographed by polarized light to show "maltese-cross" birefringence. × 350.

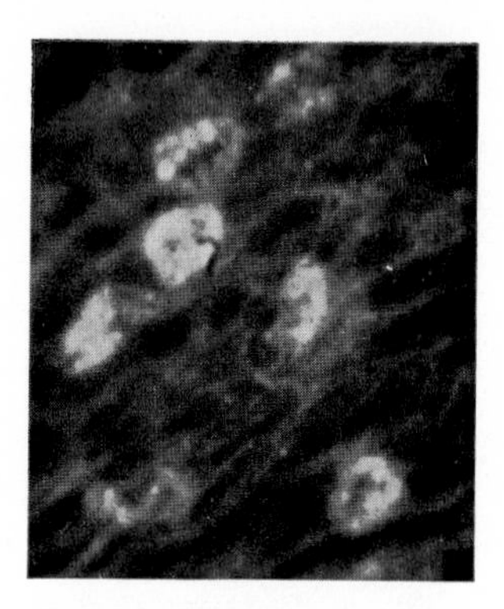

FIG. 110. Mouse myometrium. Mast cell granules showing strong fluorescence in ultraviolet light. Benzpyrene method (Berg). × 540.

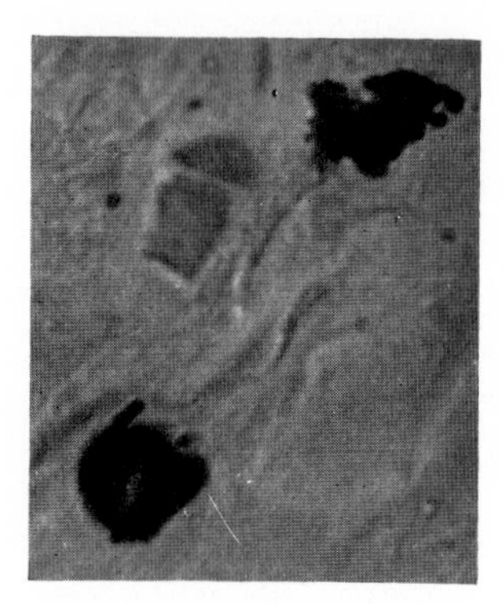

FIG. 111. Human stomach. Mast cell granules stained for lipid. Colloidal Sudan IV. × 1000.

Examination by Polarized Light

Very little assistance can be obtained, in distinguishing between neutral fats, fatty acids, phospholipids and glycolipids, from examination by polarized light. Lison (1936, p. 198) put the position very clearly. Neutral fats, examined in the fresh state, may be liquid and therefore monorefringent, but on cooling they may set in the form of birefringent crystals. Fatty acids in the crystalline state are also birefringent, but may be present in tissue sections in an amorphous isotropic state. It seems, therefore, that in the fresh state both neutral fats and fatty acids are monorefringent but that when cooled, or when fixed in formalin, both may become crystalline. The cholesterol esters and lipins behave in an identical manner. Between certain ranges of temperature they exist as liquid spherocrystals which give Maltese cross birefringence. At temperatures above their melting points they are isotropic and below a certain point they may exist as acicular crystals giving only plain refringence. Lison summed up by stating that examination by polarized light alone gave very ambiguous information regarding the chemical nature of the fats. Three conditions might be found, according to this author (1936, 1953).

(1) *Lipids Observed to be Monorefringent.* No conclusion is possible. Glycerides and fatty acids in the melted state are never birefringent. Cholesterol esters and lipins can be but are not always so, if conditions are such as to prevent the formation of liquid crystals.

(2) *Lipids Birefringent without Maltese Cross.* In this case they behave as true crystals, in the solid state. When the stage of the microscope is rotated through 360° there are four positions of extinction. No diagnostic conclusion is possible since all lipids can exist in this condition.

(3) *Lipids show Maltese Cross Birefringence* (Fig. 109). The only conclusion which can be made is that they consist either of cholesterol esters or of lipins. Glycerides and fatty acids never exist in the form of spherocrystals.

In the case of conditions 1 and 2 an attempt can be made to turn the lipid into a liquid crystalline state. This is carried out in the first case by chilling or by mounting in levulose syrup and in the second case by warming. If spherocrystals appear this confirms the presence of cholesterol esters or lipins but their absence cannot be presumed in the case of failure to observe spherocrystals.

Digitonin Reactions. Windaus (1910) showed that free cholesterol (but not its esters) treated with alcoholic solutions of digitonin, yielded a crystalline complex insoluble in water, acetone and ether but soluble in glacial acetic acid or pyridine. Digitonin precipitates in general those steroids which have a 3 β-hydroxyl group as shown here, the hydroxyl group at C(3) being on the

CH_3 10 3 HO

same side of the ring system as the methyl group at C(10). The reaction was first used in histochemistry by Brunswick (1922). He used a 0·5 per cent. solution of digitonin in 85 per cent. alcohol and examined the sections, in this medium under a coverslip, for the birefringent crystals of digitonin-cholesterol. Leulier and Revol (1930) used digitonin in 35 per cent. alcohol and distinguished the birefringence of digitonin-cholesterol from that of the cholesterol esters by subsequently staining the sections with a red Sudan dye. The esters became coloured and lost their birefringence while the crystals of the digitonin complex remained anisotropic and colourless. Bennett (1939) also employed the reaction in his studies on the adrenal cortex of the cat. Though the digitonin reaction is specific for free sterols, it is of little use for showing the presence of ketosteroids in the adrenal cortex since the birefringence due to these (if any) cannot be distinguished from that of the accompanying cholesterol esters, even by the Leulier-Revol technique.

In practice useful results are seldom obtained by this technique. In particular it is usual to find that all the lipids in a digitonin-treated section stain with the Sudan dyes and with Oil Red O and that digitonin cholesterides are not to be distinguished. Since the latter are supposedly insoluble in cold acetone, and since free cholesterol and cholesterol esters are freely soluble, a further basis for differentiation is offered. Sometimes one can observe acetone-insoluble crystals after digitonin treatment of fresh tissue sections but the doubts expressed in the previous section of this chapter do not allow the clear conclusion that these represent cholesterol digitonides.

Primary Fluorescence Methods

Examination of lipids by means of the ultraviolet microscope has been, in the past, of little importance in histochemistry, except in the case of the oxidation products of lipins which are better known as lipofuscins. These are fully considered in Chapter 26. If carotenoid pigments are dissolved in lipids they may impart to them a strong fluorescence, but no information as to the nature of the solvent lipid can be obtained therefrom. The variable autofluorescence of lysosomes and phagosomes is presumably to be attributed to their lipid or lipoprotein membranes, or to substances dissolved therein. Work on this problem will make little progress until spectral analyses can be made (see Chapter 29).

Secondary Fluorescence Methods

These methods depend on the solution of fluorescent substances, usually referred to as fluorochromes, in the lipids of the tissues. A number of dyes have been used for the purpose, including methylene blue, Thioflavin S, Rhodamine B, Rose Bengal, Magdala red and Phosphine 3R. This last dye was considered by Volk and Popper (1944) to reveal the presence of lipids other than fatty acids, soaps and cholesterol. Its great advantage over the usual staining methods for lipids lies in the use of aqueous solutions so that the smaller fat droplets are not dissolved out during staining. Phosphine 3R imparts a clear silvery-white fluorescence to lipids in which it is dissolved and this is easily distinguished from the natural fluorescence of the tissues without employing control sections. Details of this method are given in Appendix 12. It was used, in a combined method with Sudan IV, by Woolfrey and Pearson (1962) who applied it to fresh cold microtome sections.

A second method for the demonstration of lipids, depending on the development of secondary fluorescence in ultraviolet light, is the benzpyrene-caffeine method of Berg (1951). This method has the same kind of advantages as the Phosphine 3R technique and it is particularly useful as a sensitive method for demonstrating lipids in fine structures, occurring as lipoprotein complexes, or in fine disperse form. It is not suitable as a routine technique but only for application to problems of research into the finer distribution of lipids. Benzpyrene is apparently soluble in all lipid substances, including cholesterol and its esters, so that it cannot be used as the basis of a method for distinguishing the various types of lipid. Berg's method is given in Appendix 12, and the results of its application to mast-cell granules are illustrated in Fig. 110. For comparison Fig. 111 shows them stained by colloidal Sudan III.

Fat-soluble Colorant Methods for Lipids

Historical Survey

The most widely used techniques for lipids depend on the solubility of the inert bis-azo, or other, dyes in the fats themselves and for this purpose the red dyes Sudan III and Sudan IV are commonly employed. The more strongly red Sudan IV has largely replaced Sudan III as the most important single fat stain, though more recently a number of dyes, in related and unrelated groups, have been tried out. Some of these are mentioned below and in Table 27, which is a historical survey of the use of colorant dyes and other methods for staining fats.

The first group in the Table comprises the classical as well as most of the commonly employed fat-staining techniques. The third group also contains methods in common usage. The second and fourth groups are really the same, since both make use of the principle of colloidal suspension of the "dyes" employed. The fifth group is the only one using a relatively new principle, the solution of the "dyes" in hydrotropic detergents, while the sixth and last group contains the newer fluorescent techniques as well as Lorrain Smith's classical method for fatty acids.

Various Solvents for the Red Sudan Dyes

Variations in the use of the Sudan and other dyes depend largely on the type and concentration of the fluid in which they are dissolved or suspended, but in each case the solubility of the dye in fat must exceed its solubility in the solvent. Herxheimer's (1903) mixture (equal parts acetone and 70 per cent. alcohol) remains a popular solvent for the red Sudan dyes, although there are

TABLE 27

Historical Survey of Fat-staining Methods

(After Berg, 1951)

Colorant	Solvent	Remarks	Author
		Simple solutions	
Sudan III	Alcohol 70 per cent.	Adipose tissue, fat droplets	Daddi (1896)
Scharlach R and	Alcohol 35 per cent.	Adipose tissue, fat droplets	Michaelis (1901)
Sudan IV	Acetone 50 per cent.		Herxheimer (1901)
Sudan Brown	Isopropanol	Fine fat droplets, myelin sheaths;	Lillie (1944)
Oil Red 4B	Pyridine 70 per cent.	rapid intensive staining	Proescher (1927)
Oil Red 0			
Scharlach R	Diacetin 50 per cent.	Adipose tissue, fat droplets	Gross (1930)
Sudan red	Diacetin 50 per cent.	Adipose tissue, fat droplets	Domagk (1933)
Sudan black B	Alcohol 70 per cent.	Mitochondria. Leucocyte granules.	Lison and Dagnelie (1935)
		Bacterial lipids	Sheehan (1939)
			Hartman (1940)
Sudan black B	Diacetin 50 per cent.	Myelin, etc.	Leach (1938)
Sudans	Propylene glycol	Mitochondria. Finest granules	Chiffelle and Putt (1951)
		Colloidal solutions	
Sudan III	Alcohol 40 per cent.	Leucocyte granules, diffuse cytoplasmic stain	Romeis (1927)
Sudan III	Alcohol-phenol-acetic acid	Leucocyte granules, "Sudanophobic" lipids	Jackson (1944)
Sudan III	Acetic acid-gelatine	Leucocyte granules, "Sudanophobic" lipids	Govan (1944)
		"Oversaturated" solutions	
Sudan III and IV	Alcohol 70 per cent.	Improved staining of lipids	Kay and Whitehead (1935)
			Kay and Whitehead (1941)
Sudan IV	Alcohol 65 per cent.	Improved staining of lipids	Froeboese and Spröhnle (1928)
		Serum solutions	
Sudan III	Serum	Vital staining of lipochondria	Ludford (1934)
3:4-benzpyrene	Glycerine and serum	Vital staining. Bacteria	Günther (1941)
			Graffi (1939)
3:4-benzpyrene	Glycerine	Free fat and structure lipids	Graffi and Maas (1938)
		Hydrotropic solutions	
Sudan IV or	Trichloroacetic acid	Finest fat droplets. Diffuse	Hadjioloff (1938)
Scharlach R	Sulphosalicylic acid	cytoplasmic staining. Myelin sheaths	
	Caffeine solutions		
3:4-benzpyrene	Caffeine solution	Critical staining of small amounts of lipid and lipoproteins	Berg (1951)
		Aqueous solutions	
Nile blue	Water	Unsaturated fatty acids pink. Lipids blue	Smith (1907)
			Smith and Mair (1911)
Rhodamine B	Water	Vital staining of neutral fats (plant)	Strugger (1937, 1938)
Phosphine 3R	Water	Lipids except cholesterol, fatty acids and soaps	Popper (1944)
			Volk and Popper (1944)

strong objections to its use on account of the ready solubility of small fat droplets in this solution and on account of the tendency of the dye to crystallize out of solution during differentiation. Dissatisfaction with the results of staining by means of the red Sudan dyes dissolved in concentrated alcoholic solutions led to the employment of weaker alcoholic solutions for longer periods in an effort to avoid extraction of the smallest fat globules. Kaufmann and Lehmann (1926) made a series of experiments with lipid substances incorporated in elder pith in which, for staining glycerides and lipins, they used a 40 per cent. alcoholic solution of Sudan III. Romeis (1927) used a similar solution for staining fats in frozen sections. Froeboese and Spröhnle (1928) used the Romeis method and concluded that the superior staining of fat which was obtainable, though irregularly, with 40 per cent. alcoholic solutions was due not to the preservation of small lipid droplets but to the colloidal properties conferred on the stain by the increased water content. Kaufmann and Lehmann (1929) conclusively demonstrated by chemical estimation that 70 per cent. alcohol dissolves fatty substances from frozen sections at a considerable rate and Kay and Whitehead (1934) considered that this indicated that the chief merit of the Romeis technique lay in the superior fat-preserving properties of 40 per cent. alcohol. These authors (1935), having previously found Sudan III to be unsatisfactory, used saturated Sudan IV in absolute alcohol, diluted to 66 per cent. with water before use. Romeis (1929) brought out a modification of his original technique using weak alcoholic solutions of a fraction of commercial Sudan III which he called Sudan orange. Technical difficulties in the preparation of his solution and the fact that other authors could not duplicate his results, prevented the general adoption of the method.

Colloidal Suspensions

Jackson (1944) evolved a new method of Sudan staining as the result of attempts to demonstrate histochemically in the cells of a contagious venereal tumour of dogs, a lipid substance which was resistant to staining by ordinary Sudan methods but which was partly revealed by the Romeis technique. His staining solution consisted of a colloidal suspension of Sudan III produced by adding 5 per cent. aqueous phenol to an 80 per cent. alcoholic stock solution of the dye until the alcohol content was reduced to 60 per cent. After standing for a few hours, glacial acetic acid was added in the proportion of 2·5 drops per ml. of carbol Sudan, the mixture being filtered and allowed to stand 24 hours before use. Using frozen sections, Jackson stained these in his acetic-carbol-Sudan for 1½ hours or more and differentiated in acetic acid-alcohol. Besides neutral fats, with this technique other lipids, some of which were probably phosphatides, stained bright orange. Some weak staining of phosphatides occurs when the Sudan dyes are used even in the more conventional techniques and, since many superior methods for phosphatides now exist there is no point in using either the Romeis or Jackson methods for their appreciation.

Work on different solvents for the Sudan dyes continued and Govan (1944), following the method of Hadjioloff (1938), went further than previous authors in avoiding the use of fat solvents altogether. He used a colloidal suspension of either Sudan III, Sudan IV, or Sudan black B in 1 per cent. aqueous gelatin containing 1 per cent. acetic acid, staining frozen sections for 30 minutes at 37°. With this technique preservation and staining of the smallest fat particles is excellent and it can be recommended for the purpose. Govan himself regarded his method as fitted for research rather than for routine use on account of the difficulty in preparing the solutions.

Other Solvents and Dyes

Alternative solvents have been suggested for the Sudan dyes in place of alcohol which, though they are still in the class of fat solvents, have been considered to have various advantages. Gross (1930) used Sudan IV in 50 per cent. diacetin (glycerol diacetate) and found that the smaller fat particles were well preserved, and Lillie and Ashburn (1943) used Sudan IV as a saturated solution in 99 per cent. isopropanol, diluting to 50 per cent. with water immediately before staining. Subsequently Lillie (1944) found Oil Red O and Oil Red 4B to be superior to the red Sudan dyes in the supersaturated isopropanol technique and later still (1945a and b) he recommended Oil Blue NA and Coccinel Red. These last two dyes were used from saturated solutions in 60 per cent. isopropanol diluted to 40 per cent. with water immediately before use, and Lillie suggested that these more dilute isopropanol solutions should cause smaller fat loss during staining. Chiffelle and Putt (1951) recommended propylene glycol or ethylene glycol as solvents for Sudan IV or Sudan black B. According to the authors both these glycols gave stable dye solutions which did not extract lipid particles.

The Use of Dye Mixtures

The problem which besets the use of fat-soluble dyes for staining neutral fats remains a two-fold one. First, despite Jackson's views on the greater importance of the colloidal state of the dyes, the solvent in which they are dissolved must be used in such strength, and for so short a period of time, that removal of small fat particles is reduced to a minimum. Secondly, the solubility of the dyes in fat must be high and the colour developed must be similarly high. Kay and Whitehead (1941) tested a large number of Sudan dyes and concluded that impurities played a part in strengthening the colour imparted to fat in tissue sections. They showed that in mixtures of the Sudan dyes in alcohol each dissolves independently of the other, and the total amount of dye available for staining depends on the number of Sudans present. It was therefore good practice, in a routine fat stain, to use a solution saturated with both Sudan III and Sudan IV (commercial batches of which are themselves mixtures of a number of dyes), in 70 per cent. alcohol or in Herxheimer's mixture, in place of Sudan IV alone. This method, which was used for many years as a routine procedure in the Department of Pathology of the Royal Postgraduate Medical School, gave a satisfactory colour with neutral fats when used for so short a period as 30–60 seconds, and solution of smaller fat particles was avoided by the short staining time. Methods of this type should be replaced by the Oil Red isopropanol technique of Lillie for the demonstration of neutral fats in routine practice. Where the dye Fettrot is available it will be found superior to either of the red Sudans but not to the Oil Red, Oil Blue series of dyes (see Appendix 12). For research problems such techniques may be supplemented by the phosphine method of Popper and by the colloidal suspension method of Govan.

Sudan Black B

Particularly in the case of the staining of phospholipids by fat-soluble dye techniques the greatest advances have come from the use of the disazo dye

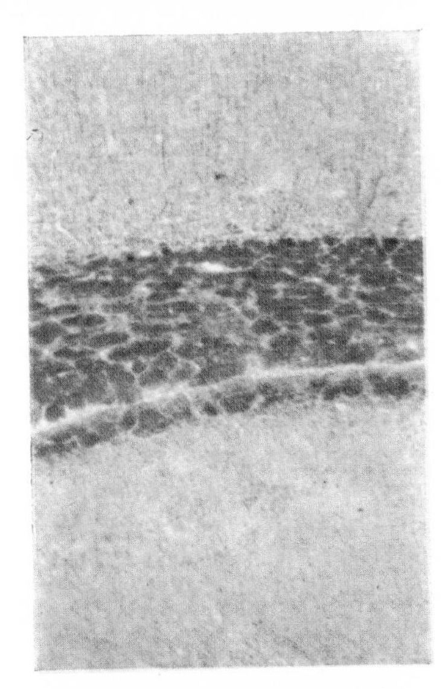

XIA. Rat Brain Formol-calcium fixed frozen section. Shows orange-red staining of myelin phospholipid. OTAN method. × 170

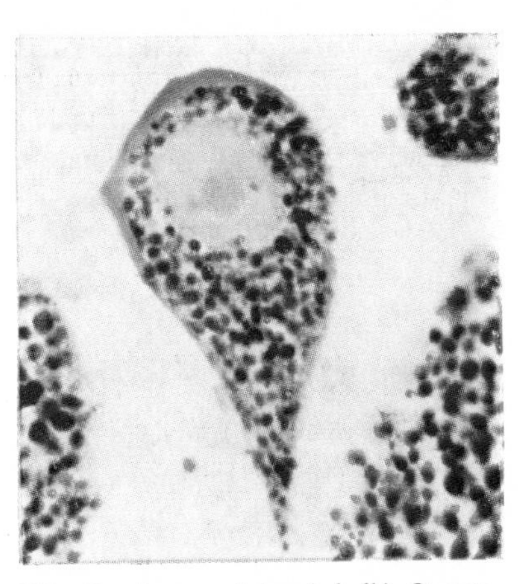

XIB. *Buccinum undatum* (whelk). Oocyte. Development of phospholipid granules in an early stage. Copper phthalocyanin —Neutral red. × 400.

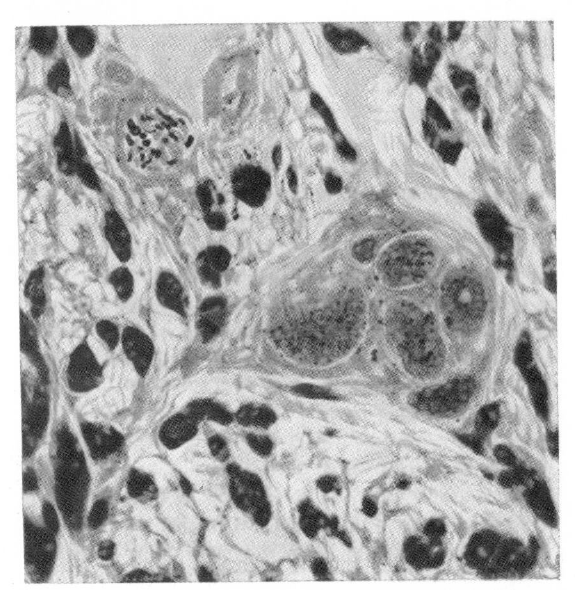

XIC. Human skin (10 μ frozen section). So-called granular myoblastoma. Myelin sheaths, fat globules and the tumour cell granules stain strongly. Sudan black B, carmalum. × 150

Sudan black B whose formula is given below. This dye was first introduced into histochemistry by Lison and Dagnelie (1935) and employed among other things, for staining the granules in leucocytes by Sheehan (1939), McManus (1945), Wislocki and Dempsey (1946) and Eränkö (1950). Sudan black B was fractionated by Bermes and McDonald (1956, 1957) using a 40 cm. celite-silicic acid column. They found no less than 10 fractions, respectively yellow, red, orange, green, blue, green-blue, blue, blue, blue, and black. All of these fractions stained the tissues and fraction 3 was fluorescent in ultraviolet light.

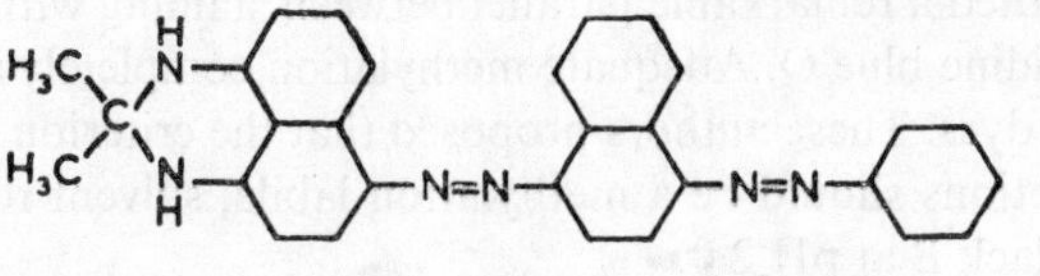

It was reported by Fredricsson *et al.* (1958) that SBB decomposed in acid solutions, and especially rapidly below pH 4. The product, which had a pK in the region of 4, was brown above this level and red below it. It stained nucleoproteins and mucopolysaccharides and the authors considered that brown staining with Sudan black B was not to be regarded as due to lipid. The point might be tested by examination of sections in citrate buffer at pH 2·0.

Further fractionation studies were carried out by Schott (1964) on commercial samples of Sudan black B. He noted that the dye had only two peaks in its absorption spectrum (417 and 592 nm in ethanol) but separated three fractions upon chromatography. The main fraction (B) was blue and the others orange (C) and red (A). Fraction B gave the typical fat staining associated with the dye. The synthesis of Sudan black B, and the possible side reactions and contaminants present in commercial preparations, were discussed by Terner *et al.* (1963).

Sudan black B is unnecessary for the demonstration of neutral fats, for which the Fettrot or Oil Red or Blue dyes are more suitable, but it possesses the advantage of being far more soluble in phospholipids and to a lesser extent in cerebrosides than the above-mentioned dyes, and it will stain them in tissue sections provided they are preserved by suitable fixation. It was therefore used in a Ciaccio type of procedure (on frozen sections) by Baker (1944) and by McManus (1946). The latter employed fixation in a formalin-calcium-cobalt solution, substituting cobalt for the cadmium of Baker's fixative for phospholipids. McManus postchromed his material where necessary and, although he used frozen sections as a routine, he pointed out that Sudan black could readily be used on paraffin sections if suitable fixatives were employed. According to Holczinger (1965) the use of Sudan black B after chromium, or osmium, fixatives is accompanied by a number of difficulties in interpretation. These are due to unspecific, non-lipid, staining.

Since formalin fixation alone preserves lipids to some extent it is always worth applying the Sudan black technique to formol-fixed paraffin sections when nothing else is available; although a negative result is of no importance, positive results signify "formol-fixed" lipoprotein whose lipid component is usually phosphatide or cerebroside.

Baker, and also McManus, used Sudan black in 70 per cent. alcohol, and this solvent is most commonly employed despite the objections of Leach (1938), who recommended 50 per cent. diacetin. After testing a large number of fat-soluble dyes for specificity Schott and Schoner (1965) concluded that the influence of solvents was particularly marked in the case of Sudan black. They recommended that it should be used only in ethanol or isopropanol. Terner *et al.* (1963) noted a remarkable parallel between staining with Sudan black B and with toluidine blue O. Adequate methylation completely abolished staining with both dyes. These authors proposed that the criterion for acidic lipids in paraffin sections should be a methylation-labile, solvent resistant staining with Sudan black B at pH 2·0.

It was suggested by Rodé (1962) that Sudan black B in alcoholic solution developed, on standing, the character of a basic dye. This, he considered, was responsible for the progressively stronger staining of pancreatic islet β granules which occurs with time. Almost certainly, however, the specific granules possess a stainable phospholipid envelope.

The Sudan black technique is not, of course, specific for phosphatides, but phosphatides are the most common lipids which it demonstrates if neutral fats are excluded. The result obtainable in frozen sections is illustrated in Plate XIc, facing p. 415, where the phospholipids present in the cells of a so-called granular myoblastoma are seen to have great affinity for the dye. The Sudan dyes gave only faint pink staining in this case. The Sudan black technique is also useful in conjunction with the various extraction methods (outlined above) and, in conjunction with acid hæmatein, in Elftman's (1954) controlled chromation technique (see p. 423).

METACHROMASIA AND SUDAN BLACK B

Following the observation by Lennert and Weitzel (1952) that Sudan black B appeared reddish-violet when dissolved in octane, nonane and decane Diezel and Neimanis (1957a and b) tested 65 different organic compounds with the dye. They observed that, in addition to the aliphatic hydrocarbon series, aromatic hydrocarbons and many other substances produced a change of colour from black to red which was described as metachromasia. When viewed by polarized light lecithin, sphingomyelin and ganglioside (in alcoholic solution) produced the metachromasia, while cerebroside and cholesterol did not. In the case of lecithin the metachromasia was so strong that it was visible by ordinary light. The authors explained the metachromasia of the dye as being due to mesomerism and claimed that lipids capable of eliciting metachromasia are those which are dipoles (zwitterions). The metachromasia was not due to the production of anomalous colours since it was not abolished by rotating the preparations.

Diezel and Neimanis tested the dye on paper chromatograms with various solvents and noted changes in the colour of the front. They did not suggest that these might be due to different fractions (cf. Bermes and McDonald). Although Sudan black B metachromasia was considered to be a useful test for the zwitterion lipids I believe that further examination of the mechanism is necessary before such an interpretation of the effect observed in tissue sections can be fully accepted.

ACETYLATED SUDAN BLACK B

Observing that ethylation of fat solvent dyes could prevent or diminish staining of the leucocyte granules (stable sudanophilia) Lillie and Burtner (1953) acetylated a number of disazo and other dyes including Sudan black B. They found that the new dye would no longer produce stable sudanophilia but that the staining of fats was unimpaired. Lillie (1954) noted that the acetylated dye gave less background staining and other workers (Casselman, 1954a) have made similar observations. This author (1954b) gave details of an improved method of acetylation using an equivalent of acetic anhydride in diethyl ether (Kaufmann, 1909) in place of Lillie's excess acetic anhydride in pyridine.

Most workers feel that under the staining conditions usually employed the results with unacetylated Sudan black are sufficiently good for most purposes. Moreover, the increased specificity of the acetylated product for lipid is open to doubt. Feagler and McManus (1964) noted very strong staining of nuclear chromatin but this was abolished by brief pretreatment of their (cryostat) sections with dilute acid.

THE PROPYLENE GLYCOL METHOD

The Sudans and other lipid-soluble dyes are far more soluble in propylene glycol than in alcohol or acetone, and staining in 0·5 per cent. solutions of dye becomes perfectly feasible (Chiffelle and Putt, 1951). Using Sudan IV, Fettrot (Ciba), or Sudan black B, as saturated solutions in propylene glycol, I was able to confirm the original authors' findings. No tendency to the extraction of finer lipid particles by the solvent was observed and the intensity of staining of neutral fat particles was far greater than with the usual alcoholic solvents. In addition, a large number of phosphatide-containing structures were often intensely stained, including the mitochondria and even small particles which were perhaps lysosomes. Structures in which lipids cannot usually be demonstrated, such as the prickles of the prickle-cell layer of the epidermis, are also stained by this method. In spite of the strictures of Schott and Schoner (1965), therefore, there may still be occasions when propylene glycol can be employed with advantage as a solvent for Sudan black B.

A modified propylene glycol method is given in Appendix 12. It is recommended for use where demonstration of the total lipid content of cells is required, but not as a routine fat stain. Sudan black B in propylene glycol is particularly good for staining chromated lipids produced by Elftman's method. Ethylene glycol may be used as an alternative solvent.

MASKED LIPIDS

A modified Sudan black B technique for masked or bound lipids in blood films was described by Ackerman (1952). This involved pretreatment of the films with various organic acids (acetic, citric, oxalic and formic) and subsequent staining with a matured 70 per cent. alcoholic solution of Sudan black B. The reaction of lymphocyte mitochondria was considerably enhanced and the platelets stained positively after acid treatment. The staining of masked lipids was also studied by Brolin (1952). A further application of Sudan black B to the staining of bound lipids in tissue sections was described by Berenbaum (1954) who found that after long washing or exposure to dry heat, *inter alia*, many presumably lipid-containing structures could be stained with Sudan black in acetone or with hot (burning) ethanolic Sudan black. The staining of lipoproteins by methods such as these was more fully considered by Berenbaum (1958), and by Schott (1962). Holczinger and Bálint (1961) found that treatment of fresh sections with 0·14M NaCl at 3° for 30 minutes, followed by 0·4M NaCl for 12 hours, 0·7M for $1\frac{1}{2}$ hours and, finally, 1·0M alternately at pH 4 and 7 for 10 hours, would unmask mitochondrial lipids and allow them to be stained with Sudan black B. Ackerman's technique was modified by Mironescu (1964) for use with smears. Details of these techniques are given in the Appendix to this chapter.

MARCHI METHODS FOR DEGENERATING MYELIN

The original method of Marchi (1892) involved fixation of blocks of tissue for 3–4 weeks in dichromate before impregnation with his OsO_4/CrO_7 mixture. Later refinements of the technique (Swank and Davenport, 1934; Glees, 1943) involved abandonment of the original fixation in dichromate and the substitution of other oxidizing agents ($NaIO_3$, $KClO_3$) in the impregnating mixture. It was usually considered (Weisschedel and Jung, 1939) that treatment with oxidizing agents prevented the unsaturated fatty acids of normal myelin from reducing OsO_4. Since degenerating myelin could still reduce OsO_4 after oxidation it was presumed either to contain more unsaturated groups than could be inactivated by the oxidant or reducing groups of a type not susceptible to oxidation. Johnson *et al.* (1950), Brante (1949), Burt *et al.* (1950) and Mannell (1952) made extensive studies of the chemical changes in degenerating myelin. All agreed that the myelin lipids (cerebroside, cholesterol and sphingomyelin) change very little during the first 8 days after section of a nerve and that they then decrease rapidly and coequally between the 8th and the 32nd day. There was thus, as pointed out by Wolman (1956a and b) no chemical explanation for the histochemical findings in degenerating myelin. Wolman therefore tested a number of pure substances with Marchi-type procedures in order to determine, if possible, the type or types of material which might be responsible for the reaction. He found reduction only with unsaturated fatty acids and with pyrogallol, cysteine, glutathione and heparin. Pretreatment with dichromate slowed the reaction with the first of these but increased the reaction with heparin. Wolman suggested, on the basis of these studies, that the normal myelin sheath contains acidic mucopolysaccharides and that these are liberated during demyelinization. Using his Bi-Col method (Chapter 10, p. 351, and Appendix 10, p. 671) he observed a marked increase in red (gold) staining which was maximal 2–4 days after transection of the nerve. This was interpreted as being due to the presence of a polyuronic polysaccharide which was weakly Marchi-positive. A strongly acidic polysaccharide such as heparin would have taken the blue (iron) stain of the Bi-Col mixture. From the 6th–8th day onwards, rising to a maximum on the 12th day, a non-acidic polysaccharide was found which Wolman considered as a derivative of the polyuronide. This was strongly Marchi-positive.

Opposition to Wolman's view was expressed by Adams (1958) who pointed out that the Marchi-positive component of degenerating myelin was soluble in acetone and behaved more like a lipid than a carbohydrate. He offered an alternative explanation for the Marchi-negative quality of normal myelin and later (1960) provided stronger evidence in favour of his hypothesis.

According to Adams normal myelin has an affinity for water, due to the presence of the polar, hydrophilic, groups of phospholipids, and it is therefore permeable to water-soluble oxidizing agents such as chlorate. Degenerating myelin has no such affinity. Adams quoted the work of Zelikoff and Taylor (1950) who showed that OsO_4 in an OsO_4/chlorate mixture acted as an oxidation catalyst rather than as an oxidizer. According to this theory it can accelerate the oxidation of various compounds without reduction to OsO_2.

Tests carried out by Adams on myelin lipids *in vitro*, and on oleic acid and cholesterol oleate, confirmed these views and suggest that the hydrophobic properties of degenerating myelin could be related to the formation of cholesterol esters during the chemical degradation of myelin. After chromatographic separation of degenerating myelin, only the esterified cholesterol fraction gave an unequivocally positive Marchi reaction. Unsaturated lipid, but not protein or polysaccharide were observed to stain after brief treatment with OsO_4. Adams thus concluded that the Marchi reaction distinguished hydrophilic from hydrophobic unsaturated lipids.

Contrary to established opinion Smith (1951, 1956), and her colleagues (Smith *et al.*, 1956), maintained that material long stored in formalin could be used for Marchi preparations, particularly if the stage of clearing was avoided. The more usual view (Swank and Davenport, 1934; Mettler and Harada, 1942) had been that prolonged formalin fixation caused the normal myelin sheaths to stain positively and that fixation in formalin alone should not be continued for over 24 hours. Smith and her colleagues found that Marchi-positive material was at first extracellular (up to 10 weeks after injury) and then progressively intracellular. The extracellular material could be stained for about 2 years but after this most of it became negative. The intracellular material stained strongly after very long periods of storage in formalin.

Marchi methods must be performed on blocks of tissues if aqueous mixtures of the Swank-Davenport type are used. Wolman (1956b) showed, however, that if staining was carried out in non-aqueous media such as CCl_4 the method could be applied to paraffin sections mounted on slides. In Appendix 12, p. 695, details of the Swank-Davenport procedure, and also of Wolman's procedure for paraffin sections, are given. The advantages of dichromate or chromic acid fixation for degenerating lipids, prior to Marchi staining, were stressed by Harman and Bernstein (1961).

Chemical and Other Methods for Lipids

Osmic Acid (Osmium Tetroxide) Methods

The reduction of osmium tetroxide, in weak aqueous solutions, has long been used as the basis of methods for demonstration of lipids. Lison (1936) made it quite clear that the conditions under which osmium tetroxide is allowed to react have a great influence on the final result. Pure OsO_4 is blackened by all tissue elements but this blackening does not occur if an oxidizing agent is used simultaneously. Primary blackening with OsO_4 is considered to be due to the presence of unsaturated fatty acids but the phenomenon of secondary blackening, which takes place when fats giving a negative primary reaction are subsequently treated with alcohol, remains inadequately explained. Lison (1936, 1953) considered that osmium tetroxide should not be used in the histochemical study of lipids, since interpretation of the results was subject to very great error, and that the reaction was unsatisfactory from the histochemical point of view. Cain (1950) did not agree with this exclusive point of view, but supported a limited use of osmium tetroxide in histochemistry. He maintained that if it was allowed to act on frozen sections, fixed in formol-calcium for 6 hours, it could be regarded as specific for reducing lipids, provided that only structures known to contain lipid were considered.

The Osmium Tetroxide-Alpha Naphthylamine Method. This method (OTAN) was introduced by Adams (1959) for the simultaneous demonstration of normal and degenerating myelin. Hydrophobic lipids, as in the latter, are stained black due to the reduction of OsO_4 to OsO_2. The hydrophilic lipids of normal myelin are stained red by the formation of an osmium-α-naphthylamine chelate. The method therefore distinguishes three types of lipid: (*a*) Unsaturated hydrophobic (black), (*b*) Unsaturated hydrophilic (red), and (*c*) Saturated (colourless). In the first class are oleic acid, triolein and cholesterol oleate. In the second are lecithin, kephalin, sphingomyelin and cerebroside, and in the third, stearic acid, tristearin, cholesterol stearate and cholesterol.

If the method is preceded by hydrolysis with 2 N-NaOH at 37° for one hour, only alkali-resistant lipids are stained. The most important of these is sphingomyelin, but kephalin B would also stain, presumably.

Both techniques are given in Appendix 12, p. 694, and the results are illustrated in Plate XIa, facing p. 415.

METALLIC MORDANT METHODS

Ciaccio Method I. Ciaccio (1909) developed a method depending on the simultaneous treatment of fresh or formol fixed tissue with formalin and dichromate, followed by prolonged postchroming and by paraffin embedding. This treatment rendered certain lipid substances resistant to extraction by the usual processes of embedding and they could be stained, in paraffin sections, with saturated 70 per cent. alcoholic solutions of Sudan III or Sudan IV. In the original technique myelin stained red and "Ciaccio lipids" orange. In his original description Ciaccio included among the positive-staining lipids both saturated and unsaturated phospholipids (lecithin, kephalin and myelin) and mixtures of cholesterol and cholesterol esters with oleic acid. Kaufmann and Lehmann (1926), however, concluded that all Ciaccio-positives were unsaturated fats and that the reaction was specific for these.

Cain (1947a and b) considered that paraffin embedding could not be regarded as an efficient procedure for the removal of non-chromated lipids and Lison (1953) could place no confidence in the reaction although he observed that it was better when carried out with Sudan black B in place of Sudan III or IV. Although the original Ciaccio method is not now used in histochemistry the most important developments are derived directly from it. I refer to procedures using direct chromation (without previous or concomitant fixation with formalin) followed by staining with Sudan black B or hæmatoxylin mixtures.

The original Ciaccio procedure had two major disadvantages; first, the necessity for prolonged treatment with dichromate and second, the poor final colour developed. Lillie and Laskey (1951), recognizing these features, developed a modified Ciaccio method using formol-calcium fixation followed by Sudan black B or Spirit blue. This method should not be called a Ciaccio procedure, but it is a useful variant of McManus' (1946) Sudan black method and it is interesting for its use of the Feulgen reaction as a nuclear counterstain.

Smith-Dietrich Type Methods. The original method of Dietrich (1910) for lipins depended on the treatment of blocks of tissue with formaldehyde-dichromate mixtures by which the lipins were oxidized and combined with chromium. Frozen sections were cut and the presence of combined chromium was subsequently shown by the black colour produced by combination with acid hæmatoxylin. The class of lipins included the phospholipins, or phosphatides, with which we are concerned in this section, and the galacto- and glycolipins, or cerebrosides, considered in the section below. Chromation was originally carried out at 37°, but Kaufmann and Lehmann (1926) considered a temperature of 60° necessary and subsequent workers have followed this precept. Lison (1936) believed that a black colour with the Smith-Dietrich test indicated the presence of lipins, but that a grey tint was given by many fats and could not be regarded as of diagnostic value. A negative test, moreover, was not considered to exclude lipins. Baker (1944), by means of *in vitro* tests, showed that kephalin and sphingomyelin both stained black by the

Smith-Dietrich procedure, as did lecithin when combined with other (Smith-Dietrich-negative) lipids. Lecithin alone did not react.

The Acid Hæmatein Method. The principle of the Smith-Dietrich method was used by Baker (1946b) as the basis of his acid hæmatein test for phospholipins. After fixation of tissue in formol-calcium, prolonged chromation was carried out (at 22° and 60°) in a dichromate-calcium mixture. Frozen sections (gelatin-embedded) were subjected to further brief chromation before staining in a freshly prepared and oxidized acid hæmatein solution. Differentiation was carried out with a borax-ferricyanide mixture. Baker gave a list of substances and their reactions by his method, of which the following is an incomplete summary:

Fibrinogen	pale dirty blue.
Collagen	negative.
Mucin	dark blue.
Hæmoglobin	grey.
Egg lecithin Brain lecithin Kephalin Sphingomyelin	dark blue or blue-black.
Nucleoprotein	dark blue.

Baker (1947) tested the reaction of purified galactolipin and obtained a negative result. He therefore concluded that, among lipid substances, the reaction was specific for phospholipids. In order to distinguish between these and other non-lipid substances giving a positive reaction, he introduced his pyridine extraction method which has already been considered in the section on extractive techniques.

Cain (1947a) conducted an examination of Baker's acid hæmatein method and concluded that the test was specific for phospholipins provided that only a definitely positive result was considered. Very pale blue and greys might be caused by other lipids which, if present in very large masses, might even show as medium to dark blue granules. Cain explained the mechanism of the test as follows:

(1) Phospholipids are not fixed by formol-calcium but are restrained from passing into solution by the calcium, which plays no other part.

(2) Phospholipids combine readily with chromium salts and are thereby rendered insoluble and mordanted as well.

(3) On staining, blue and brown colours are formed when the dye attaches itself to chromium in the various tissue components.

(4) On differentiation, some brown and most blues, particularly those in substances which contain phosphoric acid, remain nearly fast. Most browns and the weak blues are reduced or removed entirely, and for this reason **the period of differentiation must not be shortened.**

Since the specificity of the test depends on the relatively greater affinity of

From the practical point of view the method depends on the treatment of thin formalin-fixed sections, after prior extraction with acetone, with alcoholic mercuric nitrate at 0° to 4°. After washing, unbound mercury is removed with potassium iodide and the lipid-bound mercury is then demonstrated by means of diphenylcarbazone. A violet colour is given by phospholipids and cerebrosides and in order to separate these two types of material the authors proposed their cold pyridine extraction, performed after staining. Their claim that this removed only cerebrosides was not sustained by the investigations of Edgar and Donker (1957). Further claims that subsequent ether extraction would leave only sphingomyelin are also unsubstantiated. The results of the method are illustrated in Fig. 113, facing p. 422. The principle involved in this method may be an important one and efforts should be made to elucidate the mechanism. If this can be done successfully a useful technique will be added to the rather poor selection at present available for demonstrating phospholipids.

The Phosphomolybdic Acid Method. Choline-containing lipids can be demonstrated on paper chromatograms by forming insoluble choline-phosphomolybdic acid complexes which are subsequently rendered visible by conversion to molybdenum blue. This mechanism is to be distinguished from that of the method for inorganic phosphates (Chapter 27) in which water soluble phosphomolybdates are formed by treatment with ammonium molybdate.

Landing *et al.* (1952) successfully demonstrated choline-containing lipids in tissue sections in this way. They were able to stain sphingomyelin deposits in Niemann-Pick's disease and the cerebrosides of Gaucher's disease (in spite of their presumed lack of choline) and I used their method to show the gangliosides in various cases of amaurotic idiocy. Landing and his co-authors could prevent the staining of sphingomyelin and cerebroside by brief treatment of sections with a number of organic solvents, especially ethanol, pyridine and acetone-ether mixtures. Chloroform, xylene and acetone were ineffective. They considered that the results of these extractions indicated that the lipids concerned were present as lipoprotein complexes and emphasized the inapplicability in histochemistry of any differential extraction scheme based on the known solubility characteristics of purified lipids. When the phosphomolybdic acid method is applied to paraffin sections much more intense staining of various interstitial tissues is noted and a number of procedures, such as treatment with dichromate or 70 per cent. ethanol, or with acetone for long periods, will enhance this staining. Landing *et al.* believed these effects to be due to the release of potentially active compounds normally present in bound form. The phosphomolybdic acid method is especially applicable to fresh cold microtome sections, free-floating or mounted on slides, but even with these the results are difficult to interpret. Its use is particularly not recommended by Adams (1965), with whose view I concur. Notwithstanding, for those who may wish to investigate for themselves the specificity of the method, details are given in Appendix 12, p. 693.

Other Methods

The Cis-Aconitic Anhydride Method for Choline Lipids. This method (abbreviation CAZA) was described by Böttcher and Boelsma-van Houte (1964) and by the second author alone (1965). It was based on the reaction developed by Sass *et al.* (1958) for the estimation of quaternary ammonium bases. After a step described as polymerization of the lipids, with cobaltous chloride and sodium periodate, sections are reacted with *cis*-aconitic anhydride. Violet complexes are formed with lipids containing choline (lecithin, sphingomyelin and lysolecithin). The method has a high degree of specificity and details are given in Appendix 12, p. 689.

The Gold Hydroxamate Method for Phosphoglycerides. Originally, this method was introduced by Adams and Davison (1959) as the ferric hydroxamate reaction. Staining intensity was poor. This method was modified by Gallyas (1963) who used silver reduction for demonstrating the hydroxamic acid formed by the hydroxylaminolysis of fatty acid esters. Adams *et al.* (1963) added a final stage of gold toning.

In its final form (Adams *et al*, 1963) the gold hydroxamate reaction demonstrates phosphoglycerides in a stable reddish-purple colour.

The Copper Phthalocyanin Methods. Copper Phthalocyanin (CuPC) dyes were first used in lipid histochemistry by Klüver and Barrera (1953, 1954) who employed Luxol Fast blue MBS (Dupont) for staining myelin sheaths in paraffin sections. This dye is the amine salt of a sulphonated CuPC which has the same basic structure as Alcian blue (see p. 698). Dyes of this type, which can be represented by the symbols (CuPC) SO_3H. Base, may have from 1 to 4 sulphonyl groups per molecule and one of a number of different bases. Their staining characteristics with regard to lipids depend not so much on the type of base employed as on their solubility in organic solvents. The nearest British equivalent to Luxol Fast blue is Methanol Fast blue 2G (I.C.I. Ltd) and this performs in an essentially similar manner.

Klüver and Barrera (1954) carried out a number of experimental studies on formalin-fixed sections, using Luxol Fast blue in combination with various counter-stains. They found that pretreatment of fixed sections with organic solvents never abolished staining. If these solvents were applied to unfixed

Table 28

In vitro Tests with Copper Phthalocyanins

Substance	Solvent	Result	Type of Ppt.	Time
Lecithin	Warm EtOH	ppt.	Dark Blue	Rapid
Kephalin	Warm $CHCl_3$	nil	—	—
Sphingomyelin	Warm $CHCl_3$/MeOH	ppt.	Dark Blue	Rapid
Cerebroside	Cold pyridine	ppt.	Paler Blue	Slow
Phosphoryl choline	30% EtOH	ppt.	Dark Blue	Slow

blocks of brain tissue, however, myelin was effectively prevented from staining. No attempt was made by Klüver and Barrera to explain the staining of myelin in chemical terms. I therefore tested (1955) a number of lipid materials *in vitro* with Luxol, and Methanol, fast blue with the results given in Table 28.

As a result of these findings I suggested (1955) that since formation of the characteristic dark blue precipitates was apparently confined to choline-containing compounds the following mechanism of staining (in alcoholic solutions) might apply.

$$\text{R-O-P}(=\text{O})\text{-O}\text{-}OCH_2CH_2N(CH_3)_3 + (CuPC)SO_3H.\text{Base} \longrightarrow$$

$$\text{R-O-P}(=\text{O})\text{-O}\text{-}OCH_2CH_2N(CH_3)_3O_3S(CuPC) + \text{Base.}$$

Salthouse (1962a, 1963) carried out further investigations on the solubility of CuPC-lipid complexes. He concluded that CuPC would stain all phospholipids provided that a solvent for the dye was chosen in which the complex was insoluble. An alternative hypothesis for the mechanism of CuPC staining might be that an ion association reaction takes place linking the Base-CuPc. SO_3H to the free oxygen atom of the phospholipid phosphate.

Effect of Neutral Red. Although Klüver and Barrera used neutral red as one of their counterstains they did not record any unusual effects. I found, however, that the dark blue CuPC-lipid complex in the tissues was converted by aqueous neutral red into a black or blackish-purple one. This effect could be produced *in vitro* but only in aqueous media in which the insoluble CuPC-lipid complexes had to be suspended. With alcoholic solutions of the dyes neutral red formed a black complex (or salt) of similar colour to the one mentioned above. These mechanisms were explained as follows:
(1) in alcohol,

$$(CuPC)SO_3H.\text{Base} + \overset{\oplus}{\text{Neutral red}}\ \overset{\ominus}{\text{Chloride}} \rightarrow (CuPC)\overset{\ominus}{SO_3}\overset{\oplus}{\text{Neutral red}} + \text{Base HCl}$$

and (2) in water,

$$\text{R-O-P}(=\text{O})\text{-O}\text{-}OCH_2CH_2N(CH_3)O_3S(CuPC) + \text{Neutral red chloride} \longrightarrow$$

$$(CuPC)\overset{\oplus}{SO_3}\overset{\oplus}{\text{Neutral red}} + \text{R-O-P}(=\text{O})\text{-O}\text{-}OCH_2CH_2N(CH_3)_3Cl$$

While the first of these two equations is probably true, if the second were also true it should be possible to recombine the liberated phospholipid with a further phthalocyanin molecule. No such effect could be produced and I believe, therefore, that the reaction of Neutral red is with SO_3H.Base groups left uncombined after formation of the CuPC-lipid complex.

It was always evident that CuPC would stain certain non-lipid tissue components such as nuclei, nucleoli, and calcium deposits though the latter, of course, provide no problems in distinction. For this reason further studies were carried out (Pearse and Almeida, 1958) with lipid and non-lipid materials either on filter paper or on Coujard-type slides. Alcoholic solutions of Luxol Fast blue or Methanol Fast blue stained all the lipids tested, including ganglioside, with the single exception of sphingomyelin which was dissolved out. Of the non-lipid substances only protamine base and protamine sulphate were stained. Almost certainly the reaction of these last two substances was due to their arginine content. The possibility of formation of a histidine complex must also be kept in mind. Arginine gives a dark blue precipitate with $(CuPC)SO_3H$. Bases, and with various sulphonated CuPC's in solution, but it is removed by processing in the paper and Coujard tests.

My previous (1960) view of the mechanism of CuPC staining of lipids and lipoproteins was as follows: (1) In fixed tissues (especially paraffin sections) lipo-proteins rather than lipids are responsible for staining. (2) The mechanism is that of an acid-base reaction with salt formation where the base of the lipoprotein replaces the base of the phthalocyanin. (3) With the majority of lipoproteins the reaction does not take place unless the dye is dissolved in a lipid solvent. (4) Sphingomyelin stains only when it is insoluble in the dye bath, as when chloroform is used for this. (5) The chemical basis of ganglioside staining remains an enigma. The constantly observed staining of the nucleoplasm of a small percentage of nuclei in most tissues, and of the nucleoli in others, may not be due to phospholipid. However, it is worth noting that Chayen and Gahan (1958) found in calf thymus nucleohistone the equivalent of 6 per cent. sphingomyelin (as $CHCl_3$-soluble nitrogen). Since the staining of elastic tissue with CuPC (Chapter 8, p. 225) has also been ascribed to the presence of sphingomyelin there may be some link between the two. In place of item 2 I now think that the ion association mechanism offers a better explanation of the facts.

It must be conceded that in spite of the expenditure of considerable effort it has been impossible to ascribe to the CuPC technique any specificity within the broad groups of lipids which are undoubtedly stained. Among other components which are strongly and distinctly stained in the tissues we must note the basic proteins and Mallory's alcoholic hyaline (Becker and Treurnich, 1959).

From the practical point of view the CuPC method nevertheless provides an excellent stain for myelin in formalin-fixed frozen sections and in paraffin sections after a variety of fixatives. As shown by Almeida and Pearse (1958), the sensitivity of the method for myelin lipids is high, being equal to that of the acid hæmatein method except, in developing rabbit brain, at the stage where sphingomyelin alone is present in the white matter. Developing myelin sheaths are shown in Fig. 114, and in Fig. 115, facing p. 423, the abnormal type of ganglioside stored in the neurones in cases of juvenile amaurotic idiocy can be seen to stain very strongly. Plate XIb, facing p. 415, shows the application of the method to the demonstration of phospholipid stores in the egg of a marine gastropod *Buccinum undatum*. Practical details of the method are given in Appendix 12, p. 698. Alternative dyes of the Luxol series were tested by Salthouse (1962a and b). These included Luxol Fast blue ARN (C.I. Solvent blue 37), Luxol Fast blue G, and Luxol Fast black L. All these dyes formed complexes with phospholipids and the two blue dyes were recommended for use as 0·1 per cent. solutions in absolute ethanol. Later Salthouse (1964) indicated the superiority of Luxol Fast blue G (C.I. Solvent blue 34), especially when used as a 0·1 per cent. solution in isopropanol, for the demonstration of myelin.

Tetrazolium Reduction Method for Phospholipids. Noting that after treatment with hydroquinone sections containing phospholipids would reduce tetrazolium salts, particularly MTT, Carmichael (1963) suggested that his observations formed the basis for a histochemical method. Hydrogen bond formation between the nitrogen of the base (choline, ethanolamine, serine) and hydroquinone was postulated as the mechanism of the reaction. At much the same time, and independently, Tranzer and Pearse (1963) suggested a similar mechanism but attributed it to the presence, in phospholipids or other lipids, of compounds of the ubiquinone and Vitamin E series. (See Chapter 23).

Uptake of Dyes by Phospholipids. The staining of phospholipids with basic dyes *in vitro* was investigated by Byrne (1962, 1963) who found that these dyes, without exception, were taken up by lecithin and kephalin. Weakly acid dyes such as aurantia, eosin Y and erythrosin were also taken up by phospholipids but not by cerebrosides. The author suggested that a chemical reaction was responsible, in part, for the uptake of basic dyes by conjugated lipids.

The staining properties of lecithin with acid dyes was considered by Riemersma and Booij (1962). They showed that Ponceau 4 RC was bound to the positively charged choline group.

Sinapius and Thiele (1965) tested the methylene blue binding capacity of 20 lipid substances *in vitro*. Only the acidic lipids were stained and the glycerophosphatides were found to stain down to pH 2·0. These authors drew attention to the need for strict control of the pH when staining fats with basic dyes (e.g. Nile blue). Copper mordanted mitochondria were stained selectively with methylene blue at pH 3·7 by Schabadasch (1958). The mechanism of this reaction may be due to the presence of cardiolipin but no suggestions on this point have been advanced.

METHODS FOR FATTY ACIDS

The Copper Method (Fischler). Benda (1900) first demonstrated the affinity of fatty acids for heavy metals by forming green copper salts with cupric acetate, and Fischler's (1904) technique, based on this principle, depended first on the formation of calcium soaps from fatty acids. Tissues were fixed in formalin containing calcium salts, frozen sections were cut and then mordanted in a saturated copper acetate solution. They were stained with Weigert's lithium hæmatoxylin and differentiated in a dilute solution of Weigert's borax ferricyanide mixture. According to Lison (1936), iron and calcium salts and "certain non-fatty tissue elements" were also coloured grey or black. Mallory (1938) pointed out that hæmoglobin was strongly stained but that this objection could be overcome by differentiating in borax-ferricyanide until the red cells were colourless. He apparently overlooked the fact that red cells might be stained because of their lipoprotein envelopes. Mallory also observed that errors due to calcium could be avoided by pretreatment with dilute HCl and those due to iron by pretreatment with 5 per cent. oxalic acid. Lison, however, maintained that the reaction was not specific for fatty acids and that Fischler's method should not be used in histochemistry for the demonstration of these substances.

An alternative method for fatty acids, making use of the same principle, was developed by Okamoto, Ueda and Kato (1944). In this variant of Fischler's technique the copper soaps resulting from the first stage of the reaction are made visible by means of *p*-dimethylaminobenzylidene rhodanine. Details of this method are given in Appendix 12, though I am doubtful of its efficacy. In several tests carried out on normal tissues I was unable to obtain a positive reaction.

Holczinger (1959) tested a variety of procedures designed to demonstrate free fatty acids in the tissues. He recommended a method depending on the production of a copper soap, removal of non-specifically bound copper with E.D.T.A. and demonstration of the copper with rubeanic acid (Chapter 27). Adams (1965) found this method to be absolutely specific for free fatty acids although Holczinger had obtained a positive reaction with lecithin and kephalin. Adams explained the latter effect, no doubt correctly, on the basis of lyso-phosphatide formation in Holczinger's phospholipid samples. This method is given in Appendix 12, p. 699.

A rather similar principle formed the basis of the Copper-Nadi and Copper-Hæmatoxylin reactions of Ishiwatari *et al.* (1960). Mironescu (1965) found that he could obtain reactions with the latter type of technique even after methanol/ether extractions. He stated his belief that nuclear and cytoplasmic nucleic acids were responsible. While it is true that the original Fischler test can demonstrate substances other than fatty acids, the copper-rubeanic acid method of Holczinger does not do so. It is therefore the method of choice.

Methods for Cholesterol and its Esters

The Liebermann-Burchardt Reaction. This reaction, considered at one time to be specific for cholesterol and cholesterol esters, was considered by Bierry and Gouzon (1936) to be a general reaction for unsaturated steroids and by Everett (1947) to indicate the presence of diols formed by mild oxidative procedures. Boscott and Mandl (1949), however, applied the reaction to pure samples of dehydro-iso-androsterone, progesterone and deoxycorticosterone acetate without obtaining the characteristic blue-green colour. They obtained instead colours varying from yellow to orange.

Kent (1952) tested 24 compounds, including cholesterol and cholesterol esters, by the Liebermann-Burchardt test and by the Schultz histochemical variant. All the compounds tested produced a reaction with the former but only in the case of linoleic acid, carotene, vitamin A and ergosterol could the blue-green colour be confused with that given by cholesterol and its esters. With the Schultz test only carotene produced the same blue colour as cholesterol and its esters. In practice, therefore, the Schultz test can be regarded as highly specific for cholesterol and cholesterides although its specificity is not absolute.

The original histochemical modification, which was described by Schultz (1924, 1925) and by Schultz and Löhr (1925), necessitated previous oxidation of the cholesterol either by exposure to light for at least four days or by treatment with 2·5 per cent. iron alum at 37° for 2 days. Sections were then mounted on slides and treated with a mixture of equal parts of glacial acetic acid and concentrated sulphuric acid. They were examined in this medium, under a coverslip. Previous oxidation was omitted in the modification introduced by Romieu (1927). In this technique, the action of a few drops of concentrated sulphuric acid was arrested by the addition of a similar amount of acetic anhydride. After washing with the latter reagent, the preparation was examined under a coverslip. The initial blue, violet or red colour resulting from either of these modifications of the Liebermann-Burchardt reaction is not diagnostic. If cholesterol or cholesterol esters are present the colour changes after a few seconds to bluish-green, and this colour alone is to be regarded as specific. After a longer period the whole section becomes brown. It is essential that the reagents used for this test should be as pure as possible. Most authors now regard the stage of oxidation and mordanting with iron alum as essential. Baker and Selikoff (1952) carried out this stage for 3 days before proceeding to the sulphuric acid—acetic anhydride stage and Weber *et al.* (1956) mordanted their sections for 7 days, in buffered ferric ammonium sulphate. They found that a final pH of 2 gave the best results, and their modification appears in Appendix 12, p. 702.

After investigating the specificity of the Schultz reaction for steroids Lewis and Lobban (1961) devised a further modification of the reaction. Formalin-fixed sections were treated with 80 per cent. aqueous sulphuric acid containing 0·5 per cent. iron alum. Cholesterol itself gave no colour unless previously oxidized but an intense blue-green colour was produced by steroids related to testosterone and a pink or mauve colour was produced with œstrogens. Adams (1965), however, observed that even oxidized cholesterol continued to give a pink colour.

A slightly different type of reaction for cholesterol was used by Okamoto, Shimamoto and Sonoda (1944). This depended on the development of a green or bluish-green colour when formalin-fixed sections were treated with a 30 per cent. solution of sulphuric acid containing iodine and potassium iodide. Preparations made by this method are no more stable than with the various Schultz reactions but the reagent is less damaging to the sections. Its specificity has not been tested, to my knowledge.

Histochemical distinction between cholesterol and its esters was described by Feigin (1956) who designed a method based on the insolubility of the cholesterol-digitonin complex in ethanol/ether, in which the esters of cholesterol are soluble. He used a standard Schultz procedure for the demonstration of the digitonide. The reaction was thoroughly tested by Schnabel (1962, 1964) who proposed a number of modifications, observing that cholesterol digitonide was somewhat soluble in ethanol/ether, especially if excess solvent was present. He also observed that the digitonide was less amenable to subsequent oxidation with iron alum. Schnabel's modification of Feigin's method is given in Appendix 12, p. 703.

The Perchloric Acid-Naphthoquinone Method (PAN). An entirely new principle was invoked by Adams (1961) in the development of the PAN method for cholesterol. This is shown in the equation below:

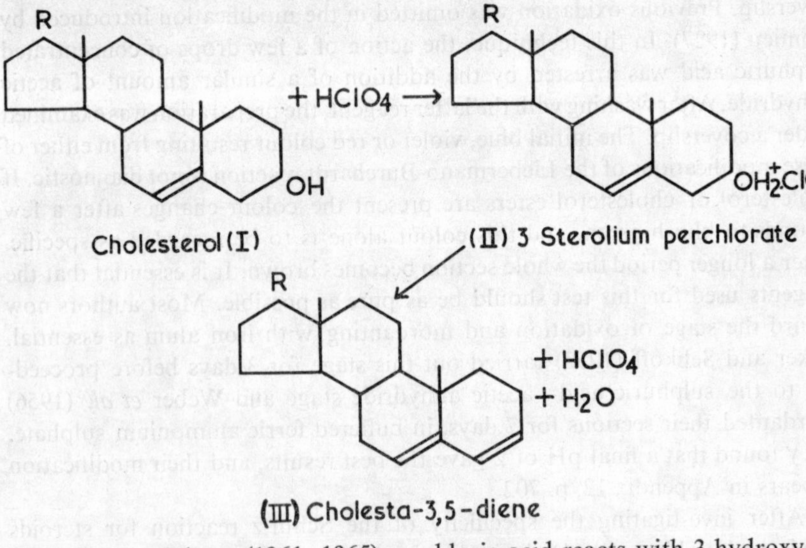

R
$+ HClO_4 \longrightarrow$
OH
Cholesterol (I)
R
$OH_2^+ClO_4^-$
(II) 3 Sterolium perchlorate
R
$+ HClO_4$
$+ H_2O$
(III) Cholesta-3,5-diene

According to Adams (1961, 1965) perchloric acid reacts with 3-hydroxy-Δ^5 steroids (I) to form 3-sterolium salts (II) which are insoluble. When excess perchloric acid is present a molecule of water is eliminated with the production of cholesta-3,5-diene (III). The subsequent reaction with naphthoquinone produces a dark blue-grey product but the mechanism is unknown (Fig. 116).

The specificity of the reaction is high. Only cholesterol and related steroids give the characteristic blue colour under histochemical conditions. A pink colour may occur with tryptophan and some mucosubstances.

Details of the PAN method are given in Appendix 12, p. 704.

The Bismuth Trichloride Method. A method for demonstrating cholesterol was reported by Grundland, Bulliard and Maillet (1949) which made use of the colour developed when tissues fixed in an alcoholic digitonin solution were treated for 24 hours with bismuth trichloride in anhydrous nitrobenzene. This was adapted from the colorimetric method of Clark and Thompson (1948) for steroids. After development of the colour the tissues were cleared in a bath o glycerol monostearate, embedded in paraffin and sections were cut in the usual manner. This method seemed to possess a definite advantage over the more commonly employed Liebermann-Burchardt procedure, in that it was far less destructive to the tissues. For this reason localization would be expected to be more accurate, but the specificity of the method (given in Appendix 12) could not be established and my own results were always patchy and lacking in clarity.

METHODS FOR ACIDIC LIPIDS

Staining with Nile Blue Sulphate. According to Lorrain Smith (1908) his Nile Blue sulphate method stained neutral fats red, fatty acids dark blue and nuclei, elastic tissue and cytoplasm pale blue. The essential part of this method was the preparatory hydrolysis of the blue dye salt with dilute sulphuric acid. This process yields a small percentage of the free base (oxazine) which is red, and of the oxazone derivative (Nile Red) which has a similar colour. Hydrolysed solutions of Nile Blue, therefore, consist partly of the blue salt and partly

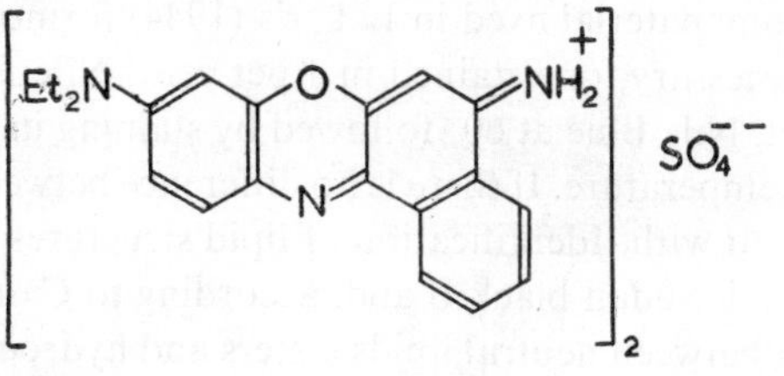

$$\left[Et_2N\text{–(benzophenoxazine)}=\overset{+}{N}H_2 \right]_2 \; SO_4^{--}$$

Nile Blue sulphate

of the red free base and the red oxazone. Lorrain Smith considered that the red base dissolved in glycerides while the blue salt dissolved in the fatty acids Lison (1936), however, concluded that a rose colour by the Nile Blue sulphate method certainly indicated the presence of unsaturated glycerides, while a blue colour had little or no significance and did not even signify the presence of fat. The method as originally proposed was not altogether satisfactory and interpretation of the results, especially of the blue staining part, was difficult. Some authors, Knaysi (1941) for instance, demonstrated that the hydrolysis of fat by bacteria could be shown equally well by other dyes such as neutral red or methylene blue.

Mechanism of Staining. The experimental basis of the Nile Blue sulphate method has been more firmly established by the work of Cain (1947b). This author discussed the results of Kaufmann and Lehmann, who used Nile Blue sulphate on various lipid substances incorporated in elder pith, and he concluded: (1) That substances in the solid state, except "greases", do not colour with aqueous Nile Blue solutions in any way whatsoever. (2) That triglycerides, or mixtures of them, free or dissolved in hydrocarbons, are coloured red by the oxazone and much less so by the free base (oxazine). (3) That fatty acids, if liquid, colour blue with solutions of the oxazine (by forming salts) and blue with dilute Nile Blue sulphate solutions. With concentrated Nile Blue solutions, on the contrary, only a slight change towards blue is seen except with oleic acid which colours fairly strongly. (4) Lecithin (perhaps all phosphatides) stains deep blue, when solid, with the oxazine or with concentrated or dilute aqueous Nile Blue solutions. It is not affected by the oxazone. (5) Cholesterol in solution does not stain at all with Nile Blue.

Cain's Technique using Nile Blue Sulphate. Cain evolved a new technique with Nile Blue sulphate based on the following observations. Neutral lipids will dissolve out of aqueous solutions of Nile Blue only the oxazone and free base (both red in colour). Acidic lipids will dissolve the oxazone and combine with the free base to form blue lipid-soluble compounds. As Lison demonstrated, 1 per cent. or stronger Nile Blue solutions are not hydrolysed and Cain found that only lecithin and oleic acid would colour blue to any extent with 1 per cent. Nile Blue. In order to make full use of the histochemical possibilities of Lorrain Smith's method one must employ both the oxazone and the oxazine and the lipids under investigation must be liquid. The new technique employs frozen sections from material fixed in Baker's (1944) formol-calcium mixture, postchromed if necessary, one stained in 1 per cent. Nile Blue at 60° and the other in 1 per cent. Nile Blue at 60° followed by staining in 0·02 per cent. Nile Blue at the same temperature. If there is no difference between the two sections the first is dispensed with. Identification of lipid structures is made in control sections stained with Sudan black B and, according to Cain, the new method allows distinction between neutral lipids (esters and hydrocarbons) and acidic lipids (lecithins). It cannot be used for the demonstration of fatty acids. Cain's Nile Blue method, which is a useful one under some circumstances, appears in Appendix 12, p. 697. Its validity is supported by the observations of Adams (1965).

Menschik's Method using Nile Blue Sulphate. Menschik (1953) introduced into the technique a series of differentiations which he claimed could make the technique specific for phospholipids. The first of these, with acetone at 50° was followed by a second stage of differentiation with weak acid since the author had observed that Nile Blue-stained phospholipids were weakly acid-fast, while similarly stained proteins were not. Menschik's tests for the specificity of the method included staining of the pure substances, with and without extraction with lipid solvents, and he concluded that it was as specific as the acid hæmatein method. It is certainly far easier to carry out. Whether the Nile Blue method is capable of demonstrating protein-bound phospholipid is open to doubt. According to Singh (1964a) Menschik's technique fails to distinguish phospholipids in avian nervous tissues. Details are nevertheless given in Appendix 12.

Acid Solochrome Cyanine. This dye has been described in Chapter 6, p. 113 (see also Chapter 27). It is an indicator dye which stains acidic proteins (nucleoproteins) blue and basic proteins red. In mammalian tissues it does not stain mucins by virtue of their acidic groups but only by virtue of the available basic groups of the protein. They therefore stain in shades of pink to red. In unfixed frozen sections, and in sections briefly fixed in ethanol or acetic-ethanol, the lipids of the myelin sheath stain dark blue in acid solutions of the dye (pH 2·1). When the sections are extracted with lipid solvents, such as light petroleum and chloroform-methanol, or if paraffin sections are stained, the myelin sheaths stain red. This they do presumably by virtue of their protein component which may or may not be neurokeratin. Fig. 117, facing p. 468, shows the staining of acidic lipid with acid solochrome.

Metachromatic Methods

Feyrter's Mounting-staining Method. Feyrter (1936) evolved his tartaric acid-thionin mixture, in the first instance, for demonstrating the metachromasia of myelin sheaths in frozen sections. The method has since become widely employed on the Continent (see Pischinger, 1943; and Pretl, 1948), where it is apparently regarded as specific for lipins in the sense described earlier in this chapter. The method is carried out by mounting formalin-fixed frozen sections on slides (with egg albumin and gentle heat if necessary) and covering them with Feyrter's acid-thionin mixture. A coverslip is then applied and the preparation is ringed with vaseline to prevent evaporation. After an interval of 24 hours or more, which may be shortened by the application of moderate heat (60°), lipid materials other than neutral fat are coloured deep rose pink. They are examined with the stain *in situ*, without removing the coverslip. The word "Einschluss", used in respect of Feyrter's method by Continental authors, refers to the process of covering and ringing and not, as might easily be supposed from the common translation "inclusion", to the particular bodies which are stained by the method. Even the translation "enclosure" is imperfectly descriptive and it has been suggested to me that "Mounting-Staining" might be a better and more acceptable term. I have therefore adopted it in the heading to this section. Hamperl (1950) gives some additional details of the application of the method particularly to the demonstration of lipid substances present in cells known as oncocytes, originally described by him in salivary, pituitary and thyroid glands, and in various other organs. There is no doubt that phosphatides are the principal lipid substances demonstrated by Feyrter's method and this is presumably on account of the capacity of the phosphoryl group to induce metachromasia. Distinction of Feyrter-positive lipid from non-lipid substances giving metachromasia with thionin is not usually a matter of difficulty. Interference due to the presence of small quantities of mucopolysaccharide may make it difficult to use the method for detection of small quantities of lipid combined, for instance, as lecithoproteins or other protein-phosphatide complexes.

No critical studies of the Feyrter technique appear to have been carried out in recent years although Brante (1957), while regarding the technique as "more or less specific for glycolipids or lipopolysaccharides" showed that gangliosides would produce a small γ-band in high concentrations with toluidine blue. He further showed that 0·5 per cent ganglioside in gelatin readily produced metachromasia with Feyrter's reagent. I certainly agree that a positive Feyrter reaction can be obtained on the stored material in most of the cerebral lipidoses whatever its supposed composition and regardless of widely differing responses to other histochemical tests. It is possible that the Feyrter metachromasia of nervous tissues does not run exactly parallel to their toluidine blue metachromasia which Landsmeer (1952), and Landsmeer and Giel (1956), regarded as due entirely to phosphate groups. Although the part played by sulpholipids is probably a small one it must not be overlooked entirely.

Heparin Precipitation Method. A method for the precipitation of serum β-lipoproteins, using heparin in the presence of calcium at low ionic concentration, was described by Burstein and Samaille (1958). This was adapted by Mustakallio and Levonen (1964) to form the basis of a histochemical method. Details are given in Appendix 12, p. 701. The method cannot be used for the investigation of lipids in tissues where large amounts of endogenous mucopolysaccharides giving pink γ-metachromasia are present.

Reduction Methods

The Iodine Cyanide Method. A somewhat complex method, stated by the author to be specific for phosphatides, was described by Alsterberg (1941). This method depended on the premise that the strong oxidizing agent iodine cyanide (cyanogen iodide, ICN) gave no precipitate with silver salts unless a reducing agent was present at the same time. Lecithins and kephalins acted as reducing agents in this respect and produced a precipitate of silver iodide and silver cyanide. Other substances were not demonstrated to possess this property. In the application of the method unfixed blocks of tissue, after incubation in an iodine cyanide-silver chlorate solution for 3 days, were embedded by the combined celloidin-paraffin method and sections brought to xylene were treated with H_2S in xylene to convert the silver precipitates to brown sulphides.

Alsterberg's work was done entirely on animal brain tissues and the histological results obtained were in some cases excellent. Although I have no practical experience of the method, it seems likely, from a study of Alsterberg's experiments, that many factors influenced the deposition of silver salts which were independent of the nature of the substances present. Differences in pH between centre and periphery of the blocks, for instance, led to staining variations in the final result. The difficulty of making both iodine cyanide and silver chlorate has prevented the use of the method, even for research on the central nervous system; it is given a place in this section solely on account of its theoretical interest, as illustrating the employment of a new principle in the histochemical demonstration of phosphatides. The application of the method to electron microscopy might, perhaps, be rewarding.

Oxidation Methods for Unsaturated Lipids

The Periodic Acid-Schiff (PAS) Reaction. The reaction of periodic acid with lipid substances has already been considered in Chapter 10 and will not be further considered at this point. Due to the very wide specificity of the reaction it cannot be used primarily for the demonstration of unsaturated lipids in the phosphatide class, but it may draw attention to the presence of these when used for other purposes.

The Performic Acid-Schiff (PFAS) and Peracetic Acid-Schiff (PAAS) Reactions. The reaction of keratin, after oxidation with performic and peracetic acids, with Schiff's reagent and, *inter alia*, with methylene blue and Alcian blue, has already been described in Chapters 6 and 8. Here the application of the performic and peracetic acid-Schiff techniques to lipids present in the tissues is considered.

It was noted in the early stages of some work on keratin (Pearse, 1951a) that the red cell envelopes constantly gave a positive reaction and, subsequently, a number of other structures were observed to react. Among these were myelin sheaths and certain intracellular granules contained usually, but not invariably, in macrophages. The latter were considered to belong to the lipofuscin group (see Chapter 26) and the presence of lipid was the factor common to all. Materials which constantly give a positive PAS reaction, such as epithelial mucins, glycogen, cartilage, pituitary β granules and the basement membranes of the kidney, were PFAS-negative.

Since Verne (1929) had shown that mild oxidation of fatty substances containing unsaturated groups, such as lecithin for instance, would produce groups having all the reactions of aldehydes it seemed probable that a similar mechanism was responsible for the positive PFAS reaction in known lipid-containing structures. Various experiments were carried out in order to test this hypothesis, and Table 29 (Pearse, 1951b) gives the reactions obtained with red cell envelopes and myelin sheaths together with those of hair shafts (keratin) for comparison. Figs. 118 and 119 (facing p. 468) illustrate some of the results obtained with performic acid oxidation of the first two structures. In the first, the groups produced in red cell envelopes have been demonstrated by means of an alkaline silver solution and in the second, the groups in myelin have been shown by performance of the NAHD routine (see Appendix 13, p. 705).

TABLE 29

Reactions of Oxidized Lipids and Keratin

STRUCTURES	TESTS				CONTROLS	
		Performic acid			Formic acid	H_2O_2
	Schiff	NAHD	Meth. blue pH 2·6	Ammon. silver	Schiff	Schiff
Red cell envelope	+	+	−	+	+	−
Hair shaft	+	+	+	−	−	−
Myelin sheath	+	+	−	+	weak +	−

The reactions used were performic acid oxidation, followed by Schiff's reagent, by the NAHD routine (2-hydroxy-3-naphthoic acid hydrazide followed by diazotized *o*-dianisidine), by methylene blue at pH 2·6 and by an ammoniacal silver solution. The first and second reagents demonstrate aldehydes and some other groups, the third demonstrates strongly dissociating acid groups and the fourth reducing groups including aldehyde.

Since formic acid is certainly, and free H_2O_2 probably, present in the performic acid solution in these tests, control sections were treated with the two reagents separately. With 10 per cent. H_2O_2 negative results were obtained, but with 98 per cent. formic acid a weak positive reaction occurred in myelin sheaths and a strong one in the red cell envelope. The reaction with myelin was entirely reversed by fixation in boiling chloroform-methanol for 8–16 hours, but traces of a positive reaction remained in the red cells even after this procedure, which is presumed to remove all but the most closely bound lipid from small blocks of tissue. Both red cell envelopes and myelin sheaths reduce ammoniacal silver solutions after performic acid oxidation, while keratin in tissue sections fails to do so. According to Cain (1949), aldehydes are the reducing groups most likely to be present in oxidized lipids and the reaction with ammoniacal silver was considered to have this significance. The peroxides and hydroperoxides, which Cain considered to be responsible for the plasmal reaction (Chapter 13, p. 460) of certain lipid structures, should give a positive reaction with the "Nadi" reagent (Chapter 19) but this effect was not seen in performic acid-oxidized sections. Lillie (1952) working independently on the PFAS reaction in the case of ceroid pigment (Chapter 26, Vol. 2) considered that the positive result with Schiff's reagent was due to the oxidation of the ethylene groups of unsaturated lipids to aldehydes. He gave the following formula to explain the reaction:

$$-HC{=}CH- + 2HCO.O.OH \rightarrow HC\overset{O}{\underset{O}{\diamond}}CH + 2HCOOH$$

and, by rearrangement, two aldehyde groups $(-HC{=}O)_2$ are produced. Lillie showed that prior acetylation, which prevents the reaction of 1:2-glycols with periodic acid, had no effect on the PFAS reaction of ceroid whereas bromination ($HC{=}CH \rightarrow BrHC.CHBr$), which has no effect on the oxidation of glycols by periodic acid, reversed the PFAS reaction completely. He concluded that ceroid contains both ethylene and 1:2-glycol groupings. According to Fieser and Fieser (1956), however, stereochemical studies indicate that after initial epoxide formation acetolysis with excess parent acid takes place to form the ester of the glycol. The formate so produced from performic acid has an aldehyde group which can react with Schiff's solution.

$$R'(H)C{=}C(H)R'' \xrightarrow{HCO_3H} R'(H)C\overset{O}{—}C(H)R'' \xrightarrow{HCO_2H} R'{-}C(OH)(H){-}C(H)(C(H){=}O){-}R''$$

The PFAS reaction cannot be considered as a routine method for the demonstration of phospholipids (and other lipids in frozen sections) containing unsaturated C=C bonds. Its chief uses, at present, are as a control method for the PAS reaction and as an instrument of research into the mechanism of fat oxidation. A suggested method for carrying out the PFAS reaction on tissue lipids, together with Lillie's method for this purpose appear in Appendix 12. If a tissue component in which the PAS reaction is positive also gives a positive PFAS reaction, it is necessary to determine whether the former is due to the 1:2-glycol grouping. According to Lillie, therefore, the PAS reaction is carried out after (1) acetylation and (2) bromination. A negative reaction in the first case, with a positive reaction in the second, indicate 1:2-glycol.

The U.V.-Schiff Reaction. Belt and Hayes (1956), after carrying out a number of experiments on the Schiff's reactivity of tissue lipids, evolved a method in which thin blocks are briefly fixed in formalin and then infiltrated with gum arabic. Frozen sections were subjected to long and short wave (254 nm) UV irradiation for 3–4 hours and then placed in Schiff's solution. The authors claimed that "there was considerable evidence that this reaction demonstrated double bonds" and I agree with this view although complete proof of specificity is difficult to obtain. Studying the release of enzymes from irradiated lysosomes Wills and Wilkinson (1966) provided evidence that irradiation (up to 750 rads/min) caused oxidation of the double bonds of unsaturated fatty acids to lipid peroxides.

Using a 250-watt B.T.H. Mercra lamp and 2–4 hours' exposure, strong staining (with negative controls) was produced in the intestine of the larva of the waxmoth *Galleria melloneila* which has been shown by Przelecka (1956) to contain large quantities of phospholipid. Other cytochemical evidence (osmium tetroxide, bromination, PFAS) suggests that these phospholipids contain a considerable quantity of unsaturated fatty acids. Details of the method of Belt and Hayes are given in the Appendix (p. 701). It is particularly suitable for application to fresh cold microtome sections.

Bromination Methods for Unsaturated Lipids

The Bromine-silver Nitrate Method. This method was described by Mukherji *et al.* (1960). It is based on the combination of the bromine with unsaturated lipids at the site of double bonds:

$$—CH{=}CH— + Br_2 \rightarrow —CHBr—CHBr—$$

In the presence of alcohol silver nitrate (as a silver ammonia complex) reacts with the brominated lipid:

$$—CHBr—CHBr— + 2AgNO_3 \rightarrow 2CHO + 2AgBr + 2NO_2$$

and

$$—CHO + AgNO_3 \rightarrow COOH + Ag + NO_2$$

The free aldehyde reduces another molecule of silver nitrate to metallic silver, as indicated. Finally the silver bromide is reduced *in situ* to metallic silver:

$$2AgBr + HCHO + H_2O \rightarrow 2Ag + 2HBr + HCOOH$$

Mukherji *et al.* used an alkaline silver nitrate solution and they claimed that neither glycogen nor DNA gave any reaction. Proteins, however, were stained to some extent.

The superiority of bromination to iodination is quite clear. The latter is too slow to be effective. According to Critchfield (1959), however, α,β-unsaturated acids and esters react slowly even with bromine. After conversion to their sodium or potassium salts the addition of bromine proceeds rapidly. Prior saponification of tissue sections has not been tested.

A modification of the bromine-silver method was given by Norton *et al.* (1962). These authors used a KBr/Br_2 solution for bromination and acidified silver nitrate. A photographic reducer was finally employed. Both methods are given in Appendix 12. According to Adams (1965) the method of Norton *et al.*

has the advantage that it does not stain proteins. It fails to react with double bonds of hydrophilic lipids (phospholipids and glycosphingosides), however, and thus has limited application.

Reactions for Sugar Components

Modified Molisch Reaction (Diezel, 1954). In his studies on the lipid storage diseases Diezel employed a modification of the Molisch reaction for carbohydrates which gave a red colour with hexose-containing compounds (cerebrosides and gangliosides). Details of the histochemical test are given in the Appendix (p. 688). This is a heroic type of procedure, reminiscent of the older microchemical tests which had to be interpreted in the few remaining fragments of the section. If Diezel's instructions are carried out, however, the sections remain surprisingly intact.

TABLE 30

Recommended Methods for Lipids and Lipoproteins

Routine (R) or Research (S)	Lipid Component	Type of Section	Fixation	Method
R	Neutral fats	Frozen	Formalin Formal-calcium	Oil Red O in isopropanol (Lillie) Fettrot in propylene glycol
S	Neutral fats	Frozen	Formalin Formol-calcium	Colloidal Sudan III (Govan) Phosphine 3 R (Popper)
R	Phospholipids	Frozen	Formalin	Sudan black B in alcohol
		Frozen	Formol-calcium	Nile Blue sulphate (Cain)
		Paraffin	Formalin	Sudan black B in alcohol
		Paraffin	Controlled chromation	Hæmatein, controlled by Sudan black B
S	Phospholipids	Cold microtome	Controlled chromation	Sudan black, Nile Blue, acid hæmatein, copper phthalocyanin
R	Phospho- and glycolipids	Paraffin	Various	Copper phthalocyanin
R	Glycolipids	Frozen or paraffin	Various	Periodic acid-Schiff
R & S	Lipoproteins	Cold microtome	Dichromate	Copper phthalocyanin
S	Lipoproteins	Frozen	Formalin	Benzpyrene-Caffeine (Berg) Performic acid-Schiff
R & S	Unsaturated fatty acids (lipid-bound)	Frozen or paraffin	Formalin	Br-Silver, OsO_4, OTAN
S	Ethylenes	Cold microtome	Nil, or brief ethanol	U-V. Schiff
R & S	Acidic lipids	Cold microtome	Nil, or brief fixation (various)	Acid solochrome cyanine. Mounting-staining (Feyrter)
S	Choline lipids	Frozen	Formalin	CAZA
R & S	Phospho-glycerides	Frozen	Formal-calcium	Gold Hydroxamate
R & S	Sphingo-myelin	Frozen	Formal-calcium	NaOH-OTAN NaOH-Hæmatein
R & S	Fatty acids	Unfixed (Cryostat)	—	Cu-Rubeanic Acid
R & S	Cholesterol	Frozen	Formalin Formal-calcium	PAN

Modified Brückner Reaction. This reaction (Brückner, 1943), which is positive *in vitro* with pentoses and hexoses, was also modified by Diezel (1954) for use on tissue sections. A red colour is given by cerebrosides and gangliosides but mucopolysaccharides do not react. Details of the histochemical test are given on p. 689.

Modified Roe and Rice Test for Pentoses. A reaction specific for pentoses was developed by Roe and Rice (1948) in which carbohydrates of different types cause little or no interference. Applied by Diezel (1954) to tissue sections the reaction was negative with cerebrosides and gangliosides but it gave a red colour with sections of plant seeds containing pentoses. Details are given in the Appendix, p. 689.

Recommended Methods for Various Types of Lipid

A number of methods for neutral fats, fatty acids, and phospholipids have been considered in this chapter. It is concluded that by staining methods alone absolute distinction between the various groups of lipids is still sometimes a matter of great difficulty. Recommended methods are given in Table 30 (p. 441).

The suggestions are given entirely without reference to the various extraction techniques,which can be used in conjunction with most of them. As stated already in this chapter, there is no reason to discontinue the use of extraction techniques. There is at present, however, a need for great caution in accepting their effects as in any way similar to those obtained *in vitro*. Until parallel chemical studies have been carried out for each histochemical extraction technique interpretations must continue to be made with reserve.

REFERENCES

Ackerman, G. A. (1952). *Science*, **115**, 629.
Adams, C. W. M. (1958). *J. Neurochem.*, **2**, 178.
Adams, C. W. M. (1959). *J. path. Bact.*, **77**, 648.
Adams, C. W. M. (1960). *J. Histochem. Cytochem.*, **8**, 262.
Adams, C. W. M. (1961). *Nature*, **192**, 331.
Adams, C. W. M. (1965). In "Neurohistochemistry," Ed. C. W. M. Adams. Elsevier, Amsterdam, p. 6.
Adams, C. W. M., and Davison, A. N. (1959). *J. Neurochem.*, **3**, 347.
Adams, C. W. M., Bayliss, O. B., and Ibrahim, M. Z. H. (1963). *J. Histochem. Cytochem.*, **11**, 560.
Adams, C. W. M., Abdulla, Y. H., Bayliss, O. B., and Weller, R. O. (1965). *J. Histochem. Cytochem.*, **13**, 410.
Almeida, D. F., and Pearse, A. G. E. (1958). *J. Neurochem.*, **3**, 132.
Alsterberg, G. (1941). *Z. Zellforsch.*, **31**, 364.
Ansell, G. B., and Hawthorne, J. N. (1964). "Phospholipids." BBA Library, Vol. 3. Elsevier, Amsterdam.
Baker, J. R. (1944). *Quart. J. micr. Sci.*, **85**, 1.
Baker, J. R. (1946a). *Ibid.*, **87**, 1.
Baker, J. R. (1946b). *Ibid.*, **87**, 441.
Baker, J. R. (1947). *Ibid.*, **88**, 463.
Baker, R. D., and Selikoff, E. (1952). *Amer. J. path.*, **28**, 573.
Becker, B. J. P., and Treurnich, D. S. F. (1959). *Stain Tech.*, **34**, 261.
Belt, W. D., and Hayes, E. R. (1956). *Stain Tech.*, **31**, 117.
Benda, C. (1900). *Virchows Arch.*, **161**.
Bennett, H. S. (1939). *Proc. Soc. exp. Biol., N.Y.*, **42**, 786.
Benson, A. A., and Strickland, E. H. (1960). *Biochim. Biophys. Acta*, **41**, 328.
Berenbaum, M. C. (1954). *Nature, Lond.*, **174**, 190.

Berenbaum, M. C. (1958). *Quart. J. micr. Sci.*, **99**, 231.
Berg, N. O. (1951). *Acta path., Scand.*, Suppl., 90.
Bermes, E. W., Jr., and McDonald, H. J. (1956). *Fed. Proc.*, **15**, 220.
Bermes, E. W., Jr., and McDonald, H. J. (1957). *Arch. Biochem.*, **70**, 49.
Bierry, H., and Gouzon, B. (1936). *C.R. Acad. Sci., Paris*, **202**, 686.
Blix, F. G., Gottschalk, A., and Klenk, E. (1957). *Nature, Lond.*, **179**, 1088.
Bloor, W. R. (1943). "Biochemistry of the Fatty Acids." New York.
Boelsma-van Houte, E. (1965). "Histochemie van Fosfolipiden in Verbrand met Atherosklerose van de Aorta." Thesis, Leiden.
Boscott, R. J., and Mandl, A. M. (1949). *J. Endocrinol.*, **6**, 132.
Böttcher, C. J. F., and Boelsma-van Houte, E. (1964). *J. Atherosclerosis Res.*, **4**, 109.
Bourgeois, C., and Hack, M. H. (1962). *Acta histochem.*, **14**, 297.
Bourgeois, C., and Hubbard, B. (1965). *J. Histochem. Cytochem.*, **13**, 571.
Brante, G. (1949). *Acta Physiol. Scand.*, **18**, Suppl. 63.
Brante, G. (1951). *Fette und Seifen*, **53**, 457.
Brante, G. (1952). *Proc. 1st Internat. Congr. Neuropath.* Elsevier, Amsterdam.
Brante, G. (1957). In "Cerebral Lipoidoses." Blackwell, Oxford, p. 164.
Brockerhoff, H. (1961). *Arch. Biochem.*, **93**, 641.
Brolin, S. (1952). *Acta Soc. med. Upsala*, **57**, 33.
Brückner, J. (1943). *Z. physiol. Chem.*, **277**, 181.
Brunswick, O. (1922). *Z. wiss. Mikr.*, **39**, 316.
Burt, N. S., McNabb, A. R., and Rossiter, R. J. (1950). *Biochem. J.*, **47**, 318.
Burstein, M., and Samaille, J. (1958). *Presse méd.*, **66**, 974.
Byrne, J. M. (1962). *Quart. J. micr. Sci.*, **103**, 47.
Byrne, J. M. (1963). *Ibid.*, **104**, 441.
Cain, A. J. (1947a). *Ibid.*, **88**, 383.
Cain, A. J. (1947b). *Ibid.*, **88**, 467.
Cain, A. J. (1949). *Ibid.*, **90**, 411.
Cain, A. J. (1950). *Biol. Rev.* (*Trans. Camb. phil. Soc.*), **25**, 73.
Carmichael, G. G. (1963). *J. Histochem. Cytochem.*, **11**, 738.
Carter, H. E., Glick, F. J., Norris, W. P., and Philips, G. E. (1947). *J. biol. Chem.*, **170**, 285.
Casselman, W. G. B. (1954a). *Quart. J. micr. Sci.*, **95**, 321.
Casselman, W. G. B. (1954b). *Biochim. Biophys. Acta*, **14**, 450.
Chayen, J., and Gahan, P. B. (1958). *Biochem. J.*, **69**, 49P.
Chiffelle, T. L., and Putt, F. A. (1951). *Stain Tech.*, **26**, 51.
Ciaccio, C. (1909). *Anat. Anz.*, **35**, 17.
Ciaccio, C. (1934). *Boll. Soc. ital. Biol. sper.*, **9**, 137.
Ciaccio, C. (1956). *Bull. Histol. Tech. micr.*, **4**, 97.
Clark, L. P., Jr., and Thompson, H. (1948). *Science*, **107**, 429.
Critchfield, F. E. (1959). *Analyt. Chem.*, **31**, 1406.
Daddi, L. (1896). *Arch. ital. Biol.*, **26**, 143.
Dawson, R. M. C. (1957). *Biol. Rev.*, **32**, 188.
Dawson, R. M. C. (1966). In "Essays in Biochemistry," Vol. 2, p. 68. Acad. Press, London.
Deuel, H. R. (1951). "The Lipids; Their Chemistry and Biochemistry." Interscience, New York.
Dietrich, A. (1910). *Ver. dtsch. path. Ges.*, **14**, 263.
Diezel, P. B. (1954). *Virchows Arch.*, **326**, 89.
Diezel, P. B. (1957). In "Cerebral Lipoidoses." Blackwell, Oxford, p. 11.
Diezel, P. B., and Neimanis, G. (1957a). *Naturwiss.*, **20**, 560.
Diezel, P. B., and Neimanis, G. (1957b). *Virchows Arch.*, **330**, 619.
Domagk, K. (1933). Quoted by Romeis, 1948.
Douglas, B. E. (1956). In "Chemistry of the Coordination Compounds." Edit. Bailar Rheinhold, New York, p. 487.
Edgar, G. W. F. (1956). "Myelination Studied by Quantitative Determination of Myelin Lipids." Thesis, Amsterdam.
Edgar, G. W. F., and Donker, C. H. M. (1957). *Acta neurol, psychiat. belg.*, **5**, 451.

ELFTMAN, H. (1954). *J. Histochem. Cytochem.*, **2**, 1.
ELFTMAN, H. (1958). *Ibid.*, **6**, 317.
ERÄNKÖ, O. (1950). *Nature, Lond.*, **165**, 116.
EVERETT, J. W. (1947). *Endocrinology*, **41**, 366.
FEAGLER, J. R., and MCMANUS, J. F. A. (1964). *J. Histochem. Cytochem.*, **12**, 530.
FEIGIN, I. (1956). *J. biophys. biochem. Cytol.*, **2**, 213.
FEYRTER, F.)1936). *Virchows Arch.*, **296**, 645.
FIESER, L. F., and FIESER, M. (1956). "Organic Chemistry," 3rd Ed. Chapman & Hall, Boston, p. 287.
FISCHLER, C. (1904). *Zbl. allg. Path. path. Anat.*, **15**, 913.
FOLCH, J., and SPERRY, W. M. (1948). *Ann. Rev. Biochem.*, **17**, 147.
FRANZ, H., and HOLLE, G. (1965). *Histochemie*, **5**, 163.
FREDRICSSON, B., LAURENT, T. C., and LÜNING, B. (1958). *Stain Tech.*, **33**, 155.
FROEBOESE, C., and SPRÖHNLE, G. (1928). *Z. mikr.-anat. Forsch.*, **14**, 13.
GALLYAS, F. (1963). *J. Neurochem.*, **10**, 125.
GLEES, P. (1943). *Brain*, **66**, 229.
GOTTSCHALK, A. (1955). *Nature, Lond.*, **176**, 881.
GOVAN, A. D. T. (1944). *J. Path. Bact.*, **56**, 262.
GRAFFI, A. (1939). *Z. Krebforsch.*, **49**, 477.
GRAFFI, A., and MAAS, H. (1938). In *Arbeiten Staatl. Institut f. exp. Therapie, Frankfurt*, **3**, 21.
GROSS, W. (1930). *Z. wiss. Mikr.*, **47**, 64.
GRUNDLAND, I., BULLIARD, H., and MAILLET, M. (1949). *C.R. Soc. Biol., Paris*, **143**, 771.
GÜNTHER, W. H. (1941). *Z. Krebsforsch.*, **52**, 57.
HACK, M. H. (1953). *Biochem. J.*, **54**, 602.
HADJIOLOFF, A. (1938). *Bull. Histol. Tech., micr.*, **15**, 81.
HAMPERL, H. (1950). *Arch. Path.*, **49**, 563.
HARMAN, P. J., and BERNSTEIN, P. W. (1961). *Stain Tech.*, **36**, 49.
HARTMAN, T. L. (1940). *Stain Tech.*, **15**, 23.
HERXHEIMER, G. W. (1901). *Dtsch. med. Wschr.*, **36**, 607.
HERXHEIMER, G. W. (1903). *Zbl. allg. Path. path. Anat.*, **14**, 491.
HOLCZINGER, L. (1959). *Acta Histochem.*, **8**, 167.
HOLCZINGER, L. (1964). *Histochemie*, **4**, 120.
HOLCZINGER, L. (1965). *Acta histochem.*, **20**, 374.
HOLCZINGER, L., and BÁLINT, Z. (1961). *Acta histochem.*, **11**, 284.
HORI, S. H. (1963). *Stain Tech.*, **38**, 221.
ISHIWATARI, Y., OSHIDA, G., and MATSUZAKI, H. (1960). *Proc. Jap. Histochem. Assoc.*, **1**, 35.
JACKSON, C. (1944). *Onderstepoort J. vet. Sci.*, **19**, 169.
JOHNSON, A. C., MCNABB, A. R., and ROSSITER, R. J. (1950). *Biochem. J.*, **45**, 500.
KANWAR, K. C. (1961). *The Microscope*, **12**, 316.
KAUFMAN, K. (1909). *Chem. Ber.*, **42**, 3481.
KAUFMANN, N. C., and LEHMANN, E. (1926). *Virchows Arch.*, **216**, 623.
KAUFMANN, N. C., and LEHMANN, E. (1929). *Z. mikr.-anat. Forsch.*, **16**, 586.
KAY, W. W., and WHITEHEAD, R. (1934). *J. path. Bact.*, **39**, 449.
KAY, W. W., and WHITEHEAD, R. (1935). *Ibid.*, **41**, 303.
KAY, W. W., and WHITEHEAD, R. (1941). *Ibid.*, **53**, 279.
KEILIG, I. (1944). *Virchows Arch.*, **312**, 405.
KENT, S. P. (1952). *Arch. Path.*, **54**, 439.
KLENK, E. (1942). *Hoppe-Seyl. Z.*, **273**, 76.
KLENK, E., and GIELEN, W. (1960). *Hoppe-Seylers Z. physiol. Chem.*, **319**, 283.
KLÜVER, H., and BARRERA, E. (1953). *J. Neuropath. exp. Neurol.*, **12**, 400.
KLÜVER, H., and BARRERA, E. (1954). *J. Psychol.*, **37**, 199.
KNAYSI, G. (1941). *J. Bact.*, **42**, 587.
KUTT, H., LOCKWOOD, D., and MCDOWELL, F. (1959a). *Stain Tech.*, **34**, 197.
KUTT, H., LOCKWOOD, D., and MCDOWELL, F. (1959b). *Ibid.*, **34**, 203.
LANDING, B. H., UZMAN, L. L., and WHIPPLE, A. (1952). *Lab. Invest.*, **1**, 456.
LANDING, B. H., and FREIMAN, D. G. (1957). *Amer. J. Path.*, **33**, 1.

LANDING, B. H., and HALL, H. E. (1956). *J. Histochem. Cytochem.*, **4**, 382.
LANDSMEER, J. M. F. (1952). *Acta physiol. pharmacol., Neerl.*, **2**, 712.
LANDSMEER, J. M. F., and GIEL, B. (1956). *J. Histochem. Cytochem.*, **4**, 9.
LEACH, E. H. (1938). *J. path. Bact.*, **47**, 635.
LENNERT, K., and WEITZEL, G. (1952). *Zeit. wiss Mikr.*, **61**, 20.
LEULIER, A., and REVOL, L. (1930). *Bull. histol. Tech. micr.*, **7**, 241.
LEWIS, P. R., and LOBBAN, M. C. (1961). *J. Histochem. Cytochem.*, **9**, 2.
LHOTKA, J. F. (1955). *Stain Tech.*, **30**, 235.
LILLIE, R. D. (1944). *Stain Tech.*, **19**, 55.
LILLIE, R. D. (1945a). *Ibid*, **20**, 7.
LILLIE, R. D. (1945b). *Ibid.*, **20**, 73.
LILLIE, R. D. (1952). *Ibid.*, **27**, 37.
LILLIE, R. D. (1954). "Histopathologic Technic and Practical Histochemistry." Blakiston, New York.
LILLIE, R. D., and ASHBURN, L. L. (1943). *Arch. Path.*, **36**, 432.
LILLIE, R. D., and BURTNER, H. J. (1953). *J. Histochem. Cytochem.*, **1**, 8.
LILLIE, R. D., and LASKEY, A. (1951). *Bull. int. Ass. med. Mus.*, **32**, 77.
LISON, L. (1936). "Histochimie Animale." Paris.
LISON, L. (1953). "Histochimie et Cytochimie Animales." Paris, p. 347.
LISON, L., and DAGNELIE, J. (1935). *Bull. Histol. appl.*, **12**, 85.
LORRAIN SMITH, J. (1908). *J. path. Bact.*, **12**, 1.
LOVERN, J. A. (1955). "The Chemistry of Lipids of Biochemical Significance." Methuen, London.
LUDFORD, R. J. (1934). In *11th Scient. Rep. Invest. Imp. Cancer Res. Fund*, p. 169.
MALLORY, F. B. (1938). "Pathological Technique." Squnders, Philadelphia.
MANNELL, W. A. (1952). *Canad. J. med. Sci.*, **30**, 173.
MARCHI, V. (1892). *Arch. ital. Biol.*, **17**, 191.
MCMANUS, J. F. A. (1945). *Nature, Lond.*, **156**, 173.
MCMANUS, J. F. A. (1946). *J. path. Bact.*, **58**, 93.
MENSCHIK, Z. (1953). *Stain Tech.*, **28**, 13.
METTLER, F. A., and HARADA, R. E. (1942). *Stain Tech.*, **17**, 111.
MICHAELIS, L. (1901). In "Einführung in die Farbstoffchemie für Histologie." Berlin.
MIRONESCU, ST. (1964). *Acta histochem.*, **18**, 106.
MIRONESCU, ST. (1965). *Acta histochem.*, **20**, 115.
MORRISON, R. W., and HACK, M. H. (1949). *Amer. J. Path.*, **25**, 597.
MUKHERJI, M., DEB, C., and SEN, P. B. (1960). *J. Histochem. Cytochem.*, **8**, 189.
MÜLLER, H. (1859). *Verh. phys. med. Ges. Wurzburg.*, **10**, 179.
MUSTAKALLIO, K. K., and LEVONEN, E. (1964). *J. Atheroscl. Res.*, **4**, 370.
NORTON, W. T. (1959). *Nature*, **184**, 1144.
NORTON, W. T., KOREY, S. R., and BROTZ, M. (1962). *J. Histochem. Cytochem.*, **10**, 83.
OKAMOTO, K., UEDA, M., and KATO, A. (1944). *Jap. J. Constitutional Med.*, **13**, 102.
OKAOMTO, K., SHIMAMOTO, H., and SONODA, H. (1944). *Jap. J. Constitutional Med.*, **13**, 113.
OKAMOTO, K., SHIMAMOTO, T., SENO, M., UEDA, M., KUSUMOTO, Y., KATO, A., and SHIBATA, D. (1947). *Trans. Soc. path. jap.*, **36**, 16.
PEARSE, A. G. E. (1951a). *J. clin. Path.*, **4**, 1.
PEARSE, A. G. E. (1951b). *Quart. J. micr. Sci.*, **92**, 4.
PEARSE, A. G. E. (1955). *J. Path. Bact.*, **70**, 554.
PEARSE, A. G. E., and ALMEIDA, D. F. (1958). In press.
PISCHINGER, A. (1943). *Z. mikr. Anat. Forsch.*, **53**, 46.
POPPER, H. (1944). *Physiol. Rev.*, **24**, 205.
PRETL, K. (1948). *Virchows Arch.*, **315**, 229.
PROESCHER, F. (1927). *Stain Tech.*, **2**, 60.
PRZELECKA, A. (1956). *Acta biol. exp.*, **17**, 231.
RIEMERSMA, J. C., and BOOIJ, H. L. (1962). *J. Histochem. Cytochem.*, **10**, 89.
ROE, J. H., and RICE, E. W. (1948). *J. Biol. Chem.*, **173**, 507.
RODÉ, B. (1962). *Nature, Lond.*, **193**, 402.

ROMEIS, B. (1927). *Virchows Arch.*, **264**, 301.
ROMEIS, B. (1929). *Z. micr. Anat. Forsch.*, **16**, 525.
ROMEIS, B. (1948). "Mikroskopische Technik." Munich.
ROMIEU, P. (1927). *C.R. Soc. Biol., Paris*, **96**, 1232.
ROSSITER, R. J. (1955). In "Neurochemistry." Thomas, Springfield, Illinois, p. 11.
SALTHOUSE, T. N. (1962a). *Nature*, **195**, 187.
SALTHOUSE, T. N. (1962b). *Stain Tech.*, **37**, 313.
SALTHOUSE, T. N. (1963). *Nature*, **199**, 821.
SALTHOUSE, T. N. (1964). *Stain Tech.*, **39**, 123.
SASS, S., KAUFMAN, J. J., CARDENAS, A. A., and MARTIN, J. J. (1958). *Analyt. Chem.*, **30**, 529.
SCHABADASCH, A. L. (1958). *Arch. Anat. Histol. Embryol.*, **35**, 3.
SCHNABEL, R. (1962). *Histochemie*, **3**, 127.
SCHNABEL, R. (1964). *Acta histochem.*, **18**, 161.
SCHOTT, H. J. (1962). *Histochemie*, **3**, 138.
SCHOTT, H. J. (1964). *Histochemie*, **3**, 467.
SCHOTT, H. J., and SCHONER, W. (1965). *Histochemie*, **5**, 154.
SCHULTZ, A. (1924). *Zbl. allg. Path. path. Anat.*, **35**, 314.
SCHULTZ, A. (1925). *Verh. dtsch. path. Ges.*, **20**, 120.
SCHULTZ, A., and LÖHR, G. (1925). *Zbl. allg. Path. path. Anat.*, **36**, 529.
SHEEHAN, H. L. (1939). *J. Path. Bact.*, **49**, 580.
SINAPIUS, D., and THIELE, O. W. (1965). *Histochemie*, **4**, 553.
SINGH, R. (1964a). *J. Histochem. Cytochem.*, **12**, 42.
SINGH, R. (1964b). *J. Histochem. Cytochem.*, **12**, 712.
SMITH, J. L. (1907). *Ibid.*, **12**, 1.
SMITH, J. L., and MAIR, W. (1911). *Skand. Arch. Physiol.*, **25**, 245.
SMITH, M. C. (1951). *J. Neurol. Neurosurg. Psychiat.*, **14**, 222.
SMITH, M. C. (1956). *Ibid.*, **19**, 67.
SMITH, M. C., STRICH, S. J., and SHARP, P. (1956). *Ibid.*, **19**, 62.
STRUGGER, S. (1937). *Arch. exp. Zellforsch.*, **19**, 199.
STRUGGER, S. (1938). *Protoplasma*, **30**, 85.
SVENNERHOLM, L. (1964). *J. Lipid. Res.*, **5**, 145.
SVENNERHOLM, L. (1966). *Biochem. J.*, **98**, 20P.
SVENNERHOLM, L., and THORIN, H. (1960). *Biochim. Biophys. Acta*, **41**, 371.
SWANK, R. L., and DAVENPORT, H. A. (1934). *Stain Tech.*, **9**, 129.
TATTRIE, N. H. (1959). *J. Lipid Res.*, **1**, 60.
TERNER, J. Y., SCHNUR, J., and GARLAND, J. (1963). *Lab. Invest.*, **12**, 405.
TRANZER, J-P., and PEARSE, A. G. E. (1963). *Nature*, **199**, 1063.
UEDA, M. (1952). *Hyoyo J. med. Sci.*, **1**, 117.
VERNE, J. (1929). *Ann. Physiol. Physicochim. biol.*, **5**, 245.
VOLK, B. W., and POPPER, H. (1944). *Amer. J. clin. Path.*, **14**, 234.
WEBER, A. F., PHILLIPS, M. G., and BELL, J. T., Jr. (1956). *J. Histochem. Cytochem.*, **4**, 308.
WEISSCHEDEL, E., and JUNG, R. (1939). *Z. Anat. Entwickl.*, **109**, 374.
WILLS, E. D., and WILKINSON, A. E. (1966). *Biochem. J.*, **99**, 657.
WINDAUS, T. (1910). *Z. phys. Chem.*, **65**, 110.
WISLOCKI, G. B., and DEMPSEY, E. W. (1946). *Anat. Rec.*, **96**, 249.
WOLMAN, M. (1956a). *J. Histochem. Cytochem.*, **4**, 195.
WOLMAN, M. (1956b). *Neurology*, **6**, 636.
WOLMAN, M. (1957a). *J. Neurochem.*, **1**, 370.
WOLMAN, M. (1957b). *Exp. Cell. Res.*, **12**, 231.
WOOLFREY, B. F., and PEARSON, H. L. (1962). *Amer. J. clin. Path.*, **37**, 437.
ZELIKOFF, M., and TAYLOR, H. A. (1950). *J. amer. Chem. Soc.*, **72**, 5039.

CHAPTER 13

ALDEHYDES AND KETONES

THE importance of aldehydes and ketones in histochemistry is related to two things. First, their presence in naturally-occurring substances, and second, their production in the tissues by a wide variety of chemical manœuvres commonly used as tests for one substance or another. Naturally-occurring *reactive* aldehydes do not greatly concern us, but *potential* aldehyde groups are widespread in two classes of material especially, namely deoxyribonucleic acids and lipids. From these they are released by very simple procedures such as mild acid hydrolysis, as in the Feulgen nucleal reaction, or by mild oxidation as in the plasma reaction considered later in this chapter.

Stronger reactions involving oxidation commonly result in the production of reactive aldehyde groups in carbohydrates and lipids as in the Bauer reaction for glycogen, the PAS reaction for 1:2-glycol groups and the PFAS reaction for unsaturated HC=CH groups in lipids. In many of these cases there is room for doubt as to whether the reactive group is aldehyde (RH.C=O) or ketone

$$\begin{array}{l}\mathrm{R.C{=}O}\\ \quad\;\;|\\ \quad\;\mathrm{R'}\end{array}$$

since the reactions of the two are essentially very similar. The only important naturally-occurring ketones with which we are concerned are present in the ketosteroids, and many of the methods described have been applied to the histochemistry of these substances.

All the reactions considered in this chapter are given first under the heading of aldehyde reactions, without prejudice to their position in relation to ketones and ketosteroids. The performance of the various reactions in the latter respect are dealt with in short separate sections. In another section full consideration is given to the plasmal reaction, not for its intrinsic importance in histochemistry but for the very great bearing which it has on the whole question of tissue aldehydes.

Positive Methods for Aldehydes

THE SCHIFF REACTION (SCHIFF, 1866)

Enough has already been said about this reaction, in Chapters 9 and 10, to establish its primary position for the demonstration of reactive aldehyde groups. The mechanism of the reaction is still not known with absolute certainty. It was considered by Wieland and Scheuing (1921) to be in the nature of an addition followed by a condensation.

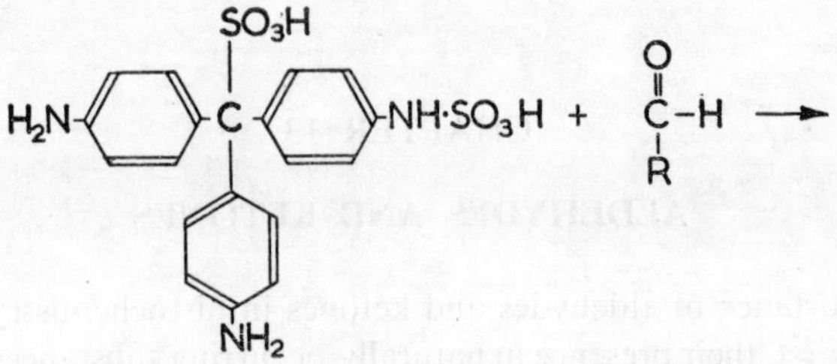

Fuchsin-sulphurous acid Aldehyde

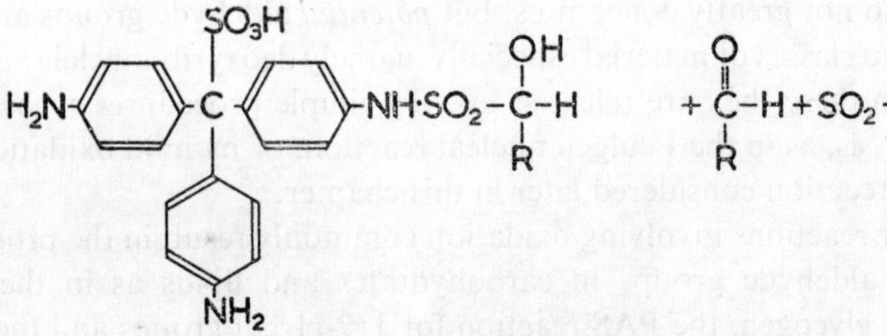

N-sulphinic acid-aldehyde
(colourless)

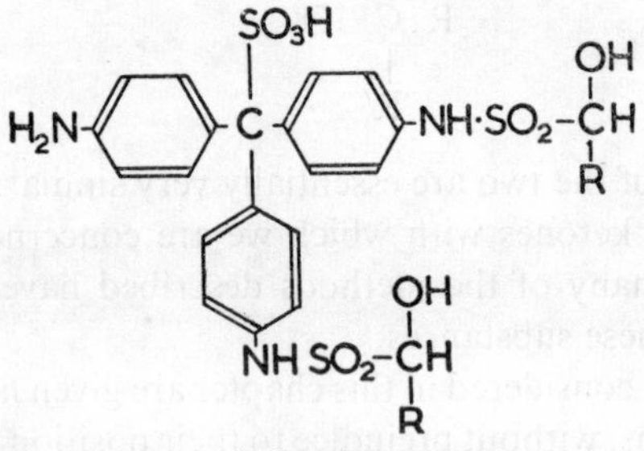

(colourless)

Molecular rearrangement then takes place to give:

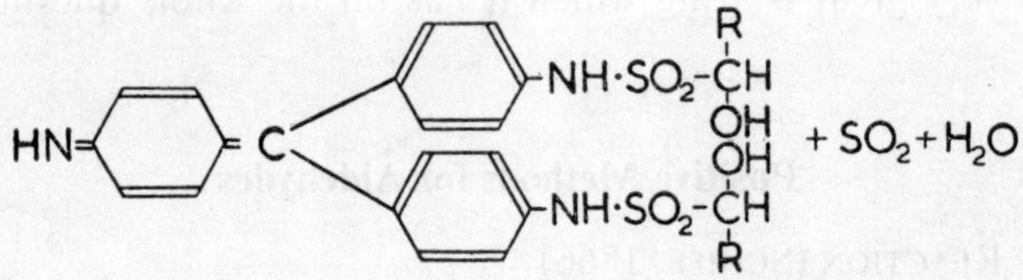

$+ SO_2 + H_2O$

and this compound, since it possesses the quinonoid grouping,

is coloured.

Although in his review of the Schiff reaction in cytochemistry Kasten (1960) strongly supported the hypothesis of Wieland and Scheuing, recent investigations have produced evidence which throws doubt on the validity of their explanation. The most popular alternative hypothesis was first put foward by Rumpf (1935) and supported by Hörmann *et al.* (1958). This suggests that an alkylsulphonic acid, rather than a sulphinic acid, is formed.

Applying the Schiff reaction to estimation of long chain aldehydes Sloane-Stanley and Bowler (1962) found that their extinctions were always less than those calculated on the Wieland and Scheuing (1921) formula for the Fuchsin-SO_2-aldehyde complex. They suggested that the molecular proportion of aldehyde to Fuchsin in the Schiff complex was not two but three. Using electrophorectic analysis Hiraoka (1960) found that the coloured product of the Schiff-formaldehyde condensation reaction consisted of several components. Hiraoka concluded that the mechanism of the reaction was more complicated than assumed by Wieland and Scheuing and with this conclusion one can only agree. A further indication that the matter was far from simple came from the work of Barka and Ornstein (1959, 1960) who indicated that from the reaction mixture of Schiff's reagent and formaldehyde 3 to 6 (sometimes even 7) compounds could be separated by paper chromatography.

The essential difference between the two hypotheses lies in the role of sulphite. This question was investigated by Hardonk and van Duijn (1964a and b), using model systems. These authors found that their spectra were identical whether staining was carried out with pararosanilin (rinsing in HCl and then water), pararosanilin (rinsing with sulphite and then water), or with Schiff's reagent (rinsing with sulphite and then water). With the second and third methods intensity of staining was identical. These findings were considered to refute the Wieland-Scheuing hypothesis and to support the view that an alkylsulphonic acid ($NH.CHR.SO_3H$) is the substituted group in the pararosanilin molecule. Supporting this hypothesis on general grounds Stoward (1966) nevertheless objected to the inferences drawn from their data by Hardonk and van Duijn. He considered that these would fit equally well if the final product was a simple azomethine (a view opposed by van Duijn and Hardonk, 1966) or even an *N*-sulphinic acid. Stoward considered that the essential chromophore was the electron-deficient central methane carbon atom (which gives rise to the 538 nm adsorption band). This is present in all three postulated final products.

The stoichiometry of Sloane-Stanley and Bowler (1962) would necessitate combination of the alkyl α-hydroxysulphonic acid with all 3 amino groups of the dye. According to observations made by Lacoste and Martell (1955) this should be possible. These authors indicated that methylene sulphonic acid groups could be added to both H atoms of an amino group. It would seem that there are many theoretically possible structural formulæ and several of these must be assumed to exist, in proportions which will vary with the concentration of SO_2 and the molar ratio of dye to aldehyde.

Shown below is the explanation presently most acceptable:

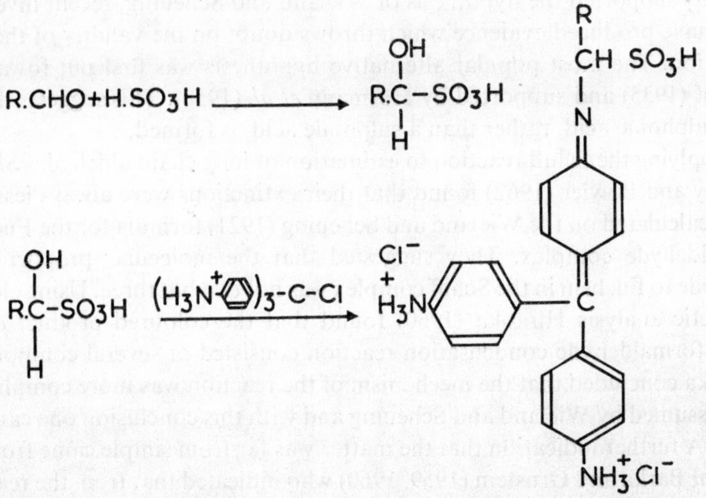

$$R.CHO + H.SO_3H \longrightarrow R.\overset{OH}{\underset{H}{C}}-SO_3H$$

$$R.\overset{OH}{\underset{H}{C}}-SO_3H \xrightarrow{(H_3\overset{+}{N}-C_6H_4)_3-C-Cl} \overset{+}{H_3}N\langle C_6H_4\rangle=C\langle C_6H_4\,\overset{+}{N}H_3\,Cl^-\rangle\langle C_6H_4=N-CH(R)SO_3H\rangle\ Cl^-$$

The colour appears in the Schiff reaction because the concentration of acid is insufficient to give the colourless anilinium-like salt of the sulphonic acid compound.

Variations in Colour of the Schiff Reaction. The possible interpretations of variation in the shade of colour developed in the Schiff reaction have been the subject of discussion by a number of authors, some of whom have suggested that the magenta shade of fuchsin-sulphurous acid recolorized by exposure to air is the only colour given by "genuine" reactions. Others have insisted that a "genuine" reaction gives a shade considerably bluer than that of naturally recolorized fuchsin-sulphurous acid. Lison (1932) was of the opinion that wide variations in final colour occurred and that the interpretation of differences in shade should be made with caution. Gomori (1950), investigating the amine oxidase reaction of Oster and Schlossman (1942), in which the final product with Schiff's reagent was stated to be bluish in colour, determined the reaction of four different aromatic aldehydes. He noted that benzaldehyde gave a purple colour, of a similar shade to that developed in the plasmal reaction, anisaldehyde a redder shade, vanillin rose red and *p*-dimethylaminobenzaldehyde a redder shade, vanillin rose red and *p*-dimethylaminobenzaldehyde scarlet. No blue shades occurred and Gomori suggested that these might be due to alterations in the plasmal-Schiff compound regenerated from its bisulphite complex, or to the acid pH of the bisulphite.

Elftman (1959) observed that the colour of the Schiff-aldehyde product could be altered by varying the SO_2 content of the reagent. High aldehyde concentration resulted in bluish and low aldehyde in reddish hues. In histochemical practice dye and SO_2 are always present in excess. Barka and Ornstein (1960) showed that when formaldehyde combined with a low SO_2 Schiff reagent the products were red in colour and had large Rf values. With a high SO_2 Schiff reagent the opposite was the case, bluish colours and smaller Rf values.

As yet no intelligent interpretation can be given for any differences in colour which may result from the use of the Schiff reagent in histochemical practice.

Ammoniacal Silver Reactions

Many reducing reagents will convert ammoniacal silver solutions to metallic silver. Their reduction has been used for the demonstration of aldehydes and for certain special types of ketone, referred to below. Although ammoniacal silver solutions can be used to demonstrate the presence of plasmal, and to block the reaction of the latter with Schiff's reagent, they are most commonly employed in histochemistry to render visible the reaction products formed from glycogen and mucin by oxidation with various reagents (see methods for glycogen, Chapter 10, p. 364). According to Lhotka and Davenport (1951) they will not react with deoxyribonucleic acid previously hydrolysed by a Feulgen type of hydrolysis, and this observation, among others, caused these authors to doubt the aldehydic nature of the substance responsible for the nucleal reaction. My experience does not entirely agree with the observations of these authors, although I agree that the usual acid hydrolysis followed by short immersion in an ammoniacal silver bath gives a negative result. I have found that after hydrolysis with acids other than HCl, formic acid for instance, followed by 18 hours in the silver solution, there is an appreciable blackening of the nuclei, absent from control sections. Cain (1949b) drew attention to the fact that ethylene oxide groups

$$-\underset{\substack{|\\H}}{C}\overset{O}{\frown}\underset{\substack{|\\H}}{C}-$$

which can be formed by oxidation of fats in the presence of sulphydryls such as glutathione, will also reduce ammoniacal silver solutions. Although the presence of such groups may have to be taken into account in order to explain the silver reactions of lipid-containing materials, aldehydes remain the most likely groups to be responsible for a positive silver reaction in oxidized tissue sections.

The Nitrohydroxylamine Reaction

From time to time references to the use of this specific test for aldehydes occur in the histochemical literature. Originally suggested by Angeli (1892) and Rimini (1901), it depended on the reaction of nitrohydroxylamine with aldehydes (but not with ketones) to give substituted hydroxamic acids. The latter give an intense red colour with ferric salts. The majority of accounts refer to the use of this test with a negative histochemical result, interpreted as indicating the absence of aldehydes. Such an interpretation is certainly not valid until a positive result has been shown to occur with known tissue aldehydes and even then it is possible that potential aldehydes may fail to react. In practice, the method can be carried out with benzene sulphonyl hydroxamic acid in dilute alkali. Under these conditions the compound behaves as though it splits to form benzene sulphinic acid and an NOH group:

$$C_6H_5—\overset{\overset{\displaystyle O}{\|}}{\underset{\underset{\displaystyle O}{\|}}{S}}—NHOH \xrightarrow{\text{ALKALI}} C_6H_5SO_2H + NOH$$

The latter combines with the aldehyde to form a substituted hydroxamic acid:

$$\underset{\underset{\displaystyle H}{|}}{R—C}{=}O + H—N{=}O \rightarrow R—\overset{\overset{\displaystyle OH}{|}}{\underset{\underset{\displaystyle H}{|}}{C}}—NO \rightarrow R—\overset{\overset{\displaystyle OH}{|}}{C}{=}NOH$$

This gives the red colour with ferric salts. I did not succeed in making the method work with known aldehyde groups such as those produced by the action of periodic acid on tissue polysaccharides, and it is doubtful whether a positive histochemical reaction has ever been obtained. (But see Chapter 12, gold hydroxamate reaction, p. 426.)

PHENYLHYDRAZINE AND DINITROPHENYLHYDRAZINE REACTIONS

Both aldehydes and ketones condense with phenylhydrazine to form substituted hydrazones which are used extensively for chemical characterization. With aldehydes the following reaction takes place:

$$R.CHO + H_2N.NHC_6H_5 \rightarrow RCH{=}N.NHC_6H_5 + H_2O$$

Bennett (1940) was the first to make histochemical use of the phenylhydrazine reaction for demonstration as opposed to blocking, noting the yellow colour which it gave with frozen sections of the adrenal gland. He suggested that the reaction was due to ketones present in the adrenal cortex presumably in the form of fat soluble ketosteroids. Dempsey and Wislocki (1946), who used 2:4-dinitrophenylhydrazine for the same purpose, were also of this opinion. The colour developed by the latter reagent is stronger than that given by phenylhydrazine, but in either case it is faint and not easily visible in small structures such as chromosomes, although Danielli (1947) claimed that the dinitrophenylhydrazine reaction of hydrolysed chromosomes was identical with that given by Schiff's reagent. This author (1949) subsequently developed a substituted azobenzenephenylhydrazine sulphonic acid which gave a purple colour with aldehydes.

NAPHTHOIC ACID HYDRAZIDE REACTIONS

A considerable technical advance on the phenylhydrazine techniques was made independently by Camber (1949) and by Ashbel and Seligman (1949) who substituted 2-hydroxy-3-naphthoic acid hydrazide (NAH) for the various phenylhydrazides. This compound condenses with both aldehydes and ketones to give a product no more highly coloured than that given by the earlier techniques. By subsequent coupling with a diazonium salt, however, the carbonyl-hydrazide compound is readily made visible, and the authors referred to above used diazotized *o*-dianisidine for the purpose. This gives a strong bluish-purple as the final colour. The reaction is considered to take place as follows:

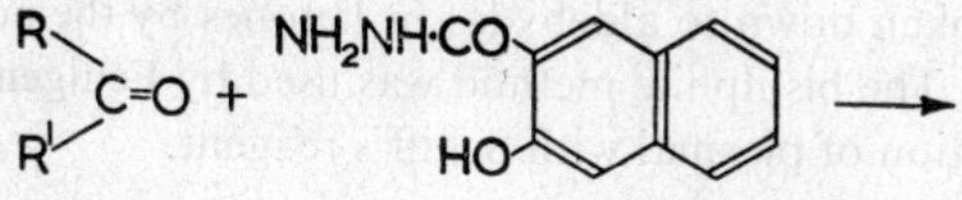

R, R' C=O + $NH_2NH{\cdot}CO$, HO (naphthalene) ⟶

NAH

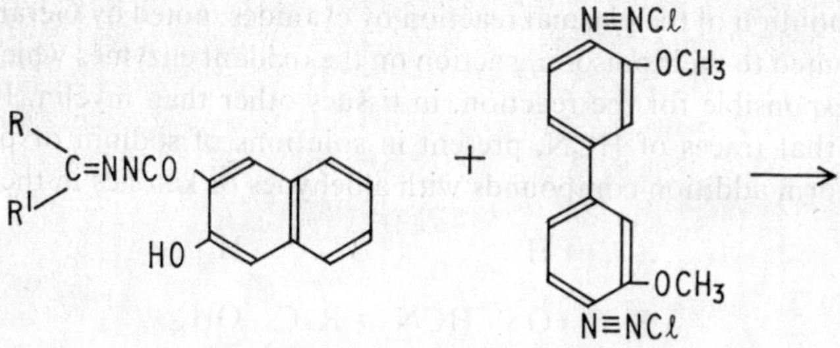

R, R' C=NNCO, HO (naphthalene) + N≡NCl, OCH_3, OCH_3, N≡NCl ⟶

Diazotized *o*-dianisidine

N=N, N=N, OC, OH, OCH_3, CH_3O, HO, CO, HN, NH, N, N, C, C, R, R', R, R'

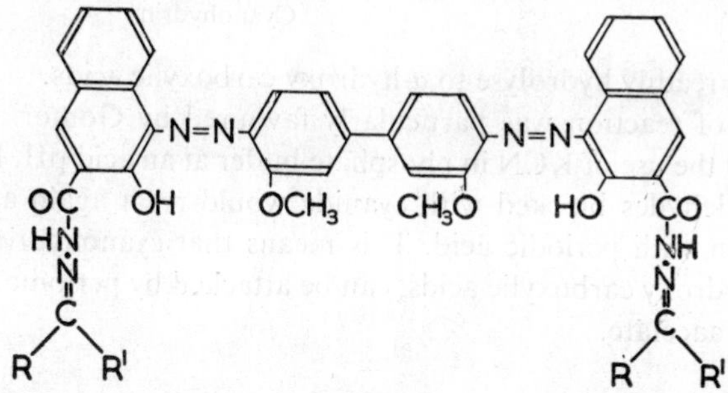

Purple dye

Blocking Reactions for Aldehydes

Aldehydes are exceedingly reactive compounds and it is not surprising that a very large number of methods exist for their chemical characterization. What is surprising, however, is the fact that so many have been successfully applied to histochemistry. Nearly all those which have been used in this way are given below, approximately in historical order, and judgment as to which of them is preferable for any given purpose may usefully be withheld until all have been considered.

BISULPHITE

Sodium bisulphite forms addition compounds with aldehydes and ketones of the type

$$\text{R.}\overset{\text{H}}{\overset{/}{\text{C}}}\text{=O} + \text{NaHSO}_3 \rightarrow \text{R.}\overset{\text{H}}{\overset{/}{\text{C}}}\text{—OH}\ \ (\searrow \text{SO}_3\text{Na})$$

which can be broken down to aldehydes or ketones by the addition of dilute acids and alkalis. The bisulphite method was used by Feulgen and Voit (1924) to block the reaction of plasmal with Schiff's reagent.

CYANIDES

The abolition of the plasmal reaction by cyanides, noted by Gérard (1935), was attributed to their poisoning action on the oxidant enzymes which he considered responsible for the reaction, in tissues other than myelin. It is more probable that traces of HCN, present in solutions of sodium or potassium cyanide, form addition compounds with aldehydes or ketones in the tissues.

$$\text{R.}\overset{\text{H}}{\overset{/}{\text{C}}}\text{=O} + \text{HCN} \rightarrow \text{R.}\overset{\text{H}}{\overset{/}{\text{C}}}\text{—OH}\ \ (\searrow \text{CN})$$

Cyanohydrin

Cyanohydrins readily hydrolyse to α-hydroxy carboxylic acids.

This type of reaction was particularly favoured by Gomori (1953) who recommended the use of KCN in phosphate buffer at an acid pH. Lillie (1956) noted that aldehydes blocked with cyanide would react again as aldehydes after oxidation with periodic acid. This means that cyanohydrins, or more probably α-hydroxy carboxylic acids, can be attacked by periodic acid as well as by lead tetraacetate.

PHENYLHYDRAZINES

The reactions of primary amines with aldehydes have long been known. Verne (1929b) noted that treatment with phenylhydrazine would reverse the plasmal reaction in the adrenal cortex and medulla and Bennett (1939, 1940) made particular use of the reaction in his studies of the adrenal cortex of the cat. The histochemical use of the phenylhydrazine reaction is now often known by his name, although this usage, if sanctioned at all, should be reserved for the colour reaction rather than for the blocking reaction. According to Wild (1947), monosubstituted and unsymmetrically disubstitued thydrazines condense with the carbonyl groups of both aldehydes and ketones to form substituted hydrazones.

$$\begin{matrix} R \\ \quad\diagdown \\ \quad C{=}O \\ \quad\diagup \\ R' \end{matrix} + H_2N.NH.C_6H_5 \rightarrow \begin{matrix} R \\ \quad\diagdown \\ \quad C{=}N.NH.C_6H_5 \\ \quad\diagup \\ R' \end{matrix} + H_2O$$

Phenylhydrazine quickly forms these derivatives, which are yellow in colour, and a reaction may be obtained in tissue sections in a few hours. Danielli (1949) used 2:4-dinitrophenylhydrazine as a blocking reagent for the plasmal reaction and Lhotka and Davenport (1951) used *p*-nitrophenylhydrazine in their study of the Feulgen technique. Naphthoic acid hydrazide, and no doubt very many other hydrazides, can be used in a similar manner. Phenylhydrazine blockade is resistant to further oxidation by periodic acid.

DIMEDONE

This agent (5:5-dimethyl-*cyclo*-hexane-1:3-dione) forms condensation products with aqueous, alcoholic or glacial acetic solutions of aldehydes.

If acetic acid solutions are employed the internal anhydride of this product is formed. The dimedone reaction was used by Verne in his extensive studies on plasmal and was presumed by Cain (1949b) to react with ketones as well as with aldehydes. Wild (1947), however, states that no condensation occurs with ketones and that the reaction is specific for aldehydes. If it is employed at room temperature the dimedone reaction is slow. Fig. 120 shows the effect of dimedone (60° for 2 hours) on a section of gastric mucosa oxidized for 10 minutes with 1 per cent. aqueous periodic acid. Fig. 121 shows the untreated control

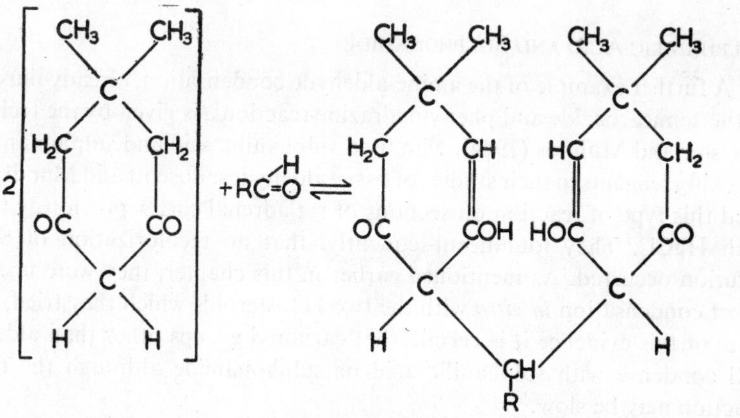

$$2\left[\begin{matrix} CH_3 \quad CH_3 \\ C \\ H_2C \quad CH_2 \\ | \qquad | \\ OC \quad CO \\ C \\ H \quad H \end{matrix}\right] + RC\overset{H}{=}O \rightleftharpoons \begin{matrix} CH_3 \quad CH_3 \qquad CH_3 \quad CH_3 \\ C \qquad\qquad C \\ H_2C \quad CH \quad HC \quad CH_2 \\ | \quad \| \qquad \| \quad | \\ OC \quad COH \; HOC \quad CO \\ C \qquad\qquad C \\ H \qquad\qquad\qquad H \\ CH \\ R \end{matrix}$$

section. The abolition of the reaction with Schiff's reagent is rapid in the connective tissues and muscle (except for glycogen and mast cell granules) and fairly rapid in free mucin in the lumen. In this short period dimedone has no effect on the staining intensity of the mucinogen granules of the mucosa.

Prolonged treatment, for 24 hours at 60°, still fails to abolish the positive reaction in the latter situation. The significance of these observations remains to be explained, though it is probable that they reflect both a quantitative difference in the number of aldehyde groups in a given structure and the degree of protection from dimedone caused by stereochemical factors.

Dimedone blocking is stable to further oxidation with periodic or peracetic acids.

Semicarbazides and Thiosemicarbazides

These very reactive compounds form insoluble products (semi-carbazones, thiosemicarbazones) with aldehydes and ketones. The reactions are reversible and the rate at which they proceed depends on the hydrogen ion concentration. Since semicarbazides are basic substances an acid pH favours the formation of semicarbazones

$$R.C(H){=}O + \underset{\text{Semicarbazide}}{H_2NNHCONH_2} \rightleftharpoons R\diagdown C(H){=}NNHCONH_2 + H_2O$$

In much of the early research on plasmal, semicarbazides were used to block the Schiff reaction of tissue aldehydes; Bennett (1940) used a 10 per cent. solution in absolute methanol, with the addition of sodium acetate, but the method given in Appendix 13 using acetate buffer at pH 4·5, is preferable. Thiosemicarbazides may be used in a similar manner, but they have no advantage in the simple blocking reaction.

Sulphanilic Acid and Sulphonamide

A further example of the amine-aldehyde condensation, already illustrated in the semicarbazide and phenylhydrazine reactions, is given by the technique of Oster and Mulinos (1944), who used sulphanilic acid and sulphonamide as blocking reagents in their studies of tissue aldehydes. Boscott and Mandl (1949) used this type of reaction on sections of rat adrenal cortex previously treated with $HgCl_2$. They found, subsequently, that no recolorization of Schiff's solution occurred. As mentioned earlier in this chapter, they were unable to effect condensation *in vitro* with the two ketosteroids which they tried, but in spite of this evidence it is certain that carbonyl groups other than aldehydes will condense with sulphanilic acid or sulphonamide although the rate of reaction may be slow.

Hydroxylamine

Yet another example of an amine-aldehyde condensation is given by the reaction with hydroxylamine which occurs readily with both aldehydes and ketones. The condensation products are known as aldoximes or ketoximes.

$$R.\overset{H}{C}{=}O + H_2NOH \rightarrow R.\overset{H}{C}{=}NOH + H_2O$$

The rate of formation of these derivatives is dependent on the hydrogen ion concentration, with an optimum at pH 4·7, decreasing on either side of this point but increasing again in strongly alkaline solutions (Ölander, 1927). In practice hydroxylamine is used in the presence of excess sodium carbonate or acetate. The reaction was used by Danielli (1949) to block free tissue aldehydes prior to the release of acetal aldehydes by treatment with either $HgCl_2$ or N-HCl.

Staple (1958) observed that blocking of aldehydes with hydroxylamine was reversed by oxidation with periodic acid. It cannot therefore be used to block existing aldehydes in frozen sections, for instance, before carrying out the PAS reaction.

Aniline Chloride and Aniline

Lillie (1954) recommended the use of this reagent to produce aldehyde blockade with relatively short incubation (15 minutes). The blockade so produced is stable to periodic acid oxidation but some tissues require much longer than the minimal incubation time. Lillie and Glenner (1957), testing aniline blockade of various carbonyl groups, found that the reaction was more rapid and complete when carried out with a molar solution of aniline in acetic acid. This last reagent has been used very extensively in applied histochemistry, in preference to the aforementioned blocking reagents.

After an extensive study of 22 different compounds Osse and Geyer (1964) concluded that aniline could with advantage be replaced by several different arylamines. They recommended especially *p*-toluidine and *m*-aminophenol. In both cases blockage was stable to periodic acid, lead tetraacetate, chromic acid, potassium permanganate, ninhydrin, *N*-bromosuccinimide and HCl. Osse and Geyer noted that after periodic acid oxidation some of the blocking reagents induced a certain degree of coloration in PAS-positive structures. Although they could not explain the phenomenon it is clearly related to the PA-diamine reactions (Chapter 10, p. 354).

Cannizzaro Reactions

Aldehydes undergo a peculiar change when treated with alkali. An intermolecular displacement of oxygen occurs, one molecule of aldehyde being reduced at the expense of another which is oxidized to an acid (cf. Karrer, 1950).

$$2R.\overset{H}{C}{=}O + NaOH \rightarrow RCH_2OH + RCOONa$$

This reaction was discovered by Cannizzaro and now bears his name. Lhotka and Davenport (1951) made use of the method in a modified form which they called a crossed Cannizzaro reaction. They considered that the supposed aldehyde radicals produced in deoxyribonucleic acid by Feulgen hydrolysis would have only limited motion and could not, therefore, be expected to react freely with each other. Additional aldehyde groups in the form of formaldehyde were therefore provided and this crossed Cannizzaro reaction was found to prevent the recolorization of Schiff's solution by the nuclei in hydrolysed blocks of tissue; sections were not used since these could not withstand the prolonged action of strong alkali.

Meerwein-Ponndorf Reaction

This reaction was also used by Lhotka and Davenport (1951), in the studies already considered above. It depends on the addition of hydrogen to the carbonyl group, the product being a primary alcohol, and may be written as follows:

$$R.\overset{H}{\overset{/}{C}}{=}O + H_2 \rightarrow RCH_2OH$$

In practice the reaction is more complicated and several views apparently exist as to the mechanism involved. Lhotka and Davenport used blocks of tissue, exposed to Feulgen hydrolysis, and refluxed these for 2–4 hours in isopropyl alcohol to which excess aluminium ethoxide was added. According to Karrer (1950) the reaction with aluminium alcoholates proceeds as follows:

$$R.\overset{H}{\overset{/}{C}}{=}O + Al(OCH_2.CH_3)_3 \rightleftharpoons RCH_2OAl(OCH_2CH_3)_2 + CH_3CHO$$

and this intermediate compound reacts with water:

$$RCH_2OAl(OCH_2CH_3)_2 + 3H_2O \rightarrow RCH_2OH + Al(OH)_3 + 2CH_3CH_2OH$$

An alternative explanation for the reaction given by Gilman (1943) is less satisfactory, in some respects. By this means Lhotka and Davenport were able to block the Feulgen reaction in previously hydrolysed tissues. They were unable to draw any firm conclusions as to the nature of the tissue group involved. The Meerwein-Ponndorf reduction cannot, in any case, be used to distinguish between aldehydes and ketones since both are able to react.

Borohydride Reduction

Searching for a reagent which would increase the specificity of the PAS, PFAS or PAAS reactions Jobst and Horváth (1961) found that previously existing oxo (aldehyde, ketone, quinone, lactone) groups in the tissues could be blocked irreversibly by reduction with sodium borohydride. According to Chaikin and Brown (1949) this reagent, in aqueous solution, does not attack carboxylic acids, esters or olefines. It was used by Jobst and Horváth in 0·2M borate buffer. Details are given in Appendix 13.

Apparently independently Geyer (1963) reported that borohydride reduction provided a convenient method for permanent blockage of aldehydes. In many cases this reaction will be found more suitable than the other blocking reactions described.

Uses of the Various Staining and Blocking Techniques for Aldehydes

There are two chief histochemical uses of the above techniques: (1) For selective blocking of aldehydes, i.e. to block free aldehydes prior to the plasmal reaction or to block aldehydes and leave ketones free to react. (2) For confirmation that a tissue group giving a positive Schiff reaction, by whatever means it may have been produced, is aldehyde rather than ketone, peroxide, hydroperoxide, ethylene oxide or some other group.

The efficiency of the blocking reagents can be assessed by their effect in preventing the recolorization of Schiff's solution by tissue aldehydes, and these can conveniently be produced by periodic acid oxidation. It must be remembered, however, that the configuration of the group produced in the tissues will have a pronounced effect on the rate of condensation with the various reagents and that the reagent, say, which condenses best with the aldehydes produced from glycogen by periodic acid will not necessarily be the best reagent to block the groups produced by treatment of acetals with $HgCl_2$. All the reactions are known to proceed faster with aldehydes than with ketones and with the aliphatic than with the aromatic series.

Leaving theoretical considerations aside, for selective blocking of tissue aldehydes six of the methods discussed are of practical utility. These are the bisulphite, hydroxylamine, semicarbazide, phenylhydrazine, aniline chloride and borohydride reduction reactions. With hydroxylamine in sodium acetate, at room temperature, condensation with tissue aldehydes of all varieties is rapid and apparently complete, but the product is susceptible to further oxidation. The semicarbazide and phenylhydrazine reactions act similarly but are considerably slower. Their products are resistant to oxidation. The best methods for routine use are probably Lillie's acetic acid-aniline and borohydride reduction.

For the accurate characterization of a tissue component as aldehyde, only three reactions out of the fourteen staining and blocking reactions listed have any claim to specificity. The first, which gives a coloured product, is the nitrohydroxylamine reaction of Angeli and Rimini. This reaction is not of practical use in histochemistry. The second, which also gives a coloured product, is the phenylhydrazine-formazan reaction and the third, used solely for blocking, is the dimedone reaction. In respect of the second, I consider that if a positive result is achieved it may be concluded that the tissue group under investigation is an aldehyde, while a negative result cannot be said to exclude this possibility.

This is because the reaction is not technically easy and because in a given section it tends to show only regions giving a strongly positive Schiff reaction. Weaker sites are often negative. There is also a strong probability that the configuration of the tissue aldehyde, even if it allows condensation with phenylhydrazine, may not permit the hydrogen substitution necessary to achieve formazan production (see Chapter 10, p. 326). The dimedone reaction, though theoretically specific for aldehydes, is sometimes found to block these exceedingly slowly in practice. It is often difficult to distinguish the reduction in recolorization of fuchsin-sulphurous acid caused by dimedone from that observed in control sections treated with the solvent (acetic-alcohol) only. The failure of dimedone to block the reaction of a given component does not therefore mean that this is not due to aldehyde. The acetic acid solvent for phenyhydrazine (used for blocking carbonyl groups) was noted by Sweat and Puchtler (1964) to intensify the resorcin-fuchsin reaction of collagen and reticulin fibres, and basement membranes.

Much more use may be made of all these reactions in the future and comparative studies should be made of their various effects. The rapid blocking by dimedone, for example, of the "depolymerized" gastric mucin of the lumen, compared with its exceedingly slow blocking of the "polymerized" product in the cell, requires further investigation and explanation. The difference between hydroxylamine and phenylhydrazine blocking on secondary tissue components (say, glycogen or mast-cell granules), when the Schiff reaction of the primary component (say, periodic-acid-oxidized mucin) has been reduced equally by both, is another example requiring further elucidation.

The Plasmal Reaction

Chemistry of the Reaction

The acetal phosphatides, for which this reaction is specific, are widely distributed in animal tissues where they commonly occur in association with ester phosphatides. In such combinations they may be present in amounts up to 12 per cent. Attention was first drawn to their existence by Feulgen and Rossenbeck (1924), who noticed, in performing the nucleal reaction (Chapter 9) now known by the first author's name, that substances occurred in the cytoplasm which gave a reaction with Schiff's reagent in sections not exposed to the hydrolysing action of N-HCl. Feulgen and Voit (1924) subsequently showed that the reaction with Schiff's solution could be abolished by previous treatment with sodium bisulphite or phenylhydrazine (see section on aldehyde-blocking techniques), and they concluded that the reacting groups were aldehydes. These cytoplasmic aldehydes they called plasmals. Using frozen sections, Feulgen and Voit demonstrated that the plasmal reaction could usually be abolished by the application of lipid solvents, but some of the lipids which gave rise to the reaction were sufficiently strongly attached in the tissues to resist the process of paraffin embedding. They noted that the plasmal reaction could be intensified by brief treatment of the tissues with mercuric chloride or by acid hydrolysis, and that the acid Schiff's reagent could effect a similar hydrolysis. Previous treatment with alcohol, if sufficiently prolonged, reversed the positive reaction obtained by the use of $HgCl_2$. In view of these findings, Feulgen and Voit considered that the aldehyde plasmal was derived from a lipid precursor and this precursor they named plasmalogen.

Feulgen and Behrens (1928) succeeded in isolating plasmalogen from the phospholipid fraction in which it occurred by saponifying the latter with alkali, and the same authors (1938) were able to oxidize plasmalogen to stearic and palmitic acids. Later Feulgen and Bersin (1939) recognized that the fatty aldehyde of plasmalogen was joined to glycerol by an acetal linkage and that the whole structure was that of an acetal phosphatide:

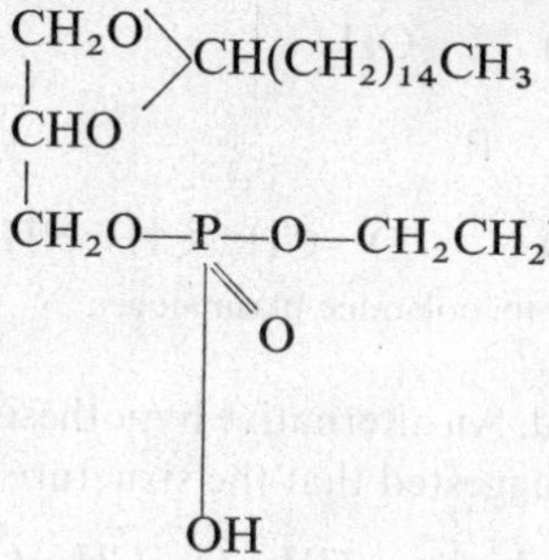

```
CH2O\
|    >CH(CH2)14CH3
CHO /
|
CH2O—P—O—CH2CH2NH2
     |\\
     |  O
     |
     OH
```

α-palmital phosphatide

This formula shows the original Feulgen-Bersin acetal structure. Note its relationship to that of *α*-kephalin (p. 399). More recent work by Klenk and Böhm (1951) and by Klenk and Debuch)1954, 1955), which is described in English by Debuch (1957), has revised our knowledge of the basic chemistry of the plasmalogens. These authors noted that in the case of brain tissue it was impossible to separate acetal phosphatide from ethanolamine although the former is insoluble and the latter soluble in ether. They obtained from an acid-treated fraction of brain lipids, in addition to phosphatidyl ethanolamine (formula on p. 399), an ethanolamine-containing lysokephalin whose structure is given below:

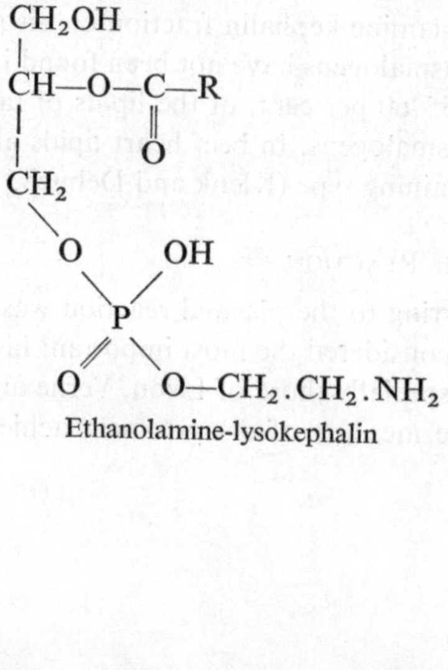

```
CH2OH
|
CH—O—C—R
|    ||
|    O
CH2
  \
   O      OH
    \    /
      P
    //  \
   O     O—CH2.CH2.NH2
```

Ethanolamine-lysokephalin

Since it was produced by acid treatment this component could only have come from the original phosphatide of the brain by splitting off of an aldehyde group. The formula of ethanolamine plasmalogen is therefore:

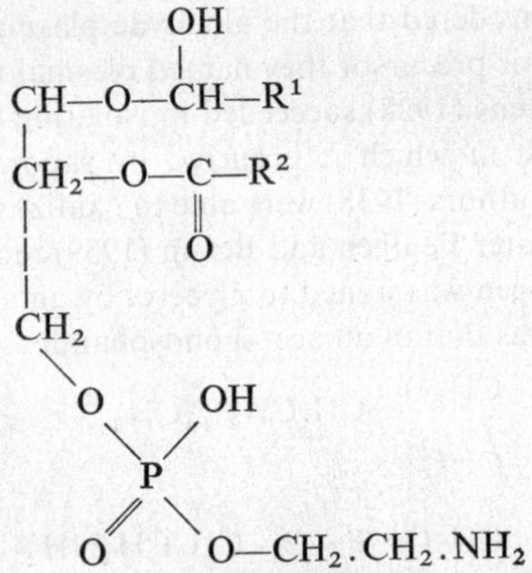

```
                OH
                |
  CH—O—CH—R¹
  |
  CH₂—O—C—R²
  |         ||
  |         O
  |
  CH₂
     \
      O      OH
       \    /
         P
       //  \
      O     O—CH₂.CH₂.NH₂
```

Ethanolamine plasmalogen

This structure is a hemiacetal. An alternative hypothesis is that of Rapport and Franzl (1957a and b) who suggested that the structure of plasmalogens might be that of an α–β unsaturated ether, $(CH—O—\overset{\alpha}{C}H{=}\overset{\beta}{C}H—R')$ in the formula above. Support for the vinyl ether structure came from observations by Blietz (1958) who hydrolyzed plasmalogen with tritiated water and obtained a radioactive aldehyde. This could only be derived from a plasmalogen with the Rapport-Franzl structure. This structure would explain the specific reaction with $HgCl_2$ (see pp. 465 and 466) and the findings of Norton (1959) who showed that $HgCl_2$ becomes an integral part of the phospholipid molecule and that its reaction is prevented by prior bromination or iodination.

Klenk and his co-workers found that two-thirds of the ethanolamine kephalin fraction of brain, obtained by Folch's method, consisted of acetal phosphatide. Stammler *et al.* (1954) have shown that the plasmalogen content of the myelin sheaths is twice that of the cerebral cortex, and in this case 55 per cent. of the ethanolamine kephalin fraction consists of the plasmalogen. Choline-containing plasmalogens have not been found in brain but Lovern *et al.* (1957) found that 55–60 per cent. of the lipids of ram spermatozoa were choline-containing plasmalogens. In beef heart lipids also, the plasmalogens are of the choline-containing type (Klenk and Debuch, 1954, 1955).

Histochemistry of the Reaction

The literature referring to the plasmal reaction was thoroughly reviewed by Cain (1949b), who considered the most important histochemical investigations into its nature, especially those of Lison, Verne and Gérard. Cain attributed the considerable measure of disagreement achieved by these authors partly to variations in the techniques employed. Both Cain (1949a) and Hayes (1949) wisely redefined the plasmal reaction in terms of a specific technique.

The prolonged investigations of Verne upon the subject of the plasmal reaction extended from 1928 to 1942 (Verne, 1928a, b and c, 1929a and b, 1936a and b, 1937, 1940; Verne and Verne-Soubiran, 1939, 1942). Verne observed (1928c) that various lipid structures could be made to give a positive plasmal reaction by oxidation with $KMnO_4$, CrO_3 or H_2O_2, and he considered that the groups responsible for the reaction were aldehydes produced by oxidation of hydroxyl groups in the phospholipin molecule, in other words that the aldehydes were derived from alcohols. Following on this reasoning Verne made the assumption that substances in the tissues which give a positive plasmal reaction without oxidative procedures must already exist in an oxidized state. In the same paper Verne also noted that prolonged oxidation with the reagents described would reverse the reaction of structures initially plasmal positive. Later (1929a), as the result of studies on the oxidation of pure fats, he decided that the aldehydes produced by oxidation were not formed from alcohols but from unsaturated (double) bonds. When they were derived from natural oxidative processes, the aldehyde groups were thought to exist in some loose combination in the tissues, which could be revealed by relatively mild procedures such as treatment with $HgCl_2$, for instance. Verne (1937) also showed that formaldehyde fixation prevented certain tissue lipids from giving a plasmal reaction in the untreated state, but that they could be made to react once more by means of $HgCl_2$.

Gérard (1935) studied the distribution of plasmal in various tissues and was the first to show that formol-fixed material could be used for such a study provided that thorough washing followed the action of the fixative. Unfortunately for our relatively simple conception of the plasmal reaction Gérard also studied the distribution of what he called "oxidases" by means of the "Nadi" reaction (Chapter 19) and found complete parallelism between the sites of plasmal and those containing "oxidase". Lison (1936a) confirmed Gérard's observations, but in addition he found that the positive "Nadi" reaction given by fats was not inhibited by cyanide and that the substance concerned would give a positive reaction with benzidine in the presence of a peroxidase. Lison therefore considered the "Nadi"-positive, benzidine-positive material to be a peroxide. Verne had concluded that the "Nadi" reaction recorded the process of autoxidation (peroxidation) of fats, at a time when it was actively proceeding, and that the plasmal reaction recorded the products of autoxidation. Later still (1940), however, he recorded his opinion that the "Nadi" reaction showed the presence of a secondarily acquired phenol oxidase system.

Oxidation of Fats. The process of oxidation of unsaturated fatty acids is now thought to proceed by the formation of a hydroperoxide (OOH) group on the carbon atom adjacent to the double bond, followed by the opening of this bond to form keto and hydroxy groups and, subsequently, by breaking of the carbon chain to form aldehydes. All the above groups may co-exist in any particular fatty acid undergoing oxidation. The initiation of lipid peroxidation by reduced metal ions was studied by Smith and Dunkley (1962) who found that the ferrous ion was much more effective than ferric ion and cuprous ion than cupric. According to their hypothesis the reduced metal ion produces a perhydroxyl radical:

$$Cu + O_2 \rightarrow CuO_2$$
$$CuO_2^+ + H^+ \rightarrow Cu^{++} + HO_2$$
$$RH + HO_2 \rightarrow R + H_2O_2$$
$$R + O_2 \rightarrow ROO$$
$$ROO + RH \rightarrow ROOH + R$$

Gomori (1942) dealt with the action of formalin on tissue lipids which he considered to be in the nature of an oxidation. He stressed the necessity of keeping tissues designed for the study of the plasmal reaction from contact with air or oxygenated water. Many samples of formalin contain appreciable quantities of reduced metal ions so that the mechanism proposed by Smith and Dunkley could clearly take place.

Some additional evidence for this explanation of Gérard's results, namely that they depend on the presence of a peroxide stage in the oxidation of fats, may be found in the following, at first sight unrelated, observations. Dam and Granados (1945a) demonstrated that the formation of peroxides preceded, and in some instances paralleled the production of a yellow-brown pigment in the adipose tissue of rats on a high cod-liver oil, low vitamin E, diet. They also (1945b) showed that the peroxidation effect of cod-liver oil was due to its highly unsaturated fatty acids. Glavind *et al.* (1949) developed a histochemical method for demonstrating these fat peroxides, using leuco-dichlorpheno-lindophenol with hæmin as catalyst, which gave a red colour at the sites of peroxidation, and using this method Granados and Dam (1950) found that in the early stages of brown pigment formation the parallelism previously observed chemically could be shown to exist histochemically (see Chapter 26, and Appendix 26). In later stages, however, a lack of parallelism was observed and these authors concluded that the insoluble yellow-brown pigment represented an abnormal product formed from highly unsaturated fatty acids beyond the fat-soluble, yellowish, peroxide stage. Beyond showing that the yellow-brown pigment was acid fast, no additional histochemical studies of its nature were attempted. Raviola and Raviola (1963) made an extensive cytochemical study of the reactions of the so-called "Gersh granules" of pituicytes. Their views, and those of the other authors referred to above are summarized in Table 31.

Original work by Lea (1956) indicated that the best naturally occurring antioxidant was vitamin E; other compounds having this property were the dihydrocaffeic acids, fluorones, tannins and catechol derivatives. Many acids, themselves possessing no intrinsic antioxidative property, considerably in-

TABLE 31

Histochemical Views on Fat Oxidation

Author	Reacting substance	Mechanism	Product	Demonstrating reaction
Feulgen and Voit (1924)	Acetal-phosphatides	$HgCl_2$	Aldehydes	Schiff
Verne (1940)	Unsaturated fatty acids	Oxidizing enzyme	Aldehydes	"Nadi" and Schiff
Gérard (1935)	Unsaturated fatty acids	Atmospheric O_2	"Oxidizer"	"Nadi"
Dam and Granados (1945a)	Unsaturated fatty acids	Atmospheric O_2	Peroxide	Leuco-indophenol
Lison (1936)	Unsaturated fatty acids	Atmospheric O_2	Peroxide	"Nadi" and Benzidine
Raviola and Raviola (1963)	Unsaturated neutral lipid	Formalin/O_2	Hydro-peroxide	Leuco-crystal violet

crease the anti-oxidative effects of phenolic substances. Among these acids are citric and phosphoric acids and ascorbic acid and kephalin. Davidow and Radomski (1953) regarded the oxidation of unsaturated compounds as a detoxication process. They suggested that the reactive C=C bond was first oxidized to an epoxide and this was then hydrolysed to a diol. Anhydrous HBr in dioxane converts epoxides into halohydrins and these can be hydrolysed back to epoxides with KOH.

$$\begin{matrix} H \\ | \\ H \end{matrix}\!\!>\!O \underset{KOH}{\overset{HBr}{\rightleftharpoons}} \begin{matrix} HOH \\ | \\ HBr \end{matrix}$$

Epoxide

This reaction has not been used histochemically. The whole subject of the autoxidation of fats and related substances was reviewed very ably by Holman (1954), by Wolman (1961), and by Harms (1965), whose articles should be consulted for further details.

Criticisms of the Plasmal Reaction. In view of more modern experience in the oxidation of fats, criticism can be levelled at the method of applying the plasmal reaction employed by Feulgen and his collaborators. They used unfixed sections and attached these to the slides by heating, a process which might well lead to initial oxidation especially when subsequently reinforced by the action of $HgCl_2$. Feulgen particularly stressed the necessity for examining control sections unexposed to $HgCl_2$ and the techniques of Verne may be criticized on the grounds that no control sections were employed and also because he occasionally used $HgCl_2$ as a fixative, which precluded the use of controls. Danielli (1949), in his critical study of the plasmal reaction, raised and answered a number of questions. Using three additional techniques for localizing aldehydes (ammoniacal silver, phenylhydrazine and a substituted phenyl-hydrazide) he found that with each of these the distribution of plasmal was the same as with Schiff's reagent. He concluded that these additional reactions confirmed the specificity of the plasmal reaction for tissue aldehydes. The possibility of diffusion of the aldehyde-Schiff compound was investigated and rejected. Cain (1949b) criticized the plasmal techniques used by Danielli because an acid fixative was used (acetals were shown by Feulgen and Bersin to be destroyed in acid media), and because of a supposed lack of control sections. Danielli (1950) replied to these criticisms with a restatement of his method of distinguishing free aldehydes from acetals and from aldehydes produced by oxidation. He regarded the Schiff-positive groups produced by $HgCl_2$ as aldehydes produced by oxidation.

The True Plasmal Reaction. After some careful experimental work Cain (1949a) drew a distinction between the true Feulgen-plasmal reaction and the Feulgen-Verne or pseudoplasmal reaction, considering Danielli's plasmal to belong to the second variety. According to his definition the true plasmal reaction was due to the release of higher fatty aldehydes from acetal phosphatides by the action of $HgCl_2$. It was progressively reduced and finally destroyed by prolonged fixation. The Feulgen-Verne or pseudoplasmal reaction, on the other hand, increased in intensity with the duration of fixation provided oxygen was not excluded. Mercuric chloride did not influence the pseudoplasmal reaction. The investigations of Waelsch (1950) showed that caution was necessary in evaluating the higher fatty aldehydes detected by Schiff's reagent. Both the age of the preparation and the accessibility of the lipids were shown to influence the amounts detected *in vitro*. After five crystallizations as the cadmium complex Waelsch found that a sample of lecithin, stored for 4 days over sodium hydroxide at 15 mm./Hg. gave a fuchsin colour value equal to 70 μg. palmitaldehyde. After 8 days the value was 185 μg. and after 14 days 235 μg. per 100 mg. These figures he attributed to the progressive oxidation of lecithin to higher fatty aldehydes. The evidence produced by Verne to show that aldehydes were responsible for his reaction was based on the assumptions that all positive plasmal reactions were due to a single component and that the techniques which he used for revealing and for blocking the aldehydes (see below) were specific for these compounds. According to Cain such assumptions were unjustified.

Modern Views on the Plasmal Reaction. The results obtained by Hayes (1947, 1949) substantially agreed with those of Cain. Hayes maintained that prolonged fixation in formalin would oxidize all acetal groups and unmask Schiff-positive material which was unaffected by $HgCl_2$ in control sections. This was certainly true in the case of the adrenal cortex with which he was mainly concerned. He regarded the reaction secondarily developed by formalin fixation as demonstrating aldehyde-containing lipids other than acetals, whereas Cain believed that groups other than aldehyde, also produced by autoxidation, were responsible. Hack (1952) made a very complete study of the distribution of plasmals in animal tissues, using freeze-dried material embedded in carbowax, and Belt and Hayes (1954) made a critical evaluation of procedures for carrying out the reaction.

A satisfactory explanation of the mechanism by which $HgCl_2$ gives rise to the plasmal reaction was provided by Terner and Hayes (1961). These authors found that only highly ionized mercury compounds, whether mono- or divalent, would produce the classical plasmal reaction at its full intensity. They showed, furthermore, that mercury was always incorporated into the plasmalogen-lipid, in agreement with Norton's (1959) findings, and postulated that it must be added to the β-carbon with concomitant addition of a hydroxyl group to the α-carbon and subsequent hydrolysis of the mercury-containing hemiacetal.

Terner and Hayes also investigated the effect of pH on the plasmal reaction and found that 6 N-HCl produced the same rapid hydrolysis as $HgCl_2$. The rapid and complete blocking of the plasmal reaction by bromination proves that tissue plasmalogens contain a double bond, and that the latter is necessary for the reaction.

In Appendix 13 the methods of Cain and Hayes are both given, but Hack's method can be used as an alternative to these.

Present Status of the Reaction. To most histologists the plasmal reaction is of little importance in itself, but it is an important complication in the interpretation of other reactions involving Schiff's reagent. The established position of the Feulgen nucleal reaction and the growing importance of the periodic acid-Schiff technique therefore necessitate a thorough understanding of the problems involved and of the nature of substances, outside the main ones with which either technique is concerned, which may give positive reactions with Schiff's solution. As a diagnostic procedure in histochemistry the true plasmal reaction may be of value in suggesting or confirming the presence of ethanolamine or choline-containing phospholipids in any particular structure, provided that neutral fats can be shown to be absent. Since the detection of these phospholipids by histochemical means, when they are present in small quantities, is made difficult by the insensitivity of the methods available, we can take advantage of the sensitivity of the true plasmal reaction to make a presumptive diagnosis.

Reactions of Aldehyde Reagents with Ketones

In the foregoing account of the various aldehyde reactions used in histochemistry references to ketones were purposely kept to a minimum to avoid confusion. Many of the reactions dealt with, however, were known to be equally positive with ketones *in vitro* and a complication in interpretation thus existed. This was largely resolved by the work of Karnovsky and Deane (1955) which brought to an end the protracted arguments of the previous decade. This work, which is considered at the end of this chapter, makes much of the intervening discussion on the reactivity of the various reagents wholly academic.

Schiff's Reagent

Many groups other than aldehydes are known to recolorize fuchsin-sulphurous acid, but the most important ones are either oxidizers or α-ketones. Dempsey and Bassett (1943) and Dempsey and Wislocki (1944, 1946) claimed that α-hydroxyketones (α-ketols) such as deoxycorticosterone would recolorize Schiff's solution after treatment with $HgCl_2$. They supposed that this treatment would convert the ketone into an aldehyde. Boscott *et al.* (1948) performed the true plasmal reaction on samples of free deoxycorticosterone but could obtain no evidence of oxidation to an aldehyde. They considered that typical hydroxyketones would not react as Dempsey had suggested but that atypical α-ketols might possibly do so. Albert and Leblond (1946), Oster and Oster (1946), Ashbel and Seligman (1949) and Nicander (1951) also failed to achieve the plasmal reaction with ketosteroids. Karrer (1950) agreed that Schiff's reagent itself would react with certain easily oxidized α-ketols but not with the majority of ketones. In histochemical practice, therefore, the presence of ketones is not regarded as responsible for a positive Schiff reaction.

A new histochemical method for α-ketolic steroids was proposed by Khanolkar *et al.* (1958). This was based on oxidation of the α-ketol group with $FeCl_3$, and subsequent demonstration of the presumed keto-aldehyde with Schiff's reagent. The theoretical specificity of this method appeared to be high and, although not fully tested in this respect, it was inserted into the Appendix of the appropriate chapter in the second edition of this book. It was subsequently modified by Penar (1961) and employed by several workers to demonstrate 20-ketosteroids in different tissues (Niwelinski, 1961; Borowicz, 1963; Romer, 1964; Mietkiewski *et al.*, 1966).

The specificity of the Khanolkar reaction was tested, *in vitro*, by Adams (1965), using paper impregnated with lipids. A positive reaction was given by plasmalogen, sulphatide, triolein, cholesterol oleate, linolenic acid, and a number of other compounds. With cortisone the reaction was negative. In view of these observations it is no longer possible to uphold the specificity of the reaction and, although details are given in Appendix 13, it cannot be recommended for applied histochemical investigations. The most likely mechanism for the reaction is plasmalogen conversion (Bergmann and Liebrecht, 1958).

Ammoniacal Silver Reagents

Many reducing reagents will convert ammoniacal silver solutions to metallic silver. Their reduction has been used for the demonstration of aldehydes and certain special types of ketones, particularly those with a carbonyl group at C^{20} and an adjacent hydroxyl at C^{21}. Both Bennett (1940) and Reichstein and Shoppee (1943) were of the opinion that the reaction could demonstrate the presence of ketosteroids in tissue sections, yet proof that aldehydes were not responsible was lacking. It is not permissible to use

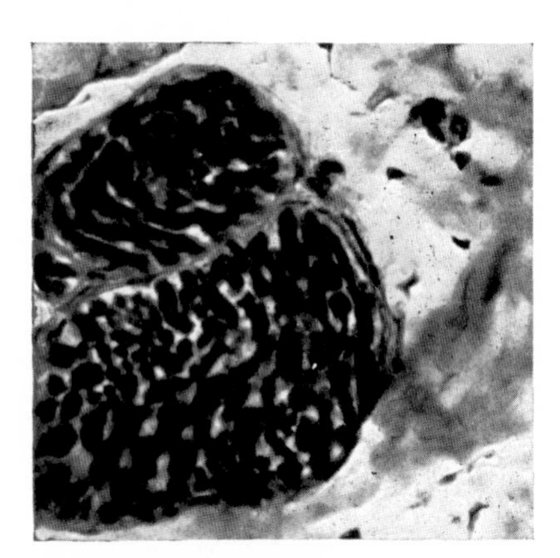

Fig. 117. Rat. Peripheral nerve. The myelin sheaths stain deep blue in this fresh frozen section. Acid Solochrome cyanine. × 360.

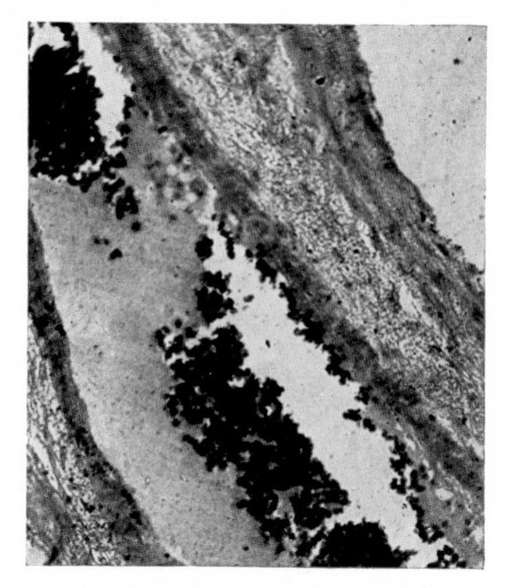

Fig. 118. Masses of black-stained red cells within a blood vessel. Performic acid —ammoniacal silver. × 100.

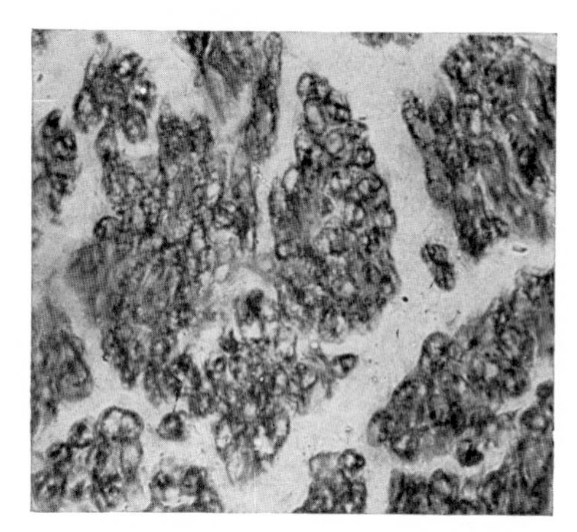

Fig. 119. Myelin sheaths stained a dark purple colour. Performic acid—naphthoic acid hydrazide—diazotized *o*-dianisidine. × 300.

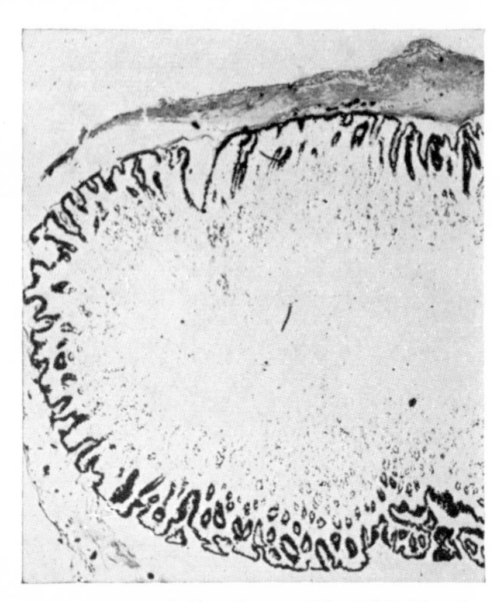

FIG. 120. Serial section to Fig. 121, identically treated except for incubation with dimedone in 5 per cent. acetic acid (60° 2 hours) after oxidation. Schiff. × 21.

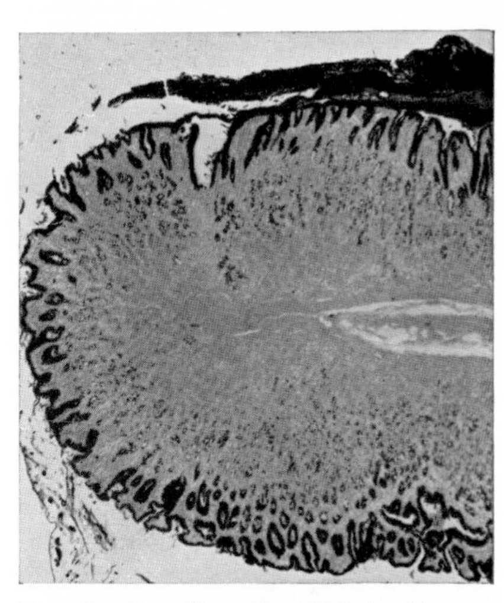

FIG. 121. Paraffin section (5·5 μ) of human gastric mucosa. Oxidized for 10 minutes at 22° with 1 per cent, aqueous HIO_4. Schiff. × 21.

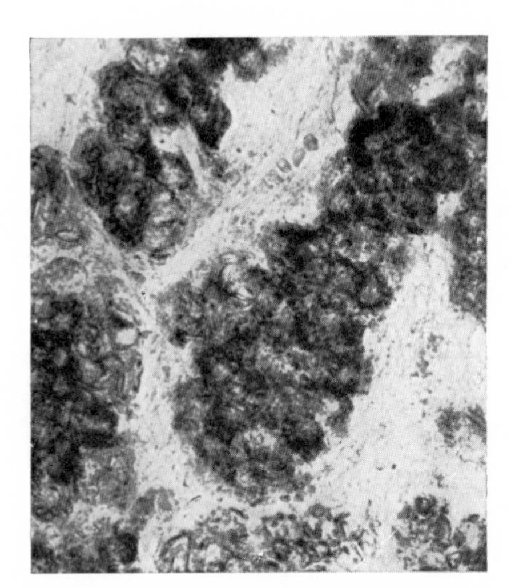

FIG. 122. Frozen section (10 μ) of human adrenal cortex, showing a strong reaction in the zona fasciculata. Naphthoic acid hydrazide—diazotized *o*-dianisidine. × 335.

ammoniacal silver reagents to demonstrate ketones histochemically, since their specificity for this purpose is altogether too low.

Phenylenediamine Reaction

Observing that according to Woker (1914) the oxidation velocity of peroxide with *p*-phenylenediamine is greatly increased in the presence of aldehydes, Scarselli (1961) introduced a new histochemical reaction for aldehydes. Details of this reaction are given in Appendix 13, p. 709. A positive result occurs only with aldehydes, nitriles, aldehyde-ammonia and aldehyde-bisulphites.

Phenylhydrazine Reagents

The specificity of the Bennett and similar reactions for ketosteroids was questioned by Gomori (1942) who pointed out that the reaction was certainly given by tissue aldehydes. Later (1950) he concluded that a positive phenylhydrazine reaction signified only an overwhelming bulk of plasmal, and later still (1953) that hydrazines absolutely did not react with ketones *under histochemical conditions*. This conclusion has not been refuted although there is plenty of evidence that steroids and hydrazides react *in vitro*. Progesterone, for instance, reacts at C3 and C20 with two molecules of hydrazides. The work of Albert and Leblond (1946), Claesson and Hillarp (1947), Rogers and Williams (1947) (and see review by Anderson, 1948), supported Gomori's opinions. Boscott and Mandl (1949) used the amine-aldehyde condensation of Oster and Mulinos (see below) to block the reaction of the adrenal cortex with 2:4-dinitrophenylhydrazine. They showed, *in vitro*, that the reagents used (sulphanilic acid or sulphonamide) would not react with either dehydroisoandrosterone or with deoxycorticosterone acetate. If either of these, or similar steroids, remained in the sections they should have given a positive phenylhydrazine reaction according to these authors. The extraction experiments of Olson *et al.* (1944) showed that the ketosteroid content of the tissues is minute, and Rogers and Williams (1947) calculated that 0·017 μg. of ketosteroid might be present in an average frozen section of adrenal. They considered that such an amount could not possibly be demonstrated by the phenylhydrazine reaction even if condensation took place. This view overlooks the fact that the local concentration of the steroid hormones and their inactive precursors may be very high. If all the steroid in the adrenal cortex was present in a single layer of cells its concentration per cell might be high enough to give a positive histochemical reaction. These arguments were considered at length by Deane and Seligman (1953) in their review of the known procedures for the cytological localization of ketosteroids.

Naphthoic Acid Hydrazide

The naphthoic acid hydrazide-dianisidine (NAHD) reaction was originally described as specific for ketosteroids and was employed for this purpose particularly upon frozen sections of adrenal glands. Camber used watery solutions, both for the hydrazide and for coupling, and inserted a short period of oxidation with iodine between cutting the sections and immersion in the hydrazide solution. He noted that this manœuvre improved the final colour considerably. In the light of our knowledge of the plasmal and pseudoplasmal reactions this point is of particular significance, suggesting the production of lipid aldehyde groups by oxidation with iodine. Ashbel and Seligman used watery solutions in their earlier experiments but rejected these later as giving inferior results. They subsequently employed 50 per cent. alcoholic solutions, buffered at pH 7·5 for the final coupling with diazotized dianisidine. Fig. 122, facing p. 469, shows the results of the NAHD reaction in human adrenal cortex.

These authors were well aware of the fact that naphthoic acid hydrazide would condense with aldehydes and they realized that the specificity of the NAHD reaction for ketosteroids depended on whether tissue aldehydes were involved or not. They maintained, however, that this point depended on whether the hydrolysis of plasmalogens to plasmal occurred under the conditions involved although, in fact, the chief aldehydes present would be those formed from lipids during formalin fixation.

In answering the question to their own satisfaction, Ashbel and Seligman made much of the statement by Hayes (1949) that exposure of tissues to formalin for more than 6 hours would hydrolyse plasmalogen to plasmal, but that further exposure would destroy all the liberated plasmal. In the case of ketosteroids, these authors claimed that the action of formalin was absolutely necessary for unmasking the reactive carbonyl group, but once this had been accomplished (1 or more hours) deterioration of the staining reaction did not occur with further exposure to formalin (weeks). Feldman (1950), however, found that 48 hours' treatment with formalin was necessary to produce the reaction in sections of adrenal cortex, and that after 3 months' exposure the cortical cells ceased to react either with fuchsin-sulphurous acid or with NAHD. My own results (1950) confirmed the gradual deterioration of the NAHD reaction with prolonged formalin treatment, but I found that 18 hours in formalin, and possibly an even shorter time, was quite sufficient to produce a strong reaction. The explanation offered by Ashbel and Seligman in their denial of the participation of plasmal in the NAHD reaction overlooked the important fact, established by Cain (1949a), that oxidation of the double bonds in unsaturated fatty acids by atmospheric oxygen produces what this author has termed the pseudo-plasmal reaction. Autoxidation of fatty acids to aldehydes proceeds quite readily in solutions of formalin and there is every likelihood that such aldehydes are responsible for the positive Schiff and NAHD reactions in sections of adrenal fixed even for short periods in formalin. This explanation is now usually regarded as correct (Adams, 1965).

Three further pieces of evidence in favour of the specificity for ketosteroids of the NAHD reaction were offered by Ashbel and Seligman, however. The first was the fact that the four monoketosteroids, which they used for *in vitro* tests, failed to recolorize Schiff's solution, though three of them (testosterone, œstrone and pregnenolone) representing 3, 17 and 20 ketosteroids, combined readily with NAH. This was interpreted to mean that while Schiff's reagent revealed aldehydes only, NAHD revealed both aldehydes and ketones. Ashbel and Seligman noted in their experiments that no reaction with Schiff's reagent occurred in control sections of adrenals fixed for a longer or shorter period in formalin. This observation was quite inexplicable; Feldman's results and my own agreed in showing that a positive Schiff reaction invariably occurred in control sections.

The second piece of evidence was the negative reaction obtained by the phenylhydrazone-formazan method. Arguments put forward in Chapter 10, p. 326, will indicate that the conditions used for this reaction at that time were far from optimal and a negative reaction clearly had no significance.

The third piece of evidence (Seligman and Ashbel, 1951) was the observation that excess sodium *o*-sulphobenzaldehyde could cause reversal of the reaction of NAH with both ketones and aldehydes *in vitro*. But whereas in the case of ketones reversal occurred rapidly, in the case of aldehydes it was relatively slow. This reaction was applied to tissue sections, and nervous tissue which had been treated with the hydrazide was incubated for 24 hours at 22° in a solution of 5 per cent. sulphobenzaldehyde. After washing in dilute sodium bicarbonate and in water for 24 hours, subsequent treatment with the diazonium salt gave an intensity of staining "strikingly less" than that observed in the controls. The authors suggested that this indicated that the carbonyl group in formalin-fixed nervous tissue was ketonic in nature.

This reversal reaction was used by Stoward and Adams Smith (1964) who observed that ketones were restored from their methyl hydrazones by 2 hours' treatment with sulphobenzaldehyde, whereas aldehyde hydrazones required 18 hours for even partial restoration. Their method is given, in full, in the Appendix.

Salicyloyl Hydrazide

Camber (1954, 1957) introduced salicyloyl hydrazide (I) as a fluorescent reagent for aldehydes and ketones. He observed that the hydrazones formed by this compound with aldehydes gave brilliant fluorescence in many colours when exposed to UV light. Its ketonic hydrazones gave a uniform dull blue fluorescence. A non-fluorescent hydrazide (β-resorcylic acid hydrazide) was prepared by Camber and Dziewiatskowski (1951) for use as an alternative to naphthoic acid hydrazide. Its hydrazones could be coupled with diazonium salts to give purple or red compounds.

Testing the fluorescent reaction Chen (1959) found that aldehyde-salicyloyl hydrazones (II) possessed the same fluorescence characteristics as the unreacted hydrazide I

$$\text{C}_6\text{H}_4(\text{OH})\text{CONH}_2\text{NH}_2 \; (\text{I}) + \text{O=C}(\text{R}^1)(\text{R}^2) \longrightarrow \text{C}_6\text{H}_4(\text{OH})\text{CONH-N=C}(\text{R}^1)(\text{R}_2) \; (\text{II})$$

It is therefore fortunate that excess salicyloyl hydrazide can be extracted from tissue sections by rinsing them briefly in a freshly prepared solution of trisodium pentacyanoammine ferroate (III) (Camber, 1957) with which it forms a deep purple water-soluble compound (IV).

$$Na_3[Fe(CN)_5NH_3] \; (\text{III}) + [\text{I}] \longrightarrow Na_3[Fe(CN)_5NH_2.NHCO\text{-}C_6H_4\text{-}OH] \; (\text{IV}) + NH_3$$

The whole sequence was used by Stoward and Adams-Smith (1964) for the demonstration of ketones regenerated from their hydrazones (see above, p. 471). Sites showing a greenish-yellow fluorescence were presumed to contain ketosteroids.

REFERENCES

ADAMS, C. W. M. (1965). "Neurohistochemistry," Elsevier, Amsterdam, p. 31.
ALBERT, S., and LEBLOND, C. P. (1946). *Endocrinology*, **39**, 386.
ANDERSON, E. (1948). *Ann. Rev. Physiol.*, **10**, 329.
ANGELI, A. (1892). *Gazz. chim. ital.*, **26**, 17.
ASHBEL, R., and SELIGMAN, A. M. (1949). *Endocrinology*, **44**, 565.
BARKA, T., and ORNSTEIN, L. (1959). *J. Histochem. Cytochem.*, **7**, 385.
BARKA, T., and ORNSTEIN, L. (1960). *Ibid.*, **8**, 208.
BELT, W. D., and HAYES, E. R. (1954). *Anat. Rec.*, **118**, 282.
BENNETT, H. S. (1939). *Proc. Soc. exp. Biol., N.Y.*, **42**, 786.
BENNETT, H. S. (1940). *Amer. J. Anat.*, **67**, 151.
BERGMANN, H., and LIEBRECHT, E. K. (1958). *Biochem. Z.*, **312**, 51.
BLIETZ, R. J. (1958). *Hoppe-Seylers Z. physiol. chem.*, **310**, 120.
BOROWICZ, J. W. (1963). *Folia histochem. cytochem.*, **1**, 253.
BOSCOTT, R. J., MANDL, A. M., DANIELLI, J. F., and SHOPPEE, C. W. (1948). *Nature, Lond.*, **162**, 572.
BOSCOTT, R. J., and MANDL, A. M. (1949). *J. Endocrinol.*, **6**, 132.
CAIN, A. J. (1949a). *Quart. J. micr. Sci.*, **90**, 75.
CAIN, A. J. (1949b). *Ibid.*, **90**, 411.
CAMBER, B. (1949). *Nature, Lond.*, **163**, 285.
CAMBER, B. (1954). *Ibid.*, **174**, 1107.
CAMBER, B. (1957). *Clin. chim. Acta*, **2**, 188.
CAMBER, B., DZIEWIATSKOWSKI, D. D. (1951). *J. Amer. chem. Soc.*, **73**, 4021.
CHAIKIN, S. W., and BROWN, W. G. (1949). *J. Amer. chem. Soc.*, **71**, 122.
CHEN, P. S. (1959). *Analyt. Chem.*, **31**, 396.

CLAESSON, L., and HILLARP, D. A. (1947). *Acta anat.*, **3**, 109.
DAM, H., and GRANADOS, H. (1945a). *Acta physiol. scand.*, **10**, 162.
DAM, H., and GRANADOS, H. (1945b). *Science*, **102**, 327.
DANIELLI, J. F. (1947). *Symp. Soc. exp. Biol.*, **1**, 101.
DANIELLI, J. F. (1949). *Quart. J. micr. Sci.*, **90**, 67.
DANIELLI, J. F. (1950). *Ibid.*, **91**, 215.
DAVIDOW, B., and RADOMSKI, J. L. (1953). *J. Pharmacol.*, **107**, 259.
DEANE, H. W., and SELIGMAN, A. M. (1953). In "Vitamins and Hormones," **11**, 173.
DEBUCH, H. (1957). In "Cerebral Lipoidoses." Blackwell, Oxford, p. 203.
DEMPSEY, E. W., and BASSETT, D. L. (1943). *Endocrinology*, **33**, 384.
DEMPSEY, E. W., and WISLOCKI, G. B. (1944). *Ibid.*, **35**, 409.
DEMPSEY, E. W., and WISLOCKI, G. B. (1946). *Physiol. Rev.*, **26**, 1.
DUIJN, P. VAN, and HARDONK, M. J. (1966). *J. Histochem. Cytochem.*, **14**, 683.
ELFTMAN, H. (1959). *J. Histochem. Cytochem.*, **7**, 93.
FEIGL, F., ANGER, V., and FREHDEN, O. (1934). *Mikrochemie*, **15**, 184.
FELDMAN, J. D. (1950). *Anat. Rec.*, **107**, 347.
FEULGEN, R., and BEHRENS, M. (1928). *Z. physiol. Chem.*, **177**, 221.
FEULGEN, R., and BEHRENS, M. (1938). *Ibid.*, **256**, 15.
FEULGEN, R., and BERSIN, K. (1939). *Ibid.*, **260**, 217.
FEULGEN, R., and ROSSENBECK, H. (1924). *Ibid.*, **135**, 203.
FEULGEN, R., and VOIT, K. (1924). *Pflüg. Arch. ges. Physiol.*, **206**, 389.
GÉRARD, P. (1935). *Bull. Histol. Tech. micr.*, **12**, 274.
GEYER, G. (1963). *Acta histochem.*, **15**, 1.
GILMAN, H. (1943). "Organic Chemistry." New York. Quoted by Lhotka and Davenport, 1951.
GLAVIND, J., GRANADOS, H., HARTMAN, S., and DAM, H. (1949). *Experientia*, **5**, 84.
GOMORI, G. (1942). *Proc. Soc. exp. Biol., N.Y.*, **51**, 133.
GOMORI, G. (1950). *Ann. N.Y. Acad. Sci.*, **50**, 968.
GOMORI, G. (1952). *J. Lab. clin. Med.*, **39**, 649.
GOMORI, G. (1953). *J. Histochem. Cytochem.*, **1**, 381, 389.
GRANADOS, H., and DAM, H. (1950). *Acta path. scand.*, **27**, 591.
HACK, M. H. (1952). *Anat. Rec.*, **112**, 275.
HARDONK, M. J., and DUIJN, P. VAN (1964a). *J. Histochem. Cytochem.*, **12**, 533.
HARDONK, M. J., and DUIJN, P. VAN (1964b). *J. Histochem. Cytochem.*, **12**, 748.
HARMS, H. (1965). "Handbuch der Farbstoffe für die Mikroscopie." Staufen Verlag, Lintfort, II, 264.
HAYES, E. R. (1947). *Anat. Rec.*, **97**, 391.
HAYES, E. R. (1949). *Stain Tech.*, **24**, 19.
HIRAOKA, T. (1960). *J. biophys. biochem. Cytol.*, **8**, 286.
HOLMAN, R. T. (1954). In "Progress in the Chemistry of Fats and Other Lipides," Vol. **12**, pl. 51.
HÖRMANN, H., GRASSMAN, W., and FRIES, G. (1958). *Liebig's Ann. Chim.*, **616**, 125.
JOBST, C., and HORVÁTH, A. (1961). *J. Histochem. Cytochem.*, **9**, 711.
KARNOVSKY, M. L., and DEANE, H. W. (1955). *J. Histochem. Cytochem.*, **8**, 85.
KARRER, P. (1950). "Organic Chemistry." 4th English Ed. Elsevier, New York.
KASTEN, F. H. (1960). *Int. Rev. Cytol.*, **10**, 1.
KHANOLKAR, V. R., KRISHNAMURTHI, A. S., BAGUL, C. D., and SAHASRABUDHE, M.B. (1958). *Indian J. Path. Bact.*, **1**, 84.
KLENK, E., and BÖHM, P. (1951). *Hoppe-Seyl. Z.*, **288**, 98.
KLENK, E., and DEBUCH, H. (1954). *Ibid.*, **296**, 179.
KLENK, E., and DEBUCH, H. (1955). *Ibid.*, **299**, 66.
LACOSTE, R. G., and MARTELL, A. E. (1955). *J. Amer. chem. Soc.*, **77**, 5512.
LEA, C. H. (1956). *Nature, Lond.*, **178**, 776.
LHOTKA, J. F., and DAVENPORT, H. A. (1951). *Stain Tech.*, **26**, 35.
LILLIE, R. D. (1954). "Histopathologic Technic and Practical Histochemistry." Blakiston, N.Y., p. 159.
LILLIE, R. D. (1956). *J. Histochem. Cytochem.*, **4**, 479.
LILLIE, R. D., and GLENNER, G. G. (1957). *Ibid.*, **5**, 167.
LISON, L. (1932). *Bull. Histol. Tech. micr.*, **9**, 77.

LISON, L. (1936a). *Bull. Soc. Chim. biol.*, **18**, 185.
LISON, L. (1936b). "Histochimie Animale." Gautier Villars, Paris.
LOVERN, J. A., OLLEY, J., HARTREE, E. F., and MANN, T. (1957). *Biochem. J.*, **67**, 630.
MIETKIEWSKI, K., LUKASZYK, A., and KOPACZYK, F. (1966). *Folia histochem. cytochem.*, **4**, 73.
NAUMAN, R. V., WEST, P. W., TROU, F., and GEAKE, G. C. (1960). *Analyt. Chem.*, **32**, 1307.
NICANDER, L. (1951). *Acta Anat.*, **12**, 174.
NIWELINSKI, J. (1961). *Folia biol.*, **9**, 263.
NORTON, W. T. (1959). *Nature*, **184**, 1144.
ÖLANDER, A. (1927). *Z. physiol. Chem.*, **129**, 1.
OLSON, R. E., JACOBS, F. A., RICHERT, D., THAYER, H., KOPP, L. J., and WADE, N. J. (1944). *Endocrinology*, **35**, 430.
OSSE, K., and GEYER, G. (1964). *Acta histochem.*, **17**, 159.
OSTER, K. A., and MULINOS, M. G. (1944). *J. Pharmacol.*, **80**, 132.
OSTER, K. A., and OSTER, J. G. (1946). *Ibid.*, **87**, 306.
OSTER, K. A., and SCHLOSSMAN, N. C. (1942). *J. cell. comp. Physiol.*, **20**, 373.
PENAR, B. (1961). *Folia morphol.*, **20**, 55.
RAPPORT, M. M., and FRANZL, R. E. (1957a). *J. biol. Chem.*, **225**, 851.
RAPPORT, M. M., and FRANZL, R. E. (1957b). *J. Neurochem.*, **1**, 303.
RAVIOLA, E., and RAVIOLA, G. (1963). *J. Histochem. Cytochem.*, **11**, 176.
REICHSTEIN, T., and SHOPPEE, C. W. (1943). *Vitamins and Hormones*, **1**, 345.
RIMINI, E. (1901). *Gazz. chim. ital.*, **31**, 84.
ROGERS, W. J., and WILLIAMS, R. H. (1947). *Arch. Path.*, **44**, 126.
ROMER, T. E. (1964). *J. Histochem. Cytochem.*, **12**, 646.
RUMPF, P. (1935). *Ann. Chim.*, **3**, 327.
SCARSELLI, V. (1961). *Nature*, **190**, 1206.
SCHIFF, H. (1866). *Liebig's Ann. Chem.*, **140**, 92.
SELIGMAN, A. M., and ASHBEL, R. (1951). *Cancer*, **4**, 579.
SLOANE-STANLEY, G. H., and BOWLER, L. M. (1962). *Biochem. J.*, **85**, 34P.
SMITH, G. J., and DUNKLEY, W. L. (1962). *Arch. Biochem.*, **98**, 46.
STAMMLER, A., STAMMLER, U., and DEBUCH, H. (1954). *Hoppe-Seyl. Z.*, **296**, 80.
STAPLE, P. H. (1958). *Nature*, *Lond.*, **181**, 288.
STOWARD, P. J. (1963). *D. Phil. Thesis, Oxford.*
STOWARD, P. J. (1966). *J. Histochem. Cytochem.*, **14**, 681.
STOWARD, P. J., and ADAMS SMITH, W. N. (1964). *J. Endocrinol.*, **30**, 273.
SWEAT, H., and PUCHTLER, H. (1964). *J. Histochem. Cytochem.*, **12**, 392.
TERNER, J. Y., and HAYES, E. R. (1961). *Stain Tech.*, **36**, 265.
VERNE, J. (1928a). *C.R. Ass. Anat.*, **23**, 465.
VERNE, J. (1928b). *Bull. Soc. Neurol.*, **1**, 722.
VERNE, J. (1928c). *C.R. Soc. Biol., Paris*, **99**, 266.
VERNE, J. (1929a). *Ann. Physiol.*, **5**, 245.
VERNE, J. (1929b). *Arch. Anat. micr.*, **25**, 137.
VERNE, J. (1936a). *C.R. Soc. Biol., Paris*, **121**, 609.
VERNE, J. (1936b). *Bull. Histol. Tech., micr.*, **13**, 433.
VERNE, J. (1937). *C.R. Ass. Anat.*, **32**, 1.
VERNE, J. (1940). *C.R. Soc. Biol., Paris*, **133**, 75.
VERNE, J., and VERNE-SOUBIRAN, A. (1939). *Ibid.*, **130**, 1232.
VERNE, J., and VERNE-SOUBIRAN, A. (1942). *Bull. Histol. Tech., micr.*, **19**, 57.
WAELSCH, H. (1950). *Ann. N.Y. Acad. Sci., Wash.*, **50**, 978.
WIELAND, H., and SCHEUING, G. (1921). *Ber. dtsch. chem. Ges.*, **54**, B, 2527.
WILD, F. (1947). "Characterization of Organic Compounds." Cambridge University Press, pp. 110 nad 136.
WOKER, G. (1914). *Ber.*, **47**, 1024.
WOLMAN, M. (1961). *Ann. Histochim.*, **6**, 329.

CHAPTER 14

THE PRINCIPLES OF HYDROLYTIC ENZYME HISTOCHEMISTRY

General Survey of Enzyme Histochemistry

In the first (1953) edition of this book the following statement appeared in the preface to the section of four chapters on enzyme methods. "Although a few years ago only 2 or 3 enzymes could be demonstrated in the tissues by histochemical means, there are now techniques for at least 18." In the next five years advances in the field of enzyme histochemistry were so great that the sentence quoted had to be amended to read "for at least 45." This rapid expansion was accompanied by improvements in technique affecting the majority of the enzymes recorded in the 1953 list. At that time it was notable that the majority of methods concerned the hydrolytic enzymes, mainly phosphatases and carboxylic acid esterases, and that other enzyme groups were poorly represented. No methods existed for the various amino-acid oxidases, urease, arginase or peptidases. Five years later, methods had been described for D-amino-acid oxidase and leucine aminopeptidase, while enzymes of the cathepsin class could be demonstrated by a combination of existing methods with suitable activators and inhibitors. Eight years later methods for about 75 enzymes have been described and at least another 30 could be localized by simple adaptations of existing techniques.

Considered against a probable total of nearly 900 enzymes (Dixon and Webb, 1958, 1964; I.U.B. Commission on Enzymes Report 1961), the histochemical total of less than 80 may not seem impressive. If, however, we separate off bacterial, fungal and plant enzymes the total number in vertebrate tissues is less than 400. If we review the distribution of this 20 per cent. histochemically covered we find there are still some large enzyme groups for which no methods exist. Chief among the latter are the *isomerases* and *ligases*. Only one example, carbonic anhydrase, is found among the carbon-oxygen lyases and only one, *aldolase*, among the aldehyde lyases. The carboxylyases (decarboxylases) are not represented by conventional methods, although no insuperable barrier to their demonstration appears to exist. Their presumptive presence can be indicated by the use of the fluorescent amine technique (Chapter 26) after administration of fluorogenic amine precursors (amino-acids).

In the large group of transferring enzymes the division of oxidizing and reducing enzymes now contains many histochemically demonstrable members. The *peroxidases* are represented by the enzyme which gives its name to the whole sub-division and the aerobic oxidases, by *cytochrome oxidase* and *catechol oxidase*. The aerobic dehydrogenases are somewhat inadequately represented by D-*amino-acid oxidase*, but other enzymes in this division will doubtless be added in the near future. In the division of flavoprotein enzymes (reduced co-enzymes dehydrogenases) the so-called *$NADH_2$ and $NADPH_2$-diaphorases* can be demonstrated and the anærobic *dehydrogenases* are probably better represented histochemically than any other class of enzyme.

We see, therefore, that histochemical coverage of the various enzyme classes continues to extend. There is, of course, no room for complacency, but certainly enzyme histochemistry has come to occupy an important place in relation to enzymology and enzyme biochemistry. Its application to pathology in particular will extend the horizons of that science far beyond the level at which they have remained fixed since Virchow published his great work, *Cellular Pathology*, 100 years ago.

A list of the available techniques in enzyme histochemistry is given in Table 32, below, and against each enzyme appears the number allocated to it

TABLE 32

1.1.1.1	Alcohol dehydrogenase	2.4.1.1	α-Glucan phosphorylase
1.1.1.1	Pyridoxal dehydrogenase	2.4.1.11	UDP-glucose-glucan glucosyltransferase
1.1.1.8	Glycerol phosphate dehydrogenase	2.4.1.18	α-Glucan branching glucosyltransferase
1.1.1.14	Sorbitol dehydrogenase	2.7.5.1	Phosphoglucomutase
1.1.1.22	UDPG dehydrogenase	2.7.7.16	Ribonuclease
1.1.1.27	Lactate dehydrogenase	3.1.1.1	Carboxylesterase
1.1.1.30	3-Hydroxybutyrate dehydrogenase	3.1.1.2	Arylesterase
1.1.1.38	Malate dehydrogenase (NAD)	3.1.1.3	Lipase
1.1.1.40	Malate dehydrogenase (NADP)	3.1.1.5	Phospholipase B
1.1.1.41	Isocitrate dehydrogenase (Bound)	3.1.1.7	Acetylcholinesterase
1.1.1.42	Isocitrate dehydrogenase (Soluble)	3.1.1.8	Cholinesterase
1.1.1.44	Phosphogluconate dehydrogenase	3.1.3.1	Alkaline phosphatase
1.1.1.49	Glucose-6-phosphate dehydrogenase	3.1.3.2	Acid phosphatase
1.1.1.50	3-α-hydroxysteroid dehydrogenase	3.1.3.5	5-Nucleotidase
1.1.1.51	β-hydroxysteroid dehydrogenase	3.1.3.9	Glucose-6-phosphatase
1.1.2.1	Glycerol phosphate dehydrogenase (Mitochondrial)	3.1.4.1	Phosphodiesterase
1.1.3.4	Glucose oxidase	3.1.4.6	Deoxyribonuclease II
1.1.99.1	Choline dehydrogenase	3.1.6.1	Arylsulphatase
1.2.1.12	Glyceraldehyde phosphate dehydrogenase	3.2.1.1	α-Amylase
1.2.3.2	Xanthine oxidase	3.2.1.2	β-Amylase
1.3.99.1	Succinate dehydrogenase	3.2.1.20	α-Glucosidase
1.4.1.2	Glutamate dehydrogenase	3.2.1.21	β-Glucosidase
1.4.3.3	D-Amino-acid dehydrogenase	3.2.1.23	β-Galactosidase
1.4.3.4	Monoamine oxidase	3.2.1.28	Trehalase
1.5.1.3	Tetrahydrofolate dehydrogenase	3.2.1.30	β-Acetylaminodeoxyglucosidase
1.6.4.3	Lipoamide dehydrogenase	3.2.1.31	β-Glycosidase
1.6.5.2	Menadione reductase	3.4.1.3	Aminopolypeptidase (s)
1.6.99.1	$NADPH_2$ diaphorase	3.5.1.5	Urease
1.9.3.1	Cytochrome oxidase	3.6.1.2	Trimetaphosphatase
1.10.3.1	Catechol oxidase	3.6.1.3	ATPase (myosin)
1.10.3.2	Diphenol oxidase	3.6.1.4	ATPase (Mg)
1.11.1.6	Catalase	3.6.1.6	Nucleoside diphosphatase (s)
1.11.1.7	Peroxidase	3.9.1.1	Phosphoamidase
		4.1.1.26	DOPA decarboxylase
		4.1.1.28	5-hydroxytryptophan decarboxylase
		4.1.2.7	Aldolase
		4.2.1.1	Carbonic anhydrase

by the I.U.B. Commission on Enzymes (1961). The name given is in all cases the trivial name. There are a few histochemical enzymes which have no equivalent in the I.U.B.C. list and these have been omitted from the Table.

Of the enzymes listed above those which concern phosphate esters, and which operate at neutral, alkaline or acid pH levels, will be considered in Chapters 15 and 16. The carboxylic acid esterases are dealt with in Chapter 17, and the glycosidases in Chapter 18. The various oxidases, irrespective of type, are described in Chapter 19 and the dehydrogenases and diaphorases in Chapters 20 and 21. The principles of dehydrogenase histochemistry, which differ in many important respects from those of the hydrolytic enzymes, are considered separately in Chapter 20, rather than in this chapter. The proteolytic enzymes are described in Chapter 22 and a rather miscellaneous remainder of enzymes in Chapter 23.

The main subject-matter of this chapter concerns the general principles of enzyme histochemistry as they appear at the present time. Since this book first appeared efforts have been made to develop various branches of enzyme histochemistry in a manner less empirical than was customary in the past and, taken together, these efforts permit a set of principles to be drawn up which can stand as a guide for future work. It is probable that for some time to come the empirical approach to the development of new methods will continue to be used in enzyme histochemistry. As soon as possible, however, all methods should be subjected to critical evaluation, and it is to be hoped that studies of the type discussed in this chapter will ultimately be universally applied.

General Principles of Hydrolytic Enzyme Histochemistry

Although many individual authors have contributed to the development and understanding of the various techniques of enzyme histochemistry the most significant contributions, from the theoretical point of view, have come from Holt (1956, 1958) and his collaborators, and from Seligman and his associates in Boston and Baltimore. Some of my own work (Pearse, 1954; Defendi and Pearse, 1955) has also been concerned with the theoretical basis of enzyme histochemistry. The whole matter was admirably set forth by Nachlas, Young and Seligman (1957) in their contribution to the symposium on *Localization in Histochemistry* held at the Eighth Annual Meeting of the Histochemical Society in Baltimore, and I have used their text as a basis for much of the discussion which follows.

Preparation of Tissues

Although the question of fixation and the preservation of enzymes has already been considered in Chapter 5, certain additional problems remain. The first of these is the question of selective loss of enzyme during processing, whether this involves fixation or not. That such selective loss in fact occurs was shown as long ago as 1950, by Mængwyn-Davies and Friedenwald, in the case of alkaline phosphatases. It may affect part of a single enzyme, less strongly bound than the part remaining in the tissues, or more selectively a single specific enzyme within a group having wider specificity for the histochemical substrate. No hard and fast rules can be laid down for the processing of tissues for different histochemical enzyme techniques but, broadly speaking, two things are of paramount importance. These are the preservation of the maximum possible amount of enzyme activity and the maintenance of the *in vivo* ocalization of the enzyme. Adherence to these two principles results essentially in some form of compromise since, to take the case of a soluble enzyme, preservation of localization by fixation must result in a reduction of enzyme activity. In dealing with soluble enzymes, fixation of some sort is absolutely necessary, although it can sometimes take the form of simultaneous fixation, with the fixative included in the incubating medium. Up to 3 per cent. of formalin can be included in this way or, under other circumstances, up to 50 per cent. ethanol or acetone. Holt (see Holt and O'Sullivan, 1958) has frequently drawn attention to the paradox that perfect preservation of morphology and enzyme activity may lead to less accurate localization of enzyme by a given cytochemical method. This must be understood and accepted, but it must also be appreciated that the converse is not necessarily true, and that the loss of accuracy may in any case be only an apparent one. Histochemists trained as histologists must always avoid the too facile description of a sharper and clearer microscopical image as *better* enzyme localization. These points will be elaborated at the end of this chapter in considering the criteria by which the histochemist may assess the accuracy of his results.

If fixation can be avoided altogether, as it can when the enzyme system concerned is localized within certain intracellular organelles, new principles for handling of the tissues can be formulated. These are included in the comprehensive term "protection" which describes procedures developed by Scarpelli and Pearse (1958) for the preservation of Krebs cycle and other enzymes in mitochondria, and later used for non-mitochondrial systems as well. Novikoff (1957) first extended the use of substances employed for preserving mitochondria in homogenates to their preservation in tissue sections, and protection of tissues *after* cold formalin fixation has been used by Holt and Withers (1958). These authors fixed their tissue blocks in cold formol-calcium solutions and then transferred them to ice-cold 0·88M-sucrose containing 9 g./litre of gum acacia. The blocks were allowed to remain in this solution at 0°–2° for 24 hours and then cut on the freezing microtome. Table 33, below, sets forth the various types of protection, any or all of which it may be necessary to apply in order to ensure preservation of a given enzyme system in tissue sections. Certain criteria here set forth are not valid for the preservation of enzymes in free floating cells or in smears.

PRESERVATION OF ENZYMES IN FRESH TISSUES

The first type of injury occurs when mitochondria are deprived of their supply of oxygen. Sensitive mitochondria show signs of increased permeability

TABLE 33

Type of Injury	Cause	Effect	Remedy
ANOXIC	Deprivation of blood supply.	Mitochondrial damage.	Minimum delay in handling fresh tissues.
THERMAL	Storage at ambient temperatures.	Reduced enzyme activity.	Maintain sections at 0°–4° until required. Blocks at −20° or better at −70°.
	Freezing and thawing.	Loss of enzyme, cofactors, etc.	None, except use of smears.
	Warming of sections drying on storage.	Loss of enzyme.	Keep sections in moist air at 0°–4°.
OSMOTIC	Swelling of mitochondria, etc. Increased permeability.	Loss of soluble enzymes. Destruction of mitochondria.	Protection by 0·88M-sucrose, or dextrin, or 7·5 per cent. PVP.
CHEMICAL	Loss of activators, co-factors, etc.	Reduced enzyme activity.	Replacement of lost factors. Alternative factors. Mg^{2+}, ATP, Mn^{2+}, Co^{2+}.

of their membranes almost at once. Insensitive mitochondria, such as those of cardiac and voluntary muscle, may show no changes for a considerable period of time. In practice this means that when intra-mitochondrial enzyme systems are being studied, *post mortem* tissues are unsuitable for anything except relatively crude studies at histological levels. Even material from the operating theatre may show severe damage to sensitive mitochondria if the interval between tying off the blood vessels and removal of the material is more than a few minutes. Needle or drill biopsy tissues are excellent.

The second type of injury, described in the table as thermal, occurs first if the tissues are left at room temperature. This can be avoided by chilling the tissues or by quenching small pieces at once at −78° by the application of dry ice or by immersion of a metal tissue holder in a dry ice-acetone mixture. Unfortunately this manœuvre also causes thermal injury to mitochondrial enzyme systems. Damage due to freezing at −78° was assessed by Porter *et al.* (1953) in the case of a number of enzyme systems. They found, for instance, that less than one-fifth of the succinoxidase activity of a tissue homogenate survived rapid freezing and subsequent thawing and maintained that slow freezing to −10° caused less damage to enzymes, although ice crystal growth was much greater. The effects noted by Porter and her associates cannot be reproduced when histochemical tests for soluble and insoluble mitochondrial enzyme systems are applied to sections from rapidly quenched (−78°) and slowly frozen (−12°) tissue blocks. Using the NAD-linked L-Malate dehydrogenase system for instance, with the substrate concentration at 10^{-2} M and NAD at 10^{-2}, 10^{-3}, and 10^{-4} M, enzyme activity decreases in the same order of decreasing NAD concentrations after quick or slow freezing. Moreover, the overall activity is less in slowly frozen material and the enzyme is localized entirely outside the mitochondria, indicating a severe degree of damage. Tissue sections cut on the cold microtome are adequately preserved for short periods at 0°–4° or at −20° in moist atmospheres. Blocks of tissue can be stored at

−20° for longer periods, perhaps up to 7 days or so, but for adequate long-term preservation temperatures as low as −70° are necessary. The design and production of an inexpensive low-temperature storage cabinet is long overdue.

Part of the damage described as due to freezing may well be due to the subsequent thawing which precedes incubation. With fresh frozen sections additional damage may be caused by the passage of the knife through the tissues, since this is preceded by a melting zone and followed by refreezing (see Chapter 2, p. 22). It is probable, therefore, that double-freezing and thawing is inevitable in the case of fresh frozen sections.

In addition to the damage caused by thawing after freezing further damage to intramitochondrial enzymes is produced by drying the sections (mounted on slides or coverslips) and allowing them to remain at room temperature before proceeding to incubation. This can be avoided by storage in a cold (0° to −20°) moist environment.

The third type of damage, to which mitochondria and other intracyto-plasmic organelles are particularly sensitive, is described in the table as osmotic. This occurs in several stages, of which the final and most severe stage involves disruption of the mitochondria. It can be avoided by the use of hypertonic media. Of these, polyvinyl pyrrolidone (PVP) with or without additional sucrose, as used by Novikoff (1956), seems to be the most suitable for general use.

The final type of injury, described in Table 33 as chemical, comprises a miscellaneous group of circumstances. During processing many smaller molecules may become dissociated from their situation, and during subsequent incubation they are lost. Metallic ions functioning as activators may come into this category and their replacement is necessary for restitution of enzyme activity. Other factors, such as ATP and Mg^{2+}, may have to be added in order to assist in the preservation of mitochondrial structure, although there is no evidence that these have been lost in processing.

The protection of cytoplasmic structures, in which the enzyme systems are localized, from damage during processing and during incubation must be given far more consideration in modern enzyme histochemistry than it has received in the past. In many cases, the application of the principles of protec-tion to histochemical practice results in enzyme localizations which agree closely with the best results obtainable by the techniques of other disciplines.

Histochemical Reactions for Hydrolytic Enzymes

The four types of histochemical reactions for hydrolytic enzymes are:

(1) Simultaneous capture (coupling, conversion, chelation).
(2) Post-incubation coupling (chelation, conversion).
(3) Self-coloured substrate (solubility change).
(4) Intramolecular rearrangement.

Of these four types the first is of outstanding importance, the second of somewhat lesser importance, and the third and fourth are confined, at present, to a few isolated examples. The simultaneous capture principle operates in the older metal salt precipitation techniques, such as those for alkaline and acid phosphatases, and in the azo dye methods for various hydrolytic enzymes (Chapters 15, 16, and 17) when the coupler may be either naphthol or naphthylamine. It operates also in the oxidation of liberated indoxyls to indigos (Chapter 17) and in the metal chelation of formazans from mono and ditetrazolium salts (Chapter 21), although in this case the kinetics are not those of the more usual capture reactions.

Simultaneous Capture

In discussing reactions based on the simultaneous capture principle it is expedient to consider first the coupling azo dye methods, since in the evolution of these have arisen most of the concepts on which modern enzyme histochemistry is based. With these methods, in which a diazonium salt present in the incubating medium combines with the coupler as soon as the latter is released from the substrate, the main obstruction to accurate localization of the responsible enzyme is diffusion of the primary reaction product (PRP). This is likely to be governed by several factors; by the rate of hydrolysis of the substrate, by the diffusion coefficient of the PRP for the buffer system and pH employed, and by the rate of coupling of PRP and diazonium salt (Defendi and Pearse, 1955). All three can be controlled to some extent; first, by modifying the substrate, which affects the rate of hydrolysis, the diffusion coefficient and the coupling rate and, secondly, by modifying the diazonium salt. This may affect the rate of hydrolysis (by inhibition) and the rate of coupling. If the PRP has an affinity for protein this quality of "substantivity" will obviously affect its rate of diffusion. Substantivity of the PRP is a quality which is certainly desirable in designing post-incubation coupling methods (see below) and, according to many authorities, it is desirable also in simultaneous capture reactions. I do not altogether agree with this latter view myself, preferring whenever possible to obtain rapid capture of a soluble non-substantive PRP. This preference is based on practical rather than on theoretical experience.

In order to assess the importance of the coupling rate in simultaneous capture azo dye reactions experiments were made (Pearse, 1954; Defendi and Pearse, 1955) to measure the rate under various conditions. Using a continuous flow apparatus and photometric control, we tested the coupling rates of α and β-naphthol, and of a number of substituted naphthols, with tetrazotized di*ortho*anisidine (Fast blue B salt) at various pH levels. The half time (t) for coupling with α-naphthol varied from 92 millisecs at pH 9·2 to 750 millisecs at pH 5·3. For β-naphthol the figures were 80 millisecs at pH 9·2 and 3 minutes 15 seconds at pH 5·8. As published, the velocity constants for the reactions were not strictly accurate since k was calculated from the usual formula for a bimolecular reaction.

$$k = \frac{1}{t(a-b)} \log_n \frac{b(a-x)}{a(b-x)}$$

where a and b are initial concentrations of the reactants and x the amount of the product formed.

This would be in order for the coupling of naphthols with monodiazonium salts, but not with bisdiazonium salts like Fast blue B, where the initial stage of the reaction may be supposed to be monocoupling followed by a second coupling at the other end of the molecule. The error entailed, however, is not very great and, since our results showed a linear relationship between pH and log k, their significance remains as stated in the original papers. For both naphthols there were critical pH levels below which the coupling rates fell off rapidly (Fig. 123). The greater efficiency of α-naphthol on the acid side of neutrality was reflected in the histochemical results which were carried out in parallel.

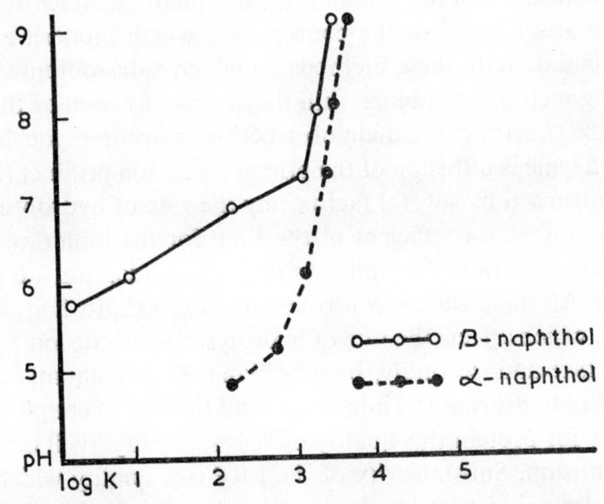

FIG. 123. Coupling of Fast Blue B with Naphthols as a function of pH.

It was suggested that the coupling rate was of importance in histochemical azo dye methods in the case of the more soluble naphthols, but of less importance with the relatively insoluble ones in producing accurate enzyme localization.

Subsequently Nachlas *et al.*, (1959) used a similar apparatus to measure the coupling rates of a series of naphthols and naphthylamines with a number of different diazonium salts, in order to test the effect of various substituent groups in both diazonium salt and coupler. Plotting log k ($\frac{1}{2}$ time) against σ, the substituent constant for the particular functional group (Hammett, 1953), they found that as substituents were added to increase coupling efficiency the rate of coupling for the naphthols continued to increase while that for the naphthylamines levelled off sharply. Fig. 124, p. 483 illustrates this effect.

This important observation explains the histochemical differences in performance of naphthol and naphthylamine primary reaction products and forms a basis for new efforts to improve existing methods in the coupling azo dye field of enzyme histochemistry.

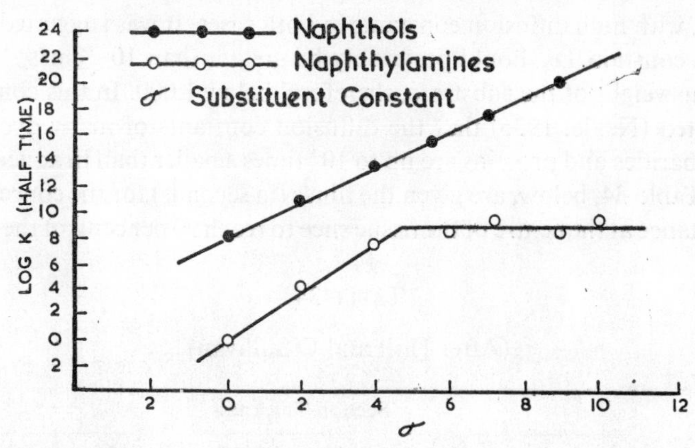

FIG. 124. Coupling Rates of Naphthols and Naphthylamines.

KINETICS OF SIMULTANEOUS CAPTURE REACTIONS

The kinetics of enzyme histochemistry, particularly as applied to the indoxyl acetate methods (Chapter 17), have been considered by Holt and his co-authors (Holt, 1956, 1958; Holt and O'Sullivan, 1958) in a comprehensive series of papers. Their findings and conclusions, which are reported extensively below, are applicable directly to simultaneous capture reactions of many different types. They have less direct application to the simultaneous chelation methods (Pearse, 1957) and may require extensive modification in respect of enzymes enclosed in such box-like structures as mitochondria or lysosomes.

For the purpose of theoretical treatment considerations of histochemical enzyme reactions were divided by Holt into those of the initial process, followed by considerations of the rate of production of the primary reaction product (PRP) the formation of the final reaction product (FRP) and the life of the PRP, and lastly, those of variations in distribution of the FRP. This order is the order in which they are here presented.

The Initial Process. It is necessary, in the first instance, that both the substrate and the capture reagent should penetrate rapidly into the tissues, and for this purpose it is desirable that both should be sufficiently soluble in water to provide high concentrations in the incubating medium. For rapid diffusion of substrate and capture reagent both should be of *low molecular weight* and also of *low polarity*, so that they can pass through lipid layers and avoid adsorption by protein. Holt and O'Sullivan (1958) made theoretical studies of simplified models in which the tissue was regarded as a homogeneous entity containing an enzyme site. Permeability barriers were ignored and depletion of substrate considered as negligible. Calculations of the time taken for the concentration of a substance at the centre of a tissue slice to reach 90 per cent. of the external value showed that for good penetration thin tissue slices were essential. High external concentrations of substrate and capture reagent were also shown to be essential, with high diffusion constants in both cases. It was suggested that the diffusion constant D′ should preferably be greater than $10^{-6}cm^2s^{-1}$ and the molecular weight of the substrate, therefore, below 1,000. In this context it is to be noted (Neale, 1936) that the diffusion constants of many dyestuffs in polysaccharides and proteins are up to 10^4 times smaller than in aqueous solution. In Table 34, below, are given the times (in seconds) for the concentration of a substance at the centre of the tissue slice to reach 90 per cent. of the external value.

TABLE 34

(After Holt and O'Sullivan)

D′	Section Thickness						
(cm^2s^{-1})	1μ	5μ	10μ	15μ	20μ	50μ	500μ
10^{-5}	0·00025	0·00625	0·025	0·056	0·1	0·625	62·5
10^{-6}	0·0025	0·0625	0·25	0·56	1·0	6·25	625
10^{-7}	0·025	0·625	2·5	5·6	10·0	62·5	6,250
10^{-8}	0·25	6·25	25·0	56·0	100·0	625	62,500

Holt and O'Sullivan commented that the desirable non-bonding substances are usually non-polar and therefore lipid soluble. This enables them to penetrate intact lipid layers in the cells and tissues. Collection of the substrate in specific sites, as by preferential adsorption, is manifestly undesirable and it is obvious that in biological systems many other (uncontrollable) factors will modify the assumptions made above.

Rate of Production of the Primary Product. When the concentration of a substrate is high in relation to that of enzyme it is removed by a reaction which obeys zero order kinetics. That is to say, the rate of the reaction is independent of the substrate concentration and related solely to enzyme concentration. If, however, the substrate is present in low concentration the kinetics may become first order. That is to say, they may be related directly to the concentration of substrate. Holt suggested that it was reasonable to suppose that rapid ingress of substrate into tissue sections gives rise to zero order production of the PRP. This being the case he considered the situation at a highly active site, into which the substrate could diffuse rapidly, and concluded that zero kinetics were in fact valid for a wide range of conditions *irrespective of the activity of the site.*

FORMATION OF THE FINAL REACTION PRODUCT AND LIFE OF THE PRIMARY REACTION PRODUCT

Continuing the foregoing investigations Holt and O'Sullivan studied the simultaneously occurring processes of PRP formation, PRP diffusion and removal of PRP by conversion to FRP. Diffusion and capture are the two methods by which the PRP is removed from the reaction and, for the purposes of calculation, it was assumed that the former obeyed Fick's law. Since the concentration of free PRP is likely to be low at any given point, if the concentration of the capture reagent is high the reaction between them should obey first order kinetics and be dependent solely on capture reagent concentration. This proposition assumes that the capture reaction is irreversible and that the FRP is the sole product. According to Holt and O'Sullivan the presence of several stages in the capture reaction introduces no complications provided the overall kinetics are first order. If the capture reagent is not present everywhere in adequate excess, however, the reaction will deviate towards second order kinetics. In this case quantitative theoretical study is not possible.

From the above considerations it may be deduced that the smaller the diffusion constant of the PRP, and the larger the velocity constant of the capture reaction, the more precise will be the localization of the FRP.

In earlier papers O'Sullivan (1955, 1956) described a model system in which the enzyme was presumed to be localized in a spherical site of known radius set in an infinite medium. It was assumed that the substrate diffused rapidly and that the FRP was completely non-diffusible. A measure of the degree of localization achieved was derived theoretically and described as the *localization factor*. This is now defined as the ratio of the mass of FRP actually deposited in the site in a given time to the mass of FRP that would be produced in the given time, if the PRP were converted *quantitatively* into FRP in that time. The localization factor thus defined is dependent on the radius of the site, the velocity constant of the capture reaction and on the diffusion coefficient of the PRP. Provided experimental values for these three can be obtained the localization factor provides a means of comparison between different cytochemical methods. For convenience Holt and O'Sullivan defined the localization factor as the percentage of total FRP which is to be found in the site of enzyme activity and their results are given in the two tables which are reproduced below. The first of these shows the relationship of the localization factor to the diffusion constant of the PRP and to the velocity constant of the capture reaction for a site of radius 1μ. This the authors took as the practical limit of resolution of the optical microscope in this context.

TABLE 35

Variation in Percentage of FRP in a Site of Radius 1μ

Diff. Constant (cm^2s^{-1})	Velocity Constant of Capture Reaction (s^{-1})							
	10^5	10^4	10^3	10^2	10	1	10^{-1}	10^{-2}
10^{-5}	85·1	57·2	19·0	3·5	1·4	10^{-1}	10^{-2}	10^{-3}
10^{-6}	95·3	85·1	57·2	19·0	3·5	1·4	10^{-1}	10^{-2}
10^{-7}	98·5	95·3	85·1	57·2	19·0	3·5	1·4	10^{-1}
10^{-8}	99·5	98·5	95·3	85·1	57·2	19·0	3·5	1·4
10^{-9}	99·9	99·5	98·5	95·3	85·1	57·2	19·0	3·5
10^{-10}	100	99·9	99·5	98·5	95·3	85·1	57·2	19

The second table (Table 36) gives the calculated values for the localization factor for sites of different size, for various combinations of velocity constant and diffusion constant.

TABLE 36

Variation in Percentage of FRP with Site Radius

Site Radius (μ)	Ratio Velocity Constant/Diffusion Constant (cm^{-2}) 10^{12}	10^{11}	10^{10}	10^{9}	10^{8}	10^{7}
0·1	85·1	57·2	19·0	3·5	1·4	10^{-1}
1·0	98·5	95·3	85·1	57·2	19·0	3·5
10·1	99·9	99·5	98·5	95·3	85·1	57·2

Holt and O'Sullivan considered that for methods depending on stain density at least 50 per cent. of the FRP should be deposited in the site of enzyme activity, and Table 36 shows that with a site of radius 10μ a velocity constant for the capture reaction of 10 s^{-1} is sufficient. With sites of radius less than 1μ, however, velocity constants of 10^3 s^{-1} and greater are necessary. According to Cotson and Holt (1958) the velocity constants of cytochemical enzyme reactions are never higher than 10 s^{-1} and the authors deduced from this that the minimum radius of sites in which *quantitative* enzyme localization could be achieved is about 10μ, when the PRP is freely diffusible. They therefore maintained that the diffusion constant of the PRP must be reduced and that this could best be brought about by arranging for the PRP to possess substantivity for protein. This concept is further discussed below in the section dealing with the properties of the PRP.

VARIATION OF STAIN DISTRIBUTION WITH THE LOCALIZATION FACTOR

Localization factors discussed above give a measure of the relative amounts of FRP within and without sites of different radii under different conditions. They do not indicate the actual distribution pattern of the FRP. This has to be considered if a full interpretation is to be made of the significance of localization factors in relation to the microscopic examination of tissues after the performance of histochemical enzyme reactions.

Holt and O'Sullivan (1958) therefore described a further factor the *specific stain density*. This can be defined as a density of FRP such that, if perfect localization is achieved, its value is unity throughout the enzyme site and zero elsewhere. The specific stain density (s) varies with the value of the localization factor (F) and with a factor p which is the ratio of the radius of an arbitrary spherical zone around the site to the radius of the site itself. These relationships are shown in Fig. 125.

It is evident from the above graph that if precise localization of an enzyme is desired the value of F should certainly be higher than 0·65. Holt considered

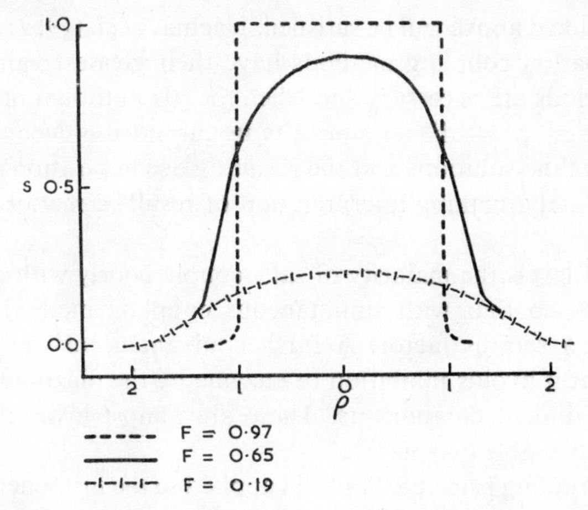

FIG. 125. Density distribution of FRP produced by a spherical enzyme site with different values for the localization factor. (After Holt and O'Sullivan.)

that a reasonable picture of the site might still be obtained with much smaller F values, but that values below 0·01 should not be tolerated with any method. His calculations showed that for a site of radius 1 μ the ratio k/D (the ratio of the velocity constant of the capture reaction to the diffusion coefficient of the PRP) should be greater than 10^6 cm^{-2}. A value greater than 10^8 cm^{-2} would give a good indication of the size of the site, and a value of 10^{10} cm^{-2} would provide excellent localization.

The numerical results considered above cannot be applied generally throughout the whole field of enzyme histochemistry. Holt pointed out that a number of things prevent this. The calcium phosphate techniques for alkaline phosphatase, for example, depend on the deposition of a PRP from a supersaturated solution and its absorption and/or crystallization on the tissues. Knowledge of the factors which influence the deposition of materials from supersaturated solutions is scanty and the effects of crystallization are difficult to assess and measure. Calcium phosphate precipitation in the alkaline phosphatase method was considered from the theoretical point of view by Danielli (1958). The kinetics of chelation are also difficult to assess and satisfactory models have not yet been constructed for systems based on capture by metal chelation.

POST-INCUBATION COUPLING

This type of reaction is based on the supposition that a PRP which is sufficiently insoluble will remain *in situ* without diffusion either during incubation or during the subsequent performance of the necessary demonstrating reaction. The post-incubation coupling (post-coupling) principle has so far been applied only to the large group of azo dye methods. If the PRP is substantive as well as insoluble there is, of course, an even greater chance that the conditions outlined above will be satisfied. Nachlas *et al.* (1957) suggested that the post-incubation coupling methods have their greatest value where long incubation periods are necessary and when the pH optimum of the enzyme is in the acid range. Most diazonium salts in current use decompose on long standing in alkaline solutions and the resulting decomposition products stain the tissues diffusely, making interpretation of results a matter of some difficulty.

At acid pH levels the majority of salts couple poorly with naphthols and naphthylamines so that with simultaneous coupling methods diffusion of PRP becomes a serious factor. A further advantage of the post-coupling principle is that it avoids inhibition of enzyme by the diazonium salt and its stabilizer and diluent components. These are almost invariably present in commercially available diazotates.

The post-coupling principle has been applied to the histochemical localization of acid phosphatase (Rutenburg and Seligman, 1955), sulphatase (Rutenburg, Cohen and Seligman, 1952), β-glucuronidase (Seligman, Tsou, Rutenburg and Cohen, 1954), and glycosidases (Cohen *et al.*, 1952), using in each case esters or glycosides of 6-bromo- or 6-benzoyl-2-naphthol. Criticisms of the principle were put forward by Burton and Pearse (1952), by Pearse (1954), and by Defendi (1957). In particular the specific binding of naphthols, and especially substituted naphthols, to various tissue components has been considered to be one of the gravest objections to the post-coupling principle. Tests carried out by Defendi showed that 6-benzoyl and 6-bromo-2-naphthol exhibited not only the well-known naphthol binding to elastic tissue but also specific binding to various epithelial cells. Using fresh frozen (cryostat) sections, binding of naphthols by tissue components was observed to be dependent on the solubility and concentration of the naphthols, on the time of contact, the pH, and the type of tissue.

The more soluble naphthols (α- and β-naphthol) proved to have little affinity for any tissue component at alkaline pH levels though at an acid pH slight binding to liver cells occurred. The two insoluble naphthols had strong affinity for particular protein structures, especially the cytoplasm of parenchymal cells. Defendi concluded his paper with the observation that the post-coupling principle should be used with caution, not because of the occurrence of false positive results but because of the probability of false negative results.

Altered Solubility Methods

At the present time there are few methods making use of this principle in current use. It is based on the employment of coloured soluble substrates which are rendered insoluble when the solubilizing group is removed by enzyme activity. The insoluble dye is then presumed to precipitate at the site of enzyme activity.

A method for alkaline phosphatase making use of the altered solubility principle was described by Loveless and Danielli (1949) and a method for β-glucuronidase was produced by Friedenwald and Becker (1948). These methods are further considered in the appropriate chapters (15 and 18). It is unlikely that methods of this type will be developed on a wide scale, since only a small number of enzyme-sensitive solubilizing groups are known and in practice the large substrate molecules produced are often hydrolysed with extreme difficulty.

Intramolecular Rearrangement

This principle depends on the development of soluble substrates which, after hydrolysis, undergo molecular rearrangement to give highly coloured insoluble products. It has so far been employed, in preliminary explorations only, by Nachlas *et al.* (1957) for the development of a method for carboxylic acid esterases using an acetic acid ester of indophenol. In this case the FRP was found to be too soluble to give accurate enzyme localization, but future studies in this direction may be expected to yield important results.

Development of Histochemical Techniques for Hydrolytic Enzymes

Properties of the Substrate. It follows, particularly from the work of Holt and his associates, that the first essential is that the substrate should be water-soluble and of low molecular weight. Unless these two criteria are satisfied the substrate will not diffuse rapidly enough into the tissues nor will it be present in sufficient concentration for the enzyme reaction to have zero order kinetics. Unless zero order kinetics are established the amount of FRP will not reflect accurately the level of enzyme activity.

Many of the substrates designed for use as esterase and phosphatase reagents are very insoluble in water and it becomes necessary to add organic solvents to the incubating medium in order to keep them in solution at all. If the PRP, as is often the case, is soluble in the chosen organic solvent serious errors of diffusion may result. It is essential that the substrate be easily hydrolysed at the pH and temperature necessary for optimum enzyme activity. This attribute must not, of course, be accompanied by any tendency to spontaneous hydrolysis. Many substrates in current use (Naphthol AS esters, halogen-substituted indoxyl esters) are hydrolysed slowly, or with difficulty, and long incubation times are necessary. Even with fixed tissues this factor can lead to serious disturbances and with fresh tissues long incubations are entirely unacceptable.

A further necessary property which all substrates should possess is lack of polarity. Possession of polar groups may lead to attachment to specific protein components in the tissues and to interference with diffusion. Diffusion through lipid or lipid protein interfaces will also occur, and interference with this is a particularly serious factor in the case of enzyme systems which are enclosed in lipid or lipoprotein envelopes.

Following brief observations on the effects of lipid solubility of substrates by Wolfgram (1961), a series of papers by Alqvist (1963a, b and c, 1964a and b) stressed the importance of tissue binding of the naphtholic substrates. Alqvist showed that these were unevenly distributed in tissue sections and, furthermore, that they were no longer adsorbed by sections which had been freed from lipid (by extraction with chloroform/methanol). He considered that differences in solubility of various substrates could influence both the speed of the reaction and the quantity of the reaction product formed. This is clearly proven by Alqvist's experiments and his findings must be taken into account when designing or performed enzyme reactions based on naphtholic substrates.

Properties of the Reaction Product. It is desirable that both the primary and final reaction products should not only be insoluble in water but also in lipid, since otherwise they may be concentrated in the latter. Such an occurrence is common with the reaction products of a large number of currently-used substrates, especially those in the azo-dye field. If the PRP is to be removed by a capture reaction it must be capable of reacting rapidly with the chosen reagent, but the question of substantivity (i.e. affinity for protein) of the reaction product is one which cannot easily be settled. Nachlas *et al.* (1957) take the view that substantivity for protein is a desirable property, but rightly stress that there should be no special affinity for *particular* protein components. Naphtholic reaction products of many current enzyme methods have a special affinity for elastic tissue. This is an undesirable quality. Strong polarity is also to be avoided, since this may lead to attachment of the reaction product to acidic or basic proteins at points distant from the site of enzyme activity. If the capture reaction is sufficiently rapid the effects of polarity will not be felt.

Modification of the substrate by the introduction of groups calculated to increase the reactivity of the PRP have been undertaken by Seligman and his associates. Further work on these lines cannot fail to be productive. The nature of affinity and substantivity, with respect to dyestuffs and textile fibres, is fully discussed by Wegmann (1959). According to this author substantivity is defined as "Die Fähigkeit eines Farbstoffes oder Textilapplikations produktes aus einem flüssigen Medium auf ein textiles Substrat aufzuziehen und sich zu fixieren."

Properties of the Capture Reagent. There are several types of capture reagent in use in enzyme histochemistry. In the case of the indoxyl substrates the oxidizing agent (ferricyanide, cupric ions) plays this role, as do Ca^{2+} and Pb^{2+} in the older phosphatase techniques. In the coupling azo dye methods the diazonium salt is the capture reagent and with the metal chelation techniques for dehydrogenases it is the divalent cobalt or nickel ions of the chelator. There is thus no similarity between the various methods and few general rules can be formulated. Cotson and Holt (1958) showed that first order kinetics obtained when indoxyls were oxidized to indigoid dyes in the presence of excess O_2 at alkaline pH levels. It has not been shown that the other types of capture reaction employed in enzyme histochemistry permit first-order kinetics, and it is probable that the diazonium salts usually employed do not do so. Nachlas *et al.* (1957, 1959) suggested appropriate modifications of diazonium salts which might be expected to increase the rate of coupling with naphthols and naphthylamines. They pointed out, however, that knowledge of the effect of various functional groups in the diazotate on its rate of coupling was extremely sparse.

Choice of diazotate for an individual coupling azo dye reaction is still largely based on empirical methods and the best-looking result is considered to indicate that the best diazotate has been selected. This is by no means necessarily true. Many diazotates strongly inhibit enzyme activity either by virtue of active groups in the molecule or because of the metal ions (Zn, Al, Mg, Cu) which are present as stabilizers. It has repeatedly been shown that partial inactivation of an active enzyme will lead to *apparently* better localization. This effect must never be overlooked and, if at all possible, it must be avoided.

Validity of Localization

With all histochemical enzyme methods, not only those for hydrolytic enzymes, the question of accurate localization has ultimately to be considered. In a few cases, where the enzyme is predominantly mitochondrial or microsomal, biochemical evidence is sufficiently strong to be considered accurate. In the majority of cases, however, there is no yardstick by which the validity of histochemical localization can be assessed.

The number of factors involved is considerable and it is almost impossible to take them all into account when designing or modifying histochemical methods for enzymes. The division of enzymes into diffusible or lyo, and fixed or desmo, forms was originally made by Willstätter and Rohdewald (1932). This conception was studied experimentally by Nachlas, Prinn and Seligman (1956) and by Hannibal and Nachlas (1959). Many important points have been raised, and answered, and it is necessary to consider some of them here.

It is clear that diffusion of enzyme occurs to different degrees in different fixatives, and that it varies for each enzyme. Moreover, the loss of activity during fixation is due not only to inactivation by the chemical reaction of the fixative on the enzyme protein but also to diffusion of the lyo component into the fixative during fixation and into the medium during incubation. Measurements made by Nachlas and his associates showed that in the absence of fixation 30 per cent. of esterase, 61 per cent. of alkaline phosphatase, 48 per cent. of acid phosphatase, 65 per cent. of leucine aminopeptidase and 5 per cent of β-glucuronidase were present in the desmo form. After formalin fixation, as expected, an increase in the proportion of desmo to lyo enzyme took place in each case. Absolute acetone prevented significant loss of activity over the whole range of hydrolytic enzymes studied by Hannibal and Nachlas, but lower concentrations allowed the lyo component to diffuse. Lyo-esterase and lyo-alkaline phosphatase were observed to diffuse into absolute alcohol, but lyo-acid phosphatase and lyo-β-glucuronidase did not do so. Earlier studies by Doyle and Liebelt (1954) had already drawn attention to the diffusion of esterases into the incubating medium after fixation.

The procedures of fixation, and fixation and paraffin embedding, do not wholly prevent the diffusion of the lyo component of hydrolytic enzymes into the incubating medium. Many factors influence this post-fixation diffusion and among those which have to be considered are the duration of incubation, temperature, pH and the concentration of electrolytes and other additives. According to Janigan (1964) the elimination or reduction of the soluble enzyme component by fixation is related in great part to inactivation and, to a lesser degree, to aqueous extraction into the fixative.

Increasing the time of incubation, and elevation of the temperature, naturally cause increased loss of diffusible enzyme. Alteration of the pH of the medium produces very considerable changes in the amount of enzyme diffusion. Alkaline phosphatase and leucine aminopeptidase activities, in particular, are noted to be much higher at the pH optimum than in distilled water. the pH at which fixation is carried out also exerts a strong effect on the amount of desmo enzyme subsequently recorded. The results obtained by Hannibal and Nachlas are shown in Fig. 126.

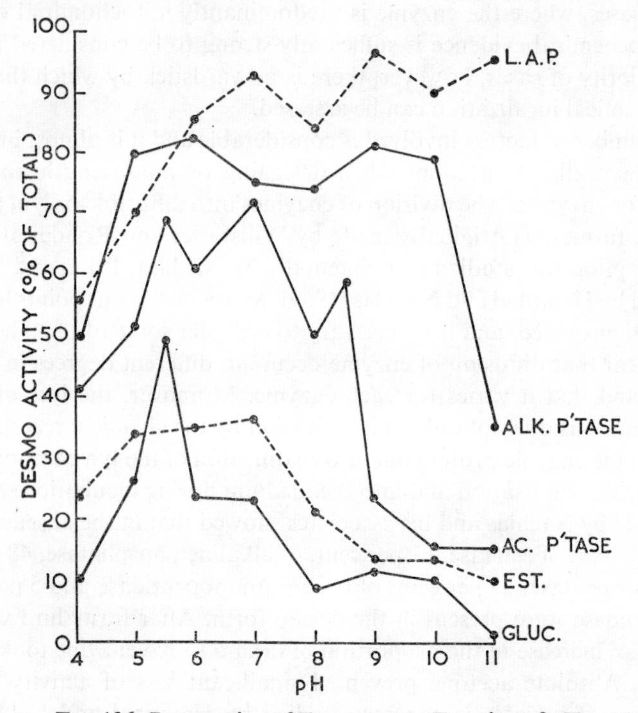

FIG. 126. Preservation of Enzymes as a Function of pH.

Additives. Improved preservation of esterase can be obtained by the addition of sodium acetate or methyl cellulose and the activity of β-glucuronidase was observed by Hannibal and Nachlas to be doubled by addition of salts and 0·88M-sucrose, or methyl cellulose.

Chloral hydrate was employed by Fishman and Baker (1956) to preserve β-glucuronidase in thin tissue blocks. It was observed that in low concentration in dilute formaldehyde (0·1 per cent. in 6 per cent.) this substance improved the preservation of enzyme, as judged by the final results in tissue sections using the ferric-8-hydroxyquinoline method (Chapter 18). Further studies on the effect of chloral hydrate on the preservation of five hydrolytic enzymes have now been made by Baker, Hew and Fishman (1958). With their chloral-formalin fixative, β-glucuronidase activity was still present after fixation for over 3 weeks, while acid phosphatase was preserved for 6 weeks and esterase and lipase for 9 weeks. Alkaline phosphatase, on the other hand, was unable to survive for more than one day in chloral-formalin.

In their series of experiments referred to above Hannibal and Nachlas used chloral hydrate as an additive (1 per cent. in 10 per cent. formalin) and found that it did not prevent loss of the lyo component of any of the enzymes tested and that in some cases definite inactivation was produced. It is difficult to compare the results of Hannibal and Nachlas with those of Baker *et al.*, since not only did they employ different concentrations of chloral hydrate but also different terms of reference (percentage of desmo and lyo forms; length of survival of enzyme).

Since the survival of alkaline phosphatase in chloral-formalin was enormously increased by buffering to pH 7·0 it is evident that chloral hydrate *per se* was not the important factor in this case, and the improved preservation of acid phosphatase and β-glucuronidase particularly might well be due to alteration of the pH of fixation so as to approach the value for optimum preservation of the lyo component (cf. Hannibal and Nachlas and Fig. 126).

Conclusion

It will obviously be necessary to consider all the methods, to be described hereafter in the enzyme sections of Vols I and II of this book, in the light of the theoretical and practical observations recorded in this chapter. Until such time as the necessary investigations have been carried out it is essential that those who use the methods in their present form should clearly understand their limitations, particularly in respect of enzyme localization.

REFERENCES

ALQVIST, J. (1963a). *Suomen Kemi. B.*, **36**, 23.
ALQVIST, J. (1963b). *Acta path. Scand.*, **57**, 353.
ALQVIST, J. (1963c). *Ibid.*, **59**, 171.
ALQVIST, J. (1964a). *Ann. Med. exp. Fenn.*, **42**, 85.
ALQVIST, J. (1964b). *Histochemie*, **4**, 273.
BAKER, J. R., HEW, H., and FISHMAN, W. H. (1958). *J. Histochem. Cytochem.*, **6**, 244.
BURTON, J. F., and PEARSE, A. G. E. (1952). *Brit. J. exp. Path.*, **33**, 87.
COHEN, R. B., TSOU, K-C., RUTENBURG, S. H., and SELIGMAN, A. M. (1952). *J. biol. Chem.*, **195**, 239.
COTSON, S., and HOLT, S. J. (1958). *Proc. Roy. Soc.*, **148**, 506.
DANIELLI, J. F. (1958). In "General Cytochemical Methods," I, 423.
DEFENDI, V. (1957). *J. Histochem. Cytochem.*, **5**, 1.

DEFENDI, V., and PEARSE, A. G. E. (1955). *Ibid.*, **3**, 203.
DIXON, M., and WEBB, E. C. (1958). "Enzymes." Longmans, London.
DIXON, M., and WEBB, E. C. (1964). "Enzymes," 2nd Ed. Longmans, London.
DOYLE, W. L., and LIEBELT, R. (1954). *Anat. Rec.*, **118**, 384.
FISHMAN, W. H., and BAKER, J. R. (1956). *J. Histochem. Cytochem.*, **4**, 570.
FRIEDENWALD, J. S., and BECKER, B. (1948). *J. cell. comp. Physiol.*, **31**, 303.
HAMMETT, S. J. (1953). *Chem. Rev.*, **53**, 191.
HANNIBAL, M. J., and NACHLAS, M. M. (1959). *J. biophys. biochem. Cytol.*, **5**, 279.
HOLT, S. J. (1956). *J. Histochem. Cytochem.*, **4**, 541.
HOLT, S. J. (1958). In "General Cytochemical Methods," **1**, 375. Ed. J. F. Danielli. Acad. Press, New York.
HOLT, S. J., and O'SULLIVAN, D. G. (1958). *Proc. Roy. Soc. B.*, **148**, 465.
HOLT, S. J., and WITHERS, R. F. J. (1958). *Ibid.*, **148**, 520.
JANIGAN, D. T. (1964). *Lab. Invest.*, **13**, 1038.
LOVELESS, A., and DANIELLI, J. F. (1949). *Quart. J. micr. Sci.*, **90**, 57.
MAENGWYN-DAVIES, G. D., and FRIEDENWALD, J. S. (1950). *J. cell. comp. Physiol.*, **36**, 421.
NACHLAS, M. M., GOLDSTEIN, T. P., ROSENBLATT, D. H., KIRSCH, M., and SELIGMAN, A. M. (1959). *J. Histochem. Cytochem.*, **1**, 50.
NACHLAS, M. M., PRINN, W., and SELIGMAN, A. M. (1956). *J. biophys. biochem. Cytol.*, **2**, 487.
NACHLAS, M. M., YOUNG, A. C., and SELIGMAN, A. M. (1957). *J. Histochem. Cytochem.*, **5**, 565.
NEALE, S. M. (1936). *J. Soc. Dy. Col.*, **52**, 252.
NOVIKOFF, A. B. (1956). *J. biophys. biochem. Cytol.*, **2**, 65.
NOVIKOFF, A. B. (1957). *Cancer Res.*, **17**, 1010.
O'SULLIVAN, G. D. (1955). *Bull. math. Biophys.*, **17**, 243.
O'SULLIVAN, D. G. (1956). *Ibid.*, **18**, 199.
PEARSE, A. G. E. (1954). *Int. Rev. Cytol.*, **3**, 329.
PEARSE, A. G. E. (1957). *J. Histochem. Cytochem.*, **5**, 515.
PORTER, V. S., DEMING, N. P., WRIGHT, R. C., and SCOTT, E. M. (1953). *J. biol. Chem.*, **205**, 883.
RUTENBURG, A. M., COHEN, R. B., and SELIGMAN, A. M. (1952). *Science*, **116**, 539.
RUTENBURG, A. M., and SELIGMAN, A. M. (1955). *J. Histochem. Cytochem.*, **3**, 455.
SCARPELLI, G. D., and PEARSE, A. G. E. (1958). *Ibid.*, **6**, 369.
SELIGMAN, A. M., TSOU, K-C., RUTENBURG, S. H., and COHEN, R. B. (1954). *Ibid.*, **2**, 209.
TAKEUCHI, T. (1958). *Ibid.*, **6**, 208.
WEGMANN, J. (1959). *Textil-Rundschau*, **11**, 1.
WILLSTÄTTER, R., and ROHDEWALD, M. (1932). *Zeit. physiol. Chem.*, **208**, 258.
WOLFGRAM, F. (1961). *J. Histochem., Cytochem.*, **9**, 171.

CHAPTER 15

ALKALINE PHOSPHATASES

The Phosphatases

THE hydrolytic enzymes responsible for the breakdown of phosphate esters can be divided into three types, the mono, di- and triphosphatases, and in histochemistry we are mainly concerned with the first of these. The phosphomonoesterases, generally speaking, are not specific in respect of the alcohol radical which is attached to the phosphoric acid group of the substrate and they will thus hydrolyse a wide variety of organic phosphates. In the equations given below, R is the alcohol radical.

$$\mathrm{R{-}O{-}\overset{\overset{\displaystyle O}{\|}}{\underset{\underset{\displaystyle OH}{|}}{P}}{-}O{-}H + HOH \rightarrow R.OH + H{-}O{-}\overset{\overset{\displaystyle O}{\|}}{\underset{\underset{\displaystyle OH}{|}}{P}}{-}O{-}H}$$

In the case of phosphatases hydrolysing esters with three phosphate groups, of which adenosine triphosphatase is the chief example, the reaction occurs thus, producing a diphosphate:

$$\mathrm{R{-}O{-}\overset{\overset{\displaystyle O}{\|}}{\underset{\underset{\displaystyle OH}{|}}{P}}{-}O{-}\overset{\overset{\displaystyle O}{\|}}{\underset{\underset{\displaystyle OH}{|}}{P}}{-}O{-}\overset{\overset{\displaystyle O}{\|}}{\underset{\underset{\displaystyle OH}{|}}{P}}{-}O{-}H + HOH \rightarrow R{-}O{-}\overset{\overset{\displaystyle O}{\|}}{\underset{\underset{\displaystyle OH}{|}}{P}}{-}O{-}\overset{\overset{\displaystyle O}{\|}}{\underset{\underset{\displaystyle OH}{|}}{P}}{-}O{-}H}$$

$$\mathrm{+\ H{-}O{-}\overset{\overset{\displaystyle O}{\|}}{\underset{\underset{\displaystyle OH}{|}}{P}}{-}O{-}H}$$

Further enzyme action produces monophosphate and another molecule of phosphoric acid. This monophosphate is a substrate for the monoesterase and if the latter is also present the reaction proceeds to completion as in the first formula given above. In the case of the diesterases the reaction is essentially similar,

$$\mathrm{R{-}O{-}\overset{\overset{\displaystyle O}{\|}}{\underset{\underset{\displaystyle OH}{|}}{P}}{-}O{-}R' + HOH \rightarrow R{-}O{-}\overset{\overset{\displaystyle O}{\|}}{\underset{\underset{\displaystyle OH}{|}}{P}}{-}O{-}H + R'{-}O{-}H}$$

and again one of the reaction products is a substrate for the monoesterase. For this reason histochemical localization of the di- and triphosphatases lacks precision unless it can be shown that the monoesterase is absent from the particular location concerned. Histochemical methods have been evolved for revealing the phosphomonoesterases which depend on the demonstration of either reaction product; those depending on the deposition of the released phosphoric acid as calcium phosphate will be considered first, since up to 1950 all the important work on phosphatases was done by their use.

The phosphomonoesterases are divided on the basis of their optimum pH levels *in vitro* (cf. Kroon, Neumann and Krayenhoff Sloot, 1944). Those which work on the alkaline side of neutrality, and particularly at pH 9·0 and above, are called alkaline phosphatases, while those which work in the region of pH 5·0 constitute the acid phosphatases. The numerous curves with which Kroon *et al.* illustrated their paper show "summits" of two types, alkaline and acid. In none of the tissues examined could these authors find evidence of a third summit at a neutral pH.

It is probable that the most important *in vivo* activity of alkaline and acid phosphatases is concerned with the transfer of phosphate from one alcohol to another (Axelrod, 1948; Appleyard, 1948; Dixon, 1949; Meyerhof and Green, 1950; Dixon and Webb, 1953; Davison-Reynolds *et al.*, 1955; Morton, 1955; London and Hudson, 1955). This has been stressed by Danielli (1958) and by Dixon and Webb (1958, 1964).

$$\mathrm{R{-}O{-}\overset{\overset{\displaystyle O}{\|}}{\underset{\underset{\displaystyle OH}{|}}{P}}{-}OH + R'OH \rightleftharpoons R'{-}O{-}\overset{\overset{\displaystyle O}{\|}}{\underset{\underset{\displaystyle OH}{|}}{P}}{-}OH + ROH}$$

Since we have, at present, few histochemical methods for the detection of phosphokinases or transphosphorylases interpretation of alkaline or acid phosphatase activity as a manifestation of phosphate transfer remains essentially speculative. The significance of much of the applied histochemistry of the phosphatases is nevertheless greatly improved if one assumes that high activity indicates increased phosphate transfer rather than hydrolysis of phosphate esters.

Techniques for the Non-specific Alkaline Phosphomonoesterases (E.C.3.1.3.1)*

Methods Depending on Deposition of $CaHPO_4$

Of all the histochemical techniques for hydrolytic enzymes the most important, historically speaking, is the method for alkaline phosphatase designed independently by Gomori (1939) and by Takamatsu (1939). Although the technique is simple to apply it is still necessary to emphasize the warning given by Danielli (1950) that from the point of view of the interpretation of results

* These are the Enzyme Commission numbers of the enzyme or enzymes which the reaction to be described are designed to demonstrate.

it is deceptively simple. The arguments and criticisms of the method which are given below will at least make this last point clear. A variation of the original technique (Gomori; 1946, 1952) is commonly used at the present time. It depends on the deposition of calcium phosphate at sites of enzyme activity when sections are incubated with an organic phosphate ester in the presence of calcium ions, usually at pH 9 or higher. Most of the phosphatases require magnesium ions as activators and these are therefore added in the form of small concentrations of magnesium sulphate or chloride. Mg^{2+} is the natural activator of most of the enzymes which act on phosphorylated substrates, especially the phosphokinases but not the phosphorylases. In many cases it can be replaced by Mn^{2+}. It had been suggested that the metal formed an essential part of the active centre of the enzyme, or that it acted as a binding link between enzyme and substrate. It was also considered possible that the metal might directly activate the enzyme by altering the surface charge on the protein and thus its electrokinetic potential.

Evidence has been presented (Mathies, 1958) that Zn^{2+} is the prosthetic group of kidney alkaline phosphatase and that the substrate for the enzyme is actually the magnesium salt of the phosphate ester. This would account for the high level of Mg^{2+} (5×10^{-3}M) needed.

Various substrates are, or have been, employed, including α and β glycerophosphate, fructose diphosphate, yeast nucleic acid, yeast adenylic acid (adenosine-3-phosphate), phenyl phosphate, glucose-1-phosphate, creatine phosphate, thiamine pyrophosphate, barium phytate, adenosine triphosphate and muscle adenylic acid (adenosine-5-phosphate). In special studies like those of Gomori (1949a) many others were employed. The most usually employed substrate, sodium-β-glycerophosphate, is probably the best for ordinary purposes, being hydrolysed at a rapid rate which is exceeded, however, by that of phenyl phosphate and the naphthyl phosphates.

Following Gomori's example, tissues for phosphatase study were usually fixed in three changes of absolute acetone at 4°, after which they were dehydrated in absolute alcohol and embedded in paraffin wax. The double embedding process, using celloidin prior to embedding in wax, has also been employed with advantage. Although it is now realized (see section on fixation for enzyme studies in Chapter 5) that alkaline phosphatases are better preserved by other methods of fixation, the convenience of handling paraffin sections, prepared as above, and the fact that they can be used for acid phosphatase, lipase and esterase methods, have ensured their practical survival. After incubation of the sections in a buffered solution of calcium nitrate, containing magnesium ions and substrate, it remains only to localize the precipitated calcium phosphate *in situ*. Two principal methods have been employed for this. Gomori, and Takamatsu, originally used the von Kóssa silver stain but Gomori (1941) later used the cobalt sulphide procedure. Both are given in Appendix 15. Bourne (1943) considered the cobalt procedure to be the more efficient, but Kabat and Furth (1941) used von Kóssa's method and Newman *et al.* (1950)

observed that if the cobalt method was used the control sections, incubated in calcium nitrate without substrate, showed a non-specific deposition of cobalt which subsequently appeared as cobalt sulphide. I agreed (1953, 1960) with Gomori and Bourne in preferring the cobalt method for the final demonstration in most cases, since it produced sharper pictures with finer granulation. The von Kóssa procedure transforms the precipitated calcium phosphate into silver phosphate which is then reduced by the action of light to metallic silver. With some substrates (adenylic acid for instance), it may be preferable to the cobalt technique but the final appearances vary considerably. Where the enzyme is present in strength, the precipitate is usually dark brown and granular, but the final colour differs widely with different substrates. With some (fructose diphosphate) it is yellowish brown, while with others (muscle adenylic acid, thiamine pyrophosphate) it is black and often macro-crystalline. With substrates giving a macro-crystalline end result high power examination is fruitless. With both methods the presence of xylene in the final mounting medium must be avoided since it tends to cause fading, rapid in the case of cobalt, slower with the silver precipitates. Experiments conducted by Ruyter (1952) suggested that the cobalt sulphide method of conversion failed to demonstrate calcium phosphate when the concentration of the latter reached a certain critical level. Thus in the striated border of the duodenum, after a long incubation period, the deposits were well shown by the silver method but were unstained by cobalt sulphide.

Another method for demonstrating calcium phosphate precipitates, which I have not found to be efficient in practice, is the use of the dye sodium alizarin sulphonate which stains calcium phosphate deposits orange red. It was used by Bourne (1943), who added it to the incubating fluid. Many alternatives have been proposed but few have been found acceptable, even for critical studies. For instance, Bélanger (1951) suggested that sections should be examined by polarized light, in which the deposits show up clearly since the crystals of which they are composed are birefringent. Hancox and Nicholas (1956) used phase contrast microscopy in order to follow the reaction during incubation. They found the method more sensitive than the CoS technique. Takeuchi and Nogami (1954) devised a method of fluorescent staining and Molnar (1952) proposed the substitution of lead for cobalt and its subsequent staining with rhodizonate (see Chapter 26). Since the conversion of calcium to cobalt or to lead is not without its own peculiar problems (see below, p. 503) it would seem that fluorescent or visible staining of the deposits by lake formation with one of the more modern methods for calcium might well be preferable. The morin method (Chapter 27) and the copper phthalocyanin method (p. 426) have been suggested for the purpose. Deimling (1964) has proposed, as an alternative, that Pb^{2+} ions should be used to capture the phosphate released by alkaline phosphatase. This procedure avoids the calcium to cobalt conversion stage, which might be a considerable advantage. The dangers of false localization, due to Pb^{2+} ion-membrane effects increase markedly at alkaline pH levels.

Factors Influencing the Result. In order that a calcium deposit may take place in the tissues the rate of production of phosphate ions must be high enough for the solubility product of calcium phosphate to be exceeded locally. The establishment of this condition is prevented by many factors. Conditions controlling the nature of the precipitated calcium phosphate, at physiological concentrations, were fully examined by Boulet and Marier (1961).

Inactivation of the Enzyme during Fixation and Embedding. Fixation of enzymes, including the phosphatases, has already been considered in Chapters 5 and 14 and will not be considered further here. The remaining processes connected with embedding have little effect on the amount of phosphatase remaining in the block, until we come to the stage of infiltration with paraffin. No loss is caused by dehydration in alcohol, and Mowry (1949) stated that, as clearing agents, light petroleum, xylene, benzene and chloroform were all satisfactory and appeared to cause no significant decrease in enzyme activity in 12–14 hours. Because of its suitability as a prelude to vacuum-embedding he expressed a preference for light petroleum. The usual methods of embedding in paraffin (m.p. 56°–58°) cause a significant but variable loss of enzyme which Danielli (1946) estimated at 75 per cent. The actual loss depends on both the duration and the temperature of embedding. If thin blocks of acetone or alcohol-fixed tissues are vacuum-embedded in low m.p. paraffin (42°–44°) enzyme activity is higher (Mowry, 1949), although it is much less than that observed in fresh frozen sections.

Inactivation during Mounting or Storage. Considerable loss of enzyme activity occurs if paraffin sections, mounted on egg-albuminized slides, are redried on hot plates at 60° or higher. On no account should temperatures higher than 37° be used for drying slides intended for use in enzyme histochemistry. Lison (1948) drew attention to the possible solubility of phosphatases in watery solutions, suggesting that only the so-called desmo-enzymes would remain in tissue sections while the soluble lyo-enzymes would disappear during processing. In the same paper Lison stated that deterioration of phosphatase occurred with storage in the paraffin block but Gomori (1950b) could not confirm this. He maintained that no deterioration could be observed in sections cut from blocks stored at room temperature for 5 years. With Lison's warning in mind, I used at one time to store blocks for phosphatase studies at 4°, but I could observe no difference between sections cut from these and from similar blocks stored in the ordinary way.

Inactivation by Departure from Optimum pH. Raising or lowering the pH of the incubating medium from the optimum pH (for alkaline phosphatases pH 9·2–9·8) interferes with the result by slowing the rate of the reaction, although within fairly wide limits it does not inactivate the enzyme completely. Much more important is the effect of lowering the pH on the solubility of calcium phosphate. At pH 9·2 this compound is extremely, though not absolutely, insoluble and it has been considered to precipitate as soon as it is produced, thereby giving a picture which should correspond to the true localization of phosphatase (but see comparison with azo dye techniques below). At pH 7·5 it is much more soluble and will not precipitate locally unless the rate of production is very rapid. False localization is invariably seen when incubation is prolonged at levels below pH 8·5 and at pH 7·5 with even short periods of incubation. Apparent inactivation may be produced by pH changes which is in fact due to loss of lyo-phosphatase into the incubating medium (see Chapter 14, p. 491). The effects of pH on alkaline phosphatase activity were considered by Robbins (1950), Cacioppo *et al.* (1953) and Sonnenschein and Kopac (1953).

Effect of Activator. The principal activator for alkaline phosphatases is the magnesium ion whose accelerating effect was first noted by Erdtman (1928). This is usually included in the incubating medium whatever the method subsequently employed. Its absence is not therefore a factor which has to be taken into account in routine employment of phosphatase methods of the type we are considering. Other substances are known to enhance the activity of alkaline phosphatase, 0·01 M-ascorbic acid, for example, or salts of the bile acids in low concentrations. Newman *et al.* (1950) found that 0·001 M-sodium taurocholate, glycocholate, or deoxycholate produced a distinct increase in staining by the Gomori alkaline phosphatase technique. Zinc salts were mentioned as activators, by Hore, Elvehjem and Hart (1940) and by Cloetens (1941). It was reported by Sadasivan (1950, 1952) that cyanide inhibition of ox kidney phosphatase could be reversed by Zn ions at alkaline pH levels and by Mg ions at acid pH levels. Hoare and Delory (1955), on the other hand, found that zinc first activates and then inhibits the cyanide-treated enzyme. They showed that zinc is not indispensable for enzyme activity and considered that Cloetens' theory must be abandoned.

A specific inhibitor of intestinal alkaline phosphatase, L-phenylalanine, was described by Fishman *et al.* (1963) and histochemical studies were carried out by Watanabe and Fishman (1964) to ascertain the precise conditions of inhibition. They reported that at a concentration of 10 mM intestinal alkaline phosphatase was 60 per cent. inhibited, other phosphatases 0 to 10 per cent. With the calcium-cobalt technique inhibition was almost complete (using either β-glycerophosphate or AS-TR phosphate as substrate). Using the second substrate in an azo dye technique (p. 517) the degree of inhibition was much less. The most likely explanation for the discrepancy is that in the presence of 10 mM L-phenylalanine the solubility product of calcium phosphate is never exceeded.

Inhibition of different types of alkaline phosphatase (serum, bone, intestine, milk) by adrenalin, adrenochrome and other *O*-dihydroxyphenols with substitution at 0–4, at a concentration of 2 to 5 mM, was described by Anderson (1961). Thomas and Aldridge (1966) found that alkaline phosphatase was inhibited by Be^{2+} ions at very low concentrations (1 μM or less). The effect of anions (acetate, chloride, fluoride, bromide, phosphate, sulphate, iodide, nitrate) on alkaline phosphatase and other enzymes was recorded by Freiman *et al.* (1962).

Unsuitable Substrates. The most commonly employed substrate for both alkaline and acid phosphatase techniques depending on the demonstration of released phosphate ions is glycerophosphate. As indicated in the diagram (Fig. 138, p. 521), this ester is less rapidly hydrolysed at an alkaline pH than either muscle adenylic acid or phenylphosphate. Substrates which are even more slowly hydrolysed have been employed in various investigations. They have two main drawbacks. First, a prolonged incubation time becomes necessary and the occurrence of diffusion and other artifacts is thereby significantly raised. These points have recently been considered by Hill (1956). Secondly, comparison of the results obtained by the use of different substrates is made difficult by the presence of widely differing rates of hydrolysis.

Interpretation of the Result. If, in spite of the pitfalls enumerated above, a successful reaction is achieved, there still remains the interpretation of the result. Gomori (1950a) drew attention to some of the difficulties of interpretation and discussed them under three headings.

False Negative Reactions. These are defined as failures to produce a positive reaction when alkaline phosphatase is known to be present in the tissue under consideration. The chief causes are the presence of the enzyme in subthreshold amounts and the operation of the factors, mentioned above, which prevent the formation of a precipitate. In the case of the alkaline phosphatase technique the limit of sensitivity is in the neighbourhood of 25 μ moles (about 1 Bodansky unit) per gram of reacting tissue if the time of incubation is 1 hour (Gomori, 1950c). A section of rat kidney, for instance, measuring $12 \times 12 \times 0{\cdot}007$ mm. and weighing about 1 mg., may contain as much as 0·02 Bodansky units or about 20 units per gram. Any factor which increases the rate of diffusion of calcium phosphate, raises its solubility or decreases its rate of production, may cause a negative reaction in the presence of large amounts of enzyme.

False Positive Reactions. These are defined as reactions similar to the genuine one but not due to enzyme contained in the tissue. Substances already present may be confused with a positive reaction, hæmosiderin if treated by the cobalt method and phosphates in preformed calcification treated by either cobalt or silver methods. The first may be distinguished by applying Perls' method for iron to control sections and the second by removal of preformed deposits with 0·1 M-citrate buffer at pH5 for 5–10 minutes. This process does not injure the enzyme in any way. Melanin may cause trouble, especially in eye tissues. It may be bleached, by exposure to 3 per cent. H_2O_2 for 24 hours for example, without greatly affecting the enzyme, but this short exposure is usually insufficient to bleach dark melanin deposits. As mentioned above, if the cobalt method is used there may be non-specific absorption on to certain tissue elements (muscle, collagen, colloid, etc.), but the colour of these is yellow-brown, easily distinguished from the dark brown or black of a genuine reaction. Two other (rare) factors may cause false positive results. These are spontaneous hydrolysis of the substrate and its hydrolysis by bacterial action. The former occurs especially with adenosine triphosphate and with some acyl phosphates but not with the usual substrates. Bacterial hydrolysis can be avoided by making fresh substrate solutions at frequent intervals or by keeping a trace of $CHCl_3$, caphor or thymol in the stored incubating medium.

False Localization. This is defined as a positive reaction, due to enzyme contained in the tissue, appearing at sites other than the primary location. It may be due to two causes: (*a*) diffusion of the enzyme from its original site and its absorption at another, and (*b*) diffusion of the products of enzymic hydrolysis from their sites of production and their deposition or absorption at other sites. Lison (1948), in a paper already referred to, stated that phosphatases were present in the form of soluble (lyo-) and fixed (desmo-) enzymes. Diffusion of the former occurred readily in his opinion. Ruyter and Neumann (1949) found that phosphomonoesterase I (alkaline phosphatase) was soluble in dilute ethanol and that 20 hours' extraction with a 30 per cent. solution would remove it from tissue sections entirely. They believed that by taking sections from xylene to water via various ethanol concentrations the true picture of localization of the enzyme was disturbed. This danger was certainly exaggerated. Newman *et al.* (1950) observed a diffuse precipitate distributed uniformly over the section and the glass slide, and thought that it was due to activity of enzyme which had diffused from the section into the incubating mixture. Other interpretations are obviously possible.

Diffusion of products of hydrolysis already briefly discussed under the heading of inactivation by departure from optimum pH, is the most important single factor not only in the alkaline phosphatase techniques but in many other techniques for the localization of enzymes. Four experiments were carried out by Gomori (1950b) in order to illustrate the importance of this factor. In the first, he showed that in sections completely inactivated by treatment with nitric acid, and then placed in a solution of purified alkaline phosphatase, an intense staining reaction could subsequently be obtained in which the distribution was entirely different from that seen in control sections. In the second, an active slide and an inactivated one were turned face to face, separated by a strip of celluloid at either end. Pairs such as these were incubated at various pH levels. Between pH 7 and 8·5 the inactivated tissue showed nuclear staining around centres of high activity. At pH 9·2 no diffusion was noted, although Jacoby and Martin (1949) had found evidence of diffusion even at this pH. Finally, inactivated sections were incubated together with active sections in a Coplin jar for 24–96 hours. Once again, between pH 7 and 8·5, nuclear staining occurred in the inactivated slides. These experiments confirmed the existence of both short and long range diffusion effects. Danielli (1946) carried out some similar experiments, studying the precipitation of $CaHPO_4$ by non-enzymatic procedures, the ability of $CaHPO_4$ to diffuse, and the sites of deposition of the alcoholic end of the phosphate ester (see below). He concluded that $CaHPO_4$ did not diffuse at pH 9 but that there was a strong tendency for it to deposit at the sites of so-called phosphatase activity. Gomori (1951a) found that no nuclear staining occurred in sections incubated with 0·01M glycerophosphate at pH 9·3 for up to $2\frac{1}{2}$ hours, provided that the concentration of calcium ions exceeded 0·4M. At lower concentrations of calcium nuclear staining became more and more apparent.

By means of a simple model Johansen and Linderstrøm-Lang (1951, 1952) investigated the liberation and diffusion of phosphate ions, and their precipitation by calcium ions as in the Gomori type of technique. These authors stressed the fact that not only must the solubility product of calcium phosphate be exceeded for a precipitate to occur locally but that "the essential condition for a local precipitation at the site of enzymatic activity is therefore that the ion product at which spontaneous crystallization occurs is reached so rapidly that the entire cell is bathed in super-saturated calcium phosphate solution which will give rise to precipitation wherever this is induced, either by stray crystal nuclei embedded in the fixed cytoplasmic material or by other constituents of the cell." From their calculations Johansen and Linderstrøm-Lang concluded that spontaneous precipitation was unlikely to occur in the Gomori system, but that the precipitate was always induced either by preformed crystal nuclei or by other elements. What was measured, they considered, was just as much the distribution of precipitation centres as that of enzyme molecules.

As a possible mechanism of precipitation it was suggested that if calcium phosphate begins to precipitate somewhere in the neighbourhood of a site of enzyme activity (due to the presence of a preformed nucleus) then the HPO_4 concentration falls to a low value at this point. More phosphate ions therefore diffuse in but the diffusion is one-sided (more phosphate comes from the side of the site of production). Hence the solid particle grows asymmetrically and if growth on one side is accompanied by dissolution on the other side, the particle will wander towards the site of production of phosphate.

In opposition to these ideas Gomori and Benditt (1953) pointed out that several of the assumptions made by Linderstrøm-Lang and Holter were not valid for the conditions existing in tissue sections. In particular, they pointed out that the local concentration of enzyme in a cell might be 1,000 times the average concentration. Experimentally Gomori and Benditt observed that when phosphate was added to solutions similar to those used histochemically there was no tendency to supersaturation. They concluded that the factors contributing to false localization with the Ca—Co technique were not amenable to mathematical analysis and (Gomori, 1952) it was believed that most of the sources of error could be rectified by raising the concentration of calcium ions in the incubating medium to 0·05 or 0·1 M.

Johansen and Linderstrøm-Lang (1953) replied to these criticisms by comparing their work with that of Gomori and Benditt. It appears that in their calculations they assumed a *turnover number* (the number of moles of substrate reacting per minute per mole of enzyme) of 1,000, equal to 40,000 Bodansky units per gram. This is very different to the average figure of 700 units used by Gomori and Benditt. It is worthwhile to note at this point that the turnover number should be expressed in terms of active centres of enzyme (Dixon and

Webb, 1958, 1964) and that the expression used above is better called the *catalytic constant*. In any case, unless the molecular weight and turnover number of alkaline phosphatase are known, theoretical calculations are necessarily arbitrary and probably inaccurate. According to Schramm and Armbruster (1954), the molecular weight of alkaline phosphatase was about 60,000 and Morton (1957) calculated the turnover number (catalytic constant) of intestinal alkaline phosphatase for phenyl phosphate at 38° as 236,000 moles per minute per mole of enzyme.

With regard to precipitation centres the observations of Danielli (1946) are still pertinent. He found that the artificial precipitation of calcium phosphate over a section of kidney resulted in a large amount of deposition in the nuclei and very little in the brush borders. Since an exactly opposite result occurs in the histochemical reaction Danielli considered that the quantitative distribution of precipitation nuclei did not affect the accurate localization of enzyme.

All the above considerations lead up to the difficult question of the staining of the nuclei by Gomori's method. We have to decide whether this is due to the presence of a phosphatase, specific or otherwise, or to diffusion artifact of one or the other variety.

Staining of the Nuclei and Nuclear Phosphatases. With the Gomori method for alkaline phosphatases it was consistently observed that the nuclei of cells in areas of intense staining were themselves deeply stained. For example, the nuclei of the proximal convoluted tubules of the kidney stained more darkly and in a shorter time than those of other portions of the nephron (see Fig. 127). This corresponded to the most intense phosphatase activity which was in the brush borders of the same tubules. Observations of this kind, and the experimental manœuvres considered above, suggested very strongly that the staining of the nuclei was due to diffusion artifact. While there can be no possible doubt that under a variety of conditions the staining of nuclei is due to this cause, we require to know whether it is invariably due to artifact or whether the nuclei in fact contain alkaline phosphatase, demonstrable by histochemical methods.

Newman *et al.* (1950), in a comprehensive study, used eleven different substrates and ten inhibitors and their results with respect to localization, substrate range, and inhibition effects suggested that the phosphate-liberating enzyme systems could be divided into three groups. **Group I enzymes** comprised the alkaline phosphatases, and these gave results with all substrates except the barium salt of inositol hexaphosphate. **Group II enzymes** were demonstrated when muscle adenylic acid or adenosine triphosphate (ATP) were used as substrates and they were regarded as specific 5-nucleotidases. (Although the latter do not hydrolyse ATP, this substrate is subject to spontaneous hydrolysis, one of the products being adenosine-5-phosphate.) **Group III enzymes** were only present in nuclei and their differentiation was based partly on the fact that with 5 of the 11 substrates nuclei were better shown than with the remainder and partly on the finding that heat, trichlor-

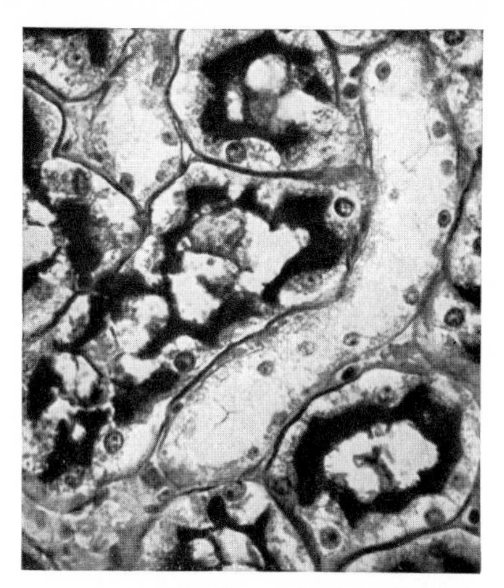

Fig. 127. Application of the Gomori Ca-Co method to a 5·5 μ paraffin section of rat kidney. Note staining of the brush borders of the primary convoluted tubules and of the nuclei. × 370.

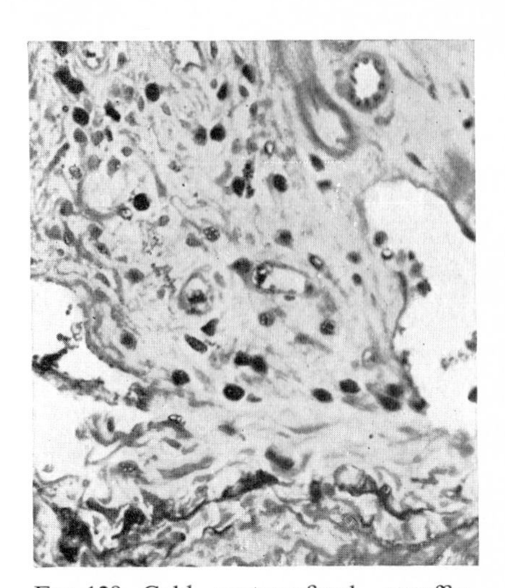

Fig. 128. Cold acetone-fixed, paraffin-embedded rat lung. Showing, *below*, alkaline phosphatase in adventitial fibres in the wall of an artery and, *above*, isolated lymphocyte and other nuclei also strongly stained. Sodium α-naphthyl phosphate—diazotized *o*-dianisidine. × 500.

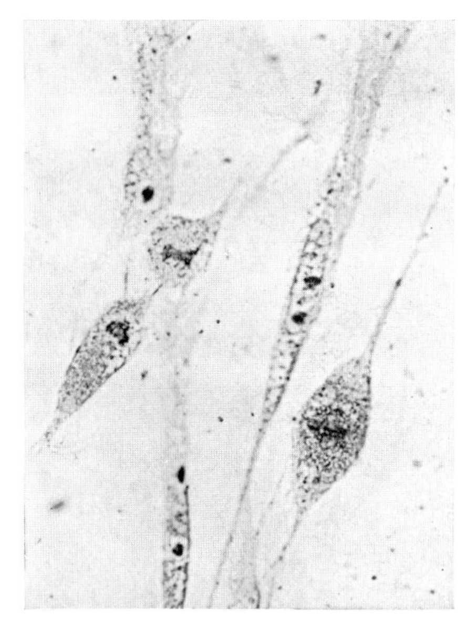

Fig. 129. Tissue culture (Chick embryo, muscle and connective tissue). Black stained regions were stained reddish-violet, grey areas orange-yellow. Coupling azo dye method for alkaline phosphatase. × 900.

acetic acid of 0·25 M to 0·125 M glycine or arginine inhibited the staining of Group I sites but allowed the nuclei to stain as before. In order to show this so-called Group III activity, the authors incubated inactivated sections with the various substrates for no less than 7 days. If, under similar conditions, phosphate was released in the incubating fluid by any means whatsoever, the staining pattern was identical with that observed in the case of active sections. These results were probably due to adsorption of enzyme from the incubating medium, derived from active sections which were incubated at the same time.

Feigin, Wolf and Kabat (1950), using the techniques detailed above, determined that the increased nuclear staining adjacent to a locus of intense enzyme activity was due to enzymes of the same group as that locus and not to the so-called nuclear enzyme of Group III. They concluded that nuclear staining was an artifact "under some circumstances" and that it might be produced by diffusion of enzyme into the nuclei, after death, from sites in which it is normally present. That this is not the whole story is shown by the fact that nuclear artifact can be produced in sections freeze-dried less than a minute after death of the animal and subsequently fixed in absolute acetone for 30 minutes prior to staining. Continuing their studies, Feigin and Wolf (1953) compared the cobalt sulphide and silver nitrate methods for "developing" the final picture and reported that with the former nuclear staining might be both enzymatic and non-enzymatic. With the silver method only enzymatic staining was revealed, though this might be partly intrinsic and partly due to absorption of enzyme. This absorption of alkaline phosphatase by nuclei, as well as by other active sites, was demonstrated by Barter (1954), and also by Yokoyama, Stowell and Matthews (1951). More recently Feigin and Wolf (1954, 1957) were able to demonstrate, in cold acetone-fixed paraffin sections, a remarkable influence of magnesium ion concentration on the final localization of enzyme. If incubation was maintained at 5 minutes, there was only light cytoplasmic staining in the absence of magnesium, very considerable staining in the presence of 10 mM Mg, and practically none in the presence of 160 mM Mg. Nuclear staining, on the other hand, was present only when the highest concentration of magnesium was employed.

This nuclear activity bore no resemblance to other types of nuclear activity since it was observed in only a few organs, notably in guinea-pig endometrium, ovary and adrenal cortex. It was completely inhibited by 10 mM cyanide. The authors considered that their results indicated the presence of a true nuclear alkaline phosphatase.

Leduc and Dempsey (1951) studied the role of activation and diffusion in the alkaline phosphatase reaction. They showed that if an active section were incubated face to face with a heated section, no alkaline phosphatase reaction appeared in the latter and there was no diffusion of calcium phosphate from the active section. Under the same conditions, nuclear staining was produced in a section inactivated by treatment with acid. The authors considered that this indicated a complete inactivation by heat but a reversible inactivation by acid, restored in this case by diffusion of an activator from the active section. Their findings might equally well be due to physical differences in the affinity of the nuclear proteins for phosphate, brought about by heat or acid treatment. The fact that nuclear staining was still obtained in sections extracted with deoxyribonuclease or 0·1 N-HCl does not prove that affinity for the reaction product ($CaHPO_4$) was not the factor responsible, since the protein structure of the chromosomes remained intact.

With the evolution of histochemical techniques for revealing the alcoholic part of the phosphate ester, the existence of nuclear alkaline phosphatase has been further called into question. These techniques are fully considered in a succeeding section and only their bearing on the present question is dealt with here. The majority of the azo dye techniques for phosphatase depend on coupling of the reaction product with a diazonium salt present in the incubating medium, and these techniques are usually stated to show no activity in the nuclei. In fact, as already reported (Grogg and Pearse, 1952), the nuclei in acetone-fixed paraffin-embedded sections are often stained by coupling azo dye techniques, although in cold formalin-fixed frozen sections they are always colourless. Fig. 128 shows the staining of lymphocyte nuclei, and adventitial fibres of an artery, in rat lung. This observation indicates either the occurrence of enzyme diffusion during acetone fixation or the activation of phosphatase in occasional nuclei by precipitation of their proteins with acetone. In the azo dye method of Loveless and Danielli (1949) in which coupling was not used, the nuclear reaction was far more intense than that given by Gomori's method and the structure of the nuclei was in most cases "completely obliterated by deposits of dye." The significance of this remains uncertain, though Feigin and Wolf (1957) claimed that their experiments with the same substrate showed that non-enzymatic staining of nuclei by the phenol did not occur.

Lorch (1947) considered that the lack of nuclear staining with the coupling azo dye techniques was due to the insensitivity of the reaction when naphthyl phosphates were employed as substrates. Yokoyama, Stowell and Tsuboi (1950), comparing Gomori's method with the azo dye method of Manheimer and Seligman (1949), found that the former showed "definite nuclear activity" in many tissues and the latter "very little". Novikoff (1951) discussed the validity of the phosphatase methods for intracellular demonstration of enzyme. He stated that nuclear staining was absent with the azo dye method and concluded that both this and the Gomori method were of little value for intracellular localization except in structures with extremely high activity. Gomori (1951a), using a newer modification of the azo dye method, also reported no nuclear staining in his preparations. Both these authors used acetone-fixed paraffin sections. On the other hand, Chèvremont and Firket (1953) invariably obtained nuclear staining in fibroblasts and myoblasts growing actively *in vitro*, by both the Ca—Co and azo dye techniques (Fig. 129, facing p. 504).

These authors observed that procedures which interfered with the activity of the enzyme would produce negative results in the nuclei but that these could be restored by prolonging the incubation period.

The bulk of the experimental work detailed above suggests that the nuclear staining observed with the Gomori technique is mainly due to artifact. This conclusion is supported to some extent by observations on isolated nuclei from mouse or rat liver cells. Yokoyama *et al.* (1950) found little or no nuclear alkaline phosphatase activity in isolated mouse liver cells and Novikoff, Podber and Ryan (1950) found only a small fraction (15 per cent.) of the total enzyme activity in rat liver cells. Dounce (1950) also reported similar results. There remain, nevertheless, legitimate doubts concerning the interpretation of all nuclear staining with the calcium phosphate techniques as due to one or other type of artifact.

Histochemical Specificity of the Phosphatases. In earlier days there was much argument as to whether there were a number of different non-specific alkaline phosphatases. It is now agreed (Moss, 1964) that alkaline phosphatases from several human tissues are distinct in electrophoretic and enzymatic properties. Apparently only one non-specific alkaline phosphatase is usually present at any given site, however. Similarly, there is now no doubt that several distinct substrate-specific alkaline phosphatases exist. Many early investigators (Glick and Fischer, 1945, 1946; Dempsey, 1949; Friedenwald and Maengwyn-Davies, 1950) stated that various alkaline phosphatases could be distinguished histochemically. Many authors, like Dempsey, Greep and Deane (1949), using various substrates, attributed the different pictures obtained to substrate specificity and referred, for instance, to fructose diphosphatase, nucleic acid phosphatase, etc. Gomori (1949a and b) studied the question of the unity or plurality of the phosphatases. He admitted, *ab initio*, that there could be no doubt that alkaline and acid phosphatases were distinctly different enzymes and that the individuality of the adenosinetriphosphatase of striped muscle, of pyrophosphatase, and hexosediphosphatase, was firmly established. More modern evidence (Eaton and Moss, 1966) indicates that intestinal alkaline phosphatase and intestinal pyrophosphatase are one and the same enzyme, perhaps having different active centres.

Zorzoli and Stowell (1947) compared the distribution of glycerophosphatase and hexosediphosphatase in varying tissues. They believed that they could localize a specific fructose-diphosphatase by histochemical means. In order to determine whether the histochemical approach could give an answer to the question, Gomori (1949a) used nineteen different substrates and found with eighteen of them no evidence, in acetone-fixed paraffin-embedded tissues, of any other than non-specific alkaline and acid phosphatases. With the nineteenth substrate, *p*-chloranilidophosphonic acid ($ClC_6H_4NHPO(OH)_2$), a different distribution was noted in the acid range which led him (1949b) to describe a specific phosphamidase (see p. 567). In further studies (1950b) he used ten additional substrates and observed intense reactions in 2 hours with all of these except the pyrophosphates, which required 24 hours' incubation. The distribution was identical with all substrates except 5-nucleotide. In this case, although in some organs the picture was identical with that produced by glycerophosphate, in others (brain, testis, smooth muscle, spleen) the difference was marked. Gomori considered that these findings indicated the presence of the specific 5-nucleotidase described by Reis (1937, 1940, 1951) and by Gulland and Jackson (1938). This specific 5-nucleotidase was later demonstrated histochemically by Newman *et al.* (1950) and by Pearse and Reis (1952). In this last study, which is referred to below and illustrated in Figs. 130, 131, 132 and 133, two additional sites of strong 5-nucleotidase were noted, the adrenal medulla (rat) and pars nervosa of the pituitary (man). Although Gomori (1950a), in a single paper, unaccountably maintained that different alkaline phosphatases could not be demonstrated histochemically, the above evidence suggested that at least one enzyme could be so distinguished in the acid range and another two on the alkaline side.

The question of distinguishing substrate-specific phosphatases was further considered by Novikoff (1958), who stressed the value of pH activity curves for various substrates when assessing the possibility of demonstrating specific phosphatases histochemically in a given tissue. He pointed out that in intestinal mucosa, for instance, the extremely high activity of non-specific alkaline phosphatase effectively prevented the demonstration of substrate-specific phosphatases such as 5-nucleotidase and adenosine triphosphatase. In other tissues, such as liver, the specific phosphatases can certainly be demonstrated histochemically (Wachstein and Meisel, 1957) owing to the small amount of alkaline phosphatase activity present. These points are further considered in the sections dealing with the substrate-specific phosphatases (pp. 521–539), and at the end of the section on non-specific alkaline phosphatase the important question of comparison between the calcium phosphate and the naphtholic substrate methods is dealt with.

In a series of papers on the subject Bourne (1954a and b) and Baradi and Bourne (1951) presented their findings on the hydrolysis of a large number of mono- and diphosphate esters by fixed tissue sections, using a Gomori type of procedure. Their substrates included dipotassium α-D-glucose-1-phosphate, barium D-fructose-1-phosphate, dipotassium D-galactose-6-phosphate, barium L-sorbose-1-phosphate, tetrasodium 2-methyl-1,4-naphthohydroquinone diphosphate, tetrasodium 3,4′-dihydroxy-4-methoxychalcone diphosphate, riboflavin-5-phosphate and pyridoxal phosphate. Comparison was made in all cases with glycerophosphate.

Bourne considered that non-specific alkaline phosphatase was not responsible for the dephosphorylation of riboflavin and pyridoxal phosphates, since marked differences were evident in the localization of the calcium phosphate deposit in nearly all tissues examined. Differences between the various sugar phosphates and glycerophosphate were less striking, but Bourne concluded that his results indicated the existence of a "spectrum of phosphate-splitting enzymes with considerable overlap at pH 9." The results obtained by Baradi and Bourne (1951) in the tongue and olfactory mucosa of the rabbit also suggested to the authors the presence of a number of specific enzymes. This view was based largely on the differential inhibitions produced by vanillin, tea infusion, quinine, capsicum, aniseed, peppermint and coffee.

I examined the reaction of a number of the above substrates (1960), together with some less likely ones, using fresh frozen cold microtome sections incubated in the acetone-containing medium of Fredricsson (1952, 1956) (Appendix 15, p. 711). Most of the substrates were hydrolysed only weakly by comparison with glycerophosphate and the principal and almost invariable difference was the presence of strong nuclear staining. This was completely lacking in glycerophosphate-incubated sections. In view of the number of alternative explanations for such findings, and in the complete absence of supporting biochemical data, Bourne's thesis must be considered unproven. If parallel biochemical and histochemical studies are carried out, strict comparison will only be obtained if the latter are done on fresh tissue (fresh frozen or freeze-dried sections). Fixation introduces too many variables into a problem already top-heavy with them.

Quantitative Estimation. Quantitative estimation of alkaline phosphatase in tissue sections was possible according to Gomori (1950b) only if laborious comparative experiments were carried out. Danielli (1950), on the other hand, suggested that incubation for a logarithmic series of times gave quantitative appreciation, the sites of highest activity appearing before the other sites. In all histochemical studies of the phosphatases, the shorter the incubation can be kept the more accurate is localization likely to be. An interesting method for the quantitative evaluation of Gomori-type histochemical preparations was evolved by Doyle (1950). This was based on the substitution of lead for cobalt (in the alkaline phosphatase method), the extraction of the final precipitate of lead sulphide with 5 N-HCl, and its conversion to methylene blue by the method of Fogo and Popowsky (1949). The amount of methylene blue was measured spectrophotometrically at 670 nm. With methods such as these (Doyle, Omoto and Doyle, 1951) only quantitation per section can be obtained. Radioactive tracer methods have been used by several workers following the early technique of Dalgaard (1948), who made use of ^{32}P-labelled glycerophosphate. This work was extended by Barka *et al.* (1952), who employed ^{212}Pb, as lead nitrate, to precipitate the phosphate ions at pH 9·4. With the aid of a perforated lead disc, and a specially designed counter-tube, measurements were made of the enzyme activity of relatively small areas of tissue. Further studies on these lines were carried out by Shugar *et al.* (1957, 1958).

There is little doubt that the most convenient and easily applicable method is that of Barter *et al.* (1955) employing the interferometer microscope. This method affords only approximate values, however, since exact knowledge of the composition of the calcium phosphate deposit is not forthcoming. In calculating the mass of the deposit, therefore, an arbitrary value is employed instead of an exact one and, according to Danielli (1958) an error of ±20 per cent. is possible.

Applications of the Gomori Method. Most of the work mentioned below was carried out by means of the calcium-cobalt or calcium-silver techniques. Now that they have been made more workable, the azo-dye techniques are being increasingly employed for demonstrating the non-specific alkaline phosphatases. There appear below references to the classical papers in phosphatase histochemistry with brief mention of a number of others which are of particular interest to pathologists.

The general distribution of alkaline phosphatase in normal tissues was described by Gomori (1941) and by Bourne (1943), and in normal and neoplastic tissues by Kabat and Furth (1941). The *kidney* was investigated especially by Kritzler and Gutman (1941), Wilmer (1943, 1944), Bunting (1948), Eränkö and Niemi (1954), Eränkö and Lehto (1954), Wachstein (1955), Longley (1955), Sachs and Dulskas (1956) and by Tapia Freses *et al.* (1957), while the *endometrium* was studied by Atkinson and Elftman (1946, 1947), Atkinson and Engle (1947), Arzac and Blanchet (1948), Atkinson and Gusberg (1948), Hall (1950), Erichsen (1953), Váczy *et al.* (1955) and by Skjerven (1956). The *ovaries* were studied by Corner (1944), and later by Leckie (1955), Ford and Hirschman (1955), Mukherjee and Banerjee (1955–56), Dhom and Mende (1956), McKay *et al.* (1961) and Arvy (1960). The *placenta* was studied by Hard (1946), Dempsey and Wislocki (1947), Strauss and Stark (1960) and Curzen (1964). The phosphatases of the *liver* were dealt with by Wachstein and Zak (1946a and b) and by Deane (1947), whose survey included also the *pancreas* and *salivary glands*. Later studies on the liver were those of Minamitani (1953), Meyner and Williams (1954), and Wajchenberg and Hoxter (1955). Jacoby and Martin (1951) compared the biochemistry of the bile with the histochemical distribution of alkaline phosphatase in the liver, in the rabbit, dog and guinea-pig. There was a rough correlation between the bile alkaline phosphatase and the histochemical demonstration of the bile canaliculi. Krugelis (1946) and Emmel (1946) considered especially the intracellular distribution of alkaline phosphatase, while Danielli (1946) declared that, apart from the nuclei and brush borders of the kidney tubules, this enzyme seldom occurred in the cytoplasm except in rapidly regenerating tissue. He connected it especially with fibre formation, but Robertson, Dunihue and Novikoff (1950), repeating and extending the work of Fell and Danielli (1943) and of Danielli, Fell and Kodicek (1945), could find no such evidence. Many other authors worked on this problem of the association of alkaline phosphatase with fibre formation (Johnson, Butcher and Bevelander, 1945; Bradfield, 1946; Jeener, 1947) and later two papers by Gold and Gould (1951) and Gould and Gold (1951) supported the relationship between the two. These last authors claimed to have confirmed their histochemical findings with the Gomori method by using the azo-dye technique of Manheimer and Seligman (see below).

For other sites a short list is appended. *Blood and bone marrow:* Wachstein (1946), Storti (1951), Wiltshaw and Moloney (1955), Kaplow (1955), Cesàro and Granata (1955–56), Boivin and Robineaux (1956), Graziadei and Zaccheo (1956), Trubowitz *et al.* (1957), Valentine *et al.* (1957), Hayhoe and Quaglino (1958), Ackerman (1962), Kaplow (1963), Wetzel *et al.* (1963), Elves *et al.* (1963), Wulff (1967). *Nervous system:* Landow, Kabat and Newman (1942), Sinden and Scharrer (1949), Shimizu (1950), Tolone and Ventra (1950), Naidoo and Pratt (1953), Chiquoine (1954), Feigin and Wolf (1955). *Skin and hair:* Johnson and Bevelander (1946), Fischer and Glick (1947), Foraker and Wingo (1955), Spier and Martin (1956), Kopf (1957). *Cartilage and bone:* Lorch (1947), Zorzoli (1948), Follis (1949, 1953), Majno and Rouiller (1951), Zorzoli and Nadel (1953), Schajowicz and Cabrini (1954). *Teeth:* Engel and Furata (1942), Bevelander and Johnson (1945, 1950), Greep, Fischer and Morse (1948), Mori *et al.* (1961). *Pituitary gland:* Abolins (1948), Abolins and Abolins (1949). *Thyroid:* McAlpine (1955), Haley *et al.* (1955). *Adrenal:* Verne and Herbert (1955), Allen and Slater (1956), Clayton and Hammant (1957) (Azo-dye method) Naik and George (1964). *Nuclei and chromosomes:* Krugelis (1942), Danielli and Catcheside (1945), de Nicola (1949), Brachet and Jeener (1946, 1948). Sulkin and Gardner (1948), Firket (1952), Chèvremont and Firket (1953). *Choroid Plexus:* Bartoníček and Lojda (1964). *Cells in tissue culture:* Chèvremont and Firket (1949), Henrichsen (1956). The distribution and relationships of alkaline phosphatase and glycogen were studied particularly by Horowitz (1942), Wislocki and Dempsey (1945), Johnson and Bevelander (1946) and Pritchard (1947), while Martin (1949), using rabbit liver, introduced a technique for demonstrating the presence of alkaline phosphatase and glycogen in the same section.

Simultaneous Coupling Azo Dye Methods (Unsubstituted Naphthols)

The first method for the histochemical demonstration of alkaline phosphatase by precipitation and demonstration of the alcoholic part of the phosphate ester used as substrate was described by Menten, Junge and Green (1944). This depended on the hydrolysis of calcium β-naphthylphosphate and the rapid reaction *in situ* of the liberated β-naphthol with diazotized α-naphthylamine at pH 9·4 to give a red precipitate at the sites of phosphatase activity. Methods of this type are referred to as simultaneous coupling methods. The same principle was used by Danielli (1946) who used phenylphosphate as well as β-naphthyl phosphate to accomplish a similar reaction. Subsequently, modifications of the technique of Menten *et al.* were introduced by Manheimer and Seligman (1949) and by Gomori (1951a).

The essentials of the procedure as detailed above, are the hydrolysis of a monoaryl phosphate, instead of an alkyl phosphate such as is usually employed in the Gomori technique, and the immediate coupling of the phenol, freed by hydrolysis, with a suitable diazotized amine to form an insoluble coloured precipitate at the site of enzymic activity. All subsequent simultaneous coupling azo dye methods, for other enzymes as well as for phosphatases, have employed this same principle. Such development was foreseen by the original authors. Menten *et al.* tried other phosphate esters before finally deciding on the use of β-naphthyl phosphate. Phenyl phosphates proved useless because the azo dye formed from the free phenol with a large number of diazotized amines was too soluble to be retained *in situ* in the tissues. β-naphthol on the other hand gave insoluble precipitates with a variety of amines, and of these α-naphthylamine was selected on account of the deep red colour and extreme insolubility of the resulting dye. The reactions involved are given below.

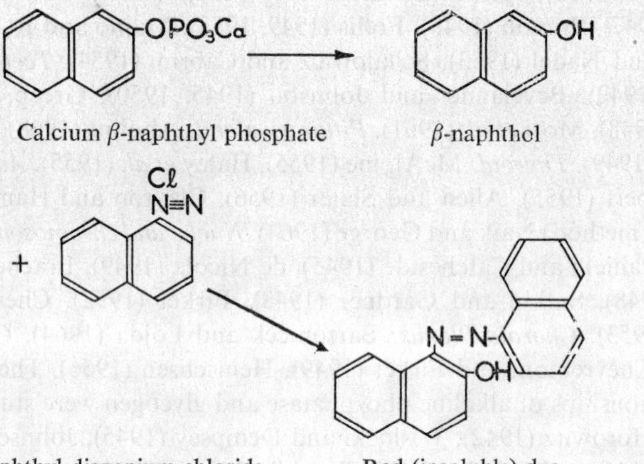

OPO_3Ca → OH

Calcium β-naphthyl phosphate β-naphthol

Cl N≡N + N=N OH

α-naphthyl-diazonium chloride Red (insoluble) dye

This ingenious method, performed on acetone-fixed paraffin sections, was technically laborious. After mixing the substrate and freshly diazotized α-naphthylamine the pH was adjusted to 9·4 with 6N-NaOH and stabilized with veronal-acetate buffer. The brown precipitate which formed was removed by filtration and the solution cooled to 6°. Slides were incubated for 15–30 minutes, washed in water, counterstained with light green and mounted in glycerol. No activator, in the shape of the magnesium ion, was employed, and the optimum incubation time was 18 minutes at pH 9·4. The extreme rapidity of the reaction was notable and a deposit of the dye could be observed after 5 minutes' incubation of active tissue. This extreme rapidity however, only exceeds the hydrolysis rate of glycerophosphate by a small margin. In active tissues the Calcium-Cobalt method gives a visible result in as short a time as this. Tests performed at 37° increased the rate of the reaction but disintegration of the diazonium salt was also more rapid. Secondary products formed by the breakdown of the latter stained the background structures a diffuse yellow colour.

Modifications of the Original Method. The results obtained by Menten *et al.* were comparable with those of the Calcium-Cobalt method but the authors did not mention the presence or absence of nuclear staining. Lorch (1947) stated that only sites of great enzyme activity could be demonstrated by this method and she attributed the observed absence of nuclear staining to the weakness of these sites in phosphatase. One of the drawbacks to the method as outlined above was the necessity for the preparation of a fresh diazonium salt for each batch of the incubating medium. Manheimer and Seligman (1949) overcame this by forming a stable diazotate of α-naphthylamine by the addition of naphthalene-1:5-disulphonic acid. They also used commercially available diazotates of *o*-dianisidine and 2-amino-4-chloroanisole (see Table in Appendix 15, p. 711). Using acetone-fixed paraffin sections and a pH of 9·4, deposition of dye in phosphatase-rich tissues was noted within 1 minute. By using the diazonium salt in low concentration, together with low temperature and short incubation, the brownish staining of the background was reduced to a minimum. Because of the solubility of the azo dye in organic solvents, the final mounting was made in an aqueous medium. Manheimer and Seligman used calcium α-naphthyl phosphate as well as the β-naphthyl salt, but found that it was unsuitable as a substrate for alkaline phosphatase because it gave relatively weak staining under optimal conditions. These authors compared their modified Menten-Junge-Green method with Gomori's calcium-cobalt method, in a wide variety of tissues, and reported a close correspondence. No specific study of the reaction of the nuclei was made, but it was noted that "some staining occurred in tumour-cell nuclei". The advantages of the method were considered to be: (1) The incubation period was short and hence the method more sensitive. (2) Control sections were unnecessary as naphthols are not normally present in the tissues. (3) The azo dye was very much less soluble than $CaHPO_4$, so that pictures were very much sharper and diffusion artifact was reduced. (4) Development of colour could be watched, and stopped at the optimal time.

Theoretical and Practical Considerations. A further modification of the Menten-Junge-Green method for alkaline phosphatase was suggested by Friedman and Seligman (1950). This involved the substitution of the relatively insoluble calcium salt by the readily soluble sodium salt. This modification was used by Gomori (1951a), who incubated acetone-fixed paraffin sections with 0·002 to 0·005 M-sodium α-naphthyl phosphate at pH 9·4 in the presence of a variety of stable diazonium salts. The buffered substrate-diazotate medium was filtered into a Coplin jar and the sections incubated at room temperature. Within a few minutes the sites of strong phosphatase activity were observed to be coloured black, blue or red, according to the diazotate employed. After 15 minutes or so the incubating medium became dark and turbid and the author advised its renewal at this stage. The background staining, due to breakdown products of the diazotates, was easily removed with 1 per cent. acid alcohol according to the author. Gomori (1951b) did not consider the reaction to be faster than that obtained in comparable sections by means of the Ca—Co technique. A comparison between his two methods, applied to acetone-fixed paraffin sections of rat kidney, is given by Figs. 127 and 134. Absence of nuclear staining with the azo dye method is the most important point to observe.

The superiority of the Gomori modification is due to two things; the substitution of the far more water soluble sodium salt, which enables the optimum concentration of substrate to be employed, and the substitution of the α-naphthyl salt. The azo dyes produced by coupling α-naphthol with diazonium salts are far more insoluble in water than those produced with β-naphthol and diffusion of the final reaction product is thus diminished. Grogg and Pearse (1952) reported attempts to improve the coupling azo dye technique for alkaline phosphatase still further, and subsequent work on these lines is discussed below.

Sodium α-naphthyl phosphate was therefore the substrate of choice until the advent of Naphthol AS substrates. King (1952) observed that the hydrolysis of α-naphthyl phosphate proceeded as rapidly as that of phenylphosphate at alkaline pH levels. Since acetone fixation and paraffin embedding destroy at least 70 per cent. of alkaline phosphatase present in the tissues, it is better to use short fixation in cold buffered formalin (4°, 8–16 hours) and frozen sections. Alternatively, one may use freeze-dried or cold microtome sections, with or without post-fixation. In sections of these types preservation of enzyme is good, fixation artifact small and enzyme diffusion also small, so that the most important single factor in the performance of the method becomes the choice of a suitable diazonium salt for coupling at an alkaline pH.

Certain properties of diazonium salts have particular bearing on the problem. Inhibition of enzyme activity is produced by all diazonium salts in varying degree, and the possibility of reducing this inhibition by lowering their concentration is prevented by the fact that for efficient coupling with the reaction product (α-naphthol) excess diazonium salt is necessary. The optimum concentration for most diazotates was found to be in the region of 1 mg./ml. of substrate solution. If the concentration was much lower than this (0·1–0·2 mg./ml.) two particular effects were noted. First, diffusion of α-naphthol occurred, and, secondly, the final colour of the azo dye precipitate was less strong. With diazonium salts like Fast blue B, which gives a black final product at the optimum concentration, for instance, the resulting dye with lower concentrations was invariably red. If the concentration of diazonium salt was higher than 5 mg./ml. inhibition was very great and no staining at all resulted except in situations of strong enzyme activity. The effect of the zinc or aluminium salts, present in most stable diazotates, on the activation or inactivation of alkaline phosphatase must also be considered. The optimum concentration of $ZnCl_2$ for activation has been given as 0·001M (0·316 g./litre) and the average concentration of $ZnCl_2$ in the diazotates is 5 per cent. At a concentration of 1 mg./ml. of the diazotate 0·05 g./litre of $ZnCl_2$ will be present, a figure about 2½ times smaller than the optimum concentration for activation.

A second property of the diazonium salts which is of great importance is the non-specific staining of the tissues resulting from their breakdown products. In extreme cases this may be mistaken for a positive reaction due to enzyme activity, and it is not always removable beyond a certain point. It is essential that the colour of the azo dye final product should be as dark as possible and that the size of the precipitate should be small. The first increases the sensitivity of the method by raising the point of minimum appreciation, the second enables accurate localization to be made on the intracellular level. Since the azo-α-naphthol dyes, and to a greater extent the azo-β-naphthol dyes, are soluble in lipids, the final product precipitated in the tissues should have particles visible under high magnification. If the final product is not particulate, staining of lipid components of the tissues is likely to be indistinguishable, if it occurs, from the specific deposition of dye.

Unpublished observations by Loustalot (1955) suggested that some of the changes occurring after storage in preparations made by simultaneous coupling azo dye techniques for alkaline phosphatase might be due to continued enzyme activity on residual substrate. He found considerable improvement after brief treatment with Lugol's iodine (Figs. 135 and 136). There is no doubt that this procedure can reduce storage artifacts, but the mechanism by which it does so is far from clear. One of the most annoying developments, which occurs after only brief storage, is the development of gas bubbles under the coverslip. According to Meier-Ruge and Meier (1963) post-incubation with 1 per cent. sulphanilic acid in 10 per cent. acetic acid for up to 10 minutes, or in 5 per cent. potassium alum, or 0·5M $AlCl_3$ in 0·1 N-HCl, prevents gas formation. Peters (1964) recommended a method of mounting smears and sections in gelatine without coverslip.

I tested a number of stable diazotates, thought likely to be suitable on theoretical grounds, as coupling agents in the alkaline phosphatase method. In each case a note was made of (1) the speed of the reaction, (2) the rate of decomposition of the diazonium salt (and hence the amount of non-specific staining), (3) the colour of the final product, and (4) the type of precipitate. The results, in the case of fifteen stable diazotates, appear in Table 37. A very thorough survey, on these lines, was also carried out by Goessner (1958). His findings, which agreed substantially with my own, are referred to more fully in succeeding paragraphs.

The question of diffusion of the reaction product was controlled by incubating two sections face to face, one inactivated by heat (15 minutes/90°) and Lugol's iodine solution (15 minutes), the other untreated. No diffusion of the reaction product was detected with any of the salts within the limits (100 μ) of the distance between the two sections. Absence of enzyme inhibition by salts 2, 6, 7, 9, 13 and 18 to 22, at the optimum concentration, was indicated by the production of a strongly positive result in 2–5 minutes in sections of rat kidney. This is shown in the case of salt 9 in Plate XIIa, facing p. 516. With all the other salts mentioned in the table inhibition was considerable. The decomposition of salts 9, 19, 20 and 22 was negligible, even after incubation for 8–16 hours; with salt 2 it was slight up to 4 hours, and with salts 7 and 16 it was slight up to 1 hour. Salt 6, on the other hand, rapidly stained the tissues dark brown and salt 14 equally rapidly stained them dark red. Precipitate

TABLE 37

Properties of Coupling Agents at pH 9·2
(Frozen Sections)

	Diazonium salt of	Inhibition	Decomposition	Colour	Diffuse or particulate
1	1-amino-anthraquinone	+++	+++	Brick red	Diffuse
2	4-benzolyamino-2:5-dimethoxy-aniline	+	+	Black	Particulate
3	*p*-nitroaniline	+++	++	Brick-red	Particulate
4	4-chloro-2-nitroaniline	++++	+	Rose-red	Particulate
5	2:5-dichloroaniline	++	++	Red-brown	Particulate +
6	*o*-dianisidine	+	++++	Black	Diffuse
7	4-chloro-*o*-anisidine	+	++	Brick-red	Particulate
8	5-nitro-*o*-anisidine	++	+++	Red-brown	Diffuse
9	5-chloro-*o*-toluidine	+	Nil to +	Brown	Particulate
10	4-nitro-*o*-anisidine	+++	+++	Brown-red	Particulate
11	3-nitro-*p*-toluidine	++++	++	Brown-red	Particulate
12	3-nitro-*p*-anisidine	++	++	Brown-red	Particulate +
13	4-amino-3:1′-dimethyl azobenzene	+	+++	Dark-brown	Particulate
14	4-amino-2:5-dimethoxy-4-nitroazobenzene	+++	++++	Purplish-black	Particulate +
15	4-amino-4′-methoxy diphenylamine	++	++	Orange-red	Particulate
16	3-nitro-*o*-toluidine	++	+	Brown	Particulate +
17	4-amino-4′-nitro-3:6-dimethoxy azobenzene	+++	++	Red-brown	Particulate +
18	4-amino-3′:1′-dimethyl-azobenzene	+	+	Brown	Particulate
19	2-methoxy-5-diethylamino-sulphaniline	++	Nil to +	Red-brown	Diffuse
20	2-benzoylamino-4-methoxy-toluidine	+	Nil to +	Brown-black	Particulate
21	4-amino-diphenylamine	+	+++	Black	Diffuse
22	*p-p*′-diaminodiphenylamine	+	Nil to +	Black	Particulate

In the above Table degrees of inhibition are registered from + to ++++, the former indicating almost complete absence. Similarly, in the case of decomposition, one + means practically no visible staining of the background structures. The reference particulate means that the dye particles were large enough to be clearly visible with the 2/3rds objective; diffuse means a particle size not appreciable at this magnification. Numbers given are Laboratory Reference Numbers given in the Table of Diazotates (Appendix 15, p. 711). Numbers in bold type are those of recommended salts. Apparently identical salts have different stabilizers.

size was largest with 2, smaller with 7, 9, 20 and 22, and very much smaller with 6. The reactions with salts 2, 9, 20 and 22 were the best obtained in respect of speed, colour and lack of decomposition. The full method, for application to both acetone-fixed paraffin sections and to formalin-fixed frozen sections, is given in Appendix 15, pp. 713 and 714.

The investigations carried out by Goessner (1958) on the choice of diazonium salt in the simultaneous coupling method, using α-naphthyl phosphate, were done with formalin-fixed frozen sections. He recommended three salts particularly for the purpose: Echtrotsalz TR (9), Echtblausalz BB, and Variaminblausalz B (21). His average incubation time was 30–90 minutes, at 22°, in

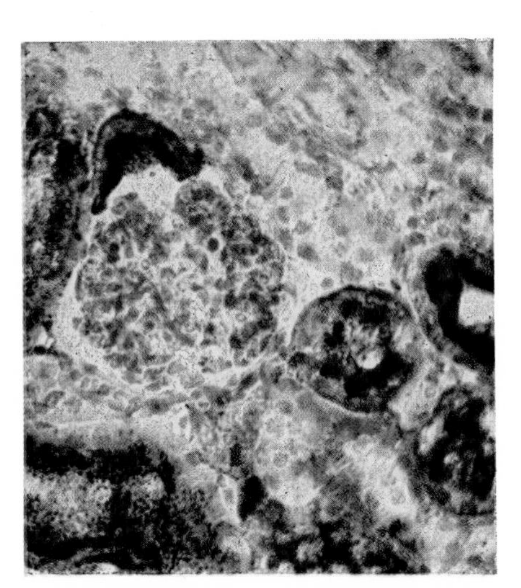

XIIA. Guinea-pig kidney (10 μ frozen section). Alkaline phosphatase in tubules and in a polymorph. Incubation 15 minutes. Coupling azo dye method, Mayer's hæmalum. × 220.

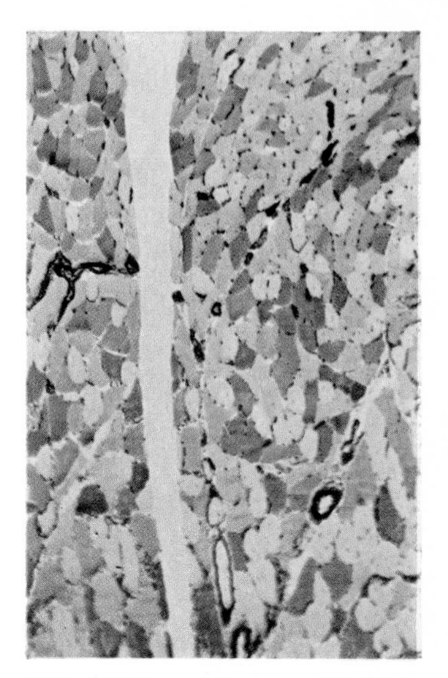

XIIB. Chinese Hamster. Skeletal Muscle. Fresh frozen (cryostat) section. Shows division of muscle fibres into 3 distinct types. ATPase preparation. × 70.

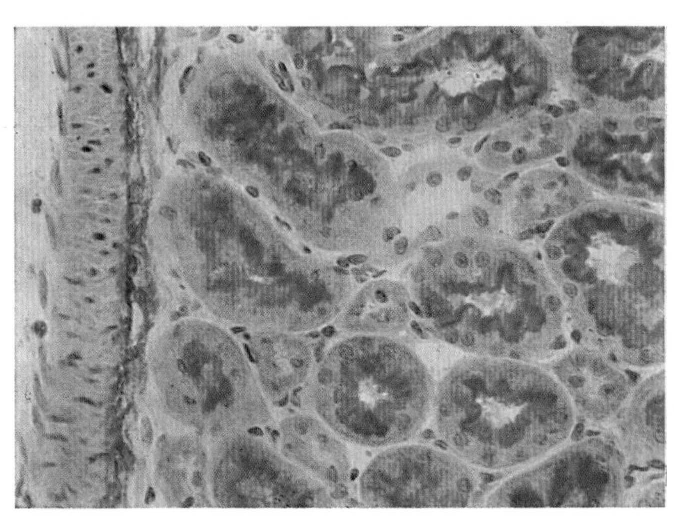

XIIC. Rat Kidney. Fresh frozen (cryostat) section. Strong alkaline phosphatase in the brush border of the proximal convoluted tubules. Reaction also in adventitial coat of the arterial walls. Naphthol AS-MX phosphate, Hexazotized Neufuchsin. × 250.

0·1 M-phosphate at pH 9·2. The final choice of diazonium salt must still, unfortunately, be governed by the results, and these will depend especially on the type of tissue preparation employed. In the case of alkaline phosphatase, a predominantly microsomal enzyme with a large lyocomponent, we can expect difficulty in obtaining precise intracellular localization. The azo dye methods with simple naphthol phosphates in fact give only rough intracellular localization of enzyme. They may be improved by faster coupling, by the use of protective media and, perhaps, by the use of simultaneous fixation. They are particularly applicable to fresh and to formalin-fixed frozen sections.

SIMULTANEOUS COUPLING AZO DYE METHODS (SUBSTITUTED NAPHTHOLS)

The first substantial work on the use of substituted naphthol phosphates in simultaneous coupling techniques for alkaline phosphatases was that of Burstone (1958a and b) though similar work with a single substitute (Naphthol AS phosphate) was reported by Rutenburg *et al.* (1958). The phosphate esters of a variety of complex arylides of 2-hydroxy-3-naphthoic acid were prepared by Burstone using modifications of the technique of Freeman and Colver (1938). In his book *Enzyme Histochemistry*, Burstone (1962) considered the whole matter much more fully than it is considered here. He gave complete lists of available bases and full details for synthesizing their phosphates.

The most suitable esters for use as substrates for non-specific alkaline phosphatase are those of Naphthol AS-MX (I), AS-TR (II) and AS-B.I (III), whose formulæ are given below:

I OPO_3Na_2 CONH— CH_3 CH_3

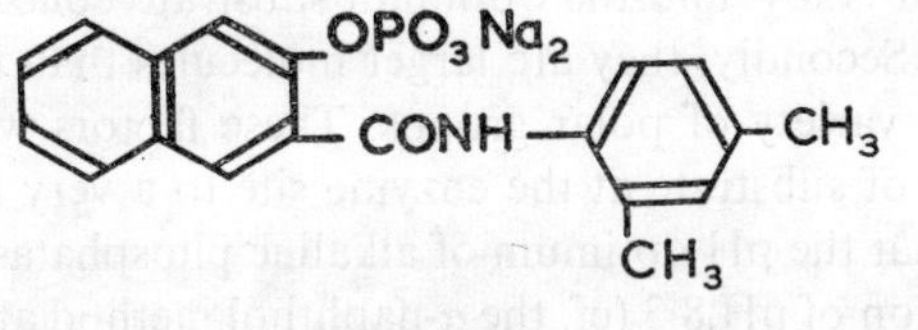

II OPO_3Na_2 CONH— Cl CH_3

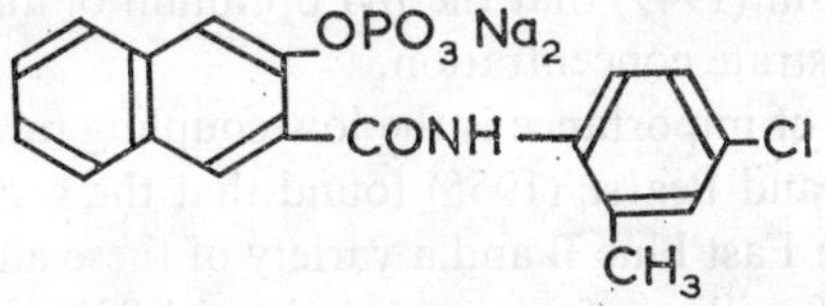

III OPO_3Na_2 CONH— OCH_3

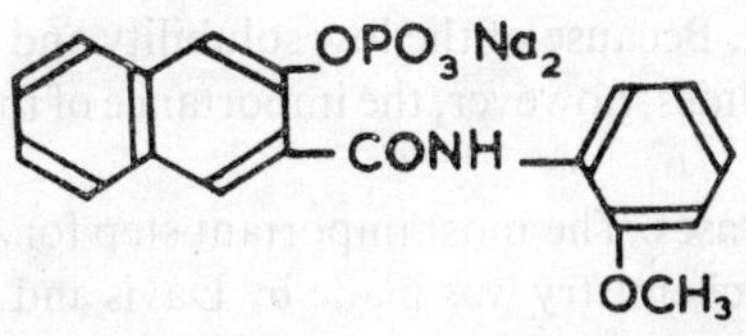

These substrates are used as saturated solutions in buffered 2 per cent. dimethylformamide at pH 8·3. The usual incubation period is 15–30 minutes and good coupling is obtained with Fast blue RR (2), Fast red RC (7), Fast red

TR (9), Fast red RL (16), and Fast violet B (20). Red violet LB (5-benzamido-4-chloro-*o*-toluidine) and Blue BBN (4-amino-2,5-diethoxy benzaniline) have also been found to be successful, but the latter has a rapid rate of decomposition

Although Burstone employed many types of tissue preparation his best results were obtained with freeze-dried sections. These were incubated without subsequent fixation. The results he obtained were "essentially similar" to those observed with the older phosphatase methods, but localizatio was claimed in all cases to be sharper and the colours more intense. There is no doubt that the localization of the final reaction product is particularly clean and precise. This is shown in Fig. 137, facing p. 535, where the localization of non-specific alkaline phosphatase in rat pancreas is illustrated.

Apart from their application to freeze-dried tissues the Naphthol AS methods are particularly applicable to blood smears (Monis and Rutenburg, 1958; Ackerman, 1962) for the demonstration of leucocyte phosphatases. A combination should be chosen which allows a stable stock solution to be prepared (e.g. AS-MX phosphate) and which produces a dark blue final reaction product (e.g. Fast blue BBN).

The precise localizations obtained with the AS-phosphates are a reflection of the low solubility of their naphtholic reaction products and of the affinity of these for protein (substantivity). Furthermore, they are highly stable esters and can be kept for long periods at alkaline pH levels in the form of stock solutions. These are undoubted advantages. On the other side, however, the AS-phosphates possess certain undesirable qualities. First, their solubility in aqueous media is low and the optimum substrate concentration is therefore unobtainable. Secondly, they are larger molecules than the simple naphthols and possess a variety of polar groups. These factors will tend to bring the concentration of substrate at the enzyme site to a very low level. Burstone's observation that the pH optimum of alkaline phosphatase with his substrates was in the region of pH 8·3 (cf. the α-naphthol method at pH 10) supports the finding of Neumann (1949) that the pH optimum of alkaline phosphatase is dependent on substrate concentration.

A third factor of importance is the low coupling rate of hydroxynapthoic anilides. Defendi and Pearse (1955) found that the average half-time of the coupling rate with Fast blue B and a variety of these anilides was 62 seconds compared with 300 millisecs for α-naphthol and 1,800 millisecs for β-naphthol, at an alkaline pH level. Because of the low solubility and high substantivity of hydroxynaphthoic anilides, however, the importance of this factor is somewhat reduced.

Hexazotized Dye Bases. The most important step forward in simultaneous coupling azo dye histochemistry was made by Davis and Ornstein (1959) with the introduction of hexazonium pararosaniline (hexazotized fuchsin, HPR). This was freshly prepared and used originally with simple naphthol substrates for the demonstration of acid phosphatases (p. 565). Subsequently, HPR was used increasingly with Naphthol-AS substrates, not only for acid phosphatase

(Barka and Anderson, 1962) but also for esterases (Thybusch *et al.*, 1966) and for alkaline phosphatases.

Alternative bases having three available amino groups for hexazotization have been tested. Lojda *et al.* (1964) synthesized and tested tetrabromfuchsin but found it less satisfactory than HPR, for a variety of reasons. Subsequently Stutte (1967) suggested the use of hexazotized neufuchsin (triamino-tritolyl-methane chloride; New Magenta; HNF; C.I.42520) with Naphthol AS-MX phosphate as substrate for alkaline phosphatase. The final reaction product is essentially non-particulate and localization is excellent (see Plate XIIc, facing p. 516).

Non-coupling Azo Dye Methods

A method was devised by Loveless and Danielli (1949) which was designed to overcome the disadvantages of the Menten-Junge-Green technique. According to the authors these disadvantages were: (1) the instability of diazonium salts and their fresh preparation; (2) the necessity for low temperatures, thus denying the enzyme its optimal conditions; and (3) the inhibitory effect of diazonium salts. It will be realized that the first of these was removed by the use of stable diazotates and the second by the successful employment of temperatures as high as 37°. The third is still a significant factor, though it can be reduced to a low value by using a suitable diazotate. Loveless and Danielli produced a coloured phosphate ester which, on hydrolysis, yielded an insoluble highly coloured base which had no tendency to crystallize in the tissues but remained amorphous. This ester, *p*-nitrobenzene-azo-4-naphthol-1-phos-

$$O_2N-C_6H_4-N=N-C_{10}H_6-O-PO_3H_2$$

phate was moderately soluble and stable in the absence of the enzyme. It had no inhibitory effects and was hydrolysed rapidly. With such properties the substrate should have been ideal for the purpose for which it was designed. Using the sodium salt, the results were studied with sections of rat kidney fixed in 80 per cent. alcohol and embedded in paraffin. By comparison with the glycerophosphate method it was seen that the dye was deposited in all the usual sites but, in addition, the nuclear reaction was far more intense. It was possible to obtain permanent sections mounted in balsam, provided that clearing in xylene was carried out rapidly, but this theoretical advantage over the coupling techniques has been offset by the application to these of diazotized *o*-dianisidine as coupling agent. The azo dye, formed by the union of this salt and α-naphthol is insoluble in alcohol and in xylene, though not in mixtures of the two, so that permanent mounts can be obtained if required.

Although the Danielli method is of considerable theoretical interest, especially in relation to non-coupling azo dye techniques which have been developed for other enzymes, its disadvantages are too great for it to be of practical use.

Post-coupling Azo Dye Techniques

Because of the rapid coupling obtainable in simultaneous coupling techniques with α- and β-naphthols, and the relatively rapid coupling obtainable with the substituted naphthols, post-coupling techniques have scarcely been used for the demonstration of alkaline phosphatases. The post-coupling technique of Danielli (1946), using sodium phenolphthalein phosphate, is of historical interest only.

The primary reaction product of the majority of post-coupling methods developed for hydrolytic enzymes was 6-bromo-2-naphthol. Its phosphate has been used as a substrate for alkaline phosphatase in both post-coupling and simultaneous coupling techniques. At alkaline pH levels the solubility of 6-bromo-2-naphthol is sufficient to allow diffusion to occur with resulting non-specific staining of many tissue components (see Pearse, 1954; Defendi, 1957). No sufficient indication exists to justify the use of a post-coupling azo dye method for alkaline phosphatase, despite their apparently successful employment by Burstone (1962).

Non-coupling Fluorescence Methods

These techniques, which were introduced by Burstone (1960, 1962), employed as substrate 5, 6, 7, 8-β-tetralol carboxylic acid-β-naphthylamide phosphate. Paraffin sections (freeze-dried) were incubated at pH 8·2 to 9·3 and subsequently mounted in glycerol and examined under the fluorescence microscope. Coupling was thus avoided altogether but there is no evidence to suggest that fluorescence techniques have any advantages over the more usual chromogenic methods.

Indoxyl Phosphate Methods

Very little work has been done on the localization of alkaline phosphatases using indoxyl substrates. The sodium salt of indoxyl phosphate was used by Seligman, Heymann and Barrnett (1954) and the result, in freeze-dried sections incubated at pH 9·5, differed to some extent from those obtained by alternative methods. Using this substrate, I found that localization was far from precise since the indigo was deposited in a macrocrystalline form. Far better results were obtained by Holt (1954), using the calcium salt of 5-bromoindoxyl phosphate in the presence of a ferro-ferricyanide oxidation catalyst as used in the indoxyl esterase method (Chapter 17). In this case the precipitate of 5,5′-dibromindigo was microcrystalline and the localization obtained was excellent. Good results were also obtained by Przelecka *et al.* (1962a and b) who used the barium salt of 5-bromoindoxyl phosphate, with $CuSO_4$ as oxidation catalyst. Investigating intestinal alkaline phosphatase in the mouse they found a considerable increase after high fat feeding.

The synthesis of halogen-substituted indoxyl phosphates is not easy and the more complex leucoindigo phosphates and thioindoxyl phosphates are not hydrolysed by tissue phosphatases. This branch of phosphatase histochemistry has remained undeveloped, therefore.

Techniques for Specific Alkaline Phosphatases

METHODS FOR 5-NUCLEOTIDASE (E.C. 3.1.3.5)

The 5-nucleotidases are enzymes which catalyse the hydrolysis of nucleoside-5-phosphates to yield nucleosides and inorganic phosphate. Their specificity is not completely established but they attack ribo- and deoxylribonucleoside-5-phosphates.

In spite of what was said earlier in this chapter it is possible, using the Gomori technique, to obtain useful results even at pH 7·5, provided that one is dealing with an active enzyme. This was shown by Pearse and Reis (1952), who used adenylic acid (adenosine-5-phosphate) as substrate in order to demonstrate the sites of activity of the specific 5-nucleotidase described earlier by Reis (1937, 1940, 1950). This enzyme has a pH optimum in the region of pH 7·8 in human tissues and pH 8·5 in the rabbit. Gomori (1949b) and Newman, Feigin, Wolf and Kabat (1950) estimated the presence of 5-nucleotidase by comparing two slides incubated at pH 9·2, one with adenylic acid, the other with β-glycerophosphate. The difference between the two was taken as corresponding to the activity of 5-nucleotidase, since adenylic acid is hydrolysed both by 5-nucleotidase and by alkaline phosphatase, whereas β-glycerophosphate is hydrolysed by non-specific alkaline phosphatase only. The result may be sufficiently clear when 5-nucleotidase is present in the examined tissue in the absence of alkaline phosphatase. If both enzymes are present, however, it is difficult to judge whether the heavier precipitates with adenylic acid are really due to 5-nucleotidase action.

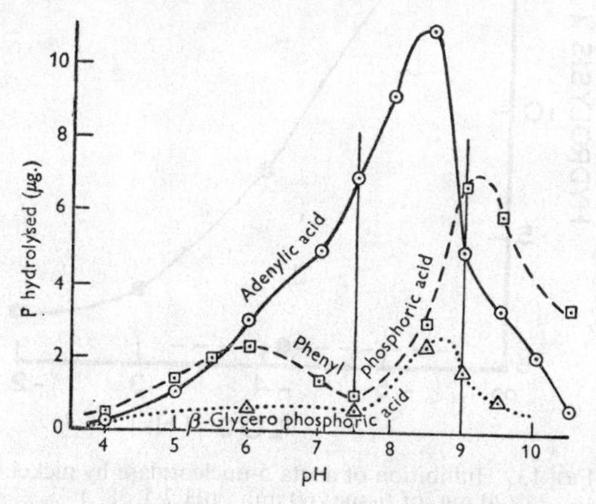

FIG. 138. Enzymic hydrolysis of phosphoric esters by an extract of rabbit lung. 5 mg. of tissue (dialysed extract), 20 min., temp. 68°, substrate concentration 0·9 mM. Details of technique as described by Reis (1951).

For some time little was known about the activation and inhibition of 5-nucleotidase, although the earlier studies of Reis (1937, 1950) had shown that magnesium activated this enzyme much less than it did non-specific alkaline phosphatase. Later work by Ahmed and Reis (1958) confirmed this and indicated that the most potent activator of 5-nucleotidase was Mn^{2+}, which produced a 60 per cent increase in activity. Confirming earlier work by Kaye (1955), these authors showed that Zn^{2+} ions strongly inactivated the enzyme, but the strongest inhibition was produced by Ni^{2+}. As indicated in Fig. 139 below, a millimolar solution of a nickel salt almost totally inhibited 5-nucleotidase activity, while non-specific alkaline phosphatase activity was unaffected. Scott (1965), however, found that mouse brain 5-nucleotidase was activated rather than inhibited by Ni^{2+}, at concentrations as low as 10^{-6}M, especially when thymidine monophosphate was used as substrate.

The different activity of alkaline phosphatase against various phosphoric esters has also to be taken into account. Fig. 138 shows the rate of hydrolysis of three phosphate esters *in vitro*, by slices of rabbit lung, as a function of the hydrogen ion concentration. At pH 9·0 it is obviously difficult to say whether the difference between adenylic acid and β-glycerophosphate hydrolysis is due to 5-nucleotidase activity or whether it is due to a different activity of alkaline phosphatase towards adenylic acid. The varying activity of alkaline phosphatase is illustrated by the hydrolysis of phenylphosphate which, as was shown by King and Delory (1939), at pH 9·0 is about four times as fast as that of β-glycerophosphate,while that of adenylic acid is between the two. At pH 7·5, on the other hand, the rate of hydrolysis of phenylphosphate and β-glycero-

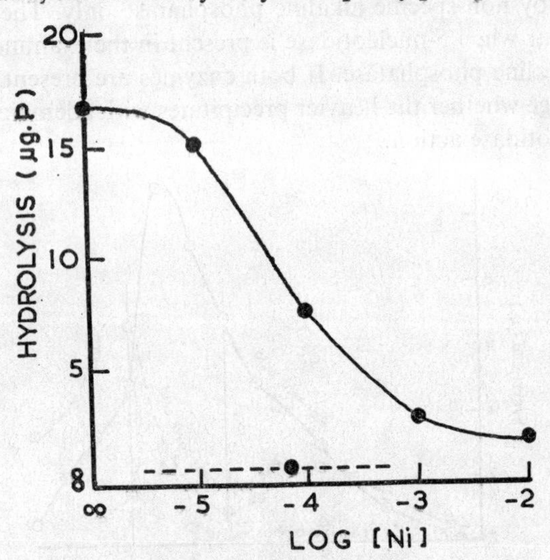

FIG. 139. Inhibition of aorta 5-nucleotidase by nickel.
(20 mg. of tissue; 60 min., pH 7·5, 38°.)

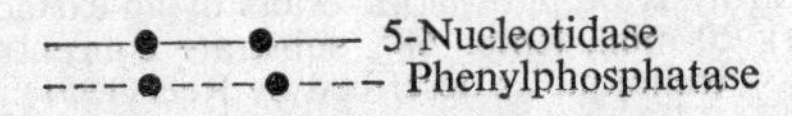

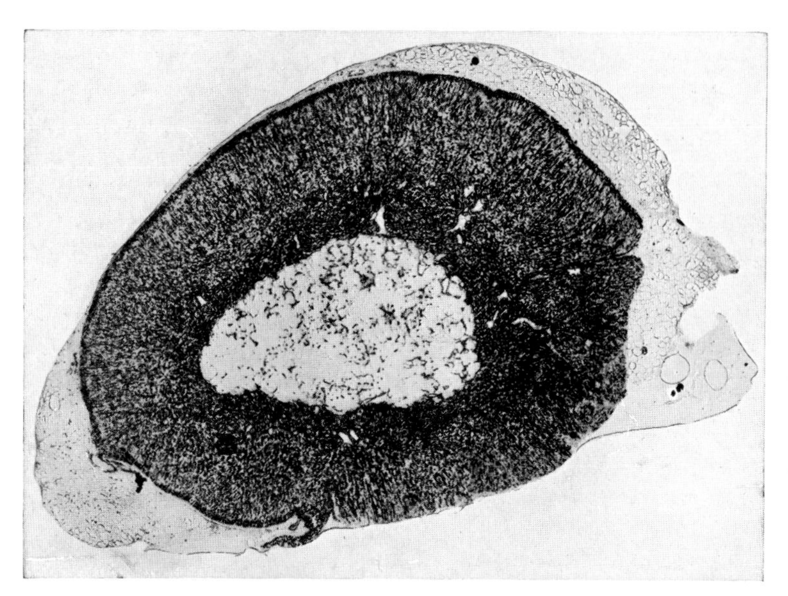

FIG. 130. Rat adrenal gland incubated for 3 hours with 5mM phenyl phosphate at pH 9·0. Disposition of alkaline phosphatase. × 19.

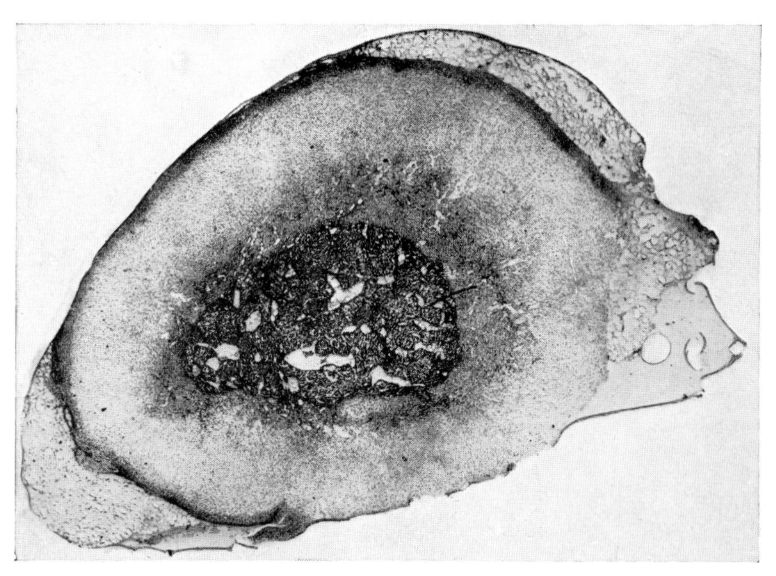

FIG. 131. Rat adrenal gland incubated for 3 hours with 5mM adenylic acid at pH 7·5. Disposition of 5-nucleotidase. × 19.

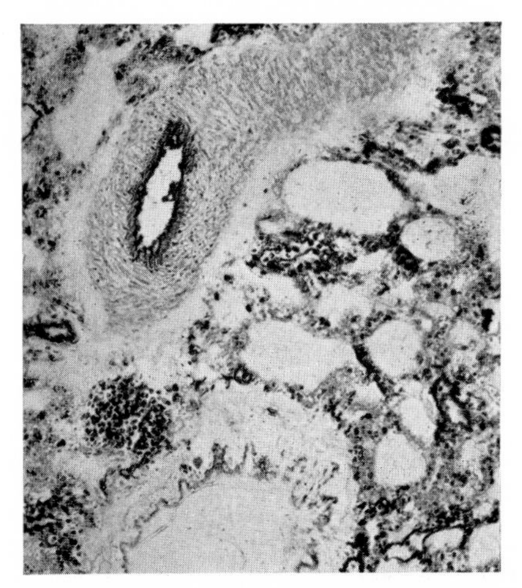

FIG. 132. Rabbit lung. Stained for alkaline phosphatase (phenyl phosphate, pH 9·0), which is present especially in the endothelium of a large artery. × 100.

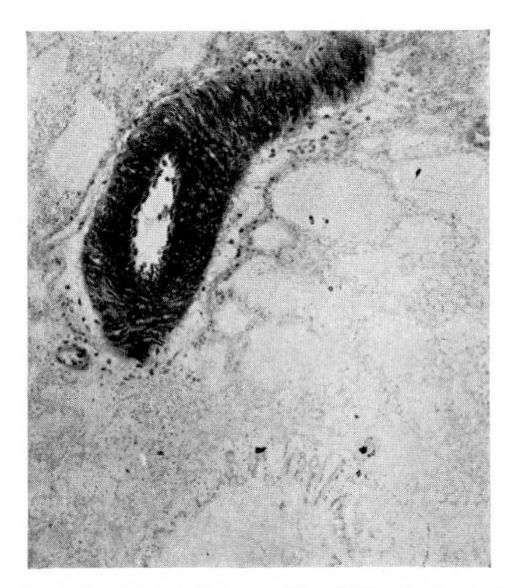

FIG. 133. Rabbit lung (serial section to Fig. 132). Stained for 5-nucleotidase (adenylic acid, pH 7·5), which is concentrated in the medial coat of the artery and absent from the endothelium. × 100.

phosphate is very low, and the rate of hydrolysis of adenylic acid by non-specific phosphatase has a similarly low value at this pH. In subtracting such a small value from the total hydrolysis of adenylic acid a much smaller error is likely to be made than in subtracting the far larger values for non-specific hydrolysis obtained at pH 9·2. Incubation was therefore carried out by Pearse and Reis at pH 7·5, in spite of the known increase in solubility of calcium phosphate at this pH and the consequent increase in diffusion which was expected.

The disadvantage of using a low pH with the calcium nitrate method of phosphate precipitation is shown by a comparison of Figs. 130 and 131. Fig. 130 shows the disposition of alkaline phosphatase in the rat adrenal gland and indicates that it is fairly sharply localized and entirely in the cortex except for the walls of small arteries in the medulla. Fig. 131 shows the localization of 5-nucleotidase in the rat adrenal. All the staining observed in the photograph is due to this enzyme since a control secion incubated with 5 mM-phenyl phosphate at pH 7·5 was blank. This indicates that alkaline phosphatase, which can also hydrolyse adenylic acid at more appropriate pH levels, was inactive under the conditions of the experiment. The most intense staining in this case is in the medulla, but the presence of diffusion from this site is revealed by the staining of the nuclei in the cortex, which is most intense in a zone immediately surrounding the medulla. With incubation at pH 7·5, although in this case the production of calcium phosphate has been sufficiently rapid to cause local precipitation, localization is confined broadly to tissue regions and intracellular localization cannot be attempted. Figs. 132 and 133 show alkaline phosphatase and 5-nucleotidase, respectively, in serial sections of rabbit lung. No counterstain was used in either case.

The Calcium-Cobalt Technique

Details of the method used by Pearse and Reis appear in Appendix 15, p. 717. By means of this technique only the sites of maximal 5-nucleotidase activity are demonstrated, even when cold formalin-fixed frozen sections are used, since for precipitation of calcium phosphate to occur at pH 7·5 production must be exceedingly rapid. The actual sites of lower 5-nucleotidase activity are not usually revealed, although diffuse staining of nuclei and other structures indicates the presence of 5-nucleotidase somewhere in the section. A positive histochemical reaction using adenylic acid at pH 7·5 cannot be considered as an artifact since it corresponds with chemical estimations and its intensity is, to a large extent, proportional to the time of incubation.

An essentially similar technique to that used by Pearse and Reis was employed by Antonini and Weber (1951) to demonstrate 5-nucleotidase in the medial coat of the aorta and other blood vessels. These authors used frozen sections after brief fixation in acetone. McManus *et al.* (1952) and McManus and Lupton (1953) used a Gomori type of procedure, with muscle adenylic acid as substrate, at a pH between 8·5 and 8·8. In the first paper a general study of human tissues was made, and the results agreed, on the whole, with those of Pearse and Reis. In the second paper a study of the enzyme in obsolescent renal glomeruli was reported. Other studies were those of Goebel and Puchtler (1954), who described the localization of the enzyme in renal infarcts, and of Otte (1958), who found a strong 5-nucleotidase in the surface layers of articular cartilage. Otte used a pH of 9·2 but observed no reaction with glycerophosphate as substrate except in the basal zone (of ossifying cartilage). High levels of 5-nucleotidase were demonstrated cytochemically, in muscle tissues from cases of human muscular dystrophy by Golarz *et al.* (1961). This finding has amply been confirmed by biochemical assays carried out in other laboratories.

The Lead-sulphide Technique

The calcium phosphate technique fails to give satisfactory results in fresh frozen or formalin-fixed frozen sections. An alternative technique using lead nitrate at pH 7·2 was employed by Naidoo and Pratt (1954), and by Naidoo (1962) for use with freeze-dried sections. This gave a cleaner result, and Wachstein (1955) found this method gave a "very satisfactory and reproducible preparation" when applied to fresh frozen kidney sections. He also used cold formalin-fixed frozen sections, however, and suggested that with these the diffuse activity of the capillaries was suppressed, making the recognition of positive-staining tubules easier. Although this last effect seems undesirable, it is probable that the lead nitrate technique is the technique of choice for the demonstration of 5-nucleotidase in all types of tissue. The method most commonly employed is that of Wachstein and Meisel (1957), which appears in Appendix 15 (p. 717). A later modification, designed by Barron and Boshes (1961) expressly for nervous tissues, is also given in the Appendix. Novikoff (1958) maintained that variations in staining patterns (in rat liver) under varying conditions might be due to different 5-nucleotidases. It may therefore be justifiable, in many instances, to use a calcium and a lead method in parallel, on both fresh and formalin-fixed frozen sections.

The Lead-sulphide technique was used by Scott (1963, 1964, 1965), who demonstrated a particularly striking distribution of the enzyme in mouse cerebellum, and by Thiery and Willighagen (1963) and Willighagen (1962) for their studies on enzymatic dedifferentiation in neoplastic cells. In their work on human thyroid nucleotidase Harcourt-Webster and Stott (1965) also used the lead-sulphate technique. Scott (1965) found that the enzyme in mouse brain would hydrolyse a number of mononucleotides, as indicated by Heppel and Hilmoe (1951) and by Herman and Wright (1959) for the clostridial enzyme. Unlike the latter, however, mouse brain 5-nucleotidase will not hydrolyse di- or trinucleotides.

Methods for Aldolase (E.C. 4.1.2.7)

This enzyme catalyses the splitting of a molecule of fructofuranose-1:6-diphosphate (hexose diphosphate) into one molecule each of 3-phosphoglyceraldehyde and α-phosphodihydroxyacetone. It is believed to play an important part in the reactions of glycolysis in yeast and in muscle. A small fraction of the total aldolase of rat liver is found in the nuclear fraction of homogenates. The remainder is present in the supernatant. Much higher specific activity occurs in skeletal muscle and somewhat less in cardiac muscle. Tung *et al.* (1954) investigated the substrate specificity of rabbit muscle aldolase and found that D-fructose-1,6-diphosphate and L-sorbose-1,6-diphosphate, which possess a *trans* configuration of hydroxyls on carbon atoms 3 and 4, were both hydrolysed more rapidly than tagatose-1,6-diphosphate which has the *cis* configuration. The affinity of crystalline muscle aldolase for fructose-1,6-diphosphate was much greater than that for fructose-1-phosphate, the respective Michaelis constants being 5×10^{-5} M and $3{\cdot}4 \times 10^{-3}$ M.

O

$H_2O_3P{\cdot}OH_2C$ $CH_2O{\cdot}PO_3H_2$

H H HO OH

HO H

$$\rightleftharpoons \begin{array}{c} H_2O_3P{\cdot}OCH_2 \\ | \\ HCOH \\ | \\ CHO \end{array} + \begin{array}{c} CH_2O{\cdot}PO_3H_2 \\ | \\ C{=}O \\ | \\ HOCH_2 \end{array}$$

CALCIUM-COBALT METHOD

Allen and Bourne (1943) used a Gomori type of procedure to demonstrate the phosphates produced by this reaction and thus to localize aldolase in the tissues. Further enzymic breakdown of the two triosephosphate reaction products was prevented by the presence of iodoacetate in the incubating medium and the reaction products were precipitated by a mixture of magnesium and ammonium chlorides in alkaline solution. Frozen sections of tissues fixed in 80 per cent. alcohol for 24 hours were used and these were incubated for 1–2 hours at 37°. The precipitated phosphates were revealed by their conversion to the cobalt salt and finally to the brownish-black sulphide.

This method, in the hands of a number of different workers, proved unsatisfactory. In my hands it produced, at best, a diffuse precipitate over the whole section which provided no cellular localization.

I suggested (1960) that an alternative method might be developed on the basis of hydrazide capture of the reaction products. A successful assay technique had been developed by Jagannathan *et al.* (1956) which was based on the hydrazino derivatives of the products of aldolase activity. Attempts, by myself and others, to develop this technique proved abortive.

Indirect Tetrazolium Method

A new method for aldolase was introduced by Nepveux and Wegmann (1962). This was based on methods of biochemical assay developed by Colowick and Kaplan (1955). Essentially the reaction depends on oxidation of the glyceraldehyde phosphate product of the initial reaction by means of added phosphoglyceraldehyde dehydrogenase and coenzyme I (NAD). The resulting $NADH_2$ transfers its electrons to the tetrazolium salt, by way of the diaphorase unless additional electron transfer agents such as phenazine methosulphate (PMS) are added (see Chapter 21).

Modifications of this principle were described by Lake (1965), who used MTT and PMS in place of Nitro-BT, and by Abe and Shimizu (1964) who relied on phosphoglyceraldehyde dehydrogenase present in their tissue sections. Both these authors used sodium arsenate, as in the biochemical assay, to produce phosphoarsenoglycerate and thus accelerate the NAD-linked secondary reaction by removal of its product.

Still later Simon *et al.* (1966) used the same principle (apparently independently to judge by their references) to produce an indirect aldolase reaction based on Nitro-BT, NAD and exogenous phosphoglyceraldehyde dehydrogenase. In technical detail the various methods differ in several respects. Although Abe and Shimizu found formalin fixation of cryostat sections strongly inhibitory to aldolase, both Lake and Simon *et al.* used it.

The indirect method for aldolase gives precise histochemical localizations in most tissues where more than a trace of the enzyme is present. As with all multistep reactions it is necessary to be cautious in interpreting any given localization. Details of the reaction are given in Appendix 15, p. 719.

Methods for Adenosine Triphosphatase (E.C. 3.6.1.3 and 3.6.1.4)

The two enzymes which appear in this heading in the Enzyme Commission list are now properly termed ATP phosphohydrolases. They are separated into the calcium-activated myosin ATPase and the magnesium-activated mitochondrial ATPase. Both catalyse the reaction:

$$ATP + H_2O \rightarrow ADP + \text{orthophosphate}$$

The first enzyme (3.6.1.3) is inhibited by Mg^{2+} ions and the second (3.6.1.4) by Ca^{2+} ions. As will be seen the above statements represent a great simplification of the situation both from the biochemical and the histochemical point of view. At least 5 enzymes have been described which can dephosphorylate ATP (Freiman and Kaplan, 1960). In addition to the two already mentioned they are adenylpyrophosphatase, apyrase and alkaline phosphatase. The so-called adenylpyrophosphatase splits ATP into adenosine-5-phosphate and 2 moles of orthophosphate, and apyrase is a plant enzyme which transfers the terminal phosphate of ATP to some other compound. With these two enzymes we are little concerned, from the histochemical point of view. In the earlier stages of development of histochemical methods for ATPase most of the trouble encountered was due to activity of non-specific alkaline phosphatase. With the advent of newer methods emphasis has shifted to the problem of non-enzymatic hydrolysis, as will be described below.

There is already a voluminous biochemical literature on the ATPases, particularly with respect to activation and inhibition studies. Since 1960, moreover, a very high proportion of the histochemical literature on phosphatases concerns the ATPases and their demonstration in the tissues. The simplest approach to what is still a very complex problem may well be to describe the development of histochemical techniques for ATPase and, at the same time, to refer to biochemical findings as and when they affect the histochemical situation.

The Calcium-cobalt Method

Using ATP as substrate Glick and Fischer (1945, 1946) claimed that they could localize the specific adenosine triphosphatase of plant tissues, and they found the enzyme especially in the nucleoli of the cells of the wheat epicotyl.

The accuracy of their method depended on the absence of activity by non-specific alkaline phosphatase, which Moog and Steinbach (1946) had shown to be able to hydrolyse ATP. These authors criticized the method of Glick and Fischer and concluded that only non-specific alkaline phosphatases were demonstrated. They stated that localization of adenosine triphosphatase could only be made by assessing the difference between ATP-incubated and glycerophosphate-incubated slides. Glick (1946), however, maintained that the method would localize the specific enzyme whenever this was predominant. Maengwyn-Davies *et al.* (1952) demonstrated the activity of myosin ATPase in rat muscle, using fresh frozen sections and a Gomori type of procedure. They attempted to control the specificity of the method by the use of a variety of activators and inhibitors and found that at pH 8·5 ATP was dephosphorylated by a specific SH-dependent enzyme. At pH 9·9, however, a different enzyme, possibly a pyrophosphatase, was concerned.

No further advances in the histochemistry of ATPase were made until Padykula and Herman (1955a and b) re-examined the whole question. These authors (1955a) compared the method of Maengwyn-Davies *et al.* with a modified Gomori type method of their own, using fresh frozen sections in both cases. They found a much higher activity with their own method and considered that this was mainly due to the much lower concentration of salts in their medium. Another factor of importance was the substrate concentration (12 mM compared with 5 mM). In this case it was observed that the higher concentration of substrate produced nuclear staining. The lower concentration was therefore subsequently adopted.

In their second paper (1955b) Padykula and Herman reported tests on the specificity of their method with a series of activators and inhibitors. They found that BAL (2,3-dimercapto-1-propanol) at about 5 mM, would inhibit non-specific alkaline phosphatase activity while it enhanced the activity of ATPase. Sulphydryl inhibitors such as *p*-chloromercuribenzoic acid, on the other hand, abolished the activity of the specific enzyme and this inhibition was reversible with BAL or cysteine.

Padykula and Herman classified three types of ATPase by means of their tests. These were sulphydryl dependent (mitochondrial), sulphydryl indifferent (endothelial, vascular smooth muscle), and sulphydryl sensitive.

The Lead-sulphide Method

Introduced originally by Naidoo and Pratt (1951, 1952), the use of Pb^{2+} ions to capture the initial reaction product of ATP hydrolysis was extensively employed by Wachstein and Meisel (1957a and b) and by Wachstein *et al.* (1960, 1962). This had previously been adapted successfully (Wachstein and Meisel, 1956) for the demonstration of 5-nucleotidase. With the lead nitrate method the bile canaliculi were particularly well shown and Novikoff *et al.* (1958) showed that fragmented canaliculi in liver homogenates would react in the same way. In this paper the authors assessed the value of the Calcium-cobalt and Lead-sulphide procedures by both biochemical and histochemical means.

It was apparent that with both methods enzyme activity was much below its maximum. In homogenates incubated with the Padykula-Herman mixture only 20 per cent. ATPase activity was recorded, and with the Wachstein-Meisel medium only 15 per cent. Provided that loss of enzyme activity is not confined to the enzyme in any particular localization, leaving others unaffected, these percentages are of course perfectly acceptable in histochemical practice. Novikoff and his co-workers found that if formol-calcium fixation was superadded an additional loss of 80–90 per cent. occurred, and this was hardly to be tolerated.

The data obtained suggested that enzyme inhibition with the Wachstein-Meisel medium was caused by the high concentration of Pb^{2+} ions, a fact already established by Pratt (1954) in his biochemical and histochemical studies of the specific phosphatases in brain.

It was evident that the ATPase localized by both methods in the bile canaliculi was different from the one found histochemically and biochemically in the mitochondria. Already Novikoff *et al.* (1952) had shown that the ATPase of their nuclear fraction in their homogenates (later shown to be canalicular rather than nuclear) was activated by calcium. The mitochondrial enzyme is not activated by calcium, of course, and in fact it is almost totally inactivated in formol-calcium fixed sections. In Fig. 140, p. 529, is recorded the activity of rat-liver homogenates in the dephosphorylation of ATP, adenylic acid, phenyl phosphate and glycerophosphate.

At pH 7·2, the pH employed in the Wachstein-Meisel procedure, although the activity of both calcium-activated and magnesium-activated ATPase is only 50 per cent. of the maximum value, both values are nevertheless

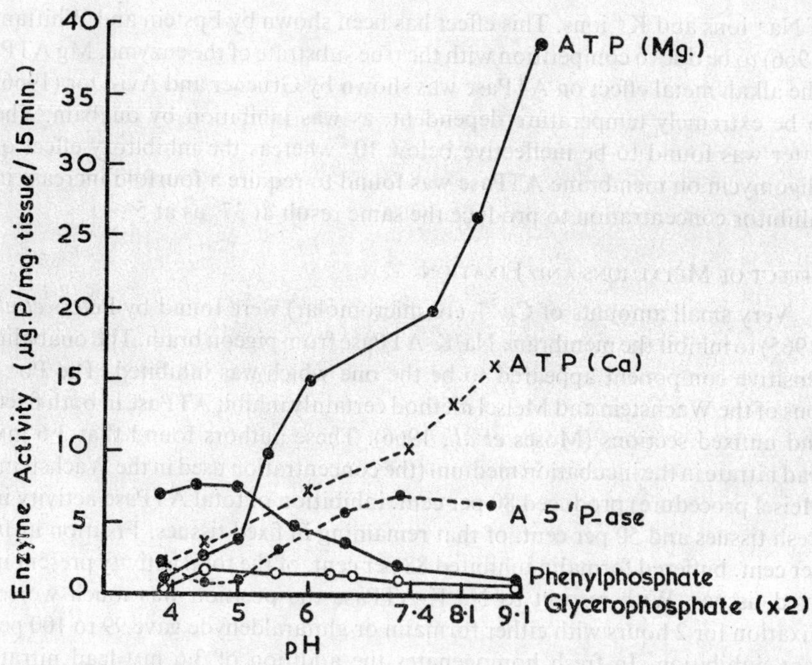

FIG. 140. Dephosphorylation by rat liver homogenates as a function of pH. (After Novikoff *et al.*)

significantly higher than the activities of the homogenate towards the other substrates. In particular, the activity of non-specific alkaline phosphatase can obviously be ignored.

ACTIVATORS AND INHIBITORS

In addition to Ca^{2+} and Mg^{2+} by means of which the two main types of ATPase are distinguished, a further type, or subtype, of the enzyme has been described on the basis of activation by Na^{+} and K^{+} ions. This ATPase is magnesium-dependent and sensitive to the concentration of Na^{+} and K^{+} ions. It is presumed to be concerned with electron transport and was first described by Skou (1957) in crab nerve. Later it was characterized in a number of mammalian tissues by Järnefelt (1961, 1962), Deul and McIlwain (1961), and Aldridge (1962). In the kidney the properties of the membrane associated Na/K-ATPase were described by Wheeler and Whittam (1962). These authors confirmed the strong inhibitory effect of ouabain (strophanthin-G), which they used in a 0·33 mM concentration. The cardiac glycosides are potent inhibitors of active cation transport.

The proportions of Mg^{2+} ions and ATP required for maximal activity of ATPase vary with pH. In the region of pH 6·9 to 7·6 the highest level of hydrolysis occurs when equimolar concentrations are present. Calcium inhibits Na/K-ATPase in low concentrations which do not inhibit in the absence of Na^+ ions and K^+ ions. This effect has been shown by Epstein and Whittam (1966) to be due to competition with the true substrate of the enzyme, Mg ATP. The alkali metal effect on ATPase was shown by Gruener and Avi-Dor (1966) to be extremely temperature dependent, as was inhibition by ouabain. The latter was found to be ineffective below 10° whereas the inhibitory effect of oligomycin on membrane ATPase was found to require a fourfold increase in inhibitor concentration to produce the same result at 37° as at 5°.

Effect of Metal ions and Fixation

Very small amounts of Cu^{2+} (10 micromolar) were found by Peters *et al.* (1965) to inhibit the membrane Na/K-ATPase from pigeon brain. The ouabain-sensitive component appeared to be the one which was inhibited. The Pb^{2+} ions of the Wachstein and Meisel method certainly inhibit ATPase in both fixed and unfixed sections (Moses *et al.*, 1966). These authors found that 3·6 mM lead nitrate in the incubation medium (the concentration used in the Wachstein-Meisel procedure) produced 80 per cent. inhibition of total ATPase activity in fresh tissues and 50 per cent. of that remaining in fixed tissues. Fixation in 10 per cent. buffered formalin inhibited 88 per cent. of the total activity present in fixed tissues. With respect to Na/K-ATPase the position was much worse. Fixation for 2 hours with either formalin or glutaraldehyde gave 99 to 100 per cent. inhibition. In fresh homogenates the addition of 3·6 mM lead nitrate abolished the effect of Na^+ and K^+ in stimulating the membrane-bound enzyme. Sodium and ouabain fail to influence ATPase staining with the Wachstein-Meisel medium (Novikoff *et al.*, 1961; McClurkin, 1964).

Observations published by Rosenthal *et al.* (1966) showed that Pb^{2+}, in the concentration indicated above, hydrolysed ATP at pH 7·2 at 37°. The rate of hydrolysis was increased with elevated temperatures. Rosenthal *et al.* suggested that the above reaction might be a source of artifact in the lead salt method for the histochemical localization of nucleoside phosphatases. Moses *et al.* (1966) found that either decreasing concentrations of Pb^{2+} or increasing concentrations of ATP completely altered the localization of the reaction product. At 0·9 mM nuclear staining was predominant, at 0·45 mM a cytoplasmic pattern was obtained which suggested mitochondrial localization. It was suggested that plasma membrane phospholipids, by acting as phosphate activators in a lead catalysed transphosphorylation of ATP, could result in local deposition of metal phosphate precipitates. Since Lowenstein (1958) and Tetas and Lowenstein (1963) had shown that other bivalent cations, including Ca^{2+}, Cu^{2+}, Mn^{2+} and Mg^{2+}, could act in the same way as Pb^{2+} it was further postulated that Ca^{2+} ions in the histochemical calcium-cobalt method for ATPase might produce a similar false localization.

The above observations constitute a serious objection to all the current methods for localizing ATPase. There was already serious reason for doubt in view of the alterations produced by relatively small changes in ionic concentration of various components in the media. It is to be expected that changes will be made in the histochemical approach to ATPase localization. These may take the form of a return to the Calcium-cobalt technique at all pH levels at which the product provides accurate localization. Long incubations may be avoided, possibly by using freeze-dried material without fixation, and activity of alkaline phosphatase may be inhibited by the use of specific inhibitors.

Three basic methods for ATPase are given in Appendix 15, p. 720 together with some minor variants. They should be used with full realization that the presence of a visible product may not indicate ATPase activity, and that the localization of the product may not be that of the enzyme. Tormey (1966), for instance, obtained experimental results which clearly indicated that the histochemical localization of Na/K-ATPase was impossible.

METHOD FOR THIAMINE PYROPHOSPHATASE

The problem of the specific organic pyrophosphatases is far from being solved. The dephosphorylation of thiamine pyrophosphate (TPP) was demonstrated in freeze-dried brain sections by Pratt (1954) using a lead nitrate type of procedure. This author found, as intimated above, that the concentration of Pb ions in the medium affected the amount of enzyme demonstrable histochemically. This effect is shown in Fig. 141, below.

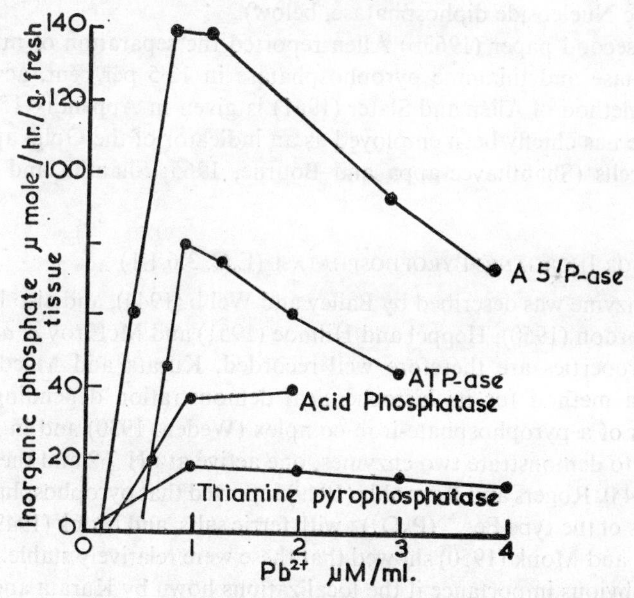

FIG. 141. Effect of Pb^{2+} ions on hydrolysis of thiamine pyrophosphate. (After Pratt, 1954.)

Pratt noted that at the lowest concentration of Pb^{2+}, where enzyme activity was at its highest, the sections stained unevenly. This was due to inefficient capture of liberated phosphate, and it serves to emphasize once again the double necessity in hydrolytic enzyme histochemistry to observe the principles set forth in Chapter 14 and to effect the best possible compromise within this framework.

A similar study of the histochemical dephosphorylation of thiamine pyrophosphate was carried out by Eränkö and Hasan (1954) on fresh frozen sections of rat kidney. At alkaline pH levels thiamine pyrophosphate and glycerophosphate gave identical pictures. In acid media, however, while the picture with glycerophosphate was that associated with non-specific acid phosphatase, the sections incubated with thiamine pyrophosphate showed strong activity, especially in the glomeruli. There was little or no activity in the convoluted tubules. The method used by Eränkö and Hasan is given in Appendix 15, p. 722.

Further studies were undertaken by Allen and Slater (1961) on the TPP-hydrolysing enzyme of mouse epididymis. This enzyme was found to have its maximum activity at pH 9·5. The work was continued by Allen (1963a and b), using two different conditions of localization. (1) pH 9·5, Tris-calcium chloride. (2) pH 7·0 Tris-maleate-lead nitrate). Using acetone-fixed cryostat sections, and the first set of conditions, only TPP was hydrolysed at a rapid rate. Inosine diphosphate (IDP) and uridine diphosphate (UDP) were hydrolysed only very slowly. With the second set of conditions all three substrates were broken down rapidly (see Nucleoside diphosphatase, below).

In his second paper (1963b) Allen reported the separation of nucleoside diphosphatase and thiamine pyrophosphatase in 12·5 per cent. acrylamide gels. The method of Allen and Slater (1961) is given in Appendix 15, p. 722.

TPPase has chiefly been employed as an indicator of the Golgi apparatus in nerve cells (Shanthaveerappa and Bourne, 1965; Shantha and Bourne, 1966).

METHOD FOR INORGANIC PYROPHOSPHATASE (E.C. 3.6.1.1)

This enzyme was described by Bailey and Webb (1944), and also by Mann (1944), Gordon (1950), Heppel and Hilmoe (1951) and McElroy *et al.* (1951), and its properties are therefore well recorded. Kurata and Maeda (1956) designed a method for its histochemical demonstration depending on the hydrolysis of a pyrophosphate-iron complex (Weden, 1930) and in fact they were able to demonstrate two enzymes, one active at pH 7·2 and the other at pH 3·7 to 4·0. Rogers and Reynolds (1949) reported that pyrophosphate forms complexes of the type Fe^{3+} $(P_2O/)^-$ with ferric salts, and Monk (1949a and b) and Jones and Monk (1950) showed that these were relatively stable. This is a point of obvious importance if the localizations hown by Kurata and Maeda are to be considered accurate. Their curves for the activity of the alkaline enzyme are presented in Fig. 142.

This shows the considerable activation produced by Mg^{2+} ions. Lohmann (1933) had already shown that inorganic pyrophosphatase of muscle and liver required Mg^{2+}, and later he extended his studies to yeast, red cells, brain, cereals, moulds and fireflies. Robbins *et al.* (1955) suggested that the need of the enzyme for Mg^{2+} was due to the fact that its true substrate is magnesium pyrophosphate. They found that pyrophosphate itself inhibited the activity of rat-brain pyrophosphatase.

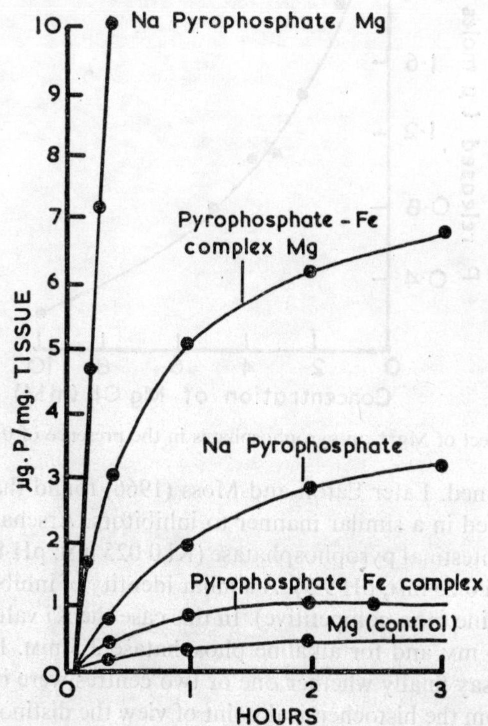

FIG. 142. Pyrophosphatase activity with various substrates at pH 7·6.

In the presence of fluoride, however, Elliot (1957) reported that the microsomal inorganic pyrophosphatase of guinea-pig liver was inhibited by magnesium rather than activated by it. Some discrepancy between these findings and those of Kurata and Maeda is obvious. The effect observed by Elliot is shown in Fig. 143, below.

Marked stimulation of enzyme activity was also noted by Elliot after the addition of 5 mM-ATP, the rate of hydrolysis being about double that found in control systems. Kunitz (1952) prepared a crystalline inorganic pyrophosphatase from bakers' yeast. This enzyme had no effect on organic esters such as ATP, ADP or thiamine pyrophosphate. It required Mg^{2+}, Co^{2+} or Mn^{2+} for activation and was antagonized by Ca^{2+}.

The position was further complicated when Heppel *et al.* (1962) showed that alkaline phosphatase from *Escherichia coli* would readily hydrolyse inorganic pyrophosphate. Fernley and Walker (1966) found that calf intestinal alkaline phosphatase had similar properties and Moss *et al.* (1966) indicated that the two types of activity might well represent functions of a single active centre. Alternatively, they suggested that two active centres on a single protein

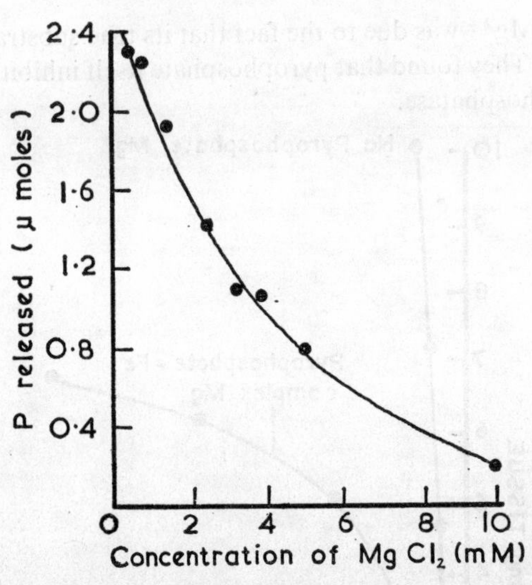

FIG. 143. Effect of Mg^{2+} on pyrophosphates in the presence of 0·5 M-fluoride.

might be concerned. Later Eaton and Moss (1966) found that both types of activity responded in a similar manner to inhibitors. Arsenate competitively inhibited both intestinal pyrophosphatase (Ki 0·025 mM, pH 8·5) and alkaline phosphatase (Ki 0·25 mM, pH 9·9). A similar identity of inhibition was shown by L-phenylalanine (non-competitive). In this case the Ki value for pyrophosphatase was 0·5 mM and for alkaline phosphatase 3·5 mM. Eaton and Moss were unable to say finally whether one or two centres were involved but it is obvious that from the histochemical point of view the distinction is too fine.

Although I have no experience of the method of Kurata and Maeda; details nevertheless appear in Appendix 15, p. 722.

METHODS FOR INORGANIC POLYMETAPHOSPHATASES (E.C. 3.6.1.2)

This enzyme, whose biochemical characteristics were described by Kitasato (1928a and b) and by Malmgren (1952), has a molecular weight of about 33,000. It has been found in a variety of genera, including moulds, yeasts and vertebrates, but its substrate, polymetaphosphate, is much more restricted in distribution. The enzyme breaks down its substrate, progressively, to the stage of pentametaphosphate.

A histochemical method for polymetaphosphatase was devised by Berg (1955), who tested homogenates of frog duodenum and frog embryos and found that they contained a magnesium-activated polyphosphatase with a pH optimum between pH 3·5 and 4·5. The enzyme was able to withstand a variety of fixatives and methods of preparation, the best being freeze-substitution (isopentane-acetone, −78°, 14 days). Preparation of the incubating medium was a somewhat tedious and critical procedure and, having no particular curiosity about this enzyme, I did not test the method.

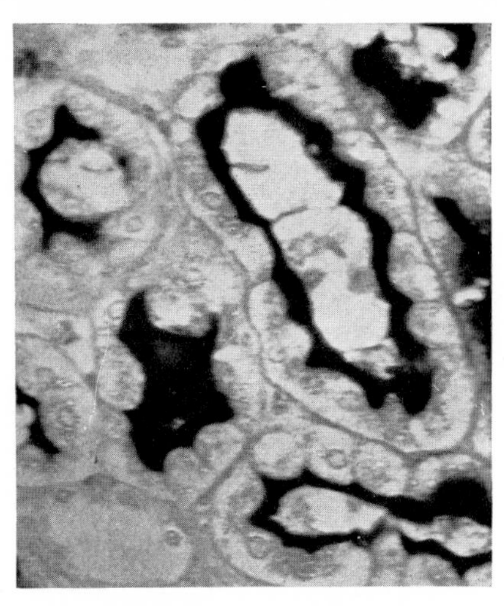

FIG. 134. For comparison with Fig. 127. Application of the coupling azo dye technique to the same material. Alkaline phosphatase is confined to the brush borders, the nuclei are unstained. × 400.

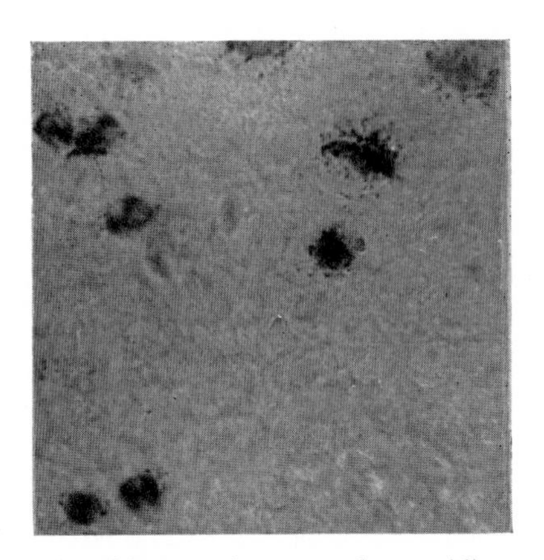

FIG. 135. Rat. Agar granuloma. Alkaline phosphatase in leucocytes photographed after mounting in pure glycerine. Coupling azo dye method. × 750.

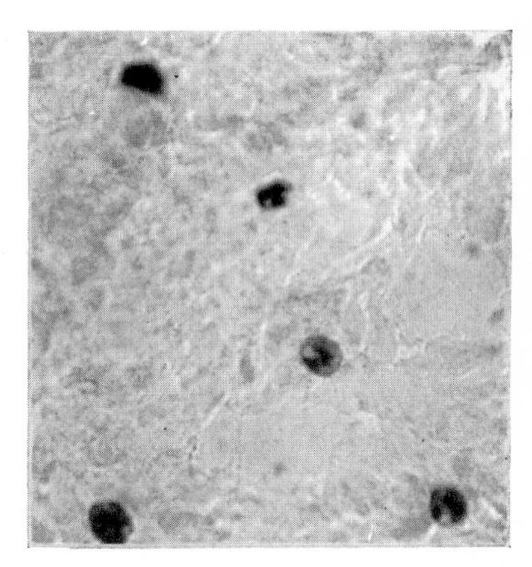

FIG. 136. As Fig. 135. Treated after incubation, and before mounting, with Lugol's iodine solution. Coupling azo dye method. × 750.

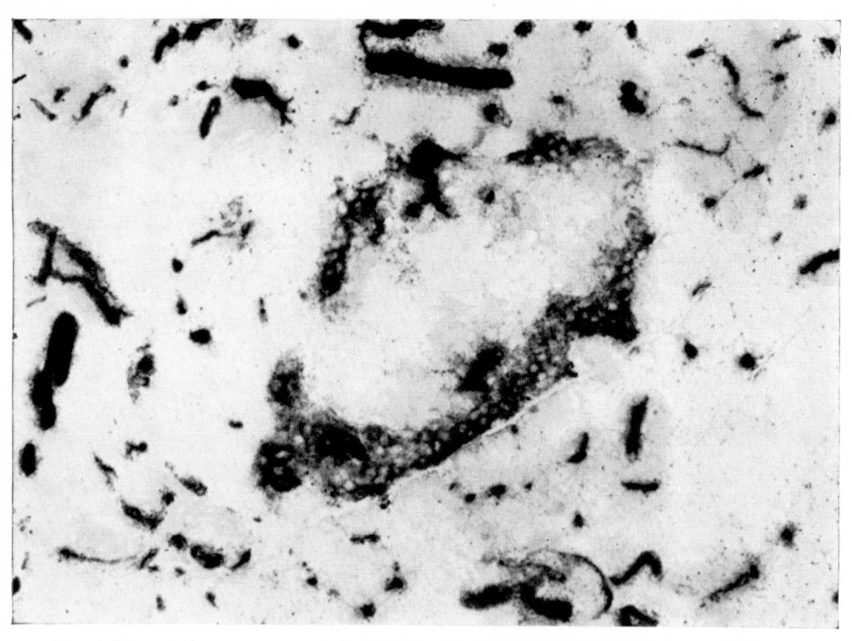

FIG. 137. Rat Pancreas. Cold Formalin-fixed, frozen section. Shows alkaline phosphatase in vessels and in some cells at the periphery of an islet of Langerhans. Napthol AS-MX phosphate-Fast red TR Salt. × 96.

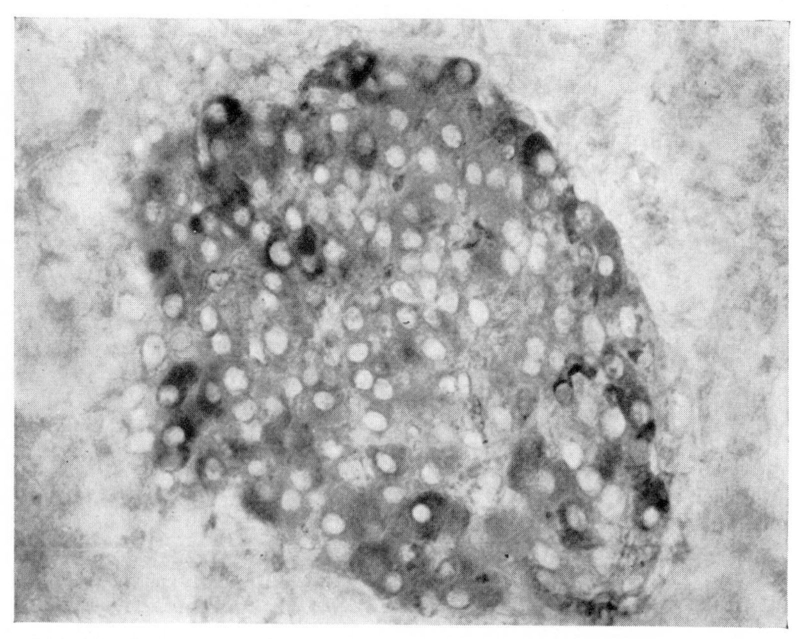

FIG. 144. Rat Pancreas. Cold formalin-fixed, frozen section. Glucose-6-phosphatase reaction strongly positive in the β-cells in an islet of Langerhans. Metal Precipitation technique. × 96.

Further communications on the subject have appeared (Berg and Gordon, 1960; Berg, 1960, 1964). In the first two papers the claim was made that the enzyme differed in distribution and properties from both acid and alkaline phosphatases. It was totally inactivated by alcohol and best preserved by acetone fixation.

The chelate-removal method for triphosphatases (Berg, 1964) depends on enzymatic hydrolysis of a stable lead triphosphate chelate, with release of Pb^{2+} ions together with pyrophosphate and orthophosphate. The solubility product of Pb^{2+} ions is exceeded and a precipitate forms. Berg's two methods are both given in Appendix 15, p. 723.

METHOD FOR GLUCOSE-6-PHOSPHATASE (E.C. 3.1.3.9)

A specific enzyme capable of dephosphorylating glucose-6-phosphate has been found in a variety of mammalian tissues (Fantl and Rome, 1945; de Duve *et al.*, 1949; Swanson, 1950). It is absent from the liver in cases of glycogen storage disease (Cori, 1952–53) but present otherwise in all mammalian organs capable of releasing glucose into the blood stream (liver, kidney, small intestine, rectum). According to Shull *et al.* (1956) the level of glucose-6-phosphatase in the livers of hereditary obese hyperglycæmic mice is higher than in control animals. After cortisone, a marked increase in the enzyme level in homogenates was noted by Weber *et al.* (1956). The properties of the enzyme in various organs of the rat were investigated by Freedland (1962). No activity was found in intestine and the liver and kidney enzymes appeared to be identical. According to Beaufay and de Duve (1955) glucose-6-phosphatase acts on a variety of substrates, including phenyl phosphate, ethyl phosphate, α-glycerophosphate and fructose-6-phosphate. It is specifically inhibited by 1,5-sorbitan-6-phosphate (Crane, 1955). The enzyme is found predominantly, if not exclusively, in the microsomal fraction of homogenates and it has therefore been used as a marker for the presence of microsomes.

It was found by Nordlie and Arion (1964) that the glucose-6-phosphatase and inorganic pyrophosphatase activities of rat liver microsomes could not be separated by fractionation and further similarities were found by Stetten and Taft (1964) and Fisher and Stetten (1965). According to Feuer *et al.* (1964) *o*-hydroxyphenylacetic acid is a potent inhibitor of both enzymes, while *o*-tyrosine activates both selectively.

The current histochemical method was introduced by Chiquoine (1953, 1955) using a Gomori type procedure with lead nitrate at pH 6·5–6·7. This pH optimum should, properly, cause it to be considered with the acid phosphatases in Chapter 16. It fits much better, however, into the sequence of enzymes discussed in this chapter. Chiquoine used fresh frozen sections in his studies, and these, or freeze-dried sections, are essential. Since both non-specific alkaline and acid phosphatase are capable of hydrolysing glucose-6-phosphate, it is obviously necessary to produce some proof of histochemical specificity. Chiquoine considered that, since formalin destroyed glucose-6-phosphatase but not acid or alkaline phosphatase, the absence of any reaction after 15 minutes' incubation of a formalin-fixed section with glucose-6-phosphate was sufficient proof of the specificity of the reaction. This is manifestly insufficient evidence, since the depressed activity of the non-specific phosphatases towards glucose-6-phosphate cannot be compared with their activity in fresh sections. The use of specific inhibitors, sorbitan-6-phosphate or *o*-hydroxyphenylacetic acid, is perhaps indicated.

Details of Chiquoine's method are given in Appendix 15, p. 724, together with instructions for the conversion of the barium salt of glucose-6-phosphate to the potassium salt. Since the latter is now commercially available these may be redundant. Wachstein and Meisel (1956) suggested a variation of the method, using the potassium salt as substrate, and this also is given in Appendix 15 and illustrated in Fig. 144, facing p. 535.

Fresh frozen sections, unless they are on the thick side (20 μ), tend to fragment during the later stages of both the above methods. Cold microtome sections mounted on coverslips are better, but the reaction is still sometimes patchy, especially when incubation is carried out with Chiquoine's original medium. Probably for this reason the method of Wachstein and Meisel is usually employed. A selective demonstration of the vascular endothelium was produced by Kazimierczak (1965) and changes in the nitrosamine damaged liver were recorded by Goessner and Friedrich-Freska (1964). The enzyme in the salivary glands of larval *Sciara coprophila* was investigated cytochemically by Terner *et al.* (1965) and the strong activity of the pancreatic islet β cells (illustrated in Fig. 144) was studied at both light and electron microscope level by Lazarus and Barden (1962). The technique used by these authors also appears in Appendix 15, p. 725.

METHODS FOR NUCLEOSIDE DIPHOSPHATASES (E.C. 3.6.1.6)

These enzymes were first demonstrated histochemically by Novikoff and Goldfischer (1961) using the diphosphates of inosine, guanosine and uridine (IDP, GDP, UDP) as substrates in a Wachstein-Meisel procedure at pH 7·2. The enzyme was found exclusively in the Golgi region of various cell types. An apparently distinct enzyme, similarly found in the Golgi region, was described by Lazarus and Barden (1962) in several rabbit tissues. This enzyme was active at pH 9·2 (Calcium-cobalt method) against the diphosphates of cytidine guanosine, inosine and uridine. It also hydrolysed uridine triphosphate (UTP). No staining of the Golgi region was observed with other nucleoside triphosphates, with adenosine diphosphate, nucleoside monophosphates or with β-glycerophosphate. TPP was not hydrolysed. The authors concluded that rabbit tissues contained a nucleoside diphosphatase which differed from the enzyme described by Novikoff and Goldfischer, and also from the TPPase of Allen and Slater (1961) and the thiamine diphosphatase of Naidoo (1962), in that it was inhibited by formalin fixation (10 per cent. overnight).

In a later paper Lazarus and Wallace (1964) described the separation of nucleoside diphosphatase activity from TPPase activity, in rabbit tissues. The TPPase of cerebral neurons (calcium-cobalt, pH 9·2) was inhibited by 25 mM sodium fluoride and a substrate concentration of 12 mM was required, six times that employed by Novikoff and Goldfischer using the Pb method at pH 7·2. With the latter method, and IDP as substrate, Braunstein and Stenger (1966) demonstrated nucleoside diphosphatase activity in hyaline bodies in carbon tetrachloride-damaged liver cells. When TPP, AMP or glycerophosphate were used in place of IDP no staining resulted.

The position with regard to nucleoside diphosphatases is not clear even if we restrict ourselves to a single animal species (an unacceptable restriction). In Appendix 15 appear details of the modified Padykula-Herman (pH 9·2) procedure used by Lazarus and Barden (1962).

Methods for Phosphodiesterases (PDases) (E.C. 3.1.4.1)

As indicated by Schmidt and Laskowski (1961) the term phosphodiesterase has undergone several changes of meaning. At present it is customary to divide these enzymes into three groups, Deoxyribonucleases (DNases), Ribonucleases (RNases) and a third group called simply phosphodiesterases. Schmidt and Laskowski prefer to use the term exonucleases for the third group.

Two new synthetic substrates for exonucleases were introduced by Razzell and Khorana (1959a and b) and further described by Razzell (1961). These were *p*-nitrophenyl-5′-thymidylic acid, and thymidylic acid-3′-nitrophenyl. According to Razzell (1961) the first was hydrolysed by PDase I and the second by PDase II. The corresponding *α*-naphthyl esters were synthesized by Sierakowska *et al.* (1963) in the hope that the two enzymes could be demonstrated and differentiated histochemically. They found that *α*-naphthyl thymidine-5′ phosphate was rapidly hydrolysed by PDase I but the 3′-phosphate ester was not hydrolysed by either enzyme. The preparation of the two substrates is given in Appendix 15, together with cytochemical directions for their use.

A histochemical method for cyclic 3′,5′-nucleotide phosphodiesterase was developed by Shanta *et al.* (1966). The substrate (cyclic 3′,5′-AMP) is split by the enzyme to 5′-AMP. The latter is then split by endogenous 5-nucleotidase, and by exogenous 5-nucleotidase added to the medium, and the free phosphate groups are captured by Pb^{2+} groups, at pH 7·5 to 7·67.

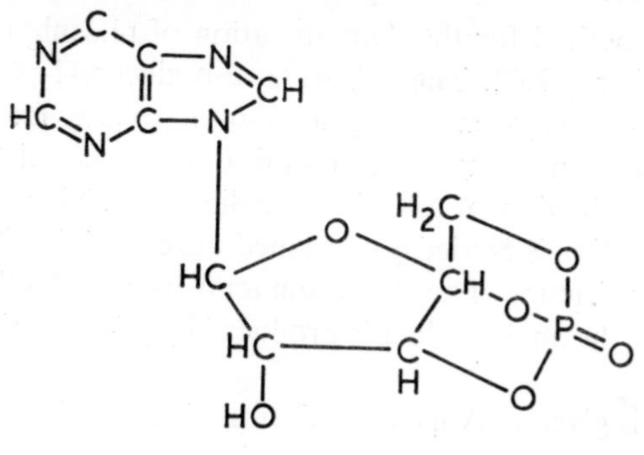

This method is thus a multistep procedure and the localization of the final product must be questioned, even if it can be shown by suitable controls that the whole reaction is dependent on the first stage. The phosphodiesterase present in snake venom certainly forms 5-nucleotides but it is commonly supposed that the spleen enzyme forms 3-nucleotides. The latter would not be susceptible to the activity of either endogenous or added 5-nucleotidase. Details of the method are given in Appendix 15, despite the above reservations.

METHODS FOR RIBONUCLEASE (E.C. 2.7.7.16)

Until recently the only accepted histochemical technique for ribonuclease was a substrate-film method (see Chapter 24). A new substrate synthesized by Shugar and his colleagues (Sierakowska *et al.*, 1965; Zan-Kowalczewska *et al.*, 1966) has been used by them for histochemical demonstration of the enzyme in mammalian tissues.

Ribonuclease acts by transferring the 3′-phosphate of a pyrimidine nucleotide residue of a polynucleotide from the 5′-position of the adjoining nucleotide to the 2′-position of the pyrimidine nucleotide itself, to form a cyclic nucleotide. It can also act by catalysing transfer of the phosphate group in the 2′-position of a cyclic nucleotide to water.

The reaction described by Shugar and his colleagues depended on the transesterification of the substrate uridine-3′-naphthylphosphate by tissue ribonucleases to give uridine-2′,3′-cyclic phosphate and free naphthol. Localization of both alkaline and acid ribonucleases was described. For the former, at pH 9, the capture reagent was Fast red TR salt and for the latter, at pH 5·7, hexazotized pararosanilin was used.

In an addendum to their second paper (Zan-Kowalczewska *et al.*, 1966) the authors indicated that, as would be expected, non-specific alkaline phosphatase could hydrolyse their substrate. Localization of specific alkaline RNase was therefore impossible without inhibition of alkaline phosphatase. On the other hand, it was claimed that acid RNase could still be demonstrated specifically. Until further information is available the use of pyrimidine naphthyl phosphates as substrates for RNase cannot freely be recommended.

METHOD FOR PHOSPHOGLUCOMUTASE (E.C. 2.7.5.1)

A multistep method for the demonstration of phosphoglucomutase was reported by Meijer (1967). The substrate, α-D-glucose-1-phosphate, is converted by the tissue enzyme to α-D-glucose-6-phosphate. Exogenous glycerophosphate dehydrogenase then converts this to the gluconolactone phosphate with coincident reduction of added tetrazolium salt. Since the enzyme is a soluble one incubation is carried out in a medium containing 2 per cent. gelatin which has the consistency of a gel at room temperature. Additional sharpness of localization of the final product is produced by prior fixation of sections in cold acetone.

The method is given in Appendix 15. If used it must be accompanied by a careful appraisal of the result, checking where possible by means of biochemical data. Specific inhibition of phosphoglucomutase (and alkaline phosphatase) is produced by beryllium sulphate at concentrations of 10 μM or lower (Aldridge and Thomas, 1964; Thomas.and Aldridge, 1966).

METHOD FOR ACETYL PHOSPHATASE (E.C. 3.6.1.7)

A magnesium-sensitive enzyme specifically hydrolysing acetyl phosphate was described in skeletal muscle by Bourne and Golarz (1959). Its pH optimum was 7·0 and a calcium cobalt method was used. In progressive muscular dystrophies there was an increase of the enzyme, particularly in the connective tissue. Although the significance of this observation is not clear there are strong indications for investigating acetyl phosphatases (and 5-nucleotidase) in wasting disorders of skeletal muscle.

REFERENCES

ABE, T., and SHIMIZU, N. (1964). *Histochemie*, **4**, 209.
ABOLINS, L. (1948). *Nature, Lond.*, **161**, 556.
ABOLINS, L., and ABOLINS, A. (1949). *Ibid.*, **164**, 455.
ACKERMAN, G. A. (1962). *Lab. Invest.*, **11**, 563.
AHMED, Z., and REIS, J. L. (1958). *Biochem. J.*, **69**, 386.
ALDRIDGE, W. N. (1962). *Biochem. J.*, **83**, 527.
ALDRIDGE, W. N., and THOMAS, M. (1964). *Biochem. J.*, **92**, 16P.
A LEN, J. (1963a). *J. Histochem. Cytochem.*, **11**, 529.
ALLEN, J. (1963b). *Ibid*, **11**, 542.
ALLEN, J. M., and SLATER, J. J. (1956). *J. Histochem. Cytochem.*, **4**, 110.
ALLEN, J., and SLATER, J. (1961). *J. Histochem. Cytochem.*, **9**, 418.
ALLEN, R. J. L., and BOURNE, G. H. (1943). *J. exp. Biol.*, **20**, 61.
ANDERSON, A. B. (1961). *Biochim. Biophys. Acta*, **54**, 110.
ANTONINI, F. M., and WEBER, G. (1951). *Arch. de Vecchi*, **16**, 985.
APPLEYARD, J. (1948). *Biochem. J.*, **42**, 596.
ARVY, L. (1960). *Zest. Zellforsch.*, **51**, 408.
ARZAC, J. P., and BLANCHET, E. (1948). *J. clin. Endoc.*, **8**, 315.
ATKINSON, W. B., and ELFTMAN, H. (1946). *Proc. Soc. exp. Biol., N.Y.*, **62**, 148.
ATKINSON, W. B., and ELFTMAN, H. (1947). *Endocrinology*, **40**, 30.
ATKINSON, W. B., and ENGLE, E. T. (1947). *Ibid.*, **40**, 327.
ATKINSON, W. B., and GUSBERG, S. B. (1948). *Cancer*, **1**, 248.
AXELROD, P. (1948). *J. biol. Chem.*, **172**, 1.
BAILEY, K., and WEBB, E. C. (1944). *Biochem. J.*, **38**, 394.
BARADI, A. F., and BOURNE, G. H. (1951). *Nature, Lond.*, **168**, 977.
BARKA, T., SZALAY, S., POSALAKY, Z., and KERTESZ, L. (1952). *Acta anat.*, **16**, 45.
BARKA, T. T., and ANDERSON, P. J. (1962). *J. Histochem. Cytochem.*, **10**, 231.
BARRON, K. D., and BOSHES, R. (1961). *Ibid*, **9**, 455.
BARTER, R. (1954). *Nature, Lond.*, **173**, 1233.
BARTER, R., DANIELLI, J. F., and DAVIES, H. G. (1955). *Proc. Roy. Soc. B.*, **144**, 412.
BARTONIČEK, V., and LOJDA, Z. (1964). *Acta histochem.*, **19**, 357.
BEAUFAY, H., and DE DUVE, C. (1954). *Bull. Soc. chim. Biol.*, **36**, 1525.
BEAUFAY, H., and DE DUVE, C. (1955). Quoted by Crane, R. K.
BÉLANGER, L. F. (1951). *Proc. Soc. exp. Biol., N.Y.*, **77**, 266.
BERG, G. W. (1955). *J. Histochem. Cytochem.*, **3**, 22.
BERG, G. G. (1960). *Ibid*, **8**, 92.
BERG, G. G. (1964). *Ibid*, **12**, 341.
BERG, G. G., and GORDON, L. H. (1960). *J. Histochem. Cytochem.*, **8**, 85.

BEVELANDER, G., and JOHNSON, P. L. (1945). *J. cell. comp. Physiol.*, **26**, 25.
BEVELANDER, G., and JOHNSON, P. L. (1950). *Anat. Rec.*, **108**, 1.
BOIVIN, P., and ROBINEAUX, R. (1956). *Le Sang, Par.*, **27**, 535.
BOULET, M., and MARIER, J. R. (1961). *Arch. Biochem.*, **93**, 157.
BOURNE, G. (1943). *Quart. J. exp. Physiol.*, **32**, 1.
BOURNE, G. (1954a). *Acta anat.*, **22**, 289.
BOURNE, G. (1954b). *Quart. J. micr. Sci.*, **95**, 359.
BOURNE, G. H., and GOLARZ, N. (1959). *Arch. Biochem.*, **85**, 109.
BRACHET, J., and JEENER, R. (1946). *C.R. Soc. Bio., Paris*, **140**, 1121.
BRACHET, J., and JEENER, R. (1948). *Biochim. Biophys. Acta*, **2**, 423.
BRADFIELD, J. R. G. (1946). *Nature, Lond.*, **157**, 876.
BRAUNSTEIN, H., and STENGER, R. J. (1966). *J. Histochem. Cytochem.*, **14**, 112.
BUNTING, H. (1948). *Proc. Soc. exp. Biol., N.Y.*, **67**, 370.
BURSTONE, M. S. (1957). *Amer. J. clin. Path.*, **28**, 429.
BURSTONE, M. S. (1958a). *J. Histochem. Cytochem.*, **6**, 87.
BURSTONE, M. S. (1958b). *J. nat. Cancer Inst.*, **20**, 601.
BURSTONE, M. S. (1960). *J. nat. Cancer. Inst.*, **24**, 1199.
BURSTONE, M. S. (1962). "Enzyme Histochemistry." Acad. Press, New York.
CACIOPPO, F., QUAGLIARIELLO, C., COLTORTI, M., and DELLA PIETRA, G. (1953). *Arch. Sci. Biol.*, **37**, 563.
CESÀRO, A. N., and GRANATA, A. (1955–56). *Acta histochem.*, **2**, 88.
CHÈVREMONT, M., and FIRKET, H. (1949). *Arch. de Boil.*, **40**, 441.
CHÈVREMONT, M., and FIRKET, H. (1953). *Int. Rev. Cytol.*, **2**, 261.
CHIQUOINE, A. D. (1953). *J. Histochem. Cytochem.*, **1**, 429.
CHIQUOINE, A. D. (1954). *J. comp. Neurol.*, **100**, 415.
CHIQUOINE, A. D. (1955). *J. Histochem. Cytochem.*, **3**, 471.
CLAYTON, B. E., and HAMMANT, J. E. (1957). *J. Endocrinol.*, **15**, 1.
CLOETENS, R. (1941). *Biochem. Z.*, **310**, 42.
COLOWICK, S. P., and KAPLAN, N. O. (1955). In "Methods in Enzymology." Acad. Press, New York.
CORI, G. T. (1952–53). *Harvey Lectures*, p. 145.
CORNER, G. W. (1944). *Science*, **100**, 270.
CRANE, R. K. (1955). *Biochim. Biophys. Acta*, **17**, 443.
CURZEN, P. (1964). *J. Obstet. Gynaecol.*, **71**, 388.
DALGAARD, J. B. (1948). *Nature, Lond.*, **162**, 811.
DANIELLI, J. F. (1946). *J. exp. Biol.*, **22**, 110.
DANIELLI, J. F. (1950). *Nature, Lond.*, **165**, 762.
DANIELLI, J. F. (1951). *Ibid.*, **168**, 464.
DANIELLI, J. F. (1958). *General Cytochemical Methods*, **1**, 423.
DANIELLI, J. F., and CATCHESIDE, D. G. (1945). *Nature, Lond.*, **156**, 294.
DANIELLI, J. F., FELL, H. B., and KODICEK, E. (1945). *Brit. J. exp. Path.*, **26**, 367.
DAVIS, B. J., and ORNSTEIN, L. (1959). *J. Histochem. Cytochem.*, **7**, 297.
DAVISON-REYNOLDS, M. M., BARRUETO, R. D., and LEMON, H. M. (1955). *Enzymol.*, **17**, 145.
DEANE, H. W. (1947). *Amer. J. Anat.*, **80**, 321.
DE DUVE, C., BARTHET, J., HERS, H. G., and DUPRET, L. (1949). *Bull. Soc. chim. Biol.*, **31**, 1242.
DEFENDI, V. (1957). *J. Histochem. Cytochem.*, **5**, 1.
DEFENDI, V., and PEARSE, A. G. E. (1955). *J. Histochem. Cytochem.*, **3**, 203.
DEIMLING, O. H. (1964). *Histochemie*, **4**, 48.
DEMPSEY, E. W. (1949). *Ann. N.Y. Acad. Sci.*, **50**, 336.
DEMPSEY, E. W., and DEANE, H. W. (1946). *J. cell. comp. Physiol.*, **27**, 159.
DEMPSEY, E. W., GREEP, R. O., and DEANE, H. W. (1949). *Endocrinology*, **44**, 88.
DEMPSEY, E. W., and WISLOCKI, G. B. (1947). *Amer. J. Anat.*, **80**, 1.
DEUL, D. H., and MCILWAIN, H. (1961). *J. Neurochem.*, **8**, 246.
DHOM, G., and MENDE, H-J. (1956). *Virchows Arch.*, **328**, 337.
DIXON, M. (1949). "Multi-enzyme Systems." Cambridge Univ. Press.
DIXON, M., and WEBB, E. C. (1953). *Brit. med. Bull.*, **9**, 110.
DIXON, M., and WEBB, E. C. (1958). "Enzymes." Longmans, London.

DIXON, M., and WEBB, E. C. (1964). "Enzymes," 2nd Ed.
DOUNCE, A. L. (1950). "The Enzymes." Ed. Sumner and Myrbach. Vol. I, Pt. 1, p. 189.
DOYLE, W. L. (1950). *Science*, **111**, 64.
DOYLE, W. L., OMOTO, J., and DOYLE, M. E. (1951). *Exp. Cell. Res.*, **2**, 20.
DE DUVE, C., BERTHET, J., HERS, H. G., and DUPRET, L. (1949). *Bull. Soc. chim. biol.*, **31**, 1242.
EATON, R. H., and MOSS, D. W. (1966). *Proc. Biochem. Soc.*, **100**, 45P.
ELLIOT, W. H. (1957). *Biochem. J.*, **65**, 315.
ELVES, M. W., ROATH, S., and ISRAELS, M. C. G. (1963). *Acta haematol.*, **29**, 141.
EMMEL, V. M. (1946). *Anat. Rec.*, **95**, 159.
ENGEL, M. A., and FURATA, W. (1942). *Proc. Soc. exp. Biol., N.Y.*, **50**, 5.
EPSTEIN, F. H., and WHITTAM, R. (1966). *Biochem. J.*, **99**, 232.
ERÄNKÖ, O., and HASAN, J. (1954). *Acta. path. scand.*, **35**, 563.
ERÄNKÖ, O., and NIEMI, M. (1954). *Ibid.*, **45**, 357.
ERÄNKÖ, O., and LEHTO, L. (1954). *Acta anat.*, **22**, 277.
ERDTMAN, H. (1928). *Z. physiol. Chem.*, **177**, 211.
ERICHSEN, S. (1953). *Acta path. scand.*, **33**, 263.
FANTL, P., and ROME, M. N. (1945). *Austr. J. exp. Biol.*, **23**, 21.
FEIGIN, I., WOLF, A., and KABAT, E. A. (1950). *Amer. J. Path.*, **26**, 647.
FEIGIN, I., and WOLF, A. (1953). *Lab. Invest.*, **2**, 115.
FEIGIN, I., and WOLF, A. (1954). *J. Histochem. Cytochem.*, **2**, 435.
FEIGIN, I., and WOLF, A. (1955). *J. Neuropath. exp. Neurol.*, **14**, 11.
FEIGIN, I., and WOLF, A. (1957). *J. Histochem. Cytochem.*, **5**, 53.
FELL, H. B., and DANIELLI, J. F. (1943). *Brit. J. exp. Path.*, **24**, 196.
FERNLEY, H. N., and WALKER, P. G. (1966). *Biochem. J.*, **99**, 39P.
FEUER, G., GOLBERG, L., and GIBSON, K. I. (1964). *Biochem. J.*, **93**, 5P.
FIRKET, H. (1952). *Bull. Microsc. appl.*, **2**, 57.
FISCHER, I., and GLICK, D. (1947). *Proc. Soc. exp. Biol., N.Y.*, **66**, 14.
FISHER, C. J., and STETTEN, M. R. (1965). *Fed. Proc.*, **24**, 474.
FISHMAN, W. H., GREEN, S., and INGLIS, N. I. (1963). *Nature*, **198**, 685.
FOGO, J. K., and POPOWSKY, M. (1949). *Anat. Chem.*, **21**, 732.
FOLLIS, R. H. R. (1949). *Bull. Johns Hopk. Hosp.*, **85**, 360.
FOLLIS, R. H. R. (1953). *Ibid.*, **93**, 386.
FORAKER, A. G., and WINGO, J. W. (1955). *Arch. Dermatol.*, **72**, 1.
FORD, D. H., and HIRSCHMAN, A. (1955). *Anat. Rec.*, **121**, 531.
FREDRICSSON, B. (1952). *Anat. Anz.*, **99**, 97.
FREDRICSSON, B. (1956). *Acta anat.*, **26**, 246.
FREEDLAND, R. A. (1962). *Biochim. Biophys. Acta*, **62**, 427.
FREIMAN, D. G., and KAPLAN, N. (1960). *J. Histochem. Cytochem.*, **8**, 159.
FREIMAN, D. G., GOLDMAN, H., and KAPLAN, N. (1962). *J. Histochem. Cytochem.*, **10**, 520.
FREEMAN, H. F., and COLVER, C. W. (1938). *J. Amer. chem. Soc.*, **60**, 750.
FRIEDENWALD, J. S., and MAENGWYN-DAVIES, G. D. (1950). *Fed. Proc.*, **9**, 44.
FRIEDMAN, O. M., and SELIGMAN, A. M. (1950). *J. Amer. chem. Soc.*, **72**, 624.
GLICK, D. (1946). *Science*, **103**, 599.
GLICK, D., and FISCHER, E. E. (1945). *Ibid.*, **102**, 429.
GLICK, D., and FISCHER, E. E. (1946). *Arch. Biochem.*, **11**, 65.
GOEBEL, A., and PUCHTLER, H. (1954). *Virchows Arch.*, **326**, 119.
GOESSNER, W. (1958). *Histochemie*, **1**, 48.
GOESSNER, W., and FRIEDRICH-FRESKA, H. (1964). *Zeit. naturforsch.*, **19**, 862.
GOLARZ, M. N., BOURNE, G. H., and RICHARDSON, H. D. (1961). *J. Histochem. Cytochem.*, **9**, 132.
GOLD, N., and GOULD, B. S. (1951). *Arch. Biochem. Biophys.*, **33**, 155.
GOMORI, G. (1939). *Proc. Soc. exp. Biol., N.Y.*, **42**, 23.
GOMORI, G. (1941). *J. cell. comp. Physiol.*, **17**, 71.
GOMORI, G. (1946). *Amer. J. clin. Path.*, **16**, Tec. Sect. **7**, 177.
GOMORI, G. (1949a). *Proc. Soc. exp. Biol., N.Y.*, **70**, 7.
GOMORI, G. (1949b). *Ibid.*, **72**, 449.

GOMORI, G. (1950a). *Ann. N.Y. Acad. Sci.*, **50**, 968.
GOMORI, G. (1950b). *J. Lab. clin. Med.*, **35**, 802.
GOMORI, G. (1950c). *Exp. Cell. Res.*, **1**, 33.
GOMORI, G. (1951a). *J. Lab. clin. Med.*, **37**, 520.
GOMORI, G. (1951b). Personal communication.
GOMORI, G. (1952). "Microscopic Histochemistry." Chicago Univ. Press.
GOMORI, G., and BENDITT, E. P. (1953). *J. Histochem. Cytochem.*, **1**, 114.
GORDON, J. J. (1950). *Biochem. J.*, **46**, 96.
GOULD, B. S., and GOLD, N. (1951). *Arch. Path.*, **52**, 413.
GRAZIADEI, P., and ZACCHEO, D. (1956). *Boll. Soc. ital. sper.*, **32**, 702.
GREEP, R. O., FISCHER, C. J., and MORSE, A. (1948). *Science*, **105**, 666.
GROGG, E., and PEARSE, A. G. E. (1952). *Nature, Lond.*, **170**, 578.
GRUENER, N., and AVI-DOR, Y. (1966). *Biochem. J.*, **100**, 762.
GULLAND, J. M., and JACKSON, E. M. (1938). *Biochem. J.*, **32**, 590, 597.
HALEY, H. L., DEWIS, G. M., and SOMMERS, S. C. (1955). *Arch. Path.*, **59**, 635.
HALL, J. E. (1950). *Amer. J. Obstet. Gynec.*, **60**, 212.
HANCOX, N. M., and NICHOLAS, E. (1956). *Acta anat.*, **26**, 302.
HARCOURT-WEBSTER, J. N., and STOTT, N. C. H. (1965). *J. roy. micr. Soc.*, **84**, 155.
HARD, W. L. (1946). *Amer. J. Anat.*, **78**, 47.
HAYHOE, F. G. J., and QUAGLINO, D. (1958). *Brit. J. Hæmatol.*, **4**, 375.
HENRICHSEN, E. (1956). *Exp. Cell. Res.*, **11**, 115.
HEPPEL, L. A., and HILMOE, R. J. (1951). *J. biol. Chem.*, **192**, 87.
HEPPEL, L. A., HARKNESS, D. R., and HILMOE, R. J. (1962). *J. biol. Chem.*, **237**, 841.
HERMAN, E. C., and WRIGHT, B. E. (1959). *J. biol. Chem.*, **234**, 122.
HILL, M. (1956). *Československ. Morphol.*, **4**, 1.
HOARE, R., and DELORY, G. E. (1955). *Arch. Biochem. Biophys.*, **59**, 465.
HOLT, S. J. (1954). *Proc. Roy. Soc. B.*, **142**, 160.
HORE, E., ELVEHJEM, C. A., and HART, E. B. (1940). *J. biol. Chem.*, **136**, 425.
HOROWITZ, N. H. (1942). *J. Dent. Res.*, **21**, 519.
JACOBY, F., and MARTIN, B. F. (1949). *Nature, Lond.*, **163**, 875.
JACOBY, F., and MARTIN, B. F. (1951). *J. Anat.*, **85**, 391.
JAGANNATHAN, V., SINGH, H., and DAMODARAN, M. (1956). *Biochem. J.*, **63**, 94.
JÄRNEFELT, J. (1961). *Biochim. Biophys. Acta.*, **48**, 104.
JÄRNEFELT, J. (1962). *Ibid*, **59**, 643.
JEENER, R. (1947). *Nature, Lond.*, **159**, 578.
JEFFREE, G. M., and TAYLOR, K. B. (1961). *J. Histochem. Cytochem.*, **9**, 93.
JOHANSEN, G., and LINDERSTRØM-LANG, K. (1951). *Acta chem. Scand.*, **5**, 965.
JOHANSEN, G., and LINDERSTRØM-LANG, K. (1952). *Acta med. Scand.*, Suppl. 266, p. 601.
JOHANSEN, G., and LINDERSTRØM-LANG, K. (1953). *J. Histochem. Cytochem.*, **1**, 442.
JOHNSON, P. L., and BEVELANDER, G. (1946). *Anat. Rec.*, **95**, 193.
JOHNSON, P. L., BUTCHER, E. O., and BEVELANDER, G. (1945). *Ibid.*, **93**, 355.
JONES, H. W., and MONK, C. B. (1950). *J. chem. Soc.*, 3475.
KABAT, E. A., and FURTH, J. (1941). *Amer. J. Path.*, **17**, 303.
KAPLOW, L. S. (1955). *Blood*, **10**, 1023.
KAPLOW, L. S. (1963). *Amer. J. clin. Path.*, **39**, 439.
KAZIMIERCZAK, J. (1965). *Acta. path. scand.*, **63**, 319.
KAYE, M. A. G. (1955). *Biochim. Biophys. Acta*, **18**, 456.
KING, E. J. (1952). Personal communication.
KING, E. J., and DELORY, G. E. (1939). *Biochem. J.*, **33**, 1185.
KITASATO, T. (1928a). *Biochem. Z.*, **197**, 257.
KITASATO, T. (1928b). *Ibid.*, **201**, 206.
KOPF, A. W. (1957). *Arch. Dermatol.*, **75**, 1.
KRITZLER, R. A., and GUTMAN, A. B. (1941). *Amer. J. Physiol.*, **134**, 94.
KROON, D. B., NEUMANN, H., and KRAYENHOFF SLOOT, W. J. A. T. (1944). *Enzymologia*, **11**, 186.
KRUGELIS, E. J. (1942). *J. cell. comp. Physiol.*, **19**, 1.
KRUGELIS, E. J. (1946). *Biol. Bull.*, **90**, 220.
KUNITZ, M. (1952). *J. gen. Physiol.*, **35**, 423.

KURATA, Y., and MAEDA, S. (1956). *Stain Tech.*, **31**, 13.
LAKE, B. D. (1965) *J. roy. micr. Soc.*, **84**, 489.
LANDOW, H., KABAT, E. A., and NEWMAN, W. (1942). *Arch. Neurol. Psychiat.*, **48**, 518.
LAZARUS, S. S., and BARDEN, H. (1962). *J. Histochem. Cytochem.*, **10**, 368.
LAZARUS, S. S., and WALLACE, B. J. (1964). *J. Histochem. Cytochem.*, **12**, 729.
LECKIE, F. H. (1955). *J. Obst. Gynecol.*, **62**, 542.
LEDUC, E. H., and DEMPSEY, E. W. (1951). *J. Anat.*, **85**, 305.
LISON, L. (1948). *Bull. Histol. Tech. micr.*, **25**, 23.
LOJDA, Z., VEČEREK, B., and PILICHOVÁ, H. (1964). *Histochemie*, **3**, 428.
LOHMANN, K. (1933). *Biochem. Z.*, **262**, 137.
LONDON, M., and HUDSON, P. B. (1955). *Biochim. Biophys. Acta*, **17**, 485.
LONGLEY, J. B. (1955). *Science*, **122**, 594.
LORCH, I. J., (1947). *Quart. J. micr. Sci.*, **88**, 159.
LOUSTALOT, P. (1955). Personal communication.
LOVELESS, A., and DANIELLI, J. F. (1949). *Quart. J. micr. Sci.*, **90**, 57.
LOWENSTEIN, J. M. (1958). *Biochem. J.*, **70**, 222.
MAENGWYN-DAVIES, G. D., FRIEDENWALD, J. S., and WHITE, R. T. (1952). *J. cell. comp. Physiol.*, **39**, 395.
MAJNO, G., and ROUILLER, C. (1951). *Virchows Arch.*, **321**, 1.
MALMGREN, H. (1952). *Acta chem. scand.*, **6**, 16.
MANHEIMER, L. H., and SELIGMAN, A. M. (1949). *J. nat. Cancer Inst.*, **9**, 181.
MANN, T. (1944). *Biochem. J.*, **38**, 345.
MARTIN, B. F. (1949). *Stain Tech.*, **24**, 215.
MATHIES, J. C. (1958). *J. biol. Chem.*, **233**, 1121.
MCALPINE, R. J. (1955). *Amer. J. Anat.*, **96**, 191.
MCELROY, W. D., COULOMBRE, J., and HAYS, R. (1951). *Arch. Biochem. Biophys.*, **32**, 207.
MCCLURKIN, I. T. (1964). *J. Histochem. Cytochem.*, **12**, 654.
MCKAY, D. G., PINKERTON, J. H. M., HERTIG, A. T., and DANZIGER, S. (1961). *Obstet. Gynæcol.*, **18**, 13.
MCMANUS, J. F. A., LUPTON, C. H., and HARDEN, A. G. (1952). *Lab. Invest.*, **1**, 480.
MCMANUS, J. F. A., and LUPTON, C. H. (1953). *Ibid.*, **2**, 76.
MEIER-RUGE, W., and MEIER, E. (1963). *Experientia*, **19**, 266.
MEIJER, A. E. F. H. (1967). *Histochemie*, **8**, 248.
MENTEN, M. L., JUNGE, J., and GREEN, M. H. (1944). *J. biol. Chem.*, **153**, 471.
MEYERHOF, O., and GREEN, H. (1950). *Ibid.*, **183**, 377.
MEYNER, R. H., and WILLIAMS, W. L. (1954). *Anat. Rec.*, **119**, 289.
MINAMITANI, S. (1953). *Okajimas Fol. anat., Jap.*, **25**, 19.
MOLNAR, J. (1952). *Stain Tech.*, **27**, 221.
MONIS, B., and RUTENBERG, A. M. (1958). *J. Histochem. Cytochem.*, **6**, 91.
MONK, C. B. (1949a). *J. chem. Soc.*, 423.
MONK, C. B. (1949b). *Ibid.*, 427.
MOOG, F., and STEINBACH, H. B. (1946). *Science*, **103**, 144.
MORI, M., TAKADA, K., and OKAMOTO, J. (1961). *J. Osaka Univ. Dent. Sch.*, **1**, 67.
MORTON, R. K. (1955). In "Methods in Enzymology," **2**, 556. Acad. Press Inc., New York.
MORTON, R. K. (1957). *Biochem. J.*, **65**, 674.
MOSES, H. L., ROSENTHAL, A. S., BEAVER, D. L., and SCHUFFMAN, S. S. (1966). *J. Histochem. Cytochem.*, **14**, 702.
MOSS, D. W. (1964). *Scientific Basis of Med. Ann. Revs.*, p. 334.
MOSS, D. W., EATON, R. H., SMITH, J. K., and WHITBY, L. G. (1966). *Biochem. J.*, **99**, 19P.
MOWRY, R. W. (1949). *Bull. int. Ass. med. Mus.*, **30**, 95.
MUKHERJEE, A. K., and BANERJEE, S. (1955–56). *Acta histochemica*, **2**, 259.
NAIDOO, D. (1962). *J. Histochem. Cytochem.*, **10**, 421.
NAIDOO, D. (1962). *Ibid*, **10**, 572.
NAIDOO, D., and PRATT, O. E. (1951). *J. Neurol. Neurosurg. Psychiat.*, **14**, 287.
NAIDOO, D., and PRATT, O. E. (1952). *Ibid.*, **15**, 164.

Naidoo, D., and Pratt, O. E. (1953). *Enzymologia*, **16**, 91.
Naidoo, D., and Pratt, O. E. (1954). *Ibid.*, **16**, 298.
Naik, D. V., and George, J. C. (1964). *J. Histochem. Cytochem.*, **12**, 772.
Nepveux, P., and Wegmann, R. (1962). *Ann. Histochem.*, **7**, 21.
Neumann, H. (1949). *Biochim. Biophys. Acta*, **3**, 117.
Newman, W., Feigin, I., Wolf, A., and Kabat, E. A. (1950). *Amer. J. Path.*, **26**, 257.
de Nicola, M. (1949). *Quart. J. micr. Sci.*, **90**, 391.
Nimmo-Smith, R. H. (1961). Personal Communication.
Nordlie, R. C., and Arion, W. J. (1964). *J. biol. Chem.*, **239**, 1680.
Novikoff, A. B. (1951). *Science*, **113**, 320.
Novikoff, A. B. (1958). *J. Histochem. Cytochem.*, **6**, 251.
Novikoff, A. B., Hausman, D. H., and Podber, E. (1958). *Ibid.*, **6**, 61.
Novikoff, A. B., Hecht, L., Podber, E., and Ryan, J. (1952). *J. biol. Chem.*, **194**, 153.
Novikoff, A. B., Drucker, J., Shin, W. Y., and Goldfischer, S. (1961). *J. Histochem., cytochem.*, **9**, 434.
Novikoff, A. B., and Goldfischer, S. (1961). *Proc. Nat. Acad. Sci. Wash.*, **47**, 802.
Novikoff, A. B., Podber, E., and Ryan, J. (1950). *Fed. Proc.*, **9**, 210.
Otte, P. (1958). *Hoppe-Seyl. Z.*, **310**, 103.
Padykula, H. A., and Herman, E. (1955a). *J. Histochem. Cytochem.*, **3**, 161.
Padykula, H. A., and Herman, E. (1955b). *Ibid.*, **3**, 170.
Pearse, A. G. E. (1951). *J. clin. Path.*, **4**, 1.
Pearse, A. G. E. (1954). *Int. Rev. Cytol.*, **3**, 329.
Pearse, A. G. E., and Reis, J. L. (1952). *Biochem. J.*, **50**, 534.
Peters, H. (1964). *Histochemie*, **4**, 345.
Peters, R. A., Shorthouse, M., and Walshe, J. M. (1965). *Biochem. J.*, **96**, 47P.
Pratt, O. E. (1954). *Biochim. Biophys. Acta*, **14**, 380.
Pritchard, J. J. (1947). *J. Anat.*, **81**, 352.
Przelecka, A., Dominas, H., and Szrzala, M. G. (1962a). *Folia morphol.*, **13**, 371.
Przelecka, A., Ejsmont, G., Sarzala, M. G., and Taracha, M. (1962b). *J. Histochem. Cytochem.*, **10**, 596.
Razzell, W. E. (1961). *J. biol. Chem.*, **236**, 3028.
Razzell, W. E., and Khorana, H. G. (1959a). *J. biol. Chem.*, **234**, 2108.
Razzell, W. E., and Khorana, H. G., (1959b). *Ibid.*, **234**, 2114.
Reis, J. L. (1937). *Enzymologia*, **2**, 110.
Reis, J. L. (1940). *Bull. Soc. Chim. biol., Paris*, **22**, 36.
Reis, J. L. (1950). *Proc. Biochem. Soc., Biochem. J.*, **46**, p. xxi.
Reis, J. L. (1951). *Biochem. J.*, **48**, 548.
Robbins, E. A., Stulberg, M. P., and Boyer, P. D. (1955). *Arch. Biochem. Biophys.*, **54**, 215.
Robbins, S. L. (1950). *Amer. J. med. Sci.*, **219**, 376.
Robertson, W. van B., Dunihue, F. W., and Novikoff, A. B. (1950). *Brit. J. exp. Path.*, **31**, 545.
Rogers, L. B., and Reynolds, C. A. (1949). *J. Amer. chem. Soc.*, **71**, 2081.
Rosenthal, A. S., Moses, H. L., Beaver, D. L., and Schuffman, S. S. (1966). *J. Histochem. Cytochem.*, **14**, 698.
Rutenburg, A. M., Barrnett, R. J., and Tsou, K-C., Monis, M., and Teague, R. (1958). *J. Histochem. Cytochem.*, **6**, 90.
Ruyter, J. H. C., and Neumann, H. (1949). *Biochim. Biophys. Acta*, **3**, 125.
Ruyter, J. H. C. (1952). *Acta anat.*, **16**, 209.
Sachs, H. W., and Dulskas, A. (1956). *Virchows Arch.*, **329**, 466.
Sadasivan, V. (1950). *Arch. Biochem. Biophys.*, **28**, 100.
Sadasivan, V. (1952). *Nature, Lond.*, **169**, 418.
Schajowicz, F., and Cabrini, R. L. (1954). *J. Bone Jt. Surg.*, **36B**, 474.
Schmidt, G., and Laskowski, M. (1961). In "The Enzymes," 2nd Ed. Eds. Boyer, P. D. Lardy, H., and Myrbäck, K. Acad. Press, New York, p. 3.
Scott, T. G. (1963). *Nature*, **200**, 793.
Scott, T. G. (1964). *J. comp. Neurol.*, **122**, 1.

SCOTT, T. G. (1965). *J. Histochem. Cytochem.*, **13**, 657.
SCHRAMM, G., and ARMBRUSTER, D. (1954). *Z. Naturf.*, **96**, 114.
SELIGMAN, A. M., HEYMANN, H., and BARRNETT, R. J. (1954). *J. Histochem. Cytochem.*, **2**, 441.
SHANTA, T. R., WOODS, W. D., WAITZMAN, M. B., and BOURNE, G. H. (1966). *Histochemie*, **7**, 177.
SHANTHA, T. R., and BOURNE, G. H. (1966). *Ann. Histochim.*, **11**, 337.
SHANTHAVEERAPPA, T. R., and BOURNE, G. H. (1965). *Act. Histochem.*, **22**, 155.
SHIMIZU, N. (1950). *J. comp. Neurol.*, **93**, 201.
SHUGAR, D., SZENBERG, A., and SIERAKOWSKA, H. (1957). *Exp. Cell Res.*, **13**, 424.
SHUGAR, D., SIERAKOWSKA, H., and SZENBERG, A. (1958). *Acta Biochim. Polon.*, **5**, 27.
SHULL, K. H., ASHMORE, J., and MAYER, J. (1956). *Arch. Biochem. Biophys.*, **62**, 210.
SIERAKOWSKA, H., SZEMPLINSKA, H., and SHUGAR, D. (1963). *Acta biochim. polon.*, **10**, 399.
SIERAKOWSKA, H., ZAN-KOWALCZEWSKA, M., and SHUGAR, D. (1965). *Biochem-Biophys. Res. Comm.*, **19**, 138.
SIMON, H., ARNOLD, F., and FINDFLEISCH, B. (1966). *Acta Histochem.*, **23**, 322.
SINDEN, J. A., and SCHARRER, E. (1949). *Proc. Soc. exp. Biol., N.Y.*, **72**, 60.
SKJERVEN, O. (1956). *Fertility and Sterility*, **7**, 31.
SKOU, J. C. (1957). *Biochim. Biophys. Acta*, **22**, 394.
SONNENSCHEIN, N., and KOPAC, M. J. (1953). *Anat. Rec.*, **117**, 611.
SPIER, H. W., and MARTIN, K. (1956). *Arch. klin. exp. Dermatol.*, **202**, 120.
STETTEN, M. R., and TAFT, H. L. (1964). *J. biol. Chem.*, **239**, 4041.
STRAUSS, G., and STARK, G. (1960). *Histochemie*, **2**, 87.
STUTTE, H. J. (1967). *Histochemie*, **8**, 327.
STORTI, E. (1951). *C.R. IIIieme Cong. Soc. int. Europ. Hematol., Rome.*
SULKIN, W. M., and GARDNER, J. H. (1948). *Anat. Rec.*, **100**, 143.
SWANSON, M. A. (1950). *J. biol. Chem.*, **184**, 647.
TAKAMATSU, H. (1939). *Trans. Soc. path., Japan*, **29**, 429.
TAKEUCHI, T., and NOGAMI, S. (1954). *Acta path., Japan.*, **4**, 277.
TAPIA FRESES, A., GARCIA LLAQUE, J., YEN MU, F., and CASTILLO, Y. (1957). *Boll. soc. quim., Peru*, **23**, 89.
TERNER, J. Y., GOODMAN, R. M., and SPIRO, D. (1965). *J. Histochem. Cytochem.*, **13**, 168.
TETAS, M., and LOWENSTEIN, J. M. (1963). *Biochemistry*, **2**, 350.
THIERY, M., and WILLIGHAGEN, R. G. J. (1963). *Nature*, **197**, 1312.
THOMAS, M., and ALDRIDGE, W. N. (1966). *Biochem. J.*, **98**, 94.
THYBUSCH, D., BROSOWSKI, K. H., and WOOHSMANN, H. (1966). *Acta Histochem.*, **23**, 127.
TOLONE, S., and VENTRA, D. (1950). *Boll. Soc. ital. sper.*, **26**, 1403.
TORMEY, J. McD. (1966). *Nature*, **210**, 820.
TRUBOWITZ, S., FELDMAN, D., BENANTE, C., KIRMAN, D. (1957). *Prod. Soc. exp. Biol., N.Y.*, **95**, 135.
TUNG, TA-CH, LING, K-H., BYRNE, W. L., and LARDY, H. A. (1954). *Biochim. Biophys. Acta*, **14**, 488.
VÁCZY, L., SANDOR, T., and JUHOS, D. (1955). *Acta endocrinol.*, **18**, 87.
VALENTINE, W. N., FOLLETTE, J. H., SOLOMON, D. H., and REYNOLDS, J. (1957). *J. lab. clin. Med.*, **49**, 723.
VERNE, J., and HÉRBERT, S. (1955). *Ann. d'Endocrinol.*, **16**, 279.
WACHSTEIN, M. (1946). *J. lab. clin. Med.*, **31**, 1.
WACHSTEIN, M. (1955). *J. Histochem. Cytochem.*, **3**, 246.
WACHSTEIN, M., and MEISEL, E. (1956). *Ibid.*, **4**, 592.
WACHSTEIN, M., and MEISEL, E. (1957a). *Amer. J. clin. Path.*, **27**, 13.
WACHSTEIN, M., and MEISEL, E. (1957b). *J. Histochem. Cytochem.*, **5**, 204.
WACHSTEIN, M., and ZAK, F. G. (1946a). *Proc. Soc. exp. Biol., N.Y.*, **62**, 73.
WACHSTEIN, M., and ZAK, F. G. (1946b). *Arch. Path.*, **42**, 501.
WACHSTEIN, M., MEISEL, E., and NIEDWIEDZ, A. (1960). *J. Histochem. Cytochem.*, **8**, 387.

WACHSTEIN, M., BRADSHAW, M., and ORTIZ, J. M. (1962). *J. Histochem. Cytochem.*, **10**, 65.
WAJCHENBERG, B. L., and HOXTER, G. (1955). *Ibid.*, **60**, 669.
WATANABE, K., and FISHMAN, W. H. (1964). *J. Histochem. Cytochem.*, **12**, 252.
WEBER, G., ALLARD, C., DE LAMIBBANDE, G., and CANTERO, A. (1956). *Endocrinol.*, **58**, 40.
WEDEN, H. (1930). *Arch. exp. Path. Pharmakol.*, **150**, 332.
WETZEL, B. K., HORN, R. G., and SPICER, S. S. (1963). *J. Histochem. Cytochem.*, **11**, 812.
WHEELER, K. P., and WHITTAM, R. (1962). *Biochem. J.*, **75**, 495.
WILLIGHAGEN, R. G. J. (1962). *Nature*, **194**, 691.
WILMER, H. A. (1943). *J. exp. Med.*, **78**, 225.
WILMER, H. A. (1944). *Arch. Path.*, **37**, 327.
WILTSHAW, E., and MOLONEY, W. C. (1955). *Blood*, **10**, 1120.
WISLOCKI, G. B., and DEMPSEY, E. W. (1945). *Amer. J. Anat.*, **77**, 365.
WULFF, H. R. (1967). *Med. dansk. (résumé) Munksguard, Copenhagen.*
YOKOYAMA, H. O., STOWELL, R. E., and TSUBOI, K. K. (1950). *Proc. Histochem. Soc., J. nat. Cancer Inst.*, **10**, 1367.
YOKOYAMA, H. O., STOWELL, R. E., and MATHEWS, R. M. (1951). *Anat. Rec.*, **109**, 139.
ZAN-KOWALCZEWSKA, M., SIERAKOWSKA, H., and SHUGAR, D. (1966). *Acta biochim. polon.*, **13**, 237.
ZORZOLI, A. (1948). *Anat. Rec.*, **102**, 445.
ZORZOLI, A., and NADEL, E. M. (1953). *J. Histochem. Cytochem.*, **1**, 362.
ZORZOLI, A., and STOWELL, R. E. (1947). *Ibid.*, **97**, 495.

CHAPTER 16

ACID PHOSPHATASES

UNTIL 1960 histochemical methods for acid phosphatases were all more or less unsatisfactory and the number of references in the literature to the use of such methods was far smaller than the number of references to alkaline phosphatases. Histochemists and histologists tended to overlook the importance of acid phosphatase in concentrating their efforts on the alkaline variety. With the advent of improved histochemical methods the position of acid phosphatase was radically altered and a glance through current bibliography shows that the gap is almost closed.

The acid phosphatases are widely distributed in animal tissues, prostate, spleen and liver being three of the richest sources. They are also found in protozoa, molds, yeasts, and in higher plants. Their presence in yeasts was noted as long ago as 1935 (Schaeffer and Bauer) and histochemical investigations have been carried out on a number of tissues and cells belonging to the above categories. Powdery mildew, for example, was shown by Atkinson and Shaw (1955) to contain a strong acid phosphatase and the protozoon *Tetrahymena pyriformis* was investigated by Klamer and Fennell (1963). In the case of higher plants the enzyme was demonstrated in the root tips of *Vicia faba* by Beneš *et al.* (1961) and later by Gahan (1965). It was shown in the meristematic tissues of *Triticum vulgare* by Poux (1963) and Beneš and Opatrná (1964) found strong activity in similar tissues from *Ricinus*, *Lupinus*, *Sinapis*, *Allium* and *Zea mays*, so that it is obviously very widely distributed (see also Avers, 1961; and Harrington and Altschul, 1963). There are a number of different acid phosphatases of animal origin besides prostatic and red cell, the two main types with which the clinical pathologist is concerned. Prostatic acid phosphatase was first described by Kutscher and Wolberg (1935) and its possible functions were extensively studied by the Gutmans (Gutman and Gutman, 1938, 1940, 1941), and Herbert (1944, 1946) first showed that it could be differentiated from the normal serum acid phosphatase on account of its lability to alcohol. The distinctive red-cell acid phosphatase was described by King, Wood and Delory (1945), and Abu'l Fadl and King (1948a and b, 1949) found that it was totally inactivated by 0·5 per cent. formaldehyde, which did not affect the prostatic enzyme at all.

These observations have been the basis for most of the clinical biochemistry of the serum acid phosphatases, but histochemically their implications have not been adequately considered. The work of Folley and Kay (1936) indicated that there were three types of acid phosphatase which could be differentiated by their pH optima, their sensitivity to Mg^{2+} ions and their relative activity towards α- and β-glycerophosphates. Further studies by Roche (1950) definitely suggested that more than one acid phosphatase was present in the liver, and later work by Goodlad and Mills (1957) produced further evidence that this was the case. Table 38, which is derived mainly from the work of the last-mentioned authors, illustrates the effect of various inhibitors on acid phosphatase from four different sources.

TABLE 38

Percentage Inhibition of Acid Phosphatases at pH 5·0–5·5

Inhibitor	Liver*	Prostate†	Erythrocytes†	Adrenal cortex‡
0·01 M-Mg^{2+}	Nil	Slight	Slight	Nil
0·0002 M-Cu^{2+}	−85	Slight	−95	−86
0·01 M-cyanide	Nil	+12	+8	—
0·01 M-DL-tartrate	−41	−95	Nil	−91
0·01 M-fluoride	−53	−96	−8	—
0·5 per cent. H.CHO	−68	Nil	−100	−30

* Goodland and Mills (1957) 0·005 M-phenylphosphate, acetate buffer.
† Abu'l Fadl and King (1949) 0·005 M-phenylphosphate, acetate buffer.
‡ Gordon (1952) 0·01M-β-glycerophosphate, acetate buffer.
Negative values indicate inhibition; positive values activation.

It must be emphasized that species differences are apparent, since Abu'l Fadl and King found almost complete inhibition of human liver acid phosphatase by 0·01 M-tartrate between pH 4 and 6, whereas in the rat only 40 per cent. inhibition occurs.

It is clear from the findings of Goodlad and Mills that rat liver contains certainly two, and possibly three, acid phosphatases. One, having a pH optimum between 3·5 and 4·0, is activated by veronal and inhibited by citrate and tartrate as well as by fluoride. The other, having a pH optimum between 5·0 and 5·5, is inhibited by veronal and fluoride but not by citrate or tartrate. The third enzyme has a pH optimum of 6·0–7·0 with glycerophosphate as substrate. Figs. 145 and 146 show the effect of the various specific inhibitors on the pH activity curves of acid phosphatase in rat-liver homogenates.

It will be observed that the curve for fluoride inhibition, in the lower graph, shows evidence of two maxima, one at pH 4·0 and the other at pH 5·5. Observations on the inhibition of particularly the prostatic acid phosphatase by α-hydroxy carboxylic acids have been made by Anagnostopoulos (1953) and by Kilsheimer and Axelrod (1957).

More modern studies continue to support the idea that a number of separate acid phosphatases exist in single situations. Differences in the rate of acid phosphatase activity in *Tetrahymena pyriformis* towards several substrates (Connor and MacDonald, 1964; Elliot and Bak, 1964) have been attributed to multiple enzyme forms. In the case of rat liver acid phosphatase Arsenis and

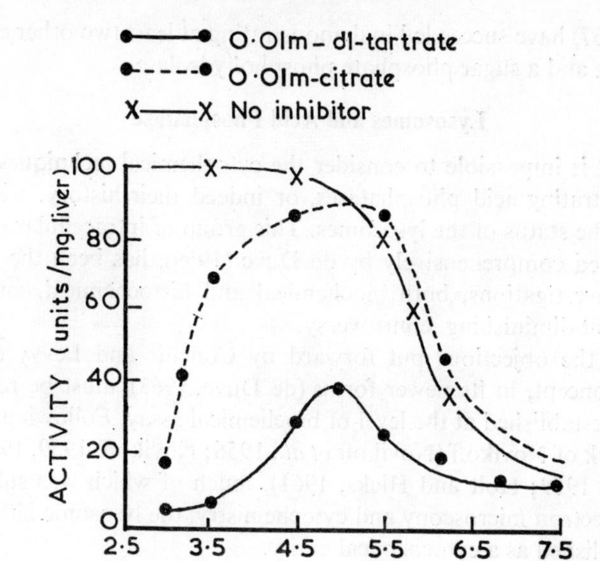

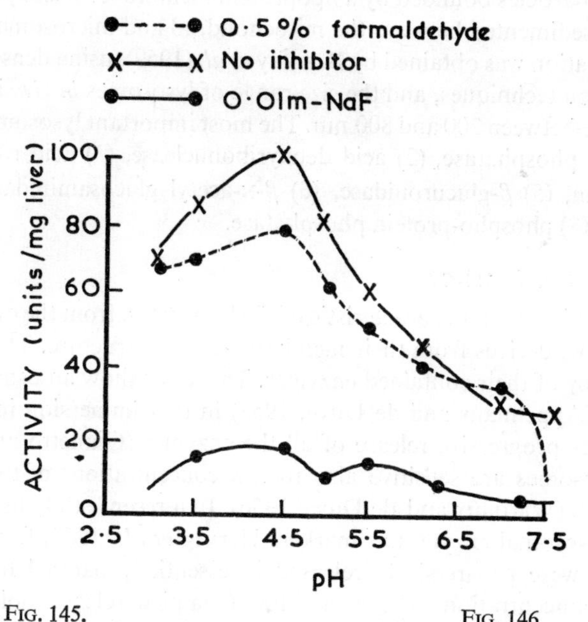

FIG. 145. FIG. 146

Effect of Inhibitors on the pH-activity curve of acid phosphatase. (After Goodlad Mills.)

(0·005 M-phenylphosphate; veronal-acetate buffer; 38°.)

Touster (1967) have succeeded in demonstrating at least two other enzymes, a nucleotidase and a sugar phosphate phosphohydrolase.

Lysosomes and Acid Phosphatase

Today it is impossible to consider the cytochemical techniques available for demonstrating acid phosphatases, or indeed their history, without first examining the status of the lysosomes. This group of intracellular organelles, first described comprehensively by de Duve (1959), has been the subject of extensive investigations, both biochemical and histochemical, and of considerable, but diminishing, controversy.

Despite the objections put forward by Conchie and Levvy (1963) the lysosome concept, in its newer forms (de Duve, 1963), must be regarded as completely established at the level of biochemical assay. Following the initial pioneer work of Novikoff (Novikoff *et al.*, 1956; Novikoff, 1959, 1961) and of Holt (Holt, 1959; Holt and Hicks, 1961), much of which was substantially based on electron microscopy and cytochemistry, the lysosome has now been firmly established as a cytochemical entity.

More or less standard techniques of differential centrifugation were employed by de Duve and his colleagues (de Duve *et al.*, 1955) to determine the intracellular localization of various enzymes. It was observed that a group of acid hydrolases (now numbering 8 or more) apparently existed in the cell in the form of particles bounded by a lipoprotein membrane. These particles, the lysosomes, sedimented between the mitochondrial and microsomal fractions. Better separation was obtained by Beaufay *et al.* (1959), using density gradient centrifugation techniques, and the size range of lysosomes *in vivo* is now considered to be between 200 and 800 nm. The most important lysosomal enzymes are (1) acid phosphatase, (2) acid deoxyribonuclease, (3) acid ribonuclease, (4) cathepsin, (5) β-glucuronidase, (6) β-N-acetyl glucosaminidase, (7) aryl sulphatase, (8) phospho-protein phosphatase.

The Concept of Latency

The most important characteristic of the lysosomes, from the cytochemical point of view, derives from their membrane-bound structure. This is the so-called latency of their contained enzymes. The latter show an osmotic activation curve (Appelmans and de Duve, 1955) in that immersion in hypotonic media causes progressive release of all the enzymes (Gianetto and de Duve 1955). Lysosomes are sensitive also to low concentrations of the detergent Triton X 100 (Wattiaux and de Duve, 1956). In a recent study using 4 of the principal lysosomal enzymes as markers Herveg *et al.* (1966) found that the 4 activities were progressively released in essentially parallel fashion with increasing concentrations of Triton X 100. Complete release took place at a concentration of about 0·05 per cent. No sign of inhibition was observed, up to 0·14 per cent. of the detergent. Unmasking of latent lysosomal enzyme activity can also be produced by freezing and thawing, provided that freezing

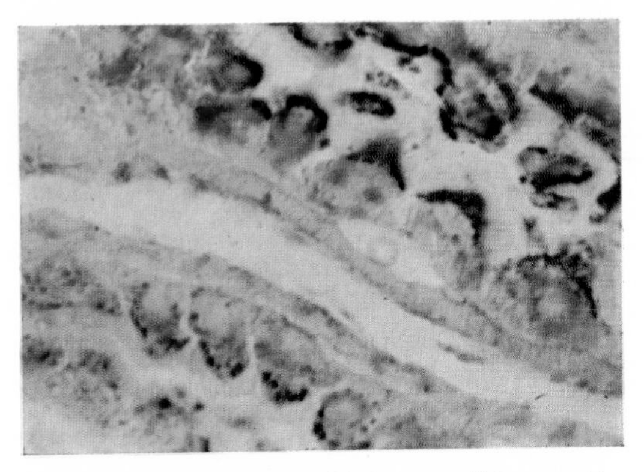

XIIA. Rat Kidney. Animal injected with horse-radish peroxidase 15 minutes before death. Shows blue stained phagosomes close to the lumen of the proximal convoluted tubules. Red stained lysosomes at deeper levels in the cell. Combined Straus technique. × 660.

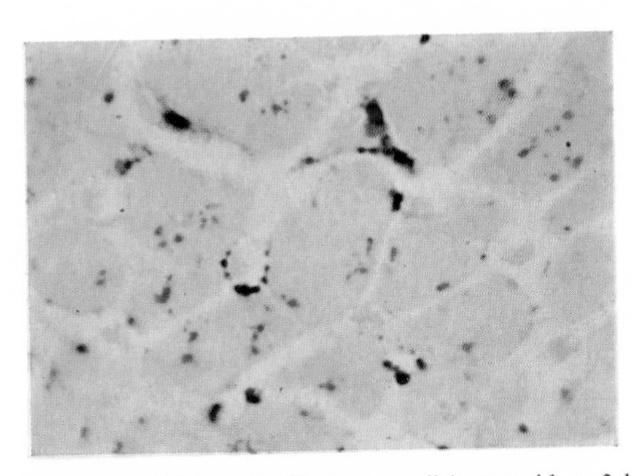

XIIB. Rat Liver. Animal injected with horse-radish peroxidase 3 hours before death. Shows purple-stained phagosomes in a Kupfer cell. Otherwise, blue phagosomes at the sinusoidal borders and red lysosomes within the cells. Combined Straus technique. × 660.

takes place at a sufficiently low temperature (Bendall and de Duve, 1960). Herveg *et al.*, for instance, found that 2 cycles of freezing and thawing would produce activations of the same order as those produced by Triton X. Lysosomal enzymes are not always released at equivalent rates by all solubilizing procedures. Decreasing the pH was found by Herveg *et al.* to release much more β-glucuronidase than acid phosphatase from calf thyroid lysosomes.

The concept is thus of a lipoprotein-bound particle containing acid hydrolases which are released progressively by procedures which alter or damage the membrane. Damage which allows escape of enzyme may be presumed at the same time to allow increased access of substrate into the interior of the particle. Work carried out at the cytochemical level supports this view (Bitensky, 1962, 1963a and b; Bitensky and Gahan, 1962; Gahan, 1965; Gahan and Maple, 1966). A comprehensive review has been presented by Gahan (1967).

Additional cytochemical information has come from studies carried out by Wolman (1965) and by Wolman and Bubis (1966). In the first of these two papers the author investigated the effect of ions ($CaCl_2$ and NaCl), lipid solvents, formalin, and freezing and thawing on lysosomal acid phosphatase, demonstrated by both metal precipitation and azo-dye techniques. The protective effects of both formalin and of half molar $CaCl_2$ find an echo in their regular use as fixatives for acid phosphatase (Chapter 5, p. 96). Wolman concluded from his experimental observations that acid phosphatase exists in nature as a protein bound to a lipid, probably a phospholipid. Amplification of this view produced the hypothesis that lysosomes are derived from endoplasmic reticulum by disintegration into corpuscles having a hydrophobic (phospholipid) layer on the outside whereas disintegration with the hydrophilic (protein) layer on the outside produces microsomes. These are most interesting concepts of which much more will be heard in the future.

Other Acid Phosphatase-containing Particles

In a comprehensive review on "Biochemical and Staining Reactions of Cytoplasmic Constituents" Novikoff (1960) described or redefined a number of terms for lipid or lipoprotein-bound cytoplasmic particles. These included pinosomes (the pinocytosis vacuoles of Holter and Marshall, 1954, and Holter, 1959), cytolysosomes, dense bodies, ferrisomes and phagosomes. The diagram (Fig. 147) gives some details of the various particles and their relationships.

The term phagosome was first proposed by Straus (1958), following earlier papers (1956, 1957) in which he described a group of particles appearing in kidney and liver cells, and in macrophages, after injection of different foreign proteins. These particles were isolated from kidney cells and shown to contain high concentrations of the same acid hydrolases as found by de Duve in lysosomes.

Straus continued his important work on phagosomes with a further series of papers (1962a and b; 1964a, b, c and d; 1967a and b) and he established clearly that, provided care was taken to avoid a number of technical artifacts,

Cytoplasmic Lipoprotein-bound Particles

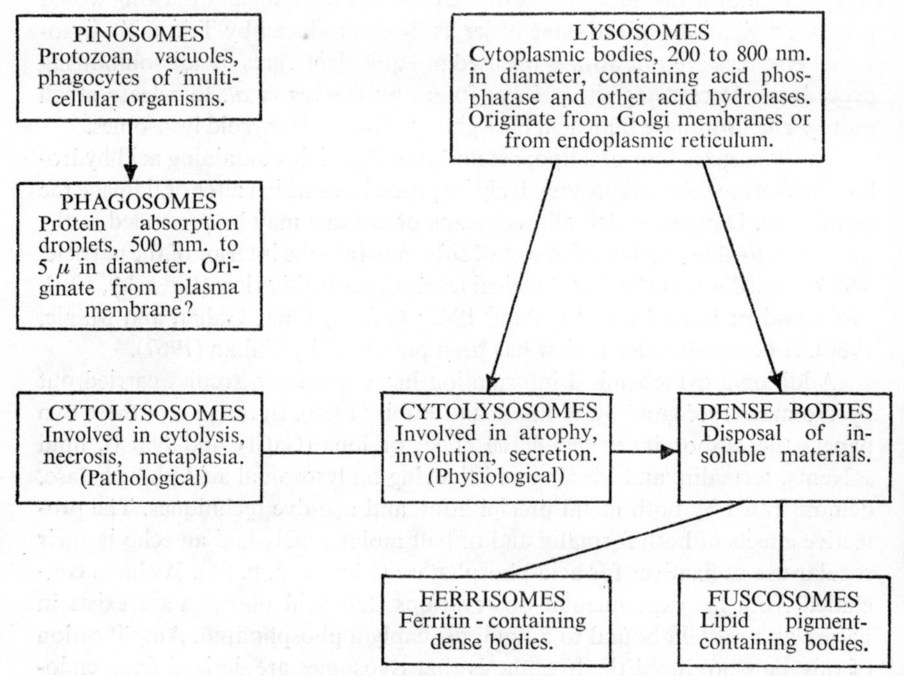

FIG. 147. Modified from Novikoff (1960).

the original benzidine-peroxidase method (Chapter 19, Vol. II) was capable of distinguishing phagosomes of large, and also of medium (800 to 400 nm), range. The smallest particles, which Straus equated with micropinocytic vesicles, were not clearly stained however. The combined technique for lysosomes (acid phosphatase) and phagosomes (benzidine-peroxidase), which was first described by Straus (1964d), can be used with confidence. The results of the combined method are illustrated in Plate XIII, a and b, facing p. 551.

Techniques for the Non-specific Acid Phosphomonoesterases (E.C. *3.1.33*)

METHODS DEPENDING ON DEPOSITION OF LEAD PHOSPHATE

The original technique for demonstrating acid phosphatases was introduced by Gomori (1941). It depended, as did his alkaline phosphatase method, on the liberation of phosphate ions from organic phosphate esters. The reaction is buffered at pH 4·7 to 5·0 and it is therefore impossible to use calcium ions to form a precipitate of calcium phosphate, since this product is completely soluble in the acid range. Lead nitrate is used instead of calcium nitrate in the incubating solution and its phosphate is insoluble at pH 5·0. The precipitated lead phosphate is converted by means of yellow ammonium sulphide into the brown sulphide of lead.

Wolf, Kabat and Newman (1943), in the course of a study on the distribution of acid phosphatases in which they used the Gomori technique with a variety of phosphate esters, noted that non-specific but localized deposits of lead salts occasionally occurred. These superficially resembled the accepted histochemical distribution of the enzyme, especially in spleen, testis and brain. Moog (1943) reported a similar capriciousness of the Gomori technique when applied to the spinal cords of chick embryos. Hard and Lassek (1946) and Lassek (1947) also criticized the technique. The latter described the persistence of staining in the axons of the brain stem and spinal cord of the cat, monkey and man, and in the sciatic nerve of the cat, after manœuvres which certainly destroyed all enzymes. He suggested that the appearances produced by the Gomori technique were artifacts due to non-specific deposition of lead.

Later studies by Newman, Kabat and Wolf (1950) threw some light on the subject. These authors re-examined the specificity of the method in view of the implication that methods using lead salts for the histochemical localization of phosphatases and other esterases demonstrated nothing other than the affinity of various tissue components for lead salts in the acid range. They compared the effect, at various pH levels, of incubation with buffered lead salts only, and of incubation with lead-glycerophosphate mixtures. Their results showed that non-specific staining by lead varied directly with the pH, occurring infrequently at pH 5 and below, increasing in intensity and regularity above pH 5·3 and decreasing again in intensity only between pH 6·0 and 6·8. At pH 5·6 to 6·8, the distribution pattern in some organs, notably kidney, intestine, testis, epididymis, heart and brain, resembled that described for alkaline phosphatase. In most tissues the lead effect was noted especially in the nuclei, in cuticular borders and at surface interfaces. It was not influenced by 0·01 M-sodium fluoride, which has been shown chemically and histochemically to inactivate acid phosphatases, so that any staining in the absence of substrate, or the persistence of staining in the presence of substrate and 0·01 M-sodium fluoride, was due solely to lead. The lead effect was also noted to be unaffected by formalin or by 95 per cent. alcohol, both of which were considered to inhibit acid phosphatases. Both the non-specific staining by lead and the true enzymic staining were prevented by heating the sections at 80° in distilled water for 10 minutes or by treatment with 5 per cent. trichloracetic acid for 10 minutes. Both these manœuvres might be expected to depolymerize nucleic acids in addition to destroying any enzyme present, thus reducing the number of acid groups available for combination with the metal.

Later still information concerning the mechanism by which lead becomes attached to tissue components came from the work of Clarkson and Kench (1958). These authors showed that when 5 μg. of lead (as $PbCl_2$) was added to 5 ml. of whole blood, over 95 per cent. rapidly became attached to the red cells. Because of the failure of a variety of potential competitors to interfere with this reaction Clarkson and Kench considered that the lead must be operating either as a metal complex or as a chelate compound or, perhaps, as a dispersed colloidal metal salt.

Millet and Jowett (1929) had showed that, in solutions having the same ionic strength and pH as plasma, lead orthophosphate was precipitated when the concentration of lead rose to about 0·0001 M. The concentration in the standard acid phosphatase medium is about 0·003 M. When, however, organic substances are present, as they are in plasma, a process of "peptization" (Lovelock, 1954) takes place and the lead phosphate sol is prevented from coagulating and forming a precipitate. Clarkson and Kench stated that metal-salt sols, except at their equivalence point, carried electrical charges and could be removed from suspension by non-specific absorbents. This process presumably occurs under histochemical conditions of incubation. A strong pH effect was noted. Uptake of lead by cells was markedly reduced when the pH was lowered from 6·6 to 6·0, in which range, according to the authors, the highly insoluble lead orthophosphate is transformed into the more soluble lead monohydrogen phosphate. This point is of obvious histochemical significance.

A further important point concerned the charge on the lead phosphate sol. If excess phosphate is present the sol bears a negative charge, but if the concentration of free phosphate ions is low, as it is under histochemical conditions, the net charge becomes positive. Thus Clarkson and Kench found that combination of lead sol with the negatively charged erythrocyte occurred more rapidly in saline than in plasma or other phosphate-containing solutions. On the above evidence one would expect histochemical non-specific absorption of lead to be confined to structures carrying a negative charge, and this is certainly observed in practice.

Newman *et al.* concluded that valuable information might still be derived from the Gomori procedure, carried out at pH 4·7 with glycerophosphate as substrate, provided that two control sections were used. It was suggested that control should be incubated with substrate to which 0·01 M-sodium fluoride has been added and the other in the buffer + lead nitrate only. The non-specific effects noted in these two sections are subtracted from the total picture to give the result due to enzymic action alone.

Gomori's Improved Technique. Gomori (1950), after examining the causes of failure of his original method, introduced an improved technique. Discussing the various factors which were responsible for such failures, he admitted that patchy fixation might account for certain differences in intensity between one part of the section and another, but agreed with Sulkin and Kuntz (1947) that this factor could not explain all the observed phenomena. Particularly difficult to explain on this basis were the failures with some but not all sections cut from the same block, and failures with a fresh batch of incubating fluid prepared in the same way as a previously successful batch. In his experimental investigations Gomori used serial sections of the same block of tissues and incubated these with glycerophosphate at pH 5·0, altering variously the concentrations of substrate, buffer and lead salt. The original substrate concentration of 0·005 M was found to be inadequate, but uniformly good results were given with sub-

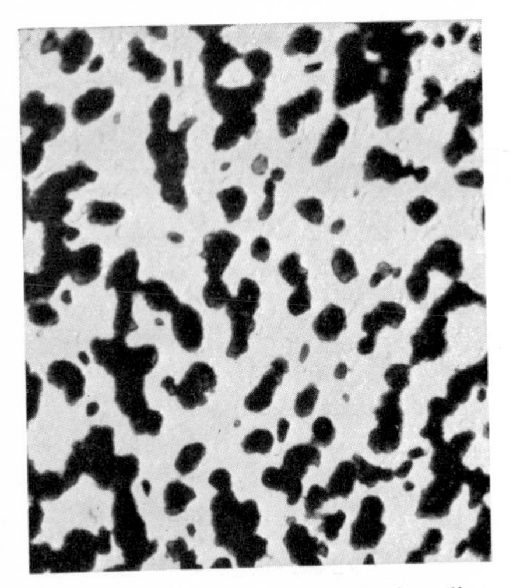

FIG. 148. Frozen section (10 μ) of rat liver after 16 hours' cold formalin fixation and 16 hours' incubation in Gomori's (1941) medium for acid phosphatase. The precipitate is chiefly present in the nuclei. × 220.

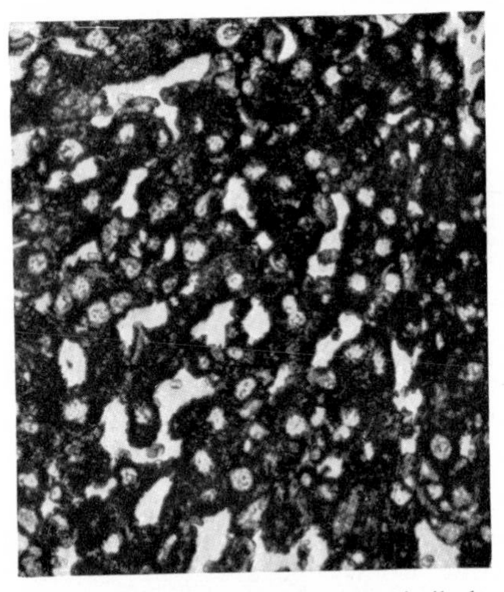

FIG. 149. As Fig. 148. Section similarly treated. The precipitate is largely absent from the nuclei. × 220.

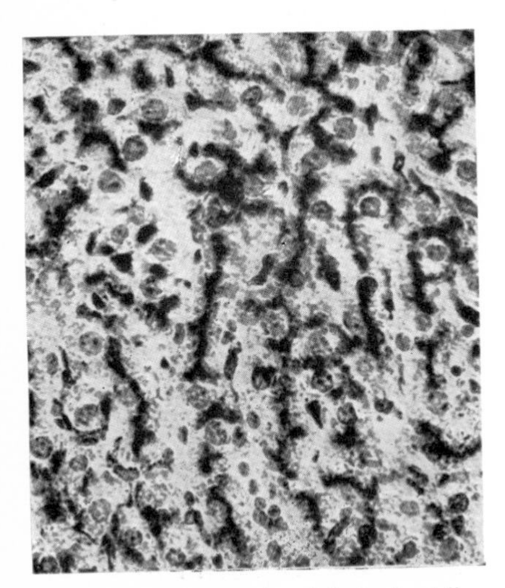

FIG. 150. Frozen section (10 μ) of rat liver after 16 hours' cold formalin fixation and 1 hour's incubation in Gomori's (1950) medium. Eosin. × 220.

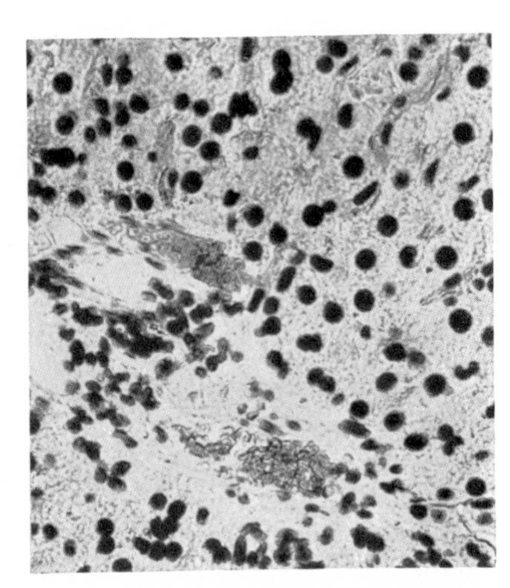

FIG. 151. Frozen section (10 μ) of rat liver after 40 hours' cold formalin fixation and 5 hours in Gomori's (1941) medium. Long fixation produces predominantly nuclear staining. Eosin. × 220.

strate concentrations between 0·014 and 0·05 M (buffer 0·08 M, lead nitrate 0·037 M in the final solution). On varying the buffer concentration Gomori found that this should not exceed that of the substrate by more than a factor of 6. A ratio of 5 or 6 to 1 was found to be optimal in the case of glycerophosphate. Increasing the concentration of lead increased the intensity and uniformity of the reaction only within narrow limits.

Gutman and Gutman (1938) showed that in the case of acid phosphatase the rate of hydrolysis was increased with rising substrate concentration up to 0·1 M, and Gomori emphasized the need to keep the substrate concentration high, to use substrates which were hydrolysed rapidly and to minimize the solubility of the raction product as far as possible. Since raising the concentration of lead depresses the concentration of substrate, a low concentration of lead was employed. Similarly, high concentrations of buffer increase the solubility of the reaction product and one should therefore use the minimal amount of buffer necessary to stabilize the pH. The final concentrations arrived at by Gomori are given in the description of his improved method in Appendix 16, p. 728.

Gomori also attempted to increase the efficiency of his method by employing substrates which were hydrolysed faster than glycerophosphate. Phenylphosphate could not be used since its lead salt is completely insoluble, but the monophosphates of catechol and resorcinol .were successfully employed as histochemical substrates. These offered no advantages over glycerophosphate, however, and their routine use was not advised. In spite of the observations of Moog (1943) that intensification of the acid phosphatase reaction could be obtained by means of manganese ions (0·01 M-manganese sulphate) or ascorbic acid (0·01 M), Gomori did not recommend the use of activators.

Other Modifications. Goetsch and Reynolds (1951) stated that two major factors radically affected the results observed when using Gomori's (1941) technique, at pH 4·3, on acetone-fixed paraffin sections. These were (1) the technique of mounting the section on the slide, and (2) the method of deparaffinization and hydration. They recommended that the temperature of the water bath used for flattening the sections should be no higher than 38°–40°, or just sufficient to melt the paraffin. Higher temperatures caused inactivation, beginning at the edges of the slide. In this context it is interesting to note that Woodard (1951) showed that at pH 8·0 or higher 1 hour's exposure to 37° was sufficient totally to inactivate prostatic acid phosphatase. Marked inactivation of the enzyme in the sections was said to be caused by the use of egg albumin as an adhesive and it was recommended that no adhesive be used. An improved version of this technique was described by Ruyter (1964), however, who found no evidence of enzyme inhibition by albumin. Using fixation in excess cold 80 per cent. acetone, and very careful embedding techniques it could be shown that the critical factor was the incubation of unmounted, undeparaffinized, sections floating on the medium. This technique has evidently a high degree of sensitivity, even with short incubation times. Details of the method are given in Appendix 16, p. 729.

Eränkö (1951b and c), observing that the results he obtained with the Gomori (1941) technique on acetone-fixed paraffin sections were entirely inadequate, used the same technique on frozen sections, cut preferably from fresh material or, failing this, from material fixed in 4 per cent. neutral formalin for 2–6 hours at room temperature or for 12–24 hours at 0°–3°. The results he obtained with these modifications were consistently reproducible and no patchy negative areas were seen.

An interesting suggestion was made by Tandler (1953), who replaced the lead ions of the usual media by 0·04 M-Co^{2+}. Unfortunately, this modification could not be employed at pH 5·0 since at this pH the precipitation of phosphate was incomplete. At pH 6·0 to 6·5, however, acetone-fixed tissues produced in 12 hours a deep blue colour at the sites of enzyme activity. This was converted into black CoS in the usual way. Investigations into the properties of other metallic ions having phosphates insoluble between pH 3·0 and pH 7·0 have so far produced no alternative phosphate precipitates for use in the histochemistry of acid phosphatase.

Another modification, which attracted less attention than it deserved, was that of Takeuchi and Tanoue (1951). This employed a somewhat higher pH than usual (up to pH 6·0 could be employed), and final development of the lead phosphate precipitate was achieved with an ammoniacal silver nitrate solution. Details appear in Appendix 16, p. 728.

Practical Considerations of the Method. In order to decide which, if any, of the reported modifications of the Gomori technique might be recommended for general use, Grogg and Pearse (1952), in a series of experiments, compared Gomori's 1941 with his 1950 technique, using (1) acetone-fixed paraffin sections, (2) cold formalin-fixed frozen sections, and (3) freeze-dried sections, and both Gomori techniques with the modification of Goetsch and Reynolds using paraffin sections only. In all cases control sections were incubated (*a*) without substrate, and (*b*) with substrate plus 0·01 M-sodium fluoride. Using acetone-fixed sections of rat liver, it was found, in agreement with Gomori (1950), that the ratio of buffer to substrate was the most important factor in obtaining uniform results. With a substrate concentration of 0·005 M either the sections were negative or precipitation of lead phosphate was patchy. Subsequent work (Holt, 1959) established the fact that the buffer to lead concentration ratio is even more important.

With formalin-fixed frozen sections of rat liver the 0·005 M substrate failed to give uniform results, and the precipitate appeared first in the nuclear membrane and nucleoli ($\frac{1}{2}$–2-hour incubation). After 4 hours' incubation the nuclei were completely black and a few fine cytoplasmic precipitates were visible, along the bile canaliculi. After prolonged incubation (16 hours) varying pictures were obtained. One section (Fig. 148, facing p. 555) showed coarse precipitation in the nuclei, but another identically-treated section (Fig. 149) showed staining of the cytoplasm only, the nuclei remaining colourless. Using the 0·01 M substrate concentration a completely different precipitation pattern was obtained. After short incubation, precipitates occurred along the bile canaliculi and the nuclei were scarcely stained escept in the Kupfer cells. After longer incubation (1–4 hours) the precipitates in the cytoplasm became coarser and the nuclei more definitely stained (Fig. 150). Storage of sections did not alter the distribution pattern of precipitate, but with prolonged formalin fixation (208 instead of the usual 16 hours) practically only the nuclei were stained. An essentially similar result to the last was obtained by using the Gomori (1941) medium on frozen sections fixed for 16–40 hours only (Fig. 151).

Using the Gomori (1950) technique on freeze-dried sections of rat liver a precipitation pattern closely resembling that seen in frozen sections was observed, but it was markedly less uniform. In some parts the precipitates were in the nuclei and in others in the cytoplasm along the bile capillaries and in the Kupfer cells.

THE ACID RINSE

This step was included in his instructions for the acid phosphatase technique by Gomori (1952) with the words "Rinse . . . for a minute or so in 1–2 per cent. acetic acid". The rinsing stage was used by Holt (1959), by Holt and Hicks (1961), and also by Rosenbaum and Rolon (1962). These last authors expressed the view that it made little or no difference to the final result. Ericsson and Trump (1964) used only a short acid wash, however.

It remained for Desmet (1962) to point out the dangers of acid differentiation when he showed that even total loss of stainable precipitate could result from brief acid washing. As indicated by Desmet, the step was omitted from the directions given in the 2nd Edition of this book, but without any explanation. (My reasons were indeed those which prompted Desmet to publish his results.) Subsequent authors have, for the most part, omitted the acid wash from their practice. Those who have done so include Bitensky (1963), Ruyter (1964), Barka and Anderson (1963), Goldfischer *et al.* (1964), Janigan (1965) and Lake (1966).

It may be concluded that the modified acid phosphatase procedure of Gomori (1950) affords very precise enzyme localization when applied to tissues which have received optimal fixation (usually cold formalin) and adequate protection (gum sucrose) before subsequent sectioning in the cryostat or on a freezing microtome. Typical results are shown in Figs. 152 and 153, facing p. 558.

Distribution of Acid Phosphatases by the Gomori Method. The classical papers dealing with the distribution of acid phosphatases in the tissues are those of Gomori (1941) and of Wolf, Kabat and Newman (1943). There are a few other papers dealing primarily with the application of the technique; with those that have already been quoted the following list is representative, if not comprehensive. *Brain and Neurones:* Bodian and Mellors (1944), Smith (1948),

Josephy (1949), Eränkö (1951a), Naidoo and Pratt (1951), Mottet and Barron (1953), Walter (1954–55). *Peripheral nerves:* Bartelmez and Bensley (1947). *Blood:* Rabinowitch (1949a and b), Woodward (1950), Weiss and Fawcett (1953). *Nuclear structures:* Wachstein (1945), Rabinowitch (1949c). *Adrenal cortex and medulla:* Soulairac, Desclaux and Tesseyre (1949a and b), Eränkö (1951b), Hillarp and Falck (1956). *Fibroblasts:* Noback and Paff (1951). *Prostate* (*Benign and Malignant*)*:* Downey *et al.* (1954), Brandes and Bourne (1954), Mathes and Norman (1956). It is notable that in a biochemical study of prostatic acid phosphatases Schwartz *et al.* (1953) showed that in the normal gland two peaks (pH 4·0 and 5·5) were present, while in the malignant gland only one peak (pH 5·5) could be found. *Skin:* Moretti and Mescon (1956), Spier and Martin (1956). *Metanephros:* Eränkö and Lehto (1954).

More modern studies of acid phosphatase, using the metal precipitation technique at optical microscope level, have largely been critical studies devoted primarily to fixation effects (Janigan, 1965) or to the behaviour of lysosomes (Bitensky, 1963; Maggi, Franks and Carbonell, 1966). Others have been purely concerned with technique (Lake, 1965) or with comparisons between metal precipitation and azo-dye techniques (Rosenbaum and Rolon, 1962). Practically no applied work on acid phosphatase has been done with the first alone whereas the azo-dye techniques are almost invariably so used.

Simultaneous Coupling Azo Dye Methods

The Seligman-Manheimer Technique. The principle of the Menten-Junge-Green coupling azo dye technique for the demonstration of alkaline phosphatase was adapted by Seligman and Manheimer (1949) to the histochemical demonstration of acid phosphatase. These authors found that α-naphthyl phosphate was hydrolysed more rapidly than the β-naphthyl ester at acid pH levels and this substrate was therefore employed, in the form of its calcium salt. Since it was found that alcohol had a greater effect than acetone in inactivating acid phosphatase, acetone-fixed paraffin-embedded sections were carried through xylene and acetone to water and incubated in 0·3 M-acetate buffer at pH 5·0 in the presence of substrate (approximately 0·0018 M), 4 M-NaCl, and anthraquinone-1-diazonium chloride. Although some text-books suggest the contrary, phenols can couple efficiently with diazonium salts in acid solution. With a number of diazonium salts, nevertheless, coupling with α-naphthol will not take place at pH 5·0, and if it does occur the resulting dye often differs from the one which is formed at an alkaline pH. After a search for a suitable diazonium salt Seligman and Manheimer selected anthraquinone-1-diazonium chloride, which they considered to be very stable in aqueous solutions, making possible the necessary prolonged incubations without renewing the medium. They also found that a temperature of 20° could conveniently be employed. The reaction, with calcium α-naphthyl phosphate and anthraquinone-1-diazonium chloride, follows the pattern of the coupling azo dye method for alkaline phosphatase, the equation for which is given on p. 512.

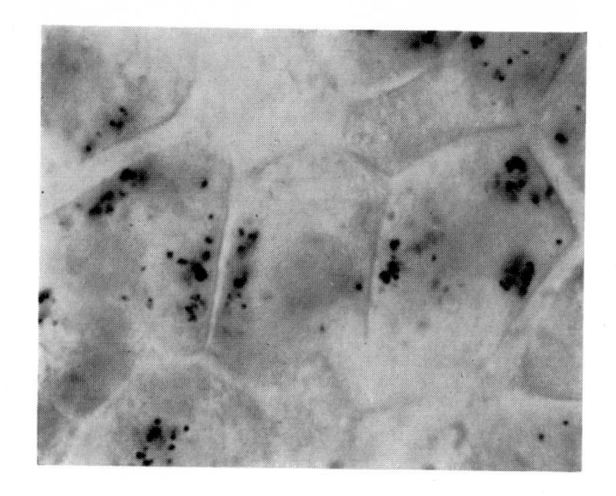

Fig. 152. Rat Liver, cold formalin-fixed (24 hours). Incubated 12 minutes in Gomori medium. ×1100.

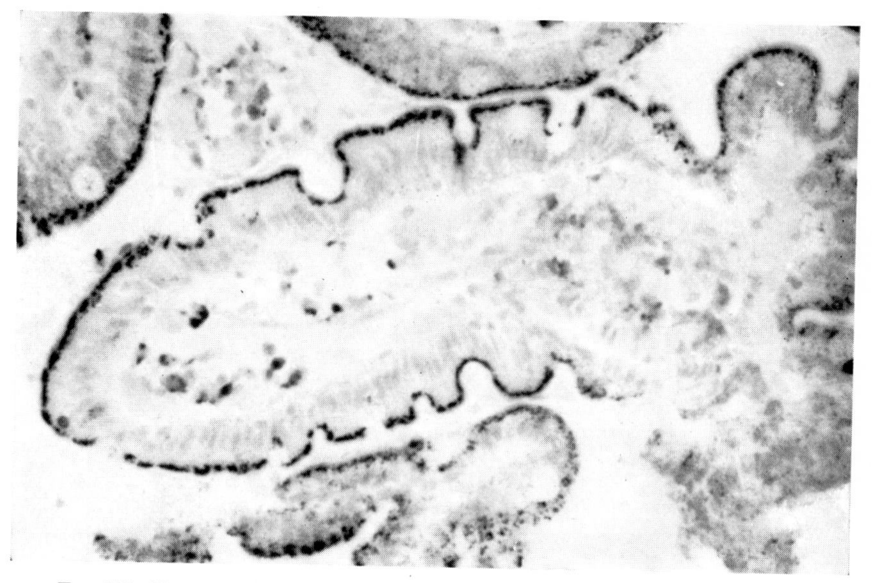

Fig. 153. Human Jejunal Biopsy, cold formol calcium-fixed. Incubated 10 minutes in Gomori medium. × 280.

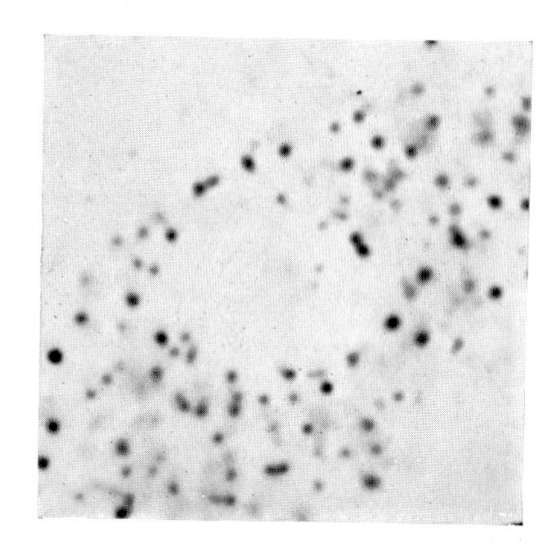

Fig. 154. Rat kidney. Proximal convoluted tubule. Osmotic protection of fresh frozen section with 7·5 per cent. PVP. Acid phosphatase is localized in round or ovoid bodies measuring 0·6–1·2 μ. Note absence of background staining. Standard coupling azo dye method. × 4200.

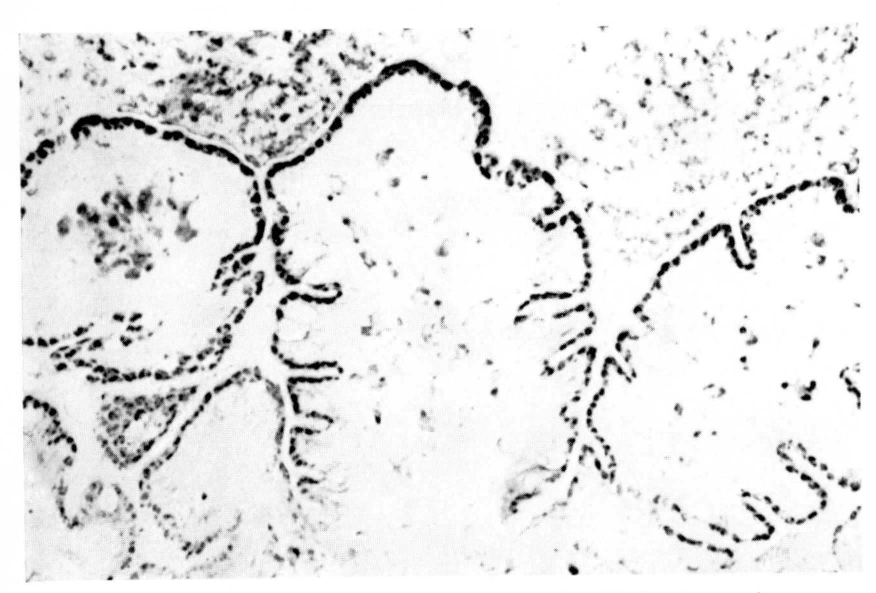

FIG. 155. Human Jejunal Biopsy, cold acetone fixed. Acid phosphatase demonstrated by use of Naphthol AS–BI phosphate and Hexazotized pararosanilin. × 280.

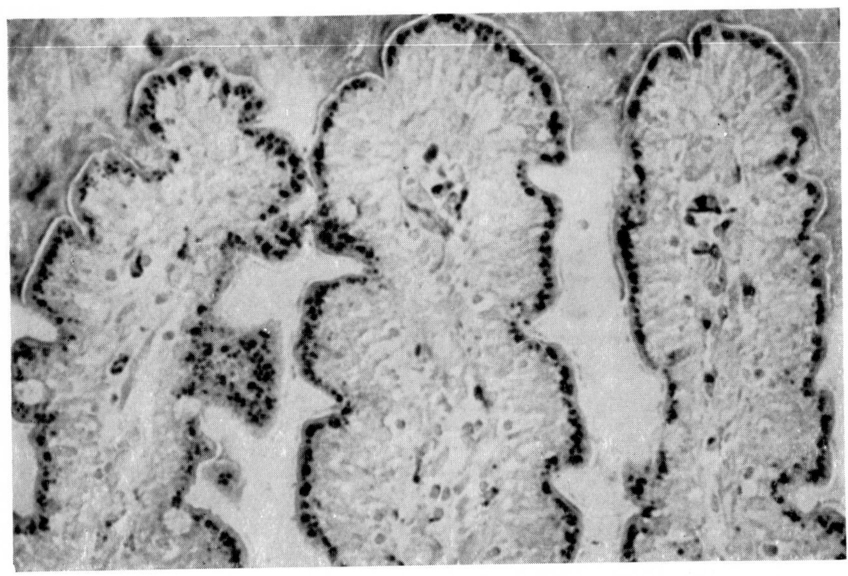

FIG. 156. Human Jejunal Biopsy, cold formol calcium-fixed. Treated as Fig. 155, for comparison. × 280.

Seligman and Manheimer found that for human and dog prostate the incubation period was as short as 1 hour, but for all other tissues it was in the region of 12–24 hours. With these longer periods they observed that the nuclei were invariably more densely stained than any other tissue components, but with the shorter incubation period I have observed that this staining does not always occur.

Using their azo dye method, Seligman and Manheimer found that the distribution of acid phosphatase in normal human tissues was the same as that reported by other observers using the Gomori method. On the basis of a selective inhibition by D-tartaric acid, following the observations of Abu'l Fadl and King (1948a and b), they described two varieties of acid phosphatase detectable histochemically. One, the type found in prostatic epithelium, was completely inhibited by 4 per cent. tartaric acid, while the other, present mainly in the nuclei and in skeletal muscle, retained its usual activity. The second variety was completely inhibited by 10 per cent. tartaric acid.

The best results which I was able to obtain (1953) with the Seligman-Manheimer method were far from satisfactory and they compared unfavourably with the results of the Gomori (1950) method. Three factors were responsible for these poor results. First, the extremely small amount of acid phosphatase remaining in paraffin sections made the long incubation period inevitable. This led not only to diffusion of enzyme and reaction product, and to non-specific staining of the background, but also to diffuse precipitation of some of the breakdown products in particulate form. The general effect, for this reason, was usually dirty. Secondly, the low solubility of calcium α-naphthyl phosphate made it impossible to obtain the substrate concentration necessary for maximum enzyme activity. This factor also prolonged the incubation period. Thirdly, anthraquinone-1-diazonium chloride, though strongly recommended by Seligman and Manheimer (1949), was found quite unsatisfactory as the coupling agent for this technique.

The necessary qualities of a diazonium salt for use in enzyme histochemistry of the type we are considering are: (1) that it should be sufficiently soluble at the pH employed; (*b*) that it should decompose slowly at this pH; (*c*) that it should not inhibit the enzyme; (*d*) that it should couple rapidly with the enzymically released α-naphthol or other phenol; (*e*) that the resulting azo dye should be absolutely insoluble in the incubating medium; (*f*) that it should be highly coloured; and (*g*) that the size of the dye particles should be small. In the case of paraffin sections, an additional desirable quality is that the dye should be insoluble in alcohol and xylene so that permanent preparations may be made. In the case of frozen sections this quality is less necessary, since it is not advisable to mount these in balsam or other permanent medium in view of disturbances caused by the necessary dehydration.

The Standard Monocoupling Technique (Simple Naphthols). Grogg and Pearse (1952), having found that only broad localization of acid phosphatase could be obtained by Gomori-type methods, modified the coupling azo dye technique in the light of the above observations. In order to overcome the first objection, cold formalin-fixed frozen sections were substituted for acetone-fixed paraffin sections. The second objection was removed by the substitution of the very soluble sodium α-naphthyl phosphate for the calcium salt, and there remained, therefore, only the selection of a suitable coupling agent for the

TABLE 39

Properties of Coupling Agents at pH 5·0
(*Frozen Sections*)
(*Coupler α-naphthol*)

	Diazonium salt of	Inhibition	Decomposition	Colour	Diffuse or particulate
1	1-amino-anthraquinone	++	++++	Brick-red	Diffuse
2	4-benzoylamino-2:5-dimethoxyaniline	++	+	Brick-red	Particulate +
3	*p*-nitroaniline	+++	+++	Orange	Particulate +
4	4-chloro-2-nitroaniline	+++	++	Orange-red	Particulate +
5	2:5-dichloroaniline	+++	+++	Orange-red	Particulate +
6	*o*-dianisidine	+	++	Dark red	Diffuse
7	4-chloro-*o*-anisidine	++++	+	Red	Particulate
8	5-nitro-*o*-anisidine	++	++	Brown-orange	Particulate +
9	5-chloro-*o*-toluidine	++	+	Dark orange-red	Particulate
10	4-nitro-*o*-anisidine	++	+	Orange-brown	Particulate +
11	3-nitro-*p*-toluidine	+++	+	Orange-red	Particulate +
12	3-nitro-*p*-anisidine	+++	++	Orange-red	Particulate +
13	4-amino-3:1′-dimethyl azobenzene	++	+	Red-brown	Particulate +
14	4-amino-2:5-dimethoxy-4-nitroazobenzene	+	++	Purplish-red	Particulate
15	4-amino-4′-methoxy diphenylamine	++++	+++	Nil	
16	3-nitro-*o*-toluidine	+++	+	Nil	
17	4-amino-4′-nitro-3:6-dimethoxy azobenzene	++	+++	Red	Particulate +
18	4-amino-3:1′-dimethyl-azobenzene	+	+	Red-brown	Particulate
19	2-methoxy-5-diethylamino-sulphaniline	++	++	Brick red	Particulate
20	2-benzoylamino-4-methoxy-toluidine	+++	+	Red	Particulate +
21	4-amino:-diphenylamine*	+	+	Red	Particulate
22	*p*-*p*′-diaminodiphenylamine	++	+	Brown	Particulate +
23	*p*-rosanilin	++	++	Red-brown	Particulate

In the above Table degrees of inhibition are registered from + to ++++, the former indicating almost complete absence. Similarly, in the case of decomposition, one + means practically no visible staining of the background structures. The reference particulate means that the dye particles were large enough to be clearly visible with the 2/3rds objective; diffuse means a particle size not appreciable at this magnification.

The addition of + means that the particles were larger than is desirable.

Numbers given are Laboratory Reference Numbers given in the Table of Diazotates in Appendix 15, p. 711. Numbers in bold type are those of salts providing most efficient coupling. Apparently identical salts have different stabilizers.

* As the diazonium sulphate (Dajac).

precipitation of α-naphthol at pH 5·0. With this object in view, the stable diazotates or bis-diazotates of most of the above compounds were tested in respect of the above criteria, using sections of human and dog prostate and mouse and rat liver.

With only six of the coupling agents listed was the final result of incubation at all satisfactory. These were salts 6, 7, 9, 14, 18 and 19. The best results were obtained with salt 18 (Fast Garnet GBC salt), which satisfied all the criteria listed above except the additional one of insolubility in alcohol and xylene. This failure was regarded as unimportant since the frozen sections, which were finally used as a routine, were always mounted in glycerine jelly. Sections prepared with salts 18 and 19 are fairly stable. No change in the distribution of the final azo dye product, or in the size of its particles, could be detected after 4 months' storage at room temperature. As first reported by Davis and Ornstein (1959), freshly hexazotized pararosanilin will couple extremely rapidly with 3 moles of α-naphthol to produce a reddish-brown azo-dye which is totally insoluble in aqueous and organic solvents. The introduction of hexazotized pararosanilin completely revolutionized the simultaneous coupling azo-dye techniques for acid phosphatase (see below, p. 565) and of many other hydrolases.

The information given in the second column of Table 39 is based on purely visual estimation of the intensity of staining in each case. More accurate information in respect of a number of diazonium salts, based on assays, was given by Lojda *et al.* (1964). This is shown in Table 40, below.

TABLE 40

Inhibitory Effect of Diazonium Salts on Acid Phosphatases

Diazonium Salt	Concn.	Unfixed		Formalin-fixed	
		Liver	Kidney	Liver	Kidney
Red Violet LB	0·6 mg/ml.	79·6	80	72·2	82·8
Fast Blue B	1 mg/ml.	83·2	88·5	65	69·3
Fast Red ITR	1 mg/ml.	82·4	88	70·9	60
Fast Red ITR	2 mg/ml.	69·9	72·5	52·7	53
Fast Garnet GBC	1 mg/ml.	84·3	82·5	74·9	71·7
Hexazotized PR	0·25 ml/10 ml.	77	80	84·1	82·6
Hexazotized PR	0·5 ml/10 ml.	59	73·5	68·2	75·5

Inhibition shown as percentage of control activity remaining after 80 mins incubation at 37° and pH 5·5.

Effect of pH on the Behaviour of Diazonium Salts. In the course of the above-mentioned investigations into the nature of the coupling azo dye techniques some interesting observations on the behaviour of diazonium and bis-diazonium salts at various pH levels were made. If diazotized *o*-dianisidine, at the optimum concentration, was coupled with enzymically (alkaline phosphatase) released α-naphthol at pH 9·2, the resulting dye was black and insoluble in alcohol and xylene as well as in water. If it was coupled with α-naphthol released enzymically (esterase) at pH 7·4, the dye was at first red and later black, as more was produced. It was only partially soluble in alcohol and xylene. When coupled with α-naphthol released by acid phosphatase at pH 5·0, as in the experiments here described, the dye was purplish-red and soluble in alcohol and xylene. These findings indicated that the mode of coupling in the case of this bis-diazonium salt depended, *inter alia*, on the pH. A possible explanation of the mechanism of the present example is that at higher pH levels

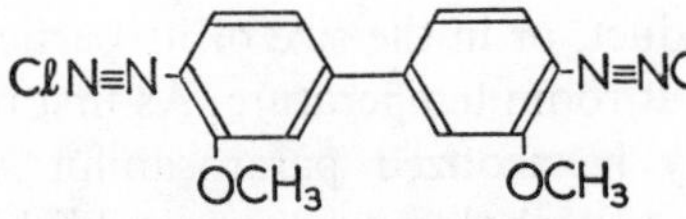

ClN≡N– –N≡NCl

OCH_3 OCH_3

Diazotized *o*-dianisidine.

both ends of the molecule are able to react, whereas at acid pH levels only one end is available for coupling. In this case the bis-diazonium salt acts like many mono-diazonium salts and produces a red dye on coupling with α-naphthol. These observations lead naturally to consideration of Burton's (1954) modification of the Seligman-Manheimer technique.

Pyridine or Quinoline-Catalyzed Coupling. When salt 6 was used in the standard type of coupling method for acid phosphatase, with high substrate concentrations and relatively long incubation periods, Burton found that the most active sites contained black pigment and the weaker sites the usual red product. When the concentration of the diazonium salt was below 1 mg./ml. the reaction was rapid and the final product always red. With high concentrations of salt 6 the reaction was appreciably slower but the product was predominantly black. Burton therefore introduced the use of catalyzers to increase the coupling rate (see Chapter 14, p. 481), and for this purpose he employed pyridine or 2-butoxy-N-(2-diethylaminoethyl) cinchonamide hydrochloride (percain). After incubation sections were treated with acid alcohol before dehydration with alcohol, clearing in xylene and mounting in a synthetic medium.

The results of the modification certainly appeared to be much sharper than those obtained by the usual technique with salt 6, in which diffusion was obvious after all but the briefest incubation periods. Unfortunately Burton's technique introduced an entirely new artifact in that the black final dye product was produced only in the most active regions. Elsewhere the dye retained its usual red colour. Furthermore the black dye, considered by Rutenburg and Seligman (1955) to be a polymerized product due to multiple coupling, was insoluble in alcohol and xylene. During the final processing in Burton's technique, therefore, differentiation was produced with the result that only areas of strong activity remained stained. The red dye in the weaker areas was entirely removed. For this reason the technique could not be recommended. Its interest is purely historical.

The standard simultaneous coupling azo dye method of the 1950s gave excellent results, particularly with salt 18 (Fast Garnet GBC) or salt 19 (Fast Red ITR). Application of the latter method to rat pancreas is shown in Plate XIVa, facing p. 564. At a later date the standard method was used with salt 23 (hexazotized pararosanilin) with even better results, in a variety of different tissue preparations. The latter included cold formol-calcium-fixed, frozen, free-floating sections, fresh unfixed cold microtome sections mounted on coverslips, and the same preparations post-fixed either in cold formalin or in cold acetic-ethanol. Incubation for 15–30 minutes was usually adequate and in the first two types of tissue preparation mentioned above the final dye product appeared to be localized in bodies resembling lysosomes (Fig. 154, facing p. 558). If protective media (containing 7·5 per cent. PVP) were employed (Scarpelli and Pearse, 1958) the reaction proceeded more slowly but the lysosomal type of localization was more easily obtained.

MONOCOUPLING TECHNIQUES (SUBSTITUTED NAPHTHOLS)

Five mono- and dihalogen-substituted naphthylphosphates (6-bromo-2-naphthyl; 1-chloro-2-naphthyl; 1,6-dibromo-2-naphthyl; 2,4-dichloro-1-naphthyl; 2,4-dibromo-1-naphthyl) were tested by Pearse (1954) as substrates in simultaneous coupling azo dye methods for both acid and alkaline phosphatases. The results obtained at pH 5·0 with 6-bromo-2-naphthyl phosphate were similar to those obtained with α-naphthyl phosphate, but with the two di-halogen-substituted α-naphthols the results were certainly superior. Although according to Fierz-David and Blangey (1949) diazonium coupling could not take place with α-naphthol derivatives substituted in the 2 and 4 positions it was observed to take place readily under histochemical conditions or in the presence of pyridine or alcohol. Presumably under suitable conditions the ejection of one of the halogen groups allowed coupling to take place.

Gomori (1956a) also tested 2,4-dichloro- and 2,4-dibromo-1-naphthyl phosphates as substrates for acid phosphatase and found that the results obtained were "essentially the same as, although æsthetically far inferior to, those obtained by the Rutenburg-Seligman method". This method is described below on p. 566. According to Gomori (1956b) the molecular extinction of azo dyes from the two halogenated α-naphthols was less than one-fifth that of the dyes from 6-bromo- or 6-benzoyl-2-naphthol. This was a powerful objection to their use in histochemistry in the absence of any supporting evidence on other counts.

The Naphthol AS Series of Substrates

An alternative approach to the problem was that of Burstone (1958), who carried out studies similar to those reported in the previous chapter (p. 517) using a variety of substituted Naphthol AS phosphates. In conjunction with these Burstone particularly recommended the use of freeze-dried sections although cold-acetone-fixed paraffin sections and cold formalin-fixed frozen sections were stated to give good results. As substrates the phosphates of Naphthol AS-BI, AS-TR and AS-MX were found to be stable at acid pH levels and readily hydrolyzed. Long incubation periods were perfectly feasible and recommended diazotates were salts 4 and 5 (pp. 711, 712) and Fast-red-violet LB.

The advantages of substituted naphthoic arylides as reaction products are due to three things. First, their insolubility in aqueous media; secondly, their substantivity for protein; and thirdly, the high molecular extinctions of azo dyes derived from them. From Table 41 it can be seen that the solubility in water of Naphthol AS and its derivatives is far less than that of any other substituted or unsubstituted naphthols.

TABLE 41

Solubilities of Naphthols (*after Gomori* (1956b) *and Burstone*)
(0·05 M-acetate; pH 5·0; 22°)

Naphthol	Solubility (micrograms/ml.)
2-Naphthol	740
1-Naphthol	303
6-Bromo-2-naphthol	195
6-Benzoyl-2-naphthol	9–27
2,4-Dichloro-1-naphthol	12
2,4-Dibromo-1-naphthol	3·5
Naphthol AS and derivatives	>1

The results of the application of Burstone's method, details of which are given on p. 731 (Appendix 16) were æsthetically satisfactory, especially in freeze-dried sections, as the many coloured illustrations in his (1962) book show. The colour and clarity of the final result varied with the method of fixation and with the choice of (stable) diazonium salt (Plate XIVb, opposite.

The first really satisfactory series of azo-dye techniques for acid phosphatase were produced by the combination of Naphthol AS phosphates as substrates with freshly hexazotized fuchsin (pararosanilin) as diazonium salt (Beneš *et al.*, 1961; Lojda, 1962; Barka, 1962; Barka and Anderson, 1962; Anderson and Song, 1962; Goldberg and Barka, 1962). Variations of the Naphthol AS-hexazotized pararosanilin (HPR) method have superseded all others for studies on the optical microscopic cytochemistry of the enzyme. Nevertheless, in view of continuing evidence (Meany, Gahan and Maggi, 1967) of the existence of more than one type of acid phosphatase, critical studies may well continue to be carried out using both the azo-dye and the metal precipitation technique.

A very large number of naphthoic arylides could probably be phosphoryl-

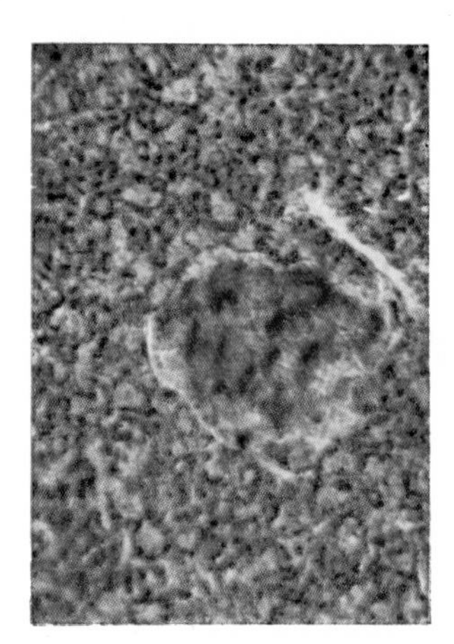

XIVA. Rat pancreas. Islet of Langerhans. Acid phosphatase in the beta cells, no staining of the peripherally situated alpha cells. α-Naphthyl phosphate method. × 60.

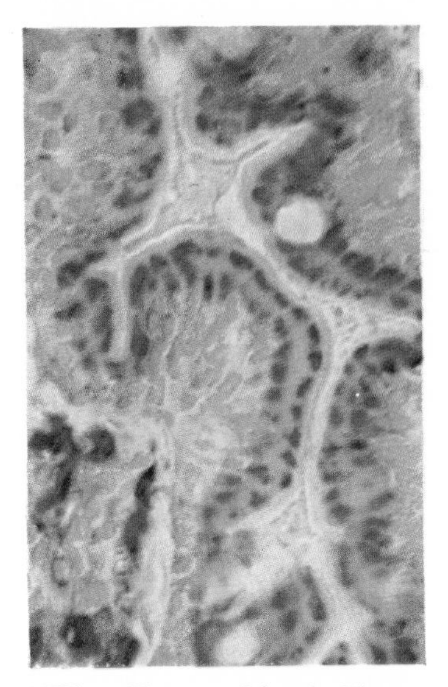

XIVB. Human Jejunal Biopsy. Case of idiopathic steatorrhoea 27 months after commencing treatment with gluten-free diet. Shows high acid phosphatase activity in the Golgi zone (reduplication of acid phosphatase zone) Naphthol AS-BI phosphate and hexazotized pararosaniline. × 240

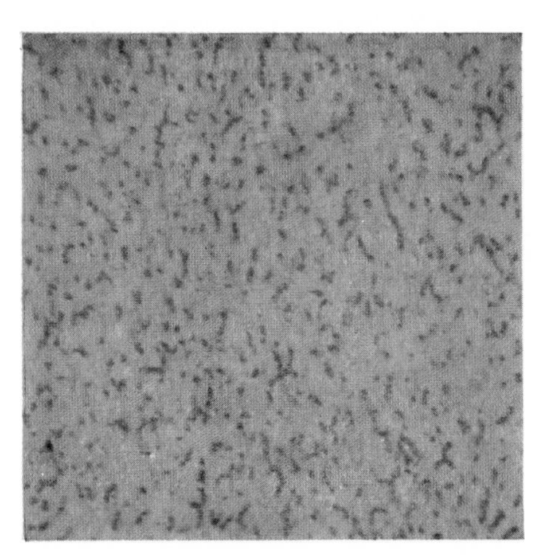

XIVC. Rat Liver, cold formol calcium-fixed. Burstone-type method for acid phosphatase using Naphthol AS-TR phosphate and Fast Dark blue R salt. × 250.

ated to give phosphatase substrates. At least 14 have been so phosphorylated and are theoretically available for histochemical use. Lojda *et al.* (1964) prepared a long series which included the phosphates of Naphthols AS, AS-D, AS-BO, AS-CL, AS-LC, AS-BI and AS-TR. Details of the preparation of AS-TR phosphate of suitable quality for histochemical use had been given by Jeffree and Taylor (1961), who observed that the method advocated by Burstone (1958) gave a product containing over 50 per cent. free naphthol. An alternative modification of Burstone's procedure was given by Chytil and Müller (1961).

As indicated by Lojda *et al.* (1964), the ease with which Naphthol AS phosphates can be prepared varies greatly with different naphthoic arylides. Their method for synthesizing AS-BO phosphate (one of the most difficult to prepare) is given in Appendix 16, Vol. 2, as are details of Jeffree and Taylor's method for preparing AS-TR phosphate.

Choice of Substrate

Commercially available Naphthol AS phosphates include those of AS-BI, AS-TR, AS-E, AS-GR, and AS-AN (see Source List, Vol. 2). Chromatographic studies carried out by Lojda *et al.* (1964) on a number of Naphthol AS phosphates showed that in every case their product was a complex mixture, varying according to the technique of preparation. Although it was possible to agree with Burstone's (1962) view that such crude products were useful histochemical substrates, they pointed out that the extraneous substances could and did affect the velocity of the histochemical reaction appreciably. If free naphthol is present it will combine with the diazonium salt in the incubation medium and filtration will effectively remove it. However, the concentration of diazonium salt will be lowered, in some cases below the level necessary for an effective capture reaction (Chapter 14, p. 481).

The chromogenicity of the various naphthoic arylides with HPR must obviously affect the choice of substrate. The deepest coloured azo-dye is given by naphthol AS-BI and, for most purposes, its phosphate is the substrate of choice for demonstration of acid phosphatase. Although Burstone (1962) recommended and used dimethylformamide (at low concentrations) in order to dissolve his AS substrates Lojda *et al.* (1967) failed to demonstrate any advantage at alkaline pH levels. They noted, however, that at acid pH levels only the sodium salts of Naphthol AS phosphates were soluble in the buffers tested. They therefore used 1 per cent. (final concentration) of dimethylformamide in order to dissolve the free acid substrates. In either case the substrates were soluble up to 4 mg./ml.

Concentration of Substrate

It was clearly shown by Lojda *et al.* (1964) that the substrate concentrations employed in "standard" incubating mixtures employing Naphthol AS esters (or glycosides) were too low. Raising the concentration from the average (0·1 mg./ml.) up to 0·5 mg./ml. resulted in a greatly increased reaction rate. This concentration, or at least nothing below 0·3 mg./ml., is therefore recommended. When the substrate concentration is raised to the levels suggested the "standard" concentration of stabilized diazonium salt (1 mg./ml.) is insufficient to provide an effective capture reaction and it must therefore be raised to 2 mg./ml. In the case of HPR the concentration originally used by Barka and Anderson (1962) is much higher than is required for efficient capture. Final concentrations of 0·05 to 0·15 ml./ml. were recommended by Lojda *et al.* (1967). The type of result which may be expected is shown in Plate XIVc, facing p. 564, and a comparison between cold formalin and cold acetone as fixatives is shown in Figs. 155 and 156.

Post-coupling Techniques for Acid Phosphatases

The demonstration of acid phosphatase by a post-incubation coupling technique using 6-bromo-2-naphthyl phosphate as substrate was proposed by Seligman *et al.* (1949). The method proved unsatisfactory, however, because of the long incubation period required and the solubility of 6-bromo-2-naphthol (see Table 41). My own observations on post-coupling techniques based on esters or glycosides of this naphthol (Pearse, 1954; Burton and Pearse, 1952) caused me to regard the post-coupling techniques in general as unreliable indicators of enzyme localization. Improvements in the post-coupling technique for acid phosphatase developed by Rutenburg and Seligman (1955) necessitated a new assessment of the situation.

The newer method was based on the use of sodium 6-benzoyl-2-naphthyl phosphate as substrate, and details of its preparation are given in Appendix 16. As described, the original incubating medium was made hypertonic with 2 per cent. NaCl with the idea of preventing diffusion of enzyme during incubation. The same result could perhaps better be achieved by the use of PVP (Scarpelli and Pearse, 1958) or methyl cellulose (Hannibal and Nachlas, 1959). Formalin-fixed free-floating sections were originally recommended, but fresh-frozen sections mounted on slides were also used. After relatively short incubation periods (½–2 hours) the sections were washed in cold saline and then treated with Fast blue B salt at pH 7 to 8. In active tissues the final product was a deep blue azo dye, but when dissolved in lipid tissue components this appeared purplish-red. Weaker areas of activity stained red in any case (mono-coupling). The tendency to spontaneous hydrolysis shown by 6-benzoyl-2-naphthyl phosphate was stressed by Lojda *et al.* (1964), who found complete decomposition after 2 hours. The incubation period was normally much shorter than this but the observation provides a further objection to the post-coupling technique, if one was needed.

In the 2nd Edition of this work (1960, p. 448) appeared the following commentary. "The view was expressed by Gomori (1956a) that the post-coupling technique was the best available at that time for the general demonstration of acid phosphatases. With the advent of simultaneous coupling methods using Naphthol AS phosphates this view may have to be modified. In making a choice between several available methods it is necessary here, as elsewhere, to ask oneself what it is that one requires." It is necessary today to conclude that post-coupling techniques for acid phosphatase have completely been supplanted by the Naphthol AS-HPR techniques.

Indigogenic Methods for Acid Phosphatase

Early attempts to use the above principle for demonstration of phosphatases were made by Seligman *et al.* (1954), using sodium indoxyl phosphate and by Holt (1958), using calcium-5-bromoindoxyl phosphate. The results achieved at both alkaline and acid pH levels, did not justify the use of either method. Later Evans *et al.* (1966) introduced an alternative substrate, disodium 5-iodoindoxyl phosphate, after noting that free 5-iodoindoxyl was oxidized to 5-5′-diiodoindigo much faster than was the 5-brom-substituted equivalent to 5-5′-dibromindigo. The ferri-ferrocyanide mixture described and used by Holt (1958) was employed as oxidation catalyst and an incubation period of only 2–5 minutes was required, with active tissues.

As reported by Evans *et al.*, their substrate appeared to be less than ideal since it provided at least two main products. The first was reddish-purple, and granular or microcrystalline and the second was blue-purple and similarly microcrystalline (Granules 0·5 to 1·5 μ). At present it must be concluded that indigogenic methods, using oxidation catalysts, do not provide localization comparable to that given by the best azo-dye techniques or, for that matter, by the Gomori method. The possibility remains that their combination with HPR, or with some other fast-coupling diazonium salts, would provide the basis for an efficient technique.

Techniques for Specific Acid Phosphomonoesterases

Lead Nitrate Technique for Phosphoamidase (E.C. 3.9.1.1)

In the course of experiments on the histochemical specificity of phosphatases, identical patterns of enzyme distribution were found with eighteen out of the nineteen substrates employed by Gomori (1949). The nineteenth substrate, described as *p*-chloranilidophosphonic acid, yielded a pattern identical with the others at pH 9 and 7 but gave very different pictures at pH 5. This finding strongly favoured the existence of a specific enzyme and a method for its demonstration was, therefore, set forth by Gomori (1948).

A specific enzyme acting on phosphocreatine was first described by Waldschmidt-Leitz and Köhler (1933) and by Ichihara (1933), and later by Bredereck and Geyer (1938). It was present in a wide variety of animal and plant tissues. Holter and Li (1950, 1951) showed that in brain extracts the enzyme had a pH optimum of 4·6 and that activity declined rapidly on either side of this point. They also showed (1950b) that *p*-chloranilidophosphonic acid monoamide was hydrolyzed by trypsin, pepsin and chymotrypsin. Several points of distinction between phosphoamidase and non-specific acid phosphatase were found and there seemed to be no doubt of the existence of the specific enzyme. Speculations about the possible functions of phosphoamidase in brain were made by Tseitlin (1952), who found a high specific activity using phospholysine as substrate and later (1953), using phosphocreatine.

It was shown by Rorig (1949) that the method used by Gomori for the synthesis of his substrate did not, in fact, yield *p*-chloranilidophosphonic acid but N-*p*-chlorophenyldiamidophosphonic acid.

The first of these two compounds can be synthesized, however, by a method given by Li (1950).

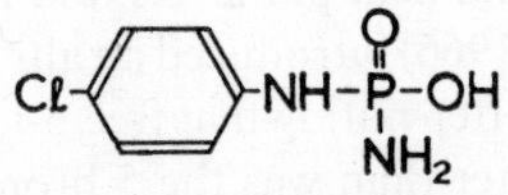

$$Cl-C_6H_4-NH-P(=O)(OH)-NH_2$$

p-chloranilidophosphonic acid.

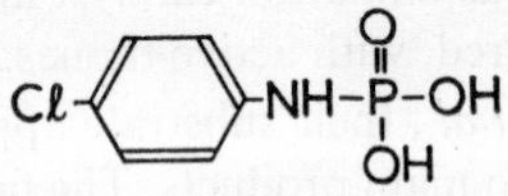

$$Cl-C_6H_4-NH-P(=O)(OH)-OH$$

N-*p*-chlorophenyldiamidophosphonic acid.

The above disclosure does not in any way invalidate Gomori's experimental observations, and these are described below. Acetone-fixed paraffin sections were used and these were incubated with 0·004 M substrate between pH 5·4 and 5·8, in the presence of 0·001 M-lead nitrate. A 0·05 M-maleate buffer was employed and manganese ions as activator. Above pH 6·2 the reaction became weaker and above pH 6·7 it vanished. Below pH 5·3 artifacts due to lead impregnation of tissues were numerous. They could be distinguished by examination of slides inactivated by treatment with Lugol's iodine for 5 minutes. Many other drawbacks were present; at pH 5·6, for instance, the substrate was not entirely stable and it was found necessary to incubate the sections face downward or facing downwards at least. After incubation a good deal of the superficial precipitate was removed by treatment with 0·1 M-citrate buffer at pH 4·5–5, followed by washing in water, and the sites of enzymatic activity were demonstrated by conversion of the lead salt to its sulphide by yellow ammonium sulphide.

Gomori did not find the method to be entirely dependable in that uniformity of results was not achieved. In two tissues particularly, however, a strong reaction invariably occurred. These were the grey matter of the central nervous system (especially the cerebellum) and malignant epithelial tumours. In the latter, the intracellular localization of the enzyme was in striking contrast to the extracellular situation of alkaline phosphatase in the majority of cases. I applied the phosphamidase technique (1953) to a variety of tissues fixed in

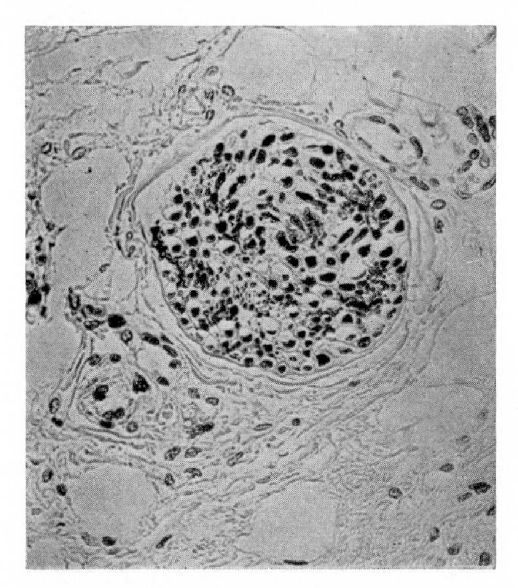

FIG. 157. Cold acetone-fixed paraffin section. Showing localization of phosphamidase, especially in the nuclei and axons of a small medullated nerve. Phosphamidase method. × 205.

cold acetone and embedded in paraffin, following the exact directions of Gomori. In the case of the majority of tissues a satisfactory result could not be obtained with less than 16–24 hours' incubation. By employing standard 16-hour cold-formalin-fixed frozen sections this time was reduced to 2–4 hours. Fig. 157 shows the result of applying the method to paraffin sections and it indicates the predominant localization of precipitate in the nuclei and in nervous tissues, in a medullated nerve fibre in this example.

The majority of workers who employed the original technique of Gomori obtained unsatisfactory results and it was usually criticized even by those (e.g. Meyer and Weinmann, 1953) who employed it with a certain degree of success. These authors used the original technique on cold ethanol-fixed paraffin sections as did Winter (1954–55) in his studies of the normal and neoplastic human cervix uteri. Eger and Schulte (1954–55), on the other hand, used fresh-frozen sections in their study of the enzyme in rat liver and kidney. The localization observed was in both cases different from that of acid or alkaline phosphatase and the authors concluded that they were dealing with a specific enzyme. There is no doubt that the localization in fresh tissues is sometimes different from that of acid phosphatase. The distinction between phosphoamidase and acid phosphomonoesterase activity still rests largely on localization differences, however, and these have often been invented where they do not in fact occur. For instance, Meyer and Weinmann (1957) gave a list of sites in the rat containing high phosphoamidase activity from which, according to their researches in the literature, acid phosphatase was absent. These included the granular cells of keratinizing epithelium, gastric parietal cells, secretory cells of the seminal vesicle, adrenal cortex and pancreatic islet (β-) cells. All these according to my (1958) observations possessed a strong or very strong acid phosphatase with almost maximal activity at pH 5·2 (and see Rutenburg and Seligman, 1955). While I was prepared to admit that the lead nitrate technique might show a separate and specific phosphoamidase, proof of this fact was not vouchsafed by inaccurate assertions like those reported above.

Notwithstanding these criticisms, it remains true that Meyer and Weinmann (1955) produced a distinct improvement on the original method of Gomori. I tested their method with fresh frozen and formalin-fixed frozen sections as well as with the conventional paraffin sections, and I concluded that the use of the latter was obligatory if worthwhile results were to be obtained. The modified method appears in Appendix 16, p. 734. Very few examples of its successful application can be quoted. It was used by Schajowicz and Cabrini (1964) on formalin-fixed bone tissues which were decalcified in citrate-formic acid mixtures or in 5 per cent. EDTA at pH 7·0. The method is, in my opinion, still too capricious for routine application to pathological and other problems. It might well be investigated again at the E.M. level, however, since there is ample precedent for the improvement of metal precipitation techniques in this way.

Coupling Azo Dye Methods for Phosphoamidase

Histochemical attempts to demonstrate the hydrolysis of phosphoamides by capture of the alcoholic reaction product have hitherto been delayed by lack of appropriate naphtholic substrates. Preliminary reports by Burstone (1958) suggested that the phosphoamides of substituted AS-naphthols which he synthesized might be useful substrates for a specific phosphoamidase. The two most promising compounds were the phosphoamides of Naphthol AS-TR (I) and AS-BI (II).

I: $OP(O)(NH_2)_2$; CONH ; Cl ; CH_3

II: $OP(O)(NH_2)_2$; CONH ; OCH_3

Details of the synthesis of Naphthol AS-phosphoamides were given by Burstone (1962). These substrates are used in the same way as similar substrates for acid phosphatase.

Lead-Nitrate Technique for Deoxyribonuclease II (E.C. 3.1.4.6)

This mainly mitochondrial enzyme breaks down desoxyribonucleate with the formation of polynucleotides (purine and pyrimidine-3′-nucleotides). The pH optimum of the enzyme is between 4·5 and 6, and Mg^{2+} ions do not, in this case, act as activators. An attempt to demonstrate the enzyme histochemically was made by Aronson *et al.* (1958), using a Gomori type lead nitrate procedure. In order to produce a reaction product capable of capture by Pb^{2+} ions the activity of other enzymes is considered to be necessary. The nucleotide reaction product of the first part of the reaction is presumed to be acted on by endogenous non-specific acid phosphatase and by 5-nucleotidase. The original authors added exogenous acid phosphatase to their medium but could not obtain any evidence that it was effective. Reliance on endogenous auxiliary enzymes obviously constituted a weakness of the method since the final localization of precipitate could at best only be that of the auxiliary enzyme.

Aronson and his co-workers use three control sections, one lacking substrate, one lacking exogenous acid phosphatase and the third containing 0·01 M-NaF. The method suffered from all the defects of the original acid phosphatase method and a few extra ones of its own. Modifications of the method were introduced by Vorbrodt (1961, 1962) which brought it within the realm of usable techniques. His most important modification was the employment of cold formol-calcium fixation combined with the use of unpolymerized DNA from herring sperm. The pH optimum of the enzyme under cytochemical conditions was found to be pH 5·9, when the localization of the product was mainly lysosomal. At pH 5·0 to 5·2, on the contrary, it was largely in the nuclei. Details of the method are given in Appendix 16, p. 735.

The enzyme(s) localized in the nuclei were considered by Vorbrodt (1961) to be nucleotide phosphodiesterases since he could obtain identical reactions at pH 5·0 with either DNA or RNA. Study brain tissues, however, Coimbra and Tavares (1964) used both DNA and RNA over the whole pH range. They selected pH 9·0 for incubation with both substrates, since this produced the maximum cytoplasmic localization, and they regarded the results as indicating two distinct enzymes (acid DNase II and acid RNase).

Both Aronson *et al.* (1958) and Vorbrodt (1961) invariably noted diminution of staining when they omitted exogenous acid phosphatase from the medium. Coimbra and Tavares, on the other hand, never found any difference from such omission in the case of tissues from the central nervous system. Investigating the presence of acid DNase II in the leucocytes of Swiss mice, Atwal *et al.* (1965) obtained results which were not exactly equivalent to those of Vorbrodt. It would appear that some aspects of the metal precipitation technique for acid DNase require further investigation.

Digestion Method for Deoxyribonuclease

An ingenious method for the histochemical demonstration of enzymes of this class was suggested by Daoust (1955, 1957). Films of gelatine containing DNA were applied to microscope slides and fixed *in situ* with formalin. Fresh frozen sections were then mounted on the films and incubated in a moist chamber for 5 hours at 20°. After fixation of the tissue, performance of the Feulgen reaction (Chapter 9, p. 254) coloured the nuclei in the section and also the underlying film of DNA magenta. Negative (unstained) images in the stained sheet of DNA indicated sites of DNAse activity. This substrate-film technique is fully considered in Chapter 24, together with other techniques using the same principle.

REFERENCES

Abu'l Fadl, M. A. M., and King, E. J. (1948a). *J. clin. Path.*, **1**, 80.
Abu'l Fadl, M. A. M., and King, E. J. (1948b). *Proc. Biochem. Soc., Biochem. J.*, **42**, xxviii
Abu'l Fadl, M. A. M., and King, E. J. (1949). *Biochem. J.*, **45**, 51.
Anagnostopoulos, C. (1953). *Bull. Soc. Chim. Biol.*, Paris, **35**, 575.
Anderson, P. J., and Song, S. K. (1962). *J. Neoropath. exp. Neurol.*, **21**, 263.
Appelmans, F., and de Duve, C. (1955). *Biochem. J.*, **59**, 426.
Applemans, F., Wattiaux, R., and de Duve, C. (1955). *Biochem. J.*, **59**, 438.

ARONSON, J., HEMPELMANN, L. H., and OKADA, S. (1958). *J. Histochem. Cytochem.*, **6**, 255
ARSENIS, C., and TOUSTER, O. (1967). *J. biol. Chem.*, **242**, 3399.
ATKINSON, T. G., and SHAW, M. (1955). *Nature*, **175**, 993.
ATWAL, O. S., FRYE, F. L., and ENRIGHT, J. B. (1965). *Nature, Lond.*, **205**, 185.
AVERS, C. J. (1961). *Amer. J. Bot.*, **48**, 137.
BARKA, T. (1962). *J. Histochem. Cytochem.*, **10**, 231.
BARKA, T., and ANDERSON, P. J. (1962). *J. Histochem. Cytochem.*, **10**, 741.
BARKA, T., and ANDERSON, P. J. (1963). "Histochemistry, Theory, Practice and Bibliography," Hoeber, New York, p. 240.
BARTELMEZ, G. W., and BENSLEY, S. H. (1947). *Science*, **106**, 639.
BEAUFAY, H., BENDALL, P., BAUDHUIN, P., WATTIAUX, R., and de DUVE, C. (1959). *Biochem. J.*, **73**, 628
BENDALL, P., and de DUVE, C. (1960). *Biochem. J.*, **74**, 444.
BENEŠ, K., LOJDA, A., and HOŘAVKA, B. (1961). *Histochemie*, **2**, 313.
BENEŠ, K., and OPATRNÁ, J. (1964). *Biol. Plant., Praha*, **6**, 8.
BERTHET, F., and DE DUVE, C. (1951). *Biochem. J.*, **50**, 174.
BERTHET, J., BERTHET, F., APPELMANS, F., and DE DUVE, C. (1951). *Ibid.*, **50**, 182.
BITENSKY, L. (1962). *Quart. J. micr. Sci.*, **103**, 205.
BITENSKY, L. (1963a). *Quart. J. micr. Sci.*, **104**, 193.
BITENSKY, L. (1963b). In "Lysosomes". Ed. de Reuck, A. V. S., and Cameron, M. P. Churchill, London, p. 362
BITENSKY, L. and GAHAN, P. B. (1962). *Biochem. J.*, **84**, 13P.
BODIAN, D., and MELLORS, R. C. (1944). *Proc. Soc. exp. Biol., N.Y.*, **55**, 243.
BRANDES, D., and BOURNE, G. H. (1954). *Brit. J. exp. Path.*, **35**, 577.
BREDERECK, H., and GEYER, E. (1938). *Hoppe-Seyl. Z. physiol. Chem.*, **254**, 223.
BURSTONE, M. S. (1958). *J. Nat. Cancer Inst.*, **21**, 523.
BURSTONE, M. S. (1962). "Enzyme Histochemistry." Acad. Press, New York.
BURTON, J. F. (1954). *J. Histochem. Cytochem.*, **2**, 88.
BURTON, J. F., and PEARSE, A. G. E. (1952). *Brit. J. exp. Path.*, **33**, 1.
CHYTIL, F., and MÜLLER, J. (1961). *Čs. Morfol.*, **9**, 173.
CLARKSON, T. W., and KENCH, J. E. (1958). *Biochem. J.*, **69**, 432.
COIMBRA, A., and TAVARES, A. S. (1964). *Histochemie*, **3**, 509.
CONCHIE, J., and LEVVY, G. A. (1963). *Biochem. Soc. Symposium No.* 23, p. 867
CONNOR, R. L., and MACDONALD, L. A. (1964). *J. cell. comp. Physiol.*, **64**, 257
DAOUST, R. (1955). *Brit. Emp. cancer Camp.*, 33rd *Ann. Rep.*, p. 67.
DAOUST, R. (1957). *Exp. Cell. Res.*, **12**, 203.
DAVIS, B. J., and ORNSTEIN, L. (1959). *J. Histochem. Cytochem.*, **7**, 297.
DAWSON, R. M. C. (1956). *Biochem. J.*, **64**, 192.
DESMET, V. J. (1962). *Stain Tech.*, **37**, 373.
DOWNEY, M., HICKEY, B. B., and SHARP, M. E. (1954). *Brit. J. Urol.*, **26**, 160.
DE DUVE, C., PRESSMAN, B. C., GIANETTO, R., WATTIAUX, R., and APPELMANS, F. (1955). *Biochem. J.*, **60**, 604
DE DUVE, C. (1959). In "Subcellular Particles." Ed. Hayashi, T. Ronald Press, New York, p. 128
DE DUVE, C. (1963). In "Lysosomes." Ed. de Reuck, A. V. S., and Cameron, M. P. Churchill, London, p. 1
EGER, W., and SCHULTE, W. (1954–55). *Acta Histochem.*, **1**, 60.
ELLIOT, A. M., and BAK, I. J. (1964). *J. cell. Biol.*, **20**, 113.
ERÄNKÖ, O. (1951a). *Acta physiol. scand.*, **24**, 1.
ERÄNKÖ, O. (1951b). *Nature, Lond.*, **168**, 250.
ERÄNKÖ, O. (1951c). *Ann. med. exp. Fenn.*, **29**, 287.
ERÄNKÖ, O., and LEHTO, L. (1954). *Acta Anat.*, **22**, 277.
ERICSSON, J. L. E., and TRUMP, B. F. (1964). *Lab. Invest.*, **13**, 1427.
EVANS, G. M., WHINNEY, C. L., and TSOU, K. C. (1966). *J. Histochem. Cytochem.*, **14**, 171
FIERZ-DAVID, H. E., and BLANGEY, L. (1949). "Fundamental Processes of Dye Chemistry." *Interscience, New York.*
FOLLEY, S. J., and KAY, H. D. (1936). *Ergebn. Eyznmforsch.*, **5**, 159.

GAHAN, P. B. (1965). *J. exp. Bot*, **16**, 350.
GAHAN, P. B. (1965). *J. Histochem. Cytochem.*, **13**, 334.
GAHAN, P. B. (1967). *Int. Rev. Cytol.*, **21**, 2.
GAHAN, P. B., and MAPLE, A. J. (1966). *J. exp. Bot.*, **17**, 151.
GIANETTO, R., and DE DUVE, C. (1955). *Biochem. J.*, **59**, 433.
GOETSCH, J. B., and REYNOLDS, P. M. (1951). *Stain Tech.*, **26**, 145.
GOLDBERG, A. F., and BARKA, T. (1962). *Nature, Lond.*, **195**, 297.
GOLDFISCHER, S., ESSNER, E., and NOVIKOFF, A. B. (1964). *J. Histochem. Cytochem.*, **12**, 72
GOMORI, G. (1941). *Arch. Path.*, **32**, 189.
GOMORI, G. (1945). *Proc. Soc. exp. Biol., N.Y.*, **58**, 362.
GOMORI, G. (1948). *Proc. Soc. exp. Biol., N.Y.*, **68**, 354.
GOMORI, G. (1949). *Ibid.*, **72**, 449.
GOMORI, G. (1950). *Stain Tech.*, **25**, 81.
GOMORI, G. (1952). "Microscopic Histochemistry." Chicago Univ. Press, p. 193.
GOMORI, G. (1956a). *J. Histochem. Cytochem.*, **4**, 425.
GOMORI, G. (1956b). *Ibid.*, **4**, 453.
GOODLAD, G. A. J., and MILLS, G. T. (1957). *Biochem. J.*, **66**, 346.
GORDON, G. B., MILLER, L. R., and BENSCH, K. G. (1963). *Exp. Cell. Research*, **31**, 440
GORDON, J. J. (1952). *Biochem. J.*, **51**, 97.
GROGG, E., and PEARSE, A. G. E. (1952). *J. Path. Bact.*, **64**, 627.
GUTMAN, A. B., and GUTMAN, E. B. (1938). *J. clin. Invest.*, **17**, 473.
GUTMAN, A. B., and GUTMAN, E. B. (1940). *J. biol. Chem.*, **136**, 201.
GUTMAN, A. B., and GUTMAN, E. B. (1941). *Proc. Soc. exp. Biol., N.Y.*, **48**, 687.
HANNIBAL, M. J., and NACHLAS, M. M. (1959). *J. biophys. biochem. Cytol.*, **5**, 279.
HARD, W. L., and LASSEK, A. M. (1946). *J. Neurophysiol.*, **9**, 121.
HARRINGTON, J. F., and ALTSCHUL, A. M. (1963). *Fed. Proc.*, **22**, 475.
HERBERT, F. K. (1944). *Biochem. J.*, **38**, xxiii.
HERBERT, F. K. (1946). *Quart. J. Med.*, **15**, 221.
HERVEG, J. P., BECKERS, C., and DE VISSCHER, M. (1966). *Biochem. J.*, **100**, 540.
HILLARP, N. A., and FALCK, B. (1956). *Acta Endocrinol.*, **22**, 95.
HOLT, S. J. (1958). *Gen. Cytochem. Meth.*, **1**, 375.
HOLT, S. J. (1959). *Exp. Cell Research, Suppl.*, **7**, 1.
HOLT, S. J., HOBBIGER, E. E., and PAWAN, G. L. S. (1960). *J. biophys. biochem. Cytol.*, **7**, 383.
HOLT, S. J., and HICKS, R. M. (1961). *J. biophys. biochem. Cytol.*, **11**, 47.
HOLTER, H., and LI, S. O. (1950). *Acta chem. scand.*, **4**, 1321.
HOLTER, H., and LI, S. O. (1951). *C. R. Trav. Lab. Carlsberg Sér. chim.*, **27**, 393.
HOLTER, H., and MARSHALL, J. M. jr. (1954). *C. R. Trav. Lab. Carlsberg. Sér. chim.*, **29**, 7
HOLTER, H. (1959). *Int. Rev. Cytol.*, **8**, 481.
ICHIHARA, M. (1933). *J. Biochem., Japan.*, **18**, 87.
JANIGAN, D. T. (1965). *J. Histochem. Cytochem.*, **13**, 476.
JEFFREE, G. M., and TAYLOR, K. B. (1961). *J. Histochem. Cytochem.*, **9**, 93.
JOSEPHY, H. (1949). *Arch. Neurol. Psychiat.*, **61**, 164.
KILSHEIMER, G. S., and AXELROD, B. (1957). *J. biol. Chem.*, **227**, 879.
KING, E. J., WOOD, E. J., and DELORY, G. E. (1945). *Biochem. J.*, **39**, xxiv.
KLAMER, B., and FENNELL, R. A. (1963). *Exp. Cell Research*, **29**, 166.
KUTSCHER, W., and WOLBERG, H. (1935). *Hoppe-Seylers Z. physiol. Chem.*, **236**, 237.
LAKE, B. D. (1965). *J. roy. micr. Soc.*, **85**, 73.
LASSEK, A. M. (1947). *Stain Tech.*, **22**, 133.
LI, S. O. (1950). *Acta chem. scand.*, **9**, 610.
LOJDA, Z. (1962). *Čs. Morfol.*, **10**, 46.
LOJDA, Z., VEČEREK, B., and PELICHOVÁ, H. (1964). *Histochemie*, **3**, 428.
LOJDA, Z., VAN DER PLOEG, M., and VAN DUIJN, P. (1967). *Histochemie*, **11**, 13.
LOVELOCK, J. E. (1954). *Biochem. J.*, **60**, 692.
MAGGI, V., FRANKS, L. M., and CARBONELL, A. W. (1966). *Histochemie*, **6**, 305.
MARPLES, E. A., and THOMPSON, R. H. S. (1960). *Biochem. J.*, **74**, 123.

MATHES, G. L., and NORMAN, T. D. (1956). *Lab. Invest.*, **5**, 276.
MEANY, A., GAHAN, P. B., and MAGGI, V. (1967). *Histochemie*, **11** (in press).
MEIJER, A. E. F. H. (1966). *Histochemie*, **6**, 317.
MEYER, J., and WEINMANN, J. P. (1953). *J. Histochem. Cytochem.*, **1**, 305.
MEYER, J., and WEINMANN, J. P. (1955). *Ibid.*, **3**, 134.
MEYER, J., and WEINMANN, J. P. (1957). *Ibid.*, **5**, 354.
MILLET, H., and JOWETT, M. (1929). *J. Amer. chem. Soc.*, **51**, 997.
MOOG, F. (1943). *Proc. Nat. Acad. Sci., Wash.*, **29**, 176.
MORETTI, G., and MESCON, H. (1956). *J. Histochem. Cytochem.*, **4**, 247.
MOTTET, K., and BARRON, D. H. (1953). *Yale J. Biol. Med.*, **26**, 275.
NAIDOO, D., and PRATT, O. E. (1951). *J. Neurol. Neurosurg.*, **14**, 287.
NEWMAN, W., KABAT, E. A., and WOLF, A. (1950). *Amer. J. Path.*, **26**, 489.
NOBACK, G. R., and PAFF, G. H. (1951). *Anat. Rec.*, **109**, 71.
NOVIKOFF, A. B. (1959). *Biol. Bull.*, **117**, 385.
NOVIKOFF, A. B. (1960). In "Developing Cell Systems and Their Control." Ronald Press, New York, p. 167.
NOVIKOFF, A. B. (1961). In "The Cell," ed. Brachet, J., Vol. 2. Academic Press, New York, p. 423
NOVIKOFF, A. B., BEAUFAY, H., and DE DUVE, C. (1956). *J. biophys. Biochem. Cytol.* (Suppl.), **2**, 179.
PALADE, G. E. (1951). *J. exp. Med.*, **94**, 535.
PEARSE, A. G. E. (1954). *Int. Rev. Cytol.*, **3**, 329.
POUX, N. (1963). *J. Microscopie*, **2**, 485.
RABINOWITCH, M. (1949a). *Stain Tech.*, **24**, 147.
RABINOWITCH, M. (1949b). *Blood*, **4**, 580.
RABINOWITCH, M. (1949c). *Nature, Lond.*, **164**, 878.
ROCHE, J. (1950). In "The Enzymes," Vol. 1, p. 275. Ed. Sumner, J. B., and Myrback, K. Acad. Press, New York.
RORIG, K. (1949). *J. Amer. chem. Soc.*, **71**, 3561.
ROSENBAUM, R. M., and ROLON, C. I. (1962). *Histochemie*, **3**, 1.
RUTENBURG, A. M., and SELIGMAN, A. M. (1955). *J. Histochem. Cytochem.*, **3**, 455.
RUYTER, J. H. C. (1964). *Histochemie*, **3**, 521.
SAUNDERS, L., and THOMAS, I. L. (1958). *J. chem. Soc.*, 483.
SCARPELLI, G. D., and PEARSE, A. G. E. (1958). *J. Histochem. Cytochem.*, **6**, 369.
SCHAEFFER, A., and BAUER, E. (1935). *Hoppe-Seylers Z. physiol. chem.*, **232**, 66.
SCHAJOWICZ, F., and CABRINI, R. L. (1964). *Acta histochem.*, **17**, 371.
SCHWARTZ, M. K., ASIMOV, I., and WOTIZ, H. H. (1953). *Cancer*, **6**, 924.
SELIGMAN, A. M., and MANHEIMER, L. H. (1949). *J. nat. Cancer Inst.*, **9**, 427.
SELIGMAN, A. M., NACHLAS, M. M., MANHEIMER, L. H., FREIDMAN, O. M., and WOLF, G. (1949). *Ann. Surg.*, **130**, 333
SELIGMAN, A. M., HEYMANN, H., and BARRNETT, R. J. (1954). *J. Histochem. Cytochem.*, **2**, 441
SMITH, W. K. (1948). *Anat. Rec.*, **102**, 523.
SOULAIRAC, A., DESCLAUX, P., and TESSEYRE, J. (1949a). *Ann. endocrinol., Paris*, **10**, 285
SOULAIRAC, A., DESCLAUX, P., and TESSEYRE, J. (1949b), *Ibid.*, **10**, 535.
SPIER, H. W., and MARTIN, K. (1956). *Arch. klin. exp. Dermatol.*, **202**, 120.
STRAUS, W. (1956). *J. biophys. biochem. Cytol.*, **2**, 513.
STRAUS, W. (1957). *Ibid.*, **3**, 933.
STRAUS, W. (1958). *Ibid.*, **4**, 541.
STRAUS, W. (1962a). *J. cell. Biol.*, **12**, 231.
STRAUS, W. (1962b). *Exp. Cell Research*, **27**, 80.
STRAUS, W. (1964a). *J. cell Biol.*, **20**, 497.
STRAUS, W. (1964b). *Ibid.*, **21**, 295.
STRAUS, W. (1964c). *J. Histochem. Cytochem.*, **12**, 462.
STRAUS, W. (1964d). *Ibid.*, **12**, 470.
STRAUS, W. (1967a). *Ibid.*, **15**, 375.
STRAUS, W. (1967b). *Ibid.*, **15**, 381.
SULKIN, N. M., and KUNTZ, A. (1947). *Anat. Rec.*, **99**, 639.

TAKEUCHI, T., and TANOUE, M. (1951). *Kumamoto med. J.*, **4**, 41.
TANDLER, C. J. (1953). *J. Histochem. Cytochem.*, **1**, 151.
TSEITLIN, L. A. (1952). *Biokhimiya, Moscow*, **17**, 203.
TSEITLIN, L. A. (1953). *Ibid.*, **18**, 311.
VORBRODT, A. (1961). *J. Histochem. Cytochem.*, **9**, 647.
VORBRODT, A. (1962). *Acta Un. int. Cancer*, **18**, 66.
WACHSTEIN, M. (1945). *Arch. Path.*, **40**, 51.
WALDSCHMIDT-LEITZ, E., and KÖHLER, F. (1933). *Biochem. Z.*, **258**, 360.
WALTER, W. (1954–55). *Acta Histochem.*, **1**, 3.
WATTIAUX, R., and DE DUVE, C. (1956). *Biochem J.*, **63**, 606.
WEISS, L. P., and FAWCETT, D. W. (1953). *J. Histochem. Cytochem.*, **1**, 47.
WINTER, G. F. (1954–55). *Acta Histochemica*, **1**, 303.
WOLF, A., KABAT, E. A., and NEWMAN, W. (1943). *Amer. J. Path.*, **19**, 423.
WOLMAN, M. (1965). *Z. Zellforsch.*, **65**, 1.
WOLMAN, M., and BUBIS, J. J. (1966). *Histochemie*, **7**, 105.
WOODARD, H. Q. (1950). *Blood*, **5**, 660.
WOODARD, H. Q. (1951). *J. Urol.* **65**, 688.

APPENDIX 1

CHAPTER 1 has no Appendix; this is a General Appendix and it contains, *inter alia*, technical details of methods of preparation, embedding, mounting, etc., which do not readily fit into the other Appendices. A selection of buffer tables, covering all the buffers in common histochemical use, is also given.

METHODS FOR THE PREPARATION OF TISSUES

(*Polyethylene Glycol Wax* (*Carbowax*) *Embedding for the Demonstration of Lipids and Enzymes*)

Blank (1949) and Blank and McCarthy (1950) described a rapid embedding technique for histological purposes using a melted mixture of 1 part Carbowax 1500 and 9 parts Carbowax 4000 (50°–54°). Other authors (Berlin and Brines, 1951; Firminger, 1950) used similar techniques and obtained thin sections (1–3 μ) in which lipids were well preserved. The method was also used for plant tissues by Van Horne and Zopf (1951) and by McLane (1951), the latter reporting preservation of enzymes in unfixed material embedded in frozen aqueous Carbowax 4000. Rinehart and Abu'l-Haj (1951), following other workers in this field, have described a method for dehydration and embedding which will be found useful for studies of lipids. For work on enzymes, where these are sufficiently stable to heat, the polyethylene glycol wax methods promised to be useful also. In practice, however, with alkaline phosphatase at all events, considerable diffusion of enzyme into the embedding medium and into the surrounding tissue structures has been observed to occur. With other enzymes (acid phosphatase, peroxidases) diffusion is much less evident. Details of the method are given below; modification of the technique by the addition of some of the higher m.p. waxes (mol. wt. 1540–4000) may be necessary in hotter climates.

PREPARATION OF SOLUTIONS

(A) STOCK SOLUTIONS OF POLYETHYLENE GLYCOL WAX 1000*

These are prepared by heating the wax until vapour appears on the molten surface. Water is then added at 50° to make a 70 per cent. and a 90 per cent. solution. These solutions, and the molten 100 per cent. wax, should be kept in an oven at 48°–52°.

(B) SOLUTION FOR FLOATING OUT

Diethylene glycol	40 parts
Distilled water	50 parts
40 per cent. formaldehyde	10 parts

(C) AFFIXATIVE MEDIUM

Gelatin (granular)	10 g.
Distilled water	60 ml.
Glycerine	50 ml.
Phenol	1 g.

Dissolve the gelatin in water at 32°–35°, add glycerine, filter through gauze and add phenol. Apply a thin film to the slides and wipe well.

* The whole range of polyethylene glycol waxes is available from the Watford Chemical Co. Ltd., Copperfield Road, London, E.3, and from General Metallurgical and Chemical Ltd., Moorgate, London, E.C.2. (Known as Carbowaxes in the United States, they are there available from the Carbide and Chemical Corporation.)

Method. (1) Fix tissues in neutral 6 per cent. formalin (24–48 hours at room temperature for lipids, 16 hours at 4° for many enzymes).

(2) Trim blocks to 3 or 4 mm. thick.

(3) Place blocks in 70 per cent. wax for 30 minutes, with occasional stirring.

(4) Transfer to 90 per cent. wax for 45 minutes, with occasional stirring.

(5) Transfer to 100 per cent. wax for 60 minutes, rotating the block and stirring.

(6) Embed in wax in paper cups, and harden at 4°.

(7) Cool blocks before cutting and use only dry implements.

(8) Cut sections 4–6 μ (or thicker or thinner as desired).

(9) Float out sections on solution B.

(10) Mount on gelatinized slides (C); allow to dry at 37° for 10 minutes.

After sectioning the block face should be coated with wax and the block stored in a dry cold atmosphere. Röhlich (1956) has used this method, embedding for 4–5 hours at 42°, for studies of acetylcholinesterase by the Koelle method (Appendix 17). He found a relatively rapid fall in the amount of enzyme demonstrable after storage, and after 1 month could scarcely obtain any reaction at all. Miles and Linder (1953), after testing various polyethylene glycol mixtures, developed a method using Nonex 63B (polyethylene glycol 1000 monostearate) at 39° after preliminary infiltration of tissues with a softer wax, polyethylene glycol 900, at 28°. They found excellent preservation of alkaline phosphatase in cold acetone-fixed tissues but noted some diffusion.

Polyethylene glycol 400-distearate (m.p. 35°) was recommended as a suitable polyester wax by Sidman *et. al.* (1961). It is a moderately hard wax with low solubility in water but with excellent properties as a histological embedding medium.

Effect of Temperature and Relative Humidity

The chief drawback to the use of water-soluble waxes is the difficulty of cutting satisfactory sections even after careful experiment to determine the optimum proportions of waxes for a particular climate. Hale (1952) investigated this problem, and found that, for sections cut at 7 μ, at 25°, ribboning ceased when the relative humidity reached 40 per cent. and that no sections could be cut above 56 per cent. humidity. At 20° ribboning continued up to 55 per cent. humidity. As a result of these experiments Hale recommended cutting in a cool room (below 18°) so as to be independent, as far as possible, of atmospheric humidity.

Gelatine Embedding for Enzyme Histochemistry

This process is usually carried out when it is necessary to support delicate tissues so that sections can be cut on the freezing microtome. When sections of delicate tissues are required for enzyme studies, and particularly for the coupling azo dye techniques (Appendices 15 and 16), a modified gelatine embedding procedure, designed to damage the enzymes as little as possible, can be employed. This is described below.

Method. (1) Fix thin slices or portions of tissue in 15 per cent. cold neutral formalin (4°) for 10–16 hours.

(2) Wash in running water for 30 minutes.

(3) Embed in gelatine* for 1 hour at 37°.

(4) Cool and harden in formalin (40 per cent. formaldehyde) for 1 hour at 15°–22°. Wash.

(5) Store at 4° or below until frozen sections are required.

* Gelatine pulv.	15 g.
Glycerine	15 ml.
Distilled water	70 ml.
Thymol	A small crystal

Provided that the gelatine block surrounding the tissues is more than a few mm. thick, the hardening process does not affect the enzyme. It should not be prolonged, however, beyond the time indicated.

MOUNTING MEDIA

This section is not meant to be comprehensive; only those media which are mentioned in the text and in the various Appendices are included. For a comprehensive review of the available permanent mounting media the reader is referred to the Interim and Final Reports of the Committee on Mounting Media (Lillie *et al.*, 1950, 1953).

DPX

(Kirkpatrick and Lendrum, 1939, 1941)

Distrene 80 (British Resin Products Ltd.)	10 g.
Dibutylphthalate	5 ml.
Xylene	35 ml.

This is a useful mountant for general purposes which does not cause fading with the majority of stains. Shrinkage is considerable and only a thin layer must separate the section from the coverslip. Thicker layers cause distortion of the microscopic image. The original DPX mountant could not be used for preparations containing silver deposits (reticulin, Gomori's acid and alkaline phosphatase methods), or for sections finished by the cobalt sulphide technique (acid and alkaline phosphatase, Gomori's cholinesterase method).

The new DPX, in which tri-orthocresyl phosphate is no longer used as plasticizer, can be employed successfully for the above-mentioned preparations.

KARO CORN SYRUP (*Corn Products Ltd.*)

(Patrick, 1936)

Corn syrup	1 part
Distilled water	1 part

Add a crystal of thymol and keep at 4°.

CHROME GLYCERIN JELLY (*Glychrogel*)

(Zwemer, 1933)

Distilled water	80 ml.
Glycerine	20 ml.
Gelatin (powdered)	3 g.
Chrome alum	0·2 g.

Dissolve the chrome alum in 30 ml. of water and the gelatin in 50 ml. of water. Mix and add the glycerine. Test pH before use and adjust to pH 7·4 with $NaHCO_3$ if necessary.

This mountant shrinks and sets very hard. Removal of the coverslip and renewal of the medium is therefore impossible.

GLYCERINE JELLY

(Roulet, 1948, after Kaiser, 1880)

Dissolve 15 g. best quality white gelatine in 100 ml. distilled water, with moderate heating. Add 100 g. glycerine and warm for 5 minutes on a waterbath, filter hot through glass wool. To each 100 ml. of the mixture add 1 drop of phenol (Phenol liquefactum B.P., U.S.P.).

This medium is kept at 37° in an incubator and applied to the wet section, mounted on its slide, with a thick glass rod. (In cold weather it is sometimes an advantage to warm the slide in running hot water before applying the jelly.) A coverslip is then lowered gently over the medium and the excess is wiped away with blotting paper. The refractive index of glycerine jelly is 1·47, and although it sets hard and has no tendency to draw up under the edges of the coverslip, it can easily be removed by dipping the slide in hot water.

APATHY'S GUM SYRUP
(Lillie and Asburn, 1943)

Dissolve 50 g. gum acacia and 50 g. cane sugar in 100 ml. water with frequent shaking at 55°. Restore original volume with water. Add 15 mg. menthiolate (sodium ethylmercurithiosalicylate) or 100 mg. thymol as a preservative.

POLYSORBITOL MEDIUM
(Karion-Merck)

Karion is a proprietary medium which can be used for mounting frozen sections. It has a tendency to crystallize on prolonged storage even when the coverslips are ringed.

DECALCIFICATION FOR HISTOCHEMICAL PURPOSES

The standard methods of decalcification using nitric, hydrochloric or acetic acids are usually quite unsatisfactory if histochemical methods of enquiry are proposed. This refers not only to enzyme techniques but also to analytical methods for nucleic acids and polysaccharides and, to a lesser extent, to other methods.

With the three standard methods given below adequate fixation is an important prerequisite. Enzymes which survive this are largely if not entirely destroyed, even when the decalcifying process is short, as when only minor calcium deposits have to be removed.

DE CASTRO'S FLUID

Chloral hydrate	50 g.
Absolute alcohol	300 ml.
Water	670 ml.
Nitric acid (conc.)	30 ml.

JENKINS' FLUID

Hydrochloric acid (conc.)	80 ml.
Acetic acid	80 ml.
Chloroform	200 ml.
Alcohol, 95 per cent.	1,460 ml.
Water	200 ml.

With the above fluids treat thin well-fixed blocks for 16–48 hours, as necessary, and post-fix for 48 hours in 10 per cent. neutral formalin. Prolonged decalcification interferes with nuclear staining and Gram staining.

CLARKE (1953)

Method. (1) Fix tissue in 10 per cent. neutral formalin for 48 hours.

(2) Decalcify in a mixture of equal parts of 45 per cent. formic acid and 6·8 per cent. sodium formate (pH 2·0). Change solution daily (7–14 days).

(3) Wash in water.

(4) Dehydrate, clear and embed as usual.

ELECTROLYTIC DECALCIFICATION
(Richman *et al.*, 1947)

With this method a 2–6 volt direct current is passed through the decalcifying solution (1 part 10 per cent. formic acid and 1 part 8 per cent. HCl). The positive electrode must be platinum, the negative electrode can be carbon. If the temperature of the bath exceeds 40° considerable damage occurs, but even at lower temperatures much protein, nucleoprotein and polysaccharide is extracted.

Decalcification takes 2–6 hours for thin slices (1 cm.) of normal bone, but may take up to 18 hours for dense bone specimens. Treatment with neutral or alkaline 10 per cent. formalin should follow decalcification by this method.

DECALCIFICATION WITH ENZYME PRESERVATION
(after Lorch, 1946)

This method was designed for the preservation particularly of alkaline phosphatases.

Method. (1) Fix in two changes of absolute acetone at 0°–4° (thin blocks).

(2) Decalcify for 3–10 days in Lorch's fluid:

Citric acid (crystalline)	14·7 g.
0·2 N-NaOH	700 ml.
0·1 N-HCl	300 ml.
1 per cent. zinc sulphate	2 ml.
Chloroform	0·1 ml.

(3) Wash in distilled water, several changes.

(4) Reactivate the enzyme by treatment with 0·1 M-veronal-acetate buffer (pH 9) for 6–8 hours.

(5) Dehydrate in alcohol.

(6) Clear in chloroform.

(7) Embed *in vacuo* in 48° wax.

DECALCIFICATION WITH ENZYME PESERRVATION
(after Greep *et al.*, 1948)

This method, like that of Lorch, was designed for the preservation of alkaline phosphatases.

Method. (1) Fix thin slices of bone or whole bones from small animals in absolute alcohol at 4° for 24 hours.

(2) Rinse briefly in 50 per cent. alcohol and in distilled water.

(3) Decalcify in a mixture of equal parts of 2 per cent. formic acid and 20 per cent. sodium citrate for 2–10 days. (Change solution every 3–4 days.)

(4) Rinse in three changes of tap water.

(5) Reactivate by immersion in 1 per cent. sodium diethylbarbiturate for 24–48 hours.

(6) Rinse in tap water.*

(7) Dehydrate in alcohol, clear in benzene and embed *in vacuo* in 48° wax.

The preservation of both alkaline and acid phosphatase during decalcification has been considered by a number of authors. Some of the most important papers are those of Greep *et al.* (1948), Zorzoli (1948), Lillie (1951), Majno and Rouiller (1951), and Schajowicz and Cabrini (1954, 1958). These last authors (1955) also tested a number of decalcifying solutions with respect to their effects on the histochemical

* Frozen sections may be cut at this stage for enzyme demonstrations.

staining of bone and cartilage. Among the solutions tested were 10 per cent. HCl and 7·5 per cent. HNO_3 (1 hour, 24 hours, 5 days), Greep's fluid and Lorch's fluid, for 1 hour, 24 hours, 5, 15 and 30 days, and lactic and perchloric acids for varying times. Sections were stained with toluidine blue, PAS, Hale's dialysed iron, aldehyde fuchsin, mucicarmine and mucihæmatein. Schajowicz and Cabrini found that with the strong acids there was a rapid loss of metachromasia in cartilage matrix, complete by the fifth day. Glycogen was quickly lost but PAS staining hardly modified. Greep's fluid produced similar changes but much more slowly, scarcely any effect being visible until after the fifth day. Lorch's fluid was observed to produce a phenomenon described as "inversion". Within 7 days an intense metachromasia appeared in bone and osteoid, while that of cartilage was abolished. These observations produced the recommendation that for mucopolysaccharide studies thin pieces of bone should be treated for up to 5 days with Greep's fluid.

For the preservation of alkaline phosphatases Schajowicz and Cabrini (1955) obtained excellent results with Greep's fluid followed by reactivation with sodium barbiturate-glycine buffer.

The Use of Chelating Agents

An ion exchange resin was used by Dotti, Paparo and Clarke (1951) for the decalcification of bone, and many other authors have followed this lead in using organic chelators for the same purpose (Hahn and Reygadas, 1951; Screebny and Nikiforuk, 1951; Birge and Imhoff, 1952; Hilleman and Lee, 1953; Freiman, 1954). The chelator universally employed is the disodium salt of ethylenediamine tetraacetic acid (Versene, Sequestrene).

The preservation of acid phosphatase in bone and teeth has been improved considerably by EDTA decalcification procedures. Reports of its successful use have come from Takada *et al.* (1960), Mori *et al.* (1962), Yoshiki (1962) and Balogh (1965).

HILLEMANN AND LEE

(1953)

(Average time for rat femur, 4 weeks.)

Method. (1) Fix in neutral formalin for 24 hours at room temperature.

(2) Transfer to 5·5 per cent. versene in 10 per cent. formalin, renewing the solution every 10 days.

(3) Dehydrate in alcohol, clear, embed in paraffin wax.

FREIMAN

(1954)

(Suitable for enzyme studies)

Method. (1) Fix thin slices (or whole bones from small animals) in 80 per cent. alcohol at 4° for 24 hours.

(2) Wash in tap water.

(3) Decalcify in 5 per cent. aqueous EDTA., adjusted to pH 6·0–6·5 with NaOH, for 24 hours at 4°.

(4) Change solution and keep at room temperature for 2–21 days, changing the solution at regular intervals.

(5) Wash in water.

(6) Reactivate enzyme by treatment with 1 per cent. sodium barbiturate for 24 hours.

(7) Dehydrate, clear and embed *in vacuo* in 48° wax.

BALOGH
(1965)
(For Acid Phosphatase)

Method. (1) Decalcify unfixed tissues at 4° in buffered 10 per cent. EDTA (dissolve in 0·1 M-phosphate buffer at pH 7·0 and adjust pH with normal NaOH, if necessary). Thin blocks of cancellous bone take 3–4 days, rat long bones about 72 hours.

(2) Blot blocks and mount on tissue holders.

(3) Cut cryostat sections 10–20 μ thick, and mount on clean coverslips. Dry at oom temperature for 30 minutes.

(4) Post-fix for 1 hour in 10 per cent. formol-calcium at 4°.

(5) Wash in distilled water and incubate as required.

BUFFERS

pH 0·65 to 5·20

Walpole (1914) 50 ml. N-sodium acetate + x ml. N-HCl, made up to 250 ml.

pH	x ml. HCl	pH	x ml. HCl	pH	x ml. HCl
0·65	100	1·99	52·5	3·79	42·5
0·75	90	2·32	51·0	3·95	40·0
0·91	80	2·64	50·0	4·19	35·0
1·09	70	2·72	49·75	4·39	30·0
1·24	65	3·09	48·5	4·58	25·0
1·42	60	3·29	47·5	4·76	20·0
1·71	55	3·49	46·25	4·92	15·0
1·85	53·5	3·61	45·0	5·20	10·0

pH 2·2 to 8·0

McIlvaine (1921). 200 ml. mixtures of x ml. 0·2 M-Na_2HPO_4 with y ml. 0·1 M-citric acid.

x ml. Na_2HPO_4	y ml. citric acid	pH	x ml. Na_2HPO_4	y ml. citric acid	pH
4·0	196	2·2	107·2	92·8	5·2
12·4	187·6	2·4	111·5	88·6	5·4
21·8	178·2	2·6	116·0	84·0	5·6
31·7	168·3	2·8	120·9	79·1	5·8
41·1	158·9	3·0	126·3	73·7	6·0
49·4	150·6	3·2	132·2	67·8	6·2
57·0	143·0	3·4	138·5	61·5	6·4
64·4	135·6	3·6	145·5	54·5	6·6
71·0	129·0	3·8	154·5	45·5	6·8
77·1	122·9	4·0	164·7	35·3	7·0
82·8	117·2	4·2	173·9	26·1	7·2
88·2	111·8	4·4	181·7	18·3	7·4
93·5	106·5	4·6	187·3	12·7	7·6
98·6	101·4	4·8	191·5	8·5	7·8
103·0	97·0	5·0	194·5	5·5	8·0

pH 3·6 to 5·6

Walpole (1914). 200 ml. mixtures of 0·1 N-acetic acid and 0·1 N-sodium acetate.

pH	0·1 N-acetic acid (ml.)	0·1 N-sodium acetate (ml.)	pH	0·1 N-acetic acid (ml.)	0·1 N-sodium acetate (ml.)
3·6	185	15	4·8	80	120
3·8	176	24	5·0	59	141
4·0	164	36	5·2	42	158
4·2	147	53	5·4	29	171
4·4	126	74	5·6	19	181
4·6	102	98			

pH 5·29 to 8·04

Sörensen (1909–12). 100 ml. mixtures of 0·06 M-Na_2HPO_4 and 0·06 M-KH_2PO_4.

pH	Na_2HPO_4 (ml.)	KH_2PO_4 (ml.)	pH	Na_2HPO_4 (ml.)	KH_2PO_4 (ml.)
5·29	2·5	97·5	6·81	50	50
5·59	5	95	9·98	60	40
5·91	10	90	7·17	70	30
6·24	20	80	7·38	80	20
6·47	30	70	7·73	90	10
6·64	40	60	8·04	95	5

pH 1·5 to 3·5

Lewis (1962). Acid Phosphate Buffer. Stock solution. 0·2 M-acid potassium phosphate, 27·22 g. potassium dihydrogen phosphate to 1 litre.

pH values at 2 concentrations

Molar Proportion of HCl	pH values at 0·02 M	0·1 M
0·05	3·55	3·35
0·10	3·2	3·0
0·15	3·0	2·75
0·20	2·85	2·55
0·25	2·7	2·4
0·30	2·6	2·25
0·35	2·5	2·1
0·40	2·4	2·0
0·45	2·3	1·85
0·50	2·2	1·75
0·60	2·1	1·55
0·75	2·0	

pH 7·8 to 10·0

Clark and Lubs (1916). 50 ml. 0·2 M-H_3BO_3 + 0·2 M-KCl + x ml. 0·2 M-NaOH, diluted to 200 ml.

pH	x ml. 0·2 M-NaOH	pH	x ml. 0·2 M-NaOH
7·8	2·65	9·0	21·40
8·0	4·00	9·2	26·70
8·2	5·90	9·4	32·00
8·4	8·55	9·6	36·85
8·6	12·0	9·8	40·80
8·8	16·40	10·0	43·90

pH 8·45 to 12·77

Sörensen-Walbum (1920). 100 ml. mixtures of 0·1 M-glycine and 0·1 M-NaCl with 0·1 M-NaOH.

pH	x ml. glycine-NaCl	y ml. NaOH	pH	x ml. glycine-NaCl	y ml. NaOH
8·45	95	5	11·14	50	50
8·79	90	10	11·39	49	51
9·22	80	20	11·92	45	55
9·56	70	30	12·21	40	60
9·98	60	40	12·48	30	70
10·32	55	45	12·66	20	80
10·9	51	49	12·77	10	90

pH 2·62 to 9·16

Michaelis (1931). Stock solution: 9·714 g. sodium acetate $3H_2O$ + 14·714 g. sodium barbiturate in CO_2 free distilled water, made up to 500 ml. 5 ml. of this solution + 2 ml. of 8·5 per cent. NaCl (may be omitted) treated with x ml. of 0·1 N-HCl and (18 − x) ml. water.

x ml. 0·1 N-HCl	pH	ml. 0·1 N-HCl	pH	x ml. 0·1 N-HCl	pH
16·0	2·62	9·0	4·93	4·0	7·66
15·0	3·20	8·0	5·32	3·0	7·90
14·0	3·62	7·0	6·12	2·0	8·18
13·0	3·88	6·5	6·75	1·0	8·55
12·0	4·13	6·0	6·99	0·75	8·68
11·0	4·33	5·5	7·25	0·5	8·9
10·0	4·66	5·0	7·42	0·25	9·16

pH 2·7 to 5·7

Lewis (1962). Stock solutions:

(1) 1·0 M-sodium acetate, 27·22 g. in 200 ml. water.

(2) 0·2 M-formic acid, 14·2 ml. 90 per cent. formic acid, 11·69 g. NaCl, to 1 litre.

pH values at 2 concentrations

Molar Proportion of Acetate	pH values at 0·02 M	pH values at 0·1 M	Molar Proportion of Acetate	pH values at 0·02 M	pH values at 0·1 M
0·00	2·72	2·38	0·48	4·11	4·07
0·04	2·85	2·55	0·52	4·25	4·21
0·08	2·96	2·72	0·56	4·39	4·35
0·12	3·06	2·87	0·60	4·52	4·48
0·16	3·15	3·02	0·64	4·65	4·61
0·20	3·25	3·17	0·68	4·78	4·74
0·24	3·36	3·30	0·72	4·91	4·87
0·28	3·48	3·42	0·76	5·04	5·00
0·32	3·59	3·54	0·80	5·18	5·14
0·36	3·71	3·66	0·84	5·32	5·28
0·40	3·84	3·79	0·88	5·48	5·45
0·44	3·97	3·93	0·92	5·71	5·68

Formic-Acetate Buffer Series

Molar Proportion of Acetate	Ml. Stock per litre 0·2 M Formic	Ml. Stock per litre 1·0 M Acetate	pH	Molar Proportion of Acetate	Ml. Stock per litre 0·2 M Formic	Ml. Stock per litre 1·0 M Actate	pH
0·1	90	2	3·0	0·55	45	11	4·35
0·15	85	3	3·1	0·6	40	12	4·5
0·2	80	4	3·25	0·65	35	13	4·7
0·25	75	5	3·4	0·7	30	14	4·85
0·3	70	6	3·55	0·75	25	15	5·0
0·35	65	7	3·7	0·8	20	16	5·2
0·4	60	8	3·85	0·85	15	17	5·35
0·45	55	9	4·0	0·9	10	18	5·6
0·5	50	10	4·2				

pH 7·4 to 9·0

Holmes. Mixtures of 0·2 M-boric acid with 0·2 M-sodium tetraborate.

pH	H_3BO_3 (ml.)	$Na_2B_4O_7$ 10 H_2O (ml.)	pH	H_3BO_3 (ml.)	$Na_2B_4O_7$ 10 H_2O (ml.)
7·4	18	2	8·2	13	7
7·6	17	3	8·4	11	9
7·8	16	4	8·7	8	12
8·0	14	6	9·0	4	16

This buffer is recommended by Burtner and Lillie (1949) for making up Gomori's hexamine-silver solution (see Appendix 26).

pH 7·19 to 9·10

Gomori. 25 ml. of 0·2 M-tris(hydroxymethyl)aminomethane, Eastman-Kodak (M.W. 121·14), $+x$ ml. 0·1 N-HCl and distilled water up to 100 ml.

pH	x ml. 0·1 N-HCl	pH	x ml. 0·1 N-HCl
7·19	45·0	8·23	22·5
7·36	42·5	8·32	20·0
7·54	40·0	8·41	17·5
7·66	37·5	8·51	15·0
7·77	35·0	8·62	12·5
7·87	32·5	8·74	10·0
7·96	30·0	8·92	7·5
8·05	27·5	9·10	5·0
8·14	25·0		

pH 5·2 to 8·6

Tris (hydroxymethyl) aminomethane-maleate buffer. Stock solutions. 0·2 M-tris-maleate (24·2 g. tris + 23·2 g. nucleic acid, or 19·6 g. maleic anhydride, in 1000 ml. distilled water.

50 ml. Stock + x ml. 0·2 M-NaOH, make up to 200 ml.

pH	x	pH	x
5·2	7·0	7·0	48·2
5·4	10·8	7·2	51·0
5·6	15·5	7·4	54·0
5·8	20·5	7·6	58·0
6·0	26·0	7·8	63·5
6·2	31·5	8·0	69·0
6·4	37·0	8·2	75·0
6·6	42·5	8·4	81·0
6·8	45·0	8·6	86·5

pH 5·0 to 7·4

Sabatini *et al.* (1963). 0·2 M-Cacodylate buffer. Stock solution: 42·8 g. $Na(CH_3)_2AsO_2.3H_2O$ in 1000 ml. distilled water. 50 ml. cacodylate + x ml. 0·2 M-HCl and make up to 200 ml.

pH	x	pH	x
7·4	2·7	6·0	29·6
7·2	4·2	5·8	34·8
7·0	6·3	5·6	39·2
6·8	9·3	5·4	43·2
6·6	13·3	5·2	45·0
6·4	18·3	5·0	47·0
6·2	23·8		

NOMENCLATURE AND DEFINITION TABLES

Table 47

Diminutives

Definition	Symbol	Abbreviation	Equivalent
deci	d	10^{-1}	0·1
centi	c	10^{-2}	0·01
milli	m	10^{-3}	0·001
micro	μ	10^{-6}	0·000,001
nano	n	10^{-9}	0·000,000,001
pico	p	10^{-12}	0·000,000,000,001
femto	f	10^{-15}	0·000,000,000,000,001
atto	a	10^{-18}	0·000,000,000,000,000,001

Table 48

Symbols and Equivalents
(Disused Symbols in Brackets)

Definition	Symbol	Quantity	Equivalent or Application
Ångstrom	Å	0·1 nm	10^{-10} metres
British Thermal Unit	BTU	1055·8 J	—
Calorie	cal	4·1855 J	—
Centigrade	° or °C	273·15° K	32° F
Centimetre	cm	—	—
Constant (velocity)	k	—	—
Curie	c	3700×10^{10} disintegrations/sec	—
Diffusion Constant	D	—	—
Dissociation Constant (minus log)	pK	—	—
Electrode potential (standard)	E_0	—	—
Electrode potential (constant pH)	E_0^1	—	—
Electron volt	eV	$1·602 \times 10^{-19}$ J	$1·59 \times 10^{-12}$ erg
Erg	erg	10^{-7} J	—
Gram	g	—	—
Hydrogen ion concn.	pH	—	—
Half wave potential	E_0	—	—
Joule	J	10^7 erg	$2·7 \times 10^{-7}$ kWh
Kilocalorie	kcal	4185·5 J	—
Kelvin	K	—	—
Median lethal dose	LD_{50}	—	—
Melting point	m.p.	—	—
Millicurie	mc	$3·7 \times 10^7$ disintegrations/sec	—
Millimeter	mm	0·3937 inch	—
Millimolar	mM	10^{-3} M	Concentration
Millimole	m-mole	—	—

Table 48—*contd.*

Definition	Symbol	Quantity	Equivalent or Application
Micron	μ, μm	$3{\cdot}937 \times 10^{-5}$ inch	Histology and E.M.
Microgram	μg (γ)	$1{\cdot}543 \times 10^{-5}$	—
Micrometer	μm(μ)	$3{\cdot}937 \times 10^{-5}$ inch	—
Molar	M	—	Concentration
Molecular weight	mol. wt.	—	—
Nanometer	nm (mμ)	10 Å	Millimicron
Normal	N	—	Concentration
Photon	ph	10^4 lux	—
Rad	rad	100 erg/lg	—
Sedimentation constant	S	—	—
Specific gravity	sp. gr	—	—
Substrate constant	K_s	—	Dissociation constant E/S
Torricelli	Torr (mmhg)	$1333{\cdot}223$ dyn/cm^2	—
Wavelength	λ	—	—

REFERENCES

BALOGH, K. Jr. (1965). *J. Histochem. Cytochem.*, **13**, 303.
BERLIN, R. B., and BRINES O. A. (1951). *Amer. J. clin. Path.*, **21**, 332.
BIRGE, E. A., and IMHOFF, C. E. (1952). *Ibid.*, **22**, 192.
BLANK, H. (1949). *J. invest. Derm.*, **12**, 95.
BLANK, H., and MCCARTHY, P. L. (1950). *J. Lab. clin. Med.*, **36**, 776.
CLARKE, G. (1953). *Amer. J. clin. Path.*, **24**, 113.
DOTTI, L. B., PAPARO, G. P., and CLARKE, B. E. (1951). *Ibid.*, **21**, 475.
FIRMINGER, H. I. (1950). *Stain Tech.*, **25**, 121.
FREIMAN, D. G. (1954). *Amer. J. clin. Path.*, **24**, 227.
GREEP, R. O., FISHER, C. J., and MORSE, A. (1948). *J. Amer. Dent. Assoc.*, **36**, 427.
HALE, A. J. (1952). *Stain Tech.*, **27**, 189.
HAHN, F. L., and REYGADAS, F. (1951). *Science*, **114**, 462.
HILLEMAN, H. H., and LEE, C. H. (1953). *Stain Tech.*, **28**, 285.
KIRKPATRICK, J., and LENDRUM, A. C. (1939). *J. Path. Bact.*, **49**, 593.
KIRKPATRICK, J., and LENDRUM, A. C. (1941). *Ibid.*, **53**, 441.
LEWIS, P. R. (1962). *Histochemie*, **2**, 423.
LILLIE, R. D., and ASHBURN, L. L. (1943). *Arch. Path.*, **36**, 432.
LILLIE, R. D., LASKEY, A., GRECO, J., BURTNER, J. H., and JONES, P. (1951). *Amer. J. clin. Path.*, **21**, 711.
LILLIE, R. D., WINDLE, W. F., and ZIRKLE, C. (1950). *Stain Tech.*, **25**, 1.
LILLIE, R. D., ZIRKLE, C., DEMPSEY, E. W., and GRECO, J. P. (1953). *Ibid.*, **28**, 57.
LORCH, I. J. (1946). *Nature, Lond.*, **158**, 269.
MAJNO, G., and ROUILLER, C. (1951). *Virchows Arch.*, **321**, 1.
MCLANE, S. R., Jr. (1951). *Stain Tech.*, **26**, 63.
MILES, A. E. W., and LINDER, J. E. (1953). *J. Roy. micr. Soc.*, **72**, 199.
MORI, M., TAKADA, K., and OKAMOTO, J. (1962). *Histochemie*, **2**, 427.
PATRICK, R. (1936). *Science*, **83**, 85.
RICHMAN, I. M., GELFAND, M., and HILL, J. M. (1947). *Arch. Path.*, **44**, 92.
RINEHART, J. F., and ABU'L HAJ, S. (1951). *Arch. Path.*, **51**, 666.
RÖHLICH, P. (1956). *Nature, Lond.*, **178**, 1398.
ROULET, F. (1948). "Methoden der Pathologischen Histologie." Springer, Wien, p. 162.
SABATINI, D. D., BENSCH, K., and BARRNETT, R. J. (1963). *J. cell. Biol.*, **17**, 19.
SCHAJOWICZ, F., and CABRINI, R. L. (1954). *J. Bone Joint Surg.*, **36B**, 474.

SCHAJOWICZ, F., and CABRINI, R. L. (1955). *J. Histochem. Cytochem.*, **3**, 122.
SCHAJOWICZ, F., and CABRINI, R. L. (1958). *Science*, **129**, 1147.
SCREEBNY, L. M., and NIKIFORUK, G. (1951). *Science*, **113**, 560.
TAKADA, K., TANI, T., and MORI, M. (1960). *Jap. J. clin. Path.*, **8**, 237.
VAN HORNE, R. L., and ZOPF, L. C. (1951). *J. Amer. pharm. Assoc.*, **40**, 31.
YOSHIKI, S. (1962). *Bull. Tokyo, dent. Coll.*, **3**, 14.
ZORZOLI, A. (1948). *Anat. Rec.*, **102**, 445.
ZWEMER, K. (1933). *Ibid.*, **57**, 41.

APPENDIX 2

MAINTENANCE OF CRYOSTAT AND MICROTOME

In general, satisfactory instructions are sent out with their instruments by the manufacturers. These normally enable the user to assemble and operate his machine without trouble. It is, therefore, unnecessary to set down here full instructions for successful cryotomy but it may be useful to emphasize one or two particular points. (A very comprehensive manual on cryostat maintenance is published by the South London Electrical Equipment Co. Ltd, Hither Green Lane, London, S.E.13, and obtainable from them.)

Certain aspects of maintenance have altered considerably since the previous (1960) edition of this book.

(1) It is now customary to lubricate the working parts of the microtome with a suitable low temperature oil, having a flow point down to $-50°$.

(2) It is not necessary to incorporate any device for absorbing humidity in the cryostat chamber.

(3) The period between defrostings will vary according to the degree of humidity of the atmosphere, the amount of use, and the extent to which the cabinet is kept open for changing blocks and other manœuvres. It may thus vary between 4 weeks and 4 months. The time of defrosting is determined rather by icing of the microtome than of the chamber. Within wide limits the latter has no significant effect on performance.

Storage of Tissues

Quenched tissues can be stored for short periods in the cryostat cabinet at its normal working temperature (usually $-20°$ in conventional, non-thermoelectric, cryostats). Unless the tissues are kept in airtight vessels, rapid drying will take place. Two popular types of container are screw-top aluminium boxes (round section) and polythene film tubing (4 cm. wide, 500 gauge).* The latter must be heat sealed at both ends. It is not impervious to gases and drying of the block may still take place.

Whichever type of container is employed it is necessary to provide storage temperatures lower than those available in the cryostat chamber, if storage for more than a few days is required. Commercial deep freeze cabinets ($-70°$, $-90°$) are often used but special library storage cabinets ($-70°$) are available commercially (Fig. 158).† Accurate information concerning the changes which take place during storage is not available. Provided that the initial quenching has been carried out rapidly, and at a sufficiently low temperature, ice crystal growth is not rapid and tissues may be kept for at least 2 years without obvious deterioration.

For really prolonged (perhaps indefinite) storage the only method is to use a liquid nitrogen refrigerator (Fig. 159). Several models are available commercially.‡

Technical Accessories

(1) **Brush.** A stiff paint brush of medium size hog bristle is essential both for dislodging sections which may have become stuck to the knife or to the anti-roll plate and for removing condensation from the knife surface. When "warm" coverslips or slides are used to pick up the sections this condensation may have to be removed between cutting of each section or pair of sections.

* British Visqueen Ltd, Six Hills Way, Stevenage, Hertfordshire.

† South London Electrical Equipment Ltd, Hither Green Lane, London, S.E.13.

‡ Union Carbide, Linde Company, 8 Grafton St., London, W.1.

FIG. 158. Library-type −70° Storage Cabinet for Tissue Samples. (Slee Co.)

FIG. 159. Liquid Nitrogen Storage Tank for Tissue Samples. (Union Carbide Co.)

(2) **Coverslip Pick-up.** As stated in Chapter 2, there are a number of ways in which the freshly cut section can be handled. If it is to be picked up on a coverslip (standard size $\frac{7}{8}$ inch square), or on a microscope slide, the provision of a coverslip pick-up is absolutely essential. A simple piece of apparatus, shown in the diagram (Fig. 160) below, consists of a round perforated sucker disc attached to a capillary glass tube 9 inches long, having at its other extremity an inclined side arm to which a rubber teat is fixed. The end beyond the side arm should be solid.

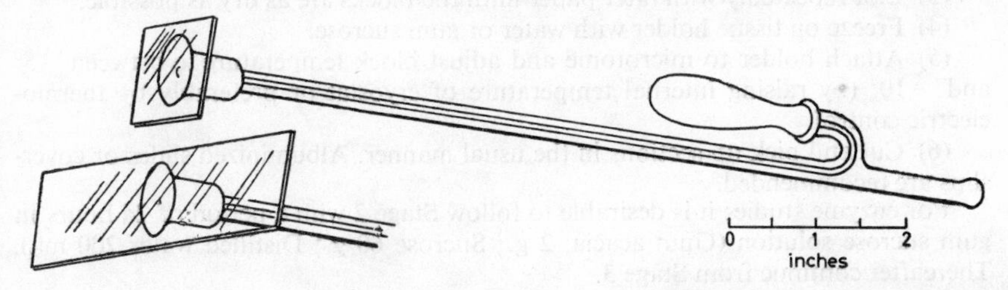

FIG. 160. Coverslip Suction Pick-up.

Clean coverslips or slides are picked up by squeezing the teat between the right thumb and forefinger, expelling air from the capillary tube. The sucker disc is then applied to the coverslip or slide and the teat released. After picking up the section by pressing coverslip or slide firmly and steadily on to the surface of the knife (having previously swung down the fluon-coated anti-roll plate), the coverslip or slide is grasped between the thumb and forefinger of the left hand. Momentary pressure on the teat will then release either.

Some workers prefer a more positive type of attachment of the holder to the coverslip. A spring-loaded holder was described by Combs (1961), designed for use with a Slee-type cryostat. For the Coons' (glove-box) instrument Lundy (1963) recommended a holder made of glossy paper and to a special design.

If a "warm" coverslip is employed for picking up the section, it is necessary to remove the ice film which forms on the knife surface as each section is picked up. This is conveniently carried out with a long artist's hog bristle brush.

(3) **Dry Ice Knife Cooler.** The standard pattern knives used with Cambridge Rocker microtomes have a conveniently long handle which projects from the side of the knife holder, to one side or other. A hemi-cylindrical box of brass or steel is provided with a slot to take the handle of the knife, and a reversible screw for maintaining its position (Fig. 11, p. 21). The screw is reversed when changing the box (and the knife) from one side of the holder to the other. The box is filled with powdered dry ice and moved in and out on the handle of the knife to adjust the temperature of the cutting zone to the desired temperature. One filling lasts 3 hours or more.

RAPID TECHNIQUE FOR DIAGNOSTIC BIOPSIES

(1) Mount tissue block (preferably not larger than 2 × 1·5 cm.) on the brass microtome chuck with a little water.

(2) Freeze in CO_2 jet and transfer to microtome.

(3) Cut two or three 4 to 8 μ sections and mount on "warm" coverslips.

(4) Fix for 1–2 minutes in absolute methanol, or in 5 per cent. acetic-ethanol.

(5) Stain one section in 1 per cent. aqueous methylene blue (30 seconds); rinse, blot, dehydrate in absolute alcohol, clear in xylene and mount.

(6) Stain another section in 1 per cent. Solochrome cyanine RS in 0·1 M-citric acid (pH 2·1) for 3 minutes. Wash in running water until differentiated and mount in glycerine jelly.

The first section should be available in less than 4 minutes from the time the tissue is received.

CRYOSTAT SECTIONS OF FORMALIN-FIXED TISSUES

(1) Fix small tissue blocks in formol saline (general studies), or in cold formol-calcium (enzymes), for the appropriate period.

(2) Wash blocks in running tap water for 5–10 minutes.

(3) Blot repeatedly with filter paper until the blocks are as dry as possible.

(4) Freeze on tissue holder with water or gum sucrose.

(5) Attach holder to microtome and adjust block temperature to between −5° and −10° (by raising internal temperature of cryostat or preferably by thermo-electric control.

(6) Cut and pick up sections in the usual manner. Albuminized slides or cover-slips are recommended.

For enzyme studies it is desirable to follow Stage 2 with a period of 24 hours in gum sucrose solution (Gum acacia, 2 g.; Sucrose 60 g.; Distilled water 200 ml.). Thereafter continue from Stage 3.

Table 42. CRYOSTAT MANUFACTURERS LIST

Manufacturer and address	Type	No. (if any)	Microtome(s)	Approx. temperature range (where known)
International Equipment Co., 300B Needham Hts., Mass 02194, U.S.A.	Open Top	CT1	International Minot. Rotary	−10° to −30°
W. Dittes, D 6900, Heidelberg, Bergstrasse 117, Germany	Glove box	—	Jung Rotary	
Lipshaw Mfg. Co., 7446 Central Avenue, Detroit, Mich. 48210, U.S.A.	Open Top	—	Lipshaw Rotary	—
Lab Teck Instrument Co., Westmont, Ill., U.S.A.	Open Top	—	Lab Teck Rotary	
De la Rue Frigistor Ltd., Canal Estate, Station Rd., Langley, Bucks, England	Open Top		Cambridge Rocker (Thermoelectric)	Block +30° to −35° Knife +30° to −50°
Bright's Refrigerator Service Ltd., Factory No. 3, Clifton Rd., Huntingdon, Hunts, England	Remote Control	BM/RS	Cambridge Rocker or Rotary Rocker	+20° to −30°
Slee Medical Ltd., Lanier Works, Hither Green Lane, London S.E. 13, England Slee International Inc., 8010 U.S. Highway 130, Pennsauken, N.J. 08110, U.S.A. Slee Mainz GmbH, Mainz-Gonsenheim, Heidesheimer strasse, 1, Germany	Open Top	C	Cambridge Rocker	+20° to −30°
	Remote Control	H	Cambridge Rocker	+20° to −30°
		HR	Slee Rotary	+20° to −30°
		HM	Internat. Minot	+20° to −30°
		HRF	Slee Rotary (Fine Feed)	+20° to −70°
	Mechanized	HRM	Slee Rotary	+20° to −30°
		MS	Leitz Sledge	+20° to −30°
Spencer (Optical), 141 Fulton Street, New York, N.Y., U.S.A.	Open Top	—	Spencer Rotary	+10° to −30°
W.K.F. Gesellschaft fur Electrophysikalischen Apparatebaue, 6101 Brandau/Darmstadt 2, Germany	Glove box	—	Jung Rotary	0° to −45°

APPENDIX 3

PREPARATION OF TISSUES BY FREEZE-DRYING

THE scheme of events detailed below refers to the use of an apparatus where the drying tissues are carried on a cold platen or thermoelectric module as in the Thieme and Edwards designs. Particularly it refers to the latter. The preliminary procedures are the same, however, whatever type of apparatus is employed.

PRELIMINARY ATTENTIONS

Before commencing operations it is advisable to check the performance of unit with a clean dry empty chamber in order to ensure that the vacuum is adequate.

(1) With the Pyrex lid in place, switch on the rotary pump (Main switch to position PUMP). Close the two air admittance valves, open the isolation valve. The chamber pressure should reach 0·5 Torr within 1 minute.

(2) Close isolation valve, open chamber air admittance and remove the lid.

(3) Fill the tray or trays provided for the purpose with P_2O_5 or Siccapent.* Replace the lid (the "O" ring should be lightly greased with silicone grease), close chamber air admittance, open isolation valve. The chamber pressure should fall to:

0·05 Torr in 1 minute
0·02 Torr in 2 minutes
0·01 Torr in 5 minutes

Assuming that the machine's performance is satisfactory:

(4) Turn main switch to position COOL. Turn on cooling water (minimum rate 5 litres per minute). Adjust the voltage regulator to obtain the lowest possible temperature on the top module.

(5) Allow the machine to operate for 10 minutes.

QUENCHING

(1) Pieces of tissue, from 1 to 6 mm. thick, are cut from the original larger specimens. In area they can be of any size up to that of the module itself. Small organs can be dried *in toto*.

(2) These are placed on narrow plates of copper or brass, or on thin strips of copper or aluminium foil.

(3) The plate or foil, with its mounted tissues, is plunged into a tube containing one of four alternatives: (1) isopentane cooled to about $-160°$ with liquid nitrogen (there is no need to measure the exact temperature; a point at which the isopentane becomes noticeably more viscous is chosen); (2) propane cooled to $-185°$ (see Stephenson, 1954, on need for caution); (3) dichlorodifluoromethane (Arcton 12,† Freon 12; CCl_2F_2) cooled to $-158°$ with liquid air or O_2; (4) monochlorofluoromethane (Arcton 22,† Freon 22; $CHClF_2$).

(4) The frozen tissues are dislodged from the plate or strip of metal with the tip of a metal rod.

(5) Excess coolant is poured off.

DRYING

(1) Close isolation valve, open chamber air admittance, remove lid, close chamber air admittance.

* The Schuchardt A.G. München, Germany.
† I.C.I. Ltd.

(2) With cold forceps transfer the specimens as rapidly as possible from the quenching bath to the module (or to the specimen holder standing thereon).

(3) Replace the lid and instantly open the isolation valve.

(4) When the vacuum reaches 0·5 Torr adjust the voltage regulator to obtain the desired drying temperature. (This may vary from −10° for rapid histological preparations, to −65° for certain critical studies. These limits refer to the standard module unit.) A suitable temperature for most histochemical purposes is −40°.

(5) No further attention to the machine is required during the drying run.

The vacuum will fall to 0·001 Torr, or lower, well before drying is complete. It will rise only if the vapour trap (P_2O_5) is completely saturated with moisture. The vacuum and module temperature are constantly indicated.

The average drying times given in the Table below will provide some indication of the probable duration of a given drying run.

TABLE 43

Drying Times for Mammalian Tissues as a Function of Thickness and Module Temperature

Thickness of tissue	Temperature maintained	Spleen	Kidney	Liver
1·0 mm.	−10° C.	55 min.	1 hr.	45 min.
2.0 mm.	−10° C.	1 hr.	2 hr.	1 hr.
3·0 mm.	−10° C.	3 hr., 30 min.	3 hr., 30 min.	3 hr., 30 min.
4·0 mm.	−10° C.			
1·0 mm.	−20° C.	1 hr.	1 hr. 10 min.	45 min.
2·0 mm.	−20° C.	2 hr., 30 min.	3 hr.	1 hr., 30 min.
3·0 mm.	−20° C.	4 hr.	4 hr.	4 hr.
4·0 mm.	−20° C.	5 hr.	5 hr., 30 min.	5 hr.
1·0 mm.	−30° C.	1 hr., 20 min.	1 hr., 20 min.	1 hr., 20 min.
2·0 mm.	−30° C.	3 hr.	3 hr.	3 hr.
3·0 mm.	−30° C.	3 hr., 45 min.	4 hr.	3 hr., 45 min.
4·0 mm.	−30° C.			
1·0 mm.	**−40° C.**	**2 hr.**	**2 hr.**	**2 hr.**
2·0 mm.	**−40° C.**	**3 hr., 45 min.**	**3 hr., 45 min.**	**3 hr., 45 min.**
3.0 mm.	**−40° C.**	**4.hr., 30 min.**	**4 hr., 30 min.**	**4 hr., 30 min.**
4·0 mm.	**−40° C.**	**6 hr., 30 min.**	**6 hr., 45 min.**	**6 hr., 30 min.**
1·0 mm.	−50° C.	3 hr.	3 hr.	3 hr.
2·0 mm.	−50° C.	4 hr., 45 min.	4 hr., 45 min.	4 hr., 45 min.
3·0 mm.	−50° C.	6 hr., 15 min.	6 hr., 15 min.	6 hr., 15 min.
4·0 mm.	−50° C.			
1·0 mm.	−60° C.	5 hr., 30 min.	5 hr., 30 min.	5 hr., 30 min.
2·0 mm.	−60° C.	8 hr.	8 hr.	8 hr.
3·0 mm.	−60° C.	Dry overnight	Dry overnight	Dry overnight
4·0 mm.	−60° C.	Dry overnight	Dry overnight	Dry overnight
2·0 mm.	−69° C.	Not tested	16 hr.*	16 hr.
3·0 mm.	−69° C.	Not tested	16 hr.*	16 hr.

* Were completely dry by the 16th hour.

(6) When drying is complete, switch off the current to the modules. (Turn voltage regulator to zero, turn main switch to PUMP.)

Allow the module to reach room temperature.

(7) Close isolation valve, open chamber air admittance, remove lid.

(8) Process tissues as required. For standard paraffin embedding:

(9) Place tissues in melted 45–56° wax.*

(10) Transfer to vacuum-embedding bath for 10 minutes.

(11) Embed in the normal manner.

Alternatively:

(9) Transfer tissues to vapour fixation chamber and treat according to individual requirements. (See Appendix 5, p. 605.)

(10) Subsequently, place tissues in 56° wax.*

(11) Transfer to vacuum-embedding bath for 10–60 minutes.

Alternatively to embed tissues within the unit, in previously degassed wax (see below):

(7) Turn main switch to HEAT position. Slowly increase voltage until the wax melts.

(8) Maintain suitable temperature for 10 minutes.

(9) Return voltage regulator to zero, turn main switch to PUMP.

(10) Allow module to cool. Proceed as above to remove tissue boat and embedded tissues.

Switching-off

When the unit is to be closed down for any length of time (e.g. overnight) the chamber should be maintained under vacuum with fresh P_2O_5 in the trays.

(1) Turn main switch to PUMP.

(2) Fill desiccant trays.

(3) Replace lid, close chamber air admittance valve, open isolation valve.

(4) Evacuate chamber to about 0·1 Torr.

(5) Close isolation valve.

(6) Switch off Pirani gauge.

(7) Open rotary pump air admittance valve.

(8) Switch off mains current.

(9) Switch off cooling water supply.

Degassing Procedure for Paraffin Wax

The tissue boat (see Fig. 26, facing p. 50), filled with wax, must have a central space entirely free of wax so that contact between the metal and the tissue is not prevented.

(1) Close rotary air pump admittance valve, close isolation valve.

(2) Turn main switch to HEAT and bring the top module to a temperature of 2–3 degrees above the melting-point of the wax selected.

(3) Place a suitably shaped block of perspex (lucite), or other clean solid material, in the centre of the tissue boat. Pour melted wax to fill the remaining space.

N.B. The bottom of the boat must be clean and dry.

(4) Transfer boat to dryer and replace lid.

(5) Open isolation valve.

(6) When degassing is complete (judged by cessation of spluttering) return voltage regulator to zero. Allow wax to solidify.

(7) Close isolation valve, open chamber air admittance valve. Remove lid, clean off wax deposits from lid and baseplate.

(8) Replace chamber lid and maintain under vacuum (see Switching-off procedure) until unit is required for use.

* For satisfactory embedding it is essential to use paraffin waxes containing fairly high amounts of plastic polymers. Suitable waxes are Ralwax (R. A. Lamb, 25 St. Stephen's Road, London, W.13) or Tissuemat (Fisher).

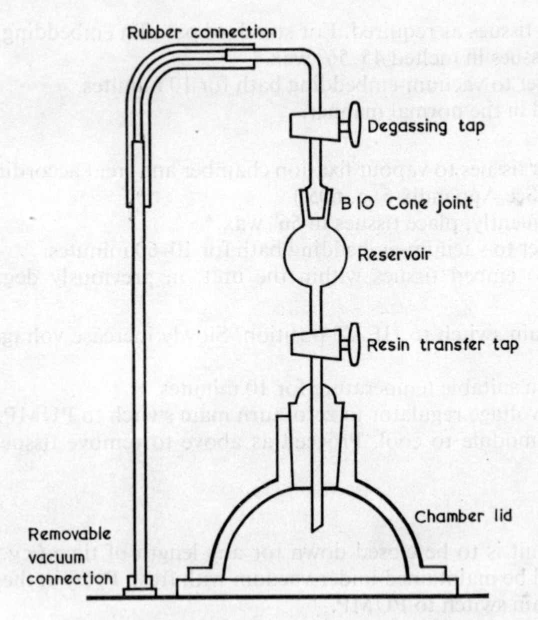

FIG. 161. Resin Embedding Accessory.

AIR BALLASTING

If the tissue dryer is used continuously it is essential that the pump be operated on full air ballast for one hour at least three times a week. This procedure cleans the pump oil and maintains the full vacuum performance of the unit.

RESIN EMBEDDING

The resin embedding accessory for the Speedivac Tissue Dryer allows a resin to be degassed at any stage of the drying run without affecting the latter. The procedure is as follows:

(1) Before commencing the drying run remove the cone joint in the chamber lid and insert the resin dropping funnel.

(2) Close the taps. Connect the removable vacuum connection.

(3) As required, fill the reservoir with resin by removing the standard B 10 cone.

(4) Replace B 10 cone, open degassing tap.

(5) When degassing is complete, close degassing tap.

(6) When tissue is dry open the resin transfer tap and allow resin to drip into the tissue boat, to cover the tissue.

(7) Close resin transfer tap.

(8) Admit air to the chamber.

(9) Remove boat and carry out appropriate hardening routine in suitable capsules.

DOUBLE-EMBEDDING PROCEDURE FOR FREEZE-DRIED TISSUES (AFTER BURSTONE (1962))

(1) Place the dried tissues in cold acetone (0°–4°) for about 4 hours.

(2) Transfer to cold absolute ethanol (0°–4°) for 16 hours.

(3) Transfer to cold alcohol-ether (1:1) at 0°–4° for 3 hours.

(4) Transfer to 5 per cent. celloidin (Necol or flexible collodion U.S.P.), diluted with an equal volume of alcohol-ether, for about 16 hours.

(5) Remove excess celloidin with blotting paper and transfer tissues to two changes of chloroform for 1 hour each.

(6) Transfer tissues to melted 56° paraffin wax and infiltrate under vacuum for 30–45 minutes.

(7) Cut sections and float out on water.

TABLE 44
Commercial Freeze Dryers

Manufacturer and Address	Type of Instrument
Canal Scientific, 5635 Fisher Lane, Rockville, Md. 20852, U.S.A.	Cold Hand.
Virtis Co., Route 200, Gardiner, N.Y. 12525, U.S.A.	Conduction Heat.
Fisher Scientific, 633 Greenwich Street, New York 14, N.Y., U.S.A.	Cold Finger.
HETO, Kirkeröd, Denmark	Thieme design Refrigerant cooled head.
Edwards High Vacuum Ltd., Manor Royal, Crawley, Sussex, England; Edwards Alto Vuoto, S.p.A., Via Piosacane arg. Via Dante, Pero (Ml), Italy; Edwards High Vacuum (Canada) Ltd., 430.S. Service Road, W. Oakville, Ontaria, Canada; Edwards Hochvakuum GMBH, Frankfurt/Main—Niederrad, Hahnstrasse 46, Germany; Edwards High Vacuum Inc., 3279 Grand Boulevard, Grand Island, N.Y., U.S.A.	Thermoelectric.*
E. Leybold, 5 Köln-Bayental, Bonner Strasse 504, Germany	(*a*) Refrigerant Cooled Conduction Type. (*b*) Thermoelectric.*
Phipps & Bird Inc., 6th and Byrd Streets, Richmond, Va., U.S.A.	Conduction Heat.

* Edwards High Vacuum Ltd. hold patents on this design.

APPENDIX 4

PREPARATION OF TISSUES BY FREEZE-SUBSTITUTION

(1) Block Substitution Techniques

(A) GLYCOL TECHNIQUE OF BLANK, MCCARTHY AND DE LAMATER

Method. (1) Cut fresh slices of tissue 2–3 mm. thick into cold propylene glycol at −20°.

(2) After 2 or more hours transfer to a mixture of Carbowax 4000, 9 parts; Carbowax 1500, 1 part, at 55°. Allow the wax to infiltrate for 2–3 hours and then prepare blocks.

This method was recommended by the original authors for lipid studies.

(B) PEYROT'S MODIFICATION OF LISON TECHNIQUE FOR GLYCOGEN

Method. (1) Immerse very small pieces of tissue briskly in a solution of absolute acetone saturated with dry ice (−80°).

(2) Transfer the frozen pieces to Rossman's fluid maintained at −60° to −65° in a container of dry ice.

(3) After 24–36 hours allow the fluid to rise slowly to room temperature.

(4) Wash two or three times in absolute alcohol.

(5) Clear and embed in the usual manner.

(C) THE WOODS AND POLLISTER TECHNIQUE FOR PLANT TISSUES

Method. (1) Quench small pieces of tissue (less than 1 $mm.^3$) in isopentane at −160°.

(2) Transfer quickly to absolute ethanol at −41° to −45° (dry ice/ethyl oxalate).

(3) After 3 days replace the cold ethanol with either 75 per cent. ethanol at −40° or with 95 per cent. ethanol containing 0·3 ml. acetic acid per 100 ml., also at −40°. After 30–60 minutes allow to rise to room temperature and maintain for the same period.

(4) Dehydrate in absolute alcohol and embed in the usual way.

(D) GENERAL TECHNIQUE FOR MAMMALIAN TISSUES (AFTER FEDER AND SIDMAN)

Method. (1) Quench small (1 to 3 mm. diameter) specimens by rapid immersion in 3:1 propane-isopentane cooled to −170° with liquid N_2. (Any equivalent system may be employed.)

(2) Transfer specimens rapidly into jars containing the chosen solvent or solvent-fixative at −70°.

The substituting fluid is usually acetone or ethanol. The former may contain 1 per cent. osmium tetroxide or 2 per cent. chromium trioxide, and the other may contain 5 per cent. picric acid or 1 per cent. mercuric chloride.

(3) Maintain the specimen jars at −70° for 2–7 days, preferably in an apparatus providing for continuous stirring.

(4) If the solvent contains any additional fixative, wash specimens *well* with fresh dry* solvent at −70°.

(5) Transfer specimens, in solvent, to a refrigerator at 0°–4°.

* Solvents should be dried over anhydrous sodium sulphate or with molecular sieves such as Linde type 4A. In the latter case subsequent ultrafiltration may be required to remove suspended particles.

(6) When the solvent reaches this temperature, transfer specimens to absolute alcohol at 0–4° (3 changes in 24 hours).

(7) Transfer to chloroform at 0°–4° for 6–12 hours.

(8) Transfer to fresh chloroform at 0° and then allow the bath to reach room temperature.

(9) Change to 52° paraffin wax (2 changes, 15 minutes each).

(10) Embed in fresh 52° wax.

Alternatively:

(6) Transfer specimens to methanol-methyl cellosolve (1:1 v/v) at 0°–4°. Change bath 3 times, 15 minutes each.

(7) Warm to room temperature and transfer to a rising series of concentrations of polyester wax* in solvent at 37°.

(8) Infiltrate with fresh wax at 37° (3 changes) for at least 24 hours.

(9) Pour wax into former and orientate specimen at room temperature. Place block in refrigerator to harden.

(10) Cut on microtome provided with thermoelectric tissue holder and knife and cut with block at −5° and knife at 0°.

(E) SUBSTITUTION TECHNIQUE FOR E.M. (AFTER FERNÁNDEZ-MORÁN)

Method. (1) If feasible, infiltrate specimen (0·5 mm. or smaller) in increasing concentrations of glycerol in Ringer's fluid, buffered to pH 7·4 (up to 60 per cent. glycerol).

(2) Quench by ultra-rapid technique using liquid helium II (1° to 2° K) if available. If not, use the most suitable alternative (as for freeze-drying).

(3) Substitute specimens at −80° to −130° in acetone-alcohol-ethyl chloride mixtures containing 1–2 per cent. platinum or gold chlorides, or osmium tetroxide. The duration of this process may be several weeks.

(4) Infiltrate specimens in (*a*) methacrylate/butyl/methyl methacrylate mixtures at −75° or (*b*) in a mixture containing 5–10 ml. methacrylate, 75–85 ml. *n*-butyl methacrylate and 10–15 ml. methyl methacrylate, for 24–48 hours at −25°.

(5) Continue infiltration at −25°, as in (*b*) above, in the same mixture with the addition of 0·2 to 0·3 per cent. benzoin and 1 per cent. benzoyl peroxide.

(6) Embed specimens at −25° in the above mixture by photopolymerization (48–72 hours).

(F) SUBSTITUTION TECHNIQUE FOR E.M. (AFTER BULLIVANT)

Method. (1) Treat small pieces of tissue successively in an ascending series of glycerol baths, made up in isotonic veronal-acetate buffer, pH 7·2 (Appendix 1, p. 584) as follows:

Glycerol (%)	3	6	10	20	30	45	60
Time (minutes)	5	5	5	10	15	20	30
Temperature (° C.)	4	4	4	4	4	4	−25

(2) Quench the glycerinated tissues in liquid propane cooled to −175° in liquid nitrogen.

(3) Transfer to dry ethanol at −75° (14 days, 3 changes of solvent).

(4) Fix according to the following schedule:

(*a*) 75 per cent. ethanol: 25 per cent. 90/10 butyl/methyl methacrylate, for 1 day at −75°.

(*b*) 50 per cent. ethanol: 50 per cent. 90/10 butyl-methyl methacrylate; for 1 day at −75°.

(*c*) 25 per cent. ethanol: 75 per cent. 90/10 butyl/methyl methacrylate, for 1 day at −75°.

* See Appendix 1, p. 576.

(*d*) 100 per cent. 90/10 butyl/methyl methacrylate, for 3 days at $-75°$.
(*e*) 100 per cent. 90/10 butyl/methyl methacrylate with 1 per cent. Lucidol, for 1 week at $-75°$.
(*f*) 100 per cent. 90/10 butyl/methyl methacrylate with 1 per cent. Lucidol, for 3 days at $-25°$.
(*g*) Polymerize by U.V. light, 3 days at $-25°$.
(*h*) Place in Durcupan mixture and polymerize at 45°.
(*i*) Leave for 1 day in a 50:50 mixture of ethanol and Epon mixture.
(*j*) Embed in Epon according to the method of Luft (1961).

Sections may be stained in 1 per cent. uranyl acetate for 1 hour, or by other metal impregnation techniques.

(2) Section Substitution Techniques

Method. (1) Cut freshly excised tissue into 1–2 mm. slices. Quench in liquid N_2 or in isopentane cooled to $-185°$.

(2) Cut 8–10 μ cryostat sections and allow them to fall into a suitable cold container. (With open top cryostats this is not possible; alternatively pick up sections on a dry brush, directly from the knife.)

(3) Transfer carefully (with brush) to a screw-cap vial (2 inch diameter) filled with dry acetone chilled to $-70°$ with powdered dry ice.

(4) Return vial to dry ice and leave for 12 hours.

(5) Open vial and insert (with forceps) a clean coverslip which has been prechilled in cold acetone. Agitate gently to float sections and guide a floating section on to the coverslip (with a glass needle). Withdraw and allow to dry.

(6 Optional) Coat with cold celloidin (1 per cent. in 30 parts ether, 30 parts ethanol and 40 parts acetone) chilled to $-70°$. Remove and allow to dry.

(Celloidin coating is essential for studies on oxidative enzymes and 5-nucleotidase.)

APPENDIX 5

FIXATIVES

FORMALIN is the most widely used fixative in histological practice and the diagnostic histochemist will probably continue to receive a great deal of material already immersed in formalin-containing fluids. Since it is difficult to arrange to receive routine surgical and autopsy specimens in the fresh condition, and to freeze selected blocks in isopentane cooled with liquid N_2 (−180°) prior to storage at −70°, it is necessary to be reconciled to the evils of formalin. After recent demonstrations of the excellent preservative qualities of formalin in respect of many common enzymes, this fixative is gradually returning to favour with histochemists.

The bad qualities of formalin from the histological point of view are: (1) it allwos considerable shrinkage during subsequent embedding; (2) it forms black acid hæmatin (formalin) pigment in many tissues; (3) it gives poor nuclear definition, especially in very cellular tissues; and (4) it acts slowly. From the histochemical, as well as the histological point of view the first objection can be overcome by the use of frozen sections. The second objection can be overcome by maintaining the pH of the formalin solution in the region of neutrality and nothing but neutral or, preferably, buffered neutral formalin should be used for histochemical work. The fourth objection becomes in many cases a histochemical advantage, especially in enzyme work, but the third remains. The presence or absence of NaCl makes some difference, according to the views of Birge and Tibbitts (1961), but it is seldom employed either for morphological preservation or for enzyme fixation. Below is given the buffered 10 per cent. neutral formalin of Lillie (1954) which is recommended for routine histochemical use:

40 per cent. formaldehyde	100 ml.
Distilled water	900 ml.
NaH_2PO_4, H_2O	4 g.
Na_2HPO_4, (anyhydrous)	6·5 g.

(Fixation time 24–72 hours.)

As an alternative I employ either a variant of Baker's (1944) formol-calcium:

40 per cent. formaldehyde	150 ml.
1·3 per cent. calcium chloride*	850 ml.

(Fixation time 16–24 hours.)

or better, perhaps, his 1946 variant:

40 per cent. formaldehyde (Analar, Reagent grade)	100 ml.
Distilled water	500 ml.
10 per cent. calcium chloride*	100 ml.

Make up to 1 litre and add marble chips.

The use of buffers may be avoided by using Amberlite IR-45 resin (Rohm and Hass Co., Philadelphia) either as a column or as a layer of resin at the bottom of the bottle.

Formalin-stable enzymes are well demonstrated in frozen sections from tissues fixed in these fixatives at 4° for 16 hours. The following routine is used for paraffin embedding after cold formalin fixation:

(1) Rinse once in cold (0°–4°) distilled water and leave in excess cold distilled water for 24 hours.

* Anhydrous, granular, technical grade.

(2) Begin dehydration in three changes of absolute acetone, each for 20 minutes at 0°–4°.

(3) Continue dehydration in two changes of absolute acetone containing 1 per cent. Necol, 30 minutes in each, at 0°–4°.

(4) Transfer to 5 per cent. Necol in acetone, 30–60 minutes (0°–4°).

(5) Harden and clear in chloroform for 90 minutes, allowing the temperature to rise, at this stage, to that of the environment.

(6) Embed quickly in 56° paraffin wax *in vacuo* or, better, in one of the lower melting-point waxes or wax mixtures.

It is now more usual to follow cold formalin fixation for enzymes with storage in cold gum sucrose, introduced by Holt *et al.* (1960). The whole procedure is as follows:

(*a*) Phosphate-buffered formaldehyde:

40 per cent. formaldehyde (Analar, Reagent grade)	100 ml.
0·067 M phosphate buffer, pH 7·2	500 ml.
Sucrose	75 g.

Dissolve the sucrose and make up to 1 litre with buffer.

(*b*) Gum-sucrose:

Mix, in the dry state, 30 g. sucrose and 1 g. gum-acacia. This disperses the gum and prevents clumping during solution. Add distilled water to 100 ml. Cold storage is advised, or a crystal of thymol may be added. The final concentration of sucrose is 0·88 M.

(1) Fix in (*a*) for 18–24 hours at 0°–4°.*

(2) Wash briefly in distilled water at 0–4°.

(3) Transfer to (*b*) for 12–24 hours, or longer, at 0–4°.

For enzyme cytochemistry (less frequently for light microscopy) other fixatives may be used such as those suggested by Sabatini *et al.* (1963). Suitable conditions are given in the Table below.

TABLE 45

Fixation for E.M. Enzyme Cytochemistry

(After Janigan)

Fixative	Amount (ml.)	Sucrose*	Buffer† to	Final Vol. & Con.
Formaldehyde 40%	5	3·76 g.	0·1 M, pH 7·4	50 ml. 4%
Glutaraldehyde 25%	13	nil	0·1 M, pH 7·2	50 ml. 6·5%
Acrolein 100%	5	nil	0·1 M, pH 7·6	50 ml. 10%
Methacrolein 100%	2·5	3·76 g.	0·1 M, pH 7·6	50 ml. 4%
Crotonaldehyde 100%	5	7·52 g.	0·1 M, pH 7·4	50 ml. 10%
Glyoxal 30%	6·7	3·76 g.	0·2 M, pH 6·5	50 ml. 4%
Hydroxyadipaldehyde 22·5%	27·7	7·52 g.	0·1 M, pH 7·5	50 ml. 12·5%

* Dissolve sucrose in buffer before bringing to final volume.
† Phosphate or cacodylate (See Appendix 1).

Fixation times for blocks 3–4mm.^{3} normally from 2–6 hours.

Fixatives for Special Purposes

Glycogen. With the possible exception of Bouin's fixative, the formalin-containing fixatives in routine use are all more or less objectionable from the histochemical

* Thin slices, 1–2 mm. thick should be cut cleanly with a clean (grease-free) razor blade. No material should exceed 3–4 mm^3.

point of view. Bouin's fluid is sometimes used as a routine histological fixative for glycogen, but for histochemical purposes it should be superseded by Lison's "Gendre fluid."

Saturated picric acid in 96 per cent. alcohol	85 parts
40 per cent. formaldehyde	10 parts
Glacial acetic acid	5 parts

This fixative is cooled to −73° before use (fixation time 18 hours), and the resulting localization of glycogen compares favourably with that shown by freeze-dried tissues. The weak Bouin fixative for use with Baker's acid hæmatein test is given in Appendix 12.

There is little advantage in using any of the above fixatives in preference to absolute ethanol, provided that very small blocks are taken.

Nucleic Acids. For routine studies of the nucleic acids in paraffin-embedded material Carnoy's fixative is recommended:

Ethyl alcohol	60 ml.
Chloroform	30 ml.
Glacial acetic acid	10 ml.

For smears and for freeze-dried sections fixation in absolute methanol is preferable.

An alternative to Carnoy, which gives excellent fixation without interfering with the action of nucleases, is Lillie's (1949) AAF (acetic alcohol formalin):

40 per cent formaldehyde	10 ml.
Acetic acid	5 ml.
Ethanol	85 ml.

(Fixation time 24 hours at 4°.)

Newcomer (1953) designed his isopropanol fixative especially for the preservation of chromosomes. It is excellent for this purpose and permits most histochemical reactions to be carried out. Feulgen hydrolysis is prolonged, however, to 10 minutes for plant tissues and to 20 minutes for animal tissues.

Isopropanol	6 parts
Propionic acid	3 parts
Light petroleum	1 part
Acetone	1 part
Dioxane	1 part

(Fixation time 12–24 hours.)

Antigen-antibody Reactions. For both freeze-dried and cryostat sections the best fixative is absolute acetone, closely followed by 95 per cent. ethanol. All aldehyde fixatives, especially glutaraldehyde and acrolein, rapidly interfere with antibody-binding sites on tissue antigens. Ball *et al.* (1964) observed that while intracellular "rheumatoid factor", γ^2 globulins, and nuclear antigens could all be demonstrated by fluorescent antibody techniques in ethanol-fixed paraffin sections, only "rheumatoid factor" was found to resist glutaraldehyde.

Mucopolysaccharides and Mucoproteins

There are several more or less satisfactory variants of the older lead subacetate mixtures for fixing acid mucopolysaccharides. Two are given below. They are Lillie's (1954) alcoholic lead nitrate and the ethanol-acetic-subacetate of Mota, Ferri, and Yoneda (1956). The latter was devised for and is particularly good for the fixation of mast-cell granules.

The formula for Lillie's fixative is as follows:

Lead nitrate	8 g.
40 per cent. formaldehyde	10 ml.
Water	10 ml.
Ethanol	80 ml.

(Fixation time 24 hours at room temperature; 2–3 days at 4°; 10–14 days at −25°.)

Mota *et al.* formula is given below:

Lead subacetate	1 g.
Ethanol	50 ml.
Water	50 ml.
Acetic acid	0·5 ml.

(Fixation time 24 hours.)

The two fixatives for acid mucopolysaccharides introduced by Williams and Jackson are as follows:

(1) 40 per cent. formaldehyde	10 ml.
Cetyl pyridinium chloride	0·5 g.
Water	90 ml.

(Fixation time 48 hours.)

(2) Ethanol	50 ml.
5-aminoacridine hydrochloride	0·4 g.
Water	50 ml.

(Fixation time 48 hours.)

For the soluble acid mucopolysaccharides of Hurler's disease Wolfe (1964) recommended the following routine:

(1) Fix in dioxane for 24–48 hours.
(2) Double embed in celloidin and paraffin.
(3) Cut sections at 6 μ and float on to albuminized slides from a warm bath of 80 per cent. ethanol.
(4) Dry on a warm hotplate.
(4) Deparaffinize in xylene and transfer through absolute to 95 per cent. ethanol.
(6) Stain in 1 per cent. toluidine blue in 80, 70, and 50 per cent. ethanol (3 minutes in each).
(7) Differentiate and dehydrate in absolute acetone.
(8) Clear in xylene and mount.

This method is to be recommended if, for any reason, freeze-drying and formaldehyde vapour fixation cannot be employed.

ENZYMES

A few special fixatives may be mentioned here. In general the best fixation for a given enzyme method will be mentioned in the appropriate appendix.

Acid Buffered Acetone (Kaplow and Burstone, 1963). For blood and bone-marrow smears, use 60 per cent. acetone buffered to pH 4·2 with citrate. Fix for 20–30 secs at room temperature.

Good preservation of phosphatases, acetyl- and chloroacetyl-naphthol esterases, peroxidase and DOPA-oxidase.

Formalin-Sucrose-Ammonia (Pearson, 1963). For cholinesterases the following gives excellent preservation.

Formalin 40 per cent. (Analar, Reagent grade)	10 ml.
Sucrose (Analar, Reagent grade)	15 g.
0·880 Ammonia	1 ml.
Distilled water to	100 ml.

Fix in the above (pH 6·7) for 18–24 hours.

LIPIDS

Baker's (1944) formol-calcium has already been mentioned (p. 601) as a general fixative which preserves phospholipids well without destroying more than afr action of the various formol-stable enzyme systems.

If the preservation of enzymes is not required Elftman's (1957) fixative may be employed.

Mercuric chloride	5 g.
Potassium dichromate	2·5 g.
Water	100 ml.

(Fixation time 3 days at room temperature.)

Fixation with osmium tetroxide is seldom required in histochemistry. Mixtures containing an oxidant as well as OsO_4 serve to distinguish normal from degenerating myelin, and they preserve lipids and lipoproteins. Swank and Davenport's (1935) fixative is a suitable one:

1 per cent. potassium chlorate	60 ml.
40 per cent. formaldehyde	12 ml.
Acetic acid	1 ml.

(Fixation time 6–10 days.)

Vapour Fixatives

The following vapours are used after freeze-drying.

Transfer the dry tissue block directly from the drying chamber to an enclosed thermostatic chamber provided with a suitable heating mechanism (Fig. 32, facing p. 51).

The apparatus illustrated provides continuous readings of the humidity and means of controlling the latter.

Formaldehyde. This vapour is obtained by heating paraformaldehyde at temperatures between 50° and 80°. For pure fixation the water vapour content of the thermostatic chamber is not critical. For fluorescent amine technique see Appendix 26, Vol 2. Optimum conditions 3 hours at 50°.

Osmium tetroxide. Crystalline osmium tetroxide is maintained at 37°. Duration of fixation, up to one hour.

Glutaraldehyde. To prepare a suitable anhydrous solution add 100 ml. 25 per cent. glutaraldehyde to 100 ml. heavy liquid paraffin (British Drug Houses Ltd). Shake automatically for several hours. Allow to stand and then remove the upper, paraffin, layer with a separating funnel.

Sediment remaining water by centrifugation. Liquid paraffin treated in this way contains about 10 per cent. glutaraldehyde.

Treat tissues for 7 hours at 60°.

Acrolein. Liquid acrolein is used, at 37° for 1 hour.

Chromyl Chloride. Liquid chromyl chloride is used, at 37° for 1–2 hours.

Double embedding is recommended:

(1) Place tissue block in alcohol: ether (equal parts) at −70° for 3 hours.
(2) Transfer to 2 per cent. celloidin in alcohol: ether at −70° for 16 hours.
(3) Remove excess celloidin by blotting.
(4) Transfer to 2 changes of chloroform, 1 hour each, at −70°.
(5) Allow to reach room tempreature.
(6) Transfer to 56° paraffin wax and impregnate under vacuum for 1½ hours.

APPENDIX 6

THE MILLON REACTION (Bensley and Gersh Modification)

(*Freeze-dried, formalin; alcohol, etc., paraffin sections*)

PREPARATION OF THE REAGENT

Add 400 ml. conc. nitric acid (Sp. G. 1·42) to 600 ml. distilled water and stand 48 hours. Dilute 1:9 with distilled water and saturate with crystalline mercuric nitrate. Filter and add to 400 ml. of the filtrate 3 ml. of the original diluted nitric acid and 1·4 g. sodium nitrite.

Method. (1) Remove wax from paraffin or freeze-dried sections with light petroleum.

(2) Rinse in absolute acetone and allow to dry in air.

(3) Cover section with Millon reagent and either stand at room temperature until the maximum colour develops or place in a covered Petri dish in the 60° incubator. The reaction proceeds slowly in the cold, but is complete in 30–60 minutes in the incubator.

(4) Rinse in cold or warm 2 per cent. nitric acid (2 ml. conc. acid Sp. G. 1·42, 98 ml. water).

(5) Dehydrate rapidly in 70 per cent. and absolute alcohol.

(6) Clear in xylene.

(7) Mount in Canada balsam.

Result (Fig. 43). Proteins containing tyrosine stain orange to rose-red. The colour is stable for over 12 months.

MILLON REACTION (Baker Modification)

(*Formalin; paraffin sections*)

The original author advises celloidin embedding, but this is not absolutely necessary.

PREPARATION OF THE REAGENT

Add 10 g. $HgSO_4$ to 100 ml. 10 per cent. H_2SO_4 and heat until dissolved. Make up to 200 ml. Add 0·5 ml. 0·25 per cent. $NaNO_2$ to 5 ml. of above solution.

Method. (1) Bring sections through 50 per cent. alcohol to water.

(2) Place sections in a small beaker containing the reagent and boil gently.

(3) Stop heating and allow the solutions to come to room temperature.

(4) Remove sections and wash three times in distilled water (2 minutes each wash).

(5) Mount in glycerine jelly, or

(6) Dehydrate, clear and mount in DPX or other suitable synthetic medium.

Result. Tyrosine-containing proteins are stained red, pink or yellowish-red.

ACROLEIN-SCHIFF METHOD FOR PROTEIN
(van Duijn, 1961)

(*Freeze-dried, Carnoy, formaldehyde; paraffin sections*)

Method. (1) Bring sections to 95 per cent. ethanol.

(2) Incubate for 15–60 minutes in 5 per cent. acrolein* in 95 per cent. ethanol.

(3) Pass through 3 changes of ethanol, 5 minutes each, to water.

* Must be fresh, do not use if discoloured.

(4) Immerse in Schiff's reagent (Appendix 9) for 10–20 minutes.
(5) Wash in water.
(6) Dehydrate, clear, and mount in a synthetic resin medium.
Result. Magenta staining indicates reactive groups (NH_2, NH, SH, imidazole) in protein.

NITROSO-NAPHTHOL METHOD FOR TYROSINE

(Udenfriend and Cooper, 1952; Stoward, 1963)

(*Fresh frozen; freeze-dried, Carnoy, paraffin sections*)

Method. (1) Treat sections for 30 minutes at 60° with a mixture of equal parts of 1 per cent. ethanolic nitroso-naphthol and HNO_3 containing a trace of $NaNO_2$.
(2) Wash in ethylene dichloride to remove excess reagent.
(3) Wash in 98 per cent. ethanol and examine dry.
Result. A greenish-yellow fluorescence (fading) under U.V. microscope indicates sites containing tyrosine.

MERCURY-BROMPHENOL BLUE METHOD FOR PROTEINS

(after Bonhag)

(*Carnoy, formalin, etc.; avoid osmium*)

PREPARATION OF SOLUTION

Two alternatives have been employed: (1) 1 per cent. alcoholic bromphenol blue saturated with $HgCl_2$ and (2) 1 per cent. $HgCl_2$ and 0·05 per cent. bromphenol blue in 2 per cent. aqueous acetic acid. The second of these is recommended.
Method. (1) Bring paraffin sections to water.
(2) Stain in one of the two alternative solutions for 2 hours at room temperature.
(3) Rinse sections for 5 minutes in 0·5 per cent. acetic acid.
(4) Transfer sections directly into tertiary butyl alcohol.*
(5) Clear in xylene and mount in a suitable synthetic medium.
Result. Proteins are stained a deep clear blue colour.

DIAZOTIZATION-COUPLING METHOD FOR TYROSINE

(Glenner and Lillie, 1959)

(*Formalin, etc., hot chloroform/methanol, freeze-dried; paraffin sections*)
Method. (1) Bring paraffin sections to water.
(2) Nitrosate (in the dark) for 18–24 hours at 3° in a mixture containing 6·9 g. $NaNO_2$, 5·8 ml. acetic acid, distilled water to 100 ml.
(3) Rinse in four changes of ice-cold distilled water.
(4) Treat for 1 hour at 3° with a mixture containing 1 g. S-acid (8-amino-1-naphthol-5-sulphonic acid), 1 g. KOH, 1 g. ammonium sulphamate and 100 ml. 70 per cent. alcohol.†
(5) Wash in three changes of 0·1 N-HCl, 5 minutes in each.
(6) Wash in running water, 10 minutes.
(7) Dehydrate, clear, and mount in neutral synthetic resin.
Result. Tyrosine-containing proteins stain purplish-red to pink.

* This step obviates the necessity for restoring the pH to the neutral point in order to produce the blue form of the dye.
† This part of the reaction should also be carried out in the dark since the reaction product is light-sensitive.

ISOELECTRIC POINT BY MINIMUM SOLUBILITY

(after Catchpole)

(*Freeze-dried, paraffin sections*)

Method. (1) Cut sections of paraffin-embedded freeze-dried material at 6–10 μ.

(2) Float out on warm mercury (40° or 58° wax).

(3) Attach section to slide and dewax with light petroleum.

(4) Dry in air.

(5) Cover with selected buffer solution, and place section in a closed Petri dish.

A 0·1 M-veronal acetate buffer in stages of about pH 0·25 can conveniently be used between pH 3·0 and pH 9·0. Alternatively, acetate buffer may be employed from pH 3·6 to 5·4, phosphate buffer from pH 5·4 to 8·0 and borate buffer from pH 8·0 to 10·0. In the case of these buffers 0·1-M solutions and steps of pH 0·2 should be used.

(6) After 1–2 hours in buffer at room temperature, wash with fresh buffer solution.

(7) Immerse section in absolute alcohol for 8–16 hours.*

(8) Bring section to water (except for Millon and DNFB reactions) and stain by the chosen method.

The following methods are recommended:

Simple Proteins—Millon, DNFB, Acrolein-Schiff, coupled tetrazonium (pp. 606, 612 and 614).

Histones—Arginine reaction (p. 618).

Nucleoproteins—Feulgen reaction (p. 647); methyl green-pyronin (p. 651).

Mucoproteins and Mucopolysaccharide-protein Complexes—Periodic acid-Schiff (p. 660).

OXIDIZED TANNIN-AZO METHOD (OTA)

(Dixon, 1962)

(*Freeze-dried, Carnoy, formalin; paraffin sections*)

Method. (1) Remove wax and bring sections through ethanol to distilled water.

(2) Immerse sections for 10 minutes in a Coplin jar filled with aqueous tannic acid-HCl (10 g. tannic acid in distilled water, add 25 ml. N-HCl and make up to 250 ml.).

(3) Wash in 3 changes of distilled water.

(4) Place sections on rack and flood with 0·5 per cent. aqueous H_5IO_6. Allow to act for 5 minutes. The section becomes pale brown.

(5) Wash in 3 changes of distilled water.

(6) Place in ice-cold distilled water.

(7) Treat for 20 minutes with ice-cold buffered (pH 4·0) diazotized *o*-dianisidine,† in the refrigerator. The sections become dark salmon-pink in colour.

(8) Wash in running tap water. Dehydrate, clear, and mount in a neutral mounting medium.

(9) Control sections as above, omitting stages 4 and 5. These should show a faint pink only.

Result. Tannophilic protein, bright salmon-pink.

OXIDIZED TANNIN-OXALINE METHOD (OTO)

(Dixon, 1962)

Method. (1) Follow stages 1–5 of the OTA method.

(2) Wash with 3 changes of glacial acetic acid.

* When mucoproteins are involved a formalin-containing fixative is preferable since some materials of this class remain soluble in water after alcoholic fixatives. In the case of nucleoproteins, 10 per cent. acetic acid should be added to the alcohol.

† 100 mg. Fast blue B salt in 82 ml. ice-cold 0·2 M acetic acid and 18 ml. 0·2 M sodium acetate.

(3) Place sections in a Coplin jar containing a freshly prepared solution of 6-amino-3-dimethyl aminophenol in glacial acetic acid, 5–20 minutes at 37°. (Dissolve 500 mg. 6-nitroso-3-dimethylaminophenol in 50 ml. glacial acetic acid. Cool till crystals begin to form and add excess zinc dust. The orange colour disappears. Filter until 34 ml. of filtrate is obtained and make up to 100 ml. with glacial acetic acid. Use immediately.)

(4) Wash successively in glacial acetic acid, ethanol and xylene.

(5) Mount in neutral synthetic medium.

Result. Tannophilic proteins stain dark slate grey.

BIEBRICH SCARLET FOR BASIC NUCLEOPROTEINS

(Spicer and Lillie, 1961; Spicer, 1962)

(Buffered $HgCl_2$, Carnoy, Bouin; paraffin sections)

Method. (1) Bring sections to distilled water. (Remove mercury if necessary.)

(2) Stain for 30–90 minutes in a freshly prepared 0·72 mM solution of Biebrich Scarlet (C.I. 26905) in glycine buffer at pH 8, 9·5, 10·5.

TABLE 46

Glycine Buffer

pH	N-NaOH ml.	N. glycine ml.	Distilled water ml.
8	0·5	20	19·5
9·5	7	20	13
10·5	17	20	3

(3) Without rinsing in water, pass directly through 95 per cent. and absolute alcohol, to alcohol/xylene, and thence to pure xylene.

(4) Mount in a synthetic resin medium.

Result. At pH 9·5 basic nucleoproteins stain strongly. Above and below this pH level the staining decreases rapidly.

ACID SOLOCHROME CYANINE METHOD FOR BASIC PROTEINS

(Freeze-dried or fresh frozen sections with brief acetic-ethanol fixation, formalin-fixed frozen and paraffin sections; Carnoy, Zenker, Bouin, etc.)

Method. (1) Bring paraffin sections to water.

(2) Stain in 1·0 per cent. solochrome cyanine R* in 1 per cent. orthophosphoric acid (pH 1·7) or in 0·1 M-citric acid (pH 2·1) for 5–20 minutes at room temperature.

(3) Wash in running *hot* water until the section changes from orange through red and begins to turn blue.

(4) Pass rapidly through 70 per cent. to absolute alcohol.

(5) Clear in xylene and mount in a suitable synthetic medium.

Result (Plate Ic). Nuclear chromatin, dark steel blue.

Acidophil tissue components, various shades of red.

The blue staining of the nuclei is completely fast, but washing may remove too much of the red complex. This is easily restored by brief immersion in the staining solution, followed by a further brief wash in water.

* I.C.I. Ltd, Blackley, Manchester.

NINHYDRIN-SCHIFF METHOD FOR PROTEIN-BOUND NH_2

(Yasuma and Itchikawa, 1953)

The original authors recommended fixation in Zenker, with and without acetic acid, or absolute alcohol. I prefer 85 per cent. ethanol, Carnoy or 5 per cent. acetic-ethanol. Alternatively, cold microtome sections may be treated directly with the reagent.

Method. (1) Bring paraffin sections to water.

(2) Treat sections with 0·5 per cent. ninhydrin* in absolute ethanol for 16–20 hours at 37°.

(3) Wash gently in running water for 2–5 minutes.

(4) Immerse sections in Schiff's reagent† for 15–25 minutes.

(5) Wash in running water for 10 minutes.

(6) Counterstain nuclei, if required, in Mayer's hæmalum, wash, and differentiate in 1 per cent. acid alcohol.

(7) Dehydrate, clear and mount in a suitable synthetic resin.

Result. Proteins stain pinkish-red to magenta if they contain a sufficient number of reactive NH_2 groups. Not all reactive amines are oxidized by ninhydrin-Schiff, since it is possible to obtain a positive König-Sassi reaction after prolonged ninhydrin treatment.

KÖNIG-SASSI METHOD FOR PRIMARY AND SECONDARY AMINES

(After Stoward, 1963)

(*Freeze-dried, Carnoy or ethanol; paraffin sections. Fresh frozen*)

PREPARATION OF CYANOGEN BROMIDE‡ SOLUTION

Dissolve 4 g. KCN in 7·5 ml. distilled water and add this dropwise to 4 ml. liquid bromine, covered with 2·5 ml. water in a reaction flask immersed in ice-salt mixture. Stir for 2 hours.

The product is an orange-coloured liquid. Store at −20° in a screw-top bottle.

Method. (1) Bring sections to methanol.

(2) Immerse in 5 per cent. v/v pyridine/methanol, containing 2–3 drops of CNBr solution, for 3–5 minutes. (Use within 24 hours.)

(3) Rinse in methanol, ethanol, xylene.

(6) Mount in fluormount.

Result. An intense green fluorescence indicates reactive primary and secondary amines.

CHLORAMINE-T SCHIFF METHOD FOR PROTEIN-BOUND NH_2

(Chu, 1953; Burstone, 1955)

(*Carnoy, cold acetone, etc.*)

Method. (1) Bring paraffin sections to water.

(2) Treat with 1 per cent. aqueous chloramine-T§ (pH 7·5) for 6 hours at 37°.

(3) Wash briefly in distilled water.

(4) Treat with 5 per cent. aqueous sodium thiosulphate for 3 minutes.

(5) Rinse in distilled water and immerse in Schiff's reagent for 20–30 minutes.

* 1·0 per cent. Alloxan may be used as an alternative.

† Full-strength Schiff reagent should be used; either the Itikawa and Ogura (1954) or the de Tomasi (1936) variants are recommended (see Appendix 9, p. 647).

‡ Obtainable commercially: see Mol. Wt. and Source List, Vol. II.

§ 10 per cent. solutions of commercial preparations of sodium hypochlorite can be used in place of chloramine-T.

(6) Rinse in 10 per cent. aqueous sodium bisulphite.
(7) Wash in tap water.
(8) Dehydrate in alcohols, clear in xylene and mount in DPX.
Result. Protein-bound NH_2 groups stain pink or reddish-magenta.

HYDROXYNAPHTHALDEHYDE METHOD FOR PROTEIN-BOUND NH_2

(Weiss, Tsou and Seligman, 1954)

(*Ethanol, Carnoy, acetone, and paraffin sections, are best for cytoplasmic NH_2. Nuclear NH_2 is most clearly demonstrated after Zenker fixation*)

Method. (1) Bring paraffin sections to water, avoiding thiosulphate and iodine treatment in the case of mercury-fixed tissues.

(2) Incubate sections for 1 hour at room temperature in a freshly prepared solution containing 20 mg. 3-hydroxy-2-naphthaldehyde in 20 ml. acetone to which 30 ml. 0·1 M-veronal acetate buffer at pH 8·5 has been added.

(3) Wash in three changes of distilled water for 5 minutes in each.

(4) Develop sections by placing them in 0·1 M-veronal acetate buffer at pH 7·4 and shaking 25 mg. tetrazotized diorthoanisidine (Fast blue B salt) on to the surface. This is then stirred into the solution and the colour reaches its maximum in 3–5 minutes.

(5) Wash in running tap water for 5 minutes.

(6) Dehydrate in alcohols, clear in xylene and mount in a suitable synthetic medium.

Result (Fig. 45). A blue colour develops in sites containing many reactive NH_2 groups. Where these are more sparsely distributed the colour is red or pink.

ALKALINE SALICYLALDEHYDE METHOD FOR NH_2

(Stoward, 1963)

(*Freeze-dried, Carnoy, Alcohol; paraffin sections. Fresh frozen*)

Method. (1) Bring sections to 95 per cent. ethanol.

(2) Treat with a fresh 1 per cent. solution of salicylaldehyde in 1 per cent. alcoholic KOH, 5–20 minutes.

(3) Wash in 95 per cent. ethanol.

(4) Dehydrate, clear and mount in fluormount.

Result. NH_2 groups give intense green fluorescence (410 nm) which is heat labile.

DIACETYLBENZENE REACTION FOR PROTEIN-BOUND NH_2

(Voss, 1940; Wartenburg, 1956–57)

(*Various fixatives; paraffin sections*)

Method. (1) Paraffin sections, preferably mounted with egg albumin, should be brought to water in the usual way.

(2) Immerse sections for 5–10 minutes in 0·1 M-veronal acetate buffer (pH 8·2).

(3) Immerse in 2 per cent. *o*-diacetylbenzene* in 70 per cent. ethanol with an equal amount of 0·1 M-veronal acetate buffer (pH 8·2).† Leave for 30–60 minutes at room temperature.

(4) Wash briefly in buffer at pH 7·0 and in three changes of distilled water.

(5) Dehydrate in alcohols, clear in xylene and mount in Canada balsam or in a suitable synthetic medium.

Result. Protein-bound amino groups are shown by the presence of a reddish colour.

* This can be prepared by oxidation of *o*-ethylbutophenone (Remschneider, 1947; Remschneider and Weygand, 1955).

† The buffer should be added shortly before the beginning of the incubation.

ACID AZIDE REACTION FOR NH_2 GROUPS

(After Geyer, 1965)

(*Carnoy, Bouin, Alcohol; paraffin sections*)

Method. (1) Bring sections to distilled water.

(2) Incubate for 1½–2 hours at 22° in the acid azide reagent (50 mg. *p*-nitrobenzoic acid azide in 10 ml. absolute ethanol, with 10 ml. veronal acetate buffer, pH 8·2, added).

(3) Wash twice, for 5 minutes, in absolute ethanol.

(4) Rinse in 70 per cent. and 50 per cent. ethanol. Bring sections to water.

(5) Reduce for 15–30 minutes at 37° in the following:

15 per cent. $TiCl_3$	2 ml.
Sodium citrate buffer (0·5 M, pH 4·5)	8 ml.

(6) Wash in citrate buffer and then in distilled water.

(7) Diazotize at 4° for 5 minutes in fresh nitrous acid. (8 ml. 5 per cent $NaNO_2$ and 1 ml. N-H_2SO_4).

(8) Wash in distilled water.

(9) Couple with H-acid at 4° for 5 minutes, pH 9. (Sat. solution in borax or veronal acetate buffer.)

(10) Wash in running water (10 minutes).

(11) Dehydrate in alcohols, clear in xylene and mount in synthetic medium.

Result. A bright red colour indicates the presence of protein-bound NH_2 groups.

THE COUPLED TETRAZONIUM REACTION

(*Formalin, alcohol, freeze-dried; paraffin sections*)

Method. (1) Bring sections to water and remove mercury if necessary.

(2) Immerse in freshly tetrazotized benzidine* at 4° for 15 minutes. Alternatively (Burstone, 1955) one can use a 0·2 per cent. aqueous solution of Fast blue B salt in tris buffer at pH 9·2, for 5 minutes at room temperature.

(3) Wash in water and in three changes of veronal acetate buffer at pH 9·2, for 2 minutes in each change.

(4) Immerse in a saturated (1 g. in 50 ml.) solution of H-acid in veronal acetate buffer at pH 9·2 for 15 minutes.

(5) Wash in water for 3 minutes.

(6) Dehydrate in alcohol.

(7) Clear in xylene.

(8) Mount in balsam or DPX.

Result (Figs. 71, 75; Plate IVd). The majority of tissue components stain in shades of reddish brown. If the coupled tetrazonium reaction is preceded by mild heat and benzoylation positive structures now stain either in shades of deep reddish brown or, as in the case of collagen, in purplish red. These colours are quite stable (over 3 years).

The use of formalin-fixed sections is probably less advisable than alcohol-fixed sections for critical work. Aromatic hydrogen groups are irreversibly blocked by formalin (see Chapter 5, p. 73) but the tetrazonium reaction is not prevented by formalin fixation, even though it is reduced.

* To about 3 ml. of a 2 per cent. suspension of benzidine, in 2 N-HCl at 4°, add 8 drops of cold fresh 5 per cent. sodium nitrite. Agitate rapidly for 10 minutes. Add 1 ml. 5 per cent. cold ammonium sulphamate and about 10 ml. of cold saturated aqueous sodium carbonate. As effervescence ceases and the solution becomes alkaline, it changes to a clear dark yellow. Add water to 50 ml. This solution deteriorates rapidly even at 4° and should not be used for more than an hour after preparation. Stable tetrazoates have not proved satisfactory substitutes for the fresh solution.

Benzoylation

(*Freeze-dried ethanol fixed; alcohol or Carnoy-fixed, paraffin sections*)

(1) Sections are immersed in light petroleum (petroleum ether) for 3 minutes.

(2) Remove and allow to dry in air.

(3) *Controlled heat stage* (optional). Apply heat at 60° in an incubator, or at 80° on the warm stage, for 5–10 minutes, to the dry sections.

(4) *Hydration Control* (optional). Treat section with water for 2–5 minutes. Remove water by blotting and allow to dry in air.

(5) Place the dry sections in 10 per cent. benzoyl chloride in dry pyridine* for 10–16 hours at room temperature. Alternatively, use Barnards (1961) mixture: dry acetonitrile 50 ml., benzoyl chloride 4·2 ml., dry pyridine 2·2 ml. Use at room temperature for 3 hours in a desiccator over $CaCl_2$. (The duration of benzoylation depends on the type of fixation employed; it is shorter for alcoholic and longer for formalin-containing types.)

(6) Rinse in absolute acetone and

(7) Immerse in absolute alcohol.

(8) Take sections to water.

(9) Proceed to the coupled tetrazonium reaction, Stage 2.

Results. Protected histidine bonds are shown by a purple to reddish-brown stain. The hydrated control should be colourless. Histidine-containing proteins, which are normally uncoloured in the benzoylation-coupled tetrazonium procedure, may become benzoylation resistant and therefore coloured after heating.

Acetylation (Drastic)

(1) Bring sections to absolute ethanol and then allow to dry in air.

(2) Immerse in 10 per cent. acetic anhydride in dry pyridine and heat for 4–8 hours at approximately 100°, under a reflux condenser.

Acetylation for reversal of the periodic acid-Schiff reaction is not carried out by this method (see Appendix 10, p. 660).

Acetylation (Mild)

(1) Bring sections to ethanol and then allow to dry in air.

(2) Immerse in acetic anhydride for 3 minutes.

(3) Immerse in a mixture containing 4 ml. acetic anhydride, 36 ml. ethyl acetate, 0·05 ml. perchloric acid.† For 36 hours.

(4) Rinse in acetic anhydride.

(5) Rinse in ethanol.

DIAZO-DEAZO PROCEDURE FOR NH_2 GROUPS

(Stoward, 1963)

Method. (1) Bring sections to water.

(2) Immerse in fresh HNO_2 solution for 48 hours at 0°–5°, in the dark (1 g. $NaNO_2$ in 30 ml. 3 per cent. H_2SO_4).

(3) Wash in distilled water.

* This solution lasts for about 1 month but rapidly loses strength if water is allowed to contaminate it. Pyridine should be dried by distillation over barium sulphide.

† $HClO_4$ ruptures $(CH_3CO)_2O$ to give acetylium, $CH_3CO^{\oplus}$. Protein-bound amino groups are completely blocked by this procedure.

(4) Treat for 4 hours at 60° in
(*a*) Water
(*b*) Absolute ethanol.

This procedure gives reliable deamination.

THE DINITROFLUOROBENZENE (DNFB) METHOD FOR TYROSINE, SH AND NH_2

(Danielli; Burstone)

(*Freeze-dried, alcohol, acetone, formalin, etc.*)

Method. (1) Bring sections to absolute acetone or alcohol, remove and allow to dry in air.

(2) Treat with a saturated solution of DNFB in 90 per cent. ethyl alcohol saturated with sodium bicarbonate, 2–16 hours, at room temperature.

(3) Wash in three changes of 90 per cent. alcohol, and finally in water.

(4) Treat with 5 per cent. sodium hydrosulphite, 30 minutes at 45°.

(5) Wash in water.

(6) Treat with cold (0°–4°) nitrous acid,* 30 minutes.

(7) Wash in water.

(8) Treat with a cold (0°–4°) saturated solution of H-acid in veronal acetate buffer at pH 9·4, 15 minutes.

(9) Wash in water, dehydrate in alcohols, clear in xylene, mount in balsam or DPX.

Result. The sites of DNFB attachment to the tissues appear in reddish-purple.

MODIFIED DNFB SCHEDULE

(Zerlotti and Engel)

(*Freeze-dried, paraffin embedded*)

PREPARATION OF REAGENT

Mix 1 ml. DNFB with 30 ml. absolute ethanol and 15 ml. aqueous 1·0 M $NaHCO_3$. Remove precipitated bicarbonate by filtration. Use only freshly prepared reagent.

Method. (1) Remove paraffin wax with light petroleum. Allow to dry.

(2) Treat with DNFB for 30 minutes or 24 hours at 25°.

(3) Rinse sequentially in 75 per cent. ethanol, warm distilled water (30°), 75 per cent ethanol, 95 per cent. ethanol, 95 per cent. ethanol/ether (1:1), absolute ethanol.

(4) Clear in xylene and mount in a synthetic resin medium.

(5) Examine with tungsten filament lamp source and interference filter (400–420 nm).

Result. Terminal α-amino groups, ε-amino groups and SH groups contribute to the reaction. The colourless DNP derivatives of imidazoles and tyrosine do not do so.

Complete blockade was produced by formaldehyde vapour treatment of sections prior to stage 2. A similar effect is produced by vapour treatment of freeze-dried blocks (Tock and Pearse, 1965).

MODIFIED DNFB SCHEDULE

(Tranzer and Pearse, 1964)

(*Fresh cryostat; Carnoy, freeze-dried, paraffin sections*)

Method. (1) Remove wax with light petroleum (if necessary) and bring sections to absolute ethanol.

(2) Incubate in 1 per cent. DNFB in absolute ethanol, alkalinized with 0·2 ml. N-NaOH for 2–20 hours at 22°.

(3) Wash 4 times in 90 per cent. alcohol and rinse in distilled water.

* Add 1 vol. of freshly prepared 5 per cent. sodium nitrite to 4 vols. of 2 N-HCl.

(4) Reduce for 15–30 minutes at 37° in the following:

15 per cent. $TiCl_3$ 2 ml.
Sodium citrate buffer 8 ml.
(0·5 M, pH 4·5)

(5) Wash in citrate buffer and then in distilled water.

(6) Diazotize at 4° for 5 minutes in fresh nitrous acid (8 ml. 5 per cent. $NaNO_2$ and 1 ml. N-H_2SO_4).

(7) Wash in distilled water.

(8) Couple with H-acid at 4° for 5 minutes, pH 9.

(9) Rinse in running water, dehydrate, clear and mount in a synthetic resin.

Result. If no preliminary blockade has been applied a deep reddish-purple colour indicates sites of DNP-amino-acids.

THE DMAB-NITRITE METHOD FOR TRYPTOPHAN
(Adams, 1957)

The original author recommended short fixation (6 hours) in neutral formalin but stated that up to 24 hours was acceptable. As alternatives, 1 per cent. trichloro-acetic-ethanol, 70 per cent. methanol or 10 per cent. aqueous sulphosalicylic acid may be employed. Sections should be mounted on albuminized slides or, preferably, on chromate gelatin, followed by 3–4 days in the 37° incubator. Fresh frozen (cold microtome) sections mounted on slides and air-dried are quite suitable.

Method. (1) Bring sections to absolute alcohol and allow them to become just dry in the air at room temperature.

Alternatively,

(2) Remove from alcohol and coat with a thin film of celloidin (0·25 per cent.); proceed to the next stage without drying.

(3) Immerse sections in 5 per cent. *p*-dimethylaminobenzaldehyde in conc. HCl* (S.G. 1·18) for 1 minute.

(4) Transfer to an approximately 1 per cent. solution of $NaNO_2$ in conc. HCl for a further minute.

(5) Wash for 30 seconds in tap water.

(6) Rinse in 1 per cent. acid-alcohol.

(7) Dehydrate, clear and mount.

Result (Fig. 47, and Plate Ia). Tryptophan-containing proteins are shown in varying intensities of deep blue. Strong reactions occur in fibrin, fibrinoid, Paneth cell granules, peptic cell granules, zymogen granules, muscle, neurokeratin and in hair root sheaths.

TRYPTOPHAN METHOD FOR FORMALIN-FIXED TISSUES
(Adams, 1960)

(*Formalin, paraffin embedded*)

Method. (1) Bring sections to absolute alcohol.

(2) Dip sections into the following solution: glycerol, 5 ml.; 60 per cent. ferric chloride, 1 ml.; conc. H_2SO_4, 5 ml.; distilled water, 9 ml.; methylated alcohol, 80 ml. (this solution keeps for several months in a stoppered bottle).

(3) Grip slide in forceps, tip off excess fluid, ignite in a small flame. While burning, hold slide horizontal with the section uppermost. Repeat process 3–6 times.

(4) Wash in absolute ethanol.

(5) Rinse until clean in glacial acetic acid/ethanol (1 : 1). Clear in xylene and mount.

Result. A mauve pigment indicates sites containing tryptophan. Examine (and photograph) within 24 hours as the pigment is not stable.

* Keeps 3–4 weeks in a Coplin jar sealed with petroleum jelly.

THE ROSINDOLE REACTION FOR INDOLES

(Glenner, 1957)

(10 *per cent. calcium acetate-formalin for* 3–6 *hours*)

Method. (1) Bring sections to absolute alcohol.

(2) Dry in air for 30 seconds.

(3) Treat for 3 minutes at 25° in a solution containing 60 ml. perchloric acid, 34 ml. acetic acid, 1 ml. conc. HCl and 1 g. *p*-dimethylaminobenzaldehyde.

(4) Transfer to a solution containing 35 ml. acetic acid and 5 ml. conc. HCl with 0·5 g. $NaNO_2$ added just before use. Allow to stand for 1 minute.

(5) Wash three times in acetic acid and carry through 50 per cent. acetic-xylene to pure xylene.

(6) Mount in Permount or other suitable synthetic medium.

Result. Tryptophan-containing proteins appear deep blue. The depth of staining is usually less than that shown by the DMAB-nitrite reaction.

THE POST-COUPLED BENZYLIDENE REACTION FOR INDOLES

(Glenner and Lillie, 1957)

(10 *per cent. calcium acetate formalin,* 3–6 *hours. Tissues fixed in* $HgCl_2$*, picrate or bichromate give a diminished reaction*)*

Method. (1) Bring paraffin sections to absolute alcohol.

(2) Dry in air for 30 seconds.

(3) Treat for 5 minutes at 25° in a mixture containing 1 g. *p*-dimethylamino-benzaldehyde, 10 ml. conc. HCl and 30 ml. acetic acid.

(4) Wash gently three times in acetic acid.

(5) Place sections in a Coplin jar containing 40 ml. acetic acid with the addition of 1 ml. of a fresh diazotate of S-acid (8-amino-1-naphthol-5-sulphonic acid).† Leave for 5 minutes at room temperature.

(6) Wash gently three times in acetic acid (1 minute in each).

(7) Clear in 50 per cent. acetic-xylene, 20 per cent. acetic-xylene and in three changes of pure xylene.

(8) Mount in a suitable synthetic medium.

Result (Fig. 48). Indole derivatives appear pale to dark blue, depending on their local concentration.

THE XANTHYDROL REACTION FOR INDOLES

(Lillie, 1957)

(10 *per cent. calcium acetate formalin,* 3, 24 *or* 48 *hours,* 6 *per cent. glutaraldehyde*)

Method. (1) Bring paraffin sections to absolute alcohol.

(2) Dry in air for 30 seconds.

(3) Immerse sections for 5 minutes in 2·5 per cent. xanthydrol in a mixture of 90 ml. acetic acid and 10 ml. conc. HCl.

(4) Wash gently in acetic acid, three times.

(5) Clear in 50 per cent. acetic-xylene, 20 per cent. acetic-xylene and in three changes of pure xylene.

(6) Mount in a suitable synthetic medium.

* Poorly reactive tissues fixed in picrate can be unmasked by treatment with saturated lithium carbonate in 70 per cent. ethanol. Mercury blockade can be reversed in the same way, or by treatment with BAL or 5 per cent. thiosulphate (37°, 5 minutes).

† Add 240 mg. S-acid to 3 ml. N-HCl and 6 ml. distilled water. Cool to 4° and add 1 ml. N-$NaNO_2$. Stir at 4° for 15 minutes.

Result. Indole-containing tissue components stain violet. Strong reactions are given by zymogen granules, and by stomach chief cells in some species. After glutaraldehyde fixation the enterochromaffin granules of the gastro-intestinal tract give a strong positive reaction.

THE NAPHTHYL ETHYLENEDIAMINE REACTION FOR TRYPTOPHAN

(Bruemmer, Carver and Thomas, 1957)

(*Carnoy, Zenker, formalin; paraffin sections*)

Method. All reactions, except where otherwise stated, are carried out at 0°–4°.

(1) Bring paraffin sections to ice-cold 50 per cent. ethanol.

(2) Place for 15 minutes in a mixture containing equal parts of 8 per cent. $NaNO_2$ and 6 N-HCl.

(3) Wash in two changes of distilled water, 5 minutes in each.

(4) Immerse for 15 minutes in a freshly prepared solution of 2 per cent. N-(1-naphthyl) ethylenediamine in 95 per cent. ethanol containing N-HCl.

(5) Dehydrate in 70 per cent. tertiary butyl alcohol to which has been added a few drops of conc. H_2SO_4.

(6) Continue dehydration at room temperature in absolute butyl alcohol containing H_2SO_4.

(7) Clear in two changes of xylene containing 2 ml. acetic acid per 100 ml.* (Room temperature.)

(8) Mount in a suitable synthetic medium.

Result. A purple colour develops in tryptophan-containing proteins.

THE SAKAGUCHI REACTION FOR ARGININE

(Baker's 1947 Modification)

(*Various; paraffin sections; freeze-dried, cold microtome*)

Baker himself advised the use of celloidin-covered paraffin sections fixed in Zenker, Bouin, Susa or formal sublimate.

Method. (1) (Optional.) Remove wax and cover with a thin film of 1 per cent. celloidin; harden.

(2) Bring sections to water.

(3) Remove from water and flick off till nearly dry.

(4) Flood with α-naphthol hypochlorite solution,† 15 minutes.

(5) Drain and blot.

(6) Immerse in a mixture of equal parts of dry pyridine and chloroform.

(7) Mount in pyridine-chloroform, or in dry pyridine.

Result (Fig. 57). Arginine-containing proteins appear in various shades of orange-red.

THE SAKAGUCHI OXINE REACTION FOR ARGININE

(Carver, Brown and Thomas' 1953 Modification)

(*Bouin, formalin, Carnoy; paraffin sections*)

Method. (1) Bring paraffin sections to 70 per cent. alcohol.

(2) Immerse sections in 0·3 per cent. 8-hydroxyquinoline in 30 per cent. ethanol for 15 minutes at room temperature.

* If an orange-red colour is preferred to purple, the acid is omitted at this stage.

† Mix 2 ml. 1 per cent. NaOH, 2 drops of 1 per cent. α-naphthol in 70 per cent. alcohol, and 4 drops of 1 per cent. Milton (proprietary brand of stable sodium hypochlorite containing about 1 g. of NaOCl and 18 g. NaCl per 100 ml.).

(3) Quickly transfer the slide, without draining, into alkaline hypochlorite solution.* Leave for 60 seconds without movement.

(4) Transfer quickly, without draining, into alkaline urea solution.† Move the slide gently for 10 seconds, then transfer to a fresh alkaline urea bath for 2 minutes.

(5) Place slide in tertiary butyl alcohol for 4 minutes.

(6) Place in aniline oil for 3 minutes.

(7) Wash in xylene, 10 seconds.

(8) Mount in Permount or other synthetic resin containing 0·025 ml. aniline per 100 ml.

Result. An orange colour indicates the presence of arginine.

THE SAKAGUCHI DICHLORONAPHTHOL REACTION FOR ARGININE

(McLeish *et al.*, 1958; Deitch, 1961)

(*Acetic-ethanol, freeze-dried, paraffin sections*)

PREPARATION OF REAGENTS

Immediately before use prepare: (*a*) 4 per cent. barium hydroxide (filtered), (*b*) 1 per cent. sodium hypochlorite,* (*c*) 1·5 per cent. 2,4-dichloro-α-naphthol in *tert* butanol.

Method. (1) Bring slides to water, using 2 changes of distilled water. Blot slides and place in an empty staining jar.

(2) Pour 5 parts $Ba(OH)_2$, 1 part NaOCl and 1 part dichloronaphthol into a flask in succession. Agitate after each addition. Pour contents of the flask into the staining jar.

(3) Allow reagent to act for 10 minutes at 22°.

(4) Transfer slides to three 5 second changes of *tert* butanol; agitate vigorously in each change.

(5) Transfer to two changes of xylene (30–60 secs) containing 5 per cent. tri-N butylamine.

(6) Drain and mount in Shillaber's oil (ref. index, 1·580) containing 10 per cent. tri-N-butylamine.

Result. An orange-red colour indicates sites containing arginine. (If a $BaCO_3$ precipitate forms on the underside of the slide it can be removed with cotton wool soaked in dilute acetic acid.)

BLOCKING AND CONVERSION METHODS FOR PROTEIN END GROUPS

DEAMINATION (NH_2)

Nitrous acid is used at 0°–4° in various mixtures. The diazo-deazo sequence (p. 613) is recommended.

* Standardize a fresh commercial preparation of sodium hypochlorite (Milton, Chlorox) against 0·1 N-sodium thiosulphate. Use for the titration 1 ml. of hypochlorite, 5 ml. N-potassium iodide, 8 ml. conc. HCl and 50 ml. water. (The average chlorine content of fresh commercial hypochlorite is 1·6 N). The fresh hypochlorite is used as a stock solution. Prepare a 0·15 N stock solution of KOH.

For use, measure appropriate quantities of the two stock solutions into a 100 ml. cylinder, to make a final concentration of 0·15 N-chlorine and 0·015 N-KOH.

† To 10 ml. 0·15 N-KOH in a 100 ml. cylinder add 15 g. urea. Dilute with water to 25–30 ml. and mix until the urea is dissolved. Add 70 ml. tertiary butyl alcohol, dilute to volume or mix.

DINITROPHENYLATION (α-NH_2, ε-NH_2, tyrosyl-OH, SH, imidazole)

Treat sections with 1 per cent. 2,4-dinitrofluorobenzene in 90 per cent. ethanol containing 0·01 M-NaOH, for 16–20 hours at room temperature.

ACETYLATION. See p. 613.

BENZOYLATION. See p. 613.

NITROBENZOYLATION (NH_2, tyrosyl-OH)

Treat sections with 5 per cent. *p*-nitrobenzoyl chloride in dry pyridine for 4 hours at 37°.

OXIDATION TO ALDOXIMES (Primary NH_2)

Treat sections with 1·5 per cent. H_2O_2 containing 0·1 per cent. sodium tungstate for 2–8 hours at 22°.

METHYLATION (COOH, NH_2, SH, etc.)

(A) **Mild Methylation** (Fisher and Lillie). Treat sections with 0·1 N-HCl in absolute methanol for 8–96 hours at 37°.

(B) **Drastic Methylation.** Use 0·1 N-HCl in absolute methanol for 8–96 hours at 60°.

Mild methylation blocks protein carboxyls (reversibly) but not primary amines. The latter are converted into secondary and tertiary amines by methylation at 60° (*N*-methylation). Drastic methylation also produces desulphation (p. 329) and a range of effects on nucleic acids and carbohydrates which are due to hydrolysis and removal of various components. O-alkylation of polysaccharides does not occur with methanolic-HCl.

(C) **Methyl Iodide** (Terner). Pretreat sections with a saturated aqueous solution of sodium carbonate (30 minutes, 22°). Follow by treating with 50 per cent. methyl iodide in absolute methanol at 45° for 18 hours. This method blocks protein and carbohydrate COOH groups, and also NH_2, SH, tyrosine and phosphate.

(D) **Diazomethane** (Peters). Treat sections with a 0·5 per cent. solution in ether, for 8–18 hours at 22°. Carboxyls and phosphates are blocked after 8–10 hours, sulphates in 18 hours.

The reagent is prepared by dissolving 2 g. tolyl sulphonyl methyl nitrosamide in 12·5 ml. ethyl ether and adding this solution slowly to 0·5 g. KOH in 2·5 ml. ethanol and 1 ml. water. This mixture is warmed and distilled into 20–30 ml. ether. The distillate is further diluted with ether until a yellow colour is no longer visible. The final concentration of diazomethane is about 0·5 per cent.

(E) **Methanolic thionyl chloride** (Bello, Geyer). Add 1 ml. $SOCl_2$ slowly to 50 ml. methanol. Allow to stand overnight before use. Sections are treated with the reagent for 4–6 hours at 22°. All tissue basophilia is abolished in this time. Cytoplasmic basophilia due to COOH and RNA disappears in 30 minutes, and that due to sulphated mucopolysaccharides in 4 hours. After 6 hours the extraction of the phosphate groups of nucleic acids begins. It is complete in 24 hours. Extraction of mucopolysaccharides is negligible up to 12 hours. Primary amines are methylated in 30–60 minutes.

DEMETHYLATION

Reversal of methylation is only possible to a limited extent in the case of the majority of tissue components. Treatment for 20 minutes with 0·5 per cent. potassium permanganate is followed by a brief rinse in 2 per cent. oxalic acid. More usually hydrolysis of methyl groups is carried out by saponification (1 per cent. KOH in 80 per cent. ethanol, 20 minutes, 22°). As an alternative, barium hydroxide can be employed.

PERFORMIC ACID OXIDATION (Tryptophan, Cystine)

Prepare fresh performic acid by adding 4 ml. of 30 per cent. H_2O_2 (100 vol.) and 0·5 ml. conc. H_2SO_4 to 40 ml. 98 per cent. formic acid. The peroxide must be reasonably fresh and stocks older than 2–3 weeks should not be used. The performic acid solution should be degassed just before use by vigorous stirring with a glass rod and used within 8 hours.

Oxidation for 15–60 minutes at room temperature is normally employed.

PERACETIC ACID OXIDATION (Tryptophan, Cystine)

Peracetic acid (40 per cent.) is usually available commercially.* If it is not, it can be made by adding 9·5 ml. acetic acid and 0·25 ml. conc. H_2SO_4 to 30 ml. H_2O_2. Stand for 1–3 days before use and store at 0°–4° (months). Commercial (40 per cent.) peracetic acid is best employed as a 2 or 4 times dilution with distilled water, and tissues are oxidized for 2–4 hours.

PERMOLYBDIC AND PERTUNGSTIC ACIDS

These acids are made by adding to 60 ml. water, 30 g. orthophosphoric acid (H_3PO_4) and 3 ml. 30 per cent. H_2O_2, 0·5 g. of either ammonium molybdate or sodium tungstate. Oxidize tissue sections at room temperature for 4–16 hours.

PERSULPHATE OXIDATION (tryptophan)

A fresh alkaline solution of persulphate is prepared by dissolving 1 g. potassium persulphate ($K_2S_2O_8$) in 40 ml. 0·5 N-KOH. Sections are incubated for 12–18 hours at 20°–22°.

IODINATION (tyrosine, possibly tryptophan)

Treat sections for 24 hours at room temperature with a mixture of 30 ml. Gram's iodine (iodine, 1 g.; KI, 2 g.; water, 300 ml.) and 2 ml. 3 per cent. ammonia. The pH of this solution should be approximately 10. Alternatively, 0·78 N-I_2 in ethanol may be used for 72 hours at 25°. Tryptophan is not iodinated by this last procedure.

IODINE OXIDATION (SH)

Treat sections for 4 hours at room temperature with an aqueous solution containing 38 mg. iodine and 33 mg. potassium iodide per 100 ml. This solution should be titrated to pH 3·2 with 0·01 N-HCl before use.

IODOACETATE BLOCK (SH)

Treat sections for 20 hours at 37° with 0·1 M-iodoacetic acid titrated to pH 8 with 0·1 N-NaOH.

MALEIMIDE BLOCK (SH)

Treat sections for 4 hours at 37° with 0·1 M-*N*-ethyl maleimide in 0·1 M-phosphate buffer at pH 7·4. Follow this by washing in 1 per cent. acetic acid and then in tap-water.

MERCAPTIDE BLOCK (SH)

The most commonly used reagents for this purpose are mercuric chloride and phenylmercuric acetate. The latter has been used as a 0·001 M-solution in *n*-propanol (72 hours, 20°). Tolyl mercuric chloride and methylmercury iodide have also been used at the same concentration. Many other organic mercurials can be employed. All require long periods for complete reaction, and the resulting mercaptides break down easily in some cases so that the blocked SH groups become free to react.

* From Laporte Chemicals Ltd., Luton, Beds.

THIOGLYCOLLATE REDUCTION (SS)

Cover sections with a thin layer of 0·5 per cent. celloidin. Incubate for 4 hours at 37° in 0·5 M-thioglycollic acid (freshly made) titrated to pH 8 with 0·1 N-NaOH. Afterwards wash in tap-water, rinse briefly in 1 per cent. acetic acid and wash again in water.

THE NITROPRUSSIDE REACTION FOR SH

(Hammett and Chapman)

(*Fresh slices*, 40–100 μ)

Method. (1) Place the section on a dry watch-glass.

(2) Cover with a small amount of distilled water (say 0·25 ml.).

(3) Add 0·05 ml. 27–29 per cent. ammonium hydroxide and 0·05 ml. 1 per cent. sodium nitroprusside.*

(4) Displace section to one side and underline with 0·25 g. crystalline ammonium sulphate.

Result. A pink colour develops which fades after a short period. It is possible to distinguish between areas containing many or few SH groups and, with practice, to estimate roughly the amount of SH present.

THE FERRIC FERRICYANIDE METHOD FOR SH

(Chèvremont and Fréderic, 1943)

(*Formalin-frozen sections; various, including fresh sections*)

The authors recommend short fixation, of a few hours only, and frozen or paraffin sections. In the case of the latter the time occupied in the embedding process should be as short as possible.

REAGENT

One per cent. ferric sulphate, $Fe_2(SO_4)_3$, 3 parts, added to a freshly prepared 0·1 per cent. potassium ferricyanide, $K_3Fe(CN)_6$, 1 part. The pH should be adjusted to 2·4.

Method. (1) Wash sections in distilled water.

(2) Immerse in three changes of the ferricyanide reagent for a total of 10–20 minutes (frozen sections) or 20–25 minutes (paraffin sections).

(3) Wash in distilled water.

(4) Counterstain nuclei with carmalum 6–18 hours (optional).

(5) In the case of frozen sections, mount in glychrogel. With paraffin sections, dehydrate in alcohols, clear in xylene, and mount in DPX.

If the method is being used as a critical histochemical test, differentiation is not permissible. It is possible, however, by brief rinsing in 2 per cent. alkaline alcohol (NaOH in 60 per cent. alcohol), to reduce the diffuse blue staining of background structures and so to emphasize the strongly positive SH-containing elements.

CONTROL SECTIONS

Yao (1949) suggested the use of control sections treated with saturated aqueous $HgCl_2$ for 1 hour. Such sections should give a negative result in all areas where the reduction of ferricyanide is due to sulphydryl. In practice neither the mercaptan-forming agent nor the time of treatment are satisfactory, particularly for frozen sections. In the latter sulphydryls continue to react freely and, in addition, reactive groups are formed by the oxidation of lipid substances. The use of saturated phenyl mercuric chloride in butanol, for 2–3 days, is recommended instead of saturated aqueous mercuric chloride, but maleimide blocking is usually preferable.

Result (Fig. 48). Sulphydryls blue, nuclei red.

* Fresh solutions must be used. No solution should be more than 1 hour old.

THE MERCURY ORANGE METHOD FOR SH

(Bennett, 1951)

(Trichloroacetic-teased; freeze-substituted, butanol; freeze-dried; cold microtome, etc.)

Preparation of the Reagent

(1-(4-chloromercuriphenylazo)-2-naphthol. Called mercury orange.

(1) Prepare *p*-aminophenylmercuric acetate by the method of Dimroth (*Ber. dtsch. chem. Ges.*, **35**, 2032, 1902).

This involves the direct mercuration of aniline which is performed as follows:

Add 31·8 g. mercuric acetate in 160 ml. of water to 18·6 g. of freshly distilled aniline. After standing for a short time the pale yellow prisms of the *p*-isomer begin to separate; after 3 hours the supernatant is decanted and the crystals are washed with water.

(2) Diazotize 35·4 g. of *p*-aminophenyl mercuric acetate, in 500 ml. 50 per cent. acetic acid with 7·0 g. of sodium nitrite, at −5°.

(3) Filter off the diazonium salt and couple with 2-naphthol (15 g. of 2-naphthol, 180 g. of NaOH in 2 litres of iced water).

(4) Stand for a few hours, filter and wash precipitate, dissolve in 200 ml. glacial acetic acid, filter and reprecipitate by dilution to 2 litres.

(5) Collect precipitate, wash and dissolve by refluxing with 3 litres of ethanol in a water bath.

(6) Filter the hot solution, bring to boil under a reflux condenser and add 5·8 g. NaCl in 150 ml. of 60 per cent. ethanol.

(7) Reflux for 30 minutes.

(8) Collect the cottony red precipitate by hot filtration and wash several times with hot 50 per cent. ethanol.

(9) Recrystallize three times from *n*-butanol. (Yield 3·6 g.)

Solvents for the Reagent

A variety of solvents can be used for mercury orange and the number of sulphydryl groups which react depends on the particular solvent used. Bennett originally recommended toluene, *n*-butanol or *n*-propanol, but later suggested that from six to eight times the number of SH groups would react if mercury orange was dissolved in dimethylformamide. Mescon and Flesch used 80–100 per cent. ethanol, and in some early studies with the method I used the same solvent. There is some evidence, however, that the increase in colour by comparison with mercury orange in other solvents is due to non-specific staining. For some time past I have used dimethylformamide exclusively, and unfixed cold microtome sections mounted on slides or coverslips.

Method (according to Bennett, with modifications). (1) Fix in 5 per cent. trichloroacetic acid and follow by washing in distilled water and dehydration in various strengths of ethanol. Immerse in *n*-propanol or *n*-butanol and tease into small fragments.

Alternatively,

(2) Freeze small strips of tissue in isopentane cooled to about −160° with liquid N_2 and follow by immersing the frozen tissue in *n*-propanol or *n*-butanol at −20°. Allow to remain for 10–12 days with occasional changes of solvent at −20°. Finally, tease into small fragments for study.

(3) Place the fragments obtained by either of the above manœuvres in a saturated solution of mercury orange in butanol (about $1{\cdot}25 \times 10^{-5}$ M), propanol (about $5{\cdot}6 \times 10^{-6}$ M), or dimethylformamide (5×10^{-5} M). Leave for 16–48 hours (until no further uptake of mercury orange occurs).

(4) Wash in several changes of pure solvent (propanol or butanol).
(5) Clear in xylene.
(6) Mount in clarite.

No fading occurs unless mercaptans are present in the mounting medium.

For formalin-fixed paraffin sections or for unfixed freeze-dried and cold microtome sections, proceed as follows:

(1) Bring sections to absolute alcohol (not cold microtome sections).
(2) Immerse in a saturated solution of mercury orange in dimethylformamide for 16–48 hours.
(3) Wash twice in absolute alcohol.
(4) Clear in xylene.
(5) Mount in DPX.

For formalin-fixed frozen sections proceed as follows:

(1) Bring sections from water to 75 per cent. alcohol.
(2) Immerse in saturated mercury orange in dimethylformamide, 16–48 hours.
(3) Wash twice in 75 per cent. alcohol.
(4) Bring to 50 per cent. alcohol and thence to water.
(5) Mount in glycerine jelly.

Result (Fig. 47). The colour of SH-containing structures varies between pale orange and a darker orange-red.

CONTROL SECTIONS

Where there is any doubt as to the specificity of the reaction observed with mercury orange, control sections should be treated for 48–72 hours with a saturated solution of phenyl mercuric chloride in butanol or alcohol, or with 0·1 M-*N*-ethyl maleimide (4 hours, 37°).

CHROMOPHORIC REAGENTS FOR PROTEIN THIOL GROUPS

(Fai Chong and Liener, 1964)

PREPARATION OF PHENYLAZO-CHLOROMERCURIPHENOL (PCMP)

To 1·14 g. *o*-chloromercuriphenol in 30 ml. 15 per cent. KOH add 2·5 g. sodium acetate. Cool to −2°. Add dropwise a solution of freshly diazotized aniline. (0·32 g. freshly distilled aniline, 4 g. crushed ice, 1·7 ml. conc. HCl, 0·25 g. $NaNO_2$).

Collect the orange-yellow precipitate by filtration. Wash thoroughly and dry over CaCl. (Yield 1·13 g.)

Purify by passing a hot benzene solution through a column of acid-washed alumina. After elution of a yellow impurity with benzene, PCMP is eluted with benzene/ethanol (100:4). After evaporation of the solvent, the residue is crystallized from hot benzene. *Product:* Orange crystals (m.p. 146°).

PREPARATION OF *p*-NITROPHENYLAZO-CHLOROMERCURIPHENOL (NPCMP)

As above using *p*-nitroaniline.

Product. Orange-red crystals (m.p. 218–220).

THE PERFORMIC ACID-SCHIFF METHOD

(Pearse, 1951)

(*Various; paraffin sections*)

REAGENTS

Performic Acid. Add to 40 ml. of 98 per cent. formic acid 4 ml. of 30 per cent. (100 vol.) H_2O_2* and 0·5 ml. conc. H_2SO_4. Allow the mixture to stand for at least an hour before using, and use preferably within 24 hours.

* Use fresh 100 vol. H_2O_2, preferably not more than 3 weeks old after opening the bottle.

Schiff's Reagent. The de Tomasi or Itikawa preparations are preferred to the Barger and DeLamater modification for the purpose of this method. (See Appendix 9, p. 647.)

Method. (1) Bring paraffin sections to water with removal of mercury precipitate where necessary. (Celloidin-coating may be employed if sections fail to remain adherent to the slides.)

(2) Treat with performic acid solution for 10–30 minutes.

(3) Wash in water 2–5 minutes.

(4) Immerse in Schiff's solution for 30–60 minutes.

(5) Wash in warm running water for 10 minutes.

(6) Dehydrate in the alcohols, clear in xylene, mount in DPX.

Result (Fig. 49). Keratin-containing structures appear bring pink to deep magenta and the amount of colour reflects the amount of available SS groups.

THIOGLYCOLLATE-FERRIC FERRICYANIDE FOR SS

(Adams, 1956)

(*Trichloroacetic-ethanol, formol-calcium, formalin, etc.*)

Method. (1) Bring paired sections to water.

(2) Immerse both sections for 30 minutes in 2·5 per cent. aqueous sodium thioglycollate adjusted to pH 8·0 (green to 9-naphthol-phthalein).

(3) Wash in weakly acidified distilled water (pH 4·0) for 3 minutes.

(4) Wash in running tap-water for 3 minutes, and rinse in distilled water.

(5) Transfer the second (control) slide to phenyl mercuric chloride in butanol (48 hours) or to 0·1 M-*N*-ethyl maleimide (4 hours, 37°). After blocking, bring again to water.

(6) Both sections should now be immersed in a fresh* solution containing 10 ml. freshly made 1 per cent. potassium ferrocyanide and 30 ml. 1 per cent. ferric chloride.† Leave for 1½ minutes.

(7) Wash in three changes of distilled water for 10 minutes.

(8) Dehydrate in alcohols, clear in xylene and mount in a suitable synthetic medium.

Result. Disulphides (stratum corneum, posterior pituitary neurosecretory material, etc.) stain Prussian blue. Any green colour may be tentatively interpreted as indicating weak reducing groups.

PERFORMIC ACID-ALCIAN BLUE METHOD FOR SS GROUPS

(Adams and Sloper, 1955, 1956)

(*Formalin, formol-calcium, etc.; paraffin sections*)

PREPARATION OF REAGENTS

Performic Acid. Prepared as described for the performic acid-Schiff method.

Alcian Blue. Three per cent. (w/v) in 2 N-sulphuric acid. Dissolve the dye by heating to 70° and filter when cool (pH 0·2–0·3).

Method. (1) Bring paraffin sections to water.

(2) Blot gently to remove excess water.

(3) Immerse in the performic acid reagent for 5 minutes after vigorous stirring to remove dissolved gas.

(4) Wash gently in tap-water for 10 minutes.

* Not more than 5 minutes old.

† Filter before use.

(5) Rinse in 70 per cent. and absolute alcohol, blot with filter paper to flatten the creases and again rinse in tap-water.

(6) Warm at 50°–60° until the section is just dry.

(7) Rinse again in absolute alcohol and finally in tap water for 1 minute.

(8) Stain for 1 hour at room temperature in the acid Alcian blue mixture.

(9) Wash for 5 minutes in tap-water.

(10) Counterstain can be applied at this point, if required.

(11) Dehydrate, clear and mount in a suitable synthetic medium.

Result (Fig. 52; Plates Va and VIa). Structures containing 4 per cent. or more of cystine appear dark steely blue in colour. Lesser amounts of cystine appear pale blue.

THE ALKALINE TETRAZOLIUM REACTION (SS AND SH)

(Formalin, trichloroacetic-ethanol, etc.; paraffin sections)

Method. (1) Bring sections to water.

(2) Incubate at 56°–60° in a solution of blue tetrazolium salt, in glycine buffer at pH 12–12·5*† for 60–90 minutes.

(3) Wash well in running cold water.

(4) Mount in glycerine jelly.

Result (Fig. 53). Thickly distributed protein-bound (i.e. non-lipid) reducing groups appear dark blue. Lipid reducing groups appear red. Weaker reduction (sparsely distributed reducing groups) gives a reddish colour in all cases. If the tetrazolium mixture of Barrnett and Seligman is used most of the colour developed can be ascribed to SS and SH groups.

MODIFIED ALKALINE TETRAZOLIUM REACTION
(Deguchi, 1964)

PREPARATION OF THE REAGENT

Prepare a 5 per cent. KCN solution and adjust the pH to 8·4 with 0·2 M-acetic acid. To 10 ml. add 10 ml. 0·2 M-borate buffer (pH 8·4) and 15 mg. nitro-BT. Keep in a covered Caplin jar. If more than a few hours' preservation is required, store at −20°.

Method. (for SS & SH)

(1) Bring sections to ethanol.

(2) Cover with 1% celloidin and allow to dry.

(3) Treat with fresh 0·5-M thioglycollic acid, adjusted to pH 8 with 1 per cent. NaOH, for 2–3 hours at 50°.

(4) Rinse in water, 1 per cent. acetic acid and, finally, water again.

(5) Treat with the nitro-BT reagent for 30–60 minutes at 37°.

(6) Wash in 1 per cent. acetic acid, then in water.

(7) Dehydrate, clear, and mount in a synthetic resin.

Result. Protein-bound SS and SH give dark blue to red colours in the section.

THE DDD REACTION FOR SH GROUPS
(Barrnett and Seligman, 1952)

(Trichloroacetic-ethanol, Carnoy, Bouin, formalin, etc.; paraffin sections)
(Unfixed, or trichloroacetic-ethanol fixed, cold microtome sections)

Method. (1) Bring paraffin sections to water.

(2) Incubate for 1 hour at 50° in a solution containing 35 ml. 0·1 M-veronal acetate

* Dissolve 10 mg. of tetrazolium salt in 40 ml. 0·1 M-glycine buffer at the required pH, heating to 70° and stirring. Filter while hot.

† Barrnett and Seligman (1956) used 25 mg. blue tetrazolium in 48 ml. aqueous KCN with 2 ml. NaOH, and incubated sections for 8–12 hours.

buffer (pH 8·5) and 15 ml. absolute ethanol in which has been dissolved 25 mg. of the DDD reagent. Some of the reagent may remain in suspension.

(3) Cool to room temperature.

(4) Rinse sections briefly in distilled water.

(5) Wash for 10 minutes in two changes of distilled water acidified to pH 4–4·5 with acetic acid. This step converts the sodium salt of the reagent, and of the unwanted reaction product (6-thio-2-naphthol), to free naphthols.

(6) Extract the free naphthols by passage through a graded series of alcohols and wash twice in absolute ether for 5 minutes in each wash.

(7) Rinse in distilled water.

(8) Stain for 2 minutes at room temperature in a freshly prepared solution of 50 mg. tetrazotized diorthoanisidine (Fast blue B salt)* in 50 ml. 0·1 M-phosphate buffer at pH 7·4.

(9) Wash in running tap-water.

(10) Dehydrate in alcohols, clear in xylene and mount in Permount, DPX, or other suitable medium.

Result (Fig. 54 and 55). Blue staining indicates a high concentration of SH groups; red staining areas contain lower concentrations.

THE DDD REACTION FOR SH AND SS GROUPS

If the reaction is required to demonstrate SH and SS groups together sections are first treated with thioglycollate (see p. 621). If SS groups alone are required it is necessary to block existing SH groups with iodoacetate or *N*-ethyl maleimide (see p. 620) and then to reduce the SS groups to SH with some reagent which will not unblock the original SH groups. Potassium cyanide is suitable for this purpose.

THE *p*-NBAF METHOD FOR SH GROUPS

(Gershstein, 1958)

(*Carnoy, Bouin; paraffin sections*)

Method. (1) Bring sections to 95 per cent. ethanol.

(2) Incubate for 1 hour at 22° in the following: 56 mg. *p*-NBAF in 20 ml. 95 per cent. ethanol with 8 ml. veronal acetate buffer, pH 7·8.

(3) Wash in acetone and proceed through descending alcohols to water.

(4) Reduce for 30 minutes at 22° in stannous chloride solution (5 ml. 10 per cent. $SnCl_2$ in 5N-HCl, with 5 ml. 10 per cent. w/v H_2SO_4).

(5) Wash in 5N-HCl and bring to water.

(6) Diazotize with the usual reagent ($NaNO_2$/HCl) at 2° for 30 minutes.

(7) Wash in water and then in veronal-acetate buffer.

(8) Couple with 2 per cent. H-acid in veronal acetate buffer, 5 minutes.

(9) Dehydrate, clear and mount in synthetic resin.

Result. Violet staining indicates protein-bound SH groups.

THE NAPHTHYL MALEIMIDE (HNI) REACTION FOR SH

(after Seligman, Tsou and Barrnett, 1954)

(*Trichloroacetic-ethanol, paraffin sections. Cold microtome sections, unfixed or TCA-fixed*)

PREPARATION OF SOLUTIONS

HNI. Dissolve 25–50 mg. *N*-(4-hydroxy-1-naphthyl) iso-maleimide in 40 ml. dimethylformamide.

* Alternatively a mono-coupling diazotate such as Fast blue salt K may be used.

Diazonium Salts. Stir 10–20 mg. of the chosen stable diazotate into 40 ml. 0·1 M-Tris buffer (pH 10). Filter through Whatman No. 40 paper and use at once.

Method. (1) Bring paraffin sections or mounted fresh frozen sections to alcohol.
(2) Incubate in the HNI solution for 4–16 hours at 37°.
(3) Bring through 70 per cent. alcohol to water.
(4) Treat with the alkaline diazotate for 5–10 minutes at room temperature.
(5) Wash in running water.
(6) Dehydrate in alcohols, clear in xylene and mount in Fluormount† or other non-fluorescent medium.

Result (Fig. 56). By visible light proteins containing reactive SH groups appear reddish-purple, occasionally blue if the concentration is sufficiently high. Under ultra-violet light a brilliant deep red fluorescence indicates the presence of SH in regions in which it is not visible with the light microscope.

The results obtained with a variety of diazonium salts are given in the table below.

TABLE 49

Results with HNI and various Diazotates at pH 10·0

No.*	Diazotate from:	Visible light	U.V. light	Coupling time (mins.)
2	4-benzoylamino-2:5-dimethoxyaniline	Purple	Red (contrast +)	3
6	*o*-dianisidine	Red	Brilliant red (contrast +)	20
7	4-chloro-*o*-anisidine	Brick red	Orange-red	20
9	5-chloro-*o*-toluidine	Orange-red (contrast +)	Orange	4
12	3-nitro-*p*-anisidine	Red (contrast +)	Fiery red	15
15	4-amino-4′methoxy diphenylamine	Brown	Pinkish-red	5
16	3-nitro-*o*-toluidine	Brown	Pinkish-red	5
18	4-amino-3,1′dimethyl azobenzene	Red	Nil	5

For visible light observations I found Salts 2, 6, 9, 12 and 18 to be excellent, while for fluorescence observation Salts 2, 6, 7, 9 and 12 were good. The relative intensity with which identical structures are stained varies with the salt used for coupling, and this suggests that stereochemical factors are operating. In doubtful cases two or more different salts should be used on serial sections.

* See list of diazotates on p. 711.

NEM-SALICYLOYL HYDRAZIDE-ZINC METHOD FOR SH

(Stoward, 1963)

(*Fresh frozen; freeze-dried, Carnoy; paraffin sections*)

Method. (1) Bring sections to water and allow to dry.
(2) Cover with fresh 1·25 per cent. *N*-ethyl maleimide in 0·1 M-phosphate buffer, pH 7·4. Incubate in small moist chamber for 4 hours at 37°.
(3) Wash in water, then for 2 minutes in 1 per cent. acetic acid, return to water.
(4) Treat with fresh 1 per cent. salicyloyl hydrazide in 5 per cent. acetic acid for 20–40 minutes.
(5) Rinse in water.

† Obtainable from G. T. Gurr Ltd, London, S.W.6.

(6) Rinse (2 minutes) in fresh dilute solution of sodium pentacyano-ammine ferroate. This extracts excess hydrazide.

(7) Wash twice in distilled water.

(8) Treat with 1 per cent. aqueous zinc acetate for 5–10 minutes.

(9) Rinse in water, dehydrate, clear in xylene and mount in Fluormount.

Result. Thiols fluoresce in green when examined by U.V. light.

HYDROGEN SELENIDE METHOD FOR SS GROUPS

(Olszewska *et al.*, 1967)

(*Formol-calcium, paraffin sections*)

Method. (1) Mount sections on slides pretreated with adhesive.

(2) Bring sections to water and dry in air.

(3) Place in a closed vessel over water for 24 hours.

(4) Treat for 2 hours at room temperature, in a closed vessel and in the absence of oxygen, with hydrogen selenide (from a Kipp generator; toxic, therefore handle with care and in a fume cupboard).

(5) Clear directly in xylene and mount in a suitable synthetic resin.

Result. Orange, red, reddish-brown and deep brown colours indicate the presence of SS groups.

MIXED ANHYDRIDE METHOD FOR SIDE-CHAIN COOH

(after Barrnett and Seligman, 1958)

(*Formalin, alcohol, paraffin sections; freeze-dried/formaldehyde vapour*)

Method. (1) Remove wax from paraffin sections with light petroleum.

(2) Allow to dry and then wash in glacial acetic acid, 2 minutes.

(3) Incubate sections for 1 hour at 60° in a mixture of equal parts of acetic anhydride and anhydrous pyridine (redistilled over barium oxide).

(4) Rinse in glacial acetic acid and wash in absolute alcohol.

(5) Incubate for 2 hours at 22° in 0·1 per cent. 2-hydroxy-3-naphthoic acid hydrazide. (50 mg. hydrazide in 2·5 ml. warm glacial acetic acid, to which is added 47·5 ml. of 50 per cent. ethanol).

(6) Wash in 3 changes (10 minutes each) of 50 per cent. ethanol.

(7) Incubate for 30 minutes in 0·5 N-HCl at 22°.

(8) Rinse in distilled water and then in 3 changes of 1 per cent. sodium bicarbonate.

(9) Rinse in several changes of distilled water.

(10) Transfer slides to a solution containing equal parts of 0·06 M Sorensen's phosphate buffer (pH 7·6) and absolute ethanol, to which has been added 1 mg./ml. tetrazotized diorthoanisidine (Fast blue B salt). Colour develops in 2–5 minutes.

(11) Wash in distilled water.

(12) Dehydrate, clear, and mount in a synthetic medium.

Result. Protein-bound side-chain carboxyl groups give a deep reddish-purple colour.

FLUORESCENT METHYL KETONE METHOD FOR C-TERMINAL COOH

(after Stoward, 1963; Stoward and Burns, 1967)

Method. Proceed to Stage 3 of the preceding method and then carry on as follows:

(4) Wash in two changes of 95 per cent. ethanol, 2 minutes each, and take down to water.

(5) Treat with 1 per cent. salicyloyl hydrazide* in 5 per cent. acetic acid† (pH 3) for 20–40 minutes at 22°.

* Sigma or Eastman-Kodak.

† If this solution is stored, a brown resinous precipitate is deposited after a few weeks. This may be ignored.

(6) Rinse in deionized or distilled water.

(7) Rinse (2 minutes) in a fresh dilute solution of pentacyano-ammine ferroate (tri-sodium salt).* This extracts excess hydrazide.

(8) Wash twice in distilled water.

(9) Treat with 1 per cent. aqueous zinc acetate for 5–10 minutes.

(10) Rinse in deionized or distilled water.

(11) Dehydrate in alcohols, clear in xylene and mount in Flourmount.

Result. C-terminal COOH groups give an intense blue fluorescence (366 nm).

EPOXYETHER REACTION FOR CARBOXYL GROUPS

(After Geyer, 1962)

(*Formalin, paraffin sections*)

Method. (1) Bring sections to absolute alcohol. Dry at 37° in an incubator.

(2) Prepare a 2 per cent. solution of *p*-biphenyl-methoxyethylene oxide in dry xylene. Boil sections under reflux for at least 10 hours.

(3) Wash in 3 changes of xylene at room temperature.

(4) Wash in absolute and then in 50 per cent. ethanol.

(5) Incubate for 2 hours at 22° in 0·1 per cent. naphthoic acid hydrazide in acetic ethanol (p. 649).

(6) Wash in 3 changes of 50 per cent. ethanol.

(7) Incubate for 30 minutes in 0·5 N-HCl at 22°.

(8) Rinse in distilled water and then in 3 changes of 1 per cent. sodium bicarbonate.

(9) Rinse in several changes of distilled water.

(10) Transfer slides to a solution composed of equal parts of 0·06 M Sorensen's phosphate buffer (pH 7·6) and absolute ethanol, to which has been added 1 mg./ml. Fast blue B salt. Leave for 5–10 minutes.

(11) Wash in distilled water.

(12) Dehydrate, clear, and mount in a suitable synthetic medium.

Result. A grey-blue colour indicates protein-bound carboxyl groups.

CARBODIIMIDE REACTION FOR CARBOXYL GROUPS

(After Geyer, 1964)

(*Formalin, paraffin sections*)

PREPARATION OF REAGENT

Dissolve 100 mg. 1-cyclohexyl-3-(2-morpholinyl-(4)-ethyl)-carbodiimide-*p*-toluolmethyl sulphonate and 50 mg. naphthoic acid hydrazide in 5 ml. tetrahydrofuran and add 5 ml. 70 per cent. alcohol with 10 ml. distilled water.

Method. (1) Bring sections to 70 per cent. alcohol.

(2) Incubate for 2–3 hours at 37° in the reaction mixture.

(3) Wash in 3 changes of 50 per cent. alcohol.

(4) Incubate in 0·5 N-HCl for 30 minutes.

(5) Wash several times in distilled water.

(6) Treat sections with 1 per cent. aqueous sodium bicarbonate for 5 minutes.

(7) Wash in distilled water.

(8) Treat for 5 minutes with a solution of Fast blue B salt (1 mg./ml.) in phosphate buffer at pH 7·6.

(9) Wash in water. Dehydrate in alcohols, clear in xylene and mount in a synthetic resin medium.

Result. Strongly reacting proteins may appear deep purple. Other sites are stained reddish-purple.

* Koch-Light, Hopkin Williams.

APPENDIX 7

PREPARATION OF FLUORESCEIN ISOCYANATE AND ISOTHIOCYANATE

(Coons and Kaplan, 1950; de Repentigny and James, 1954; Riggs *et al.*, 1958)

Preparation of Nitrofluorescein

100 g. 4-nitrophthalic acid and 100 g. resorcinol are intimately mixed in a beaker and heated on an oil bath at 195°–200° until the mass is dry (12 hours). When cool, the melt is chipped from the beaker, ground in a mortar, and boiled with 1,600 ml. 0·6 N-HCl for 1 hour. After washing by decantation with 3 × 300 ml. hot HCl solution it is collected on a Buchner funnel. The resulting brown paste is washed with 5 litres of water and dried.

Two isomers exist. According to Borek (1961) the isomer described by Coons and Kaplan (1950) as nitrofluorescein I has the nitro group *para* to the carboxyl group (5′-nitrofluorescein). Nitrofluorescein II, with the nitro group *meta* to the carboxyl group is therefore 4′-nitrofluorescein.

Reduction to Aminofluorescein

Suspend 2 g. crude nitrofluorescein in 100 ml. absolute ethanol and shake with 1·5 g. Raney nickel in an atmosphere of hydrogen at room temperature and pressure. After 90 minutes remove the nickel by centrifugation, wash with 15 ml. ethanol and add the washings to the main lot. Dilute with an equal quantity of water and allow to stand. Colourless needles form which are separated by filtration. The two isomers, aminofluorescein I and aminofluorescein II, are present. (See also McKinney *et al.* (1962) for additional details.)

Separation of the Isomers by Chromatography

Coons and Kaplan separated the isomers by fractional crystallization of nitrofluorescein diacetates. This method is laborious and, provided that small quantities (up to 400 mg.) are acceptable, the alternative method of separation of the two aminofluoresceins by chromatography is preferable.

Stationary Phase. Kieselguhr (Hyflo-Supercel) column with 0·2 M-phosphate buffer at pH 8·0 (3 ml. buffer to 5 ml. kieselguhr).

Mobile Phase. A mixture of *n*-butanol and *cyclo*hexane; de Repentigny and James recommend the attachment of an automatic fraction collector to a column 70 × 3 cm., containing 100 g. kieselguhr.

Preparation of Fluorescein Isocyanates

The required amount (10–60 mg.) of isomer I or isomer II, or of a mixture of the two if the pure isomers are not available, is added to 5 ml. of dry acetone. This is added through a dropping funnel to 15 ml. acetone previously saturated with phosgene. As each drop of the amine solution enters the flask a yellow precipitate forms and dissolves. Allow the reaction to proceed for a further 30 minutes after adding all the amine. Then remove the flask from the phosgene train and add some anthracite chips (to prevent bumping). Take the solution to dryness *in vacuo* over a water bath at 45°. Dissolve the resulting greenish-brown gum in 2 ml. acetone and 1 ml. dioxan and use this solution for adding to the chilled protein solution.

PREPARATION OF FITC I

(McKinney *et al.*, 1964)

Dissolve 5·6 g. aminofluorescein I in 200 ml. dry acetone and add dropwise, over a 2- to 5-hour period to a stirred solution of 16 ml. thiophosgene in 250 ml. acetone. Maintain constant atmosphere of nitrogen throughout. Continue stirring for 2–3 hours, using a magnetic stirrer. Collect the acetone-insoluble product on a sintered glass filter and wash with 250 ml. acetone. Dry under reduced pressure over anhydrous calcium sulphate. (Yield 5·4 g.) Transfer to a sintered glass suction filter and wash with 500 ml. distilled water to remove as much HCl as possible. Dry again as above; (Yield 4·6 g.).

PREPARATION OF FITC II

(McKinney *et al.*, 1964)

Dissolve 2·8 g. aminofluorescein II in 150 ml. acetone and add dropwise, over a 3-hour period to a stirred solution of 8 ml. thiophosgene in 350 ml. acetone. Continue stirring for a further 3 hours, maintaining an atmosphere of nitrogen throughout the whole period. Collect the insoluble product by suction filtration (Yield 3·55 g.). Dissolve 1 g. of the product in 200 ml. 0·05 M Na_2HPO_4 and precipitate the FITC II by adding 4 ml. acetic acid. Collect on suction filter and wash thoroughly with water. Dry to constant weight under reduced pressure (Yield, 0·73 g.). The product is almost free from HCl.

PREPARATION OF 1-DIMETHYLAMINONAPHTHALENE-5-SULPHONIC ACID

(Mayersbach, 1957b)

Suspend 10 g. 1-aminonaphthalene-5-sulphonic acid (Laurent's acid, technical grade) in 100 ml. water in a 250 ml. stoppered flask. Add 30 ml. of 0·5 N-KOH and warm to 40°; the Laurent's acid dissolves.

Add 10–20 ml. of dimethyl sulphate in small (2 ml.) portions, shaking the flask constantly. The addition of each portion of $(CH_3)_2SO_4$ causes the temperature to rise, and shaking should continue until it falls. A further portion of $(CH_3)_2SO_4$ is then added until the point is reached at which there is no longer a rise of temperature in the flask. On cooling, the colourless crystals of the dimethyl compound begin to separate.

Add sufficient conc. HCl to neutralize (2 ml.); a mass of crystals separates.

Filter through a Buchner funnel and wash with water.

Recrystallize the product from 300 ml. boiling water (twice) and dry at 80°.

Store under usual conditions. (Yield 60 per cent.)

PREPARATION OF 1-DIMETHYLAMINONAPHTHALENE-5-SULPHONYL CHLORIDE

(Mayersbach, 1957; Laurence, 1957)

Place 1 g. 1-dimethylaminonaphthalene-5-sulphonic acid and 2 g. phosphorus pentachloride (PCl_5) in a wide-mouthed Pyrex tube. Mix and rub together with a stirring rod until a greyish-yellow melt is produced; gentle heating will accelerate this process. Smear the melt over the walls of the tube. Add 50 ml. ice-cold 1·0 M-disodium phosphate and suspend the melt in this by stirring. Add 30 ml. ether and shake. The sulphonyl chloride dissolves entirely to form a dark yellow solution. Evaporate the ether under vacuum until the product crystallizes out.

Keep over $CaCl_2$, in the cold.

PREPARATION OF LISSAMINE RHODAMINE SULPHONYL CHLORIDE

(Chadwick *et al.*, 1958)

As an alternative to DANSYL the sulphonyl chloride of the red fluorescent dye Lissamine Rhodamine B 200 (RB 200) can be employed. When conjugated with proteins this gives an orange-red fluorescence (maximum emission 610 nm) which shows up clearly against the usual autofluorescence of the tissues.

METHODS OF PREPARATION

Grind 1 g. of RB 200 and 2 g. PCl_5 in a mortar for 5 minutes (in a fume cupboard). Add 10 ml. *dry* acetone and stand for 5 minutes with occasional stirring. Filter and use the resulting solution for conjugation.

METHOD OF CONJUGATION

Dilute each ml. of serum with 1 ml. physiological saline and 1 ml. carbonate-bicarbonate buffer (0·5 M, pH 9·0). Add 0·1 ml. RB 200 solution drop by drop with constant stirring and continue at 0°–4° with constant stirring for 12–18 hours. Dialyse against saline for 5–7 days, until the dialysate is non-fluorescent.

PREPARATION OF TRIHYDROXY-AMINOAZOBENZENE

(After Dowdle and Hansen)

Dry 3·5 g. sodium nitrite at 110° for 24 hours. Add this slowly with constant stirring, to 50 ml. conc. H_2SO_4 (sp. gr. 1·84). Maintain the mixture at 1 to 10° by means of an ice-bath. Heat slowly, while stirring, to 70° and then cool to room temperature. Add 9 g. 2,4-dinitroaniline and stir until completely dissolved. The resulting dark brown syrup should be allowed to stand for 2 hours in the dark.

Dissolve 60 g. anhydrous Na_2CO_3 in 1250 ml. distilled water and cool to 2° by adding 750 g. of ice. Add the diazotized dinitroaniline and stir vigorously. When the evolution of gas ceases remove the yellowish-brown flocculent precipitate by filtration and collect the clear yellow alkaline solution. To the latter, at 10°–15°, add 6 g. resorcinol in 25 ml. 20 per cent. NaOH. Continue stirring for several hours.

The resulting solution of 2,4,2′-trihydroxy-4-′aminoazobenzene is purplish-red in colour. Warm the solution to 75° on a water-bath and add 15 g. sodium sulphide monohydrate, dissolved in 30 ml. water, dropwise over a 10-minute period with constant stirring. Continue for 20 minutes then cool to room temperature. Adjust pH to neutrality with glacial acetic acid. Collect the reddish-brown precipitate and dry in a vacuum desiccator.

Dissolve the powdered precipitate in 250 ml. benzene and boil for 10–15 minutes. Filter while hot and reduce the volume to one-half by boiling. When cool a brick-red precipitate forms. (M.P. 205°–207°.)

PREPARATION OF FLAZO ORANGE

Dissolve 15·1 g. 4-chloro-2-aminophenol in 60 ml. water to which 10·2 ml. of 10 N sodium hydroxide has been added. Cool the solution to 0° by adding ice and then acidify with 24·2 ml. 10 N HCl. To this acid solution, under stirring, add within 30 seconds 25 ml. 5 N sodium nitrite. Continue stirring for 45 minutes at 5–10°.

Dissolve 15·8 g. 2-naphthol in 80 ml. water to which 13 ml. of 10 N sodium hydroxide has been added. To this add within 2 minutes the diazo solution obtained as above. Stir for 1 hour and then acidify with glacial acetic acid. Collect the precipitated dye by filtration. Dry under vacuum and recrystallize from a hot solution in methyl cellosolve and nitrobenzene. (Yield approximately 50 per cent.)

PRESERVATION OF FLUORESCEIN ISOCYANATE
(Goldman and Carver, 1957)

Although sealed ampoules containing an acetone solution keep for "several months" they are not really convenient either for storage or use.

Goldman and Carver suggest the following procedure:

(1) Dissolve 20 mg. per ml. of the isocyanate in acetone-dioxan.*

(2) Place 1 ml. in a flat dish and soak up with a 20 mm. square of chromatographic paper.

(3) Dry in front of a fan (*N.B.*—dioxan is toxic) and store over $CaCl_2$ on a desiccator.

(4) For labelling. Cut a strip of the paper to give the calculated amount of isocyanate and add to the protein solution with enough 0·5 M-carbonate to produce pH 9·0. The volume of buffer should be about 10 per cent. of that of the protein solution.

With this procedure denaturation of protein is avoided.

PREPARATION OF FLUORESCENT CONJUGATES
(After Marshall *et al.* and Cochrane)

The following details may be used for the preparation of conjugates with FIC, FITC, TMRIC or other isocyanates or isothiocyanates, with modifications. As given they refer to the preparation of FITC conjugates only.

CARBONATE BUFFER STOCK SOLUTION

(*a*) Na_2CO_3, 13·25 g. in 250 ml. distilled water.

(*b*) $NaHCO_3$, 10·5 g. in 250 ml. distilled water.

When required for use mix 8 parts of (*b*) with 1 part of solution (*a*) to give a 0·5 M buffer, pH 9·0.

Method. (1) Decide on a convenient final volume for the conjugating solution (say 50 ml.).

(2) Calculate the amount of antibody globulin solution which contains 500 mg (To provide a final protein concentration of 1 per cent.)

(3) Prepare sufficient carbonate buffer (pH 9·0) to constitute 15 per cent. of the final volume (7·5 ml.). Keep at 0°–4°.

(4) Dilute the antibody globulin solution with cold 0·15 M NaCl so that when the buffer is added the final volume is 50 ml.

(This volume is 50 ml. minus 7·5 ml. minus the volume of the original antibody solution which contains 500 mg.)

(5) Add cold buffer solution (0°–4°).

(6) Slowly add the calculated amount of FITC powder. (This will be in the ratio of 1:40 up to 1:20 of the antibody protein.) Mix and maintain at 0°–4° for 12 to 18 hours with constant stirring. The resulting solution is ready for purification (see below). This is usually carried out by passage through a DEAE-cellulose column.

PREPARATION OF FITC-CONJUGATE
(After Wood *et al.*, 1965)

This method differs from the foregoing in that the pH of the coupling solution is maintained at above 9·0 during the early phase of the reaction.

Method. (1) Standardize a pH meter (with probe electrodes and temperature compensator set at 10°) using standard pH 10·2 buffer at 4°.

(2) Adjust the pH of the cold antibody globulin solution (20 mg./ml. in 0·15 M-NaCl) to 9·5 by adding 0·1 N-NaOH.

* Both the acetone and the dioxan should be dried by distillation over sodium (metal) and by keeping over $CaSO_4$.

(3) Add to a dry weighing bottle (of sufficient volume to accommodate the volume of globulin solution to be added) the calculated amount of FITC (25 μg./mg. protein).

(4) Add the antibody solution and stir constantly at 0°–4°.

(5) During the first hour maintain the pH above pH 9·0 by adding 0·1 N-NaOH.

(6) After one hour, seal the reaction vessel and continue stirring at 0°–4° for 18–24 hours.

(7) Dilute the reaction mixture with 10 times its volume of 0·01 M sodium phosphate buffer, at pH 7·2 to 7·5.

(8) Apply to a DEAE-cellulose column (see below) equilibrated with the same buffer.

PURIFICATION OF CONJUGATES

(Wood *et al.*, 1965; Cebra and Goldstein, 1965)

FITC conjugates are usually purified by step-wise elution on DEAE-cellulose columns. Other conjugates, especially those of TMRITC, must first be passed through a column of Sephadex G 50 which retains the uncoupled dye and other low mol. wt. components.

PROCEDURE (FITC CONJUGATES)

(1) Equilibrate a DEAE-cellulose column (17 × 1·7 cm.) with 0·01 M-sodium phosphate buffer, pH 7·2. Some authors, e.g. Hamashima *et al.* (1964) recommend pH 6·4.

(2) Elute with 0·01 M-phosphate buffer at the chosen pH. This fraction will contain lightly coupled antibody globulin (which is not bound to the column).

(3) Elute with 0·03 M-phosphate buffer at the same pH. This fraction should contain the correctly coupled globulin.

(4) Elute successively with 0·05 and 0·1 M-buffer (pH 7·2). This removes overcoupled globulin and some unreacted fluorescent materials. The correct fraction may be obtained at a different molarity if another pH is used.

PROCEDURE (TMRITC CONJUGATES)

(1) Prepare a 20 × 2·5 cm. column with Sephadex G 50.*

(2) Wash the column with physiological saline buffered to pH 7·1 with 0·01 M-phosphate.

(3) Apply a small volume of the conjugate (say 5 ml.) to the column.

(4) Maintain the flow through the column with buffered saline.

(5) Collect fractions of 3 to 4 ml. The earliest fractions contain no conjugate. Fractions 10 to 20 contain most of the labelled globulin.

(6) Apply the pooled fractions containing labelled globulin to a DEAE-cellulose column equilibrated as indicated above.

(7) Carry out step-wise elution with solutions of NaCl in phosphate buffer at pH 7·5, in ascending order of molarity. (0·02, 0·04, 0·06, 0·1, 1·0.)

(8) Collect fractions of about 6 ml. Monitor the procedure by examining the fluorescence of the eluate. Increasing NaCl concentrations yield fractions with increasing fluorochrome content. Fractions eluted at 0·04 and 0·06 M should contain the correctly labelled globulins.

PREPARATION OF ACETONE-DRIED TISSUE POWDERS FOR ABSORPTION OF FLUORESCEIN CONJUGATES

(Coons, Leduc and Connolly)

Place 25–50 gm. of fresh or frozen tissue in a Waring blendor with an equal volume of 0·15 M-NaCl solution. Homogenize with short repeated runs to avoid

* As described in the maker's instructions.

heating. Pour the homogenate into a beaker and add 4 volumes of acetone, with stirring. Allow to stand for a few minutes, decant and discard the supernatant. Pack the precipitate by centrifugation, wash in the centrifuge with several changes of saline until the supernatant is free from hæmoglobin. Suspend the washed precipitate in an amount of saline about equal to the volume of the precipitate and add 4 vols. acetone and harvest on a Buchner funnel. Wash the precipitate with acetone and allow to dry on the funnel. Dry overnight at 37° and store at 4° in stoppered containers.

Absorption of Fluorescein Conjugates

For use, stir the powder obtained as above into an aliquot of the fluorescent antibody solution (5 ml.) at 100 mg./ml. After standing for 1 hour at room temperature with occasional stirring the supernatant is harvested by centrifugation in the cold at 18,000 r.p.m. in an angle head. The process is then repeated, or it may be replaced for the second absorption by DEAE-cellulose chromatography.

Suitable general tissues for absorption are mouse liver, pig liver and, for rabbit tissues especially, to prevent non-specific staining of eosinophils, rabbit-bone marrow.

Table 50

Absorption and Emission of Fluorescent Compounds
(After Hansen, 1964)

Compound	Absorption nm	Emission nm
Fluorescein isothiocyanate		520
Fluorescein isothiocyanate conjugate	280, 495	520
Fluorescein isocyanate		520
Fluorescein		527
Fluoresceinamine II	325, 490, pH 7·6	
Fluorescein isocyanate conjugate	275, 325, 490, pH 7·6	550, pH 7·6
Fluorescein		530
Tetramethylrhodamine isothiocyanate		
isomer R	545	580
isomer G	550	585
Tetramethylrhodamine isothiocyanate conjugate isomer G	550	585
4-Aminorosamine B		620–630, pH 3
4-Aminorosamine B conjugate		620–630, pH 3
Lissamine rhodamine B 200 conjugate		590–610, 650–700
Lissamine rhodamine B 200 free dye	560, 570, considerable absorption below 300	610
Lissamine rhodamine B 200 albumin conjugate	570, slightly higher than free dye	610
Lissamine rhodamine B 200 free dye	350	
Lissamine rhodamine B 200 conjugate	280, 360, 575	595, 710
Dimethylamino-5-sulphonic acid free dye	310–370	
Dimethylamino-5-sulphonic acid conjugate	340, considerable absorption below 300	525, broad
3-Hydroxy-5,8,10-pyrene-sulphonic acid		520

METHOD FOR CONCENTRATION OF PROTEIN-CONTAINING SOLUTIONS

(After Kohn, 1959)

This method is based on dialysis against substances of high molecular weight.

Concentrating Procedure

Place the dilute protein-containing solution in a glass tube, or small beaker, having a protuberance or sump at the bottom end.

Break up a sufficient quantity of polyethylene glycol (Carbowax 20 M) or, alternatively, use powdered polyvinylpyrrolidone. Inflate a section of dialysis tubing and fill with PEG or PVP. Moisten with a small quantity of water. Insert the filled tubing into the protein solution adjusting its level to the degree of concentration required. (When the level of fluid in the glass tube falls below the bottom of the dialysis tube the concentration process ceases.)

At least 1 part of PEG should be used per 10 parts of the original solution.

PREPARATION AND TESTING OF IMMUNE SERA

(Kabat and Mayer; Cochrane, Boyd)

Animal Immunization

(1) *Viruses.* Inoculate rabbits intramuscularly with 1 ml. of virus suspension in appropriate dilution. After two weeks a second inoculation should be given and after a further two weeks, a third. Ten days after the last inoculation the animals should be bled. To the pooled sera add 1:10′000 merthiolate. Store at −20°.

(2) *Killed antigens.* The method of inactivation must assure retention of antigenicity. Give three doses of 0·1 ml. intravenously on three successive days (rabbits). Ten days later repeat with increased dose (0·5 ml.). Ten days later bleed the animals and treat sera as above.

(3) *Soluble protein antigens.* These include albumins, γ-globulins, fibrinogen, prothrombin, and such products as purified protein hormones. There are a number of techniques available. The best are those which tend to diminish the production of antibodies against trace contaminants in the antigen. For this purpose the alum precipitation method (Proom, 1943) is increasingly employed.

For each 100 mg. protein add 2·5 to 4·5 ml. 10 per cent. (w/v) aluminium potassium sulphate ($AlK(SO_4)_2 . 12H_2O$). Adjust the pH to 6·5 with 5 N-NaOH. Wash deposit twice with 100 to 200 ml. 1:10′000 merthiolate saline. Make up final precipitate to a suitable volume with merthiolate-saline.

(4) *Soluble polypeptide antigens.* Proteins having a molecular weight below 10′000 fail to stimulate antibody production. A method of immunization developed by Berglund (1965) employs polymethylmethacrylate particles (0·6 μ diameter).* The peptide is mixed with these, in 0·05 M-citrate buffer, pH 4·1. Subsequently the mixture is injected, alternatively, intravenously and subcutaneously. In the latter case it is emulsified with incomplete adjuvant (Freund).

Determination of Antibody Concentration

A rough determination of the level of precipitating antibody in a given serum can be made by adding increasing amounts of soluble antigen to a constant amount of the antiserum, as shown in the table below. This illustrates the case of anti-human gamma globulin (HGG). The ratio of anti-HGG to HGG at equivalence is 3·3:1. The figures in column 3 (†) are obtained by dividing those in column 2 by 3·3. The ratio of most serum γ-globulins is the same as that of HGG at equivalence. For serum albumin it is 5·6:1 and for ovalbumin 9:1.

* AB. Bofors, Sweden.

TABLE 51

Tube	Anti HGG ml.	µg. AB protein	µg. HGG† to add	ml. HGG†† to add	Max. Ppt.
1	0·5	150	45·5	0·05	
2	0·5	300	91·0	0·09	
3	0·5	600	182·0	0·18	
4	0·5	1800	545·0	0·55	+
5	0·5	3000	910·0	0·91	

The values in column 4 (††) represent the amounts of a stock solution of antigen (HGG) containing 100 mg./ml. which must be added to each tube. After filling the tubes are incubated at 37° for 30 minutes and then placed in a cold room overnight. The tube with maximal precipitate is the one closest to equivalence.

In the example given the antibody level is 1800 µg. per 0·5 ml. This is more than adequate for immunofluorescent studies.

If quantitative precipitin tests are unsuitable the passive hæmagglutinin test can be substituted. Tests for purity are performed by gel diffusion in Ouchterlony plates or by immunoelectrophoresis.

CONCENTRATION OF THE GLOBULIN FRACTION OF ANTISERA

This is usually carried out by means of ammonium or sodium sulphate precipitation.

Method. (1) Dilute the antiserum with an equal volume of isotonic saline and cool to 0°–4°.

(2) Add slowly an equal volume of cold saturated ammonium sulphate, buffered to pH 7·0, with constant stirring. Continue stirring for 30 minutes.

(3) Centrifuge at 4° and 2500 r.p.m. for 30 minutes.

(4) Resuspend the precipitate in cold half-saturated ammonium sulphate and wash twice in this medium.

(5) Dissolve the precipitate in the minimum amount of cold saline.

(6) Dialyze against buffered saline, in the cold, until the dialyzing solution is free of sulphate.

(7) Dialyze for 10–24 hours against unbuffered saline.

(8) Add merthiolate to a concentration of 1:10′000 and store at 4°.

PREPARATION OF SATURATED AMMONIUM SULPHATE

Add reagent grade crystalline $(NH_4)_2SO_4$ to about 500 ml. distilled water while heating over a burner. A considerable amount is required but it is not necessary to obtain saturation at boiling point. Test by cooling small amounts of the solution. After cooling to 0–4°, bring to pH 7·0 by adding dilute NaOH.

DETERMINATION OF PROTEIN CONCENTRATION

Either the Micro-Kjeldahl or Biuret methods are suitable. The latter is probably the easier of the two for performing in general laboratories.

PREPARATION OF BIURET REAGENT

Dissolve 9 g. sodium potassium tartrate in just under 400 ml. 0·2 N-NaOH in a volumetric flask. Add 3 g. copper sulphate ($CuSO_4 . 5H_2O$) and dissolve completely with stirring. Add 5 g. potassium iodide and make up to 1 litre with 0·2 N-NaOH (0·8 g./litre of distilled water).

PREPARATION OF STANDARD CURVES

A standard curve should be prepared for each protein solution using either a solution of known protein content obtained by micro-Kjeldahl or other determination. Several dilutions should be used and the reference curve constructed by plotting the absorbance at 555 nm against protein concentration.

Method. Take duplicate 100 ml. samples and add 1·5 ml. biuret reagent. Mix and incubate the tubes at 37° for 30 minutes. Read in the spectrophotometer at 555 nm against a reagent blank. Compare reading with standard reference curve to obtain a value for N content of the sample.

ALCOHOL-FIXATION AND PARAFFIN-EMBEDDING METHOD
(Sainte-Marie)

Method. (1) Remove tissue from animal as soon as possible. Cut small blocks not more than 5 mm. thick. Drop these into pre-cooled 95 per cent. ethanol at 0°–4°. Leave at this temperature for 1 hour.

(2) Remove and trim blocks to 2–4 mm. thick.
(3) Fix in 95 per cent. ethanol at 0°–4° for 15–24 hours.
(4) Dehydrate in 4 changes of pre-cooled alcohol 1–2 hours in each with agitation.
(5) Clear in three changes of pre-cooled xylene for 1–2 hours in each.
(6) Allow tissues, in final xylene bath, to come to room temperature.
(7) Embed *via* 4 changes (1–2 hours) in 56° paraffin wax.
(8) When sectioning avoid floating-out on water. Use saturated sodium sulphate solution.
(9) Dry at 37° for 30 minutes on slide.
(10) Use promptly for immunofluorescent studies.

PRESERVATION OF CRYOSTAT SECTIONS FOR IMMUNOFLUORESCENCE
(George and Walter, 1962)

Method. (1) Dip mounted section (on coverslip) into polyethylene glycol 300 (mol. wt. 285 to 315) for 5 secs.
(2) Transfer to −20° storage cabinet.
(3) For use, warm coverslip and wash 3 times in physiological saline.

APPENDIX 8

THE ALDEHYDE-FUCHSIN STAIN FOR ELASTIC TISSUE
(Gomori, 1950)

(Formalin, Bouin, etc.; paraffin sections; avoid dichromate)

PREPARATION OF THE STAIN

Add 1 ml. of conc. HCl and 1 ml. of paraldehyde to 100 ml. of a 0·5 per cent. basic fuchsin in 60–70 per cent. alcohol. Keep at room temperature until the shade of the mixture darkens to a deep violet (about 24 hours). The aldehyde-fuchsin stain gradually alters its properties with age, staining more rapidly and strongly when fresh.

Electrophoretic analyses carried out by Sumner (1965) showed that solutions of aldehyde-fuchsin of different ages had different components. In the case of the precipitated dye the bluish-violet component was more soluble in acidified 70 per cent. ethanol than the reddish-violet component of the young dye.

Method. (1) Bring sections to water.

(2) Oxidize sections in Lugol's iodine for 10 minutes to 1 hour.

(3) Remove iodine with 5 per cent. thiosulphate for 1 minute.

(These two steps will remove mercury precipitate, if present.)

(4) Immerse in aldehyde-fuchsin 5 minutes to 2 hours depending on the tissue component it is desired to stain.

(5) Rinse the slide in several changes of 60–70 per cent. alcohol.

(6) Counterstain as desired—Gomori recommended hæmatoxylin-orange G, Masson's trichrome or the Mallory-Heidenhain azocarmine method, replacing aniline blue with light green or fast green in the last two cases.

(7) Dehydrate in alcohol, clear in xylene, mount in balsam or DPX.

Result. All components which take the stain appear in shades of deep purple. The optimum staining times are given in brackets. Elastic tissue (5 minutes), β-cells of the pancreas (15–30 minutes, pituitary β-granules (20 minutes to 2 hours), mast cell granules (5–10 minutes). Certain other substances and structures are stained, such as mucins and gastric chief cells.

PREPARATION OF ALDEHYDE-FUCHSIN IN DRY FORM
(Rosa, 1953)

One of the main difficulties experienced in using the aldehyde-fuchsin stain is due to its instability. Especially when used for the demonstration of cells in the anterior pituitary gland, it may be found adequate for only a few days between ripening and deterioration. The dry form of the stain has overcome this difficulty. It can be made up as required and is stable for a much longer period than the dye as originally used, even for "difficult" components like pituitary β-cells.

(1) Prepare aldehyde-fuchsin according to the (original) directions given above.

(2) Allow to ripen at room temperature for 3 days.

(3) Add 100 ml. of the mixture to 50 ml. chloroform in a separating funnel and add 200 ml. distilled water.

(4) Shake briefly and allow the formed precipitate to settle.

(5) Drain off the contents of the separating funnel containing the suspended precipitate and filter *without* suction.

(6) Dry at 50° and store in a stoppered bottle.

For use add 0·5 g. to 100 ml. 70 per cent. ethanol containing 1 ml. conc. HCl.

WEIGERT'S ELASTIC STAIN

(Moore, 1943)

(*Various fixatives, paraffin sections*)

This method depends for its success on the making of a satisfactory batch of the stain. A good batch will keep for about a year.

Preparation of the Stock Solution

Heat 500 ml. distilled water *nearly* to boiling in a large evaporating basin. Mix 2·5 g. crystal-violet (C.I. 42555), 2·5 g. basic fuchsin and 1·0 g. dextrin and dissolve them in the hot water. Add 10·0 g. resorcinol and bring to the boil. When boiling add slowly 62·0 ml. of a freshly prepared 30 per cent. aqueous solution of ferric chloride (British Drug Houses, anhydrous), stirring continuously with a glass rod. The mixture must be kept boiling, but not too vigorously. Continue boiling for a further 2 minutes to coarsen the precipitate. Cool, filter through a Buchner funnel. Wash the deposit with water (8–10 litres) until no further colour is removed. Dry the precipitate overnight at 56°. Remove from the filter paper and dissolve in 550 ml. of absolute ethanol to which has been added 1 ml. conc. HCl by simmering on a water bath for 30 minutes. Cool and filter; add 19 ml. conc. HCl and allow to stand 24–48 hours before use. The colour should be dark greenish-blue.

Preparation of the Stain

Add 35 ml. stock solution to 30 ml. 70 per cent. alcohol. If necessary these proportions can be varied.

Method. (1) Bring sections to water.

(2) Treat for 2–5 minutes with acidified permanganate (47·5 ml. of 0·5 per cent. aqueous potassium permanganate with 2·5 ml. of 3 per cent. H_2SO_4 added).

(3) Wash in water.

(4) Bleach in 1 per cent. oxalic acid for 1 minute.

(5) Wash in water and rinse in 70 per cent. alcohol.

(6) Stain in Weigert's stain for 8–24 hours (the time must depend on the results obtained since it varies considerably with each batch of the stain.

(7) Wash in 1 per cent. acid alcohol until only elastic tissue is stained.

(8) Wash in water.

(9) Counterstain lightly in carbol-safranin or neutral red.

(10) Rinse in water, dehydrate, clear and mount in a suitable synthetic resin.

Result. Elastic fibres appear dark blue-black; nuclei, red.

VERHOEFF'S ELASTIC STAIN

(*Various fixatives, paraffin sections*)

Preparation of the Stain

Mix, in the order stated, 10 ml. of 5 per cent. *fresh*, unripened, alcoholic hæmatoxylin, 4 ml. of 10 per cent. *fresh* aqueous ferric chloride and 4 ml. of Lugol's iodine.

Method. (1) Bring sections down to water.

(2) Stain in Verhoeff's stain until black (15 minutes).

(3) Rinse in water.

(4) Differentiate in 2 per cent. ferric chloride until only elastic fibres and nuclei are stained.

(5) Rinse in distilled water.

(6) Counterstain with Van Gieson (10 ml. 1 per cent. acid fuchsin, 90 ml. saturated aqueous picric acid, 100 ml. distilled water, boiled for 3 minutes to ripen) for 30–60 seconds.

(7) Rinse in distilled water, dehydrate, clear and mount.

Result. Elastic fibres, black; nuclei, grey; collagen, red.

Background structures and cells, yellow.

The period of counterstaining must not be prolonged since the picric acid differentiates the stain further. Fine elastic fibrils are usually not stained by this method.

ORCINOL-NEW FUCHSIN METHOD FOR ELASTIC TISSUE

(Fullmer and Lillie, 1956)

(*Various fixatives, paraffin sections*)

This is a modification of Weigert's technique.

PREPARATION OF STAINING SOLUTION

Add 2 g. new fuchsin (C.I. 42520) and 4 g. orcinol (highest purity) to 200 ml. distilled water and boil for 5 minutes. Add 25 ml. 30 per cent. ferric chloride (U.S.P. IX or freshly prepared from analytical reagent grade anhydrous $FeCl_3$) and boil for a further 5 minutes. Cool, collect the precipitate and dissolve this in 100 ml. 95 per cent. ethanol. In contrast to Weigert's procedure this solution is used for staining.

Method. (1) Bring sections to water.

(2) Stain for 15 minutes at 37°.

(3) Differentiate in three changes of 70 per cent. alcohol for 15 minutes in each.

(4) Dehydrate, clear and mount in synthetic resin.

Result. Elastic fibres, deep violet.

Counterstaining may be carried out with 1 per cent. aqueous safranin or with Van Gieson. The latter may be applied before or after differentiation.

MODIFIED OXYTALAN METHOD

(After Gawlik and Jarocińska)

(*Formalin, alcohol, Zenker, Orth; paraffin sections*)

PREPARATION OF STAIN

Dissolve 20 mg. cresylechtviolet in 100 ml. distilled water. Filter before use and dilute with an equal part of 96 per cent. ethanol.

Method. (1) Bring sections to water.

(2) Treat with peracetic acid (Greenspan: see p. 620) for 20 minutes.

(3) Rinse in running water for 3 minutes.

(4) Stain for 5–10 minutes.

(5) Rinse quickly in distilled water.

(6) Dehydrate in alcohols.

(7) Clear in xylene and mount in a synthetic resin.

Result. Collagen grey-blue or grey-violet. Oxytalan fibres violet. Nuclei blue.

SILVER IMPREGNATION FOR RETICULIN

(Gordon and Sweet, 1936)

(*All ordinary fixatives, including Helly; paraffin sections*)

Method. (1) The sections should be firmly attached to the slides (egg albumin followed by heating).

(2) Bring sections to water and remove mercury deposits if necessary.

(3) Oxidize for 1–7 minutes in acidified permanganate (47·5 ml. 0·5 per cent. aqueous $KMnO_4$ with 2·5 ml. 3 per cent. H_2SO_4) for 1 minute.

(4) Wash in water.

(5) Blanch until white in 1 per cent. oxalic acid or 10 per cent. HBr for 1 minute.

(6) Wash in two changes of *glass* distilled water.

(7) Mordant for 2–15 minutes in 2 per cent. aqueous iron alum $[(NH_4)_2SO_4 . Fe_2(SO_4)_3 . 24H_2O]$.

(8) Wash in two or three changes of distilled water.

(9) Impregnate for 5–7 seconds in Wilder's silver bath. (To 5 ml. 10 per cent. $AgNO_3$ add ammonia (28 per cent. ammonia water) drop by drop until the brown precipitate which forms is *nearly* dissolved. Add 5 ml. 3 per cent. NaOH and add ammonia drop by drop until the solution is clear. Make up to 50 ml. with glass distilled water.*)

(10) Wash briefly in distilled water.

(11) Reduce with 10 per cent. *neutral* formalin for 30 seconds.

(12) Wash in water. (If sections appear over-impregnated repeat the process from stage 7.)

(13) Tone in 0·2 per cent. yellow gold chloride for 1–3 minutes (optional).

(14) Wash in tap-water.

(15) Fix in 5 per cent. sodium thiosulphate for 5 minutes.

(16) Wash well in tap-water.

(17) Dehydrate, clear and mount.

Result. Reticulin, black; collagen fibres, yellow to brown.

If desired a light counterstain can be employed between stages (16) and (17). The sensitivity of the method is said to be improved if stages (3)–(5) are repeated.

MALLORY'S P.T.A.H. METHOD FOR FIBRIN

(*Various fixatives, paraffin sections*)

Method. (1) Bring sections down to water. Remove mercury deposits with iodine and remove the iodine with alcohol, *not thiosulphate*.

(2) Postchrome for 30 minutes in a mixture of 3 parts of 3 per cent. aqueous $K_2Cr_2O_7$ and 1 part of 10 per cent. HCl.

(3) Wash in water.

(4) Differentiate for 1 minute in acid permanganate (see reticulin stain, above).

(5) Wash in water.

(6) Bleach in 1 per cent. oxalic acid until white.

(7) Rinse in water and transfer to P.T.A.H.† for 12–24 hours.

(8) Shake off excess stain (water removes the red component).

(9) Dehydrate in 99 per cent. alcohol (this differentiates the blue component).

(10) Clear and mount in a synthetic resin.

Result. Fibrin, neuroglia, red cells: dark blue. Nuclei: light blue. Collagen: rose red.

The balance of red and blue depends on the postchromating and removal of the chromate by the acid permanganate. The more chromate left in, the darker the blue.

* All silver solutions must be made up in chemically clean glassware. The Wilder's silver bath will keep for 3–6 months.

† Hæmatoxylin (or hæmatein), 0·1 g.; phosphotungstic acid, 2·0 g.; distilled water, 100 ml. Dissolve separately and mix. Leave to ripen for several months. When ripe the stain keeps for years.

PICRO-MALLORY V FOR FIBRIN

(After Lendrum *et al.*, 1962)

(*Formalin-mercuric chloride, formalin; avoid chromium; paraffin sections*)

PREPARATION OF REAGENTS

(1) Yellow mordant. To 200 ml. 80 per cent. ethanol, saturated with picric acid, add 0·4 g. Orange G. and 0·4 g. Lissamine Fast Yellow 2 G.*

(2) Acid fuchsin 1 per cent. solution in 1 per cent. aqueous acetic acid.

(3) Differentiator. To 30 ml. of yellow mordant add 70 ml. 80 per cent. ethanol.

(4) 1 per cent. aqueous phosphotungstic acid.

(5) 1 per cent. Soluble blue (acid blue 22), or 2 per cent. light green SF, in 1 per cent. acetic acid.

Method. (1) Stain nuclei by the celestin-blue hæmalum sequence (Appendix 10, p. 659). Rinse in tap-water.

(2) Apply yellow mordant for 3–5 minutes, followed by 1 minute in tap-water.

(3) Stain with fuchsin for 5 minutes. Rinse in tap water.

(4) Differentiate for 10–15 seconds. Rinse in tap water.

(5) Apply phosphotungstic acid (5 minutes).† Rinse in tap water.

(6) Stain with soluble blue or light green SF for 2 minutes. Rinse in tap water.

(7) Dehydrate rapidly, clear in xylene, mount in DPX.

Result. Nuclei, blue-black; basement membranes and collagen, pale bluish-grey; red cells, yellow; fibrin, red.

MARTIUS-SCARLET-BLUE METHOD FOR FIBRIN

(After Lendrum *et al.*, 1962)

(*As for previous method*).

Method.(1) Stain nuclei by the celestin-blue hæmalum sequence (Appendix 10, p. 659). Rinse in tap water.

(2) Differentiate nuclei in 0·25 per cent. HCl in 70 per cent. ethanol.

(3) Wash well in running water.

(4) Rinse in 95 per cent. ethanol and stain with 0·5 per cent. Martius yellow in 95 per cent. ethanol containing 2 per cent. phosphotungstic acid (2 minutes).

(5) Rinse in water and stain with 1 per cent. Brilliant Crystal Scarlet 6 R‡ in 2·5 per cent. aqueous acetic acid (10 minutes).

(6) Rinse in water and treat with 1 per cent. aqueous phosphotungstic acid for 5 minutes.

(7) Rinse in water and stain with 0·5 per cent. soluble blue in 1 per cent. aqueous acetic acid (10 minutes).

(8) Rinse in water, blot, dehydrate in running absolute ethanol, clear in xylene, mount in DPX.

Result. Nuclei, blue-black; red cells, yellow; fibrin, red.

MASSON 44/41 METHOD FOR FIBRIN

(After Lendrum *et al.*, 1962)

(*As for previous two methods*).

Method. (1) Remove wax with xylene, rinse with trichloroethylene and immerse in a closed jar of trichloroethylene for 48 hours.

* Acid yellow 17, I.C.I. Ltd.

† Dense tissues such as hyaline collagen may still retain red. In this case continue for 10 minutes and follow with red differentiator (McFarlane, 1944). This is made by diluting a stock solution (2 parts) with 95 per cent. ethanol (2 parts) and distilled water (1 part). The stock contains 25 g. phosphotungstic acid and 2·5 g. picric acid in 100 ml. 95 per cent. ethanol.

‡ Acid red 44. L. B. Holliday Ltd, Huddersfield, Yorks.

(2) Rinse in absolute ethanol and immerse for 3 hours in a closed jar of absolute ethanol saturated with picric acid and containing 3 per cent. mercuric chloride.

(3) Bring to water, remove mercury by treatment with Lugol's iodine followed by thiosulphate. Wash until yellow colour is no longer visible.

(4) Stain nuclei with the celestin blue-hæmalum sequence. Rinse in tap water.

(5) Differentiate with 0·25 per cent. HCl in 70 per cent. ethanol.

(6) Wash well in running water.

(7) Stain with Brilliant Crystal Scarlet 6 R in 1 per cent. aqueous acetic acid for 5 minutes.

(8) Rinse in water and treat with 1 per cent. aqueous phosphotungstic acid for 5 minutes.

(9) Rinse in water. Stain with 1 per cent. Naphthalene Blue Black CS in 1 per cent. aqueous acetic acid, for 30 minutes.

(10) Rinse in water, dehydrate in 95 per cent. and absolute ethanol, clear in xylene and mount in DPX.

Result. Nuclei, greyish black or black; red cells, red; fibrin, deep black.

BARGMANN'S CHROME HÆMATOXYLIN FOR NSS

(Bouin, Susa; paraffin sections)

NSS can be demonstrated in alcohol-fixed sections if these are floated out on Bouin's fluid instead of water.

PREPARATION OF THE STAIN

1 per cent. aqueous hæmatoxylin, 50 ml.

3 per cent. aqueous chrome-alum-[$Cr_2(SO_4)_3(NH_4)_2SO_4 . 24H_2O$], 50 ml.

5 per cent. aqueous potassium dichromate, 2 ml.

5 per cent. aqueous H_2SO_4, 1 ml.

Allow to ripen for 48 hours before use. Keeps for several weeks at 0°–4°. Filter before use.

Method. (1) Bring sections down to water.

(2) Mordant in a solution of Bouin's fixative containing 3–4 per cent. chrome-alum for 12–24 hours at 37°.

(3) Wash in running tap-water until the sections are colourless.

(4) Oxidize for 2–3 minutes in a mixture of 2·5 per cent. $KMnO_4$ (1 part), 5 per cent. H_2SO_4 (1 part), distilled water (6–8 parts).

(5) Wash in distilled water.

(6) Bleach in 1 per cent. oxalic acid for 1 minute.

(7) Wash in running tap-water for 5 minutes.

(8) Stain for 10 minutes.

(9) Differentiate for 30 seconds in 0·5 per cent. acid alcohol.

(10) Wash in running tap-water for 2–3 minutes.

(11) Stain for 2–3 minutes in 0·5 per cent. aqueous phloxine.

(12) Rinse in 5 per cent. aqueous phosphotungstic acid (2 minutes).

(13) Wash in running tap-water for 5 minutes.

(14) Differentiate in alcohols, clear and mount.

Result (Figs. 82 and 83). NSS: deep purple. Nuclei: purple. Background: pinkish-red.

PSEUDOISOCYANIN METHOD FOR INSULIN, ETC.

(Schiebler and Schiessler, 1958; Wolff, 1965)

(Bouin, Susa, Zenker, Formalin, Carnoy, Ethanol; paraffin sections)

Method. (1) Bring sections to water.

(2) Oxidize with performic acid (p. 620) for 1 hour or with acid permanganate

(10 ml. 2·5 per cent. $KMnO_4$, 10 ml. 5 per cent. H_2SO_4, 70 ml. distilled water) for the same period.

(3) Stain for 2–15 minutes in a freshly prepared aqueous solution of pseudo-isocyanin.

(Dissolve 8·6 mg. N,N′-diethyl-6,6′-dichloropseudoisocyanin chloride in a few drops of methanol and add to 100 ml. hot distilled water. Heat for 5 minutes, cool and filter.)

(4) Rinse in ammonia water (1 drop ·880 ammonia in 100 ml. water).

(5) Examine in the above medium or in a watery alkaline mounting medium. A green filter facilitates appreciation of the metachromatic complex. Alternatively use monochromatic light at 578 nm., or examine by U.V. microscopy.

Result. β-granules of the pancreatic islets, pituitary S-cell (thyrotroph) granules, and other cystine-containing tissue components exhibit a strong red (fluorescent) metachromasia.

HALMI METHOD FOR PITUITARY BASOPHILS

(Halmi, 1952)

The original author recommends the use of Bouin's fluid, modified by substitution of 0·5 per cent. trichloroacetic acid in place of acetic acid. As reproduced below, the method works satisfactorily only in the case of the rat hypophysis.

Method. (1) Bring paraffin sections to water.

(2) Oxidize in Lugol's iodine for 30 minutes.*

(3) Remove iodine with 5 per cent. aqueous sodium thiosulphate, 2 minutes.

(4) Rinse slides thoroughly in distilled water.

(5) Stain in aldehyde-fuchsin (3–10 days old) for 2–10 minutes.

(6) Rinse in two changes of 95 per cent. alcohol and leave in 95 per cent. alcohol for 5–10 minutes.

(7) Rinse in 70 per cent. alcohol and then in distilled water.

(8) Stain in Ehrlich's hæmatoxylin (3–4 minutes).

(9) Rinse in distilled water and differentiate in 0·5 per cent. acid alcohol.

(10) Blue in running tap-water.

(11) Counterstain for 45 seconds in 0·2 per cent. aqueous light green S.F. yellowish with 1 per cent. orange G, 0·5 per cent. phosphotungstic acid, and 1 per cent. acetic acid.

(12) Rinse briefly in 0·2 per cent. acetic alcohol.

(13) Dehydrate in two changes of absolute alcohol.

(14) Blot and transfer to xylene.

(15) Mount in a suitable synthetic resin.

Result. β-granules of the rat pituitary gland stain deep purple; δ-granules, green; α-granules, orange.

THE PFAAB, PAS, ORANGE G METHOD FOR THE HUMAN HYPOPHYSIS

(After Adams, 1956)

(Preferred fixation is formol-mercury; paraffin sections)

Method. (1) Bring sections to water and remove mercury deposits.

(2) Blot carefully.

(3) Apply performic acid (see p. 701) for 5 minutes.

* Alternative oxidations have been employed, e.g. performic and peracetic acids, periodic acid, acid permanganate, bromine, etc. With suitable oxidation the method can be made to work on pituitary glands of other species besides the rat. Performic acid, used as on p. 620, gives good results.

(4) Rinse in tap-water for 10 minutes.
(5) Rinse in 70 per cent. and absolute alcohol.
(6) Blot, to flatten section on slide.
(7) Rinse in tap-water and dry section (just dry) at 60°.
(8) Rinse in absolute alcohol.
(9) Rinse in tap-water for 1 minute.
(10) Stain in acid Alcian blue (p. 624) for 1 hour.
(11) Rinse in tap-water for 5 minutes.
(12) Stain by the PAS, hæmalum, orange G sequence (p. 659).
(13) Dehydrate, clear and mount in synthetic resin.

Result (Plate Id). R-type mucoid granules stain magenta red.
S-type mucoid granules stain blue.
α-granules of the acidophils stain orange.
Nuclei, blue-black.

THE WILSON-EZRIN METHOD FOR THE HUMAN HYPOPHYSIS

(Wilson and Ezrin, 1954)

(*Formalin, formol mercury; paraffin sections*)

Method. (1) Bring sections to distilled water.
(2) Oxidize in 0·5 per cent. aqueous periodic acid for 5 minutes.
(3) Treat with Schiff's reagent (p. 647) for 15 minutes.
(4) Wash in running water for 10 minutes.
(5) Stain in 1 per cent. aqueous orange G for 10–15 seconds.
(6) Mordant in 5 per cent. phosphotungstic acid for 15 seconds.
(7) Rinse in running tap-water for 15 seconds.
(8) Stain in 1 per cent. aqueous methyl blue for 1 minute.
(9) Rinse briefly in 1 per cent. acetic acid.
(10) Dehydrate in alcohols, clear and mount in a synthetic medium.

Result. The pituitary mucoid granules are divided into two types: β-granules, magenta-red; γ-granules, purple. The α-granules are stained orange.

This method works on both rat and human hypophyses. Examination of sections by daylight with the aid of a blue filter (Ansco No. 45) brings out the contrast between red and purple.

APPENDIX 9

THE FEULGEN REACTION

(Feulgen and Rossenbeck, 1924, modified)

(*various*)

PREPARATION OF SCHIFF'S REAGENT

Although Schiff's reagent is spoken of as if it were an analytical reagent of uniform purity and constitution it is necessary to point out that many factors in its production are not well understood and are therefore not easy to control. Firstly, not all batches of basic fuchsin will produce a satisfactory reagent and solutions made from two batches of the dye will often differ considerably in performance. Secondly, the different methods of preparation quoted in Chapter 9 do not produce a single uniform reagent. Thirdly, different conditions of use produce very different results. (See Atkinson (1952) and Longley (1952) for information on these points.)

Three methods of preparation only are given here in detail. With the first method choice of a suitable batch of basic fuchsin is especially important, and the best procedure is to obtain a number of samples of the dye and to make up a number of different solutions. When a satisfactory sample has been found a large quantity should be put into store. Thus uniform results over a long period can be attained.

Schiff's Reagent (de Tomasi, 1936). Dissolve 1 g. of basic fuchsin in 200 ml. of boiling distilled water. Shake for 5 minutes and cool to exactly 50°. Filter and add to the filtrate 20 ml. of N-HCl. Cool to 25° and add 1 g. of sodium (or potassium) metabisulphite ($Na_2S_2O_5$). Stand this solution in the dark for 14–24 hours. Add 2 g. of activated charcoal and shake for 1 minute. Filter. Keep the filtrate in the dark at 0°–4°. Allow to reach room temperature before use.

Schiff's Reagent (Barger and De Lamater, 1948). Dissolve 1 g. of basic fuchsin in 400 ml. of boiling distilled water. Cool to 50° and filter. To the filtrate add 1 ml. of thionyl chloride ($SOCl_2$). Stand in the dark for 12 hours. Clear by shaking for 1 minute with 2 g. activated charcoal. Filter. Store in the dark at 0°–4°. Use in the dark at room temperature.

Schiff's Reagent (Itkawa and Oguru, 1954). Boil 200 ml. distilled water and add 1 g. basic fuchsin. Shake until dissolved, cool and filter into a flask. Bubble SO_2 gas slowly through the solution through fritted glass, shaking occasionally. The SO_2 is most conveniently supplied from an SO_2 syphon.* When the solution becomes a clear transparent red colour the gas is turned off. The flask should now be stoppered and left overnight in the dark at room temperature. The pale red fluid can then be decolourized by adding 1 g. activated charcoal,* shaking for 1 minute and filtering. Store at 0°–4°.

Purification of Parafuchsin for Use as Schiff's Reagent (Gabler, 1965). (1) Dissolve 4 g. parafuchsin (pararosanilin, C.I. 42500) in 800 ml. ethanol at room temperature. Treat 4 times with 8 g. activated charcoal (powder). After adding the first portion of charcoal bring quickly to the boil and filter while hot through a hard (Whatman No. 1) filter. Carry out this procedure a further 3 times. Evaporate the final solution, under vacuum, warming gently to assist the process.

(2) Dissolve the purified dye in ethanol/chloroform (1:1) and pass through an aluminium oxide column. A relatively broad dark red zone on the column indicates the region to be eluted. After elution the product is dried under vacuum. This procedure gives a dye which is completely decolorized by the addition of metabisulphite, thionyl chloride or SO_2.

* British Drug Houses Ltd, Poole, Dorset, England.

TIMES OF HYDROLYSIS IN N-HCl AT 60° (CONVENTIONAL HYDROLYSIS)

The duration of hydrolysis varies with the fixative employed. The figures below, given in minutes, are mainly derived from K. Bauer (1932).

Apathy	5	Formalin	8
Bouin	not recommended	Formol-sublimate	8
Bouin-Allen	22	Helly	8
Bouin-Allen-sublimate	14	Newcomer (Plants)	10
Carnoy 3:1	6	Newcomer (Animals)	20
Carnoy 6:3:1	8	Petrunkevitch	3
Carnoy-Lebrun	6	Regaud	14
Champy	25	Regaud-sublimate	8
Chrome-acetic	14	Serra	20
Flemming	16	Susa	18
Formaldehyde vapour		Zenker	5
(freeze-dried)	30–60	Zenker-formol	5

Standard Feulgen Method. (1) Bring sections to water and remove mercury if necessary.

(2) Rinse briefly in cold N-HCl.

(3) Place in N-HCl at 60° for the optimum time of hydrolysis.

(4) Rinse briefly in cold N-HCl and then in distilled water.

(5) Transfer to Schiff's solution for the optimum time (½–1 hour with de Tomasi, longer with Barger and De Lamater).

(6) Drain and rinse in three changes of freshly prepared bisulphite solution (5 ml. 10 per cent. $K_2S_2O_5$, 5 ml. N-HCl, water to 100 ml.).

(7) Rinse in water.

(8) Counterstain if desired (1 per cent. aqueous light green, 1 minute, or 0·5 per cent. alcoholic fast green, ½–1 minute).

(9) Dehydrate in alcohol.

(10) Clear in xylene and mount in balsam or DPX.

Result (Fig. 85, facing p. 262). DNA appears in shades of reddish-purple.

TIMES OF HYDROLYSIS IN 5 N-HCl AT 20°–22°

Alcoholic fixatives	20 minutes to 2 hours
Formalin-containing fixatives	35 minutes to 4 hours
Formaldehyde Vapour (freeze-dried)	2 to 8 hours

Modified Feulgen Method. (1) Bring sections to water.

(2) Treat with 5 N-HCl at 20–22° for the optimum time of hydrolysis.

(3) Rinse in distilled water.

(4) Transfer to Schiff's reagent for 30–60 minutes.

(5) Continue as in stage 6 of the standard method.

Result. DNA appears in shades of reddish-purple.

THE FEULGEN-NAPHTHOIC ACID HYDRAZIDE REACTION

(Pearse, 1951)

(*various*)

PREPARATION OF 2-HYDROXY-3-NAPHTHOIC ACID HYDRAZIDE (NAH) (SELIGMAN, FRIEDMAN AND HERZ, 1949)

(1) Prepare the acid chloride by warming 20 g. of 2-hydroxy-3-naphthoic acid with 25 ml. of thionyl chloride until the solid has dissolved and HCl is no longer evolved.

(2) Remove excess thionyl chloride by distillation at reduced pressure. The acid chloride solidifies on cooling and is dissolved in 50 ml. of methanol (pure).

(3) Remove the methanol by distillation at low pressure.

(4) Add 35 ml. of 85 per cent. hydrazine hydrate and heat on a steam bath for 3 hours. The hydrazide separates as a crystalline mass on cooling.

(5) Add water and collect the crystals with suction. Wash in water.

(6) Dissolve crystals in 500 ml. of hot ethanol and treat with activated charcoal.

(7) Chill the solution, collect the crystals and wash with ether.

Yield. 16 g. (75 per cent.). Pale yellow platelets, M.P. 203°–204°.

Method. (1) Bring sections to water.

(2) Rinse briefly in cold N-HCl.

(3) Place in N-HCl at 60° for the optimum time of hydrolysis (as for Feulgen reaction).

(4) Rinse briefly in cold N-HCl, in distilled water, and finally in 50 per cent. alcohol.

(5) Treat sections at about 22° for 3–6 hours with 0·1 per cent. NAH in 50 per cent. ethanol with 5 per cent. acetic acid.

(6) Rinse in three changes of 50 per cent. alcohol, 10 minutes in each change.

(7) Rinse in water.

(8) Transfer to a freshly prepared solution of diazotized *o*-dianisidine,* at 0° and pH 7·4, for 1–3 minutes.

(9) Wash in water.

(10) Dehydrate in alcohol, clear in xylene and mount in DPX.

Result (Fig. 86). DNA bluish-purple. Cytoplasmic and other proteins, especially if strongly basic, may be stained pinkish red.

FEULGEN-SILVER METHENAMINE

(After Korson, 1964)

(*Neutral formalin, Carnoy, Formol sucrose; fresh sections*)

Method. (1) Place slides in 1 M-Citric acid, preheated to 60°, and allow to remain at this temperature for 30 minutes.

(2) Wash in distilled water for 5 minutes.

(3) Immerse in Gomori's methenamine-silver bath (App. 26), preheatet do 60° for 30 minutes. After immersion maintain temperature at 60° and incubate slides for 1 hour.

(4) Wash briefly in distilled water.

(5) Treat with 0·2 per cent. gold chloride for 5 minutes.

(6) Rinse in distilled water, dehydrate, clear in xylene and mount in a permanent resin medium.

Result. Chromatin (DNA) black.

FEULGEN-SILVER HEXAMETHYLENETETRAMINE

(After Martino *et al.*, 1965)

(*Methanol-acetic acid*, 3:1; *Carnoy, ethanol*)

This method was designed for squashes and smears.

Method. (1) Wash in distilled water.

(2) Hydrolyze for 5 minutes at room temperature in 1 N-HCl.

(3) Hydrolyze for 7–10 minutes at 58°.

(4) Hydrolyze for 1–2 minutes at room temperature.

(5) Rinse thoroughly in distilled water.

* The stable diazotate, Fast Blue B salt (I.C.I. Ltd.) may be used as an alternative.

(6) Stain for 5–12 hours at 45° in silver-hexamethylenetetramine borax solution.
(7) Rinse thoroughly in distilled water.
(8) Place in 0·2 per cent. aqueous gold chloride for 10–60 minutes.
(9) Rinse in distilled water.
(10) Wash in 1 per cent. aqueous sodium thiosulphate for 1–2 minutes.
(11) Dehydrate in alcohol.
(12) Clear in xylene and mount. (Resin medium.)
Result. Chromatin (DNA) black.

MODIFIED TURCHINI METHOD FOR DNA AND RNA

(After Blackler and Alexander, 1952)

(*Formol-mercury, Zenker, Bouin, paraffin sections*)

PREPARATION OF THE REAGENT (9-METHYL-2,3,7-TRIHYDROXYFLUORONE)

Mix 1 mole of 1,2,4-triacetyl-trioxybenzene, 1·25 moles of paraldehyde in a quantity of reagent ethanol 5 times the weight of the two reactants. Add 5–10 per cent. (v/v) sulphuric acid. Allow to stand for 18–24 hours at room temperature. Add 30 volumes of distilled water. Stand for 24 hours. A reddish-orange precipitate of the dye settles out. Filter and discard the filtrate. Dry the product at 37–40°. Redissolve the dye in the minimum volume of ethanol and filter. Add 30–60 volumes of water to the filtrate and stand for 24 hours. Filter and dry as above. The reddish-orange powder decomposes at 319°. It is moderately soluble in alcohol and practically insoluble in water.

Method. (1) Bring sections to water as usual.
(2) Hydrolyse in N-HCl for 6–12 minutes at 60°.
(3) Transfer 80 per cent. ethanol for 15 seconds.
(4) Immerse in the fluorone solution for 4–14 hours. (Dissolve 0·5 g. of the dye† in 100 ml. 95 per cent. ethanol containing 1 ml. H_2SO_4; filter before use.)
(5) Transfer to 1 per cent. aqueous Na_2CO_3 for 2 minutes.
(6) Immerse in distilled water for 2 minutes.
(7) Dehydrate in 50 per cent. acetone, and in 100 per cent. acetone.
(8) Clear in acetone-xylene and in xylene.
(9) Mount in a suitable synthetic resin.
Result. DNA in chromatin bluish-purple; RNA reddish-orange.

If alcoholic differentiation is used some of the dye is extracted from the nucleic acids.

HYDRAZINOLYTIC METHOD FOR DNA PYRIMIDINES

(After Smith and Anderson, 1960)

(*Formalin, ethanol, acetic-ethanol, chloroform-acetic-ethanol* 3:1:6, *Carnoy; paraffin sections*)

Method. (1) Remove wax and bring sections to 80 per cent. alcohol.
(2) Transfer through 3 changes of 100 per cent. isopropanol.

* (*a*) Dissolve 3 g. $(CH_2)_6N_4$ in 100 ml. distilled water.
(*b*) Dissolve silver nitrate in distilled water to produce a 5 per cent. solution.
(*c*) Dissolve borax in distilled water to produce a 5 per cent. solution.
Add 5 ml. of silver nitrate solution to 100 ml. of (*a*). Mix thoroughly until complete solution of the precipitate is obtained. This stock solution keeps for one month in the dark. Immediately before use add 50 ml. stock solution to 50 ml. solution (*c*). The final pH should be 8·8. Renew the staining solution every 5–6 hours.
† Obtained from Koch Light Co., Colnbrook, Bucks, England.

(3) Treat for 4 hours at 50° in the following preheated mixture in a screw-top Coplin jar with a polythene cap liner.

Absolute isopropanol	37·5 ml.
Distilled water	4·5 ml.
Anhydrous hydrazine*	8·0 ml.

(4) Transfer through 3 changes of fresh isopropanol, 5 minutes in each.
(5) Transfer to 1:1 benzaldehyde-isopropanol for 1 hour.
(6) Transfer through 3 changes of fresh isopropanol, 20 minutes in each. Rinse twice in distilled water.
(7) Treat with Schiff's reagent for 1 hour.
(8) Rinse 3 times in a sulphurous acid bath (p. 648).
(9) Wash in running water.
(10) Dehydrate, clear and mount in synthetic resin medium.
Result. DNA stains bluish-purple.

THE METHYL-GREEN-PYRONIN AND RIBONUCLEASE METHOD FOR RNA

(Brachet, 1942)

(*various*)

Short fixation in 10 per cent. formalin, at pH 7·0 ± 0·2 for 4–16 hours, is recommended.

The use of ribonuclease is considered in Chapter 25, and of the various solutions for the extraction of RNA (and DNA) in Chapter 9. Any of these methods can be used in conjunction with methyl green-pyronin staining.

Preparation of Methyl Green Solutions

Samples of methyl green are always mixtures of this dye and methyl violet, from which the former dye differs only in possessing 7 instead of 6 methyl groups. Before use as a histochemical agent methyl green should always be freed from the violet component by shaking an aqueous solution with excess of chloroform or amyl alcohol, both of which dissolve methyl violet. Preferably after two or three days' standing the aqueous supernatant is removed for use. Methyl green purified in this way tends to break down slowly, by loss of the seventh methyl group, into methyl violet. The rate of conversion over a period of up to 5 years is nearly negligible.

Preparation of Methyl Green-pyronin Solutions

Pyronin G or pyronin Y are usually recommended but many samples of these dyes are not suitable for the present purpose. On the whole the bluish shades perform most satisfactorily but it is often necessary to test a number of samples.

The older methyl green-pyronin solutions were made up in a dilute alcoholic solution containing 0·5 per cent. phenol. These are not satisfactory in practice and the method given below, modified from Trevan and Sharrock (1951), can be substituted. Alternatively, and especially if spectrophotometric measurements are intended, the method of Kurnick, also given below, is recommended.

Solution A. Five per cent. aqueous pyronin 17·5 ml., 2 per cent. aqueous methyl green (chloroform-washed) 10 ml., distilled water 250 ml.

* Hydrazine which has absorbed water can be dried by refluxing 1 mol over 5 mols solid NaOH for 15 to 30 minutes under an efficient moisture trap. Heat slowly and add 1 to 2 parts of light petroleum per 30 parts of hydrazine to serve as an inert vapour padding. Cool the flask and add a further 5 parts of light petroleum (B.P. 30–60°). Collect the distillate (BP. 113° to 114°) in a small container. Add a small volume of light petroleum and keep tightly closed.

Solution B. 0·2 M acetate buffer, pH 4·8. In the original directions this solution contains 30 ml. of 1 per cent. orange G, but for purely histochemical work this should be omitted. Murgatroyd (1963) used instead an acid citrate buffer (pH 5·0), prepared by adding 51·5 ml. 0·2 M-di-sodium hydrogen phosphate to 48·5 ml. 0·1 M-citric acid.

For use mix equal volumes of A and B in a Coplin jar. The mixture keeps for about a week, it should not be used for longer.

Method. (1) Bring sections to water.

(2) Stain in methyl green-pyronin solution 10 minutes to 24 hours.

(3) Rinse in distilled water for a few seconds. (Some pyronin is removed at this stage, which must be kept short.)

(4) Blot dry.

(5) Dehydrate rapidly in absolute acetone.

(6) Rinse briefly in equal parts of acetone and xylene.

(7) Rinse briefly in 10 per cent. acetone in xylene.

(8) Clear in two changes of clean xylene.

(9) Mount in DPX.

Result. Nuclear chromatin: green, bluish-green or purplish-green.

RNA: red.

The purplish-red staining of the nuclei, so often obtained with the older methods, was partly due to the admixture of methyl violet and partly to the low pH of the staining solutions. At low pH levels the pyronin component stains more strongly and at higher levels, from pH 4·5 to 5·5, the effect of methyl green is accentuated. Before extraction of RNA the majority of nuclei contain some pyronin; after extraction with ribonuclease the nuclei are green but with other methods, due to depolymerization of DNA, they may take up even more pyronin than before extraction.

METHYL-GREEN-PYRONIN Y METHOD FOR DNA AND RNA

(Kurnick, 1955)

(*Carnoy, Wolman 22°; cold microtome or paraffin sections*)

PREPARATION OF STAIN

Make up a 2 per cent. aqueous solution of Pyronin Y.* Extract with $CHCl_3$ by shaking in a separating funnel until the chloroform layer becomes colourless. Make up a 2 per cent. aqueous solution of methyl green and extract with $CHCl_3$, as above until the chloroform layer is no longer violet coloured. For use mix 12·5 ml. Pyronin Y solution and 7·5 ml. of methyl green with 30 ml. distilled water.

Method. (1) Bring paraffin sections to water; cold microtome sections can be immersed directly in the staining mixture.

(2) Stain for 6 minutes in methyl green-pyronin.

(3) Blot with filter paper.

(4) Immerse in 2 changes of *n*-butyl alcohol,† 5 minutes in each.

(5) Immerse in xylene for 5 minutes.

(6) Immerse in cedar oil for 5 minutes.

(7) Mount in Permount.

Result (Fig. 89 and Plate IVb). Chromatin clear green, nucleoli bright red, cytoplasmic RNA bright red. Eosinophil granules and osteoid also stain bright red.

* Kurnick found oniy Pyronln Y (05564, G. T. Gurr) or Pyronin Y (Edward Gurr) suitable for the purpose. He considered that Pyronin B, which differs from Pyronin Y in being tetraethyl instead of tetramethyl compound, gave only non-specific counterstaining of cytoplasmic components.

† Tertiary butyl alcohol cannot be used for differentiation.

THE GALLOCYANIN-CHROMALUM METHOD FOR NUCLEIC ACIDS

(Einarson, 1951)

(*various*)

PREPARATION OF THE STAINING SOLUTION

Dissolve 5 g. of chromalum ($K_2SO_4 . Cr_2(SO_4)_3 . 24H_2O$) in 100 ml. of distilled water. Add 0·15 g. gallocyanin and mix by shaking. Warm gradually and bring to the boil. Boil for 5 minutes. Cool to room temperature, filter and add distilled water through the filter paper until the volume of the filtrate reaches 100 ml. The pH of this stock solution, which is ready for use, is 1·64 (lasts 4 weeks).

The pH of the staining solution may be altered by adding given amounts of 1·0 M-HCl and 1·0 M-NaOH, as indicated in the Table below, to 40 ml. of the stock solution.

TABLE 52

ml. 1·0 M HCl	pH	ml. 1·0 M NaOH	pH
10	0·83	0	1·64
9	0·90	1	1·84
8	0·92	2	2·16
7	0·94	3	2·90
6	1·02	4	3·42
5	1·10	5	3·76
4	1·14	6	3·98
3	1·18	7	4·07
2	1·29	8	4·18
1	1·44	9	4·27
0	1·64	10	4·35

The addition of strong acid or base to the stock solution does not cause any abrupt change in pH. After the addition of NaOH a precipitate of dye-lake-hydroxide occurs and the solution should not be used for longer than 7 days.

Method. (1) Bring sections to water.

(2) Stain in gallocyanin-chromalum for 48 hours at room temperature.

(3) Wash briefly in water.

(4) Dehydrate in the alcohols.

(5) Clear in xylene and mount in DPX.

Result. Nucleic acids stain deep blue. Depending on the pH of the staining solution, other structures may be stained. At pH 1·64 cartilage gives a fine red metachromasia. At low pH values (0·83 to 0·94) staining of structures other than the nucleic acids is very slight. From pH 1·1 to 2·9 this non-specific staining increases and it reaches a maximum between pH 3·3 and 3·5, falling abruptly at pH 4·07. Staining of the nucleic acids does not vary in this way.

MODIFIED GALLOCYANIN METHOD FOR NUCLEIC ACIDS

(de Boer and Sarnaker, 1956)

(*Formalin, frozen sections; various fixatives, paraffin sections*)

PREPARATION OF THE DYE SOLUTION

Shake up 600 mg. of gallocyanin in 200 ml. distilled water for one minute. Filter and discard the filtrate. Transfer the filter paper and residue to 200 ml. of 5 per cent. chrome alum solution in distilled water. Place on a water bath and boil for 30 minutes. Cool, filter and adjust filtrate to pH 1·6 with 1 per cent. HCl.

ALTERNATIVE PREPARATION OF DYE SOLUTION (Berube *et al.*, 1966)

Mix 150 mg. gallocyanin, 15 g. chromalum and 100 ml. distilled water. Boil for 10–20 minutes. Cool, filter and restore the volume of the filtrate to 100 ml. by washing the precipitate. The filtrate can be used as a stain. If, however, the chelate is to be separated bring the filtrate to pH 8–8·5 with ammonia (dilute). Filter, with suction, through a medium-porosity fritted glass funnel. Wash the precipitate with a small amount of anhydrous ethyl ether. Dry and store in a tightly capped container. For use, make a 3 per cent. solution in 1 N-H_2SO_4.

Method. (1) Bring sections to water.

(2) Stain in gallocyanin-chromalum for 24 hours or longer (depending on the age of the stain).

(3) Rinse for 1 minute in distilled water at pH 1·6. (Acidified with HCl.)

(4) Repeat rinse until no further dye can be removed.

(5) Dehydrate in alcohol, clear and mount in DPX.

Result. Nucleic acids stain deep blue. Nissl granules are especially well shown. The non-specific staining of the original method is substantially reduced by the procedures outlined above.

FLUORESCENT PSC REACTION FOR DNA

(Sterba, 1965)

(*Carnoy; paraffin sections*)

Method. (1) Bring sections to absolute alcohol.

(2) Treat with acid methanol (H_2SO_4, 3 ml.; MeOH,100 ml.) for 6–7 hours at 40°.

(3) Transfer directly to acid permanganate solution (Distilled water, 80 ml.; 2·5 per cent. $KMnO_4$, 5 ml.; 5 per cent. H_2SO_4, 5 ml.) for 3 minutes at room temperature.

(4) Bleach in 3 per cent. oxalic acid for 30 seconds and rinse in distilled water.

(5) Wash in running water for 30 minutes and blot dry with filter paper.

(6) Stain in N,N′-Diethylpseudoisocyanin chloride (PS)C (29 or 50 mg. per 100 ml. distilled water) for 12–24 hours at room temperature.

(7) Wash briefly in distilled water; mount and examine in this medium by fluorescence microscopy (Filters; Schott UG 1/15, OG 1).

Result. Bright yellow fluorescence on a dark red ground indicates sites of DNA.

THE EOSIN-GRAM-WEIGERT METHOD

(*Formalin and various other fixatives; paraffin sections*)

A large number of variations of this method exist. If it is proposed to employ the method, in comparative studies, as an index of the physical structure of protein components of the tissues, it is essential to standardize every detail of the technique.

Method. (1) Fix tissues in 10 per cent. formol saline for 4 days.

(2) Wash for 12–24 hours in running water.

(3) Embed in paraffin by any standard technique. (The period in hot wax should be constant in each case.)

(4) Cut sections 5–6 μ, float on water, mount on egg-albuminized slides.

(5) Dry, preferably in paraffin-oven at 60°, for a constant time ($\frac{1}{2}$–1 hour).

(6) Remove wax by dipping in light petroleum (avoid hot stage).

(7) Transfer to absolute acetone, absolute alcohol, 70 per cent. alcohol and finally to water.

(8) Stain in 1 per cent. aqueous Eosin for 10–15 seconds.

(9) Wash briefly in water and blot dry.

(10) Stain in freshly filtered aniline-Gentian violet, 2 minutes. (Saturated alcoholic crystal oivlet, 16 ml., 2 per cent. aniline water to 100 ml.)

(11) Wash briefly and flood with Gram's iodine, 2 minutes. (Iodine 1 g., KI 2 g., distilled water 300 ml.)

(12) Wash briefly and blot well.

(13) Differentiate, with frequent blotting, in aniline-xylene (equal parts). A suitable time must be arrived at by experiment (say that necessary to give positive staining in normal collagen), and this time must be used as a constant for any one series of slides.

(14) Blot.

(15) Wash in 2 changes of xylene.

(16) Mount in DPX.

Result (Fig. 94). Gram-positive structures stain purple in varying degrees.

TOLUIDINE BLUE-MOLYBDATE METHOD FOR RIBONUCLEOPROTEIN

(Love, 1962; Love and Walsh, 1963)

(*Tissue cultures or smears, wet fixation, TCA and formol sublimate*)

Only certain batches of the dye give satisfactory results with the standard technique. The authors listed Coleman Bell CU-3 and National Aniline NU-2. Others which could be used with the modified methods included National Aniline NU-17, Harleco NU-14, Matheson-Coleman Bell CU-9 and Biological Stain Commission NU-19.

The dye samples are dissolved in McIlvaine's buffer at pH 3·0 and 20°.

Standard Method. (1) Rinse tissue cultures in 0·85 per cent. saline for 10 seconds before fixation.

(2) Wet fix two slides in 5 per cent. aqueous trichloroacetic acid for 10 minutes.

(3) Rinse slides briefly in distilled water.

(4) Fix one slide for 5 minutes and the other for 10 minutes in formol sublimate (1 part 40 per cent. formaldehyde, 9 parts 6 per cent. aqueous $HgCl_2$).

(5) Wash in tap water and treat with Lagol's iodine for 5 minutes.

(6) Treat with 5 per cent sodium thiosulphate for 5 minutes and then wash in water.

(7) Stain for 30 minutes in Toluidine blue.

(8) Immerse in 4 per cent. aqueous ammonium molybdate for 15 minutes.

(9) Rinse briefly in tap water.

(10) Dehydrate in *tert*-butyl alcohol, clear in xylene and mount in a synthetic resin.

Result. The pars amorpha of the nucleolus is stained bluish-green. The nucleolini are bright purple.

Modified Methods. According to the authors, 9 types of RNP can be distinguished with modifications of the original method. The Tables below give details:

TABLE 53

Types of RNP Stained by the TBM Method

Type	Site	Morphology
I. Nucleolini, granular parachromatin	(1) in pars amorpha (2) in nucleoplasm	Nucleolini solid or hollow spheres
II. Chromosomal RNP (Type A)	Late prophase, metaphase, anaphase, early telophase	As chromosomes

TABLE 53 *contd.*

Type	Site	Morphology
III. Chromosomal RNP (Type B)	Interphase chromatin	As chromatin
IV. Diffuse RNP (Cytoplasmic Type A)	Cytoplasm in interphase	Diffuse
V. Diffuse RNP (Cytoplasmic Type B)	Cytoplasm	Diffuse
VI. Perichromosomal	Nucleoplasm	Amorphous
VII. Cytoplasmic (granular)	Cytoplasm	Coarse granules
VIII. Pars amorpha	Nucleolus	Amorphous
IX. Parachromatin (amorphous)	Increased in prophase. Extruded into spindle	

TABLE 54
Methods for Nine Types of RNP

Method	Types of RNP shown	Fixation		Toluidine blue		Ammonium molybdate	
		TCA	$HgCl_2$	Conc^n	Time	Conc^n	Time
A	I, II	10′	5′	S*	30′	4	15′
B	I, II, IV	10′	5′	S	2 hrs.	15	7′
C	II, V, VIII and DNP	10′	5′	0·5 S	2 hrs.	15	15′
D	VI, VII	10′	45′	S	1 hr.	4	40′
E	IX	omit	3′ to 5′ 37°	0·8 S	2 hrs.	4	30′

* S is the standard concentration (4 to 9 mg./100 ml.).

ACROLEIN-TOLUIDINE BLUE METHOD FOR DNA AND RNA
(After Feder and Wolf, 1965)
(*Acrolein; polyester wax embedding*)

Method. (1) Fix small pieces of tissue in 10 per cent. aqueous acrolein with 0·5 per cent. calcium acetate.

(2) Dehydrate in 1:1 methanol/methoxyethanol, transfer to ethanol, then to *n*-propanol and finally to polyester wax.

(3) Cut 5 μ sections and mount on slides in the usual manner.

(4) Bring sections to water and stain for 8 minutes in 0·1 per cent. toluidine blue in phosphate buffer at pH 4·2 and 22°.

(5) Dehydrate in *tert*-butyl alcohol for 5 minutes.

(6) Clear in xylene and mount in a synthetic resin.

Results. The two nucleic acids are distinguished by sharp colour differences. DNA, deep (orthochromatic) blue; RNA pale (metachromatic) purple.

ALKALINE FAST GREEN FOR BASIC PROTEINS OF NUCLEI

(Alfert and Geschwind, 1953)

(10 *per cent. neutral formalin, paraffin sections*)

Method. (1) Fix tissues for 3–6 hours in 10 per cent. neutral formalin, wash overnight in running water, dehydrate, clear and embed in paraffin.

(2) Bring sections to water.

(3) Immerse for 15 minutes in 5 per cent. aqueous trichloroacetic acid at 100°.

(4) Wash in 3 changes of 70 per cent. alcohol, 10 minutes in each.

(5) Wash in distilled water.

(6) Stain at 22° for 30 minutes in 0·1 per cent. aqueous fast green F.C.F. adjusted to pH 8·0 to 8·1 with the minimum amount of NaOH.

(7) Wash in distilled water for 5 minutes.

(8) Dehydrate in 95 per cent. alcohol, clear and mount in a suitable synthetic resin.

Result (Fig. 92). Basic proteins stain bright green, due to their content of arginine and lysine.

NAPHTHOL-YELLOW S METHOD FOR BASIC PROTEINS OF NUCLEI

(Deitch, 1955)

(*Freeze-dried, alcohol post-fixed; cold microtome, acetic-ethanol-fixed; Formalin, Carnoy etc., paraffin sections*)

Method. (1) Bring sections to water.

(2) Stain for 4–6 hours in 0·5 per cent. aqueous Naphthol Yellow S at pH 2·7.

(3) Rinse in distilled water.

(4) Dehydrate in 95 per cent. alcohol, clear and mount.

Result. The amount of yellow staining (λ^{max}= 435 nm) reflects the number of available basic groups of the protein.

In the case of the nuclei, if the nucleic acid is not previously removed, only those basic groups not blocked by DNA will be demonstrated.

METAPHOSPHORIC ACID-GALLOCYANIN FOR BASIC PROTEINS

(Jobst and Sandritter, 1964)

(*Air-dried smears; alcohol or formalin fixation*)

Method. (1) Treat smears for 15 minutes with 5 per cent. aqueous trichloracetic acid at 95°, to remove DNA.

(2) Rinse 3 times in 70 per cent. alcohol, then in water.

(3) Dissolve 0·8 g. crystalline metaphosphoric acid (HPO_3) in 100 ml. distilled water by shaking at 20° until the crystals disappear.

(4) Immerse smears for 1 hour at 20° in a freshly prepared solution made as above.

(5) Wash 3 times for 1 minute each in distilled water.

(6) Stain for 48 hours at 20° in gallocyanin-chromalum (150 mg. gallocyanin boiled for 10 minutes with 5 g. chrome alum in 100 ml. distilled water and restored to a final volume of 100 ml.).

(7) Wash in running water for 5 minutes.

(8) Dehydrate in alcohol; clear in xylene; mount in a synthetic resin.

Result. Metaphosphoric acid combined with NH_2 and guanidino groups imparts a deep blue stain to basic proteins.

EXTRACTIVE METHODS FOR NUCLEIC ACIDS

(Methods using enzymes are given in Appendix 25)

PERCHLORIC ACID (AFTER ERICKSON *ET AL.*, 1949)

(*formalin, formol sublimate; paraffin sections*)

For removal of RNA alone, bring sections to water, after removal of mercury where necessary, and treat with 10 per cent. perchloric acid at 4° for 12–18 hours.

For removal of both nucleic acids treat with 5 per cent. perchloric acid at 60° for 20–30 minutes.

Neutralize sections in 1 per cent. sodium carbonate (1–5 minutes), wash, and stain with 1 per cent. aqueous toluidine blue.

BILE SALTS (AFTER FOSTER AND WILSON, 1952)

(*preferably formalin-fixed paraffin sections*)

Bring sections to water and treat for 24–48 hours at 60° in a 2 per cent. aqueous solution of sodium tauroglycocholate. It is essential that this extractive solution be kept continuously oxygenated by means of a stream of oxygen. After incubation, wash the sections in water and stain with toluidine blue or by Gram's method. Only RNA is released by bile salt extraction.

TRICHLOROACETIC ACID (AFTER SCHNEIDER, 1945)

(*various, paraffin sections*)

Bring sections to water and treat with 4 per cent. trichloroacetic acid at exactly 90° for 15 minutes. Wash and stain with toluidine blue or in White's (1950) orange G-aniline blue mixture. Both types of nucleic acid are extracted by this procedure.

HYDROCHLORIC ACID (AFTER DEMPSEY *ET AL.*, 1950)

(*Zenker, etc., paraffin sections*)

Bring sections to water and treat with 1 N-HCl for 3 hours at 37°. Wash, and stain with 2 mM methylene blue at pH 5·7 for 12–24 hours, or with any other suitable basic dye for shorter periods. This method removes both types of nucleic acid.

APPENDIX 10

THE PERIODIC ACID-SCHIFF TECHNIQUE

(after Hotchkiss, 1948)

(*most types of fixation; frozen or paraffin sections*)

PREPARATION OF THE SOLUTIONS

(1) **Periodic Acid.** Dissolve 0·4 g. of periodic acid (HIO_4, $2H_2O$) in. 35 ml. of reagent ethyl alcohol and add 5 ml. of 0·2M-sodium acetate (27·2 g. of the hydrated salt in 1,000 ml.) and 10 ml. of distilled water. This solution should be kept in the dark at 17°–22° and used at this temperature. It should be discarded if a brown colour appears.

(2) **Reducing Bath.** Dissolve 1 g. potassium iodide and 1 g. sodium thiosulphate ($Na_2S_2O_3$, $5H_2O$) in 30 ml. of reagent ethyl alcohol and 20 ml. of distilled water. Add 0·5 ml. of 2 N-HCl (20 per cent. conc. HCl). A deposit of sulphur forms which can be ignored. Keep between 17°–22°; the solution lasts for about 14 days, not longer.

(3) **Schiff's Reagent.** For the preparation of this reagent see Appendix 9, p. 647. For most purposes the reagent of Barger and DeLamater is recommended.

(4) **Celestin Blue Solution.** Dissolve 2·5 g. iron alum in 50 ml. distilled water by standing overnight at room temperature. Add 0·25 g. Celestin blue B (C.I. 51050) and boil for 3 minutes. Filter when cool and add 7 ml. glycerol.

(5) **Orange G Solution.** Dissolve 2 g. orange G (C.I. 16230) in 100 ml. of 5 per cent. aqueous phosphotungstic acid. Stand for 24 hours and use the supernatant.

Method (short title PARS). (1) Bring sections down to water and remove mercury.

(2) Rinse in 70 per cent. alcohol.

(3) Immerse in periodic acid solution for 5 minutes.

(4) Rinse in 70 per cent. alcohol.

(5) Immerse in the reducing bath for 1 minute.

(6) Rinse in 70 per cent. alcohol.

(7) Immerse in Schiff's solution (Barger and DeLamater) 20 minutes.

(8) Wash in running water for 10 minutes.

(9) Stain nuclei lightly with celestin blue 2–3 minutes, followed by Mayer's hæmalum 2–3 minutes.

(10) Differentiate in 1 per cent. acid alcohol.

(11) Wash in running water for 30 minutes.

(12) Counterstain with orange G for about 10 seconds.

(14) Wash in water till sections are pale yellow (about 30 seconds).

(14) Dehydrate in alcohol, clear in xylene, mount in DPX.

Stage 14 may be completed after stage 8, or after stage 11, as well as in its final position. The resulting variations (PAS, Dipas, Tripas) will be found useful for different purposes. Tripas or Trichrome-PAS (Pearse, 1949, 1950b) has been used as a routine stain for the human and many animal hypophyses. Dipas is useful in cases where orange G overstaining of PAS-stained components is not desired. Alternatively, the PAS stain may be followed by orange G, omitting the nuclear stain altogether. This variant is useful for differential cell counts in the animal hypophysis. For all other purposes the method given below gives more satisfactory results.

Result. Hexose-containing mucosubstances (*vic*-glycols) deep purplish-red. In the full method (Trichrome-PAS) the nuclei are blue-black and RBC's and acidophil proteins are yellow.

THE PERIODIC ACID-SCHIFF TECHNIQUE

(after McManus)

(*various*)

Method (short title PAS). (1) Bring sections to water and remove mercury, if necessary.*

(2) Oxidize for 10 minutes in 1·0 per cent. aqueous periodic acid.

(3) Wash in running water for 5 minutes.

(4) Immerse in Schiff's reagent 10 minutes.

(5) Wash in running water for 5 minutes.

(6) Counterstain nuclei with the celestin blue-hæmalum sequence (Stage 9 of the previous method) if required.

(7) Differentiate if necessary in 1 per cent. acid alcohol and follow by thorough washing in running water.

(8) Dehydrate in alcohol, clear in xylene and mount in a suitable synthetic medium.

Result. Hexose-containing mucosubstances stain in various shades of purplish-red. Glycogen stains deeply.

PREVENTION OF PAS REACTION BY ACETYLATION

(Lillie, 1954a)

Control sections are treated at 22° for 1–24 hours in 16 ml. acetic anhydride in 24 ml. of dry pyridine, washed in water and then subjected to the PAS reaction as above. A negative reaction after acetylation indicates that 1:2-glycol groups were responsible for the original reaction observed. The following double procedure follows that recommended by McManus and Cason (1950).

(1) Bring three sections to water.

(2) Leave one section (A) untreated until stage 7.

(3) Treat the second and third sections (B and C) in the acetic anhydride-pyridine mixture.

(4) Wash sections B and C in water.

(5) Treat section C with 1 per cent. KOH in 70 per cent. alcohol for 20 minutes at room temperature. Sections may have to be protected with a film of celloidin (collodion). Alternatively treat with 20 per cent. ammonia in 70 per cent. alcohol at 37° (Lillie) for 24 hours.

(6) Wash this section in water.

(7) Treat all three sections by the PAS routine.

A positive result in a given structure in sections A and C with a negative result in section B confirms the carbohydrate nature of the reacting groups.

THE LEAD TETRAACETATE-SCHIFF METHOD

(Shimizu and Kumanoto, 1952)

(*paraffin sections after various fixatives*)

PREPARATION OF SOLUTION

Dissolve 1 g. $Pb(O.CO.CH_3)_4$ in 30 ml. glacial acetic acid. Immediately before use add 70 ml. saturated sodium acetate solution.

Method. (1) Bring sections to water.

* When using fresh or formalin-fixed frozen sections it has been customary to block existing aldehydes by one of the techniques for this purpose given in Appendix 13, pp. 706 and 707. In some cases the blockade may be reversible by subsequent procedures. Lillie's acetic-aniline is particularly recommended, as an irreversible blocking reagent.

(2) Immerse in 1·0 M-sodium acetate, 5 minutes.
(3) Treat with fresh tetraacetate solution, 10 minutes.
(4) Immerse in 1·0 M-sodium acetate, 5 minutes.
(5) Wash in running water, 10 minutes.
(6) Treat with Schiff's reagent, 15 minutes.
(7) Treat with sulphite water (as in Feulgen reaction, Appendix 9, p. 647).
(8) Wash in running water, 10 minutes.
(9) Dehydrate, clear and mount in Canada balsam.

Result. Glycogen and various mucosubstances are stained reddish-purple by this method, as with the PAS methods.

SODIUM BISMUTHATE-SCHIFF METHOD

(after Lhotka)

(*paraffin sections after most fixatives*)

Method. (1) Bring sections to water.

(2) Oxidize for 3 minutes in 1 per cent. sodium bismuthate in 20 per cent. orthophosphoric acid (mix 40 ml. 1·25 per cent. bismuthate with 10 ml. of acid and use immediately).

(3) Wash in running tap-water for 1 minute.

(4) Rinse for 15 seconds in 1 N-HCl to remove adherent bismuth pentoxide and rinse in distilled water.

(5) Immerse in Schiff's reagent for 10 minutes.

(6) Treat with sulphite water (as in Feulgen reaction, Appendix 9, p. 647).

(7) Wash in running water.

(8) Dehydrate, clear and mount in a suitable resin.

Result. Staining is similar to that obtained with the PAS and $PbAc_4$-S methods.

PERIODIC ACID-SALICYLOYL HYDRAZIDE

(after Stoward, 1963, 1967d)

(*paraffin sections after most fixatives; FDFV*)

Method. (1) Bring sections to water.

(2) Oxidize with 1 per cent. periodic acid for 10 minutes.

(3) Wash in running water for 2 minutes.

(4) Treat with 0·5 per cent. salicyloyl hydrazide* in 5 per cent. acetic acid† (pH3) for 20–40 minutes.

(5) Rinse quickly in distilled water.

(6) Rinse in a freshly prepared dilute (0·5 per cent.) solution of trisodium pentacyanoammineferroate‡ (see Appendix 6, p. 627).

(7) Wash twice in distilled water.

(8) Counterstain in 0·5 per cent. Solochrome black/aluminium (p. 325) for 10 minutes.

(9) Rinse briefly in water.

(10) Dehydrate in alcohols, clear in xylene and mount in Fluormount, or equivalent medium.

Result. A bright blue fluorescence (366 nm.) indicates sites containing PA-reactive glycols.

* Sigma or Eastman-Kodak.
† A brown resinous deposit forms in this solution after a few weeks. It may be ignored.
‡ Koch-Light, Hopkin Williams.

ARYLAMINE-ALDEHYDE CONDENSATION

(Lillie, 1962)

(*paraffin sections after formalin; FDFV*)

Method. (1) Bring sections to water.

(2) Oxidize for 10 minutes in 1 per cent. periodic acid.

(3) Wash for 10 minutes in running water.

(4) Transfer to glacial acetic acid for 1 minute.

(5) Treat with 11 per cent. *m*-aminophenol in glacial acetic acid for 1 hour.

(6) Wash for 10 minutes in running water and rinse in distilled water.

(7) Azo couple for 2 minutes at 3° in a solution of Fast black salt K (3 mg./ml.) in 0·1 M-veronal-HCl buffer at pH 8·0.

(8) Wash in 3 changes of 0·1 N-HCl, 5 minutes in each.

(9) Wash in running water for 10 minutes.

(10) Dehydrate in alcohols, clear in xylene and mount in synthetic resin.

Result. Sites containing PA-reactive glycols are coloured black. Nuclei, muscle, cell cytoplasm, greyish red to pink.

PERIODIC ACID-PARADIAMINE METHOD (PAD)

(Spicer, 1966)

(*Paraffin sections, most fixatives; FDFV*)

REAGENTS

Add 50 mg. N,N-dimethyl-*p*-phenylenediamine HCl, just before use, to 50 ml. citrate-phosphate buffer (0·1 M-Citric acid, 4·8 ml., 0·2 M disodium phosphate, 7·2 ml., distilled water 38 ml.).

Alternatively, dissolve 100 mg. *para*diamine in 50 ml. distilled water and adjust pH to 5·0 with 0·4 M-Na_2HPO_4.

Method. (1) Bring slides to water.

(2) Oxidize in 1 per cent. periodic acid for 10 minutes.

(3) Rinse in running water for 10 minutes.

(4) Immerse in *para*diamine solution for 7, 24, or 48 hours.

(5) Differentiate in 1 per cent. HCl in 70 per cent. alcohol for 8 seconds (after 24 hours' stain) or for 10 seconds (after 48 hours' stain).

(6) Wash in water for 5 minutes.

(7) Dehydrate through the alcohols, clear in xylene and mount in a synthetic medium.

Result. Neutral mucopolysaccharides, brown; periodate-reactive polymers, purple or grey-brown; periodate unreactive mucosubstances, black.

MIXED DIAMINE STAIN

(Spicer, 1966)

(*Paraffin sections, most fixatives; FDFV*)

REAGENT

Prepare a fresh solution by dissolving 30 mg. N,N-dimethyl-*m*-phenylenediamine dihydrochloride and 5 mg. N,N-dimethyl-*p*-phenylenediamine hydrochloride in 50 ml. distilled water. Adjust pH to 3·4 to 4·0 with 0·2 M-Na_2HPO_4 (0·15 to 0·65 ml.).

Method. (1) Treat duplicate slides. Remove paraffin wax in xylene and bring through alcohols to water.

(2) Hydrolyse both sections for 10 minutes in 1 N-HCl at 60° (to remove interference by nucleic acids).

(3) Wash in running water for 5 minutes.

(4) Oxidize one of the two sections for 10 minutes in 1 per cent. aqueous periodic acid. Rinse again for 5 minutes.

(5) Stain both sections for 24–48 hours in the mixed diamine solution.

(6) Proceed directly to 95 per cent. alcohol (2 changes) and absolute alcohol (2 changes).

(7) Clear in xylene-alcohol and then in xylene.

(8) Mount in synthetic resin.

Result. Many sulphated and non-sulphated acid mucosubstances are stained black or brown. Periodate-reactive substances (in the PA-oxidized section) fail to stain.

MIXED DIAMINE-SODIUM CHLORIDE METHOD

(Spicer, 1966)

(*Paraffin sections, most fixatives*)

Method. (1) Bring duplicate slides to water.

(2) Feulgen hydrolysis (10 minutes, 60°).

(3) Wash in running water for 5 minutes.

(4) Oxidize one of the two sections in 1 per cent. aqueous periodic acid for 10 minutes; rinse for 5 minutes in running water.

(5) Stain for 24–48 hours in the mixed diamine reagent (above) but replace 3–7 ml. of water with 1·0 M-NaCl when making up.

(6) Proceed as in previous method.

Result. Sulphated acid mucopolysaccharides in cornea, ovarian follicles, and certain connective tissues, are well stained. Sialomucins are unstained.

LOW IRON DIAMINE-ALCIAN BLUE STAIN (LID)

(Spicer, 1966)

(*Paraffin sections, formalin; FDFV*)

REAGENT

Prepare a fresh solution containing 30 mg. N,N-dimethyl-*m*-phenylenediamine dihydrochloride and 5 mg. N,N-dimethyl-*p*-phenylenediamine hydrochloride in 50 ml. distilled water. When the reagents are dissolved, pour this solution into a Coplin jar containing 0·5 ml. 10 per cent. $FeCl_3$.

Method. (1) Bring duplicate sections to water.

(2) Oxidize one section for 10 minutes in 1 per cent. aqueous periodic acid (if demonstration of neutral, as well as acidic, mucosubstances is desired).

(3) Rinse for 5 minutes in running water.

(4) Stain for 18 hours in the low iron diamine solution.

(5) Rinse very rapidly in water and stain in 1 per cent. alcian blue in 3 per cent. acetic acid for 30 minutes.

(6) Dehydrate, clear and mount in synthetic resin.

Result. Sulphated and uronic acid-containing mucosubstances (and elastic fibres) stain grey to purple-black. PA-oxidation renders neutral mucopolysaccharides purple-grey.

If Alcian blue staining is superadded, LID-negative non-sulphated acidic mucosubstances are blue and LID-reactive non-sulphated and sulphated acidic mucosubstances are black.

HIGH IRON DIAMINE-ALCIAN BLUE STAIN (HID)

REAGENT

Prepare as for the LID method but substitute 120 mg. *meta*diamine and 20 mg. *para*diamine in 50 ml. distilled or tap water. Add this solution to a Coplin jar containing 1·4 ml. 10 per cent. $FeCl_3$.

Method. Proceed as in LID method (stain for 18 hours).

Result. Sulphated mucosubstances, purple-black. Uronic acid-containing mucosubstances and sialomucins unstained (blue after Alcian blue).

PERIODIC ACID-PSEUDOSCHIFF METHODS

(after Stoward, 1963, 1967c)

(*Paraffin sections, most fixatives; FDFV*)

REAGENTS

These are 0·01 to 0·1 per cent. aqueous or 30 per cent. ethanolic solutions of various aminoacridines (Acridine yellow, Acriflavine, Coriphosphine) saturated with SO_2 by adding 2–3 drops of thionyl chloride to 100 ml. of dye solution. The pH should be adjusted to 2·5 to 3·5 with dilute NaOH.

Method. (1) Bring sections to water.

(2) Methylate for 4 hours at 37° with 2 per cent. thionyl chloride in methanol.

(3) Wash in running water for 2 minutes.

(4) Oxidize with 1 per cent. aqueous periodic acid for 10 minutes.

(5) Rinse in 0·1 per cent. thionyl chloride.

(6) Rinse in water.

(7) Immerse in (*a*) Pseudo-Schiff reagent, or (*b*) Dye solution prepared as above, acriflavine is recommended, for 20–90 minutes.

(8) Wash twice in distilled water.

(9) Mount in water and examine by UVL. Alternatively

(10) Blot dry; dehydrate quickly in 3 changes of isopropyl alcohol, clear through 50:50 isopropyl alcohol/xylene, to xylene. Mount in DPX and examine by UVL.

Result. PA-reactive mucosubstances fluoresce bright yellow.

METACHROMATIC METHODS

(*Fresh frozen or formalin-fixed frozen; various paraffin*)

Schmorl's Thionin Method

Method. (1) Bring paraffin or frozen sections to water.

(2) Treat with saturated aqueous mercuric chloride, 30 seconds.

(3) Wash quickly in water.

(4) Stain in dilute aqueous thionin (2 drops of a hot-saturated solution in 5 ml. of distilled water) for 5–15 minutes.

(5) Wash briefly in distilled water.

(6) Mount frozen and paraffin sections preferably in glycerine jelly.

Result. Metachromatic substances red, nuclei blue.

Maxilon Blue RL Method

(de Almeida, 1960)

(*Formol-calcium, paraffin sections*)

Method. (1) Bring sections to water.

(2) Stain in 0·5 per cent. aqueous Maxilon blue RL, for 30–60 seconds.

(3) Wash in distilled water.

(4) Blot to remove excess water.
(5) Dehydrate in *tert*-butanol for 1 to 2 minutes.
(6) Clear in xylene.
(7) Mount in a synthetic medium.

Result. Strongly acidic mucosubstances stain red (γ-metachromasia) to violet (β-metachromasia). Other basophilic structures stain in various shades of blue.

Toluidine Blue Method (Standard)

Different samples of this dye behave differently in respect of metachromasia and particularly in respect of the resistance of this metachromasia to the usual process of dehydration in alcohol. Selection of a suitable sample of the dye is, therefore, of paramount importance.

Method. (1) Bring sections to water.
(2) Stain in 0·5 per cent. aqueous toluidine blue for 4–6 hours.
(3) Rinse in distilled water.
(4) For histochemical purposes examine immediately in water.
(5) Mount in glycerine jelly.

Result (Plate IVa). Metachromatic substances red or pink (γ-metachromasia) or purple (β-metachromasia). Nuclei blue (usually). An orange-yellow filter alters the shift from blue to reddish-purple to one from green to red.

Toluidine Blue Method (Kramer and Windrum)

(*Various fixatives, paraffin sections; FDFV*)

Method. (1) Bring sections to water.
(2) Stain in 0·1 per cent. toluidine blue in 30 per cent. ethanol for 5–20 minutes.
(3) Rinse in 95 per cent. alcohol.
(4) Dehydrate in absolute alcohol.
(5) Clear in xylene and mount in a suitable synthetic resin.

Result. γ-metachromasia red or pink; β-metachromasia purple.

Toluidine Blue Method for Permanent Preparations

(after Hess and Hollander, 1947)

(*Zenker-fixed paraffin sections*)

Method. (1) Bring sections to water.
(2) Cover for 30 seconds with toluidine blue solution (0·25 g. toluidine blue in 100 ml. 0·25 per cent. borax in distilled water).
(4) Apply a solution of 0·5 per cent. potassium dichromate, previously saturated with mercuric chloride and filtered, for 15 seconds.
(5) Add fresh dichromate-sublimate solution and leave for 2 minutes.
(6) Blot dry.
(7) Immerse in absolute alcohol for 10 seconds with constant agitation.
(8) Immerse in clean xylene for 30 seconds.
(9) Rinse in absolute alcohol and drain.
(10) Cover with the aqueous dichromate-sublimate solution for 30 seconds.
(11) Rinse in colophony-alcohol (1 g. colophonium resin, 100 ml. alcohol) and drain.
(12) Clear in two changes of benzene.
(13) Mount in Canada balsam dissolved in benzene.

Result. The red (γ) metachromasia is preserved and in some cases even enhanced by the treatment with resin-alcohol.

Toluidine Blue Method for Permanent Preparations

(after Bélanger and Hartnett, 1960)

(*Formaldehyde or alcohol; paraffin sections*)

Method. (1) Bring sections to distilled water.

(2) Stain for 10 minutes in 0·5 per cent. toluidine blue in potassium acid phthalate-tartaric acid buffer (1·0 M) at pH 4·40. (The dye takes 24 hours to dissolve and the solution must be filtered before use.)

(3) Rinse for 2 minutes in buffer alone.

(4) Dehydrate in 4 changes of *tert*-butyl alcohol, 2 minutes in each.

(5) Clear in 3 changes of xylene.

(6) Mount in synthetic resin.

Result. γ-metachromatic tissue components retain their red colour. Orthochromatic materials are well stained.

Resin Mounting Medium for Toluidine Blue Metachromasia

(Izard, 1964)

Method. (1) Proceed as in any of the toluidine blue staining methods as far as the stage of rinsing after staining.

(2) Mount directly in prepolymerized butyl methacrylate (butyl methacrylate monomer, 1 per cent. Luperco CDB, prepolymerized for 45–60 minutes at 57° and cooled to 22°.

SULPHATION METHODS FOR INDUCING METACHROMASIA

Methods. (1) Conc. H_2SO_4. Treat sections for 60–70 seconds. Remove and wash in running tap water. Stain.*

(2) Chlorosulphonic acid in pyridine. Care must be taken in the preparation of this reagent and the pyridine must be dry. Add 11 ml. chlorosulphonic acid to 100 ml. pyridine contained in a flask immersed in an ice-water-salt mixture. The reaction product is solid at room temperature and is used at 65°–70° for 5 minutes. Sections are then washed in water and stained.*

(3) Sulphuric-acetic anhydride. Equal volumes of sulphuric acid and acetic anhydride mixed slowly in a cooled vessel. Immerse sections for 3 minutes, then wash and stain.*

(4) Chlorosulphonic-acetic-chloroform. Mix 5 ml. chlorosulphonic acid, 10 ml. acetic acid and 12 ml. chloroform. Use immediately. Immerse sections for 5 minutes at 22°, wash and stain.*

(5) Fuming sulphuric-acetic anhydride-ether. Mix 35 ml. acetic anhydride and 15 ml. ether. Add 50 ml. of fuming H_2SO_4, drop by drop. Immerse sections for 2–3 minutes only. Wash and stain.*

(6) Sulphuric ether (Gomori). Add 80 ml. fuming H_2SO_4 carefully to 20 ml. dry ether. Immerse sections for 2–6 minutes. Wash and stain.*

(7) Sulphuryl chloride (Grillo and Lewis, 1959). Expose dry dewaxed sections to vapour for 20 minutes. (Place a small amount of $SOCl_2$ in a Coplin jar and warm on a hot plate to 40°.) Stain without intermediate washing. This method sulphates glycogen but not other carbohydrates.

(8) Dioxane/SO_3 Complex. Preparation: Distil SO_3 from fuming oleum over P_2O_5 into a cold 25 per cent. mixture of dry dioxane and CCl_4. Continue until precipitation of a white solid is complete and excess fumes appear. Filter the adduct

* 0·01 per cent. Azure A (Hopkin Williams) in 60 per cent. ethanol.

and wash 4 times in cold CCl_4. Sulphate for 3 minutes using a 5 per cent. solution in dimethyl formamide.

(9) Dimethyl formamide/SO_3 Complex. Preparation: Add 5·5 ml. chlorosulphonic acid to 50 ml. ice-cold dry DMF with vigorous stirring. Remove the adduct by filtration. Alternatively, distil SO_3 from 65 per cent. oleum over P_2O_5 into cold DMF until a permanent precipitate just persists. Treat sections with this preparation for 3 minutes, wash in DMF and then in acetone.

METHYLENE BLUE EXTINCTION

(after Dempsey and Singer, 1946)

(*Various fixatives, paraffin sections*)

PREPARATION OF SOLUTIONS

Because of the differing extractive tendencies, and the different effects on staining affinities, of the various buffer solutions it is preferable to prepare the whole range of methylene blue solutions with a single buffer. For this purpose the veronal-acetate buffer of Michaelis (1931) can be used. Details are given in Appendix 1, p. 584. A series extending from pH 8·18 to pH 2·62 is adequate for most investigations. Lewis (1962) advised the use of a formic acid-sodium acetate buffer between pH 3·0 and 5·6, and an acid phosphate buffer down to pH 2·0, for use with methylene blue extinction. Details are given in Appendix 1, p. 585.

In the chosen buffer solution dissolve methylene blue to give a concentration of 5 mM. Place the solutions in stoppered tubes rather than in Coplin jars.

Method. (1) Bring sections to water, removing mercury precipitate if necessary.

(2) Immerse sections in the buffered methylene blue solutions for 24 hours at room temperature (17°–22°, avoid great variation).

(3) Wash quickly in distilled water and examine in water.

(4) Mount in glycerine jelly.

Result. The intensity of staining varies directly with the pH. Colorimetric measurements were made by the original authors to compare the binding capacity for methylene blue of various basophilic substances over a wide range of pH.

The point at which virtual extinction of staining occurs is taken as the methylene blue extinction of the tissue component concerned.

AZURE A STAINING PROCEDURES

PREPARATION OF SOLUTIONS

Dissolve Azure A 1:5000 in one of the above recommended buffer series. Alternatively, use the following mixtures, to obtain a full pH range.

pH 0·5 1:5000 azure A in 0·5 N-HCl.

pH 1·0 1:5000 azure A in 0·1 N-HCl.

pH 1·5 1:5000 azure A in 50 ml. buffer (30 ml. 0·1 N-HCl, 20 ml. 0·1 M-KH_2PO_4)

pH 2·0 1:5000 azure A in 50 ml. buffer. (20 ml. 0·1 N-HCl, 30 ml. 0·1 M-KH_2PO_4)

pH 2·5 1:5000 azure A, 48 ml.; 2 ml. 0·1 M citric acid.

pH 3·0 1:5000 azure A, 48 ml.; 1·65 ml. 0·1 M-citric acid; 0·35 ml. 0·2 M-Na_2HPO_4.

pH 3·5 1:5000 azure A, 48 ml.; 1·4 ml. 0·1 M-citric acid; 0·6 ml. 0·2 M-Na_2HPO_4.

pH 4·0 1:5000 azure A, 48 ml.; 1·25 ml. 0·1 M-citric acid; 0·75 ml. 0·2 M-Na_2HPO_4.

pH 4·5 1:5000 azure A, 48 ml.; 1·1 ml. 0·1M-citric acid; 0·9 ml. 0·2 M-Na_2HPO_4.
pH 5·0 1:5000 azure A, 48 ml.; 1·0 ml. 0·1M-citric acid; 1·0 ml. 0·2 M-Na_2HPO_4.

Method. (1) Bring sections to water.
(2) Stain in selected pH solutions for 30 minutes.
(3) Dehydrate in alcohols and in alcohol-xylene.
(4) Clear in xylene and mount in synthetic resin.

Result. Below pH 2·0 only sulphated mucosubstances retain the dye in dehydrated sections. Above pH 3·0 many sialomucins stain metachromatically.

ACRIDINE ORANGE-CTAC METHOD

(after Saunders, 1964)

(*Newcomer; paraffin sections*)

PREPARATION OF TISSUES

(1) Fix small pieces of tissue in Newcomer's solution for 12–24 hours. (Isopropanol 6, propionic acid 3, acetone 1, dioxane 1, light petroleum 1.)
(2) Process in 1:1 Newcomber/*n*-butanol for 30 minutes.
(3) Process in 3 changes of *n*-butanol for 30 minutes each.
(4) Transfer to 1:1 *n*-butanol/56° wax.
(5) Three wax changes, 30 minutes each. Cut 5 μ paraffin sections and mount on slides.

Method. (1) Treat slides in triplicate with 1 per cent. cetyltrimethylammonium chloride (CTAC) for 10 minutes.
(2) Wash for 10 minutes in running water.
(3) Treat all 3 slides with ribonuclease (Appendix 25, Vol. 2) for 2 hours at 45°.
(4) Treat slide 1 with CTAC again, and wash.
(5) Treat slide 1 with 0·1 per cent. aqueous acridine orange (pH 7·2) for 3 minutes.
(6) Wash slide 1 in running water for 10 minutes.
(4a) Treat slide 2 with 0·1 per cent. acridine orange in 0·01 M-acetic acid (pH 3.2).
(5b) Treat slide 2 in running water and then differentiate in 0·3 M-NaCl in 0·01 M-acetic acid.
(6b) Wash in running water.
(4c) Treat slide 3 as slide 2 but substitute 0·6 M-NaCl.
(7) Air dry all three slides.
(8) Mount in Fluor-free (Gurr).

Result. Slide 1; Red fluorescence due to hyaluronic acids.
Slide 2; Red fluorescence due to chondroitin sulphates and heparin.
Slide 3; Red fluorescence due to heparin.

ION ASSOCIATION METHODS FOR POLYANIONS

(Zugibe, 1966)

(*F.D. unfixed, Newcomer, paraffin or carbowax*)

PREPARATION OF TISSUES

As an alternative to FD (or FDFV) fix small pieces of tissue in modified Newcomer's fluid (Isopropanol 5, propionic acid 2, acetone 1, dioxane 2, light petroleum 1) for 12–24 hours. Dehydrate and embed as in previous method.

REAGENTS

(1) 0·1 per cent. cetyltrimethylammonium bromide (CTAB) in 0·03 M-NaCl or 0·05 per cent. cetylpyridinium chloride (CPC) in 0·03 M-NaCl.

(2) Stock ferric thiocyanate. 67·58 ml. $FeCl_3 . 6H_2O$ added to 300 ml. ammonium thiocyanate (170 g/.100 ml.).

(3) Working solution of (2). 1 part stock to 10 parts distilled water.

Method. (1) Bring sections to water.

(2) Treat with CTAB or CPC at 32° for 30 minutes.

(3) Wash for 1 minute in lukewarm tap-water.

(4) Process slides one at a time as follows: stain for 30 seconds in ferric thiocyanate working solution.

(5) Dip into distilled water.

(6) Blot with fine filter paper (Whatman No. 50). (The rapidity of this step is critical.)

(7) Place two drops of Dow-Corning No.710 silicone oil on the section and apply cover slip.

Control sections should be processed as above, omitting the initial CTAB or CPC treatment.

Results. Sites of high polyanion concentration stain reddish brown.

ION ASSOCIATION AND C.E.C. METHOD

(Zugibe and Fink, 1966)

(*FD unfixed, paraffin or carbowax; fresh frozen; Newcomer*)

Method. This method incorporates four separate procedures:

I. Total Polyanion Staining

(1) Treat sections with 0·1 per cent. CPC or 0·05 per cent. CTAB in 0·03 M-NaCl at 32° for 30 minutes.

(2) Wash for 1 minute in tap water.

(3) Proceed to treat *one slide at a time* as follows:

(4) Stain for 30 seconds in ferric thiocyanate (working solution).

(5) Dip in distilled water and blot.

(6) Apply silicone oil (2 drops) and cover slip.

II. Elimination of Hyaluronate

(1) 0·1 per cent. CPC or 0·05 per cent. CTAB in 0·2 M-KCl at 32° for 30 minutes.

(2) Proceed as for Sequence I.

III. Elimination of Hyaluronate, CS, Heparitin S, RNA, DNA

(1) 0·1 per cent. CPC or 0·05 per cent. CTAB in 1·0 M-KCl at 32° for 30 minutes.

(2) Proceed as for Sequence I.

IV. Elimination of Hyaluronate, CSA, B, C° HS, RNA, DNA, Heparin

(1) 0·1 per cent. CPC or 0·05 per cent. CTAB in 2·0 M-KCl at 32° for 30 minutes.

(2) Proceed as for Sequence I.

Results. As for previous method; positive reddish-brown staining in reactive sites.

ALDEHYDE FUCHSIN FOR SULPHATE AND CARBOXYL GROUPS

(after Halmi and Davies, 1953)

FDFV; formalin or alcohol; paraffin sections)

PREPARATION OF SOLUTION

Dissolve 0·5 g. basic fuchsin in 100 ml. 60 per cent. ethanol. Add 1 ml. paraldehyde and 1·5 ml. conc. HCl. Ripen for 24 hours at room temperature. Store in refrigerator. Use for 6–7 days only before making fresh solution.

Method. (1) Bring sections to water.

(2) Rinse in 70 per cent. alcohol.

(3) Stain for 10–30 minutes.
(4) Rinse in 70 per cent. alcohol.
(5) Dehydrate in alcohols, clear in xylene and mount in synthetic resin.

Result. Sulphated mucosubstances (and elastic tissue) are the only strongly staining tissue components in unoxidized sections (see Chapter 8, p. 225 and Appendix 8, p. 639). Non-sulphated acidic mucosubstances stain weakly to moderately.

DIALYSED IRON METHODS FOR ACID MUCOPOLYSACCHARIDES

(Hale, 1946)

(*Unfixed fresh frozen sections or fresh frozen sections briefly fixed in Wolman's fixative are excellent. Carnoy or formalin-fixed paraffin sections usually recommended.*)

Method. Sections should be mounted on the slides without any adhesive.
(1) Bring sections to water, removing mercury precipitate where necessary.
(2) Flood with dialysed iron (1 vol. dialysed iron, B.D.H.,* 1 vol. 2 M-acetic acid) for 10 minutes. Alternatively, use colloidal iron reagent, in the same manner.†
(3) Wash well with distilled water.
(4) Flood with acid ferrocyanide solution (0·02 M-$K_4Fe(CN)_6$, 0·14 M-HCl, equal parts), leave for 10 minutes.
(5) Wash in water.
(6) Counterstain nuclei with Mayer's carmalum (6–18 hours) or with 1 per cent. aqueous neutral red (1 minute). Counterstaining is not recommended in the case of frozen sections.
(7) Dehydrate in alcohol, clear in xylene and mount in DPX or any suitable synthetic resin.

Result (Figs. 100 and 105). Strongly acidic mucosubstances are stained blue, nuclei red.

COMBINED DIALYSED IRON AND PAS STAIN FOR POLYSACCHARIDES

(after Ritter and Oleson, 1950)

(*Paraffin sections after formol-alcohol, formalin or Helly*)

Method. Fix tissues (as fresh as possible) in 10 per cent. formalin in 90 per cent. alcohol. The other fixatives mentioned above are adequate. Embed in paraffin wax.
(1) Bring sections to water.
(2) Perform Hale's method, stages 2–5 inclusive.
(3) Wash in water.
(4) Apply the PAS reaction, stages 2–5 inclusive.
(5) Counterstain nuclei if desired, lightly, with Mayer's hæmalum, 4–6 minutes.
(6) Wash well in water.
(7) Dehydrate in alcohol, clear in xylene, mount in Canada balsam or DPX.

Result. Acid mucosubstances, blue. Most protein structures, pale blue. Nuclei, pale blue or dark blue if counterstained. Neutral mucosubstances, purplish red.

In the kidney the visceral glomerular basement membrane is blue, the parietal glomerular, and the tubular, basement membranes are red. The glomerular intercapillary space is red.

* British Drug Houses Ltd, Poole, England.

† The colloidal iron reagent of Rinehart and Abu'l Haj (1951) is made by dissolving 75 g. of $FeCl_3$ in 250 ml. distilled water, adding 100 ml. glycerol and then, gradually, 55 ml. 28 per cent. ammonia. This last should be added with constant stirring. The mixture is dialysed against distilled water for three days with regular changes of the water. The staining time with this mixture is also about 10 minutes.

COLLOIDAL IRON-HÆMATOXYLIN FOR ACID MUCOSUBSTANCES

(after Müller, 1955; Mowry, 1958, 1963)

PREPARATION OF SOLUTIONS

Stock Colloidal Iron (Krecke). Bring 250 ml. distilled water to the boil. Pour in 4·4 ml. 29 per cent. $FeCl_3$ and stir. Keep the solution boiling the whole time. When the solution turns dark red, allow it to cool.

To remove acid formed by hydrolysis, and non-colloidal iron, dialyse for 24 hours against 3 changes of distilled water of a volume 5 to 10 times that of the stock solution.

Store at room temperature and filter if any precipitate develops.

Working Colloidal Iron. Distilled water 18 volumes; glacial acetic acid, 12 volumes; stock colloidal iron, 10 volumes. (Remains usable for 24 hours but should preferably be made fresh every day.) The pH of this solution is 1·8. It does not require constant checking.

OTHER REAGENTS

(1) Acetic acid rinse, 30 per cent.
(2) Potassium ferrocyanide, 2 per cent.
(3) HCl, 2 per cent.
(4) Harris' hæmatoxylin (optional nuclear stain).
(5) Acid alcohol (1 per cent. HCl in 70 per cent. ethanol).
(6) Picric acid 0·5 per cent. (optional counterstain).

Method. (1) Bring sections to water.
(2) Place sections in colloidal iron solution for 2 hours.
(3) Rinse in 3 changes of 30 per cent. acetic acid, 10 minutes each.
(4) Wash in running water for 5 minutes.
(5) Rinse in (deionized) distilled water.
(6) Treat for 20 minutes in a freshly prepared mixture of equal volumes of 2 per cent. HCl and 2 per cent. potassium ferrocyanide.
(7) Wash for 5 minutes in running water.
(8) Stain in Harris' hæmatoxylin for 5 minutes.
(9) Rinse in water and differentiate (10 to 20 seconds) in acid alcohol.
(10) Wash briefly (2–3 minutes) in water.
(11) Stain in picric acid solution for 60 seconds; rinse for a few seconds in water.
(12) Dehydrate, clear and mount.

Result. Acidic mucosubstances are coloured Prussian blue. These include sialomucins, goblet cell mucins, some connective tissue mucins and mast cell granules. Staining resembles that given by Alcian blue at pH 2·5, except for weaker staining in some epithelial sites. Strongly acidic, non-Alcianophil, mucins do not stain with colloidal iron.

MODIFIED BI-COL METHOD

(Wolman, 1956, 1961)

(*Carnoy-fixed paraffin sections*)

PREPARATION OF SOLUTIONS

(1) *Working colloidal iron* according to the method of Müller (Krecke) given above.

(2) *Colloidal gold.* This solution must be prepared with special care. Glass-distilled water must be used throughout. Glassware must be acid washed and rinsed in glass-distilled water. The following stock solutions are used:

(1) Reagent formaldehyde (40 per cent.).

(2) Aqueous KOH, 30 per cent., kept in well stoppered bottle.

(3) 0·2 per cent. aqueous gold chloride ($HAuCl_4$ or $AuCl_3HCl$).

Remove stock solutions with a pipette so as to avoid disturbing any precipitates which may have formed.

Testing Colloidal Gold Solutions. Pipette 0·4, 0·6, 0·8 and 1·0 ml. of KOH stock into 4 graduated cylinders (50 ml.). To each cylinder add 3 ml. formaldehyde and bring volume to 30 ml. with distilled water.

Place 48 ml. of water into each of 4 beakers. Add to each 1 ml. 0·2 per cent. gold chloride and then add 1 ml. of each of the 4 alkaline formalin solutions, with vigorous shaking. The pH of the solutions should be between 4 and 6. Each solution should be boiled for 20–30 minutes, after which the volume should be restored with distilled water. A "good" solution is red or purple-red, and orange-red or rose-red after boiling. Use whichever preparation provides the best solution.

An alternative colloidal gold solution may be prepared according to Módis *et al.* (1964), as follows: To 50 ml. distilled water add 0·1 ml. freshly prepared 0·1 M-ascorbic acid. To this, with continuous shaking, add dropwise 0·5 ml. 1 per cent. gold chloride. The resulting solution, which is red in colour, does not require boiling for stabilization. The pH is between 4·8 and 5·0.

Method. (1) Bring sections to distilled water.

(2) Immerse in the colloidal iron reagent for 10 minutes.

(3) Develop with acid ferrocyanide as in Hale's method after washing in distilled water.

(4) Treat with colloidal gold for 12 24 hours at 37°.

(5) Rinse in tap-water.

(6) Dehydrate, clear and mount in Canada balsam.

Result. Chromatin (nuclei) blue or brownish-red. Compounds containing free phosphoric or sulphuric radicals stain blue. Those containing weak acids stain reddish-brown. Hyaluronic acid stained blue *in vitro* but, according to Wolman, reddish-brown in the tissues. Sulphated hyaluronic acid would of course stain blue

THE ALCIAN BLUE METHOD FOR ACID MUCOPOLYSACCHARIDES

(after Steedman, 1950)

(*Suitable for all types of material, fixed and unfixed; for paraffin sections fixation in Bouin, Carnoy, formalin or formol-mercury is recommended*)

Method. (1) Bring sections to water, removing mercury precipitate where necessary.

(2) Stain in a freshly filtered 1·0 per cent. solution of Alcian Blue 8GX in 3 per cent. acetic acid for 10–30 minutes.

(3) Rinse in distilled water.

(4) Stain in Ehrlich's hæmalum 5–10 minutes, or in 1 per cent. Neutral red,* 30 seconds, or in 0·5 per cent. Chlorantine fast red 5B (Lison, 1954b) 10–15 minutes.

(5) Differentiate in 1 per cent. alcohol.
(6) Wash in running water 10–20 minutes.
{Omit if using Neutral red or Chlorantine fast red.

* If connective tissue mucins (hyaluronic acid, dermatan sulphate) are to be demonstrated the following variation of the technique is recommended:

(4) Stain in 1 per cent Neutral Red, 60 seconds.

(5) Rinse quickly in water.

(6) Blot dry and allow the slide to dry completely in air (22° to 37°).

(7) Clear in xylene and mount in DPX.

With this variation the Neutral Red stains cytoplasmic as well as nuclear structures, concealing the tendency of Alcian Blue to stain connective tissues but leaving the acid mucopolysaccharide of the ground substance a deep blue-green.

(7) Dehydrate in alcohol, clear in xylene and mount in Canada balsam or in DPX.

If the use of stains in acid solution is to follow staining with Alcian Blue, the following manœuvre may be performed after stage 3: Immerse in 1 per cent. alkaline alcohol (pH 8 or over) for 2 hours or more to convert the dye into the insoluble pigment Monastral Fast Blue.

Result (Plates Va, VIa). Acid mucopolysaccharides, clear blue-green. Nuclei, dark blue or dark red. In place of Alcian blue 8GX, Alcian green 3BX or Alcian green 2GX can be employed.

ALCIAN BLUE 2·5 PROCEDURE

Method. (1) Bring sections to water.

(2) Stain in freshly filtered Alcian blue 8GX, 1 per cent. in 3 per cent. acetic acid, 30 minutes (pH 2·5).

(3) Wash in running water for 5 minutes.

(4) Dehydrate in alcohols, clear in xylene and mount.

Result. Weakly acidic sulphated mucosubstances, hyaluronic acids and sialomucins, dark blue. Strongly acidic sulphated mucins are stained weakly or not at all.

ALCIAN BLUE 1·0 PROCEDURE

Method. (1) Bring sections to water.

(2) Stain for 30 minutes in 1 per cent. Alcian blue 8GX in 0·1 N-HCl (pH 1·0).

(3) Blot dry with fine filter paper (Whatman No. 50).

(4) Dehydrate in alcohols, clear in xylene and mount.

Result. Only sulphated mucosubstances stain. The most acidic mucins, paradoxically, stain the least strongly.

ALCIAN BLUE-CEC PROCEDURES

Method. (1) Bring 6 sections to water.

(2) Stain for 30 minutes in 0·1 per cent. Alcian blue 8GX in 0·05 M-sodium acetate buffer at pH 5·7 adding, respectively, 0·1 M, 0·2 M, 0·5 M, 0·6 M, 0·8 M, and 1·0 M-$MgCl_2$.

(3) Wash in running water for 5 minutes.

(4) Dehydrate quickly in alcohols, clear in xylene and mount in synthetic resin.

Result. Hyaluronic acid, sialomucins and some weakly acidic sulphomucins are not stained at or above 0·1 M-$MgCl_2$. Most sulphated mucosubstances, including those which stain metachromatically with azure A at pH 0·5 (p. 667), stain strongly at 0·2 M levels. The various sulphated mucosubstances lose Alcianophilia at different levels with increasing $MgCl_2$ concentration. Those in mast cells, cornea, and some epithelia persist at the 1·0 M level.

ALCIAN BLUE-PAS PROCEDURE

(after Mowry, 1963)

(*FDFV; various fixatives, paraffin sections*)

Method. (1) Bring sections to water.

(2) Rinse briefly in 3 per cent. aqueous acetic acid.

(3) Stain for 2 hours in 1 per cent. Alcian blue 8GX in 3 per cent. acetic acid.

(4) Rinse briefly in water and then in 3 per cent. acetic acid, running water and distilled water.

(5) Oxidize for 10 minutes in 1 per cent. periodic acid (aqueous) at room temperature.
(6) Wash in running water for 5 minutes.
(7) Immerse in Schiff's reagent for 10 minutes.
(8) Wash in running water for 2 minutes.
(9) Rinse in 3 changes of 0·5 per cent. sodium bisulphite, 1 minute in each.
(10) Wash in running water for 5 minutes.
(11) Dehydrate, clear and mount.

Result. This procedure stains periodate-unreactive, Alcinophilic mucosubstances blue; periodate-reactive and Alcianophilic components are bluish-purple and periodate-reactive, non-Alcianophilic, components are red. Acid mucosubstances stained blue by the above procedure include hyaluronic and sialomucins, and all but the most strongly acidic sulphated mucosubstances stain blue or bluish-purple.

THE PHENYLHYDRAZONE-FORMAZAN REACTION (PHF)

(Stoward, 1963, 1967e)

(*Formol-calcium, paraffin sections; FDFV*)

Method. (1) Bring sections to water.
(2) Oxidize for 10 minutes in 1 per cent. aqueous periodic acid (pH adjusted to pH 3–5 with 1 N-NaOH).
(3) Rinse briefly in water.
(4) Immerse in a freshly prepared 2·5 per cent. solution of phenylhydrazine hydrochloride in 5 per cent. mixed phosphate buffer (2·5 g. each of sodium dihydrogen phosphate and disodium monohydrogen phosphate in 100 ml. of distilled water) for 3 minutes.
(5) Rinse in 3 changes of water.
(6) Flood sections with a freshly made solution of 0·1 to 1·0 per cent. tetrazotized 3,3′-dimethoxybenzidine fluoroborate (TDMBF)* in 25 per cent. aqueous pyridine (pH 9·3–10·4) for 3 minutes.
(7) Rinse briefly in water. Counterstain nuclei, if required, in 2 per cent. $CHCl_3$-washed methyl green, 30 seconds.
(8) Dehydrate in alcohols, clear in xylene, mount in synthetic resin (DPX or Permount).

Result (Plate VIc). The phenylazo-dimethoxybiphenyl formazans formed from periodate-oxidized mucosubstances vary in colour from brown, through red, to yellow. With the exception of certain colonic mucins all PAS-prositive mucins react.

MODIFICATIONS OF THE PHF REACTION

(Stoward, 1967e)

(*a*) *Methylation*

Before stage 2, bring sections to methanol and methylate in 2 per cent. thionyl chloride in methanol for 4 hours at room temperature. Follow with a double rinse in methanol and return sections to water.

Result. In some sites (notably glycogen) formazans no longer form. Colour is always less than in unmethylated sections.

(*b*) *ANSA coupling*

Between stages 7 and 8, immerse sections in a freshly made solution of 1-amino-8-naphthol-4-sulphonic acid† in 25per cent. pyridine for 3 minutes at room temperature.

* For details of preparation see below.

† This compound is often described as 8-amino-1-naphthol-5-sulphonic acid. The designation used here is based on the I.U.P.A.C. rules of nomenclature; Rules A23.3, A12.3 and A2.4 (Handbook for Chemical Society Authors, 1960).

Result (Plate VId). The original colour of the phenylhydrazone formazan is changed to purple. Depth of colour varies between one site and another. Some mucins (lingual serous glands) only show visible colour after ANSA coupling.

(*c*) *Methylation—PHF-ANSA Sequence*

Stages (*a*) and (*b*) above, with the PHF routine between.

Result. Deep colonic mucous cells contain a pale purple product.

PREPARATION OF TDMBF

(Stoward, 1967e)

Tetrazotized 3,3′-dimethoxybenzidine fluoroborate is synthesized as follows:

A solution of sodium nitrite (3 g.) in water is slowly added, with constant stirring (keep temperature below 5°), to a suspension of 3,3′-dimethoxybenzidine (5 g.) obtained by adding conc. HCl (30 ml.) and water (70 ml.) to a filtered solution of the amine in dimethylformamide (50 ml.) and cooling to below 5°. Stirring must be continued for 30 minutes after the last addition of sodium nitrite solution.

After this procedure, 40 ml. of 50 per cent. fluoroboric acid is then added to the diazotization mixture and stirring is continued for a further 3 minutes. The buff-coloured crystals of TDMBF are collected, washed successively with small quantities of cold fluoroboric acid, cold absolute ethanol, and ether twice. The crystals are then dried at room temperature and stored at 0° to 5°.

Yield: 76 per cent. M.P. of crystals obtained by cooling solution in water at room temperature to 0°: 181–182° (decomp). Found: N, 13·6; F, 33·7 per cent.

FERRIC ALUM—CORIPHOSPHINE FOR SULPHOMUCINS

(Stoward, 1967b)

(*FDFV, formalin; paraffin sections*)

Method. (1) Bring sections to water.

(2) Immerse in 4 per cent. aqueous ferric alum, 10 minutes.

(3) Wash twice in distilled water.

(4) Stain in 0·01 per cent. coriphosphine (pH 6·0), 20 minutes.

(5) Rinse in distilled water, blot, dehydrate in 3 changes of isopropyl alcohol.

(6) Clear in xylene and mount in DPX.

Result. Sulphomucins fluoresce green or dull red.

CORIPHOSPHINE-THIAZOL YELLOW FOR SULPHOMUCINS

(Stoward, 1967b)

FDFV, formalin; paraffin sections)

Method. (1) Bring sections to water.

(2) Stain in 0·01 per cent. aqueous coriphosphine, 20 minutes.

(3) Rinse in 2 changes of distilled water.

(4) Stain in 0·001 per cent. thiazol yellow at pH 2·0 to 2·1, for 1 minute.

(5) Proceed as in previous method.

Result. Sulphomucins fluoresce red or orange-yellow. Nuclei green (UVL) or greenish-yellow (blue light).

Purification of Coriphosphine

Coriphosphine O (Basic yellow. C.I. 46020) can be purified from commercial preparations as its dichloride, as follows:

Add 150 ml. diethyl ether to a filtered solution of 0·5 g. coriphosphine in 50 ml. absolute ethanol and 0·15 ml. conc. HCl. Discard most of the ether-alcohol layer but keep the precipitated coriphosphine covered (otherwise it is oxidized to a tar). Wash the precipitate 3 times with light petroleum (b.p. 60°–70°) by decantation and, finally, evaporate a slurry of the dye in light petroleum *in vacuo*. Yield of dull orange powder, approximately 0·1 g. (20 per cent.). Check the purity by ascending paper chromatography using 38:38:24 amyl alcohol:ethanol:water mixture as developing solvent. Repeat the precipitation procedure until a single spot is obtained (1 to 3 precipitations may be required).

BENZIDINE REACTION FOR SULPHATE GROUPS

(after Bracco and Curti, 1953)

(*FDFV, formalin; paraffin sections*)

REAGENTS

(1) 0·1 N-HCl.

(2) Dissolve 1 g. benzidine* in 100 ml. saturated aqueous boric acid, with constant shaking. Filter before use. This solution keeps for a few days in the dark. It should be renewed when it turns brown.

(3) 1 per cent. potassium dichromate in saturated aqueous boric acid.

(4) Rinsing solution (2 per cent. aqueous boric acid).

Method. (1) Bring sections to water.

(2) Immerse in benzidine solution for 5 minutes.

(3) Wash 3 times in 2 per cent. boric acid.

(4) Treat with dichromate for 30–50 minutes.

(5) Wash in distilled water. (Counterstain nuclei red, if required, with carmalum or neutral red.)

(6) Dehydrate in alcohols, clear in xylene, embed in synthetic resin.

Result (Fig. 98). Sulphated mucosubstances stain dark blue. (Since the sensitivity of this method is not high it may be convenient to use 8–10 μ sections.)

TETRAZONIUM METHOD FOR SULPHATE GROUPS

(after Geyer, 1962)

(*F, FDFV, paraffin; fresh, unfixed*)

Method. (1) Bring paraffin sections to water.

(2) Treat with acidic Fast blue B solution for 10–30 minutes (50 mg. Fast blue B salt in 10 ml. 5 per cent. acetic acid).

(3) Rinse for 30–60 seconds in cold (0°–5°) distilled water.

(4) Treat for 2–5 minutes in a cold (0°–5°) saturated solution of 1-naphthol in borax buffer at pH 9·4.

(5) Wash in distilled water and mount in glycerine jelly, or other watery medium.

Result. Tissue sulphate esters, and sulphonic acid groups induced by oxidation or sulphation, are stained reddish violet.

ACRIFLAVINE METHODS FOR SULPHATE ESTERS

(Hollander, 1964)

(*Formalin, frozen sections; FDFV, formalin, paraffin*)

STOCK DYE SOLUTION

Dissolve 500 mg. purified Acriflavine (Trypaflavin) in 100 ml. distilled water at 80°. Cool and store in the dark.

Method I. (1) Bring paraffin sections to water.

(2) Stain in 1:20′000 Acriflavine in 0·1 M-citrate buffer (citrate-HCl, pH 2·5).

* This compound is a carcinogen and should be used with care

The staining solution is made by adding 1 ml. stock dye to 99 ml. buffer.

(3) Differentiate for 1 minute in 70 per cent. isopropanol.

(4) Proceed through 90 per cent. to 100 per cent. isopropanol.

(5) Clear in xylene and mount in a fluorescence-free medium. Examine by UVL.

Result. Clear yellow fluorescence marks the sites of sulphated mucosubstances or sulpholipids.

Method II. (1) Perform stages 1, 2 and 3 as above.

(2) Treat with DMAB solution for 30–45 seconds. (30 parts 2 per cent. dimethylaminobenzaldehyde in 20 per cent. HCl, 70 parts isopropanol.)

(3) Rinse in distilled water for 2–3 minutes.

(4) Counterstain nuclei in Mayer's hæmalum for 3–4 minutes.

(5) Wash in running water for 10 minutes.

(6) Dehydrate in ascending concentrations of isopropanol.

(7) Clear in xylene, mount in synthetic resin.

Result. Sulpho-mucosubstances and sulpholipids stain reddish-brown.

BIAL REACTION FOR SIALIC ACIDS

(after Ravetto, 1964)

(*F, frozen; FDFV; F, paraffin sections*)

REAGENT

Dissolve 200 mg. orcinol (5-methylresorcinol) in 80 ml. conc. HCl and add 0·25 ml. 0·1 M-$CuSO_4$. Make up the volume to 100 ml. with distilled water. (Allow to stand for 4 hours before use. If stored at 0° the solution is stable for one week.)

Method. (1) Spray sections with the above solution.

(2) Place sections, face downwards, on a glass frame in a preheated container, which has on the bottom a thin layer of conc. HCl, at 70° for 5–10 minutes.

(3) Dry the sections in air.

(4) Clear in xylene and mount in Canada Balsam.

Result (Fig. 101). Sites containing (high levels of) sialic acids stain red or reddish-brown. (The colour fades rapidly.)

DISCHE REACTION FOR *N*-ACETYLAMINO HYDROXYLS

(after Stoward, 1963)

(*FDFV; formalin, paraffin sections*)

Method. (1) Bring sections to water.

(2) Rinse in 30 per cent. orthophosphoric acid, 1 minute.

(3) Treat with 60 per cent. orthophosphoric acid, 90 minutes (to complete hydrolysis of the *N*-acetyl groups).

(4) Immerse in Warren's reagent (5 per cent. HIO_4 in 60 per cent. H_3PO_4) for 20 minutes.

(5) Wash in 60 per cent. and then in 30 per cent. H_3PO_4, 1 minute in each.

(6) Wash in water, 1 minute.

(7) Wash in 10 per cent. aqueous ethylene glycol.

(8) Wash in water, 1 minute.

(9) Treat with 0·6 per cent. 2-thiobarbituric acid at 100° for 15 minutes.

(10) Wash and mount and examine, in water, by UVL.

Result. The resulting diffuse blue fluorescence may indicate sites of sialic acid-containing mucosubstances.

PARA-ANISIDINE REACTION FOR SIALIC ACIDS

(after Cerbulis and Zittle, 1961)

(*FDFV; Cryostat-FV; Formalin; paraffin*)

REAGENT

Dissolve 0·5 g. *p*-anisidine in 3 ml. H_3PO_4 in 100 ml. 80 per cent. methanol.

Method. (1) Bring sections to water.

(2) Rinse in 30 per cent. H_3PO_4.

(3) Treat with 60 per cent. H_3PO_4 for 90 minutes.

(4) Treat with *p*-anisidine reagent for 30 minutes at 95°.

(5) Wash in 60 then in 30 per cent. H_3PO_4.

(6) Wash, and then mount in water.

Result. A brownish colour indicates sites with high concentration of sialic acid.

THE EVAN'S BLUE METHOD FOR CONNECTIVE TISSUE GROUND SUBSTANCE

(Perl and Catchpole, 1950)

Method. (1) Administer to selected animals, by the intravenous route, a suitable amount of 1·25 per cent. aqueous Evan's blue (4 ml. in the case of the guinea-pig).

(2) After 10 minutes kill the animals, preferably with chloroform or ether.

(3) Remove small blocks of tissue and immerse directly in isopentane at −160°.

(4) Freeze-dry the tissues (see Appendix 3).

(5) Embed in soft paraffin (M.P. 40°).

(6) Cut thick sections (120 μ).

(7) Mount on albuminized slides.

(8) Remove wax with several changes of hot liquid paraffin.

(9) Mount in liquid paraffin and seal the edges of the coverslip with varnish.

Result. According to the authors, areas in which depolymerization has occurred, making the mucopolysaccharides of the ground substance more water soluble, are coloured blue by the dye.

OKAMOTO METHOD FOR GLUCOSE

(after Müller, 1955–56)

Method. (1) Fix thin (2 mm.) tissue slices in methanol saturated with barium hydroxide for 24 hours at −10°.

(2) Dehydrate in three changes of absolute alcohol.

(3) Clear and embed in methyl benzoate, benzene, paraffin series.

(4) Cut 10μ sections and mount on albuminized slides without contact with water.

(5) Place in paraffin oven at 56–58° until the wax melts.

(6) Remove wax with chloroform and bring to absolute alcohol.

(7) Immerse in alcoholic silver nitrate* for about 30 minutes.

(8) Wash repeatedly in 96 per cent. alcohol.

(9) Reduce in alcoholic formalin.†

(10) Bring through descending strengths of alcohol to water.

(11) Fix for 30 seconds in 5 per cent. sodium thiosulphate.

(12) Wash in distilled water, dehydrate, clear and mount in a suitable (non-reducing) resin.

Result. Black silver deposits indicate sites containing glucose.

* Add 90 ml. ethanol to 10 ml. of a 20 per cent. aqueous $AgNO_3$.

† Add 90 ml. ethanol to 10 ml. of 40 per cent. formaldehyde.

METHODS FOR PECTIN

The Ruthenium Red Method

(*Paraffin sections; various fixatives*)

(1) Bring sections to water.
(2) Treat with 25 per cent. alcoholic HCl for 24 hours at 22°.
(3) Immerse in dilute (2 per cent.) ammonia for 4–6 hours.
(4) Transfer to 0·05 per cent. $RuCl_3$ and store in the dark until an adequate colour develops.

Result. A deep red colour indicates materials containing pectin. A similar colour is given by cellulose starch, inulin and lignin. The method is therefore quite unspecific.

The Krajčinovič Amine Reaction

(*Paraffin sections; various fixatives*)

(1) Bring sections to water.
(2) Treat for 2 minutes in 1 per cent. HCl.
(3) Wash in three changes of absolute alcohol.
(4) Treat with 0·1 M-benzidine* for 45 minutes.
(5) Wash in four changes of absolute alcohol.
(6) Diazotize by immersion for 1 minute in 0·1 N-HCl with 0·1 N-$NaNO_2$ (3:1).
(7) Wash in water.
(8) Couple with 0·1 M-β-naphthol.
(9) Wash in water and mount in glycerol.

Result. Pectin stains red. According to the author the method does not stain other polysaccharides.

FERRIC HYDROXAMIC ACID FOR ACETYLATED POLYSACCHARIDES

(after McComb and McCready, 1957)

This method is best carried out on fresh frozen cold microtome sections which have been fixed briefly in alcohol or in Wolman's 5 per cent. acetic-ethanol.

Method. (1) Bring sections to absolute ethanol, remove and dry in air for a few minutes.
(2) Actylate for 6 hours at 60° in 16 ml. acetic anhydride and 24 ml. dry pyridine.
(3) Wash in water.
(4) Treat for 5–10 minutes in hydroxylamine† (Appendix 13, p. 706).
(5) Wash in water.
(6) Treat with ferric perchlorate solution‡ for 5 minutes. (Colour should appear in ½ to 1 minute.)
(7) Wash in water, dehydrate, clear and mount in DPX.

Result (Fig. 104). Acetylated materials give an orange-red colour whose intensity is related to the number of reacting groups and therefore in some measure to the amount of polysaccharide glycol. The specificity of this method has not been tested.

* This substance is a carcinogen and should be used with care.

† Dissolve 3·75 g. hydroxylamine hydrochloride in 100 ml. water and make alkaline with NaOH.

‡ Dissolve 1·96 g. of ferric chloride ($FeCl_3 . 6H_2O$) in 5 ml. conc. HCl. Add 5 ml. 70 per cent. perchloric acid and evaporate almost to dryness. Dilute to 100 ml. with water. This is the stock solution (stable 4 weeks at 0°–4°). For use, add 8·3 ml. 70 per cent. perchloric acid to 60 ml. stock solution. Make up to 500 ml. with reagent grade methanol.

CARMINE STAIN FOR GLYCOGEN

(after Best, 1906)

(*Bouin, Carnoy, alcohol, formalin etc.; paraffin sections*)

The best fixatives for the preservation of glycogens are either the ice-cold picro-alcohol formalin of Deane, Nesbett and Hastings (1946) or the mixture of Lison and Vokaer (1949) which is used at −73°. The second is composed of saturated picric acid in absolute alcohol (85 parts), formalin (10 parts) and acetic acid (5 parts). An acetone/CO_2 snow mixture is employed to cool the fixative and the smallest pieces of tissue possible are immersed therein.

PREPARATION OF SOLUTIONS

Carmine Stock Solution. Add 2 g. carmine, 1 g. potassium carbonate and 5 g. potassium chloride to 60 ml. distilled water. Boil gently for 5 minutes, cool and filter. Add to the filtrate 20 ml. of ammonia (Sp. G. 0·880). This solution lasts 3 months at 0°–4°.

Carmine Staining Solution. Dilute 15 ml. of stock solution with 12·5 ml. of ammonia (Sp. G. 0·880), and 12·5 ml. of methyl alcohol. This solution lasts for 2–3 weeks.

Best's Differentiator.	Absolute alcohol . .	8 ml.
	Methyl alcohol . .	4 ml.
	Aq. dest. . . .	10 ml.

Method. (1) Bring sections to absolute alcohol.

(2) Place sections in 1 per cent. celloidin in absolute alcohol/ether (equal parts) for 2 minutes.

(3) Dry in air.

(4) Pass through alcohol to water.

(5) Stain in Ehrlich's hæmalum for 5 minutes.

(6) Rinse and differentiate rapidly in 1 per cent. acid alcohol.

(7) Rinse in water.

(8) Stain in Best's carmine solution for 15–30 minutes.

(9) Differentiate in Best's differentiator, without previous rinsing, 5–60 seconds

(10) Wash in 80 per cent. alcohol.

(11) Dehydrate in absolute alcohol, clear in xylene and mount in DPX.

Result (Figs. 34 and 35). Nuclei, dark blue; glycogen, red.

SILVER METHOD FOR ASCORBIC ACID

(after Bourne, 1935, and Barnett and Bourne, 1941)

The original author advised the use of thin pieces of fresh tissue. If total ascorbic acid is required, small pieces of tissue are exposed to the vapour of acetic acid (5 minutes) and to H_2S (15 minutes), subsequently removing the latter by exposure to a moderate vacuum for 10–30 minutes. This converts the oxidized to the reduced form, which alone reduces acid silver solutions.

Method. (1) Treat slices with acid silver nitrate solution (5 ml. glacial acetic acid to 100 ml. 5 per cent. aqueous silver nitrate) 5–10 minutes.

(2) Treat in the dark with 5 per cent. ammonia for 10–15 minutes.

(3) Wash in distilled water.

(4) Mount in glycerine.

Result. Black silver granules indicate the presence of reduced ascorbic acid.

SILVER METHOD FOR ASCORBIC ACID

(after Bacchus, 1950)

Method. (1) Place tissues within a short while of removal in a dark vial containing 5 per cent. silver nitrate at pH 2–2·5. Incubate at 56° for 45–60 minutes.

(2) Pour off silver solution and add distilled water. Wash for 10–15 minutes.

(3) Pour off distilled water and add 5 per cent. sodium thiosulphate. Leave for 30–45 minutes.

(4) Dehydrate in dioxan.

(5) Embed in paraffin wax.

(6) Cut sections 6–10 μ thick and mount on slides.

(7) Bring to water, counterstain with hæmalum.

(8) Wash, dehydrate in alcohol, clear in xylene, mount in Canada balsam.

Result. Black granules indicate ascorbic acid.

SILVER METHOD FOR ASCORBIC ACID

(after Jensen and Kavaljian, 1956)

(*Freeze-dried; paraffin sections*)

Method. (1) 10 μ sections, dry mounted on to albuminized slides, are incubated at 37°–40° overnight.

(2) Without removing paraffin place sections in 10 per cent. $AgNO_3$ in 3 per cent. acetic acid for 8–14 hours.*

(3) Wash in 95 per cent. alcohol and dehydrate in absolute alcohol.

(4) Remove paraffin wax with xylene.

(5) Mount in Permount or suitable non-reducing resin.

The finished preparations should be stored in the dark.

Result. Reduced silver granules indicate ascorbic acid.

SILVER METHOD FOR GLYCOGEN AND MUCIN

(Gormori, 1946, modified by Arzac and Flores, 1949)

(*Bouin or alcohol; paraffin sections*)

PREPARATION OF REAGENTS

Lithium-silver Solution

10 per cent. aqueous $AgNO_3$ . .	4 ml.
Sat. aqueous Li_2CO_3 . . .	16 ml.

Add conc. ammonium hydroxide (Sp. G. 0·880) drop by drop until the lithium silver complex is almost dissolved, leaving a light turbidity.

Add sat. aqueous Li_2CO_3 to make up 100 ml.

Filter and keep in well-stoppered dark bottles at 4°.

Use at full strength (if formalin reduction is employed) or diluted 1:10 (if no formalin reduction).

Piperazine-silver Solution

0·5 per cent. piperazine . . .	100 ml.
10 per cent. aqueous $AgNO_3$ . .	1 ml.

For use, dilute 1:4 with distilled water and add 2 drops of sat. aqueous Li_2CO_3 to each 50 ml.

* A modified method for reduction of dehydroascorbic acid was used before this stage, when appropriate, as follows:
Place sections in an atmosphere of H_2S for 15 minutes. Pass O_2-free nitrogen over the sections for a further 15 minutes.

Method. (1) Bring sections to alcohol and cover with 1 per cent. celloidin, dry and pass through alcohol to water.

(2) Immerse in 10 per cent. chromic acid, 20–30 minutes.

(3) Wash in running water and then in distilled water.

(4) Treat with dilute lithium-silver or piperazine-silver at 45° for 15–60 minutes. Alternatively, use stock lithium-silver at 17°–22° for the same period.

(5) After the first two silver solutions wash in water; after the third in 2 per cent. neutral formalin for 30–60 seconds.

(6) Tone with 1:500 gold chloride if necessary.

(7) Rinse in 5 per cent. sodium thiosulphate, 3–5 minutes.

(8) Wash in water and remove celloidin with acetone.

(9) Dehydrate, clear and mount as usual.

Result. Glycogen and mucin, brown to black. Background colourless.

APPENDIX 11

METACHROMATIC METHODS FOR AMYLOID

(I) Dahlia Method

(Lendrum, 1951)

(*Formalin or alcohol; paraffin sections, fresh frozen, alcohol-fixed sections*)

Method. (1) Bring paraffin sections to water.

(2) Stain in 1 per cent. aqueous methyl violet or 1 per cent. aqueous dahlia, 3 minutes.

(3) Differentiate in 70 per cent. formalin (commercial formalin, 7 parts; water, 3 parts).

(4) Rinse well and flood with saturated aqueous sodium chloride, 5 minutes.

(5) Rinse well and mount in corn syrup.

Result. Amyloid, pink or red. Other structures blue or violet. According to the author this method gives preparations which remain stable for a number of years.

(II) Modified Methyl Violet Method

(*Various types of preparation can be employed; fresh frozen sections briefly fixed in 5 per cent. acetic-ethanol are recommended*)

Method. (1) Stain mounted sections for 12 or more hours at 22° in 40 ml. distilled water containing 1 drop of freshly filtered 1 per cent. aqueous methyl violet.

(2) Differentiate quickly in dilute acetic acid (4 drops per 100 ml.).

(3) Immerse in saturated aqueous sodium chloride, 5 minutes.

(4) Mount in glycerine jelly.

Result. Amyloid deposits stain pink or red.

(III) Modified Methyl Violet Method

(Fernando, 1961)

(*Formol saline, formol mercury, alcohol, Carnoy; paraffin sections*)

Method. (1) Bring sections to 95 per cent. alcohol and stand for 5 minutes.

(2) Blot carefully.

(3) Stain for 10 minutes in 1 per cent. crystal violet in 3 per cent. formic acid. (Filter immediately before use.)

(4) Pour off stain and blot.

(5) Differentiate (2–3 minutes) in 1 per cent. formic acid.

(6) Blot after each immersion in formic acid.

(7) Cease when amyloid is bright pink.

(8) Rinse in distilled water.

(9) Blot carefully.

(10) Mount in the following medium (pH 3·75; R.I. 1·54).

Dextrin	16·7 g.
Sucrose	16·7 g.
NaCl	10 g.
Sodium methiolate	10 mg.
Distilled water	100 ml.

Heat the above substances in a porcelain dish and stir until all are dissolved. Cool and store in airtight bottles.

(11) Seal the edge of the cover slip immediately with asphalt varnish.

Result. Amyloid bright pink. Other tissue components blue.

(IV) Modified Methyl Violet Method

(Bancroft, 1963)

(*Unfixed cryostat sections*)

Method. (1) Stain in 1 per cent. aqueous methyl violet, 5 minutes.

(2) Wash in distilled water, 1 minute.

(3) Partially differentiate in 1 per cent. acetic acid (15–20 seconds).

(4) Wash in distilled water.

(5) Stain and differentiate in 2 per cent. methyl green (chloroform-washed), 5–15 minutes.

(6) Wash in distilled water.

(7) Mount in Apathy's medium or blot dry, rinse in xylene and mount in synthetic resin.

Result (Plate Xc). Amyloid red or pink, mast cell granules purple, nuclei green, background clear.

(V) Modified Methyl Violet Method

(Bancroft, 1967)

(*Unfixed cryostat sections*)

Method. (1) Stain in 2 per cent. aqueous methyl green (unwashed) for 15–60 seconds.

(2) Wash in tap water.

(3) Blot gently.

(4) Flood section with triethylphosphate.

(5) Clear in xylene and mount in DPX.

Result. Amyloid, pink. Nuclei, green. Elastic fibres, blue.

MODIFIED CONGO RED METHOD

(*All types of material; fresh frozen sections briefly fixed in acetic-ethanol are recommended*)

Method. (1) Stain in 1 per cent. aqueous Congo red for 1–6 hours.

(2) Treat with 1 per cent. KI for 60 seconds.

(3) Differentiate in 70 per cent. alcohol until only amyloid deposits are stained.

(4) Wash in water.

(5) Counterstain nuclei with Mayer's hæmalum 1½ to 3 minutes.

(6) Differentiate for a few seconds in 1 per cent. acid alcohol.

(7) Blue in dilute buffer (pH 7–8).

(8) Dehydrate in alcohols, clear in xylene and mount in a suitable synthetic resin.

Result (Fig. 108). Amyloid stains brick-red, nuclei blue. (Examination by polarized light (Plates IXa and b) must always be carried out as well as optical microscopy.)

HIGHMANS CONGO RED METHOD

(Modified from Highman, 1946)

(*All types of material; frozen sections recommended*)

Method. (1) Stain in 0·5 per cent. Congo red in 50 per cent. ethanol, 5 minutes.
(2) Wash in tap water.
(3) Differentiate under microscope in 0·2 per cent. KOH.
(4) Wash in tap water.
(5) Stain in Mayer's haemalum, 2 minutes.
(6) Wash in tap water.
(7) Differentiate, if necessary, in 1 per cent. acid alcohol.
(8) Dehydrate in alcohols, clear in xylene, mount in DPX.
Result. Amyloid, orange-red, birefringent by polarized light. Nuclei, blue.

THE THIOFLAVINE T METHOD (SECONDARY FLUORESCENCE)

(after Vassar and Culling, 1959)

(*Formalin etc.; paraffin sections. Unfixed cryostat; FDFV*)

Method. (1) Bring paraffin sections to water.
(2) Stain in alum hæmatoxylin, 10 minutes, and differentiate in acid alcohol for about 5 seconds. (This blocks nuclear fluorescence.)
(3) Stain sections in 1 per cent. aqueous Thioflavine T for 3 minutes. (Filter stain before use.)
(4) Differentiate in 1 per cent. aqueous acetic acid for 10 minutes.
(5) Mount in glycerol-saline (9 volumes glycerol; 1 vol. 0·9 per cent. saline).
(6) Examine under UVL (350 nm.).
Result (Plate Xa). Amyloid deposits fluoresce bright greenish-yellow.

REDUCING METHOD FOR AMYLOID

(King, 1948)

(*Formalin; frozen sections*)

PREPARATION OF SILVER SOLUTION

Add ammonia drop by drop to 5 ml. of 10 per cent. $AgNO_3$ to produce and dissolve the precipitate. Add 6·8 ml. 3·5 per cent. sodium carbonate (anhydrous) and 75 ml. distilled water. For use take 10 ml. of this solution and add a few drops of pyridine.

Method. (1) Cut thin frozen sections (10–15 μ or less).
(2) Warm 10 ml. of silver solution to about 40° and immerse sections after brief rinsing to remove excess formalin. Leave until sections are brown.
(3) Wash in 5 per cent. sodium thiosulphate, 3 minutes.
(4) Rinse in water.
(5) Dehydrate in 50 per cent. alcohol for 5–10 minutes.
(6) Pass rapidly through 70–95 per cent. alcohol.
(7) Clear in carbol xylene (phenol, 1 part; xylene, 3 parts).
(8) Lift sections on to clean slides and mount in Canada balsam.

Result. Amyloid blackish-brown. Some other substances may be stained brown, particularly those containing reducing lipids.

PERMANENT GOLD STAIN FOR AMYLOID

(Lynch and Inwood, 1963)

(*Formalin; paraffin sections*)

Method. (1) Bring sections to distilled water.

(2) Treat for 2½ to 5 minutes in iodine solution (I_2, 1·0 g.; KI, 2·0 g.; distilled water, 100 ml.).

(3) Rinse, 5–10 seconds each, in 3 changes of distilled water.

(4) Immerse in gold chloride solution for 2½ to 5 minutes. (1 per cent. w/v aqueous brown-gold chloride.) The reddish-brown colour quickly fades and the sections become almost colourless.

(5) Rinse in 3 changes of distilled water.

(6) Treat with 3 per cent. H_2O_2 (freshly prepared from 30 per cent. stock) at 37° for 6–36 hours.

(7) Dehydrate in alcohols, clear in xylene and mount in a synthetic resin.

Result. Amyloid; golden yellow or, occasionally, faint purple. R.B.C.'s, muscle, liver cells; blue. Other cell cytoplasm and nuclei; brown. Connective tissue; grey.

PEPSIN DIGESTION TECHNIQUE

(McAlpine, Radcliffe and Friedmann, 1963)

(*All types of material, including fresh and fixed frozen sections*)

Method. (1) Incubate sections in porcine pepsin (5 mg./ml.) in 0·2 N-HCl at pH 1·6 for 4 hours at 37°. (Control sections in pepsin-free acid.)

(2) Wash very carefully in distilled water.

(3) Stain with mercury-bromphenol blue or eosin.

(4) Dehydrate, clear, and mount in DPX.

Result. Amyloid alone resists extraction.

APPENDIX 12

EXTRACTIVE METHODS FOR LIPIDS

Baker's Pyridine Extraction Test

Method. (1) Fix in a weak solution of Bouin's fixative (sat. aqueous picric acid 50 ml., commercial formalin 10 ml., glacial acetic acid 5 ml., distilled water 35 ml.) 20 hours.

(2) Wash in alcohol to remove picric acid.

(3) Immerse in pyridine at 17°–22°, 30 minutes.

(4) Immerse in pyridine at 60°, 24 hours.

(5) Wash in running water, 2 hours.

(6) Transfer to the dichromate-calcium mordant of the acid hæmatein method (q.v.).

Result. All lipids should be removed by the pyridine extraction procedure which can be followed by Sudan black B or other colorant methods as well as by the acid hæmatein method.

Keilig's Extractions

Thin, unfixed, blocks of tissues are treated with the four solvents, cold acetone, hot acetone, hot ether and hot chloroform/methanol (equal parts). Three changes of each of the solvents should be used, for 3, 3 and 12 hours respectively. For the hot solvents a Soxhlet extractor may be used, or a simpler apparatus if desired. After extraction, the blocks are washed in water and frozen sections are stained in Sudan black B in 70 per cent. alcohol. Dispersion effects are marked so that localization is inaccurate. Cold acetone, for instance, disperses certain phospholipids from the cells containing them and hot acetone invariably does so. They remain in the section, however, and stain with Sudan black.

BROMINATION TECHNIQUES

(Lillie, 1954)

PARAFFIN SECTIONS

(1) Remove wax with xylene.

(2) Wash in two changes of carbon tetrachloride.

(3) Brominate for one hour in 39 ml. CCl_4 containing 1 ml. of bromine. (Control sections should be exposed to the solvent alone.)

(4) Wash in two changes of carbon tetrachloride.

(5) Transfer to 95 per cent. alcohol, 70 per cent. alcohol and bring to water.

(6) Carry out selected technique.

FROZEN SECTIONS

(1) Attach sections to slides, blot lightly and dry for 10–15 minutes.

(2) Immerse in freshly made bromine water (1 ml. bromine, 39 ml. water) for 1–6 hours.

(3) Wash in water.

(4) Treat with 0·5 per cent. sodium metabisulphite for 1–2 minutes, to remove free bromine.

(5) Wash in running water.

(6) Carry out selected technique.

This method is applicable to studies of solvent-labile fats.

Result. Ethylene groups (C=C, double bonds) no longer react with oxidizing agents to form aldehydes.

SECONDARY FLUORESCENCE METHODS

Aqueous Phosphine 3 R

(Popper, 1944)

(*Formalin or formol-calcium; frozen sections*)

Method. (1) Formalin or formol-calcium fixation is preferable. Frozen sections 10 μ thick are cut.

(2) Stain sections for 3 minutes in 0·1 per cent. aqueous Phosphine 3 R.

(3) Rinse briefly in water.

(4) Mount in 90 per cent. glycerine.

(5) Examine by ultraviolet light.

Result. Fatty acids, soaps and cholesterol negative. Other lipids, including cholesterol esters, give a silvery-white fluorescence.

3:4-Benzpyrene

(Berg, 1951)

(*Formalin or formol-calcium; frozen sections*)

PREPARATION OF THE STAINING SOLUTION

Prepare a saturated solution of caffeine in water at room temperature (about 1·5 per cent.). To 100 ml. of filtered solution add 0·002 g. 3:4-benzyprene and incubate at 37° for 2 days. Filter off excess benzpyrene and dilute with an equal volume of distilled water. Allow to stand for 2 hours and refilter. The solution contains about 0·00075 g./100 ml. and it can be used for some months if stored in the dark in a well-stoppered bottle.

Method. (1) Cut frozen sections at 10 μ and stain for 20 minutes in the benzpyrene solution.

(2) Wash in water and mount the sections on slides.

(3) Examine in water by ultraviolet light.

Result (Fig. 110). Lipids, including the finest lipid granules, give a blue or bluish-white fluorescence. Since this fades relatively rapidly, permanent sections cannot be obtained.

METHODS FOR SUGAR-CONTAINING LIPIDS

Modified Molisch Reaction

(after Diezel, 1954)

(*Formalin; frozen or paraffin sections*)

Method. (1) Apply to the dry section, mounted on a slide and after removal of wax if necessary, 2–3 drops of a mixture containing 0·5 ml. 5 per cent. alcoholic α-naphthol and 10 ml. of 2N-H_2SO_4.*

(2) Place section in an incubator at 90°–100° (without coverslip), or on a hot stage at the same temperature.

Result. Cerebrosides and gangliosides stain violet-red. The colour is not stable and fades to grey after 1 hour.

* Diezel recommends the addition of one drop of some convenient wetting agent (Wasa, Teepol) to the section to ensure proper wetting of the tissues.

Modified Bruckner Reaction

(after Diezel, 1954)

(Formalin; frozen or paraffin sections)

Method. (1) Apply to mounted, dry sections, after removal of wax if necessary, 2–3 drops of a mixture of 1 ml. 2 per cent. aqueous orcinol (3,5-dihydroxytoluene) and 10 ml. 2N-H_2SO_4.

(2) Place section in an incubator or on a hot stage at 90°–100°.

Result. Cerebrosides and gangliosides stain red to bluish red.

Modified Roe-Rice Method for Pentoses

(Formalin; frozen or paraffin sections)

Method. (1) Immerse mounted sections for 5–10 minutes at 70°–85° in acetic-bromoaniline.*

Result. Pentoses stain red within a few minutes.

CAZA METHOD FOR CHOLINE LIPIDS

(After Boelsma-van Houte, 1965)

(Fresh cryostat; postfixed formol-calcium)

Method. (1) Treat sections with 10 per cent. aqueous cobalt chloride for 24 hours at room temperature. The sections must be mounted on quartz† slides.

(2) Rinse thoroughly.

(3) Treat with 1 per cent. aqueous sodium periodate for 1 hour at 37°.

(4) Rinse and take through graded alcohols to absolute ethanol.

(5) Remove ethanol by transferring to toluene, 3 changes.

(6) Transfer to acetic anhydride/toluene‡ (1:1 v/v).

(7) Develop colour by treatment for 30–60 minutes at room temperature with a 2·5 per cent. solution of *cis*-aconitic anhydride in acetic anhydride/toluene (2:3 v/v), which should be prepared 24 hours before use. This reaction must be carried out in quartz† vessels.

(8) Remove excess reagent by rinsing in acetic anhydride/toluene (1:1 v/v).

(9) Clear in toluene and cover with an ordinary glass coverslip.

Result. Choline-containing lipids stain pinkish-red. The colour is stable for only a few hours.

THE ACID HÆMATEIN METHOD

(Baker, 1946)

(Formol-calcium, post-chromation; frozen sections)

PREPARATION OF SOLUTIONS

(1) **Formol Calcium Fixative**

40 per cent. formalin . . .	10 ml.
10 per cent. $CaCl_2$ (anhydrous) aqueous solution . . .	10 ml.
Distilled water	80 ml.

* Saturate 100 ml. of acetic acid with 4 g. of thiourea. Allow sediment to settle and pour off supernatant. Add to this 2 g. bromoaniline. Stored in a dark bottle this solution is stable for about 7 days.

† Normal glassware is contaminated with Lewis bases which interfere with complex formation.

‡ It is essential to remove both water and ethanol completely.

To this is added a piece of chalk to maintain neutral pH.

(2) **Dichromate-calcium**

Potassium dichromate . . .	5 g.
$CaCl_2$ (anhydrous) . . .	1 g.
Distilled water	100 ml.

(3) **Acid Hæmatein.** Place 0·05 g. hæmatoxylin in a flask, add 48 ml. distilled water and exactly 1 ml. of 1 per cent. $NaIO_4$. Heat until the water begins to boil, cool and add 1 ml. of glacial acetic acid. (The author recommended B.D.H. reagent hæmatoxylin.) This solution should be used on the day of preparation.

(4) **Borax-ferricyanide Differentiator**

Potassium ferricyanide . . .	0·25 g.
Sodium tetraborate, 10 H_2O . .	0·25 g.
Distilled water	100 ml.

This solution should be kept in the dark.

Method. (1) Fix small blocks of tissue in formol-calcium, 6–18 hours.
(2) Transfer to dichromate-calcium for 18 hours at 22°.
(3) Transfer to dichromate-calcium for 24 hours at 60°.
(4) Wash well in distilled water.
(5) Cut frozen sections at 10 μ (embed in gelatine if necessary).
(6) Mordant in dichromate-calcium for 1 hour at 60°.
(7) Wash in distilled water.
(8) Transfer to acid hæmatein solution for 5 hours at 37°.
(9) Rinse in distilled water.
(10) Transfer to borax-ferricyanide for 18 hours at 37°.
(11) Wash in water.
(12 Mount in glycerine jelly.

Result (Fig. 112). Nucleoprotein and phospholipins dark blue or blue-black. Mucin dark blue, fibrinogen pale blue. Cytoplasm pale yellow.

N.B. A positive result for phospholipins with this technique should be accompanied by a negative reaction after pyridine extraction. In the absence of such a negative extraction test the dark blue or blue-black staining material is not phospholipin.

SIMPLIFIED ACID HÆMATEIN METHOD

(after Hori, 1963)

(*Formal-calcium or formal-calcium-cadmium; cryostat sections*)

Method. (1) Chromate sections in 5 per cent. $K_2Cr_2O_7$ containing 1 per cent. $CaCl_2$ at 60° for 2–24 hours. (Usually 4 hours.)
(2) Wash in water.
(3) Stain for 5 hours in standard acid hæmatein at 37°.
(4) Differentiate with standard borax-ferricyanide for 18 hours.
(5) Wash in water.
(6) Dehydrate in alcohols, clear, and mount in Canada balsam.

Result. Mitochondrial lipids are especially well seen.

CONTROLLED CHROMATION

(Elftman, 1954)

This method is normally applied to thin blocks of tissue which, after chromation is completed, are dehydrated and embedded in paraffin. These blocks are brittle and sections must usually be cut at 8 μ. Elftman also applied the method to frozen sections, remarking that the inevitable increase in fragility presented a problem. The use of fresh, cold microtome, sections mounted on slides or coverslips entirely overcomes this problem and the sections can be as thin as 2 μ if required.

Method. (1) Treat blocks of fresh tissue (or fresh frozen sections) with 2·5 per cent. potassium dichromate adjusted to pH 3·5 with 0·2 M-acetate buffer. The average time of chromation is 18 hours at 56°–60°.

(2) Wash in water. Sections are ready for subsequent treatment at this stage.

(3) Dehydrate blocks in ascending strengths of alcohol, clear in benzene and embed in paraffin. Cut sections 8–10 μ thick.

(4) Stain pairs of sections, one in Sudan black B in triethyplhosphate (p. 693) and the other in 0·1 per cent. hæmatoxylin in 0·2 M-acetate buffer at pH 3·0, containing 0·005 per cent. potassium ferricyanide (2 hours, 56°).

Result. Chromated lipids are stained by both Sudan black and hæmatoxylin; non-chromated lipids only by Sudan black; non-lipids only by hæmatoxylin.

Many other methods given in Appendix 12 can be applied to sections after controlled chromation. Especially useful is the copper phthalocyanin technique (p. 698).

MODIFIED ACID HÆMATEIN-OIL RED METHOD

(after Bourgeois and Hubbard, 1965)

(*Fresh frozen cryostat sections*)

Method. (1) Fix sections in formal-calcium for 1–2 hours at room temperature

(2) Rinse and mordant for 12–16 hours at 60° in dichromate-calcium.

(3) Wash in three changes of distilled water.

(4) Stain for 2 hours at 37° in freshly prepared acid hæmatein. (50 mg. hæmatoxylin, 1 ml. 1 per cent $NaIO_4$ in 48 ml. water; heat to boiling, cool, add 1 ml. glacial acetic acid.)

(5) Rinse and differentiate in borax-ferricyanide for 1–2 hours at 37°.

(6) Rinse and mount in glycerine; *or* proceed to stage 7. Control sections before stage 1, are treated with pyridine for 2 hours at 60°, washed in water and transferred to mordant.

(7) Rinse and transfer to 60 per cent. triethyl phosphate for 2 minutes.

(8) Stain for 15 minutes at room temperature in 0·5 per cent. Oil Red O in triethyl phosphate.

(9) Dip briefly in 60 per cent. triethyl phosphate.

(10) Wash in distilled water.

(11) Mount in glycerine or P.V.P.

Result. Choline-containing lipids, black. Triglycerides, red.

SUDAN BLACK B STAINING FOR LIPIDS IN PARAFFIN SECTIONS

(after McManus, 1946)

(*Various fixatives; paraffin sections*)

McManus advised fixation of tissues for 1–5 weeks in a solution of 1 g. cobalt nitrate in 80 ml. distilled water with 10 ml. 10 per cent. $CaCl_2$* and 10 ml. commercial (40 per cent.) formalin. Subsequently, as an optional step, he recommended post-chroming for 24–48 hours in 3 per cent. $K_2Cr_2O_7$.

The blocks were dehydrated in 3 changes of acetone, each of ½ hour duration, and placed directly in molten paraffin wax. This procedure seems to have no advantage

* Lillie advised calcium acetate in place of the chloride, to make a final strength of 2 per cent.

over the routine methods of embedding, especially if post-chroming has been carried out.

Method. (1) Bring sections to 70 per cent. alcohol.

(2) Stain for 30 minutes at room temperature in saturated Sudan black B in 70 per cent. alcohol (longer times are advisable in the case of tissues fixed other than in formol-calcium and for tissues which have not been post-chromed. I employ up to 16 hours at 60° for resistant cases).

(3) Remove excess dye by rinsing quickly in 70 per cent. alcohol.

(4) Wash in running water.

(5) Counterstain in Mayer's carmalum 16 hours, or in 1 per cent. aqueous neutral red, 1 minute.

(6) Wash in water and mount in glycerine jelly.

Result. Lipids stain black or blue if present in sufficient quantity. Even a brownish-black stain may be an indication of the presence of lipid or lipoprotein. Nuclei, red.

SUDAN BLACK B METHOD FOR MASKED LIPIDS

(Ackerman, 1952)

(*Blood smears*)

Method. (1) Fix in formalin vapour for 2–5 minutes.

(2) Immerse films in 25 per cent. acetic acid (or 5 per cent. citric, 10 per cent. oxalic or 20 per cent. formic acids) for 2 minutes.

(3) Wash thoroughly in tap water, then in distilled water and allow films to dry.

(4) Stain in saturated Sudan black B in 70 per cent. ethanol (this solution must be at least one week old).

(5) Differentiate in 70 per cent. alcohol.

(6) Blot dry and mount in glycerine jelly.

Result. Bound lipids and lipids surviving the staining procedure stain black.

ACETONE-SUDAN BLACK METHOD FOR BOUND LIPIDS

(after Berenbaum)

(*Formalin, Zenker, Carnoy, etc.; paraffin sections*)

Method. (1) Wash sections overnight in running water after first bringing to water in the usual manner.

(2) Rinse in absolute acetone.

(3) Stain in 2 per cent. Sudan black B in acetone at 37° for 1–24 hours.

(4) Wash in xylene till no further stain comes away (5–10 minutes).

(5) Mount in Canada balsam or in a suitable synthetic medium.

Result. Chromatin, nucleoli and various bound lipids stain black. Neutrophil granules, platelets, amyloid and fibrin also stain.

BURNT SUDAN BLACK METHOD

(after Berenbaum)

This method is carried out as above except that, instead of stage 3, sections are placed on a rack and covered with saturated Sudan black B in 70 per cent. alcohol. This is set alight and allowed to burn off. The process is repeated 6–12 times. Sections are then washed in 70 per cent. alcohol, cleared in xylene and mounted.

Result. As with the acetone-Sudan black method except that red cells, eosinophil granules and myelin sheaths also stain.

TRIETHYLPHOSPHATE METHODS

(Gomori, 1952)

(*Formol calcium-fixed, frozen sections*)

PREPARATION OF STAINING SOLUTIONS

(1) Dissolve 1 g. Oil red O in 100 ml. 60 per cent. triethylphosphate. Heat to 100° for 5 minutes with constant stirring. Filter when hot and again when cool. Filter immediately before use, also.

(2) Dissolve 1 g. Sudan black, as above.

Method. (1) Bring sections to 60 per cent. triethylphosphate.

(2) Transfer to dye solution, 1 or 2, for 10 minutes.

(3) Rinse in 60 per cent. triethylphosphate.

(4) Wash in water.

(5) Stain nuclei in Mayer's hæmalum, 2 minutes.

(6) Wash in running tap water.

(7) Mount in glycerine jelly.

Result. Triglycerides and cholesterol, esters red or black. Cerebrosides and phospholipids, pink or black. Nuclei, blue.

SALT EXTRACTION FOR RESIDUAL LIPOPROTEINS

(after Holczinger and Bálint, 1961)

(*Unfixed cryostat sections*)

Method. (1) Treat with 0·14 M-NaCl at 3° for 30 minutes.

(2) Agitate and pour off saline solution.

(3) Run in 0·4 M-NaCl and treat for 12 hours at 3°.

(4) Proceed, in turn, to 0·7 M-NaCl for 1½ hours, and to 1·0 M-NaCl (at pH 7 and pH 4 alternately) for 4–10 hours.

(5) Rinse in water.

(6) Stain in saturated Sudan black B in 70 per cent. ethanol.

Result. Residual lipoproteins stain black.

THE PHOSPHOMOLYBDIC ACID METHOD

(Landing *et al.*, 1952)

(*Formalin or formol-calcium; frozen sections*)

Method. (1) Place section on a gelatinized slide, drain, blot, and expose to formalin vapour for 10–15 minutes.

(2) Allow the section to dry thoroughly and dip into 50/50 acetone-ether.

(3) Transfer slide to 1 per cent. phosphomolybdic acid in 50/50 ethanol-chloroform for 15 minutes.

(4) Rinse in ethanol-chloroform, then in chloroform and dry.

(5) Dip slide into 1 per cent. aqueous stannous chloride in 3-NHCl (freshly made).

(6) Wash in water.

(7) Counterstain with 1 per cent. aqueous eosin.

(8) Dehydrate, clear and mount in Canada balsam.

Result. Choline-containing lipids stain in molybdenum blue shades. This method can be applied to fresh, cold microtome, sections.

OKAMOTO'S MERCURY DIPHENYLCARBAZONE METHOD

(after Ueda)

(*Formalin; frozen sections*)

Method (so-called Common Method). (1) Cut 6–10 μ sections and place in acetone containing 0·65 per cent. magnesium chloride for 48 hours at 22°.

(2) Rinse thoroughly in water.

(3) Place sections in mercuric nitrate solution at 0°–4° for 24 hours. (Saturated mercuric nitrate in 60 per cent. alcohol, 10 ml.; 0·2 per cent. NaCl, 0·2 ml.).

(4) Wash thoroughly in water.

(5) Treat with 5 per cent. KI for 4–5 minutes. (The colour of the sections changes to reddish-yellow; after 2–3 minutes this colour is discharged.)

(6) Wash in water for 10 minutes.

(7) Treat with 2 per cent. sodium acetate for 10 minutes.

(8) Place in saturated diphenylcarbazone in 90 per cent. alcohol, 10 minutes.

(9) Wash thoroughly in water and mount in glycerine jelly.

Result (Fig. 113). Phospholipids appear bluish-violet. Cerebosides also stain.

The original authors stated that not only phospholipids and cerebrosides were positive but also "some other materials besides lipids." They endeavoured to distinguish between phospholipids and cerebrosides by means of cold pyridine extraction for 48 hours, carried out *after* the staining procedure.

The so-called separation test methods were designed to demonstrate sphingomyelin.

Method A. (1) Rinse sections (formalin-fixed, frozen) in acetone.

(2) Place in ether for 48 hours at room temperature.

(3) Rinse thoroughly in acetone.

(4) Apply stages 2–9 of the common method.

Result. Sphingomyelin only remains positive.

Method B. (1) Apply stages 1–3 of the common method.

(2) Rinse in acetone.

(3) Treat with ether for 48 hours.

(4) Rinse in acetone and wash in water.

(5) Apply stages 5–9 of the common method.

Result. Sphingomyelin and lecithin are positive.

These extractions have been shown by Edgar and Donker (1957) to produce much less clear cut results when tested by chemical procedures than the original authors suggest. They should only be used, with extreme caution in interpretation, as pointers to the development of more satisfactory methods.

OSMIUM TETROXIDE-ALPHA NAPHTHYLAMINE METHOD

(Adams, 1959)

(*Formal-calcium; frozen sections*)

Method. (1) Treat free-floating sections for 18 hour in a mixture of 1 part of 1 per cent. OsO_4 and three parts of 1 per cent. $KClO_3$. Fill container and stopper tightly.

(2) Wash for 10 minutes in distilled water and then mount on slides.

(3) Treat with saturated aqueous α-naphthylamine*† at 37° for 10–20 minutes.

(4) Wash sections in distilled water for 5 minutes.

* Add α-naphthylamine to warm (40°) distilled water and filter.

† Carcinogenic β-naphthylamine may be a contaminant.

(5) Counterstain, if required, in 2 per cent. Alcian blue in 5 per cent. acetic acid, for 15–60 seconds.

(6) Mount in glycerine jelly.

Result (Plate XIa). Phospholipids, orange red. Cholesterol esters and triglycerides black.

NaOH-OTAN FOR SPHINGOMYELIN

(Adams, 1965)

As above method except:

(1) Treat free-floating sections with 2 N-NaOH at 37° for 1 hour.

(2) Gently wash in water, 1 per cent. acetic acid, and again in water.

Result. Sphingomyelin and alkali-resistant lipids, orange-red.

MARCHI METHOD FOR DEGENERATING MYELIN

(after Swank and Davenport)

(*Formalin*)

Method. (1) Fix thin (3 mm.) tissue slices in 10 per cent. formalin for 48 hours.

(2) Transfer to solution containing 1 per cent. aqueous potassium chlorate, 60 ml., 1 per cent. aqueous osmium tetroxide, 20 ml., acetic acid, 1 ml., formalin, 12 ml. (About 15 volumes of this mixture to 1 volume of tissue is adequate.)

(3) Shake every day and turn the tissues over. Keep in the OsO_4 solution for 7–10 days, in the dark, in a vapour-tight container.

(4) Wash in running water for 48 hours.

(5) Dehydrate in 70 per cent. and 95 per cent. alcohol, transfer via absolute alcohol and alcohol/ether (1:1) to 1 per cent. celloidin in alcohol/ether.

(6) Cut sections 30–60 μ thick and mount in Canada balsam.

Glees (1943) recommends in place of stage 5 that 60–90 μ sections should be cut on the freezing microtome and Smith (1956) states that all kinds of artifacts are much less in frozen sections.

Result. Degenerating myelin products black; background yellow or brown.

MODIFIED MARCHI METHOD FOR PARAFFIN SECTIONS

(Wolman, 1957)

(*Alcohol or formalin; paraffin sections*)

Method. (1) Fix small pieces of tissue in 95 per cent. alcohol for 4 days.

(2) Dehydrate in absolute alcohol, clear in xylene and embed in paraffin.

(3) Cut 8 μ sections and transfer to 1 per cent. OsO_4 in carbon tetrachloride for 24 hours.

(4) Wash in three changes of carbon tetrachloride.

(5) Transfer to xylene (two changes).

(6) Mount in a suitable synthetic resin.

Result. Degenerating myelin brown to black.

BROMINE-SILVER METHOD I

(Mukherji *et al.*, 1960)

(*Formalin; frozen sections*)

Method. (1) Wash well in running water.

(2) Blot dry and expose to bromine vapour in a closed jar for 2 hours at 37°. (Saturated aqueous bromine.)*

* As an alternative, freeze-dried blocks may be treated with bromine vapour at 30° for 6 hours, then embedded in carbowax and cut at 8–10 μ before treatment with silver solution.

(3) Treat with 5 per cent. sodium thiosulphate to remove yellow colour.

(4) Wash in several changes of distilled water.

(5) Place in dark coloured glass vessel containing 10 per cent. silver nitrate in 50 per cent. ethanol at 37° until sections turn yellow or faint brown.

(6) Wash in subdued light and reduce in 5 per cent. methyl hydroquinone to the desired depth of colour.

(7) Wash in water and fix in 5 per cent. sodium thiosulphate for 2 minutes.

(8) Wash in distilled water, dehydrate in alcohols, clear in xylene and mount in Canada balsam.

Result. Unsaturated lipids, brown to black.

BROMINE-SILVER METHOD II

(Norton *et al.*, 1962)

(*Formal-calcium; frozen sections*)

Method. (1) Mount sections on slides and dry in air.

(2) Treat with bromine-KBr solution (1 ml. Bromine in 390 ml. 2 per cent. KBr) for 1 minute.

(3) Wash in water.

(4) Rinse in 1 per cent. sodium bisulphite for 5 minutes.

(5) Rinse in several changes of distilled water.

(6) Treat with 1 per cent. $AgNO_3$ in 1 N nitric acid for 18 hours.

(7) Rinse in several changes of distilled water.

(8) Reduce for 10 minutes in Kodak Dektol developer diluted 1:1 with water.

(9) Wash well and mount in glycerine jelly.

Result. Unsaturated lipids, brownish-black or black.

METHODS USING COLORANT DYES IN SIMPLE SOLUTION

Sudan III and IV Method for Neutral Fats

(after Kay and Whitehead, 1941)

(*Formalin; frozen sections*)

PREPARATION OF THE STAINING SOLUTION

Mix equal quantities of dry Sudan III and Sudan IV and place in a clean dry bottle. Fill bottle with Herxheimer's mixture (equal parts of acetone and 70 per cent. alcohol) and shake well. Leave solution for a few days until saturated. For use, pipette off some of the supernatant. Used in a closed vessel to avoid evaporation and precipitation of the stain.

Method. (1) Cut frozen sections at 10 μ into distilled water, or into 1 per cent. formalin if storage is necessary.

(2) Rinse in 70 per cent. alcohol.

(3) Stain for not more than 1 minute in the Sudan mixture, taking care to avoid letting the section fold over.

(4) Rinse in 50 per cent. alcohol to remove excess stain.

(5) Rinse in water, holding section under until all alcohol has diffused out.

(6) Counterstain in dilute (1:4) Ehrlich's hæmatoxylin for 1–3 minutes.

(7) Differentiate, if necessary, in 0·5 per cent. HCl in 50 per cent. alcohol.

(8) Wash in distilled water to which a few drops of ammonia have been added.

(9) Mount in glycerine jelly.

Result. Lipids (neutral fats), orange-red to orange. Nuclei, blue.

The main disadvantages of this method are: (1) there is a tendency to extraction of fine lipid droplets; (2) the colour imparted to the remaining fine lipid droplets is not sufficiently strong; and (3) the dyes have a marked tendency to form crystalline precipitates on storage.

The Oil Red O Method for Neutral Fats

(after Lillie, 1944)

(*Formalin; frozen sections*)

PREPARATION OF THE STAIN

Prepare the stock solution by adding about 0·5 g. Oil Red O to 100 ml. 98 per cent. isopropanol. For use, dilute 6 ml. with 4 ml. of water, stand for 24 hours and filter. Use this filtrate as a stock staining solution, filtering through Whatman No. 42 paper, as necessary, a sufficient amount for immediate use.

Lillie (1945a and b) also recommended the dyes Oil Blue N and Coccinel Red for fat staining from isopropanol solutions.

Method. (1) Stain frozen sections, after rinsing in water and then in 60 per cent. *iso*propanol, in freshly filtered Oil Red solution for 10 minutes.

(2) Differentiate briefly in 60 per cent. isopropanol (keep tightly stoppered or make up freshly).

(3) Wash in water.

(4) Stain for 5 minutes in Mayer's hæmalum.

(5) Wash in running water for at least 10 minutes.

(6) Mount in glycerine jelly or gum syrup.

Result. Lipids, red. Nuclei, blue.

This method has two advantages over the Sudan III and IV method. First, the colour is deeper and the staining of smaller droplets can therefore be appreciated; secondly, there is less tendency to formation of dye precipitates.

The Fettrot Method for Neutral Fat

(*Formalin; frozen sections*)

Method. (1) Cut frozen sections 10–15 μ, mount on egg-albuminized slides and dry thoroughly in air to ensure adherence.

(2) Stain in Meyer's hæmalum 4–6 minutes.

(3) Wash in running water for 30 minutes.

(4) Rinse quickly in 50 per cent. alcohol.

(5) Stain in a saturated solution of Fettrot 7B (Ciba) 10–15 minutes.

(6) Wash in water.

(7) Mount in glycerine jelly.

Result. Lipids, pinkish-red. Nuclei, dark blue.

Fettrot 7B or Technical Fettrot are available from Ciba Ltd. This method gives very beautiful and permanent results, since unlike the red Sudan dyes and Oil Red O, Fettrot has no tendency to crystallize when used from alcoholic solutions, and mounted specimens retain their clarity over a long period.

The Nile Blue* Method for Neutral and Acidic Lipids

(after Cain, 1947)

(*Formol-calcium; frozen sections*)

Method. (1) Fix small blocks of tissue in Baker's formol-calcium fixative.

(2) Treat with dichromate-calcium (Baker) in the two stages of the acid hæmatein method. This step is optional.

(3) Cut frozen sections at 10 μ.

* The term Nile Blue is used to mean the commercial dye which is a mixture of the oxazine sulphate (true Nile Blue) and the oxazone (Nile Red).

(4) Stain one section (A) in saturated Sudan black B in 70 per cent. alcohol as a control for lipids.

(5) Stain one section (B) in 1 per cent. aqueous Nile Blue at 60°, 5 minutes.

(6) Wash quickly in water at 60°.

(7) Differentiate in 1 per cent. acetic acid at 60°, 30 seconds.

(8) Stain another section (C) as section (B) and restain in 0·02 per cent. Nile Blue at 60°.

(9) Wash and differentiate section (C) as section (B) (stages 6 and 7).

(10) Mount all sections in glycerine jelly.

Comment. Only purely lipid inclusions can be considered. An inclusion can be established with certainty as purely lipid only by staining a pyridine-extracted control section with Sudan black B and comparing the result with formol-calcium-fixed, unextracted, sections stained in the same way. If there is no difference between sections (B) and (C) the first may be discarded, since what will stain with 1 per cent. Nile Blue will stain with 0·02 per cent.

Result. Neutral lipids, red or pink. Acidic lipids, blue.

NILE BLUE METHOD FOR PHOSPHOLIPIDS

(Menschik, 1953)

(*Formol-calcium; frozen sections*)

Method. (1) Fix for 6–12 hours and cut frozen sections without embedding in gelatin or carbowax.

(2) Stain for 90 minutes at 60° in saturated aqueous Nile blue sulphate 500 ml. with 50 ml. 0·5 per cent. H_2SO_4 (boil for 2 hours before use).

(3) Rinse in distilled water.

(4) Place in acetone heated to 50°.

(5) Remove acetone from source of heat and allow sections to remain in it for 30 minutes.

(6) Differentiate in 5 per cent. acetic acid 30 minutes.

(7) Rinse in distilled water.

(8) Differentiate again in 0·5 per cent. HCl for 3 minutes.

(9) Wash in distilled water and mount in glycerine jelly.

Result. Phospholipids blue.

THE COPPER PHTHALOCYANIN METHOD

(after Klüver and Barrera, 1953)

(*Formalin or formol-calcium; frozen sections; various; paraffin sections*)

Method. (1) Bring sections to absolute alcohol.

(2) Stain in 0·1 per cent. Luxol Fast blue MBS, Luxol Fast blue G (Matheson, Coleman), (Dupont) or Methasol Fast blue 2 G (I.C.I. Ltd) in 95 per cent. alcohol for 6–18 hours at 56°–60°.

(3) Rinse in 70 per cent. alcohol and wash in water.

(4) Differentiate in 0·05 per cent. aqueous lithium carbonate for $\frac{1}{2}$–2 hours (90 minutes for formalin-fixed frozen sections).

(5) Rinse in water.

(6) Counterstain if required either in Mayer's carmalum 10–30 minutes or in 1 per cent. aqueous Neutral red (1–30 minutes).

(7) Rinse in water and dehydrate in 70 and 95 per cent. alcohol.

(5) Clear in xylene and mount in a suitable synthetic resin.

Result (Figs. 114 and 115). Phospholipids except sphingomyelin stain blue. So do gangliosides and probably cerebrosides.

Sphingomyelin can be stained by substituting chloroform for ethanol as the dye solvent (stage 2) but many other tissue components also stain if this is done.

The method is suitable for application to fresh, cold microtome, sections, treated by Elftman's controlled chromation (p. 423).

METHOD USING COLORANT DYES IN COLLOIDAL SUSPENSION

The Gelatine-Sudan Method

(after Govan, 1944)

(*Formalin; frozen sections*)

PREPARATION OF THE SUSPENSION

Make a saturated solution of Sudan III and Sudan IV in acetone and add, drop by drop, to a 1 per cent. aqueous gelatine solution containing 1 per cent. acetic acid until a brick-red milky fluid is obtained after constant stirring. Place the mixture in the 37° incubator to evaporate the acetone (2 hours). Filter through coarse paper.

Method. (1) Cut frozen sections at 10 μ and transfer to 1 per cent. gelatine 2–3 minutes.

(2) Stain for 30 minutes in the suspension.

(3) Wash in 1 per cent. gelatine for 2–3 minutes.

(4) Stain in Mayer's hæmalum 4–6 minutes.

(5) Blue in water with added ammonia.

(6) Mount in glycerine jelly or in Karo corn syrup.

Result (Fig. 111). Lipids, including the smallest droplets, orange-red. Nuclei, blue.

FISCHLER'S METHOD FOR FATTY ACIDS

(*Formalin; frozen sections*)

PREPARATION OF SOLUTIONS

(1) **Weigert's Lithium-hæmatoxylin.** *Solution A*. Ten per cent. hæmatoxylin in absolute alcohol.

Solution B. Sat. aqueous lithium carbonate 10 ml., distilled water 90 ml.

For use mix one part each of solutions A and B.

(2) **Weigert's Borax-ferricyanide**

Borax ($Na_2B_4O_7$, $10H_2O$) . .	20 g.
Potassium ferricyanide . . .	25 g.
Distilled water . . .	1,000 ml.

Method. (1) Mordant 10 μ frozen sections for 24 hours at 37° in saturated aqueous cupric acetate.

(2) Wash in distilled water.

(3) Stain for 20 minutes in Weigert's lithium-hæmatoxylin.

(4) Differentiate in borax-ferricyanide until the red cells are no longer dark blue.

(5) Wash in distilled water and mount in glycerine jelly.

Control sections should be stained which have previously been extracted for 24 hours with 3 changes of alcohol/ether (equal parts).

Result. Tissue components staining dark blue in the unextracted and colourless in the extracted sections are presumed to contain fatty acids. For interferences see text.

HOLCZINGER'S FATTY ACID TECHNIQUE

(*Unfixed cryostat or formalin; frozen*)

Method. (1) Treat sections with 0·005 per cent. aqueous copper acetate for 3–5 hours at room temperature.

(2) Wash twice, for 10 seconds, with 0·1 per cent. disodium—EDTA (pH 7·1).
(3) Wash twice in distilled water.
(4) Treat with 0·1 per cent. rubeanic acid in 70 per cent. alcohol for 30 minutes. (Dissolve rubeanic acid in ethanol, warm slightly, dilute with distilled water.)
(5) Wash for a few minutes in 70 per cent. ethanol.
(6) Wash in water.
(7) Mount in glycerin jelly.

Result. Fatty acids stain greenish-black.

OKAMOTO'S METHOD FOR FATTY ACIDS

(after Ueda)

(10 *per cent. formol-saline; frozen sections*)

Method. (1) The fixative should contain at least 20 per cent. NaCl.
(2) Cut frozen sections and wash in 20 per cent. NaCl.
(3) Transfer sections to copper solution for 24 hours (saturated aqueous $Cu(NO_3)_2 . H_2O$, 0·5 ml., distilled water 100 ml.).
(4) Alternatively, transfer to an alcoholic copper solution (sat. aq. $Cu(NO_3)_2 . H_2O$, 0·2 ml., 75 per cent. ethanol, 100 ml.) for 2–3 hours.
(5) Wash thoroughly in water.
(6) Treat with alcoholic *p*-dimethylaminobenzylidene rhodanine* for 24 hours at 58°.
(7) Washing in water.
(8) Mount in glycerine jelly.

Result. Fatty acids and their alkaline earth salts stain reddish-brown.

FEYRTER'S "MOUNTING-STAINING" METHOD FOR ACIDIC LIPIDS

(*Formalin; frozen or cold microtome sections*)

PREPARATION OF STAINING SOLUTIONS

Dissolve 1 g. of either thonin or cresyl violet in 0·5 per cent. aqueous tartaric acid. Not all samples of these dyes given equally good staining.

Method. (1) Cut frozen sections (10–15 μ) and wash these thoroughly in water until free from all but closely bound formalin.
(2) Mount on clean slides from distilled water and blot dry.
(3) Cover with staining solution and apply a coverslip.
(4) Remove excess stain with filter paper until the edges of the coverslip are dry
(5) Seal the coverslip with molten paraffin wax.
(6) Allow to stand at room temperature and examine from time to time.

(If rapid results are required, ring the coverslip with petroleum jelly and incubate the slide at 60° for 15–30 minutes.)

These preparations, sealed with paraffin wax, are stable for years.

Result. A fine red or pink metachromasia is given by epithelial mucins containing acid mucopolysaccharide and by connective tissue mucin and mast cell granules. Myelin sheaths are bright rose-red in colour and phosphatides and cerebrosides are a similar, usually less intense, rose-red. This method must always be controlled by the use of other tests, Baker's acid hæmatein and the Sudan black methods for instance.

* Saturated alcoholic *p*-dimethylaminobenzylidene rhodanine 5 ml., absolute ethanol containing 2 per cent. sodium acetate, 95 ml.

HEPARIN PRECIPITATION METHOD

(after Mustakallio and Levonen, 1964)

(*Unfixed cryostat sections*)

Method. (1) Mount 10 μ sections on coverslips.

(2) Carry through three sections: (*a*) dried in air, (*b*) treated for 20 minutes with absolute acetone at room temperature, (*c*) treated with 2:1 boiling $CHCl_3$/methanol for 20 minutes.

(3) Treat for 1 hour with the following solution: 0·01 M-$CaCl_2$, 50 ml.; heparin 5 mg. (500 IU). Omit either calcium or heparin for controls.

(4) Wash in 2 changes of 0·01 M-$CaCl_2$.

(5) Stain for 15 minutes in 0·1 per cent. toluidine blue in 15 per cent. ethanol.

(6) Dehydrate, clear and mount in synthetic resin.

Result. Possible phospholipoproteins are shown by heparin metachromasia (γ-metachromasia).

THE U-V SCHIFF METHOD

(after Belt and Hayes)

(*Cold formalin; frozen sections. Cold microtome*)

Method. (1) Fix sections for 12–18 hours in cold 10 per cent. neutral formalin.

(2) Cut sections and place under a sourse of ultraviolet light for 2–4 hours.

(3) Treat with Schiff's reagent for 15 minutes.

(4) Rinse in three changes of sulphurous acid water.

(5) Rinse in distilled water.

(6) Mount in glycerine jelly.

Result. A magenta-red colour absent from unirradiated control sections indicates unsaturated lipids.

This method produces less damage to tissues than the performic acid-Schiff method given below.

THE PERFORMIC ACID-SCHIFF METHOD FOR LIPIDS CONTAINING UNSATURATED BONDS

(Lillie, 1952)

(*Formalin, Helly, Zenker; paraffin sections. Formalin; frozen sections*)

PREPARATION OF SOLUTIONS

(1) **Performic Acid.** Add 4·5 ml. H_2O_2 (30 per cent or 100 vol.) and 0·5 ml.conc. H_2SO_4 to 45 ml. of 98 per cent. formic acid. Stand for 1 hour. Stir well before use. Use for 24 hours only. Alternative use Lillie's prescription (below).

(2) **Performic Acid (Lillie).** To 9·2 g. 90 per cent. formic acid (8 ml.) add 33·7 g. 30 per cent. H_2O_2 (31 ml.) and 0·4 g. conc. H_2SO_4 (0·22 ml.). Keep temperature below 25°. 4·7 per cent. performic acid is formed within 2 hours and the solution deteriorates after 2 hours.

(3) **Peracetic Acid.** This reagent (40 per cent.) is available commercially. A similar solution may be made (Pearse, 1951) by adding 20 ml. of acetic anhydride to 5 ml. of 30 per cent. H_2O_2. After thorough mixing, and standing for 24 hours, the solution is diluted to twice its volume before use.

(4) **Schiff's Reagent.** Any except the thionyl chloride modification may be used.

Method. (1) Bring sections to water, removing mercury precipitate if necessary.

(2) Blot dry.

(3) Treat with performic or peracetic acids for 1–2 hours.

(4) Wash in running water for 10 minutes.
(5) Immerse in Schiff's solution for 10 minutes.
(6) Wash in 3 changes of sulphite water for 2 minutes each.
(7) Wash in warm running water for 10 minutes.
(8) Mount in glycerine jelly without counterstaining.

Lillie recommends counterstaining in Weigert's acid hæmatoxylin and in saturated aqueous picric acid, dehydration in alcohol, clearing in xylene and mounting in a synthetic medium. He particularly recommends the nuclear counterstain for covering up the Feulgen-stained nuclei. This practice is not recommendable histochemically. The increased basophilia of the tissues after performic acid oxidation (see Chapter 6, p. 151) causes basic dyes to overlay a number of structures in which a positive PFAS reaction is visible. The red compound formed in the tissues is partially soluble in alcohol and dehydration is therefore best avoided.

Result. Structures which stain red may be regarded as containing lipids (phospholipid or cerebroside) with unsaturated bonds.

Lillie regards the reaction as specific for the ethylene group HC=CH and recommends controlling this point by means of control sections treated with bromine (directions on p. 696). The ethylene groups are blocked by this procedure and the PFAS reaction becomes negative. The PAS reaction, however, where it is due to the 1:2-glycol group, remains positive and is reversible by acetylation (Appendix 10, p. 660).

METHODS FOR CHOLESTEROL AND ITS ESTERS

Schultz Method

(after Weber *et al.*)

(*Cold calcium-cadmium-formalin; frozen sections*)

Method. (1) Cut frozen sections at 20–30 μ.

(2) Wash for 24 hours in distilled water (several changes).

(3) Treat with 2·5 per cent. ferric ammonium sulphate in 0·2 M-acetate buffer at 37° for 7 days. It is recommended that the buffer should be adjusted to pH 3 by mixing 2 ml. of 0·2 M-sodium acetate with 98 ml. of 0·2 M-acetic acid. The final pH is about 2.

(4) Wash sections for 1 hour each in 3 changes of acetate buffer.

(5) Rinse in distilled water, transfer to 5 per cent. formalin for 10 minutes.

(6) Mount sections on slides and remove excess water by blotting the edges. Do not dry the sections.

(7) Place one drop of a mixture of equal parts of sulphuric and acetic acids on a coverslip. Invert the slide and apply it to the coverslip. Turn right way up and apply even pressure to the coverslip so as to flatten the section. Grasp the coverslip at the corners and oscillate several times.

Result. Cholesterol and cholesterol esters appear pale violet or red, turning rapidly to green within a few seconds. The colour remains stable for 30–60 minutes.

Okamoto Method

(after Ueda)

(*Formalin; frozen sections*)

Method. (1) Mount the section on a slide and blot thoroughly.

(2) Cover with 2 drops of freshly made sulphuric-iodine (conc. H_2SO_4, 10–12 ml., alcoholic iodine,* 20 ml.).

* Iodine 3·5 g.
KI 2·5 g.
95 per cent. alcohol 180 ml.

(3) Apply a coverslip and examine under the microscope.

Result. Cholesterol and its esters give a blue-green colour stable at most for 2–3 hours.

Bismuth Trichloride Method

(Grundland, Bulliard and Maillet, 1949)

PREPARATION OF THE REAGENT

Bismuth trichloride	0·2 g.
Acetyl chloride	1·0 ml.
Anhydrous nitrobenzene .	100·0 ml.

Method. (1) Fix small pieces of tissue for 24–48 hours in 70 per cent. alcohol saturated with digitonin.

(2) Allow the alcohol to evaporate and immerse in molten paraffin wax containing 5 per cent. glyceryl monostearate for 12 hours.

(3) Embed in paraffin.

(4) Cut sections (6–10 μ) and mount on slides. Do not remove wax.

(5) Expose to the bismuth trichloride reagent for 15–45 minutes.

(6) Rinse rapidly in 10 per cent. acetyl chloride in nitrobenzene.

(7) Clear and remove wax in benzene.

(8) Mount in liquid paraffin.

Result. Cholesterol appears brown, its esters are not revealed. (The preparations are stable for a long period.)

Further treatment can be employed to accentuate the colour of the reaction. For this purpose proceed thus after stage 6:

(1) Wash in absolute alcohol/conc. nitric acid (1:3) to remove excess bismuth.

(2) Wash rapidly in absolute alcohol.

(3) Immerse in a dilute solution of yellow ammonium sulphide.

(4) Wash in absolute alcohol.

(5) Clear in xylene and mount in Canada balsam.

DIFFERENTIATION OF CHOLESTEROL FROM ITS ESTERS

(after Feigin, 1956; and Schnabel, 1964)

Method I. (1) Immerse formalin-fixed frozen sections in a 0·5 per cent. solution of digitonin in 40 per cent. ethanol for 3 hours at room temperature.

(2) Drain and immerse in a mixture of equal parts of absolute ethanol and ether for 3 hours at room temperature.

(3) Drain and immerse, together with an untreated control section, in 2·5 per cent. aqueous iron alum at 37° for 2–4 days.

(4) Drain dry and carry out stage 7 of the Schultz method.

Method II. (1) Treat sections for 3–4 days in 2·5 per cent. aqueous iron alum solution.

(2) Wash in water.

(3) Treat with 0·5 per cent. digitonin in 40 per cent. ethanol for 3 hours at room temperature.

(4) Extract for 1 hour with absolute acetone at room temperature.

(5) Carry out stage 7 of the Schultz method.

Result. I. Cholesterol esters are extracted by the alcohol-ether mixture and therefore stain only in the control section. The free cholesterol-digitonin complex is not extracted and therefore stains in both test and control sections.

Result. II. Only cholesterol esters are extracted by the acetone. Free cholesterol-digitonin complex alone is stained.

PERCHLORIC ACID-NAPHTHOQUINONE METHOD

(after Adams, 1961)

(*Formal-calcium; frozen sections*)

Method. (1) Cut sections and leave in formalin for 1 or more weeks, to allow oxidation of cholesterol to occur.

(2) Mount on slides and dry in air.

(3) Paint with the following reagent: 0·1 per cent. 1,2-naphthoquinone-4-sulphonic acid in ethanol-perchloric acid (60 per cent.)-formaldehyde (40 per cent.)-water (2:1:0·1:0·9 v/v).

(4) Heat slides on hotplate at 60°–70° for 5–10 minutes (until original red colour turns dark blue).

(5) Place a drop of 60 per cent. perchloric acid on the section and apply coverslip.

Result (Fig. 116). Cholesterol, demosterol and cholesterol esters, dark blue. (The colour is not stable in water or glycerine jelly.)

MODIFIED HALE METHOD SUITABLE FOR NERVOUS TISSUES

(Seitelberger, Vogel and Stepan, 1957)

Method. (1) Mount formalin-fixed frozen sections on slides. Bring these, or paraffin sections to 50 per cent. alcohol.

(2) Stain for 10 minutes in Sudan IV or Herxheimer's solution.

(3) Pass rapidly through 70 per cent. alcohol and wash twice in distilled water.

(4) Transfer for 2 minutes to the colloidal iron solution (see p. 671).

(5) Wash in distilled water.

(6) Treat for 10 minutes with acid ferrocyanide (1 per cent. HCl, 2 parts; 2 per cent. potassium ferrocyanide, 1 part).

(7) Wash twice in distilled water.

(8) Fix in alum solution for ten minutes. (Aluminium ammonium sulphate 24 H_2O, 26 g.; 28 per cent. ammonia 5 ml.; distilled water 450 ml.)

(9) Wash in distilled water.

(10) Mount in glycerine jelly.

Result Acidic polysaccharides, free or bound, stain deep blue. Neutral fats red.

APPENDIX 13

REACTIONS FOR ALDEHYDES

THE method of using Schiff's reagent for this purpose does not differ from its use in the Feulgen reaction, already described in Chapter 9 and in Appendix 9, p. 647.

Ammoniacal Silver Reaction

Gomori's hexamine-silver solution is recommended.

PREPARATION OF THE SOLUTION

Add 5 ml. 5 per cent. silver nitrate to 100 ml. 3 per cent. aqueous hexamine. A precipitate forms which redissolves. Add 5 ml. of borate buffer (approximately pH 8) made by adding a drop of alcoholic phenolphthalein to 3 per cent. boric acid and then adding 1 N-NaOH until a pink colour appears. Finally, make up to 200 ml. with distilled water.

Method. (1) Bring paraffin sections to water.

(2) Oxidize sections (for instance with 5 per cent. chromic acid for 1 hour, or with 1 per cent. periodic acid for 5 minutes).

(3) Wash in running water for 10 minutes.

(4) Immerse in the hexamine-silver solution for 1–3 hours at 37°.

(5) Wash in distilled water.

(6) Rinse in 5 per cent. sodium thiosulphate, 30 seconds.

(7) Wash and counterstain nuclei, if desired.

Result (Fig. 118). Aldehyde groups are revealed by a brown or black deposit of silver.

The Naphthoic Acid Hydrazide Reaction

(after Ashbel and Seligman)

The synthesis of this reagent is carried out according to details given in Appendix 9, p. 648. It can be applied to tissues fixed in any kind of fixative, and to frozen or paraffin sections, after the application of any reagent known to produce aldehyde groups. Besides these the reaction is given by sulphonic and sulphinic acids (see Chapter 6, p. 151) and by ketones if these are present in sufficient amounts.

PREPARATION OF SOLUTIONS

(1) 0·1 per cent. 2-hydroxy-3-naphthoic acid hydrazide in 50 per cent. reagent ethanol containing 5 per cent. acetic acid. Solution is effected by warming to 70° and shaking vigorously.

(2) Alcoholic phosphate buffer (pH 7·2–7·5) prepared by mixing equal quantities of 0·1 M buffer and absolute reagent ethanol.

Method. (1) If frozen sections of formalin-fixed tissues are used, formaldehyde may still be present. This may be removed either by thorough washing or, preferably, by blocking treatment with hydroxylamine (p. 706). In the case of paraffin sections such reatment is unnecessary.

(2) Immerse sections in the fresh hydrazine solution at 60° for 1–3 hours.

(3) Wash in three changes of 50 per cent. ethanol with 5 per cent. acetic acid.

(4) Wash in three changes of 50 per cent. ethanol.

(5) Wash in distilled water.

(6) Transfer to the alcoholic buffer solution in a Coplin jar.

(7) Add either 5–10 ml. of freshly diazotized *o*-dianisidine or 50 mg. Fast blue B salt (I.C.I.) in 0·5 ml. distilled water. Agitate the sections to effect mixture. Leave the sections for 2–5 minutes.

(8) Wash in running water.

(9) Mount frozen sections in glycerine jelly, paraffin sections in DPX or Canada balsam after rapid dehydration in alcohol and clearing in xylene.

Result (Figs. 119 and 122). Tissue components in which aldehydes are present, or in which they have been produced, bluish to reddish-purple. Various cytoplasmic proteins and lipids may be coloured pink or red, only blue shades constitute a positive reaction.

BLOCKING REAGENTS FOR ALDEHYDES AND KETONES

(*Various*)

(1) **Bisulphite**

Add 10 ml. ethyl alcohol to 40 ml. 50 per cent. aqueous sodium bisulphite. Expose tissues for 2–8 hours at room temperature.

(2) **Phenylhydrazine**

Phenylhydrazine	5 ml.
Glacial acetic acid . . .	10 ml.
Distilled water	to 50 ml.

Expose tissues for 2–3 hours at 60°.

(3) **2:4-Dinitrophenylhydrazine**

A saturated solution in 1 N-HCl at 0°. Expose tissues for 20 hours at 0°, wash in cold 1 N-HCl and then in water.

(4) **Semicarbazide**

Semicarbazide hydrochloride . .	2 g.
Sodium acetate (crystalline) . .	5 g.
Distilled water	40 ml.

Expose tissues for 2–3 hours at 60°.

(5) **Thiosemicarbazide**

Thiosemicarbazide	1 g.
Glacial acetic acid	2 ml.
Distilled water	40 ml.

Expose tissues for 2–3 hours at 60°.

(6) **Hydroxylamine**

Hydroxylamine hydrochloride . .	10 g.
Sodium acetate (crystalline) . .	20 g.
Distilled water	40 ml.

Expose tissues for 1–3 hours at 22°.

(7) **Amine-aldehyde Condensation**

Sulphanilic acid	50 mg.
0·1 M-KH_2PO_4	40 ml.

Incubate sections for 24 hours at 37°.

(8) **Amine-aldehyde Condensation** (Lillie, 1954)

Aniline	9 ml.
Conc. HCl	8 ml.

Shake during addition and dilute with

Distilled water	100 ml.

Incubate sections for 1–6 hours at 22°.

Amine-aldehyde Condensation (Lillie and Glenner, 1957)

Aniline	10 ml.
Acetic acid	90 ml.

Incubate sections for 20–30 minutes at 22°.

(9) **Dimedone** (Fig. 120)

A saturated solution in 5 per cent. acetic acid. Treat sections for 1–16 hours at 60° or for 2–3 days at 22°.

(10) **Cannizzaro Reaction**

Treat blocks of tissue with reagents necessary to produce aldehyde groups (HCl, CrO_3, HIO_4, etc.). Treat with 50 per cent. KOH in 50 per cent. ethanol for 1–4 hours, and wash thoroughly in water.

(11) **Meerwein-Ponndorf Reaction**

Treat blocks of tissue as for the Cannizzaro reaction to produce aldehydes. Heat for 2–4 hours under a reflux condenser in isopropanol containing excess aluminium isopropoxide.

(12) **Borohydride Reduction**

Dissolve 5 mg. $NaBH_4$ in 10 ml. 0·2 M borate buffer (pH 7·6). Treat sections of tissue for 1 hour at room temperature.

THE PLASMAL REACTION

(after Hayes, 1949)

(*Unfixed, cold microtome sections or smears*)

Method. (1) Cut frozen sections 10–15 μ. Mount on coverslips or slides as desired.

(2) Wash in several changes of distilled water.

(3) Place one section in 1 per cent. aqueous $HgCl_2$ for 2–10 minutes.

(4) Place this section and an untreated control section in Schiff's reagent for 5–15 minutes.

(5) Wash in three changes of bisulphite water (5 ml. 10 per cent. $K_2S_2O_5$, 5 ml. 1 N-HCl, 100 ml. distilled water), 2 minutes in each change.

(6) Wash in water.

(7) Mount in glycerine jelly, or in DPX, after mounting the section on a slide, dehydrating in gradual alcohols and clearing in xylene.

(Glass section lifters must be used throughout.)

Result. Acetal lipids, reddish-purple.

THE PLASMAL REACTION

(after Cain, 1949)

(*Fresh tissues*)

Method. (1) Drop pieces of fresh tissue into a mixture of equal parts of (A) Schiff's solution half diluted with bisulphite water and (B) saturated (7 per cent.) aqueous $HgCl_2$. Leave for 15 minutes.

(2) Wash in several changes of bisulphite water.

(3) Treat with 10 per cent. formalin for 2 hours.

(4) Wash in running water for 2 hours or until control pieces become pale lilac.

(5) Cut frozen sections.

(6) Mount on slides and enclose in glycerine jelly or dehydrate, clear and mount in DPX.

Cain used control sections stained with Sudan black B to indicate that the plasmal formed was derived from lipid sources.

Result. Acetal lipids, reddish-purple.

PHOSPHOLIPIDS BY THE PSEUDO-PLASMAL REACTION

(*Cold formalin; frozen sections*)

The plasmal reaction may be used, on frozen sections of tissues fixed in formalin for 4–16 hours at 4°, as an indication of the presence of associated phospholipids.

Method. (1) Treat a number of sections with hydroxylamine (see Blocking Reagents, above) for 1–3 hours at 22°.

(2) Wash the sections in three changes of distilled water.

(3) Treat some of the sections with 1 per cent. $HgCl_2$ for 10 minutes.

(4) Wash in distilled water.

(5) Immerse treated sections, and untreated control sections, in Schiff's reagent for 15–20 minutes.

(6) Wash in bisulphite water, 10 minutes.

(7) Wash in water.

(8) Mount in glycerine jelly.

Result. Control hydroxylamine-treated sections unexposed to $HgCl_2$ should be absolutely colourless. Lipids containing acetals and easily-revealed aldehyde groups appear reddish-purple in sections treated with $HgCl_2$.

METHOD FOR α-KETOL GROUPS OF CORTICOIDS

(after Khanolkar *et al.*, 1958)

(*Fresh, cold microtome, sections; mounted on slides or coverslips*)

This method is based on the oxidation of α-ketols to keto-aldehydes by ferric chloride. Its theoretical specificity is high but *in vitro* tests fail to support the authors' views.

Method. (1) Block existing aldehydes by immersing sections in Lillie's acetic-aniline (p. 707) for 20 minutes at room temperature.

(2) Rinse gently in distilled water for 1–3 minutes.

(3) Incubate in 5 per cent. ferric chloride at 50° for 30 minutes.

(4) Rinse gently in distilled water.

(5) Immerse in Schiff's reagent for 20 minutes.

(6) Wash in 3 changes of bisulphite water (see above).

(7) Mount in glycerine jelly.

Result. Corticoids are revealed by magenta staining. Plasmalogens stain similarly.

Androgens, œstrogens, pregnandiol and cholesterol do not react.

HYDRAZONE REGENERATION METHOD FOR KETOSTEROIDS

(after Stoward and Adams-Smith, 1964)

(*Formalin-hydroquinone; carbowax embedding*)

Method. (1) Fix thin fresh tissue samples in neutral 10 per cent. formalin containing hydroquinone, overnight at 5°.

(2) Wash in running water for 10 minutes.

(3) Embed in carbowax and cut 5–10 μ sections on the freezing microtome.

(4) Treat mounted sections on slides for 8–10 hours at 22° with fresh 1 per cent. methyl hydrazine (sulphate) whose pH has been adjusted to 4–6 with dilute NaOH.

(5) Wash sections in 1 per cent. acetic acid and then in two changes of distilled water.

(6) Immerse in 5 per cent. aqueous sulphobenzaldehyde, for 2 hours at room temperature.

(7) Wash in distilled water.

(8) Treat for 20 minutes with 0·5 per cent. salicyloyl hydrazide in 5 per cent acetic acid.

(9) Rinse successively in (*a*) water, (*b*) fresh 1 per cent. (approximately) trisodium pentacyanoammine ferroate,* 5 minutes, and (*c*) water.

(10) Mount in water and examine under UVL (exciting light 366 nm.).

Result. Ketosteroids fluoresce greenish-yellow.

PHENYLENEDIAMINE REACTION FOR ALDEHYDES

(after Scarselli, 1961)

(*Various fixatives; paraffin sections*)

REAGENTS

(1) Stock *p*-phenylenediamine, 1 per cent. aqueous. Made up in boiled distilled water and kept, well-stoppered, in the dark.

(2) 1·5 per cent. H_2O_2.

(3) 1 per cent. gold chloride.

Method. (1) Bring sections to water.

(2) Apply selected aldehyde-generating reaction.

(3) Wash in distilled water.

(4) Mix equal parts of reagents 1 and 2 and treat sections for 15 minutes at room temperature, in the dark.

(5) Wash thoroughly in distilled water.

(6) Apply reagent 3 for 5 minutes.

(7) Wash in water, dehydrate, clear and mount in synthetic resin.

Result. Aldehydes stain bluish-violet.

*(Feigl, Anger and Frehdeu, 1934)

APPENDIX 15

THE CALCIUM-COBALT METHOD FOR ALKALINE PHOSPHATASE

(after Gomori)

(*Cold acetone, paraffin sections; cold formalin, frozen sections*)

Paraffin Sections

THIN blocks of tissue should be fixed, according to Gomori's (1946) method, in two or three changes of cold absolute acetone at 4° for 24 hours. There are many subsequent ways in which the tissues may be embedding in paraffin wax, most of them being minor variations designed to minimize destruction of the enzyme. As a routine procedure the following method gives good results.

(1) Transfer the blocks progressively at half-hourly intervals to absolute ethanol, absolute ethanol-ether with one or two changes, and thence to 1 per cent. celloidin.

(2) Drain off excess celloidin and harden in chloroform.

(3) Clear in benzene.

(4) Embed in paraffin wax, avoiding prolonged exposure to the high temperature of the wax bath.

(5) Cut sections at 5 μ and mount on albuminized slides.

(6) Dry the slides for 3 hours at 37°.

(7) Store at 4° until required for incubation.

Method Proper. (1) Remove wax from the slides by brief immersion in light petroleum.

(2) Pass to water via absolute acetone.

(3) Incubate for ½–16 hours, at 37°, in the following medium:

(10 ml. 3 per cent. sodium β-glycerophosphate.
10 ml. 2 per cent. sodium diethyl barbiturate.
5 ml. distilled water.
20 ml. 2 per cent. calcium chloride.
1 ml. 5 per cent. magnesium sulphate.

(4) Rinse in running water.

(5) Treat with 2 per cent. cobalt nitrate or acetate, 3–5 minutes.

(6) Rinse in distilled water.

(7) Treat with a dilute solution of yellow ammonium sulphide, 1–2 minutes.

(8) Wash in water, counterstain in 1 per cent. eosin, 5 minutes, if desired.

(9) Dehydrate in alcohol, clear in xylene and mount in Canada balsam.

Result (Figs. 127 and 132). Various structures are stained black or brownish black in tissues possessing alkaline phosphatase activity.

Frozen Sections

Method. (1) Cut sections 10–15 μ thick and mount on clean glass slides without any adhesive.

(2) Dry in air at room temperature for 1–2 hours.

(3) Incubate in the substrate solution for ½–4 hours.

(4) Wash in water, treat with 2 per cent. cobalt solutions, wash, treat with dilute yellow ammonium sulphide.

(5) Counterstain in 1 per cent. aqueous eosin, 5 minutes.

(6) Wash in running water, 5 minutes.

(7) Mount in glycerinejelly.

MODIFIED GOMORI METHOD FOR ALKALINE PHOSPHATASE

(after Fredricsson, 1952, 1956)

(*Alcohol; paraffin sections*)

Method. (1) Fix small blocks of tissue in 90 per cent. alcohol for 24 hours at 22° (two changes).

(2) Transfer to 96 per cent. alcohol for 1 hour.

(3) Transfer to two changes of absolute alcohol, 15 minutes each.

(4) Two changes in benzene, 30 minutes each.

(5) Embed in paraffin wax at 56° (3 changes, 20 minutes each).

(6) Cut sections 3–10 μ and mount on slides without floating out on water.

(7) Remove wax with xylene and pass through absolute acetone, and 40 per cent. acetone, to water.

(8) Incubate for 10–60 minutes in a closed Coplin jar at 37° in the following mixture:

2 per cent. sodium-β-glycerophosphate .	25 ml.
2 per cent. sodium veronal . . .	25 ml.
2 per cent. calcium nitrate . . .	5 ml.
0·8 per cent. magnesium chloride . .	5 ml.
Acetone	40 ml.

(9) Rinse in 40 per cent. acetone.

(10) Treat with 2 per cent. cobalt nitrate in 40 per cent. acetone for 5 minutes.

(11) Rinse in 40 per cent. acetone.

(12 Treat with dilute yellow ammonium sulphide in 40 per cent. acetone for 3 minutes.

(13) Rinse in acetone, continue dehydration in alcohol.

(14) Clear in xylene and mount in Canada balsam.

Result. Clean black deposits of cobalt sulphide indicate sites of enzyme activity.

TABLE 55

DIAZONIUM SALTS IN ENZYME HISTOCHEMISTRY

No.	Salt Name	Chemical Composition	Formula (Stabilizer may vary)
1	Fast Red AL	1-amino-anthraquinone	[O, N≡N⁺, O]₂ $ZnCl_4^{=}$
2	Fast Blue RR (Echtblausalz RR)	4-benzoylamino-2:5-dimethoxyaniline	[CONH, OCH_3, CH_3, N⁺≡N]₂ $ZnCl_4^{=}$
3	Fast Red GG	*p*-nitroaniline	O_2N–N⁺≡N BF_4^-
4	Fast Red 3GL (Echtrotsalz 6GL)	4-chloro-2-nitroaniline	[Cl, NO_2, N⁺≡N]₂ $ZnCl_4^{=}$

TABLE 55

DIAZONIUM SALTS IN ENZYME HISTOCHEMISTRY—*continued*

No.	Salt Name	Chemical Composition	Formula (Stabilizer may vary)
5	Fast Scarlet GG	2:5-dichloroaniline	Cl, Cl, $N^+ \equiv N$; $[\ ]_2$ $ZnCl_4^=$
6	Fast Blue B (Echtblausalz B)	*o*-dianisidine	$ZnCl_4^=$ OCH_3, OCH_3, $N^+ \equiv N$, $N^+ \equiv N$; $[\ ]_2$ $ZnCl_4^=$
7	Fast Red RC (Echtrotsalz RC)	4-chloro-*o*-anisidine	OCH_3, Cl, $N^+ \equiv N$; $[\ ]_2$ $ZnCl_4^=$
8	Fast Red B (Echtrotsalz B)	5-nitro-*o*-anisidine	OCH_3, O_2N, $N^+ \equiv N$; ^-O_3S, SO_3H
9	Fast Red TR (Echtrotsalz TR)	5-chloro-*o*-toluidine	CH_3, Cl, $N^+ \equiv N$; ^-O_3S, SO_3H
10	Fast Scarlet R	4-nitro-*o*-anisidine	OCH_3, NO_2, $N^+ \equiv N$; $[\ ]_2$ $ZnCl_4^=$
11	Fast Red GL (Echtrotsalz 6 GL)	3-nitro-*p*-toluidine	NO_2, H_3C, $N^+ \equiv N$; ^-O_3S, SO_3H
12	Bordeaux GP	3-nitro-*p*-anisidine	NO_2, CH_3O, $N^+ \equiv N$; $[\ ]_2$ $ZnCl_4^=$
13	—	4-amino-3:1′-dimethyl azobenzene	CH_3, CH_3, N = N, $N^+ \equiv N$
14	Echtschwarzalz K	4-amino-2:5-dimethoxy 4′-nitro azobenzene	O_2N, N = N, CH_3O, CH_3O, $N \equiv N^+$
15	Variamine Blue B (Variaminblausalz B) Fast blue VB	4-amino-4′-methoxy diphenylamine	H_3CO, NH, $N^+ \equiv N$, Cl^-

TABLE 55

DIAZONIUM SALTS IN ENZYME HISTOCHEMISTRY—*continued*

No.	Salt Name	Chemical Composition	Formula (Stabilizer may vary)
16	Fast Red RL	3-nitro-*o*-toluidine	$O_2N-C_6H_3(CH_3)-N^+\equiv N$ BF_4^-
17	Fast Black K	4-amino-4′-nitro-3:6-dimethoxy-azobenzene	$[O_2N-C_6H_4-N=N-C_6H_2(OCH_3)_2-N^+\equiv N]_2$ $ZnCl_4^{=}$
18	Fast Garnet GBC (Echtgranatsalz GBC)	4-amino-3:1′dimethyl azobenzene	$CH_3C_6H_4-N=N-C_6H_3(CH_3)-N^+\equiv N$ SO_4H^-
19	Fast Red LTR (Echtrotsalz ITR)	2-methoxy-5-diethylamino sulphaniline	$[N\equiv N^+-C_6H_3(OCH_3)-SO_2-N(C_2H_5)_2]_2$ $ZnCl_4^{=}$
20	Fast Violet B (Echtvioletsalz B)	2-benzoylamino-4-methoxy toluidine	$[C_6H_5-CONH-C_6H_2(OCH_3)_2-N^+\equiv N]_2$ $ZnCl_4^{=}$
21	Fast Blue VRT (Variaminblausalz B)	4-amino-diphenylamine	$C_6H_5-NH-C_6H_4-N^+\equiv N$ Cl^-
22	Fast Black B (Echtschwarz B)	*p, p′*-diaminodi-phenylamine	$ZnCl_4^{=}$ $[N\equiv N^+-C_6H_4-NH-C_6H_4-N^+\equiv N]_2$ $ZnCl_4^{=}$

A MODIFIED COUPLING AZO DYE METHOD FOR ALKALINE PHOSPHATASE

(*Cold formalin or fresh frozen sections; freeze-dried paraffin sections*)

Frozen Sections

Method. (1) Fix thin slices of tissues in 10 per cent. neutral formalin at 4° for 10–16 hours. Alternatively use fresh frozen cold microtome sections, mounted on cover slips.

(2) Cut frozen sections 10–15 μ thick and mount on clean slides without adhesive.

(3) Allow to dry in air for 1–3 hours to ensure adherence.

(4) Dissolve 10–20 mg. sodium α-naphthyl phosphate in 20 ml. 0·1 M stock "tris" buffer (pH 10).* Add 20 mg. of the stable diazotate of 5-chloro-*o*-toluidine (Salt 9, Table 55) and stir well. Filter on to the slides sufficient to cover each section

* The pH of the medium is lowered both by the addition of the substrate and also by the various diazotates.

adequately and incubate at room temperature (17°–22°)* for 15–60 minutes.† (Alternatively, use the same quantity of the stable diazotate of 2-benzoylamino-4-methoxy toluidine (Salt 20, Table 55), or of *p-p′*-diaminodiphenylamine (Salt 22, Table 55), and proceed in the same manner.)

(5) Wash in running water for 1–3 minutes.

(6) Counterstain in Mayer's hæmalum, 1–2 minutes.

(7) Wash in running water for 30–60 minutes.

(8) Mount in glycerine jelly.

Result (Plate XIIa). The sites of alkaline phosphatase activity are coloured brown with Salts 9 and 20 or black with Salt 22. Nuclei, dark blue.

Paraffin Sections

(*Cold acetone-fixed; paraffin-embedded*)

Method. (1) Bring sections to water via absolute acetone after removing the paraffin with light petroleum.

(2) Cover with freshly made and filtered substrate-diazonium salt mixture as above.

(3) Incubate for 30 minutes to 4 hours (Salt 2) or for up to 2 hours (Salt 7) or for up to 12 hours (Salt 9).

(4) Wash in water, counterstain as above and blue in running water.

(5) Mount in glycerine jelly.

(The use of Salt 9 is particularly recommended.)

Result (Figs. 134 and 136). With Salt 9, the sites of alkaline phosphatase activity appear dark reddish-brown, localization is excellent. Nuclei, blue.

AZO DYE METHODS FOR LEUCOCYTE PHOSPHATASE

(after Kaplow, Monis, Hayhoe)

(*Formalin-methanol*)

Method. (1) Prepare smears from capillary or sequestrene-venous blood.

(2) Allow the freshly made smears to dry in air for a few minutes.

(3) Fix for 30 seconds in ice-cold 10 per cent. formalin in methanol (40 per cent. formaldehyde, 10 ml.; methanol, 90 ml.).

(4) Prepare incubating solution A‡ or B§ and filter directly on to the smears. Incubate 5–10 minutes at 22° (A) or 30 minutes (B).

(5) Wash in running water.

(6) Counterstain with Mayer's hæmalum for 1½–3 minutes or with 2 per cent. chloroform-washed methyl green for 2 minutes.

(7) Wash in running water.

(8) Mount in glycerine jelly.

Result. Neutrophil granulocytes (40–90 per cent.) show alkaline phosphatase activity as a brown, red or blue deposit.

* With temperatures above 22°, diffusion and abnormal crystallization effects may be noted.

† It is possible to remove the aluminium sulphate, present in most stable diazotates, by the addition of a few ml. of a saturated aqueous solution of purpurin. The red aluminium lake is retained by a Whatman No. 40 paper. This procedure appreciably increases the speed of the reaction. Removal of inhibitory zinc salts by means of chelators such as dithizone unfortunately removes the diazotates also.

‡ To 25 ml. 0·1 M-"tris" buffer (pH 9·2) add 20 mg. sodium α-naphthyl acid phosphate and 30 mg. Fast Garnet GBC salt (No. 18).

§ To 5 ml. "Stock" Naphthol AS-MX phosphate add 10 mg. Fast violet LB salt or Fast blue RR salt (No. 2).

PREPARATION OF SODIUM α-NAPHTHYL ACID PHOSPHATE

(Friedman and Seligman, 1950)

(1) α-NAPHTHYLPHOSPHORYL DICHLORIDE

Heat 25 g. α-naphthol and 26·5 g. (15·5 ml.) phosphorus oxychloride in 90 ml. dry benzene under a reflux condenser and add 13·7 g. dry pyridine over a period of 30 minutes. Heat for a further 15 minutes, cool and remove precipitate of pyridine hydrochloride by filtration. Distil off the solvent and redistil to leave a clear colourless syrup (b.p. 199°–201°).

(2) α-NAPHTHYL ACID PHOSPHATE

Place a thin layer of α-naphthylphosphoryl dichloride over a layer of aqueous KOH in a Petri dish in a partially evacuated dessicator. Quantitative conversion to α-naphthyl phosphate occurs with formation of a white crystalline solid (m.p. 155°–157°). The monosodium salt is precipitated from a solution of the latter in methanol by the addition of an equivalent of sodium methoxide.

PREPARATION OF NAPHTHYL AS-PHOSPHATES

(Burstone, 1958)

A suspension of 5 gm. of the appropriate hydroxy-naphthoic arylide (Naphthols AS, AS-MX, AS-TR, AS-D, AS-CL, etc.) in 50–70 ml. of tetrahydrofuran is treated with excess $POCl_3$ (4 ml.) in the presence of sodium chloride (1 gm.), sodium sulphate (1 gm.) and tetrasodium ethylene diamine tetra-acetic acid (150 mg.). Pyridine (2 ml.) is then added to the mixture with constant stirring. After standing for 18 hours the acid phosphate is prepared by evaporating the solvent in a stream of air and adding to the resulting slush one or two cubes of ice. The reaction mixture is then added to 500 ml. cold water, stirred, and collected under suction.

A stock solution ready for histochemical use can be prepared from the majority of AS acid phosphates by dissolving 50 mg. in 20 ml. N,N′-dimethyl formamide, adding 20 ml. water and sufficient molar Na_2CO_3 to bring the pH to 8·0. After the addition of 600 ml. water the volume is brought up to 1 litre by adding 0·2 M-"tris" buffer (pH 8·3). The opalescent solution is stable at room temperature for several months. The disodium salts of the AS phosphates are prepared, if required, by treatment with sodium methylate in methanol or with sodium cabonate.

PREPARATION OF AS-TR PHOSPHATE

Using the phosphorus oxychloride method Jeffree and Taylor (1961) failed to obtain better than 50 per cent. conversion of the naphthol. They recommended a technique, using phosphorus pentachloride in dioxan, for preparing AS-TR phosphate (see p. 731).

Purification (Nimmo-Smith, 1961)

Dissolve the product in warm dioxan (20 ml. per 5 g.) with the aid of a few drops of conc. HCl. Treat with charcoal, cool and filter. To the filtrate, after re-warming, add toluene (slowly) until after thorough mixing a slight turbidity results. After standing overnight a crop of fine needles results. These are washed with a little cold dioxan. Further crops of crystals are obtained by adding more toluene to the mother liquor. Recrystallize the combined crops (about 1 g.) by solution in 15 ml. warm dioxan, cooling, and filtration, followed by the addition of 20 ml. toluene to the filtrate. Dry *in vacuo* at room temperature (m.p. 169°–170°).

NAPHTHOL AS-PHOSPHATE AZO DYE METHOD

(after Burstone)

(*Freeze-dried or cold acetone-fixed; paraffin sections*)

Method. (1) Bring sections to water via xylene and acetone.

(2) Incubate for 5–30 minutes at 22° in 40 ml. Stock Solution* (*vide supra*) containing 40 mg. of the chosen diazonium salt (Fast red-violet LB salt, Fast blue RR, red RC and red TR are recommended).

(3) Wash briefly in water and mount in PVP medium (Burstone, 1957).†

Result. Intensely coloured red or blue azo dyes indicate sites of alkaline phosphatase activity.

NAPHTHOL AS-PHOSPHATE HNF METHOD

(after Stutte, 1967)

(*Freeze-dried paraffin sections; fresh cold microtome sections*)

Reagents

Naphthol AS, AS-BI, AS-MX, AS-CL or AS-TR phosphates may be used.

HNF (Hexazotized New Fuchsin) is prepared as follows:

Solution A. Dissolve 4 g. New Fuchsin in 100 ml. 2 N-HCl at 60°, filter. This solution is stable for some weeks.

Solution B. Dissolve 4 g. $NaNO_2$ in 100 ml. distilled water. Should be freshly prepared every 2 days.

Immediately before use mix equal volumes of A and B.

Preparation of Incubating Medium

Add 0·2 ml. freshly prepared HNF solution to 40 ml. 0·2 M Tris-HCl buffer and adjust the pH to 8·8 to 9·2 by adding 1 N-NaOH. To this add 5–10 mg. of the chosen substrate in 0·2 ml. dimethylformamide. Filter and use immediately.

Method. (1) Bring paraffin sections to water via xylene and acetone.

(2) Incubate sections in the freshly prepared medium for 2–60 minutes.

(3) Wash briefly in water and mount in a suitable watery medium.

Result (Plate XIIc). Alkaline phosphatase appears in different shades of red. With the same substrate the colour may differ appreciably in different sites.

PREVENTION OF GAS BUBBLES IN AZO DYE METHODS

(after Peters, 1964)

Method. (1) Following incubations by any of the recognized techniques, immerse the slides in a dilute solution of gelatine (the original author used Gelatinol-Chroma, 3 parts; distilled water, 1 part) for 2 minutes, with gentle movement.

(2) Remove slides and stand them on their long edge to drain.

(3) Leave to dry for 20–60 minutes.

(4) If required for critical microscopy or photomicrography apply a coverslip after application of a few drops of a suitable aqueous mounting medium.

* According to Burstone AS-D, AS-OL, AS-BS and AS-MX phosphates are rapidly hydrolysed at alkaline pH levels.

† Dissolve 50 gm. polyvinyl pyrrolidone in 50 ml. distilled water. Stand overnight. Add 2 ml. glycerol and a crystal of thymol and stir. (Refractive Index 1·46, increasing as evaporation of water takes place.)

CALCIUM METHOD FOR 5-NUCLEOTIDASE

(after Pearse and Reis)

(*Cold acetone, double embedded sections; cold formalin, frozen sections*)

Method. (1) Prepare acetone-fixed paraffin sections or formalin-fixed frozen sections as for the methods described for alkaline phosphatase.

(2) Incubate one section (with glycerophosphate or phenyl phosphate) as for alkaline phosphatase at pH 9·2, another similarly at pH 7·5, a third in the following medium at pH 7·5 and a fourth in the same medium containing water instead of substrate, all for 3–18 hours at 37°.

5 vol. barbiturate buffer (pH 7·5)
(sodium diethyl barbiturate, 0·1 M, 3 vol.; HCl, 0·1 N, 2 vol.; distilled water, 1 vol.).
1 vol. 12 per cent. (w/v) $Ca(NO_3)_2$.
1 vol. 2 per cent. (w/v) $MgCl_2$.
1 vol. 0·04 M-adenylic acid (adenosine-5-phosphate).

(3) Wash with 2 per cent. $Ca(NO_3)_2$ of pH about 8·0 and then in distilled water.
(4) Immerse in 1 per cent. $AgNO_3$ and expose to daylight for 1 hour.
(5) Rinse in distilled water and treat with 5 per cent. $Na_2S_2O_3$ for 10 minutes.
(6) Wash, dehydrate in alcohol, clear in xylene and mount in Canada balsam.

The Co and $(NH_4)_2S_2$ method for revealing the precipitate of calcium phosphate may be used as an alternative.

Result (Figs. 131 and 133). Various structures in the neighbourhood of weak or moderately strong sources of 5-nucleotidase are stained black. In the case of very strong sources of enzyme, localization to a group of cells may be possible. (The glycerophosphate control at pH 7·5 must be subtracted from the total result given by adenylic acid at pH 7·5 to obtain a true estimate of 5-nucleotidase. This control is often negative and the substrate control must be negative.)

LEAD METHOD FOR 5-NUCLEOTIDASE

(after Wachstein and Meisel)

(*Free-floating fresh frozen sections*)

Method. (1) Cut 10–15 μ sections from the cold microtome directly into the following medium:

10 ml. of 1·25 per cent. adenosine-5-phosphate.
5 ml. of 0·2 M-"Tris" buffer at pH 7·2.
30 ml. of 0·2 per cent. $Pb(NO_3)_2$.
5 ml. of 0·1 M-$MgSO_4$.

(2) Remove from cryostat and incubate sections for ½–2 hours at 37°. (Immerse any floating sections with the aid of a glass needle.)
(3) Stop reaction with 2 ml. of 40 per cent. formaldehyde, 30 minutes.
(3) Remove sections with section lifter and rinse in distilled water.
(4) Treat with dilute yellow ammonium sulphide, 2 minutes.
(5) Rinse in distilled water.
(6) Mount on slides and allow to dry (almost).
(7) Apply glycerine jelly and coverslip.

Result. Brown deposits of lead sulphide indicate sites of enzyme activity. Studies on Ni^{2+} inhibition must be carried out with acetate instead of Tris buffer. 100 mM nickel completely inhibits 5-nucleotidase.

LEAD METHOD FOR 5-NUCLEOTIDASE

(after Barron and Boshes, 1961)

(*Short formol-calcium; frozen sections*)

SUBSTRATE SOLUTION

25 ml. 0·1 M-Tris-maleate, pH 6·5.
3 ml. 2 per cent. lead nitrate.
5 ml. 0·1 M-manganous acetate.
20 mg. adenosine-5-phosphate.
17 ml. distilled water.

Add reagents and filter immediately before use.

Method. (1) Incubate sections for 15–30 minutes.
(2) Wash briefly in distilled water.
(3) Treat with dilute yellow ammonium sulphide, 2 minutes.
(4) Rinse in distilled water.
(5) Mount sections on slides and continue as in previous method.

Result. As with previous method.

A CALCIUM-COBALT METHOD FOR ALDOLASE

(after Allen and Bourne)

(*Alcohol; frozen sections*)

The authors used tissues fixed in 80 per cent. alcohol for 24 hours, from which frozen sections were cut after removal of the alcohol by washing in water.

PREPARATION OF SOLUTIONS

(1) **The Precipitant Mixture.** Dissolve 5·5 g. $MgCl_2.6H_2O$ and 7·0 g. NH_4Cl in 35 ml. 5N-NH_4OH. Stand for 1 hour, filter and add 60 ml. 4N-NH_4OH to the filtrate.

(2) **Purified Substrate Solutions A and B.** (*A*) Mix 40 ml. 4 per cent. sodium hexose diphosphate with 20 ml. of the precipitant mixture. Stand for 30 minutes and filter.

(*B*) Mix 20 ml. of 4 per cent. sodium hexose diphosphate with 20 ml. of the precipitant mixture and 20 ml. distilled water. Stand for 30 minutes and filter.

(3) **Final Substrate Solutions A and B.** (*A*) Add 10 ml. purified substrate solution A to 1·7 ml. of 0·1 M-sodium iodoacetate and 5 ml. distilled water.

(*B*) Add 15 ml. purified substrate solution B to 2·5 ml. of 0·1 M-sodium iodoacetate and 7·5 ml. distilled water.

Method. (1) Mount frozen sections on clean slides without adhesive and dry at room temperature for 1–3 hours.
(2) Incubate sections in final substrate solutions A and B for 1–2 hours at 37°.
(3) Immerse in 2 per cent. cobalt acetate for 5–10 minutes.
(4) Wash and immerse in dilute yellow ammonium sulphide solution 1–2 minutes.
(5) Dehydrate in alcohol, clear in xylene, mount in Canada balsam.

Result. A brownish-black precipitate in either section indicates the presence in the section of aldolase.

MULTISTEP TETRAZOLIUM METHOD FOR ALDOLASE

(after Abe and Shimizu, 1964)

(*Fresh frozen; cold microtome sections*)

INCUBATING MEDIUM

10 ml. 0·02 M sodium fructose-1,6-diphosphate.
5 mg. NAD.
10 mg. Nitro-BT.
10 ml. 0·05 M-arsenate-HCl buffer, pH 7·6.

Mix and use as soon as possible.

Method. (1) Post-fix sections for 20 minutes at 0–4° in 80 per cent. ethanol.
(2) Allow to dry for 20 minutes.
(3) Incubate for 10–30 minutes at 37°.
(4) Fix in neutral formalin for 1 to 12 hours.
(5) Dehydrate in alcohols and xylene; mount in a suitable synthetic medium.

Result. Purple staining indicates aldolase activity.

ADDED ENZYME-TETRAZOLIUM METHOD FOR ALDOLASE

(after Lake, 1965)

(*Fresh frozen cold microtome; post-fixed in cold acetone*)

PREPARATION OF STOCK SOLUTIONS. (Prepare in bulk, store in deep freeze)

Solution A

25 ml. 0·2 M Tris buffer pH 7·4.
35 ml. distilled water.
25 ml. Tetrazolium salt (MTT, 1 mg./ml.).
5 ml. 0·5 M-$CoCl_2$.

Solution B

8·5 ml. Solution A.
1·0 ml. 2 per cent. glycine.
1·0 ml. 5 per cent. sodium arsenate.
0·5 ml. 0·05 M-$MgCl_2$.

Solution C

8·5 ml. Solution A.
1·0 ml. 2 per cent. glycine.
0·5 ml. 0·05 M-$MgCl_2$.
0·5 ml. menadione bisulphite (2 mg./5 ml.).
2 mg. phenazine methosulphate.

Incubating Media. (1) For aldolase via glyceraldehyde phosphate dehydrogenase (GAPD).

1·1 ml. Solution B.
0·1 ml. 0·5 M fructose-1-phosphate or 1,6-fructose diphosphate.
0·1 ml. Dilute enzyme (GAPD) solution (0·02 ml. to 0·5 ml. with distilled water).
2 mg. NAD.

(2) For Aldolase via α-glycerophosphate dehydrogenase (α-GPD).

1·0 ml. Solution C.
0·1 ml. 0·5 M Substrate.
0·1 ml. αGPD (diluted 0·02 ml. to 0·5 ml.).

Method. (1) Incubate sections for 30 minutes to 4 hours at 37°.
(2) Wash briefly and fix in 10 per cent. formalin for 15 minutes.
(3) Wash, and mount in glycerine jelly.
Result. Black deposits indicate sites of aldolase activity.

CALCIUM METHOD FOR ADENOSINE TRIPHOSPHATASE

(after Maengwyn-Davies *et al.*)
(*Unfixed frozen sections*)

PREPARATION OF INCUBATING MEDIUM

This must be prepared immediately before use, heated to 37° for 1 hour filtered, and finally adjusted to pH 9·9 with molar KOH.

0·1 M-glycine, 0·4 M-KCl } in saturated sodium acetate*	12 ml.
Saturated sodium acetate	12 ml.
0·36 M-$CaCl_2$	3·6 ml.
1 M-KOH	0·6 ml.
0·04 M-sodium adenosine triphosphate†	6 ml.
Distilled water	13·8 ml.
Saturated sodium phosphate ($NaPO_4$)	0·3 ml.

Method. (1) Incubate free-floating sections (10–15 μ) for 5 minutes to 3 hours.
(2) Wash in three changes of 1 per cent. $CaCl_2$ in 75 per cent. ethanol.
(3) Transfer to 2 per cent. $CoCl_2$ for 3 minutes.
(4) Develop in 1 per cent. yellow ammonium sulphide.
(5) Wash, dehydrate and mount in a suitable synthetic medium.
Result. ATPase activity shows as a blackish-brown deposit.

CALCIUM METHOD FOR ADENOSINE TRIPHOSPHATASE

(after Padykula and Herman)
(*Unfixed frozen sections; free-floating or attached to cover slips*)

PREPARATION OF INCUBATING MEDIUM

This solution must be absolutely freshly prepared.

0·1 M-sodium barbiturate (2·062 g./100 ml.)	20 ml.
0·18 M-$CaCl_2$ (1·998 g./100 ml.)	10 ml.
Distilled water	30 ml.
Adenosine triphosphate (disodium salt)‡	152 mg.

As soon as the ATP is dissolved, adjust the pH to 9·4 with 0·1 M-NaOH and make up to 100 ml. with distilled water. Filter if turbid.

Method. (1) Incubate sections for 5 minutes to 3 hours at 37°.
(2) Wash in three changes of 1 per cent. $CaCl_2$.
(3) Transfer to 2 per cent. $CoCl_2$, 3 minutes.
(4) Wash in distilled water for 1 minute.
(5) Develop in dilute yellow ammonium sulphide.
(6) Wash, counterstain nuclei if desired (carmalum) or background (eosin). Dehydrate, clear, and mount in a suitable synthetic medium.

* Made up by dissolving the glycine and KCl in a small amount of water and adding saturated sodium acetate to make up the volume.
† Final concentration, 0·005 M.
‡ Final concentration, 0·005 M.

Result (Plate XIIb). Black deposits indicate ATPase. Localization is good. Whenever formalin fixation is employed, in conjunction with this method, it is advisable to use formaldehyde made by dissolving paraformaldehyde in appropriate concentration. Methanol (contaminant in formalin) inhibits ATPase.

LEAD METHOD FOR ADENOSINE TRIPHOSPHATASE

(after Wachstein and Meisel)

(*Unfixed frozen sections, free-floating or mounted; formalin-fixed frozen sections*)

PREPARATION OF THE INCUBATING MEDIUM

	Final concentration
20 ml. 125 mg. per cent. ATP (disodium salt) .	$8{\cdot}3 \times 10^{-4}$ M
20 ml. 0·2 M-"Tris"-maleate buffer (pH 7·2) .	8×10^{-2} M
3 ml. 2 per cent. $Pb(NO_3)_2$	$3{\cdot}6 \times 10^{-3}$ M
5 ml. 0·1 M-$MgSO_4$	1×10^{-2} M
2 ml. Distilled water	

Add constituents in the above order and adjust the pH if necessary.

Method. (1) Incubate fresh or formalin-fixed sections for 5–60 minutes.

(2) Rinse in distilled water.

(3) Develop in 1 per cent. yellow ammonium sulphide, 1 minute.

(4) Rinse again in water.

(5) Mount in glycerine jelly.

Result. Brownish-black deposits indicate ATPase activity.

METHOD FOR SODIUM ACTIVATED ADENOSINE TRIPHOSPHATASE

(after McClurkin, 1964)

(*Cold formaldehyde; frozen sections*)

PREPARATION OF SOLUTIONS

0·01 M ATP. Dissolve the barium salt of ATP in distilled water acidified to pH 3·1 with 0·1 N-HCl.

PREPARATION OF MEDIUM

	Final Concentration
1·5 ml. 0·01 M ATP (Ba salt)	3 mM
1·4 ml. distilled water	
0·6 ml. Tris buffer, pH 7·8, 0·2 M	24 mM
0·5 ml. 1 per cent. lead nitrate (0·03 M) . . .	3 mM
1·0 ml. magnesium sulphate (0·015 M) . . .	3 mM

The final pH should be 7·2. Centrifuge to remove precipitate. Place half of the medium into each of two containers.

To one, add 15 mg. NaCl	100 mM
and 5 mg. KCl	30 mM.

Method. (1) Incubate free floating sections in medium containing Na and K and in control medium (Na/K free), for 10–15 minutes at 37°.

(2) Rinse briefly and treat with yellow ammonium sulphide.

(3) Rinse and mount in glycerine jelly.

Result. Black deposits indicate enzyme activity.

METHOD FOR THIAMINE PYROPHOSPHATASE

(after Eränkö and Hasan)

(*Cold microtome or cold-knife sections; free-floating or mounted*)

PREPARATION OF THE MEDIUM

Dissolve 60 mg. $Pb(NO_3)_2$ in 45 ml. 0·005 M-acetate buffer (pH 5·0). Add 5 ml. 5 per cent. thiamine pyrophosphate. Incubate at 37° for 24 hours, filter and store in the cold, if necessary. Filter again before use.

Method. (1) Incubate slides at 42° for 5–60 minutes.
(2) Rinse in 1 per cent. acetic acid.
(3) Blot and dry at 60° for 5 minutes.
(4) Cover sections with a thin film of celloidin.
(5) Treat with dilute yellow ammonium sulphide, 2 minutes.
(6) Rinse in water.
(7) Mount in glycerine jelly.

Result. A brownish-black deposit indicates sites of thiamine pyrophosphatase activity.

METHOD FOR THIAMINE PYROPHOSPHATASE

(after Allen and Slater, 1961)

(*Unfixed cryostat sections, mounted on slides*)

PREPARATION OF INCUBATING MEDIUM

Make up a medium containing (final concentrations):
33 mM-sodium veronal, pH 9·5.
15 mM-$CaCl_2$.
5 mM-thiamine pyrophosphate chloride.
4 mM-cysteine hydrochloride.
Adjust the pH before the final dilution.

Method. (1) Incubate sections for 2 to 4 minutes at 37°.
(2) Rinse briefly.
(3) Treat with 2 per cent. cobalt nitrate, 3–5 minutes.
(4) Rinse well in distilled water.
(5) Treat with dilute yellow ammonium sulphide, 1–2 minutes.
(6) Wash in water, dehydrate, clear and mount in a suitable synthetic medium.

Result. Brown or black deposits indicate sites of TPPase activity. Control sections incubated without cysteine may show non-specific alkaline phosphatase activity.

METHOD FOR INORGANIC PYROPHOSPHATASE

(after Kurata and Maeda)

(*Cold microtome or cold-knife sections, free-floating or mounted; cold formalin-fixed frozen sections*)

PREPARATION OF SUBSTRATE SOLUTIONS

Alkaline Pyrophosphates. Dissolve 1·088 g. sodium pyrophosphate ($Na_4P_2O_7$ $12H_2O$) in 20 ml. distilled water. Prepare a 6 per cent. solution of ferric chloride ($FeCl_3 . 6H_2O$) and add 10 ml. to the above solution. A white precipitate forms. Add 10 per cent. sodium carbonate drop by drop until the precipitate just dissolves. Adjust the pH of the medium to 7·2 to 7·3 with 1 N-HCl. Dilute to 100 ml. with distilled water and make hypertonic by adding 0·9 g. NaCl. This solution must be freshly prepared and, just before using, about 1 ml. of 10 per cent. magnesium chloride is added, drop by drop. Slight clouding of the medium can be ignored.

Acid Pyrophosphatase. Prepare the pyrophosphate-ferric chloride solution, as above, and centrifuge. Add the precipitate to a small amount (10 ml.) of distilled water and dissolve by adding 10 per cent. Na_2CO_3. Adjust the pH to 4·7 to 6·0 by adding 1 N-HCl. Dilute to 100 ml. Allow to stand at 22° for 2–3 hours and readjust pH to 3·7 to 4·0 by adding 100 ml. 0·5 M-acetate buffer (pH 3·5).

Method. (1) Incubate 20 μ sections at 37° for 3 hours (alkaline medium) for for 1 hour (acid medium).

(2) Wash twice (10–15 minutes) in 0·9 per cent. NaCl.

(3) If free-floating sections have been used they should be mounted at this point.

(4) Treat with dilute yellow ammonium sulphide, 2 minutes (sections turn deep or light green).

(5) Dehydrate in alcohols, clear in xylene and mount in a suitable synthetic medium or in Canada balsam.

Result. A green deposit indicates sites of pyrophosphatase activity. The colour is not stable.

METHOD FOR INORGANIC POLYMETAPHOSPHATASE

(after Berg, 1955)

(*Freeze-substitution, Wolman; paraffin sections*)

PREPARATION OF TISSUES

Quench small pieces of tissue in cold isopentane at −160° Transfer to acetone at −78°, without thawing. Allow 14 days for dehydration. Subsequently the dehydrated specimens are brought to acetone at room temperature, cleared in benzene (30 minutes) and vaccuum embedded in 56° paraffin wax. Tissues fixed in Wolman's 5 per cent. acetic-ethanol at −4° for 1 hour can be used as an alternative.

PREPARATION OF SUBSTRATE SOLUTION

Stock Solution I. 70 mM-lead acetate, 180 mM-acetic acid and 45 mM-nitric acid.

Stock Solution II. Freshly prepared 5·1 per cent. sodium metaphosphate.

Stock Solution III. 0·3 N-nitric acid.

Add 10 ml. Stock I and 1 ml. Stock II to 75 ml. water. Slowly add 4 ml. Stock II with constant stirring. Throughout the procedure the pH must be maintained at below pH 3·9 by addition of small amounts of Stock III. Finally filter with suction through a Whatman No. 50 paper.

Method. (1) Cut sections at 6 μ. Pass through xylene and dilutions of dioxane to water (exposure to alcohol must be avoided).

(2) Incubate mounted or unmounted sections for 18–24 hours at 37°.*

(3) Rinse in 50 mM-acetate buffer at pH 3·9.

(4) Develop with dilute yellow ammonium sulphide.

(5) Wash in water.

(6) Dehydrate in alcohol, clear with tetrachloroethylene and mount in a medium made up with this solvent in place of xylene.

Control sections inactivated by exposure to 90° for 10 minutes should be used.

Result. Brownish-black deposits indicate sites of polymetaphosphatase activity.

METHOD FOR ALKALINE POLYMETAPHOSPHATASE

(after Berg, 1960)

(*Fresh frozen, cryostat. Cold acetone post-fixation*)

PREPARATION OF STOCK SOLUTIONS

(1) 6 mM lead acetate adjusted to below pH 6 with acetic acid.

* The original author recommends that a filtration accelerator pad (Fisher) should be placed in the incubating solution.

(2) Freshly prepared substrate 2 mM sodium polyphosphate (96 mg. anhydrous sodium tripolyphosphate to 100 ml. water).

(3) 0·2 M veronal buffer, pH 8·5 to 9·0.

To 100 ml. substrate add alternatively small amounts of 1 and 2 until the mixture becomes slightly opalescent. Adjust pH to 8·7 with NaOH or HNO_3. If opalescence disappears add 1 or 2 drops of Solution I, to restore it.

Method. (1) Incubate for 5–30 minutes.
(2) Wash briefly.
(3) Treat with dilute yellow ammonium sulphide.
(4) Wash briefly.
(5) Dehydrate, clear and mount.

Result. Brown or black deposits indicate enzyme activity.

CHELATE REMOVAL METHOD FOR TRIPHOSPHATASES

(after Berg, 1964)

(*Formalin, frozen sections; Acetone or cold acetone, paraffin*)

PREPARATION OF SUBSTRATE MEDIA FOR ALKALINE TRIPHOSPHATASE

Prepare a solution containing the following:
(1) Substrate (Sodium tripolyphosphate) 2·6 mM.
(2) Buffer (Veronal-Na) 20 mM.

Add slowly and in small amounts, lead acetate powder until the solution remains opalescent on stirring. Adjust pH to 8·7 with NaOH or HNO_3.

PREPARATION OF SUBSTRATE MEDIA FOR ALKALINE PYROPHOSPHATASE

Prepare a solution containing the following:
Sodium pyrophosphate 1 mM.
Veronal-Na 50 mM.

Add 50 mM lead acetate in 50 mM acetic acid until opalescence. Adjust pH, as above, to 9·4.

PREPARATION OF SUBSTRATE MEDIA FOR NEUTRAL TRIPHOSPHATASE

Prepare a solution containing the following:
Sodium tripolyphosphate 2·6 mM
Tris-HCl buffer, pH 6·3 50 mM.
$Co(NO_3)_2$ 6·5 mM.

Add lead nitrate powder to point of opalescence.

Method. (1) Incubate sections for 2 to 10 minutes.
(2) Wash briefly.
(3) Treat with dilute ammonium sulphide.
(4) Wash, dehydrate, clear and mount.

Result. Black deposits indicate enzyme activity.

METHOD FOR GLUCOSE-6-PHOSPHATASE

(after Chiquoine)

(*Cold microtome or cold-knife sections; free-floating or mounted*)

CONVERSION OF BARIUM SALT OF GLUCOSE-6-PHOSPHATE TO POTASSIUM SALT

Dissolve 250 mg. of the barium salt in 10 ml. distilled water containing 2 drops of 2 N-HCl. Add 120 mg. potassium sulphate. Allow to stand for 2 hours with frequent stirring. Centrifuge and test the supernatant for barium with a pinch of potassium sulphate. If no precipitation occurs dilute to 30 ml. with distilled water and adjust pH to 6·7 with 1 N-KOH.

PREPARATION OF THE INCUBATING MEDIUM

Dilute 1 part of the above solution with 2 parts of 6 mM-lead nitrate (0·2 per cent). Filter before use.

Method. (1) Incubate sections for 5–15 minutes at 32°.
(2) Wash in distilled water.
(3) Develop with dilute yellow ammonium sulphide.
(4) Wash in water, dehydrate, clear in xylene and mount in Clarite. (Alternatively mount in glychrogel without dehydration.)

Result. Brownish-black deposits indicate sites of glucose-6-phosphatase activity.

METHOD FOR GLUCOSE-6-PHOSPHATASE

(after Wachstein and Meisel)

(*Cold microtome, free-floating sections*)

Method. (1) Incubate 10–15 μ sections for 5–15 minutes at 32° in a substrate mixture consisting of 20 ml. of a 125 mg. per cent. solution of potassium glucose-6-phosphate, 20 ml. of 0·2 M-"Tris" buffer (pH 6·7), 3 ml. of 2 per cent. lead nitrate, and 7 ml. distilled water.

Alternatively (Lazarus and Barden, 1964) (1) Incubate 10 μ cryostat sections, mounted on slides and post-fixed in 3 per cent. neutral formalin for 1 hour at 0°, for 15 minutes at 40° in a solution containing:

20 mM Tris buffer, pH 6·7.
2 mM Lead nitrate.
6·4 mM Glucose-6-phosphate (K salt).

(2) Wash in distilled water.
(3) Develop in dilute yellow ammonium sulphide.
(4) Wash in water.
(5) Post-fix in 6 per cent. neutral formaldehyde.
(6) Mount in glycerine and ring the coverslip with nail polish.

Result (Fig. 144). Brownish-black deposits indicate sites of glucose-6-phosphatase activity.

METHOD FOR ALKALINE NUCLEOSIDE DIPHOSPHATASE

(after Lazarus and Barden, 1962)

(*Cryostat sections, briefly post-fixed in formalin at* 0°)

PREPARATION OF INCUBATING MEDIUM

Make up a solution containing 5 ml. 100 mM sodium barbital, 3 ml. 180 mM $CaCl_2$, 7 ml. water, 5 mM uridine diphosphate or inosine diphosphate or uridine triphosphate, 5 ml. 0·1 N-NaOH. Adjust pH with 1 N-NaOH or 1 N-HCl to 9·2 and add water to bring volume to 25 ml.

Method. (1) Incubate sections for 5–20 minutes at 40°.
(2) Wash briefly in water.
(3) Apply Cobalt-Sulphide procedure (Gomori).
(4) Rinse, dehydrate, clear and mount.

Result. A black or brown deposit indicates enzyme activity.

METHOD FOR PHOSPHODIESTERASE

(after Sierakowska *et al.*, 1963)

PREPARATION OF SUBSTRATES

(A) *α*-Naphthyl thymidine-3′-phosphate (Tp-Naphthyl)

To a magnetically stirred solution containing 1·98 m-moles 5′-tritylthymidine and 3·92 m-moles *α*-naphthyl phosphoryl dichloride in 5 ml. anhydrous dioxan at 22°, add dropwise a mixture of 7·84 m-moles dry pyridine in 5 ml. anhydrous dioxan. When complete (1 hour) continue stirring for a further 2 hours. Then add, with stirring, 7·84 m-moles pyridine in 3 ml. water. Evaporate the reaction mixture under reduced pressure and take up the residue in chloroform. Decant to get rid of insoluble material. Extract the $CHCl_3$ solution with 1/5 vol. water and again with 1/5 vol. 1M pyridine hydrochloride at pH 5·5. Finally take to dryness under reduced pressure. Dissolve the resulting gum in 2·5 ml. 80 per cent. acetic acid and heat under reflux for 20 minutes. Take to dryness under reduced pressure, suspend the residue in 15 ml. water and leave for 18 hours. Filter off the crystalline precipitate of triphenylcarbinol and bring the aqueous solution to dryness under reduced pressure. Redissolve and dry, once more.

The product still contains naphthyl phosphate and this is removed in one of two ways: (*a*) enzymatically, or (*b*) by paper chromatography.

(*a*) Concentrate solution to 40 mg./ml., neutralize with 1 N-NaOH and add acetate buffer (pH 5·5) to a concentration of 20 mM. Add purified acid phosphatase (free from diesterase) and incubate at 37°.

Test aliquots with Fast Garnet GBC salt until no more free naphthol is liberated. Extract with 2 ml. ether. Pass through Dowex 50 (H^+) column, evaporate to dryness, take up the residue in 2 ml. anhydrous methanol. Filter off the insoluble enzyme protein and remove methanol under reduced pressure. Dissolve the residue in water to about 40 mg./ml. and bring to pH 7·5 with hot saturated barium hydroxide. Centrifuge off the barium phosphate, concentrate the supernatant to a small volume and precipitate the Ba salt of Tp-naphthyl by adding excess cold acetone. Separate the flocculent precipitate from the sticky yellow oil at the bottom of the tube, centrifuge down and dissolve in water. Remove Ba with Dowex 50 (H^+), neutralize with 1 N-NaOH, concentrate to a small volume and precipitate Tp-naphthyl-Na by adding excess cold acetone.

(*b*) Ascending chromatography on Whatman No. 3 paper with the following solvent:

Sat. $(HN_2)SO_4$, 80
Isopropanol, 2
1·0 M-Sodium acetate, 18.

Elute the slower migrating band ($R_F = 0{\cdot}1$) with slightly acidified methanol. Evaporate to dryness, dissolve in water, pass through Dowex 50 (H^+) column and take twice to dryness. Finally, dissolve in water at about 50 mg./ml. and neutralize sulphate ions with hot saturated $Ba(OH)_2$. Centrifuge off precipitate and add excess cold acetone to supernatant. The Ba salt of Tp-naphthyl which is the product can be converted, in the usual manner, to the sodium salt.

(B) *α*-Naphthyl thymidine-5′-phosphate (naphthyl-pT)

Prepare, as above, with appropriate modifications. The gum resulting from the first chloroform extraction is evaporated twice from ethanol, dissolved in a small quantity of ethanol and 60 ml. ether added to precipitate an oil. Dissolve the latter in acetone, filter, and evaporate to a product consisting mainly of 3′-acetylthymidine-5′-naphthylphosphate. Remove acetyl groups by heating with 10 mM NaOH on a boiling water bath for 10 minutes. Maintain pH at 12. Pass solution through Dowex 50 H^+) column. Take to dryness twice. Dissolve in water and remove contaminating naphthylphosphate as for previous synthesis.

Incubation Medium. Naphthyl-pT, 2 mg./ml.; Fast red TR salt 4 mg./ml.; 100 mM Tris-HCl buffer pH 9.

Method. (1) Incubate, preferably on coverslips, for 10–30 minutes.
(2) Wash briefly.
(3) Counterstain nuclei with hæmalum sequence.
(4) Mount in glycerine jelly.

Result. Red precipitate indicates sites of phosphodiesterase I activity.

METHOD FOR CYCLIC 3′,5′-NUCLEOTIDE PHOSPHODIESTERASE

(after Shanta *et al.*, 1966)

(*Cryostat sections, air dried*)

PREPARATION OF INCUBATING MEDIUM

Prepare 10 ml. solution containing:

Cyclic 3′-5′-AMP	1·44 mM
Tris-maleate (pH 7·62)	50 mM
Magnesium chloride	10 mM
Lead acetate	2 mM
Snake venom*	1 mg.

Check and adjust pH to 7·50 with Tris base or maleate.

Method. (1) Incubate for 30 minutes to 3 hours at 37°.
(2) Rinse in 3 changes of distilled water.
(3) Treat with yellow ammonium sulphide (dilute) for 2 minutes.
(4) Wash and amount in watery medium.

Result. Black or brown deposits indicate sites of enzyme activity.

METHOD FOR PHOSPHOGLUCOMUTASE

(after Meijer, 1967)

(*Fresh frozen sections*)

PREPARATION OF INCUBATING MEDIUM

Disodium glucose-1-phosphate	50 mg.
NADP	1·2 mg.
ATP	2·5 mg.
$MgCl_2 . 6H_2O$	12·5 mg.
Nitro-BT	2·0 mg.
Imidazole buffer (40 mM (pH 7·4)	1·5 ml.
3 per cent. gelatin	3·5 ml.
G-6-P dehydrogenase solution†	0·02 ml.

Adjust pH to 7·4.

Method. (1) Post-fix mounted sections in acetone at −25° for 30 minutes.
(2) Incubate at 22° for 1–2 hours.
(3) Wash briefly.
(4) Fix in 10 per cent. formalin for 30 minutes.
(5) Wash and mount in glycerine jelly.

Result. Purple deposits indicate sites of enzyme activity.

* *Crotalus atrox* Sigma. † Boehringer.

APPENDIX 16

PREPARATION OF TISSUES FOR ACID HYDROLASE STUDIES

(after Holt, 1959; Holt *et al.*, 1960; Holt and Hicks, 1961)

ALTERNATIVE FIXATIVES

(A) 4 per cent. formaldehyde containing 7·5 per cent. sucrose and buffered at pH 7·2 with 0·067 M-phosphate buffer.

(B) 4 per cent. formaldehyde containing 1 per cent $CaCl_2$, pH adjusted to 7·0 with additional $CaCl_2$.

GUM-SUCROSE SOLUTION

(0·88 M-Sucrose containing 1 per cent. gum acacia.)

Mix 2 g. dry gum acacia powder with 60 g. dry sucrose. Add distilled water to make 200 ml. Allow to dissolve and store at 4°.

Method. (1) Cut small (5 mm^3) blocks and transfer to ice-cold fixative (A or B) at 0° to 4°. Leave for 18–24 hours.

(2) Blot gently to remove excess fixative and transfer blocks to gum-sucrose solution at 0° to 4° for 24 hours or longer.

(3) Wash blocks in distilled water for 1–2 minutes and again blot dry.

(4) Mount blocks on cryostat tissue holder and cut sections 8–10 μ thick.

(5) Pick up sections on slides or coverslips, preferably pretreated with 1 per cent. gelatin and 2 per cent formaldehyde (equal parts) and dried. Allow sections to dry on the slides or coverslips for at least 1 hour at 22° to 37° before incubation.

THE LEAD NITRATE METHOD FOR ACID PHOSPHATASE

(after Gomori, 1950)

(*Cold acetone, paraffin sections; cold phosphate-buffered formalin or formol-calcium, frozen sections; cold microtome—post-fixed Wolman—sections*)

Method. (1) For preference, use the preparative technique described above. The method is still applicable, however, to tissues prepared in the alternative ways listed.

(2) Incubate at 37° for 15–30 minutes (up to 4 hours if necessary) in freshly prepared 0·01 M-sodium β-glycerophosphate in 0·05 M-acetate buffer (pH 5·0), containing 0·004 M-lead nitrate.

(3) Wash briefly and immerse in dilute yellow ammonium sulphide, 1–2 minutes.

(4) Wash and counterstain with 1 per cent. aqueous eosin, 5 minutes.

(5) Wash well, mount in glycerine jelly.

Result (Figs. 148, 151, and 152). The presence of acid phosphatase in the sections is indicated by a black precipitate of lead sulphide.

MODIFIED LEAD NITRATE METHOD FOR ACID PHOSPHATASE

(after Takeuchi and Tanoue)

(*Cold acetone, paraffin sections; cold microtome, mounted sections; post-fixed*)

Method. (1) Incubate sections for $\frac{1}{2}$–2 hours in the following:

2 vols. 2 per cent. sodium-β-glycerophosphate.
1 vol. 0·1 M-acetate buffer (pH 5·0–6·0).
1 vol. 2 per cent. lead acetate.
0·3 vol. 1–5 per cent. $MgCl_2$.

(2) Rinse in distilled water.

(3) Develop in ammoniacal silver nitrate solution for 30 minutes. (Add 28 per cent. ammonia water drop by drop to 5 per cent. aqueous $AgNO_3$ until the precipitate just dissolves.)

(4) Rinse in 5 per cent. sodium thiosulphate for 5 minutes.

(5) Dehydrate, clear and mount in a suitable synthetic medium or mount directly in glycerine jelly.

Result. A brownish precipitate indicates sites of acid phosphatase activity.

MODIFIED LEAD NITRATE METHOD FOR ACID PHOSPHATASE
(after Weissenfels, 1967)

This method was designed particularly for tissue cultures grown on coverslips or glass slides.

FIXATION (Modified from Gordon, Miller and Bensch, 1963)

Fix for 12 minutes at 4° in 2 per cent. glutaraldehyde in 0·05 M-cacodylate buffer at pH 7·4, containing 0·5 per cent. $CaCl_2$.

WASHING

(1) Wash 6 times, for 20 minutes each, in distilled water.

(2) Leave overnight at 4° in 6·5 per cent. sucrose in 0·05 M-cacodylate buffer at pH 7·4.

(3) Transfer to same medium at 37° for 30 minutes.

INCUBATION MEDIA

Solution A. Dissolve 120 mg. sodium β-glycerophosphate in 30 ml. 0·05 M-cacodylate buffer, containing 6·5 per cent. sucrose, at pH 5·2.

Solution B. Dissolve 60 mg. lead nitrate in 20 ml. 0·05 M-cacodylate buffer, containing 6·5 per cent. sucrose, at pH 5·0.

Add Solution B dropwise to Solution A. Allow to stand at room temperature overnight, then for one hour at 37°. Filter before use to remove turbidity.

INCUBATION AND DEVELOPMENT

(1) Incubate preparations for 1–4 hours at 37°.

(2) Wash briefly, twice, in distilled water.

(3) Wash three times, ½ hour each, in cacodylate-sucrose at pH 5·0 and 4°.

(4) Treat with 0·5 per cent. yellow ammonium sulphide for 3 minutes.

(5) Wash 6 times, 5 minutes each, in distilled water.

(6) Counterstain nuclei, if required, in Meyer's hæmalum.

(7) Wash in water and mount in glycerine jelly.

Result. Black deposits of lead sulphide indicate sites of acid phosphatase activity.

PARAFFIN SECTION TECHNIQUE FOR ACID PHOSPHATASE
(after Ruyter, 1964)

FIXATION

Wrap small pieces of tissue (2–3 mm^3) in cotton wool and fix in a large volume of 80 per cent acetone at 0°–4° for 12–24 hours. Renew the fixative at least once during this period.

EMBEDDING

(1) Transfer blocks to absolute acetone, at room temperature, two changes, 6 hours in each.

(2) Place for 1–2 hours in a mixture of equal parts of acetone and benzene and then in 2 changes of pure benzene, 1 hour each.

(3) Transfer to 5:1 benzene-paraffin wax for 30 minutes at 37°.

(4) Impregnate in 3 changes of 56° wax, 1 hour each.

(5) Embed in 56° wax.

INCUBATION PROCEDURE

(1) Cut 5 μ sections and, with the aid of a spatula, float them on stock 80 per cent. acetone,* for 3–5 minutes or longer. During this manœuvre the upper face of the sections must remain completely dry.

(2) Remove sections and float on to Gomori-type medium (0·002 M Substrate; 0·001 M-lead nitrate; 0·05 M-acetate buffer, pH 5·0 to 5·2). Incubate at 37° for 30 minutes to 4 hours.

(3) Transfer to distilled water and wash twice.

(4) Transfer to 2 per cent. sodium sulphide in 0·1 N-acetic acid, in a sealed container, for 3–5 minutes.

(5) Wash well (still floating) on two changes of tap water, 2–3 minutes each.

(6) Mount on slides, blot, and dry in oven at 60°, standing upright.

(7) Remove wax with xylene.

(8) Bring to water if nuclear or other counterstain is required, and return *via* alcohols to xylene. Otherwise mount directly in synthetic resin.

Result. Sites of acid phosphatase appear black.

RHODIZONATE METHOD FOR LEAD PHOSPHATES

(After Meijer, 1966)

Method. (1) Interrupt any of the lead sulphide techniques before the stage of conversion to sulphide.

(2) Wash in distilled water, 3 minutes.

(3) Immerse for 1 minute in a freshly prepared solution containing 30 mg. disodium rhodizonate in 40 ml. tartrate buffer (pH 3·0). This buffer contains 3·5 g. sodium hydrogen tartrate and 1·2 g. tartaric acid per 100 ml. (Keeps indefinitely.)

(4) Wash in distilled water.

(5) Dehydrate, clear and mount in synthetic resin. (Store sections in the dark.)

Result. Scarlet-red deposits of lead rhodizonate indicate sites of enzyme activity.

LEAD ACETATE METHOD FOR ACID PHOSPHATASE

(after Lake, 1965)

Lake advises substitution of lead nitrate by the acetate in all procedures in which it is employed for enzyme demonstration.

Method. (1) Air dry cryostat sections, mounted on slides or coverslips for 2–3 minutes.

(2) Fix in 4 per cent. formaldehyde in Holt's gum-sucrose† at 4° for 20 minutes.

(3) Wash in running water for 5 minutes.

(4) Incubate in standard Gomori medium‡ at 37° for 15–30 minutes.

(5) Wash well in tap water.

(6) Develop in buffered (pH 7–8) dilute ammonium sulphide.

(7) Wash well.

(8) Counterstain nuclei, if required, in Meyer's hæmalum.

(9) Wash well.

(10) Mount in glycerine jelly.

Result. Black deposits indicate enzyme activity.

* Freshly made acetone-water mixtures develop bubbles.

† 50 ml. 40 per cent. formaldehyde, 5 g. $CaCl_2.6H_2O$, 5 g. gum acacia, 150 g. sucrose. Water to 500 ml.

‡ 31·5 mg. sodium β-glycerophosphate in 5 ml. 0·1 M-acetate buffer at pH 5·0. Add 5 ml. 0·008 M-lead acetate before use.

STANDARD COUPLING AZO DYE TECHNIQUE FOR ACID PHOSPHATASE

(*Cold microtome, unfixed or post-fixed; cold formalin, frozen sections*)

Method. (1) Fix thin (2–4 mm.) slices of tissue in 10 per cent. neutral formalin at 4° for 10–16 hours.

(2) Cut frozen sections 10–15 μ thick and mount them on slides, without adhesive, drying for 2–3 hours at room temperature to ensure adherence. Alternatively use cold microtome sections, mounted on coverslips, fresh, or post-fixed in cold acetone or cold Wolman's fixative.

(3) Incubate the sections at 37° in the following mixture: Dissolve 10–20 mg. of sodium α-naphthyl phosphate in 20 ml. 0·1 M-veronal acetate buffer (Michaelis, 1931) at pH 5·0. (Alternatively 0·1 M-acetate buffer (Walpole) may be used.) Add 1·5 g. polyvinyl pyrrolidone and allow to dissolve. Add approximately 20 mg. of the stable diazotate of *o*-amino azotoluene (Fast Garnet GBC salt, I.C.I. Ltd.) or of diethylsulphamino-*o*-anisidine (Fast red ITR). Shake well and filter the mixture on to the dry sections. The incubation times vary: ½–1 minute for dog prostate, 30–60 minutes for rat liver.

(4) Wash in running water for 2 minutes.

(5) Counterstain in Mayer's hæmalum, 1–2 minutes.

(6) Wash in running water.

(7) Mount in glycerine jelly.

Result (Plate XIVa). Sites of acid phosphatase activity are coloured reddish-brown with salts 18 or 19, p. 713. With longer incubation periods the result is invariably crystalline. Nuclei, deep blue.

STANDARD NAPHTHOL AS PHOSPHATE METHOD

(after Burstone)

(*Freeze-dried or cold acetone fixed; paraffin sections*)

PREPARATION OF INCUBATING MEDIUM

Dissolve 4 mg. Naphthol AS-BI phosphate* in 0·25 ml. dimethyl-formamide† and add 25 ml. 0·2 M-acetate buffer, pH 5·2–5·6. Add 35 mg. of a suitable diazonium salt (Red-violet LB salt, or salts 2, 4 or 5) and 2 drops of 10 per cent. $MnCl_2$. Shake and filter.

Method. (1) Incubate sections, after removal of wax, for ½–6 hours at 37°.

(2) Wash in running water.

(3) Counterstain with Mayer's hæmalum, 1½ minutes or with carmalum for a similar time.

(4) Wash in running water.

(5) Mount in glycerine jelly or in Burstone's PVP mounting medium (see p. 716).

Result (Plate XIVb). Sites of acid phosphatase activity appear in various shades of red or blue. Nuclei blue or red.

PREPARATION OF NAPHTHOL AS-TR PHOSPHATE

(after Jeffree and Taylor, 1961)

SYNTHESIS

Suspend one equivalent of Naphthol AS-TR (C.I. 37525) and 5 equivalents of PCl_5 in dioxan. Add 5 equivalents of pyridine and stir for 30 minutes. Pour slowly into a large volume of ice-cold water, with constant mechanical stirring. A yellow

* Alternative substrates are AS-TR and AS-MS phosphates.
† If frozen sections are being used, reduce this to 0·1 ml.

precipitate forms. Wash precipitate with fresh distilled water on a Buchner funnel and dry in a vacuum desiccator.

The product contains approximately 60 per cent. Naphthol AS-TR phosphate.

PURIFICATION

Prepare a concentrated solution in dimethyl formamide and add sufficient hot 0·1 N-sodium carbonate to convert the acid phosphate to the sodium salt. Cool, and filter off insoluble material. Acidify the clear filtrate with normal HCl. Collect and dry the fine precipitate.

An alternative synthetic procedure, using suspensions of $POCl_3$ and naphthoic arylide in tetrahydrofuran, was offered by Burstone (1962, p. 259).

PREPARATION OF NAPHTHOL AS-BO PHOSPHATE

(Lojda *et al.*, 1964)

SYNTHESIS

Combine 5 g. Naphthol AS-BO (C.I. 37560) with 8 ml. pure $POCl_3$ in a glass-stoppered 100 ml. flask. Heat under a reflux condenser for 24 hours at 110°.The product is transformed into a honey-like mass. Cool and hydrolyse with 100 ml. ice-cold water in a water bath. Stand for 4 hours, then grind in a mortar and discard the liquid. Wash the product with 50 ml. water and allow to dry.

Yield: 6·1 g.; m.p. 156°.

PURIFICATION

Dissolve 0·5 g. of the product in boiling methanol (8 ml.). After 5 minutes, while still hot, filter off the insoluble residue and add to the filtrate, dropwise, sodium methylate until the solution turns yellow. Filter off turbidity. Acidify with excess 0·1 N-HCl. Discard supernatant and wash precipitate. Dry the resulting cream-coloured product.

Yield: 0·23 g.; m.p. 224°.

PREPARATION OF HEXAZOTIZED PARAROSANILIN

(Davis and Ornstein, 1959)

SOLUTION A

Dissolve 1 g. pararosanilin hydrochloride* in 20 ml. distilled water and add 5 ml. conc. HCl. Warm gently, cool and filter. Store in the dark, preferably at 4°.

SOLUTION B

A freshly prepared 4 per cent. aqueous solution of sodium nitrite. (Keep 48 hours at 4°.) For use mix equal parts of A and B, shake for a few seconds, until the colour becomes amber. Adjust to pH 5·0 by the addition of a few drops of normal NaOH.

As an alternative Lojda *et al.* (1964) recommended the following procedure: add 1 ml. freshly prepared hexazotized pararosanilin to 10 ml. 0·1 M-sodium acetate and adjust to pH 5·0 with 2 N-NaOH. For use equal parts of this solution and the buffered substrate solution are mixed to give a final HPR concentration of 0·05 ml./ml.

NAPHTHOL AS AND HPR TECHNIQUE FOR ACID PHOSPHATASE

(after Lojda, 1962; Barka and Anderson, 1962)

PREPARATION OF STOCK SUBSTRATE SOLUTIONS

If the sodium salt of the chosen naphthol AS phosphate is available it may be dissolved directly in 0·1 M-acetate buffer at pH 5·0 (Concentration 1 mg./ml.). This forms Stock Substrate A.

* Unsuitable fuchsins will yield reddish-brown solutions. These are useless. The final HPR solution must be pale yellow or amber in colour.

Alternatively the free acid may be dissolved at 10 mg./ml. in dimethylformamide to form Stock Substrate B. The latter can be added to 0·1 M-acetate buffer to produce a concentration of 1 mg./ml. (Stock Substrate C).

PREPARATION OF INCUBATING MEDIA

(1) (Barka and Anderson)

Stock Veronal Acetate Buffer (Appendix 1)	5 ml.
Distilled water	12 ml.
Stock Substrate B	1 ml.
Fresh HPR Solution (full strength)	1·6 ml.

Adjust to pH 5·0 with 1 N-NaOH.

(2) (Lojda *et al.*)

Buffered dilute HPR solution	5 ml.
Stock Substrate Solution A or C	5 ml.

Method. (1) Incubate sections at room temperature for 30–90 minutes (or shorter times at 37°).

(2) Rinse in distilled water.

(3) Counterstain nuclei, if required, in hæmalum.

(4) Wash well.

(5) Dehydrate rapidly in alcohols and clear in xylene.

(6) Mount in synthetic resin.

Result. Bright red deposits indicate acid phosphatase.

POST-COUPLING METHOD FOR ACID PHOSPHATASE

(after Rutenburg and Seligman)

(*Cold formalin; free-floating frozen sections; cold microtome sections, mounted, unfixed*)

PREPARATION OF SODIUM 6-BENZOYL-2-NAPHTHYL PHOSPHATE*

Dissolve 14 g. 6-benzoyl-2-naphthol and 8·6 g. phosphorus oxychloride in 100 ml. dry benzene. Heat under reflux for a few minutes. Add 5 ml. dry pyridine and continue refluxing for 30 minutes. Cool and remove precipitate of pyridine hydrochloride by filtration. Remove the solvent by distillation on a water pump until a thick syrup is obtained. Do not allow the temperature to exceed 70–80°. Pour the warm syrup into a Petri dish in a dessicator over saturated aqueous KOH. Close the dessicator and apply low vacuum. In the course of several days crystals of the acid phosphate appear. Precipitate the monosodium salt by dissolving the acid phosphate in methanol and adding an equivalent of sodium methoxide in methanol. Yield 11·2 g. (60 per cent.). To remove sodium chloride and sodium phosphate suspend the solid in a few ml. of cold water, filter, wash with methanol and ether, collect and dry.

PREPARATION OF SUBSTRATE SOLUTION

Dissolve 25 mg. in 80 ml. distilled water and add 20 ml. 0·5 M-acetate buffer pH 5·0). Make this solution hypertonic by adding 2 per cent. of solid NaCl.

Method. (1) Incubate sections, after brief washing in 0·8 per cent., 1·0 per cent. and 2·0 per cent. NaCl, for ½–2 hours (fixed sections) or for 10–60 minutes (fresh sections).

(2) Wash fresh sections in three changes of cold saline. Fixed sections can be washed in water.

* This substrate can be used with HPR in a simultaneous coupling procedure. The results are less precise than with Naphthol AS substrates.

(3) Transfer to a cold aqueous solution of diazonium salt* (1 mg./ml.), made alkaline with sodium bicarbonate, for 3–5 minutes.

(4) Wash in three changes of cold saline.

(5) Fix (unfixed sections) in cold formalin (10 per cent., 2 hours).

(6) Mount in glycerine jelly.

Result. A fine precipitate, blue or reddish-blue (Salt 6) or red (Salt 18), indicates sites of acid phosphatase activity.

THE LEAD NITRATE METHOD FOR PHOSPHOAMIDASE

(after Gomori, 1948)

(*Cold acetone; double-embedded sections*)

PREPARATION OF SOLUTIONS

(1) **Stock Substrate.** Dissolve sufficient *p*-chloroanilidophosphonic acid to make a 0·1 M solution in an excess of 10 per cent. NH_4OH. Adjust the pH to about 8 by adding dilute acetic acid and make up the volume with distilled water. This solution is stable at 4° for at least a month.

(2) **Maleate Buffer.** Dissolve 5·8 g. of maleic acid in 500 ml. distilled water. Add 62 ml. 1 N-NaOH and make up to 1,000 ml. The pH should be in the neighbourhood of 5·6.

(3) **0·1 M-Lead Nitrate.**

(4) **Ten per cent. Manganese Chloride.**

(5) **The Incubating Medium.** Add 2 ml. of 0·1 M stock substrate solution to 50 ml. maleate buffer with 1·5 ml. of 0·1 M-lead nitrate and a few drops of $MnCl_2$. Incubate at 60° for 30 minutes and filter into a Coplin jar.

Method. (1) Incubate mounted frozen sections for 2–4 hours, keeping the Coplin jar at an angle and the slides face downwards.

(2) Rinse in distilled water, wipe precipitate from around sections and from the backs of the slides.

(3) Rinse in 0·1 M-citrate or acetate buffer at pH 4·5 until the diffuse white precipitate covering the section disappears.

(4) Rinse in running water.

(5) Treat with dilute yellow ammonium sulphide solution, 1–2 minutes.

(6) Wash in running water.

(7) Counterstain in 1 per cent. aqueous eosin, 3–5 minutes.

(8) Wash well and mount in glycerine jelly.

Result (Fig. 157). A black precipitate is presumed to indicate phosphoamidase activity.

MODIFIED PHOSPHOAMIDASE TECHNIQUE

(after Meyer and Weinmann)

(*Cold acetone; double-embedded sections*)

PREPARATION OF INCUBATING MEDIUM

(1) **Stock Solution A.** Dissolve 2·08 g. *p*-chloroanilidophosphonic acid in 15 ml. 1 N-NaOH and make up to 100 ml. with distilled water.

(2) **Solution B.** Dissolve 534 mg. maleic acid in 5 ml. 1 N-NaOH and make up volume to 100 ml. with distilled water. Add 175 mg. NaCl and 94 mg. $Pb(NO_3)_2$. The mixture becomes turbid and should be heated gently until it clears. Add 4·5 ml. of Stock Solution A and heat to 44°. Filter. The filtrate should be clear.

* The authors recommended Fast blue B salt (6), but if a single-coloured end product is desired, salt 18 may be used instead.

Method. (1) Mount two (serial) sections, one at either end of the slide.

(2) Immerse one of the sections in 10 per cent. nitric acid for 90 minutes, without removal of paraffin wax.

(3) Wash this section and blot dry.

(4) Remove wax from both sections witn xylene.

(5) Incubate slides for 1½–4 hours at 42°, in a horizontal position with the sections facing downwards. The authors recommend a rectangular bath and 7 mm. solution below and 4 mm. above.

(6) Differentiation. Sections incubated for over 3 hours require treatment with 0·1 M-citric acid until the inactivated section is clear of precipitate.

(7) Wash in water and treat with dilute yellow ammonium sulphide for 2 minutes.

(8) Wash in water, dehydrate, clear and mount in a suitable synthetic medium.

Result. A brownish-black precipitate indicates sites of phosphoamidase activity.

METHOD FOR ACID DEOXYRIBONUCLEASE

(after Vorbrodt, 1961)

(*Cold formol-calcium, cryostat sections*)

INCUBATION MEDIUM

DNA	10 mg.
Acid phosphatase	5 mg.
0·2 M-Acetate buffer (pH 5·0)	12·5 ml.
0·4 M-Lead acetate or nitrate	0·25 ml.
Distilled water to	50 ml.

Method. (1) Incubate for 30 minutes to 10 hours in above medium.

(2) Wash briefly in distilled water.

(3) Treat with buffered dilute yellow ammonium sulphide, 3–5 minutes.

(4) Wash in distilled water.

(5) Mount in glycerine jelly.

Result. Black deposits (lysosomal) indicate acid DNase II.

COMBINED TECHNIQUE FOR LYOSOMES AND PHAGOSOMES

(Straus, 1967)

INJECTION OF PEROXIDASE

Dissolve horseradish peroxidase (12·5 or 6·2 mg./100 g. body weight) in 0·3 to 0·5 ml. saline. Inject into suitable vein.

RECOMMENDED FIXATION

After initial perfusion with warm saline, perfuse with warm (37°) 2–4 per cent. glutaraldehyde in 0·8 per cent. NaCl with 0·02 M-cacodylate buffer at pH 7·3 (final pH 6·8 to 7·0).

This procedure is followed by further fixation of blocks in 4 per cent. formaldehyde or 4 per cent. glutaraldehyde in 30 per cent. sucrose with 0·02 M-cacodylate (pH 6·6–7·0), for 3 hours at 4°.

After washing in 30 per cent. sucrose blocks are quenched in isopentane at −78° and mounted on tissue holders for cryotomy. Sections 4–8 μ thick are mounted on coverslips and allowed to dry for several hours.

Method. (1) Stain for acid phosphatase by the Naphthol AS-HPR technique.

(2) Wash in 5 per cent. sucrose for 5 minutes.

(3) Wash in ice-cold distilled water for 5–10 minutes.

(4) Immerse in 25 per cent. ethanol at 25°, 30 seconds.

(5) Immerse in ice-cold 0·2 per cent. benzidine* with 0·045 per cent. H_2O_2 in 40 per cent. ethanol, for 10–60 seconds.

(6) Wash three times in 35 per cent. ethanol at 10°.

(7) Transfer to 4·5 per cent. Sodium nitroprusside in 2mM-acetate buffer (pH 4·0) containing 25 per cent. ethanol, cooled to 10°.

(8) Wash twice in 25 per cent. ethanol and then once in 50 per cent. ethanol, 5–10 seconds in each.

(9) Dehydrate in absolute ethanol at −10°.

(10) Transfer to two changes of 1:1 alcohol-xylene, thence to xylene.

(11) Mount in a synthetic resin.

Result (Plates XIIIa and b). Phagosomes appear blue, lysosomes red. Intermediate forms may appear purple but the possibility of superimposition of the two types of granule must be kept in mind.

* This compound is a carcinogen.

AUTHORS' INDEX

SUBJECT INDEX